NATIONAL GEOGRAPHIC

COMPLETE
BIRDS
OF NORTH AMERICA

Yellow Warbler (NL, July)

COMPLETE
BIRDS
OF NORTH AMERICA

THIRD EDITION
FULLY REVISED and UPDATED

EDITED BY **JONATHAN ALDERFER**
AND **JON L. DUNN**

MAPS BY PAUL LEHMAN

CONTRIBUTING AUTHORS

Jessie H. Barry, Edward S. Brinkley, Steven W. Cardiff, Allen T. Chartier,
Cameron D. Cox, Donna L. Dittmann, Jon L. Dunn, Ted Floyd,
Kimball L. Garrett, Matthew T. Heindel, Paul Hess, Steve N. G. Howell,
Marshall J. Iliff, Alvaro Jaramillo, Ian L. Jones, Tony Leukering, Mark W. Lockwood,
Guy McCaskie, Bill Pranty, David E. Quady, Kurt Radamaker, Gary H. Rosenberg,
Brian Sullivan, Clay Taylor, and Chris Wood

NATIONAL GEOGRAPHIC
WASHINGTON, D.C.

CONTENTS

Mexican Violetear
adult ♀

Broad-tailed Hummingbird
adult ♀

Calliope Hummingbird
adult ♀

INTRODUCTION by Jonathan Alderfer and Jon L. Dunn

Since the publication of the first edition of this work in 2005 the birding landscape in North America has changed in many ways. In this new edition of *Complete Birds*, you will find information on the identification of 1,035 species.

COVERAGE

This book includes all species of native birds reliably recorded in North America, which we describe as the continent north of Mexico, plus adjacent islands and seas within 200 miles of the coast, excluding Greenland. We also include exotic species that are established or regularly observed. At the end of the book (pages 734–735) are short sections describing the avifauna of Greenland and Bermuda. With a cutoff date of August 2018, the book's last included species was the Pied Wheatear, a stray from Eurasia in the summer of 2018.

Text The text for the first edition was written in 2005 by 24 experienced field ornithologists in North America (see Contributors, page 736). Edward S. Brinkley, Jon L. Dunn, Paul Hess, Kimball L. Garrett, and Paul Lehman provided additional editorial work and writing in previous editions. Added text for the second edition, including 12 new family accounts and over 60 new species accounts, was written by Edward S. Brinkley, Paul Hess, and Kimball L. Garrett, with updating and editorial work by Jonathan Alderfer and Jon L. Dunn. For this third edition, we were assisted by Kimball L. Garrett (editor of family taxonomy sections and general consultant) and Paul Lehman (consultant on status and distribution).

Art and Photographs Most of the artwork in this edition comes from the *National Geographic Field Guide to the Birds of North America*, seventh edition (2017), including 357 art figures new to that edition. Since the publication of the seventh edition, 13 new species have been added: Kermadec Petrel, Juan Fernandez Petrel, Tahiti Petrel, Black Kite, Great Black Hawk, Red-backed Shrike, Thick-billed Warbler, River Warbler, European Robin, Pied Wheatear, Mistle Thrush, Cassia Crossbill, and Black-backed Oriole. Illustrations of those species were painted for this volume by Jonathan Alderfer, David Quinn, N. John Schmitt, and Thomas R. Schultz. Illustration credits appear on page 739.

Maps Paul Lehman has been the map editor for all three editions of this work. The range maps in this edition are from the seventh edition field guide with updates to reflect new information. A map key is located below. Note that boundaries are drawn where a species ceases to be regularly seen in its proper habitat. The 825 range maps in this edition include 55 large-scale subspecies maps; many of the regular range maps

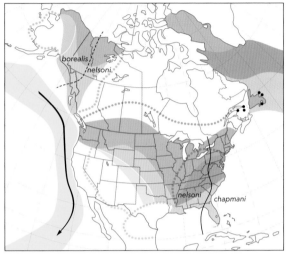

Range Map Symbols

Breeding range, generally in spring and summer

Year-round range

Winter range (*if no winter or year-round range is shown, bird winters outside North America*)

Migration range in spring and autumn

Migration range primarily in spring

Migration range primarily in autumn

Principal direction of migration

Selected breeding colonies

Extent of irregular breeding range

Extent of irregular year-round range

Extent of irregular or irruptive range in some winters

Extent of irregular migration in spring and autumn

Extent of irregular migration primarily in spring

Extent of irregular migration primarily in autumn, or post-breeding dispersal

fulva Subspecies name and its boundary

Subspecies boundary where only approximate

Subspecies Maps Only
(*these symbols not shown on sample map above*)

Subspecies boundary during particular seasons

Subspecies group boundary

Zone of intergradation between two subspecies

also delineate subspecies ranges. Nine specialty maps show long-range migration routes, patterns of rare occurrence, or historical data.

TAXONOMIC ORGANIZATION

With its contents organized by order, family, subfamily, genus, and species, this edition follows the taxonomic sequence and naming conventions adopted by the American Ornithological Society (AOS) as of July 2018 (59th Supplement). The taxonomy of birds is currently in a period of flux as genetic research uncovers evolutionary relationships. These changes have greatly affected the linear sequence of families and of species within families, as well as the splitting and occasional lumping of species. Getting to know the latest taxonomic sequence is part of a birder's education.

Family A short overview introduces each of the 95 families. The scope is worldwide and includes information on structure, plumage, behavior, distribution, taxonomy, and conservation.

Genus Genera (pl. of *genus*) are groups of closely related species. The genus name is the first part of a species' scientific name: i.e., *Spizella* in *Spizella passerina,* for Chipping Sparrow. Determining a bird's genus is often useful in identifying it. A brief text describes many of the genera, but where distinctions are less useful, genera are grouped together or have no text.

Species A species is identified by its scientific **binomial** name—its genus and specific epithet. Species accounts start with an overview and follow a standard layout; some accidental species have abbreviated accounts.

Subspecies (Geographic Variation) Some species show distinctive geographical variation. This book attempts to give information on many field-recognizable subspecies or groups of subspecies. A species that has no recognized subspecies is known as **monotypic** (e.g., Cerulean Warbler). When within-species variation forms two or more distinctive subspecies, the species is known as **polytypic**. Those subspecies are each given a **trinomial** scientific name: for example, *Zonotrichia leucophrys gambelii* is the trinomial for "Gambel's" White-crowned Sparrow, a distinctive, field-recognizable subspecies of the White-crowned Sparrow, which breeds in the taiga region west of Hudson Bay. In polytypic species, the first type of the species described is the **nominate subspecies**, with a name that repeats the second part of its scientific species name (e.g., *Zonotrichia leucophrys leucophrys* is the nominate subspecies of the White-crowned Sparrow). Many subspecies have no commonly used English name, so we use their scientific subspecies epithet throughout. When different subspecies interbreed, the result is known as an **intergrade**. When a species or subspecies changes gradually across a geographical region, it is said to form a **cline**, or to vary clinally. In most cases, our authority for subspecies names and ranges is the *Howard and Moore Complete Checklist of the Birds of the World,* fourth edition.

PLUMAGE VARIATION

In many species the male and female look quite different and are **sexually dimorphic**. In some species the sexes look alike but can be separated by size, particularly when they are together (e.g., raptors). This book illustrates and describes most of those differences. **Age** determination ranges from simple to complex, depending on the species and how quickly or slowly it reaches adult plumage. **Molt** describes the replacement of a bird's feathers: A new feather growing from the same follicle pushes out the old feather. All birds molt, and a molt produces a plumage. Birds with old, abraded feathers are referred to as **worn**; birds with new feathers are **fresh**. Most birders have adopted an amalgam of terms for molt and plumage to describe what they are seeing—not a rigorous or scientific use of terminology, but a practical one. Some of the terms used in this book are defined below.

Juvenile Juveniles (birds in **juvenal** plumage) are wearing their first true coat of feathers, the ones in which they usually fledge (leave the nest). Some young birds hold their juvenal plumage only a few weeks, while others—such as hawks, loons, and many other waterbirds—hold their juvenal plumage into the winter or even the following spring.

First-Fall or First-Winter Most species replace their juvenal plumage in late summer or early fall with a partial molt into this plumage, usually while retaining the major juvenal wing and tail feathers. All plumages between juvenal and adult are **immature**, an imprecise term. The sequence of immature plumages may continue for some years until adult plumage is attained.

First-Summer, Second-Winter, etc. These terms describe immature plumages between first-winter and adult. **Life-year** terms may be used, such as first-year, second-year, etc. These are age (not molt or plumage) terms; that is, a first-year bird is in its first 12 months of life. **Calendar-year** refers to the bird's actual calendar-year age; that is, a first-calendar-year bird becomes a second-calendar-year bird after December 31.

Adult Birds are adult when their molts produce a stable cycle of identical (or **definitive**) plumages that repeats for the rest of their lives. Most adult birds have either one complete molt per year *or* one partial and one complete molt per year. Adult birds with a single complete molt per year (usually after breeding) look the same all year, except for the effects of wear. Adults with two annual molts generally undergo a partial (**pre-breeding** or **prealternate**) molt in late winter or early spring—often involving the head, body, and some wing coverts—

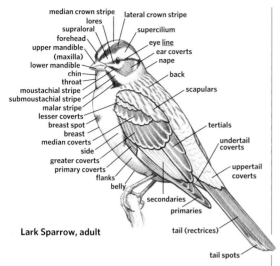

Lark Sparrow, adult

into **breeding** (or alternate) plumage. Most of these species later undergo a complete (**postbreeding** or **prebasic**) molt after breeding into their nonbreeding or winter (or **basic**) plumage. **High breeding** refers to plumage or bare-parts colors evident only during the brief period of courtship, such as the rich colors of the bill, lores, legs, and feet of many herons. Most male ducks in late summer acquire a briefly held, female-like **alternate** plumage (formerly known as "eclipse").

Other Variations Two species may interbreed, producing **hybrids** that may look partly like one parent, partly like the other, or may show unexpected characteristics. For a few species, hybridization is frequent; we describe and illustrate some of these plumages. **Morph** describes a regularly occurring variation, usually in plumage coloration, such as the gray-morph Eastern Screech-Owl. Other terms describe abnormal plumages: **Albino** birds are pure white with red eyes; **leucistic** birds have unusually pale plumage or a piebald appearance with scattered white feathers, sometimes in a symmetrical pattern; and **melanistic** birds have excessively dark pigmentation.

Art Labels Art labels point out distinguishing characteristics. Male (♂) and female (♀) symbols are used where the sexes are discernibly different. Many figures are labeled with an age and subspecies designation. If a species shows little or no difference between ages or sexes (e.g., storm-petrels), there are no age or sex labels.

FEATHER TOPOGRAPHY

Knowing the names of the feather groups and how they are arranged is central to becoming a better birder. The color and pattern of feathers as well their physical proportions are important (e.g., the primary extension past the tertials). Most of the bolded terms below are labeled on the illustrations here.

Head A careful look at a bird's head can often provide the best clues to make a correct identification. At the top of the Lark Sparrow's head (left) there is a pale **median crown stripe**, bordered by dark **lateral crown stripes**. The line running from the base of the bill up and over the eye is the **eyebrow** or **supercilium**, and the part in front of the eye is the **supraloral area**. The area between the eye and the bill is the **lores**. Note eye color: The pupil is always black, but the **iris** (pl. *irides*) can be colored. The **orbital ring** (below) is naked flesh around the eye; when contrastingly feathered, the area forms an **eye ring** or **eye crescents**. A dark stripe extending back from the eye is a **postocular stripe**; when it extends through the eye it is an **eye line**. The Lark Sparrow shows a distinctive brown patch behind and below the eye known as the **ear coverts**, or **auricular**. Its lower border is the **moustachial stripe**; below that is the **submoustachial stripe**, and below that is the **malar stripe**. Most gulls (below) and some other species have a prominent ridge on their lower mandible, the **gonys**, which forms the **gonydeal angle**. The top of the bill is the **culmen**. Raptors have a bare patch of skin over part of the upper mandible, the **cere** (below). Some species, such as pelicans and cormorants, have a fold of loose skin hanging from the throat, the **gular pouch**.

Wings Folded wings and extended wings look very different, and it's useful to study both aspects. The longest, strongest feathers of the wing, the **flight feathers** or **remiges**, are covered on both their upper and lower surfaces by shorter, protective feathers called **coverts** (opposite). Each feather's shaft divides its **inner web** from its **outer web**. The outermost 9 or 10 tapering flight feathers are the **primaries (P1–P10)**; inward from the primaries are a variable number of **secondaries**. The innermost secondaries are a group of feathers known as the **tertials**. On the folded wing, the distance the primaries extend past the longest tertial is the **primary tip projection**. Covering the area where the wing joins the body are the **scapulars**. Rows of wing coverts cover the bases of the flight feathers. From **leading edge** to **trailing edge** they are the **marginal, lesser, median,** and **greater wing coverts**. The greater and median coverts form single rows; the lesser and marginal coverts consist of multiple rows of small feathers. **Wing bars** are formed by the pale tips of the greater and median wing coverts. On long-winged species (e.g., albatrosses, above right), the **humerals** are well developed. **Mantle** refers to the overall plumage of

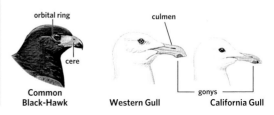

Common Black-Hawk   **Western Gull**   **California Gull**

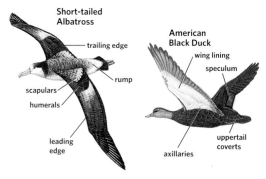

Short-tailed Albatross

trailing edge

rump

scapulars

humerals

leading edge

American Black Duck

wing lining

speculum

uppertail coverts

axillaries

the back and scapulars. The uniquely colored secondaries of some ducks are referred to as the **speculum** (above). A contrasting bar on the leading edge of the wing is a **carpal bar**; a bar that cuts diagonally across the inner wing is an **ulnar bar**. Dark primaries, ulnar bars, and rump may form an **M pattern**, prominent on some seabirds. On the underwing, **wing linings** refer to the underwing coverts as a whole; **axillaries** are the feathers of the bird's "armpit."

Tail Most species have 12 tail feathers, also called **rectrices**. On the folded tail from above, only the two central tail feathers are visible; from below only the two outermost tail feathers are visible. Note the pattern of the undertail and length of the tail projecting past the undertail coverts. Tail pattern and shape is often best viewed in flight.

Size All species accounts include average length, **L**, from tip of bill to tip of tail; some large or soaring species also have wingspan, **WS**, measured from wing tip to wing tip. The perception of size is fraught with variables. Lighting, background coloration, and posture affect size perception, as does the use of birding optics.

ABUNDANCE TERMS AND CODES

Abundance must be considered in relation to habitat. Some species are highly local, found only in specialized habitats. Ranges and abundance are detailed in the text, as well as extralimital records and population trends. The terms *abundant, common, fairly common,* and *uncommon* are self-explanatory. For more unusual occurrences, we use the following terms. **Rare**: Occurs in low numbers but annually (e.g., visitors or rare breeding residents). **Casual**: Not recorded annually but with six or more total records, *including* three or more in the past 30 years, reflecting a pattern of occurrence. **Accidental**: Five or fewer records *or* fewer than three records in the past 30 years.

Banding Codes These four-letter codes, used by bird banders, are usually derived from the first two letters of a bird's two-word name (e.g., TUVU is the code for Turkey Vulture), but this varies to avoid duplication. Banding codes are found here after each species' scientific name.

ABA Abundance Codes Number codes developed by the American Birding Association (ABA) are found after the banding codes on the opening line of each species account. Recently the ABA added Hawaii to the ABA List and changed the codes for some species: Those changes were not adopted in this book since it does not cover Hawaii. **1**: Common (usually widespread). **2**: Uncommon (or with a limited range). **3**: Rare. **4**: Casual. **5**: Accidental. **6**: Not found (extirpated, extinct, or exists as a released population not yet naturally reestablished).

Conservation Codes BirdLife International codes categorize the conservation status of the world's birds on the International Union for Conservation of Nature (IUCN) Red List and are regularly updated. In this book, they appear in the Conservation section of each family introduction and in the Population section for species under some level of threat. **LC** (Least Concern, not used); **NT** (Near Threatened); **VU** (Vulnerable); **EN** (Endangered); **CR** (Critically Endangered), **CR (PE)** (Critically Endangered, Possibly Extinct); and **EX** (Extinct). For more information see *datazone.birdlife.org*. These codes are different from the Endangered Species List of the U.S. Fish and Wildlife Service (USFWS).

Abbreviations In general, we use US and Canadian postal codes to identify states and provinces. We also use N.A. for North America, C.A. for Central America, M.A. for Middle America, and S.A. for South America.

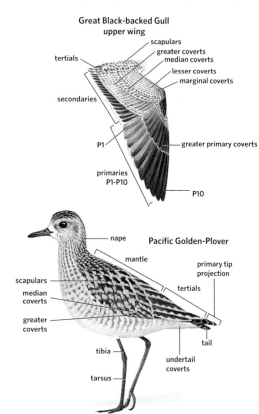

Great Black-backed Gull upper wing

scapulars
greater coverts
median coverts
lesser coverts
marginal coverts

tertials

secondaries

P1

greater primary coverts

primaries P1–P10

P10

Pacific Golden-Plover

nape

mantle

primary tip projection

scapulars

tertials

median coverts

greater coverts

tibia

tail

undertail coverts

tarsus

DUCKS, GEESE, AND SWANS   Family Anatidae

Black Scoter, female (second from left) with males (NJ, Mar.)

These species all share some distinctive traits. They have fully webbed feet. Most species have a standard duck bill that is flattened and broadened at the end; the major exception is the mergansers, which have long, thin, serrated bills. Anatids usually have short tails and pointed wings. The placement of the legs differs considerably depending on whether the species is a surface swimmer or a diver. The legs of surface species are more centrally placed and allow for a horizontal body stance, while those of divers tend to be set far back on the body, making it difficult for these ducks to walk on land. Whistling-ducks, an exception, are slim and have legs placed far back, allowing them to stand more upright than other ducks.

Plumage Variation is huge; many are very attractive and are kept in captivity. Whistling-ducks are attractively patterned and often show characteristic markings on the flanks; they are sexually monomorphic. Geese tend to be brownish or subdued colors, but often with attractive patterns, and there is no sexual dimorphism. Swans are often white, although two are black or black necked. Ducks are incredibly variable, but are usually sexually dimorphic, with males showing bright drake plumage for much of the year, except during late summer when they acquire a briefly held dull female-like alternate (formerly known as eclipse) plumage; females tend to cryptic brown plumage. Some species of dabbling ducks, particularly those on islands, have henlike male plumages. Dabbling ducks and some other species show a colorful, and often iridescent, color patch on the secondaries known as the speculum.

Behavior This family is aquatic. Some birds feed from the surface of the water, either filtering food or tipping down to feed on vegetation, while other species are active divers ranging from piscivorous predators to species specializing on shellfish or submerged vegetation. Geese graze on open fields. Courtship behavior is well marked in ducks and is usually easy to observe during the winter and early spring. Geese and swans tend to mate for life, while ducks find new mates each year.

Distribution Anatids occur worldwide, breeding in all continents except Antarctica.

Taxonomy A large family with 53 genera and 159 species that is divided into four subfamilies, all of which—whistling-ducks (Dendrocygninae), true geese and swans (Anserinae), shelducks (Tadorninae), and ducks (Anatinae)—are found in North America. The family is most closely related to the magpie-geese (Anseranatidae) and screamers (Anhimidae). These families and the galliform families form an old lineage (Galloanserae) within modern birds (Neognathae). Apart from the basal Paleognathae (Ratites and Tinamous), all modern birds are grouped as the Neognathae. The Galloanserae is the oldest Neognathae lineage and sister to all remaining groups (the Neoaves). Species level taxonomy within North America has been quite stable, although currently there is interest in determining adequate species level divisions within the geese and certain dabbling ducks. One confounding factor in waterfowl is that they readily hybridize with each other, and even between genera. The genus *Anser* now includes all the geese formerly placed in genus *Chen*, and the large dabbling duck genus *Anas* has recently been split into several genera.

Conservation BirdLife International codes: 11 NT, 18 VU, 8 EN, 6 CR, and 7 EX.

WHISTLING-DUCKS Genus *Dendrocygna*

The eight species of whistling-ducks are found mainly in tropical regions. Slim, long necked, and long billed, they stand rather upright. Their whistling calls inspired their name; some perch on trees, giving them the alternate name tree-ducks. They fly with shallow wingbeats on rounded wings. The whistling-ducks comprise a subfamily of the Anatidae.

BLACK-BELLIED WHISTLING-DUCK *Dendrocygna autumnalis* BBWD ■ 1

This gorgeous duck is often found perching on trees or shrubs. Polytypic (2 sspp.; *fulgens* in N.A.). L 21" (53 cm) **Identification** A long-legged, broad-winged, and long-necked duck. **ADULT:** Bright red bill obvious on gray face; the eye is surrounded by a bold white eye ring. Lower neck and breast tawny, with contrasting black belly and flanks. Upperparts warm brown,

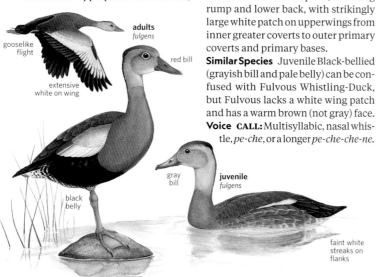

gooselike flight

adults
fulgens

red bill

extensive white on wing

gray bill

juvenile
fulgens

black belly

faint white streaks on flanks

folded wings show whitish shoulder/forewings, and dark primaries and tertials. Legs bright pink. **JUVENILE:** Similar to adult but bill grayish and body plumage much duller. Flanks brownish and belly off-white, not black. Wing pattern similar to that of adult, although duller and less extensively white. **FLIGHT:** A warm brown with black hindparts, including rump and lower back, with strikingly large white patch on upperwings from inner greater coverts to outer primary coverts and primary bases.
Similar Species Juvenile Black-bellied (grayish bill and pale belly) can be confused with Fulvous Whistling-Duck, but Fulvous lacks a white wing patch and has a warm brown (not gray) face.
Voice CALL: Multisyllabic, nasal whistle, *pe-che*, or a longer *pe-che-che-ne*.

Status & Distribution Fairly common, widespread in Neotropics south to central Argentina. **YEAR-ROUND:** Wetlands, those with overhanging woody vegetation preferred. **MIGRATION:** A short-distance migrant, traveling not far into northern Mexico. Arrival on breeding grounds by Apr., southbound movements noted from Aug. to Oct. **RARE STATUS:** Casual (mostly spring and summer) north of regular range as far north as MN, MI, ON, QC, and NS: also to Southern CA, southern NV, and NM. Often appears in small flocks. **Population** Population increasing. Range has expanded over the last several decades; now found over much of the Southeast.

FULVOUS WHISTLING-DUCK *Dendrocygna bicolor* FUWD ■ 1

The Fulvous Whistling-Duck is a long-legged and long-necked largely tawny duck. When swimming feeds by up-ending and dabbling, but also frequently dives. Monotypic. L 20" (51 cm). **Identification** Bill gray, legs blue-gray. Face, neck sides, breast, and belly are tawny. Throat and foreneck buffy white, speckled with black on lower neck. Top of crown brown, continues as a brown line (broken in male, continuous in female) down the back of the neck to the upper back. Upperparts black with broad cinnamon edges.

Tawny underparts accented by crisp white stripes on flanks, vent white. **FLIGHT:** Tawny with blackish wings. Lower back and tail black, contrasting with white rump band.
Similar Species Juvenile Black-bellied has a gray face and large white wing patch. Fulvous is always more tawny, with white flank stripes. In flight, Fulvous shows a strong contrast between blackish underwings and tawny body, and a white rump band.
Voice CALL: A two-note, high-pitched, nasal, whistled call, *pt-TZEEW*.
Status & Distribution Fairly common. Widely distributed in New World tropics south to central Argentina, as well as in Africa and the Indian subcontinent. **YEAR-ROUND:** Freshwater grassy wetlands, rice fields commonly used. **MIGRATION:** Many move south into Mexico. Arrival on breeding grounds in Feb.–Mar., southbound movements Aug.–late Sept. **DISPERSAL:** Regular northward dispersal after breeding. **RARE STATUS:** Casual throughout continent, north to BC, AB, and NS. These

blackish rounded wings

white U-shaped rump band

white lines on sides and flanks

records occur in years of increased postbreeding northward wandering. Records are concentrated in Mississippi Basin and Atlantic coast.
Population Numbers have fluctuated; however, CA population and range has seriously decreased and contracted since the mid-1900s; it was extirpated as a breeder shortly after 2000 and now is only of casual occurrence.

Genus *Anser*

Our domesticated goose is derived from a member of this genus, the Graylag *(Anser anser)*. There are nine species now in this genus; recently the three *Chen* geese have been merged into *Anser*. These geese are variable in color from brownish to largely white. Bill color, pattern, and shape, as well as leg color, aid in identification.

EMPEROR GOOSE *Anser canagicus* EMGO ▪ 2

The Emperor Goose is a gorgeous marine goose of the Bering Sea area. Monotypic. L 26" (66 cm)
Identification A small stocky goose with thick and strong legs. Bill pink with black cutting edges and nostril, and legs bright orange. **ADULT:** Distinctive, gray bodied, with each feather neatly tipped dark subterminally and white terminally, giving a characteristic scaly look. Head white, continuing down the back of the neck and contrasting with black throat and foreneck. Sometimes the head may be stained rust with oxides. The white tail contrasts strikingly with all-dark upper- and undertail coverts. **IMMATURE:** Browner than adult, but shows scaly appearance and white tail although it has an entirely brown head and neck; head and neck become largely white by late fall.
Similar Species Distinctive, although blue-morph Snow or Ross's Geese show similar dark body with white face or neck but have white undertail coverts. Emperor Goose has a diagnostic pattern of white head and hind neck contrasting with black throat and foreneck. Barnacle Goose shares grayish scaly appearance, but it has an entirely black breast and neck

with only face white and a dark tail.
Voice CALL: A hoarse and repeated *kla-ha*, although often remains quiet.
Status & Distribution Uncommon; breeds in AK and northeastern Russian Far East. **BREEDING:** In coastal wetlands with tidal influence. **MIGRATION:** Molt migration starts as early as late June, with dispersal south to the Aleutian Is. by Nov. Wintering areas largely in Aleutian Is. and south coast of AK Peninsula. Birds depart Mar.–early Apr. for staging lagoons on the north shore of the AK Peninsula. Birds stage until May, and then fly to nesting areas, largely in the Yukon-Kuskokwim Delta. **WINTER:** Intertidal habitats, from rocky coasts to eelgrass beds. **RARE STATUS:**

Casual south to CA on the Pacific coast, with a few inland records in the Sacramento Valley. Accidental HI and Japan.
Population Near Threatened. Appears to have decreased between 1960 and the 1980s, but stable since then.

SNOW GOOSE *Anser caerulescens* SNGO ▪ 1

This species is the larger and more common white Arctic goose. There is also a blue morph, which commonly occurs in the mid portion of the continent and is called the "Blue Goose." Polytypic (2 sspp.; both in N.A.). L 26–33" (66–84 cm)
Identification Medium size, with a pink bill and distinctive "grinning patch" along the cutting edge of the bill. Adult's legs pink, immature's legs dusky pink. **ADULT WHITE:** White, with black primaries. Often face is stained rust from oxides in water. **JUVENILE WHITE:** Similar to adult, but gray-brown wash on head, neck, and upperparts; dark centered tertials. In flight shows dusky secondaries. **ADULT BLUE:** Brown body, white head and neck. Wings with gray-blue coverts, long tertials, inner greater

coverts black with bold white fringes. Tail gray with white border. **JUVENILE BLUE:** Dull brown, including head and neck. Pattern of wing and tertials less well developed.
Geographic Variation Interior and western populations known as "Lesser Snow Goose," *caerulescens*; eastern population is the "Greater Snow Goose," *atlantica*. Size overlaps and perhaps better considered monotypic. Blue morph lacking in Pacific and Atlantic wintering populations.
Similar Species Most similar to Ross's Goose, but adult Snow Geese of both morphs have black "grinning patch" on bill, are larger, longer necked, and longer billed. Blue morph has white head and neck and blue wing panel;

compare to Ross's. Immature white-morph Snow Geese are extensively brownish washed above, and in flight shows brownish secondaries.
Voice CALL: Call is a barking *whouk*, or *kow-luk*.
Status & Distribution Abundant, casual to Europe, now rare to Japan where formerly numerous. **BREEDING:** Moist tundra. **MIGRATION:** Southbound movements begin late Aug.–early Sept., arriving in wintering areas by early Oct., peaking late Oct. and Nov. Northbound movements begin in Feb., peak early Apr. in southern Canada and reach breeding areas in late May. **WINTER:** Wetlands and agricultural fields.
Population Both subspecies growing

exponentially since the 1960s; they are exceeding the carrying capacity at breeding sites, overgrazing sensitive grassy tundra.

ROSS'S GOOSE *Anser rossii*   ROGO ■ 1

The smallest of the white Arctic geese, Ross's Goose has a short, triangular bill. Monotypic but polymorphic. L 23" (58 cm)
Identification Stocky and short necked, with high forehead and short legs. The bill is pink and shows grayish "warts" on the base of the upper mandible, most developed on older males. Legs pink. **ADULT:** Entirely white, with black primaries. **JUVENILE:** Similar to adult but variably washed dusky, particularly on the nape and back of neck. The tertials show dark shafts or centers. **HYBRIDS:** With the Snow Goose, Ross's Goose is intermediate in size and structure and "grinning patch," best identified by comparing with the two parental species.
Geographic Variation Monotypic, but a rare blue morph exists. The origin of this blue morph is controversial and is thought to be due either to introgression with blue-morph Snow Geese or a recurrent mutation of genes control-

more rapid wingbeats than Snow

paler than juvenile Snow

rounder head than Snow

adult

white-morph juvenile

stubby bill lacks "grinning patch"

white-morph adult

dark neck

short neck

blue-morph adult

darker than blue-morph Snow

ling feather color. Blue-morph Ross's Geese are most frequent in wintering flocks in CA's Central Valley. They are structurally like typical Ross's Geese.
Similar Species Like the Snow Goose but smaller, shorter necked, smaller billed, and rounder headed. Ross's lacks "grinning patch" but some show gray warts at bill base. In flight, smaller and shorter necked with more rapid wingbeats than Snow Goose. Blue-morph Ross's Geese differ from blue-morph Snow Geese in showing only a white face, with dark hind crown, nape, and neck; and extensive white on the lower breast and belly.
Voice CALL: A high, nasal *hawhh*.
Status & Distribution Common. **BREEDING:** Wet tundra. **MIGRATION:** Little is known about boreal staging areas in this species. However, the western population stops over in eastern AB, western SK largely in Sept. From there they move through

western MT, northern ID, eastern OR into the wintering areas in CA by late Oct. The eastern population moves through Hudson Bay and stages in eastern SK and MB, before moving south through the Dakotas to wintering areas in central Mexico, eastern TX, and LA. Spring migration begins as early as Feb. and retraces the fall route, following the advance of the snowmelt. Staging in Canadian prairies takes place mid-Apr.–mid-May, and arrival in breeding grounds from end of May to early June. Lingers later in wintering grounds than the Snow Goose. **WINTER:** Wetlands and adjacent agricultural areas. **RARE STATUS:** Rare but regular east of the Mississippi River; increasing.
Population Since the mid-1950s Ross's Goose populations have increased dramatically; some estimates give a 10 percent annual increase from the 1950s to the 1990s.

GRAYLAG GOOSE *Anser anser*   GRGO ▪ 5

This goose is the progenitor of most domestic geese. Often not shy, the domestic forms are larger and heavier bodied, with deeper bellies than their wild ancestors. Many show variable amount of white at base of bill, like the Greater White-fronted. When swimming, the domestic's rear end floats higher than rest of its body. Polytypic (2 sspp., N.A. records likely nominate *anser*). L 29–32" (74–81 cm)

Identification Wild-type Graylag is large, with thick neck, large head, heavy, pinkish orange to pink bill, and dull pink legs. The dumpy domestic Graylag is a very large, heavy-bodied, thick-necked goose. Domestic forms range from gray-brown to entirely white or multicolored. Hybridization between domestics and sometimes with Canada Goose is frequent, producing a wide array of structures and plumages. Presumed Canada x domestic Graylag often has a large, buffy cheek patch, white eye ring, brown crown and neck, and some white at base of bill, suggestive of Canada x Greater White-fronted hybrids.

Similar Species Wild form compared to juvenile White-fronted Goose is stockier, thicker necked, and grayer overall. Compare domestics carefully to Greater White-fronted Goose. Graylag is larger, stockier, with deep belly, often practically dragging on the ground. Domestic Graylag bills are usually larger and thicker at base than those of Greater White-fronted Goose. The Graylag is larger and stockier and lacks dark markings on bill of the two species of bean-geese or Pink-footed Goose.

Voice Loud, raw, deep *honks*. Wide repertoire of calls.

Status & Distribution Native to Europe (west to Iceland) and northern Asia. **RARE STATUS:** Accidental. Two records of single birds well off NL; records from NS, QC, CT, and RI are of unknown origin. Also recorded Greenland, Azores, and Canary Is.

Population Stable.

head, neck, and body uniformly gray-brown

large, thick, orangish bill

adult
anser

striking pale gray forewing visible in flight

GREATER WHITE-FRONTED GOOSE *Anser albifrons*   GWFG ▪ 1

Colloquially known as the "Specklebelly" due to variable black barring on adults' underparts, this is a medium-size grayish brown goose. Polytypic (5 sspp.; 4 in N.A.). L 28" (71 cm)

Identification Legs and feet orange, bill pink, more orange on *flavirostris*. **ADULT:** Brownish with a distinctive white band around the bill base, most obvious on the forehead. The underparts are brown with variable (individual and geographical) black barring from lower breast to upper belly; belly and vent are white. The flanks are bordered by a white line. Above brown

dark neck

orangish bill

"Greenland" adult
flavirostris

heavier black bars on underparts

long, attenuated, thick-based bill

dark head and long, thick dark neck

"Tule Goose" taiga adult
elgasi

thicker legs than *sponsa*

western AK *sponsa* is small and pale; birds breeding from northern AK (sometimes recognized as *gambelli*) are also pale, but larger

adult
sponsa

gray on upperwing

adult
sponsa

white rump band with dark gray tail base

high-pitched laughing calls heard in flight

immature
sponsa

bill color of immature varies from orangish to pinkish

some white at base of bill by end of first winter

juvenile
sponsa

fewer black bars (variable)

largely unmarked belly

with paler feather tips. **JUVENILE:** Similar to adult, but lacks white forehead and speckling on underparts; bill has a dark nail. **FLIGHT:** Brown showing a gray wash to the coverts and a narrow white rump band.

Geographic Variation Five subspecies recognized, three breeding in N.A. The two most distinctive subspecies are *flavirostris* and *elgasi* and could perhaps be treated as distinct species. The smaller *sponsa* (breeds in western AK and perhaps the Russian Far East; winters in Pacific states and western Mexico); larger, but similarly colored *gambelli* (breeds northern AK and northwest Canada, winters from southern Midwest to northeast Mexico). The declining "Greenland White-fronted Goose," *flavirostris* (breeds Greenland, winters primarily in Ireland and western Scotland, a few to Wales and rarely to northwestern Europe and northeastern N.A.), is barely larger than *gambelli*. It is much darker than *sponsa* and *gambelli* with an orange bill (can be pale orange or peach colored in *sponsa* and *gambelli*) and a blacker tail with a thinner whitish terminal band. It is blacker below, the black often coalescing on the belly into solid patches. Taiga-breeding *elgasi* breeds in wooded marshlands between Cook Inlet and the beginning of the AK Range, and winters mainly in the western Sacramento Valley, CA. It is the largest subspecies (averages 25 percent larger than *sponsa*) and has a thick-based and long attenuated bill. It has a long and thick neck. The head and neck are a dark brown, and this

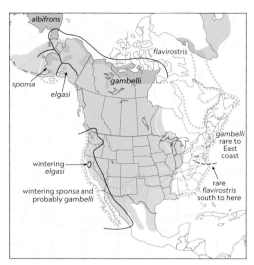

contrasts with a whitish belly. The dark belly barring is sparser than *sponsa*. Many have yellow eye rings, but other species can have pale eye rings too. Old World birds are assigned to nominate *albifrons*, the smallest and palest subspecies; northeast Asian populations are closer to *sponsa*. The nominate subspecies breeds in northern Eurasia.

Similar Species White front and barred belly are diagnostic in adult, although some adult Pink-footed Geese have limited white feathering at the bill base. Compare immature with other brown geese and with accidental Lesser White-fronted Goose.

Voice **CALL:** A laughing two- to three-syllable *ka-yaluk* or *kaj-lah-aluk*, somewhat grating and high-pitched. Calls are lower pitched in *elgasi*.

Status & Distribution Common, also in Eurasia. **BREEDING:** Tundra; *elgasi* breeds in more wooded wetlands. **MIGRATION:** Mid-continent populations move south from Aug.–Sept. to stopover areas from eastern AB to western MB; most depart by mid-Oct. for coastal TX, then to sites from Mexico to LA. Northward movements begin in late Jan., ending by early May; most arrive in central AK late Apr.–early May. Pacific Population of

central to southwestern AK, including "Tule Goose," fly south along coast eventually moving inland to stop in Klamath region of southern OR, northern CA in Sept.–Oct. More southern breeders and "Tule Goose" continue to winter in Central Valley of CA, while others continue to Mexico. **WINTER:** Wetlands, agricultural areas, short-grass fields; *elgasi* largely remains in wetlands, not going much like other subspecies to feed in grain fields. **RARE STATUS:** Rare east of range in migration and winter.

Population Generally increasing, Pacific populations declined strongly in the 1970s and the 1980s, but are regaining in numbers; Greenland *flavirostris* is declining (2011 population about 24,000). "Tule Goose" numbers estimated between 5,000 and 10,000.

LESSER WHITE-FRONTED GOOSE *Anser erythropus*   LWFG ▪ 5

This medium-size brown goose is accidental from the Palearctic. Monotypic. L 22–26" (55–66 cm)

Identification The Lesser White-fronted is extremely similar to the White-fronted Goose but is smaller and stockier in structure, shorter necked, and smaller billed, with longer and narrower wings. The wings extend beyond the tail when folded. The legs and feet are deep orange, bill is pink, orbital ring is yellow. **ADULT:** White "front" extends back onto the crown. Brown below with minimal black barring on the belly. **JUVENILE:** Similar to the adult, but fresh juveniles lack the white forehead and speckling on the underparts.

Similar Species Smaller, darker, longer winged, shorter necked, shorter

legged, and smaller billed than Greater White-fronted. In adults the extensive white on the forehead and the bright yellow orbital are diagnostic. Juveniles may be separated from the Greater White-fronted by darker body, size and structure, yellow orbital, and bill with pale nail.

Voice Yelping and high-pitched calls.

Status & Distribution Accidental, breeds from Scandinavia to the Russian Far East, wintering in a few midlatitude sites. **RARE STATUS:** One specimen from Attu I., AK (5 June 1994), and a recent record from St. Paul I., Pribilofs, AK (21 June 2013). Some records involve obvious escapes.

Population Vulnerable. There has been a significant decline over the

last century and a half with perhaps a 90 percent reduction, and a collapse of the breeding population in Scandinavia. The current world population is estimated at about 30,000.

conspicuous yellow orbital ring

stubby pink bill

folded wings extend past tip of tail

adult

TAIGA BEAN-GOOSE *Anser fabalis* TABG ■ 3

Taiga Bean-Goose of northern Eurasia has been recorded only a handful of times in N.A. It was formerly combined with Tundra Bean-Goose as a single species known as Bean Goose, *A. fabalis*. The recent (2007) split of Bean Goose has prompted extensive discussions about N.A. records of bean-geese, most of which have not yet been identified to species despite good photographs or even specimens. Polytypic (3 sspp.). L 30–35" (76–90 cm)

Identification A large brownish goose with long, almost swanlike neck and long, black bill with distinctive pale orange subterminal band. Legs orange. **ADULT:** Brown with head usually noticeably darker and white belly. **FLIGHT:** Coverts show gray cast, contrasting with

long sloped bill with little or no "grinning patch"

adults
middendorffii

overall coloration and marking like Tundra Bean-Goose but much larger and longer necked

much darker remiges; dark brown rump separated from dark rectrices by white coverts, recalling uppertail pattern of white-fronted geese.

Geographic Variation Subspecies size increases from west to east from smaller *fabalis* west of the Urals to largest *middendorffii* mainly in the Russian Far East. The westernmost subspecies, nominate *fabalis*, breeds from northern Scandinavia east to the Urals. The medium-size *johanseni* (sometimes synonymized with *fabalis*) breeds in Siberia from the Urals to Lake Baikal, and the large *middendorffii* breeds from Lake Baikal to the Russian Pacific coast and south to northern Mongolia and the Altay. Intergradation between subspecies apparently occurs, but more research is needed. Some authorities continue to treat Taiga and Tundra Bean-Geese as a single species; ongoing field and biochemical studies present a complex situation with no clear taxonomic solutions. It does appear that at least some populations of *middendorffii* are genetically the most distinct. They winter in Japan, but even there, wintering populations at different locations in western and eastern portions of the country differ by size likely reflecting different breeding grounds, the smaller ones breeding farther west. It might be better to just treat the bean-goose complex as a single polytypic species until there is a careful study on the breeding grounds across northern Siberia and the Russian Far East, a most difficult project to accomplish given the logistics. Although they are still treated by AOS as two species, many birds should simply just be identified as bean-geese.

Similar Species Although measurements overlap between the smallest Taiga Bean-Geese (*fabalis*) and the largest Tundra Bean-Geese (*serrirostris*), most Tundras are considerably smaller than Taigas, with much shorter necks, stubbier bills, and darker heads. Some, however, appear intermediate. Taiga typically shows longer bill with minimal or no "grinning patch," unlike Tundra, and usually has more extensive area of orange in bill, but this feature is highly variable. At least some Taiga (*middendorffii*) are paler headed than Tundra Bean-Geese. Bill color and pattern eliminate Graylag and Greater White-fronted Geese; pale belly (lacking speckles or bars) also unlike white-fronted geese.

Voice CALL: A nasal, sonorous, somewhat grating *gang gang*, which is usually lower in pitch and more resonant than Tundra.

Status & Distribution BREEDING: Breeds in taiga and tundra-forest from northern Scandinavia east to western Anadyrland. **WINTER:** From northeastern Europe to eastern China and Japan. **RARE STATUS:** AK has certain records from St. Paul I. (19 Apr. 1946), Shemya I. (3 birds; fall/winter 2007–2008), and Adak I. (May 2009), but other AK records of bean-geese are still being evaluated. Records of single bean-geese from IA/NE (1984–1985), QC (1987), NE (1998), and WA (2002) appear to be of Taiga, with some authorities identifying them as the largest subspecies *middendorffii*. A single bean-goose at the Salton Sea, CA, in fall 2010 was not identified conclusively.

Population All populations are in decline.

TUNDRA BEAN-GOOSE *Anser serrirostris* TUBG ■ 3

Tundra Bean-Goose is a smaller version of the closely related Taiga Bean-Goose and nests farther north, as its English name suggests. Polytypic (2 sspp.). L 28–33" (71–84 cm)

Identification A medium-size to moderately large brownish goose, comparable in size to *gambelli* Greater White-fronted Goose. Relatively compact appearance, a product of the short neck and short, stout bill (especially in smaller western subspecies *rossicus*). Bill blackish with pale orange or yellow subterminal band. Legs orange.

ADULT: Brown with head usually considerably darker and white belly. **JUVENILE:** As adult, but pattern in flanks and coverts more muted. **FLIGHT:** Impressions of plumage as described for Taiga Bean-Goose, but head often appears contrastingly darker, neck much shorter.

Geographic Variation The easterly nominate subspecies *serrirostris* breeds in the Russian Far East from Khatanga to Anadyrland and Kamchatka; it is relatively larger and longer billed than subspecies *rossicus*, which

breeds in areas west of the nominate, in northwestern Siberia from the Taymyr Peninsula west to the Kanin Peninsula.

Similar Species From Taiga Bean-Goose, note smaller size (a bit larger than Greater White-fronted), darker head and more stubby bill. Some (many?) encountered appear intermediate and those should simply be left as bean-geese. See Taiga Bean-Goose account for more details. Tundra Bean-Goose could be confused with Pink-footed Goose, but the latter

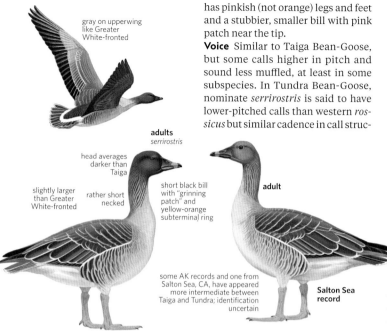

gray on upperwing like Greater White-fronted

adults
serrirostris

head averages darker than Taiga

slightly larger than Greater White-fronted

rather short necked

short black bill with "grinning patch" and yellow-orange subterminal ring

adult

some AK records and one from Salton Sea, CA, have appeared more intermediate between Taiga and Tundra; identification uncertain

Salton Sea record

has pinkish (not orange) legs and feet and a stubbier, smaller bill with pink patch near the tip.

Voice Similar to Taiga Bean-Goose, but some calls higher in pitch and sound less muffled, at least in some subspecies. In Tundra Bean-Goose, nominate *serrirostris* is said to have lower-pitched calls than western *rossicus* but similar cadence in call struc-ture. Obviously, more study needed.

Status & Distribution BREEDING: Tundra north to the Arctic across northern Russia. **WINTER:** From Europe to eastern China and Japan. **RARE STATUS:** In AK, where rare, this species is recorded mostly in spring, with verified specimens and photographs from Bering Sea and Aleutian Is. and probably the Seward Peninsula. Away from AK, there are apparent records from QC (possibly subspecies *rossicus*; 1982) and YT (2000). In autumn 2013, the first Tundra Bean-Goose for CA turned up at the Salton Sea in Oct. (at the site of the 2010 bean-goose), and first Tundra Bean-Goose for NS was found at Yarmouth in Nov. OR has recent confirmed winter records from the northwest coast and the Willamette Valley.

Population Western *rossicus*, the most numerous of all the bean-geese, appears to be stable. Eastern *serrirostris* appears to be declining.

PINK-FOOTED GOOSE *Anser brachyrhynchus* PFGO ■ 4

A midsize brownish gray goose similar in structure to the Greater Whitefronted. Pink-footed Goose is now nearly an annual visitor from Greenland to the Northeast. Monotypic. Although Greenland and Svalbard populations appear identical, mitochondrial DNA genetic studies show distinct differences. L 26" (66 cm)

Identification Legs and feet pink, bill black with extensive pink on terminal half. Juveniles may show dull or even orange tone to legs. **ADULT:** A brownish goose with darker head and white belly. Upperparts tipped white, creating neatly barred appearance. Upperparts show gray wash, contrasting with browner underparts. Tertials crisply fringed white. Flanks edged with narrow white line, which forms a separation with the folded wings. Hind flanks dark barred. **JUVENILE:** As adult, but lacking white edge to flanks and barring on flanks. Legs often orange tinged. **FLIGHT:** A brown goose showing obvious gray wash to the upperwings. The upperparts contrast with the narrow white rump band; the tail is brown with a comparatively wide white border.

Similar Species Juvenile Whitefronted Goose similar, but it has longer pink (or orange in *flavirostris*) bill lacking black base. In addition, Pink-footed Geese show obvious white fringes on the tertials. Also very similar to both bean-geese, but the Pink-footed has pink on bill and legs. Bean-geese are browner, lacking grayish wash to upperparts of most Pink-footed Geese. Furthermore, Pink-footed is much smaller than Taiga Bean-Goose with stockier shape; closer in size to Tundra Bean-Goose. In flight, Pink-footed Goose has a more prominent white edge to the tail and more gray on wings.

Voice CALL: A high-pitched *ayayak*.

Status & Distribution Rare, but with records increasing in northeastern N.A. Breeds in Greenland, Iceland, and Svalbard, winters in the British Isles and Low Countries of Europe. In N.A., casual to Atlantic Canada and northeastern US south to the mid-Atlantic states. Records from upper Midwest, CO, and BC are of uncertain origin.

Population Both the Svalbard and the larger Greenland populations are increasing, a trend likely reflected in the more numerous records from northeastern N.A.

more extensive bluegray on upperwing than Greater White-fronted

small, stubby, dark-based bill with pinkish band

comparatively small head, dark neck

some have white rim at bill base

warm cinnamon-buff tinge at base of neck

light silvery or "frosty" gray upperparts on most adults; some are browner

broad white tip to tail

broad white tail tip with paler gray base

pinkish legs and feet

CANADA GOOSE AND ALLIES Genus *Branta*

This is a genus of six species found throughout the Northern Hemisphere, including HI. The birds have distinctive patterns on the face, neck, or breast. Most show a black neck "sock," with contrasting white face or neck patches. All have black bills and legs. Several are high-Arctic breeders; most forage on short grass during the nonbreeding season.

BRANT *Branta bernicla*   BRAN   ■ 1

A sea goose that forages largely on marine grasses, the Brant is known as the "Brent Goose" in the UK. Polytypic (3 sspp.; 2 in N.A.). L 25" (64 cm)
Identification A smallish, short-billed, short-necked stocky goose. Bill and legs black. **ADULT:** All subspecies show a black head, neck, and breast; a white neck collar varies in extent geographically. Note contrasting white rump and vent and dark tail. Underparts variable depending on subspecies, varying from pale brown to blackish brown with contrasting white flanks. **JUVENILE:** Similar to adult but lacks or largely lacks neck collar and shows white tips to wing coverts.

Geographic Variation There are three named subspecies—nominate *bernicla*, "Dark-bellied Brant," breeding in western and central Russian Arctic and wintering in western Europe; *hrota*, "American Brant" (also known as "Pale-bellied Brant"), breeding in eastern Arctic of N.A., Greenland, and Svalbard and wintering on the Atlantic coast of US and Europe (mainly Ireland and Denmark); and *nigricans*, "Black Brant" (includes *orientalis*, recognized by some authorities), breeding from eastern Russia through western Arctic N.A. to Victoria I., NU, and wintering on Pacific coast south to Mexico. There is a fourth, as yet unrecognized population known as "Gray-bellied Brant," breeding in the Arctic Archipelago, mainly on Melville and Prince Patrick Is. and wintering in the Puget Sound area and Boundary Bay, BC. The "American Brant" has a small neck collar, divided at front, pale underparts contrasting with black neck "sock," and white belly behind legs. "Black Brant" is darker above and below. It

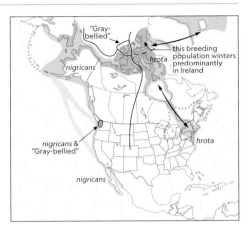

shows little contrast between the dark underparts and the neck sock; pale flanks contrast with the dark underparts; the neck collar is broad and full and meets in the front of the neck. "Dark-bellied Brant" is dark below, but paler than "Black Brant," lacks the contrasting white flanks, and has a small and broken neck collar. "Gray-bellied

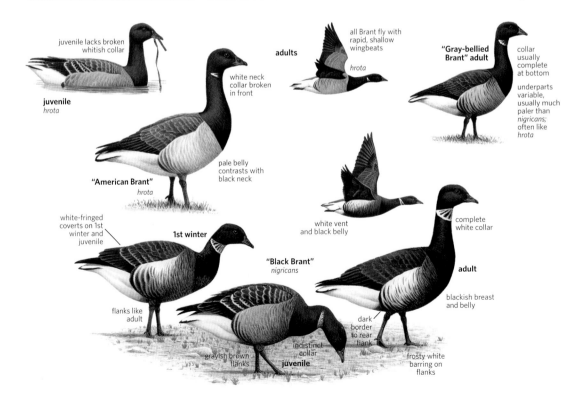

juvenile lacks broken whitish collar

juvenile
hrota

adults

all Brant fly with rapid, shallow wingbeats

hrota

"Gray-bellied Brant" adult

collar usually complete at bottom

underparts variable, usually much paler than *nigricans*; often like *hrota*

white neck collar broken in front

"American Brant"
hrota

pale belly contrasts with black neck

white-fringed coverts on 1st winter and juvenile

1st winter

white vent and black belly

complete white collar

adult

"Black Brant"
nigricans

blackish breast and belly

flanks like adult

grayish brown flanks

indistinct collar

juvenile

dark border to rear flank

frosty white barring on flanks

Brant" appears to be variable, ranging from pale birds similar to "American Brant," to birds with gray underparts and contrasting paler flanks; they have smaller, broken neck collars than "Black Brant."

Similar Species Entirely black head, neck, and breast; white collar diagnostic. Cackling and Canada Geese have white face patch.

Voice CALL: A quavering *crrr-oonkkk.*

Status & Distribution Common. **BREEDING:** All subspecies breed in salt marshes in low Arctic and moist tundra in high Arctic. **MIGRATION:** Highly migratory. **"AMERICAN BRANT":** Most migrate to James Bay, ON, in Sept. and fly nonstop to coastal NY and NJ, arriving in late Oct.–Nov. They move north in Apr. Some use a coastal route to the St. Lawrence River Estuary; others move inland over Lake Ontario and up the Ottawa River Valley (late May). Both routes converge in James Bay, where they stage before heading to the breeding grounds by early June. Historically more birds moved up the coast, with important staging areas in the Maritimes; these routes have shifted westward. The population breeding in the northeast Arctic Archipelago leaves the breeding grounds in late Aug. and stages in western Greenland in Sept. before continuing to Iceland (mid-Sept.) and wintering sites mainly in Ireland by late Sept. or Oct. Spring migration begins in late Apr.; they stage for several weeks in Iceland and fly over the Greenland ice cap directly to breeding grounds by mid-June. Svalbard breeders mainly winter in Denmark. **"BLACK BRANT":** N.A. breeders make their way from the breeding areas to stage at Izembek Lagoon on the AK Peninsula as early as mid-Sept. Victoria I. breeders fly a coastal route around AK and arrive at the staging grounds by mid-Oct. They then fly directly to Haida Gwaii and the southern BC coast before continuing south, well off the Pacific coast to wintering sites, arriving late Oct.–Nov. Most Russian breeders winter along the coasts of northern China, Korea, and Japan. Northbound migration in N.A. peaks in late Mar. and Apr. along CA coast, with staging areas on east coast of Vancouver I. before flight to AK Peninsula, peaking there in mid-May, and arriving at breeding grounds mid–late June. **WINTER:** Coastal, primarily in large estuaries where beds of eelgrass and other intertidal plants are available for forage. **RARE STATUS:** "American Brant" is casual in the eastern interior, away from eastern Great Lakes. "Black Brant" is casual in the interior of the West, except at Salton Sea where regular in spring, and some summer; also casual in winter on mid-Atlantic coast. "Dark-bellied Brant" has been reported but not conclusively documented from the East; it should be looked for in wintering concentrations of "American Brant."

Population "Black Brant" populations appear to be stable; "American Brant" have fluctuated markedly and are now stable or increasing. The "Gray-bellied Brant" population is small and of conservation concern.

BARNACLE GOOSE *Branta leucopsis*   BARG ▪ 4

Before bird migration was accepted, legends told that when Barnacle Geese disappeared from wintering sites in western Europe, they turned into barnacles. Monotypic. L 27" (69 cm)

Identification A small, short-necked goose with a stubby bill. Bill and legs black. **ADULT:** White face and black lores contrast with black neck and breast. Underparts white, gray barring on flanks. Upperparts gray, with scaled appearance created by dark subterminal band and white terminal band on each feather. **FLIGHT:** Gray wings and back, with black neck; lower back and tail contrasting with white rump band.

Similar Species Distinguishable from Cackling Goose and Brant by white

usually seen with Canada Geese

bluish gray upperparts

barred upperparts

white face

sharp contrast between breast and belly

underparts and grayish wings and back; white face a further distinction from Brant. Hybrids with Cackling or Canada geese are known.

Voice CALL: A barking *kaw.*

Status & Distribution Breeds on tundra in eastern Greenland, Svalbard, and Novaya Zemlya; winters largely in Ireland and Scotland and the Netherlands. **RARE STATUS:** Numerous sightings throughout continent, mostly from the East; many of these, especially from coastal regions, are likely of wild origin; others, particularly well away from East Coast, are more likely escapes. Most records from the East Coast are from Atlantic Canada south to the mid-Atlantic region from Oct. to Apr.; most are mixed with other geese, especially Canada Geese, and some involve family groups.

Population Has rapidly increased in last couple of decades. The Greenland breeding population (including some from Iceland) was estimated in the early 2000s at 70,500 birds.

CACKLING GOOSE *Branta hutchinsii* CACG ■ 1

The Cackling Goose was split from the Canada Goose in 2004. Polytypic (4 sspp.; all in N.A.). L 23–33" (58–84 cm). **Identification** A small, stubby-billed goose with short legs, stocky body, and short neck. Long winged, with primaries extending noticeably past the tail. **ADULT:** Black neck and head with a contrasting white cheek patch that wraps around the throat. Often shows a white neck ring.

Geographic Variation Four subspecies recognized. "Richardson's Goose" (*hutchinsii*) breeds from Mackenzie Delta, Northwest Territories, east to western Baffin I. and south to Southampton I. and McConnell River, Hudson Bay, NU; may breed in western Greenland, winters from eastern NM and northern TX south into highlands of Mexico, also coastal TX to western LA and south to northern Veracruz, Mexico. "Taverner's Goose" (*taverneri*) breeds in coastal wetlands of Seward Peninsula and North Slope, AK, and winters in Columbia River Valley, WA and OR, south to the Central Valley of CA. "Cackling Goose" (*minima*) breeds in the Yukon-Kuskokwim Delta, AK, and winters largely from Willamette Valley, OR, south to Central Valley of CA, previously wintered largely in CA but has shifted north; "Aleutian Goose" (*leucopareia*) breeds on Aleutian and Semidi Is., AK, and winters in Central Valley, CA, but expanding. "Richardson's" is the palest, with a pale breast and intermediate in size but with a disproportionately long bill. Dark-breasted *minima* is the smallest, with a stubby short bill. "Taverner's" is the largest, with a longer neck and rounder head; intermediate in darkness, but variable. "Aleutian" is larger than *minima*, but smaller than "Taverner's"; a moderately dark form; it typically has a complete white neck collar.

Similar Species Due to the split of Cackling and Canada Geese, we have a vexing problem in field identification, mainly that the largest Cackling Goose (subspecies *taverneri*) is close in size to the smallest Canada Goose (subspecies *parvipes*). In the past, *taverneri* and *parvipes* were grouped together as a single subspecies; however, recent studies clearly show them to be genetically different. On average, *parvipes* Canada is larger than *taverneri* Cackling and has a relatively longer neck and longer bill. In general, the Cackling Goose has longer and more pointed wings than the Canada Goose, and this difference shows up as a longer primary extension (past tertials) and wing extension (past tail) in the Cackling. The wintering range of *taverneri* is not clearly known, but it is not expected to be regular east of the Cascade Range/Sierra Nevada. So, a troubling intermediate goose in the East is more likely *parvipes*, which migrates and winters largely east of the mountains. The dark-bodied *parvipes* from Anchorage, AK, winters in the Pacific Northwest.

Voice CALL: Higher pitched and more cackling than Canada Goose.

Status & Distribution Common. **BREEDING:** Moist coastal tundra. **MIGRATION:** Richardson's flies south from late Aug., staging in central and eastern Prairie Provinces, peaking in Oct. Northward movements begin in Feb.–Mar. "Cackling" (*minima*) and "Aleutian" stage in AK Peninsula into Oct., then perform nonstop flight (two to four days) to OR and CA; northbound migration continues until early May. **WINTER:** Short-grass fields, including agricultural areas and wetlands. **RARE STATUS:** Rare away from regular migratory areas.

Population Subspecies *minima* is stable; *taverneri* thought to be declining; *hutchinsii* has increased; *leucopareia* has increased dramatically over the last four decades and has been removed from the endangered species list.

"Aleutian" *leucopareia*

"Cackling" *minima*

"Richardson's" *hutchinsii*

flies with rapid, shallow wingbeats

short neck especially noticeable in flight

calls higher pitched than Canada

rounded head

short neck

stubby bill

"Richardson's" *hutchinsii*

overall pale

like *parvipes* Canada but with rounder head and stubbier bill

"Taverner's" *taverneri*

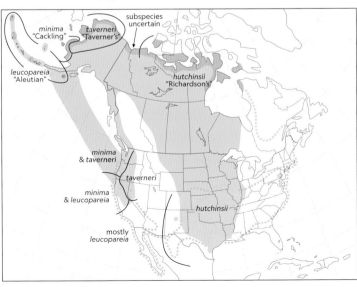

minima "Cackling"

taverneri "Taverner's"

subspecies uncertain

leucopareia "Aleutian"

hutchinsii "Richardson's"

minima & taverneri

taverneri

minima & leucopareia

hutchinsii

mostly leucopareia

CANADA GOOSE *Branta canadensis*   CANG ■ 1

The "honker" is the common goose in most of N.A. Polytypic (7 sspp.; all in N.A.). L 30–43" (76–109 cm)

Identification A large and long-necked goose; legs and bill black. **ADULT:** Black neck and head with a contrasting white cheek patch. Body brown, paler below, often darker on rear flanks. Belly and vent white. **FLIGHT:** Brown above including wings, lower back blackish, as is the tail, contrasting with a white rump band.

Geographic Variation Seven subspecies in N.A. Subspecies *canadensis* breeds Ungava Bay east to Newfoundland, wintering on Atlantic seaboard; *interior* breeds west of Ungava Bay through Hudson Bay lowlands to northern MB (Manitoba) and north to southern Baffin I. and southwestern Greenland, wintering throughout the East; *maxima* ("Giant Canada Goose") breeds central MB and MN south to KS and western KY some resident, others short-distance migrants; *moffitti* breeds south-central BC to western MB south to CO and OK, wintering southern portion of breeding range to Southern CA, northern Mexico, and TX; *parvipes* ("Lesser Canada Goose") breeds in boreal forest zone

from Cook Inlet and central AK to northwestern Hudson Bay, wintering eastern WA and eastern OR to northeastern Mexico and eastern TX; *fulva* ("Vancouver Canada Goose") breeds in Copper River Delta and Prince William Sound, AK, wintering in Willamette Valley, OR; *occidentalis* ("Dusky Canada Goose") breeds from Glacier Bay, AK, to northern Vancouver I., many resident, others winter in Willamette and Columbia Valleys, OR. These subspecies intergrade to various extents and can be thought of as fitting three general groups. The subspecies *canadensis*, *interior*, *maxima*, and *moffitti* are very similar. The "Lesser Canada Goose" (*parvipes*) varies in size, but on average is smaller than all other Canada Geese. The "Vancouver" and "Dusky Canada Geese" are very dark and color saturated, the "Dusky" being smaller than the "Vancouver."

Similar Species See Cackling Goose.

Voice CALL: Males give a lower pitched *hwonk*, females a higher *hrink*.

Status & Distribution Abundant, the birds have been introduced to western Europe. **BREEDING:** Various freshwater wetlands, golf courses. **MIGRATION:** Complex due to number of different populations and staging areas; however, most wild populations of the Canada Goose are migratory. In modern times feral Canada Geese have been introduced in various parts of the continent, many of these being primarily stocks descended from mixes of "Giant Canada Goose" and others. Many of these birds are residents in urban and suburban areas, or they perform only minor migrations. Migrant geese tend to leave breeding grounds in Aug.–Sept., peak in Oct., and arrive at wintering areas from mid-Oct.–Nov. Spring movements begin in Feb., peaking in Mar. **WINTER:** Various grassy habitats, from urban parks and golf courses to native wetlands; also agricultural fields.

Population This species has dramatically increased in number since the 1940s. Some subspecies however (*canadensis*, interior, and *occidentalis*) have declined. It is now estimated that nearly six million Canada Geese (mainly *moffitti* and *maxima*) live in N.A., and the birds are considered pests in some regions.

"Lesser Canada Goose,"
B. c. parvipes

long, black neck

honking calls lower pitched than Cackling

"Atlantic"
canadensis

flies with slower, deeper wingbeats than Cackling

smallest Canada; compare to *taverneri* Cackling

"Lesser"
parvipes

long, black neck

"Dusky"
occidentalis

darkest Canada Goose

moffitti

widespread, large western subspecies

SWANS Genus *Cygnus*

Of the seven species of swans, six constitute the genus *Cygnus*. The genus has a worldwide distribution, but only in temperate regions. All *Cygnus* are huge birds, with long necks, long bills, and strong legs. Southern Hemisphere species are black or black necked, but all N.A. species are entirely white, differing from each other in subtle structural and bill pattern variations.

TRUMPETER SWAN *Cygnus buccinator* TRUS ■ 1

The long-necked Trumpeter is our native swan of forested habitats. Monotypic. L 60" (152 cm)
Identification Huge swan, with a long, sloping head and bill profile, as well as a relatively thick-based neck. The bare loral skin is as wide as the eye and narrowly encircles the eye. Shape of upper edge of bill, where it meets forehead, comes forward to create a V-shaped point along the bill's midline. **ADULT:** White, often stained yellowish on head and upper neck. Bill black with orange stripe on lower mandible along cutting edge. Legs black. **JUVENILE:** Appears dirty, pale brownish gray throughout. Bill dull pink with dark base, dark loral skin, and dark on tip and cutting edge.
Similar Species Most likely to be confused with the Tundra Swan. The

Trumpeter is larger, has a longer, sloping bill with a V-shaped upper edge. The upper mandible of the Trumpeter is always black, but a few Tundras may lack yellow bill spot. The loral skin of the Trumpeter is wide, but narrow on the Tundra.
Voice CALL: A bugling *oh-OH* like an old car horn, second syllable emphasized.
Status & Distribution Uncommon to rare, currently being stocked in various states and provinces in the East. **BREEDING:** Various freshwater wetlands, at least 300 ft long, needed for takeoff. **MIGRATION AND WINTER:** Migrants leave north by mid-Oct., arriving in south by early Nov. or later; northbound movements begin late Feb., arriving on breeding grounds

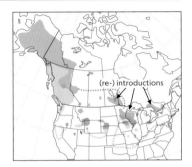

(re-) introductions

from Apr. AK population winters mainly in coastal BC and western WA. YT and NT population migrates largely east of Rockies to winter in tristate area of MT, WY, and ID. Southern breeding populations resident or make only local movements. Newly stocked populations in central and eastern N.A. largely resident, but some move as far as AR and the mid-Atlantic region. Winter birds are most often found in freshwater wetlands and grassy and agricultural fields near water bodies. **RARE STATUS:** Rare to CA; casual to AZ, NM, and TX. Some strays likely involve escapes.
Population Historically suffered a huge population decrease, but conservation efforts have allowed native western populations to increase. A somewhat controversial reintroduction to the central part of the continent has been successful, although former range in eastern N.A. is very unclear.

white forms V-shape on forehead; black facial skin includes eye

retains darker plumage later in winter than Tundra

adult

juvenile

TUNDRA SWAN *Cygnus columbianus* TUSW ■ 1

The Tundra is the widespread and more highly migratory swan in the continent. Polytypic (2 sspp.; both in N.A.). L 52" (132 cm)
Identification Large, shorter-necked swan with rounded forehead, clearly setting off bill profile. The bare skin on the lores narrows to a point before the eye. Upper edge of bill, where it meets forehead is smoothly curved—shallow U shape when seen from front.
ADULT: White throughout. Bill black

with variable yellow patch at bill base, extending forward from eye; patch rarely absent. Yellowish stripe on lower mandible along cutting edge. Legs black. **JUVENILE:** Dull whitish gray throughout, but some become largely white by midwinter. Bill pink to base, including loral skin; this darkens from base outward as bird ages. Obvious black nostril on pink bill.
Geographic Variation Subspecies *columbianus* widespread, *bewickii*

("Bewick's Swan") of Asia and Europe is rare to western N.A. It is differentiated from *columbianus* by having much more extensive yellow on bill base, covering more than a third of the bill. Hybrid family groups have been seen in N.A.
Similar Species Most likely to be confused with the larger Trumpeter Swan and identification often difficult, especially on silent birds at distance. Tundra has shorter bill, steeper forehead,

and rounder crown. When observed from the front, the upper edge to the bill, where it meets the forehead, is shaped like a shallow U, not the sharp V of the Trumpeter. Yellow at bill base is characteristic of Tundra, and absent on Trumpeter; the loral skin of Tundra pinches in before the eye. Immatures begin with a pink base to bill and quickly become much whiter-plumaged than most Trumpeters.

Voice CALL: A loud barking and somewhat gooselike *kwooo*.

Status & Distribution Common. **BREEDING:** Tundra lakes and ponds. **MIGRATION & WINTER:** Western popu-lation stages on Great Salt Lake, then moves to more coastal wintering sites. Eastern population stages from ND to MN and flies to mid-Atlantic coast wintering areas. Staging areas used in late Oct., arrival in wintering sites by late Oct. to mid-Nov. Northbound by late Feb. and early Mar. Eastern population stages around Lake Erie, continuing to MN and Canadian prairies, arrives in Arctic breeding grounds by mid-May. Western population retraces route through Great Salt Lake, or through Klamath region, arrives in breeding grounds of AK in Apr. Winter birds and migrants usually found in wetland habitats and agricultural fields.

RARE STATUS: Casual to Maritimes and Gulf Coast; rare throughout interior away from regular migration routes. **Population** Appears stable now, but is thought to have doubled between the 1960s and the 1990s.

white on forehead forms a V

white on forehead cuts more straight across

bill has more concave shape than Trumpeter

eye stands out

juvenile
columbianus

size of yellow lore spot variable, absent on some

extensive yellow squared off at base of bill

"Bewick's Swan"
adult
bewickii

adult
columbianus

Trumpeter   **Tundra**

WHOOPER SWAN *Cygnus cygnus*   WHOS ■ 3

This Old World species occurs in N.A., mainly on the Aleutians in winter. Monotypic. L 60" (152 cm)

Identification Structurally similar to Trumpeter Swan. Bill long; sloping forehead accentuates bill length. Adults have extensive yellow on bill, at least to the level of the nostril. Much of the underside of the lower mandible is also yellow. The shape of the yellow comes forward to a point. **ADULT:** Body entirely white, sometimes stained yellowish on head and neck. **JUVENILE:** Body grayish brown throughout, bill whitish gray shaped like yellow pattern of adult; nostril, cutting edge, and nail black.

Similar Species In shape and size resembles Trumpeter Swan, although large amount of yellow on bill easily separates the Whooper. Tundra Swan smaller, with shorter bill and steeper forehead as well as small yellow patch

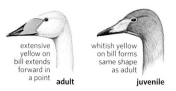

extensive yellow on bill extends forward in a point   **adult**

whitish yellow on bill forms same shape as adult   **juvenile**

on bill base. However, *bewickii* subspecies, "Bewick's Swan," has more extensive yellow on bill, more like Whooper Swan. Note that yellow on Whooper comes forward to a point and extends past nostril; on "Bewick's" yellow ends more abruptly and does not reach the nostril. Immatures have a bill pattern that suggests that of the adult, helping to identify them.

Voice CALL: Bugling; often gives three or four calls in a group *kloo-kloo-kloo*.

Status & Distribution Widely distributed from Europe (including Iceland) to Asia, breeding in higher latitudes, wintering at midlatitudes. **BREEDING:**

Has bred on Attu I. **WINTER:** Large wetlands and agricultural fields. **RARE STATUS:** Rare in winter on central and outer Aleutians and casual elsewhere in Bering Sea region. Casual south to OR and northern CA. Many records in eastern N.A. likely pertain to escapes. **Population** Increasing in Europe; declining in Siberia; no information for East Asia population.

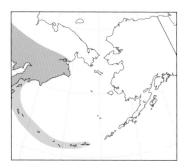

MUTE SWAN *Cygnus olor*   MUSW ▪ 1

This is the quintessential swan, a species introduced from Europe no doubt to give our local parks and ponds a royal touch. However, this species is a great ecological problem here. Originally native to central Asia; widely introduced to Europe, as early as the 12th century in the UK. Monotypic. L 60" (152 cm)

Identification A large, long-necked swan that holds its neck in a characteristic S-shape. The wings are often held partially raised, arched over the back. The bill has a prominent black knob at the base, larger in males than females. **ADULT:** Entirely white body. Bill bright orange to reddish with black base, cutting edge, and nail, as well as black bulbous knob at base of culmen. Bill color of females duller. Legs black. **JUVENILE:** Entirely brownish gray; bill gray with black base and black around nostril. Older immature becomes progressively whiter, and bill becomes pale pink; black knob slowly enlarges as immature ages. The pale morph ("Polish Swan") with paler coloration and duller soft parts of juveniles is also found in N.A.

Similar Species Both Tundra and Trumpeter Swans may be found in established range of the Mute Swan; black-and-orange bill with black knob diagnostic for Mute Swan. Immatures of all swan species more similar; Mute has narrow black base to bill; bill begins grayish and turns pinkish orange later on in life, by which time growth of bill knob has started. Swans in city parks are likely Mute Swan, although reintroduced Trumpeter Swans in the East may also frequent urban parks.

Voice CALL: As its name suggests, this swan is usually silent. However, it is not mute. Usually calls heard are hisses and other soft alarm calls. Rarely it gives a resonant bugle.

Status & Distribution Common, introduced from Europe; native to central Asia. **YEAR-ROUND:** From small urban ponds and lakes to large wetlands. **WINTER:** May flock together in larger bodies of water, including the Great Lakes. **RARE STATUS:** Rare in MS watershed and the southern Atlantic coast; at least some of these birds are dispersing from more northerly established sites. Small to moderate populations now locally established in Pacific states. Most birds elsewhere are either feral or local escapes.

Population Booming in northeastern states, with substantial increases in Chesapeake Bay population and continuing expansion of the range particularly to New England. Other populations variable, depending on management efforts. Mute Swans can greatly alter the ecology of wetlands by uprooting emergent vegetation as they feed. Most Old World populations are increasing.

dark border at base of bill

dark knob

sometimes arches wings

adult

juvenile

darker than other juvenile swans

long tail

> Genera *Alopochen* and *Tadorna*

EGYPTIAN GOOSE *Alopochen aegyptiacus*   EGGO ▪ 2 ▪ EXOTIC

This native of tropical Africa south of the Sahara is usually found inland on freshwater. It feeds by grazing but also dabbles, swimming with rear end floating higher than rest of the body. Monotypic. L 27" (68 cm)

Identification ADULT: Head and neck buffy with dark brown patch around eye, base of bill, and breast. Body buffy to gray-brown, paler on flanks, whitish belly. **JUVENILE:** Duller than adult, with dark crown and hind neck, lacking brown patch around eye and on breast. **FLIGHT:** Strong and quick flier with slow wingbeats, white wing coverts, green to purple speculum.

Similar Species Likely to be confused only with Ruddy Shelduck (p. 52), which is smaller, shorter legged, with orange-chestnut body, and no dark eye patch.

Voice Male gives a harsh, breathing sound; female a loud, harsh quacking.

Status & Distribution An established, introduced population is found in southeastern FL; becoming established in southwestern Los Angeles Co. and Orange Co., CA; sightings

dark reddish around eye

strikingly white wing coverts

adults

brown spot on breast

in other locations are considered escapes. Historic records of wild strays in the Middle East and north in Europe to the Danube. Introduced into western Europe in the late 1960s and now widely established (tens of thousands), including the UK. Introduced populations also locally present in the Middle East (Israel and UAE). Control measures are used in western Europe, less so in the Middle East and N.A. **Population** Apparently stable in most of native African range. Feral birds greatly increasing in northwest Europe and N.A.

COMMON SHELDUCK *Tadorna tadorna*   COMS ▪ 2 ▪ EXOTIC

This native to Europe, Asia, and northern Africa is often known as the Shelduck. Commonly kept in collections and reported as an escape in N.A., Common Shelduck is as at home on land as it is in water; favors salt or brackish water. Monotypic. L 25" (64 cm)
Identification ADULT: Red bill; green head; broad chestnut band across white breast and black center to belly. Sexes are similar, but much smaller females have a duller red bill with dark tip, lack the knob at the base of the culmen, lack the green head sheen, and have variable whitish on the anterior portion of the face and chest. **JUVENILE:** White anterior portion to face and brown scapulars; lacks chestnut band on breast; soft parts pinkish. **FLIGHT:** White wing coverts prominent.
Similar Species Generally distinctive.
Voice Loud calls. Males give musical whistles in flight and perched females give a low and nasal *ak-ak-ak-ak*.
Status & Distribution Abundant in western Europe; casual in N.A. Large numbers (100,000) gather in huge flocks to molt on the North Sea mudflats at Helgoland Bight, Germany. Records have substantially increased from Iceland, perhaps explaining the scattering of records from Atlantic Canada and New England. Scattered records elsewhere from N.A. are more problematic, especially well away from Atlantic coast.
Population Stable.

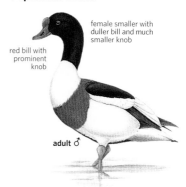

female smaller with duller bill and much smaller knob

red bill with prominent knob

adult ♂

PERCHING DUCKS Genera *Cairina* and *Aix*

These genera are represented by three species worldwide. One species from each genus is present in N.A., the Muscovy (*Cairina*) and the Wood (*Aix*). Both are at home in slow-moving waters surrounded by large trees. They nest in cavities or use nest boxes and have dark glossy plumage. Otherwise, the two species are very different.

MUSCOVY DUCK *Cairina moschata*   MUDU ▪ 2

This large, widespread, tropical duck is found in the wild near Falcon Dam in southern TX. Wild birds are shy; most active at dusk and dawn. Monotypic. L 26–33" (66–84 cm)
Identification Adults blackish with iridescent greenish and purple highlights. **MALE:** Glossier than females, with dark reddish knob above bill. **FEMALE:** Duller and smaller; lacks knob on bill. **JUVENILE:** Duller, slowly acquires white wing patch during first winter. **FLIGHT:** Appears massive, with very broad wings with large white patches.
Similar Species Domestic Muscovies can vary from having white blotches on head and body to being almost entirely white.
Status & Distribution Domestic birds are in parks throughout N.A. and are now widely established and locally common in the southern half of FL. Locally uncommon to common from Mexico to northern half of S.A. **YEAR-ROUND:** Wild birds present near Falcon Dam; uncommon, and difficult to find. Most active shortly after dawn and before dusk when most often seen in flight. Populations downriver are believed to consist largely, if not entirely, of feral birds. **BREEDING:** Nests in holes, hollow trees, or earth banks.
Voice Generally silent.
Population Declining over much of range due to hunting; nest box program in northern Mexico is believed to have led to increases and enabled them to reach the Rio Grande in the vicinity of Falcon Dam, south TX.

introduced

knob

slow gooselike flight

female much smaller than male

juvenile ♀

highly variable

domestic variety ♂

extensive white underwing

large white wing patch

adult ♂

long tail

adult ♂

WOOD DUCK *Aix sponsa*   WODU ◼ 1

Wood Ducks are usually found in heavily wooded swamps, often detected by the loud squeal of the female when taking flight. Monotypic. L 18" (47 cm) **Identification** Both sexes have a bushy crest that makes the head appear large and rounded. **MALE:** Stunning, intricate plumage is distinctive. **ALTERNATE MALE:** As female, but retains distinctive bill, eye, colors, and throat pattern. **FEMALE:** Distinguished by large white eye-patch, white throat, and head shape. **JUVENILE:** Like female but duller, with spotted belly. **FLIGHT:** Dark wings with white trailing edge; uplifted head, broad wings, and long rectangular tail create distinctive flight profile.

Similar Species The female is similar to female Mandarin (p. 52), a closely related Asian species common in waterfowl collections and parks, which lacks eye-patch of the Wood Duck.

Voice CALL: Female flight call a rising squeal, *ooEEK*. Male infrequently gives a high, thin *jeeee*.

Status & Distribution Widespread and common, generally in low densities. Feral populations established in a number of areas (e.g., southwestern CA). **BREEDING:** Nests in cavities in hollow trees or nest boxes in wet woods. **MIGRATION:** Spring: Begins moving north in Feb.; peaking in New England in late Mar.; in the Great Lakes in mid-Apr. Fall: Many Wood Ducks disperse widely after breeding and prior to fall migration. Most begin moving south in late Sept.; numbers in southern states increase steadily through Dec. **WINTER:** Found in bottomland forests and swamps. Rather rare in much of Southwest. Rare to casual in southeastern AK.

Population Currently fairly stable, expanding range across northern plains in the last 50 years.

white around and behind eye

♀

dark wings

♂

longish tail

♀

juvenile ♂

♂

unique pattern

Genera *Sibirionetta* and *Spatula*

BAIKAL TEAL *Sibirionetta formosa*   BATE ◼ 4

The Baikal Teal is an East Asian species only slightly larger than the Green-winged Teal. Most Baikal Teal winter in eastern China and South Korea. The bird favors several lakes, where counts can reach hundreds of thousands. Although Baikal Teal is rare in captivity, most sightings, especially away from the West, are possibly escapes. Monotypic. L 17" (43 cm) **Identification** Small dabbler with a disproportionately large, square head and small bill. **MALE:** Its intricate face pattern and long, ornate scapulars are exquisite. **FEMALE:** Tawny brown with two short stripes on the face; pale loral spot with a brown outline and an extensive pale throat that extends up onto the face.

Similar Species Female Baikal is similar to female Green-winged Teal; look for the pale throat, face pattern, and a loral spot stronger than on most Green-winged Teal. However, some female Green-wingeds have strongly patterned faces that appear more similar to that of the Baikal. The upperwing pattern is much like the Green-winged, but the narrow cinnamon-buff upper border to the green speculum is straight, not wedge shaped; the white trailing edge is broader than on the Green-winged; and the underwing is more extensively gray.

Voice Male frequently gives a deep, repeated *wot-wot-wot*. Female gives a soft *quack*.

Status & Distribution Locally common in East Asia. **RARE STATUS:** Casual spring (May–June) and very rare fall (Sept.–Oct.) migrant to western and northern AK. Fall migrants recorded on Seward Peninsula, St. Lawrence I., western Aleutians, and Pribilof Is. Casual in winter to Pacific states; accidental MT and AZ.

Population Recent increases followed sharp declines in the 1960s and the 1970s. The Korean winter population was estimated at over one million in 2009 with declines thereafter. The scattering of recent records in AK and elsewhere in the West likely reflects the overall recent population increase.

white on throat sweeps up to eye

dark crown

prominent white spot bordered with dark

♀

green on speculum usually hidden on swimming or standing bird

distinctive head pattern

♂

pale sides to undertail coverts

long rufous-edged scapulars

more extensive dark on leading edge of underwing than Green-winged

BLUE-WINGED TEAL *Spatula discors*   BWTE ■ 1

This small duck is largely a summer resident. Monotypic. L 15" (39 cm) **Identification** Both sexes have all-black bills and a pale area at the base of the bill that is crescent shaped in males and oval shaped in females. **MALE:** Violet-gray head with large white facial crescent and white flank patch; the body is bronze, covered with black spots. **FEMALE:** Grayish brown with paler face, a distinct dark eye stripe, white eye arcs, and a pale loral spot.

FLIGHT: Powder blue wing coverts and green speculums are visible; the wingbeats are rapid. Male has a broad white border to blue forewing. **Similar Species** Larger and longer billed than Green-winged, with a pale loral spot; lacks whitish streak on sides of undertail coverts and longer bill. **Voice** Males give a high, whistled *peew*; females give a shrill *quack*. **Status & Distribution BREEDING:** Nests in grassy clumps near small ponds and small flooded fields. Most dense in the prairie pothole region, where it is the most abundant duck. **MIGRATION:** Relative to other waterfowl, is a late spring and early fall migrant. **SPRING:** Begin arriving along the Gulf Coast in Feb.; peaking in Apr. in the Midwest; migration

is largely over by the end of May. **FALL:** Peaks in the Midwest and Southwest in late Sept.–early Oct.; peaks in the Gulf states in early Oct., but quickly declines later in the month. **WINTER:** The majority of birds winter in S.A. At all seasons uncommon and somewhat local on West Coast. Rare in British Isles; casual from western Europe and northwestern Africa. **Population** Stable.

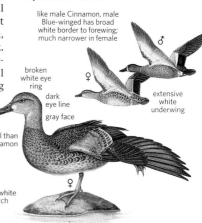

like male Cinnamon, male Blue-winged has broad white border to forewing; much narrower in female

♂

♀

broken white eye ring

dark eye line

gray face

extensive white underwing

white face crescent

smaller bill than Cinnamon

vertical white flank patch

♀

♂

CINNAMON TEAL *Spatula cyanoptera*   CITE ■ 1

In N.A., this colorful teal is exclusively a western species. Polytypic (5 sspp.; *septentrionalium* in N.A., 4 others in S.A., *borreroi* of Colombian Andes is possibly extinct). L 16" (41 cm) **Identification** The head is large and rounded, grading into a long bill reminiscent of Northern Shoveler. Males have intense red eyes in all but first few months of life. **MALE:** Distinctive bright cinnamon head and body. **FEMALE:** Rusty or golden brown overall with faintly scalloped flanks; variable, but usually small, pale loral spot and faint eye stripe. **FLIGHT:** Powder blue upperwing coverts contrast sharply with bright cinnamon plumage of male. Male has a broad white border to blue coverts. Flight is swift with rapid wingbeats. **Similar Species** The female and summer/early fall males can be very similar to the female Blue-winged Teal. It differs by having a warmer brown color (less grayish) with more subdued

mottling, a more blended face with no whitish patch at base of bill, and less of an eye line and a broken white eye ring. On dull-plumaged adult males and older young males, the red eye is distinctive. The bill of a Cinnamon Teal is larger than Blue-winged and more spatulate shaped. Female is smaller and darker than female Northern Shoveler with less distinctly spatulate bill. **Voice** Male makes dry series of *click* notes; female gives shrill *quack*. **Status & Distribution** Locally common in the West. **BREEDING:** Unlike its close relative, the Blue-winged Teal, the Cinnamon largely avoids prairie potholes; most common in the Great Basin and other interior regions of the western US. Nests on marshes, ponds, and shallow lakes; often uses highly alkaline water. **MIGRATION:** In spring, begins in mid-Jan. in South-

west; peaks in late Apr. in UT; most breeders are in place by mid-May. In fall, numbers diminish sharply after early Sept. **WINTER:** Uncommon to rare in Southwest. **RARE STATUS:** Rare east to eastern Great Plains. Casual to AK, eastern N.A., and C.A.; some records may involve escapes. **Population** Stable.

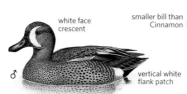

underwing of Cinnamon and Blue-winged Teal more extensively white than Green-winged Teal

♂

buffier face, less evident eye line and eye ring than female Blue-winged

♀

larger spatulate bill than Blue-winged

all but youngest males have red-orange eyes

♂

GARGANEY *Spatula querquedula*   GARG ▪ 4

The Garganey is the highly migratory Eurasian counterpart of the Blue-winged Teal. Monotypic. L 15" (39 cm) **Identification** Structure is very similar to that of Blue-winged. **MALE:** Bold white supercilium, brown breast, drooping tertials, and gray flanks. **FEMALE:** Pale face with strong dark eye stripe and additional diffuse dark stripe across the face starting at the lores. Pale loral spot and white throat. **ALTERNATE MALE:** Like female, but retains the paler gray upperwing of the breeding male. The Garganey holds its alternate plumage much longer than

most ducks, so that from midsummer to late Jan. or Feb., males look much like females. **FLIGHT:** The wing pattern is distinctive. The male has pale blue forewing; the female's forewing is gray-brown. Silvery gray inner primary webs. The underwing is pale with a strongly contrasting dark leading edge. **Similar Species** Compare female and alternate plumaged (fall) male to female Blue-winged Teal. Garganey's wing pattern is diagnostic. On the water look for its double-striped face, pale loral spot separated from the white throat, and sharp white fringes

on the tertials. Garganey is larger, with a heavier bill than Green-winged Teal, has a pale throat, and lacks a pale streak on sides of undertail coverts. **Voice** Male gives a series of dry clicking noises; female utters a harsh *quack* like Green-winged Teal. **Status & Distribution** The Garganey is rather common across Europe and Asia. **MIGRATION:** Highly migratory, completely vacating its breeding range. **RARE STATUS:** Casual in western AK; one nest found on Attu I. Also casual in the Pacific states; additional records are widely scattered throughout N.A. Most often found in migration. **WINTER:** Central Africa and southern Asia. **Population** Common but declining in recent decades in eastern Asia, likely resulting in fewer records from western N.A.

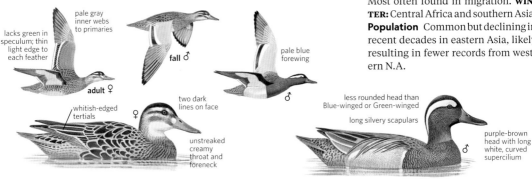

pale gray inner webs to primaries

lacks green in speculum; thin light edge to each feather

fall ♂

pale blue forewing

adult ♀

whitish-edged tertials

♀

two dark lines on face

♂

unstreaked creamy throat and foreneck

less rounded head than Blue-winged or Green-winged

long silvery scapulars

♂

purple-brown head with long white, curved supercilium

NORTHERN SHOVELER *Spatula clypeata*   NSHO ▪ 1

This is an odd-looking, Holarctic species with a distinct spatulate bill. It obtains most of its food from the surface, straining it through its large bill. Partial to sewage ponds, often in large flocks. Groups of Northern Shovelers often bring food to the surface by swimming rapidly in a circle while swinging their bills side to side. Monotypic. L 19" (48 cm) **Identification** On the water, shovelers look front heavy and short necked, with the bill angled downward. Adult males have bright golden eyes, while eye color in young males and females ranges from dull yellow to brown. Upperwing coverts powder blue in both sexes. **MALE:** Distinctive, with bright green head, white breast, reddish

brown flanks and belly. **FEMALE:** Mottled light brown head and body; some have a faint rusty wash; whitish fringed tertials. Most have dark bills with orange edges, but a few have all-dark bills. **ALTERNATE MALE:** As female, but has golden eyes and bright orange legs. **FALL MALE:** An intermediate plumage in early fall, with heavily spotted breast and flanks, dull head, and pale facial crescent. **Similar Species** The long, spatulate bill is diagnostic. **Voice** Generally silent, courting males give a repeated low, nasal *erp-EERP*.

Status & Distribution Common. **BREEDING:** Uses a wide variety of shallow wetlands for nesting, such as saline ponds and sewage-treatment plants. **MIGRATION:** In spring, a late migrant: begins migration in late Mar., peaks in the southern Great Lakes in early Apr., arriving on the prairie breeding grounds mid-Apr.–early May. Fall migration begins in late Aug.; peaks in BC in late Sept. or early Oct.; Great Lakes peaks mid-Oct. **WINTER:** Primary wintering areas CA and the Gulf Coast. Found in marshes, ponds, and bays. **Population** Population has increased in recent decades. Breeding range expanding eastward.

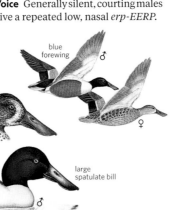

blue forewing

♂

white facial crescent, compare to male Blue-winged Teal

fall ♂

♀

orange line on bill

♀

large spatulate bill

♂

Genus *Mareca*

A 2009 study comparing mtDNA sequences resulted in the split of the polyphyletic genus *Anas* into four monophyletic genera. The genus *Mareca* was introduced in 1824; the type species for the genus is the European Wigeon. Currently, *Mareca* includes five extant species, the four described in this book and the Chiloé Wigeon of southern S.A.

GADWALL *Mareca strepera*   GADW ■ 1

The Gadwall is generally thought to be nondescript, but the male's intricate plumage shows a subtle beauty. Monotypic. L 20" (51 cm)
Identification From Mallard, has a steeper forehead and a narrow, shallow-based bill. Both sexes show a mixture of chestnut and black on the upperwing and a white square on the inner secondaries. Males have a more colorful wing pattern, often concealed on swim-ming birds; in flight, the wing pattern is diagnostic. **MALE:** Tan face, black-and-white scalloped breast that appears gray at a distance, gray flanks, black rump, and long scapulars with faint reddish fringes. **FEMALE:** Dull, mottled brown head and body, with a black bill evenly edged with orange, and a white belly.
Similar Species Female similar to female Mallard in plumage; separated by smaller bill and high, steep forehead; in flight by wing pattern and white belly. Female wigeon has shorter gray bill, rounded head, brighter flanks.
Voice CALL: Female gives a low, harsh *aack*. Male utters a soft, burping *meep*.
Status & Distribution Common and widespread across Northern Hemisphere. **BREEDING:** Nests on ponds, lakes, and marshes. **MIGRATION:** In spring, a late migrant; many do not begin to move north until well into Apr. Numbers peak in the Great Lakes and western plains in late Apr. Fall migration begins in Sept.; Great Lakes peak occurs in early Nov.; OR peak mid- to late Oct. **WINTER:** Marshes of LA are the most important wintering area for Gadwall. Widespread in shallow lakes and marshes.
Population Significantly increased in recent decades. Breeding range expanding in the East.

small white speculum patch ♂

white belly

chestnut-edged scapulars

mostly gray body ♂

black undertail coverts

steep forehead

even line of orange on sides of bill ♀

FALCATED DUCK *Mareca falcata*   FADU ■ 4

This stunning East Asian species is closely related to the Gadwall. Individuals that reach N.A. often mix with flocks of Gadwall or wigeon. Monotypic. L 19" (48 cm)
Identification Structure like that of the Gadwall but with a smaller thin bill and a shaggy crest coming to a point on males. **MALE:** Elongated tertials trail in the water. Green-and-white striped throat often hidden, visible when the head is lifted. **FEMALE:** All brown with tan face, pale belly, and dull green speculum with narrow white upper border.
Similar Species Female separated from female Gadwall by longer, thinner, dark bill; crested nape, uniform head color, and subtle dark fringes to the body feathers. Dark thin bill combined with the large head eliminate all other female dabblers.
Status & Distribution Uncommon to fairly common in eastern Asia. Rare migrant to western Aleutians; casual migrant on the Pribilofs and in winter from the Pacific region south to CA. The developing pattern of records suggests that likely most records away from AK likely involve wild birds.
Voice Generally silent.
Population Near Threatened. Declining in China.

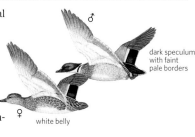
dark speculum with faint pale borders ♂

white belly ♀

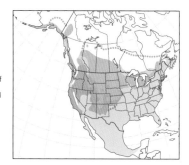

slight hint of crest ♀

all-gray bill

distinctive head shape

white throat with blackish band ♂

long sickle-shaped tertials

EURASIAN WIGEON *Mareca penelope* EUWI ■ 2

This Old World counterpart of the American Wigeon is regular in small numbers in N.A., mainly in the Pacific states and BC. It is typically found in flocks of American Wigeon; individuals often return to the same location for several years. Monotypic. L 20" (51 cm)

Identification Structure is almost identical to that of the American Wigeon. In any plumage the gray axillaries and underwing coverts of the Eurasian Wigeon are diagnostic. **MALE:** Distinctive, with bright chestnut head, cream-colored crown, and vermiculated gray back and sides. Some males may show small patches of green around the eye, which falls within the normal variation of the Eurasian Wigeon and does not necessarily indicate hybridization. **FEMALE:** Very similar to female American, but usually has warmer brown head; some gray-morph females have dull gray heads. Females often have pale, unmarked throats and are uniformly dull with no contrast between the head and the breast. Female American

shows contrast between gray throat and warmly colored breast. This is the single best field mark on females, unless the underwing can be seen. **IMMATURE MALE:** Adult male plumage acquired during first winter. Most individuals closely resemble adult male by Jan., but retain brown wing coverts. **FLIGHT:** Gray axillaries and underwing coverts are usually visible; white ovals on upperwings of adult males like American Wigeon. Females have uniform brown upperwing, lacking white covert bar of the female American.

Similar Species Compare males to hybrids with the American Wigeon. Most hybrids show obviously intermediate characteristics. Typical hybrids show an American Wigeon head pattern with rusty wash; flanks are a mix of gray and pink, usually more gray.

Voice CALL: Male whistles a sharp, high-pitched, single note, wiry

WHEEOOO, with a distinct similarity to the Gray-cheeked Thrush. The call is more attenuated than that of the American Wigeon and is clearly audible above a noisy group of American Wigeon. Female gives a harsh call like that of the American Wigeon.

Status & Distribution Uncommon to fairly common migrant and rare winter visitor on the western and central Aleutians; uncommon migrant elsewhere in western AK. Small numbers winter along and near both coasts; much more numerous in the Pacific states and BC; generally rare to casual elsewhere, with records from almost every state and province. Counts at favored wigeon sites in the Pacific states can reach double figures. At such sites, paired off Eurasian Wigeon can be encountered in late winter among large flocks of American Wigeon, indicating at least some degree of assortative mating. In the lower 48, the earliest arrivals are seen in late Aug.; some individuals linger into May.

Population Overall stable in Eurasian range.

rufous morph ♀
rusty cast to head
head and chest uniform in color
white forewing
dusky underwing and axillaries
adult ♂
cream crown
gray morph ♀
rufous head
gray morph ♀
adult ♂
adult ♂
plumage intermediate but always with rusty cast to head
American x Eurasian hybrid
immature ♂
mottled coverts; white on adult male
adult ♂

AMERICAN WIGEON *Mareca americana* AMWI ■ 1

The American Wigeon is a noisy dabbler equally comfortable on land as in the water. Monotypic. L 19" (48 cm)

Identification The American Wigeon's large head, chunky body, short legs, and fairly long, pointed tail are characteristic of wigeon. Males have a small blue bill and females are gray. Either sex can have a black ring at the base of the bill that the Eurasian Wigeon lacks, but many female and immature American Wigeons also lack this ring.

Axillaries and underwing coverts are white, which, if visible, is a diagnostic difference from the Eurasian. Another characteristic is the color of the innermost secondary; visible just below the tertials, which is gray in the American, usually white in the Eurasian. **MALE:** Sports a white crown, gray face, glossy green eye stripe, pinkish breast and flanks. One common variation, known as the "storm wigeon," has variable amounts of cream on the throat and

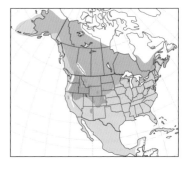

face. **FEMALE:** Gray-brown overall with orange or pinkish tinged flanks; brown upperwing with a white greater covert bar. **FIRST-WINTER MALE:** Like adult male, but more mottling on the upperwing coverts. **FLIGHT:** White ovals on the upperwings of males are visible from a great distance.

Similar Species Males are distinctive, but see Eurasian Wigeon account for separation of hybrids. The female is separated with caution from the female Eurasian Wigeon by gray head and speckled throat that contrasts rather sharply with much more warmly colored breast and flanks. The Eurasian shows uniform coloration on throat and breast. Note American's gray throat contrasts with its warm-colored breast. Note grayish, not white, underwings.

Voice CALL: Male gives two- or three-part whistle, *whee-WHOO* or *whee-WHOO-who*, given in flight or on the water; call is heard constantly from any large flock. Call of the female, given infrequently, sounds like a harsh, nasal cough.

Status & Distribution Common and widespread, particularly in the West. **BREEDING:** Nests on tundra pools, river deltas, boreal lakes, and prairie potholes. A small isolated population

breeds in the Russian Far East (Anadyr Valley). **MIGRATION:** Wintering birds begin leaving in early Feb.; in most states bordering Canada peak arrival is in the first two weeks of Apr., with birds arriving on the northern breeding grounds in the first two weeks of May. Early fall migrant, beginning in mid-Aug.; Great Lakes peak is the end of Oct., and migration is largely over by late Nov. It peaks in Nov. on coast

of BC. **WINTER:** Widespread in winter, with scattered areas of high concentrations. Found in open marshes, coastal estuaries, wet farm fields; flocks may even gather at parks and golf courses, where they become fairly tame. Often associate with geese and swans. Rare to British Isles; very rare to the rest of Europe and the Russian Far East. **Population** Stable.

white or buffy white crown

white forewing

white underwing coverts and axillaries

adult ♂

alternate (late-summer) adult ♂

bluish gray bill with dark tip

adult ♂

gray face contrasts with cinnamon-buff chest

♀

Genus *Anas*

Many species previously placed in this genus were recently moved to three newly created genera (*Sibrionetta*, *Spatula*, and *Mareca*). Seven *Anas* species occur in N.A.; of these, five species breed, and two are visitors from the Old World or the Caribbean. Males have colorful plumages that are held for most of the year, except in midsummer, when bright plumage is replaced by a dull, alternate plumage, previously called the eclipse plumage. It is usually worn until fall and camouflages males as they replace flight feathers. Females are mottled brown and show little seasonal variation. Most males and some females show a colorful speculum, an iridescent patch on the secondaries.

EASTERN SPOT-BILLED DUCK   *Anas zonoryncha*   ESBD   4

This robust, blackish brown duck is an East Asian species and casual stray to AK. Its structure and habits are very similar to the Mallard. It is found on freshwater marshes and lakes, lagoons, and rivers. Despite its name, this recently split species lacks the red spots at its bill base that the two related, more westerly taxa exhibit (now known as the Indian Spot-billed Duck, *A. poecilorhyncha*, which lack the dark lower facial bar and have much more white in the tertials). Monotypic. L 22" (56 cm)

Identification The diagnostic, sharply defined yellow tip on the black bill is visible at great distances. Sexes are similar; female slightly paler. Head is

pale with light gray throat and cheeks, dark crown, dark stripe through the eye, and shorter, dark stripe beginning at the base of the bill. Body is blackish brown; speculum is bluish with white border. Diagnostic white-edged tertials are readily visible when at rest. **FLIGHT:** Large and dark with striking white underwing coverts, and dark upperwing with narrow white borders to the blue speculum.

Similar Species Like American Black Duck and female Mallard; separated by distinctive bill coloration, paler head, and pale-fringed tertials.

Voice CALL: Females give a loud *quack*. Males give raspy *kreep*.

Status & Distribution Generally widespread and locally common throughout its range in eastern Asia. **RARE STATUS:** Casual to western and central Aleutians. Accidental to Kodiak I. **Population** Declining.

distinct yellow tip to black bill

dark facial stripes

mostly dark body

distinct white tertial edges

pale neck

MALLARD   *Anas platyrhynchos*   MALL ■ 1

Probably the most widely recognized and widespread duck in N.A. Mallards make use of almost any body of water. Feral populations in various plumage patterns are widely established in city parks. Polytypic (3 sspp.; 2 in N.A.). L 23" (58 cm)

Identification Mallards are large with rounded heads and bright blue speculums with bold white borders. The central uppertail coverts of the males are curled upward, a characteristic frequently passed on to hybrid offspring. **MALE:** Distinctive, with bright yellow bill and green head, white neck ring, and chestnut breast. **FEMALE:** Brown overall, with an indistinct brown eye line and orange bill with a black saddle. The outer tail feathers are whitish. **ALTERNATE MALE:** Like female but the bill remains bright yellow and the crown and breast are darker than female. **FLIGHT:** Wingbeats steady and shallow, slower than other ducks.

Geographic Variation Two subspecies in N.A., nominate and the southwestern subspecies *diazi*, "Mexican Duck," which is found in pockets along the Rio Grande River in TX, NM, and southern AZ. "Mexican" is 10 percent smaller, and the sexes are similar, resembling a female Mallard. Its plumage is darker than female northern Mallards, and it has a brown tail. "Mexican Duck" resembles Mottled Duck, but has a grayer more distinctly patterned face, streaked throat, and stronger white borders to the speculum.

Similar Species Female paler than Mottled (especially western subspecies) and American Black Duck, with distinct white borders to the blue speculum. From the smaller female Gadwall by rounded head shape, lacking a distinct forehead; larger, deeper based, and differently patterned bill; and speculum pattern.

Voice Female gives series of loud, descending *quacks*. Male less vocal; shorter, soft *quack* is often drowned out by the calls of the females.

Status & Distribution Abundant and widespread across Northern Hemisphere. **BREEDING:** Any place that has both water and cover. Typically nests on the ground in grass or shrubs, but also nests in hollow trees. **MIGRATION:** An early spring migrant, along with Northern Pintail and Cinnamon Teal, appearing north of its winter range as soon as there is open water. Begins in early Feb.; peaks in the Midwest and Great Lakes region in late Mar.; northernmost breeders arrive in AK in early May. In the fall it is among the last to leave the north, retreating only as water freezes. Some begin moving south in Sept.; however, most northern breeders do not move south until Oct.; peak numbers arrive in the Great Lakes and across the northern plains in early Nov. **WINTER:** Spends the winter in a variety of habitats, mostly shallow freshwater.

Population The most abundant wild duck in N.A., population stable.

alternate ♂

dark saddle on orange bill

blue speculum bordered broadly with white

underwing contrasts less with body than Black Duck

mostly white tail

male with yellow bill, female similar; no dark saddle

diazi ♂

body and tail darker than female Mallard

yellow bill ♂

curled tail feathers

"Mexican Duck"
soon to be restored as a full species

AMERICAN BLACK DUCK   *Anas rubripes*   ABDU ■ 1

The American Black Duck is most numerous in northeastern coastal salt marshes. Monotypic. L 23" (58 cm)

Identification Sexes similar and structurally identical to the Mallard. The two species are often found in mixed flocks and frequently hybridize. **MALE:** Body blackish with a dark crown, sharply contrasting gray and streaked face and yellow-olive bill. **FEMALE:** Like male but slightly paler body with duller olive bill. **FLIGHT:** Speculum is deep purple and can show thin white borders. Dark body contrasts with white underwing.

Similar Species Resembles female

Mallard but darker, showing strong contrast between body and face, bill yellow to olive, speculum purple. Hybrids with Mallard may be paler in coloration overall, with partially green head on males, brighter yellow or orange bill, white bordering the speculum, curved central tail feathers or paler tail feathers. See Mottled Duck.

Voice Similar to Mallard.

Status & Distribution Locally common. **BREEDING:** Nests in a wide variety of wetlands. **MIGRATION:** Spring migration begins in early Feb. and proceeds gradually following the appearance of open water. Arrives in ME in early Apr., northern QC in late May. Fall migrants peak in mid-Atlantic in early Nov., peak in the western Great Lakes in mid-Nov. **WINTER:** Primarily in mid-Atlantic salt marshes. Also use a variety of wetlands throughout the Northeast; some wintering as far north as NL. **RARE STATUS:** Casual to western Gulf States and western N.A. Casual to British Isles.

Population Apparently stable in recent decades; replacement by Mallards in deforested portions of its range and overhunting in past decades are risk factors.

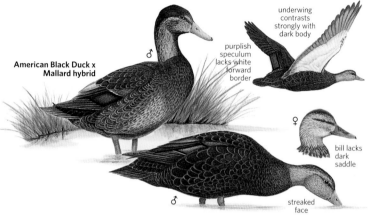

American Black Duck x Mallard hybrid

♂

underwing contrasts strongly with dark body

purplish speculum lacks white forward border

♀

bill lacks dark saddle

♂

streaked face

MOTTLED DUCK *Anas fulvigula*   MODU   1

This southern look-alike of the American Black Duck is primarily a resident of the coastal marshes of the Gulf of Mexico and FL. Mottled Ducks retain their pair bond for most of the year, so are most often seen in pairs and do not form large winter flocks. Polytypic (2 sspp.; both in N.A.). L 22" (56 cm)

Identification Overall color is close to American Black Duck, especially darker *maculosa*. The blue-green speculum has very narrow white borders. Noticeable black spot at gape. **ADULT MALE:** Dark brown body with golden brown, V-shaped marks on interior of flank feathers. Buffy brown face with an unmarked throat; yellow bill. **ADULT FEMALE:** Like male, with dull orange or olive-yellow bill with black marking on the culmen.

FLIGHT: Upperwing appears all-dark; underwing coverts flash bright white. **Geographic Variation** Two subspecies, nominate in the FL Peninsula and along the coast of GA and SC; *maculosa* found, primarily coastally, from LA and coastal MS to northern Mexico. Though similar, *maculosa* is darker and more coarsely marked than nominate *fulvigula*.

Similar Species Similar to female Mallard and American Black Duck. Separated from female Mallard by darker overall color; buffy, unstreaked throat; all brown tail; and narrower white border to the speculum. Separated with great care from American Black Duck by buffier face, unstreaked throat, paler brown V-shaped markings on the interior of the flank feathers, and a blue-green speculum with narrow white borders. Black Duck has essentially all-dark flanks, a grayer face, streaked throat, and lacks white borders on the speculum.

Voice Similar to Mallard.

Status & Distribution Closely tied to the Gulf Coast, where it is fairly common but rarely found in large numbers. **YEAR-ROUND:** Essentially nonmigratory; however, individuals move in response to food resources and water levels. Large numbers move from coastal marshes to interior rice fields in fall to feed on the ripening rice. Rare away from coastal areas but known to wander and occasionally nest in the interior. Introduced to SC coast. **BREEDING:** Nests in cordgrass meadows, fallow rice fields, and grassy islets in marshes. **RARE STATUS:** Rare breeder in northeastern TX, southern AR, and OK. Very rare north to KS. Casual to ND, SD, KY, IL, and mid-Atlantic states; accidental to southern ON.

Population Apparently stable, but trend unknown.

maculosa

fulvigula

unmarked buffy face and throat

greenish yellow bill

black spot at gape

♂

western Gulf Coast
maculosa

mainly FL
fulvigula

yellower bill

averages paler overall than *maculosa*

black spot at gape

♂

WHITE-CHEEKED PINTAIL   *Anas bahamensis*   WCHP ■ 4

This Neotropical duck is a casual visitor to southern FL. The White-cheeked Pintail is found on shallow-water habitats in salt water or freshwater, where it feeds by dabbling and tipping up. The large, white patch on the side of the White-cheeked's head and throat, for which it is named, is distinctive. Polytypic (3 sspp.; *bahamensis* in N.A.). L 17" (43 cm)

Identification Sexes of this pintail are similar; the female is paler than the male, with a slightly shorter tail. White cheeks and throat contrast with a dark forehead and cap; the blue-

red at bill base

sharply delineated white cheek and throat

bahamensis

buffy pointed tail

♂

gray bill has a red spot near the base. The long, pointed tail is buffy; the underparts are tawny or reddish and are heavily spotted. **FLIGHT:** Slender body and long, pointed tail are evident. Both sexes of White-cheeked Pintail show brown forewing, green speculum bordered on each side with broad buffy edge. Pale, buffy rump and tail contrast with the darker brown body.

Geographic Variation Nominate *bahamensis* from the West Indies and northeastern S.A. occurs in N.A. Subspecies *bahamensis* is brighter and larger than *rubrirostris*, which is from mainland S.A., and *galapagensis* of the Galápagos Is. is the palest.

Similar Species Generally distinctive, not likely to be mistaken. Prominent white cheek patch readily distinguishes the White-cheeked Pintail from the Northern Pintail.

Voice Generally silent. Female gives weak descending *quack* notes. Male produces low whistle.

Status & Distribution Found in the West Indies and S.A., including

the Galápagos. Now largely rare to uncommon in West Indies; small movements between islands occur.

RARE STATUS: Casual to southern FL, majority of records mid-Dec.–late Apr. Accidental to NC; one accepted record from coastal southern TX, perhaps an escape. Sightings even from FL are often of uncertain origin as the White-cheeked Pintail is frequently kept in captivity; those away from FL (recorded coast to coast) are most likely birds that have escaped from captivity.

Population Moderate to severe declines in the West Indies due to habitat loss, hunting, and introduced predators.

NORTHERN PINTAIL   *Anas acuta*   NOPI ■ 1

This elegant species is as easily identified by structure as by plumage. The birds frequently gather in huge flocks to feed in grain fields. Monotypic. L 20–26" (51–66 cm)

Identification The thin grayish bill; long, slender neck; slim body; and long, wedge-shaped tail give pintails a diagnostic shape. **MALE:** Combination of brown head with white neck stripe, white breast, and spikelike tail is distinctive. **FEMALE:** Mottled brown, with a plain, warm buff head. **FLIGHT:** Looks long and lean, with

slender neck extended. The wings are thin and pointed, with bold white trailing edges to the secondaries.

Similar Species Female's combination of structure and unmarked buffy head is unlike any other female dabbler.

Status & Distribution This numerous duck is most common in the West. **BREEDING:** Holarctic species; nests on marshes, lakes, and tundra pools. **MIGRATION:** In spring, early migrant relative to other dabbling ducks. Mid-Atlantic peak late Feb.; northern plains peak in early Apr.; Arctic breeders arrive in mid- to late May. An early fall migrant, arrives in Southwest mid-Aug. Peak in the mid-Atlantic and Great Lakes late Oct.–early Nov. Pacific Northwest peak mid- to late Oct. **WINTER:** Prefers open marshes, agriculture fields, and tidal flats. Much of the population winters in CA; also, rarely, in south-

ern AK and the Great Lakes region.

Voice Male gives a high, whining *mee-meee* during courtship and short, mellow *proop-proop*, like the Green-winged Teal but lower pitched and more musical; female gives a hoarse, weak *quack*.

Population Stable, but population fluctuates; low during prairie droughts.

grayish bill

♀

♂

long neck

long pointed tail

dark speculum with narrow white trailing edge

white stripe extends up neck

long pointed tail

♂

GREEN-WINGED TEAL *Anas crecca*   GWTE ◼ 1

A tiny, Holarctic dabbler, the Green-winged Teal is frequently seen foraging on mudflats. Polytypic (2 sspp.; both in N.A.). L 14" (37 cm)

Identification Small, with rounded head; short, thin bill. **ADULT MALE:** Chestnut head with green ear patch and a vertical white bar on each side. **ADULT FEMALE:** Brown body and head, black bill with dull orange edges, a pale line on the sides of the undertail coverts, and white belly. **FLIGHT:** Flies with very rapid wingbeats, forming tight bunches that twist and turn erratically. White restricted to central part of underwing, unlike the Blue-winged and Cinnamon. **Geographic Variation** Subspecies *carolinensis* is widespread in N.A.; the Eurasian subspecies nominate *crecca*, split by some, is regular in the Aleutians and the Pribilofs, but is rare or casual elsewhere in N.A. It has a white horizontal bar (not vertical) above the flanks and more white on the face. Females very similar, but inner wing bar on *crecca* is typically broader than white on trailing edge, more equal width in *carolinensis*.

Similar Species Small bill and white line on sides of undertail coverts separate female from other teal.

Voice Males give a sharp, whistled *kreek* (like the Northern Pintail); females give a high, thin *quack*.

Status & Distribution Common and widespread. **BREEDING:** Wooded ponds, potholes, and tundra pools. **MIGRATION:** In spring, mid-Atlantic peak late Mar. In fall, mid-Atlantic peak mid- to late Oct. Peak late Oct.–early Nov. on BC coast. **WINTER:** Shallow wetlands and marshes, often in large flocks. Eurasian *crecca* is regular to western AK, uncommon to fairly common on Aleutians and Pribilofs, rare to St. Lawrence I. and mainland western AK, very rare on West Coast, regular to NL in winter; casual south to the mid-Atlantic. Accidental elsewhere in N.A. **Population** Stable.

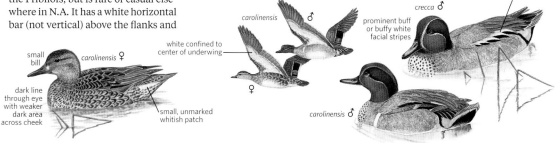

"Eurasian Teal" *crecca* ♂   white horizontal bar

prominent buff or buffy white facial stripes

carolinensis ♂

white confined to center of underwing

♀

carolinensis ♀

small bill

dark line through eye with weaker dark area across cheek

small, unmarked whitish patch

BAY DUCKS Genus *Aythya*

This genus is made up of 12 species worldwide. Five species breed in N.A., and two occur as rarities from Eurasia. *Aythya* are medium to medium-large diving ducks that often gather together in impressive rafts on coastal bays and large lakes in the winter. *Aythya* ducks regularly hybridize with each other.

COMMON POCHARD *Aythya ferina*   COMP ◼ 3

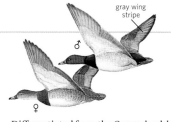

gray wing stripe

♂

♀

This Old World counterpart of the Canvasback, also similar to the Redhead, is a Eurasian species rare in western AK. Monotypic. L 18" (46 cm)

Identification Heavily built *Aythya* with a thick neck and a long sloping forehead grading smoothly into the bill. Common Pochard's bill has a dark base and a black tip separated by a thick whitish blue stripe. **MALE:** Coloration similar to Canvasback with chestnut head, black breast, and light gray body. **FEMALE:** Light brown head with pale throat, lores often paler than the face, fairly prominent pale eye ring, brown breast, mottled gray and brown body. **FLIGHT:** Wings uniform gray with no wing stripe. Appears chunkier than the Canvasback with short, rounded wings.

Similar Species Males separated from the Canvasback by shorter bill with more concave profile, more compact build, shorter neck, slightly grayer back, and lack of black crown; from the Redhead by sloping forehead, bill pattern, darker eye color, and paler back. Females separated from Redheads and Ring-necked Ducks by structure and the presence of gray on the flanks and back.

Differentiated from the Canvasback by structure, bill pattern, pale loral spot, and darker color. Hybrids between Canvasback and Redhead can appear very similar to Common Pochard but lack its unique bill pattern.

Voice Generally silent.

Status & Distribution Widespread in Europe and Palearctic Asia. **RARE STATUS:** Rare migrant to the western and central Aleutians; casual to the Pribilofs; records from St. Lawrence I., Seward Peninsula, and coastal southern AK and CA.

Population Vulnerable. Declining.

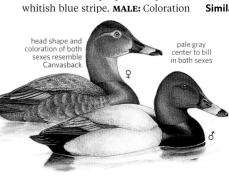

head shape and coloration of both sexes resemble Canvasback

pale gray center to bill in both sexes

♀

♂

CANVASBACK *Aythya valisineria*   CANV ■ 1

The Canvasback is one of the more elegant ducks; its distinctive structure makes identification simple. Monotypic. L 21" (53 cm)

Identification Forehead slopes straight down to a long, dark bill. The largest *Aythya* species, it has a thick neck and long body. **MALE:** Chestnut head with a dark crown and forehead, red iris. The flanks, back, and tertials are very pale, almost white. **FEMALE:** Light brown head and breast with a slightly darker crown; contrasting pale gray flanks and back. **FLIGHT:** Wings are uniform, with only a faint gray wing stripe.

Similar Species Compare to Redhead and Common Pochard. Canvasback can be distinguished by its sloping forehead and long, uniformly colored bill. The forehead is straight, not concave as in similar species. In all plumages, Canvasback has a dark bill and pale back unlike any other *Aythya* species.

Voice Generally silent.

Status & Distribution Uncommon to common. **BREEDING:** Uses a wide variety of freshwater wetlands, from lakes and ponds to large marshes; also found on alkali lakes. **MIGRATION:** Spring migration begins in early Feb.; peaks in the Midwest and Great Plains in early Apr.; most breeders have arrived by mid- to late May. Fall migration is fairly late. Most begin moving south in Oct.; peaking in the Great Lakes in early Nov.; peaks in early Dec. on the Gulf Coast. **WINTER:** Ranges widely in freshwater lakes and brackish bays and estuaries; south to central Mexico. **RARE STATUS:** Casual on the Aleutian and Pribilof Is., and HI. Also casual south to Costa Rica, northwest Europe, and East Asia.

Population Stable with a slight increase over last two decades.

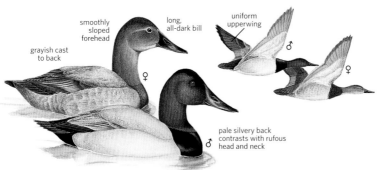

grayish cast to back

smoothly sloped forehead

long, all-dark bill

uniform upperwing

♂

♀

♀

pale silvery back contrasts with rufous head and neck

♂

REDHEAD *Aythya americana*   REDH ■ 1

An attractive duck usually found in small groups except in massive rafts along the barrier islands of southern TX in winter. Monotypic. L 19" (48 cm)

Identification Large *Aythya*, slightly larger than Greater Scaup, with a short bill, rounded head. **MALE:** Tricolored bill with a black tip, white subterminal ring, and pale blue base. The head is uniformly rufous with a yellow eye. Back and flanks appear smoky gray. **FEMALE:** Brown overall, with paler face and dark crown. Bill is tricolored with a slate gray base. **FLIGHT:** Brown upperwing, with a paler gray stripe on the flight feathers.

Similar Species Male separated from Canvasback by short, tricolored bill and darker back. Female from Canvasback by structure and browner color.

Voice Generally silent.

Status & Distribution Fairly common, south to central Mexico: casual to Honduras. **BREEDING:** Highest concentrations in the western prairies; nesting on prairie potholes, seasonal pools, and lakes. **MIGRATION:** In spring, begins leaving the wintering grounds in late Jan., peaking in the southern prairies in mid-Mar. In fall, migration begins in late Aug., peaking in the southern prairies in mid- to late Oct., arriving on the Gulf Coast in numbers in early Nov. **WINTER:** Widespread but fairly sparse except along the western Gulf Coast, where most of the population winters, on the Laguna Madre of southern TX. **RARE STATUS:** Accidental in Bermuda, HI, and the UK.

Population Stable, but eastern breeders are declining.

♂

♀

gray wing stripe

more rounded and uniform buffy head than female Ring-necked

head and back uniform in color

♀

rounded head

pale blended band on bill

darker gray back than Canvasback

♂

RING-NECKED DUCK *Aythya collaris* RNDU ■ 1

This distinctive duck is usually found in marshy pools and small ponds. Monotypic. L 17" (43 cm)
Identification Angular head, peaking near the rear. Both sexes have a white ring adjacent to a black-tipped bill; male has an additional thin ring at base of bill. **ADULT MALE:** White wedge between the breast and flanks, a unique feature; dark purple head, gray flanks, and black back. **ADULT FEMALE:** Gray face with a white eye ring, darker crown, and indistinct pale area at the base of the bill; some show a pale thin stripe behind the eye. **FLIGHT:** Rapid; gray stripe on secondaries.
Similar Species Male similar to Tufted Duck, but lacks tuft and gray flanks separated from breast by a white wedge, and lacks wing stripe; bill pattern differs. Female similar to female Redhead, but with an angular head, darker back and crown. From female scaup by gray face contrasting with peaked crown and white eye ring, and lacks white wing stripe in flight.
Voice Generally silent.
Status & Distribution Fairly common. **BREEDING:** Nests on freshwater marshes and shallow ponds and lakes. **MIGRATION:** In spring, begins early Feb., peaking in Great Lakes in early Apr.; most breeders in place by late Apr. In fall, migration begins in mid-Sept., peaking on Great Lakes in late Oct., Pacific Northwest in late Nov. **WINTER:** Sometimes uses coastal marshes, but usually in shallow freshwater. Winters south through West Indies and C.A. **RARE STATUS:** Rare to western Europe. Casual to Aleutians and Bering Sea Is.
Population Stable.

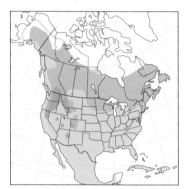

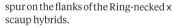

adult ♂

grayish stripe

♀

peaked head

white eye ring and grayish face ♀

black back

adult ♂

prominent white ring near tip of bill

cinnamon collar on lower neck hard to see

vertical white bar on side

TUFTED DUCK *Aythya fuligula* TUDU ■ 3

An Old World species, regular in western AK and rare to casual to the rest of N.A., the Tufted Duck is usually found in the company of scaup or sometimes Ring-necked Ducks. Individuals often return to the same locations year after year. Monotypic. L 17" (43 cm)
Identification **ADULT MALE:** A variably long crest is the most striking feature. The bill is similar to that of a scaup but has more black on the tip. From a distance, black back and gleaming white flanks distinctive. **FEMALE:** Dark brown head, back, and breast, with paler mottled brown flanks. The female shows a variably sized crest and a black tipped bill. **FIRST-WINTER MALE:** Like the adult male, but has a shorter crest; flanks not as white. **FLIGHT:** Shows an extensive white stripe on the wing, broader than that of the Greater Scaup.
Similar Species First-winter males are similar to hybrids between a scaup and a Tufted Duck or a scaup and a Ring-necked Duck. Scaup x Tufted Duck hybrids are regular, almost as frequent as pure Tufted Ducks; hybrids tend to show a short crest, reduced black on the tip of the bill, a pale gray wash on the flanks, and grayish barring on the upper back; they lack the faint white spur on the flanks of the Ring-necked x scaup hybrids.
Voice Generally silent.
Status & Distribution Abundant and widespread, breeding across northern Palearctic Europe and Asia. **WINTER:** Ponds, bays, and rivers; a number of records have come from ponds or lakes in city parks. In southern Europe and Asia, south to Africa. **RARE STATUS:** Regular migrant in western AK; rare to very rare winter visitor on West Coast south to Southern CA. Casual on East Coast to mid-Atlantic. Accidental elsewhere.
Population Stable and widespread with a very large world population.

some females whitish around bill

♀

crest

♀

broad black bill tip

dark brown back

1st winter ♂

black back

adult ♂

extensive and bold white wing stripe

adult ♂

crest length variable on males and females

♀

pure white sides

GREATER SCAUP *Aythya marila*   GRSC ▪ 1

Very similar to the Lesser Scaup, the Greater Scaup often forms mixed flocks. In coastal bays and inlets or other areas of deep water, the Greater Scaup generally dominates these flocks, while the Lesser Scaup is more numerous in the interior and on shallower lakes and ponds. Polytypic (2 sspp.; *nearctica* in N.A.). L 18" (46 cm)

Identification Structure is the best way to separate and identify scaup. The larger Greater Scaup has a larger and more rounded head, thicker neck, and larger, thicker bill. The Greater Scaup also has more black on the bill, covering much of the bill tip. **ADULT MALE:** Dark head that shows a bright green sheen in bright light. The breast is black; the back is pale gray covered with fine vermiculations; the flanks are

white with very light to moderate gray vermiculations. **FEMALE:** Dark brown head with a slight reddish tone. There is an extensive white ring encircling the base of the bill. Females in worn plumage show a variably sized pale ear patch. The back of the female is brown, and the flanks are mottled gray-brown. **FIRST-WINTER MALE:** Like females in fall, but with less white surrounding the bill; acquires plumage like adult males by spring, but with more vermiculations on the flanks and heavily worn brown tail feathers. **FLIGHT:** White wing stripe extends almost to the tip of the wing.

Similar Species See above for separating the two scaup species. Compare females to other female *Aythya*. Smaller female Ring-necked shows a more angular head, ringed bill, and more defused pale face pattern. Female Redhead is more evenly and paler brown, with more black on the bill, and lacks the white ring surrounding the bill.

Geographic Variation N.A. and East Asian birds are of the subspecies *nearctica*. There is one specimen record of the nominate *marila* from St. Paul I. in AK in late June.

Voice Generally silent.

Status & Distribution Widespread and abundant in the Northern Hemisphere. **BREEDING:** Holarctic breeder; nests on tundra pools and lakes. **MIGRATION:** In spring, begins gradually in early Mar. Mid-Atlantic peak mid-Mar to mid-Apr. In the West, most have left the southern part of their winter range by late Mar. Peak in BC the last two weeks of Apr. In fall, departs breeding grounds mid-Sept.–late Oct. The bulk of the eastern population migrates to the Atlantic coast via the Great Lakes. Peak in Great Lakes the second week of Oct.; late Oct.–early Nov. in the mid-Atlantic. Peak in southern BC mid- to late Oct. **WINTER:** Uses bays, estuaries, and deep inland lakes and rivers.

Population Stable.

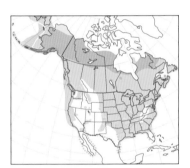

extensive white stripe

adult
nearctica ♂

many females with prominent whitish ear patches

nearctica ♀

nearctica ♀

larger, rounder head than Lesser

1st winter
nearctica ♂

nearctica ♀

more black on bill tip than Lesser

adult
nearctica ♂

LESSER SCAUP *Aythya affinis*   LESC ▪ 1

The most abundant, frequently encountered *Aythya* in N.A., Lesser Scaup is smaller than Greater Scaup and usually more common in freshwater. Monotypic. L 16.5" (42 cm)

Identification Lesser Scaup are medium size with small, narrow heads; high, peaked crowns; and thin necks. Their bills are small and thin, with only a small amount of black surrounding the nail. **ADULT MALE:** Dark head has a purple gloss in good light, but may also appear green. It has a vermiculated, gray back and lightly vermiculated flanks that may appear light gray or white. **FIRST-WINTER MALE:** Like female in fall, but with no white surrounding the bill; acquires plumage like adult male by spring, but with more vermiculations on the

flanks. **FEMALE:** Dark brown head with a moderate white ring enclosing the base of the bill. A few females, in worn plumage, show a small, pale ear patch; back is dark brown, and flanks are a mottled mix of gray and brown. **FLIGHT:** White wing stripe is strong on the secondaries and very faint on the primaries.

Similar Species Compare Lesser Scaup females to other female *Aythya*. Female Ring-necked has a paler head and a sharp white eye ring and white ring at the base of the bill. Female Lesser Scaup is smaller and darker than female Redhead or Canvasback. Separate with caution from female Greater Scaup by noticeably smaller size in direct comparison, angular head shape, usually with a slight bump

(topknot) at the peak of the nape. Usually shows noticeably less white on the face enclosing the bill; also most have little or no pale ear patch. On average appears slightly more mottled and "messier" than Greater Scaup.

Voice Generally silent.

Status & Distribution Abundant and widespread. **BREEDING:** Nest near lakes and pools and large, permanent potholes in the western prairies. **MIGRATION:** In spring, some Lesser Scaup begin to leave their winter range in early Feb., while large numbers remain through Apr. Mid-Atlantic and Great Lakes peak early Apr., late Mar in the Pacific Northwest. In fall, they remain on the breeding grounds until Sept. South of its breeding range, in the lower 48, the fall peak occurs in the first two weeks of Nov. virtually everywhere. **WINTER:** Found on lakes, reservoirs, and coastal lagoons. The bulk of the population winters near the Gulf of Mexico in FL, LA, and TX; south through West Indies and C.A. **RARE STATUS:** Casual to Bering Sea, north to Greenland, and Europe. Accidental in S.A.

Population Our most numerous *Aythya*; but declines noted over last decade.

adult ♂

white restricted to secondaries

♀

smaller size; head with slight peak at rear

less black on bill tip than Greater

♀

adult ♂

EIDERS Genera *Polysticta* and *Somateria*

Each of the four species worldwide are found in N.A.. These large, bulky diving sea ducks have dense down feathers that help insulate them from cold northern waters. Females pluck their own down to line nests. Eiderdown is harvested and sold for pillows and blankets. Eiders generally migrate in large flocks. Spectacular movements can be witnessed at points in coastal northwestern AK. Most eiders head north to tundra nesting grounds as soon as sea ice breaks up. In years of late breakup in combination with severe storms, many eiders, mainly King, starve to death. In spring 1964, an estimated 100,000 King Eiders, some 10 percent of the population migrating on the Arctic coast, died.

STELLER'S EIDER *Polysticta stelleri*   STEI ■ 3

The Steller's Eider lacks the highly developed frontal lobes and bill feathering characteristic of other eiders. Monotypic. L 17" (43 cm)

Identification Head is rectangular, with a flat crown and sharply angled nape. Its long, pointed tail is generally held out of the water. **ADULT MALE:** Distinctive. White head with small green patch on back of head and black spot on breast sides; cinnamon buff underparts; appears mostly white at a distance. **ALTERNATE MALE:** Like adult female, but darker, retaining white in the wing and broad pale tertial tips. **FEMALE:** Easily overlooked in mixed flock. Plain dark brown face, with pale eye ring, warm cinnamon brown coloration overall. Blue speculum and white-tipped tertials diagnostic when present. **JUVENILE:** Gray-brown, plumage wears and fades quickly; thin white speculum borders, tertials dull and short. **FLIGHT:** Flies swiftly in tight flocks, with a long-winged, short-necked appearance.

Similar Species Blue speculum with broad white borders, like a Mallard, in combination with long, curved white-tipped tertials separate Steller's from all other sea ducks. Female Steller's Eider distinguished from Harlequin Duck by pale eye ring on warm brown face and lack of white spots on head and shape.

Voice Noisy in winter flocks; females giving rapid guttural call, inciting females give loud *qua-haa* or *cooay*. Growling and barking noises given by both sexes; does not produce cooing calls like other eiders.

Status & Distribution **BREEDING:** Northern Russia and AK. Arrive on

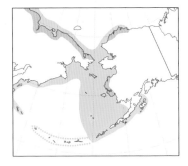

AK Arctic tundra late May–early June. **WINTER:** To southern Bering Sea, arrive late Oct.–Nov. Steller's Eider favors shallow, sheltered coastal lagoons and bays. **RARE STATUS:** Casual outside AK waters to BC, WA, OR, CA. Casual to accidental on East Coast; records from QC, ME, MA, MD, and Baffin I.

Population Vulnerable. Marked decline in AK and Russia since 1960s. Listed as Threatened by the USFWS.

white forewing

adult ♂

blue speculum bordered by white

♀

white underwing contrasts with dark body

dark cinnamon-brown color overall

♀

rather small unfeathered, dark bill

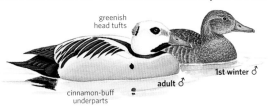

greenish head tufts

1st winter ♂

adult ♂

cinnamon-buff underparts

SPECTACLED EIDER *Somateria fischeri*   SPEI ■ 3

This hardy sea duck is appropriately named for its bold, pale "spectacles," which are apparent in all plumages. The Spectacled Eider has a gradually sloping forehead, with feathers extending down the bill. It lacks the highly developed frontal lobes of Common and King Eiders. Otherwise, Spectacled is structurally similar to Common Eider in flight but significantly smaller, with a shorter neck. Monotypic. L 21" (53 cm)

Identification ADULT MALE: Bold head pattern with big white, black-bordered goggles and orange bill; generally distinctive. **ALTERNATE MALE:** Resembles breeding plumage, but white and green areas replaced with dusky gray feathers. **FEMALE:** Head appears light overall with dark forehead, due to pale "spectacles" and cheek. Bill is dark, blue-gray. Vertical barring on the flanks, paler body coloration overall than other eiders. **FLIGHT:** Male is only eider with sooty gray belly extending to upper breast. Upperwing coverts all-white, like on Common Eider. Underwing duskier than on other eiders. Female difficult to identify, but dark forehead contrasts with pale spectacles and cheek and can be useful in combination with structural characteristics.

Similar Species Spectacled Eider is smaller than Common and King Eiders. Similar to Common in flight, but more extensive dark breast on males and distinctive head pattern. Female Spectacled Eider is separated from Common and King Eiders by pale cheek and "goggles"; feathering extends down the smaller bill, and lacks the chevron pattern on the flanks of King Eider.

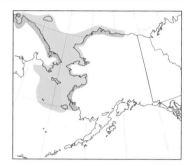

green head with white "goggles"

feathering on all birds extends well out on bill

adult ♂

sooty gray extends into breast

female has pale "goggles," but pattern is subdued

adult ♀

adult ♂

pattern of white on upperparts similar to Common Eider

♀

Voice Typically silent. Male display call is a very soft *hoo-hoo*. Female gives a short, guttural *croak* and a two-syllable clucking call.

Status & Distribution BREEDING: Return to AK breeding grounds, coastal tundra near lakes and ponds, first week of May. **MIGRATION:** Depart late June for molting areas. **WINTER:** Bering Sea in large flocks in openings of pack ice (polynyas), nearly all south of St. Lawrence I. to St. Matthew I. **RARE STATUS:** Casual on Aleutian and Pribilof Is. Not prone to wander far from breeding and wintering grounds.

Population Near Threatened. Three distinct breeding populations in western and northern AK and in Russia. Russian population is currently much larger than AK populations, which have experienced dramatic declines in the last 50 years. Listed as Threatened by the USFWS.

KING EIDER *Somateria spectabilis*   KIEI ■ 2

The King Eider undergoes spectacular migrations in large flocks to arrive on the breeding grounds with enough time to breed and molt in the short Arctic summer. Monotypic. L 22" (56 cm)

Identification ADULT MALE: Very distinctive plumage and bare parts coloration, with bright orange frontal lobes, pinkish red bill, baby blue head, and a green wash on the side of the face. **FIRST-WINTER MALE:** Has a brown head, pinkish or buffy bill, and lacks white wing patches. Full adult plumage is attained by the third winter. **ALTERNATE MALE:** Dark overall, with only a bit of white mottling on mantle and in wing coverts; pinkish bill; and a distinct head shape with thin white postocular stripe. **FEMALE:** Brown overall with dark chevron markings on flanks and scapulars. Bill is dark, appears stubby. **FLIGHT:** Large rect-

angular head and short neck; stockier than the Common Eider, with faster wingbeats. The Common is larger, with a more horizontal body posture and triangular head. King Eider has its crest extending below its rectangular shaped head and a shorter neck. Male shows a partly black back and black wings with white patches. Young males have white flank patches.

Similar Species Males in flight have more black on upperparts than Common Eider. Females can be difficult to separate from Common Eider. Structurally King is smaller, with a flatter crown, rounded nape, and slightly bulging, shorter forehead that does not slope evenly into its bill, which is relatively smaller than Common. At rest the bill of King is held more horizontally than that of Common Eider, which is often angled down a

bit. Also, note that female Kings have a crescent or V-shaped markings on the flanks and scapulars as opposed to the vertical barring on Common Eider. **Voice** Males give a low, dovelike *urr urr urr*. Various low croaks given in flight.

Status & Distribution Fairly common Holarctic species. **BREEDING:** Common on tundra and coastal waters in northern part of range. **MIGRATION:** Move in impressive large flocks at all hours. In spring, begin departing Bering Sea in Apr. and early May. Undergo molt migration three to four weeks after the end of spring migration. **FALL:** Generally not seen well south of breeding range before late Oct. **WINTER:** Remains as far north as pack ice allows. **RARE STATUS:** Rare on East Coast south to mid-Atlantic. Very rare on the Great Lakes, except rare on Lake Ontario. Casual to FL and West Coast to Southern CA. Some southern winterers linger into summer. Accidental to Gulf Coast.

Population Sharp declines in late 20th century in N.A. population; more stable over last two decades.

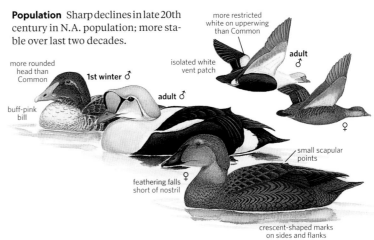

more restricted white on upperwing than Common

more rounded head than Common

1st winter ♂

buff-pink bill

isolated white vent patch

adult
adult ♂

♀

small scapular points

feathering falls short of nostril ♀

crescent-shaped marks on sides and flanks

COMMON EIDER *Somateria mollissima* COEI ▪ 1

This most common and widespread eider shows extensive geographic variation. Polytypic (6 sspp.; 4 in N.A.). L 24" (61 cm)
Identification MALE: Distinctive head pattern; frontal lobes and bill color vary with respect to subspecies. **ALTERNATE MALE:** Dark with white mottling overall. **IMMATURE MALE:** Dark with white breast. Full adult plumage attained by third year. **FEMALE:** Evenly barred flanks and scapulars. "Pacific Eider" (*v-nigrum*) is colder brown and duller overall compared to eastern subspecies, which is warm reddish brown. In the East, *dresseri* is dark and richly colored; *borealis* is reddish with dark markings; *sedentaria* palest.
Geographic Variation Four N.A. subspecies: *dresseri*, *borealis*, *sedentaria*, and *v-nigrum*. See head details below.
Similar Species See King Eider entry.
Voice Male gives a ghostly *ahoooooo* during courtship. Female gives a grating *krrrr* and other guttural calls.
Status & Distribution Locally abun-

dant. **BREEDING:** Nests in colonies along marine coasts. In Atlantic N.A. south to ME, some to MA. **MIGRATION:** Some populations sedentary, others long-distance migrants. In spring, Mar.–mid-June for Arctic breeders. Molt migration in June and July. In fall, peaks in MA Oct.–Nov.; late Nov.–early Dec. in NJ. **WINTER:** "Pacific Eider" winters in Bering Sea to south coastal AK in N.A. **RARE STATUS:** In East *dresseri* and *borealis* rare in winter on coast south to NC, casual south to FL, casual on Great Lakes, accidental northwest CA (*dresseri*). "Pacific Eider" (*v-nigrum*) rare in BC; accidental to Pacific states south to central CA; rare east to Greenland; *sedentaria* accidental outside Hudson and James Bays.
Population Near Threatened. Subspecies *v-nigrum* and *borealis* declining; *dresseri* relatively stable.

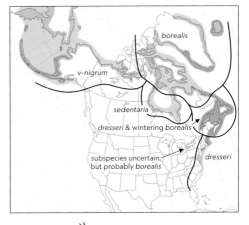

borealis

v-nigrum

sedentaria

dresseri & wintering borealis

subspecies uncertain, but probably borealis

dresseri

adult ♂

extensive white on back

v-nigrum

♀

feathering extends to nostril

♀ v-nigrum

overall color variable

♀ dresseri

white back

flat forehead

1st winter ♂

dresseri

evenly barred on sides and flanks

adult ♂

greenish tint under cap in *dresseri* and *sedentaria*

broadly rounded frontal lobe

greenish bill

"Atlantic Eider"
♂ *dresseri*

frontal lobe very similar to *dresseri*

"Hudson Bay Eider"
♂ *sedentaria*

lacks greenish under cap; face entirely white

more tapered frontal lobe

dull yellow bill

"Northern Eider"
♂ *borealis*

frontal lobe tapers to a fine point

orange bill

"Pacific Eider"
♂ *v-nigrum*

HARLEQUIN DUCK   *Histrionicus histrionicus*   HADU   ■   1

This small diving duck is structurally unique, with a steep rounded forehead, stubby bill, and chunky body. Males have a white crescent in front of the eye, a small white circular patch near the ear, and a white vertical stripe on the hind neck; these create the clown-like appearance for which it is named. Monotypic. L 16.5" (42 cm)

Identification ADULT MALE: Distinctive. Dark blue-gray overall with chestnut flanks; appears dark at a distance. Scapulars and tertials white, white band on breast and neck. **ALTERNATE MALE:** Dark sooty brown, but still shows male characteristics. **FIRST-WINTER MALE:** Like adult, but duller overall. Attains adultlike features throughout winter. **FEMALE:** Dark brown overall, pale belly, variably sized white patch in front of the eye, white spot behind eye. **FLIGHT:** Small head, steep forehead, plump body with long tail. Wings are entirely dark, but adult male has small white spots on a few coverts; whitish belly

of female is not very noticeable, and it appears dark overall. Flies swiftly, low over water with rapid wingbeats; usually seen in small numbers.

Similar Species Female, compared with female Bufflehead in flight, has darker body and longer, pointed tail. Smaller overall size, smaller bill, steeper forehead than female scoters. Chunky body with disproportionately tiny head and bill, round spot on face, and longer tail separate it from female Black Scoters. Darkest of female Long-tailed Ducks are similar, but have different head pattern and more white on body.

Voice High-pitched, nasal squeaking; commonly a mouselike squeak, *gia*.

Status & Distribution Locally fairly common from N.A. to Iceland and Eastern Asia. **BREEDING:** Nests on ground, near fast-flowing rivers, streams, lakes. **MIGRATION:** Not seen moving in large concentrations, short- to intermediate-distance migrant. In spring, moves inland from the coasts,

departing East Coast Apr.–mid-May. West Coast departs late Mar. and largely gone by mid-May. Immature and injured birds may stay on winter grounds. In fall, males undergo molt migration starting in late June. Rare in MA before Nov. **WINTER:** Rocky coastlines; south to NJ and central CA. **RARE STATUS:** Rare, but increasing on Great Lakes in winter. Rare to coastal Southern CA; casual northwest Mexico. Casual on Atlantic coast south to FL. Casual to accidental elsewhere in N.A. well away from mapped range. **Population** Stable. Much smaller Atlantic population is increasing.

round head with small bill

three white head spots

unique face pattern

adult ♂

♀

long tail

chestnut sides

adult ♂

adult ♂

head markings develop over winter

1st winter ♂

LABRADOR DUCK   *Camptorhynchus labradorius*   LABD   ■   6

This sea duck was the first N.A. endemic bird species to reach extinction. Very few reports after about 1860. The last reliable report was a specimen collected on Long I., New York, in the fall of 1875, possibly later at Elmira, NY, on 12 Dec. 1878 (the salvaged head was subsequently lost). Best known in

winter from Long I. Bill shape (wide toward tip, soft lateral flaps that hang down from bill, and a high number of lamellae inside the bill) suggests Labrador Duck was a food specialist, but very little is known about the species' natural history. It is believed to have foraged near sandbars by diving for

shellfish. Monotypic. L 22.5" (57 cm)

Identification ADULT MALE: White head, neck, and chest; the crown has a black stripe. **FEMALE:** Plumage is gray-brown overall, with throat whiter than head, mantle and scapulars slaty blue, and white speculum. **IMMATURE AND ALTERNATE MALE:** Resembles female.

Similar Species Confusion unlikely.
Voice Unknown. Reports of whistling sound likely produced by wings during swift flight.
Status & Distribution Extinct; some 60 specimens preserved. **BREEDING:** Probably remote coastal areas, likely Labrador or farther north. There are no nesting records or even definite records for Labrador. **WINTER:** Atlantic coast from NS south possibly to Chesapeake and Delaware Bays, estuaries, and sandy bays. **RARE STATUS:** Accidental to Montreal, QC, spring 1862.
Population Extinct. Explanations for extinction are largely speculative. The population already was believed to be small in the 1800s and intense market hunting was likely the most significant final factor.

white head and breast

dark band across neck

unique bill shape

adult ♂

SCOTERS Genus *Melanitta*

The AOS currently recognizes four species of scoters. Europeans split White-winged Scoter into two species for a total of five; AOS likely soon to split White-winged into three species, for a total of six species. They are medium to large, with black or dark brown plumage and orange or yellow on the bill. They nest in northern Canada and AK as well as in northern regions of the Old World. While scoters migrate nocturnally when crossing land, along the coast they migrate diurnally, often in mixed flocks. During peak migration multiple successive passing flocks may consist of several thousand in total. Scoters winter along seacoasts and on the Great Lakes. Very small numbers of all three N.A. breeding species are found throughout the interior every year, mostly in fall and most often on lakes and rivers. Generally Surf Scoter is the most numerous scoter inland.

SURF SCOTER *Melanitta perspicillata*   SUSC ▪ 1

Surf Scoters are the most common scoter in many areas. Male Surfs have a distinctive clownlike face. Monotypic. L 20" (51 cm)
Identification Forehead and bill are smoothly merged, forming a wedge-shaped head. The bill meets the face in a straight, vertical line; some feathers extend down the culmen toward the nares. Diving birds flick their wings open just as they submerge. **ADULT MALE:** Black with white patches on the nape, forecrown, and base of the bill. Eye pale and bill orange with a yellow tip. **ADULT FEMALE:** Uniform blackish brown with diffuse pale loral and postocular spots; many also have pale napes. **FIRST-WINTER MALE:** As female in fall, but with a pale belly; acquires much of the black-and-white plumage and orange bill during the first winter, but retains the pale belly through the spring. **FIRST-WINTER FEMALE:** Similar to female but slightly paler brown with a pale belly; never has pale nape as with many adult females. **FLIGHT:** Large, wedge-shaped head, sleek body, and pointed wings give the Surf Scoter a unique silhouette. Forms dense flocks.
Similar Species First-winter females can have pale cheeks like Black Scoter, but can be separated by head shape and in flight by less contrasting flight feathers.
Voice Generally silent.
Status & Distribution Common; most numerous scoter along Pacific coast. **BREEDING:** Nesting on shallow lakes, often with rocky shores. **MIGRATION:** In spring, begins in Mar.; peaking in CA and OR in second half of Apr.; mid-Atlantic peak in late Mar. or early

Apr.; movements less noticeable in spring than fall on Atlantic coast. In fall, begins in late Sept., peaking on US coasts mid-Oct. to early Nov. **WINTER:** In large flocks along both coasts, usually close to shore. **RARE STATUS:** Rare in the interior, where it is primarily a fall migrant; rare on the Gulf Coast. Casual to central Mexico, HI, Bermuda, Greenland, Europe (mostly northwest Europe, annual UK) and northeast Asia.
Population Fairly stable; possibly declining.

adult ♂

more uniformly dark wings than Black

immature ♀

whitish belly

feathering out culmen

white head patches

forward whitish patch vertically shaped

white nape on adult female

adult ♀

forward whitish patch vertically shaped

1st winter ♀

adult ♂

bill shows color by Dec.

1st winter ♂

immatures of all scoters have pale bellies

WHITE-WINGED SCOTER *Melanitta fusca*   WWSC ■ 1

When migrating, these—the largest and most distinctive of the four scoters—form smaller flocks. Polytypic (3 sspp.; 2 in N.A.). L 21" (53 cm)

Identification The white secondaries are obvious on birds in flight. White secondaries may be visible on swimming birds, but are often hidden. When diving, they flip their wings open even more distinctly than Surf Scoters. **ADULT MALE:** Head, breast, and back are black; sides are dark brown. A small white comma mark surrounds the eye; orange tip on bill. **ADULT FEMALE:** Dark brown with two faint round spots on the face. **FIRST WINTER:** Paler than adult female with distinct head patches and pale belly; male bill color develops by midwinter. **FLIGHT:** Broad wings and heavy flight are reminiscent of eiders; white secondaries diagnostic.

Geographic Variation Three subspecies: widespread *deglandi* in N.A.; Asian subspecies *stejnegeri* breeding from east of the Urals to Kamchatka has in the adult male a more obvious nasal protuberance, different bill color pattern, longer white post

ocular stripe and black flanks. It is of nearly annual occurrence in spring from St. Lawrence I. and is recorded from Nome and the Pribilof Is.; a winter record from Santa Cruz, CA, will soon be evaluated. Western Palearctic adult male *fusca* ("Velvet Scoter," Vulnerable), casual to Greenland, also has black flanks and largely lacks a bill knob, but note different bill coloration. Females of all subspecies very similar to one another, but with subtle differences in head shape and facial feathering around the bill. All three subspecies will likely soon be recognized by the AOS as distinct species.

Similar Species Compare females to Surf Scoter. Face pattern less distinct, feathering extends farther down bill. White secondaries distinctive, if visible.

Voice Generally silent.

Status & Distribution Uncommon to fairly common. Significant numbers now in late fall and winter on the Great Lakes, especially Lake Ontario. **BREEDING:** Taiga ponds. **MIGRATION:** Early Mar.–May. In fall, peaks in the

mid-Atlantic in late Nov. **RARE STATUS:** Rare to casual in the interior and the Gulf Coast. Nominate *fusca* casual to Greenland, but no N.A. records yet.

Population N.A. subspecies *deglandi* has declined in many regions, including both coasts; breeding range has contracted northward.

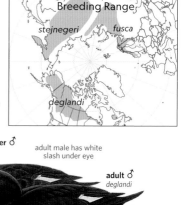

Breeding Range

stejnegeri   *fusca*

deglandi

adult ♂

deglandi

only scoter with white in wings

1st winter *deglandi* ♀

juveniles of both sexes show whitish patches; forward patch horizontally shaped

1st winter ♂ *deglandi*

adult male has white slash under eye

adult ♂ *deglandi*

adult ♀

feathering extends out bill

adult ♀ *deglandi*

brownish flanks

adult ♂ *stejnegeri*

"Velvet Scoter" *fusca*

small knob

yellow on sides of bill

recorded from Greenland

adult ♂

black flanks like *stejnegeri*

white not always visible on folded wing

bill color and white behind eye differs from *deglandi*; note bulbous nostrils

black flanks

COMMON SCOTER *Melanitta nigra*   COSC ■ 5

Common Scoter, the Eurasian counterpart to Black Scoter, was until recently considered conspecific with that species. Even before the two taxa were elevated to full species in 2010, seawatching birders in coastal areas from Labrador to Massachusetts had kept a keen eye out for Common Scoter, which breeds as near as Iceland and has been documented as close as Greenland. Monotypic. L 18–19" (45–49 cm)

Identification ADULT MALE: Uni-

formly black in plumage with a distinctive bill. The base of the maxilla appears swollen at the base, then appears oddly indented across the culmen; most of the bill is black, the central portion of the maxilla yellow-orange, including (usually) a variably narrow stripe in the middle of the swollen portion. **FEMALE:** Closely resembles Black Scoter, being dark brown with contrasting pale cheek and throat. **FIRST WINTER:** Resembles

black knob

female like Black Scoter

small yellow area on bill and thin yellow eye ring

thinner neck than Black Scoter

adult ♂

females but has paler belly. Young males acquire a bill pattern similar to the adult during the first winter.

Similar Species Adult male most likely to be confused with Black Scoter, but differences in bill shape and pattern are diagnostic. In addition, there are some very subtle structural differences: The wings of Common are more pointed than those of Black, and Common has a slightly thinner and more rounded head than Black. In these closely related species, females and immatures are not known to be distinguishable in the field; however, the nostrils are set closer to the base of the bill in Common.

Voice Male's call is very similar to Black Scoter's mournful whistle in quality, but male Common's is lower in pitch and usually (by a factor of four) much shorter than Black, though Black sometimes gives shorter calls.

Status & Distribution BREEDING: Adjacent freshwater pools, lochs, lakes, and rivers in tundra and heath habitats from Iceland across much of northern Eurasia, east to the Olenek River in the Russian Far East. **WINTER:** Coastal (saltwater) areas from Norway to the British Isles, sparingly south to Portugal and Gibraltar. **RARE STATUS:** Accidental, recently recorded twice in winter from Pacific coast in OR and northwest CA (adult males).

Population Possibly declining, but insufficient data.

BLACK SCOTER *Melanitta americana*   BLSC ▓ 1

The handsome Black Scoter is often located in winter and spring by the male's plaintive, whistling calls, given incessantly, particularly during foggy or calm conditions, as flocks are resting in rafts. Black Scoters migrate in flocks, often mixing with other scoters. Monotypic. L 18–19" (45–49 cm)

Identification Chunky body with small head and bill; at a great distance the bill seems to disappear, leaving the impression of a flat profile. The tail may be held flat but is often held cocked above the water, like a Ruddy Duck. **ADULT MALE:** Uniformly black with a bright orange knob at the base of the maxilla. **FEMALE:** Dark brown with contrasting pale cheek and throat. First-year birds resemble females but have paler belly. **FIRST-WINTER MALE:** Young males slowly acquire the orange knob during the first winter.

Similar Species Distinctive at close range. Much smaller winter male Ruddy Duck, mostly brownish with a pale cheek, is superficially like female Black Scoter. Distant Black Scoters in flight can be difficult to separate from Surf Scoter, but some are identifiable with careful study of head, wing, and body shape. Black has a shorter "hand" with more rounded wing tips, and stronger silver flash on the underside of the primaries (in optimal light). Its rounded head, with steep forehead, differs from the long, sloping profiles of White-winged and Surf. As a Black Scoter's primaries wear, they become strongly translucent, a feature that is most useful in the spring, especially for adult males, and an attribute that also separates Black from the two other N.A. scoters. Unlike White-winged and Surf, Black dives with a small leap, wings held closed. Also see Common Scoter.

Voice Males produce low, mellow whistle and less often a rattling call. Females are said to give squealing and growling calls on the breeding grounds. Adult males also produce wing noise when flying, like other scoters.

Status & Distribution Much more numerous on Atlantic coast than Pacific. **BREEDING:** On tundra pools, lakes, and rivers in two disjunct parts of N.A.; the western population nests mostly in AK and adjacent Russia, while the eastern population is confined mostly to northern QC and Labrador. Breeds west in northern Russian Far East to Yana River. **MIGRATION:** Usually in large flocks, most visible in

fall. In spring, peak in the mid-Atlantic in early Apr., in Canadian Maritime Provinces in early May, and arrival on the breeding grounds in northern QC occurs in the third week of May. In CA, migration begins in early Mar. Southward migration in fall begins in Sept., peaking in the third week of Oct. in the mid-Atlantic, and in early Nov. in BC. Smaller numbers pass through the Great Lakes, and the species can be common on Lake Ontario, where some winter. **WINTER:** Found on inshore bays and inlets and offshore along both Atlantic and Pacific coasts; very uncommon to very rare along Gulf Coast and in the interior.

Population Near Threatened. Possibly declining, but insufficient data.

1st winter ♂

dull yellow at bill base

yellow-orange knob at base of bill

round head

adult ♂

no feathering out on bill

pale face contrasts sharply with dark cap

adult ♀

adult ♂

adult ♀

secondaries and primaries paler than rest of wing, especially on adult males

Genus *Clangula*

LONG-TAILED DUCK *Clangula hyemalis*   LTDU ■ 1

Unique sea duck, formerly known as the Oldsquaw, has three plumages a year. Monotypic. L 6–22" (41–56 cm) **Identification** All males have pink band on bill, lacking on females. **ADULT MALE WINTER:** Long tail conspicuous. White head with gray patch around eye, black band across breast, long gray scapulars, light gray flanks. **EARLY SUMMER MALE:** Head black, still with eye patch, scapulars buffy with black centers. By late summer, head and neck become whiter, flanks darker gray. **FIRST-YEAR MALE:** Lacks long tail. Gradually attains adultlike features by second fall. **FEMALE:** Crown and nape blackish brown, breast grayish, flanks whitish, scapulars buff with black centers. By fall whiter head and neck. In summer head and neck dusky overall. **FLIGHT:** Appears very white, with uni-

formly dark underwings. Identifiable by swift, careening flight. **Similar Species** Female similar to female Harlequin, but Harlequin's dark flanks, steep forehead, rounded head, and facial pattern are distinct. Female Steller's Eider is darker overall with plain face, white-edged speculum, curved tertials. In winter flight, guillemots can cause confusion; they have a bold white wing patch, unlike the black wings of the Long-tailed. **Voice** Highly vocal. Male utters nasal, loud, yodeling, three-part, *ahr-ahr-ahroulit*. Both sexes give soft *gut* or *gut-gut* call while feeding. **Status & Distribution** Much more numerous on Atlantic coast than Pacific. **BREEDING:** Holarctic. Arctic tundra ponds. **MIGRATION:** In spring, East Coast and Great Lakes in late Mar.–

early Apr., some into May; in West and Great Lakes in late Feb.–May. In fall, US coasts and Great Lakes peak numbers in late Nov. and Dec. **RARE STATUS:** Away from Great Lakes and Northeast lakes, rare from interior N.A. **Population** Vulnerable. Significant recent global declines.

winter adult ♂ · dark wings · 1st fall ♀ · variably white on face · winter ♀ · winter ♀ · males have pink band on bill · small bill · early summer adult ♂ · long pointed tail · 1st winter ♂ · females have grayish bills · winter adult ♂

Genus *Bucephala*

BUFFLEHEAD *Bucephala albeola*   BUFF ■ 1

This diving duck has adapted to a wide variety of habitats from sheltered bays, rivers, to flooded fields. It is one of the smallest ducks, with a large puffy head, a steep forehead, and a short stubby bill. Its plumages are suggestive of miniature goldeneyes. It flies fast and with such rapid wingbeats that the wings blur; usually seen only in small flocks. Monotypic. L 13.5" (34 cm)

Identification ADULT MALE: Glossy black back, white below, with large white patch on iridescent purple-green head. **ALTERNATE MALE:** Like female, but darker head with larger white patch; retains white in wing.

FIRST-WINTER MALE: Similar to female. **FEMALE:** Dark gray-brown above, dusky white below. Dark head and elongated white patch under eye. **Similar Species** Female's very rapid wingbeats and small size can resemble

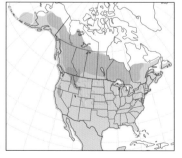

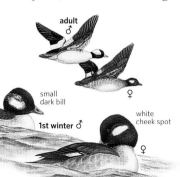

extensive white head pattern · adult ♂ · adult ♂ · small dark bill · ♀ · white cheek spot · 1st winter ♂ · ♀

some alcids. Compare female Harlequin; Bufflehead has larger head, shorter tail, and faster wingbeats. Female and juvenile Long-tailed Ducks are larger, with slower wingbeats, longer tail, whiter head. **Voice** Generally silent; most vocal during breeding season displays. **Status & Distribution** Common and widespread. **BREEDING:** Nests in woodlands near small lakes, ponds. **MIGRATION:** In spring, begins in Feb., arrives on breeding grounds early Apr.–early May. In fall, begins late Oct., with winter numbers in place by early Dec. **WINTER:** South to central Mexico. A variety of aquatic habitats including sheltered bays, rivers, and lakes. **RARE STATUS:** Casual to Europe and northeast Asia. **Population** Stable in recent decades.

COMMON GOLDENEYE *Bucephala clangula* COGO ▪ 1

This common and widespread goldeneye is usually found in or near deep, clear water. Polytypic (2 sspp.; *americana* in N.A.). L 18.5" (47 cm) **Identification** Large triangular head, sloping forehead, and longer bill than the Barrow's, with shallower base. **ADULT MALE:** White circular spot on dark head distinctive. **FIRST-WINTER MALE:** Like female; transition to more like adult male throughout first winter. White loral spot develops slowly, can appear crescent shaped, causing confusion with the Barrow's. **FEMALE:** Dull brown head, pale gray body, and black bill with yellow tip. Rarely can have mostly yellow bill. **FLIGHT:** Heavy and muscular appearance. Show more white on the upperwing than the Barrow's, but difficult to judge. **Geographic Variation** Subspecies *americana* is found in N.A.; smaller nominate *clangula* is found across the Old World.

Similar Species Males have more white on the scapulars than Barrow's and lack the black spur on the breast of Barrow's. Loral spot round and forehead not as steep as Barrow's. Hybrid Common x Barrow's are rare; male hybrids (that can be identified) typically look almost exactly intermediate between the two species (see p. 48). **Voice** Male gives a soft, short *preent* during display. Female occasionally gives a harsh croak: *gack*. **Status & Distribution** Common Holarctic species. **BREEDING:** Open lakes with nearby boreal woodlands where nest holes are available. **MIGRATION:** In spring, departs southern part of winter range in late Feb. Peak on Atlantic coast late Mar.–early Apr. In fall, New England and Atlantic coast peaks early Dec. Pacific coast migration late Oct.–early Dec.; CA peak third week of Nov. **WINTER:** Coastal areas, inland lakes, and rivers. **Population** Stable.

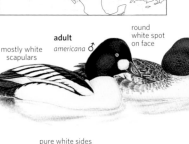

adult ♂

americana

♀

round white spot on face

1st winter *americana* ♂

americana ♀

adult *americana* ♂

mostly white scapulars

pure white sides

BARROW'S GOLDENEYE *Bucephala islandica* BAGO ▪ 1

Breeds in N.A. and Iceland. Monotypic. L 18" (46 cm) **Identification** Puffy head, steep forehead, and stubby bill. **ADULT MALE:** White crescent; black spur separates white breast and flanks. **ADULT FEMALE:** Dark brown head, gray body, yellow-orange bill that becomes black in summer. **FIRST-WINTER MALE:** Facial crescent evident on many by late Dec. **FIRST-WINTER FEMALE:** Like adult female, but with largely black bill. **Similar Species** Adult female from Common by steep forehead and shorter bill. Bill usually all yellow-orange; some female Commons have yellow bill, but color is paler, not as orange. Separation

adult ♂

slightly less white on wing coverts than Common

♀

many immature males have a thin white facial crescent by Dec.

1st winter ♂

steeper forehead and stubbier bill than Common

black scapulars with white spots

1st winter ♀

white crescent on face

vertical black spur on side

adult ♂

orange-yellow bill

adult ♀

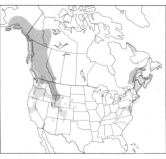

often difficult; hybridization complicates identification.

Voice Generally silent.

Status & Distribution Fairly common in West; mostly rare in East; found northeast to Greenland (one historical breeding record) and Iceland. **BREEDING:** Open lakes and small ponds.

MIGRATION: In spring, mostly late Mar.–early Apr. In fall, late Oct.–late Nov. **WINTER:** Sheltered coastal areas, lakes, and rivers. **RARE STATUS:** In East, casual away from Great Lakes, Northeast, and mid-Atlantic; casual Bering Sea Is. and TX. Accidental to Southeast. **Population** Generally stable.

Barrow's x Common hybrid adult ♂

hybrid adult males have intermediate amount of white on scapulars and white facial mark

---

MERGANSERS Genera *Mergellus, Lophodytes,* and *Mergus*

The long, thin, serrated bills and slender, long-necked bodies of mergansers aid them in catching fish, crustaceans, and aquatic insects. There are six species worldwide, four in N.A. Fast and direct in flight, mergansers show pointed wings, shallow wingbeats, long neck and bill. They nest in tree cavities, nest boxes, and on ground sheltered by vegetation.

SMEW *Mergellus albellus*   SMEW ■ 3

This Eurasian species is rare throughout N.A. Most records are western and central Aleutians. The Smew is kept in captivity in US, therefore origin is often questioned for sightings away from AK and the West Coast. Monotypic. L 16" (41 cm)

Identification ADULT MALE: White with black markings, black-and-white wings conspicuous in flight. **FIRST-**

SPRING MALE: Dark rufous crown and nape, dark around eye, white cheek, gray body; resembles female in winter. **FEMALE:** White throat and lower face contrast with reddish head and nape. **FLIGHT:** Male and female have white patches on upperwing.

Similar Species Adult male unmistakable. Brown crown and white face diagnostic on female and immature male.

Voice Generally silent, except during courtship.

Status & Distribution Rare migrant to west and central Aleutians mid-Mar.–May and primarily Oct. to early winter, where it has been observed in small flocks. **BREEDING:** Northern Eurasia near rivers and small lakes in forested areas. **RARE STATUS:** Casual elsewhere in AK and in the Pacific states; accidental in the East, including three records from ON. Most records away from AK Nov.–Mar. **Population** Stable.

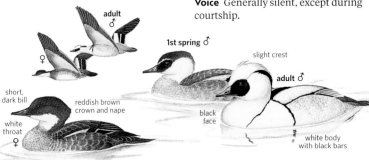

adult ♂

1st spring ♂

slight crest

adult ♂

short, dark bill

reddish brown crown and nape

black face

white throat ♀

white body with black bars

HOODED MERGANSER *Lophodytes cucullatus*   HOME ■ 1

This small, oddly shaped diving duck is frequently found on ponds, wooded sloughs, and streams in small groups or pairs. Monotypic. L 18" (46 cm)

Identification MALE: Black-and-white fan-shaped crest, black back, chestnut flanks, white breast with black band, black bill. **FIRST-YEAR MALE:** Like female, but by late winter white in crest, yellow eye, sometimes dark feathers on head, bill blackens. **ALTERNATE MALE:** Dusky-brown crest, dark bill, yellow eye. **FEMALE:** Brownish overall, paler breast, upper mandible dark, lower yellowish. **FLIGHT:** Very quick wingbeats, head usually held low, crest flattened, prominent tail.

Similar Species Female can be confused with female Red-breasted. Much smaller; thinner dark bill; lower man-

dible yellow. In flight like Wood Duck, but more slender shape and faster, shallower wingbeats.

Voice Generally silent; froglike growl often given during courtship displays.

Status & Distribution Uncommon in West, common throughout much of East. **BREEDING:** Forested areas with nest cavities. Widely scattered nesting records south of mapped range. **MIGRATION:** In spring, begin departing Southeast early Feb.; remain on coastal areas in Pacific Northwest until mid-Apr. In fall, peak mid-Nov. in New England; numbers increase by late Dec. in FL. Arriving on wintering areas late Oct. in Pacific Northwest. **RARE STATUS:** Rare to Bering Sea region and northern Mexico. **Population** Stable.

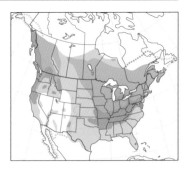

crest sometimes raised

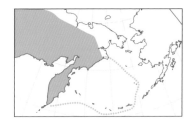

very similar to
female but pale
iris by midwinter

1st spring ♂

crest often
flattened

brownish
bushy crest

adult ♂

♀

dull
yellow bill

chestnut
sides

long
tail

adult
♂

very fast
wingbeats

♀

COMMON MERGANSER *Mergus merganser* COME ■ 1

This hardy merganser is regularly seen along raging streams or loafing on chunks of ice in winter lakes. It forms large single-species flocks in winter, often tightly packed and facing the same direction. Obtains food by diving in deep water; also forages in shallow water by swimming with its face submerged looking for prey. Polytypic (3 sspp.; 2 in N.A.). L 25" (64 cm)

Identification One of the largest ducks, with a heavy body, thick neck, and long, slender, hooked bill. Sits low in the water, holding its large head erect. Slight crest on nape can be held close to the head so it almost disappears. **ADULT MALE:** Dark head with green gloss, white breast and flanks, generally distinctive. **FEMALE:** Bright chestnut head and neck contrast with white chin; bill bright red and body uniformly pale gray. **ALTERNATE MALE:** Resembles female, but retains adult male's wing pattern. **FIRST-YEAR MALE:** Resembles adult female, but with paler body and with variable amounts of white feathers on scapulars appearing during the first winter. **FLIGHT:** Shows large white wing patches with

a single black bar across the median coverts, larger on males than females. Bulkier body and thicker neck than Red-breasted. Flight is extremely swift, with shallow, rapid wingbeats.

Geographic Variation Three subspecies worldwide; only *americanus* regular in N.A. Nominate *merganser* ("Goosander") migrant to western Aleutian Is., found across Eurasia. Males of nominate subspecies have a shallower midsection to the bill with a longer hook, and lack visible black bar across the white wing patch; head shape in display very different from *americanus* (see illustration). Largest subspecies, *comatus*, is found in Central Asia.

Similar Species Female can be confused with the smaller female Red-breasted, but has sharply delineated white chin surrounded by chestnut head and throat, unlike diffused pale throat and neck of Red-breasted. Heavier bill than Red-breasted that meets the head on a straight line. Tends to look paler and cleaner than Red-breasted, which appears more mottled. Robust build with a thick neck and large head, unlike slender Red-breasted.

Voice Generally silent, except during courtship or when alarmed.

Status & Distribution Common Holarctic species. **BREEDING:** Nests in woodlands near lakes and rivers.

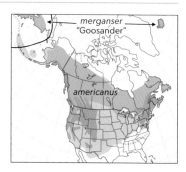

merganser
"Goosander"

americanus

MIGRATION: Short- to intermediate-distance migrant, nocturnal overland movements and diurnal migration along coasts. In spring, leave southern part of wintering range in mid-Feb.; most birds gone by Apr. Great Lakes region, arrive late Mar., peak late Apr. In fall, on Atlantic coast, arrive mid-Nov., peak Dec.–Feb. in VA. On West Coast peak in BC in Nov. **WINTER:** Found as far north as there is open water; large lakes, reservoirs, rivers, sometimes on coastal bays, estuaries, and harbors. In eastern N.A., regular in winter south to Ohio River region. **RARE STATUS:** Casual to Gulf Coast. **Population** Stable or increasing in N.A. and Europe.

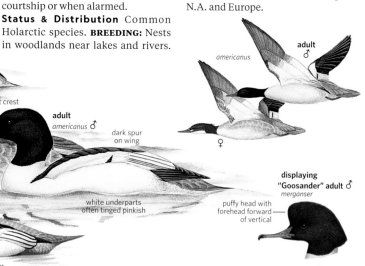

thick base
to bill

1st spring
americanus ♂

americanus ♀

short crest

rich rufous head and
lower throat contrast
sharply with white chin

adult
americanus ♂

dark spur
on wing

adult
americanus

♀

thinner midsection
to bill and more
distinct hook
than *americanus*

"Goosander"
adult ♂
merganser

smaller black spur
on wing concealed

white underparts
often tinged pinkish

**displaying
"Goosander" adult** ♂
merganser

puffy head with
forehead forward
of vertical

RED-BREASTED MERGANSER *Mergus serrator*　RBME ▪ 1

This species is most often seen in flocks in coastal wintering areas. Monotypic. L 23" (58 cm)

Identification Medium size with slim neck, thin bill, and ragged, two-pronged crest. **ADULT MALE:** Green head; white collar; brown breast; gray flanks; distinctive. **ADULT FEMALE:** Gray-brown body with a dull reddish brown head; slender reddish bill; all-white throat and breast. **FIRST-YEAR MALE:** Similar to female, but by midwinter often shows blackish around eye. **FLIGHT:** White wing patches divided by two black lines on male, one on female.

Similar Species Female similar to Common, but Red-breasted is slender, with thinner neck, much thinner bill, full ragged crest, uniformly pale throat.

Voice Generally silent, except during courtship. Male gives catlike *yeow-yeow*. Female produces raspy croaking during displays.

Status & Distribution Common Holarctic species. **BREEDING:** Arrive mid- to late May. Nests in woodlands near freshwater or sheltered coastal areas. **MIGRATION:** Abundant on Great Lakes, where some winter. In fall, pass through Great Lakes first two weeks of Nov., with numbers increasing in mid-Atlantic in early Nov. Prolonged fall migration, principally from mid-Sept.–early Nov. in BC. Uncommon in interior, more common in East (as a migrant). **WINTER:** Favors salt water more than other mergansers. Common in coastal regions and Great Lakes, except Lake Superior.

Population Perhaps stable, but recent declines from some regions, e.g., coastal New England.

adult ♂

♀

1st winter ♂

thin base to bill

head more brownish than Common; crest larger but appears wispier

cinnamon-brown chest

adult ♂

throat and chest more uniformly colored than Common ♀

STIFF-TAILED DUCKS Genera *Nomonyx* and *Oxyura*

Long, narrow, stiff tail feathers serve as a rudder for these diving ducks. Males have a bright blue bill in breeding season. They are found in fresh and brackish water; Masked Duck prefers ponds with dense emergent vegetation. When actively swimming and diving, the tail is frequently held flat to the water's surface, not cocked and spread as it is held when resting.

MASKED DUCK *Nomonyx dominicus*　MADU ▪ 3

This small tropical stiff-tail most frequently appears in southern TX and north to the upper TX coast region. Shy and difficult to see even when present, it spends most of its time hidden in dense aquatic vegetation. Monotypic. L 13.5" (34 cm)

Identification Slightly smaller than Ruddy Duck with a smaller head and shorter bill. Like Ruddy, it has a long tail that can be cocked above the body or trail behind on the water. **BREEDING MALE:** Black face with sky blue bill. Nape and body are dark reddish brown with black-centered back and flank feathers. **FEMALE:** Warm buffy brown, with a buffy supercilium and two dark stripes on the cheek. **WINTER MALE:** Like female, but with stronger stripes on the face. **FLIGHT:** Large white patches on inner wing. Unlike Ruddy, it takes to flight easily off the water.

Similar Species Duller plumages are similar to female and winter male Ruddy Duck, but overall coloration warmer (buffier) and with two distinct dark stripes on face. Also shows a pale supercilium, which Ruddy lacks. Ruddy Ducks prefer more open water with less vegetation. Any stiff-tail with white wing patch can instantly be identified as a Masked Duck. Male Ruddy can show an entirely black face or heavily flecked cheek patch during molt to breeding plumage.

Voice Generally silent.

Status & Distribution Rare; sometimes occurs in small numbers after invasions to TX, but then it will go for many years without being recorded at all. Casual to LA and FL; accidental to OK, TN, WI, NC, MD, PA, MA, and VT. **BREEDING:** Nests on lowland ponds, marshes, and rice fields with much floating vegetation. **Population** Probably stable. Numbers fluctuate in TX, more likely in wet years. Recent invasions in the 1930s, late 1960s, early 1970s, and 1990s. Due to the species' secretive nature, preference for highly vegetated ponds, and fact that it is often on private land, it is difficult to assess the population and perhaps not as rare as is perceived.

breeding ♂
long pointed tail
white wing patches, but species seldom seen in flight
♀

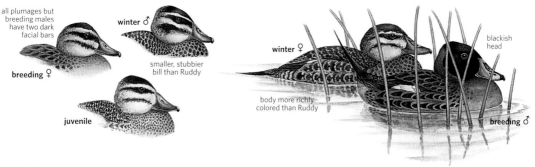

all plumages but breeding males have two dark facial bars
winter ♂
smaller, stubbier bill than Ruddy
breeding ♀
juvenile
winter ♀
blackish head
body more richly colored than Ruddy
breeding ♂

RUDDY DUCK *Oxyura jamaicensis*   RUDU ■ 1

The Ruddy Duck is the only widespread stiff-tail in N.A. Bright breeding plumage is held during summer and dull plumage during winter, unlike other ducks of N.A. Polytypic (3 sspp.; *rubida* in N.A.). L 15" (38 cm)
Identification Small, chunky duck with disproportionately large head and bill. Frequently cocks long tail in a fan shape. **BREEDING MALE:** Bright baby blue bill, black crown contrast with white cheeks and chestnut body. Often with black flecking or almost entirely black cheeks when molting to breeding. **WINTER MALE:** Dark bill; retains a white cheek and black crown but body becomes dull gray-brown. **FEMALE:** Fairly dull gray year-round. Dark crown; a single dark line crosses buffy cheek. **FLIGHT:** Prefers to dive or run on water to avoid danger. When they do fly, wingbeats are rapid, usually barely clearing the water.
Similar Species Ruddy Duck has a unique structure and plumage; it is most likely to be confused with the similarly shaped but smaller Masked Duck. Female Bufflehead is superficially similar to the Ruddy Duck, but has a rounded head with a high forehead and smaller bill.
Voice Generally silent, except for males during spectacular courtship display; females also give various sounds when being pursued by males.
Status & Distribution Common. **BREEDING:** Nests in dense vegetation of freshwater wetlands. **MIGRATION:** Almost exclusively nocturnal. In spring, begins in early Feb. in south; ends third week of May in northern parts of range. In fall, earlier than other diving ducks, beginning in late Aug., peaks mid-Sept.–late Oct., ending in Dec. **WINTER:** Lakes, bays, and salt marshes.

Population Stable or increasing in N.A. Released and escaped from wildfowl collections in the UK in the 1950s; its feral population had increased to 6,000 by the early 2000s with spread to Europe (even strays to Iceland). Eradication efforts in Europe are ongoing.

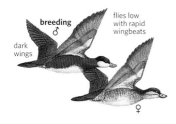

breeding ♂
flies low with rapid wingbeats
dark wings
♀

dark cap contrasts sharply with white face
winter *rubida* ♂
long pointed tail often raised
winter *rubida* ♀
raised tail is spread in display
bright blue bill
breeding *rubida* ♂
rufous body
single blurry line on face
breeding *rubida* ♀

EXOTIC WATERFOWL

Many waterfall species are brought to N.A. from other continents for zoos and private collections. Some escape from captivity. The ones illustrated are the ones most frequently encountered in the wild. In assessing origin, look especially for a leg band, clipped wing tips, and a missing hind toe, or notched web.

SWAN GOOSE *Anser cygnoides*   SWGO ■ EXOTIC

Native to eastern Asia, the domestic form is known as the "Chinese Goose," the only domestic goose not of Graylag ancestry. There are two domestic types: the slimmer-bodied "Chinese Goose" and the larger, heavy-bodied "African Goose." Color varies from all-white with orange bill to brown and white with black bill. The wild form is slim; its long, swanlike bill lacks knob at base. It could potentially reach western AK as a stray. L 45" (114 cm)

Identification Crown, nape, hind neck dark brown; cheek and sides of neck white.
Similar Species White form could be confused with other domestic white geese of the Graylag-type, but note prominent orange knob and bill of white types.
Status & Distribution Domestic forms found regularly in N.A.
Population Vulnerable in its native range of eastern Asia.

"Chinese Goose" domestic type

BAR-HEADED GOOSE *Anser indicus*   BHGO ■ EXOTIC

This native to central Asia, known for its record high-altitude flights over the Himalaya, is common in captivity in N.A. and Europe, and escapes are frequent. L 30" (76 cm)
Identification ADULT: Two black bars on rear of white head; white line runs down side of gray neck. Back is pale gray with white-edged feathers. Bill is yellow with a black nail; orange-yellow

legs. **JUVENILE:** Less distinctive than adult, with white face, dark gray crown and hind neck, paler gray body.
Similar Species Distinctive.
Voice Honking flight call.
Status & Distribution Escapes across N.A. and western Europe. **BREEDING:** High mountain lakes of central Asia. **MIGRANT:** Flies over the Himalaya to reach wintering grounds. **WINTER:**

Mountain rivers, lakes, and grassy wetlands primarily in northern half of Indian subcontinent, west to Pakistan and east to northern Myanmar.
Population Now stable.

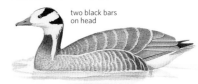

two black bars on head

RUDDY SHELDUCK *Tadorna ferruginea*   RUSH ■ EXOTIC

This Afro-Eurasian species (popular in zoos and private collections) is a nomadic species. Monotypic. L 26" (66 cm)
Identification Prominent white wing coverts and ruddy plumage. **MALE:** Head and neck buffy, with thin black band around neck, ruddy body. **FEMALE:** Like male, but lacks neck collar and has white face patch. **FLIGHT:**

Upperwing coverts white, primaries black. Underwing coverts white, contrasting with dark flight feathers.
Similar Species Unmistakable except for other shelduck species.
Voice Frequently gives a loud, nasal trumpeting call in flight.
Status & Distribution Frequent in captivity. A record of a flock of six on 23 July 2000 at Southampton I., NU,

likely wild birds from the Old World; accidental in Iceland and Greenland.
Population Declining in some regions.

in flight, note extensive white in wings

female lacks neck collar and has white face

adult ♂

MANDARIN DUCK *Aix galericulata*   MADU ■ EXOTIC

Beautiful native duck of eastern Asia that is commonly kept in waterfowl collections. Like its relative the Wood Duck, it favors wooded lake and rivers and is often seen perching in trees. Monotypic. L 16" (41 cm)
Identification MALE: Ornate plumage. Reddish bill, white band sweeps behind eye, orange "mane" and orange tertial "sails." **ALTERNATE MALE:** Resembles female, but bill is reddish with less distinct eye ring, shaggier

crest, and glossy upperparts. **FEMALE:** Resembles female Wood Duck, but paler and grayer overall, white eye ring with thin white postocular stripe. Bill grayish with pale nail. Wing coverts dull brown (glossy on Wood Duck).
Voice Most vocal in breeding season.
Status & Distribution Eastern Asia. Established in Britain. Feral population in central Sonoma Co., CA. Escapes widely seen from coast to coast. **BREEDING:** Russian Far East, China, Japan;

nests in tree cavities. **WINTER:** Mainly lowlands of eastern China and southern Japan.
Population Declining in Asia from destruction of forests, and previous export from China in large numbers.

long, thin white postocular stripe

compare to female Wood Duck

♂   ♀

CURASSOWS AND GUANS Family Cracidae

Plain Chachalaca (TX, Nov.)

Cracids are a diverse family of primarily arboreal game birds, many of which have elaborate wattles and knobs. They generally feed on leaves and fruits of trees, with some species foraging on the ground for fallen fruits.

Structure They range in size from medium (chachalacas) to large (curassows) and have heavy bodies. Common characteristics include longish necks with small heads, long broad tails, and rounded wings. Their bills are rather chickenlike and their legs are fairly long and strong.

Behavior In general, cracids live in groups; the chachalacas are the most sociable. Most species are found moving through the canopy of tropical forests, but chachalacas generally stay closer to the ground.

Plumage Overall plumage varies from olive brown to black; only the curassows, in some species, show strong sexual dimorphism.

Distribution A New World family, cracids are found from south TX and northwestern Mexico through the Neotropics to northern Argentina. The family reaches its greatest diversity in northern South America. The chachalacas are a uniform group; they appear to occupy similar niches to the point that there is virtually no overlap in distribution between the different species.

Taxonomy The family can be informally divided into three major groups: curassows, guans, and chachalacas—54 species in 11 genera. The 15 species of chachalaca—characterized by small size (for the family) and plain coloration—are all in the genus *Ortalis*.

Conservation Most chachalaca species are common; but among the guans and curassows, close to 70 percent are at risk. Small clutch sizes do not allow their populations to recover quickly from hunting pressures. Many of the species at risk depend on primary forest, seriously threatened by deforestation. BirdLife International codes: 5 NT, 8 VU, 8 EN, 5 CR, and 1 EW (Alagoas Curassow, ±130 individuals, all in captivity).

CHACHALACAS Genus *Ortalis*

PLAIN CHACHALACA *Ortalis vetula*   PLCH ■ 2

Although generally secretive, this south TX bird is very vocal and has become habituated to feeding stations. The Plain Chachalaca is always found in small social groups—numbering 10–15 individuals—as it moves through the understory and on the ground. It often hops from branch to branch without the use of its wings. The English name of the genus, chachalaca, is derived from the loud chorus given by groups of these birds. Polytypic (4 sspp. in C.A.; *mccallii* in N.A.). L 22" (56 cm)

Identification Large with a long tail and small head. **ADULT:** Brownish olive above, buffy brown below. Gray head and neck. Long dark tail with a green sheen and broad white tip. Patch of bare skin on throat gray to dull pink, becoming bright pinkish red during breeding season. **JUVENILE:** Similar to adult, but duller and often mottled brown above; tail feathers tipped with pale brown and less rounded.

Voice Noisy, especially at dawn and dusk; much more vocal during the breeding season. **CALL:** A raucous *cha-cha-lac* given in very loud chorus. Female voice is higher pitched than male. Various other guttural chatter given, including *krrr* notes.

Status & Distribution Common in limited N.A. range. **YEAR-ROUND:** Resident in brushy areas and riverine woodlands (even some well vegetated

white tail tips

urban yards) of south TX. Successfully introduced to Sapelo I., GA. World range extends to Costa Rica along the Pacific coast.

Population The population appears stable throughout its range.

♂

carmine pink wattle, more pinkish in female, is often hidden

long blackish tail

♀

NEW WORLD QUAIL Family Odontophoridae

Gambel's Quail, male (center) and females (AZ, May)

This family composed primarily of terrestrial game birds is nearly restricted to the New World; it includes both temperate and tropical members. Omnivorous, they feed on insects as well as mast from a variety of plants.

Structure Of small to medium size, these birds have heavy bodies and rather small heads. All have stout bills with slightly serrated cutting edges, and long toes and short tarsi, which do not develop spurs. A long tail separates the wood-partridges from the rest of the family.

Behavior In general, New World quail are found in pairs during the breeding season and in groups otherwise.

Plumage They generally exhibit sexual dimorphism, although it ranges from strong in some quail to weak in wood-quail to virtually nonexistent in the wood-partridges. Although the overall plumage of most species is dominated by grays and browns, many species are quite ornate. Most species of quail have crests or head plumes as well as ornate body plumage.

Distribution Found from southwestern Canada to Paraguay, the family reaches its greatest diversity in western North America and Mexico. Stone partridges are restricted to Africa.

Taxonomy This family informally divides into four groups: wood-partridges, quail, wood-quail, and stone partridges. Worldwide, there are 33 species in 10 genera.

Conservation Poorly studied. Hunting and habitat loss have led to declines. More than half of the 15 species of tropical wood-quail (genus *Odontophorus*) are at risk. BirdLife International codes: 6 NT and 7 VU.

Genera *Oreortyx* and *Colinus*

MOUNTAIN QUAIL *Oreortyx pictus* MOUQ ■ 1

Although fairly common in much of its range, the Mountain Quail is usually very shy and hard to find. This bird is perhaps best seen when males call from rocks or in late summer, when noisy family groups often forage along roadsides near dense brush. It seems to occur with greater frequency in mixed evergreen woodlands on mountain slopes than in chaparral or riparian corridors. As its name suggests, it can be found on mountain slopes at elevations as high as 10,000 ft. Polytypic (4 sspp.; 3 in N.A.). L 11" (28 cm)

Identification Plump, short-tailed quail with gray and brown plumage. Two long, thin head plumes present in both sexes. **ADULT MALE:** Brown or grayish brown above; slate blue chest and chestnut belly barred with white on the flanks; slate blue crown with chestnut throat bordered in white. **ADULT FEMALE:** Very similar to adult male, but normally has shorter, brownish plumes and brown mottling on the hind neck in interior populations. **JUVENILE:** Dull version of the adult. Lacks bold flank pattern and the rich coloration of the face and throat pattern. Body more mottled with dark and light edging.

Geographic Variation Subspecies differ in general coloration of the back and breast. The interior subspecies—*plumifer* and *russelli*—are grayer above, generally lacking brown mottling in the nape and breast. The nominate subspecies (now includes *palmeri*) is found in more mesic habitats and is brown on the nape with some brown mottling on the sides of the breast.

Similar Species Distinctive; no similar species within its range. Gambel's and California Quail share some characteristics, but a reasonable view of the bird should provide a straightforward

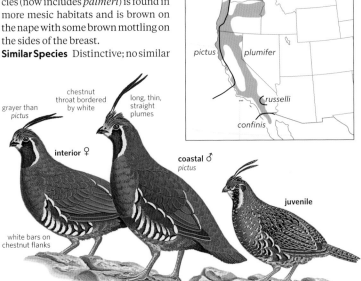

grayer than *pictus*

interior ♀

chestnut throat bordered by white

long, thin, straight plumes

coastal ♂ *pictus*

juvenile

white bars on chestnut flanks

pictus plumifer

russelli

confinis

identification; long head plumes and chestnut throat eliminate both species. The prominent white bars on chestnut flanks are often a key mark in distinguishing this species when seen in heavy cover. A juvenile Mountain Quail is more like a California in plumage, but note its straight head plumes and less intricate belly plumage.
Voice The male's advertising call is a clear, descending *quee-ark* that can carry up to a mile. The covey call that is frequently given is an extended series of whistled *kow* or *cle* notes.
Status & Distribution Uncommon to fairly common in evergreen woodlands, chaparral, brushy ravines, and mountain slopes. Nominate subspecies introduced on Vancouver I., possibly also western WA. **WINTER:** Some descend to lower elevations, usually on foot.
Population Declining in some areas in US: significant declines in Idaho, where only three small populations persist; likely extirpated from south-central WA. Heavy agricultural use of prime habitat in northern Baja California is affecting the only population (*confinis*) in Mexico.

NORTHERN BOBWHITE *Colinus virginianus*   NOBO ■ 1

slight crest

white supercilium

white throat

buffy supercilium

virginianus ♂

buffy throat

virginianus ♀

juvenile
virginianus

The Northern Bobwhite, found in coveys except during breeding season, is the most widespread quail in N.A. Polytypic (22 sspp.; 7 in N.A.). L 9.7" (25 cm)
Identification Plump, short-tailed quail with reddish brown plumage. **ADULT MALE:** Intricate body plumage of chestnut, brown, and white; blackish plumage on head; white throat and eye line. **ADULT FEMALE:** Similar to adult male but throat and eye line buffy and plumage on head brown. **JUVENILE:** Similar to adult female but body plumage browner, less rufous, and eye line less prominent. **RUFOUS MORPH:** Very rufous body plumage masks markings. Black face superficially resembles the "Masked Bobwhite." Extremely rare. **"MASKED BOBWHITE":** Similar to eastern birds, but it has a black face with the white eye line greatly reduced and flecked with black. Underparts are entirely rufous with white markings on the flanks; the upperparts have prominent rufous markings, becoming browner on the lower back and rump.
Geographic Variation Complex, with differences sometimes striking in plumage of adult male. The spe-

cies is often broken into three distinct groups, two of which occur in the US. The northern group occurs in eastern N.A.; it is made up of four similar subspecies differentiated by changes in overall coloration and the width of barring on the underparts. Birds from peninsular FL (Gainesville and south), *floridanus*, are smaller and darker than the widespread nominate subspecies. The subspecies *taylori* (SD to northern TX; introduced locally in WA, OR, and ID) and the more southerly *texanus* (southwestern TX to northern Mexico) are paler. The second group is the "Masked Bobwhite" (*ridgwayi*) found in southern AZ and Sonora; at times it has been considered a separate species.
Similar Species Distinctive; no similar species within its range. Montezuma Quail is superficially similar, but its range overlaps with Northern Bobwhite only in south-central TX.
Voice Male's advertising call is a whistled *bob-white* or *bob-bob-white*. Other calls include a low whistled *ka-lo-kee* and a variety of clucks.
Status & Distribution Uncommon to fairly common but declining resident of open habitats with sufficient brushy cover. Introduced to northwestern N.A., the Bahamas, several

ridgwayi
"MASKED BOBWHITE" group
virginianus
"EASTERN NORTH AMERICA" group
texanus
floridanus
maculatus & aridus
multiple subspecies

Caribbean islands, and New Zealand. The "Masked Bobwhite" (Endangered, USFWS) is rare in the Altar Valley of southern AZ as a result of transplanted individuals from remaining populations in Sonora. Historically common in southern AZ in tall grass-mesquite plains until about 1890; overgrazing led to extirpation by the early 1900s.
Population Near Threatened. Northern Bobwhite has declined significantly in the US. Extirpated over much of northern range from Midwest to New England; recent sightings often involve released birds. Significant declines even within its core range. Manipulation of habitat and high populations of the introduced fire ant are believed to be key factors in the decline.

FL ♂
floridanus

smaller and darker

♂ *taylori*

black throat

paler overall

cinnamon underparts

"Masked Bobwhite" ♂
ridgwayi

Genera *Callipepla* and *Cyrtonyx*

SCALED QUAIL　*Callipepla squamata*　SCQU ■ 1

A bird of the desert grasslands, mesquite savanna, and thorn-scrub of the Southwest, the Scaled Quail is generally sedentary, moving short distances to form winter coveys. Fall and winter coveys can have as many as 200 individuals. It prefers to run to escape, but it will fly when startled. Hybrids with Gambel's Quail (see opposite) and Northern Bobwhite have been documented. Polytypic (4 sspp.; 3 in N.A.). L 10" (25 cm)

Identification This quail is plump and short-tailed and has grayish plumage. The bluish gray breast and mantle feathers are edged in black or brown giving it a scaled appearance. **ADULT:** Sexes are similar. The male has a prominent white crest; the female's is smaller and buffier in coloration. **JUVENILE:** Similar to adults but more mottled above and less conspicuous scaling on the underparts.

Geographic Variation Two similar subspecies, *pallida* and the more northerly *hargravei*, inhabit most of Scaled Quail's range in N.A.; *castanogastris* occurs in southern TX, below the Balcones Escarpment. A chestnut patch on the belly and darker overall plumage, particularly on the upperparts, separates *castanogastris* from *pallida* and *hargravei*.

Similar Species Distinctive; unlikely to be confused with any other quail in N.A.

Voice During the breeding season, both sexes give a location call when separated, a nasal *chip-churr*, accented on the second note. The male's advertising call is a rhythmic *kuck-yur*, often followed by a sharp *ching*.

Status & Distribution Uncommon to locally fairly common year-round resident.

Population Scaled Quail has declined

dramatically in the US since the 1940s. In TX, the species has become increasingly scarce in the Hill Country and Rolling Plains; reasons for the decline are unknown, but they are probably attributable to habitat degradation; still numerous in the Trans-Pecos northward to the TX Panhandle.

pale top to crest

♂ pallida

south TX *♂*
castanogastris

scaly breast

dark chestnut belly patch

juvenile

GAMBEL'S QUAIL　*Callipepla gambelii*　GAQU ■ 1

Although a common quail of the desert Southwest, Gambel's Quail requires a lot of water. Generally sedentary, it moves short distances in the late summer to form coveys, usually comprising several family groups. It hybridizes with both California and Scaled Quails where their ranges overlap. Polytypic (5 sspp.; 2 in N.A.). L 11" (28 cm)

Identification Plump, short-tailed quail with gray plumage and prominent teardrop-shaped head plumes (two feathers) in both sexes. **ADULT**

MALE: Chestnut crown with black forehead and black throat; chestnut sides with whitish underparts with a black belly; gray upperparts. **ADULT FEMALE:** Similar to adult male but muted and lacking distinctive facial pattern. Head plumes smaller. **JUVENILE:** Grayish brown overall and heavily mottled. Usually has a short head plume.

Similar Species California Quail is similar in structure and size, but the chestnut crown and sides and lack of scaling on Gambel's easily separate them.

Voice CALL: A plaintive *qua-el*; and a loud *chi-ca-go-go* similar to California Quail, but higher pitched and usually four notes, sometimes shortened on only one or two syllables. Also a variety of clucking and chattering calls.

Status & Distribution Common year-round in desert shrublands and thickets, usually near permanent water sources. Introduced to HI, ID, and San Clemente I., CA.

Population Numbers appear to be stable over the past 60 years.

chestnut crown

teardrop-shaped crest

black belly surrounded by white *♂*

juvenile

chestnut sides

♀

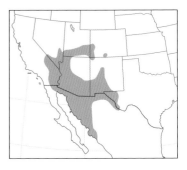

CALIFORNIA QUAIL *Callipepla californica*   CAQU ■ 1

The resident California Quail congregates in large coveys during the fall and winter. It hybridizes with the Gambel's Quail where their ranges overlap. Polytypic (6 sspp.; 4 in N.A.). L 10" (25 cm)
Identification Plump with gray and brown plumage; prominent teardrop-shaped head plumes (two feathers) in both sexes. **ADULT MALE:** Pale forehead with brown crown and black throat, scaled belly with a chestnut

patch. **ADULT FEMALE:** Similar to male but muted and lacking distinctive facial pattern; head plume smaller. **JUVENILE:** Grayish brown to brown overall and heavily mottled. Belly pale with less scaling than on adults.
Geographic Variation Differences are based on coloration and size, more pronounced in females. Adult female *canfieldae* (in east-central CA) and *californica* (the most widespread subspecies) have grayish upperparts; adult female *brunnescens* (found in the wetter coastal mountains) has brown upperparts. The *catalinensis* subspecies is endemic to Santa Catalina I., CA.
Similar Species Gambel's

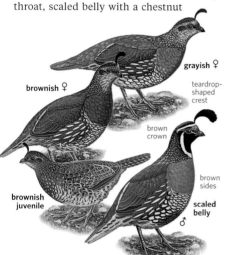

grayish ♀

brownish ♀

teardrop-shaped crest

brown crown

brown sides

scaled belly ♂

brownish juvenile

Scaled x Gambel's hybrid ♂

lacks scaled underparts, has rufous cap.
Voice CALL: An emphatic *chi-ca-go*; sometimes shortened on only one or two syllables. Territorial breeding season call is a single *cah*. Also makes a variety of grunts and sharp cackles.
Status & Distribution Common year-round in open woodlands with brushy areas, usually near permanent water sources. Introduced locally within the general boundaries of the mapped range, including UT, AZ, and Vancouver I., BC.
Population Overall population declining in the US due to urbanization.

MONTEZUMA QUAIL *Cyrtonyx montezumae*   MONQ ■ 2

The elaborate plumage of the male Montezuma Quail cryptically blends into tall grasses or in dappled sunlight under a shrub. It has highly specialized long claws, used for digging up bulbs and tubers. It remains in pairs all summer, although nesting often does not begin until the monsoon rains of July and Aug. It is most often seen cautiously crossing roads, along roadsides, or at feeders. When encountered, it prefers to crouch and remain still. In winter, it forms into small coveys. Populations appear to be as much as 60 percent male, enhancing mate selection in females. Polytypic (4 sspp.; *mearnsi* in N.A.). L 8.7" (22 cm)
Identification Plump and short-tailed. **ADULT MALE:** Distinctive facial pattern with a rounded dark buffy brown crest. Dark, intricately patterned body. **ADULT FEMALE:** Brown overall. **IMMATURE MALE:** Pale gray face; black sides and flanks, heavily spotted with white. Distinctive facial pattern appears in late fall. **JUVENILE:** Like female, but lighter in overall color and more heavily mottled with black.

Similar Species Unmistakable; no similar species exist within its range. Northern Bobwhite is superficially similar, but their ranges overlap only in south-central TX.
Voice Male's advertising call is a descending whistled *vwirrrrr.* Covey call, which is given by both sexes, is a loud descending whistle. Contact calls are a rather quiet *whi-whi.*
Status & Distribution Uncommon; secretive and local in open grassy juniper-oak or pine-oak woodlands on semiarid slopes.
Population Formerly much more widespread in N.A. In TX, it was found throughout all mountain ranges and in the central portion of the state east to San Antonio. It was also more widespread in the mountains of southern AZ and southern NM. By 1950, it had shrunk to its current range. Tall grasses in woodlands are believed to be an important part of their habitat.

slightly crested look ♀

exotic head pattern

dark buffy crest laid over nape

juvenile

solid blackish brown chest ♂

round white spots on sides and flanks

PARTRIDGES, GROUSE, AND TURKEYS Family Phasianidae

Wild Turkey (ON, Jan.)

Phasianidae is arguably one of the more spectacular families in the world. Many species have elaborate courtship rituals. Most exhibit complex plumage variation that is still poorly understood. Many species require large tracts of quality habitat.

Structure Well adapted for life on the ground, all family members have strong legs with three long toes and a small hind toe. Adults of many species have one or more sharp spurs on the back of the leg to rake opponents during quarrels. The relatively short wings allow birds to explode off the ground.

Behavior Phasianids are unobtrusive. Most will fly only when pressed; they spend their lives with minimal movement, both in their daily activities and seasonally. During the mating season, the males of most species, particularly those with polygynous systems where males compete for females at leks, engage in a barrage of displays that include drumming, wing whooshing, maniacal cackles, foot stomping, and flight displays.

Plumage Most species in this family are sexually dimorphic, dramatically so in many pheasants. Males are often larger, and in many species of grouse have special adornments on their head and neck. Otherwise, most species are cryptically colored. Most appear very similar throughout the year, although ptarmigan change dramatically from their cryptic lichen-covered-rock summer plumage to nearly completely white in winter.

Distribution Members of this family are distributed across North America. Most species are sedentary. Widespread in the Old World, with diversity peaking in South Asia; absent from the Neotropics.

Taxonomy Phasianidae includes five distinct subfamilies: Hill partridges (Rollulinae, not in North America), pheasants and partridges (Phasianinae, introduced in North America), grouse (Tetraoninae), turkeys (Meleagridinae), and guinea fowl (Numidinae, some small feral populations in North America). Of the approximately 178 species worldwide (in 52 genera), 12 grouse and the Wild Turkey are native to North America. Attempts have been made to introduce several species, but only the Chukar, Gray Partridge, Himalayan Snowcock, and Ring-necked Pheasant are considered established. The sedentary nature of phasianids with little genetic mixing has given rise to extensive regional variation and subspecies. Taxonomy continues to undergo revision: The Gunnison Sage-Grouse was described as a new species in 2000 and the Blue Grouse was split into the Dusky Grouse and Sooty Grouse in 2006.

Conservation In North America, three native species are at risk. Worldwide, many phasianids are at risk, especially Asian pheasants. BirdLife International codes: 31 NT, 26 VU, 11 EN, 3 CR, and 1 EX.

Genus *Alectoris*

CHUKAR   *Alectoris chukar*   CHUK ■ 2

This introduced Old World species can be challenging to find and is best located by call or by searching near water sources. By late summer it is not unusual to see a covey with one to three adults and 30 to 50 young. Polytypic (14 sspp.). L 14" (36 cm)

Identification Resembles an overgrown short-tailed quail. **ADULT:** Cream-colored face and throat broadly outlined in black. Flanks boldly barred with black. **JUVENILE:** Simi-

lar but smaller, mottled, and no bold black markings; usually with adults. **FLIGHT:** It explodes into the air and then glides away; look for chestnut on spread tail just before landing.

Geographic Variation Complex; most birds in N.A. believed to be nominate subspecies, which is the darkest and brownest subspecies.

Similar Species No similar species established in N.A.; however, two released game birds are similar: Redlegged Partridge (*A. rufa*) has a white face and throat; its neck is conspicuously streaked with black, giving the appearance of having a "neck-shawl." Rock Partridge (*A. graeca*) has more black (less white) between the bill and the eye and a whiter throat.

Voice CALL: A calm *chuck, chuck, chuck* that often becomes progressively louder and drawn out, culminating in a voluminous eruption of *chuckara-chuckara-chuckara*. The alarm call of flushed birds is a loud piercing squeal followed by *whitoo* notes.

Status & Distribution Uncommon to fairly common; now established in steep, rocky arid terrain in much of interior West. Released and escaped birds are possible almost anywhere, even in the East.

Population Widely released in N.A. starting in the 1930s, the Chukar became established in much of the West by the late 1960s. The largest populations are in WA, OR, NV, ID, UT, and CA.

adults

extensive rufous on outer tail feathers

buffy throat outlined by black

black-and-white barred flanks

juvenile

Genus *Tetraogallus*

HIMALAYAN SNOWCOCK   *Tetraogallus himalayensis*   HISN ■ 2

The Central Asian Himalayan Snowcock has established a toehold in N.A., but seeing this exotic is no small feat. Its numbers are small (only about 1,000 birds) and it lives only in NV's high, rugged Ruby Mts. At dawn, the Himalayan Snowcock calls from its roost site and flies downhill. It spends the day walking back up, feeding, rest-

white areas bordered by chestnut bands

♂

dark belly

ing, and preening. Birds congregate just before dark, then fly or walk to roost. Polytypic (6 sspp.; nominate in N.A.). L 28" (71 cm)

Identification Himalayan Snowcock is large (nearly 2.5 times the size of the Chukar). It is generally grayish overall with rusty-brown streaks on upperparts. Two prominent chestnut stripes on each side of the head outline the whitish face and throat. The predominantly white primaries are conspicuous in flight. **FEMALE:** Smaller and duller than the male with a buff forehead; the area around the eye is grayer; lacks spurs. **JUVENILE:** Similar to the adult female, but smaller and duller and lacks rusty tones to the spotting above. Very young birds lack the chestnut markings on the face.

Similar Species None within restricted range.

Voice CALL: Generally similar to Chukar; other calls surprisingly reminiscent of a Long-billed Curlew. Clucks and cackles persistently while foraging.

Status & Distribution Successfully introduced to the high elevations of the Ruby Mts. of northeast NV, where it resides in subalpine and alpine habitats. **MOVEMENTS:** Generally sedentary; may move slightly downslope during severe winters.

Population The Himalayan Snowcock was first released into the Ruby Mts. in 1963; the first brood was seen in 1977. By the 1980s the species was self-sustaining. Limited threats include overgrazing by sheep and pressure from hunters and birders.

Genus *Perdix*

GRAY PARTRIDGE *Perdix perdix*   GRAP ■ 2

Established as a game bird in N.A. since the early 1900s in agricultural areas, this often evasive Old World species is best found at dawn or dusk and when there is snow cover. In fall it forms coveys of 12 to 15 birds. Polytypic (8 sspp.). L 12.5" (32 cm)

Identification Mostly streaked brownish above, grayish below. Broad rufous bars and narrow cream-colored streaks on flanks. Contrasting rufous outer tail feathers often revealed as bird flicks its tail open or flies. **MALE:**

Conspicuous chestnut patch on upper belly and bright rufous-orange head and throat. **FEMALE:** Belly patch absent or reduced; head and throat more buff. **IMMATURE:** Duller in both sexes. **FLIGHT:** Explodes into the air with rapid wingbeats and then alternates glides and rapid wingbeats; flights are typically low to the ground and of short distance.

Geographic Variation Clinal. In its Old World range the western subspecies are darker gray and more rufous than subspecies farther east. Little information on introductions, but most believed to be western subspecies (including nominate *perdix*).

Similar Species None. The Gray Partridge's broad rufous bars on the flanks and face pattern are distinctive. Compare female to smaller female California Quail, which overlaps with Gray Partridge in the western part of its range.

Voice CALL: A harsh *kee-uck*, likened to a rusty gate. Alarm call a rapidly

repeated *kuta-kut-kut-kut*. Also various clucks.

Status & Distribution Uncommon in most areas; it inhabits farmland and grassy fields.

Population Gray Partridge is thought to be declining in much of its range in N.A., largely due to changes in farming practices.

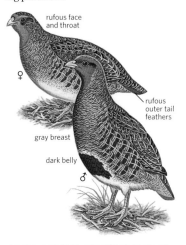

rufous face and throat

♀

rufous outer tail feathers

gray breast

dark belly

♂

PHEASANTS Genus *Phasianus*

RING-NECKED PHEASANT *Phasianus colchicus*   RNEP ■ 1

Native to Central and East Asia, the Ring-necked Pheasant has been widely introduced as a game bird in N.A. It is frequently seen along roadsides in agricultural areas. Polytypic

(±30 sspp.). L ♂ 33" (84 cm) ♀ 21" (53 cm)

Identification Large. Very long, pointed tail; short, rather rounded wings. **MALE:** Iridescent bronze overall, mottled with brown, black, gray, and rufous. Head a dark glossy green to purplish, with reddish eye patches and iridescent ear tufts. Most show a wide white neck ring. **FEMALE:** More buff colored with sparse dark spots and bars on breast and flanks. **FLIGHT:** When flushed, it erupts with powerful

loud whirring wingbeats. The long tail is distinctive even in flight.

Geographic Variation Extensive subspecies variation in males. Many subspecies have been introduced in N.A.; intergrades are numerous. The male "White-winged Pheasant" (not illustrated) with distinctive white upperwing coverts has become established in parts of the West. The Japanese Green Pheasant (*P. versicolor*), now usually considered a separate species, is largely extirpated from the areas where it was introduced, mainly the Tidewater region of VA and southern DE.

♀

white neck ring

♂

long tail; compare carefully with female Sharp-tailed Grouse

♀

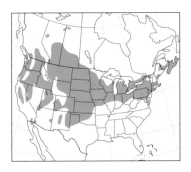

Green
Pheasant

Similar Species Females resemble Sharp-tailed Grouse, but the latter is smaller, with heavily scalloped and spotted underparts and has a shorter tail with white outer tail feathers. **Voice CALL:** Male territorial call is a loud, penetrating *kok-cack*. Both sexes give hoarse, croaking alarm notes. **Status & Distribution** Locally common to uncommon resident of open country, farmlands, brushy areas, and edges of woodlands and marshes. Local hunting releases account for the presence of some individuals outside their normal range. **Population** While populations fluctuate with weather changes, pheasants are declining in parts of the East. Some populations are preserved through continual introductions.

FOREST GROUSE Genus *Bonasa*

The sole member of this monotypic genus lives in relatively forested landscapes. The species shows considerable geographic variation and also occurs in different color morphs. Its closest relatives are Northern Hazel Grouse *(Tetrastes bonasia)* from Eurasia and Severtsov's Hazel Grouse *(T. sewerzowi)* from China.

RUFFED GROUSE   *Bonasa umbellus*   RUGR ■ 1

This generally solitary species is the most widespread grouse in N.A. In spring, the male's deep accelerated drumming often resonates in open woodlands. Displaying males also raise their ruffs and crest. Most active in morning and evening, it may be seen along edges of woodland road, or in winter, perched high in deciduous trees feeding on buds, twigs, and catkins. It sometimes visits feeding stations. Polytypic (14 sspp.; all in N.A.). L 17" (43 cm)
Identification Slim, long-tailed species with a crest (most apparent when bird is alarmed). Black ruffs on side of neck often hard to see. Most populations have two color morphs, red and gray, most apparent in coloration of tail; overall coloration varies region-ally. Barred underparts and tail with many narrow bands and a wide dark subterminal band (center of which is typically broken or blotchy on female). **JUVENILE:** Similar to female, but less well marked and lacks subterminal band; usually seen with the adult female. **FLIGHT:** When alarmed, it explodes into the air with a roar of the wings, but even then tail pattern is usually seen.
Geographic Variation Complex, with 14 recognized subspecies. Birds in the Pacific Northwest average darker and richer brown; in interior West, paler gray; in the Northeast, generally grayish; in the Southeast, brownish.
Similar Species Spruce, Dusky, and Sooty Grouse lack crests. Larger Dusky

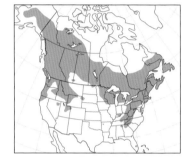

and Sooty Grouse are more uniformly colored, without the barring on the underparts. Spruce Grouse has a shorter tail and more noticeable white spotting below. Note the tail pattern.
Voice CALL: Both sexes may give nasal squeals and clucks, particularly when alarmed.
Status & Distribution Fairly common resident of deciduous and mixed woodlands.
Population Populations rise and fall in regular cycles. Significant declines in upper Midwest, Appalachian, and mid-Atlantic regions.

male with solid
dark tail band

displaying
gray-morph ♂

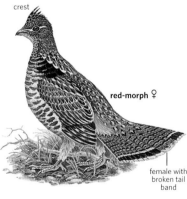

crest

red-morph ♀

female with
broken tail
band

red-morph ♂

SAGE-GROUSE Genus *Centrocercus*

GREATER SAGE-GROUSE   *Centrocercus urophasianus*   GRSG ■ 1

The largest N.A. grouse, the declining Greater Sage-Grouse is a highly social denizen of aromatic sagebrush flats and foothills mainly centered in the Great Basin. Where the species is still common, it may form flocks of several hundred birds. Impressive lekking displays begin in late Feb. or early Mar. and continue into early summer.

Monotypic. L ♂28" (71 cm) ♀22" (56 cm)
Identification Distinctive dark belly and long pointed tail feathers. **MALE:** Yellow eye combs, black throat and bib, large white ruff on the breast. **FEMALE:** Similar to male but smaller. Brown throat and breast, showing only a hint of the male's pattern. **JUVENILE:** Resembles the female but with upperparts finely streaked. **FLIGHT:** Superficially suggests waterfowl, but note dark belly, long tail, and grouse shape.
Geographic Variation Some consider birds in the interior Northwest a separate subspecies (*phaios*), but differences are minor and validity questionable. It hybridizes occasionally with Sharp-tailed and Dusky Grouse.
Similar Species See Gunnison Sage-Grouse. Distinguished from all other grouse by black belly and larger size.
Voice CALL: Courting male fans tail

rather uniform-colored tail feathers

displaying ♂

graduated, pointed tail

black belly patch

displaying males on lek

♂

♀

and rapidly inflates and deflates air sacs emitting two popping sounds, at times audible up to two miles away; also a variety of wing swishes, coos, whistles, and tail rattles. When flushed, it gives a rapid cackling *kek-kek-kek*.
Status & Distribution Uncommon to fairly common but local. **MOVEMENTS:** Generally resident, but may engage in local movements, particularly to find sage in times of deep snow. Movements of more than 100 miles have been noted in ID and WY.
Population Near Threatened. The species declined during the late 19th and 20th centuries; hunting probably reduced numbers early on, but degradation and fragmentation of habitat pose the biggest risks today.

GUNNISON SAGE-GROUSE *Centrocercus minimus*　　GUSG ▪ 2

This bird was only recognized as a separate species by scientists in the 1990s when differences in size, plumage, and lekking displays were described. Only 2,000 to 4,000 birds remain in less than 193 square miles of land in southwestern Colorado and extreme southeastern Utah. Gunnison Sage-Grouse require large expanses of sage with a diversity of forbs and grasses and a healthy riparian habitat, interspersed with grassy open areas, where the males perform elaborate courtship

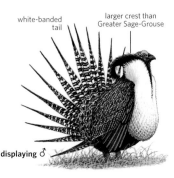

white-banded tail

larger crest than Greater Sage-Grouse

displaying ♂

displays. Monotypic. L ♂22" (56 cm) ♀18" (46 cm)
Identification Roughly two-thirds the size of the Greater Sage-Grouse. Identify by range.
Similar Species Greater Sage-Grouse is very similar. During display, the Gunnison has pronounced, even white bands on the tail, whereas the Greater's tail feathers are more mottled; it also appears noticeably paler at a distance, almost blond in color. Gunnison males display less frequently, and produce different sounds (typically nine air sac plopping sounds during each display, compared with two for Greaters). Gunnison's longer, thick filoplumes are thrown forward over the top of the head, and then flop back down (unlike Greater, where the sparse filoplumes are held erect). Female Gunnison is nearly identical to female Greater in plumage and behavior.
Voice CALL: Similar to Greater.
Status & Distribution Most in CO in Gunnison Basin of Gunnison Co. and Saguache Co.; six smaller disjunct

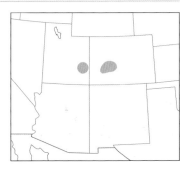

populations, each with fewer than 300 birds. Fewer than 150 birds in UT. Formerly likely more widespread; 19th-century populations in NM and KS were perhaps Gunnison, but no extant specimens. Unsuccessful 20th-century releases in NM involved Greater Sage-Grouse.
Population Endangered. Habitat loss, degradation, and fragmentation have reduced the populations. In recent years this has been exacerbated by drought. Listed as Threatened by the USFWS.

PTARMIGAN Genus *Lagopus*

All three ptarmigan species are found in N.A.: The White-tailed is endemic to the mountains of the West; the Rock and Willow Ptarmigan are circumpolar. They have thickly feathered toes and tarsi, short tails, and cryptic plumages. All but the "Red Grouse" (*scotica* subspecies of Willow Grouse) of the British Isles molt to a near completely white plumage in winter.

WILLOW PTARMIGAN *Lagopus lagopus*   WIPT   ■   2

The largest ptarmigan, the circumpolar Willow is characteristic of willow flats and low dense vegetation near tree line. It has the most widespread distribution of any grouse. Polytypic (16 sspp. worldwide; 6 in N.A.). L 15" (38 cm)

Identification Both sexes have a black tail, predominantly white wings, and white-feathered legs year-round. The red eye combs are more pronounced in males; they may be largely concealed or inflated during courtship or aggression. Both sexes are white in winter, except for black tails. **SUMMER MALE:** Mottled, but predominately rufous. **BREEDING FEMALE:** Warm brown; very similar to Rock Ptarmigan. **JUVENILE:** Heavily barred black and buffy yellow. Dark wings for a short period in summer.

Geographic Variation Extensive. Differences in both sex and molt complicate subspecies recognition.

Similar Species Willow Ptarmigan is larger than Rock Ptarmigan and has a bigger bill. Vocalizations and habitat are helpful in identification. While Willow will move upslope in mountains, it is restricted to relatively lush patches of vegetation. In winter, male Willow has a relatively plain face and head lacks the distinctive dark eye line of the male Rock.

Voice In displays, male utters a loud raucous *go-back go-back go-backa go-backa go-backa*. **CALL:** Low growls and croaks; noisy cackles.

Status & Distribution Widespread sometimes abundant tundra species. Fond of moist locations with dwarf willow and alder thickets. **BREEDING:** Generally prefers wetter, brushier habitat than Rock Ptarmigan. **MOVEMENTS:** Vary between largely resident populations with only short shifts between breeding and wintering ranges, to relatively long-distance migrants. Thousands have been recorded in a day moving through Anaktuvuk Pass, AK. Flocks of up to 2,000 individuals have been recorded in migration. **MIGRATION:** Fall peak at Anaktuvuk in mid-Oct. Spring migration begins as early as mid-Jan., peaking late Apr. **RARE STATUS:** Casual in spring and winter to southern ON, southern SK, and northern tier states (e.g., MT, ND, MN, WI, NY, and ME).

Population Substantial cyclic ups and downs in populations over several-year periods. Little data on long-term trends, but apparently stable.

dark rufous head and neck

molting spring ♂

summer ♀

thick bill

winter

summer ♂

summer ♂

black tail

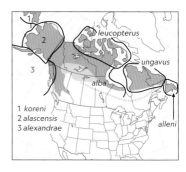

leucopterus

ungavus

alba

1 *koreni*
2 *alascensis*
3 *alexandrae*

alleni

ROCK PTARMIGAN *Lagopus muta*   ROPT   ■   2

The Rock Ptarmigan is the archetypal bird of the arctic. Found in cold and barren windswept tundra, it has been recorded as far north as 75° N in winter, when the region is enshrouded in 24 hours of darkness. Like other ptarmigan it has an unusual sequence of three body molts each year, generally allowing it to remain well camouflaged at all times of the year. Males molt later than females in spring. By early June many are still mostly white and can be spotted from up to a mile away; at the same time the females have completed their

molts and are difficult to spot from a few feet away. Polytypic (23 sspp.; 8 in N.A.). L 14" (36 cm)

Identification Like the Willow Ptarmigan, both sexes have a black tail, predominantly white wings, and white-feathered legs year-round. The mottled summer plumage is black, dark brown, or grayish brown; the exact patterning varies with subspecies. **ADULT MALE:** White winter plumage is similar to the Willow Ptarmigan but the black eye line is prominent. **ADULT FEMALE:** Very difficult to distinguish from Willow Ptar-

migan, except by overall size and bill size. **JUVENILE:** Heavily barred black and buffy yellow; has dark wings for a short period in summer.

Geographic Variation Extensive regional variation. Individual variation and complex molt complicates subspecies identification. Subspecies on the Aleutians are larger with heavier bills. Males on westernmost Aleutians are mostly blackish in breeding plumage (see illustration of *evermanni* from Attu); central Aleutian males more ochre-toned.

Similar Species Rock Ptarmigan is smaller than Willow Ptarmigan, and with a smaller bill. Use caution when estimating bill size as the dark feathers between the eye and bill on birds in transitional plumages can give the impression that the bill is larger. The presence of dark lores is diagnostic for male Rock Ptarmigan (their absence is not diagnostic). Male Rock Ptarmigan holds much of its winter plumage through the spring, unlike male Willow. **Voice** In courtship and territorial flight displays, male utters a low rattling *ah-aah-ah-aaaah-a-a-a-a*. **CALL:** Low growls and croaks; noisy cackles. **Status & Distribution** Common. **BREEDING:** Generally prefers higher and more barren habitat than the Willow Ptarmigan. **MOVEMENTS:** May withdraw from northernmost summer range, but only irregularly south of the southern boundary of the breed-ing range. Most make only limited movements, which are largely altitudinal and likely influenced by weather and snow depth. Populations on the Aleutian Is. are resident. **RARE STATUS:** Accidental on Haida Gwaii, BC, and in northeastern MN. **Population** Numbers fluctuate. Some southern populations may have contracted. In Aleutians, on islands that maintained populations, numbers substantially increased after the removal of non-native arctic fox introductions.

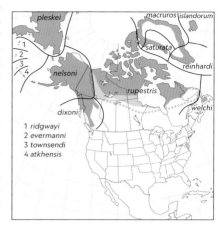

1 ridgwayi
2 evermanni
3 townsendi
4 atkhensis

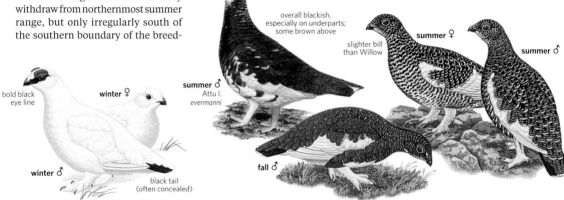

overall blackish, especially on underparts; some brown above

slighter bill than Willow

summer ♀

summer ♂

bold black eye line

winter ♀

summer ♂
Attu I.
evermanni

winter ♂

fall ♂

black tail (often concealed)

WHITE-TAILED PTARMIGAN *Lagopus leucura* WTPT ■ 2

The White-tailed Ptarmigan is the smallest grouse in N.A. and the only ptarmigan regularly found in the lower 48. Its ability to blend in perfectly with its remote alpine tundra environment makes it among the most challenging resident species to find. Although territorial during the breeding season, it forms flocks of up to 80 birds from late Oct. to late Apr. During winter, it roosts under the snow. White-tailed Ptarmigan forage largely on buds, stems, and seeds year-round, but they augment this diet with insects in the summer. Males do not have a flight display. Polytypic (5 sspp.; all in N.A.). L 12.5" (32 cm)
Identification WINTER: Completely white except for eye combs, dark eye and bill, and fine dark shaft streaks to primaries. **SUMMER MALE:** More white on belly than female; conspicuous coarsely barred brown and black breast feathers form a necklace. **SUMMER FEMALE:** More evenly patterned with barring of buffy yellow and black. **JUVENILE:** Heavily barred black and buffy yellow and have dark wings for a short period in summer.
Geographic Variation Clinal among five subspecies; average larger and grayer in the south, smaller and darker in the north.
Similar Species The white tail distinguishes it from both Willow and Rock Ptarmigan in all seasons. However, the tail feathers of all ptarmigan are typically concealed by the uppertail coverts, making it difficult to judge tail color.
Voice Males give a loud raucous flight call with four distinct syllables, *ku-ku-kii-kiieur*; also a variable chattering clucking series of calls given on the ground, *duk-duk-daaak-duk-duk*; *duk-dak-dadaak-duk*. **CALL:** Clucks, churrs, and high-pitched chirps.
Status & Distribution Uncommon to locally common. Endemic to western mountains, where it is found at or

winter

all plumages have all-white tail

above timberline in rocky alpine areas. Small numbers have been introduced in the Wallowa Mts., Sierra Nevada, Pike's Peak, and the Uinta Mts. Re-established into northern NM. **MOVEMENTS:** Depends on severity of winter and snowfall, but usually moves to slightly lower elevations; during heavy snow years moves to streambeds and avalanche chutes dominated by willows and alders.

Population Numbers fluctuate widely from year to year, but there is little data on overall population changes. Local populations may be affected by natural resource extraction, road construction, off-road vehicles, and overgrazing by domestic livestock and elk; these activities have the greatest effect on reducing winter forage, principally willow. The population on Vancouver I. (*saxatilis*) is considered vulnerable.

slighter bill than Willow

summer ♀

molting fall ♂

summer ♂

Genus *Falcipennis*

SPRUCE GROUSE *Falcipennis canadensis*   SPGR ■ 2

While this species may be seen along roadsides, particularly in fall, the Spruce Grouse is often challenging to find. Typically very tame, this species is so well camouflaged that by remaining still it can remain undetected even when observers are only a few feet away. Spruce Grouse are usually solitary, but will gather in small loose flocks in winter. Polytypic (6 sspp.; all in N.A.). L 16" (41 cm)

Identification Short-tailed compact grouse. Over most of range, both sexes have black tail with chestnut tail tip. Male has dark throat and breast, edged with white and red eye combs. Female occurs as red and gray color-morphs; in all females note the heavy dark barring and white spots below. **JUVENILE:** Similar to red-morph female.

Geographic Variation Two groups of subspecies: "Taiga Grouse" (*canadensis, atratus, osgoodi, canace,* and *isleibi*) and "Franklin's Grouse" (*franklinii*). Some argue that "Franklin's Grouse" of the northern Rockies and Cascades should be considered a separate species from "Taiga Grouse" found throughout the rest of the range. They intergrade across central BC. "Franklin's" has bold white spots on the uppertail coverts, and on males the square-tipped tail feathers are uniformly dark. On male "Taiga" subspecies, tail feathers are more rounded and tipped rufous (*isleibi* from Alexander Archipelago, southeast AK, lacks rufous tail tips and has narrow white tips on uppertail coverts). Females are not safely distinguishable. Courtship displays also differ: Both give strutting displays where the male spreads his tail, erects the red combs above his eyes, and rapidly beats his wings; some give a series of low hoots. In territorial flight display, "Taiga" male flutters upward on shallow wingstrokes; "Franklin's," however, ends the performance by beating his wings together, which produces a loud clapping sound.

Similar Species Compared with Dusky and Sooty Grouse, Spruce Grouse appears to have a shorter tail. Female Spruce is similar to both Dusky and Sooty Grouse, but it has more heavily marked underparts. The crested Ruffed Grouse has a longer tail with a broad black subterminal band and has broad dark bars on the flanks.

Voice Relatively quiet outside the breeding season. **CALL:** Birds in foraging groups give a *sreep*. Female gives a high-pitched call that is thought to be territorial.

Status & Distribution Rare to fairly common resident of conifer forests, particularly early successional stages with young dense trees. Usually found near conifers, but when dispersing in fall may be seen in deciduous woods. "Franklin's Grouse" (*franklinii*) is also found in more open subalpine habitats.

Population Population declines in the southern perimeter of the species' range are possibly due to changes in habitat; populations are generally stable in the northern part of the range. Fires are important to renew the species' favored habitat.

red comb

displaying ♂

red-morph ♀

gray-morph ♀

rufous tail band

white tips on uppertail coverts

"Franklin's Grouse" ♂
franklinii

canadensis group

isleibi

franklinii "Franklin's Grouse"

canace

| Genus *Dendragapus*

DUSKY GROUSE *Dendragapus obscurus*   DUGR ■ 2

Birding visitors to the Rocky Mts. prize the swashbuckling spectacle of a displaying male Dusky Grouse—circling in fluttering flight, strutting with tail fanned, body tipped forward, head drawn in, wings splayed, and neck feathers fanned open, revealing brilliant maroon patches of skin on the sides of the neck. Polytypic (4 sspp.; L 20" (51 cm)

Identification A relatively large, long-tailed grouse. **MALE:** Medium brownish gray, mottled plumage overall. On each side of the neck, white-based feathers cover patches of reddish bare skin, called cervical apteria, exposed during displays. Prominent yellow eye combs, called superciliary apteria, become orange during display. **FEMALE:** Similar to male but

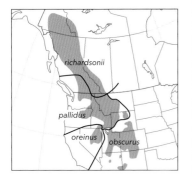

upperparts browner, with a more uniformly grayish belly. **JUVENILE:** Similar to female but with fine pale streaks above. Tail is heavily mottled brown.

Geographic Variation Subspecies *oreinus* resembles *obscurus* but is much paler, and females have more pronounced white edgings in scapulars and wing coverts. The two subspecies found farther north in the Rockies, *richardsonii* and *pallidus*, are sometimes referred to together as "Richardson's Grouse." Both essentially lack the gray terminal tail band found in the other subspecies; *richardsonii* is distinguished from *pallidus* by its darker, blacker white-tipped undertail coverts.

Similar Species See Sooty (below). Dusky is less tied to trees than Sooty and is often found foraging far out into open areas (e.g., in sagebrush). From smaller female Spruce by less heavily patterned underparts and a longer tail. Smaller Ruffed has a broad black subterminal tail band, more heavily barred underparts, and a crest.

Voice CALL: Males in display give ultra-low-pitched, short hoots, audible only at close range and usually delivered from the ground. Both sexes give various soft clucking sounds; male gives low, nasal, rapid *gr-gr-gr-gr*; female gives odd braying yelp.

northern subspecies lack gray tail band

purple air sac

displaying northern Rockies ♂ *richardsonii*

southern Rockies ♀ *obscurus*

Status & Distribution Fairly common resident of open coniferous and mixed woodlands, brushy lowlands, and mountain slopes. **MOVEMENTS:** Early spring movements are often from lower to higher elevations, into alpine/subalpine habitats, especially for northern populations. Later in the breeding season, some move from relatively open areas to dense coniferous forest. In some parts of range, these grouse move downslope into sagebrush habitats after breeding. **Population** Historical declines with some extirpations.

SOOTY GROUSE *Dendragapus fuliginosus*   SOGR ■ 2

Sooty Grouse was until recently combined with Dusky Grouse as a single species, Blue Grouse. Sooty inhabits primarily coastal Pacific ranges Polytypic (4 sspp.). L 20" (51 cm)

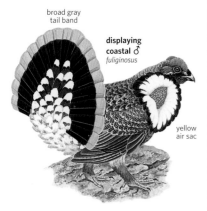

broad gray tail band

displaying coastal ♂ *fuliginosus*

yellow air sac

Identification Plumages generally very like Dusky Grouse overall, and in most places, range will determine the identification; hybridization is very limited. Sooty is darker than Dusky, and all subspecies of Sooty have a gray tail band (absent in "Richardson's" Dusky). Compared to Dusky Grouse, Sooty's plumage averages darker. During displays, the male's bare yellow (over most of range) neck skin is exposed; in southeast AK and north coastal BC, the apteria is red like Dusky. Both sexes of Sooty Grouse have 18 rectrices (versus 20 in Dusky), and in adults, the rectrix tips are rounder (versus more squared in Dusky). In Sooty, the rectrices are also more graduated in length, so that the tail overall looks more rounded than in Dusky. Even the downy chicks are different: grayish in Dusky but red-

dish above, yellowish below in Sooty. Behavior also differs: Sooty is much more arboreal and spends most of the year foraging on conifer needles except when females with broods forage on the ground; Dusky forages largely on the ground. Male Sooty mainly hoots from conifers; Dusky hoots from ground.

Geographic Variation The northernmost *sitkensis* is very dark. Nominate *fuliginosus*, found from BC south to Sonoma and Trinity counties, CA, is rather similar; males of both have fairly narrow gray terminal tail band, but female *sitkensis* are more rufous above than *fuliginosus*. Subspecies *sierrae* is found on the eastern slope of the Cascades in Washington south to Mendocino and Fresno counties, CA, while *howardi* is found south of there, in the southern Sierra Nevada.

The latter two subspecies are similar, but *howardi* is paler, with longer, more graduated tail.

Similar Species Very similar to Dusky;

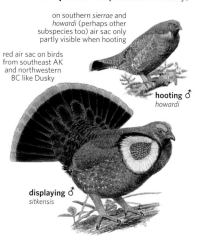

on southern *sierrae* and *howardi* (perhaps other subspecies too) air sac only partly visible when hooting

red air sac on birds from southeast AK and northwestern BC like Dusky

hooting ♂
howardi

displaying ♂
sitkensis

see Identification section for distinguishing characters of both species.

Voice CALL: Male Sooty in display gives six short hoots, higher in frequency than Dusky's and thus audible about one-half mile away; Dusky's hooting is hard to hear more than 200 ft away. Sooty usually displays in tree rather than on ground. Both sexes have repertoire of clucks and wails, similar to Dusky.

Status & Distribution Uncommon, locally declining resident of coniferous forests, from sea level (in northern portions of range) to high elevations (12,000 ft) near tree line. **MOVEMENTS:** Apparently travel by walking rather than flying between breeding and nonbreeding areas. Males begin trek to breeding grounds in early spring and vacate them beginning in June.

Population Sooty Grouse popula-

tions fluctuate, especially on the coast, but long-term declines have clearly occurred in CA, where the species is no longer found in the southern portion of its historical range (Mt. Pinos and the Tehachapi Mts.—populations likely extirpated in the 1930s). These extirpations and declines elsewhere likely due to logging operations.

sitkensis

?

fuliginosus

fuliginosus

sierrae

howardi

PRAIRIE GROUSE Genus *Tympanuchus*

Endemic to N.A., the three species of *Tympanuchus* are found in open grasslands and brushy habitats. They are generally barred or mottled in shades of brown, black, and pale buff. The males have colorful neck sacs and tufts (pinnae) on their head that they inflate and raise during their foot-stomping springtime displays.

SHARP-TAILED GROUSE *Tympanuchus phasianellus*   STGR ■ 2

Early each spring, Sharp-tailed Grouse gather at leks located on rather open elevated sites. The male inflates his purplish neck sacs, rapidly stomps his feet, bows his wings, and shakes his tail rapidly from side to side, producing a distinctive rattling sound. A variety of cackles and low coos accompany the performance. In other seasons, the Sharp-tailed is more difficult to see. During winter, look for it feeding on buds and catkins in deciduous trees early and late in the day. Where ranges overlap, it occasionally hybridizes with Greater

Prairie-Chicken and Greater Sage-Grouse; at least one record of hybridization with Dusky Grouse. Polytypic (6 sspp.; all in N.A.). L 17" (43 cm)

Identification Similar to prairie-chickens, but underparts scaled and spotted. **ADULT:** Pointed tail, with dark central (others white) rectrices extending far beyond the others. Yellow eye combs (sometimes difficult to see). Small erectile crest most noticeable when bird is agitated. Female smaller and with a less contrasting facial pattern. **JUVENILE:** Like female, but shorter central tail feathers and outer tail more brownish (not white).

Geographic Variation Four northern subspecies (*caurus, kennicotti, phasianellus,* and *campestris*) are darker, with heavy markings below and bold pale spotting on upperparts. Two southern subspecies (*columbianus* and *jamesi*) are paler brown and more uniform.

Similar Species Greater Prairie-Chicken is always heavily barred below (not spotted or scaled); it also has a short, rather square, mainly dark tail. Female Ring-necked Pheasant is superficially similar but is larger with a much longer tail, unfeathered legs, and

much less heavily marked underparts.

Voice CALL: Various clucks and peeps; also a three-note *whucker-whucker-whucker* given in flight.

Status & Distribution Fairly common in most of range; rare in western portions of range. **BREEDING:** Open and brushy habitats including grasslands, sagebrush, woodland edges, and river canyons. **MOVEMENTS:** Sedentary, but snow cover may induce movements to more wooded areas.

Population Habitat destruction and fragmentation led to declines throughout much of its southern range; populations generally stable in north. Extirpated from CA by about 1915 and OR by the 1970s; releases are underway in some places.

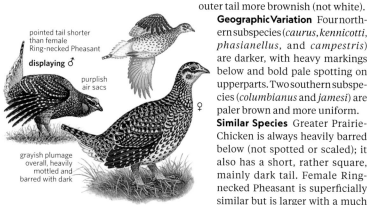

pointed tail shorter than female
Ring-necked Pheasant

displaying ♂

purplish air sacs

♀

grayish plumage overall, heavily mottled and barred with dark

GREATER PRAIRIE-CHICKEN *Tympanuchus cupido*   GRPC   ▪ 2

The once widespread Greater Prairie-Chicken is now very local in mixed-grass and tallgrass prairie. It is best found and appreciated in early spring, when males gather at hilltop leks; the sounds of their dramatic booming display may carry for nearly a mile. Polytypic (3 sspp.; all in N.A., but 1 is extinct). L 17" (43 cm)

Identification Heavily barred above and below with dark brown, cinnamon, and pale buff. Barring on underparts broader. **MALE:** Uniformly dark, short, rounded tail; yellow-orange neck sacs (inflated during the breeding season); and long pinnae, erected in displays to form "rabbit ears." **FEMALE:** Subtle version of the male, with smaller pinnae, indistinct eye combs, and barred tail. **JUVENILE:** Resembles female but is smaller with white shaft streaks on upperparts.

Geographic Variation The two coastal subspecies are darker: "Attwater's Prairie-Chicken" (*attwateri*, Endangered per USFWS) of southeastern TX is nearly extinct, and the "Heath Hen," nominate *cupido*, is extinct but was once found on the coastal plain from MA to VA—the last record was on Martha's Vineyard in 1932. The widespread *pinnatus* is found in the rest of the species' range.

Similar Species Lesser Prairie-Chicken is best separated by range. Greater Prairie-Chicken is slightly larger, with darker, broader, and more uniform barring on underparts. Sharp-tailed Grouse is similar but is more heavily scalloped and spotted below with a pointed, mostly white tail.

Voice Courting males make a deep low *ooah-hooooom* sound known as booming, often likened to the sound of blowing across a bottle, and also give a variety of wild frenzied cackles and yelps. **CALL:** Both sexes give a variety of clucking notes.

Status & Distribution Uncommon and local; "Attwater's Prairie-Chicken" is endangered. **MOVEMENTS:** Poorly understood; show no clear relationship with weather or food availability. Typically less than 25 miles between breeding and wintering areas.

Population Vulnerable. Greater Prairie-Chickens have experienced enormous declines in much of their range. The species was extirpated from TN (1850); KY (1874); AR (1913); LA (*attwateri*, 1919); TX (*pinnatus*, 1925); OH (1934); IN (1972); IA (1984); AB (1965); MB (1970); ON (1975); and SK (1976). Some of the first conservation legislation in the US was passed in 1791 to protect the "Heath Hen" from market hunting—efforts that ultimately failed. Market hunting was a major component in historic declines. Today, habitat destruction and fragmentation pose threats. Small isolated populations are especially vulnerable. Conservation and restoration efforts exist in many areas. Some, such as efforts by the CO Division of Wildlife and local communities in northeastern CO, have been extremely successful in rebuilding Greater Prairie-Chicken numbers.

pinnatus

attwateri

short, square black tail

displaying ♂

golden orange air sacs

stronger barring below than Lesser

♀

LESSER PRAIRIE-CHICKEN *Tympanuchus pallidicinctus*   LEPC   ▪ 2

This shortgrass prairie grouse has a very limited range in the southern Great Plains. It is best looked for during early spring at leks on relatively open sites on exposed hills and ridges. During this time, the differences between Lesser and Greater Prairie-Chicken are best appreciated. Note Lesser Prairie-Chicken's dull orange-red neck sacs, its more maniacal wails and cackles, and slight differences in displays. Monotypic. L 16" (41 cm)

Identification Very similar to Greater Prairie-Chicken; see above for separation from that species. As with the Greater, the Lesser male has a uniformly dark tail while that of the female is barred. Juvenile Lesser is also similar to a Greater but is slightly paler.

Similar Species Best separated from Greater Prairie-Chicken by range; the two species come very close in central Kansas and Oklahoma, but do not overlap. Lesser Prairie-Chicken is slightly smaller than Greater, with paler barring on the underparts, which is narrower and less uniform in width.

The barring typically becomes paler on the center of the belly. On Greater, the barring is bold throughout the breast and belly; the upperparts are more finely barred with black edging.

displaying ♂

smaller reddish orange air sacks

♀

weaker barring below than Greater Prairie-Chicken

There is much individual variation.
Voice CALL: Similar to Greater Prairie-Chicken; male's courtship calls are higher, more piercing, and even more demonic than the Greater.
Status & Distribution Uncommon and local; found in sand sagebrush-bluestem and shinnery oak-bluestem ecosystems. **MOVEMENTS:** Sedentary, although more likely to be found in agricultural fields during winter.
Population Vulnerable. This species has declined by an estimated 97 percent since the 1800s, including a 92 percent reduction in range (78 percent since 1963). Conversion of native habitat to cropland, excessive grazing by livestock, and herbicide control of sand sagebrush and shinnery oak all contributed to habitat loss. Almost extirpated from CO. Hunting likely played some role in early declines. Droughts have exacerbated all these factors.

Genus *Meleagris*

WILD TURKEY *Meleagris gallopavo*   WITU ■ 1

Writing to his daughter, Benjamin Franklin observed that the Wild Turkey would have made a better national symbol for the US than the Bald Eagle, proclaiming it "a bird of courage." The Wild Turkey is among our best known birds. After major declines, the Wild Turkey is once again regularly encountered in open woods bordered by clearings, particularly where oaks are prevalent. At night it roosts in trees. There is one other turkey in the world, the Ocellated Turkey (*M. cellata*), found on the Yucatán Peninsula and in Belize and Guatemala, but the Wild Turkey is the parent stock from which all domesticated turkeys are descended. Polytypic (7 sspp.; 4 in N.A.). L ♂ 46" (117 cm) ♀ 37" (94 cm)

Identification The Wild Turkey is the largest game bird in N.A., albeit smaller and more slender than its domesticated cousin. **MALE:** Dark, iridescent body, flight feathers barred with white, red wattles, blackish breast tuft, spurred legs; bare-skinned head is blue and pink. **FEMALE AND IMMATURE:** Smaller, duller, and often lack the breast tuft of the male.
Geographic Variation The widespread "Eastern Turkey" (*silvestris*) has tail, uppertail coverts, and lower rump feathers tipped with chestnut. These feathers are tipped a buff white in "Merriam's Turkey" (*merriami*), found in the Great Plains and Rockies. The "Rio Grande Turkey" (*intermedia*), found from Kansas south to Mexico, is intermediate in plumage; in fall and winter it forms huge flocks of up to 500 birds (40–50 typical of other subspecies). The "Peninsular Florida Turkey" (*osceola*) is similar to the "Eastern Turkey," but smaller. Introductions of subspecies outside of native range and presence of escaped and released domestic birds and resultant interbreeding greatly complicate

regional variation. Two other subspecies occur in Mexico.
Similar Species None.
Voice CALL: The distinctive gobble given by males in spring may be heard a mile away. Other calls include various yelps, clucks, and rattles.
Status & Distribution Restocked in much of its former range and introduced in other areas; now found in each of the lower 48 states and southernmost Canada.
Population The Wild Turkey declined throughout its range in the 19th and early 20th centuries; the loss was generally blamed on overhunting and habitat loss. Since then, the species has rebounded dramatically and continues to grow in some areas, with an estimated population of about six million (2009). Numbers in Mexico remain extremely low.

unfeathered reddish head

eastern *silvestris*

breast tuft

eastern *silvestris* ♀

unfeathered gray head

rufous tips to tail feathers

buffy white tips to tail feathers

western *merriami*

displaying ♂

FLAMINGOS Family Phoenicopteridae

American Flamingo (Bahamas, Mar.)

Extremely long necks, a unique bill shape, and pink plumage make flamingos among the most recognizable birds in the world. Nonetheless, flamingos are rare in the US, occurring regularly at only one site. They are common in captivity, however, which casts doubt on many extralimital reports, and even induces identification issues—not all escapes are species native to North America.

Structure Proportionately, flamingos are the longest necked and longest legged birds in the world. The neck is so long that it must be draped over one side of the body when the bird is at rest. The bill bends down in the middle, giving the bird a distinctive "broken nose." Webbed feet provide stability on mudflats and allow the flamingos to swim in water too deep for wading.

Plumage All species are unmistakably pink or red with black flight feathers. The color of the feathers is obtained from carotenoid pigments in the food supply; it takes three or more years to reach full adult plumage. Immatures are gray or brown.

Behavior Highly gregarious, flamingos nest, forage, and roost in colonies apart from other birds. Breeding takes place on mud- or marlflats near water; flamingos build unique volcano-shaped nests of mud at the top of which they lay a single egg. They use their curiously shaped bill to strain algae, diatoms, and aquatic invertebrates; when foraging, the head faces directly downward so that the distal half of the bill is parallel with the water. They usually prefer to forage in hypersaline estuaries and lakes; they feed by swinging their bill from side to side as they walk, and may even dabble like ducks in deep water. Their flight is swift and direct, with quick wingbeats and with the neck fully extended. Flocks fly in lines or in a V formation.

Distribution Flamingos are found on all continents except Antarctica and Australia. In North America, only the American Flamingo occurs naturally in extreme southern FL. Elsewhere along the Gulf and south Atlantic coasts it is a casual visitor; however, many of the records may be of escaped captives. Adding to the confusion, escapes of other flamingo species are regularly observed continent-wide.

Taxonomy Recent genetic studies have revealed a close relationship to grebes. Flamingos are placed in their own order, Phoenicopteriformes. There are six species worldwide in three genera. American Flamingo was formerly classified as a subspecies of Greater Flamingo, which included birds from the Old World. The two subspecies were split into two species in 2008: American Flamingo (*Phoenicopterus ruber*) in the New World and Greater Flamingo (*P. roseus*) in the Old World.

Conservation Flamingos have suffered from hunting and disturbance at nesting colonies, but they were largely spared the ravages of the late 18th- to early 19th-century millinery trade because their feathers fade once plucked from their bodies. Populations are recovering now that colonies are protected, but they remain at risk from wetland loss, pollution, disturbance, and other factors. BirdLife International codes: 3 NT, 1 VU.

Genus *Phoenicopterus*

AMERICAN FLAMINGO *Phoenicopterus ruber*   AMFL ■ 3

This well-known species is common in exhibits, but it has an extremely limited native range in the US. In Florida Bay—an area within Everglades NP—one small, mostly wintering flock offers the thrill of seeing flamingos in their natural habitat. At least some of these birds originate from Mexico—one fledgling color-banded in a colony at the Yucatán Peninsula was discovered recently at Florida Bay. Monotypic. L 46" (117 cm) WS 60" (152 cm)

Identification Unmistakable (except for escaped flamingos of other species). **ADULT:** The sexes look similar. Plumage entirely pink (darker below); black flight feathers often not visible at rest. Pink orbital ring and yellow eyes. Tricolored bill—gray base, pink mid-section, black tip. Pink legs and feet. **JUVENILE:** Grayish white overall with blackish streaking on wing coverts. Eyes dark; mid-section of bill gray; gray or grayish pink legs and feet.

Transition to adult plumage occurs gradually over three years; intermediate plumages not well known. **FLIGHT:** All ages show black flight feathers on both upper and lower wings.

Similar Species Roseate Spoonbill shares the same coastal habitats and pink plumage, but its body and bill shapes, proportions, and foraging behavior differ considerably. Escaped flamingos of other species are rare but present identification problems:

Be particularly circumspect with extralimital reports. Greater Flamingo (*P. roseus*) from the Old World is mostly whitish with reddish flight feathers, and the base of its bill is pink. Chilean Flamingo (*P. chilensis*) has very pale pink plumage except it is darker on the neck and has reddish flight feathers; its bill lacks any pink and its legs are gray with pink ankle joints. Lesser Flamingo (*P. minor*) is smaller and has a shorter, thicker neck; the basal half of its bill is dark maroon, the distal half is red with a black tip.

Voice Low guttural sounds given when feeding. **FLIGHT CALL:** A goose-like honking.

Status & Distribution Rare in the US;

restricted to southern FL. Locally common in Mexico, Cuba, and the Bahamas. **BREEDING:** Builds a volcano-shaped mound of mud in shallow coastal estuaries. No certain US nesting records. **DISPERSAL:** Prone to postbreeding and nonbreeding dispersal from colonies in Yucatán Peninsula, Cuba, and Great Inagua, Bahamas. **RARE STATUS:** Individual flamingos casually move north to Gulf or southern Atlantic coastal states, but many extralimital reports may refer to escaped captives. Reported north to KS, MI, ON, QC, NB, and NS. Flamingos in CA, NV, and WA presumably refer to escapes (a variety of species).

Population The species was formerly a locally abundant postbreeding visitor to Florida Bay from Mexican or Greater Antillean breeding grounds; only a small flock remains. Some populations now are increasing with protection. The species numbers some 80,000 in the West Indies; about 40 nonbreeders are semiresident at Florida Bay.

black remiges

fast, ducklike wingbeats

largely pinkish, strongly decurved bill is whitish at base, blackish at tip

adult

immature

very long, pink legs

mostly pink bill with small black tip

large size, whitish overall

dark bill with red area near tip

small size, pink overall

pink spotted wing coverts

whitish bill with large dark tip

from Old World

mainly from Africa

from South America

pink legs

reddish legs

pale pink overall

Lesser Flamingo
P. minor

Greater Flamingo
P. roseus

grayish legs with red "knees"

Chilean Flamingo
P. chilensis

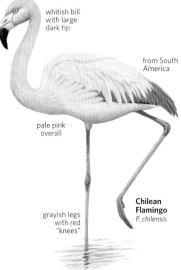

GREBES Family Podicipedidae

Least Grebe, adult with chicks (TX, Apr.)

Grebes are fairly small to medium-large diving birds of freshwater and inshore habitats. A tail-less appearance and a fairly long neck distinguish them from most swimming birds; relative to loons, note their fluffy rear ends. Rarely seen in flight, most grebes are best identified by head shape, face and neck patterns, and bill details.

Structure Grebes are heavy bodied with longish necks and lobed feet set far back on the body. Their beaks vary from somewhat long and pointed to stout and chicken-like. The wings are relatively small and pointed, with 11 primaries and 17 to 22 secondaries; vestigial tail feathers are difficult to distinguish from body feathers.

Behavior Grebes dive for food. Smaller species are mostly found singly or in small groups, but Eared and *Aechmophorus* grebes occur in flocks of hundreds, sometimes thousands. Flight is direct with steady, fairly quick wingbeats, the neck held outstretched and feet trailing.

Plumage Seasonally dichromatic plumage, but sexes look similar; juveniles resemble winter adults. Generally, plumages are dark above and light below; two North American species have golden head plumes in breeding plumage. Adult plumage aspect attained at about one year of age.

Distribution Cosmopolitan. In North America, grebes occur in all but the farthest northern latitudes.

Taxonomy Twenty-two species worldwide in six genera, with seven species in four genera in North America. Some authorities unite grebes and flamingos into a single order, Phoenicopteriformes.

Conservation Pollution, droughts, and disturbance at lakes and reservoirs affect breeding and wintering populations. Oil spills threaten nonbreeding grebes on marine waters. BirdLife International codes: 2 NT, 1 VU, 2 EN, 2 CR, and 3 EX.

Genera *Tachybaptus* and *Podilymbus*

LEAST GREBE *Tachybaptus dominicus*   LEGR ▮ 2

This well-named, diminutive, and somewhat retiring grebe of southern TX often holds its neck retracted and its rear end puffed out. Polytypic (5 sspp.; 2 in N.A.). L 8.5–9.5" (21–24 cm).

Identification Bill slender and slightly uptilted. **BREEDING ADULT:** Black face and throat; slate gray sides of neck. Black bill finely tipped white, eyes golden. **WINTER ADULT:** A paler face with a whitish throat; lower mandible mostly pale grayish to horn. **JUVENILE:** Head sides striped dark gray and whitish; lower mandible mostly pale horn; eyes paler and duller. Eyes develop golden color

over first year. First-summer bird has throat mottled sooty gray, not solidly black as in breeding adult.

Geographic Variation Subspecies *brachypterus* occurring in TX averages slightly larger and longer billed than *bangsi*, which is a stray to the Southwest from western Mexico.

Similar Species No other species should be confused with Least Grebe in its limited N.A. range; Pied-billed Grebe is nearly twice its size and has a stout, mostly pale bill.

Voice An overall descending, purring trill, *pc, pc, purrrrrrrrrrrrr*; often

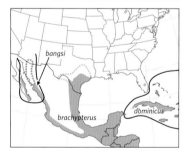

starts hesitantly. Also a quacking *kwrek*, and a slightly shrill, emphatic *ehkehr*.

Status & Distribution Southern TX to S.A. **BREEDING:** Fairly common, but local on vegetated ponds and lakes; nests year-round in southern TX (mainly Apr.–Aug.). **RARE STATUS:** Very rare to southern AZ (has bred) and upper TX coast. Casual to north-central TX and south FL. Accidental to southeastern CA, OK, and AR.

Population No data.

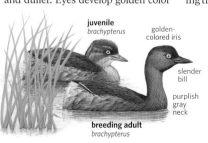

juvenile
brachypterus

golden-colored iris

slender bill

purplish gray neck

white throat

winter
brachypterus

breeding adult
brachypterus

PIED-BILLED GREBE   *Podilymbus podiceps*   PBGR   ■   1

This widespread chunky grebe is usually found singly or in small groups on small ponds. It can hide by sinking until only its head is above water. Polytypic (3 sspp.; nominate in N.A.). L 11–13.5" (28–34 cm)
Identification Thick bill diagnostic; fairly large head, and, unlike other N.A. grebes, lacks distinct white upperwing patches, visible only in flight. **BREEDING ADULT:** Face and throat black, neck sides grayer; bill pale blue-gray with sharply defined black medial

band. **WINTER ADULT:** Face paler and throat whitish, neck sides warmer and browner; bill pale horn to grayish, without black band. **JUVENILE:** Head striped dark gray and whitish, and bill pale horn. Some birds retain vestiges of dark head striping through winter. First-summer plumage resembles breeding adult, but it can have a less solidly black throat and the bill band may be narrower and less complete.
Similar Species Eared and Horned Grebes have slenderer bills, and a more capped appearance in nonbreeding plumages; also see Least Grebe.

Voice CALL: Includes single clucks and a rapid-paced, slightly nasal, chatter in interactions. **SONG:** Loud series of gulping notes that can fade away or run on into more prolonged variations.
Status & Distribution N.A. to southern S.A. **BREEDING:** Lakes, small ponds with emergent vegetation. **MIGRATION:** Mainly Mar.–Apr., late Aug.–Oct. in interior regions. In migration and winter also on sheltered coastal waters. **RARE STATUS:** Very rare to southeast AK; casual farther north and to Europe.
Population Declining locally, especially in the East.

white rear end

white eye ring surrounds dark eye

ring on thick, whitish bill

striped head

breeding adult
podiceps

black chin

juvenile
podiceps

white throat

downy young
podiceps

winter
podiceps

tawny brown neck

Genus *Podiceps*

RED-NECKED GREBE   *Podiceps grisegena*   RNGR   ■   1

This grebe often forms flocks at favored coastal locations in winter. Polytypic (2 sspp.; *holboellii* in N.A. and eastern Asia). L 17–20" (43–51 cm)
Identification Thick neck and blocky head. White secondaries and white patagial panel visible in flight. Medium-long, pointed bill with yellow at base. **BREEDING ADULT:** Neck chestnut; throat and auriculars smoky gray. **WINTER ADULT:** Neck dusky; face dusky gray with white extending up behind the auriculars. **JUVENILE:** Head sides striped dark gray and buffy to whitish, neck sides washed rufous. First-summer like breeding adult, but crown browner and foreneck duller.

Similar Species Eared Grebe in winter plumage is somewhat similar, but is distinctly smaller and has a slender dark bill, steep forehead, usually whiter and more contrasting auricular patches, and bright reddish eyes. Eared also often has a puffier rear end and shows more whitish along waterline. A molting Horned Grebe can suggest a breeding Red-necked, but is smaller and has a smaller blackish bill, bright-red eyes, and some indication of golden supercilium. In flight, Eared and Horned Grebes lack a clean-cut white forewing patch.
Voice Mainly vocal when breeding. Low, slightly gull-like wailing cries often run into slightly metallic, pul-

sating bickering or chattering series. Also a clipped, harsh note, usually repeated, *kerk! kerk!*
Status & Distribution Holarctic breeder. **BREEDING:** Favors shallow freshwater lakes. **MIGRATION:** Mainly Mar.–early May, Sept.–Nov. **WINTER:** Coastal regions and Great Lakes. When Great Lakes freeze, appears in numbers south to Ohio Valley and mid-Atlantic. **RARE STATUS:** Rare in interior south of northern tier of states; casual to the Southwest and Gulf Coast.
Population Declining.

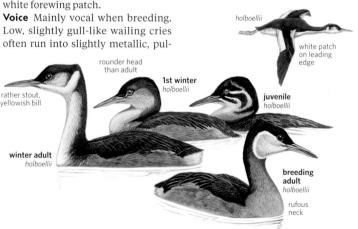

holboellii

white patch on leading edge

rounder head than adult

1st winter
holboellii

rather stout, yellowish bill

juvenile
holboellii

winter adult
holboellii

breeding adult
holboellii

rufous neck

HORNED GREBE *Podiceps auritus*   HOGR   ▇   1

This fairly small but stocky grebe is generally less familiar than the Eared Grebe on the breeding grounds, but it is more widespread in N.A. in winter, mainly on salt water, but also on freshwater lakes, where locally numerous. Overall scarcer in the interior West than interior eastern N.A. Polytypic (2 sspp.; *cornutus* in N.A.). L 11.5–13" (29–33 cm)

Identification Medium-length straight bill with fine white tip in all ages, shallow-angled forehead, and flat crown. Undertail coverts rarely fluffed into a high rear end. White secondaries visible in flight. **BREEDING ADULT:** Black head with golden bushy postocular "horns," black bill, and red eyes; neck and side of body chestnut. **WINTER ADULT:** White head sides and foreneck with blackish cap extending down to eye level; white on face nearly meets on back of head; bill grayish. **JUVENILE:** Head sides striped

dark gray and whitish, lower mandible mostly pale horn. First-summer plumage averages duller, with paler and less extensive horns.

Similar Species Eared Grebe is more lightly built with a thinner neck and steep forehead that create a pointier, less blocky head profile; its slimmer, upturned bill lacks a white tip. Eared often rides higher on the water, with a fluffed-up rear end. Most winter Eared Grebes are typically much dingier, with the black cap extending down well into the cheeks. Some Eareds, though, have surprisingly bright whitish necks and faces; check bill shape and depth of cap. Breeding Eared has a shaggy set of golden postocular plumes and black lores. The bold black-and-white pattern of the Clark's and Western Grebes can suggest a winter Horned, but the former are much

larger and longer necked with long and thin yellowish bills. Also see the larger Red-necked Grebe.

Voice Vocal mainly when breeding. A fairly high-pitched, rapid, pulsating whinny often given in duet, and a slightly hoarse, whistled *híki-sheíhr*, repeated.

Status & Distribution Holarctic breeder (known as Slavonian Grebe in the UK). **BREEDING:** Favors small to medium-size lakes with emergent vegetation. **MIGRATION:** Mainly Mar.–early May, Sept.–Nov. In migration and winter, prefers coastal waters and larger lakes. Casual in NL.

Population Vulnerable. Numbers slowly declining with a steady northwestward contraction of breeding range; also declines on winter grounds.

"horns" raised

breeding adult
cornutus

golden "horns"

chestnut neck

breeding adult
cornutus

molting birds in spring can be confused with Red-necked Grebe

adult in spring molt
cornutus

flat crown

darker winter
cornutus

white cheeks

small pale lore spot

pale tip to straight bill

winter
cornutus

EARED GREBE *Podiceps nigricollis*   EAGR   ▇   1

This fairly small, mostly western grebe is famed for its fall molt migrations, when tens of thousands of birds swarm at Mono Lake, CA, and Great Salt Lake, UT. It also nests colonially. Known as Black-necked Grebe in the UK. Polytypic (3 sspp.; *californicus* in N.A.). L 11–12.5" (28–31.5 cm)

Identification Distinctive steep forehead and relatively thin neck. Over-

all dusky to dark head and neck, and slender, slightly upturned bill. White secondaries visible in flight. **BREEDING ADULT:** Black head and neck, expansive postocular fan of golden plumes, black bill, and red eyes; body sides chestnut. **WINTER ADULT:** Dusky head sides and neck with a whitish throat and postauricular patches; gray bill with dark culmen. **JUVENILE:** Resembles a winter adult, but its head sides and foreneck are washed buff and the eyes are duller, orangey. By fall, resembles winter adult but browner

buffy wash to neck

1st fall
californicus

some quite pale-necked like Horned

often raises and fluffs out rear end

paler winter
californicus

peaked or rounded head

upturned all-dark bill

dark cheek

dusky neck

winter
californicus

downy young
californicus

golden "ears"

breeding adult
californicus

black neck

puffier cheeks, and white almost meets on back of head

winter

Horned   Eared

above; eyes are orange to orange-red. First-summer plumage resembles a breeding adult, but the head and neck average duller, with paler and less extensive head plumes. **Similar Species** Horned is more thickset and has a shallow-sloping forehead, flatter crown, and a straight, slightly thicker bill tipped white. Horned often rides flatter on the water, without the pronounced fluffed-up rear end of the Eared. Winter Horned Grebe is typically a much cleaner black-and-white,

with a shallow black cap (beware of nonbreeding Eared with bright whitish neck; check bill shape and depth of cap). Molting birds of the two species can look quite similar (mainly Apr. and Sept.): Note bill shape differences and the position of yellow on the head sides. Also see Red-necked Grebe. **Voice** An up-slurred, slightly reedy or plaintive *hoor-eep* or *eeihr-ek!* Rapid-paced, high piping chittering in interactions. The two Old World subspecies are similarly plumaged to *californicus*,

but they appear to have very different vocalizations. **Status & Distribution** Holarctic and African breeder. **BREEDING:** Common. Favors shallow ponds with high macro-invertebrate productivity. **MIGRATION:** Mainly Mar.–May, Aug.–Nov. In migration and winter, found on a wide variety of aquatic habitats, including salt water. **RARE STATUS:** Rare in East, a few winter regularly in the Southeast. **Population** No demonstrable trend apparent in N.A. over recent decades.

Genus *Aechmophorus*

WESTERN GREBE *Aechmophorus occidentalis*   WEGR ◾ 1

Western Grebe looks similar to Clark's Grebe, and the two were once considered conspecific under the name Western Grebe. Flocks in winter can number hundreds, often with Clark's Grebes mixed in. Polytypic (2 sspp.; nominate in N.A.). L 19–25" (48–64 cm) **Identification** Extent of white on face varies with age and season. Intermediate (and unidentifiable) birds are not rare; some may be hybrids. **BREEDING ADULT:** Dusky gray to blackish lores and supraorbital region, yellow bill with dark culmen, and orange-red to red eyes. **WINTER ADULT:** Dusky lores and supraorbital region, often with whitish loral patch. **FIRST-YEAR:** Aging criteria as Clark's. First summer may have pale loral spot. **Similar Species** Bill color best separates Western Grebe (greenish yellow to yellow) from Clark's Grebe (orange-yellow). See Clark's for other differences. Also see the much smaller winter Horned Grebe. **Voice** Varied screechy and scratchy calls, usually a far-carrying two-note *kree-kreet*, different from the single drawn-out note of Clark's. Begging

calls sound much like Clark's Grebe. **Status & Distribution** N.A. to central Mexico. **BREEDING:** Favors freshwater lakes with large areas of open water and bordered by reeds or other emergent vegetation. **MIGRATION:** Mainly Apr.–May and Sept.–Nov. **WINTER:** Large numbers winter on inshore Pacific coast waters (Clark's prefers more protected coastal bays); many nonbreeding birds oversummer. Also winters on large lakes. **RARE STATUS:** Casual in the East. **Population** See Clark's Grebe.

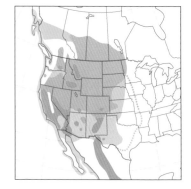

slightly paler faced in winter

darker face

greenish yellow bill

occidentalis

winter adult *occidentalis*

darker downy young than Clark's

breeding adult with downy young *occidentalis*

courtship display *occidentalis*

slightly darker sides than Clark's on average

CLARK'S GREBE *Aechmophorus clarkii*   CLGR ◾ 1

A large, striking black-and-white grebe once considered conspecific with the generally commoner Western Grebe. The two species associate readily and hybridize. Monotypic. L 18–24" (45–61 cm) **Identification** This grebe has a long, slender neck and a long, slender, pointed bill (noticeably larger on males) with a distinct gonydeal angle. **BREEDING ADULT:** White lores and supraorbital region, bright orange-

yellow bill with dark culmen, orange-red to red eyes. **WINTER ADULT:** White to mostly dusky lores and supraorbital region with whitish loral patch. **FIRST-YEAR:** Juvenile similar to winter adult but crown washed grayish and eyes duller and paler, orangey. After molt, rarely distinguishable from adult, although upperparts can be mixed with worn, brownish feathers. First-summer may have dusky face, lacking bright white surround to eye,

which is paler red through first year. **Similar Species** Western Grebe is best separated by its greenish yellow to yellow bill, but also note its more extensive black crown (extending down to the eyes), broader black hind neck stripe, and overall grayer flanks. Breeding Western Grebe has blackish to dusky-gray lores and surround to eye, but these areas are paler on winter birds, which can have a loral pattern similar to Clark's. Although the upperwing of Clark's averages more extensive white on primaries, it is not diagnostic. The range of variation within each species is not well understood, compounded by frequent hybridization; some birds, mainly in winter, may not be identifiable. Note different calls of Clark's and Western. **Voice CALL:** A rather drawn-out, slurred single-note *kreeeih* or *kreeet* among others. Juveniles solicit adults with high-pitched whistles that can still be given in midwinter.

Status & Distribution N.A. to central Mexico. **BREEDING:** Fairly common. **MIGRATION:** Mainly Apr.–May and Sept.–Nov. A few nonbreeding birds oversummer on inshore waters and bays along the Pacific coast. **WINTER:** Fairly common to uncommon on West Coast. (Only 10–20 percent of *Aechmophorus* flocks in CA are Clark's.)

RARE STATUS: Accidental in the East. **Population** From the 1890s to about 1906, tens of thousands of Western/Clark's Grebes were shot for their silky white feathers; today, oil spill mortality, disturbance, habitat degradation, and pesticides continue to keep present populations below historic levels.

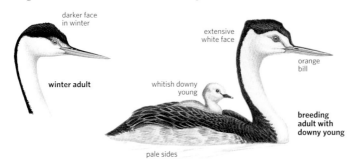

PIGEONS AND DOVES Family Columbidae

Inca Dove (TX, Mar.)

The larger species in this family are usually called pigeons, the smaller ones doves; there is no scientific or taxonomic distinction between the two groups. Pigeons and doves feed chiefly on grains, seeds, and fruits. Nests are usually flimsy structures, and most clutches are limited to two white eggs.

Structure The family varies in size from the huge crowned-pigeons of New Guinea to the towhee-size ground-doves of the Americas. Most are plump with short, round-tipped wings and have tails varying from short and square-tipped to long and pointed. All have small rounded heads, and virtually all have short legs and similarly shaped short bills.

Behavior Pigeons and doves are strong, fast fliers (although a few species fill a terrestrial role in dense jungles and fly less frequently); only a few species are long-distance migrants, but many disperse widely in search of food. While most eat various forms of vegetable matter—primarily seeds and fruit, but also leaves and flowers—some species have been known to eat invertebrates, such as worms or insect larvae. North American species feed primarily on the ground; many species in other parts of the world are arboreal. Their nest-building skills seem poor, but such untidy nests, usually appearing to be little more than several sticks connecting tree branches, might allow them to rebuild quickly in the event of a nesting failure.

Plumage Pigeons and doves vary in color from the near

patternless grays of some Old World pigeons to the highly patterned oranges and greens of fruit-doves. Many species show iridescence on the hind neck; many also show a pinkish suffusion to the underparts. Most juveniles have pale-fringed feathers and lack the neck markings of adults. Some species have spectacular head plumage, such as elaborate crests and long, pointed feathers.

Distribution Except for Antarctica, pigeons and doves are present on all continents, with the greatest variety in the tropics. Most are found in a variety of woodlands, but some members of this family have evolved to exploit habitats from rather barren deserts to sea cliffs and areas near the tree line in high mountain ranges.

Taxonomy About 313 species in 46 genera are recognized worldwide, with 20 species in eight genera recorded in the US and Canada. The taxonomy of this family continues to evolve; AOS recognized the splitting of the American pigeons, *Patagioenas*, from the Old World pigeons, *Columba*, in 2003.

Conservation More than a third of the world's pigeons and doves are threatened, with 16 already extinct (including the iconic Dodo and Rodriguez Solitaire). Predators, particularly rats, threaten island species, such as some of the fruit-doves or large terrestrial pigeons, which evolved with little competition and few natural enemies. Wholesale clearing and fragmentation of forests eliminates cover and food for many species, though a few may benefit. Overhunting has played a major role in some extinctions, such as the extermination of the Passenger Pigeon. Surprisingly, trapping for the pet trade appears to have little or no impact. BirdLife International codes: 47 NT, 40 VU, 20 EN, 11 CR, 1 EW, and 16 EX.

OLD WORLD PIGEONS Genus *Columba*

ROCK PIGEON *Columba livia*   ROPI   ■   1

This highly variable city pigeon is familiar to all urban dwellers. Many color variations were developed over centuries of near domestication. Polytypic (9 sspp.; nominate in N.A.). L 12.5" (32 cm)

Identification A medium-size compact pigeon with long wings and a short tail. Birds most closely resembling their wild ancestors are gray with head and neck darker than back, and a prominent white rump. Black tips on greater coverts and secondaries form bold black bars on inner wing, and there is a broad black terminal band on tail. **ADULT MALE:** Metallic green and purple iridescence on neck and breast; iris orange to red; orbital skin blue-gray; bill grayish black; and feet dark red. **ADULT FEMALE:** Like male, but iridescence on neck and breast more restricted and subdued. **JUVENILE:** Generally browner, lacks iridescence; orbital skin and feet gray. **Voice CALL:** Gives a soft *coo-cuk-cuk-cuk-cooo*.

Status & Distribution The Rock Pigeon was introduced from Europe by early settlers; it is now widespread and common throughout the US and southern Canada, particularly in urban settings and around grain elevators; locally on cliffs as in its native Old World range. Gregarious and forming large flocks, it feeds on handouts and grains during the day in city parks and open fields and roosts on buildings at night. Passing flocks can be found far from civilization. **BREEDING:** Nest is loosely constructed of twigs and leaves, primarily on structures such as window ledges, bridges, and in barns; female lays two white eggs.

Population Primarily associated with human development and dependent on people for food and shelter.

in flight, often gives a loud wing flapping noise

ancestral natural coloration

white rump

color variations
many other variations are seen

AMERICAN PIGEONS Genus *Patagioenas*

SCALY-NAPED PIGEON *Patagioenas squamosa*   SNPI   ▪   5

orange orbital ring

dark maroon head and neck, but in most lights appears all-dark

adult ♂

A large, dark, shy pigeon of the West Indies, the Scaly-naped Pigeon has been recorded only twice (old historical records in N.A.); larger than Rock Pigeon. Monotypic. L 13.8" (35 cm)
Identification This species appears entirely dark slate-gray, or even blackish, when seen perched or in flight

from a distance; there are no obvious markings on wings or tail. **ADULT MALE:** Head and upper breast dark maroon; feathers on sides of neck are more reddish, tipped black, forming diagonal lines and creating the scaled appearance that gives this species its name. Iris orange-red; prominent reddish orbital ring, with variable yellow-orange subocular skin below, sometimes much broader under eye forming a rectangular bar. Bill is dark red at base with extensive pale yellowish tip; feet dark red. **ADULT FEMALE:** Nearly identical; soft parts slightly duller. **JUVENILE:** Duller with rusty-brown fringes on scapulars.
Similar Species White-crowned Pigeon is slightly smaller, a little darker, normally shows white on the head, and lacks red orbital ring and yellow-orange subocular skin. However, on young White-crowned, white is reduced and washed with gray-

ish brown, so could be overlooked.
Voice CALL: An emphatic *cruu, cruu-cru-cruuu*, the first syllable soft with a pause before the last three syllables, which sound like "who are you?"
Status & Distribution Resident, primarily arboreal, found most frequently in primary upland forest. Found throughout most of the West Indies, plus some other islands off Venezuela; notably absent from the Bahamas and Jamaica. **BREEDING:** Nest is loosely built of twigs lined with grass and found in trees, including palms, but occasionally on the ground. Usually lays one white egg. **RARE STATUS:** Single birds collected at Key West, FL, 24 Oct. 1896 and 6 May 1929 are the only two recorded in N.A. Also recorded from Jamaica.
Population While some have adapted to more urban areas, the overall population is declining, with threats of local extirpations, due to deforestation and intensive hunting.

WHITE-CROWNED PIGEON *Patagioenas leucocephala*   WCPI   ▪   2

White-crowned Pigeon is the large, dark pigeon seen in the FL Everglades and FL Keys. Flocks of varying sizes

♂

white crown

pale eye

red bill with pale tip

dark slaty gray body

♀

commute daily from nesting colonies in coastal mangroves to feed inland, at times flying many miles. Monotypic. L 13.5" (34 cm)
Identification A large, square-tailed slate black pigeon with a conspicuous white crown. **ADULT MALE:** Crown is pure white; sides of neck are iridescent green merging into an iridescent purple on hind neck. Iris is whitish, with pale blue-gray orbital skin. Bill is dark reddish with a white tip; feet are dull red. **ADULT FEMALE:** Like male, but slightly less iridescent. **JUVENILE:** Browner; rusty brown fringes on coverts and scapulars. White on crown is greatly reduced and washed with grayish brown; it can be difficult to see at times. Iris is brownish, with brownish gray orbital skin.
Similar Species Some Rock Pigeons can appear black, but will normally show white on rump and lack white on crown. Check all birds carefully for the slim possibility of an accidental Scaly-naped Pigeon from the West Indies.
Voice CALL: A loud, deep *coo-cura-coo*, or *whoo-ca-cooo* repeated several times; a soft *coo-crooo* is believed to be given from the nest.

Status & Distribution Found throughout West Indies and on islets off east coast of Yucatán Peninsula. In FL, a fairly common resident in the Keys, north to Miami-Dade Co. Casual north to St. Lucie Co. and Lee Co. **BREEDING:** A colonial nester. Nest is a loose platform of twigs; in FL, normally in mangroves and other wooded areas; sometimes visits feeders. **RARE STATUS:** Accidental coastal TX (Oct. records of juveniles at Galveston and South Padre I.).
Population Near Threatened. Declines in parts of its native range due to clearing of hardwood forests, overhunting, harvesting of nestlings for food, and introduced predators.

RED-BILLED PIGEON　*Patagioenas flavirostris*　RBPI ▪ 2

The Red-billed Pigeon is a large, all-dark pigeon seen perched conspicuously (especially early and late in the day) in tall trees above brushy understory or flying in pairs or in small groups along the Rio Grande in southern TX, but in recent decades, this is a diminishing sight. Polytypic (2 sspp.; nominate in N.A.). L 14.5" (37 cm)

Identification The species generally appears entirely dark when seen from any distance. **ADULT MALE:** Head, neck, most of underparts, and lesser wing coverts are dark maroon; rest of plumage is dark blue-gray, with a blackish tail. Paler gray area on wing coverts forms a bar that might be visible in flight in good light. Iris is reddish orange, with bordering red orbital ring. Bill is pale yellow, or whitish with dark red at base. Feet are dark red. **ADULT FEMALE:** Female's coloring is similar to male but generally duller. **JUVENILE:** Like female but its maroon feathering is rustier and paler.

Similar Species Distinctive within its expected area of occurrence, although darker Rock Pigeons should always be considered, especially when only seen in flight.

Voice CALL: A distinctive long, high-pitched *cooooo* followed by two to five loud *up-cup-a-coo*'s given in the early spring and summer; also a single, swelling *whoo* often repeated several times.

Status & Distribution Widespread and still numerous throughout the lowlands of Mexico and south through Costa Rica. Locally uncommon to rare in TX; declining along the lower Rio Grande (mostly in Starr Co. and Zapata Co., fewer upriver to northern Webb Co. and southern Maverick Co. and to western Hidalgo Co.); rare and very local in winter. **RARE STATUS:** Casual farther downriver and north to Nueces Co., Kerr Co., and Real Co., TX. **BREEDING:** Nest is a loose platform of twigs built well above the ground; normally lays one white egg.

Population Deforestation along the Rio Grande in the 1920s greatly reduced the Red-billed Pigeon's numbers and range there, and the species is now protected in TX.

pale eye with red orbital ring

red bill with pale yellow tip

maroon head, neck, and wing coverts; otherwise slate gray

flavirostris

flavirostris

BAND-TAILED PIGEON　*Patagioenas fasciata*　BTPI ▪ 1

The Band-tailed is a large, heavily built gray pigeon that frequents the forests and woodlands of the West. It is most often seen in flocks of varying sizes in rapid flight. Its wings make a loud clapping sound when flushed. Polytypic (6 sspp.; 2 in N.A.). L 14.5" (37 cm)

Identification Larger, with a longer tail than a Rock Pigeon; blue-gray on upper parts with contrasting blackish gray flight feathers; a blackish gray tail with a broad, pale gray terminal band. Paler gray greater coverts show as a broad wing stripe when in flight. **ADULT MALE:** Gray on head, and breast tinged pinkish; narrow white half-collar across upper hind neck, with iridescent greenish below. Narrow orbital skin purplish; bill yellow with a black tip; and feet yellow. **ADULT FEMALE:** Like male, but pink color somewhat subdued, and with less iridescent green. **JUVENILE:** Paler than adults, with narrow whitish fringes on breast and coverts; half-collar reduced or obscured; bill and feet duller.

white band at top of nape

yellow-based bill

lacks white band on nape

duller bill

juvenile

Geographic Variation Nominate subspecies *fasciata* breeds in the Southwest from UT and CO south into Mexico; *monilis* (sometimes not recognized) breeds in the Pacific states from BC (uncommonly in southeast AK) to northern Baja California, Mexico. They are not separable in the field.

Four other subspecies are found in M.A. and S.A., the southern group from Costa Rica south being distinc-

rather long tail with broad pale tail band

tive with a richer coloration overall and an all-yellow bill.

Similar Species Rock Pigeons have blackish tails; most have black markings on wings and conspicuous white rumps; at close quarters, bill lacks yellow; and feet are reddish rather than yellow.

Voice CALL: A low-pitched *whoo-whoo* delivered several times.

Status & Distribution Locally common in low- to mid-elevation coniferous forests in the Pacific Northwest, and in oak or oak-conifer woodlands in the Southwest; presence dependent on availability of food; increasingly common in suburban gardens and parks. **BREEDING:** Nest is a platform of twigs lined with grasses placed in a tree well above the ground; bears one white egg. **MIGRATION & WINTER:** Most birds breeding in the Southwest winter in Mexico, and most breeding in the Pacific Northwest move south into CA in winter. Large flocks are often seen during migration. **RARE STATUS:** Casual across southern Canada east to NS and New England, and on Gulf Coast east to western FL. Accidental in western AK.

Population The Pacific population was formerly threatened by overhunting, but with controls, the population is recovering.

OLD WORLD TURTLE-DOVES Genus *Streptopelia*

ORIENTAL TURTLE-DOVE *Streptopelia orientalis* ORTD ■ 4

This large, heavy dove of Asia, appearing almost pigeonlike when in flight, has strayed to N.A. Polytypic (5 sspp.; likely nominate in N.A. but no specimen). L 13.5" (34 cm)

Identification Warm brown scaly pattern above, formed by broad chestnut fringes on blackish tertials and wing coverts. Flight feathers blackish; underside of wings dark gray; blunt-tipped tail is blackish with broad gray terminal band. **ADULT MALE:** Crown pale gray; underparts grayish brown merging into gray on undertail coverts; conspicuous black-and-white stripes on sides of neck; iris orange; orbital skin dark purplish; bill blackish with a trace of dark purple at base; feet dark reddish. **ADULT FEMALE:** As male, but slightly duller and browner on underparts. **JUVENILE:** Paler; fringes on tertials and coverts narrower; neck markings obscured.

Geographic Variation Two subspecies groups, the more easterly, larger and duller *orientalis* group with gray tail band, and the smaller, more west-erly and brighter *meena* group with white tail band. All N.A. records pertain to the former group.

Similar Species European Turtle-Dove is smaller, brighter, with more rufous on upperwing coverts; terminal tail band whitish unlike eastern subspecies.

Voice CALL: Nominate subspecies gives a four-phase *deh-deh co-co*, with the last two notes lower-pitched.

Status & Distribution Present throughout much of Asia; northern birds highly migratory. Eastern nominate subspecies, reaches to Southeast Asia in winter. **RARE STATUS:** Casual to Aleutians and Bering Sea region in spring, summer, and fall (recorded north to St. Lawrence I.). Accidental YT, Vancouver I., BC, and CA (twice).

Population Stable over its wide Asian range.

tail tips vary from whitish to pale gray

orientalis

rufous fringes give scaly pattern above

black-and-white streaked neck patch

orientalis

orientalis

fringes more grayish below

orientalis

gray rump

EUROPEAN TURTLE-DOVE *Streptopelia turtur* EUTD ■ 5

orange eye

black-and-white stripes on neck

scapulars and wing coverts with black centers and orange-brown edges

turtur

From biblical verses in the Song of Songs to Shakespeare's poems, folk music, Christmas carols, and spirituals, turtle-doves have been celebrated as symbols of devotion. Through much of its range, it is also a heavily hunted species. Polytypic (4 sspp.; L 10" (26 cm)

Identification In N.A. context, this slim dove is very distinctive. **ADULT:**

Vivid orange-brown fringes to scapulars and coverts; black-and-white-striped patch on side of neck. Pale orange eye, surrounded by red orbital skin, stands out starkly in pale gray head. In flight, wedge-shaped tail is strikingly patterned, with grayish brown coverts over black rectrices with broad white tips. **JUVENILE:** Paler, and colors far more muted than

in adult; lacks neck patch of adult.
Geographic Variation Four subspecies recognized: The range of nominate *turtur* extends from Europe to western Siberia; three others occur from northern Africa to western China.
Similar Species In N.A., beware escaped doves and pigeons from aviary collections. Most similar to Oriental Turtle-Dove, recorded only in the West, but that species is larger overall, and has grayer fringing on upperwing coverts.
Voice SONG: A rather unmusical, insistent purring *trrrrrrrr*, repeated in series, with some shorter than others and some rising slightly in tone.
Status & Distribution Highly migratory. **BREEDING:** From the British Isles and Europe to northern Africa and southwestern Asia; winters in sub-Saharan Africa. **RARE STATUS:** Records of single birds from Monroe Co., FL (9–11 Apr. 1990)—this record origin questioned by some; St. Pierre I., France (15–20 May 2001); and Tuckernuck I., MA (19 July 2001). Iceland has more than 200 records, most from autumn, so Atlantic Canada would be a likely place for an autumn record.
Population Vulnerable. Has declined in parts of Europe, especially in the UK.

AFRICAN COLLARED-DOVE *Streptopelia roseogrisea* AFCD ■ EXOTIC

A domesticated form of the African Collared-Dove, also known as Ringed Turtle-Dove, can be encountered as an escaped cage bird almost anywhere. Monotypic. L 10.5" (27 cm)
Identification A very pale Mourning Dove–size bird with a blunt-ended tail. It has a whitish head and neck, and is pale buff above with an obvious black collar on hind neck. Whitish below, including undertail coverts. There are many color variations in adults; a pale buff variant is commonly encountered, but others, including peach-colored and rather uniform whitish birds, are also seen. Primaries are only slightly darker than rest of wing, providing a more uniform look. Tail is black at base, visible from below; black typically falls short of longest undertail coverts. Tail has an entirely white outer web on outermost rectrices, and a broad whitish terminal band.
Similar Species Eurasian Collared-Dove is superficially similar, is larger with gray (not white) undertail coverts (beware of pale-morph Eurasian Collared-Dove). Given the huge range expansion of Eurasian Collared-Dove, combined with the generally poor ability of African Collared-Dove to survive in the wild, most encounters away from known African Collared-Dove populations will likely pertain to Eurasian Collared-Doves. Beware of hybrids between the two species. African Collared-Doves are smaller and paler. Their primaries are paler, contrasting less with rest of wing and adjacent coverts. In addition, inner wings, visible in flight, are rather uniform in African Collared-Dove, versus contrasting gray secondaries of Eurasian Collared-Dove and outer webs of outer rectrices are white in African Collared-Dove, versus black in Eurasian Collared-Dove. Black at base of tail, visible from below, falls short of the longest undertail coverts and is generally hard to see in African Collared-Dove, whereas it extends beyond the longest undertail coverts and is quite evident.
Voice CALL: A rolling bisyllabic *kooeek-krrrooooo*, noticeably softer than that of Eurasian Collared-Dove.
Status & Distribution A frequent escape. Small populations have persisted where fed, but they do not do

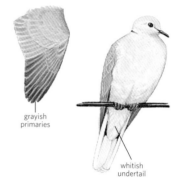

grayish primaries

whitish undertail

well in the wild; no known viable wild populations are now known in N.A. (Formerly a tiny established population was in Los Angeles.) They can be encountered well away from cities, but these are presumably birds that have recently escaped or been released. In its native range, a common species in the Sahel, from Senegal to the Red Sea, including southwestern Saudi Arabia and Yemen.
Population Stable in native range. Various N.A. populations (e.g., Los Angeles) are extirpated.

SPOTTED DOVE *Streptopelia chinensis* SPDO ■ 2

This Asian dove was introduced into Southern CA in the early 1900s; now virtually extirpated. Polytypic (5 sspp.; nominate in N.A.). L 12" (31 cm)
Identification This is a large and dark dove with broad, rather round-tipped wings. Tail is long, dark, and broad at the end with white tips on outer three to four rectrices. **ADULT MALE:** Gray on top of head bleeds into face. Dark brown upperparts have feathers finely fringed with pale brown, and flight feathers are black. Underparts are a warm dark pinkish brown, merging into gray on undertail coverts. Upperwing in flight has gray secondaries; below, underside of wings are dark. Broad black collar has prominent white spots. Iris is reddish orange, bordered by narrow, dark red orbital skin. Bill is blackish, and feet are reddish. **ADULT**

chinensis

black-and-white spotted collar

pinkish underparts

chinensis

chinensis

juvenile chinensis

rectangularly shaped tail shows extensive white from below

FEMALE: Same as male. **JUVENILE:** Browner with broader, buffer fringes on feathers of upperparts; collar obscured or missing.

Similar Species Mourning Dove, and it differs in the following ways: Spotted is a large, chunky dove, with broad, round wings and a broad, squared-off tail. In contrast, Mourning Dove has a more slender look, with a paler coloration and more pointed wing and a thinner tail, with an obviously pointed central tail. If Spotted Dove's distinctive black nape, pitted with white spots, is visible, identification is simple.

Voice CALL: A rather harsh *coo-coo-croooo* and *coo-crrooo-coo*, with the emphasis respectively on the middle and last notes of the calls.

Status & Distribution Spotted Dove occurs naturally throughout Southeast Asia and India; it was introduced into Los Angeles in the early 1900s, then spread throughout much of urban Southern CA from Santa Barbara and Bakersfield south to Baja California, peaking in the 1980s. In 2019, a very few survive in a small section of south Los Angeles and a few may survive in Bakersfield and Fresno. A small stable population is found at Avalon (Santa Catalina I.). **BREEDING:** The nest is a flimsy platform of twigs, usually built in a tree or shrub, but occasionally found on buildings or even on telephone poles; lays two white eggs.

Population Stable and common in Asia. Definitive reasons for this introduced species' decline and extirpation from Southern CA are not known for

certain, but predation from Cooper's Hawks (nesting widely in urban areas in recent decades) is a likely factor. Its continued presence with a stable population on Santa Catalina I. could be due to the absence of Cooper's Hawks there.

EURASIAN COLLARED-DOVE *Streptopelia decaocto*　EUCD　■　1

A fairly recent arrival to N.A., this large pale dove can now be found across the US. It flaps on broad wings, and often soars briefly, with wings extended slightly above horizontal as it seemingly floats down to a landing. Polytypic (2 sspp.; nominate in N.A.). L 12.5" (32 cm)

Identification A large, pale gray-buff dove with a black collar, noticeably larger than Mourning Dove. There is also a naturally occurring cream-colored variant, and this species is known to hybridize with African Collared-Dove, so plumage variation will occur. Tail is fairly long and blunt-ended. **ADULT MALE:** Head is an unmarked, pale buff-gray, while upperparts are a darker buff-brown, tinged gray; a conspicuous black collar can be seen on hind neck. The primaries are noticeably darker than rest of wing, appearing blackish; secondaries are gray and contrast with blackish primaries and brown wing coverts in flight. Undersides of wings are pale. Underparts are a paler buff-gray merging into gray on undertail coverts. A dark gray tail has obvious black at base when seen from below; black extends beyond undertail coverts. This black includes outer webs of outer rectrices, and tail has a broad, pale buff-gray terminal band. A reddish brown iris borders narrow grayish white orbital skin. Blackish bill has gray at base, and feet are dull reddish. **ADULT FEMALE:** Similar. **JUVENILE:** Paler; buff fringes on feathers of upperparts; black collar obscured or missing.

black collar

grayish undertail

stockier overall than Mourning Dove

three-toned wing

Similar Species Domesticated (in US) African-Collared Dove—feral birds were formerly known as "Ringed Turtle-Dove"—is smaller, shorter-tailed, and noticeably paler; it has far less contrast between flight feathers and rest of wing; undertail coverts are white with black at base of tail more restricted, and outer webs of outer rectrices white. In addition, call is distinctly different.

Voice CALL: A monotonous repeated, trisyllabic *kuk-koooo-kook*, slightly nasal, with emphasis on middle note is most habitually given during the long breeding season; also a harsher *kwurrr* is given frequently in flight throughout the year.

Status & Distribution This Eurasian species was accidentally introduced first in Nassau, Bahamas, when an aviary was burglarized in 1974; it spread to FL by the early 1980s. Explosive expansion in N.A. to west and to northwest

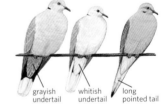

grayish undertail

whitish undertail

long pointed tail

Eurasian Collared-

African Collared-

Mourning

in the late 20th and early 21st century. Interestingly, has not spread very far to the north in the East. The species is now resident in much of the US and western Canada. Its westward expansion follows a similar expansion from its original range in Asia all the way to the Atlantic coast of Europe, including the British Isles. **BREEDING:** Nest is a flimsy construction of twigs in trees; two white eggs, up to six broods a year. **MIGRATION:** Not a migrant in the true sense, but individuals move great distances. Frequent at sea well off West Coast.

Population Continues to expand in both numbers and range in N.A., except in upper Midwest and Northeast. Small numbers present in some areas may have escaped or been released from captivity by dove breeders, but most birds are thought to represent genuine colonizers.

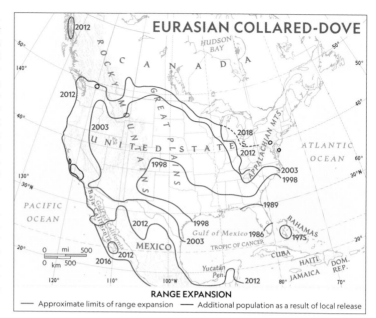

EURASIAN COLLARED-DOVE

RANGE EXPANSION
—— Approximate limits of range expansion —— Additional population as a result of local release

---

Genus *Ectopistes*

PASSENGER PIGEON *Ectopistes migratorius* PAPI ■ 6

Extinct. Formerly the most abundant bird in N.A., it occurred in huge flocks said to "blacken the sky" in the early 1800s. By the 1870s breeding numbers had been reduced to small scattered colonies, and the last wild bird was taken in OH in 1900; a reliable sight record for MO in 1902. The species became extinct when the last remaining bird in captivity died on 1 Sept. 1914 at the Cincinnati Zoo. Monotypic. L 15.8" (40 cm)
Identification Resembled the Mourning Dove, but had long, broad wings and a very long graduated tail and was substantially larger. **ADULT MALE:** Generally blue-gray above and pinkish below, brightest on breast, with black markings on coverts, and contrasting darker primaries and secondaries; white in tail restricted to outer pair of remiges; bill black; and tarsi and feet coralred. **ADULT FEMALE:** Browner above and paler below than adult male, lacking pinkish coloration on breast; bare-part colors as in male, but duller. **JUVENILE:** Similar to female, but black wing markings are obscured or lacking, with feathers of head, breast, and mantle fringed with buff-

gray. Genetically, Band-tailed is its closest relative.
Voice CALL: Variously reported to give a series of harsh *keek* notes, often ending with a *keooo*.
Status & Distribution Extinct. **BREEDING:** Formerly in vast colonies in the deciduous woodlands stretching across the northern US and southern Canada from the Great Plains east to the Atlantic. **WINTER:** The woodlands of the southeastern US, south along the Gulf Coast of Mexico. **RARE STATUS:** Casual in Cuba during the middle of the 19th century; also north to northern Canada, and west to BC and NV; records from Great Britain and France possibly involved genuine accidental visitors.
Population Extinct. Destruction of old-growth deciduous forest and overhunting led to rapid declines and the eventual extinction of this once-abundant species.

male bluish above, pinkish below

female duller

adult ♂

long tail

GROUND-DOVES Genus *Columbina*

INCA DOVE *Columbina inca* INDO ■ 1

The conspicuously long-tailed, small dove is found in urban areas of the Southwest; resident south to Costa Rica. Monotypic. L 8.3" (21 cm)
Identification A small gray dove with

black fringes on feathers of both upper and underparts forming an obvious scaled appearance. In flight, shows chestnut on upper and underside of wing like Common Ground-Dove;

however, also shows prominent white edges on long tail. **ADULT MALE:** Crown and face pale blue-gray and breast tinged lightly with pink; iris reddish; narrow eye ring blue-gray; bill blackish; and feet bright red. **ADULT FEMALE:** Similar, but duller on head and breast. **JUVENILE:** Duller with a slight brownish tinge overall, and scaling less noticeable.

long tail with white edges

rufous primary patches

scaly pattern on head and body

Similar Species Its long tail provides a very different shape from that of the ground-doves, but beware of the possibility of a shorter-tailed Inca Dove regrowing a lost or molted tail. Scaling of upperparts is unique.

Voice CALL: A long series of disyllabic *kooo-poo* that can be interpreted as "no hope."

Status & Distribution Uncommon to fairly common resident; primarily around human habitation and in city parks. Terrestrial, feeding on the ground; fairly tame and easily observed. **BREEDING:** Nest is a small fragile floor of twigs placed in a low bush or shrub; lays two white eggs; two or three broods each year. **RARE STATUS:** Casual to southwest CA, MT, southern UT, ND, NE, WI, MI

(Upper Peninsula), ON, TN, KY, AR, MS, AL, GA, western NY, and MD.

Population Some urban populations have declined sharply (e.g., Tucson, AZ), and northern range has contracted. Trends unknown in Mexico and C.A.

COMMON GROUND-DOVE   *Columbina passerina*   CGDO   ■   1

This very small, short-tailed dove lives throughout most of the southern US. It is also found in the West Indies and from Mexico to S.A. Polytypic (18 sspp.; 2 in N.A.). L 6.5" (17 cm)

Identification The smallest of the doves, with short rounded wings and a short, square-ended tail; generally grayish brown above, with underparts paler and washed with pink; prominent blackish spots on coverts and tertials; dark centers to feathers on head and breast create a scaled effect; short, square-ended tail blackish; fine white tips on outer two or three rectrices; bright chestnut panel in primaries and entirely rufous underwing all visible in flight. **ADULT MALE:** Crown paler and grayer than rest of upper parts, with iridescent blue-gray on nape and hind neck; face decidedly pinkish, with pinkish tone extending down

onto underparts; dark comma-shaped spots on wings show iridescent violet sheen; iris reddish brown; orbital ring blue-gray; bill blackish with bright pink on basal third; and feet bright red. **ADULT FEMALE:** Similar to adult male, but lacking iridescence on hind neck, and pale gray-brown below instead of pinkish; dark marks on wings more brownish and colors on soft parts less intense. **JUVENILE:** Noticeably browner; scapulars and wing coverts fringed buff-gray; black spots on wings obscured or missing.

Geographic Variation Nominate *passerina* is found in the Southeast, and *pallescens* in the West; pinkish on underparts and pink at the base of the bill more intense on nominate subspecies, but generally not separable in the field.

Similar Species Similar to Ruddy

bahamensis

pallescens   *passerina*

jamaicensis   *insularis*

Ground-Dove (and see that account), but note Common's scaled head and breast, pale base to bill, and unmarked scapulars.

Voice CALL: A repeated soft, drawn-out, and ascending *wah-up* double note given every two to three seconds.

Status & Distribution Rare to fairly common resident in open areas,

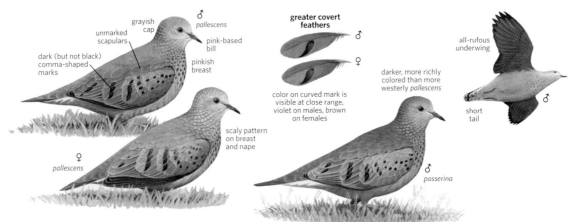

grayish cap

unmarked scapulars

dark (but not black) comma-shaped marks

♂ *pallescens*

pink-based bill

pinkish breast

scaly pattern on breast and nape

♀ *pallescens*

greater covert feathers

♂

♀

color on curved mark is visible at close range, violet on males, brown on females

darker, more richly colored than more westerly *pallescens*

♂ *passerina*

all-rufous underwing

short tail

♂

including coastal dunes in the East. Feeds on the ground. **BREEDING:** Nest is a flimsy collection of twigs on the ground or low in dense shrubs; normally two white eggs; two or three broods each year. **MIGRATION:** Although considered resident, some disperse northward in fall. **RARE STA-** **TUS:** Casually north to OR, SD, MI, ON, MA, and NS.

Population Significant declines along Gulf and south Atlantic coasts.

RUDDY GROUND-DOVE   *Columbina talpacoti*   RGDO ▨ 3

This small, short-tailed dove is widespread throughout the lowlands of Mexico, C.A., and much of S.A., and is expanding its range northward. It often associates with Inca Doves. Polytypic (4 sspp.; 2 in N.A.). L 6.8" (18 cm)

Identification Like Common Ground-Dove, but slightly larger and longer tailed and underwing coverts blackish; tail black with narrow whitish tips on outer two to three rectrices; black spots on wings and scapulars that form vertical lines. Sexually dimorphic. **ADULT MALE:** Head unmarked pale grayish with blue-gray on crown and hind neck; upper parts rufous; underparts unmarked rufous; iris reddish brown; bill gray with black tip. **ADULT FEMALE:** Head unmarked pale gray-brown; upper parts gray-brown, and underparts paler and grayer; some with whitish fringes on tertials and greater coverts; soft parts as on male. **JUVENILE:** Similar to female, but males show varying amounts of rufous; black spots on wings obscured or missing.

Similar Species Both sexes of Ruddy Ground-Dove have black linear markings on upperwing coverts and scapulars, lack scaly pattern on head and breast, and have no reddish coloration at base of bill; some, especially females, show white fringes on tertials (unpatterned on Common Ground-Dove). Underwing coverts are mostly blackish. Common Ground-Dove has

comma-shaped dark marks on the upperwing coverts, plain scapulars, dark scaling on the head and breast, and a pink-based bill; underwings coverts extensively rufous. The male Ruddy Ground-Dove with extensive rufous body coloration (especially rich in *rufipennis*), and contrasting pale gray crown, is distinctive, but rosy Commons are frequently confused with Ruddy.

Geographic Variation The two subspecies found sparingly in N.A. are *rufipennis,* found from eastern Mexico to northern S.A., and *eluta,* found from western Mexico south to Chiapas. Adult males of *rufipennis* are a richer rufous above and below, more so than the adult males of *eluta,* and tips of their outer rectrices are more rufous instead of whitish. Females of *eluta* are paler and grayer but may not be separable in the field.

Voice CALL: A monotonous series of evenly pitched disyllabic *ca-whoop* notes given at one-second intervals, suggestive of Common Ground-Dove but lower pitched and more rapidly delivered.

Status & Distribution Common to abundant throughout the lowlands of Mexico south to northern Argentina, frequenting open woodlands. The western Mexican population (*eluta*)

eluta   *rufipennis*

has expanded northward, in the 1980s, and very small numbers were found annually, primarily in fall, in southeastern CA and southern AZ with scattered records to southern NM, and West TX east to Big Bend NP. Many of these birds associated with Inca Doves, not Common Ground-Doves. During this time it nested at isolated locations within this area. More recently, many fewer to the Southwest though still of nearly annual occurrence (mostly southern AZ). **RARE STATUS:** Subspecies *eluta* is casual west to coastal Southern CA, southern NV, southwestern UT, and NM. Accidental central TX and MS. In south TX, *rufipennis* is casual.

Population Common in Mexico and C.A.; adapts well to human-altered habitats.

greater covert feather

♀

black linear mark on both sexes

♂

short tail

blackish wing lining

males with overall rufous coloration, females grayish

grayish cap

♂
eluta

♀
eluta

no scaling on head and breast

some females more richly colored

dark bill in all sexes

black linear marks on scapulars, greater and median wing coverts, and some tertials

♀
eluta

deeper rufous color than *eluta*

♂
rufipennis

no scaling on breast in all plumages

QUAIL-DOVES Genus *Geotrygon*

KEY WEST QUAIL-DOVE *Geotrygon chrysia*   KWQD ■ 4

Although named Key West, this most widespread of the quail-doves in the West Indies is now a casual stray to FL. It is rather secretive and is usually found singly, even where it is common, although pairs are also encountered. It feeds on the forest floor, primarily on seeds and fruit. Quail-doves will tend to walk from danger, as opposed to taking flight. The male will perch on low branches, from which it makes its territorial calls. Monotypic. L 12" (31 cm)

Identification Key West Quail-Dove is one of the larger quail-doves. It is bicolored with a prominent white facial stripe; rich chestnut-brown above, glossed with purple and green, and grayish white below. **ADULT MALE:** Crown, nape, and hind neck washed with iridescent blue-green. A white face is bordered below by a narrow dark chestnut malar stripe. Pale underparts. An orange iris is bordered by red orbital skin. Dark red bill has a brownish tip. Feet are bright coral red. **ADULT FEMALE:** Duller brown above than adult male, with less iridescence and with a less obvious facial stripe. **JUVENILE:** Duller still than adult female,

generally lacking rich chestnut tones, and lacking any hint of iridescence. Scapulars and wing coverts are fringed cinnamon; feet are a much duller red than on adults.

Voice CALL: A low moaning ventriloquial *ooooo* or *oooowoo*, with the second part accentuated and slightly higher; very similar to the call of White-tipped Dove.

Status & Distribution Uncommon to fairly common in the West Indies; primarily resident in arid and semi-arid woodlands and scrub thickets on the Bahamas, Cuba, and Hispaniola, and locally on Puerto Rico. **BREEDING:** The nest is a fragile platform of twigs lined with dead leaves built in low undergrowth or on the ground; normally lays two creamy-buff eggs. **RARE STATUS:** Possible former resident in the early 1800s on the FL Keys. Now casual mid-Oct.–mid-June to the FL Keys and southeastern mainland FL, with over 15 records since 1964, the northernmost being near Daytona Beach, Volusia Co.

Population Stable, but in West Indies vulnerable to hunting pressure on some islands.

prominent white line under eye

chestnut upperparts glossed with purple and green

♂

whitish below

RUDDY QUAIL-DOVE *Geotrygon montana*   RUQD ■ 5

As is typically the case for this genus, the Ruddy Quail-Dove is secretive and quite terrestrial, frequently running rather than flying when alarmed. It is normally encountered on the forest floor in the West Indies and in C.A. and S.A. Males readily fly to low branches to vocalize, but otherwise flight is infrequent, except during the courting process. Ruddy Quail-Dove feeds on seeds, fruits, and more so than most congeners, on a regular diet of invertebrates, such as beetles.

Polytypic (2 sspp.; nominate in N.A.). L 9.8" (25 cm)

Identification One of the smaller and most widespread of the quail-doves. A plump, short-tailed dove. Shows strong sexual dimorphism. **ADULT MALE:** Rich rufous on upperparts is washed with iridescent purple. Chin, throat, and sides of face are pale buff; cheeks are separated from throat by a broad, reddish brown malar stripe. Iris is yellow-orange; orbital skin red; bill dark reddish with dusky tip; feet pink-

ish red. **ADULT FEMALE:** Smaller; dark olive-brown above and below, lacking iridescence; facial stripe less obvious. **JUVENILE:** Coloration like female, but has rufous fringes on feathers.

Voice CALL: A series of very deep, resonant, monosyllabic coos, *waooo* or *wooo* repeated every three to four seconds, trailing off at the end. In the classic *Birds of the West Indies*, James Bond described the call as "reminiscent of the doleful sound of a fog buoy."

Status & Distribution Uncommon to locally common; largely resident in lowland humid forests of the West Indies and Mexico southward through

both sexes with buffy line under eye

♀
montana

♂
montana

male rich rufous above, female largely brownish

C.A. to southern Brazil. **BREEDING:** Ruddy Quail-Dove's nest is a platform of twigs lined with dead leaves, built low in a bush or tree, but sometimes on the ground; normally lays two creamy-buff eggs. **MIGRATION:** Much remains to be learned about the movements of this species. It is partially migratory or nomadic, but specific patterns and causes of migration are undiscovered. Numbers vary from year to year, even where the species is common; but the extent and regularity of the bird's movements remain unknown. As an example of the significant travel that can take place, this species has been recorded at sea in the Caribbean. **RARE STATUS:** Eight records from the FL Keys, including the Dry Tortugas, and once at Bentsen-Rio Grande Valley SP in TX.

Population Not threatened, but sensitive to habitat fragmentation and hunting pressure.

NEOTROPICAL FOREST DOVES Genus *Leptotila*

WHITE-TIPPED DOVE *Leptotila verreauxi* WTDO ■ 2

The White-tipped Dove, a plump, short-tailed terrestrial dove with broad, rounded wings, is found in forested areas of southern TX, often on the ground. It jerks its tail when alarmed. Polytypic (13 sspp.; 1 in N.A.). L 11.5" (29 cm)

Identification Head gray, palest on face; upperparts unmarked, faintly bronzed olive-brown, and underparts unmarked pale gray; short, square-tipped tail blackish with prominent white tips on outer three rectrices; underwings rufous. **ADULT MALE:** Iridescent green and maroon wash on hind neck; iris yellow; orbital skin dull red; bill entirely dark; and feet coral red. **ADULT FEMALE:** Like male, but iridescence on hind neck reduced. **JUVENILE:** Duller; fine buff fringes on scapulars and coverts; lacks iridescence.

Similar Species Most of the doves in this genus are similar, but differ in the amount of white in the tail; any *Leptotila* found away from southern TX should be identified with care.

Geographic Variation Depending on the authority: one species with 13 subspecies or two species with multiple subspecies. The subspecies *angelica* is resident in southern TX.

Voice CALL: A low-pitched *waa-*

woooo, like the sound produced by blowing across the top of a bottle.

Status & Distribution Common throughout lowlands of C.A. and S.A.; resident in southern TX and expanding northward; now found to Uvalde Co. and Refugio Co.; casual farther north. **BREEDING:** The nest is a bulky flat of twigs lined with grasses, normally in dense brush close to the ground; typically two creamy-white eggs. **RARE STATUS:** Two records from Dry Tortugas, FL (2–8 Apr. 1995; 19 Apr.–3 May 2003).

Population Appears to adapt well to fragmentation of forests, so is increasing in numbers.

chunky body shape

short, broad slightly rounded tail

angelica

whitish belly

white forehead and throat

pale yellow eye

angelica

white tail tips

AMERICAN DOVES Genus *Zenaida*

WHITE-WINGED DOVE *Zenaida asiatica* WWDO ■ 1

Flocks of these doves, with large white wing patches, and short square-ended tails, are a common sight in summer in many areas near the Mexican border. Polytypic (3 sspp.; 2 in N.A.). L 11.5" (29 cm)

Identification A little larger than Mourning Dove; generally brownish gray above and a paler gray below; white on the outer wing coverts forms a white line on the lower edge of the folded wing that, in flight, appears as a prominent white crescent; blackish square-ended tail with prominent white terminal band. Prominent black crescent framing lower edge of auricular. **ADULT MALE:** Grayish or grayish brown head and neck, with neck and breast washed lightly with pink; iris reddish brown; orbital skin bright blue; bill black; and feet bright red. **ADULT FEMALE:** Similar to male, but duller. **JUVENILE:** Paler on head; narrow pale gray fringes on scapulars and wing coverts.

Geographic Variation Most recent treatment recognizes three subspecies, but there have been varied interpretations. Until recently, it was considered conspecific with West Peruvian Dove (*Z. meloda*) of S.A. The two subspecies in N.A. are nominate *asiatica*, from eastern West TX eastward along the

Gulf Coast to FL, and *mearnsi* in the Southwest, east to NM. Differences are weak and clinal, with *asiatica* on average showing a less grayish tone to the brown plumage; but the two subspecies are likely not separable in the field. The third subspecies (*australis*) is found from Costa Rica to Panama.

Voice CALL: A drawn-out *who-cooks-for-you* cooing; has many variations.

Status & Distribution Breeds in south-ern tier of the US, from southeastern CA to the Gulf Coast of TX, and from southern FL south through C.A. to Panama. **BREEDING:** Nest is a fragile platform of twigs in medium-height brush; nominate birds typically nest in colonies, while western birds are more solitary; lays two white eggs. **MIGRATION:** Primarily a summer resident in the US, with most migrating into

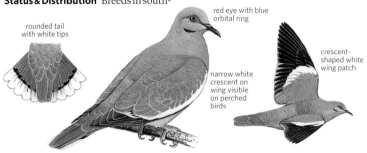

rounded tail with white tips

red eye with blue orbital ring

narrow white crescent on wing visible on perched birds

crescent-shaped white wing patch

Mexico in winter. Increasing numbers are remaining through the year, establishing isolated resident populations scattered across the US from southeastern CA to FL; regular visitor to the Gulf Coast from LA to FL. **RARE STATUS:** Very rare in the East, north to southern Canada, along the West Coast north to extreme southeastern AK.

Population Increasing in US.

ZENAIDA DOVE *Zenaida aurita* ZEND ■ 4

This shorter-tailed version of the Mourning Dove is found throughout forested regions of the West Indies. Polytypic (3 sspp.; *zenaida* in N.A.). L 10.5" (27 cm)

Identification About the same size as a Mourning Dove with similar black spots on scapulars and tertials and a small black crescent below the auriculars. Flight feathers are blackish; secondaries are tipped white, creating a white trailing edge on inner wing in flight, showing as a white patch on inner secondaries on folded wing, the single best field mark; short, squarish tail. **ADULT MALE:** Body a rich cinnamon brown with bronze iridescence on hind neck; iris dark brown; orbital skin pale blue; bill black; and feet bright red. **ADULT FEMALE:** Grayer than male and with much less iridescence on hind neck. Most females are a brownish, but some are grayer (similar to Mourning Dove). **JUVENILE:** Duller with buff fringes on both scapulars and coverts.

Similar Species Differs from Mourning Dove in having white on trailing edge of secondaries, and in having a shorter, rounded, gray-tipped tail; male is much more richly colored. In West Indies where both species occur, Zenaida prefers more forested habitats.

Voice CALL: A gentle cooing, similar to that of Mourning Dove but slower:
coo-oo, coo, coo, coo, with second syllable rising sharply.

Status & Distribution Uncommon (Bahamas) to fairly common resident of woodlands throughout the West Indies and east coast of the Yucatán Peninsula. **BREEDING:** May have up to six broods per year. **RARE STATUS:** Said to be resident on small islands off the FL Keys during the early 1800s; two extant 19th-century specimens. Strictly casual in the US (all south FL) since 1900, with just a handful of records.

Population Stable.

salvadorii
zenaida

zenaida ♂
zenaida ♀

shorter tail than Mourning

zenaida ♂

white trailing edge to secondaries

zenaida ♂

squarish white patch formed by white tips to secondaries

MOURNING DOVE *Zenaida macroura* MODO ■ 1

This familiar dove, with its slim body and tapered tail, is the most common and widespread dove in most of N.A. The wings make a fluttering whistle when the bird takes flight. Polytypic
(5 sspp.; 3 in N.A.). L 12" (31 cm)

Identification Head and underparts unmarked pale pinkish brown, but with black crescent framing lower edge of auricular; upper parts darker
and grayer brown; prominent black spots on coverts and tertials, and flight feathers contrasting darker; long pointed tail dark, with black subterminal spots and bold white tips on all

but the central rectrices. **ADULT MALE:** Iridescent blue and pink on hind neck, with pinkish bloom extending onto breast; iris blackish; orbital skin pale blue; bill dark; and feet red. **ADULT FEMALE:** Similar to adult male, but with reduced iridescence and pinkish bloom. **JUVENILE:** Generally darker and browner; pale buff-gray fringes on most of the feathers give the bird a scaly appearance; dark crescent below auricular extends forward toward base of bill; cheek area pale.

Geographic Variation Subspecies *carolinensis* breeds in the East, *marginella* breeds in the West, and nominate *macroura* from the West Indies recently invaded the FL Keys; these three are not separable in the field.

Voice CALL: A mournful *oowoo-woo-woo-woo.*

Status & Distribution Common throughout the US and southern Canada south through C.A. Prefers open areas, including rural and residential areas, avoiding thick forests; normally feeds on the ground. **BREEDING:** Nest is a loose platform of twigs, flimsy

long, pointed tail

tapered tail with white tips

black spotting on upperwing

♂

♀

juvenile

scaly appearance

enough that the eggs are frequently visible from below. **MIGRATION:** Highly migratory, with birds breeding at the northern limit of the range believed to winter in Mexico, but those breeding farther south moving less, with birds present all year in most of the US. **RARE STATUS:** Very rare to AK (mainly fall); casual to northern Canada, the Azores, and Great Britain.

Population The species is a well-managed game bird, with about 350 million birds in N.A. and 20 million killed annually by hunters.

CUCKOOS, ROADRUNNERS, AND ANIS Family Cuculidae

Almost all of the species in the Cuculidae family are arboreal, inhabiting forests and well-wooded areas. Cuckoos form a remarkably variable family, which is well known for having members that are brood parasites. Various species are known as "rain birds" throughout the world because they are very vocal at the beginning of the rainy season.

Structure Cuckoos have slender bodies and long tails. All species in the family have zygodactyl feet, where the outer toes point backward and the two inner toes point forward. The bill is usually long with a curved culmen. The wing length varies depending on whether the species is a long-distance migrant or a more sedentary species. Terrestrial species—such as the Greater Roadrunner—have long, strong legs.

Behavior Many cuckoos are solitary and are more often heard than seen. Although mostly diurnal, many species call at night. Up to 53 species of cuckoos are brood parasites; their cryptic plumage allows the females to surreptitiously approach the nests of hosts. Black-billed and Yellow-billed Cuckoos are nest builders, but they have been documented as brood parasites. Unlike cuckoos, anis live in noisy social groups and are often cooperative breeders with several females using the same nest and multiple adults feeding the nestlings. Because of the texture of their plumage, many cuckoos, when wet or in the early morning, dry their bodies by sitting on an open perch with their wings and tail spread, often with their back feathers raised to expose the skin to the sun.

Groove-billed Ani (left) and Smooth-billed Ani (FL, Mar.)

Plumage Most Cuculidae species' plumage is soft and is a brown, gray, or black color, often streaked or barred. A few species exhibit bright green, rufous, or even purple

plumage. The tail is generally tipped in a different color, often white. The sexes are similar in plumage in almost all species, with some size dimorphism, females being larger.
Distribution The species of this widespread family occur in tropical and temperate regions worldwide; however, the family reaches its greatest diversity in the Old World. Most species are nonmigratory or short-distance migrants; a few species are long-distance migrants, including the Yellow-billed and Black-billed Cuckoos of North America. Strays may turn up far out of range.
Taxonomy The family divides into six subfamilies: Old World cuckoos, coucals, malkohas and couas, American cuckoos, New World ground-cuckoos, and anis. Some

taxonomists suggest that these groups represent distinct families. Worldwide, there are 140 species in 36 genera recognized. In North America, six species regularly occur, while two additional species are rare visitors, primarily to western AK.
Conservation Twenty-one species in the family are at risk. Most of them occur in tropical forests and on islands; the populations of the Smooth-billed Ani in FL and the Yellow-billed Cuckoo in western North America are also of concern. Pollution could be a contributing factor—it appears that cuckoos are heavily impacted by pollutants. BirdLife International codes: 8 NT; 7 VU; 2 EN; 2 CR; and 2 EX.

OLD WORLD CUCKOOS Genus *Cuculus*

In this genus composed of 10 species—with most species found in southern Asia and Africa—the most well-known and widespread species is the Common Cuckoo. All species share the same basic plumage of uniform-colored upperparts and paler, often barred, underparts. They have slender bodies and long, graduated tails.

COMMON CUCKOO *Cuculus canorus*   COCU ▪ 3

An Old World species, the Common Cuckoo is well known as a brood parasite, affecting a very wide variety of hosts and occurring in numerous habitats ranging from woodlands to farmlands. Hepatic-morph birds are females. Polytypic (4 sspp.; likely *canorus* in N.A.). L 13" (33 cm)
Identification Common Cuckoo has a slender body, a long rounded tail, and pointed wings. In flight, it has a falconlike appearance; its wings rarely rise above the body. It often perches in the open; when perched, it often droops its wings. It closely resembles Oriental Cuckoo. **ADULT MALE:** Pale gray above and paler below with a white belly narrowly barred with gray.

Undertail coverts are white and lightly barred. He has a prominent yellow orbital ring. **GRAY-MORPH FEMALE:** Difficult to separate from adult male, but often has a rusty buff tinge on breast. **HEPATIC-MORPH FEMALE:** Rusty brown above and on breast and heavily barred—black bars are narrower than brown ones. Lower back and rump are a lighter brown and either unmarked or lightly spotted.
Geographic Variation Variation in subspecies is slight and possibly clinal. All records in N.A. likely refer to nominate *canorus*, although many seen in AK have been paler above with finer barring below than European birds, and perhaps represent *telephonus*, though that subspecies is not recognized by most. Subspecies identification is difficult and is based on variation in plumage coloration.

Similar Species Common Cuckoo is very similar in appearance to Oriental Cuckoo, and the identification of silent birds is difficult. Common Cuckoo is slightly larger with a more delicate bill and is generally slightly lighter in overall color. Oriental Cuckoo has a more contrasty greater covert bar

northeast Asian birds, recognized by some as *telephonus*, are paler gray above and lightly barred below

adult ♂   delicate bill

pale on underwing more blended than Oriental

adult ♀

adult ♂   plain rump

whitish undertail coverts

rusty on sides of breast as in female Oriental

dark bands on tail narrower than Oriental

hepatic morph ♀

on the underwing and buffy undertail coverts. Hepatic-morph Common Cuckoo can be identified by its paler unmarked rump.

Voice Male's call is a disyllabic *cuc-oo*, with emphasis on the first syllable. The female utters a loud bubbling trill.

Status & Distribution BREEDING: Palearctic, ranging from western Europe to eastern Russia. WINTER: Primarily in southern Africa. RARE STATUS: Very rare spring and summer visitor to central and western Aleutian, St. Lawrence, and Pribilof Is. in

AK. Casual to mainland. Accidental to Martha's Vineyard, MA (May); Watsonville, Santa Cruz Co., CA (Sept.–Oct.); Blanc Sablon, QC (Sept.); and the Lesser Antilles.

Population Significant declines since the 1980s.

ORIENTAL CUCKOO *Cuculus optatus*   ORCU ▪ 4

As with other Old World *Cuculus* species, the Oriental Cuckoo is a brood parasite. It is encountered in N.A. only as a casual visitor to western AK. As in Common Cuckoo, hepatic-morph birds are females. Monotypic. L 12.5" (32 cm)

Identification Oriental closely resembles the Common Cuckoo. It has a slender body, long rounded tail, and pointed wings. In flight it resembles a small falcon. **ADULT MALE:** Medium gray above and paler below with a white belly barred with gray. Buffy undertail coverts are generally sparsely barred or unmarked. Prominent unmarked whitish wing linings are visible in flight. Orbital ring is bright yellow. **GRAY-MORPH FEMALE:** Difficult to separate from adult male, but often has a rusty buff tinge on breast. **HEPATIC-MORPH FEMALE:** Rusty brown above and on breast and heavily barred overall, including rump—black bars are wider than brown bars. Most authorities now treat the much smaller resident bird (*C. lepidus*) of Malaysia and Indonesia as its own species, the Sunda Cuckoo, though some lump it with *C. lepidus*. Northern *C. optatus* is now treated by most as a monotypic species from the very similar but smaller Himalayan bird (*C. saturatus*). Both

of these species are highly migratory, but it appears that *C. optatus* migrates much farther south and east, as far as Australia.

Similar Species Very similar in appearance to Common Cuckoo and the identification of silent birds is difficult. Oriental Cuckoo is slightly smaller, and gray-morph birds are generally somewhat darker in overall color with slightly broader barring on belly and buffy undertail coverts. Oriental shows more contrast between the whitish and grayish areas of its underwing than Common Cuckoo. Hepatic-morph Oriental is best separated from Common Cuckoo based on its barred rump that is evenly colored with back, not paler.

Voice The male's call is a series of hollow notes usually given on the same pitch and in groups of four. The female's call is very similar to the trill of a Common.

Status & Distribution BREEDING:

Primarily Palearctic, ranging from northeastern Europe to Korea, Japan, and the Russian Far East. WINTER: Primarily from Indonesia to Australia. RARE STATUS: Casual spring, summer, and fall visitor to western Aleutians, St. Lawrence, Pribilof Is., and once from mainland AK.

Population Insufficient data, possibly declining.

adult ♂

more contrast between whitish and slate-gray areas on underwing than Common Cuckoo

adult ♂

hepatic morph ♀

bill slightly thicker and more curved than Common

barred rump

ventral barring averages stronger

buffy undertail coverts

nape spot

juveniles of both species are blackish brown with whitish feather fringes above

juvenile

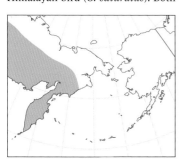

NEW WORLD CUCKOOS Genus *Coccyzus*

Thirteen species, primarily Neotropical, make up this genus. The two most common species in N.A. are long-distance migrants to S.A. Most New World Cuckoos share a basic plumage pattern that consists of brown or gray upperparts with buffy or white underparts. They have slender bodies and long, graduated tails.

YELLOW-BILLED CUCKOO *Coccyzus americanus*   YBCU ■ 1

adult

rufous primaries striking in flight

yellow orbital ring

thick, extensively yellow bill

adult

white outer web

juvenile

bold white spots

spots less bold than adult

Generally shy and elusive, Yellow-billed Cuckoo is easily overlooked, except for its loud vocalizations. Polytypic (2 sspp.; both in N.A.). L 12" (31 cm)
Identification Slender body and long tail. **ADULT:** Grayish brown above, whitish below; crown noticeably grayer. Rufous primaries especially visible in flight. Long tail is graduated with brown central rectrices tipped in black; remainder are black and broadly tipped with white. Yellow orbital ring. Bill curved with black culmen; lower mandible is yellow with a black tip.

JUVENILE: Similar to adult but has buffy undertail coverts. Undertail pattern muted, and tips of rectrices not as prominent. Orbital ring is dull yellow.
Geographic Variation Subspecies *americanus* found in eastern N.A.; *occidentalis* in the Southwest. Differences between subspecies are weak.
Similar Species From Black-billed Cuckoo, distinguished by whiter underparts, yellow orbital ring, rufous primaries, more prominently white-tipped tail, and slightly thicker bill with yellow lower mandible.
Voice SONG: A rapid staccato *kuk-kuk-kuk* that usually slows and descends into a *kakakowlp-kowlp* ending; also a loud, braying *coo, coo, coo*, ending slower and softer.
Status & Distribution Common in eastern N.A.; scarce and local in much of the West. **BREEDING:** Open woodlands with dense undergrowth, riparian corridors, and parks. Southwestern populations increasingly limited to riparian corridors. **MIGRATION:** Trans-Gulf migrant as well as over Caribbean Is. In spring, the Gulf Coast peak occurs ±1 May; southern Great Lakes ±15 May. Southwestern

population arrives primarily in June. Southern Great Lakes peak ±20 Aug; Gulf Coast peak ±10 Sept. Rare in the US after Nov. 1. **WINTER:** Casual to accidental along Gulf Coast and south FL perhaps late fall migrants. Most winter in S.A., as far south as northern Argentina.
RARE STATUS: Rare over much of West and in fall to Atlantic Canada; casual to UK; accidental to southeast AK.
Population Declining and listed as Endangered in western US by USFWS.

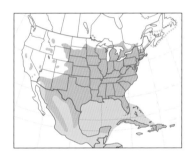

MANGROVE CUCKOO *Coccyzus minor*   MACU ■ 2

The Mangrove Cuckoo is more often heard than seen. It is widespread in the Caribbean and in coastal habitats from northern Mexico to northern S.A. Monotypic. L 12" (31 cm)
Identification It has a slender body and long tail. **ADULT:** It has grayish brown upperparts with a grayer crown. Throat and upper breast are white or buffy with remainder of underparts buffy. Long tail is graduated with brown central rectrices tipped in black, remaining tail feathers are black and broadly tipped with white.

adults

black mask

uniform brown wings

thick bill; lower mandible has yellow base

buffy underparts

black outer web

large white spots on black tail

Bill is strongly curved with a black upper mandible and a black-tipped yellow lower mandible. **JUVENILE:** Black mask is faint to near absent.
Geographic Variation Up to 14 subspecies have been described, but the species is now considered monotypic. Previously, the FL population was placed in *maynardi*. Strays reaching the western Gulf Coast tend to be more richly and extensively colored below (formerly *continentalis*) and likely originate from Mexico.
Similar Species Most closely resembles Yellow-billed Cuckoo, but it is distinguished by its black mask, buffy underparts, lack of rufous primaries,

and dark, not white, outer web to outer primary; voice is also very different.
Voice SONG: A very guttural *gaw gaw gaw*.
Status & Distribution Uncommon. **BREEDING:** Nests in mangrove swamps and low canopy tropical hardwood forests. **WINTER:** Withdraws from northern portions of its range in FL. **RARE STATUS:** Casual, presumably from northern Mexico, along the Gulf Coast from TX to northwestern FL.
Population Local declines in FL.

BLACK-BILLED CUCKOO   *Coccyzus erythropthalmus*   BBCU ▪ 1

The Black-billed Cuckoo regularly feeds on caterpillars, and it also greedily consumes tent caterpillars and gypsy moth larvae during outbreaks of those insects. Monotypic. L 12" (31 cm)
Identification Slender body and long tail. **ADULT:** Grayish brown upperparts and whitish underparts, tinged buff on throat. Long tail is graduated; it is predominantly brown above, narrowly tipped with white, while undertail is patterned in gray. It has a prominent red orbital ring. Bill is strongly curved and all-dark. **JUVENILE:** It looks similar to adult, but it has a buffy throat and undertail coverts. Tips of rectrices are buffy and not as prominent. Orbital ring is buffy.
Similar Species Most closely resembles Yellow-billed Cuckoo, but it is distinguished by its red orbital ring, lack of rufous primaries, different undertail pattern, and narrower dark bill.
Voice SONG: A monotonous *cu-cu-cu* or *cu-cu-cu-cu* phrase.
Status & Distribution Uncommon. **BREEDING:** Deciduous and mixed forests to open woodlands and brushy habitats. **MIGRATION:** Trans-Gulf migrant, rare in the US after mid-Oct. **WINTER:** Northern S.A. **RARE STATUS:** Casual visitor west to the Pacific coast and to the UK.
Population Has shown substantial declines, considered a species of conservation concern in many areas of N.A.

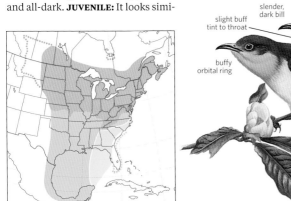

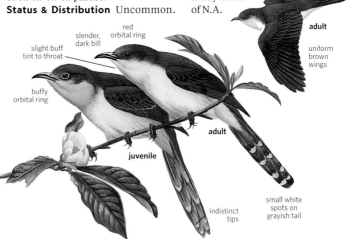

slender, dark bill
red orbital ring
slight buff tint to throat
buffy orbital ring
adult
uniform brown wings
adult
juvenile
indistinct tips
small white spots on grayish tail

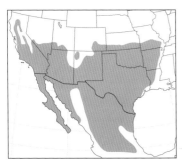

ROADRUNNERS   Genus *Geococcyx*

Worldwide there are two species in this genus, the Greater and Lesser Roadrunners. These large terrestrial cuckoos are similar in appearance and are found primarily in arid and semiarid habitats. Roadrunners can run at speeds of up to 15 miles an hour pursuing prey or to escape predators.

GREATER ROADRUNNER   *Geococcyx californianus*   GRRO ▪ 1

The Greater Roadrunner is omnivorous, although the majority of its diet includes insects, birds, reptiles and rodents. It is a fierce predator and runs at good speed to obtain prey or avoid predators, like approaching cars! Monotypic. L 23" (58 cm)
Identification Greater Roadrunner is a large ground-dwelling (including roadsides) cuckoo with a bushy crest that can be raised and lowered. Brownish overall with darker brown and white streaking on upperparts. It has short rounded wings with a white crescent in primaries. Its long tail is edged and tipped in white. Note long, heavy bill; and bare blue skin around and behind its eye with a terminal red spot, often obscured. Juvenile similar but bare facial skin grayish.

very long, graduated tail with white tips
crest can be raised or flattened
long legs

Similar Species Unmistakable. No real contenders for misidentification.
Voice SONG: A descending series of low coos. Often sings from well up in trees and bushes.

Status & Distribution Uncommon to fairly common. Resident south to central Mexico. **BREEDING:** Found in a variety of different desert types within open scrubby habitats; also in chapar-ral, and in open woodlands. **WINTER:** Some local movements away from nesting habitat; sometimes individuals turn up far from mapped range.
Population Stable.

ANIS Genus *Crotophaga*

The three species in this genus—Greater, Groove-billed, and Smooth-billed Anis—exhibit complex social behavior and are cooperative breeders. They share a loose black plumage and a laterally compressed bill.

SMOOTH-BILLED ANI   *Crotophaga ani*   SBAN ■ 3

The Smooth-billed Ani colonized FL in the late 1930s from the West Indies. The population expanded and the species was locally fairly common until about 1990. Sharp declines followed, and by 2000 only a few individuals remained (e.g., at Ft. Lauderdale); by about 2010 they too had disappeared. Monotypic. L 14.5" (37 cm)

Identification Black, but with some purplish or greenish iridescent edges to feathers. Long tail is frequently held down. Large, laterally compressed bill often has a high curve to culmen, a feature that easily separates it from Groove-billed; others are more difficult. Flight often looks labored and weak.
Similar Species Quite difficult to distinguish from Groove-billed Ani; it is best identified by voice and close examination of bill shape (see Groove-billed Ani). Grackles are the only other species likely to cause confusion with Smooth-billed Ani; both anis fly with quick awkward flaps with glides. Shaggy plumage and large bill are key features to use in identification.
Voice CALL: A whining, rising *quee-lick*, quite distinct from Groove-billed.
Status & Distribution Widespread in the Neotropics; occurs in brushy or weedy habitats, often close to water. Now very rare from Ft. Lauderdale to the FL Keys; casual to Orlando area, these may be strays from the Bahamas. **RARE STATUS:** Accidental north along Atlantic coast to NC; also LA and OH.
Population Stable within core Neotropical range.

some show raised base to culmen, but many others do not

bill has no distinct grooves

slightly larger than very similar Groove-billed; most birds best distinguished by voice

GROOVE-BILLED ANI   *Crotophaga sulcirostris*   GBAN ■ 2

The gregarious Groove-billed Ani is most often found in vocal groups of four to 10 birds. It typically roosts with other anis in tight bunches on vines and other protected perches. It (and the Smooth-billed Ani) has a very distinctive flight—weak flaps on rounded wings interspersed with glides. It feeds by gleaning insects off leaves but is often seen chasing grasshoppers on the ground through high grass. Nests may be single to fairly large communal nests shared by up to four pairs. Both males and females defend the territory and care for the young. Throughout most of its range, the Groove-billed Ani is frequently seen around cattle where it feeds on insects stirred up by the livestock. This aspect of its life history is rarely observed in TX. Within their range,

Groove-billed and Smooth-billed Anis are often found in small flocks. Monotypic. L 13.5" (34 cm)
Identification Plumage is entirely black with iridescent purple and green overtones; overall appearance is often shaggy or disheveled. Its long tail appears loosely joined to its body and is often held down. Groove-billed Ani frequently wags its tail when on open perches. Large, laterally compressed bill has a curved culmen. Adults normally have easily observed grooves on upper mandible; on juveniles upper mandible might be unmarked. Groove-billed's bill shape is similar to Smooth-billed Ani, but curvature of culmen does not extend above crown.
Similar Species Both anis are rare to very rare in FL and silent birds are difficult to distinguish. Both ani spe-

long tail

sulcirostris

bill grooves visible at close range

sulcirostris

cies have occurred far out of range. Species identification between the two anis is best accomplished by voice, but if silent, a close examination of bill shape (particularly the bill ridge) is necessary. Groove-billed is also slightly smaller. Grackles are the only other species likely to cause confusion with Groove-billed Ani. Shaggy plumage and large bill are key features to use in identification.

Voice CALL: A liquid *tee-ho*, with the accent on the first syllable. This call is given raucously in quick succession at dawn when groups leave the roost. **FLIGHT CALL:** Soft clucking or chuckling notes.

Status & Distribution Widespread in the Neotropics from southern TX to northern S.A., and south along the Pacific coast to northern Chile. **BREEDING:** Uncommon to locally common late Apr.–late Sept. Nests in scrub or low-canopied woodlands and in riparian corridors. Occasionally is found in more open habitats. Breeding also occurs in more open scrub, not necessarily near water. **WINTER:** Rare and local as majority of population has withdrawn, presumably into northeastern Mexico. Wintering populations very localized and primarily found along the TX coast and in the lower Rio Grande Valley. **RARE STATUS:** Wanders regularly, most often in fall, east along the Gulf Coast to FL. Casual throughout remainder of TX and north to CO and the Midwest. Accidental elsewhere in eastern N.A., and has occurred as far north as ON

and northern NJ. Casual to southern AZ (most in summer and fall, a few in winter) and in fall and winter to southern NV and Southern CA.

Population Stable. Populations from islands off southern Baja California, Mexico, extirpated.

GOATSUCKERS Family Caprimulgidae

Mexican Whip-poor-will (AZ, May)

Caprimulgids occur nearly worldwide and occupy a wide variety of habitats. They collectively are referred to as goatsuckers, the translation of the family name. According to folklore, nightjars suckled on nursing goats during the night. In reality, most species are insectivorous, but because of their nocturnal and secretive habits, strange vocalizations, huge mouth, and, in some species, tendency to forage in areas where livestock occur, this superstition developed.

Structure Goatsuckers are superficially reminiscent of owls because of their large head, large eyes that shine when illuminated at night, cryptic plumage, and largely nocturnal behavior. But instead of capturing and eating prey with sharp bills and talons, goatsuckers are the nocturnal counterpart of swifts and swallows. The bill is tiny, but the gape is exceptionally wide and reveals a cavernous mouth. Insect prey are literally engulfed and swallowed whole rather than being seized and manipulated with the bill. Most nightjars have elongated rictal bristles along the edges of the mouth that presumably help a nocturnal species detect and guide an insect into its mouth; these are not elongated in the more diurnal

nighthawks. Caprimulgids have disproportionately long wings; nighthawk wings are more pointed and falconlike, nightjar wings more rounded. In most species the tail is usually fairly long and provides added maneuverability. Short legs and small feet reflect mostly aerial habits, but all can grip perches or walk short distances if necessary. Virtually all species have cryptically colored plumage with complex patterns to provide camouflage in daylight. Most species are small to medium in size.

Behavior Nightjars roost by day on ground, rocks, or branches, usually in dense vegetation, and are most frequently observed during twilight hours or after dark. All species are ground nesters. Loud, distinctive calls during the breeding season usually alert observers to their presence. Nighthawks forage during long sustained flights; nightjars make short sallies from an exposed perch or ground. Most species in North America are highly migratory, but some can survive periods of cold weather by lowering their metabolism and going into torpor—the Common Poorwill, for example.

Distribution North American species occupy a diversity of habitats. All but the Common Pauraque are migratory to some extent, but the most widespread species, the Common Nighthawk, is the champion long-distance migrant, traveling from as far north as northern Ontario to as far south as central Argentina.

Taxonomy Worldwide the family is composed of 92 species in 20 genera; relationships within the family are not well established by phylogenetic analyses. Closely related families are the potoos (Nyctibiidae), frogmouths (Podargidae), and Oilbird (Steatornithidae); all are, in turn, closely related to owlet-nightjars, swifts, and hummingbirds.

Conservation Nocturnal habits of most species make study difficult. BirdLife International codes: 7 NT, 4 VU, 3 EN, 2 CR (possibly extinct).

NIGHTHAWKS Genus *Chordeiles*

A New World genus, three of the six species occur in N.A. Nighthawks are partially diurnal, open-country birds, thus easier to observe and more frequently encountered than other nightjars. All species have relatively long, pointed wings, notched tails, conspicuous pale wing patches, and a patch of puffy white feathers at the bend of the wing when perched.

LESSER NIGHTHAWK *Chordeiles acutipennis* LENI ■ 1

Found primarily in arid regions, this species often congregates at water sources morning and evening, rarely active during midday. Where overlap occurs with the Common Nighthawk, the two species must be separated with care. Polytypic (7 sspp.; *texensis* in N.A.). L 8–9.2" (20–23 cm)

Identification Generally dark gray to brownish gray above mottled with black, grayish white, or buff; paler markings more concentrated on upperwing coverts, scapulars, tertials. Secondaries and primaries dark brownish gray; secondaries and basal portions of primaries spotted with buff. Underparts generally buffy, finely barred dark brown; chest and malar area darker and grayer with whitish to buff spotting. Underwing coverts buff

mottled with brown. **ADULT MALE:** White throat, subterminal tail band, and primary patch about two-thirds out from bend of wing to tip. **ADULT FEMALE:** Buff throat and wing patch; tail band reduced or absent. **JUVENILE:** Upperparts uniformly buffy gray with fine dark markings and spots; faint primary patch.

Similar Species Some Commons very similar in general coloration and size, but have disproportionately longer, more slender, more pointed wings, and primary patch is farther from wing tip. Most Commons lack buff spotting on secondaries and primaries.

Voice CALL: (Males only) Whistled trill in variable bursts. Flight display a bleating *bao-b-bao-bao.*

Status & Distribution Common. **BREEDING:** Arid lowland scrub, farmland. **MIGRATION:** In spring, arrives early Mar.–mid-May; departs early Aug.–late Oct., peak mid-Aug.–mid-Sept. **WINTER:** Northwestern and central Mexico south to northern S.A.; rare in extreme southern US, including south FL. **RARE STATUS:** Very rare migrant along Gulf Coast east to LA. Casual/accidental to northwest AK, CO, OK, ON, WV, NJ, and Bermuda. **Population** Stable.

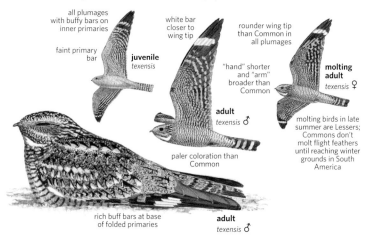

all plumages with buffy bars on inner primaries

faint primary bar

juvenile *texensis*

white bar closer to wing tip

"hand" shorter and "arm" broader than Common

adult *texensis* ♂

rounder wing tip than Common in all plumages

molting adult *texensis* ♀

molting birds in late summer are Lessers; Commons don't molt flight feathers until reaching winter grounds in South America

paler coloration than Common

rich buff bars at base of folded primaries

adult *texensis* ♂

COMMON NIGHTHAWK *Chordeiles minor* CONI ■ 1

This goatsucker performs flight displays and roosts conspicuously. Normally solitary, it sometimes forages or migrates in loose flocks. It is often out hunting or migrating during daylight hours. Polytypic (9 sspp.; 7 in N.A.). L 8.8–9.6" (22–24 cm)

Identification Varies geographically.

Black to paler brownish gray above; crown and upper back darkest; paler markings concentrated on upperwing coverts, scapulars, tertials. Below, barred blackish brown and white or buff. **ADULT MALE:** White throat patch, subterminal tail band, and primary patch halfway between bend of wing and wing tip.

juveniles show distinct geographic color variations; *sennetti* is silver-gray

juvenile *sennetti*

minor

sennetti

hesperis

henryi

howelli

chapmani

aserriensis

neotropicalis

subspecies uncertain

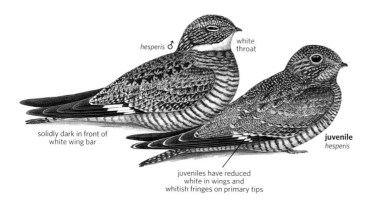

hesperis ♂

white throat

solidly dark in front of white wing bar

juvenile hesperis

juveniles have reduced white in wings and whitish fringes on primary tips

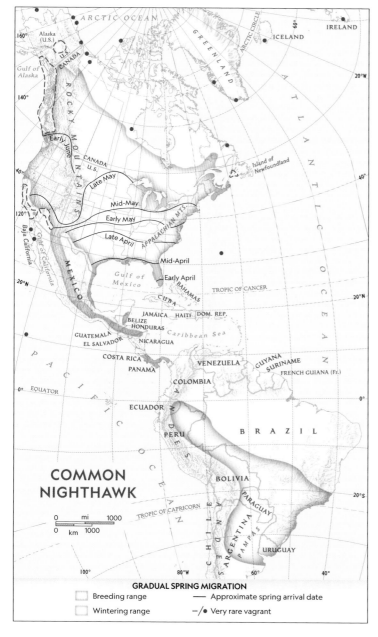

Early June

Late May

Mid-May

Early May

Late April

Mid-April

Early April

COMMON NIGHTHAWK

mi 1000
km 1000

GRADUAL SPRING MIGRATION

☐ Breeding range —— Approximate spring arrival date
☐ Wintering range —/● Very rare vagrant

ADULT FEMALE: Throat patch buffy, primary patch smaller, tail band reduced or lacking. **JUVENILE:** Generally paler, more uniform above with finer spotting and vermiculations.

Geographic Variation Eastern birds darkest, blackish above, less mottling on back; nominate *minor* (large), *chapmani* (smaller). Great Plains, Great Basin, and Southwestern birds paler and grayer; *henryi* (medium size), *howelli* (large), *sennetti* (large), and *aserriensis* (small). Western *hesperis* relatively dark, grayer, and large. Geographical color variation is most evident in juvenal plumage: juvenile *minor* and *chapmani* blackish; *sennetti* palest; *hesperis*, *howelli*, *aserriensis* intermediate; *henryi* rusty.

Similar Species Position of wing patch, lack of buff spotting on primaries, narrower and more pointed wing, and darker underwing coverts eliminate Lesser. Separation from Antillean problematic; best told by voice, but Antillean also usually smaller, shorter winged, and buffier on belly and undertail coverts.

Voice CALL: Nasal *peent* by male in flight; multiple-syllable variation may suggest Antillean. Male courtship dive vibrates primaries, producing "boom."

Status & Distribution Uncommon to fairly common. **BREEDING:** Open habitats, including towns. **MIGRATION:** Birds initiate migration by late afternoon. In spring, arrives early Apr.–mid-June, peak May (much later in West); departs Aug.–Oct., peak late Aug.–early Sept., when moderate to even large numbers can be seen throughout eastern N.A. (birds initiate migration by late afternoon); stragglers into Nov. **WINTER:** S.A.; casual Gulf Coast. **RARE STATUS:** Casual migrant to coastal central and Southern CA, HI, northern Canada, and UK.

Population Declines over much of range, especially in the East.

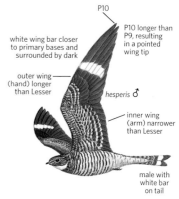

P10

P10 longer than P9, resulting in a pointed wing tip

white wing bar closer to primary bases and surrounded by dark

outer wing (hand) longer than Lesser

hesperis ♂

inner wing (arm) narrower than Lesser

male with white bar on tail

ANTILLEAN NIGHTHAWK *Chordeiles gundlachii*   ANNI ■ 2

Formerly considered a Caribbean subspecies of the Common Nighthawk, this FL Keys specialty was elevated to full species status following evidence there that both Commons and Antilleans nested without interbreeding. Monotypic. L 8.6" (21.5 cm)

Identification Relatively small nighthawk. Upperparts generally blackish, darkest on crown and upper back, varyingly mottled with whitish gray to buff, pale markings more concentrated on upperwing coverts, scapulars, and tertials. Upper breast and malar region blackish spotted with buff or grayish white; lower breast barred dull black and grayish white, belly with pale buff. Undertail coverts buffy. **ADULT MALE:** White throat, white subterminal tail band, and uniformly dark primaries except for white patch about halfway between bend of the wing and tip on five outer primaries. **ADULT FEMALE:** Similar but throat patch buffy, white primary patch smaller, lacks tail band.

Similar Species Voice is only certain identification character. Sight identification of silent, out-of-range individuals is likely impossible. Antillean Nighthawk is virtually identical to southeastern *chapmani* Common Nighthawk but averages smaller, shorter winged, and buffier on belly and undertail coverts. Some Commons have buffy underparts. Lesser overall very similar and buffy below, but has generally paler upperparts, breast, and primaries; paler, buffier underwing coverts; slightly more rounded wing tips. Antillean and Common have more pointed wing tips. Lesser has primary patch positioned about two-thirds of the way out from bend of wing to wing tip; and buffy spots on primaries and secondaries, usually visible when perched. Female Lesser has buff-and-white or completely buffy primary patch.

Voice CALL: (Males only) *pity-pit*, *chitty-chit*, or *killikadik*; also a Common Nighthawk–like nasal *penk-dik*. During male courtship dive, air rushing through primaries at the terminus of a dive produces hollow, roaring "boom."

Status & Distribution BREEDING: Uncommon in lower FL Keys (north to about Marathon) and on mainland to Homestead, in open or semi-open habitats. **MIGRATION:** Arrives after mid-Apr.; departs Aug.–Sept; rare on Dry Tortugas. **WINTER:** Unknown, presumably S.A. **RARE STATUS:** Accidental LA, NC.

Population Stable; expanded breeding range to FL Keys beginning in the 1940s.

slightly shorter wing than Common and pale bar slightly closer to tip

overall similar to Common

quite buffy overall, but so are some Commons

NIGHTJARS Genera *Nyctidromus*, *Phalaenoptilus*, *Antrostomus*, and *Caprimulgus*

Nyctidromus (2 sp.) and *Phalaenoptilus* (1 sp.) are small New World genera. *Phalaenoptilus* is characterized by its short tail. All New World goatsuckers formerly placed in *Caprimulgus* (previously with 45 sp.; now 36 sp.) are currently placed in their own genus, *Antrostomus* (11 sp.); they have disproportionately long tails.

COMMON PAURAQUE *Nyctidromus albicollis*   COPA ■ 2

This beautifully patterned tropical nightjar is often detected by its golden-red eyeshine along roadsides at night. By day it roosts on the ground in thickets, perfectly camouflaged in leaf litter. Polytypic (6 sspp.; *merrilli* in N.A.). L 11.2–12" (28–30 cm)

Identification Upperparts brownish gray finely vermiculated with black, crown coarsely streaked with black, back and upperwings crisply and complexly mottled with black and buff spots. Auriculars chestnut. Underparts buffy with narrow dark bars. **ADULT MALE:** Throat patch and large patch across outer primaries white, extensive white on third and fourth pairs of tail feathers. **ADULT FEMALE:** More diffusely marked above; reduced, buffier throat and wing patches, and less

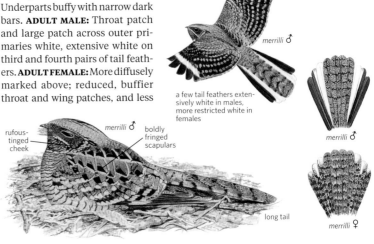

rounded wings

wing patch; white in males, buffy in females

merrilli ♂

a few tail feathers extensively white in males, more restricted white in females

rufous-tinged cheek

merrilli ♂   boldly fringed scapulars

long tail

merrilli ♂

merrilli ♀

white in tail (restricted to feather tips). **JUVENILE:** Similar, but browner, less heavily patterned; lacks throat patch, wing patches smaller and buffier. **Similar Species** Chuck-will's-widow and whip-poor-wills lack white primary patch, have shorter tail and longer wings, and are more uniformly colored. **Voice CALL:** *Whip* or *wheeeeeeer.* **SONG:** Buzzy whistle, *pur pur perp pur-wheeeeer.* **Status & Distribution** Common.

YEAR-ROUND: Woodland clearings and scrub in southern TX. **RARE STATUS:** Casual in TX north of breeding range to northern Maverick Co., Bastrop Co., Grimes Co., and Calhoun Co. **Population** Stable.

COMMON POORWILL *Phalaenoptilus nuttallii* COPO ■ 1

Our smallest nightjar is more typically heard giving its forlorn song or detected by its eyeshine at night (typically on roads). Makes short flights. Polytypic (6 sspp.; 4 in N.A.). L 7.6–8.4" (19–21 cm) **Identification ADULT MALE:** Large head; short, rounded tail. Paler (gray) and darker (browner) morphs occur throughout range. Upperparts dark brown to pale grayish brown, mottled pale gray and black. White throat bordered by blackish brown. All but central tail feathers tipped white. **ADULT FEMALE:** Similar but can have narrower or buffier tail tips. **Geographic Variation** Subspecies are distinguished by size and dorsal coloration (beware of color morphs). Two low-desert subspecies palest: *hueyi* of southeastern CA smallest, *adustus* of southern AZ larger. Widespread northern nominate *nuttallii* and *californicus* of western CA relatively large, darker brown above. **Similar Species** Whip-poor-will and Buff-collared Nightjar larger with longer tail, less white on throat. **Voice SONG:** Plaintive whistled *poor will up,* last note heard at close range. **CALL:** A soft *cluck.* **Status & Distribution** Uncommon to fairly common. **BREEDING:** Low-middle elevation dry, rocky, open, shrubby habitats. **MIGRATION:** Northern populations migratory, southern populations partially migratory or move to lower elevations in winter. Migrants typically noted in Oct. Arrives Feb.–Mar. south, late Apr.–

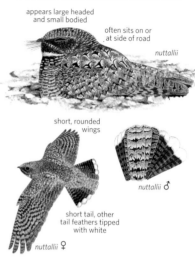

appears large headed and small bodied

often sits on or at side of road

nuttallii

short, rounded wings

nuttallii ♂

short tail, other tail feathers tipped with white

nuttallii ♀

late May north. Departs Sept. north, Oct.–Nov. south. **WINTER:** Southern areas of breeding range (CA, AZ, TX) to central Mexico. **RARE STATUS:** Accidental southwestern BC, southern MB, ON, MN, MO. **Population** Stable.

BUFF-COLLARED NIGHTJAR *Antrostomus ridgwayi* BCNI ■ 3

The Buff-collared Nightjar is primarily a west Mexican species. Polytypic (2 sspp.; nominate in N.A.). L 8.8–9.2" (22–23 cm) **Identification ADULT:** Upperparts brownish gray with black blotches on crown and scapulars, head and back separated by distinct pale buff collar. Upperwing coverts spotted with buffy white, primaries brownish black banded with cinnamon-buff. Upper throat dark grayish brown, breast somewhat paler, and belly paler tan barred with dark brown; throat and breast separated by whitish buff to buff foreneck collar. **JUVENILE:** Similar. **Similar Species** Mexican Whip-poor-will similar but slightly larger, darker, and browner, and lacks distinct pale buff hind neck collar. Common Poor-will smaller, shorter tailed. Lesser Nighthawk has pale primary patches, different proportions and behavior. **Voice CALL:** Series of *tuk* notes. **SONG:** Rapid, accelerating series of five to six *cuk* notes, concluding with rapid *cuk-a-chee-a,* somewhat reminiscent of the dawn song of Cassin's Kingbird. **Status & Distribution** Rare and irregular in southeastern AZ; once to Upper Guadalupe Canyon, NM. **BREEDING:** Local in desert canyons dominated by mesquite, acacia, and hackberry. **MIGRATION:** In US extreme dates Apr. 17–Aug. 28; presumably migratory based on lack of winter records, but no specific information. **WINTER:** Western Mexico. **RARE STATUS:** Accidental to coastal Southern CA (Ventura Co.). **Population** Stable. In the US, the species was first detected in 1958–1960; it has been designated by state wildlife agencies as endangered (NM) and of special concern (AZ).

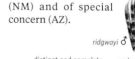

ridgwayi ♂

male with less white than whip-poor-wills

distinct and complete buff collar

ridgwayi

overall paler than Mexican Whip-poor-will

CHUCK-WILL'S-WIDOW *Antrostomus carolinensis* CWWI ■ 1

This species is heard much more often than seen. Widespread in open forest and woodlands of the Southeast, this is our largest nightjar. During the day, it roosts on the ground or on a branch. It can be quite startling when flushed at close range. It flies up in a burst of brown, sometimes accompanied by loud grunting notes. Monotypic. L 11.2–12.8" (28–32 cm)

Identification ADULT: Large nightjar, males are larger than females. Large, flat-topped head. Generally dark grayish brown, but considerable individual variation exists, from paler and rustier to darker and browner. Crown and upper back darkest; scapulars palest, with heavy black streaks or spots on these areas. Wings blackish and heavily barred rufous. Tail long, broad, rounded, buffy to rufous, and with black bars and vermiculations. Chin and upper throat brown with fine dark bars. Whitish buff to buff patch on lower throat, contrasting with dark brown to blackish malar area and breast. Belly and undertail coverts pale buff to dark rufous, barred with blackish brown. **MALE:** Extensive subterminal whitish buff inner webs of outer three pairs of tail feathers. **FEMALE:** Buffy tips to tail feathers.

Similar Species Eastern Whip-poor-will is smaller, has shorter, narrower, more round-tipped wings, averages grayer overall, has dark throat, and male has different tail pattern.

Voice CALL: A single or a series of grunting notes, *quaah*. **SONG:** Whistled *chuck-weo-wid-ow*; and the song is con-cluded by a rapid *wid-ow*; at a distance, often just the last two notes are audible.

Status & Distribution BREEDING: Deciduous, evergreen, or mixed woodlands with edges or gaps. **MIGRATION:** Spring migration mid-Mar.–mid-May, a few to late May. Depart mid-Aug.–Oct., stragglers into Nov. **WINTER:** Southern FL, northern West Indies, and east-central Mexico south to Colombia; rare in northern FL and along Gulf Coast. **RARE STATUS:** Rare, mainly spring and summer, north of breeding range as far as SD, MN, WI, MI, and southeastern Canada. Casual or accidental west to NM, NV, and coastal northern CA.

Population Near Threatened.

much larger than Eastern Whip-poor-will with big, flat head

gray morph

some are grayer

more restricted white than both whip-poor-will species ♂

rufous-brown throat

rufous morph

lacks white in tail ♀

most are rufous overall in coloration

EASTERN WHIP-POOR-WILL *Antrostomus vociferus* EWPW ■ 1

On still summer nights, the song of the "Whip" rings in eastern forests. Daytime encounters with this seldom seen creature are usually by chance when a nesting bird is flushed, but many will permit close study when found roosting. Monotypic. L 8.8–10" (22–25 cm)

Identification ADULT: Relatively rounded wings. Generally grayish brown above with black, pale gray, and buff spotting and vermiculations; thicker black streaks and blotches on crown and scapulars. Wings heavily barred with buff. Blackish brown throat and dark brown upper chest separated by white (male) to buff (female) foreneck collar; rest of underparts buffy with fine dark bars. Male has thick, white tips on outer three pairs of tail feathers; female has smaller, buffy tail tips.

Similar Species Chuck-will's-widow is much larger, with bigger, flatter head, broader and more pointed wings; it averages darker and browner, has buffy brown throat, and male has different tail pattern. Buff-collared Nightjar

blackish throat and white collar on males ♂

overall darker and less rufous than Chuck-will's-widow ♀

female with buffy tips to outer tail feathers

male with much more white in outer tail than Chuck-will's Widow ♂

is slightly smaller, paler, and grayer, with a more prominent pale buff collar across hind neck. Common Poorwill is notably smaller and has a shorter tail. Common Pauraque and all nighthawks have pale primary patch, different tail pattern and shape, and pale throat. See Mexican Whip-poor-will.
Voice SONG: Melodious, repeated *whip-poor-will*, usually delivered at night. **CALL:** Year-round, a subdued Hermit Thrush-like *tuk* or *quirt*, usually delivered at dusk and dawn.
Status & Distribution Formerly common but now absent or distinctly uncommon across most of range. **BREEDING:** Dry deciduous or mixed forest with open understory and partly open canopy. **MIGRATION:** Arrives in spring mid-Mar.–mid May, peak Apr., stragglers to late May–early June. Departs late Sept.–Oct., peak mid-Sept.–mid-Oct., stragglers to Nov.–early Dec. **WINTER:** FL and Mexico south to Honduras; rare elsewhere along southern Atlantic and Gulf Coasts. **RARE STATUS:** In the East, casual to NL and Bermuda; in the West, casual to CO and NM; accidental to southeast AK (late fall), OR (May), and CA (late fall); a territorial singing bird near Santa Cruz (May–June).
Population Near Threatened. Widespread declines (by over 60 percent between 1970 and 2014) and associated breeding-range contractions.

MEXICAN WHIP-POOR-WILL *Antrostomus arizonae* MWPW ■ 2

Until recently considered conspecific with Eastern Whip-poor-will, Mexican Whip-poor-will is a far less familiar bird in the US, where it is found in rugged and remote country, mostly in the isolated mountain ranges of the Southwest. Polytypic (5 sspp., nominate in N.A.). L 9–10.4" (23–26 cm)
Identification Very similar to slightly smaller Eastern Whip-poor-will; best distinguished by song. Mexican has longer rictal bristles that appear buffy near the base (perhaps only on adults), rather than blackish; and more black and less white (or, in females, less buff) in outer rectrices. It roosts mainly on the ground and rarely tolerates a close approach, unlike Eastern, which roosts in trees and bushes and permits close approach.
Geographic Variation Nominate *arizonae* is the only subspecies recorded north of Mexico. Subspecies *setosus*, which could reach TX, is resident from central Tamaulipas to Veracruz. Three other subspecies occur from central Mexico to El Salvador.
Voice SONG: Similar to Eastern but burrier (more modulated), lower in pitch, and slightly slower, a husky *whirr-p-wiirr*. **CALL:** A soft *tuk*, similar to Eastern.
Status & Distribution Locally fairly common. **BREEDING:** In mixed coniferous-oak or dry coniferous woodland with brushy understory, in mountains. **MIGRATION:** Some may winter in or near breeding areas in border areas, but most appear to withdraw southward from the US. Arrive in spring Apr.–mid May; departure may start by early Aug. **RARE STATUS:** Rare in Southern CA and southern NV. Casual or accidental with confirmed records as far out of range as northern CA, UT, MT (singing, which whip-poor-will species uncertain), CO, NE, exceptionally BC (specimen, Vancouver I.). The easternmost record thus far is from the NE Panhandle.
Population In contrast to Eastern Whip-poor-will, Mexican Whip-poor-will appears stable and may have expanded its range in recent decades.

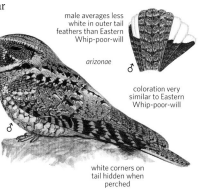

male averages less white in outer tail feathers than Eastern Whip-poor-will

arizonae ♂

coloration very similar to Eastern Whip-poor-will

longer and browner rictal bristles

throat blackish with white collar on males ♂

white corners on tail hidden when perched

GRAY NIGHTJAR *Caprimulgus jotaka* GRNI ■ 5

The Gray Nightjar has been recorded only once in N.A. This species has recently been split from three other more southerly and sedentary taxa from southern Asia and Palau. Polytypic (2 sspp.; N.A. record is of northernmost and highly migratory *jotaka*). L 11–12.8" (28–32 cm)
Identification Upperparts generally brownish gray, patterned with black, buff, and grayish white spots, streaks, and bars. Whitish submoustachial stripe. Underparts grayish brown, barred with paler gray, buff, or brown. **ADULT MALE:** Large subterminal white patch on outer primaries and white tips to all but central pair of tail feathers. **ADULT FEMALE:** Tawny primary patch, brownish white or brownish buff tail tips.
Similar Species N.A. goatsuckers either lack primary patches or have different wing, tail, and throat patterns.
Voice SONG: Loud, rapid series of down-slurred *schurks*.
Status & Distribution BREEDING: From southeast Siberia and Russian Far East to the Himalaya in a variety of forest, woodland, and scrub habitats. **WINTER:** Southern breeding range south to East Indies. **RARE STATUS:** Accidental. A female was found dead and desiccated on Buldir I., AK (31 May 1977). A record from Ashmore Reef, Australia, indicates how migratory the species is.
Population Stable.

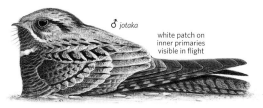

♂ *jotaka*

white patch on inner primaries visible in flight

SWIFTS Family Apodidae

White-throated Swift (BC, June)

Superficially swallowlike with their aerial habits, small bills, and long, narrow wings, swifts differ from swallows in their wing structure (the wrist joint is closer to the body), stiff wingbeats (the wings are not swept back against the body), and more rapid flight. Swifts are usually observed in flight because most species only nest, perch, and roost in hollows and crevices. Watch for swifts over freshwater, along bluffs and ridgetops, and near roost and nest sites. For species identification, concentrate on overall size, wing and tail shape, dark-and-light patterning on the body, and flight style.

Structure Swifts are small to medium-size, long-winged aerialists. The largest (up to 6.3 ounces) are 20 times the weight of the smallest. *Apodidae* literally means "without feet," a reference to the swift's tiny (but strong) feet. The "hand" portion of the wing (carpal bones and digits) is quite long relative to the "arm" (humerus and forearm bones); thus the primaries dominate the flight feathers, with the secondaries short and bunched together. There are 10 rectrices; spinelike protrusions are formed by the tip of the rachis (a feather's central shaft) in some species and are used as props. The strongly clawed toes assist in clinging to vertical surfaces, the only kind of perching of which true swifts are capable. The bill is small but with a large gape.

Behavior The most aerial of birds, swifts feed throughout the day and sometimes even "roost" on the wing. Some species can attain very high speeds as they cut through the air. The swift's nest is usually a shallow open cup, with materials glued together and attached to a vertical substrate with saliva; some extralimital species build a larger nest with plant fibers. Their diet consists of aerial arthropods, principally small insects and spiderlings, and they may forage dozens of miles from the nest site. Large numbers of swifts may gather together before dusk around roost sites, often coming together in very vocal, swirling flocks.

Plumage Swifts are generally clad in blackish to gray-brown plumage, usually with contrasting paler and darker regions and sometimes with bold white markings. Most species show short, dense blackish feathering in front of the eyes. Tail shape varies greatly among genera and can also vary with the degree of closure or spreading of the tail: For example, swifts with deep tail forks may hold the tail closed so that it appears as a single point. Sexes are similar, but juveniles are usually distinguished by duller patterning, pale feather fringes on the body, and pale tips to the secondaries and the tertials.

Distribution True swifts are found worldwide in temperate and tropical regions; many species are highly migratory. Being wide-ranging, strong fliers, swifts have a considerable potential for vagrancy: Five of our nine species occur only as rare visitors, and Alpine Swift, a European/African species, has occurred twice in the Lesser Antilles.

Taxonomy Swifts comprise 99 species, in 20 genera. These are often placed in four subfamilies: the more primitive Cypseloidinae (now limited to the New World), the widespread Chaeturinae and Apodinae, and the Hemiprocninae (four Southeast Asian tree-swifts formerly placed in their own family). Species limits in some large genera (e.g., *Chaetura*) are problematic. The swifts are grouped with hummingbirds in the order Apodiformes (this order, in turn, is sometimes placed within the Caprimulgiformes).

Conservation Some swift species have increased with the availability of human-built nest substrates, whereas other species are highly specialized in their nesting habits and have suffered local declines. BirdLife International codes: 5 NT, 6 VU, and 1 EN.

CYPSELOIDINE SWIFTS Genera *Cypseloides* and *Streptoprocne*

These are primitive swifts, medium-small to very large in size, with all modern species restricted to the New World. Unlike other swifts, they do not use saliva in their nest construction. Most species nest near waterfalls or other damp, shaded sites, but they can cover huge distances while foraging.

BLACK SWIFT *Cypseloides niger* BLSW ■ 2

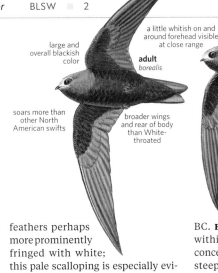

a little whitish on and around forehead visible at close range

large and overall blackish color

adult *borealis*

white scaling visible at close range; retained through fall migration

juvenile *borealis*

soars more than other North American swifts

broader wings and rear of body than White-throated

Our largest regularly occurring swift, this species is associated with waterfalls and other damp cliff habitats in western mountains and rugged northwestern coastlines. Even when on nesting grounds, adults often travel great distances to forage, returning near dusk, or later, to feed their young. It is generally rare as a migrant. Recent published research revealed that sexing of *borealis* is possible for over 95 percent of birds in hand. Polytypic (3 sspp.; *borealis* in N.A.). L 7.3" (19 cm)
Identification A large, blackish swift with long, broad-based wings that show a distinct angle at wrist joint and a rather long, broad, shallowly forked tail. **ADULT MALE:** Sooty black throughout except for frosted whitish chin, forehead, lores, and thin line over eye; tail moderately forked. **ADULT FEMALE:** Slightly smaller and typically frosted with more white scaling below than the large majority of adult males; tail fork very shallow. **JUVENILE:** Resembles adult female, but body plumage, coverts, and flight feathers perhaps more prominently fringed with white; this pale scalloping is especially evident on belly and undertail coverts. **FLIGHT:** Wingbeats can appear relatively languid, but its flight can nevertheless be exceedingly fast; it usually flies with short bursts of wingbeats interspersed with long glides. Individuals are wide ranging and fly high; and most likely observed lower on overcast days and near nest sites.
Geographic Variation West Indian nominate is smaller; might reach FL; *costaricensis* breeds from central Mexico to Costa Rica. It is slightly smaller and distinctly darker, and the adult female has broader white fringes on the underparts.
Similar Species Vaux's is much smaller, paler, has shorter wings, with rapid wingbeats and little gliding. White-throated can appear all blackish in poor light, but has a thinner and

disproportionately longer tail (often carried in a point), slimmer wings with less of an angle at wrist joint, and a more rapid wingbeat.
Voice Relatively silent.
Status & Distribution Generally uncommon; locally much more numerous in coastal BC. **BREEDING:** Locally distributed within its general breeding range, concentrating near nest sites around steep mountain cliffs with a damp microclimate (often from waterfall spray). Also nests locally in sea cliffs in central CA, northern WA, and BC. **MIGRATION:** Spring migrants usually arrive after mid-May, exceptionally the end of Apr., and migration continues into mid-June; fall migrants occur mainly Sept.–early Oct., exceptionally into early Nov. Generally rare away from nesting sites. **WINTER:** Wintering range of *borealis* recently discovered through geolocator data to be in western Amazon Basin. **RARE STATUS:** Casual in the interior West as a migrant. Accidental in East.
Population Vulnerable.

WHITE-COLLARED SWIFT *Streptoprocne zonaris* WCSW ■ 4

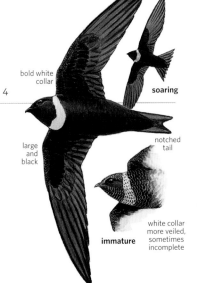

bold white collar

soaring

large and black

notched tail

white collar more veiled, sometimes incomplete

immature

This Neotropical species has occurred casually but widely in N.A. Our largest swift, it has a distinct white collar and soaring flight. Polytypic (9 sspp.; 2 in N.A.). L 8.5" (22 cm)
Identification Very large swift, with slightly forked tail. Prolonged soaring flight. **ADULT:** Blackish throughout, with a bold white collar, broader across chest and narrower on hind neck. **JUVENILE:** Resembles adult, but plumage is sooty with indistinct pale

pallidifrons

mexicana

fringes and the collar is less distinct. **Geographic Variation** Most records in N.A. are of the large Mexican subspecies *mexicana*, but a FL specimen (Lauderdale Lakes, Broward Co., 15 Sept. 1994) is of the smaller West Indian *pallidifrons*, which has more white on forehead, shows pale on chin, and has a pale streak above the eye. **Similar Species** Black Swift is smaller and lacks a white collar. **Status & Distribution** Casual visitor to N.A. Widespread from eastern (north to south Tamaulipas) and southern Mexico to northern Argentina. **RARE STATUS:** Eight records from across N.A. The four TX records are from Mar., May, and Dec.; the two FL records are from Sept. and Jan. There are also single records for CA and MI, both in May. **Population** Stable.

SPINE-TAILED SWIFTS Genera *Chaetura* and *Hirundapus*

The subfamily Chaeturniae includes small to large spine-tailed swifts found in both the New and Old Worlds. It also includes the familiar Chimney Swift, the needletails, and the southeast Asian and Pacific swiftlets, which can echolocate in caves and whose saliva is the source of bird's nest soup.

CHIMNEY SWIFT *Chaetura pelagica*   CHSW ■ 1

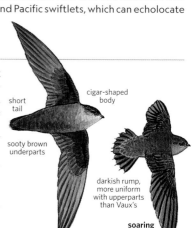

cigar-shaped body

short tail

sooty brown underparts

darkish rump, more uniform with upperparts than Vaux's

soaring

A small, dark "cigar with wings," this is the common swift of the eastern half of N.A. Its original nest sites (hollow trees, cliffs) have largely been substituted with human-built structures such as chimneys or building shafts. Monotypic. L 5.3" (13 cm)
Identification Small, dark; squared, spine-tipped tail; narrow-based wings often appear markedly pinched at the base during secondary molt in late summer, early fall. **ADULT:** Brownish black overall; paler chin, throat; slightly paler rump. **JUVENILE:** Nearly identical to adult, but with whitish tips to the outer webs of the secondaries, tertials. **FLIGHT:** Usually rapid, fairly

shallow wingbeats, including quick turns, steep climbs, short glides. V display of pairs involves long glides with wings raised in a V pattern and some rocking from side to side.
Similar Species Vaux's Swift is very similar but slightly smaller, paler; differs subtly in shape; has higher-pitched calls.
Voice CALL: Commonly heard; quick, hard chippering notes, sometimes run together into rapid twitter.
Status & Distribution Common. **BREEDING:** Widespread in variety of habitats; most abundant around towns, cities. Possibly breeds north to NL. Very small numbers summer regularly in Southern CA (though fewer since the 1990s), with breeding documented; possibly also bred in AZ. **MIGRATION:** Migrates in flocks during the day, mainly along the Atlantic coastal plain, Appalachian foothills, and Mississippi River Valley. Large concentrations may appear during inclement weather; hundreds may roost in chimneys. First spring arrivals are in mid-Mar. in southern states; peak arrivals in northernmost breed-

ing areas are late Apr.–mid-May. Most have departed breeding areas by late Sept.–mid-Oct; latest fall migrants occur in early Nov. **WINTER:** Most or all winter in Upper Amazon Basin of S.A.; unrecorded in N.A. in midwinter, but records as late as Dec. **RARE STATUS:** Casual away from CA in West, mainly May–Sept.; accidental on Pribilof Is., AK, and in western Europe.
Population Vulnerable. Population declines have been noted since the 1980s, perhaps a result of fewer nesting sites (e.g., chimneys).

VAUX'S SWIFT *Chaetura vauxi*   VASW ■ 1

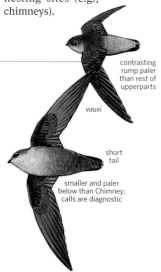

contrasting rump paler than rest of upperparts

vauxi

short tail

smaller and paler below than Chimney; calls are diagnostic

This is the counterpart of the Chimney Swift in western N.A., occurring almost exclusively west of the Rocky Mts. Only rarely does it overlap with the Chimney Swift; identifications of silent out-of-range-birds should be made very carefully, if at all, and be substantiated with photos. Vaux's also appears to be the only *Chaetura* to winter north of Mexico, though it is casual at that season. Polytypic (7 sspp.; 2, nominate nests in N.A.). L 4.8" (12 cm)
Identification Our smallest regularly occurring swift, it is very similar to

Chimney Swift in all respects. Distinctions from Chimney are fairly evident on the rare occasions the two species can be directly compared, but lone birds are very difficult to identify. **ADULT:** Brownish overall, paler on throat, breast, and rump. **FLIGHT:** Very rapid, twinkling wingbeats are interrupted by only brief glides; many quick turns and climbs when foraging, as in Chimney. Rocking V display also similar to Chimney Swift.
Geographic Variation Our birds are nominate *vauxi*, the northernmost

subspecies. Five more subspecies are found from Mexico to Venezuela, one of which (*tamaulipensis* of eastern Mexico) has been collected once in AZ (14 May 1950 at Ft. Hauchuca); it is slightly darker and glossier black than *vauxi*. Dark birds from southern Mexico to Venezuela (*richmondi* group) and small birds on the Yucatán Peninsula (*gaumeri* group) are sometimes treated as separate species.

Similar Species Chimney Swift has a darker throat and breast and is more uniform above, with a gray-brown rump only slightly paler than back; Vaux's rump and uppertail coverts

are paler, showing more contrast to back. Structural characters are subtle but important: Chimney is larger and appears relatively longer winged with a narrower wing base; its rear body appears slightly longer, thus wing placement appears slightly farther forward. Flight differs subtly, with Chimney gliding more and flapping slightly more slowly.

Voice High insectlike chipping and twittering notes, often run together into a buzzy trill. Vocalizations resemble those of Chimney, but are considerably higher pitched, more rapid, and more trilled.

Status & Distribution Common. **BREEDING:** Heavily forested lowland and lower montane areas; old standing snags generally required, but Vaux's will also nest in chimneys. **MIGRATION:** As with migrant swallows, most frequently encountered and largest numbers recorded during inclement weather. Spring migrants arrive in the Southwest by early Apr.

and peak there and in CA from late Apr. to early May, a few into late May. The northernmost breeding areas are occupied by late May; fall migration is from late Aug.–early Oct. (to late Oct. in south) with a peak in mid- to late Sept. Fall migrants occur rarely east to UT, eastern AZ, and NM. Huge postbreeding and migrant roosts noted from WA to Southern CA. **WINTER:** Individuals or small flocks rarely and irregularly winter in coastal central and Southern CA. A few late fall and winter records for LA and FL, including some small flocks, suggest occasional wintering in the Southeast. Most Vaux's winter from central Mexico south through C.A., but details of winter range are poorly known because of the presence of resident subspecies or closely related species.

Population Although still common, especially at urban roost sites, overall numbers have declined with loss of old growth forests in the Northwest.

WHITE-THROATED NEEDLETAIL *Hirundapus caudacutus* WTNE ▪ 5

Needletails are large, spectacular Asian swifts with broad triangular wings (including full secondaries) and short square or rounded tails that have moderate to long rectrix spines. All four species in this genus have a white horseshoe-shaped patch on the rear of the underparts, and glossy upperparts. Polytypic (2 sspp.; nominate in N.A.). L 8" (20 cm)

Identification Large swift with stubby spine-tipped tail, broad-based wings, and large head. Remarkable high-speed flight consists of rapid and powerful wingbeats and then long glides on bowed wings. **ADULT:** White throat sharply contrasts with dark breast and belly; extensive white undertail coverts and flanks form a large U-shaped patch; white spot on forehead, lores; white on inner webs of tertials. Upperparts show silvery white saddle, contrasting with green-glossed black crown and nape, wings, and tail. **JUVENILE:** Similar to adult, but lores grayish, white of lower underparts scalloped with dark.

Geographic Variation Two sub-

species, the more northerly and highly migratory subspecies *caudacutus* has been recorded in N.A. Himalaya *nudipes*, perhaps a separate species, is similar but lacks white on forehead and lores; saddle a little browner.

Similar Species Fork-tailed has a long notched tail, a white rump, and dark underparts. Another migratory Asian species, Silver-backed Needletail (*H. cochinchinensis*), is quite similar. It is best separated from White-throated Needletail by pale brown throat, not a well-delineated pure white throat; also tertial edges are less distinct; and light head markings are lacking, unlike nominate *caudacutus*.

Voice Common call is a soft, rapid insectlike chattering.

Status & Distribution **BREEDING:** Nests in hollow trees in eastern Asian forests from southern Siberia to northeastern China, Korea, Sakhalin I., and northern Japan. **WINTER:** The nominate subspecies migrates to New Guinea, eastern Australia. **RARE STATUS:** Casual in spring on the outer

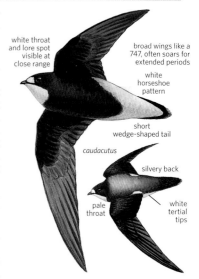

white throat and lore spot visible at close range

broad wings like a 747, often soars for extended periods

white horseshoe pattern

short wedge-shaped tail

caudacutus

silvery back

pale throat

white tertial tips

Aleutian Is., AK (five records from late May); recorded once (early summer) on St. Paul I., Pribilofs. Also casual to western Europe (most from UK) and Seychelles.

Population Stable.

TYPICAL SWIFTS Genera *Apus, Aeronautes,* and *Tachornis*

This widespread group of typical swifts ranges in size from tiny to quite large. Most have forked tails, and many species show white patterning on the rump, throat, or belly. Nests, which are generally made of feathers and plant material, are cemented with saliva to a vertical wall, tree hollow, palm frond, or other protected site.

COMMON SWIFT *Apus apus* COSW ▪ 4

The Common Swift is a familiar and well-studied swift of Eurasia. Casual in N.A., with late June records in the far northwestern and northeastern corners of the continent. It is a moderately large and dark swift, with a strongly forked tail. Polytypic (2 sspp., both likely to have occurred in N.A.). L 6.5" (17 cm)

Identification Dark nearly throughout, with a long and obviously forked tail. In flight the rapid, frenzied wingbeats alternate with long glides. **ADULT:** Plumage is blackish brown, with slight scaled effect to body feathering. Pale throat contrasts to remaining underparts. Upperwings are uniformly dark throughout; on underwings, flight feathers are only slightly paler than wing linings. **JUVENILE:** Similar to adult, but its plumage is blacker and more scaly, with whitish fringes on forehead and pale tips to flight feathers.

Geographic Variation Subspecies *pekinensis*, of the eastern half of the breeding range, is slightly paler and browner, with a more extensive whitish throat patch; AK specimen is of this subspecies. Nominate European birds

are the likely source of eastern records and reports.

Similar Species See Fork-tailed, which is casual in western AK. Compared to Black Swift, Common Swift has a narrower and more deeply forked tail, narrow-based and more pointed wings, and a pale throat patch. Adult Common Swifts have dark foreheads, but pale fringes on forehead of juveniles recall Black Swift. (Note pale throat and long, deeply forked tail of juvenile Common.) Black's flight is more leisurely; Common with rapid wingbeats; interspersed with glides. Several related Old World species (e.g., Pallid Swift) are closely similar but not likely to occur in N.A.

Voice A wheezy, screaming *sreeee* or *shreeee*; especially vocal in pre-roosting gatherings.

Status & Distribution Casual to N.A. Common breeder from Europe eastward through northern China, migrating south to winter in the southern half of Africa. **RARE STATUS:** Three late June records: a specimen from the Pribilof Is., AK (28 June 1950); one photographed there 28–29 June 1986; and one photographed on Miquelon I.,

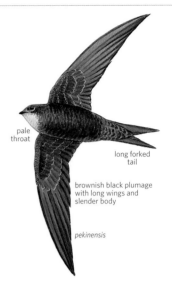

pale throat

long forked tail

brownish black plumage with long wings and slender body

pekinensis

off NL (23 June 1986). Also a single fall record from CA: one photographed near Desert Center, Riverside Co. (30 Oct. 2013), There have been other likely sightings along the Atlantic coast; recorded in Bermuda.

Population Large world population, but declines noted in some areas of Europe.

FORK-TAILED SWIFT *Apus pacificus* FTSW ▪ 4

This large Asian swift with a long forked tail has been recorded casually in western AK. It is widely known in the Old World as Pacific Swift. Polytypic (5 sspp.; nominate in N.A.). L 7.8" (20 cm)

Identification This is a large but slender, blackish brown swift with a broad (20 mm) white rump that contrasts with dark uppertail coverts and tail; it has a long, forked tail, but the tail's deep fork is often not apparent (e.g., when closed) and appears instead as a long extended point. **ADULT:** Underparts appear scaly, with dusky black feathering fringed with well-defined but narrow off-white fringes; throat is whitish and well-separated by dark breast. Underwing is uniform. **JUVENILE:** Similar to adult, but its flight feathers are fringed with white.

Geographic Variation The nominate is the largest and northernmost breeder (to Russian Far East) and highly migratory, moving as far south as New Zealand in winter. The other three, more southerly subspecies are slightly to significantly smaller

and darker; they are short-distance migrants and possibly resident in some areas. Some recent treatments suggest that this complex might consist of four species; perhaps Cook's Swift, nesting in limestone caves of Southeast Asia, might be the most distinct.

Similar Species Common Swift, also recorded in western AK, is smaller and darker, and lacks the white fringing below and the white rump. White-throated Needletail has a different wing shape, a short, rounded tail, and a distinct white horseshoe pattern on the rear of the underparts; it also has a whitish saddle on the back and lacks the white rump patch.

Voice Screaming *sre-eee* resembles calls of the Common Swift, but it is slightly softer and more disyllabic.

Status & Distribution Breeds in eastern Asia, northeast to the Kamchatka Peninsula; southern Asian populations are sedentary, whereas northern birds winter from southeast Asia to Australia. The northern limits of *pacificus* are uncertain due to identification difficulties with other southern

subspecies. **RARE STATUS:** Casual in AK, mainly in the western Aleutian and Pribilof Is., but also on St. Lawrence I., Middleton I., and mainland AK. Accidental YT. Most records are mid-May–June and Aug.–Sept.

Population Stable.

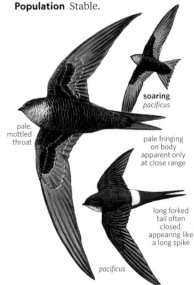

soaring
pacificus

pale, mottled throat

pale fringing on body apparent only at close range

long forked tail often closed, appearing like a long spike

pacificus

WHITE-THROATED SWIFT *Aeronautes saxatalis*   WTSW   ▪ 1

The White-throated Swift is found in western canyons, cliffs, and even urban areas. Its bold patterning, streaking flight, and staccato vocalizations attract attention wherever it occurs. Polytypic (2 sspp.; nominate in N.A.). L 6.5" (17 cm)

Identification White-throated Swift is a fairly slender swift with long, scythelike wings. Its moderately long, notched tail is usually held in a tight double-point but can be widely fanned in maneuvering birds. Bold black-and-white patterning is unique among our regularly occurring swifts. **ADULT:** Blackish to blackish brown on crown, upperparts, flanks, undertail coverts, and flight feathers; paler gray on forehead, lores, and narrow supercilium. White chin, throat, and chest; white continues more narrowly to belly. There is also a large white patch on flanks, extending up toward sides of rump, and distinct white tips to secondaries and tertials. Sexes are similar, but females have a slightly shallower tail fork and less white on

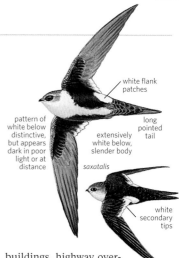

pattern of white below distinctive, but appears dark in poor light or at distance

white flank patches

long pointed tail

extensively white below, slender body

saxatalis

white secondary tips

secondary and tertial tips. **JUVENILE:** Closely similar to adult, but it appears paler and browner due to indistinct pale fringing on forehead, crown, and undertail coverts. Flight feathers are narrowly edged with white, but white tertial tips are indistinct.

Similar Species Beware that with distant views or poor lighting, White-throated Swifts can often appear all-dark. Smaller and more uniformly colored Vaux's and Chimney Swifts have shorter wings and shorter, squared tails; compared with these *Chaetura*, White-throated's long wings appear to be attached farther forward (probably due to longer rear body and tail). See also Black Swift, which is distinguished by its broader, more shallowly forked tail, crook at the wrist joint of the wings, and generally more languid flight with long periods of soaring. In distant birds, the white rump, sides, and throat may suggest a Violet-green Swallow, but White-throated Swift's long, stiff wings, blackish sides and undertail coverts, and long pointed tail make this distinction straightforward.

Voice Often quite vocal, giving a loud, rapid, shrill *tee-dee, dee, dee, dee* ... or *jee-jee-jee-jee* ... series that drops slightly in pitch.

Status & Distribution Locally common. **BREEDING:** In the interior West, mostly found in canyons, river gorges, and other areas of high relief, with records as high as 13,000 ft. It is also found on cliffs and lowlands in southern and central CA and increasingly in urban regions where crevices in

buildings, highway overpasses, and other human-built structures provide nest and roost sites. Foraging birds can wander widely over lowland areas, and like most wide-ranging swifts, their local abundance can shift with changes in wind patterns, clouds, and weather fronts. **MIGRATION:** Withdraws from northern part of breeding range, where present mainly Apr.–Sept. **WINTER:** Most birds winter from central CA, southern NV, central AZ, and west TX southward. Numbers often more concentrated in winter when large communal roosts may be used. On colder winter days, activity may be curtailed; some birds may even become torpid. **RARE STATUS:** Casual east to central and coastal TX, OK, KS, and Vancouver I., BC. Accidental MO (Nov.), AR (May, Dec.), MI (Aug.), and west FL (Apr.).

Population Stable.

ANTILLEAN PALM-SWIFT *Tachornis phoenicobia*   ANPS   ▪ 5

A tiny Caribbean swift with a forked tail, a dark-capped appearance, and a distinctive white rump and belly patch. Polytypic (2 sspp.; L 4.3" (11 cm)

Identification Very small size, white rump, and shallowly forked tail are distinctive. **ADULT:** Brown above with broad white rump; white throat and center of belly; brownish breast band (narrowest in center), sides, flanks, vent, and undertail coverts. **IMMATURE:** Similar to adult, but duller; white areas of underparts buffy. **FLIGHT:** Rapid wingbeats and short glides; generally flies low.

Geographic Variation Cuban (including Isla Juventud) subspecies *iradii* is likely the source of the FL records. The

subspecies *iradiiphoenicobia* from Hispaniola and Jamaica is smaller and darker above with a shallower tail fork and much darker flanks; sides of head more extensively grayish brown.

Similar Species Very small size and coloration distinctive. The much larger and black-and-white White-throated Swift, which lacks the dark breast band, is also accidental in the Southeast.

Voice Weak twittering calls.

Status & Distribution RESIDENT: Greater Antilles (excluding Puerto Rico), including Cuba. Found in lowlands, especially around palms where it builds its nest among dead fronds. **RARE STATUS:** Accidental in south

FL, two at Key West, FL (7 July–13 Aug. 1972) and one at Marathon (July–Dec. 2019); accidental Puerto Rico.

Population Stable, has adapted to urban habitats.

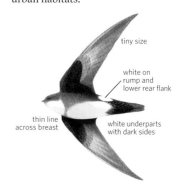

tiny size

white on rump and lower rear flank

thin line across breast

white underparts with dark sides

HUMMINGBIRDS Family Trochilidae

Black-chinned Hummingbird, male (AZ, Apr.)

Hummingbirds, or "hummers" as they are often known, are the smallest of all birds. Solely inhabiting the New World, they are very tiny and have incredible flight powers and brilliant, iridescent colors. These characteristics make field identification problematic, but hummers are readily attracted to sugar-water feeders, where they can be studied. In general, head and tail patterns, bill shape and color, and vocalizations are the best markers. Some birds will remain unidentified. Relatively frequent hybridization among species adds a cautionary dimension.

Structure Hummingbirds have relatively long wings, with 10 primaries but only six secondaries; their usually ample tails have 10 rectrices; and their feet are tiny. Bills are slender, pointed, and disproportionately long, varying from straight to distinctly decurved or arched in profile. The male Bee Hummingbird from Cuba is the smallest bird in the world.

Behavior Flight is fast and acrobatic; the wings beat so fast that they are a blur to the naked eye. Hummingbirds can reverse their primaries while hovering, in effect rotating their wings through 180 degrees and enabling them to fly backward. Aggressive for their size, several species have spectacular dive displays related to the defense of feeding territories and perhaps to courtship. They hover at flowers (or feeders) where they probe for nectar (or sugar water); they dart, spritelike, in pursuit of flying insects, which they sometimes pirate from spider webs. All hummers lay two unmarked white eggs; the male plays no part in nesting. Voices are unmusical and include the simple chip call (given by perched and feeding birds), the warning call, and the flight chase call.

Some species are migratory (including most populations in North America).

Plumage North American hummingbirds have dichromatic plumage related to both age and sex; adult males are more brightly colored than females. Juveniles typically resemble adult females, but their upperparts have buff tipping in fresh plumage, and the sides of the upper mandible have tiny scratches, or grooves (adult bills are smooth), that banders use for in-hand aging. Almost all species are metallic green above, but underparts vary greatly in color and pattern. The most common and widespread North American species are often termed the small gorgeted hummingbirds, for the iridescent throat panels, or gorgets, of the adult males, which vary from ruby red and magenta rose to violet and flame orange. Some of the large species have emerald green or royal blue gorgets. Molt occurs mostly on nonbreeding grounds. Thus, immature males of the small gorgeted species migrate south in female-like plumage and return in adultlike plumage. Because flight is integral to their existence, hummingbirds need to molt their wing feathers gradually; primary molt may require four to five months. The ninth (next-to-outermost) primary is molted last, rather than the straight P1–P10 sequence typical of most birds. Iridescence is due to the interference of reflected light; a gorget can change in appearance from blackish to flame orange in a split second. Many species show discolored (usually yellow or whitish) throats or crowns from pollen gathered during feeding.

Distribution Hummingbirds achieve their greatest diversity in the Andes of South America, from Colombia to Ecuador. At least one species can be found almost anywhere between southern AK and Tierra del Fuego. The birds occur in basically all habitats that support flowering plants; 17 species have nested in North America. Seven other species have occurred in North America as rare to accidental visitors from Mexico and the Caribbean.

Taxonomy Species-level and especially genus-level taxonomy, which is under constant revision, is improving. Worldwide, 338+ species are recognized in up to 105 genera, but much remains to be learned about many taxa in South America. Within Middle and North America, there are 105 to 115 species depending on taxonomy, with 24 species in 14 genera reported north of Mexico. Many species described in the past have proven to be unique hybrids.

Conservation Long-term data on population sizes and trends are sparse, especially where rainfall and flower abundance are highly variable. BirdLife International codes: 20 NT, 11 VU, 19 EN, 10 CR, 1 CR (PE), and 2 EX.

Parts of a Hummingbird

Hummingbirds have 10 primary feathers (P1–P10). The inner six primaries (P1–P6) of some species have distinctive tip shapes. Note the shape of the tip of P10 (the longest primary) and its relation to the tip of the tail. Hummingbirds have 10 tail feathers (rectrices), R1–R5, on each side. Differences in the shape and pattern of those feathers are best observed in photos of the spread tail. Bill length and curvature are important structural field marks.

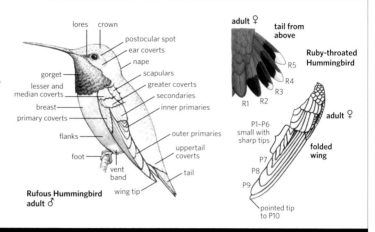

lores   crown
postocular spot
ear coverts
nape
gorget
scapulars
lesser and median coverts
greater coverts
breast
secondaries
primary coverts
inner primaries
flanks
outer primaries
foot
uppertail coverts
vent band
tail
Rufous Hummingbird adult ♂   wing tip

adult ♀   tail from above
Ruby-throated Hummingbird
R5
R4
R3
R1   R2
P1–P6 small with sharp tips
adult ♀
folded wing
P7
P8
P9
pointed tip to P10

Genus *Colibri*

MEXICAN VIOLETEAR   *Colibri thalassinus*   MEVI   ■ 3

This fairly large and overall green hummingbird is a rare visitor to the central and eastern US from the pine-oak highlands of Mexico. The AOS recently split the Green Violetear into this species and Lesser Violetear (*C. cyanotus*, Costa Rica to Argentina). Monotypic. L 4.2–4.7" (11–12 cm) Bill 18–22 mm
Identification Mexican Violetear has a short and slightly decurved bill and a fairly long, broad tail that is slightly notched. Overall dark green color, with distinctive violet-blue auricular and breast patches. **ADULT MALE:** Brighter overall with an extensive, purplish blue chest patch and auricular patches, a golden green crown, a longer, more strongly notched tail. **ADULT FEMALE:** Duller overall than male, with a smaller and less purplish

chest patch and smaller auricular patches, a bronzy green crown, and a shorter, less notched tail. **IMMATURE:** Distinguished from adult by its duller upperparts (with fine cinnamon tips when fresh) and dull bluish green throat and chest with patchy iridescent green and blue feathers.
Similar Species Unlikely to be confused with other N.A. hummingbirds but, as with all extralimital hummers, beware the possibility of hybrids that might resemble violetears, escapes of the southern subspecies, or similar species. Distinguished from Lesser Violetear (*C. cyanotus*) by larger size and violet-blue chest patch.
Voice CALL: Hard short rattles and clipped *chip* notes, both given from a perch and in flight. **SONG:** A metallic, mostly disyllabic chipping, often repeated monotonously with a jerky cadence from an exposed perch; immatures may give more varied

series, including buzzes and rattles.
Status & Distribution Central Mexico to north-central Nicaragua. **RARE STATUS:** Very rare but annual to TX, mainly in late spring and summer. Casual elsewhere (mainly May–Aug., extremes Apr.–Dec.) in central and eastern N.A., north to MI and ON. Accidental in NM, CA, and AB.
Population Apparently stable; may benefit from some habitat disturbance.

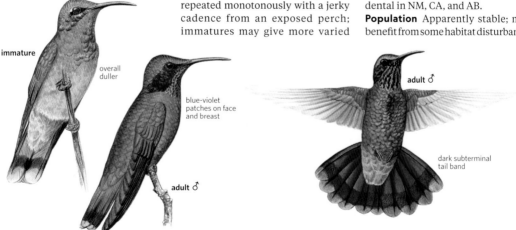

immature
overall duller
blue-violet patches on face and breast
adult ♂
adult ♂
dark subterminal tail band

Genus *Anthracothorax*

GREEN-BREASTED MANGO *Anthracothorax prevostii* GNBM ■ 4

This striking and ostensibly unmistakable large hummingbird is a casual visitor to the US from eastern Mexico. Polytypic (4 sspp.; nominate in N.A.). L 4.5–4.8" (11.5–12 cm) Bill 24–30 mm
Identification Green-breasted Mango has a thick and arched black bill and a broad tail. This species' size, bill shape, and overall plumage patterns are unlike any other N.A. hummer.

long decurved bill

rufous vertical streak borders green sides and flanks

immature
prevostii

ADULT MALE & SOME FEMALES: Plumage distinctive: deep green overall with a black throat and mostly purple tail.
ADULT FEMALE: Note blackish median throat stripe, becoming deep green on white underparts and purple tail base.
IMMATURE: Resembles female but has a thin strip of cinnamon mottling bordering the green sides of the chin down though the flanks.
Similar Species Unlikely to be confused, but beware the possibility of other mango species wandering from the Caribbean or from M.A., including the similar Black-throated Mango (*A. nigricollis*; western Panama to S.A.) and the Veraguan Mango (*A. veraguensis*; southwest Costa Rica and southern Panama).
Voice Not very vocal. **CALL:** A high, sharp *sip* and a buzzy *pzzt*, as well as fairly hard, ticking *chip* notes and

shrill tinny twitters in interactions.
Status & Distribution Northeast Mexico to northwest Panama; also found in northern S.A. **RARE STATUS:** Casual visitor (mainly in autumn and winter) to coastal lowlands of south TX (12+ records); accidental in fall and winter in GA, NC, and WI.
Population Stable.

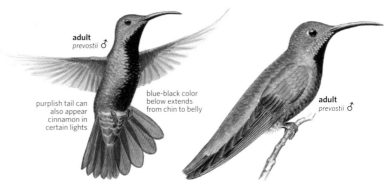

adult
prevostii ♀

dark iridescent vertical stripe down center of underparts

white tail tips

adult
prevostii ♂

purplish tail can also appear cinnamon in certain lights

blue-black color below extends from chin to belly

adult
prevostii ♂

Genus *Eugenes*

RIVOLI'S HUMMINGBIRD *Eugenes fulgens* RIHU ■ 2

This large hummer of Mexico and northern M.A. is found in the southwestern mountains of the US. In 2017, the AOS split Magnificent Hummingbird into this species and Talamanca Hummingbird (*E. spectabilis*), found in Costa Rica and western Panama. Monotypic. L 4.7–5.3" (12–13.5 cm) Bill 27.5–30 mm
Identification Rivoli's has a long black bill and a slightly notched tail.
ADULT MALE: Distinctive; often looks all-dark, since violet crown and green gorget rarely seem to catch the light.
ADULT FEMALE: Throat and underparts pale gray, mottled green on sides. **IMMATURE MALE:** Resembles female but throat and underparts

strongly mottled green, white malar streak more distinct, central throat usually has a large, iridescent green blotch. **IMMATURE FEMALE:** Resembles immature male but throat lacks iridescent green patches.
Similar Species Note large size, long bill, and face pattern. The female resembles the comparably sized Blue-throated Hummingbird. Blue-throated is easily separated by the black uppertail coverts and tail with sharply contrasting white tail tips and by a stronger face pattern and a slightly longer bill; calls also strongly differ.
Voice CALL: A fairly loud, sharp *chik* or *tsik*, given from perch and in flight. Also a higher *piik*, and males, at least,

give fairly hard, short rattles, *trrirr* and *trrrr ch-chrr* that may run into prolonged squeaky chattering. Aggression call a squeaky, slightly bubbling, accelerating chatter. **SONG:** A fairly

soft, slightly buzzy gurgling warble.
Status & Distribution Southwest US to northern Nicaragua. **BREEDING:** Fairly common (Mar.–Oct.) in pine-oak and oak highlands. **WINTER:** Rare, mainly southeast AZ. **RARE STATUS:** Casual or accidental west to CA and BC, north (mainly late summer) to NV, UT, CO (has bred), WY, KS, AR, MN, and MI, and eastward (mainly winter) to Gulf states, GA, FL, and VA.
Population Reported as increasing in its native range.

purple crown

adult ♂

blue-green throat

immature ♂

limited adult coloration by late summer

white spot

adult ♀

extensive black below

greenish back and tail

mottled underparts

large size

tail from below

often looks dark overall

adult ♀

juvenile ♀

scaly appearance

limited white on tips of extensively green-based outer tail feathers

---

Genus *Heliomaster*

PLAIN-CAPPED STARTHROAT *Heliomaster constantii* PCST ▪ 4

This large, aggressive hummingbird from Mexico often makes prolonged flycatching sallies over streams. Like many tropical hummers it shows little age/sex variation. Polytypic (3 sspp.; *pinicola* in N.A.). L 4.7–5" (12–13 cm) Bill 33–37 mm
Identification Very long, black bill; note throat and face pattern, white rump patch, and Black Phoebe-like call. **ADULT:** Throat patch is dark sooty with iridescent reddish mottling (often hard to see). Tail has bold white tips to all but central pair of rectrices; tip to R5 is broadest. **JUVENILE:** Resembles adult but throat patch has little or no red.
Similar Species Blue-throated Hummingbird has white rump patch and black uppertail coverts and tail; face pattern less striking.
Voice CALL: A sharp, fairly loud *peek!*, given from perch and especially in flight. **SONG:** A series of *chip* notes interspersed with varied notes.
Status & Distribution Northwest Mexico to Costa Rica. **RARE STATUS:** Casual visitor (May–Nov., mainly late summer) to southern AZ.
Population Stable; fairly common to common in Mexico.

large size

some red on throat

distinct white postocular and malar stripes and white on lower back

very long bill

pinicola tail from above

pinicola

grayish brown throat with little red present

juvenile *pinicola*

prominent white stripe

secondaries old and worn

P1–P6 new

molting adult *pinicola* (July–Sept.)

P7 growing

P8 missing

P9–P10 old, pale brown

prominent white tail tips

Genus *Lampornis*

AMETHYST-THROATED HUMMINGBIRD   *Lampornis amethystinus*   ATHU ■ 5

This large *Lampornis* hummingbird from montane forests from central Mexico to Honduras closely resembles its relative, the Blue-throated Hummingbird. Polytypic (2 sspp.; *amethystinus*, male with rose red gorget, from most of range; male *margaritae* from southwest Mexico [Michoacán to Oaxaca] has bluish purple throat). L 4.7–5" (12–13 cm) Bill 20–23 mm

Identification Large size and white facial stripes similar to Blue-throated. Sexes similar except adult males have a rose red gorget, buffy in females; immature male buffy with some rose pink. Underparts are dark gray with some greenish on the sides and flanks. Uppertail coverts and tail blackish like Blue-throated, but pale tail tips narrow and grayish.

Similar Species Like Blue-throated Hummingbird but slightly smaller with darker gray underparts. Best separated by smaller grayish, not white, tail tips. Rose pink gorget on male distinctive from the blue gorget on an adult male Blue-throated, but often appears dark.

Voice CALL: A high sharp *tsip tsip.* **SONG:** A long, dull staccato rattle.

Status & Distribution In native range, found in mesic montane forests from nearly 3,000 ft to over 9,000 ft. **RARE STATUS:** Accidental to N.A. with two records (*amethystinus*) of males: Le Fjord-du-Saguenay, QC, 30–31 July 2016 and Davis Mts., West TX, 14–15 Oct. 2016.

Population Considered stable.

pink gorget (often appears dark)

buffy gorget

amethystinus ♀

amethystinus ♂

white tail tips smaller than Blue-throated

BLUE-THROATED HUMMINGBIRD   *Lampornis clemenciae*   BTHH ■ 2

The largest hummingbird in N.A., the Blue-throated Hummingbird favors shady understory and edge in watered pine-oak and oak-sycamore canyons. This and Amethyst-throated—along with five C.A. species—are in the same genus; most are named mountain-gems and soon likely all will be. Polytypic (3 sspp.; 2 in N.A.). L 4.8–5.3" (12–13.5 cm) Bill 21.5–26 mm

Identification Whitish face stripes, bold white corners on tail, and high-pitched call are good field marks. **ADULT MALE:** Blue gorget can be hard to see. Outer two pairs of rectrices have large white tips. **ADULT FEMALE:** Throat plain dusky gray, white tips to outer three pairs of rectrices; some birds have a few blue throat feathers. **IMMATURE:** Resembles female but upperparts have buffy cinnamon edgings in fall, and lower mandible base often pinkish on younger birds. Males have irregular blue gorget patch restricted to central throat.

Geographic Variation TX subspecies *phasmorus* is slightly greener above and on average shorter billed than *bessophilus* from AZ.

Similar Species Female and immature Rivoli's Hummingbird longer billed with a greenish tail that has only small white corners; more mottled underparts; and a smacking chip call. Rivoli's face pattern typically dominated by a bold white postocular spot, but some can be similar to poorly marked Blue-throated. See the similarly plumaged

phasmorus

bessophilus

clemenciae

Amethyst-throated Hummingbird.
Voice CALL: A high, penetrating *siip* or *siik*, given from perch and in flight; less often a fuller *tsiuk.* **SONG:** Apparently a repetition of call, *siip siip siip,* though more complex vocalizations have been reported.

Status & Distribution Mexico to southwest US. **BREEDING:** Uncommon to fairly common (mainly Apr.–Sept.) in humid montane forested canyons. **WINTER:** Mexico. Casual in southeastern AZ. **RARE STATUS:** Casual north (mainly late summer) to UT and CO, and eastward (mainly winter) to southern LA and CA.

Population Stable.

blue throat often looks dark

adult *bessophilus* ♂

adult *bessophilus* ♂

all plumages with prominent white facial stripes

adult *bessophilus* ♀

large size

blackish uppertail coverts and tail

extensive white on tips of black-based tail feathers in all plumages

smooth gray underparts in all plumages

Genus *Calliphlox*

BAHAMA WOODSTAR   *Calliphlox evelynae*   BAWO ■ 5

This species is an endemic resident of the Bahamas. The birds on Great Inagua are now a separate species, Inagua Woodstar (*C. lyrura*). Monotypic. L 3.4–3.7" (8.5–9.5 cm) Bill 16–17 mm **Identification** Bill slightly arched; fairly long, forked tail, projecting past wing tip on perched birds. **ADULT MALE:** Unmistakable. **ADULT FEMALE:** Outer rectrices have cinnamon bases and tips. **IMMATURE MALE:** Similar to female, but throat usually has some purplish spots; tail slightly longer, more forked.

Similar Species Female and immature Rufous Hummingbirds have straighter bills and shorter tails with bold white tips to outer rectrices.

Voice CALL: A high, fairly sharp chipping *tih* or *chi*, often doubled.

Status & Distribution Bahamas. **RARE STATUS:** Casual to south and central FL (five records; Jan, Apr.–Oct); extraordinary was an adult male (photographed) at Denver, Lancaster Co., PA (20–24 Apr. 2013).

Population Stable.

adult ♂
magenta gorget
deeply forked tail with cinnamon stripes
rufous
adult ♂
older immature ♂
throat dusky with some cinnamon and a few magenta spangles
R1 very short
long forked tail, extensive cinnamon on inner webs
unspotted pale throat
bold white collar
adult ♀
cinnamon-buff tips and base
adult ♀
white collar and rufous on sides and flanks very suggestive of Rufous/Allen's
tail similar to adult male but shorter, and has a longer R1
gorget often appears dark
cinnamon underparts, mottled green on sides
long tail extends beyond primary tips
adult ♂

Inagua Woodstar

Genus *Calothorax*

LUCIFER HUMMINGBIRD   *Calothorax lucifer*   LUHU ■ 2

This small hummer is a summer inhabitant of mountain desert canyons, especially in West TX. Monotypic. L 3.5–4" (9–10 cm) Bill 19–23 mm **Identification** Bill is long and arched; fairly long, forked tail projects past wing tip on perched birds. Pale postocular stripe that broadens toward sides of neck with dusky auricular stripe of females is diagnostic. **ADULT MALE:** Expansive magenta gorget; longer tail than female. **ADULT FEMALE:** Throat is pale buff to whitish; outer rectrices have rufous bases and white tips. **IMMATURE MALE:** Like female, but throat usually has more magenta-rose spots; tail longer and more forked with narrower outer rectrices.

Voice CALL: A fairly hard, slightly smacking *chih* or *chi*.

Status & Distribution Mexico to southwest US. **BREEDING:** Uncommon and local (mainly Apr.–Oct.) in arid mountain canyons. **RARE STATUS:** Casual visitor away from traditional sites in AZ, NM, and TX.

Population Small but stable in US.

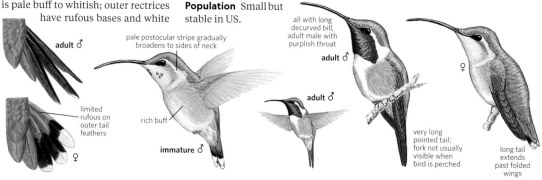

adult ♂
pale postocular stripe gradually broadens to sides of neck
all with long decurved bill, adult male with purplish throat
adult ♂
♀
limited rufous on outer tail feathers
♀
rich buff
adult ♂
immature ♂
very long pointed tail; fork not usually visible when bird is perched
long tail extends past folded wings

Genus *Archilochus*

Eastern and western counterparts are the Ruby-throated and Black-chinned. Diagnostic of the genus are the relatively narrow inner six primaries. Adult males have shield-shaped gorgets with black chins. Females/immatures have mostly plain underparts with little or no buff wash, no rufous in tail. Primary molt typically starts in fall or later (Sept.–Jan.).

RUBY-THROATED HUMMINGBIRD *Archilochus colubris* RTHU ■ 1

The only hummer seen regularly in much of the East. Monotypic. L 3.2–3.7" (8–9 cm) Bill 14–19 mm
Identification Primaries tapered. Tail forked to double-rounded. **ADULT MALE:** Solid ruby red gorget with black chin and face. **ADULT FEMALE:** Throat whitish. Sides and flanks often washed buff. **IMMATURE MALE:** Resembles adult female but throat usually with one or more ruby spots; tail slightly longer and more forked. Complete molt in winter produces plumage like adult male. **IMMATURE FEMALE:** Resembles adult female but fall plumage fresher.
Similar Species Black-chinned has blunter primaries. Adult male has black throat with violet lower band, shorter tail. Females and immatures similar to Ruby-throated but generally duller green above (especially crown) and dingier below (Ruby-throated bright emerald above and whiter below). Black-chinned tail less forked; often pumps tail strongly while feeding (Ruby-throated usually holds tail fairly still); also note triangular dark patch on lores of female and immature male Ruby-throats. Problem identifications best confirmed by checking details of primary shape. See Costa's and Anna's Hummingbirds.
Voice CALL: Slightly twangy or nasal chips, *chih* and *tchew*, given in flight and perched. Also varied twittering series. Indistinguishable from Black-chinned, but generally lacks strongly buzzy or sharp, smacking quality of Anna's and *Selasphorus*, and distinct from high, tinny chips of Costa's.
Status & Distribution Breeds N.A., winters Mexico to western Panama. **BREEDING:** Common in woodland, gardens, etc. **MIGRATION:** Mainly Mar.–May, Aug.–Oct. **WINTER:** Rare (mainly Nov.–Mar.) in the Southeast. **RARE STATUS:** Casual in the West and to AK.
Population Breeding Bird Survey data suggest increases in some areas.

adult ♀

R5 rounded tip

adult ♀

P1–P6 small with sharp tips

P7
P8
P9

pointed tip to P10

bright green above, including on crown

blackish primaries

forked tail

ruby throat

black under bill and eye

adult ♂

adult ♂

adult ♀

throat often looks blackish

immature males often develop a few red spots by Sept.

immature ♂

BLACK-CHINNED HUMMINGBIRD *Archilochus alexandri* BCHU ■ 1

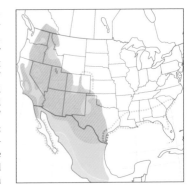

The western counterpart of Ruby-throated, Black-chinned regularly pumps its tail. Monotypic. L 3.3–3.8" (8.5–9.5 cm) Bill 16–22 mm
Identification Best marks for all ages are narrow inner primaries and blunt primary tips. Age/sex differences as Ruby-throated except as noted. **ADULT MALE:** Black chin with violet purple throat band. **ADULT FEMALE:** Black-violet spots are occasionally present on throat. **IMMATURE MALE:** Lower throat usually has black-violet spots.
Similar Species Black-chinned is often confused with Anna's and Costa's, which are chunkier and disproportionately bigger headed, shorter billed, and shorter tailed; lack narrow inner primaries of *Archilochus*; and molt wings in summer. Larger female/immature Anna's has more-mottled underparts; throat often with rose-red spots; wags tail infrequently. Female/immature Costa's slightly smaller; face often plainer but shows pale narrowly coming up from the sides of the neck to the eye; wing tips often fall beyond tail tip at rest (shorter than

tip on Black-chinned). Anna's call is a smacking chip, Costa's is a high series of tinny *tink* notes, both very unlike low-pitched notes of Black-chinned. See Ruby-throated.

Voice CALL: Indistinguishable from Ruby-throated, but distinct from high tinny chips of Costa's; distinctive male wing buzz in flight.

Status & Distribution Western N.A. to northwest Mexico. **BREEDING:** Common in riparian woodlands, foothills. **MIGRATION:** Mainly Mar.–May, Aug.–Sept. **WINTER:** Mexico. Rare (mainly Oct.–Mar.) in Southeast; casual in coastal Southern CA in winter; casual in Northeast (fall).

Population Stable.

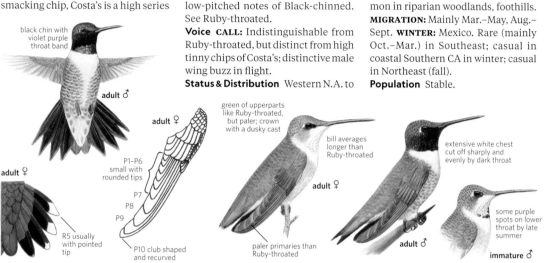

black chin with violet purple throat band

adult ♂

adult ♀

adult ♀

P1–P6 small with rounded tips

P7
P8
P9

R5 usually with pointed tip

P10 club shaped and recurved

green of upperparts like Ruby-throated, but paler; crown with a dusky cast

bill averages longer than Ruby-throated

adult ♀

paler primaries than Ruby-throated

extensive white chest cut off sharply and evenly by dark throat

adult ♂

some purple spots on lower throat by late summer

immature ♂

HELMETED HUMMINGBIRDS Genus *Calypte*

This genus comprises two western species of scrub habitats: Anna's and Costa's. Small, fairly chunky hummers with relatively large heads and short tails; adult males have an iridescent helmet (i.e., gorget and crown), with gorgets elongated at corners. Females and immatures have dingy underparts with no buff wash on the sides, no rufous in tail.

ANNA'S HUMMINGBIRD *Calypte anna*   ANHU ▪ 1

This hummer is a familiar species in West Coast gardens, where it is present year-round. Monotypic. L 3.5–4" (9–10 cm) Bill 16–20 mm
Identification Tail slightly rounded to double-rounded. **ADULT MALE:** Rose (fresh) to orange-red (worn) gorget and crown. **ADULT FEMALE:** Throat and underparts spotted and mottled dusky to bronzy green, median throat blotched rose red. **IMMATURE MALE:** Resembles adult female but throat and crown usually with more scattered rose spots; white tail tips narrower. Complete summer molt produces plumage like adult male by late fall. **IMMATURE FEMALE:** Resembles adult female but throat often lacks rose spots.
Similar Species Costa's smaller (obvious in direct comparison), and males readily identified (beware occasional hybrids, which look more like male Costa's, sound more like Anna's). Female/immature Costa's disproportionately

juvenile ♂

black spike on R5

adult ♀

head often looks blackish

shining rose red crown and gorget

dark gray tail often looks translucent

inner vane of R5 vibrates in display dive, producing a loud, sharp *pop*

adult ♂

adult ♂

scattered rose red on head

adult ♀

very short greater coverts

P1
P2
P3
P4
P5
P6
P7
P8
P9
P10

primary tips evenly graduated and spaced

usually with some rose red in center of throat

adult ♀

dusky green spots on throat

immature ♀

dingy gray below with extensive green flanks

immature ♂

new inner primaries blackish, outer primaries old and brownish

longer billed but shorter tailed, often best told by call: high, tinny *tink* calls and twitters distinct from Anna's. Costa's generally plainer on throat and underparts, without dusky throat spotting. See Black-chinned.

Voice CALL: A slightly emphatic to fairly hard *tik* or *tih* and a more smacking *tsik*, in flight and perched. In flight chases, a rapid-paced, slightly buzzy twittering, *t-chissi-chissi-chissi*, and variations. **SONG:** A high-pitched, wiry to lisping squeaky warble from perch, often prolonged and repeated with pulsating succession. Year-round.

Status & Distribution Western N.A. to northern Mexico. **BREEDING:** Common (Dec.–June) in scrub, gardens, etc.

DISPERSAL/MIGRATION: Some late summer movement upslope to mountains. Local movements complex.
RARE STATUS: Rare north to southeastern AK and casual to southern AK (mainly fall). Casual to the East (fall and winter); accidental NL.
Population Dramatic range increase since the mid-1930s.

COSTA'S HUMMINGBIRD *Calypte costae*   COHU ■ 1

Costa's is a spectacular small hummer known primarily from southwest deserts. Monotypic. L 3–3.4" (7.5–8.5 cm) Bill 16–20 mm
Identification Tail rounded to slightly double-rounded. **ADULT MALE:** Violet gorget and crown, tail lacks white tips. **ADULT FEMALE:** Throat and underparts dingy white to pale gray, throat sometimes blotched violet. **IMMATURE MALE:** Like adult female but upperparts fresher Feb. to July, with fine buff tips; auriculars darker; white tail tips narrower; usually soon develops some purple on throat that is framed broadly in white. Complete summer molt produces plumage like adult male by early winter. Like Black-chinned, but unlike Anna's, this species frequently pumps tail.

Similar Species Anna's larger (obvious in direct comparison); males readily identified (beware occasional hybrids; see Anna's). Female/immature Anna's disproportionately shorter billed but longer tailed; often best told by call, a tinny *tink* and buzzy twitters. Anna's generally more spotted on throat and underparts. See Black-chinned. Problem identifications best confirmed by using primary shape and voice.

Voice CALL: A high, slightly tinny, fairly soft *tink* or *ti*. **SONG:** A very high-pitched, thin, drawn-out, whining whistle, *tsi ssiiiiiu*, usually given in looping display dives.

Status & Distribution Western US to northwest Mexico. **BREEDING:** Fairly common (Feb.–July) in desert washes, dry chaparral. **DISPERSAL/**

MIGRATION: Mainly Feb.–May and June–July; local movements complex.
WINTER: Uncommon to rare and local north to southern NV. **RARE STATUS:** Casual north to southern AK and MT, east to KS and TX.
Population Stable.

adult ♂
extended gorget often looks black
adult ♀
adult ♀
purple head and long extended gorget
adult ♂
adult ♀
pale postocular stripe connects to side of neck
P1
P2
P3
P4
P5
P6
P7
P8
P9
P10
some purple on throat and head
overall very similar to female Black-chinned, often appears chunkier
immature ♂
primary tips evenly graduated and spaced

Genus *Atthis*

BUMBLEBEE HUMMINGBIRD *Atthis heloisa*   BUHU ■ 5

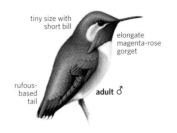

tiny size with short bill
elongate magenta-rose gorget
rufous-based tail
adult ♂

This tiny hummingbird is endemic to montane forests of Mexico, where it feeds inconspicuously with a relatively slow and deliberate, insect-like flight. Closely related to the Wine-throated Hummingbird (*A. ellioti*) of southern Mexico and northern C.A. Polytypic (2 sspp.; N.A. specimens *margarethae*). L 2.7–3" (7–7.5 cm) Bill 11–13 mm
Identification Bill is medium-short,

and the rounded to double-rounded tail projects beyond wing tips on perched birds. Both sexes have a femalelike tail with rufous bases and white tips to outer rectrices. **ADULT MALE:** Distinctive, with an elongated magenta-rose gorget. **ADULT FEMALE:** Throat whitish with lines of bronzy-green flecks, and a whitish forecollar contrasts variably with cinnamon

sides. White tail tips are bolder than male. **IMMATURE MALE:** Resembles adult female but throat flecked more heavily with bronzy green, often with rose-pink spots or streaks. **IMMATURE FEMALE:** Resembles adult female but throat has finer dusky spots. Tips of outer tail feathers washed buffy cinnamon when fresh.
Similar Species Female and immature *Selasphorus* hummingbirds are larger

with longer bills, more graduated tails, quick flight, and louder and harder calls. Calliope is slightly larger with a short, mostly black tail that falls equal with or shorter than wing tips at rest, and a quicker, darting flight.
Voice Quiet high chips. Adult males have strong insectlike wing buzz, especially in display flights.
Status & Distribution Endemic to Mexico. **RARE STATUS:** Accidental in

southeastern AZ (two female specimens collected on an ornithological expedition to the Huachuca Mts. in 1896: one in Miller Canyon on 29 June, the other on 30 June in Brown's Canyon, a side canyon off Ramsey Canyon). The specimens are still extant but the date (2 July) and location (Ramsey Canyon) on the two tags are in error.
Population Appears to be stable within its large Mexican range.

SELASPHORUS HUMMINGBIRDS Genus *Selasphorus*

This genus comprises seven species, four in western N.A. and two or three more in C.A. These small hummers have fairly long, graduated tails that project beyond the wing tips at rest (except Calliope). Males' power-dive displays from heights of 10 to 30 ft are species-specific. Females are green and rufous, with cinnamon body sides and tail bases.

BROAD-TAILED HUMMINGBIRD　*Selasphorus platycercus*　BTHU ◾ 1

An uncommon to common species of western mountains, where the male's diagnostic, loud, cricketlike wing trill is a characteristic sound. Monotypic. L 3.5–4" (9–10 cm) Bill 16–20 mm
Identification Tail weakly graduated. Pale eye ring in all plumages. **ADULT MALE:** Rose red gorget with pale chin and face. Often detected by wing trill. **ADULT FEMALE:** Throat whitish with variable lines of bronzy-green flecks, sometimes one or more rose spots; sides of neck and underparts variably washed warm buff. **IMMATURE MALE:** Resembles adult female but upperparts fresher in fall, with fine buff tips; throat usually flecked fairly heavily with bronzy green, often with rose-pink spots; tail averages more rufous at base. Complete molt in winter and spring produces plumage like adult male. **IMMATURE FEMALE:** Resembles adult female, but upperparts fresher in

fall, with fine buff tips; tail averages less rufous at base.
Similar Species Chronically confused with female/immature Rufous/Allen's Hummingbirds, which are slightly smaller (noticeable in comparison) with more strongly graduated tails that have a more tapered tip. Rufous/Allen's typically have a whiter forecollar contrasting with brighter rufous sides, and lack whitish eye ring often shown by Broad-tailed. The upperparts of Broad-tailed appear a little more bluish green, and uppertail coverts and tail base have less rufous (adult female Rufous can be all green); and their chip calls are slightly higher pitched and more metallic. Also see female and immature Calliope Hummingbird.
Voice Generally higher pitched than Rufous/Allen's. **CALL:** A slightly metallic, sharpish *chip* or *chik*, often dou-

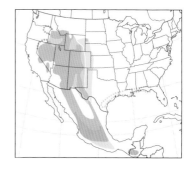

bled, *ch-chip* or *chi-tik*. Adult male's wing trill diagnostic.
Status & Distribution BREEDING: Western US to Guatemala. Uncommon to common (Apr.–Aug.) in mountains. **MIGRATION:** Mainly Apr.–May, Aug.–Sept. (a few to western Great Plains). **WINTER:** Mainly Mexico. Casual to very rare (mainly Nov.–Apr.) in the Southeast. **RARE STATUS:** Casual north to BC and west to coastal Southern CA. Accidental NJ.
Population Difficult to evaluate, but slight declines reported in some areas.

adult ♀ — less rufous at tail base than Rufous/Allen's

thin rufous edges to R2–R3 show on folded tail

adult ♀

long tail extends beyond wing tips in all plumages

bluish tint to green upperparts

adult ♂

evenly spotted throat

blended buffy underparts and usually buffy collar

loud metallic wing trill produced by modified tips to P9 and P10

adult ♂

molting immature ♂ (Mar.)

rose red gorget with narrow white line extending under bill

retained juvenal R5

new R4 (growing)

new R1–R3 (adult)

RUFOUS HUMMINGBIRD *Selasphorus rufus* RUHU ■ 1

This common summer hummingbird of the Northwest is the western species most often found in the East (in fall and winter). Note that all except (most) adult male Rufous Hummingbirds are rarely separable in the field from Allen's Hummingbird, so many observations are best termed Rufous/Allen's. In dive display (also given in migration and by immatures) male climbs to a start point, then dives with a slanted J-form trajectory, typically followed by a short, horizontal fluttering flight before climbing to repeat the dive. Aggressive at feeding locations. Hybridizes to an unknown degree with Allen's in southwest OR and northwest CA. Monotypic. L 3.2–3.7" (8–9 cm) Bill 15–19 mm

Identification Adult males often detected by wing buzz, which, like other *Selasphorus*, is produced only in direct flight, not when hovering. **ADULT MALE:** Flame orange gorget; rufous back often has some green spotting, can rarely be solidly green. **ADULT FEMALE:** Throat whitish with lines of bronzy-green flecks strongest at corners, and typically a central splotch of red. **IMMATURE MALE:** Resembles adult female but upperparts fresher in fall, with fine buff tips; throat usu-

ally flecked fairly heavily, and with red spots. Rectrices average narrower and with more rufous at bases, and white tips to outer rectrices narrower. Complete molt in winter produces plumage like adult male. **IMMATURE FEMALE:** Resembles adult female but upperparts fresher in fall, with fine buff tips; rectrices average broader; throat evenly flecked and with no (rarely a few) red spots.

Similar Species Female and immature Allen's Hummingbird safely distinguished only in the hand by narrower outer rectrices relative to age and sex. Adult male Allen's has green back (like very small percentage of Rufous). Male Allen's display dives are U-shaped, not J-shaped, and can be given by immatures in fall and winter. Females of *sedentarius* Allen's may have less rufous on flanks and instead have more green mottling there. See female and immature Broad-tailed Hummingbird.

Voice CALL: A fairly hard ticking or clicking *tik* or *chik*, often doubled or trebled, *ch-tik* or *ch-ti-tik*. **ALARM CALL:** A slightly squeaky buzz, *tssiur* or *tsirr*, and squeaky chippering in interactions. Adult male's wing buzz often draws attention; stuttering *ch-ch-ch-*

ch-chi at pullout of dive is diagnostic. Immature males make species-specific dives without the sound effects.

Status & Distribution BREEDING: Northwestern N.A. Common (Mar.–July) in open woodlands and parks. **MIGRATION:** Mainly late Feb.–early May, late June–Sept. Rare (mainly July–Nov.) in the East. **WINTER:** Mexico. **RARE STATUS:** Rare (mainly Oct.–Mar.) in the Southeast and casual in southwestern CA.

Population Near Threatened. Breeding Bird Survey data from Pacific Northwest indicate statistical, albeit unexplained, declines even though still abundant.

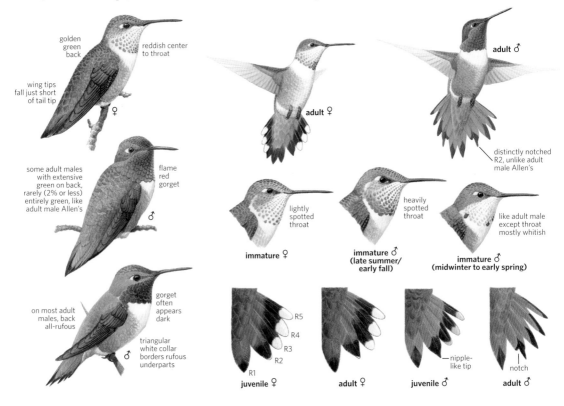

golden green back

reddish center to throat

wing tips fall just short of tail tip

♀

some adult males with extensive green on back, rarely (2% or less) entirely green, like adult male Allen's

flame red gorget

♂

on most adult males, back all-rufous

gorget often appears dark

triangular white collar borders rufous underparts ♂

adult ♀

lightly spotted throat

immature ♀

heavily spotted throat

immature ♂ (late summer/ early fall)

adult ♂

distinctly notched R2, unlike adult male Allen's

like adult male except throat mostly whitish

immature ♂ (midwinter to early spring)

R5

R4

R3

R2

R1

juvenile ♀

adult ♀

nipple-like tip

juvenile ♂

notch

adult ♂

ALLEN'S HUMMINGBIRD   *Selasphorus sasin*   ALHU ■ 1

Virtually endemic to CA as a breeding bird, this species is one of the earliest migrants in N.A.: Males return from Mexico in Jan. and head south in June! In dive display (also given in migration and by immatures) male climbs to start point and then makes repeated, often fairly low, U-shaped dives followed by a single high climb and steep dive. Polytypic (2 sspp.; both in N.A.). L 3.2–3.5" (8–9 cm) Bill 15–21 mm
Identification Rarely separable in field from Rufous except by breeding range (in which Rufous is a common migrant), and many birds are best termed Rufous/Allen's. Age-related variation and plumage sequences are as Rufous (see account) with some exceptions. **ADULT MALE:** Back always green, with rufous uppertail coverts. **ADULT FEMALE:** Uppertail coverts typically rufous (ironically, often green on Rufous). Flanks paler and mottled bronzy green on *sedentarius*.
Geographic Variation Migratory subspecies *sasin* breeds from southwestern OR to Southern CA. It may intergrade with expanding population of slightly larger and longer-billed resident subspecies *sedentarius*, originally confined to Channel Is., CA, but

now breeding on mainland. Males of the two subspecies are identical in the field, but female *sedentarius* has paler rufous and extensively green-mottled sides that create a less-demarcated vest (*sasin* females have brighter rufous, well-demarcated sides).
Similar Species See Rufous Hummingbird. Very fine, almost wirelike outer rectrices of adult male Allen's may be appreciated in good views (e.g., when aggressive males spread their tail at a feeder), but identification best confirmed by photo or in-hand examination. See female and immature Broad-tailed Hummingbird (which rarely overlap in range with Allen's).
Voice CALL: Like Rufous. However, high, drawn-out, shrieky whine at pullout of adult male Allen's display dive is diagnostic (unlike stuttering of Rufous); immature males make species-specific dive displays but without the sound effects.
Status & Distribution Breeds western N.A., winters Mexico (nominate *sasin*). Subspecies *sedentarius* is a fairly common but local resident in Southern CA (north to Ventura Co.) and on the Channel Is. **BREEDING:** Common (Feb.–July) in open woodlands of

coastal belt. **MIGRATION:** Mainly mid-Jan.–early Mar., June–Aug., in fall, ranging east to southeastern AZ. **RARE STATUS:** Casual (mainly July–Jan.) in the East, mostly the Southeast.
Population Nonmigratory *sedentarius* is expanding its breeding range.

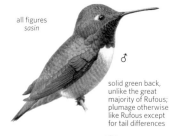

all figures
sasin

♂

solid green back, unlike the great majority of Rufous; plumage otherwise like Rufous except for tail differences

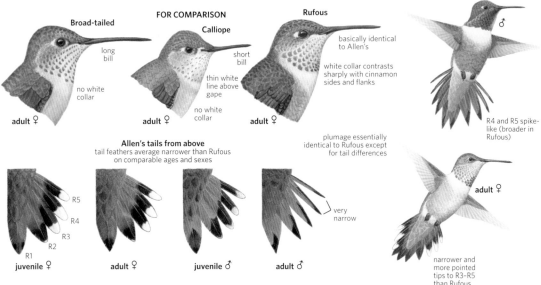

FOR COMPARISON

Broad-tailed

long bill

no white collar

adult ♀

Calliope

short bill

thin white line above gape

no white collar

adult ♀

Rufous

basically identical to Allen's

white collar contrasts sharply with cinnamon sides and flanks

adult ♀

plumage essentially identical to Rufous except for tail differences

♂

R4 and R5 spike-like (broader in Rufous)

Allen's tails from above
tail feathers average narrower than Rufous on comparable ages and sexes

R5
R4
R3
R2
R1
juvenile ♀

adult ♀

juvenile ♂

adult ♂

very narrow

adult ♀

narrower and more pointed tips to R3–R5 than Rufous

CALLIOPE HUMMINGBIRD   *Selasphorus calliope*   CAHU ■ 1

The smallest breeding bird in N.A., the Calliope Hummingbird is a summer resident of western mountain meadows, especially along wooded streams. Monotypic. L 3–3.2" (7.5–8 cm) Bill 12.5–16 mm

Identification Note relatively short and squared tail, which means that Calliope's wing tips often project slightly beyond its tail when perched; bill is medium-short. Adult males are distinctive; females and immatures are

buff-and-green and best identified by their small size, and relatively short, mostly black tail. Note whitish lores on many birds. **ADULT MALE:** Elongated gorget of rose stripes rarely looks solid; note white lores and eye ring. **ADULT**

FEMALE: Throat is whitish with variable lines of bronzy-green flecks and rarely one or more rose spots. Sides of neck and underparts are washed warm buff and lack a distinct white forecollar. Tail is slightly rounded (although it appears squared when closed). **IMMATURE MALE:** Resembles adult female but has fresher upperparts in fall, with fine buff tips. Throat is usually flecked more heavily with bronzy green and often has one or more rose pink spots or streaks; white lore area is often reduced or absent. A complete molt in winter produces an adultlike male plumage. **IMMATURE FEMALE:** Resembles adult female but upperparts are fresher in fall, with fine buff tips, and white lore area is reduced or absent. **Similar Species** Note short tail. Other female/immature *Selasphorus* are larger (obvious in direct comparison), with longer and slightly thicker

bills, longer and distinctly graduated tails that project beyond wing tips on perched birds. They are more aggressive, often uttering their louder calls and chasing other hummingbirds. However, the female Broad-tailed is remarkably similar in plumage to the Calliope; other than in size and structure and bill length, note Broad-tailed's longer tail, which is mostly green above. Rufous/Allen's Hummingbirds have brighter and more contrasting rufous body sides and much more rufous in the tail than Calliope.
Voice Often fairly quiet and inconspicuous. **CALL:** A relatively soft, high *chip*, often doubled, *chi* and *chi-ti*; sometimes given in a series. High, slightly buzzier chippering during interactions. Male's wing buzz can attract attention during flight displays.
Status & Distribution Breeds western N.A., winters southwest Mexico.

BREEDING: Fairly common (Apr.–Aug.) in mountain meadows and open forest. **MIGRATION:** Mainly Apr. (few late Mar.) to early May, July–Sept.; in fall, ranges east to western Great Plains and west TX. **WINTER:** Rare (mainly Oct.–Mar.) in the Southeast. **RARE STATUS:** Casual in late fall as far northeast as MN and ME.
Population Stable.

purple-carmine gorget stripes

adult ♂

gorget feathers flared in flight display

displaying adult ♂

smallest North American hummingbird

adult ♂

dark crescent in front of eye

thin white line above gape

adult ♀

blended buffy underparts and buffy collar

wing tips extend slightly beyond short tail

white gape line often absent; throat more heavily spotted than female

juvenile ♂

adult ♀   short tail lacks prominent rufous at base

R1 is short

Genus *Cynanthus*

BROAD-BILLED HUMMINGBIRD   *Cynanthus latirostris*   BBIH   ▪ 2

This medium-size hummingbird of desert canyons and low mountain woodlands is an inveterate tail wagger. It often attracts attention through its dry chattering calls. Polytypic (5 sspp.; *magicus* in N.A.). L 3.5–4" (9–10 cm) Bill 18.5–23.5 mm
Identification Broad-billed has a cleft tail and a medium-long bill that is reddish basally. **ADULT MALE:** Distinctive, with deep green plumage overall, a blue throat, white undertail coverts, a cleft blue-black tail, and a bright red bill, tipped black. **ADULT FEMALE:** Note broad dark auricular mask offset by a whitish to pale-gray postocular stripe; gray below with limited green spotting on sides. Blackish

tail has variable greenish basally and distinct white tips to outer rectrices. Lower mandible is reddish basally. **IMMATURE MALE:** Resembles female but plumage fresh in fall with buff-tipped upperparts. Throat usually has some blue blotching or a solid blue patch, and older birds are extensively mottled green on underparts. Red on bill base often extends to upper mandible, and tail is more extensively blue-black, with smaller white corners. **IMMATURE FEMALE:** Resembles adult female but plumage fresh in fall, buff-tipped on upperparts; underparts have reduced or no green mottling on sides, and lower mandible base averages paler, more pinkish.

Geographic Variation N.A. breeding birds are of *magicus* in nominate subspecies group, *latirostris*. Two other subspecies groups in Mexico (*lawrencei* and *doubledayi*), the latter taxon treated as a distinct species by some. **Similar Species** The female is most easily confused with a female White-eared Hummingbird, which is stockier and disproportionately shorter billed, with a bold black auricular mask, an even more flagrant white postocular stripe, and a hard, loud, metallic chipping call. Dullest females could suggest a female or immature Black-chinned Hummingbird, but note their

reddish bill base, different face pattern, broad inner primaries, and call. **Voice CALL:** Often gives a dry *cht*, singly or in short series, *ch-ch-cht*, etc., when chattering cadence suggests the common call of Ruby-crowned Kinglet; also a high squeaky chippering in interactions. **SONG:** A high, sharp song, repeated from a perch. **Status & Distribution** Mexico to southwest US; winters mainly in Mexico.

BREEDING: Fairly common (mainly Mar.–Sept.), in brushy woodland, desert washes, gardens. **WINTER:** Uncommon and local in southern AZ. **RARE STATUS:** Rare in Southern CA and TX, mainly in fall and winter. Casual or accidental to OR, ID, CO, and the East; recorded as far northeast as NS. **Population** Scant data; considered common throughout much of its range in Mexico.

white postocular stripe never as bold as White-eared

pinkish base to bill

adult ♀ *magicus*

notched tail, grayish white tips to R4–R5

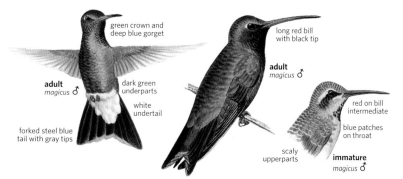

green crown and deep blue gorget

adult ♂ *magicus*

dark green underparts

white undertail

forked steel blue tail with gray tips

long red bill with black tip

adult ♂ *magicus*

scaly upperparts

red on bill intermediate

blue patches on throat

immature ♂ *magicus*

AMAZILIA HUMMINGBIRDS Genus *Amazilia*

Of the 29 species in this genus, four have occurred north of Mexico: two regular breeders, a casual breeder, and an accidental visitor. The sexes look generally similar. *Amazilia* may represent two genera: a slender-billed group with dark upper mandibles, including Berylline (subgenus *Saucerottia*) and typical *Amazilia*, with broader red bills.

BERYLLINE HUMMINGBIRD *Amazilia beryllina*   BEHU ▪ 3

This common Mexican hummingbird of foothill oak woodlands ranges into the US on an annual basis. Polytypic (5 sspp.; *viola* in N.A.). L 3.7–4" (9.5–10 cm) Bill 18.5–21 mm
Identification This medium-size hummer has a slender, medium-length bill with red on at least base of lower mandible. Diagnostic and extensive bright-rufous wing patch of all plumages noticeable in flight. **ADULT MALE:** Beryl green throat and chest contrast with dusky buff belly; crown green; coppery purple tail often looks rufous. **ADULT FEMALE:** Similar to male but duller, with spotted

green throat and more grayish belly. **IMMATURE:** Resembles female; dingier below.
Similar Species Buff-bellied Hummingbird (normally no overlap in range) lacks rufous in wings. **Voice CALL:** A fairly hard, buzzy *dzirr* or *dzzrit*. **SONG:** Short jerky and squeaky phrases; squeaky warbling. **Status & Distribution** Mexico to Honduras. **RARE STATUS:** Rare to casual visitor (Apr.–Oct., mainly June–Aug.) in oak zone of southeastern AZ mountains. Casual in summer to southwestern NM and West TX.
Population Common in Mexico.

spotted green throat

adult *viola* ♀

underparts paler

less rufous in wings than adult male

dull red on lower mandible, often obscure

adult *viola* ♂

adult *viola* ♂

rufous on wing obvious in flight

glittering green gorget and breast

buffy brown lower belly

tail violet, becoming cinnamon on outer tail feathers

BUFF-BELLIED HUMMINGBIRD *Amazilia yucatanensis*   BBEH ■ 2

This fairly large hummingbird is found in south TX. Polytypic (3 sspp.; *chalconota* in N.A.). L 3.8–4.3" (10–11 cm) Bill 19–22 mm
Identification Medium-length bill; broad, cleft tail; buffy belly; pale eye ring; and mostly rufous tail in all plumages. **ADULT MALE:** Throat and chest iridescent green, contrasting with buffy cinnamon belly. Mostly rufous tail tipped bronzy green, most broadly on central rectrices. Bright red bill tipped black. **ADULT FEMALE:** Similar to male but duller, with whitish mottled chin, a shallower tail cleft, and mostly green central rectrices. **IMMATURE:** Like female but throat and chest dingy buff to whitish, mottled iridescent green, belly paler buff, upper mandible mostly blackish, developing red over first year.
Similar Species Berylline Hummingbird (no overlap in range in US) has

rufous wings, a darker bill, and a different voice. The Rufous-tailed Hummingbird (*A. tzacatl*), erroneously reported from south TX, is similar but has a more square (less forked) tail with more extensively rufous central tail feathers and contrasting dark cinnamon undertail coverts. It is found north to southern Veracruz, Mexico.
Voice CALL: A clipped to slightly smacking chip, *tik* or *tk*, at times run into a rolled *tirr* or *tsirrr*. In warning, a slightly buzzy *sssir*. In chases, variable and usually fast-paced series of buzzy to lisping calls. **SONG:** A varied arrangement of chips alternating with slurred whistles and wheezy notes.
Status & Distribution: South TX to Guatemala. **BREEDING:** Fairly common (mainly Apr.–Aug.). **WINTER:** More widespread but less common in TX,

rare to casual (mainly Oct.–Mar.) along Gulf Coast to north FL. **RARE STATUS:** Casual to AR, coastal GA, and south FL.
Population Little studied; brush clearance in south TX (1950s to 1970s) likely had a negative impact.

adult
chalconota ♂

emerald green
throat and breast

buff belly

tail extensively
rufous with
dark tips

bill more extensively black
and with duller red than
adult male

gorget
feathers
with
pale
edges

adult
chalconota ♀

adult
chalconota ♂

bill mostly
black

immature
chalconota

variable green
on throat and
breast, paler
below than
adult

VIOLET-CROWNED HUMMINGBIRD *Amazilia violiceps*   VCHU ■ 2

An unmistakable hummer of western Mexico, with local summer populations in southeastern AZ and southwestern NM. Polytypic (2 sspp.; *ellioti* in N.A.). L 4–4.5" (10–11.5 cm) Bill 21–24 mm
Identification Bill is medium-long and straightish, and tail is broad and squared to slightly cleft (averag-

ing more so on males). Bright white underparts are unique among N.A. hummingbirds. **ADULT:** Sexes similar. Crown and auriculars are violet-blue (rarely appearing turquoise) and tail bronzy greenish to brownish. Bright lipstick red bill is tipped black. **IMMATURE:** Resembles adult, but throat

bright violet
crown, more
bluish in rear

lipstick red
bill with
black tip

adult
ellioti

duller crown;
crown feathers
tinged gray

bill color
variable, darker
than adult

pale
fringing
above

juvenile
ellioti

adult
ellioti

striking snowy
white underparts

and underparts are dingier whitish. Crown is oily bluish (can look dark) with buff feather tips in fresh plumage; upperparts and tail are tipped buff in fresh plumage; and upper mandible is mostly blackish, developing red over the first year.
Geographic Variation N.A. birds are of northern subspecies, *ellioti*.

Nominate from central Mexico has a brighter, coppery-bronze tail.
Voice CALL: A hard chip, *stik* or *tik*, often run into rattled short series; single plaintive chip, repeated from perch, *chieu chieu chieu*.
Status & Distribution Mexico to southwest US, winters Mexico. **BREEDING:** Uncommon (mainly Apr.–Sept.)

in arid to semiarid scrub, riparian woodland, and gardens. In US, most numerous in AZ, especially at Patagonia. **WINTER:** Rare in southeastern AZ. **RARE STATUS:** Casual northwest to northern CA, east to TX.
Population Range expansion in AZ and NM since the 1950s. Common in most of Mexican range.

CINNAMON HUMMINGBIRD *Amazilia rutila*   CIHU ▪ 5

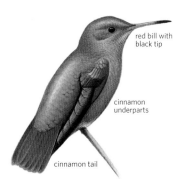

red bill with black tip

cinnamon underparts

cinnamon tail

This medium-large hummer from the tropical lowlands of west Mexico (north to central Sinaloa) and Belize south to Costa Rica has been recorded in AZ and NM. Polytypic (4 sspp.; N.A. records probably *diluta*). L 4–4.5" (10–11.5 cm) Bill 20.5–23.5 mm
Identification All plumages solidly rufous on throat and underparts with mostly cinnamon tail. **ADULT:** Sexes similar. Black-tipped, bright-red bill. **IMMATURE:** Resembles adult but upperparts tipped cinnamon when fresh; upper mandible mostly blackish, developing red over the first year.
Similar Species Should be unmistakable. See Xantus's Hummingbird.
Voice CALL: A hard, clipped tick with a buzzy or rattled quality, *tzk* or *dzk*, can be run into buzzy rattles. In interactions, high squeaks run into excited chatters. **SONG:** A short, varied arrangement of slightly squeaky chips.
Status & Distribution Mexico to Costa Rica. **RARE STATUS:** Accidental to Patagonia, AZ (21–23 July 1992), and Teresa, south-central NM (18–21 Sept. 1993).
Population Stable.

Genus *Hylocharis*

WHITE-EARED HUMMINGBIRD *Hylocharis leucotis*   WEHU ▪ 3

This medium-size, stocky hummer is common in Mexico's pine-oak highlands. Polytypic (3 sspp.; *borealis* in N.A.). L 3.5–4" (9–10 cm) Bill 15–18.5 mm
Identification Bill is medium-length and red basally. In all plumages, bold pure white postocular stripe contrasts with broad blackish auricular mask. **ADULT MALE:** Head often looks black with flagrant white stripe. **ADULT FEMALE:** Throat and underparts whitish, extensively spotted green; lower mandible reddish basally. **IMMATURE MALE:** Resembles female; throat usually has some blue and green. Red on bill extends to upper mandible. **IMMATURE FEMALE:** Resembles adult; throat

has sparser and more bronzy spotting.
Similar Species Unlikely to be confused. See Xantus's Hummingbird and Broad-billed Hummingbird.
Voice CALL: A clipped, fairly hard, metallic ticking chip, singly or often doubled or trebled, *ti-ti-tik* or *chi-tik chi-tik*. **SONG:** Rapid, rhythmical chipping, sometimes interspersed with quiet squeaks and gurgles.
Status & Distribution Visits feeders but also readily found in stands of montane wildflowers. Found northern Mexico to northern Nicaragua. **RARE STATUS:** Rare and local breeder (mainly Apr.–Oct.) in pine-oak of southeastern AZ mountains. Very rare or casual to southwestern NM and

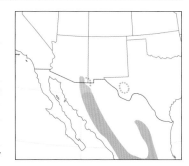

West TX; accidental elsewhere in TX and to CO, MS, AL, and MI.
Population No trends known; common and widespread throughout native range.

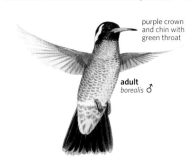

purple crown and chin with green throat

adult
borealis ♂

short, mostly black bill

spotted throat and flanks

adult
borealis ♀

rather short tail with white tips on R3–R5

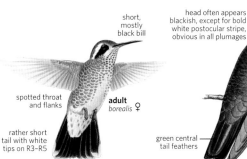

head often appears blackish, except for bold white postocular stripe, obvious in all plumages

red bill with black tip

adult
borealis ♂

green central tail feathers

XANTUS'S HUMMINGBIRD *Hylocharis xantusii*   XAHU ■ 5

This unmistakable hummingbird shares the congeneric White-eared Hummingbird's bold face pattern. Monotypic. L 3.3–3.8" (8–9 cm) Bill 16–19 mm

Identification Bill is medium length and red basally. In all plumages a bold white postocular stripe contrasts with a broad blackish auricular mask. **ADULT MALE:** Forehead and chin black, lower throat gorget glittering green. **ADULT FEMALE:** Throat and underparts cinnamon, lower mandible reddish basally. **IMMATURE:** Resembles adult female but fresh plumage has broadly buff-tipped crown feathers. Males usually have some blue-green throat spots; red on bill base extends to upper mandible.

Similar Species Shape and head pattern suggestive of White-eared (separate range in Mexico); most easily separated by rich buff belly on all birds.

Voice CALL: Most often gives a low, fairly fast-paced rattled *trrrrr* or *turrrt*. **SONG:** A quiet, rough, gurgling warble interspersed with rattles and high squeaky notes.

Status & Distribution Endemic to Baja California, Mexico, where found from about 28 degrees north, then south through the Cape Region. **RARE STATUS:** Accidental in winter to Southern CA (Ventura, 30 Jan.–27 Mar. 1988; built two nests) and southwestern BC (Gibsons, 16 Nov. 1997–21 Sept. 1998).

broad, bold white supercilium

♂

rich buff underparts

extensive rufous on tail

♀

short tail

Both birds were females.
Population Restricted-range species, but not considered at risk.

RAILS, GALLINULES, AND COOTS Family Rallidae

Purple Gallinule (FL, Mar.)

One of the most widely distributed families of terrestrial vertebrates, this group includes some of our most familiar birds as well as some of our most mysterious and secretive species. Gallinules and particularly coots tend to be less secretive, more duck-like, and often frequent open water or marsh edge.

Structure All species have disproportionately short and rounded wings, short tails, and strong legs and feet. Laterally compressed bodies allow rails and gallinules to move effortlessly, almost rodentlike, through dense vegetation. The more aquatic coots are heavier, wider bodied, with lobed toes for swimming. Bill shapes vary from short, stubby, and chickenlike to long and slender, mainly reflecting diet: more herbivorous/omnivorous.

Behavior Most occur in association with water: freshwater or brackish, salt marshes, mangrove swamps, bogs, wet meadows, or flooded fields. All species can swim. Most can dive and use their wings underwater to escape predators. Coots and gallinules are the most aquatic, forage while swimming, and must patter across the water to get airborne. Rails rarely forage in the open, but when

they do they can be surprisingly tame and oblivious. Most rails and gallinules prefer to run to avoid danger, but if startled will flush, fly a short distance and then drop back into cover. Once airborne, most species are remarkably strong fliers, and many are capable of sustained long-distance migration or vagrancy. As a group rails are very vocal, especially during the breeding season and at night. Vocalizations are usually loud and diagnostic. Calls vary from soft cooing to harsh and monotonous series of mechanical sounds; some species engage in duets.

Plumage Sexes are usually similar, and plumage generally does not vary seasonally, although some species have a slightly brighter breeding plumage. Chicks are covered in black down, some adorned with colorful plumes or naked skin on the head. Down is replaced by juvenal plumage while birds are still substantially smaller than adults, and juvenal plumage is fairly quickly replaced by first basic plumage. These immatures resemble adults, but typically have duller plumage and soft parts. The post-breeding molt is complete; the rapid loss of flight feathers results in temporary flightlessness in some species.

Distribution Worldwide except polar regions and waterless deserts. Many oceanic islands have been colonized by rails, many of which evolved into new species; some became flightless.

Taxonomy About 140 species in 40+ genera. Species and generic relationships are undergoing considerable reclassifications with ongoing genetic studies.

Conservation Many island species, especially flightless ones, are extinct or seriously threatened. Researcher David Steadman estimated that 500–1,600+ species of rails endemic to oceanic islands became extinct after human colonization, the vast majority never scientifically described. BirdLife International codes: 17 NT, 19 VU, 9 EN, 5 CR, 24 EX, 1 EW (Guam Rail, captive breeding program has resulted in releases on nearby Cocos I.).

Genera *Coturnicops, Laterallus,* and *Crex*

YELLOW RAIL   *Coturnicops noveboracensis*   YERA ■ 2

The Yellow Rail is secretive and difficult to see. If flushed the wings show a conspicuous white secondary patch. Polytypic (2 sspp.; nominate in N.A.). L 7.3" (18 cm)

Identification ADULT: Small. Sexes similar. Back feathers black broadly edged yellow-buff with narrow white terminal crossbars. Crown and nape blackish, dark stripe below eye. Throat and belly white, sides and flanks black with narrow white bars. Tips of greater coverts and trailing edge of inner wing extensively white. Nonbreeding males and most females have blackish olive bill; breeding males, some breeding females have paler yellow bills; legs brownish to greenish. **IMMATURE:**
Darker; face, neck, breast, and sides less buffy, with heavier blackish brown and white barring, white spots on crown.
Similar Species Larger Sora similar in shape (juvenile Sora similar in coloration); no white in secondaries. Corn Crake much larger, no white in secondaries. Black Rail smaller, more uniformly dark, grayer, no white in wing.
Voice Sounds like two stones tapped together, *tic-tic tic-tic-tic,* four or five notes repeated. Usually vocalizes only on breeding grounds, mainly at night.
Status & Distribution Uncommon, local. **BREEDING:** Sedge- or grass-dominated freshwater or brackish marsh. **MIGRATION:** Nocturnal. Spring: Apr.–May. Fall: late Sept.–Nov. **WINTER:**
Marsh, grassy fields, and rice fields along south Atlantic and Gulf coastal plains, NC to TX; also a few in coastal Bay Area marshes on central CA. **RARE STATUS:** Casual or accidental from southeastern AK (heard only), BC, AZ, NM; also NL, Bermuda, and Bahamas.
Population Probably declining; secretive behavior complicates estimates.

adult
noveboracensis

extensive white
secondary patch

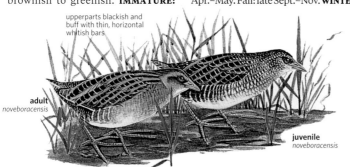

upperparts blackish and
buff with thin, horizontal
whitish bars

adult
noveboracensis

juvenile
noveboracensis

BLACK RAIL   *Laterallus jamaicensis*   BLRA ■ 2

Our smallest rail is rare, local, and usually only heard. If flushed, it flies a short distance then drops out of sight. Best chance of viewing our most secretive species is during unusually high tides in some CA salt marshes, when flooding forces rails to marsh edge. Polytypic (5 sspp.; 2 in N.A.). L 6" (15 cm)
Identification ADULT: Generally dark gray, with short black bill, red eyes, dark brown hind crown, dark chestnut nape, white speckling on upperparts, white barring on flanks, lower belly, under-
tail coverts, and underwing coverts. Female paler gray throat and breast.
IMMATURE: Similar to female but more brownish, less dorsal spotting.
Geographic Variation Two subspecies breed in N.A., three others resident in S.A. Smaller, more brightly colored *coturniculus* occurs year-round in CA, southwestern AZ, and northwestern Baja California. Larger, duller nominate occurs in eastern US, eastern Mexico, and Caribbean slope of C.A.
Similar Species Downy chicks of other rail species are small and uniformly black, but have disproportionately large legs and feet. Yellow Rail, almost as small, is dark above but more patterned and paler below, with extensive white on wing. Sora larger, paler, browner, with stouter, paler bill, and extensive white on leading edge of wing and under wing.
Voice *Kik-kee-do, kik-kee-derr,* or *kik-kik-kee-do,* mainly at night during breeding season; also gives a series of *grr* notes, heard year-round.
Status & Distribution Rare to locally uncommon. **BREEDING:** Shallow freshwater or salt marshes, wet meadows. **MIGRATION:** In spring, arrives mid-Mar.–mid-May. Departs early Sept.–early Nov., peak mid-Sept.–mid Oct. **WINTER:** Local along southern Atlantic and Gulf Coasts, NC to TX, casual north to NJ and south to Greater Antilles (may breed Cuba), C.A., S.A. **RARE STATUS:** Rare to casual inland in eastern N.A. north to MN, Great Lakes region, ON, QC, ME, CT, RI; accidental to Bermuda.
Population Near Threatened. Severe declines in mid-Atlantic; endangered in AZ.

maroon
nape

dark red
iris

tiny size with
white speckling

dark
bill

extremely
secretive

slaty gray face
and underparts

jamaicensis

CORN CRAKE *Crex crex*   CORC ▪ 4

Once a regular visitor, the Corn Crake has been recorded only six times in N.A. since 1928, most recently on Long I. in 2017. Very secretive, it prefers dense grass, runs to avoid danger, and is not easily flushed. Monotypic. L 10.8–12" (27–30 cm)
Identification Relatively large. Bill

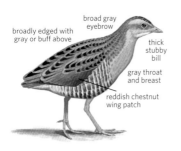

broadly edged with gray or buff above

broad gray eyebrow

thick stubby bill

gray throat and breast

reddish chestnut wing patch

stubby and pale. Back feathers blackish brown broadly edged buff, flanks banded reddish brown and buff. Some chestnut visible on folded wing; in flight wings extensively rufous. **MALE:** Gray breast and supercilium. **FEMALE AND IMMATURE:** Similar, browner, less gray on face, breast.
Similar Species Yellow Rail smaller with white-tipped secondaries, lacks chestnut shoulders. See Sora.
Voice Loud, rasping *crex crex*.
Status & Distribution Common but increasingly local in Old World. **BREEDING:** Meadows; northern Europe to Siberia. **MIGRATION:** In fall, Aug.–Nov. In spring, late Mar.–May. **WINTER:** Eastern Africa. **RARE STATUS:** Casual/accidental (fall), northeastern N.A.

extensive chestnut on wings

coast (Baffin I. to MD); most records are historical. Also recorded from Bermuda and Guadeloupe.
Population Long-term declines in some European countries have been reversed, and the species' core population in Russia is stable or possibly increasing.

TYPICAL RAILS Genus *Rallus*

Considerable size variation exists among the six New World species, but all have relatively long, slender, slightly decurved bills and rusty breasts; most have barred flanks. All have adapted to a semiaquatic existence and are often seen foraging along habitat edge.

RIDGWAY'S RAIL *Rallus obsoletus*   RIRA ▪ 2

Formerly considered part of the Clapper Rail complex. Polytypic (4 sspp., 3 in N.A.). L 13.4–17.3" (34–44 cm)
Identification ADULT: Rich buffy to almost cinnamon colored (sspp. dependent) on neck and breast with whitish barring against dark gray-brown sides and flanks. Dark, but not black, centers to feathers on upperparts with buffy gray edges. **JUVENILE:** Darker than adult above, duller below.
Geographic Variation Nominate *obsoletus* found around the Bay Area, formerly to northern Monterey Co., and tentatively and formerly to Morro Bay; Southern CA and northwestern Baja California *levipes* is similar but is a richer and darker cinnamon below; interior Southwest and west Mexican *yumanensis* is smaller and paler than

the coastal CA subspecies. Another subspecies, *beldingi*, is found in Baja California Sur.
Similar Species Ranges of Clapper and Ridgway's Rail are distinct and do not overlap. The underparts of Ridgway's are richly colored and more suggestive of King Rail, but the feathers on the upperparts with paler centers and duller fringes are more like Clapper. The cheeks are browner than Clapper. King could occur as an accidental visitor to the Southwest, where the mainly resident *yumanensis* is locally found, so any interior bird, especially away from known locations for Ridgway's, should be examined critically.
Voice All vocalizations closely resemble King and Clapper.
Status & Distribution Mainly resident,

obsoletus

yumanensis

levipes

beldingi

especially coastal subspecies, but some local movement within both *obsoletus* and *levipes*, including recent records of individuals from locations where they had been extirpated. Interior

dark and rich rusty neck

coastal southern CA
levipes

Bay Area
obsoletus

like *levipes* but underparts slightly duller and buffier

slightly duller than the two CA coastal subspecies

interior Southwest
yumanensis

yumanensis, known mainly in the US from freshwater marshes along Lower Colorado River and Salton Sea region, is believed to be less sedentary; northern populations are believed to be migratory. **RARE STATUS:** Casual to coastal regions away from known occurrence and to slightly inland regions; and to northwestern CA, Desert Center (Riverside Co.), CA, and south-central AZ. Accidental to the Farallones, CA (*obsoletus* on 18 Nov. 1886). **Population** Near Threatened. Coastal development and introduced preda-tors (e.g., Red Fox) imperil coastal CA populations. USFWS lists subspecies *levipes*, *obsoletus*, and *yumanensis* as Endangered. Populations present in Monterey Co., San Luis Obispo Co., Santa Barbara Co., and Los Angeles Co. have been extirpated.

CLAPPER RAIL *Rallus longirostris*   CLRA ■ 1

Recent genetic studies on the Clapper/King Rail complex resulted in recognition of five species. These include the King, Ridgway's, and Clapper Rails in N.A.; and two extralimital species, the Aztec Rail (*R. tenuirostris*, formerly treated as a subspecies of King Rail) and the Mangrove Rail (*R. longirostris*, 7 sspp.) of southern C.A. and S.A. Under this configuration, the Clapper Rail is restricted to coastal regions in eastern N.A. and the West Indies. There it is a familiar sound of our coastal salt marshes, and this large rail can occasionally be observed foraging for crabs, other invertebrates, and seeds along the edges of tidal channels, mudflats, or ditches. It prefers to run into vegetation rather than fly when startled, but will readily flush, fly a short distance with legs dangling, then drop back into cover. It swims across deeper tidal channels. Polytypic (±8 sspp.; 5 in N.A.). L 13.4–16.9" (34–43 cm)

Identification **ADULT:** Sexes similar in plumage, but males about 20–25 percent larger. Bill long, dark, with paler reddish to brown base of lower mandible (brighter on male), legs long and brownish red to gray, brown to blackish back feathers edged with grayish. Cheek grayish. Color of breast varies geographically from rich buff to deep rufous to olive-gray. Flanks and undertail are dark gray and heavily barred with white. **JUVENILE:** Like adult but center of breast more extensively white contrasting with darker gray sides, and underparts varyingly mottled with blackish gray feather tips.

Geographic Variation Eight sub-species, five of which are found on the Atlantic and Gulf Coasts; three others occur in the West Indies and Yucatán. These five subspecies are variably colored saltwater birds that intergrade with each other along the Atlantic and Gulf Coasts: "Northern Clapper Rail," *crepitans*, of the northern Atlantic coast from New England to NC (grayest on chest and back); "Wayne's Clapper Rail," *waynei*, of the southern Atlantic coast from NC to northeastern FL (slightly smaller and darker); "Florida Clapper Rail," *scottii*, along most of the mainland FL coast (darkest upperparts); "Mangrove Clapper Rail," *insularum*, of the FL Keys (smaller, paler); and "Louisiana Clapper Rail," *saturatus*, of the Gulf Coast from western FL to northern Tamaulipas, Mexico (more rufescent below and most similar to King Rail). Remaining subspecies found outside of the US include *pallidus* (Yucatán), *coryi* (Bahamas), and *caribaeus* (Greater and Lesser Antilles).

Similar Species The largely freshwater Virginia Rail is much smaller and more delicate, even compared with half-size juvenile Clapper, and is more richly colored on the breast. Some Clapper subspecies very similar to King Rail, but the slightly larger King Rail is overall more richly colored with browner cheeks, black-centered back feathers with rich buff edgings, and brighter chestnut wing coverts. The flanks are

blacker with more clear-cut white bars.

Voice **CALL:** Most typically a series of 10 or more loud *kek* notes, accelerating and then slowing (King Rail similar but series usually deeper, shorter, slower, more evenly spaced); also shorter *kek-kek-kek*, other squawks, grunts. Calls day or night, but most vocal at dawn and dusk during breeding season.

Status & Distribution Uncommon to common. **YEAR-ROUND:** Coastal salt marsh, mangroves. **MIGRATION:** Northern Atlantic coast populations (*crepitans*) at least partially migratory; generally scarce in winter north of southern NJ. Arrives north Atlantic coast mid-Mar. Departs late Aug.–Nov. Other populations considered non-migratory, but individuals will undertake erratic postbreeding movements. **RARE STATUS:** Casual to NE, TN, WV, central VA, PA, central NY, NH, VT, ME, NB, PE, NS, NL, and Bermuda. Accidental NM.

Population Populations in eastern US are largely stable but have experienced declines. Vulnerable to wetlands loss; a game bird in many states.

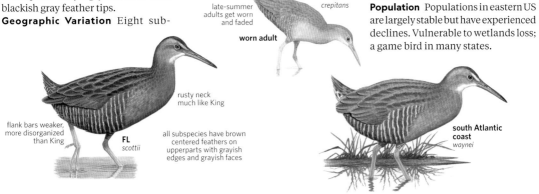

late-summer adults get worn and faded

worn adult

crepitans

rusty neck much like King

flank bars weaker, more disorganized than King

FL *scottii*

all subspecies have brown centered feathers on upperparts with grayish edges and grayish faces

south Atlantic coast *waynei*

KING RAIL *Rallus elegans*   KIRA ■ 1

King Rails occasionally forage in the open along marsh edges or in roadside ditches; in spring and summer parents venture into the open with their black downy chicks in tow. If startled, individuals run with head lowered in line with the body and disappear into vegetation. A recent genetic paper found that the former King Rail subspecies, known mainly from the interior of central Mexico, is more closely related to the western subspecies of Ridgway's Rail than it is to the two other subspecies of King Rail. It was subsequently split as the Aztec Rail (*R. tenurostris*). In areas of brackish water near the coast, hybridizes regularly with the Clapper Rail. Polytypic (2 sspp.; nominate in N.A.). L 15.0–17.7" (38–45 cm)

Identification ADULT: Sexes similar in plumage; males average larger. Very large, heavy rail, but considerable individual size variation. Bill long with yellowish orange base, eye orange-brown, legs and feet grayish brown to grayish olive. Plumage overall rusty-brown in coloration, mantle heavily streaked brownish black and edged with rich buff. Upperwing coverts are conspicuously chestnut on folded wing. Crown is dark brown, cheek mostly brown or with a small patch of gray. Throat white contrasting with bright rusty breast, belly buffy white, flanks blackish to dark brown barred with white, undertail coverts mottled dark brown and white. Plumage is brightest in fall, and becomes much duller, paler, and abraded by spring and summer. **JUVENILE:** Heavily mottled with gray below and is darker and less conspicuously

striped above. Downy chick like the Clapper Rail, black with a bicolored bill, pale at tip and around nostril.
Geographic Variation Nominate *elegans* is closely related to the subspecies *ramsdeni* from Cuba.
Similar Species Virginia Rail superficially similar but much smaller and has pure gray face. Briefly glimpsed chicks might be confused with the Black Rail, but have large legs and feet and downy plumage.
Voice CALL: Most typically a series of 10 or fewer loud *kek* notes, fairly evenly spaced; also shorter *kek-kek-kek*, other squawks, grunts. Calls day or night, but most vocal at dawn and dusk during breeding season.
Status & Distribution Rare to fairly common. **BREEDING:** Freshwater and brackish marshes, rice fields, even ditches. **MIGRATION:** Nocturnal, partial migrant. Spring: Apr.–May. Fall: Sept.–Oct. **WINTER:** Mainly near Gulf Coast, coast of eastern Mexico (separate population in Veracruz), and Cuba, occasionally farther north

within the breeding range; typically absent from areas with regular freezes. **RARE STATUS:** Casual/accidental to north TX, west to West TX, southeastern NM, CO, ND, MN, ME, MB, ON, QC, NB, NS, NL, and PE.
Population Near Threatened. Serious declines in northern breeding range since the mid-1900s. Vulnerable in Canada; considered threatened or endangered in 12 states. Despite declines, it is a game bird in 13 states.

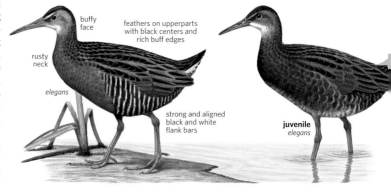

VIRGINIA RAIL *Rallus limicola*   VIRA ■ 1

Almost as widespread as Sora, Virginia Rails are seen less frequently as they forage in dense vegetation. This is our hardiest rail wintering in many areas that have cold winters. Polytypic (4 sspp.; nominate in N.A.). L 8.8–10.8" (22–27 cm)

Identification ADULT: Sexes similar. Relatively small with reddish brown eye, long reddish bill, reddish legs, and rusty chest and belly. Crown and nape brown, contrasting with gray cheek. Upperparts brown edged with rufous. Throat white, flanks black barred white, undertail coverts mottled black,

rusty, and white. **JUVENILE:** Similar but duller above, underparts extensively mottled blackish or brownish, legs and iris brown. **IMMATURE:** Like adult but buffy white on lower breast and belly.
Similar Species King and Clapper Rails somewhat similar but much larger overall. Blackish chicks could be confused with Black Rails.
Voice Year-round, the most frequent and easily recognized vocalization is a descending series of *oink* notes. Breeding calls for the male include *kid-kid-kidik-kidik-kidik-kidik* and for the female *tic-tic-tic-mcgreer*;

also a variety of squealing and other notes that are given by both sexes.

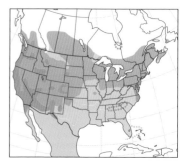

Status & Distribution Common. **BREEDING:** Primarily freshwater marsh, less frequently brackish or salt marsh. Other populations resident in southern Mexico, C.A., northwestern S.A. **MIGRATION:** Mostly migratory. Arrives in spring Mar.–May, peak early Apr.–early May. Departs mid-Aug.–Nov., peak mid-Sept.–mid-Oct. **WINTER:** Western and southern US south to central Mexico; regular in salt marshes; also local at hot springs well north of winter range. **RARE STATUS:** Rare to Bermuda, casual or accidental to AK, Cuba, Greenland.

Population Historic declines are described, but has a large range and is now considered stable. This species is a game bird in many states.

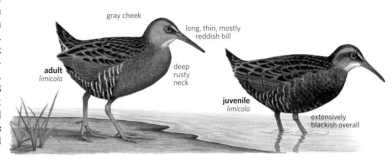

gray cheek

long, thin, mostly reddish bill

adult
limicola

deep rusty neck

juvenile
limicola

extensively blackish overall

Genus *Porzana*

SORA *Porzana carolina*   SORA   ■ 1

The calls of the Sora are some of the most familiar sounds of the freshwater marsh. With patience or luck one can frequently be observed as it ventures into the open to forage along marsh edges or the shoreline of a ditch, walking deliberately with bobbing head and cocked tail. Sora runs to avoid danger, but will also flush when nearly underfoot, fly a short distance, and then drop from sight into dense vegetation. Soras commonly swim across deeper channels and will occasionally even forage coot-style in the water. Monotypic. L 7.6–10" (19–25 cm)

Identification ADULT: Stubby yellow bill with black around base and on throat to center of upper breast. Breast gray barred white, belly white, sides barred white, brown, and black, undertail coverts buffy white. Crown and nape rich brown with black center, contrasting with gray face and neck, and white spot behind eye. Rest of upperparts brown, back and tail with black feather centers and white fringes. In flight, leading edge of inner wing conspicuously white, underwing

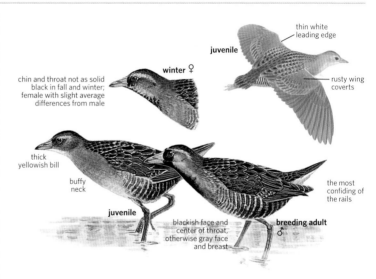

thin white leading edge

juvenile

rusty wing coverts

winter ♀

chin and throat not as solid black in fall and winter; female with slight average differences from male

thick yellowish bill

buffy neck

juvenile

the most confiding of the rails

blackish face and center of throat, otherwise gray face and breast

breeding adult ♂

coverts barred gray and white. Legs and feet are yellowish green, toes long. Female similar but bill duller, black on face and throat less extensive, slightly smaller. **JUVENILE:** Paler with white throat, buffier face and breast, lacks black on face, bill darker; immature gradually acquires black on face and throat, and gray on neck and breast. **Similar Species** Duller, buffier juvenile superficially similar to Yellow Rail. Sora is larger; upperparts are darker, less buffy, and have white streaks (not buffy stripes and white bars); and in flight, lacks conspicuous white on trailing edge of wing. Black Rail smaller, uniformly darker, with all-dark bill.

Voice Year-round gives a loud and distinctive descending whinny, *whee-hee-hee-hee-hee-hee*, and a sharp *keek*. The distinctive up-slurred and repeated incessantly *soo-rah* is most often heard on the breeding grounds. **Status & Distribution** Common. This is our commonest, most widespread, and most migratory rail. **BREEDING:** Shallow freshwater marsh. **MIGRATION:** In spring, late Mar.–mid-May, peak mid-Apr.–early May. In fall, late July–early Nov., peak Sept.–Oct. **WINTER:** Freshwater wetlands, also brackish and salt marshes, in CA and southern Atlantic and Gulf Coast regions to northern S.A. Regular migrant and winter in small numbers to Bermuda. **RARE STATUS:** Casual/accidental to east-central AK, Haida Gwaii, southern Labrador, Greenland, western Europe, and Morocco. **Population** Generally stable, but declining in central US from habitat loss. This species is a game bird over much of the East.

Genera *Aramides, Neocrex,* and *Pardirallus*

RUFOUS-NECKED WOOD-RAIL   *Aramides axillaris*   RNWR ▪ 5

Though gaudy in color, this species blends remarkably well into its shadowy environment and is most easily detected by its incisive calls. Monotypic. L 11–12.5" (28–32 cm) **Identification** A brightly colored rail unlikely to be mistaken. **ADULT:** Underparts a velveteen chestnut except for whitish throat and black or slaty gray vent. Mantle a grayish blue contrasting with olive coverts and tertials and blackish rump. Legs and iris coral red. Heavy, rather long bill green-yellow, becoming orange-yellow at base. **FLIGHT:** Seldom seen in flight. Brilliant rufous flight feathers contrast with black-and-white barred underwing coverts. **JUVENILE:**

Underparts mostly grayish, upperparts olive, with buffy gray head, darker gray crown; bill mostly dusky.
Similar Species None.
Voice A short series of high-pitched, sharp *pik!* calls, delivered mostly at night or very early or late in the day.
Status & Distribution Uncommon permanent resident of mangrove swamps, marshes, lagoons, but also in forest undergrowth to 6,000-ft elevation, from western Mexico and the Yucatán Peninsula south to Panama, along the Caribbean coast of S.A. to Suriname, and south along the Pacific coast of S.A. to northern Peru. Also on Trinidad and larger islands off Honduras. **RARE**

STATUS: One at Bosque del Apache NWR, NM (7–18 July 2013).
Population Insufficient data, but destruction of mangrove habitat imperils some populations.

bluish gray upper back

heavy pale bill with yellower base

rufous neck and breast

red legs

PAINT-BILLED CRAKE   *Neocrex erythrops*   PBCR ▪ 5

This secretive and poorly known species from S.A. and southern C.A. has been encountered only twice in N.A., both times in the 1970s during winter. Both records are substantiated by specimens. Polytypic (2 sspp.; possibly both in N.A.). L 7.2–8" (18–20 cm) **Identification ADULT:** Superficially like a small, uniformly colored Sora. Most of upperparts are dark olivebrown. Forecrown, supercilium, and face are gray. Throat whitish, rest of underparts dark gray; sides, flanks, undertail coverts, and underwing coverts barred white. Bill yellowgreen with bright orange base, legs red. **JUVENILE:** Duller.

Geographic Variation Two subspecies: Nominate *erythrops* of coastal Peru and the Galápagos Is. is accidental to TX. A VA specimen is tentatively identified as *olivascens*, patchily distributed in S.A. east of the Andes; darker, with less white on throat.
Similar Species Sora is more patterned above and below, lacks orange base to bill, has yellow-green legs. Black Rail is smaller, has dark bill and legs, chestnut nape, white spots on upperparts.
Voice Guttural, froglike *qur'r'r'r'rk,* or *pip* notes.
Status & Distribution Secretive; abundance varies locally from rare to common. Rare and local in Costa Rica

and Panama. **YEAR-ROUND:** Marsh, pasture, rice fields, overgrown drainage ditches, damp thickets in scrub or woodland. **MIGRATION:** No clear migratory pattern. **RARE STATUS:** Accidental in TX (south of College Station; 17 Feb. 1972) and VA (near Richmond; 15 Dec. 1978).
Population Insufficient data.

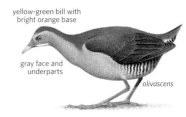

yellow-green bill with bright orange base

gray face and underparts

olivascens

SPOTTED RAIL   *Pardirallus maculatus*   SPRA ▪ 5

This striking tropical rail was detected in the US twice in the 1970s. Polytypic (2 sspp.; *insolitus* in N.A.). L 10–11.2" (25–28 cm) **Identification ADULT:** Larger and heavier than Virginia Rail. Upperparts are blackish brown with white spots;

black head and body with white speckling

long, slender greenish yellow bill with red spot at base

adult
insolitus

head and breast blacker with white spots; underparts black banded with white with white undertail coverts. Bill is long, slender, greenish yellow with red at base of lower mandible; eyes, legs, and feet red. **JUVENILE:** Polymorphic, with dark, pale, and barred morphs. All have duller legs and bill, brown eyes.
Geographic Variation M.A. *insolitus* is darker above than West Indian and S.A. *maculatus* and is spotted rather than streaked with white.
Similar Species The Spotted Rail is virtually unmistakable based on size, plumage pattern, soft parts colors. Beware juvenile Virginia, King, and Clapper Rails, which have varying

amounts of blackish or gray plumage.
Voice Sharp *geek,* a screech preceded by a grunt, and in breeding season a series of pumping *wumph* sounds, like a starting motor.
Status & Distribution Common. **YEAR-ROUND:** Vegetated wetlands and rice fields; patchily distributed southern Mexico to Costa Rica (*insolitus*); and Greater Antilles (except Puerto Rico), Trinidad, eastern Panama, and south to northern Argentina (*maculatus*). **RARE STATUS:** Accidental in Beaver Valley, PA (12 Nov. 1976), Brownwood, TX (9 Aug. 1977), Juan Fernández Is. off Chile, and at sea in South Atlantic off Brazil.
Population Unknown.

Genus *Porphyrio*

PURPLE GALLINULE *Porphyrio martinicus*   PUGA ■ 1

This colorful rail prefers floating or emergent vegetation, using long toes for balance or climbing. More terrestrial than the Common Gallinule. In flight dangles long legs and raises tail, revealing the extensive white undertail coverts. Monotypic. L 13" (33 cm)
Identification ADULT: Purplish blue head and underparts. Upperparts green, wing coverts turquoise. Undertail coverts bright white. Bill red with yellow tip; frontal shield pale blue; legs, feet bright yellow. **JUVENILE:** Upperparts brownish to greenish, underparts buffy, whiter on throat, belly, undertail. Bill is greenish brown; legs, feet dull greenish. Downy chick largely black. Full adult plumage is acquired during the second winter.
Similar Species Coots and Common

Moorhen are blackish gray and larger, and spend more time on water. Purple Swamphen is much larger with red shield and legs.
Voice A variety of calls including a sharp *puckk*; also cackling and grunting notes.
Status & Distribution Common. **BREEDING:** Freshwater marsh with dense floating or emergent vegetation. Breeds casually north to IL, OH, MD, DE. **MIGRATION:** Primarily a trans-Gulf spring migrant, late Mar.–May, peak mid-Apr.–early May;

occurs rarely north of breeding range. Migrates in fall late Aug.–Oct., peak Sept., stragglers into Nov. **WINTER:** Primarily areas of year-round occurrence, peninsular FL, eastern and central-western Mexico to S.A. Rare Gulf Coast. **RARE STATUS:** Many records well north of eastern breeding range at all seasons to southern Canada; casual to Southwest and CA; regular to Bermuda; accidental to Europe, South Africa, Galápagos Is., and South Georgia I.
Population Stable.

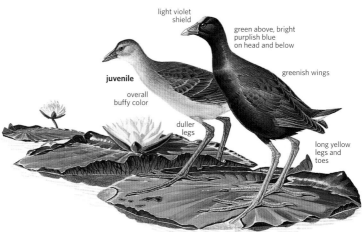

light violet shield

green above, bright purplish blue on head and below

greenish wings

juvenile

overall buffy color

duller legs

long yellow legs and toes

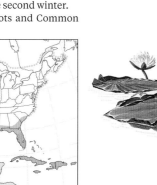

PURPLE SWAMPHEN *Porphyrio porphyrio*   PUSW ■ EXOTIC

First detected in 1996, this nonnative species quickly colonized southern FL after escaping from captivity. Eradication attempts failed. Polytypic (13 sspp.). L 18–20" (45–50 cm)
Identification ADULT: Sexes similar, females somewhat smaller. Unmistakable. Suggestive of a huge Purple Gallinule, with thick, reddish bill,

broad reddish frontal shield, red eye, massive reddish legs. Body plumage generally purplish blue; wings, lower throat, upper breast paler greenish blue; sides of head, neck, upper throat gray, tinged light blue. Undertail coverts white. **JUVENILE:** Duller, grayer plumage; bill, frontal shield, and legs are darker grayish.
Geographic Variation Most FL individuals resemble gray-headed *poliocephalus* group, occurring from Turkey to southern Asia, which includes subspecies *poliocephalus*, *caspius*, and *seistanicus*, sometimes recognized as a full species, Gray-headed Swamphen (*P. poliocephalus*).
Similar Species Adult Purple Gallinule smaller with azure frontal shield and yellow legs.
Voice Dull moans or mooing.
Status & Distribution Common in

native range. Locally common in southern FL. A possible disperser from FL was photographed in GA (Nov. 2009). **YEAR-ROUND:** Shorelines of artificial lakes, rice fields, and natural marshlands. A Purple Swamphen on Bermuda (fall 2009) was not one of the "Gray-headed" subspecies and may have been a stray from Africa.
Population Increasing in FL.

light violet-gray neck

thick red bill

large size

poliocephalus

Genus *Gallinula*

COMMON GALLINULE *Gallinula galeata*   COGA   ■ 1

This species forages while swimming, picking at floating vegetation. In protected areas where not hunted, gallinules are often tame and confiding, sometimes entering backyards to feed and rest. Polytypic (7 sspp.; *cachinnans* in N.A.). L 12.8–14" (32–35 cm)
Identification ADULT: Head, neck, and breast are blackish or gray, underparts charcoal gray with white stippling down side to flanks. Upperparts olive brown. Central undertail coverts black but otherwise white. Bill lipstick red, neatly tipped with yellow; frontal shield and eyes red. Legs bright yellow, cherry red at base, with very long toes. **JUVENILE:** Paler gray below, throat and belly white, bill dusky, and legs greenish. Downy chick is black, with yellow and orange head plumes;

frontal shield is red, orange-yellow bill is red tipped.
Geographic Variation Seven subspecies named, but only *cachinnans* recorded in N.A., where it ranges south to western Panama. Subspecies *pauxilla* nests from eastern Panama to northern S.A., and farther south, subspecies *galetea* and *garmani* have been recognized. The Caribbean has two endemic subspecies, *barbadensis* on Barbados and *cerceris* elsewhere. The very dark subspecies *sandvicensis* is endemic to HI; crimson wash on tarsus is unique.
Similar Species See American Coot, Purple Gallinule, Common Moorhen.
Voice Most commonly heard is a loud, trumpeting cackle call, a series of harsh, staccato *ka* notes, rising in pitch, followed by falling, slower series of *kreh* notes. This laughing vocalization is unknown in any subspecies of Common Moorhen; also gives clucking calls and screeches.
Status & Distribution BREEDING: Fresh to slightly brackish marshes with emergent vegetation; range highly fragmented, especially in northern regions. **MIGRATION:** Partial migrant,

mainly in East. In spring, mid-Mar.–mid-May. In fall, mid-Aug.–early Nov.
WINTER: Generally vacates eastern N.A. to southeastern coast and south.
RARE STATUS: Casual in southern Canada beyond breeding range, and to Greenland (six records!).
Population Generally stable, but considered threatened or a species of concern in some northern states due to local declines and extirpations.

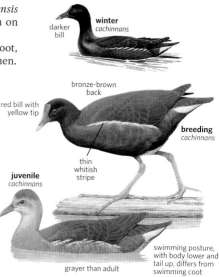

darker bill
winter *cachinnans*

bronze-brown back

red bill with yellow tip

breeding *cachinnans*

thin whitish stripe

juvenile *cachinnans*

swimming posture, with body lower and tail up, differs from swimming coot

grayer than adult

COMMON MOORHEN *Gallinula chloropus*   COMO   ■ 5

The Common Moorhen, widespread in the Old World, has recently been recognized as a species distinct from Common Gallinule of the New World. Polytypic (5 sspp.). L 10.5–12" (27–31 cm)
Identification Plumages very similar to Common Gallinule. **ADULT:** Adult gallinules show neat yellow tips to the red bill. Gallinules have a reddish brown tone to upper parts, especially distinct in the mantle; moorhens show more olive tones in the mantle. The two have differently shaped frontal shields: Moorhen's is rounded at top and widest in middle, whereas gallinules' are widest at the top and often look squared off or even notched (with two lobes) at the top. The condition of the frontal shield, however, relates to breeding condition, so a narrow shield rounded at the top is not diagnostic of a moorhen, especially in autumn, but a notched shield probably rules out a moorhen.

In adult moorhens, the iris tends to be deep ruby red, while gallinules usually have maroon irides, but this feature is variable. **JUVENILE:** Very similar to Common Gallinule.
Geographic Variation Nominate *chloropus*, likely involved in the AK record, is found throughout Eurasia and northern Africa.
Voice CALL: Lacks the trumpeting cackle of Common Gallinule. Most commonly heard calls are a rapid staccato *krrrrr!* and a rough, sharp *krek*, frequently repeated in series. Many other abrupt clucking sounds, shrieks, and grating sounds, some similar to Common Gallinule.
Status & Distribution Widespread and quite common across much of Europe, tropical Africa, and much of Asia, including oceanic islands. **RARE STATUS:** Has reached Iceland, the Faroe Is., Svalbard, and Tristan da Cunha. An immature male at Shemya I., AK (12–

14 Oct. 2010) was collected and determined by DNA analysis to represent a Common Moorhen.
Population Stable.

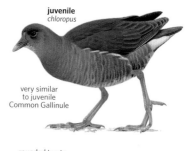

juvenile *chloropus*

very similar to juvenile Common Gallinule

rounded top to frontal shield (squared off on Common Gallinule)

adult *chloropus*

COOTS Genus *Fulica*

EURASIAN COOT *Fulica atra*   EUCO   5

This accidental visitor to the far northern corners of N.A. looks and acts very much like our American Coot. Polytypic (4 sspp.; nominate in N.A.). L 14.4–15.2" (36–38 cm)

Identification ADULT: Larger than American Coot with all-white bill and frontal shield. The all-black—rather than white-sided—undertail coverts are diagnostic. Legs and feet greenish. **JUVENILE:** Similar, but much paler on sides of face and below.

Similar Species The American Coot averages smaller, is less uniformly blackish with more head-body contrast, has white on undertail coverts, partial dark ring on bill, and usually has a dark spot on its frontal shield.

Voice *Kowk* and double *kek-wock*.

Status & Distribution Common and widespread in Old World. **BREEDING:** Freshwater marshes. **MIGRATION:** Northern Eurasian populations migratory. In fall, migrates mid-Aug.–Nov. In spring, migrates late Feb.–May. **RARE STATUS:** Casual to Greenland (11 records); accidental in spring, fall, and winter in AK, QC, and NL.

Population Stable.

all-white frontal shield and bill

atra

black undertail

AMERICAN COOT *Fulica americana*   AMCO   1

Whether swimming on a lake, grazing on a golf course, diving for submerged vegetation, or wading along a pond, this comical, chickenlike bird is one of our most familiar waterbirds. Coots have to run along the surface to take flight, but once in the air flight is strong and fast. Can occur in large flocks. Polytypic (2 sspp.; nominate in N.A.). L 12.8–17.2" (32–43 cm)

Identification ADULT: Sexes similar, males slightly larger. No other species has a short, whitish, chickenlike bill, overall gray coloration, and, in flight, white trailing edge to secondaries. Head and neck black, body somewhat paler blackish gray, center of belly usually paler, undertail coverts black with white outer patches. Bill and frontal shield mostly white with brownish red partial ring on bill and spot in center of frontal shield. Size and coloration of frontal shield variable: Some individuals have more yellowish white shield without darker center. Eyes are red, legs and lobed toes greenish yellow. **JUVENILE:** Duller and paler, with more whitish underparts. Bill dusky, legs more olive. Downy chick is black above, dark gray below, with stiff curly orange and yellow fluff on forehead, chin, and lores; dark red skin on crown, blue skin above eye, and bright red bill with black tip.

Geographic Variation Nominate occurs in N.A. The "Caribbean Coot" (*caribaea*), formerly treated as a separate species, is mainly found in the southern West Indies. It is more uniformly blackish below and has larger, broader, yellowish white frontal shield. No N.A. records are satisfactory.

Similar Species Immature Common Gallinule similar but smaller, more slender, has white stripe along sides, browner back, and more slender bill; swimming posture differs; coots are more horizontal, gallinules tip up in rear. Eurasian Coot very similar but averages larger,

overall blacker, with an all-white frontal shield and bill, and has entirely black undertail coverts.

Voice An assortment of grunting and cackling notes, including an emphatic *krrp* (lower than Common Gallinule), crowing *croooah*, and *punk-unk-punk-uh-punk-unk-uh*. Vocal day or night.

Status & Distribution Abundant. **BREEDING:** Freshwater wetlands with emergent vegetation, occasionally slightly brackish marshes. Breeds south through most of Mexico and locally in C.A., Colombia, and West Indies. **MIGRATION:** Northern populations migratory; others resident, may migrate during severe winters. Spring: Feb.–May. Fall: Late Aug.–Nov. **WINTER:** Northern interior populations move to coasts and southern half of continent. Occurs on brackish estuaries in winter. **RARE STATUS:** Rare in HI; casual to western AK, northern Canada, Greenland, Iceland, and Ireland.

Population Currently stable, but has declined historically. Considered a game bird in many states.

whitish bill with subterminal band

adult
americana

juvenile
americana

some lack dark top to shield

americana

lobed toes

swims like a duck most of the time

SUNGREBES Family Heliornithidae

Sungrebe, female (Costa Rica, Apr.)

These aquatic and generally unfamiliar birds are found in the New World tropics (the Sungrebe) and in tropical Africa and southeastern Asia (the two finfoots).

Structure These are superficially grebelike swimming birds, with lobed feet; slender bodies; long necks; and sharp, pointed bills. In size, they are roughly like gallinules or coots, but more slender. Unlike grebes or coots, they have a long, broad tail (which is stiffened in the finfoots but not the Sungrebe).

Behavior These shy swimming birds inhabit vegetated swamps, ponds, and rivers, where they are found in low densities. They are more prone to flee by moving into vegetation than by flying. Though principally aquatic, they can run on land and climb into trees. All three species are generally solitary or found in pairs during the breeding season. The nest of twigs and reeds is built within streamside vegetation. Sungrebes have a very short incubation period (about 11 days), and the altricial young can be transported by the adult male in flaps of skin under his wing—even in flight. Young finfoots are more precocial, capable of swimming when very young. Sungrebes and finfoots feed mainly on aquatic invertebrates and to some extent on small amphibians and fish.

Plumage All are generally gray-brown to brown above and pale below (barred in the African Finfoot); the outstanding visual characters include longitudinal neck stripes in all species, colorful beaks, and colorful legs and feet (orange or green in finfoots, banded yellow and black in the Sungrebe). Males and females differ slightly in plumage.

Distribution Sungrebes occur from northeastern Mexico to northern Argentina (once to NM). The African Finfoot is found widely in wetter areas of sub-Saharan Africa, while the endangered Masked Finfoot is restricted to Southeast Asia.

Taxonomy Worldwide there are three species, each in a monotypic genus. The family belongs within the "core" Gruiformes (along with cranes, rails, limpkins, and trumpeters), but its exact placement is still debated; the most recent molecular work suggests a sister relationship to the rails, and particularly to the rails known as flufftails (which are now given family rank, Sarothruridae).

Conservation All species are declining from loss of forested wetlands. Birdlife International codes: 1 EN.

Genus *Heliornis*

SUNGREBE *Heliornis fulica*   SUNG ■ 5

A small and somewhat secretive aquatic species of tropical America has occurred just once in N.A. Monotypic. L 11–12" (28–31 cm)

Identification Unlikely to be confused with any other species. Resembles some combination of large rail and duck, but with long tail and moderately long bill. Toes lobed, banded black and yellow. Usually observed on the water, swimming with slow, jerking movements of head and neck. Body plumage a rich auburn brown; tail blackish. The head and neck have a black-and-white pattern of stripes that recall a grebe chick. Female has rufous cheek, white in male. **FLIGHT:** Rapid and direct, with the long tail extending far past the trailing edge of the wing.

Voice A deeply resonant, nasal, monkey-like *eeyaaa*, repeated in short series; both softer and harsher *coo* and *cooo-ah* calls; and clucking sounds.

Status & Distribution Permanent resident in its range in coastal lagoons, slow-moving rivers, lakes, and ponds, usually with forested or vegetated banks as far north as central Veracruz (a few to southern Tamaulipas) and through lowland S.A. as far south as northern Argentina. Migration is unknown in the species, but Sungrebes do make movements in search of appropriate habitat and are strong fliers, and as with other secretive waterbirds, movements are likely poorly known.

RARE STATUS: One female at Bosque del Apache NWR, NM (13–18 Nov. 2008) was most unexpected. The species has been found on Trinidad, 12 miles off the coast of S.A.

Population Insufficient data.

ear coverts tawny-buff
(white on male)

distinctive
head pattern

♀

LIMPKIN Family Aramidae

Limpkin (FL, Mar.)

Structure Long neck, bill, and legs and rounded wings.
Behavior Loosely colonial, perhaps in part due to clumped distribution of prey abundance or availability. Active throughout the day and night, it forages by slowly walking in water; prey is detected either visually or by contact with bill. It nests in a variety of freshwater sites, ranging from mounds of vegetation built in marshes to stick nests built high in trees. When foraging, it walks slowly and flicks its tail. It flies with its neck fully extended and held below the body plane, creating a distinctive hunchbacked profile. Its flight style—quick upstrokes followed by slower downstrokes—is similar to cranes. When flying short distances, the Limpkin dangles its legs below the body, but in sustained flight it draws the legs up.
Distribution Restricted to the New World tropics.
Taxonomy This monotypic family is placed within the order Gruiformes, being most closely related to cranes and trumpeters. The Limpkin was formerly considered two species: one found in North America, Central America, and the Caribbean and the other in South America.
Conservation Overall stable in its Neotropical range, estimated at more than one million birds.

This family consists of a single species, limited in North America to FL. The Limpkin superficially resembles herons and ibises, but it is only distantly related to them—its closest relatives are the cranes.

Genus *Aramus*

LIMPKIN *Aramus guarauna*   LIMP   ▪ 2

This FL species feeds nearly exclusively on freshwater mollusks, primarily apple snails and clams. The tip of the Limpkin's lower mandible is curved to the right to facilitate extracting snails from their shells. Polytypic (4 sspp.; *pictus* in N.A.). L 26" (66 cm)
Identification Large, brown, heronlike bird with dense streaking over much of its body. Long, slender, slightly decurved bill is yellow with a blackish culmen and tip. **ADULT:** Wholly dark brown, paler on face, chin, and throat. Most feathers, especially those on the neck, have white markings; those on the upper back and upperwing coverts show large white triangles. Dark eyes; blackish legs and feet. **JUVENILE:** White

markings are narrower, appearing more streaked than spotted.
Similar Species American Bittern adults have a conspicuous dark malar streak and more pointed wings in flight. Immature night-herons have orange or red eyes, shorter, straight bills, shorter necks, and yellow legs. *Plegadis* ibises have wholly darker plumage.
Voice Loud, varied, and distinctive; mostly by male except when breeding pairs duet. Most common calls are a drawn-out *kreow* and a *kow*.
Status & Distribution Uncommon and local; in the US now restricted to peninsular FL, where generally sedentary. The N.A. subspecies (*pictus*) is also found in the Bahamas, Cuba, and Jamaica.
RARE STATUS: Casual throughout Southeast; accidental along the Atlantic and Gulf states from TX to NJ and NS.
Population Overhunting nearly extirpated the FL population by the early 1900s. The population has since recovered to about 5,000 pairs and is apparently stable, despite continuing wetland loss. No longer a game bird.

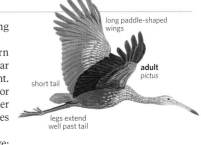

long paddle-shaped wings

short tail

adult
pictus

legs extend well past tail

large, slightly decurved, extensively pale bill with darker tip

striped neck

large white spots on upperparts

adult
pictus

long, dark legs

CRANES Family Gruidae

Sandhill Cranes (NM, Oct.)

Cranes are evocative of wild and open habitats in temperate regions around the world. In North America they present limited identification challenges, although distance and lighting can play into determining features accurately, and one should always be looking through migrating and wintering Sandhill Cranes for a stray Common Crane from Eurasia. Note overall plumage coloration and contrasts, details of head pattern, and the extent and contrast of any black on the remiges (best seen in flight); with experience, voice can be useful, but be aware that vocal variation can also be related to age and, in Sandhill Cranes, to subspecies.

Structure Cranes are large, heavy-bodied, and long-necked. They have a distinct tertial bulge, or "bustle," which is conspicuous when they stand. Their wings are long and broad, with 10 primaries and 18 to 21 secondaries (including four to five elongated tertials); their short, square tails have 12 rectrices. Their bills are moderately long, straight, and pointed, and their legs long and sturdy, with unwebbed feet. Males average 5 to 10 percent larger than females. Juveniles do not attain full growth until nearly a year old; they often look noticeably smaller than adults.

Behavior Cranes are terrestrial and social birds of open habitats, where they pick and probe for food in soil and marshes; they are not known to perch in trees. Their flight is graceful. With their necks and legs outstretched and holding their wings mostly above the body plane, they fly with stiff wingbeats with a relatively quick and slightly jerked upstroke. Migrating cranes are diurnal and often fly high, typically in V formations or well-spaced lines, and they frequently glide and sail on fairly flat wings, without flapping. Cranes are perennially monogamous; courtship involves spectacular leaping and "dancing" by pairs and groups, accompanied by much calling. Extended parental care means that families remain together for nine to 10 months, through migration and winter. Although cranes nest in territorial pairs, they form large flocks in migration and winter, when they commute from safe roost sites to feeding areas. They

often feed in agricultural land, and roost on islands or gravel bars in lakes and rivers. Their far-carrying calls are loud cries with a distinctive rolled or throaty quality, sometimes likened to bugling. Cranes show strong site fidelity to their migration staging grounds and to their wintering grounds.

Plumage Their plumage is gray to white overall, usually with darker to blackish primaries, and often with variable rusty blotching on the neck and upperparts. Adults have bare forecrowns patterned red and black; juveniles have mostly cinnamon-brown feathered heads and necks. Adult postbreeding molts occur mostly on or near the breeding grounds, whereas post-juvenal molts take place on nonbreeding grounds. Cranes have high wing loading and shed their primaries synchronously to become flightless for a short period—adults do this while nesting; pre-breeding immatures also molt their flight feathers during the summer months, in remote areas with adequate food. Adult plumage aspect is attained by the second postbreeding molt at about one year of age, although second-year birds are often distinguishable by their less developed and duller bare crowns and the presence of a few retained and worn juvenal feathers, such as secondaries.

Distribution Cosmopolitan, with both resident and migratory species.

Taxonomy Worldwide, there are 15 species in four genera; in North America, three species (two breeding, one rare stray from Eurasia) in two genera. Two other species, Demoiselle Crane (*Anthropoides virgo*) and Hooded Crane (*Grus monacha*) have been recorded in North America but with uncertain provenance.

Conservation Most crane species are at risk—several rank among the world's most threatened birds—due to habitat modification, pollution, and hunting. Their faithfulness to traditional sites in migration and winter makes them particularly vulnerable to localized habitat loss (e.g., 80 percent of mid-continent Sandhill Cranes stage in spring along NE's Platte River). Birdlife International codes: VU 7, EN 3, and CR 1.

Genus *Antigone*

SANDHILL CRANE *Antigone canadensis* SACR ■ 1

This crane is the only one seen in most of N.A. The sight of several hundred thousand staging in spring along the Platte River in NE is one of the great wildlife spectacles in N.A. Polytypic (5 sspp.; 4 in N.A.). L 34–48" (86–122 cm) WS 73–90" (185–229 cm)

Identification **ADULT:** Gray overall with variable rusty blotching on neck, body, and upperwing coverts. Naked forecrown is dark red. **JUVENILE/FIRST-WINTER:** Feathered head, neck, and upperparts cinnamon to rusty. Post-juvenal molt changes over first winter, with forecrown becoming unfeathered midwinter through summer; plumage becomes variegated with newer gray and rusty feathers. After second postbreeding molt, resembles an adult.

Geographic Variation Five subspecies, four in N.A. A fifth, *nesiotus* (Endangered), is an endemic resident to Cuba, including Isla de la Juventud. The smallest is nominate *canadensis*, the "Lesser Sandhill Crane" or the "Little Brown Crane." It breeds from northeast Russian Far East, AK, and across northern Canada to Baffin I. It winters from CA to TX and northern Mexico. The largest numbers migrate north of the AK Range and through the central Great Plains. It is these birds that form the great spectacles in migration (e.g., along the Platte River, NE). At the opposite end of the size extreme is *tabida*, the "Greater Sandhill Crane" (now includes intermediate-size "*rowani*" breeding at the southern edge of range of *canadensis*) that breeds from southeast AK and across southern Canada and the northern US and winters from CA, the Southwest, and the Southeast. Eastern populations of *tabida* are a little smaller and darker than western ones. Two other large resident subspecies are pale *praetensis* from GA and FL, and darker *pulla* from southern MS. An additional local and resident subspecies (*nesiotes*) is found in Cuba. Identification to subspecies is best done by range, although some subspecies overlap in migration and winter and can be found in mixed flocks. Identification to subspecies by size is very tricky. Males are 5 to 10 percent larger than females, but size extremes can still be striking in a mixed subspecies flock.

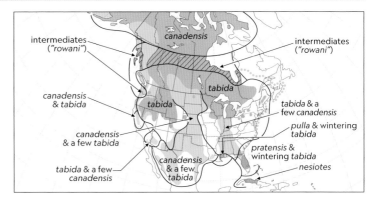

Similar Species Whooping Crane is larger and mostly white with contrasting black wing tips. See Common Crane. Great Blue Heron, often referred to by non-birders as a "crane," is rare in groups of more than 10 to 20 birds and usually flies with its neck pulled in with steady downbeats of arched wings. A distant resting heron has a longer neck, often kinked; longer bill; and lacks a bustle.

Voice Deep, far-carrying, rolled or rattled honking cries, *k'worrrh*, *grrrowh* or *grrrah-uu*, and *ah grruu*. In duets of mated pairs, female calls higher and shriller. "Lessers" average higher-pitched calls. Immatures have very different, high, slightly reedy trills through at least first year.

Status & Distribution **BREEDING:** Common to locally rare (Apr./May–Sept.) in a variety of open, undis-turbed habitats in N.A., Russian Far East, and Cuba (where resident). **MIGRATION:** Mainly Feb.–Apr./May and Sept.–Nov. **WINTER:** Farmland and open country, usually near marshes and shallow lakes or playas in N.A. and south to northern Mexico. **RARE STATUS:** Rare along West and East Coasts. Accidental in Europe and HI; very rare in Japan.

Population MS subspecies *pulla* is Endangered; its population is 75 to 80 percent captive-produced. Most populations are stable or increasing; some states permit hunting.

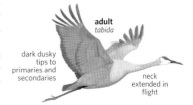

adult
tabida

dark dusky tips to primaries and secondaries

neck extended in flight

red crown

adult
tabida

juvenile
tabida

all cranes have tertial "bustles"

stained adult
tabida

COMMON CRANE *Grus grus*   CCRA ■ 4

This Old World species, a casual visitor to N.A., is usually found with migrating Sandhill Cranes. Polytypic (2 sspp.; sspp. in N.A. unknown). L 44–51"

juvenile
lilfordi

yellowish bill

black neck with white stripe

adult
lilfordi

(112–130 cm) WS 79–91" (201–231 cm)
Identification Contrasting black flight feathers on all ages. **ADULT:** Gray overall. Naked crown is black with red band behind eye; black foreneck contrasts with white auriculars and hind neck. **JUVENILE/FIRST-WINTER:** Upperparts extensively cinnamon; head and neck develop muted adult pattern.
Similar Species Larger than Sandhill Crane; easily separated by head and neck pattern and in flight by more extensively dark remiges.
Voice Similar to Sandhill Crane, a far-carrying, trumpeting *krooh* and *krooah*, and a harsher *kraah*. Immature gives high reedy *peep* or *cheerp* calls through its first year.

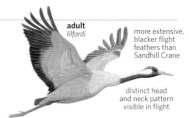

adult
lilfordi

more extensive, blacker flight feathers than Sandhill Crane

distinct head and neck pattern visible in flight

Status & Distribution Eurasia. **RARE STATUS:** Casual in migration and winter from AK to Great Plains and to CA, BC, NV, IN, QC, and recently AZ (origin uncertain); of about 20 records, some may pertain to escapes. Mixed Common (at least one a known escape) and Sandhill Crane pairs with hybrid offspring have been seen in eastern N.A. Also recorded from Greenland.
Population Extirpated as a breeding species from most of western Europe.

WHOOPING CRANE *Grus americana*   WHCR ■ 2

This large endangered species is best known on its winter grounds in coastal TX. Monotypic. L 50–55" (127–140 cm) WS 87" (221 cm)
Identification Black wing tips, usually concealed at rest, contrast strikingly in flight. **ADULT:** Large, white overall with naked red crown, black face, dark red malar. Has dark greenish bill with orange-yellow base; pale yellow eyes. **JUVENILE/FIRST-WINTER:** Feathered head, neck, and upperparts are extensively cinnamon. **SECOND-YEAR:** After second postbreeding molt in first summer, the Whooping more closely resembles an adult. Occasional cinnamon-tipped underwing coverts may be shown by birds of any age.
Similar Species Distinctive. Leucistic Sandhills have been confused with Whoopings; note wing-tip color.

(re-)introduction program

Voice Loud bugling that carries more than a mile: *k'raah-hu k'raah*; higher pitched and less throaty than Sandhill Crane. In duets of mated pairs, female calls are higher and shriller. Immatures have strikingly different, reedy whistles through at least their first winter.
Status & Distribution N.A.; formerly wintered south to central Mexico. **BREEDING:** Rare (May–Sept.), in freshwater marshes found in boreal forests. **MIGRATION:** In spring, mainly late Mar–early May; in fall, mid-Sept.–mid-Nov. Flight path along 50- to 100-mile-wide corridor in fairly direct route across Great Plains between breeding and wintering grounds. Includes brief fall staging in southern SK, irregular spring and fall stopovers in central NE, KS, and OK. **RARE STATUS:** Casual recently in migration west to southeast BC and CO, south to south TX, and east to IL and AR. **WINTER:** Restricted to estuarine marshes and shallow coastal bays in southeast TX, arriving late Oct.–mid-Nov., departing late Mar.–Apr.
Population Endangered. Resident population in LA was extirpated in 1950. The main population breeds in and around Wood Buffalo NP, AB, and winters in coastal TX in and around Aransas NWR. Some 100 were introduced to states farther east (WI and

black wing tips

adult

FL); most are migratory (to FL) and are occasionally seen in other eastern states. About 150 are in captivity. Conservation measures have brought it back from only 15 to 16 birds in the 1940s to about 600 individuals today.

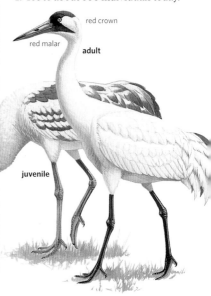

red crown

red malar

adult

juvenile

THICK-KNEES Family Burhinidae

Double-striped Thick-knee (Mexico, Mar.)

Thick-knees are large shorebirds with a unique cryptic plumage. Their crepuscular and nocturnal habits are also different from other shorebirds. They earned their name for rather thick intertarsal joints (knees).

Structure Thick-knees are large and heavily built like plovers. The head and eyes are large and the bill is stout and straight. Wings and tail are fairly long and the legs are exceptionally long.

Behavior Thick-knees are terrestrial and can run quite fast. Their typical gait is ploverlike with runs punctuated by abrupt stops. Flight is typically low over the ground with long legs projecting well beyond the tail. Wingbeats are silent and rapid. During the day they often sit or stand in the shade of a bush or tree, often with other thick-knees. Most of their activity is between dusk and dawn. They feed on insects, reptiles, and amphibians. Their nests are simple scrapes on the ground.

Plumage The sexes look alike, although females average slightly smaller than males. Their highly cryptic plumage camouflages them while they rest during the day. Juveniles are duller and can be separated from adults; adult plumage is achieved in approximately a year.

Distribution This family is composed of nine species that are found primarily in the tropics and the Old World. The two *Esacus* species live along rivers or coastlines in south Asia and Australia, while *Burhinus* thick-knees typically nest in savanna or other grassland habitats.

Taxonomy Nine species in two genera. Seven species are in the genus *Burhinus*; two species are from the New World, but only one is found in the Northern Hemisphere. Thick-knees were once thought to be more closely related to the bustards (Otididae) of the Old World, and indeed they superficially resemble bustards, but skeletal, biochemical, and other studies have proved thick-knees to be relatives of the shorebirds in order Charadriiformes.

Conservation More information is needed, although some species might actually benefit from forest destruction as it opens habitats they could not otherwise use. BirdLife International codes: 2 NT.

Genus *Burhinus*

DOUBLE-STRIPED THICK-KNEE *Burhinus bistriatus*   DSTK ▪ 5

The large head and eye, cryptic plumage, and unique behavior separate this accidental species from all other shorebirds. The terrestrial Double-striped Thick-knee can run quite fast; indeed, it seems to prefer running over taking flight. When it does fly, it is fast and low to the ground, with legs stretched behind the tail, showing white in the wings. The Double-striped Thick-knee becomes most active after sundown; its large eyes enable it to hunt in the dark for insects and other small animals. Polytypic (4 sspp.; TX specimen is nominate *bistriatus* from Mexico and C.A.). L 16.5" (42 cm).

Identification Brown upperparts; a gray-brown face, neck, and chest; a white belly; and a brown tail. A dark lateral crown stripe borders the top edge of its bold white supercilium. **ADULT:** The brown upperparts are edged cinnamon or buff in fresh plumage. Its face, neck, and chest are gray-buff and streaked darker. Its yellow bill has a dark tip. The legs and eyes are bright yellow. **JUVENILE:** It is duller and grayer overall than an adult with greenish yellow legs. The streaking along the neck is narrower and darker. In addition to the white supercilium and dark lateral crown stripe, a dark stripe below the supercilium extends to the nape. **FLIGHT:** Striking. The upperwing is dark with the outer primaries marked with a short white bar. A second white bar covers the base of the inner primaries. The underwings are white.

Similar Species Unique and offers no identification challenge.

Voice CALL: Loud and far-carrying barking or cackling *kah-kah-kah*.

Status & Distribution Fairly common to common resident in semiarid savannas and grasslands from southern Mexico to northern S.A.; also on Hispaniola. **RARE STATUS:** One record for US, King Ranch, TX (5 Dec. 1961, specimen). A tame individual in AZ was determined to have been imported from Guatemala.

Population Large geographic range with apparently stable population.

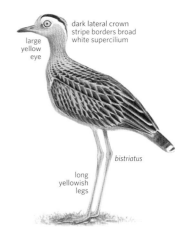

dark lateral crown stripe borders broad white supercilium

large yellow eye

long yellowish legs

bistriatus

STILTS AND AVOCETS Family Recurvirostridae

Black-necked Stilt, female (CA, Apr.)

Stilts and avocets are found around the world, primarily in warmer climates. Typically pied black-and-white, they walk gracefully on long legs, particularly the stilts.

Structure These large shorebirds are slender. The bill is long and thin, and typically black; the bill of the avocet is thicker at the base and upturned. The neck is often held in a gentle curve or pulled into the shoulders. Wings are long and pointed; in flight, the neck is outstretched and the legs trail behind the tail. Males are larger than females in stilts and equal in size in avocets.

Behavior Typically feed along the water's edge or in deeper water beyond the reach of other shorebirds. They pick at the surface, but they can also grab invertebrates well below the surface. On land, their gait is typically graceful when walking (it can be brisk with long strides), but they do appear gangly when they run. Their flight is direct, and avocets appear to be stronger fliers than stilts. Both genera swim; avocets do so well and more frequently than stilts. These are highly social birds, particularly so in the nonbreeding season, when they might form large flocks. They are usually colonial nesters and mob predators with a zeal beyond most shorebirds. Unlike most shorebirds, they are quite conspicuous when breeding, out in the open and very noisy instead of the more common cryptic strategy.

Plumage Both stilts and avocets have a striking black-and-white pattern. Differences between the sexes are slight, but noticeable. Seasonal differences are marked in avocets, but slight in stilts. The wings are dark in stilts and mixed black-and-white in avocets.

Distribution Species are scattered around the world, primarily in warmer climates. Most breeding populations withdraw to (or close to) the coasts in winter.

Taxonomy The species-level taxonomy of this family continues to be debated. The stilts (*Himantopus*) have been placed into as few as one, and as many as six, species. Avocets (*Recurvirostra*) have also been subject to taxonomic revision, but the recognition of four species spread across the world is generally accepted. The stilts and avocets share behavioral traits, and the third genus in the family (*Cladorhynchus*), the Banded Stilt from Australia, is intermediate between the two genera. Stilts and avocets have hybridized in captivity and in the wild in CA.

Conservation In North America, hunting and trapping sharply reduced populations in the 19th century, but populations are reclaiming some of the former breeding areas along the mid-Atlantic coast. Pollutants and habitat loss remain threats. The Black Stilt (*Himantopus novaezelandiae*) from New Zealand is critically endangered, with 106 adults (2017). BirdLife International codes: 1 CR.

STILTS Genus *Himantopus*

BLACK-WINGED STILT   *Himantopus himantopus*   BWST ▪ 5

An accidental visitor from the Old World, the Black-winged Stilt picks at the water with its needlelike bill for prey. Polytypic (2 sspp.; nominate in N.A.). L 13" (33 cm)

Identification Black back and wings; white underparts; long, bright pink legs. **ADULT:** Head and neck vary from entirely white to extensively black, although a portion of the hind neck will remain whitish or pale gray-brown. Extent of black on head and neck not diagnostic of sex, although males average more white headed. **BREEDING MALE:** Back glossy, pink flush to underparts at onset of breeding season. **FEMALE:** Duller than male, lacking gloss and pink flush. **JUVENILE:** Gray-brown back; thin, white trailing edge to wing (visible in flight); legs duller and paler.

Similar Species Black-necked Stilt always has entire hind neck black and has white spot above and behind eye.

Voice CALL: A repeated *kik, kik, kik* like Black-necked Stilt.

Status & Distribution Accidental. **BREEDING:** Nominate subspecies breeds in Europe and locally in Asia and the Middle East. **RARE STATUS:** Three spring records for western Aleutians and Pribilof Is. This species is becoming more frequent in Japan and Korea, with an increasing number of breeding records; there have been a few extralimital records in the Russian Far East. **Population** Stable.

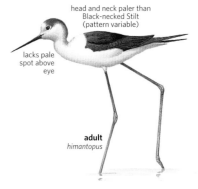

head and neck paler than Black-necked Stilt (pattern variable)

lacks pale spot above eye

adult
himantopus

BLACK-NECKED STILT *Himantopus mexicanus*   BNST ▪ 1

This graceful wader with an elegant posture is highly social. Polytypic (2 sspp.; nominate in N.A.). L 14" (36 cm)

Identification Boldly pied, black above and white below; long pink or red legs. The dark head has a white spot above and behind the eye. **BREEDING MALE:** Glossy black back, red legs, and often with a pink flush at beginning of breeding season. **NONBREEDING MALE:** Duller legs; gloss and pink flush absent. **FEMALE:** Back has brown tones. **JUVENILE:** Paler than adult. Buffy edges to upperparts; grayish pink legs. Inner primaries and secondaries tipped white, visible in flight. **Similar Species** See accidental Black-winged Stilt.

Voice CALL: A loud, piercing *kek kek kek*, particularly when disturbed while breeding. A loud *keek*, recalling a Long-billed Dowitcher, is a common contact note.

Status & Distribution Locally common. **BREEDING:** Marshy areas and shallow ponds with emergent vegetation. Range is spreading north in interior states and along Gulf and Atlantic coasts. **MIGRATION:** Short to medium-distance migrant, usually away from interior areas during winter. Spring migrants primarily Mar.–Apr., but locally as early as late Feb., and into May. Fall migration protracted with many areas seeing movement in July, but most areas peak Aug.–Sept., trickling into early Nov. **WINTER:** Central and Southern CA, southern AZ, Gulf and Atlantic coasts, and southern FL. **RARE STATUS:** Rare or casual across southern Canada and various places in the Northeast and Great Lakes.

Population Despite recent expansions, the species has not recovered from 19th-century losses.

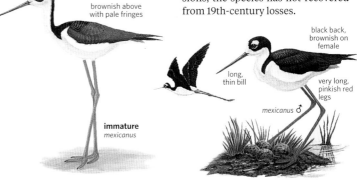

brownish above with pale fringes

black back, brownish on female

long, thin bill

very long, pinkish red legs

mexicanus ♂

immature *mexicanus*

AVOCETS Genus *Recurvirostra*

AMERICAN AVOCET *Recurvirostra americana*   AMAV ▪ 1

Avocets feed in water by sweeping their bills back and forth, and on mudflats, where they peck at the surface for prey. Monotypic. L 18" (46 cm)

Identification Black scapular stripes; white underparts; black-and-white wings. Long, thin bill is recurved; male's bill is longer and straighter, female's more recurved. Bluish gray legs. **BREEDING:** Rusty colored head and neck, with white eye ring and white at base of bill. **NONBREEDING:** Pale gray head and neck. **JUVENILE:** Pale cinnamon wash on head and neck; pale fringes on tertials and coverts.

Voice CALL: Loud, piercing *wheet* or *kleep*, usually repeated when agitated, higher pitched than the call of a stilt.

Status & Distribution Fairly common. **BREEDING:** Shallow ponds, marshes, and lakeshores. **MIGRATION:** Anywhere in the West and locally along the Atlantic (mostly DE and south) and Gulf Coasts. Spring migration primarily mid-Mar.–May. A few found in the Great Lakes region with arrivals mid-Apr.–mid-May, dozens in exceptional years. Fall migration July–Oct. with adults through early Aug., followed by juveniles mid-Aug.–mid-Oct. Lingerers into Nov. **WINTER:** Coastal and locally in the interior of CA; locally along Gulf Coast, and southern FL. Very rarely inland near more northerly breeding sites (e.g., UT, NV, OR). **RARE STATUS:** North to southern AK and southeast Canada.

Population Largely stable.

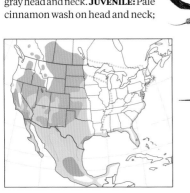

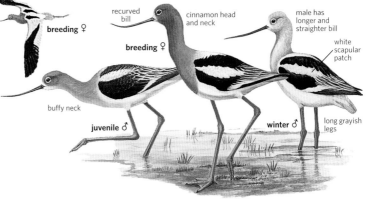

recurved bill

cinnamon head and neck

male has longer and straighter bill

breeding ♀

breeding ♀

white scapular patch

buffy neck

juvenile ♂

winter ♂

long grayish legs

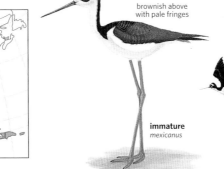

OYSTERCATCHERS Family Haematopodidae

American Oystercatcher, adult (left) with juvenile (NY, July)

Oystercatchers are large, chunky birds usually tied to the coast. Their bills, which they use like a chisel during feeding, are long and bright orange or red.

Structure These are bulky shorebirds, larger than most species. The stout and straight bill flattens laterally toward the tip. Oystercatcher legs are short and thick, and typically pink or red.

Behavior Oystercatchers frequent sandy beaches or rocky shores. They tend to run rather than take flight to escape danger. When they stand upright, they often tuck the neck in and they hold the bill below horizontal. Flight, usually low to the surface, is powerful and direct on deep, rapid wingbeats with rounded wings. To feed,

the oystercatcher plunges its bill into the sand for prey or uses its bill to chisel mollusks off surfaces and pry them open. The birds feed singly, or in small, noisy flocks on coastal beaches and mudflats. Unlike most shorebirds, they feed their young, and the young often stay with the parents through the first winter.

Plumage Oystercatchers are black or pied (black and white). Closer views show substantial brown in the upperparts plumage, with only the Eurasian Oystercatcher adult being truly black. Juveniles are paler, with duller soft part coloration.

Distribution Most oystercatchers live on shores around the world's coasts. The highly migratory Eurasian Oystercatcher, breeding in the interior of western and central Asia, is an exception.

Taxonomy The family is composed of up to 12 species all in the genus *Haematopus,* but some authorities recognize as few as four species. The pied species and the black species look very different, but hybridization occurs, resulting in intermediate plumages. In the US, two species are mostly resident and one is accidental. The Ibisbill (*Ibidorhyncha struthersii*) of mountain riverbeds in south Asia is included in the oystercatcher family by some authorities.

Conservation Beach development impacts oystercatchers, as do spills of harmful pollutants (e.g., the *Exxon Valdez* spill in AK); yet, they quickly colonize dredged areas. BirdLife International codes: 1 NT, 1 EN, 1 EX.

Genus *Haematopus*

EURASIAN OYSTERCATCHER   *Haematopus ostralegus*   EUOY   ■   5

This visitor is from the Old World. Polytypic (3 sspp.; probably 2 in N.A.). L 16.5" (42 cm)

Identification Black head, neck, and upperparts contrast with the white rump, wing stripe, and remainder of the underparts. Eyes, eye rings, and bill

white extends up back

bold white wing stripe

breeding adult
ostralegus

are bright red. **ADULT BREEDING:** Legs pinkish red. **ADULT NONBREEDING:** A white bar extends across the lower throat. **JUVENILE:** Browner above, with a narrower tail band, than the adult. Dark-tipped bill, brown eyes, and grayish legs; bare parts get brighter over first year. **FLIGHT:** White wing stripe extends almost to tip, with pale primary shafts on outer primaries.

Geographic Variation Subspecies include nominate *ostralegus* of western Europe (likely the subspecies of birds seen in NL) and *osculans* of the Russian Far East (likely the subspecies of the bird seen in AK). The subspecies *osculans* has a slightly shorter wing stripe and dark outer primary shafts.

Similar Species Similar to American Oystercatcher but adult black, not brown, above; eyes red. In flight, white wing bar bolder and more extensive.

Voice CALL: Loud, piercing whistles.

Status & Distribution BREEDING: Eurasia, from Iceland to Kamchatka, more local in Asia. **MIGRATION:** Southern populations largely resident, northern populations migratory. **WINTER:** Southern Europe, North Africa, Arabian Peninsula, coastal Asia. **RARE STATUS:** Accidental. Three records from Newfoundland (two in spring, one in fall) and one in spring from Buldir I., Aleutians, AK.

Population Declines; Near Threatened (BirdLife International).

AMERICAN OYSTERCATCHER *Haematopus palliatus* AMOY ■ 1

Found along sandy beaches, oyster bars, bays, and mudflats. Polytypic (5 sspp.; 2 in N.A.). L 18.5" (47 cm)

Identification Striking. Black head with contrasting brown back, white underparts, large red-orange bill, pinkish legs, and yellow eyes. **JUVENILE:** Whitish fringes on upperparts give a scaly appearance. Duller bill with a dusky tip, dusky eyes, and grayish pink legs. **FLIGHT:** Bold white wing stripe and white base of tail.

Geographic Variation Five subspecies, including nominate *palliatus* (along the Atlantic coast) and *frazari* (casual in Southern CA from nearby breeding grounds in Baja California). Some *frazari* appear identical to *palliatus*, but most show some dark mottling on lower breast and uppertail

coverts (white on *palliatus*); white wing stripe shorter. This subspecies is perhaps best considered as an intermediate population.

Similar Species See Eurasian Oystercatcher. Hybridization between *frazari* and Black Oystercatcher creates a variety of plumages.

Voice CALL: Loud *wheep* or *whee-ah*, often rising into an excited chatter.

Status & Distribution Fairly common. **BREEDING:** Populations extend south through both coasts of Mexico, C.A., and S.A. Expanding northward in the East. **MIGRATION:** Poorly understood but some northern Atlantic popula-

tions withdraw in Sept., returning Mar.–Apr. **RARE STATUS:** Rare coastal Southern CA. Accidental at Salton Sea, CA (flock of three) and ON.

Population Recent expansion in the Northeast, into southern ME and southern NS.

adult
frazari

limited black flecking on lower breast reflecting intergradation with Black Oystercatcher; some are more extensively marked below

limited flecking on undertail coverts

bold white wing stripe

adult
palliatus

gold eye

long red bill

black neck contrasts sharply with white belly

brown back

darker bill

adult
palliatus

juvenile
palliatus

BLACK OYSTERCATCHER *Haematopus bachmani* BLOY ■ 1

long red bill

overall blackish coloration; back browner

adult

The Black Oystercatcher is a dark, large, chunky shorebird likely to be seen only along the rocky coastline of the Pacific. It is typically found hunting for food on tide-exposed rocks, where it blends into the dark rocks, only to be revealed by its large, red-orange bill. It demonstrates tremendous site fidelity, with pairs typically returning to the same territory for many years. Monotypic. L 17.5" (45 cm)

Identification All-dark body washed with brown on the upperparts and

belly. Yellow eyes with a red eye ring; legs pinkish. Sexes similar, except females are slightly larger and bill is slightly more orange than male. While subtle, these differences might be evident in pairs. **JUVENILE:** Outer half of bill is dusky, legs are grayish pink, and eyes are dusky. Fine, pale edges to upperparts in fresh plumage.

Similar Species Unmistakable in N.A., but beware of hybrids between Black and American Oystercatchers in northern Baja California and Southern CA: The resulting offspring can look similar to either parent. See American Oystercatcher account for more detail.

Voice CALL: Similar to American Oystercatcher. Loud, piercing *wheep* or *whee-ah* notes can accelerate into an excited chatter.

Status & Distribution Common. **BREEDING:** Rocky shores and islands along the Pacific coast from the Aleutians to Baja California. **MIGRATION:** Largely resident, but some local move-

ments, not all of which are understood. Many northern breeding areas are abandoned as some birds move south. Autumn numbers in BC peak in Nov., with vacated territories reoccupied in Mar.–Apr. **WINTER:** Territories abandoned and flocks formed during winter, but range largely within breeding range. **RARE STATUS:** Casual Pribilofs, AK. Jan. record of a "distressed" bird found at 3,400 ft in the Washington Cascades is hard to explain.

Population Human disturbances have eliminated local populations in the Pacific Northwest.

LAPWINGS AND PLOVERS Family Charadriidae

Black-bellied Plover (NY, Nov.)

Birds of wetlands, grasslands, and shorelines, plovers are gregarious. They are known for undertaking long migrations, but some species are sedentary.

Structure Plovers have medium to long legs, a rather sturdy frame, a short neck, and a round head. The bill is short; thick at the base, it usually narrows and then expands at the tip. Most plovers have long, pointed wings, excellent for rapid flight and energy-efficient for long migrations. The less migratory lapwings have more rounded wings. Plovers' pectoral muscles are relatively large; this fact and their ability to store large amounts of fat allow them to migrate long distances over the ocean.

Behavior Plovers hunt by running and abruptly stopping. After a few seconds, if no prey is taken, they run again. They pick small invertebrates from either wet or dry surfaces. They rarely perch; most plovers lack the hind toe that would help them do so. Flight is generally rapid for "true" plovers, but floppy in the broad-winged lapwings. The nest is typically a scrape, and the precocial young feed themselves. Some species have exaggerated "broken-wing" displays to distract potential predators. Quite gregarious, they form flocks for migration and roosting; however, they are very territorial during breeding season. Some species will defend feeding territories in winter and during migration.

Plumage Sexual differences are slight, with males usually brighter than females. Plovers typically molt their contour/body feathers in spring. Most species are in breeding plumage when they migrate, a few start in winter plumage and molt on the way north. Many spe-

cies have black facial markings, which are boldest during breeding season and may play a role in territorial displays. Most Arctic breeders start molting head feathers when incubation begins. Fall migrants usually show some signs of molt, but some species start migrating before much molt is evident. Juveniles do not distinctly differ from winter-plumaged adults, other than look far more crisp in early fall. Some first-summer birds look similar to breeding adults, whereas others look like winter-plumaged birds. First-years have varying migration strategies, sometimes remaining on the wintering grounds, moving north to an intermediate point, or moving all the way to the breeding grounds.

Distribution Worldwide, plovers inhabit the tropics, mountains, deserts, and a variety of wetlands. They are both resident and some of the longest-distance migrants in the avian world. Seventeen species have been recorded in North America.

Taxonomy Generally the family of 11 genera and 67 species is divided into lapwings and "true" plovers, but some work suggests that the four Arctic-breeding *Pluvialis* species belong in their own subfamily. The 11 genera include some rather unique plovers; however, North American birds are quite similar apart from one rare Old World visitor, Northern Lapwing.

Conservation Historically, plovers were hunted in large numbers (with tens of thousands killed on a spring migration day); they are now protected. Today, loss or degradation of habitat and human disturbance are the biggest threats. Birdlife International codes: 10 NT, 4 VU, 1 EN, and 3 CR.

LAPWINGS Genus *Vanellus*

Lapwings, 24 species worldwide, are generally more strikingly colored than other plovers. Their large, boldly marked, broad wings often have spurs at the carpal joint. While most lapwings are sedentary and found in the tropics, some species live at higher latitudes and are migratory. They are quite noisy when approached, particularly when on the breeding grounds.

NORTHERN LAPWING *Vanellus vanellus*   NOLA ■ 4

long rounded wings with white tips

long wispy crest

winter

white wing linings and tail base

broad black breast band

rich buff undertail coverts

winter

A visitor from Eurasia, the Northern Lapwing is closely tied to wet or dry fields. Monotypic. L 12.5" (32 cm)
Identification Black and white from a distance, it is actually quite colorful. The face, throat, and upper breast are black; the back has a glossy green sheen. Most of the auriculars and adjacent areas are gray; remaining underparts are white, with pale cinnamon undertail coverts. Wings are broad with white tips, obviously rounded; wing linings white. Flight is slow and floppy. White rump and base of tail contrast with the tail's black tip. Wispy but prominent crest is unique. **NONBREEDING ADULT:** Most of the gray near the face of a breed-ing adult is replaced with buffy tones; black is reduced and throat is white. Upperparts have pale buff edges to the feathers. **JUVENILE:** Duller than adult, with a shorter crest, and bolder edges to upperparts. Legs are gray.
Similar Species None. Unique.
Voice CALL: Whistled *pee-wit*.
Status & Distribution Eurasian species. **RARE STATUS:** Casual in the Northeast and Atlantic Provinces in late fall and winter. Accidental elsewhere in the East and at other seasons; recorded south to FL, west to OH; and at Shemya I., AK (12 Oct. 2006).
Population Near Threatened. Declining in UK and elsewhere in Europe.

Genus *Pluvialis*

The largest N.A. plovers, members of this tundra-breeding group are plump and tall. Their breeding plumage is impressive—black bellies (less black in females) with white- or gold-spangled upperparts—whereas their winter plumage is dull (usually grayish or dingy brown). They feed in the typical plover fashion and usually fly and roost in flocks.

BLACK-BELLIED PLOVER *Pluvialis squatarola*   BBPL ■ 1

While lacking the more colorful tones of the related golden-plovers, the Black-bellied Plover is a handsome bird in breeding plumage. It feeds singly or in loose flocks, but collects into tight flocks for roosting. Polytypic (3 sspp.; 2 in N.A., not field-identifiable). L 11.5" (29 cm)
Identification Largest *Pluvialis*. Bill bulky. **BREEDING MALE:** Frosty crown and nape, black-and-white barred upperparts and tail. Black face, throat, and breast to belly; remainder of underparts white. **BREEDING FEMALE:** Less black; varies from rather dull to almost as much black as males. **NONBREEDING ADULT:** Drab gray-brown above, with only a hint of a supercilium; dingy below. **JUVENILE:** Like adult, but upperparts spotted whitish, with buff tone; fresh juvenile can have pale buff spots. Breast more heavily streaked. **FLIGHT:** White uppertail coverts, barred white tail, and bold white wing stripe from above; black axillaries from below.

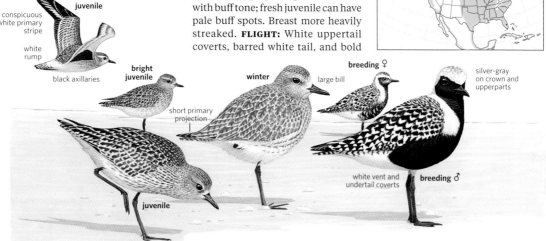

conspicuous white primary stripe

juvenile

white rump

black axillaries

bright juvenile

short primary projection

winter

large bill

breeding ♀

silver-gray on crown and upperparts

white vent and undertail coverts

breeding ♂

juvenile

Similar Species Winter and juvenile American Golden-Plover and, to a lesser extent, Pacific Golden-Plover are similar. See those species. Structurally, Black-bellied's bill is larger, with a bulbous tip, and its overall size is larger, more robust.
Voice CALL: A drawn out, three-note whistle, *wee-er-ee*; the second note is lower pitched.
Status & Distribution Common.

BREEDING: Dry Arctic tundra. **MIGRATION:** In spring, southern states (where they also winter) peak mid-Apr.–early May; late Apr. peak in NW; Great Lakes peak mid to late May; lingerers into June, nonbreeders linger all summer. In fall, adults early July, typically late July–Aug., some linger to late Sept.; juveniles first arrive late Aug. (exceptionally earlier), and peak mid- to late Sept., lingerers to mid-Nov., excep-

tionally later. Overall, an uncommon migrant in the interior away from the Great Lakes. **WINTER:** Common on coasts to S.A., Central and Willamette Valleys in CA and OR, and Salton Sea. Otherwise, rare to accidental elsewhere in the interior in winter.
Population Nineteenth-century hunting did not victimize the Black-bellied Plovers as much as it did the American Golden-Plovers.

EUROPEAN GOLDEN-PLOVER *Pluvialis apricaria*   EUGP ■ 4

This species is a rare visitor from the Old World. Polytypic (2 sspp., presumably *altifrons* in N.A.). L 11" (28 cm)
Identification Bright white underwings diagnostic in all plumages. Short tertial extension emphasizes long primaries. **BREEDING MALE:** Broad pure white on sides of underparts; flanks and undertail coverts more purely white, and gold spotting denser than on Pacific Golden-Plover. **BREEDING FEMALE:** Less black below. **NONBREEDING ADULT:** Brownish, with yellow-gold spangles above; dingy whitish

below. **JUVENILE:** Supercilium and auriculars rather indistinct; heavily barred below. **FLIGHT:** Uniform brown upperparts, including rump and tail.
Similar Species See other golden-plovers.
Voice CALL: Plaintive even whistle.
Status & Distribution Eurasian species. **BREEDING:** Greenland to western Siberia. **WINTER:** Europe to North Africa. **RARE STATUS:** Irregular spring migrant to NL; casual elsewhere in Atlantic Canada and along the East Coast of US, south to DE. Accidental AK.
Population Decreasing.

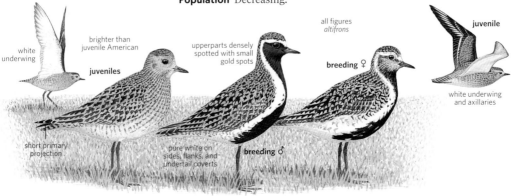

white underwing

brighter than juvenile American

juveniles

short primary projection

upperparts densely spotted with small gold spots

pure white on sides, flanks, and undertail coverts

breeding ♂

breeding ♀

all figures *altifrons*

juvenile

white underwing and axillaries

AMERICAN GOLDEN-PLOVER *Pluvialis dominica*   AMGP ■ 1

Golden-plovers are most frequently seen as migrants, often in flocks. Monotypic. L 10.3" (26 cm)
Identification American Golden-Plover is slimmer and longer winged than Black-bellied Plover. Early spring arrivals are in nonbreeding plumage; they molt over the course of spring. Four or five primaries extend beyond the tertials. **BREEDING MALE:** Gold spotting on most upperpart feathers. White forecrown extends into the supercilium and then broadens into a neck stripe, which wraps to the sides of the breast, extending slightly toward the breast's center, but goes no farther down the

lower sides and flanks. Underparts are usually all black. **BREEDING FEMALE:** Like male, but duller; white flecks may be present throughout the underparts. **NONBREEDING ADULT:** Molts on its way south, but early arrivals in southern Canada and the lower 48 are likely to be in partial breeding plumage; later in fall, it usually retains some black flecks below. Full nonbreeding plumage is dull brown above and dingy below, with minimal marking, except for the supercilium. **JUVENILE:** Grayish brown overall, with a bold whitish supercilium set off by dark cheek. Upperparts are gray with gold-notched feathers. Under-

parts dingy, with pale brown barring or mottling on sides of breasts and flanks.
FLIGHT: Rather uniform brown above,

including the rump; indistinct white wing stripe. From below, underwings pale gray with no black in axillaries.

Similar Species American Golden-Plover is similar in all plumages to the Pacific Golden-Plover (see that species). In nonbreeding and juvenal plumage, American Golden-Plover is also similar to Black-bellied Plover in those plumages. Compared to Black-bellied, the smaller American has a more slender bill, bolder supercilium, and darker crown; in flight, it appears uniform brown above, and the pale gray underwings lack black axillaries.

Voice CALL: Shrill two-note *queedle* or *klee-u*, with the second note shorter and lower pitched.

Status & Distribution Fairly common. **BREEDING:** Dry tundra. **MIGRATION:** In spring, primarily through the Midwest where partial to moist margins of flooded fields, some to Atlantic coast. Early arrivals late Feb., but more typically Mar. in southern states, late Mar. in southern Great Lakes, with peak late Apr.–mid-May. Fall migration south to Midwest but most then move to off Atlantic coast for migration to S.A. A few adults first detected early July, but

more typically after mid-Aug, peak into Sept. Juveniles peak mid-Sept.–mid-Oct., smaller numbers into Nov., and exceptionally Dec. **WINTER:** Primarily in central S.A., winter status in US uncertain; some records likely pertain to very late migrants or misidentifications. Pacific and European Golden-Plovers should always be considered. **RARE STATUS:** Casual spring migrant and rare fall migrant in the West; these fall records are nearly all juveniles.

Population Historic declines from hunting with subsequent 20th-century increases; current studies needed.

compare juvenile carefully to juvenile Black-bellied

conspicuous supercilium with dark auricular

bright juvenile

Apr. ♂

breeding ♀

slender bill

more uniform upperwing

breeding ♂

juvenile

brown rump

long primary projection (usually four primary tips)

juvenile

black undertail coverts

bulging white patches on sides of breast; pattern of female similar, but duller

grayish underwing

PACIFIC GOLDEN-PLOVER *Pluvialis fulva* PAGP ■ 2

Pacific Golden-Plover, a powerful flier, is known for long, transoceanic migrations. It prefers sandy beaches more than the American Golden-Plover does. Monotypic. L 9.8" (25 cm)

Identification Primaries do not extend much beyond the tertials; the latter are long and often extend near the tip of the tail. Adults in spring are usually in breeding plumage; only second-year birds are expected in nonbreeding plumage in spring. Migrant fall adults in July already show signs of molt. **BREEDING MALE:** White forecrown extends into the supercilium and then becomes a stripe that extends to the neck and along the flanks, where

black spots or stripes mix with the white. Underparts are black from face and throat through to belly, but not the undertail coverts. **BREEDING FEMALE:** Similar to male, but duller, with black more limited and mottled. **NONBREEDING ADULT:** Grayish brown above with yellowish edges to feathers, and pale yellow supercilium; dingy whitish or with buff tones below. **JUVENILE:** Rather bright, with distinctly buffy yellow tones. Spots on upperparts; brown streaks on breast and neck. Bold, dark postocular spot. **FLIGHT:** From above, uniform brown, including rump and tail, and indistinct wing stripe; pale gray underwings from

below. Feet usually project noticeably beyond tail.

general coloration of juveniles and winter birds brighter than American

ear coverts paler than American; often with postocular spot

winter

breeding ♂

breeding ♂

white doesn't bulge out on sides of breast and usually continues to vent

juveniles

breeding ♀

upperwing like American

short primary projection

mostly white undertail coverts

longer legs than American

brown rump

slight foot projection

juvenile

grayish underwing

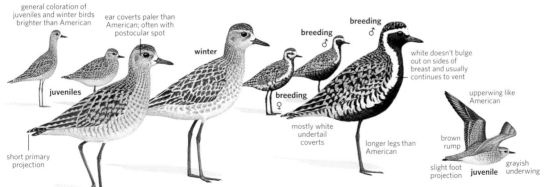

Similar Species American Golden-Plover is similar in all plumages. In breeding plumage, male American has a broad white stripe that stops at the sides of the breast, and wholly black underparts—a pattern unmatched by Pacific. However, a breeding-plumaged, female American has white flecks along the flanks, more similar to male Pacific. Juvenile and winter Pacifics typically appear brighter than juvenile and winter Americans, which are grayish brown with a bold white supercilium and a more extensively dark cheek. American has dusky barring or mottling on dingy white underparts; this barring can extend well past the legs. Pacific is usually buffier, almost yellow in some cases, with a buffy supercilium and a typi-cally more isolated postocular spot. Proportions in golden-plovers are subtly different and can be useful field marks. The tertials are long in Pacific and the primary extension is shorter, whereas the tertials are shorter in American, with a longer primary projection. Pacific typically shows three primary tips extending beyond the tertials, whereas American shows four or five. Fresh-plumaged juveniles are easiest to judge. Pacific Golden-Plover also looks to have longer legs (they project noticeably behind the tail in flight) and a longer and heavier bill than American.

Voice CALL: A loud, rich *chu-weet* with the last note higher pitched, like a Semipalmated Plover.

Status & Distribution Common breeder, rare transient. **BREEDING:** Prefers moister habitats than the American Golden-Plover, including the coast and river valleys. Russian Far East to western AK. **MIGRATION:** Movement from wintering grounds typically mid-Apr.–mid-May, with arrivals on breeding grounds mid- to late May, or June where snowmelt is later; arrivals into southwestern AK as early as late Apr. In fall, adults migrate primarily July–Aug., with lingerers into Sept. Juveniles are most numerous Sept.–Oct., lingerers into Nov. **WINTER:** Primarily southern Asia to Pacific islands; small numbers on West Coast and in central CA. **RARE STATUS:** Casual to NV and the Southwest; casual in late summer on East Coast. **Population** Possibly declining.

RINGED PLOVERS Genus *Charadrius*

Charadrius contains the most species of plovers: 30. Ringed plovers are brown or sandy-colored above and white below, and often have one or two breast bands. Many species have a white collar and/or black face markings during the breeding season. Many species reside in the tropics, while others breed at high latitudes and migrate long distances.

LESSER SAND-PLOVER *Charadrius mongolus*   LSAP   ■   3

The Lesser Sand-Plover's breeding plumage differs substantially from its nonbreeding plumage—quite unlike most of the smaller plovers. This Asian visitor was formerly called "Mongolian Plover." It is likely found on a beach or mudflat, as opposed to an irrigated field, like *Pluvialis* plovers. Polytypic (5 sspp.; 1 in N.A.). L 7.5" (19 cm)

Identification Midsize plover with a rather short stubby bill. Lesser lacks a pale collar, and has a dark bill and dark legs. **BREEDING MALE:** Bright rufous breast extends more broadly at the sides, usually bordered above by a narrow black line. Facial pattern is bold—black mid-crown, cheek, and lores contrast with a white forecrown and throat. **BREEDING FEMALE:** Duller than male, but still colorful. The facial markings are brown instead of black. **NONBREEDING ADULT:** Lacks the rufous breast band and black facial markings. Most birds show a white supercilium and have broad grayish patches on the sides of the breast. **JUVENILE:** Broad, buffy wash across the breast, sharply contrasting with the white throat. Upperparts are edged with buff. Legs usually dark grayish.

Geographic Variation N.A. records believed to be *stegmanni*, which has a white forehead and breeds farther northeast than others.

Similar Species The accidental Greater Sand-Plover presents the only identification challenge when Lesser Sand-Plover is in breeding plumage. In other plumages it might recall a Wilson's Plover, but note Wilson's pale legs, pale collar, and longer, heavier bill. The smaller Semipalmated Plover can appear to lack a collar when it is hunched over; its soft parts often retain some orange color, it is smaller, and it lacks the broad patches at the sides of the breast. Greater Sand-Plover, an accidental visitor, can be very similar to Lesser Sand-Plover in nonbreeding plumages. Most Lessers have dark legs, but juveniles can be quite pale-legged;

pattern similar to male, but duller

breeding *stegmanni* ♀

breeding *stegmanni* ♂

black forecrown bar

white forehead

white throat

rusty red breast

thick bill

grayish breast patch

winter *stegmanni*

winter *stegmanni*

buffy breast band

juvenile *stegmanni*

the thicker and longer bill of the Greater always the best distinguishing feature. **Voice CALL:** A short, rapid trill. **Status & Distribution** Asian visitor.

RARE STATUS: Rare but regular in migration in western and northwestern AK, casual in summer and it has bred. Casual along West Coast in fall.

Accidental elsewhere, primarily eastern N.A. (recorded in IN, ON, RI, NJ, VA, FL, LA); also AZ. **Population** Stable.

GREATER SAND-PLOVER *Charadrius leschenaultii*   GSAP ■ 5

Closely related to Lesser Sand-Plover, the Greater Sand-Plover is an accidental visitor from the Old World. Polytypic (3 sspp.). L 8.5" (22 cm)
Identification BREEDING MALE: Similar to Lesser Sand-Plover, but its rufous breast band is paler and not as broad. It also has black facial markings, but it lacks the dark border to the upper breast, and there's not as much contrast with the crown and throat. **BREEDING FEMALE:** Similar to male but duller and lacks black. **NONBREEDING ADULT:** Rufous and black colorations are absent. Brown breast band can be complete or incomplete. **JUVENILE:** Like a nonbreeding adult, but it has pale fringes on the upperparts. **FLIGHT:** Broad wing stripe is visible, and the toes project beyond the tail tip.
Similar Species Can be difficult to sep-

arate Greater and Lesser Sand-Plovers. Greater is a larger version of the Lesser: Its bill is up to 50 percent larger; its body is 10 percent larger than the average Lesser, with legs 20 percent longer. In combination, it looks more elongate, with a long bill. Greater's legs are pale greenish gray, sometimes yellowish green; some juvenile Lesser will show pale legs. Greater's larger bill, broader white wing stripe, and toes projecting beyond the tail in flight are the best features. Lastly, Greater's underwings are a cleaner white; Lesser has a dark mark in the under primary coverts. The long, thick bill closely suggests Wilson's, but Wilson's has a prominent pale collar.
Status & Distribution Old World species breeding from Middle East to Central Asia and wintering in coastal areas of eastern Africa and southern

Asia to Australia. **RARE STATUS:** Accidental. Two records: Marin Co., CA (29 Jan.–8 Apr. 2001) and Duval Co., FL (14–26 May 2009).
Population The nominate subspecies appears stable, while other subspecies have shown local declines due to wetlands destruction and runoff from irrigation impacting breeding grounds.

no white collar

long stout bill

long legs with yellow tint

winter

SNOWY PLOVER *Charadrius nivosus*   SNPL ■ 1

The Snowy Plover might be the most overlooked shorebird. In addition to being rather pale, it blends in with its surroundings, and on beaches, it often rests in furrows, cryptically hidden in the sand. A fast runner, it will run to escape disturbances before it will take flight. It generally eats invertebrates that it picks off the sand. Polytypic (3 sspp.; 2 in N.A.). L 6.3" (16 cm)
Identification Pale sandy brown above, with dark ear patch, and partial breast band that extends from the sides of the breast. White collar on nape is complete. Bill is thin and dark; legs are dark or grayish. **BREEDING MALE:** Ear patch, partial breast band, and forehead are black. Crown and

nape occasionally has a buffy orange tint. **BREEDING FEMALE:** Where the breeding male is black, the breeding female is mostly brown. **NONBREEDING ADULT:** Like a dull breeding female, with no black in the plumage. **JUVENILE:** Like a nonbreeding adult, with pale edges to the upperparts; it quickly fades and looks like an adult. Legs are pale and have a greenish hue. **FLIGHT:** White wing stripe is often conspicuous. Outer tail has bold white edges, and a dark subterminal bar on the tail contrasts with a paler base and rump.
Geographic Variation Gulf Coast birds from FL, *tenuirostris*, are paler dorsally (also show genetic differences) compared with

nivosus, the darker nominate subspecies from the interior West to the Pacific. Those from farther west on the Gulf Coast, from TX for example, while classified as *tenuirostris*, look more like *nivosus*. Some authors consider these the same subspecies, but differences are usually apparent. Snowy Plover was recently split from the Kentish Plover (*C. alexandrinus*) of the Old World based on strong genetic differences; the two also show distinct vocal differences.
Similar Species Females and juveniles resemble Piping,

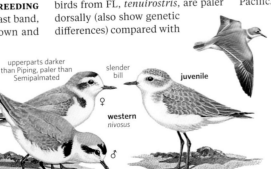

very pale upperparts, like Piping

upperparts darker than Piping, paler than Semipalmated

slender bill

juvenile

♀

western *nivosus*

♂

Gulf Coast ♂ *tenuirostris*

dark lateral patch

dark legs

the only plover that shares Snowy's pale coloration, although Piping Plover is paler than western Snowies. Note Snowy's thinner, black bill and darker legs.

Voice CALL: Low *krut* or *prit*, which can be rolled into a trill. Also a soft, whistled *ku-wheet*, second note higher. **Status & Distribution** Uncommon and declining on Gulf Coast. **BREEDING:** Barren, sandy beaches and, inland, alkali playas and marsh edges. **MIGRA-**TION: Some populations resident, others migrate short to medium distances. In spring, arrive early Mar.–Apr. in CA, slightly later on southern Great Plains. In fall, adults move from July, more frequently Aug.; migration completed by early Sept., with stragglers into Oct. **WINTER:** Primarily coastal, but irregularly in winter in southeastern CA, southern AZ, NM, western TX. Recently found locally in the Rio Grande Valley of south TX in relatively large numbers. **RARE STATUS:** Rare to casual outside of mapped range across much of N.A., including the Prairie Provinces, Great Lakes and along the Atlantic coast.

Population Near Threatened. The USFWS lists western *nivosus* as Threatened; at risk in the Southeast. Beach disturbance, as well as habitat loss and degradation, play a role in the diminished numbers of this species.

COLLARED PLOVER *Charadrius collaris*   COPL ▪ 5

The Collared Plover is found from northeast and western Mexico to S.A. Monotypic. L 5.5" (14 cm)

Identification This small plover is pale brown above and white below. It lacks the white collar that most small

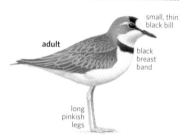

adult

small, thin
black bill

black
breast
band

long
pinkish
legs

plovers have. Black bill is small and notably thin; legs are pinkish and disproportionately long. Sexes similar. **ADULT:** Black breast band is complete and narrow. Forecrown is dark, often with a rusty border; auriculars, nape, and sides of breast often have distinct rusty tinges, especially in males. **JUVENILE:** Breast band is incomplete. Upperparts have pale rusty edges. **FLIGHT:** White wing stripe and white edge to outer tail are visible.

Similar Species Collared Plover is most similar to Snowy Plover. Collared is slightly darker above and has disproportionately longer, pinkish legs and lacks the white collar. An adult Collared has a complete breast band, whereas a Snowy has a partial breast band and dark legs.

Voice CALL: A sharp *pit*, sometimes repeated.

Status & Distribution Generally resident from southern S.A. to Mexico, as far north as southern Tamaulipas and Sinaloa, where it inhabits coastal and inland regions. **RARE STATUS:** Accidental, with three interior records (May–early Nov.) from south TX as far north as Uvalde.

Population Stable, although the population appears to be expanding in S.A.

WILSON'S PLOVER *Charadrius wilsonia*   WIPL ▪ 1

In N.A., generally only seen on breeding grounds where it prefers beaches, often on barrier islands. Polytypic (4 sspp.; 2 in N.A.). L 7.8" (20 cm)

Identification Forecrown is white, with a narrow supercilium. Brown above, with a broad, white collar; white below, except for a broad breast band. Bill is long, thick, and black; dull pinkish legs. **BREEDING MALE:** Breast band is black; some black in lores and on crown; ear patch cinnamon-buff. **BREEDING FEMALE:** Black parts of male are replaced by brown. **NONBREEDING ADULT:** Similar to a dull breeding female. **JUVENILE:** Resembles female, but note scaly-looking upperparts; breast band is usually incomplete. **FLIGHT:** White wing stripe is indistinct.

Geographic Variation Nominate *wilsonia*, occurring on the Atlantic and Gulf Coasts, has a broader white forehead and narrower breast band compared with *beldingi* of Mexico's Pacific coast (casual to the Pacific coast states).

Similar Species Smaller Semipalmated has a much smaller bill. See the sandplover species.

Voice CALL: Sharp *whit*.

Status & Distribution Uncommon and declining. **BREEDING:** Arrives on breeding grounds generally in Mar., but as early as Feb. Fall departures usually in Sept., but lingerers into Nov. Sandy beaches and mudflats; locally breeds inland up to 60 miles in south TX. **WINTER:** FL, otherwise rare along southern Atlantic coast and Gulf Coast; most likely in southernmost TX. **RARE STATUS:** Casual to coastal (and

wilsonia

♀

juvenile
wilsonia

thick
breast
band

breeding
wilsonia ♂

long
thick
bill

♀
wilsonia

dull pinkish legs

wilsonia

beldingi

one nesting record at Salton Sea) CA, OR, WA, on Atlantic coast north to the Maritimes, and inland to the Great Lakes and Great Plains.

Population Declining on the East Coast due to nesting area disturbance and habitat destruction; bred north to NJ until the 1960s. Afforded special protection by six states.

COMMON RINGED PLOVER *Charadrius hiaticula* CRPL ■ 3

Common Ringed Plover is an Old World species, but its breeding range extends into N.A. Its behavior and preferred habitat are like those of the Semipalmated Plover, to which it is closely related. It frequents beaches and mudflats, and runs in typical plover fashion in search of food. Polytypic (3 sspp.; 2 in N.A.). L 7.5" (19 cm)

Identification Brown upperparts from crown to back, except for white collar. Dark lores and auriculars offset by white forecrown and throat, the latter continuous with the white collar. Distinct white supercilium. Pale yellow eye ring incomplete, or missing entirely. White underparts, except for a broad breast band. **BREEDING MALE:** Mid-crown bar, lores, auriculars, and breast band black. Black of lores joins bill at or just below gape. Bill has extensive orange to the base. **BREEDING FEMALE:** Generally paler than male with black parts often replaced partly with brown, but can be difficult to sex. **NONBREED-ING ADULT:** Like a dull adult female, with no black on the head. Breast band possibly incomplete. Supercilium broader and more likely to extend

in front of the eye than in breeding plumage. Duller legs; bill mostly black. **JUVENILE:** Like nonbreeding adult, except duller. Obvious pale edges and darker submarginal lines create a scaly appearance above through, at least, early fall. **FLIGHT:** Prominent white wing stripe quite obvious.

Geographic Variation Subspecies *psammodroma* breeds in northeastern Canada; *tundrae* in western AK.

Similar Species Very similar to Semipalmated. Common Ringed is larger with a more prominent white wing stripe. The bill is longer and in breeding adults has more extensive pinkish at the base. The orbital ring is thinner and usually confined to the bottom of the eye. Distinctive is the more prominent white supercilium on Common Ringed, the difference being particularly useful in comparing breeding plumaged males; female Semipalmated usually has a more prominent supercilium, on some approaching Common Ringed in prominence. On Common Ringed the dark meets bill at the gape line, not above. Common Ringed has a broader breast band, but posture changes the

perceived width. Webbing, while difficult to assess in the field, is important: Common Ringed has no webbing between the outer two toes, and only slight webbing between the inner toes; Semipalmated shows obvious webbing between the two outer toes (a little webbing between the other toes). Identification of stray Common Ringed should be substantiated by vocalizations.

Voice CALL: A soft, fluted *pooee*, lower pitched and less strenuous than Semipalmated and suggestive of Gray-tailed Tattler; repeated in a series of notes during display flight.

Status & Distribution BREEDING: An almost-annual breeder on St. Lawrence I., AK; regular on Baffin I., as well as throughout Eurasia. **MIGRATION:** Long-distance migrant, wintering to Africa. **RARE STATUS:** Casual migrant on other AK islands and also recorded on Seward Peninsula and at Utqiagvik (Barrow). Casual to eastern Canada (possibly annual NL) and New England; accidental CA (Marin Co. and Yolo Co.) and NC. Given the identification issues, no doubt some strays are overlooked. **Population** Stable.

bold white supercilium

orbital ring absent or incomplete

bill slightly longer than Semipalmated with more extensive orange base

breeding ♂

breeding ♀

juvenile

SEMIPALMATED PLOVER *Charadrius semipalmatus* SEPL ■ 1

One of our most common plovers. Prefers mudflats and beaches; sometimes also in freshly irrigated fields that attract large numbers of shorebirds. Often flocks when roosting, but the flock spreads out when the birds feed. Monotypic. L 7.3" (18 cm)

Identification Very similar to Common Ringed Plover. Brown upperparts with white collar, supercilium, forecrown, and throat. Bill is short and has an orange base; legs are orange, and feet show extensive webbing between the toes. **BREEDING MALE:** Lores, auriculars, and mid-crown bar are all black, as is the breast band; soft part

colors are bright orange. The dark of the lores joins above the gape. An eye ring is usually obvious, orange or yellow in color. Supercilium usually not present during this season. **BREEDING FEMALE:** Similar to male, but facial markings and breast band might be brown; more likely than male to have a partial supercilium. **NONBREEDING ADULT:** Like a dull female with no black on facial markings; breast brown and sometimes incomplete; dark bill, duller legs. **JUVENILE:** Pale fringes and dark subterminal marks give bird a scaly appearance; darker legs than in adults.

Similar Species Semipalmated's dark back separates it from Piping and Snowy Plovers; note also complete dark

breast band in all plumages, unlike Snowy and nonbreeding Piping. Semipalmated has a much smaller bill than Wilson's Plover. Very similar to Common Ringed Plover (see that species).

Voice CALL: Distinctive. Whistled, up-slurred *chu-weet*, the second syllable higher pitched and emphatic; repeated in a series during breeding.

Status & Distribution Common. **BREEDING:** Beaches, lakeshores, rivers, tidal flats. Several nesting attempts on coasts well south of mapped range.

MIGRATION: A spring migrant on both coasts and through the center of the continent. Migration generally starts mid-Mar., but usually early Apr. before southern states see a lot of migration. Most migration is mid-Apr.–mid-May in southern and middle latitudes; peaks at Great Lakes mid- to late May. Lingers to June and some nonbreeders stay south of the breeding range, complicating departure dates. In fall, much of the eastern movement takes place to the northeast Atlantic and then south.

First arrivals usually detected in mid-July, but can be earlier. Most adults move through late July–mid-Aug., when the juveniles start migrating in numbers. Most states have passed the peak by early Sept., with mostly juveniles through that month. Lingerers occur into Oct., and in some cases Nov. **WINTER:** A coastal bird, with only a few inland locales where it is rare; accidental elsewhere. Winters to S.A. **MIGRATION:** Casual northwest Europe. **Population** Numbers are stable.

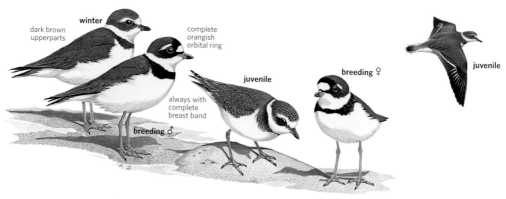

dark brown upperparts

winter

complete orangish orbital ring

juvenile

always with complete breast band

breeding ♂

breeding ♀

juvenile

PIPING PLOVER *Charadrius melodus*   PIPL   ■ 2

The beautiful and endangered Piping Plover is primarily found on beaches. Pale above, it blends well with its sandy home. Polytypic (2 sspp., both in N.A.). L 7.3" (18 cm)

Identification Pale sandy brown above, with white collar and forecrown. Orange legs. **BREEDING MALE:** Black breast band wraps around the back and black bar borders forecrown. Obvious orange base to short black bill. **BREEDING FEMALE:** Like male, but collar and bar on crown are paler, usually not black. **NONBREEDING ADULT:** Bill is black, and black from breeding plumage is lost. **JUVENILE:** Like nonbreeding adult, but with pale edges to the upperparts. **FLIGHT:** Distinct

white wing stripe and white rump. **Geographic Variation** Breast band is variable, can be incomplete, especially in eastern populations. More westerly *circumcinctus* has darker facial features than the nominate.

Similar Species In nonbreeding plumage, distinguished from Snowy Plover by its thicker bill and orange legs. **Voice CALL:** A clear *peep-lo*; also piping notes during flight display.

Status & Distribution Endangered, generally uncommon. **BREEDING:** Sandy beaches, lakeshores, dunes, and river islands. Rare and declining breeder around the Great Lakes (32 pairs in 1996 census). **MIGRATION:** Rare between breeding and winter-

ing ranges. Departures from winter range peak early Apr.–early May. Arrivals in north Atlantic and Great Plains mid-Apr.–mid-May. Departures from breeding ground as early as July, but mostly Aug.–early Sept., lingerers to Nov. **WINTER:** Atlantic and Gulf Coasts. **RARE STATUS:** Accidental in fall to OR and south end of Salton Sea, CA, and in winter to coastal Southern CA.

Population Near Threatened. In 1986 the USFWS listed the Great Plains and Atlantic coast populations as Threatened and the Great Lakes populations as Endangered. A total of fewer than 3,000 pairs are thought to remain. Development and habitat degradation have led to their decline.

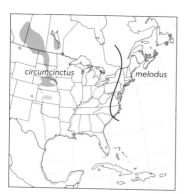

circumcinctus

melodus

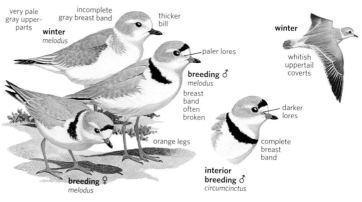

very pale gray upperparts

incomplete gray breast band

thicker bill

winter
melodus

paler lores

breeding ♂
melodus

breast band often broken

orange legs

breeding ♀
melodus

winter

whitish uppertail coverts

darker lores

complete breast band

interior breeding ♂
circumcinctus

LITTLE RINGED PLOVER *Charadrius dubius*   LRPL   5

This rather solitary Old World plover has been recorded as a casual stray in N.A. Polytypic (3 sspp.; northernmost breeder *curonicus* in N.A.). L 6" (15 cm) **Identification** Little Ringed is a small plover, notably smaller than Semipalmated Plover. Brown upperparts; white collar. Lores, forehead, auriculars, and breast band dark, contrasting with white forecrown and white bar behind the dark crown. Conspicuous yellow orbital ring on adults in all seasons. Rather dull yellow legs. Note the long tertials, which almost reach the tail tip. **BREEDING MALE:** Facial markings black; pale base to the lower mandible. **BREEDING FEMALE:** Like male, but auriculars brown. **NON-BREEDING ADULT:** Dull brown above; no black facial markings. **JUVENILE:** Often a yellow-buff tint to paler areas

on head and throat. Scaly appearance due to pale edges and dark subterminal marks. Dull yellow orbital ring. **FLIGHT:** Uniform brown above; no wing stripe.
Similar Species Little Ringed Plover most closely resembles Semipalmated Plover, but it is smaller with a slender bill and has a much more prominent yellow orbital ring. In flight, note that it lacks a white wing stripe.
Voice CALL: A descending *pee-oo* that carries a long way.
Status & Distribution Old World plover, breeds from UK to Russian Far East and south to southern Asia and Australasia. Winters from Africa to southern Asia, as far north as coastal eastern China. **RARE STATUS:** Casual in spring to western Aleutians, AK.
Population Population is generally

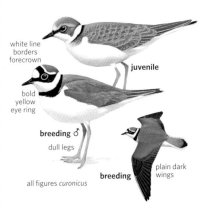

white line
borders
forecrown

juvenile

bold
yellow
eye ring

breeding ♂

dull legs

breeding

plain dark
wings

all figures *curonicus*

expanding as Little Ringeds are opportunistically able to take advantage of human-made developments—more than offsetting declines resulting from habitat loss or agricultural runoff.

KILLDEER *Charadrius vociferus*   KILL   1

The Killdeer is N.A.'s most well known plover. It can be common around human developments, frequently seen on playing fields, parking lots, and other unnatural habitats. The Killdeer's "broken-wing" display is famous and known by many non-birders. It feeds in fields and in a variety of wet areas, but rarely along the ocean shore, and in general, it is not too prevalent on mudflats. The Killdeer often forms flocks after breeding in late summer. It is noisy and reacts quickly to any perceived disturbance.

long tail and
reddish orange rump

vociferus

vociferus

two
breast
bands

vociferus

Polytypic (3 sspp.; nominate in US). L 10.5" (27 cm)
Identification Brown upperparts turn orange at the rump and uppertail coverts, but this only shows in flight or during its "broken-wing" display. Forecrown is white, as is a short but distinct supercilium, and there is a white collar. Lores are dark; this coloration continues and broadens at the auriculars. Underparts are white, except for two bold black breast bands. Legs are pale, usually dull pinkish, but sometimes with a yellow-green cast. Orbital ring is narrow, but a bright orange-red. The Killdeer is the largest of the "ringed" plovers and looks long-tailed. Sexes generally look similar in adult plumage, and there is little seasonal change. Males average more black on the face than females, but this is variable. Fresh feathers, typically in late summer, have rusty edges to them. **JUVENILE:** Plumage is paler than adult, with pale edges to the upperparts. Downy young have only one breast band, but they quickly grow out of this plumage. **FLIGHT:** A particularly bold white wing stripe marks the inner wing. In addition, the bright reddish orange rump is visible and the tail has obvious white corners, with dark central rectrices.
Similar Species Double breast band is distinctive; as is its loud, piercing call. A downy young Killdeer has one

breast band and might be identified as Wilson's Plover by overeager birders.
Voice CALL: Loud, piercing *kill-dee* or *dee-dee-dee*.
Status & Distribution Common. **BREEDING:** Open ground, usually on gravel, including in cities. **MIGRATION:** Early spring migrants show up in the middle latitudes with the first bit of warmth after mid-Feb. Peak in Great Lakes mid- to late Mar., with most migrants having passed through by mid-Apr. In fall, numbers build July–Aug., sometimes as early as late June. Migration peaks Aug., with numbers decreasing during Sept. Many birds linger until Nov., or later if warm weather persists. **WINTER:** Northern winter range varies depending on the extent of snow cover. **RARE STATUS:** Casual well north of breeding range and to western Europe.
Population Stable. Has adapted well to human encroachment.

MOUNTAIN PLOVER   *Charadrius montanus*   MOPL ■ 2

The Mountain Plover's tan coloration blends in with its barren environment. Most frequently first detected when it moves. Gregarious in winter, when it is usually found in short grass or bare dirt fields. Migrants are sometimes present on edges of saline lakes, on turf farms and even occasionally on coastal beaches. Monotypic. L 9" (23 cm)

Identification Uniform brown upperparts from collar through rump. White forecrown and narrow, somewhat indistinct whitish supercilium. Underparts generally whitish, brightest near belly; breast has a dingy brown wash. Black bill; pale pinkish to gray legs. **BREEDING ADULT:** Black lores and forecrown. **NONBREEDING ADULT:** Black on head replaced by brown; buffy tinge

on breast more extensive at this season. **JUVENILE:** Paler than adult with buff edges to the scapulars and coverts. **FLIGHT:** Brown above, with a hint of a white wing stripe, particularly toward the middle of the wing. Underwing white, visible as bird banks low above the ground. Tail has a pale tip and a dark subterminal bar.

Similar Species Except for the closely related Caspian (*C. asiaticus*) and Oriental (*C. veredus*) Plovers from Asia, no other species quite looks like a Mountain Plover, but inexperienced birders might mistake it for one of the more numerous American Golden-Plovers when that species arrives in early spring. In nonbreeding plumage in early spring, American Golden-Plovers share a rather brown-and-white plumage, but they are darker above, with pale spots, are more mottled below, have a bolder supercilium, and have gray underwings and dark legs.

Voice CALL: Harsh *krrr* note.

Status & Distribution Local and declining. **BREEDING:** Plains and shortgrass prairies. **MIGRATION:** Wintering populations thin out late Feb.,

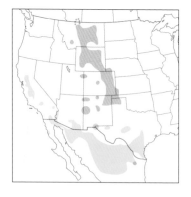

and withdrawal usually complete by late Mar., early Apr. in TX and CA. First CO arrivals early Mar., those in MT to mid-Apr. Fall dispersal starts July and ends in late Nov. (peak mid-Oct.–early Nov.). Winter populations generally do not arrive until late Oct. **WINTER:** Local. Primarily Southern CA, southern AZ, southern and central TX, and Mexico. **RARE STATUS:** Very rare to casual to Pacific coast; casual in Pacific Northwest; accidental in eastern N.A.

Population Near Threatened. Numbers have declined severely with the conversion of native grasslands to farmland.

EURASIAN DOTTEREL   *Charadrius morinellus*   EUDO ■ 4

A unique visitor from Eurasia, this species is distinct for its beauty and biology. Rather tame, it usually allows close approach. Unlike other plovers, female Eurasian Dotterels have brighter plumage and males tend the nest. Typically, the female only helps rear the young when gender ratios are

out of balance. Monotypic. L 9" (23 cm)

Identification Midsize plover with a small bill; bold white supercilium in all plumages. **BREEDING ADULT:** Gray upperparts have buff edges. Dark crown sets off a bold white supercilium that starts above and extends well behind the eye, wrapping around to the back of the head; gray above with rich buff fringes on scapulars, tertials, and

coverts. **NONBREEDING ADULT:** Similar to breeding bird, but underparts paler, grayish throughout; white line at the breast is less distinct. **JUVENILE:** Like nonbreeding adult, but has bold, colorful edgings to many feathers above and is buffy below; bold white supercilium extends around the entire head.

Similar Species Recalls goldenplovers, but supercilium that wraps around the head and white line on the breast are diagnostic; also has pale legs.

Voice CALL: Soft *put, put*, repeated.

Status & Distribution Eurasian species; formerly rare, sporadic breeder in northwestern AK on open, rocky tundra and rare late spring migrant to St. Lawrence I., AK. None in two decades. **RARE STATUS:** Until about 2000, casual to Aleutian Is. and along West Coast in the fall. Accidental in winter in CA, one record in adjacent Baja California, and one fall record in ON.

Population Declining.

JACANAS Family Jacanidae

Northern Jacana (Costa Rica, Aug.)

Jacanas are atypical shorebirds; they look more like rails. They are easy to identify; the only difficulty posed is pronouncing the name. Most people prefer ZHA-sah-na, from a local tribe in Brazil.

Structure Their long legs and very long toes and toenails are unique. The wings are rounded; some species have a pointed spur at the carpal joint, used for displays and fighting. The bill has a frontal shield.

Behavior Jacanas walk with a high-stepping gait, and frequently do so across floating vegetation. Their long toes and toenails distribute their weight over the vegetation so that they do not sink. They can also swim. Jacanas are quite conspicuous and do not hide in vegetation. Weak fliers, jacanas usually fly only for short distances; in flight, their necks are outstretched and their legs trail behind their tails. They often raise their wings upon landing or in various displays. Jacanas are fiercely territorial. In most species, the female is polyandrous, taking several males in her territory. While the female may help with nest building (on floating vegetation), the males sit on the eggs and care for the young. Jacanas usually feed on aquatic insects, but they also eat small fish and plants.

Plumage The sexes look alike; the females are larger.

Distribution Eight species in six genera are found worldwide, primarily in tropical regions.

Taxonomy The two New World jacanas are in the genus *Jacana*. Some authors consider the Northern Jacana to be conspecific with the Wattled Jacana (*Jacana jacana*) of southern Central and South America; they overlap and hybridize in Panama and southern Costa Rica. Five other genera (totaling six species) occur in Asia, Africa, and Australia.

Conservation The destruction of wetlands through drainage or overgrazing has eliminated populations. One species from Madagascar is Near Threatened.

Genus *Jacana*

NORTHERN JACANA *Jacana spinosa*   NOJA ■ 4

This conspicuous bird walks on long legs and is as adept at walking along ditches and grassy lakeshores as across lily pads. It often raises its wings, revealing yellow flight feathers; a spur sticks out from the middle of the wing

yellow flight feathers

yellow forehead shield

long bill

blackish neck

adult

immature

long toes

white supercilium and underparts

and is occasionally visible during these displays. Monotypic. L 9.5" (24 cm)

Identification ADULT: Chestnut with a glossy black head, neck, breast, and upper back. A pale blue cere separates the yellow frontal shield from the yellow bill. **JUVENILE:** White below and brown or olive-brown above with cinnamon edges in fresh plumage. Dark hind neck and crown and dark postocular stripe contrast with whitish buff supercilium. Frontal shield and spur are tiny. One-year-olds are mottled with chestnut, brown, and black; dark bill has a yellow base.

Similar Species Unique and unmistakable, except in New World from Wattled Jacana (*J. jacana*).

Voice CALL: A cackling, harsh *ka-ka-ka-ka* or *jik-jik-jik-jik*; similar to a large rail.

Status & Distribution Rare and irregular visitor. **BREEDING:** Common nesting species in Mexico and C.A., also in Cuba, Jamaica, and Hispaniola; has bred in TX, but not recently. **MIGRATION & WINTER:** Over 30 records, most from freshwater marshes and ponds near the coast of southern TX, primarily Nov.–Apr. A few records are well inland. **RARE STATUS:** Casual in southern AZ. Accidental near Marathon, TX. **Population** There was a small resident population in Brazoria Co., TX, from 1967 to 1978.

SANDPIPERS, PHALAROPES, AND ALLIES
Family Scolopacidae

Long-billed Dowitchers (far right and lower center) and Short-billed Dowitchers (NJ, May)

The large, diverse Scolopacidae family represents most of the shorebirds. Many species look confusingly similar, especially in the genus *Calidris*.

Structure This group has a great variety of shapes, from the large, long legs and long bill of a curlew to the short bills and short legs of sandpipers. Scolopacids are distinct from plovers in having longer bills. Most sandpipers forage primarily by feel; their bills have many receptors to aid in the location of prey. Species that are transequatorial migrants have long wings and primary projection.

Behavior Most sandpipers feed and roost in or near water. They often migrate and winter in mixed-species flocks. During breeding season, however, they are territorial with little tolerance for perceived intruders; many sandpipers are territorial in migration also. Many species perform breeding displays that seem unusual for shorebirds. These displays might include elaborate flights with rather well-developed songs and perching in tops of trees. Feeding behavior varies, depending on shape and length of bills and legs. Most scolopacids pick at or near the water's edge, some feed in the water, and others forage in forest litter. Most sandpipers eat invertebrates, such as worms or small bugs, and some swallow mollusks whole—their strong gizzards break up the shell—and yet others eat small fish. The family as a whole is highly migratory; some species migrate long distances over the ocean, others over land.

Plumage Understanding molt and plumages is the key to identifying difficult shorebirds. Shorebirds have at least three distinct plumages: breeding, nonbreeding, and juvenal. Transitional or second-year plumages are more complex, but these three plumages provide a good baseline. Some species show sexual dimorphism, but many show little or none. Various species have different molt strategies. After breeding, some adults molt into a winter plumage, some do so on or near the breeding grounds; most migrate south in breeding plumage and then molt into a duller plumage as they reach or near their winter grounds. By early winter most species will be paler and more unmarked below and plainer above. As spring approaches, many species begin to molt into a more colorful plumage. Birds less than a year old molt into their first breeding plumage, which is quite variable; sometimes birds look like breeding adults, but more frequently they have a dull, incomplete plumage, with a mix of new and old feathers, closer to nonbreeding plumage. In most species, juveniles leave their natal grounds in a fresh plumage that is quite distinct from either adult plumage, but a few species molt out of many juvenal feathers before moving south. Most sandpiper species retain a few juvenal feathers for months, allowing birds to be aged well into late fall or winter; seeing these feathers can be accomplished under good viewing conditions.

Distribution Of more than 90 species worldwide, most breed in the Northern Hemisphere and migrate to temperate or tropical areas or, often, well into the Southern Hemisphere. Species nest in a variety of habitats, such as marsh, prairies, tundra, and boreal forests; in winter, most species take advantage of exposed mudflats at intertidal wetlands, although a few use bogs and marshes.

Taxonomy A diverse family of some 94 species in 16 genera. Recently, genetic studies have resulted in the merging of several one- or two-species genera into genera *Calidris* and *Tringa*. Conversely, some recent authorities place the Wilson's Phalarope in its own genus, *Steganopus*.

Conservation All shorebirds face the continuing loss of wetlands habitats. Many species have narrow breeding ranges, and many use key staging areas as feeding stops on migration. The main risk to these sites is the destruction of tidal mudflats. BirdLife International codes: 18 NT, 5 VU, 3 EN, 2 CR, 1 CR (PE), and 5 EX.

CURLEWS Genera *Bartramia* and *Numenius*

These waders have long legs and, usually, long bills. Most plumages are brown with little change between seasons, ages, or sexes. Females tend to be larger than males in most species, with larger bills. The bills grow over the first year. Most species stay near water (less so during breeding season), feeding on crabs, worms, and more, but insects are also a key part of the diet for some species.

UPLAND SANDPIPER *Bartramia longicauda*   UPSA ■ 1

The Upland Sandpiper is typically found in fields, where its head and neck are visible above the grass; on the breeding grounds, it perches on posts. It often bobs the rear portion of its body, and, upon landing, will typically hold its wings up momentarily. It is one of few shorebirds that seems disinterested in water. Monotypic. L 12" (30 cm)

Identification Small, aberrant curlew is an elongate species with a long, thin neck; long legs; long tail; and long wings. Bill is short and straight; bill is yellow with dark at the tip and atop the ridge. Legs yellowish. The head is small and dove-like, with large, dark eyes. Generally brown overall; dark upperparts with buff edges; brown streaks on the foreneck that turn into chevrons on the breast and flanks. Belly and undertail coverts white. **ADULT:** Upperparts look barred, whereas the tertials are entirely barred. **JUVENILE:** Very similar to adult. Upperparts look less barred due to pale fringes on dark feathers; tertials are edged in buff. **FLIGHT:** Dark primaries contrast with mottled brown upperparts.

Similar Species Unique shape and plumage eliminate confusion with other waders.

Voice CALL: A rolling, bubbling *pulip, pulip.* A territorial "wolf whistle," given in flight on breeding grounds, sounds distinctly human; also a long guttural alarm call given at rest.

Status & Distribution Rare to fairly common. **BREEDING:** Tallgrass prairies (less common in short-grass prairies); increasingly restricted to airports in the East. **MIGRATION:** More likely found in dirt or short grassy fields, including sod farms. In spring, the first arrivals on Gulf Coast appear mid-Mar., with peak late Mar.–mid-Apr. in south, mid-Apr.–early May farther north. Fall movement begins in July, with peak late July–mid-Aug. in north, Aug.–mid-Sept. farther south, stragglers to Oct., rarely to Nov. **WINTER:** Primarily southern S.A. **RARE STATUS:** Casual on West Coast and in the Southwest in migration; casual in Europe in fall. Accidental in Australia and Guam.

Population Seriously declining over much of its breeding range; extirpated from OR and WA. The loss of habitat due to grassland conversion to agricultural fields has had a greater impact on the population than the hunting pressures of the 19th century.

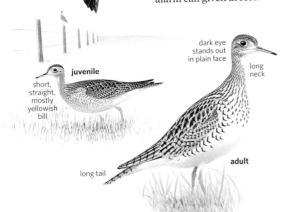

long tail

juvenile

dark outer half of wing

juvenile

short, straight, mostly yellowish bill

dark eye stands out in plain face

long neck

adult

long tail

BRISTLE-THIGHED CURLEW *Numenius tahitiensis*   BTCU ■ 3

This uncommon species is mainly seen on its limited western AK breeding grounds or its winter home on South Pacific Ocean islands. The species' English name derives from the bare shafts on the thigh feathers; reasons for this characteristic are not clear. Monotypic, although there are mensural differences between the two breeding populations. L 18" (46 cm)

Identification Medium-size curlew, like a bright Whimbrel. Bold head stripes include a dark eye line that connects to the bill and a pale supercilium, bordered above by a broad, dark lateral crown stripe. Upperparts are dark brown with bold, buff edges and notches on the scapulars and coverts. Underparts are buff, rather bright, with streaking on the neck and breast, and bars on the sides. Tail and rump vary from rusty to almost blonde, with dark bars on the tail. Worn adults can lose much of the buff edges to the upperparts, and the underparts can become faded; however, rump and tail remain distinctly buffy. Bill is dark, but much paler on the lower mandible than on most curlews, extending almost to the tip; legs are blue-gray. Stiff feathers on the thighs and flanks are hard to see except at close range and in good light. **JUVENILE:** Like adult except in early fall, when plumage is uniform and fresh, compared to more worn adult. As winter progresses, juveniles look darker than adults, retaining buff coloration on only rump and tail. Immatures generally do not return to the breeding grounds for three years. **FLIGHT:** Brown above, with noticeable buff edges. More important, the bright buff rump and tail are usually evident. **Similar Species** Possible confusion with Whimbrel, but Bristle-thighed is brighter buff with larger spots above and is buffier below, and it has a bright buff (varies from cinnamon to almost blond) and contrasting rump and tail. Calls are totally different.

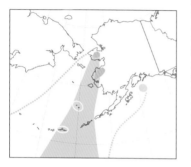

Voice CALL: Loud whistled *chu-a-whit*, somewhat recalling a Black-bellied Plover, but louder, with more-distinct syllables; somewhat humanlike. On the breeding grounds, loud, sharp whistles accompany impressive display flights.

Status & Distribution Local and uncommon. **BREEDING:** Endemic AK breeder on Seward Peninsula and in southwest AK; nests on hills with uneven, hummock tundra. **MIGRATION:** Spectacular flights covering 2,000–4,000 mi with minimal or no stops. In spring, through the Pacific islands in Apr., departing HI early to mid-May on average. Arrive in AK early to mid-May, but as early as late Apr. In fall, adults stage July–Aug. in western AK, where they fatten up on berries prior to moving south. Juveniles move south mid-Aug.–early Sept. **WINTER:** South Pacific islands from Caroline east to Fiji and east Polynesia to Pitcairn and Ducie Is.; birds from the two AK breeding populations do not winter on separate island groups. **RARE STATUS:** Casual to the Pacific coast. A weather-related occurrence accounted for 13 birds found 6–25 May 1998 in northern CA, OR, and WA. Prior to this event, there were eight records (six spring, two fall) for OR, WA, and BC; the two fall records (OR) are sight records.

Population Vulnerable. The population is at risk because of its relatively small size and restricted range.

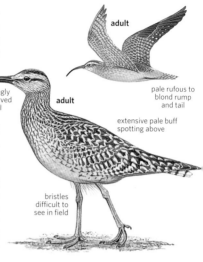

adult

strongly decurved bill

adult

adult

pale rufous to blond rump and tail

extensive pale buff spotting above

bristles difficult to see in field

WHIMBREL *Numenius phaeopus*   WHIM ■ 1

The Whimbrel is most frequently encountered along the coasts during migration where it forms large flocks. Whimbrels pick at the ground for food while walking and often form large flocks with other large shorebirds, such as godwits and curlews. Polytypic (7 sspp.; 4 or 5 in N.A.). L 17.5" (45 cm)

Identification Medium-size curlew, rather cool brown in color, with bold head stripes. Black bill, with a little pale at the base, is moderate in size and curvature; legs are blue-gray. Crown has a broad dark lateral stripe, with a pale mid-crown stripe. Dark eye line,

variegatus

phaeopus

hudsonicus

including the lores, sets off the paler supercilium. Upperparts are dark brown with pale notches along many of the scapulars and coverts. Neck and breast are lightly streaked, with barring on sides; remainder of underparts are dingy whitish buff. **JUVENILE:** Can be difficult to age this species, but juveniles typically have more extensive pale edging and notches on the upperparts and a shorter bill. There is often a greater contrast between the scapulars and coverts on young birds compared to adults. **FLIGHT:** Uniform cool brown above, including dark rump; dark underwings. Feet do not project beyond the tail.

Geographic Variation Formerly only three subspecies recognized, now up to seven recognized by some: five in the Old World, two breeding in N.A. Eastern N.A. subspecies *hudsonicus* is described above. The well-isolated breeding population from AK (to western NT) has been recognized as *rufiventris*. It is the subspecies present in the Pacific region. Coloration and pattern is like *hudsonicus*, but *rufiventris* is larger in all measurements. European nominate *phaeopus* and

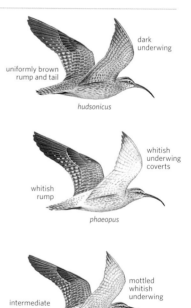

dark underwing

uniformly brown rump and tail

hudsonicus

whitish underwing coverts

whitish rump

phaeopus

mottled whitish underwing

intermediate rump

variegatus

islandicus (if recognized) are casual to East Coast. Both have a white rump and underwings and coarser dark markings on the breast. Asian *varie-*

gatus (rare in Bering Sea region) has a whitish, variably streaked rump and underwings. The three Old World subspecies share a whiter ground color to the underparts, compared to the browner ground color of *hudsonicus*. Many characters of *phaeopus*, *islandicus*, and *variegatus* overlap; identification should be made carefully.

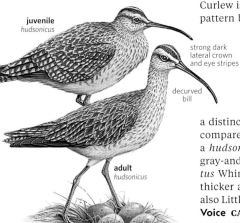

juvenile
hudsonicus

strong dark lateral crown and eye stripes

decurved bill

adult
hudsonicus

It is likely that Old World birds on the East Coast are *phaeopus* (or *islandicus*) and on the West Coast are *variegatus*. **Similar Species** Compared to Long-billed Curlew, Whimbrel is smaller, with a smaller bill and bold head stripes, and its plumage never matches the warmth of Long-billed (differences are most evident on the wings in flight). In AK, Bristle-thighed Curlew is similar. It has a bold head pattern but, when worn, might lose some of its buff coloration on the body as well as the buff notches to the upperparts, increasing the potential for confusion with the Whimbrel. However even a worn Bristle-thighed has a distinctively buffier tail and rump compared to the uniform brown of a *hudsonicus* Whimbrel or the dull gray-and-white barring of a *variegatus* Whimbrel. It also has a slightly thicker and blunter-tipped bill. See also Little Curlew.
Voice CALL: Series of hollow whistles on one pitch, *pi-pi-pi-pi*.
Status & Distribution Fairly common.

BREEDING: Open tundra. **MIGRATION:** Starts mid-Mar. or (rarely) early Mar. Peak in most southern states from mid- to late Apr.; peak in mid-Atlantic and Great Lakes states mid-May; western states slightly earlier. In fall, adults move first, typically early to mid-July. Juveniles peak mid-Aug.–mid-Sept., stragglers in Great Lakes and mid-Atlantic to mid- or late Oct. In spring migrates along and off West Coast, but thousands also move up through Salton Sea region (some through southwest AZ), Antelope Valley, and the Central Valley. In fall they are mostly rare from the interior, and the few recorded are mostly adults in July; the majority migrate along and off coast. **WINTER:** Beaches and coastal wetlands from CA and VA south to S.A. In N.A. almost unknown inland in winter. **RARE STATUS:** European *phaeopus* is casual to East, mainly Atlantic coast. Asian *variegatus* is a rare migrant on islands in western AK, casual elsewhere in Pacific region. N.A. *hudsonicus* is casual in Europe, Africa, and Australasia.
Population Stable.

LITTLE CURLEW *Numenius minutus* LICU 5

This aptly named, diminutive curlew is a casual stray from Asia. Monotypic. L 12" (30 cm)
Identification Small curlew with a short, slightly decurved, dark bill with a pale base. Legs are grayish but variable; they can be dull pinkish or greenish. The dark crown (occasionally showing pale mid-crown) is offset by a pale buff supercilium. A dark line or patch behind the eye is interrupted in front of the eye, leaving the lores

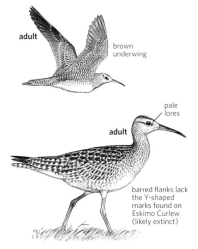

adult

brown underwing

pale lores

adult

barred flanks lack the Y-shaped marks found on Eskimo Curlew (likely extinct)

at least partially pale. Upperparts are composed of dark scapulars and coverts that are broadly edged buff. Below, there are faint streaks on a buff breast, and slight barring on the sides and flanks. It is very difficult to age this species other than in fall, when assessing the degree of feather wear will reveal juveniles, who look fresh. On adults, the tertials are usually distinctive, showing brown with narrow, dark bars, whereas juveniles show dark feathers, almost black, with buff notches along the edge of the feather. However, this feature appears sufficiently variable to limit its use as a diagnostic tool. **FLIGHT:** It appears uniform brown above, with only slightly darker primaries. The underwings are brown. The toes barely extend beyond the tip of the tail.
Similar Species Similar to the larger Whimbrel, but with a shorter and only slightly decurved bill, mostly pale lores, and buffy tones. Size and short bill of Little Curlew are like Upland Sandpiper, but note the curve to Little Curlew's bill and its stronger head pattern. Any Little Curlew found in N.A. should be scrutinized for the likely extinct Eskimo Curlew. Eskimo is slightly

smaller and longer winged (wing tips extend beyond, not to, tail tip), but shorter legged and has a bolder head pattern, including dark lores, more heavily barred breast and flanks, with bold Y-shaped marks on the flanks (just slight brown barring on Little Curlew); also has overall more cinnamon coloration, including the wing linings.
Voice Mostly silent in N.A. **CALL:** Musical *quee-dlee* and a loud *tchew-tchew-tchew*.
Status & Distribution BREEDING: Eastern Siberia and Russian Far East in burns and grassy openings within larch forests, mainly along river valleys. **WINTER:** Mainly in northern Australia; the distribution is highly dependent on rainfall. More rarely visits southern Australia and Tasmania; some winter in southern New Guinea. **RARE STATUS:** Casual in fall to coastal CA (four accepted records from early Aug. to mid-Oct. that involve both juveniles and adults, the last of which was Sept. 1994); two spring records at Gambell, St. Lawrence I., AK. Casual to northern Europe in fall.
Population Small and disjunct; threatened due to habitat loss in migration and winter ranges.

ESKIMO CURLEW   *Numenius borealis*   ESCU   ■ 6

The Eskimo Curlew is probably extinct; the last confirmed record was of an adult female shot by a hunter at a site used for hunting shorebirds in Barbados on 4 Sept. 1963 (specimen at the Academy of Natural Sciences of Philadelphia). There have been several sightings in the intervening years, but none with definitive documentation. Given the species' remote northern summer haunts, it is remotely conceivable that this species still exists. Monotypic. L 14" (36 cm)

Identification Like Little Curlew or a small Whimbrel, but its head pattern is more muted than the Whimbrel and the central crown stripe is indistinct or lacking entirely. The lores are entirely dark, like Whimbrel, but unlike Little Curlew. Upperparts are dark with buff edges; underparts are buff, with streaks on the neck and dark chevrons down the sides. The wings are long and project past the tail. **FLIGHT:** Entirely brown above, with slightly darker primaries. The underwing is cinnamon but looks darker in flight.

Similar Species Differs from Whimbrel and a buffy Bristle-thighed Curlew in its smaller size, smaller, less decurved bill, more muted head pattern, and cinnamon underwings. (Bristle-thighed Curlew has a cinnamon-rust rump and tail.) This identification would be more problematic with a fresh juvenile Whimbrel, as its bill can be substantially shorter than adult. In comparison with the Little Curlew, Eskimo is slightly larger,

with a more decurved bill, completely dark lores, more heavily marked flanks, and darker underwings.

Voice CALL: Poorly known, but includes clear whistles.

Status & Distribution This species is likely extinct, although it is still federally listed as Endangered. **BREEDING:** Nested on Arctic tundra and wintered on pampas of Argentina, but its full, former breeding range is not known. Given the historical records from AK and the Russian Far East, it likely extended farther west than NT. The only nests were found in the vicinity of Ft. Anderson, NT, during the summers from 1862 to 1866. **MIGRATION:** Formerly found in large flocks, with northbound migration primarily through the mid-continent, southbound birds were largely transoceanic" migrants, taking the Atlantic Ocean to S.A. Their stopover points hosted large numbers. During spring movements, migrants were found Mar.–Apr. In fall, Eskimo Curlews would stage along Canada's Atlantic coast (especially Labrador), fattening up on berries before their transoceanic flights, typically Aug.–Sept. Records on the US Atlantic coast were rare, particularly south of New England. **RARE STATUS:** Records from AK, Russian Far East, and the British Isles in the 19th century.

Population Likely Extinct. This species, once numerous, declined precipitously during the two decades following the American Civil War. The last definite record (specimen) in Can-

pale cinnamon wing linings

adults

like a small Whimbrel and closely resembles Little Curlew, but note Y-shaped marks on flanks

ada was at Battle Harbour, Labrador, on 29 Aug. 1932. By 1940, it was thought possibly extinct, but then up to two were photographed on Galveston I., TX, in Mar. and Apr. between 1959 and 1962. The two primary reasons for its decline were intense late-19th-century market hunting pressures, which impacted almost all shorebirds, and, uniquely for the Eskimo Curlew, the extinction of a key, once abundant prey item (the Rocky Mountain locust, *Melanoplus spretus*) on the Great Plains. The former (known and presumed) breeding grounds are hard to access. The two most likely places to search for this species during migration are along the upper TX coast from late Mar. through mid-Apr. and along the Labrador coast Aug. through Sept.

LONG-BILLED CURLEW   *Numenius americanus*   LBCU   ■ 1

barred remiges

cinnamon underwing coverts and flight feathers

adult

Long-billed Curlews are typically seen in large roosting flocks along the coast or feeding in fields. They often roost with other large shorebirds, such as godwits and Willets. In the breeding season they inhabit grasslands and shrub-steppe and eat insects and worms; however, during the nonbreeding season many move to water areas and eat crabs and worms. Polytypic

(2 sspp.; both in N.A.). L 23" (58 cm)
Identification Large, warm brown shorebird—not well marked, but distinctive. Upperparts are cinnamon-brown, with a slightly darker crown due to faint streaking. Upperpart feathers have dark stripes or bars. The face is rather uniform with only a hint of a buff supercilium and slightly darker lores. Underparts are buffy, with streaking on neck and breast. Wing coverts are warm cinnamon with fine bars. Very long bill (longer in females) curves strongly down and the basal half is pale; long legs are gray. **JUVENILE:** Shorter, less decurved bill; acquiring adult bill length can take up to a year. Dark wing coverts with broad buff edges result in a striped look.

FLIGHT: Cinnamon-buff wing linings are distinctive in all plumages. The toes extend slightly beyond the tail.

Geographic Variation Nominate *americanus* breeds in the southern part of the species' range, north to NV, ID, WY, and SD. The slightly smaller *parvus* breeds north of *americanus*; these size differences are slight and likely clinal and, given sexual (females larger and longer billed) and age variation, are not discernible in the field. Many authorities now regard the species as monotypic.

Similar Species Most likely to be confused with a Whimbrel, but Long-billed is larger, warmer brown, with buff and cinnamon tones, and lacks the dark head stripes of a Whimbrel. The bill of

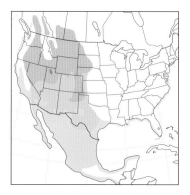

Long-billed Curlew is longer and more decurved, but be aware that juveniles have shorter bills. Eurasian Curlew has white on the rump, tail, and lower back, in addition to white underwings. Far Eastern Curlew is even larger and plain brown above, lacking any cinnamon tones. The plain brown underwings of Far Eastern are duller than Long-billed, and its bill is long, even in comparison to Long-billed Curlew. A Marbled Godwit, roosting with its bill tucked, will look smaller with unmarked or barred underparts, and has dark legs, unlike the paler, gray legs of Long-billed.

Voice CALL: Loud, musical, ascending *cur-lee*. On breeding grounds it repeats *cur-lee*, followed by sharp, descending whistles.

Status & Distribution Fairly common. The southernmost breeding curlew, with the northernmost winter range. **BREEDING:** Nests in wet and dry uplands. **MIGRATION:** Found on wetlands, grainfields, and coasts. In spring, migrants as early as mid-Feb., but more typically Mar. Peak mid- to late Apr. for UT and the Northwest. Most migrants leave wintering grounds by early May. In fall, first arrivals occur in July, occasionally late June. In most areas they peak in Aug., and almost all birds are gone from their breeding grounds by this time, peaking in southern states in Sept. **WINTER:** Pacific, Gulf, and Southeast coasts, Central and Imperial Valleys of CA, southern AZ, central TX, and Mexico. **RARE STATUS:** Casual to East in late spring (late May–mid-June) and fall (mid-July–Oct.). Accidental southeast AK.

Population Like all curlews, except perhaps Whimbrel, the Long-billed Curlew has declined, particularly in eastern parts of its range, due to habitat loss.

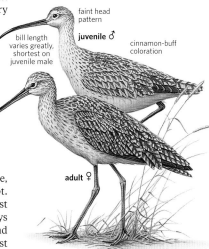

faint head pattern

bill length varies greatly, shortest on juvenile male

juvenile ♂

cinnamon-buff coloration

adult ♀

FAR EASTERN CURLEW *Numenius madagascariensis* FECU ■ 4

The Far Eastern Curlew, a casual visitor from Asia, is the largest curlew. The specific epithet does not reflect its range, as it has never occurred anywhere near Madagascar. Monotypic. L 25" (64 cm)

Identification Dull coloration, generally brown with streaks, but its size, and extremely long, decurved bill are noteworthy. Face pattern is indistinct, with slightly darker lores being the most noticeable feature. Upperparts are grayish brown with warmer tints on occasion; the neck and crown are streaked. Underparts are pale brown with dark streaks on the neck and breast, extending down the sides. **JUVENILE:** Fresh birds differ from adults by their buffier breast and belly, with less streaking, and they are also more boldly marked above, with buffy fringes to all feathers. As feathers wear, juveniles can be difficult to separate from adults. Juveniles have much shorter bills, and full length can take up to a year. **FLIGHT:** Wing linings are dingy with dark barring, but in many lights look gray or gray-brown. Upperparts are uniform brown with the rump the same color as the back.

Similar Species Large size and lack of head pattern leave the Long-billed and Eurasian Curlews as the only identification challenges. Long-billed is slightly smaller, with warm cinnamon tones on the body and, in flight, on the underwings. Far Eastern has heavier streaking than Long-billed. Eurasian and Far Eastern Curlews look remarkably similar at rest (Eurasian has slightly whiter ground color to underparts and is shorter billed), but are easily separated in flight; note Eurasian's white rump, lower back, and underwings.

Voice CALL: A strident *coour-leee* often given in a series, similar to Eurasian Curlew, but lower pitched and with second syllable distinctly longer.

Status & Distribution BREEDING: Marshes and wet meadows of Russian Far East. **WINTER:** Southeast Asia (rare); most winter from Borneo to Australasia. **RARE STATUS:** Very rare in spring and early summer on the western and central Aleutians; casual to the Pribilofs. Accidental coastal BC (south Vancouver, 24 Sept. 1984).

Population Endangered. Numbers are declining due to loss of habitat on migration routes and on winter grounds.

adult

slightly browner but otherwise extremely similar to Eurasian Curlew; best told in flight

uniform brown back and rump

brownish underwing

adult

SLENDER-BILLED CURLEW *Numenius tenuirostris* SBCU ■ 6

This Old World species is likely extinct. Monotypic. L 15" (39 cm)
Identification Smaller than a Whimbrel, but its plumage is more like a Eurasian Curlew. Rump is white; tail is white with dark bars. White underwings are visible in flight. Bill is slender, even at the base.
Similar Species Slender-billed has a short bill and bright white underparts, similar to Eurasian Curlew, but Eurasian has a longer and thicker bill, and has chevron-shaped marks rather

than round black spots on its flanks.
Status & Distribution BREEDING: The only nests were found in southwestern Siberia near Tara, north of Omsk, between 1914 and 1924. **WINTER:** Thought to winter in parts of western and northwestern Africa. **RARE STATUS:** A specimen record from Crescent Beach, Lake Erie, ON, "about 1925."
Population Critically Endangered. Likely extinct, due to habitat loss and hunting. Up to five wintering birds found at one coastal lagoon in Morocco

in 1987 gradually dwindled; the last fully credible record was in 1995.

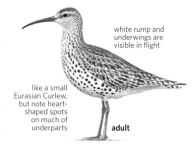

white rump and underwings are visible in flight

like a small Eurasian Curlew, but note heart-shaped spots on much of underparts

adult

EURASIAN CURLEW *Numenius arquata* EUCU ■ 4

This species is a casual visitor from Eurasia. Polytypic (2 sspp.; nominate in N.A.). L 22" (56 cm)

faint head pattern

adult
arquata

very long decurved bill

strongly patterned underparts

Identification BREEDING ADULT: Dark upperparts with black-centered feathers, edged with buff. Heavily streaked below on a whitish background; streaks extend to flanks. Long, strongly decurved bill. **NONBREEDING ADULT:** Grayer than breeding. **JUVENILE:** Smaller bill, buffier above.
Similar Species Distinguished from Long-billed Curlew by white rump, white wing linings, and cooler brown color. Separation from a stray *phaeopus* Whimbrel is more difficult; Eurasian Curlew is larger, has a longer bill, and lacks dark head stripes. Far Eastern Curlew is very similar, but Eurasian is slightly paler above and

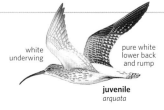

white underwing

pure white lower back and rump

juvenile
arquata

obviously different in flight.
Voice CALL: A sharp *cur-lee.*
Status & Distribution BREEDING: Eurasia. **WINTER:** Southern Europe, Africa, and Asia. **RARE STATUS:** Casual on East Coast (about seven records, none since 1991), primarily fall and winter, from NL to NY. One summer record from NU.
Population Near Threatened. Declines mostly a result of habitat loss.

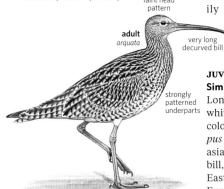

GODWITS Genus *Limosa*

Godwits are large with long legs and long, usually upturned bills. Most godwits undergo a substantial change from breeding to nonbreeding plumage, with some sexual dimorphism. They feed by probing and stitching while standing in water or along its edge; they roost with other large shorebirds. All four godwits occur in N.A.

BAR-TAILED GODWIT *Limosa lapponica* BTGO ■ 2

The Bar-tailed is primarily an Old World godwit. It has an impressive migration: The AK-breeding subspecies *baueri* stages near breeding grounds and flies nonstop, thousands of miles, to its wintering grounds in New Zealand. Polytypic (5 sspp.; 2 in N.A.). L 16" (41 cm)
Identification Slightly recurved, long, bicolored bill; brown crown and hindneck, with dark streaks; pale supercilium and dark eye line. Distinctive black-and-white barred tail. **BREEDING MALE:** Reddish brown underparts with some white near the belly, no barring; streaks limited to sides of breast. Dark scapulars with rufous edges. **BREEDING FEMALE:** Larger and paler than male;

some females mostly whitish below. **NONBREEDING ADULT:** Gray-brown upperparts with paler buff edges, giving a striped appearance. Grayish brown on foreneck and breast. **JUVENILE:** Like nonbreeding, but more obviously marked: buffy spots and notches on upperparts; wing coverts with dark streaks. Neck and breast with buff tones and light streaks; barring might be present on sides. **FLIGHT:** Weak wing stripe; tail barring visible. White or barred underwings.
Geographic Variation Nominate *lapponica*, a casual stray on the Atlantic coast, has a white rump, white wing linings, and brown-barred axillaries;

AK-breeding *baueri* has brown-and-white barring on rump and underwing.

Also, *baueri* is more boldly marked.
Similar Species Resembles Marbled Godwit but lacks cinnamon tones. (Some worn Marbleds can be more problematic, but they still retain cinnamon flight feathers.) Note Bar-tailed's shorter bill, shorter legs, bolder face pattern, white or barred underwing coverts, and barred tail. Bar-tailed's rufous breeding plumage might recall Hudsonian Godwit, but Hudsonian's underparts are more heavily barred and males are deeper chestnut; wing and tail pattern totally different.
Voice CALL: Generally silent.
Status & Distribution BREEDING: Tundra from Scandinavia to AK. **MIGRATION:** In spring, AK arrivals mid- to late May, with earliest arrivals late Apr., and arrivals to north AK into early June. In fall, adults typically depart late July–mid-Aug.; juveniles by Sept. **WINTER:** Europe, Africa, Asia, and Australasia.
RARE STATUS: Rare to casual in spring and, mostly, fall on Pacific coast, particularly in Pacific Northwest; also southern YT (spring). Casual along Atlantic coast at all seasons. Accidental IA (spring) and coastal TX.
Population Near Threatened. The subspecies *baueri* is particularly at risk in spring due to destruction of mudflats on China coast.

pale notched tertials • rather short legs • **juvenile** *baueri* • streaked above • uniform reddish brown underparts • **breeding ♂** *baueri* • recurved bill • **breeding ♀** *baueri* • female larger and much grayer • no rufous on flight feathers • grayish underwing • barred rump • **winter** *baueri* • whitish rump • whitish underwing • **winter** *lapponica*

BLACK-TAILED GODWIT *Limosa limosa*   BLTG ■ 3

Black-tailed Godwit is a rare to casual visitor from the Old World. Polytypic (3 sspp.; 2 in N.A.). L 16.5" (42 cm)
Identification Straight, or only slightly recurved, long, bicolored bill. Tail mostly black to tip, white uppertail coverts. Buff or whitish supercilium and dark loral line. Diagnostic flight pattern. **BREEDING MALE:** Chestnut head, neck, and breast; black upperparts with warm rufous edges to scapulars and coverts. White belly and undertail coverts; heavily barred sides and flanks. **BREEDING FEMALE:** Similar to male but paler, sprinkled with white. **NONBREEDING ADULT:** Gray or grayish brown above; grayish breast but otherwise whitish below, lacking barring. **JUVENILE:** Brown upperparts, with warm cinnamon edges to feathers; pale orangish hindneck and streaked crown. The dark tertials are notched with buff. Pale cinnamon neck and breast; remainder of underparts white. **FLIGHT:** Broad wing stripe across all secondaries and most primaries; conspicuous white wing linings.
Geographic Variation Three subspecies. Subspecies *melanuroides* (from Asia) and *islandica* (breeds in Iceland) are similar in breeding plumage, but an *islandica* breeding male has deeper and more extensive reddish color below; *melanuroides* is darker above in nonbreeding plumage, with a shorter bill and tarsus. AK and West Coast birds are likely *melanuroides*; Atlantic birds are likely *islandica*.
Similar Species Black-tailed shares a bold white wing stripe and black-and-white tail with Hudsonian Godwit; however, note Black-tailed's more extensive wing stripe on the inner portion of the wing and white underwings (black in Hudsonian). In breeding plumage, Hudsonian has dark streaks and lacks reddish color on the hindneck. On juveniles, Black-tailed is warmer colored, with an orange-reddish tint to the hindneck.
Voice CALL: Generally silent.
Status & Distribution BREEDING: Wet meadows from Iceland to Russian Far East. **WINTER:** Southern Europe to Australia. **RARE STATUS:** Very rare spring migrant on western and central Aleutians; casual to Pribilofs (twice to St. Lawrence I.). Casual on Atlantic coast in migration and winter. Accidental inland to ON and IN and south to LA and TX.
Population Near Threatened.

juvenile *melanuroides* • bold and extensive white wing stripe • white wing lining • white base to black tail • **juvenile** *melanuroides* • **winter** *melanuroides* • long straight bill; all godwits have extensively pink-based bills, but only Black-tailed has a straight bill • pale chestnut neck • barred sides and flanks • **breeding** *melanuroides* ♂ • **breeding** *melanuroides* ♀ • **breeding** *islandica* ♂ • extensively reddish below

HUDSONIAN GODWIT *Limosa haemastica* HUGO ■ 1

The striking Hudsonian Godwit has breeding grounds that are patchily distributed. Monotypic. L 15.5" (39 cm) **Identification** Long, slightly recurved, bicolored bill. Black tail, tipped white, with white uppertail coverts. Pale supercilium and dark loral line. **BREEDING MALE:** Dark upperparts, with buff edges; dark chestnut underparts, finely barred. **BREEDING FEMALE:** Larger and grayer, with reduced chestnut below. Tertials less patterned than on male, often with small notches. **NONBREEDING ADULT:** Grayish above, lighter on the breast. Some adults seen in fall migration are in a transitional plumage. **JUVENILE:** Like nonbreeding, but darker above, with buff feather edges to mantle and upper scapulars; lower scapulars and tertials with buff notches. **FLIGHT:** Bold white wing stripe, thin on the secondaries and broader on inner primaries; black underwing coverts and black tail with white base. **Similar Species** Nonbreeding adult resembles Black-tailed Godwit but has dark wing linings and a narrower white wing stripe. Breeding Hudsonian is chestnut below with barring, whereas Black-tailed has chestnut on the hindneck, with minimal

barring below. Barred underparts on breeding Marbled Godwit might recall Hudsonian. In flight, Hudsonian's striking wing pattern might recall a Willet at first glance, but the Willet has a shorter bill and a paler tail.
Voice CALL: Often silent. Calls strident and include rising doubled and single notes.
Status & Distribution Uncommon and local. **BREEDING:** From AK to the Hudson Bay, but in very disjunct pockets. **MIGRATION:** Spring route primarily through central and eastern Great Plains. Migrate generally Apr.–May. Peak migration in TX and KS second half of Apr.–early May. Arrives on AK breeding grounds by early May; in Churchill, ON, by early to mid-June. In fall, migrates much farther east; large staging areas from SK to James and Hudson Bays, with most birds flying southeast to the Atlantic and then to S.A. Great Lakes and mid-Atlantic coast to New England; adults arrive by late June and juveniles by late Aug.; overall peak late Aug.–early Oct., lingerers to Nov., exceptionally to Dec.

WINTER: Southern S.A., unexpected in N.A.; one sight record for GA. Any winter records should eliminate Black-tailed Godwit, which is more likely in midwinter. **RARE STATUS:** Rare to Great Lakes and mid-Atlantic in spring (mid-May–early June); casual in western states in spring (early May–mid-June) and fall (nearly all juveniles, mid-Aug.–mid-Oct.); annual in New Zealand; casual HI, Australia. Casual in Europe and South Africa.
Population Stable.

recurved bill

juvenile

fainter white wing stripe than Black-tailed

white base to black tail

black wing linings

female larger and grayer

juvenile

molting fall adult ♂

patchy chestnut bars below

breeding ♂

deep chestnut underparts

breeding ♀

MARBLED GODWIT *Limosa fedoa* MAGO ■ 1

The Marbled Godwit, our largest and most widespread godwit, is numerous along winter beaches, where it feeds by probing into sand and mud. At this time of year it roosts in coastal wetlands with other large shorebirds, typically Willets and curlews. Polytypic (2 sspp.; both in N.A.). L 18" (46 cm)
Identification Large, tawny-brown shorebird with a long, slightly recurved, bicolored bill. Upperparts have dark feather centers, with buff bars and spots; breast and sides variably barred. **NONBREEDING ADULT:** Bill with brighter, more extensive pink base. Much less barring on underparts. **JUVENILE:** Like nonbreeding, but wing coverts less heavily marked, contrast-

ing with upperparts. **FLIGHT:** Distinctive cinnamon inner primaries and wing linings.
Geographic Variation Nominate *fedoa* is the interior breeder; the slightly smaller *beringiae* population breeds on the AK Peninsula and winters south to northwest CA.
Similar Species A worn Marbled that has lost some of its color could be confused with the smaller and shorter-legged Bar-tailed Godwit. It is easily separated in flight by the lack of cinnamon in the wing. A roosting Long-billed Curlew with its bill tucked has a similar coloration, but Marbled is smaller, has darker legs, and is either unmarked or has some barring on the sides

(versus the breast streaking of a curlew).
Voice CALL: Loud *ker-ret*; also *widica widica widica*.

Status & Distribution Common. **BREEDING:** Grassy meadows, near lakes and ponds. **MIGRATION:** In spring, peak mid-Apr.–mid-May in most areas, a little earlier farther south. In fall, adults first appear mid-June, but more typically during July.

Most areas peak Aug.–mid-Sept., lingerers annually to Oct., exceptionally to Nov. or even Dec. Rare to locally

uncommon in the East, most late July–late Sept. **WINTER:** South to C.A., rarely S.A. Casual to southwest BC and Central Valley of CA. **RARE STATUS:** Casual HI and Galápagos.
Population Stable.

winter
fedoa

rufous remiges

rufous underwing

overall tawny brown mottled with black

breeding ♂
fedoa

long recurved bill; females have longer bills

winter ♀
fedoa

plain buffy below

blackish legs; sleeping birds with head tucked look very similar to Long-billed Curlew, which has grayer legs

---

TURNSTONES Genus *Arenaria*

Turnstones are named for their habit of flipping over stones and other material with their short, pointed bills in search of food. These short-legged, almost chunky birds are quite social, chattering as they feed and flock along the coasts. In flight, the black-and-white pattern on the back, wings, and tail is impressive.

RUDDY TURNSTONE *Arenaria interpres*   RUTU ■ 1

Ruddy Turnstones frequent shorelines, picking through rocks, shells, seaweed, or other flotsam in search of food. Polytypic (2 sspp.; both in N.A.). L 9.5" (24 cm)
Identification Dark breast and neck on otherwise white underparts; demarcation from black to white on bib uneven, more extensive toward the sides; white throat. Median coverts quite long, hanging down over the wing, like long scapulars. **BREEDING MALE:** Striking, bold black-and-chestnut upperparts; mostly white head, with dark streaks on crown. Bright orange legs. **BREEDING FEMALE:** Duller and browner than male; less rufous in the upperparts. **NONBREEDING ADULT:** Dull brown above and relatively unmarked. **JUVENILE:** Like nonbreeding, but with pale edges on upperparts that give the back a scaly appearance. Once edges wear, remaining upperparts look darker than nonbreeding. Legs paler orange,

some near pinkish. **FLIGHT:** Complex pattern on back and wings: white wing stripe, white back, and white humeral bar separated by rufous or brown patches. Dark tail has white base.
Geographic Variation Nominate *interpres* breeds from Greenland eastward to northwest AK; it averages more black and less rufous than *morinella*, which breeds from northeast AK through Canada.
Similar Species Black Turnstone is similar, particularly in flight, but it has more contrast, lacking brown or rufous patches. Bright orange legs of breeding Ruddy are diagnostic but are duller in winter and can be similar to Black. The brownish plumage and uneven dark breast bib still easily identify a Ruddy. Some juvenile Ruddies look fairly dark overall and have been misidentified as Blacks.

striking wing pattern

harlequin head and breast pattern

both turnstone species have slightly recurved bills

breeding ♂

extensive rufous above; female duller

Voice CALL: A low-pitched, guttural rattle.
Status & Distribution Fairly common. **BREEDING:** Coastal tundra. **MIGRATION:** Rare to casual inland, except Great Lakes region and Salton Sea. In spring, western populations (primarily *interpres*) peak mid-Apr.–mid-May, with some lingering to June. Eastern populations (*morinella*) move primarily from late Apr., with peak mid-May–early June along mid-Atlantic and Great Lakes. In fall, adults seen in late July, primarily Aug. Juveniles arrive late Aug., peak in Sept., with most gone by early or mid-Oct., depending on region, lingerers to Nov., exceptionally later. **WINTER:** Along the coasts, south to S.A. Accidental to Great Lakes.
Population Declining.

juvenile

winter

breeding ♂
orange-red legs

BLACK TURNSTONE *Arenaria melanocephala*   BLTU ■ 1

nearly solid slate black head and chest

winter

dark red legs, much darker than Ruddy

A dichromatic bird from the rocky shores of the Pacific coast, the Black Turnstone is social outside of the breeding season, wintering in small to moderate-size flocks. It has a restricted range, in contrast to the circumpolar Ruddy Turnstone. Monotypic. L 9.3" (24 cm)

Identification Dark upperparts, dark throat and upper breast, and white

underparts found in all plumages. Legs are reddish brown; some birds show pinkish brown. **BREEDING ADULT:** Head and entire upperparts are black; eyebrow and lore spot are white. White spotting is visible on sides of neck and breast. Black median wing coverts are long, hanging over the edge of folded wing, edged with white. **NONBREEDING ADULT:** Browner, but generally looks black. Some scapulars and coverts are edged white; the white on the breast, neck, and head is absent. **JUVENILE:** Like nonbreeding; median coverts are not as large and are edged with white. Legs are often paler, with more of a dull pinkish or orange tone. **FLIGHT:** Striking black-and-white pattern: white wing stripe, white back, and white humeral bar separated by black. Tail is black, with white base and tips.

Similar Species Wing pattern and habits recall Ruddy Turnstone, but Ruddy has brown or rufous upperparts, an uneven black bib, a white throat, and, usually, bright orange legs. Surfbird, with which Black Turnstone is often seen, shares the Black's general dark coloration above, white below, with black-and-white tail; however, it is paler and grayer above and on head and breast, lacks white patches on the back and humerals, and has a thicker bill and yellow legs.

Voice CALL: Includes a guttural rattle, higher than the call of the Ruddy.

Status & Distribution Locally fairly common. **BREEDING:** Coastal tundra.

MIGRATION: In spring, movement on the Pacific coast Mar.–early May; peak in southern AK mid-May, arriving on breeding grounds mid-May–early June. First fall migrants appear early July in BC, generally mid-July in OR and WA; adults move earlier, with juveniles arriving in early Aug., peaking late Aug.–Sept. **WINTER:** On rocky coasts from southern AK south. **RARE STATUS:** Casual in eastern WA, eastern OR, and interior CA, primarily May and late Aug.–early Sept.; rare at Salton Sea. Accidental: interior AK, YT, AB, MT, NV, AZ, NM, TX, and WI. Recorded from Nayarit (Mexico) and Russian Far East.

Population Stable.

striking wing pattern

winter

white lore spot

white spots on sides breast

breeding

CALIDRINE SANDPIPERS Genus *Calidris*

Calidris sandpipers—the smallest are commonly referred to as peeps or stints—make up this varied group. Plumages tend to be bright in breeding and juvenal birds, rather colorless and unmarked in nonbreeding birds. Most species are rather small with little difference in size or plumage between the sexes. Identifications within this group can be difficult. Most species pick along the shore and mudflats or in shallow water; some species feed in slightly deeper water. Although many species migrate and roost in large flocks, they can be very aggressive and territorial. Most species undergo long migrations, breeding in the Arctic and wintering to S.A.

GREAT KNOT *Calidris tenuirostris*   GRKN ■ 4

The Great Knot is a visitor from Asia. Monotypic. L 11" (28 cm)

Identification Large and chunky. Generally gray above; white and spots below. **BREEDING ADULT:** Black breast with large spots extending to flanks. Extensive rufous on scapulars. **NON-BREEDING ADULT:** Gray-brown above; white with limited streaks or spots below. **JUVENILE:** Dark back feathers

edged with brown and white; buffy wash and distinct spotting below. **FLIGHT:** Faint wing stripe; dark primary coverts and white uppertail coverts.

Similar Species Larger than Red Knot, with longer bill; less rufous on back, none on head and breast on breeding birds. In flight, wing stripe is fainter, and uppertail coverts are whiter than

in Red Knot. Remarkably similar in breeding plumage to Surfbird (with which it has hybridized, CA record), but has a longer, thinner black bill, as well as dark gray or green legs, not yellowish as in Surfbird.

Voice CALL: Generally silent.

Status & Distribution Breeds in Russian Far East. **WINTER:** Primarily southeast Asia to Australasia. **RARE**

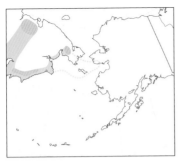

extensive rufous
on scapulars

larger and
longer-billed
than Red Knot

breeding plumage
suggests breeding
plumage of Surfbird

bold arrow-shaped
spots below

breeding

upperparts much
more patterned than
Red Knot

juvenile

extensive
spotting
below

STATUS: Casual spring migrant in western AK, mostly St. Lawrence I. and Nome; accidental in fall. Accidental in fall to OR, WV, and ME; also to western Palearctic.

Population Endangered. There is major concern over hunting pressure and especially habitat loss at migratory stopovers in China and the Korea Peninsula.

RED KNOT *Calidris canutus*   REKN ■ 1

Hundreds, formerly thousands, of Red Knots collect at staging areas during migration. The Red Knot feeds along sandy beaches and mudflats, often with dowitchers. Polytypic (6 sspp.; 3, possibly 4, in N.A.). L 10.5" (27 cm)
Identification Chunky and short-legged for a *Calidris*. Medium-length black bill. Legs vary from black to olive or gray. **BREEDING ADULT:** Dappled brown, black, and chestnut above, with buffy chestnut face and breast; entirely or mostly chestnut below. Males average brighter and females average more white on the belly. Worn adult looks darker above, once paler chestnut edges wear. **NONBREEDING ADULT:** Pale gray upperparts and head; white underparts, except breast, which has grayish wash or spots, and fine markings along sides. **JUVENILE:** Like nonbreeding, but with more distinct spotting below; pale fringes and dark subterminal lines on scapulars and coverts create pattern above. Fresh juveniles often have buff wash on breast. **FLIGHT:** White wing stripe, paler rump with gray barring.
Geographic Variation There is uncertainty over the winter distribution of some subspecies and the subspecific limits in the AK populations. Of six subspecies, *islandica*, which breeds in Greenland and northern Canadian islands and winters in Europe, is more chestnut above and below. Breeding south of *islandica* in Canada, *rufa* has less

chestnut above and is paler below. It is believed to winter in S.A. and possibly the southeast US. AK and Wrangell I. (Russia) *roselaari* is similar to *rufa*, only darker. It winters from the southern US to S.A. Subspecies *rogersi* breeds on Russia's Chukchi Peninsula and may occur in nearby western AK.
Similar Species In breeding plumage, the Red Knot's uniform color is similar to a dowitcher, but the knot has a shorter bill, paler crown, and (in flight) a whitish rump, finely barred with gray. Nonbreeding Red Knot also differs by structure, bill, and tail pattern; also see Great Knot.
Voice CALL: Generally silent, but *cur-wit* sometimes given when taking flight.
Status & Distribution Uncommon to fairly common. **BREEDING:** Dry tundra. **MIGRATION:** Rare to casual migrant in the interior. Often missed in coastal areas not used as staging grounds; may be common in one place, but only in small numbers between the staging area and the next stop.

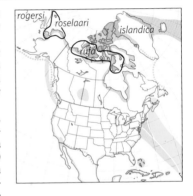

In spring, earlier in the West than in the East, peaking on Pacific coast late Apr.–mid-May. Mid-Atlantic peak mid-May–early June. In fall, adults seen mid-July, peak late July–mid-Aug. Juveniles peak late Aug.–Sept. Smaller numbers to mid-Oct.; lingers into Nov. **WINTER:** Coasts, south to Argentina. **RARE STATUS:** HI.
Population Near Threatened. Severe declines in *rufa* (Threatened, USFWS) from mid-Atlantic migration. Species is at risk at staging areas where it relies on a single food source (e.g., horseshoe crab eggs at Delaware Bay).

whitish supercilium
and dark eye line

plain gray
back

winter

spotted
chest

chunky
body with
short,
straight bill

rufous
breast
and
belly

breeding

pale vent
and undertail
coverts

subterminal dark
edges with pale
fringes on upperparts

pale buff
wash
below
when
fresh

barred
whitish
rump and
faint wing
stripe

juvenile

SURFBIRD *Calidris virgata* SURF ■ 1

A Pacific coast rocky shorebird, the rather chunky Surfbird is often seen in the company of Black Turnstones. It picks invertebrates from rocks, although rarely it will feed on mud-flats and sandy beaches. Monotypic. L 10" (25 cm)

Identification Generally gray above and on the throat, white below. Stout dark bill with a yellow base. Short yellowish green legs. **BREEDING ADULT:** Head and underparts heavily streaked and spotted with dusky black. Upperparts edged with white and chestnut; scapulars rufous and black. **NON-BREEDING ADULT:** Gray spots along sides. White edges to wing coverts. **JUVENILE:** Like nonbreeding, but upperparts have pale edges and darker subterminal markings, creating scaled appearance; chest paler and mottled. **FLIGHT:** White wing stripe and conspicuous black band at end of white tail and rump.

Similar Species Breeding Surfbird could be taken for a Great Knot; how-

ever, Great Knot has a longer, more typical sandpiper bill and dark legs. Surfbird's short, thick bill and bold flight pattern are unique.

Voice CALL: Generally silent, but chatter notes, like those of a turnstone, often given in flight.

Status & Distribution Uncommon to fairly common. **BREEDING:** Mountain tundra in AK and YT. **MIGRATION:** Spring northward movement noted Mar. to early May along coast, with most birds departed by the end of Apr. Prince William Sound, AK, is a major staging area early to mid-May. In fall, adults first noted south of breeding areas in late June or early July; most adult movement July–early Aug. Juveniles start moving early Aug.; most movement from BC south is over by late Sept. **WINTER:** Reefs and rocky beaches from southern AK to Chile. **RARE STATUS:** Casual in spring on TX coast, Salton Sea, and points east from FL to AB; accidental inland in fall from southern BC, southern YT, PA (e.g.,

sight record 18 Aug. 1979, Presque Isle, PA), ME (21 Mar.–18 Apr. 2015, Biddeford), and HI.

Population Stable, but vulnerable since a high percentage of the population uses Prince William Sound, AK, as a staging area.

winter

short, sturdy bill, always with greenish yellow at base

juvenile has scaly pattern above and below

arrow-shaped streaks on lower sides and flanks

juvenile

bold white wing stripe

winter

white base to tail

extensive rufous on scapulars

breeding

bold arrow-shaped spots below

RUFF *Calidris pugnax* RUFF ■ 3

The male Ruff has a colorful and varied breeding plumage. They gather in leks of five to 20 birds (sometimes in larger numbers) on open grassy areas or bare soil and display to females. Monotypic. L 10–12" (25–31 cm); females smaller.

Identification Tall with long legs, plump body, small head, and small bill. Back feathers often raised when feeding. White underwings with dark crescent at base of primary coverts. **BREEDING MALE:** Neck ruff colors black, rufous, or white. Legs may be yellow, orange, or red; bill has pale base, black tip. Some males are dull and small, resembling females. **BREEDING FEMALE ("REEVE"):** Lacks the ruff, is smaller, and has a variable

amount of black below. **NONBREEDING ADULT:** Legs and bill usually duller than in breeding, but some birds still with bright tones. Scapulars and coverts dark, with thin, pale edges; tertials often with bars or internal markings. Underparts dingy whitish with mottling sometimes present on the neck and sides of the breast. Typically with white feathering at base of bill and on forehead. **JUVENILE:** Like nonbreeding, but bolder edges to the upperparts, buffy breast, and tertials lack internal markings. **FLIGHT:** Distinctive U-shaped white rump band in all plumages; white underwings.

Similar Species Combination of small head, plump body, short bill, and long

legs is distinctive. Compare juvenile Ruff with Sharp-tailed Sandpiper, which has a different face pattern and

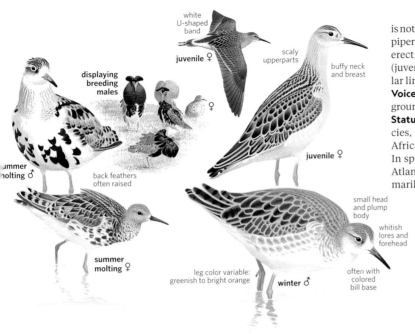

white U-shaped band

juvenile ♀

scaly upperparts

displaying breeding males

♀

summer molting ♂

back feathers often raised

buffy neck and breast

juvenile ♀

small head and plump body

whitish lores and forehead

often with colored bill base

summer molting ♀

leg color variable: greenish to bright orange

winter ♂

is not scaly above. Buff-breasted Sandpiper is also scaly above, is smaller, less erect, and lacks any facial markings (juvenile Ruffs usually have a postocular line) and white rump ovals.

Voice CALL: Silent away from breeding grounds where only grunts are given.

Status & Distribution Old World species, breeds Eurasia, most winter in Africa; has bred in AK. **MIGRATION:** In spring, rare in western AK and on Atlantic coast, casual elsewhere, primarily late Mar.–late May, into June in AK. In fall, as above, but more regular in West, primarily late July–Aug. (adults), and Sept.–early Nov. (juveniles); juveniles more numerous than adults in West, opposite in East. Casual in winter from OR, CA, AZ, Gulf Coast, and southern Atlantic coast. **Population** Stable.

Calidris Plumage and Structural Details

Three critical aspects in identifying *Calidris* are structure (especially bill length and shape), feather topography, and correct ageing. First, it is essential to learn to identify the various feather groups (especially the scapulars, the various groups of upperwing coverts, tertials, and the primaries, if visible) and to assess their pattern. There are three main groups of wing coverts that cover the secondaries: lesser, median, and greater. Lesser coverts consist of multiple rows; median and greater coverts are represented by a single row each. There are five rows of scapulars: The upper rows are smaller; the lower rows are larger and often hang down over the wing coverts. The lower scapulars are usually marked slightly differently from the uppers, especially in breeding adults. Look for any pale fringes or dark subterminal bars on the scapulars and coverts. The tertials (middle one is longest) are easily confused with the primaries. Locating the primary projection past the tertials (or lack of projection in a few species) is essential for identifying some species. The primaries are dark, so the primary tip projection area contrasts as dark and uniform to the paler and more patterned tertials. Leg color, often hailed as the best feature, can be misinterpreted due to light or when they are covered in mud. Juvenile *Calidris* look fresh and colorful and stand out among worn or molting breeding adults or plain winter-plumaged adults. They are often more approachable, sometimes within a few feet. Juveniles in most species typically migrate later in the fall than the adults, which in a number of species start their southward migration by very late June.

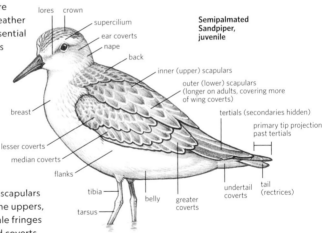

lores crown

supercilium

ear coverts

nape

back

Semipalmated Sandpiper, juvenile

inner (upper) scapulars

outer (lower) scapulars (longer on adults, covering more of wing coverts)

breast

tertials (secondaries hidden)

primary tip projection past tertials

lesser coverts

median coverts

flanks

tibia

tarsus

belly

greater coverts

undertail coverts

tail (rectrices)

Baird's Sandpiper, juvenile (CA, Sept.)

BROAD-BILLED SANDPIPER *Calidris falcinellus* BBIS ◼ 4

Broad-billed Sandpiper is a casual Old World species. Polytypic (2 sspp.; probably both in N.A.). L 7" (18 cm)
Identification Plump body, short legs, and long bill form a distinctive profile. Thick bill droops noticeably at the tip, at times looking like it has an unnatural kink. The Broad-billed is more likely to probe for food than peck, but it does both. Split supercilium is present in all plumages. Legs vary from olive to almost black. **BREEDING ADULT:** Black scapulars edged buff or rufous; coverts typically edged buff. Underparts white except for the brownish and heavily streaked breast; streaks extend down the sides. **NONBREEDING ADULT:** Gray-brown above with some scapulars having dark centers, giving a somewhat mottled look. White below with variable breast streaking. **JUVE-** **NILE:** Scapulars and coverts edged buff or rufous, with bold mantle and scapular lines. Breast lighter, more lightly streaked than breeding plumage.
Geographic Variation Nominate *falcinellus* from Europe and west Russia perhaps pertains to the Atlantic record. AK records are of Asian *sibirica* (more extensive rufous fringing on breeding adult and juvenile).
Similar Species While Dunlin is superficially similar—it shares a rather plump appearance, with a decurved bill—no species matches Broad-billed's combination of plumage and structure.
Voice **CALL:** Ascending high pitched, buzzy trill, *brreeet*.
Status & Distribution Breeds in Eurasia. **WINTER:** Arabian Peninsula to Australia. **RARE STATUS:** Casual fall migrant on western and central Aleutians and Pribilofs, AK; accidental in fall in coastal NY. All N.A. sightings have been of juveniles.
Population Stable.

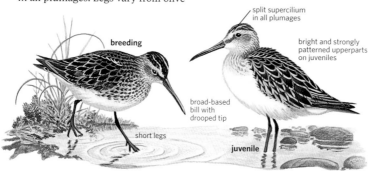

breeding

split supercilium in all plumages

bright and strongly patterned upperparts on juveniles

broad-based bill with drooped tip

short legs

juvenile

SHARP-TAILED SANDPIPER *Calidris acuminata* SPTS ◼ 3

The Sharp-tailed Sandpiper is an Asian species that principally visits coastal western N.A. in fall. Nearly all records are of juveniles, reflecting their more eastern and coastal fall migration route in Asia. Monotypic. L 8.5" (22 cm)
Identification Similar to Pectoral Sandpiper, and often seen with that species during migration. **BREEDING ADULT:** Rufous crown and edges to scapulars and tertials. Buffy breast, spotted below with dark chevrons on flanks, usually extending to undertail coverts. Distinct white eye ring, rather indistinct supercilium, and dark lores; legs greenish. **NONBREEDING ADULT:** Paler than breeding adult, lacking most rufous and buff tones. Some spots and chevrons on underparts. **JUVENILE:** Like breeding adult above, with rufous cap and edges to scapulars and tertials; bold supercilium; bold white mantle and scapular lines. Rich buff breast lightly streaked on upper breast and sides; underparts more lightly marked; streaked undertail coverts.
Similar Species Most sightings in N.A. are of juveniles, which can be distinguished from juvenile Pectoral Sandpipers by a bolder white supercilium that broadens behind the eye; bright buffy breast lightly streaked on upper breast and sides only; streaked undertail coverts; brighter rufous cap and edging on upperparts; and by call.
Voice **CALL:** A distinctive, mellow two-note whistle, *to-wheet*.
Status & Distribution Breeds in Russian Far East, where adults migrate inland and juveniles move to the coast. Casual spring, and irregularly fairly common fall migrant (nearly all juveniles) in western AK. Rare in fall (juveniles) along Pacific coast, mostly BC and northwest CA, with a few interior records (including from spring) from YT south to CA. Accidental in spring and casual in fall across rest of continent. **WINTER:** Australia. Accidental CA in winter.
Population Stable.

breeding adult

ruddy crown and bold white eye ring

streaks on undertail coverts

extensive dark chevrons on sides and flanks

broader rufous tertial edges than Pectoral

juvenile

bold supercilium

extensive buff on breast below streaking

juvenile Sharp-tailed (center) with juvenile Pectorals

streaks on sides continue faintly across upper breast

STILT SANDPIPER *Calidris himantopus*   STSA ■ 1

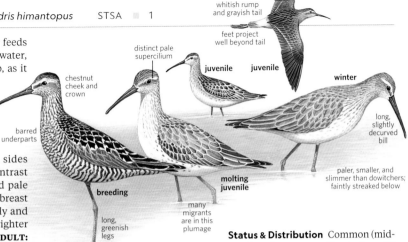

whitish rump and grayish tail

feet project well beyond tail

distinct pale supercilium

juvenile   juvenile

winter

chestnut cheek and crown

long, slightly decurved bill

barred underparts

breeding

long, greenish legs

molting juvenile

many migrants are in this plumage

paler, smaller, and slimmer than dowitchers; faintly streaked below

The Stilt Sandpiper frequently feeds with dowitchers in belly-deep water, its head down and rear end up, as it probes into the mud for food. Monotypic. L 8.5" (22 cm) **Identification** Long-legged with a long, slightly decurved bill; legs are yellow-green. **BREEDING ADULT:** Chestnut patches on sides of head and rear of crown contrast with dark gray upperparts and pale eyebrow. Heavy barring on the breast thins out somewhat on the belly and lower flanks. Males average brighter than females. **NONBREEDING ADULT:** Gray-brown above including crown, auriculars, and lores, contrasting with distinct white supercilium. Underparts are mostly whitish, with some gray on the breast. **JUVENILE:** Plumage similar to nonbreeding plumage, but scapulars and coverts have broad

pale edges. When plumage is fresh, a buff wash might be present on the breast; dusky breast streaking, and some spotting on the flanks. Migrants are often seen with molted scapulars. **FLIGHT:** White uppertail coverts (except in breeding plumage), pale tail, and faint white wing stripe; legs extend noticeably beyond tail. **Similar Species** Curlew Sandpiper is similar, but Stilt Sandpiper has a somewhat straighter bill and yellow-green legs that are noticeably longer. Also, the underparts are always more marked, and in flight, Stilt lacks a prominent wing stripe. See dowitchers and Lesser Yellowlegs. **Voice CALL:** A low, hoarse *querp*, but often silent.

Status & Distribution Common (mid-continent). **BREEDING:** Tundra from AK to Hudson Bay. **MIGRATION:** Primarily migrates mid-continent in spring; first arrivals late Mar., but typically peak in southern states mid-Apr.–early May; closer to mid-May in KS, and the latter half of May in the Prairie Provinces. Very rare in the West, except at south end of Salton Sea where it can be fairly common. Fairly common in fall on East Coast (rare spring); rare (mostly fall) in West. Adults first seen early or mid-July, peak in KS late July, generally completing migration in early Aug. Juveniles first arrive in mid-Aug., peaking in Sept. Small numbers into Oct., rarely into Nov. **WINTER:** Scattered areas in southern US most from Mexico to S.A. **RARE STATUS:** Europe, Asia, and Australia. **Population** Stable. Possible recent western shift of breeding range.

CURLEW SANDPIPER *Calidris ferruginea*   CUSA ■ 3

Curlew Sandpiper is a rare visitor from the Old World. Monotypic. L 8.5" (22 cm) **Identification** More elegant than somewhat similar Dunlin. Long, black legs. Rather long , black, evenly

decurved bill. **BREEDING MALE:** Rich chestnut underparts and mottled chestnut back; grayish wing coverts. Scattered bars on sides and across uppertail coverts; white where bill joins face. **BREEDING FEMALE:** Slightly

paler. Many birds show patchy spring plumage, showing grayer upperparts and partly white underparts. **NON-BREEDING ADULT:** Gray-brown above, whitish below. White uppertail coverts unbarred. **JUVENILE:** Appears scaly

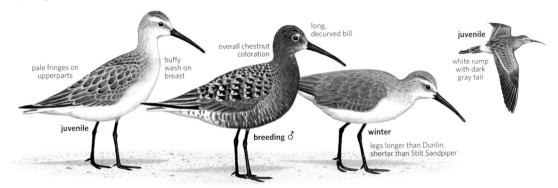

pale fringes on upperparts

buffy wash on breast

overall chestnut coloration

long, decurved bill

juvenile

white rump with dark gray tail

juvenile

breeding ♂

winter

legs longer than Dunlin, shorter than Stilt Sandpiper

above; shows rich buff across breast. **FLIGHT:** White rump conspicuous. **Similar Species** In fall, young birds are in full juvenal plumage; compare with young Dunlins, which are mostly in winter plumage in US and southern Canada. Dunlin has a dark rump, shorter legs, shorter and less bold supercilium, and its bill is more abruptly decurved at the tip. Stilt Sandpiper has longer, greenish legs that in flight extend farther beyond its paler tail. Also note Stilt's fainter wing stripe and faint streaking on underparts.

Voice CALL: A soft, rippling *chirrup*. **Status & Distribution BREEDING:** Siberia and Russian Far East; has bred in northern AK. **WINTER:** Africa, Asia, Australasia. **RARE STATUS:** Rare on East Coast (primarily mid-Atlantic), elsewhere casual continentwide. Spring records primarily late Apr.–late May. In fall, adults most likely encountered mid-July–mid-Aug., juveniles mid-Aug.–mid-Oct., a few records into Nov.; winter records FL and Southern CA. **Population** Near Threatened from wetlands loss in migration and winter.

TEMMINCK'S STINT *Calidris temminckii* TEST ■ 3

Temminck's Stint, an Old World visitor, has a preference for freshwater habitats. It is usually found singly or in small numbers. Monotypic. L 6.3" (16 cm) **Identification** Rather brown, horizontal-looking peep. White outer tail feathers are distinctive in all plumages, but hard to see at rest. Bill is dark and short, with a slight droop; legs are yellowish, sometimes with a green tint. Tail is quite long, extending past the folded wing tips. **BREEDING ADULT:** Generally brownish upperparts. Scapulars and coverts are dark with rufous or buff edges; although appearing uniform from a distance, they can be colorful. There is little of a face pattern, with no supercilium and only slightly darker lores; white eye ring is usually rather distinct. Breast is brown with dark streaks. Worn adults are brown with dark splotches above. **NONBREEDING ADULT:** Upperparts and breast are plain gray-brown; the remainder of underparts are white. Some shaft streaks are visible on the scapulars. **JUVENILE:** Like nonbreeding, but feathers of upperparts have diagnostic dark, subterminal edges and buffy fringes. **Similar Species** Breeding Temminck's resembles in plumage and shape the larger Baird's Sandpiper; however, note

Temminck's yellow or greenish yellow legs and tail that extends beyond the primaries, whereas Baird's has dark legs and primaries that project well past the tertials and tail tip. While the white outer tail feathers (best seen in flight) are diagnostic, they must be used with caution as the outer tail feathers of other *Calidris* can appear pale. **Voice CALL:** Repeated, fast, dry rattle, almost recalling a longspur; routinely calls when flushed. Call might be the first clue to its identity. **Status & Distribution BREEDING:** Breeds in wet tundra from Scandinavia to northeast Russian Far East. Rare spring and fall migrant on the Aleutians and Pribilofs; rare in spring on St. Lawrence I. (also twice in fall). One Sept. record of a juvenile

from BC and a late fall record from WA. **WINTER:** Mediterranean, Africa, east to southeastern Asia. **Population** Stable.

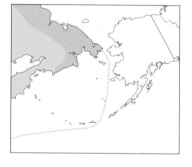

juvenile

white outer tail feathers

breeding

scattered black-centered scapulars

greenish yellow legs

dark breast

juvenile

dark subterminal fringes

short primary projection

white outer tail feathers

LONG-TOED STINT *Calidris subminuta* LTST ■ 3

This Asian visitor prefers freshwater with vegetation and is usually seen singly or in small numbers, even where common. It often appears erect, high-stepping on mudflats. Its long legs and toes, however, are of marginal use for identification. Monotypic. L 6" (15 cm) **Identification** Brown peep with a reduced wing stripe. Short and slightly

drooped black bill, usually with a pale greenish base. Yellow legs, with either brown or green tones at times; relatively long, so toes extend beyond tail in flight. Dark underwings, unique among stints. **BREEDING ADULT:** Bright above, recalling a miniature breeding Sharp-tailed Sandpiper. Crown, auriculars, mantle, scapulars,

and tertials with rufous edges; lower scapulars often edged white. Greater wing coverts edged buffy. Breast usually buff with fine dark streaks. Rather bold face pattern, with dark lores and a split supercilium. In early fall, when feather edges are worn, bird looks dark brown above. **NONBREEDING ADULT:** Gray-brown above, with dark centers to

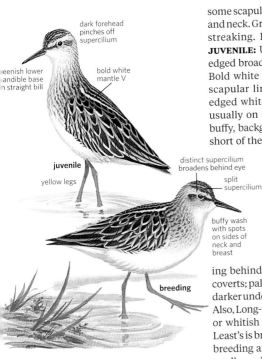

dark forehead pinches off supercilium

bold white mantle V

greenish lower mandible base on straight bill

juvenile

yellow legs

distinct supercilium broadens behind eye

split supercilium

buffy wash with spots on sides of neck and breast

breeding

some scapulars; streaked on the crown and neck. Gray-brown breast with faint streaking. Face pattern stays bold. **JUVENILE:** Upperparts and scapulars edged broadly with rufous and white. Bold white mantle line; less distinct scapular line. Median wing coverts edged white. Breast is streaked, but usually on a grayish, less frequently buffy, background. Supercilium stops short of the bill, resulting in a noticeably dark forehead.

Similar Species Most likely to be confused with the much more expected Least Sandpiper. Distinguished in all plumages by its dark forehead and region above lores; bolder, split eyebrow, broadening behind eye; white-edged median coverts; pale base to the mandible; and darker underwing coverts (hard to see). Also, Long-toed's breast is often grayish or whitish with fine streaks, whereas Least's is brown or buff with streaks. In breeding and juvenal plumage, Long-toed's median upperwing coverts are edged whitish and contrast with the rest of the bright upperparts.

Voice CALL: *Purp*, lower pitched than Least Sandpiper.

Status & Distribution Asian breeder. Rare to fairly common in spring (casual fall) on western Aleutians. Very rare in spring and casual in fall to Pribilofs and St. Lawrence I.; can be fairly common on western Aleutians, but casual in fall. **WINTER:** Southeast Asia to Australia. **RARE STATUS:** Accidental elsewhere on West Coast in fall (single records of juveniles from coastal OR and CA). **Population** Probably stable.

SPOON-BILLED SANDPIPER *Calidris pygmeus* SBSA ■ 4

This critically endangered Asian sandpiper has a unique bill shape. Peeplike in behavior, it often forages in shallow water. Monotypic. L 6" (15 cm)

Identification The spoon-shaped bill is diagnostic, but shape hard to see at some angles; bill is longer than most peeps. **BREEDING ADULT:** Plumage acquired late. Rufous head, throat, and upper breast; reddish edges above. Dark spotting along lower edge of orange breast color, spreading onto the white lower breast and sides. **NONBREEDING ADULT:** Pale grayish above. White supercilium forked and quite broad in front of eye; darker auriculars. **JUVENILE:** Supercilium forked, as in nonbreeding, leaving white forehead that contrasts boldly with dark cheek patch and lores. Upperparts edged white or buff, sometimes rufous. Mantle and scapular lines indistinct. White underparts, although a buff wash across the breast expected in fresh plumage.

Similar Species Similar in breeding plumage to Red-necked Stint, which usually has a hint of a white supercilium although Spoon-billed averages more spotting on the underparts. Juvenile Spoon-billed has a darker cheek patch than juvenile Red-necked,

distinct supercilium

dark cheek

juvenile

breeding plumage similar to breeding plumage of Red-necked Stint

breeding

very distinctive spoon-shaped bill harder to discern at a distance

emphasized by its forked supercilium. While no other sandpiper has a similar-looking bill, beware of mud creating a misleading blob.

Voice CALL: A quiet *wheet*.

Status & Distribution Endangered. **BREEDING:** Coastal tundra in Russian Far East. **WINTER:** Coastal southeast Asia. **RARE STATUS:** Casual migrant in AK (no records since 1993); one record of a breeding adult from Vancouver, BC (30 July–3 Aug. 1978).

Population Critically Endangered and still declining. In the 1970s, the population was estimated at 2,000–2,800 pairs; by 2010 it was only 120–200 pairs. A small captive breeding program from eggs collected and hatched in Russia (13 birds) was established in England in 2012. The wild population appears to be slowly increasing (about 210–228 breeding pairs in 2014) but still faces multiple threats: disturbance on breeding grounds, hunting on winter grounds, and habitat loss at migration sites. Collecting eggs from wild nests, incubating the eggs, and protecting the chicks until fledging ("head-starting") has yielded excellent results and is ongoing. Recent positive developments in China have halted the further reclamation of mudflats there; and it appears restoration of reclaimed mudflats may be in the works.

RED-NECKED STINT *Calidris ruficollis* RNST ■ 3

Red-necked Stint is an Asian visitor. Monotypic. L 6.3" (16 cm)
Identification Small peep. Black legs, feet unwebbed. Rather short black bill. Primaries extend slightly beyond tail tip. **BREEDING ADULT:** Rufous throat, upper breast, head, and mantle can be bright; paler in some birds, with throat entirely white. A necklace of dark streaks extends below the throat onto the white lower breast, with a few markings on sides. **NONBREEDING ADULT:** Gray-brown above with dark shaft streaks on scapulars. White below, with some streaks at sides of breast. **JUVENILE:** Scapulars vary from bright to pale rufous, with white tips.

White scapular and mantle lines. Drab tertials, edged in pale buff, some with a faint rust tinge. Plain brown wing coverts with dark anchor-shaped shaft streaks. Crown seems darkest in the center; partial, indistinct forked supercilium.

Similar Species Breeding Little Stint has breast spotting contained within the color of the breast and always has a white throat, usually with extensive rufous-edged tertials and upperwing coverts. Juvenile Red-necked differs from juvenile Little by its plainer gray wing coverts and tertials, dusky streaks rather than darker spots at breast sides, and less distinct mantle and scapular lines; and from juvenile Semipalmated and shorter billed Western by its plainer coverts and tertials; Semipalmated also has a darker

crown and less contrast between the scapulars and wing coverts. Away from AK, only two certain records of juveniles (CA). See Sanderling.
Voice CALL: A squeak somewhat like that of Western Sandpiper.
Status & Distribution BREEDING: Russian Far East; very small numbers to west AK. **WINTER:** Indian subcontinent to Australasia. **MIGRATION:** Regular migrant on western AK islands and Seward Peninsula. **RARE STATUS:** Breeding adults casual on both coasts and interior late June–late Aug.; lack of juvenile records in this area, although some probably pass through undetected. Accidental in CA and DE in spring, and Southeast Asia.
Population Near Threatened.

tertials and coverts rather dull in juvenal and breeding plumage

juvenile

breeding

breeding

rich brick red throat

dark streaks on breast

some have duller rufous that is confined to lower throat

SANDERLING *Calidris alba* SAND ■ 1

The Sanderling is the familiar wave runner of sandy beaches. Monotypic. L 8" (20 cm)
Identification Larger than other peeps. Black bill and legs. Lacks hind toe. Prominent white wing stripe in flight; black primary coverts contrast with remainder of wing. **BREEDING ADULT:** Not acquired until late Apr. or later; head, mantle, and breast are rusty. Tertials often with markings inside the edge, unlike any

other peep. **NONBREEDING ADULT:** Palest sandpiper in winter. **JUVENILE:** Blackish above, with pale edges near feather tips; looks checkered. Breast washed buff when fresh.
Similar Species Breeding-plumaged Red-necked Stint is smaller, with fainter wing stripe and dark necklace of streaks below rusty breast. Nonbreeding birds are paler than any congener.
Voice CALL: A sharp *kip*.
Status & Distribution Common.
BREEDING: Dry tundra of Canadian islands, Eurasia, and, rarely, AK.

MIGRATION: Spring peak on Pacific coast late Apr.–late May; Great Lakes and mid-Atlantic two weeks later. In fall, adults arrive south mid-July, peaking late July–Aug.; juveniles late Aug.–Sept.; stragglers into Nov. In interior, generally rare away from Great Lakes and northern Great Plains. **WINTER:** Coastlines to southern S.A.; accidental interior N.A. away from Salton Sea.
Population Declining.

winter

dark leading edge

bold white wing stripe

strongly patterned with black and white above

short, straight bill

dark leading edge often visible at rest

juvenile

very pale above

breeding

variably rufous upperparts, head, and breast

runs on sand

winter

no hind toe

DUNLIN *Calidris alpina* DUNL ■ 1

Holarctic species. Dunlins frequently flock in large groups on the coast during winter and migration. They often feed in water, although they will also feed along the shore, probing into the mud for prey. Polytypic (10 sspp.; 6 in N.A.). L 8.5" (22 cm)

Identification Midsize sandpiper with a short neck; it appears hunchbacked. It has a sturdy, black bill, decurved at the tip, and black legs. **BREEDING ADULT:** Reddish back is distinctive. Whitish, finely streaked underparts with a conspicuous black belly patch. **NONBREEDING ADULT:** Gray-brown above and on the breast; remainder of underparts are whitish; winter Dunlins on West Coast appear darker and browner. Molt takes place in N.A. (but not Greenland *arctica*) on or near breeding grounds. **JUVENILE:** Like nonbreeding plumage, but with rufous on crown and edges of upperparts. Neck and breast are buffy with dark streaks that extend to the belly, where there is often a partial belly patch. This plumage is not expected south of the breeding range in N.A. subspecies. **FLIGHT:** Note dark center to rump and white wing stripe.

Geographic Variation Subspecies *hudsonia* breeds in the Canadian Arctic and winters in eastern N.A.; it is marked by streaks on its flanks. Subspecies *pacifica* breeds in western AK and winters primarily on the West Coast. Asian *arcticola* breeds in northwestern AK, and Asian *sakhalina* migrates regularly through the central

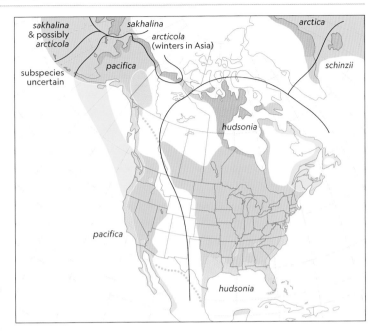

and western Aleutians. Greenland *arctica* and west Palearctic *alpina* have been recorded casually or accidentally on the East Coast.

Similar Species Rock Sandpiper in breeding plumage has a similar black patch, but it is on the chest, not the belly, and different facial pattern. In nonbreeding plumage, see Curlew and Western Sandpipers.

Voice CALL: A harsh, reedy *kree.*

Status & Distribution BREEDING: Abundant. **MIGRATION:** Peak in spring early to mid-May in mid-Atlantic and Great Lakes, one to two weeks earlier in the West; lingering birds into early June. Generally a late fall migrant, perhaps in part due to its molt prior to migration. Early arrivals usually in mid-Sept., only rarely earlier, with peak numbers mid-Oct.–mid-Nov. Rare on Great Plains. **WINTER:** Locally inland. Winters coastally south to Mexico; a few to West Indies. **RARE STATUS:** Casual C.A., accidental S.A.

Population Local declines suggested.

ROCK SANDPIPER   *Calidris ptilocnemis*   ROSA ■ 2

This western counterpart of the Purple Sandpiper is usually seen on rocky shores, often with Black Turnstones and Surfbirds. Polytypic (4 sspp.; 3 in N.A.). L 9" (23 cm)

Identification Shape like Purple Sandpiper; dark bill with pale yellow-green at base; yellow-olive legs. **BREEDING ADULT:** Rufous crown and edges to mantle and scapulars. Black patch and gray streaks on lower breast; spots on flanks and belly. **NONBREEDING ADULT:** Dark gray upperparts and breast, with distinct spotting on flanks. **JUVENILE:** Like breeding, but paler, with buffy edges to scapulars and coverts; buff color to breast. Molts prior to reaching wintering grounds. **FLIGHT:** White wing stripe and all-dark tail.

Geographic Variation Largest subspecies, nominate *ptilocnemis* (Pribilofs), has paler chestnut above, less black below, and a bolder white wing stripe and whiter underwing. Aleutian *couesi* has a darker head and breast and more obscure auricular patch. The brightest breeding bird is *tschuktschorum*, from St. Lawrence I., the Seward Pen-

insula, and elsewhere. Asian *quarta* (breeds Commander and north Kurile Is.; casual western Aleutians) is dark gray, like *couesi*, but in breeding plumage edges above are broader and paler orange than *couesi*.

Similar Species In winter plumage, best separated from Purple Sandpiper by range and duller bill base and legs. Rock Sandpiper is almost unknown inland, while Purple is casual in the Midwest and with now scattered western records, so any interior western record is more likely Purple. Still any identification of any fully winter plumaged bird is problematic. Separated easily from Surfbird, by longer, thinner bill and very different tail pattern.

Voice CALL: Rough *kreet*, like Purple Sandpiper.

Status & Distribution Fairly common. **BREEDING:** Tundra from west AK to Russian Far East. **WINTER:** Arrives WA,

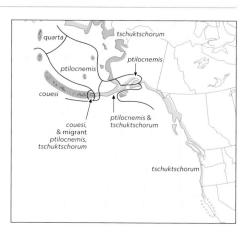

OR, and northern CA late Oct.–late Nov. (recent declines). Leaves breeding grounds Sept.–Oct. Wintering birds depart by late Apr.–mid-May. Nominate subspecies winters Cook Inlet, AK; *tschuktschorum* to West Coast. **RARE STATUS:** Casual coastal Southern CA. Accidental interior BC (Atlin, 29 Oct. 1932; specimen).

Population Declines, particularly in *tschuktschorum*.

winter
tschuktschorum

dark gray
above

thin, slightly decurved
bill with greenish
yellow base

spots on
flanks, often
spade-shaped

yellowish
legs

juvenile
tschuktschorum

nominate subspecies
larger and paler than
other subspecies

dark
postocular
spot

dark slate
above

winter
tschuktschorum

breeding
tschuktschorum

blackish
belly patch

Pribilofs
breeding
ptilocnemis

PURPLE SANDPIPER   *Calidris maritima*   PUSA ■ 1

The stocky Purple Sandpiper, primarily noted in winter along rocky coastlines in the Northeast, is the North Atlantic relative of the Rock Sandpiper. Monotypic. L 9" (23 cm)

Identification Long, slightly decurved bill, with an orange-yellow base. Yellow legs. **BREEDING ADULT:** Tawny crown streaked with black. Mantle and scapulars edged from white to

tawny-buff; rufous usually restricted to scapulars. Breast and flanks spotted with blackish brown. **NONBREEDING ADULT:** Dark gray upperparts, throat, and breast; blurred spots along sides and flanks; white belly. Purple tinge to upperparts difficult to see and also shared by Rock Sandpiper. **JUVENILE:** Like breeding, but less streaked on head and neck, more gray. Upperparts

edged buffy or white. Molts prior to arrival on wintering grounds.

Similar Species In flight and in winter, Purple very similar to Rock Sandpiper. Normally separated by range; bill base on Rock averages duller.

Voice CALL: A low, rough *kweet* or similar note, given in flight.

Status & Distribution Fairly common. **BREEDING:** Tundra, often with gravel.

MIGRATION: Reaches winter grounds in the Northeast in late Oct. (exceptionally mid-Sept.). Rare in fall on Great Lakes, especially eastern lakes in late fall. Can linger in spring, well into May. Return to breeding grounds late May–early June. The few Great Lakes spring records range late Mar.–mid-May. **WINTER:** Rocky shores and jetties, often with Ruddy Turnstones and Sanderlings. **RARE STATUS:** South of normal winter range along the coast, and annual to scattered interior areas of the eastern US. Very rare in winter to FL and casual to Gulf Coast, some have lingered through Apr. Accidental in fall at Pt. Barrow, AK, MT, CO, southwest UT, and CA; also west Mexico. **Population** Declines have been noted in Quebec.

breeding

heavily and extensively spotted below

juvenile

bright orange base to bill

overall dark slate gray

winter

winter

streaked sides and flanks

orange legs

winter

dark slate above

BAIRD'S SANDPIPER *Calidris bairdii*   BASA   ◼ 1

This long-winged and highly migratory sandpiper often feeds up from the shoreline, on drier surfaces. Monotypic. L 7.5" (19 cm)
Identification Long primary tips, projecting beyond the tail and well past tertials when standing. Brown upperparts; breast a lighter brown, with light streaks. Thin dark bill; dark legs. **BREEDING ADULT:** Mixture of brown and buff on dark scapulars and coverts. **NONBREEDING ADULT:** Paler than breeding plumage; rather nondescript light brown on upperparts, slightly paler brown (or gray-brown) on breast. **JUVENILE:** Pale fringes above give a distinctly scaled appearance; breast buffy and finely streaked. No mantle or scapular lines.
Similar Species Frequently confused with much smaller and darker Least Sandpiper. Long wings of Baird's extend past tail and note long primary projection past tertials (Least shows very short or no primary projection past tertials). Baird's also has a longer, straighter bill and black legs. Scaly upperparts on juvenile Baird's is unlike any plumage of Least. White-rumped has very similar structure to Baird's. White-rumped is grayer overall; breeding adults and some juveniles show markings down flanks; juveniles show chestnut on inner scapulars. Easily distinguished in flight by rump pattern; vocalizations completely different. Also see Pectoral Sandpiper.
Voice CALL: A low, raspy *kreeep*.
Status & Distribution Fairly common. **BREEDING:** Dry tundra from northeast Russian Far East and west AK to Baffin I. and Greenland. **MIGRATION:** In spring, through midcontinent, from TX north through the Great Plains;

rare in spring west of the Rockies, east of western Great Lakes. Early birds documented in early Mar. (exceptionally Feb.), but most pass late Mar.–mid-May. Lingerers found to early June. In fall, most migration, especially adults, is mid-continent, but uncommon (mainly juveniles) at this season to both coasts. Adults first noted early to mid-July, a few in late June. In fall, most observers see juveniles, which peak mid-Aug.–mid-Sept. Numbers drop until rare after early Oct., lingerers to Nov. or even later on occasion. **WINTER:** S.A., casual N.A. **RARE STATUS:** HI, Europe, Australasia, and South Africa. **Population** Stable.

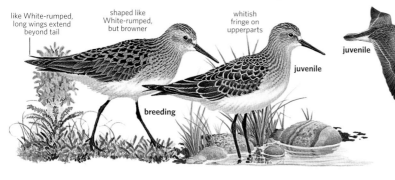

like White-rumped, long wings extend beyond tail

shaped like White-rumped, but browner

whitish fringe on upperparts

juvenile

juvenile

breeding

breeding

LITTLE STINT *Calidris minuta* LIST ■ 4

This Old World peep is likely found with other small sandpipers during fall migration. Monotypic. L 6" (15 cm) **Identification** Short, black bill has a fine tip. Legs are black; feet do not have webbing. Primaries clearly extend beyond the tail tip. **BREEDING ADULT:** Bright, with dark upperparts brightly fringed with rufous. Yellowish mantle lines appear whitish when worn. Throat and underparts are white; a warm brownish orange wash on the breast contains bold spotting, particularly on the sides of the breast. **NONBREEDING ADULT:** Gray-brown above, with some scapulars having dark feather centers. White below, with some light streaks on the sides of the breast. **JUVENILE:** Most juveniles average brighter than other peeps. Upper scapulars are edged rufous, lower scapulars are dark, usually black, with paler edges. Tertials are dark, typically with bright rufous edges. Both mantle and scapular lines are bold. Crown is dark with a distinctly forked supercilium. Streaking and spotting on the sides of the breast is distinct.
Similar Species See Red-necked Stint for separation. Breeding Sanderling is superficially similar and the only N.A.

peep that would have such a colorful throat and breast combination; since this plumage is seen only briefly away from the breeding grounds, unsuspecting birders might jump to an identification as Red-necked or Little Stint, but a Sanderling is bigger, bulkier, with a longer bill and bolder white wing bar. It also lacks a hind toe. Juveniles must also be distinguished from Semipalmated Sandpiper. Semipalmated has a darker crown, unforked supercilium, and darker wing coverts and tertials. On a bright Semipalmated, also note its more tubular bill; in comparison, Little Stint has sharper and darker streaks on the sides of the breast, and

bolder mantle and scapular lines. Although not as similar, a juvenile Least Sandpiper can often confuse birders due to its bright plumage (see Least Sandpiper).
Voice CALL: Single *kip*, often given when flushed.
Status & Distribution BREEDING: Eurasia. **MIGRATION:** Very rare on islands in western AK in spring and fall; casual to AK mainland. **WINTER:** Africa to Indian subcontinent; fewer southeast Asia. **RARE STATUS:** Casual, primarily in fall and from Pacific states (annual in recent years) and on East Coast. Accidental in CA in winter.
Population Stable.

LEAST SANDPIPER *Calidris minutilla* LESA ■ 1

A common peep throughout the continent, the Least Sandpiper prefers freshwater habitats with vegetation. It is our hardiest peep, wintering in numbers well inland. Monotypic. L 6" (15 cm) **Identification** This brown peep is smaller and darker than a Western or Semipalmated, the other two common peeps in N.A. Bill is dark, short, thin, and slightly drooped. Yellow legs can be dull or obscured by mud. **BREEDING ADULT:** Black and brown above, with rufous and buffy markings to the scapulars and all coverts; some feathers are tipped white. Breast is brown with conspicuous streaks. Supercilium is bold and meets the base of the bill, so forehead white. **NONBREEDING ADULT:** Upperparts and breast are brown. Supercilium, while present, is less distinct than in breeding plumage. **JUVENILE:** Like breeding plumage, but very crisp; upperparts are edged rufous and buff, some feathers are tipped white. White mantle line is bold; scapular line less distinct. Usu-

ally has a strong buffy wash across the finely streaked breast.
Similar Species Least Sandpiper is not too similar to Western and Semipalmated Sandpipers; however, these three species are the common peeps and the first step in learning sandpiper identification. Beginning birders often emphasize the yellow legs of the Least, compared to the darker legs of the other two species, but the identifications are sufficiently straightforward to eliminate the need to base an identification on a mark that can be altered by mud or lighting. Least Sandpiper is darker and browner overall than Western and Semipalmated and noticeably smaller when seen with either. In nonbreeding plumage, Least has a brown breast. Long-toed Stint, from Asia, is similar; see that species account. In fall, the first arrival of juveniles and their crisp, bright plumage might suggest other stints to the over-optimistic birder, but their upperparts pattern, drooped bill, yellow legs, and buffy, streaked breast

should confirm the identification.
Voice CALL: A high *kree* or *jeet*.
Status & Distribution Common. **BREEDING:** Moist tundra and wet habitats in boreal forest; breeds farther south than other peeps. **MIGRATION:** Spring movement in the West starts in late Mar., peaks in Apr. (CA) to as late as mid-May (WA). Migration in the East averages one to two weeks later. In fall, adults move as early as late June, and are common on both coasts and the Great Lakes by mid-July. Most have

migrated by the end of Aug. Juveniles first appear in late July, but more typically there is a push in mid-Aug.; they can remain common through Sept.

Where they do not winter, stragglers seen into Nov., rarely later. **WINTER:** Across southern US, as far south as Chile. This is the hardiest and most

numerous and widespread winter peep in the interior of N.A. **RARE STATUS:** Europe, Russian Far East, Azores, HI. **Population** Stable.

breeding

thin, slightly decurved bill

juvenile

streaked brownish breast

winter

yellowish legs

distinct streaks on sides of breast

WHITE-RUMPED SANDPIPER *Calidris fuscicollis* WRSA ■ 1

The White-rumped Sandpiper's larger size and longer wings separate it from the smaller peeps. It is often seen along shore edges or on mudflats. Monotypic. L 7.5" (19 cm)

Identification Long wings, with primaries extending noticeably beyond the tertials and tail. Black bill, noticeably pale reddish at the base, and dark legs. White, unmarked rump typically only visible during flight; sometimes holds wings lower, allowing a view of the rump on a perched bird. **BREEDING ADULT:** Gray-brown upperparts with pale edges on the scapulars and coverts. Distinct whitish supercilium. Crown, auriculars, mantle, and upper scapulars might have some rufous tones. Mantle and scapular lines, if present, are indistinct. Black streaks on breast extend to the flanks. Worn fall birds are overall plain; they usually show streaking along the flanks and a scattering of worn breeding-plumaged feathers above. **NONBREEDING ADULT:** Gray-brown above and on neck, giving it a hooded look; upperparts have shaft streaks or dark inner markings. Whitish below, with indistinct breast streaks

that extend along flanks. **JUVENILE:** Upperparts brighter than breeding adult, with rather broad rufous edges to crown, mantle, scapulars, and tertials. White mantle and scapular lines rather distinct. Rather bold white supercilium, contrasting with dark lores.

Similar Species Similar to Baird's Sandpiper in structure, but is grayer overall, has an entirely white rump, a bolder, whiter supercilium, streaking extends down flanks, and a pale base to the lower mandible. In flight, Curlew and Stilt Sandpipers both show a white rump; however, they show longer legs sticking out past the tail and larger, more decurved bills.

Voice CALL: A very high-pitched *jeet*.

Status & Distribution Fairly common. **BREEDING:** Moist tundra from north AK to Baffin I. **MIGRATION:** Late spring migrant, with majority of the population moving through the middle of the US and southern Canada. Although first arrivals reach southern states after mid-Apr., they peak late May–early June; small numbers are still moving through southern states in early June, the Great Lakes area in mid-June, with a few to late June. In fall, most birds take an Atlantic route to their winter-

ing grounds, moving east to Atlantic Canada or the northeast US. Adults are found south of the breeding grounds in late July, but most adults migrate Aug.–early Sept. Juveniles are very late migrants, rarely recorded south of Canada before late Sept.; they peak in Oct. and are seen regularly into Nov., strays into Dec. Juveniles yet to be documented in the western states and perhaps the Great Plains as well. **WINTER:** Southern S.A. **RARE STATUS:** Casual in the West, late May–mid-June; a few fall adults early July–Oct. Rare on the Azores and in Europe; ironically much more regular there than from Pacific states. Accidental to Galápagos, Australia, and New Zealand. **Population** Stable.

juveniles not yet recorded in West, not seen in Northeast until late Sept.

juvenile

fall migrant adults often in transitional plumage; adults appear in Northeast by early Aug.

fall-molting adult

breeding

dull rufous edges to scapulars

streaking extends to flanks

long wings and primary projection

juvenile

white rump normally only visible in flight

BUFF-BREASTED SANDPIPER   *Calidris subruficollis*   BBSA   ▪   1

Males of this elegant species join leks in search of mates, using bold wing displays. Monotypic. L 8.3" (21 cm)
Identification Dark eye prominent on buffy face; underparts paler buff. Pale orange-yellow legs. Dark bill. **ADULT:** Dark scapulars and coverts, with rather broad buff edges when fresh. **JUVENILE:** Like adult, but scapular edges duller and more narrow, look distinctly scaled. Wing coverts with dark subterminal marks, in addition to the pale fringe. Paler below than adult. Legs not as bright. **FLIGHT:** Bright white wing linings; darker underwing primary coverts.

Similar Species Unique, but see Ruff.
Voice CALL: Generally silent, but utters a low *tu*.
Status & Distribution Generally uncommon. **BREEDING:** Tundra. **MIGRATION:** Usually found at drying wetland edges and sod farms. Migrates through TX and central Great Plains in spring. Arrives TX early Apr., with peak late Apr.–early May, peak in northern Great Plains mid-May; few records to Great Lakes and Atlantic, or in the West. In fall, adults reverse migration through mid-continent; juveniles more likely to wander, very rare on the West Coast,

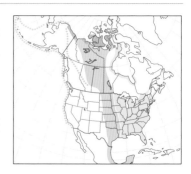

very uncommon in East. Migrate from late July, but peak (mostly juveniles) late Aug.–late Sept., lingerers through Oct., a few even later. **WINTER:** Southern S.A. **RARE STATUS:** Europe, Africa, Asia, and Australia.
Population Near Threatened; global population estimated at 56,000.

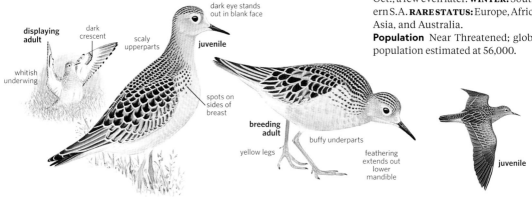

displaying adult
dark crescent
scaly upperparts
dark eye stands out in blank face
juvenile
whitish underwing
spots on sides of breast
breeding adult
yellow legs
buffy underparts
feathering extends out lower mandible
juvenile

PECTORAL SANDPIPER   *Calidris melanotos*   PESA   ▪   1

The Pectoral Sandpiper is found in wet fields, marshy ponds, and similar wetlands. It is often seen in small groups, and its migration is quite long. Monotypic. L 8.8" (22 cm)
Identification Very unusual within the genus *Calidris*, the male Pectoral (like male Ruff) is notably larger than the female. Bill has pale base; legs yellow. **BREEDING ADULT:** Upperparts are brown, with rufous or buff edges. Below, prominent streaking on breast,

darker in male, contrasts sharply with white belly. **NONBREEDING ADULT:** Duller than breeding plumage. **JUVENILE:** Brighter above than the breeding plumage, with bold white mantle and scapular lines; scapulars and coverts are broadly edged in rufous and white. Buffy breast is lightly streaked. **FLIGHT:** Weak wing stripe.
Similar Species Especially compare with juvenile Sharp-tailed Sandpiper (see that account). Pectoral is larger (especially males) than Baird's, with stronger breast streaking and yellow,

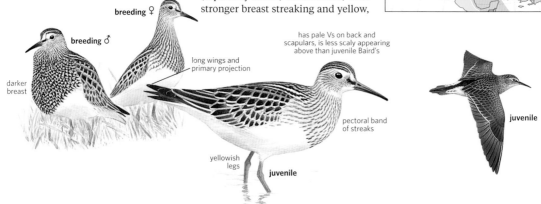

breeding ♀
breeding ♂
darker breast
long wings and primary projection
has pale Vs on back and scapulars, is less scaly appearing above than juvenile Baird's
pectoral band of streaks
juvenile
yellowish legs
juvenile

not dark, legs. Also more streaked above; the scaly juvenal plumage of Baird's being particularly distinct.
Voice CALL: A rich, low *churk*. **SONG:** A deep hoot, with chest distended.
Status & Distribution Uncommon to common (Midwest). **BREEDING:** Moist tundra from Hudson Bay west to Siberia. **MIGRATION:** Common in Midwest; fairly common on East Coast; scarcer from western Great Plains to West Coast. In spring, most arrive in Apr. in Midwest, but common by late Mar. on Gulf Coast, a few as early as late Feb. Peak migration late Apr.–

early May in Great Lakes and mid-Atlantic coast (where less numerous), with lingering birds to late May. Very rare west of Rocky Mts. in spring. In fall, they are more numerous along both coasts than in spring, with juveniles accounting for the great majority in West. Adults first arrive early to mid-July but have been seen in late June. Adult numbers decline by late Aug., when the first juveniles arrive, peaking mid-Sept.–mid-Oct. Some linger into early Nov. and later, even to Dec. **WINTER:** Casual in the southern states; most winter in southern S.A.;

juvenile Sharp-tailed (center) with juvenile Pectorals

very small numbers in Australasia.
RARE STATUS: Rare to Europe and Africa.
Population Stable.

SEMIPALMATED SANDPIPER   *Calidris pusilla*   SESA ▪ 1

Semipalmated Sandpiper is common to abundant. During migration it congregates in large flocks at wetland habitats in the eastern half of the continent. Monotypic. L 6.3" (16 cm)
Identification Short to medium length, tubular-looking bill with rather blunt tip; female's bill longer than male. Black legs; webbing on toes visible under ideal conditions. **BREEDING ADULT:** Dark upperparts, a mixture of black and brown, with only a tinge of rust on crown, auriculars, and scapular edges; streaked breast. **NONBREEDING ADULT:** Gray-brown above; largely white below. **JUVENILE:** Strong supercilium; dark crown and ear coverts; uniform brown upperparts, typically edged buff, but variably brighter, edged with rufous.

Similar Species Quite similar to Western, which has a blockier, less round head and a longer, less tubular bill (thins towards tip) that is slightly decurved. However, females of both species have longer bills (bill shape closest between a female Semipalmated and a male Western). In addition, bill length in Semipalmated shows a cline of increasing bill length from west to east. Semipalmated and Western are the only peeps/stints with partial toe webbing. The breeding adult Western is brighter rufous above and on the crown and cheek, and has arrow-shaped spots that extend down the flanks. On Semipalmated, the streaks are more confined to the breast. Juvenile Western has rich chestnut on back and inner scapulars, with plainer lower scapulars and wing coverts; on juvenile Semipalmated, the pattern is more even and uniform above, and chestnut edges to the mantle and inner scapulars are lacking, though some are brighter with rusty edges. Juvenile Semipalmated also has a darker cheek and lores with a more contrasting supercilium. For winter-plumaged birds, use bill shape and the more prominent supercilium of Semipalmated; at close range, note

Semipalmated lacks distinct breast streaks, shown by many Westerns. Calls differ. See Least Sandpiper, Little and Red-necked Stints.
Voice CALL: A short, low-pitched *churk*, very different from Western Sandpiper's raspy *jeet*.
Status & Distribution Abundant. **BREEDING:** Tundra from west AK to Atlantic. **MIGRATION:** In spring, first arrivals mid-Mar., but most in southern states Apr.–mid-May. Peak in mid-Atlantic and Great Lakes mid-May–early June. Very rare in West late Apr.–late May. In fall, adults as early as late June, peaking mid-July–mid-Aug.; rare in West south of northwest WA (mostly juveniles). Fall juveniles peak in Aug., with numbers declining by early Sept., and largely departed by early Oct., a few into early Nov. **WINTER:** Mexico (a few) to S.A. Casual in southern FL. **RARE STATUS:** Casual in Europe.
Population Near Threatened, declining overall, although it is still numerous in its range.

breeding

short, straight, tubular bill

partially webbed toes (hard to see)

female's bill averages larger than male's; birds breeding in eastern Canada have longest bills

dark ear coverts

more uniform upperparts than juvenile Western

pale supercilium

breeding ♀

rounder head than Western

juvenile

winter

darker above with more distinct head pattern than Western; breast sides unstreaked

WESTERN SANDPIPER   *Calidris mauri*   WESA ▪ 1

Western Sandpiper is the abundant western counterpart of the Semipalmated Sandpiper. It is found in large flocks during migration. Monotypic. L 6.5" (17 cm)

Identification Black legs; tapered bill, of variable length, with a slight droop at the tip. **BREEDING ADULT:** Bright rufous wash on crown and auriculars; rufous base to scapulars. White underparts, with streaks and arrow-shaped spots across the breast and along sides and flanks. **NONBREEDING ADULT:** Gray-brown above and white below, with faint streaks on breast. **JUVENILE:** Extensive rufous edges on upper scapulars contrast with gray-brown wing coverts with pale edges.

Similar Species See Semipalmated Sandpiper. In winter, Western might recall Dunlin as both share a bill that droops toward the tip; however, larger Dunlin has a longer bill with a more pronounced droop, is browner, and is more heavily marked on the throat and breast. A winter Least Sandpiper is browner and smaller, with darker throat and upper breast and yellowish or greenish legs.

Voice CALL: A high, raspy *jeet*.

Status & Distribution Abundant. **BREEDING:** Wet Arctic tundra from Russian Far East to northwest AK. **MIGRATION:** Throughout N.A., except very rare to casual in ND, MN, and eastern Canada. Movement in spring by late Mar., peaking in WA early Apr.–early May, with lingerers to late May. Casual on East Coast in spring away from wintering sites. In fall, first adults late June, most July. First juveniles arrive late July (later in the East), peak Aug.–early Sept., lingerers into Nov. **WINTER:** Primarily coastal to S.A. **RARE STATUS:** Casual Europe, Japan, and Australasia. **Population** Stable.

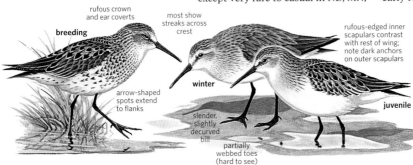

rufous crown and ear coverts

breeding

most show streaks across crest

arrow-shaped spots extend to flanks

winter

slender, slightly decurved bill

partially webbed toes (hard to see)

rufous-edged inner scapulars contrast with rest of wing; note dark anchors on outer scapulars

juvenile

DOWITCHERS   Genus *Limnodromus*

The dowitchers are medium-size shorebirds, with long, straight bills. They feed in mud or shallow water, probing with a rapid jabbing motion. In flight they show a white wedge from barred tail to middle of back. Separating the species is easiest with juveniles, difficult with breeding adults, and very difficult in winter.

SHORT-BILLED DOWITCHER   *Limnodromus griseus*   SBDO ▪ 1

Found continentwide, but scarcest from western Great Plains to Great Basin; there are three distinct subspecies. Polytypic (3 sspp.; all in N.A.). L 11" (28 cm)

Identification Rather long, gray bill (shorter in eastern birds); females have longer bills. Alternating light and dark bars, usually more light than dark, on tail. **BREEDING ADULT:** Dark upperparts with orange feather edges. Underparts a mixture of orange and white, with spots and bars (vary by subspecies). **NONBREEDING ADULT:** Uniform brownish gray above, feathers without darker center. White below with barring on the sides that contrasts with upperparts; gray breast with fine speckling visible at close range. **JUVENILE:** Dark scapulars and coverts, boldly edged orange; tertials, inner scapulars, and inner greater coverts have intricate internal markings and loops. Orange-buff neck and upper breast with dark spots. **FLIGHT:** White wedge from tail through lower back. Tail barred dark gray and white.

Geographic Variation Three subspecies: nominate *griseus* (breeds northeast Canada), *hendersoni* (central and western Canada), and *caurinus* (AK). In breeding plumage, Most Short-billed Dowitchers show some white on the belly, especially *griseus*, which also has a heavily spotted breast and may have densely barred flanks. The Pacific *caurinus* is intermediate, but more similar to *griseus*. In *hendersoni*, which is mostly cinnamon

griseus & hendersoni intermediates

caurinus

hendersoni

griseus

hendersoni

subspecies uncertain

griseus & hendersoni

hendersoni

mostly griseus

caurinus

buff below, the foreneck is much less heavily spotted than in the other subspecies. In juveniles, *caurinus* averages duller, with more narrow markings, although *griseus* is quite similar. Both are duller than *hendersoni*, which can be quite rufous, with broad edges and more buffy below.

Similar Species See Long-billed Dowitcher; also see Red Knot, which often associates with dowitchers.

Voice CALL: A mellow *tu tu tu*, repeated in a rapid series as an alarm call. More likely to remain silent than Long-billed Dowitcher. **SONG:** A rapid *di-di-da-doo* year-round.

Status & Distribution Common. **BREEDING:** Muskegs in boreal forest. **MIGRATION:** Common along the Atlantic coast (*griseus*); from the eastern Plains to Atlantic coast from NJ south (*hendersoni*); and along the Pacific coast (*caurinus*); a few *griseus* are seen on eastern Great Lakes in late spring, and a few *hendersoni* move through some interior western states and north on coast to New England. Spring movement for *griseus* begins mid-Mar., peaks late Apr.–early June, particularly mid-May. In the interior, *hendersoni* peaks on the Great Lakes and Great Plains mid-May; *caurinus* a bit earlier, with CA peak mid-Apr., WA

late Apr., and southern AK ±10 May. Fall migration begins earlier than for Long-billed; in three distinct waves, led by adult females, as early as late June; then adult males (adults complete most of migration by mid-Aug.), followed by juveniles, typically arriving in early Aug., peaking through early Sept., with small numbers into early Oct.; lingerers to early Nov.

They molt when they reach wintering grounds; most late juveniles are still in full juvenal plumage, or at least retain extensive juvenal feathering. **WINTER:** Coastal habitats (partial to tidal mudflats) to S.A., almost unknown away from the coasts. **RARE STATUS:** A few records for Europe.

Population There are indications of a decline in *griseus*.

Illustration labels:
pattern of breeding *caurinus* intermediate between *griseus* and *hendersoni*
breeding *caurinus*
worn breeding *griseus*
numerous black spots on sides of breast
some white on lower belly
extensive buffy markings above
spots on sides of breast
white belly
breeding *griseus*
breeding *hendersoni*
more extensively colored underparts than *griseus* with fewer dark markings
faint spots on breast
molting juvenile
winter
bold internal marks on tertials and greater coverts
juvenile
both dowitcher species have white stripe up back
internal bars and stripes
griseus
winter *hendersoni*
juvenile tertials

LONG-BILLED DOWITCHER   *Limnodromus scolopaceus*   LBDO ■ 1

The Long-billed Dowitcher, like the Short-billed Dowitcher, can be seen in large flocks in winter and on migration. It has a preference for a wider variety of fresh- and saltwater habitats than Short-billed. Monotypic. L 11.5" (29 cm)

Identification Long bill (longer in females). The tail is barred dark and white, with the dark bars wider than the white bars; the primaries do not extend beyond the tail. **BREEDING ADULT:** The underparts are entirely reddish below, except for whitish edges to fresh feathers that wear off in early spring. The foreneck is heavily spotted and the sides are usually barred. The black scapulars have

rufous markings and show white tips in spring. As this plumage wears, usually in July and Aug., the bird looks uniform rufous below and a mixture of rufous and black above, making identification difficult. **NONBREEDING ADULT:** The scapulars have dark centers, giving a more mottled appearance; the upperparts are gray-brown. The breast is rather dark gray and is unspotted. **JUVENILE:** It is dark above; the dark crown moderately contrasts with the supercilium. The tertials and greater wing coverts are plain, with thin gray edges and rufous tips; some birds show two pale spots near the tips. The underparts are gray, with only hints of warmth on the breast.

Similar Species Separating the Long-billed from the Short-billed Dowitcher ranges from the straightforward to difficult. A familiarity with variation, molt, vocalizations, and distribution is needed and will result in a high percentage of correct identifications. Start with the age, and consider which Short-billed subspecies might be present. In full breeding plumage, the rufous underparts of a Long-billed are easy to distinguish from those of a breeding *griseus* and many *caurinus*, but they are similar to those of *hendersoni*. Look for white tips to the scapulars of the Long-billed and small dots restricted to the sides of breast of *hendersoni*, whereas Long-billed has spots and bars on the underparts. In flight, note the lighter tail of Short-billed, owing to the wider white bars but *griseus* has wider dark bars than the other Short-billed subspecies. Breeding Long-billeds have rufous on their central tail feathers, unlike any Short-billed Dowitcher. Adult Long-billeds go to favored locations (often gathering in large numbers) in late summer to molt; Short-billeds molt when they reach winter grounds. In nonbreeding plumage, Long-billeds are darker, with a darker and browner breast that lacks the fine spotting shown by Short-billeds. Juvenile dowitchers are the easiest to identify. Long-billeds are darker above and grayer below than Short-billeds; they also lack the internal feather markings found on the tertials and greater coverts of Short-billeds. One other species is somewhat similar and often associates with dowitchers: the

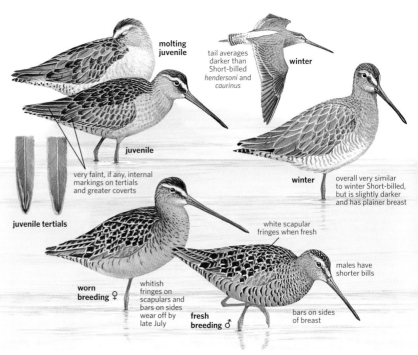

molting juvenile

tail averages darker than Short-billed *hendersoni* and *caurinus*

winter

juvenile

very faint, if any, internal markings on tertials and greater coverts

winter

overall very similar to winter Short-billed, but is slightly darker and has plainer breast

juvenile tertials

white scapular fringes when fresh

males have shorter bills

worn breeding ♀

whitish fringes on scapulars and bars on sides wear off by late July

fresh breeding ♂

bars on sides of breast

Red Knot in breeding plumage. It has a shorter bill and shorter, darker legs, and it lacks the white wedge up the back. In breeding plumage, the white lower belly and undertail coverts easily distinguishes it from a breeding Long-billed.

Voice CALL: A sharp, high *keek*, given singly or in a rapid series. **SONG:** Like the Short-billed Dowitcher.

Status & Distribution Common. **BREEDING:** Tundra from northeast Russian Far East to northwest Canada; a restricted and more northerly range than Short-billed. **MIGRATION:** Common in western half of continent. In the East, generally uncommon in fall, very rare or even casual in spring.

Spring migration is earlier than the Short-billed, arriving in nonwintering areas late Feb.–mid-Mar., peaking Great Plains and Great Lakes late Mar.–early May, before most Short-billeds. Arrive breeding grounds mid- to late May. Fall migration begins later than Short-billed, in mid-July (West) or late July (East). Juveniles migrate later than adults; very rare in lower 48 before Sept. Dowitchers seen inland after mid-Oct. are almost certainly Long-billed. **WINTER:** South to C.A., casual to northern S.A. **RARE STATUS:** Europe (where annual), Asia, and HI. **Population** Population is likely stable; appears to be expanding west in Russian Far East.

Genus *Lymnocryptes*

JACK SNIPE *Lymnocryptes minimus*   JASN ■ 4

The small and short-billed Jack Snipe, a Eurasian visitor, is quite secretive; a number of the records for N.A. were procured by hunters. Monotypic. L 7" (18 cm)

Identification The Jack Snipe is small and chunky. Secretive, it is reluctant to flush. Its flight is low, short, fluttery, on rounded wings; it is somewhat rail-like. It bobs constantly while feeding. The short bill has a pale base. The tail is dark and wedge-shaped. It has a bold head pattern with a split supercilium and, unique among

snipe, lacks median crown stripe. The broad, buffy gold back stripes are striking, and there is some dark, oily green iridescence on the scapulars. The underparts are pale, with streaking on both the breast and flanks; there is no barring.

Similar Species The Jack Snipe has the general coloration and shape of other snipes, but it is much smaller.

Voice CALL: Generally silent when

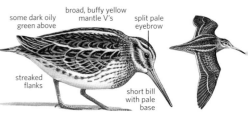

broad, buffy yellow mantle V's

some dark oily green above

split pale eyebrow

streaked flanks

short bill with pale base

bobs while feeding

flushed, occasionally a subdued note, softer than Wilson's Snipe.

Status & Distribution BREEDING: Forest bogs of northern Eurasia. **WINTER:**

Europe, Africa, and southern Asia. **RARE STATUS:** A number of recent fall records for the Pribilofs, AK. Three late-fall records for CA, two for OR, and singles for Kenai Peninsula, AK, and Labrador. Single spring records from St. Paul I. and St. Lawrence I. Some records pertain to birds shot by hunters; no doubt many are overlooked. **Population** Declines in the 20th century that resulted from habitat degradation appear to have stabilized.

WOODCOCKS Genus *Scolopax*

Six species. Two are widespread: the American in eastern N.A. and the Eurasian in the Old World (casual to N.A.). Four others are range-restricted and are found on islands from southern Japan to Indonesia. Chunky and long-billed forest dwellers, all species probably have elaborate flight displays at dawn and dusk, though the island-restricted Asian species are poorly known.

EURASIAN WOODCOCK *Scolopax rusticola*   EUWO ■ 5

A casual visitor to N.A. from Eurasia. Monotypic. L 13" (33 cm)
Identification Chunky, large-headed bird, with a long, dull pinkish bill that has a darker tip. Like all woodcocks, it has short round wings and a short tail. It is cryptically colored: earthy brown, combined with bars and stripes that allow it to blend into the forest floor. The hindcrown is barred with black; the lores are dark. The upperparts are dark brown, with paler brown markings and tips; there are pale lines on the mantle. The underparts are heavily barred, particularly the breast and flanks. The wing coverts and neck are also barred.
Similar Species The Eurasian Woodcock is similar to the American Woodcock, which is the expected species.

The Eurasian Woodcock is considerably larger, browner, and darker with heavily barred underparts against a much less buffy ground color, and with less contrasty pale scapular lines. **Voice CALL:** N.A. birds have been silent.
Status & Distribution Eurasian breeder. **WINTER:** South to North Africa and Southeast Asia. **RARE STATUS:** A few records, nearly all old (six specimens taken 1859–1890) primarily from the Northeast, inland to QC, PA, and, remarkably, AL. The only N.A. record since 1890 is one photographed at Goshen, NJ (2–9 Jan. 1956). All dated N.A. records are between early Nov. and Mar. Casual to Greenland, including a specimen as recent as 1974.
Population Millions of Eurasian Woodcocks are hunted annually in Eurasia, yet there is no clear trend in the population. Still, the lack of recent records from N.A. in recent decades may be indicative of a reduction in the European population.

larger and darker than American Woodcock

heavily barred underparts

AMERICAN WOODCOCK *Scolopax minor*   AMWO ■ 1

A bird of moist woodland floors and brushy marsh and field edges, usually found singly as it probes and picks through soil and leaves for worms or other prey. It is secretive and seldom seen until flushed; it flies up abruptly and the wings make a twittering sound. Males have a spectacular display flight, typically performed at dawn and dusk. Monotypic. L 11" (28 cm)

Identification The American Woodcock is a chunky, atypical shorebird with rounded wings and a short tail. The long bill is dull pink with a darker tip. The large eyes stand out on a rather buffy face. Bold crossbars mark the hindcrown. Upperparts are dark with obvious pale mantle lines and upper scapulars with pale pinkish or buff edges. Underparts are buffy and unmarked. **JUVENILE:** It is very much like an adult; it is only separable early in its plumage, when the juvenile has a gray-brown throat and neck that contrasts with the white chin. **FLIGHT:** A stubby bird in flight with a long bill held downward, round wings, and a short tail. It looks dark brown above and warm buffy orange below. **Similar Species** Wilson's Snipe shares the same cryptic and somewhat secretive behavior, often not noted until flushed at close range. Both species are chunky, with long bills, and share a generally brown coloration. Both also feed by probing in wet, vegetated areas, but woodcocks favor wet woods while snipe are found in marshes or wet fields. The snipe has stripes along its head as opposed to the bars that cross the head of a woodcock, and the woodcock's eyes are isolated on its pale face. The woodcock is also more colorful with rich buffy, unbarred underparts. In flight, the woodcock's rounder wings are noticeable.

chunky, with short, rounded wings

dark bands on crown and nape

large eyes

rich buffy underparts

Voice CALL: Males vocalize during spring courtship, mostly at dawn and dusk. A loud buzzy *peent* is repeatedly given from the ground between display flights. When airborne, a long series of twittering notes is given in zigzagging flight, louder just before descent. Wings also produce a twittering noise when flushed.

Status & Distribution Common but local. **BREEDING:** Moist woodlands and boggy thickets. In courtship, field edges and in flight display over open areas. **MIGRATION:** In spring, woodcocks move early; they are the earliest arriving and breeding northern shorebirds, with northern arrivals by late Feb. to early Mar. The first fall migrants are detected late Aug.–early Sept., but most migration is noted in Oct.–Nov. **RARE STATUS:** Casual in West to MT, WY, CO, and NM. Accidental in southeastern CA, two fall records.

Population Recent declines might be a result of loss of second-growth forests.

SNIPES Genus *Gallinago*

Snipes are stout, long-billed shorebirds that frequent bogs, marshes, and other similar wetlands. Some of the 17 species are poorly known, isolated on islands around the world. Snipes are generally solitary, but they can collect into flocks during migrations. Their heavily striped, cryptic patterns allow them to blend into their surroundings. They probe the mud for food, and the males' display flights during breeding season are spectacular.

PIN-TAILED SNIPE   *Gallinago stenura*   PTSN ■ 5

The Pin-tailed Snipe is a casual visitor from Asia. Monotypic. L 10" (26 cm)
Identification The Pin-tailed's bill is relatively short and thick for a snipe. It has a bold head pattern with pale mid-crown stripe; expansion of the buff supercilium in front of the eye is broad, leaving a small dark lore stripe and a small dark stripe at the top of the bill. Many of the scapulars and coverts have white around the tip of the feathers, extending to both sides, leaving a slightly more scaled look. In hand, the razor-thin outer tail feathers are diagnostic. **FLIGHT:** There is a fairly obvious buffy secondary covert patch; unlike that of the other two snipe species. The secondaries lack pale tips. The underwings are uniformly dark, and the feet distinctly project past the tail. **Similar Species** From Common Snipe, Pin-tailed is chunkier, shorter billed, and shorter tailed with barred secondary coverts and pale edges on the inner and outer webs of the scapulars, giving it a scalloped look. On Common and Wilson's, the white line on the outer webs of the scapulars is much broader

than along the inner web. Also, the Pin-tailed's face pattern subtly differs with a particularly broad supercilium in front of the eye. The underwing is the same between the Pin-tailed and Wilson's, but Wilson's lacks toe projections in flight; the noticeable buffy patch on the upperwing of the Pin-tailed is unmatched by any snipe, except the larger Swinhoe's Snipe (*G. megala*), which is unrecorded in N.A. but breeds in nearby Russian Far East. No reliable field marks are known that separate these two species in the field.
Voice CALL: A high *squak*, sounding like a duck or pig.
Status & Distribution The Pin-tailed breeds across much of Russia. **WINTER:** Primarily Southeast Asia. **RARE STATUS:** Accidental, two certain late May specimen records from Attu I., AK. Other sightings are from the Pribilofs and St. Lawrence I. and are likely Pin-tailed on probability, but Swinhoe's not eliminated.
Population Stable.

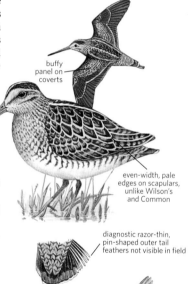

buffy panel on coverts

rather short bill

even-width, pale edges on scapulars, unlike Wilson's and Common

diagnostic razor-thin, pin-shaped outer tail feathers not visible in field

tail

darker underwing than Common with no pale trailing edge, much like Wilson's

underwing

SOLITARY SNIPE   *Gallinago solitaria*   SOSN ■ 5

Solitary Snipe nests in remote, inhospitable habitats and is very little known. Nesting above tree line in bogs, swamps, and river valleys, the species moves into foothills and lowlands in winter. Polytypic (2 sspp.; likely *japonica* in N.A.). L 12" (29–31 cm)
Identification In a genus of cryptic birds, Solitary Snipe is distinctive. Its large size, potbellied shape, and long bill recall a large woodcock as much as a snipe, but the plumage pattern is

clearly that of a snipe. **ADULT:** Upperparts overall quite dark, with dark brown and vivid red tones in mantle, scapulars, tertials, and upperwing coverts; mantle bordered by thin whitish "back braces." Underparts paler, with much barring in the flanks and a distinct brownish ginger tone in breast. Single black eye line; brown line lower on face. **FLIGHT:** Dark upperwings lack white trailing edge or other pattern.
Geographic Variation Two similar

subspecies: *japonica* and nominate *solitaria*. Subspecies *japonica*, presumed to be responsible for N.A. records, is more rufous above, with narrower pale back braces (and generally more uniformly toned upperparts) and longer bill.
Similar Species Compare to other medium-size snipe: Wilson's, Common, and Pin-tailed. Larger Asian snipe, even though unrecorded in N.A., should also be considered carefully,

including Latham's (*G. hardwickii*) and Swinhoe's (*G. megala*). Solitary's single dark eye line extending from the base of the bill to the nape appears to be

overall dark coloration, including near solid area on breast sides

very long bill

distinctive; other snipe show a variably thin dark line through the lores that either expands into two parallel lines behind the eye or broadens into a dark streak. **Voice SONG:** Displaying birds give a long series of *choka* calls in flight over territory. **CALL:** An annoyed, low, nasal *kentsh*.

Status & Distribution BREEDING: High-elevation wetlands of Kazakhstan, Kyrgyzstan, and Mongolia east through northern India and China and parts of Russia east to Kamchatka.

The breeding range of *japonica* is not well known but apparently includes part of the Kamchatka Peninsula; the breeding range of nominate *solitaria* covers much of Central Asia. **WINTER:** Mostly in wetland areas near breeding range, at lower elevations; regularly found in winter in Japan, North and South Korea, northern India, Pakistan, Iran, and eastern China; *japonica* appears to winter largely in Japan. **RARE STATUS:** Accidental: three AK records, two in Sept. from St. Paul I. and a May specimen from Attu I. **Population** Stable.

COMMON SNIPE *Gallinago gallinago*   COSN ■ 3

The Common Snipe is a visitor from the Old World. Polytypic (2 sspp.; nominate in N.A.). L 10.5" (27 cm)
Identification Long bill; legs are greenish gray. Bold, striped head pattern. The upperparts are dark, with buffy lines running along the back from the head toward the tail. The tertials are heavily barred, all the way to the base. The tail is composed of 14 feathers, the outer feather on each side is broad, and the inner web of these feathers has an orange coloration. Below, the breast is brown with dark streaks; the flanks are barred. **FLIGHT:** Like Wilson's, the Common is likely to explode up from the ground and fly straight up in the air. As it flies overhead, it gives the impression of a dark-breasted bird with a white belly. The underwing has substantial white, although juveniles have more gray than adults. There is a rather broad white trailing edge to the secondaries, easily visible in flight from above.
Similar Species Very similar to Wilson's; close scrutiny is needed. There are features that average different between the two species. Common has a paler, buffier color overall than the Wilson's and slightly fainter flank markings; however, sufficient variation renders these marks suggestive, not diagnostic. The broader white trailing edge to the secondaries and paler white-striped underwings are

consistent features separating the two species. The outer tail feathers are also diagnostic, but you'll need great views of a preening bird or a well-timed photograph; Common has one broad feather, with orange on the inner web, versus the Wilson's two narrower feathers, the outer of which lacks orange.
Voice CALL: A two-note *ski-ape*, similar to Wilson's. **SONG:** Male flight display notes distinctly lower pitched than Wilson's.
Status & Distribution BREEDING: Throughout northern Eurasia; it has bred in western Aleutian Is., AK. **MIGRATION:** Regular migrant to the western Aleutian Is. (where found casually in winter); rare to central Aleutian and Pribilof Is.; casual to St. Lawrence I. **WINTER:** Southern Europe, Africa, and Southeast Asia. **RARE STATUS:** Accidental winter records for Newfoundland (photograph), Labrador (specimen), and Riverside, CA (photograph). Given the difficulty in identifying this species,

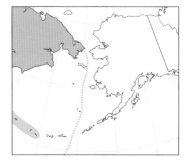

a review of hunters' snipe collections might prove valuable to assess whether this species is more numerous in N.A. **Population** Stable.

bolder white trailing edge

gallinago

bold pale stripes on upperparts

flank barring fainter than Wilson's

underwing

whitish panel on underwing coverts

warmer background color to head, breast, and sides

WILSON'S SNIPE *Gallinago delicata*   WISN ■ 1

The cryptic Wilson's Snipe is seen in various wetland habitats. On breeding grounds, often perches on fence posts. Dozens can be encountered during migration and winter. It probes the mud with its long bill, much like a dow-

itcher. Was called Common Snipe until the decision to re-split Wilson's from that species. Monotypic. L 10.3" (26 cm)
Identification Stocky shorebird with a very long dark-tipped bill, pale at the base; the legs are greenish gray. The

head is boldly striped with pale lines above and below the auriculars and in the middle of the crown, all separated by dark brown. Upperparts are dark brown with pale buff or white lines. Breast brown with dark streaks, and flanks heavily barred. Outer rectrices are barred black-and-white, with no trace of orange. **FLIGHT:** It often explodes from the ground straight up into the air. The general impression is a brown breast and white underparts, along with dark gray underwings. Toes do not extend beyond the tail in flight. **Similar Species** Separated with difficulty from a Common Snipe. Can be confused with both dowitcher species, as they share a chunky, long-billed profile. Dowitchers, however, do not share the extensive head stripes or light buffy

lines on the upperparts, and snipes lack the dowitcher's white wedge up the back. See also American Woodcock. **Voice CALL:** A raspy, two-note *ski-ape*, given in rapid, zigzagging flight when flushed. On nesting grounds, the male delivers loud *wheet* notes from perches. In swooping display flight, vibrating outer tail feathers make quavering hoots, commonly referred to as "winnowing," similar to song of Boreal Owl. **Status & Distribution** Common, but overlooked. **BREEDING:** Marshes and wet meadows. **MIGRATION:** In spring, southern states see movement in late Feb.–early Mar. Peak Pacific Northwest

to Great Lakes late Mar.–late Apr. Fall migration begins as early as mid-July, but peak mid-Sept.–late Oct. **WINTER:** To northern S.A. **RARE STATUS:** Rare or casual to central Aleutians, Pribilofs, and St. Lawrence I. Casual in UK and the Azores. **Population** Overall stable.

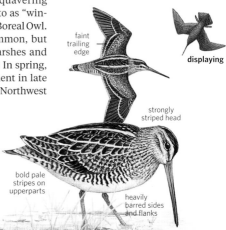

faint trailing edge

displaying

strongly striped head

dark underwing

bold pale stripes on upperparts

heavily barred sides and flanks

underwing

TRINGINE SANDPIPERS Genera *Xenus*, *Actitis*, and *Tringa*

This tribe of the sandpiper family includes the genera *Xenus*, *Actitis*, and *Tringa*. Most of these birds are of medium size, although they range from small to rather large. Most of them feed by picking, but those with longer legs will wade in water to probe or take small fish. Many of these sandpipers have gray or gray-brown as their primary color, but breeding and juvenile plumages can be attractive. A majority of these birds breed in the northern latitudes, typically in marshy openings within boreal forests; only the Willet is a common breeder in the lower 48. Most *Tringa* sandpipers migrate long distances, wintering as far south as S.A. and Australasia.

TEREK SANDPIPER *Xenus cinereus*   TESA ▪ 3

The Old World Terek Sandpiper has a frenetic personality. It teeters and makes jerky, start-and-stop runs when foraging. Monotypic. L 9" (23 cm) **Identification** Grayish brown upperparts; pale below. Long upturned bill, usually with dull orange at base. Short, orange-yellow legs. **BREEDING ADULT:** Black stripe on scapulars; faint streaking on the breast. **NONBREEDING ADULT:** Reduced black line on

scapular and breast streaking. Dark lores and faint supercilium both bolder. **JUVENILE:** Like nonbreeding, but browner above, with buffy edges to most of the upperparts. **FLIGHT:** Distinctive wing pattern with dark leading edge, grayer median coverts, dark greater coverts, and contrasting whitish secondaries. **Similar Species** Short legs and thick-based upturned bill unmistakable. **Voice CALL:** A short series of even, whistled notes. **Status & Distribution** Breeds along rivers and shores of lakes from Finland to Russian Far East; winters from Africa to Australasia. Rare migrant on western and central Aleutians, Pribilofs, and St. Lawrence I., casual mainland western AK, and Cook Inlet; accidental to coastal BC, CA, MA, VA, and Baja California Sur, Mexico. **Population** Stable.

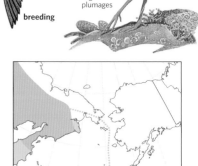

juvenile

whitish secondaries

short orangish legs in all plumages

breeding

thin dark scapular stripe

long recurved bill with orange base

breeding

COMMON SANDPIPER *Actitis hypoleucos*   COSA ■ 3

Old World equivalent to Spotted Sandpiper, Common Sandpiper has not been recorded away from western AK. Monotypic. L 8" (20 cm)

Identification Small, short-legged, with a long tail, which it bobs. Its flight is low over the water, with shallow, flicking wingbeats. **BREEDING ADULT:** Brown above, with dark streaks on the upperparts and dark spots along the edge of the tertials; distinct streaked dark patches on sides of breast. **JUVE-** **NILE:** Upperparts have pale fringes and dark subterminal lines give a scaled effect; tertials have barring around the entire feather. **FLIGHT:** Broad white wing stripe extends to the base of the wing and is adjacent to a bold white trailing edge to the inner secondaries.

Similar Species Closely resembles nonbreeding and juvenile Spotted. Note Common's longer tail; on juvenile, barring on the edge of the tertials extends along the entire feather. In flight, Common shows a longer white wing stripe, longer white trailing edge.

Voice CALL: A clear whistled even or descending *twee-wee-wee*, distinct from Spotted Sandpiper.

Status & Distribution Breeds Eurasia. Winters Europe and Africa to Australasia. Rare migrant, mostly spring, on western Aleutians (nested Attu I. in 1983); fewer to central Aleutians, Pribilofs, and St. Lawrence I. Casual to Seward Peninsula, AK, and HI.

Population Stable.

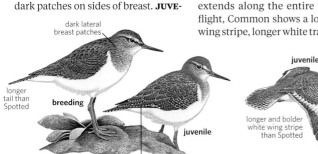

dark lateral breast patches

longer tail than Spotted

breeding

juvenile

juvenile

longer and bolder white wing stripe than Spotted

barred tertials

SPOTTED SANDPIPER *Actitis macularius*   SPSA ■ 1

This widespread species is found in freshwater habitats on the breeding grounds but uses many habitats in migration and winter. It flies low to the water, with shallow, flicking wingbeats that seem jerky. It bobs and teeters constantly. Monotypic. L 7.5" (19 cm)

Identification Short, greenish to yellow legs. **BREEDING ADULT:** Unmistakable; bold spotting throughout underparts, and dark bars above. Bill pinkish with dark tip; legs bright yellow. **NONBREEDING ADULT:** Dark patches on sides of breast; white wedge above dark wing. Spots are absent or reduced; barred wing coverts contrast with the rather uniform brown back. Bill and legs duller. **JUVENILE:** Similar to nonbreeding. Barring on the edge of the tertials, absent on some juveniles, extends no farther than halfway along each feather. **FLIGHT:** White wing stripe is evident but does not extend to body.

Similar Species Common Sandpiper shares many behavioral traits with the similar-appearing and closely related Spotted. In N.A., Common has only been recorded from western AK. Compared to Common, Spotted has a shorter tail, and in flight shows a shorter white wing stripe and a shorter white trailing edge; Common's wingbeats not as exaggerated as Spotted. On juveniles, Common's tertials are barred; bars, if present on Spotted, do not extend around the entire feather.

Voice CALL: Shrill *peet-weet*; an ascending series of *weet* notes in flight.

Status & Distribution Common and widespread. **BREEDING:** Sheltered streams, ponds, lakes, or marshes. **MIGRATION:** In spring, peak late Apr. in the southern states (e.g., FL, TX), first three weeks of May in Great Lakes, and after mid-May in the West. Females arrive mid-May in MN, a week before males. In fall, some adults leave late June, peak mid- to late July through Aug. Juveniles predominate in Sept., with stragglers Oct.–mid-Nov. **WINTER:** Most winter in C.A. and S.A. Uncommon to fairly common in coastal southern states; rare to southern edge of breeding range. **RARE STATUS:** To Russian Far East and Europe.

Population Stable.

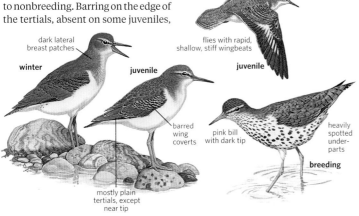

dark lateral breast patches

winter

juvenile

flies with rapid, shallow, stiff wingbeats

juvenile

barred wing coverts

pink bill with dark tip

heavily spotted underparts

breeding

mostly plain tertials, except near tip

GREEN SANDPIPER   *Tringa ochropus*   GRSA   ■   4

The Green Sandpiper, a casual visitor from Eurasia, is the Old World equivalent of the Solitary Sandpiper. Monotypic. L 8.8" (22 cm)

Identification Dark above with variable spotting; white below, with breast streaking. Straight, dark bill of medium length; greenish legs. **BREEDING ADULT:** Dark olive-gray tone to upperparts, with white spots on many scapulars and coverts. **NONBREEDING ADULT:** Brown tone to upperparts, less spotting above; neck and breast brown with indistinct streaks. **JUVENILE:** Like nonbreeding adult, but more heavily spotted above, streaks on breast on white

background. **FLIGHT:** Blackish wing linings, white rump, with barring primarily near tips of tail; outer rectrices with little or no barring.

Similar Species Like Solitary Sandpiper in plumage, behavior, and calls. Note Green Sandpiper's white rump and uppertail coverts, with less extensively barred tail; darker wing linings; and lack of solidly dark central tail feathers. The similar Wood Sandpiper has more spotting above, more barring on tail, and paler wing linings.

Voice CALL: *Peet-o-weet*. Like Solitary

Sandpiper's call, but mixes in notes with different pitch.

Status & Distribution Eurasian breeder; winters to central Africa, Indian subcontinent, and Southeast Asia. **RARE STATUS:** Casual in spring on outer Aleutian Is., Pribilofs (also one fall record), and St. Lawrence I. **Population** Stable.

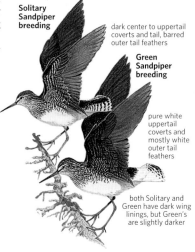

Solitary Sandpiper breeding

dark center to uppertail coverts and tail, barred outer tail feathers

Green Sandpiper breeding

pure white uppertail coverts and mostly white outer tail feathers

both Solitary and Green have dark wing linings, but Green's are slightly darker

breeding

juvenile

plumage very similar to Solitary

short olive legs

SOLITARY SANDPIPER   *Tringa solitaria*   SOSA   ■   1

This midsize sandpiper is usually seen singly or in small groups. Typically encountered around small pools, wet grassy areas, or creeks, it bobs its entire body or, more rarely, just the tail. It calls frequently as it flies overhead and is reliably vocal when flushed. Upon landing it often holds its wings up, as if getting its balance, allowing looks at the underwings. Polytypic (2 sspp.; both in N.A.). L 8.5" (22 cm)

Identification Dark brown above, with spots, and white below; lower throat, breast, and sides streaked with blackish brown; undertail coverts might be streaked. Dark bill, often with a pale base. Yellowish to greenish legs. White eye ring in all plumages; bold white supraloral line does not extend beyond the eye. **BREEDING ADULT:** Dark brown, with extensive spotting on upperparts and streaks on the neck and breast. Pale base to bill can make the bill look two-toned. Legs yellowish. **NONBREEDING ADULT:** Duller, with streaking more obscure, and fewer pale spots above. Bill usually dark; greenish legs. **JUVENILE:** Dark upperparts, with bold white spots, more marked than nonbreeding

adult. Plain brown head and breast, with minimal or no streaking. **FLIGHT:** Dark central tail feathers, white outer feathers barred with black, darkish underwing.

Geographic Variation Nominate *solitaria* breeds from interior BC to Labrador; more common in the East as a migrant. Adult has dark lores and juvenile's upperparts are spotted white. Subspecies *cinnamomea* breeds from AK to Hudson Bay; more common in the West. It is larger and paler; adult has streaked lores and juvenile's upperparts are spotted buffy.

Similar Species Most similar to Green and Wood Sandpipers. Bolder white eye ring, shorter olive legs, tail pattern,

and loud, shrill call distinguish it from Lesser Yellowlegs.

Voice CALL: Shrill, loud *peet-weet*,

juvenile *solitaria*

white eye ring

dark grayish brown breast

breeding *solitaria*

legs often yellowish in breeding plumage

short greenish legs

sometimes three notes, higher pitched and more emphatic than Spotted. **Status & Distribution** Fairly common. **BREEDING:** Shallow backwaters, taiga pools, and bogs. Uses abandoned passerine nests in small trees, unique for N.A. shorebirds. **MIGRATION:** Journey in spring starts late Mar., but migrants have been noted late Feb. Peak mid-

Apr. in TX, late Apr.–mid-May in Great Lakes, mid-Atlantic, etc. Stragglers seen late May south of Canada. In fall, adults arrive late June–early July in Great Lakes and mid-Atlantic, mid-July in TX, etc. Juveniles arrive early Aug., numbers drop in Sept., with most gone by early Oct., stragglers to Nov. Generally rare in West Coast

and Southwest in spring; uncommon in fall, juveniles predominate. **WINTER:** Primarily the tropics but rarely in the southern US, primarily south TX. **RARE STATUS:** Western Europe, Bering Sea, the Azores, and South Africa. **Population** Little data, perhaps owing to the largely remote breeding grounds and the bird's solitary nature.

GRAY-TAILED TATTLER *Tringa brevipes*   GTTA ■ 3

This visitor from Asia is very similar to the Wandering Tattler in behavior but is more catholic in its habitat choices. Monotypic. L 10" (25 cm) **Identification** Similar to Wandering Tattler in plumage and structure, although it has slightly shorter wings, with the primaries barely extending beyond the tail. Whitish supercilium is more distinct and extends across forehead. The nasal grooves usually do not extend more than halfway to the tip. **BREEDING ADULT:** Upperparts are pale gray and unmarked. Underparts

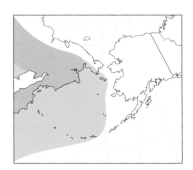

have fine barring that does not extend beyond the flanks; often there are a few dark marks on the undertail coverts. **NONBREEDING ADULT:** Pale gray above and unmarked white below, except for grayish on the sides and breast. **JUVENILE:** Similar to nonbreeding plumage, but with a white belly and sides, and pale spots on the upperparts. **Similar Species** Closely resembles Wandering Tattler but upperparts are slightly paler and the barring on underparts is finer and less extensive in a breeding adult. Juvenile Gray-tailed has less extensive gray on the sides and belly and is more heavily marked above with whitish spots and notches to the scapulars, wing coverts, and tertials. The shorter nasal groove of the Gray-tailed can be difficult to accurately assess in the field. Best distinction is voice. **Voice CALL:** Most frequent vocalization an ascending whistled *too-weet*. **Status & Distribution** Breeds northern Asia, winters Malay Peninsula to Australasia. Rare spring and uncom-

mon fall migrant on western Aleutians; fewer to rest of Aleutians, Pribilofs, and St. Lawrence I. **RARE STATUS:** Casual to northern and south-central AK; accidental in fall to WA, CA, ME, MA. **Population** Near Threatened.

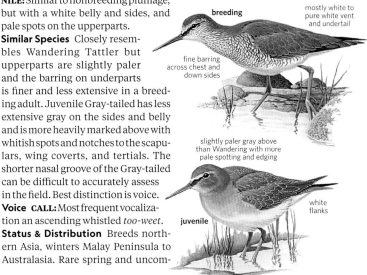

breeding

mostly white to pure white vent and undertail

fine barring across chest and down sides

slightly paler gray above than Wandering with more pale spotting and edging

juvenile

white flanks

WANDERING TATTLER *Tringa incana*   WATA ■ 1

The Wandering Tattler favors rocky Pacific coast habitats in migration and winter. A chunky gray bird, it can easily go unnoticed as it feeds on rocks exposed by the tide, bobbing as it moves. Its call can be heard over crashing waves and is often a clue to detecting its presence. It is generally seen singly or in small groups. Monotypic. L 11" (28 cm) **Identification** Long wings and short legs contribute to a horizontal look. Wings extend noticeably beyond the tail tip. Bill is rather long and black, at times showing a pale base; the nasal groove is often visible, extending more than halfway to the tip.

Legs are dull yellow, sometimes with a greenish tone. Overall, uniformly dark gray above and white below. **BREEDING ADULT:** Underparts heavily barred, all the way through the sides to the undertail coverts on most birds. White supercilium is flecked with gray and is

breeding

breeding

heavily barred undertail

extensively barred below

dark gray flanks

winter

juvenile

gray flanks

rather short, dull yellow legs

most prominent in front of the eye. Legs at their brightest, usually yellow. Some first-years, with an incomplete plum-age, can be difficult to age. **NONBREEDING ADULT:** Unbarred below with a gray wash on breast, sides, flanks; uniform gray above. Legs duller, more greenish yellow. **JUVENILE:** Like nonbreeding, but pale fringes on upperparts feathers or alternating light and dark spots along edges of those feathers.

Similar Species Closely resembles Gray-tailed Tattler and is best distinguished by voice. See Gray-tailed Tattler account for plumage details.

Voice CALL: Rapid series of clear, hollow whistles, *twee, twee, twee*; all on one pitch.

Status & Distribution Uncommon. **BREEDING:** On or near gravelly stream banks. **MIGRATION:** Mostly along the coast or over the ocean. In spring, peaking early Apr.–mid-May in Southern CA, mid-Apr.–early May on Pacific islands. Reaches southern AK by early May, on breeding grounds by late May. In fall, adults depart July, peak late July–mid-Aug. in OR. Juveniles predominate late Aug.–Sept. Smaller numbers into Oct. **WINTER:** CA to Ecuador, Pacific islands, and in Australasia. **RARE STATUS:** Casual inland in the West during migration; accidental farther east (MB, MN, TX, IL, IN, and ON). **Population** Small world population.

LESSER YELLOWLEGS *Tringa flavipes*   LEYE ■ 1

This dainty version of the Greater Yellowlegs looks almost delicate in comparison. It picks its way along the edge of the water in a deliberate fashion, taking prey with its thin, straight bill. Monotypic. L 10.5" (27 cm)

Identification Long yellow legs and a short, thin, dark bill. Generally gray above, white below. **BREEDING ADULT:** Breast is finely streaked; sides and flanks show fine, short bars. Legs can have an orange tone in high breeding plumage. **NONBREEDING ADULT:** Upperparts are grayish, with alternating dark and light notches on the feather edges; these are difficult to see from a long distance or when worn. Underparts are white, except for the dingy breast, which has obscure streaks. **JUVENILE:** Similar to nonbreeding, except upperparts have bold white spots on coverts and scapulars. **FLIGHT:** Gray wings are rather uniform; white rump and whitish tail. Legs and feet extend well beyond the tail.

Similar Species Similar to Greater Yellowlegs but smaller—a comparison easily made when the two are together. Lesser's all-dark bill is shorter, thinner, and straighter. The calls are different: Lesser utters fewer notes, which are, more importantly, not as long or as piercing as Greater and do not descend in pitch. Greater is a much more frantic, less deliberate feeder. Stilt Sandpiper can look similar in flight, but its bill is longer and decurved. They usually feed differently, especially when with their frequent companions—dowitchers. Wilson's Phalarope can be superficially similar outside of breeding plumage, but feeds differently; it has smaller, much shorter legs, is paler above, and has a white breast. Solitary Sandpiper is smaller, with shorter greenish (not yellow) legs; in flight, has darker underwings and different tail pattern.

Voice CALL: *Tew* notes, higher and shorter than Greater; notes usually given singly or double, but can be given in threes. **SONG:** *Wheedle-ree*, given with an undulating flight display.

Status & Distribution Common. **BREEDING:** Tundra or taiga woodland openings, generally farther north than the Greater Yellowlegs. **MIGRATION:** Common in the Midwest and fairly common in the East; uncommon in far West. More likely to be seen in large numbers (e.g., hundreds) than Greater Yellowlegs, especially during spring. Spring migration is usually a week later than for Greater Yellowlegs. Southern states peak in Apr.; Great Lakes and mid-Atlantic peak late Apr.–mid-May. In fall, the first adults move south in mid-June and peak late July–mid-Aug. Juveniles generally first arrive in early Aug. and peak late Aug.–Sept. Much more numerous in fall than in spring along the Pacific coast. Lessers have a more defined peak than Greaters, and have generally left by mid-Oct., with lingering birds into Nov., exceptionally later. **WINTER:** Most winter in S.A. or C.A. Local near the coasts, especially in TX, but always outnumbered by Greaters. Away from the coast, use caution when identifying winter Lessers. **RARE STATUS:** Rare to casual north and west of breeding range, including Bering Sea Is., primarily May–June; annual to Europe, primarily Aug.–Oct. Casual to Russian Far East, Japan, China; winter records from New Zealand and Australia. **Population** Stable.

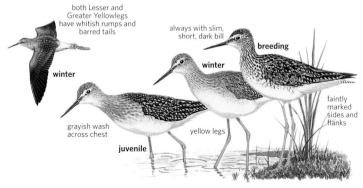

both Lesser and
Greater Yellowlegs
have whitish rumps and
barred tails

winter

grayish wash
across chest

juvenile

always with slim,
short, dark bill

breeding

winter

yellow legs

faintly
marked
sides and
flanks

WILLET *Tringa semipalmata* WILL ■ 1

The Willet's shape—large, plump, rather long legs, a thick bill—is unique for a shorebird. It picks and probes for prey, feeding along the shore, sometimes in the water, and often bobs its body like a yellow-legs. It often occurs in small flocks during migration or while feeding; it roosts in large groups with other large shorebirds. Polytypic (2 sspp.; both in N.A.). L 15" (38 cm)

Identification Generally gray or brownish gray above and whitish below. Stout, straight, dark bill, gray or brown at base. Gray legs, with possible dull olive or blue tones. **BREEDING ADULT:** Heavily streaked on neck, barred on breast and sides; white belly. Upperparts might have extensive black feathers. **NONBREEDING ADULT:** In general, unmarked gray above and white below. **JUVENILE:** Brownish upperparts, with obvious white or buff spots and fringes. **FLIGHT:** Striking black-and-white wing pattern is diagnostic in all plumages.

Geographic Variation Two distinct subspecies; nominate breeds along Gulf and Atlantic coasts and winters outside N.A.; western *inornata* breeds Great Plains and Great Basin and winters on all coasts. Nominate is unknown from interior, even in migration. It is smaller and has shorter legs and a shorter bill with a pale base, and is darker in all plumages; in breeding plumage, it is more heavily marked below; juveniles are browner. Western *inornata* is more inclined to feed in deeper water. These two subspecies are often seen near each other in spring and late summer on the Gulf and southern and mid-Atlantic coasts. They may represent distinct species.

Similar Species Unmistakable in breeding plumage and in flight. A roosting nonbreeding bird might recall a yellowlegs, but the Willet's legs are always dull, the bill larger, and the shape plumper.

Voice CALL: In nesting areas, *pill-will-willet*, higher and faster in eastern birds. Also *kip* or *yip*, sometimes repeated and other harsh alarm calls.

Status & Distribution Fairly common. **BREEDING:** Freshwater and alkaline marshes in the West; salt marshes in the East. **MIGRATION:** In spring, "Western Willet" (*inornata*) occurs (rare to uncommon) by late Mar. in interior parts of southern states; peak late Apr.–early May at Great Lakes. "Eastern" (*semipalmata*) generally returns to breeding areas along Gulf Coast and FL in Mar.; in mid-Atlantic mid- to late Apr. In fall, *inornata* starts moving in mid-June; adults peak mid-July–mid-Aug., juveniles numerous by late July; most migrants through by early Sept., with stragglers into Oct., rarely later. Most *semipalmata* depart early, and arriving *inornata* outnumber them by Aug. **WINTER:** Subspecies *inornata* winters on coastal beaches (Pacific, Atlantic, and Gulf Coasts and Caribbean islands) and inland at the Salton Sea and Owens Lake, CA. Winters south through C.A. and in S.A., including Galápagos, to northern Peru. Nominate winters on Caribbean islands and northern S.A. It is absent in winter from N.A. **RARE STATUS:** Casual or accidental to AK, YT, northern BC, NT, Hudson Bay, and southern QC. Accidental to Europe (Azores and Finland).

Population After historic declines, both subspecies now appear stable.

short pinkish-based bill

early Mar. in molt

western

eastern

heavily barred below

brownish cast to upperparts

eastern *semipalmata*

breeding

contrasting scapulars

juvenile

striking wing pattern in flight

coverts average more white on *inornata*

winter *inornata*

western *inornata*

western Willets are larger and have a longer, darker bill

paler above

less heavily marked below

worn breeding

breeding

plainer and paler upperparts

juvenile

winter

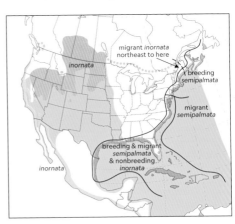

migrant *inornata* northeast to here

inornata

breeding *semipalmata*

migrant *semipalmata*

inornata

breeding & migrant *semipalmata* & nonbreeding *inornata*

SPOTTED REDSHANK   *Tringa erythropus*   SPRE   ▪   4

The Spotted Redshank, an Old World visitor to N.A., is unique among *Tringa* for its blackish breeding plumage and for its bill that droops toward the tip. Its behavior involves the expected picking and probing near the water's edge, although this species is an impressive wader and seems to take to swimming more readily than its congeners. Monotypic. L 12.5" (32 cm)

Identification Long, mostly dark bill droops at the tip. **BREEDING ADULT:** Blackish plumage overall (females more spotted with white below) with white spots on dark brown upperparts; black below. Legs are long and dark red, although they can be almost black. Bill is long and black, with red at the base. **NONBREEDING ADULT:** Upperparts a rather uniform gray with a dingy whitish breast; the remaining underparts are white. Red-orange to red legs are much brighter than in a breeding bird. **JUVENILE:** Darker upperparts than nonbreeding bird and somewhat variable. Most juveniles are dark brown with white spots on the upperparts and underparts that are brownish gray; some juveniles can be quite dark, almost approaching breeding females in coloration. Supercilium is bold and the legs are paler than on a nonbreeding bird. Base of the mandible is orange-red, not as bright as in a nonbreeding bird. **FLIGHT:** Dark above, with a white wedge extending up the back, and a barred rump and tail; light or dark below, depending on the plumage; and white wing linings. Legs and feet extend well beyond the tail.

Similar Species Drooping bill, red legs (orangish in some yellowlegs), and white wedge up the back are unlike any other species. Common Redshank, other than having red legs, is not very similar: It is shorter, smaller billed, and browner above, and has a bold white secondary patch in flight.

Voice CALL: Loud rising *chu-weet*. Similar to Pacific Golden-Plover; more strident than Semipalmated Plover.

Status & Distribution Breeds on tundra and northern boggy or marshy openings within open forests in Eurasia, winters from southern Europe and Africa to southeast Asia. **RARE STATUS:** Casual in migration to Aleutian and Pribilof Is. and on both coasts in migration and winter; accidental elsewhere with scattered records across N.A.

Population Stable, but fewer records in recent decades in N.A. perhaps indicate population declines in Old World.

juvenile

extensively barred below

extensive pale spotting above

long red-based bill in all plumages with slight droop near tip

breeding
black overall

very dark red legs

white wedge up back

winter

winter

orange legs

GREATER YELLOWLEGS   *Tringa melanoleuca*   GRYE   ▪   1

The tall and somewhat robust Greater Yellowlegs is a common bird. It feeds along the water's edge, although it will partially submerge itself and feed in belly-deep water. On shore, it may walk and pick at the surface, but often runs short distances when in an active feeding mode. It often bobs the front part of its body like Lesser Yellowlegs. Monotypic. L 14" (36 cm)

Identification Long yellow legs, long neck, and a rather thick and often upturned dark bill. **BREEDING ADULT:** Throat and breast with heavy, blackish streaks; sides and belly with blackish spots and bars. On the upperparts, many feathers are black centered, and the tertials have dark barring. Legs tend to bright yellow with orange tones. Bill is usually dark, but it can have a slightly paler base to the lower mandible. **NONBREEDING ADULT:** Rather pale and uniform, being gray above and white below. Upperparts have dark notches on the outer edges of feathers and pale fringes; neither are seen easily from a distance or if worn. Legs are paler and the bill usually has an obvious pale base. **JUVENILE:** Like nonbreeding, but upperparts have bold white spots on scapulars and coverts and lack the dark marks of the adult. **FLIGHT:** Rather plain upperparts with gray wings and white rump and whitish gray tail; legs and feet extend beyond the tail.

Similar Species Similar to Lesser Yellowlegs, but they are usually easy to distinguish when together. Greater has 50 percent more bulk, and is taller and more robust. Its bill is also thicker and longer (longer than the head) and often upturned, compared to a narrower, shorter, and straighter bill of Lesser Yellowlegs. Adult Greater molts earlier in fall, so early migrants might have dropped flight feathers; Lesser molts after migration. Greater is more heavily and extensively marked below in breeding plumage and is streaked on

neck (not washed) in juvenal plumage. Grayish-based bill differs from Lesser's all-dark bill, except in breeding plumage. Structure and differing calls are the best methods of identifying these species. Greater Yellowlegs usually gives three or more shrill and descending *tew* notes, Lesser more commonly gives two evenly pitched notes; the tone is more important than

the number of notes. Common Greenshank, a Eurasian visitor, is similar in size and shape.
Voice CALL: Loud, slightly descending series of three or more *tew* notes. This ringing call can be heard for long distances. **SONG:** Rolling *whee-oodle*.
Status & Distribution Common. **BREEDING:** Muskeg and open or sparsely wooded areas. **MIGRATION:** Migrates on a broad front, but in most areas, in smaller flocks than Lesser Yellowlegs. In spring, usually a week

or so earlier than Lesser Yellowlegs. More southerly states see peak passage late Mar.–mid-Apr., but numbers seen into early May. Great Lakes and mid-Atlantic peak mid-Apr.–mid-May, might be seen as early as early Mar. First breeding arrivals late Apr. or early May, depending on latitude. A few nonbreeders remain well south and summer. In fall, the migration is protracted late June–late Nov. First adults arrive in mid-June, but most adults arrive July. Juveniles typically

arrive by late Aug., but arrivals can be as early as early Aug, peak in most areas Sept. and into early Oct. Juvenile Lessers arrive in force a few weeks earlier, by early Aug. Some remain into winter if conditions are mild enough; this is one of our hardiest shorebirds. **WINTER:** Coastal and interior regions. **RARE STATUS:** Rare north of breeding grounds; casual western AK (primarily spring). Casual in Europe, primarily in fall, Japan, and South Africa.
Population Numbers appear stable.

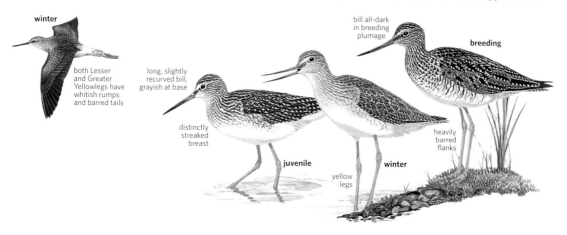

winter

both Lesser and Greater Yellowlegs have whitish rumps and barred tails

long, slightly recurved bill, grayish at base

distinctly streaked breast

juvenile

yellow legs

bill all-dark in breeding plumage

breeding

winter

heavily barred flanks

COMMON GREENSHANK *Tringa nebularia*   COMG ■ 3

An Old World visitor, the Common Greenshank is closely related to the Greater Yellowlegs, which it acts and sounds like. Monotypic. L 13.5" (34 cm)
Identification Rather large and robust *Tringa* with a long neck. Black tip of the relatively long, slightly upturned bill contrasts with a gray base. Greenish legs. **BREEDING ADULT:** Brownish gray above, with black centers to many scapulars. White underparts with light to moderate streaking on the breast and neck. **NONBREEDING ADULT:** Streak-

ing absent or restricted, with a rather pale head and foreneck. Upperparts lack black; feathers have dark and light notches on outer feather edges. **JUVE-NILE:** Like nonbreeding, but boldly marked above; white edges create a distinctly striped appearance. **FLIGHT:** A white rump with a wedge continuing up into the lower back is distinctive and is the most definitive way to separate it from either yellowlegs species.
Similar Species Similar to Greater Yellowlegs, but paler. Adult is less

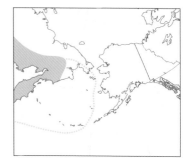

heavily streaked and the juvenile has striped, not spotted, upperparts. In all plumages, the legs are greenish, and in flight, a white wedge extends up to middle of back.
Voice CALL: A loud *tew-tew-tew* on same pitch is very similar to Great Yellowlegs, but does not descend in pitch.
Status & Distribution BREEDING: Northern Eurasia. **WINTER:** Southern Europe, Africa, southern Asia, and to Australia. **MIGRATION:** Rare spring and casual fall migrant to western and central Aleutians and to the Pribilofs; very rare in spring to St. Lawrence I., casual to CA, eastern Canada, NY, and NJ.
Population Stable.

juvenile

long and rather stout, recurved bill; paler at base

breeding

white wedge up back

winter

not as heavily marked below as Greater Yellowlegs

greenish legs

COMMON REDSHANK *Tringa totanus*   CREH ■ 5

This species is a visitor from the Old World, where it is one of the most numerous and familiar shorebirds of northwestern Europe. Multiple birds have occurred at the same location and time, suggesting that the right weather pattern (a nor'easter) is likely to deposit more Common Redshanks in the future. Polytypic (6 sspp.; probably *robusta* in N.A.). L 11" (28 cm)

Identification Slightly smaller than its relative, the Spotted Redshank. Bill is short and straight, almost stout, and red at the base. Legs are of medium length and vary from orange to red. Common Redshank is generally brownish above and white with streaking below. **BREEDING ADULT:** Brown above with heavy streaking; whitish background below, usually with streaks on the breast that turn into barring on the flanks. White eye ring is usually evident, but dark lores are obscured. **NONBREEDING ADULT:**

Rather plain brown above, with some feathers having dark notches on the feather edges; these notches are less evident than on other *Tringa* and might not be visible without good views. Breast and sides are brownish, with some spots or streaks; belly and undertail coverts white, with some barring. **JUVENILE:** Like nonbreeding, but more heavily marked above, with scapulars and coverts having bold white spots. The breast is whiter and streaked; remainder of underparts paler. Legs paler, with a yellow tone to the orange; bill has minimal color to the base of the mandible. **FLIGHT:** A white dorsal wedge is visible; particularly striking is a distinctive broad white trailing edge to the secondaries and inner primaries that looks like a white patch.

Similar Species Given a good view, no other species looks like a Common Redshank. The larger, longer-legged, and longer- and thinner-billed Spotted Redshank is grayer and paler in nonbreeding plumage (less brownish). Lesser and Greater Yellowlegs, with reddish orange legs, are sometimes confused with Redshanks.

Voice CALL: Loud *twek-twek* and a mournful, liquid whistle, *teu*, with a distinctly lower ending.

Status & Distribution Breeds in Eurasia, as close as Iceland where common. Breeding range is more southerly than its congeners, and wintering range is more northerly; the combination makes the Common less likely to occur in N.A. away from Atlantic Canada where it is casual in spring to NL (five records; four in spring 1995). Numerous records for Greenland. Even though it breeds from southeastern portions of the Russian Far East, unrecorded for AK.

Population Stable.

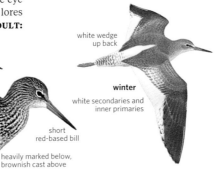

white wedge up back

winter

white secondaries and inner primaries

juvenile

short red-based bill

heavily marked below, brownish cast above

breeding

bright orange legs

WOOD SANDPIPER *Tringa glareola*   WOSA ■ 2

The Wood Sandpiper is a visitor from the Old World, somewhat intermediate between Lesser Yellowlegs and Solitary Sandpiper. Along with Green and Solitary Sandpipers, Wood Sandpiper is a more compact *Tringa*, with

obviously shorter legs. It frequents edges of wetlands, often where there is grass. It methodically picks food from the surface as it ambles along. It is often first detected when flushed from cover: It rises up steeply while giving

its shrill calls; in spring, it frequently breaks into the full display vocalizations uttered in a hovering display flight. Monotypic. L 8" (20 cm)

Identification Dark upperparts are heavily spotted, and underparts

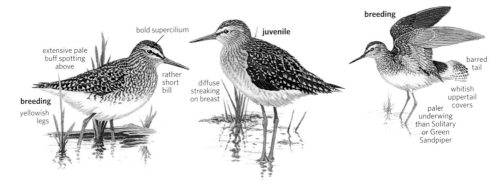

breeding

bold supercilium

juvenile

extensive pale buff spotting above

rather short bill

diffuse streaking on breast

breeding

yellowish legs

barred tail

whitish uppertail covers

paler underwing than Solitary or Green Sandpiper

are white, with brownish streaks on breast and neck. Has prominent whitish supercilium, particularly bold in front of the eye, which contrasts with a dark lore. An eye ring is usually obvious. Bill is rather short and dark; legs vary from yellow to greenish yellow. **BREEDING ADULT:** Upperparts are boldly checked with white; the breast and neck are heavily streaked. Bill often shows a pale base; legs more often with yellow tone. **NONBREEDING**

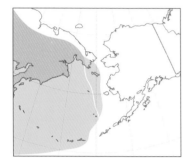

ADULT: Duller than breeding bird, with upperparts lacking the black-centered feathers. In fresh plumage, there are pale fringes on most scapulars and white spots on the coverts. **JUVENILE:** Like nonbreeding adult, but upperparts are darker, with more extensive spotting and more streaking on the breast. Crown is distinctly dark, set off from the supercilium. Legs are paler, more greenish. **FLIGHT:** Brown above and whitish below, with plain wings, a white rump, and a whitish tail, moderately barred, including the outer tail feathers. Wing linings are pale gray.

Similar Species Wood Sandpiper resembles a Lesser Yellowlegs in plumage coloration, but its shape is more similar to Green and Solitary Sandpipers. Solitary Sandpiper has dark central rectrices and is also darker, more gray, or grayish olive, as opposed to brown. Wood Sandpiper is distinguished from a Green Sandpiper by its

paler wing linings, smaller white rump patch, and more densely barred tail. Compared to Lesser Yellowlegs, note Wood Sandpiper's eye ring, shorter bill, and the well-defined supercilium, which also separates it from Solitary and Green Sandpipers. In addition, Lesser Yellowlegs is taller, with more leg extending beyond the tail in flight. **Voice CALL:** Loud, sharp whistling of three or more notes. Similar to Long-billed Dowitcher.

Status & Distribution BREEDING: Northern Eurasia. **WINTER:** Central and southern Africa, and southern Asia to Australia. **MIGRATION:** Fairly common spring migrant in AK to the western and central Aleutians (rare fall); uncommon on the Pribilofs (rare fall). Rare spring and casual fall migrant to St. Lawrence I. Casual breeder on the outer Aleutian Is. **RARE STATUS:** Casual to BC, OR, CA, NY, RI, and DE.
Population Stable.

MARSH SANDPIPER *Tringa stagnatilis*   MASA ▪ 5

The Marsh Sandpiper is a visitor from the Old World. It is a small and rather delicate *Tringa*, with very long legs and a long, thin bill. Its gait is rather deliberate as it picks its way around watery locales, and it is more stilt-like in many ways. Monotypic. L 8.5" (21 cm)
Identification Gray-brown above and white below; greenish legs; long, thin, dark bill, almost needlelike. Face pattern different from most *Tringa*, lacking a distinct loral stripe and having a supercilium that extends beyond the eye. **BREEDING ADULT:** Irregular dark markings on plain brown scapulars create somewhat mottled upperparts, alternating black and brown. Tertials and many wing coverts are brown, with black barring. Head, neck, and breast are liberally streaked or spotted with brown, largely eliminating the supercilium. Remaining underparts are white, although some light spotting or barring might be present on the flanks and undertail coverts. **NONBREEDING ADULT:** Plain gray-brown on the upperparts, with the head and back slightly paler than the wings. Scapulars and wing coverts have thin, whitish borders when fresh. Underparts are white below, generally lacking streaks. **JUVENILE:** Similar to nonbreeding, except for darker upperparts, with bold white or buff spots on

scapulars and coverts. **FLIGHT:** Gray upperparts, with slightly darker wings, contrast with the white tail, rump, and lower back. Legs extend well beyond the tail in flight.
Similar Species Only vaguely similar to Lesser Yellowlegs, but the two species share long legs and straight, thin bills. Marsh Sandpiper has a thinner, longer bill, giving it a more dainty look. In flight, the white on the lower back is more like that of Common Greenshank, whereas Lesser Yellowlegs has white restricted to the rump and tail. Somewhat similar to both Stilt Sandpiper and Wilson's Phalarope in nonbreeding plumage. Stilt Sandpiper has a slightly decurved bill and, in flight, a

white wing stripe and no white on the back. Wilson's Phalarope has a rather needlelike bill and no wing stripe, but it has a dark back. It also looks more compact as its legs are much shorter than those of Marsh Sandpiper; as a result, its legs do not extend as far beyond the tail in flight.
Voice CALL: *Tew*, repeated like a Lesser Yellowlegs.
Status & Distribution RARE STATUS: Casual from western and central Aleutians and the Pribilofs, AK; casual to CA in fall and spring; accidental northwest Baja California, Mexico, and HI.
Population Stable.

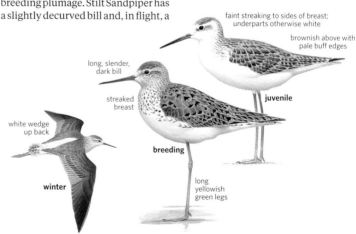

faint streaking to sides of breast; underparts otherwise white

brownish above with pale buff edges

long, slender, dark bill

streaked breast

juvenile

white wedge up back

breeding

winter

long yellowish green legs

PHALAROPES Genus *Phalaropus*

These elegant shorebirds have partially lobed feet and a dense, soft plumage. Feeding on the water, phalaropes often spin like tops, stirring up larvae, crustaceans, and insects. Females are larger and more brightly colored than males, and the sexes reverse roles: Females do the courting and males incubate the eggs and care for the chicks. In fall, they rapidly molt to winter plumage (especially Wilson's and Red); many are seen in transitional plumage farther south. Red-necked and particularly Red Phalaropes spend the winter months on the open ocean (pelagic). Although some phalaropes stay farther north, most migrate to the tropics, and many winter well south of the Equator.

WILSON'S PHALAROPE   *Phalaropus tricolor*   WIPH   ■   1

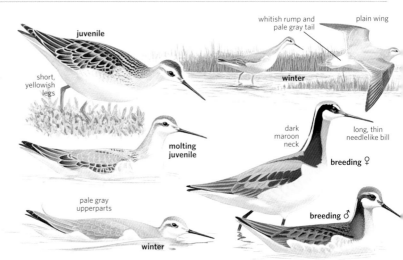

juvenile

short, yellowish legs

molting juvenile

pale gray upperparts

winter

whitish rump and pale gray tail

plain wing

winter

dark maroon neck

long, thin needlelike bill

breeding ♀

breeding ♂

Wilson's Phalarope is a splendid species with beautiful breeding plumage and an impressive biology. Large flocks are encountered during migration. It is the only phalarope that breeds south of Canada. Largest and most terrestrial of the phalaropes, it swims in circles on shallow ponds; this spinning creates a vortex that delivers food to the surface. It also feeds on mudflats, especially juveniles, where it walks in circles in a hunched-forward, almost awkward gait. Distinct genetic, behavioral, and vocal differences from other two species of phalaropes indicate that it probably belongs in its own genus, *Steganopus*. Monotypic. L 9.3" (24 cm)

Identification Wilson's long, thin bill is black and almost needlelike. **BREEDING FEMALE:** Has a bold black and rufous stripe on the face and neck. The warm orange neck contrasts with the white throat. The mantle and lower scapulars are rufous, separated by gray upper scapulars. The crown is whitish or pale gray; the whitish supercilium usually does not extend beyond the eye. **BREEDING MALE:** Like the female, only duller; the crown is dark gray. The upperparts have less contrast and are more mottled, having often dark-centered feathers with pale or rufous edges. Some males are quite dull, almost like nonbreeding plumage, except for the mottled upperparts and the dark, almost black legs. **NONBREEDING ADULT:** The legs are paler, olive to yellow. It is gray above, including gray crown and gray postocular stripe. The upperparts are edged white in fresh plumage. **JUVENILE:** The back is brown with buffy feather edges and the breast is buffy. It quickly molts out of this plumage; most juveniles seen south of breeding grounds are gray backed and identifiable as juveniles by the broad buffy edges to the dark wing coverts and tertials. **FLIGHT:** White uppertail coverts, whitish tail, and the absence of a white

wing stripe are important characters. **Similar Species** The lack of the "phalarope mark" through the eye; the long, thin bill, and the white uppertail coverts, whitish tail, and absence of white wing stripe distinguish juvenile and winter Wilson's from other phalaropes. They are more easily confused with a nonbreeding Stilt Sandpiper, but phalaropes have a straight (not curved) bill and less of a wing stripe, and their short legs do not extend well beyond the tail, as they do on Stilt Sandpiper. Lesser Yellowlegs, occasionally noted swimming, is larger and its bill is thicker, and Wilson's has a notably whiter breast. Marsh Sandpiper, a visitor from Eurasia, has a longer bill and longer legs; in flight it shows a white wedge up the back.

Voice CALL: A hoarse *wurk* and other low, croaking notes.

Status & Distribution Common. **BREEDING:** Grassy borders of shallow lakes, marshes, and reservoirs. **MIGRATION:** Common to abundant in western N.A.; uncommon to rare in the East. Late Mar.–mid-May, spring migrants pass through CA, peak late Apr.–early May. Farther north and east, extending to the Great Lakes, peak is mid-May, although they are rare, and uncommon at best. Adult females are the earliest

fall migrants, with the first arrivals, typically adult females, in early June. Large flocks stage at key areas (e.g., Mono Lake, CA, and Great Salt Lake, UT), where hundreds of thousands molt into nonbreeding plumage. Juveniles, seen as early as early July, peak mid-July–Aug., lingerers into Oct., exceptionally later. **WINTER:** Casual in Southern CA, AZ, and FL; recently found in southern TX, sometimes in small flocks. Most winter in S.A., on alkaline lakes in the Andes. **RARE STATUS:** Annual in western Europe; casual to South Africa, Australasia, the Galápagos, and the Falklands.

Population Declines due to habitat loss and drought.

RED-NECKED PHALAROPE *Phalaropus lobatus* RNPH ▪ 1

This species winters at sea, but migrating flocks occur at inland locales. Monotypic. L 7.8" (20 cm)
Identification Small, relatively thin, dark bill. **BREEDING ADULT:** Dark above with buff stripes. Chestnut on front and sides of neck, less prominent in male. More prominent supercilium on male. **NONBREEDING ADULT:** Grayish upperparts with whitish stripes. Prominent dark eye patch. **JUVENILE:** Like winter adult but darker above, with bright buff stripes. **FLIGHT:** White wing stripe, dark central tail coverts. **Similar Species** Red Phalarope is larger and thicker-billed. In winter plumage it has a plain, light gray back, lacking pale stripes. In fall, young Red-neckeds retain their dark juvenal plumage much longer than young Red Phalaropes.
Voice CALL: A high, sharp *kit*, often given in a series.
Status & Distribution Common. **BREEDING:** Tundra. **MIGRATION:** Common inland in West and off

West Coast; rare in Midwest and East; uncommon off East Coast. In spring, migration peaks first half of May, although might appear by early Apr., rarely late Mar., and linger to June. In fall, in the lower 48, adults are seen end of June–Aug., juveniles from early Aug.–late Oct. (mostly in full juvenal plumage). **WINTER:** Chiefly at sea in Southern Hemisphere, but records in Southern CA and southern TX.
Population Recent declines from the fall staging areas in Bay of Fundy.

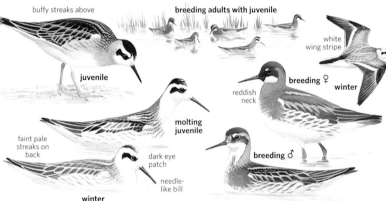

buffy streaks above

breeding adults with juvenile

white wing stripe

juvenile

breeding ♀

winter

reddish neck

molting juvenile

faint pale streaks on back

dark eye patch

breeding ♂

needle-like bill

winter

RED PHALAROPE *Phalaropus fulicarius* REPH ▪ 1

Usually highly pelagic. Monotypic. L 8.5" (22 cm)
Identification BREEDING FEMALE: Black crown; white cheek patch. Underparts chestnut. Upperparts black with buff lines. Bill yellow, with a dark tip. **BREEDING MALE:** Duller. **NONBREEDING ADULT:** Pale, uniformly gray above. Dark eye patch. Dark bill sometimes with small pale base. **JUVENILE:** Dark upperparts with buffy edges and streaks, peachy buff wash on neck. Most juveniles seen in southern Canada and US are largely in winter plumage. **FLIGHT:** White wing stripe, bolder than Red-necked.
Similar Species Compare to the Red-necked Phalarope.
Voice CALL: A sharp *keip*.
Status & Distribution BREEDING: Tundra ponds. **MIGRATION:** Fairly common to common off West Coast;

uncommon off East Coast. In spring, migration from mid-Apr., but peaks mid- to late May, some into June. In fall, adult females as early as late July, stages Aug.–Sept. (e.g., Bay of Fundy); juveniles migrate into Nov. Irregular movements in late fall and early winter off both coasts; "wrecks" sometimes occur in late fall, particularly along West Coast, some found well inland. **WINTER:** At sea to S.A. **RARE STATUS:** Very rare to casual inland and in Gulf of Mexico, mostly in late fall.
Population Stable.

molting fall adults

bolder wing stripe than Red-necked

juvenile

winter

white cheek

breeding ♀

molting juvenile

red underparts

plain gray upperparts

dark eye patch

thicker bill than Red-necked

breeding ♂

winter

COURSERS AND PRATINCOLES Family Glareolidae

Oriental Pratincole, nonbreeding (Japan, Sept.)

Although ternlike in many aspects, pratincoles are indeed shorebirds. This family also includes the terrestrial coursers.

Structure Pratincoles have long, pointed wings; long, forked tails; and short legs, which gives them a very horizontal, elongated look when perched. The bill is short. Coursers are upright birds with long legs and short toes adapted for running in deserts and brushlands.

Behavior Pratincoles are highly aerial, much more so than other shorebirds; unusual for waders, they rarely enter the water. Their flight is ternlike and can be high and fast. Quite gregarious, they often migrate in large flocks and nest colonially. They nest on bare ground, frequently in plains or desert-like habitats. They feed on insects (their main source of food) in the air, although they will chase insects on the ground. Their gait is plover-like, alternating running and abruptly stopping.

Plumage In both subfamilies the sexes are alike. Pratincoles molt prior to fall migration, resulting in a more muted plumage, blurring the crisper pattern present in spring. Plumages are mostly cryptic combinations of browns, gray, chestnut, black, and white, but flight patterns can be striking.

Distribution This Old World family is found primarily in southern Europe, Africa, and Asia, with one aberrant pratincole in Australia. There are only three records of pratincoles (of two different species) from the Americas and no records of any coursers.

Taxonomy The 17 species in this family are placed in two subfamilies, the coursers (Cursoriinae, with nine species) and the pratincoles (Glareolinae, with eight species). The enigmatic Egyptian Plover is now placed in its own family, Pluvianidae.

Conservation Land reclamation, pesticides, and development have had an impact on some members of this family. BirdLife International codes: 1 NT, 1 VU, and 1 CR.

Genus *Glareola*

ORIENTAL PRATINCOLE *Glareola maldivarum*   ORPR ▪ 5

breeding adult

no white trailing edge to secondaries

chestnut underwings

short tail

The Oriental Pratincole is an unexpected visitor from the Old World. Its graceful and unique flight would be a welcome sight to any N.A. birder, particularly given the few records to date. Where common, it is often seen in flight, as it does most of its feeding on the wing; its flight is powerful with erratic swooping, very much like a tern, or even a large swallow or martin.

On the ground, it might chase insects in ploverlike fashion, but more likely will sit or stand rather motionless until taking flight and is easily overlooked. Monotypic. L 9" (23 cm) WS 23.5–25.5" (60–65 cm)

Identification Adults are brownish above and on breast, with white rump and deeply forked black tail. When there is adequate light, chestnut underwing coverts may be viewed, but when shaded they look dark. **BREEDING:** Both males and females usually have a warmer orange tint to the brown. A black line extends below eye and frames buffy throat with a semicircle. Bill is black, with red restricted to base. **NONBREEDING:** Breast is a duller brown. Bill base is a duller red. Black frame to throat is more diffuse and streaky. **JUVENILE:** Similar to winter adult, but it is duller still, with an all-dark bill, a whitish throat, and a scaly pattern to the upperparts.

Similar Species Oriental Pratincole is unlike any bird expected in this hemisphere, but there is one record of the similar Collared Pratincole (*G. pratincola*), a bird that wintered on Barbados. Adult Oriental lacks white trailing edge on its secondaries, has darker upperparts (so outer primaries do not boldly contrast), less red on bill, and a shorter tail with a shallower fork.

Voice CALL: Gives a harsh *kik-kik-kik* or *chik* notes that recall a *Sterna* tern.

Status & Distribution BREEDING: Much of Asia to northern and northeastern China, and southeastern Russian Far East. **WINTER:** Southern part of breeding range south to northern Australia. **RARE STATUS:** Two spring records from islands in western AK: Attu I. (19–20 May 1985) and St. Lawrence I. (5 June 1986). Also recorded from northwestern Europe.

Population Fairly common, but some local populations declining.

SKUAS AND JAEGERS Family Stercorariidae

Long-tailed Jaeger, adult (AK, June)

Skuas and jaegers are superficially gull-like birds and were, until recently, united with them (along with the terns and skimmers) in the family Laridae.

Structure All are long-winged birds capable of rapid and powerful flight. The four skuas are heavily built with relatively broad wings and a short, squared tail with very slightly projecting central rectrices; the three jaegers are more slender in build and as adults have long, projecting central tail feathers. Bills are strongly hooked with a slight to pronounced gonydeal angle. The legs are short and stout compared with those of gulls, with relatively strong claws. Females are slightly larger than males.

Behavior Often solitary, but Long-tailed and Pomarine may migrate in small groups, and multiple individuals may be found in prime foraging areas. When nesting in tundra regions jaegers take small rodents, birds, nestlings, and even berries. Skuas are predatory, feeding on chicks, eggs, and adults of a variety of seabirds. Nonbreeding jaegers scavenge prey, "pirate" prey from terns or other foraging seabirds, and catch their own fish or other prey. All species are monogamous, with long-term pair bonds in some species. The nest is a simple scrape; from the usual clutch of two eggs, often only the first-hatched chick survives.

Plumage Generally clad in variations of browns and grays. Skuas are entirely dark on the head and body, some species showing reddish or golden tones or pale shaft streaks. Jaegers are mostly pale below, although all three species are polymorphic and may show entirely dark body plumages. All flash white at the bases of the primaries, which are conspicuous in flight. Age

variation is pronounced in jaegers but subtle in skuas.

Distribution The three jaegers breed in high-latitude northern regions, on tundra or coastlines; they migrate southward, mainly over the open ocean or coastlines, but with lesser movements over continental regions, and spend the winter in marine habitats from north temperate regions (a minority of birds) to subtropical, south temperate, or southern high-latitude regions. Three of the four skuas breed in islands and coastlines in the cold southern oceans south to Antarctica; the fourth (Great Skua) breeds in the North Atlantic. Many Great Skuas winter south into the subtropical Atlantic, but Brown and Chilean Skuas appear to remain within the southern oceans for the winter. The spectacular migration of South Polar Skua brings it from Antarctic nesting areas northward into the North Pacific and North Atlantic. All three jaegers and two skuas occur in North America.

Taxonomy The seven species of jaegers and skuas are closely related to the superficially somewhat similar gulls and terns; recent molecular work confirms that the skuas are the sister group to the diving auks, murres, and puffins (alcids), necessitating their removal from the family Laridae. The largest jaeger, Pomarine, shares some structural, plumage, and behavioral traits with the larger skuas, and in fact it is more similar in its mitochondrial DNA to Great Skua than to other jaegers; it appears to be the sister to the clade of larger skuas (sometimes separated in the genus *Catharacta*).

Conservation Populations of most species appear stable, although populations of some skua species on certain islands have been eliminated or reduced.

Genus *Stercorarius*

GREAT SKUA *Stercorarius skua*   GRSK ■ 3

This visitor to the North Atlantic is often associated with winter mid-Atlantic pelagic trips (where it associates with feeding assemblages around fishing boats), but it can also be found in the western North Atlantic in summer, especially off Canada. Monotypic. L 21–24" (53–61 cm) WS 51–55" (130–140 cm)

Identification Large, bulky; stout bill, broad wings, barely projecting central rectrices. **ADULT:** Ginger-brown overall; buff streaking on face and neck; rufous to buff mottling and streaking on back and upperwing coverts; often slightly darker cap or face. Underwing coverts mottled dark brown; white flash across primary bases usually bold and striking in flight. **JUVENILE & FIRST-YEAR:** More uniform overall and often darker. Head and body vary from dark rufous-brown (as shown) to fairly cold brown (suggesting juvenile South Polar, but fresh in fall), without pale neck streaking. Note pale tips or U-shaped subterminal markings on scapulars and upperwing coverts. Grayish bill base. First-year birds in May–June can be largely in bleached juvenile plumage and look paler and more uniform than winter birds, inviting confusion with South Polar.

Similar Species South Polar averages smaller and more lightly built, with more slender bill and narrower wings. Differences difficult to judge at sea, as all skuas are "big" and are not always

typical adult

all Great and South Polar Skuas have broad wings and short tails with bold white flash at base of primaries

dark adult

buffy rufous edges on upperparts

thick bill

dull cinnamon below

juvenile

short, rounded tail points

bold white primary flash

pale adult

reddish brown below

seen well. Adult and older immature South Polars differ from adult Greats in uniformly dark upperwings and cold-brown, usually fairly dark head and body, with paler hind neck. Main problem is separating bleached first-summer Great from South Polar: Note Great's evenly worn (and more tapered) juvenal outer primaries, which can be very bleached at tips; dark brown (vs. blackish) underwing coverts, and any remaining juvenal upperwing coverts, with bleached U-shaped markings. Adult South Polars generally are colder brown on head and body; contrasting pale hackles on neck (less uniformly bleached on head and neck than on Great). Many birds not seen well should be recorded as "unidentified skua species." With aseasonal, atypical-looking individuals, consider the possibility

of other skua species from Southern Hemisphere, especially Brown Skua (a few records from Maritimes and mid-Atlantic coast may be this species). Also see smaller Pomarine Jaeger.

Voice Likely to be silent in N.A.

Status & Distribution BREEDING: Breeds in northwestern Europe including Iceland. **MIGRATION & WINTER:** Uncommon. Found off Atlantic Canada, mainly Sept.–Mar., fewer June–Aug. Farther south, off East Coast of US, recorded mainly Dec.–Feb./Mar., south regularly to NC, casually to FL. Very rarely seen from shore. Winters in North Atlantic and south to off West Africa. Northbound first-summers from subtropical winter grounds probably also occur from May–June, the same time as migrant South Polar Skuas are seen off the East Coast.

Population Increasing in Iceland; global population numbers ±30,000 mature individuals.

SOUTH POLAR SKUA *Stercorarius maccormicki*   SPSK ■ 2

This polymorphic bird breeds closer to the South Pole than any other bird species and is a frequent predator on penguins. It is highly migratory, however,

and visits both of our coasts. Monotypic. L 21" (53 cm) WS 52" (132 cm)

Identification The smallest skua, its small size can allow for confusion with

larger female Pomarine Jaegers, particularly dark juveniles. Stocky, thick necked, potbellied. Thick, strong, hook-tipped bill. In many respects, intermediate between Great Skua and Pomarine Jaeger, although clearly a skua, with characteristic short-tailed appearance, broad wings, and large white primary flash visible from above and below. Typically, older ages show a contrasting pale nape; paler birds have a small pale blaze at base of upper mandible. **LIGHT ADULT:** Distinctive, pale golden brown ("blond") on head and body, contrasting with dark brown wings and mantle. Narrow golden streaks on back. Face darker, contrasting with pale golden nape. Often strongly developed pale blaze at bill base. Black bill and legs. In flight, pale body contrasts strongly with dark underwings. From above, pale nape and upper back contrast with darker rear quarters and wings. This look (pale in front and dark at back) is characteristic of this morph. **DARK ADULT:** Dark brown throughout; uniform, lacking much streaking on upperparts. Cold brown, lacks rufous or warm tones. Dark, unicolored face contrasts with paler nape, which is narrowly streaked golden. Body dark, so there is no contrast with dark underwing; paler nape most obvious area of any contrast, other than the wing flash. **INTERMEDIATE ADULT:** Intermediate between the two extremes. Body somewhat paler than dark underwings. **JUVENILE:** Similar to dark adult, but gray bill base contrasts with blackish tip; tarsus also gray. Plumage crisp and evenly worn;

lacks streaking or other markings. Nape paler but lacks golden streaking. Unicolored, as in dark adult; lacks warm tones; evenly colored. Body slightly paler than dark underwings. **HYBRID:** Frequently hybridizes with Antarctic (Brown) Skua in Antarctic Peninsula. Distribution of these hybrids during nonbreeding season is not known.

Similar Species The Great Skua is extremely similar to juvenile and dark-morph adult South Polar Skuas. South Polars lack rufous or cinnamon tones on plumage and look more uniform in general pattern, but they show a pale nape. Paler South Polars show contrast between pale body and darker underwings. While in our waters, South Polars tend to be in obvious molt, whereas adult Great Skuas molt later (starting Sept.). However, young Great Skuas molt in spring to summer, so lack of midsummer molt is useful only for identification of adult Great.

Voice Silent while in our region.

Status & Distribution Uncommon in Atlantic and Pacific Oceans; most numerous in North Pacific. Strongly pelagic, extremely unlikely to be seen from shore. **BREEDING:** Antarctica. **MIGRATION:** Hypothesized clockwise migration: arriving earliest on west side of northern oceans, moving later to the east side, before moving south. In Pacific Ocean, rare in spring; numbers increasing through summer, peak Aug.–Oct. In Atlantic, most common

May–July, numbers dropping later on in summer and autumn. Unclear whether all age groups move into Northern Hemisphere; evidence suggests that only younger birds arrive here. Light-morph adults, or similarly plumaged birds, are extremely rare in N.A., most being darker individuals. **WINTER:** Breeds during our winter (austral summer), so they are present in N.A. waters during our summer months. Even so, South Polars always seem to be on the move when present in our waters. **RARE STATUS:** Casual in AK. Accidental in interior: GA, TN, OK, and ND.

Population Stable; global population numbers ±10,000 mature individuals.

bold white primary flash

uniform dark upperparts

intermediate-morph adult

short, rounded tail points

light-morph adult

dark-morph adult

slight golden cast on nape

juvenile

most look quite dark, except for white wing patches

thick bill averages slightly smaller than Great Skua

light-morph adult

light-morph adult

juvenile

POMARINE JAEGER *Stercorarius pomarinus*   POJA ■ 1

This bulky, thick-billed species is the largest of the three jaegers. Large individuals recall skuas, while smaller ones are deceptively similar to Parasitic Jaeger. Monotypic. L 21" (53 cm) WS 48" (122 cm)

Identification A large, broad-winged, potbellied, polymorphic jaeger with a thick neck. Adult central tail extensions are broad and twisted, with diagnostic spoon-shaped tips. Outer 4–6 primaries have white shafts. Upperwings unicolored. As in Parasitic Jaeger, underwings show a white flash at base of the primaries, but a second flash on underside of primary coverts ("double wing flash") is unique to Pomarine Jaeger. This "double flash" is present in juveniles and most adults and is caused by broad dark tips to white-based underwing primary coverts. **SUMMER LIGHT-MORPH ADULT:** Dark brown above; wings with variable pale flash at base of primaries. Black cap contrasts with yellow neck; breast white with broad, dark mottled breast band and dark mottled flanks. Males show reduced breast band, rarely lacking it altogether. White underparts and dark vent. Bicolored bill with orange-pink base and dark tip. **SUMMER DARK-MORPH ADULT:** Chocolate throughout; slightly blacker cap, pale wing flash. **WINTER ADULT:** Pale birds show dark cap; variably barred underparts; coarsely black-and-white barred rump. Tail streamers lacking or less well developed than in summer. **JUVENILE & FIRST-YEAR:** Variable. Brownish, barred below and on underwings. Lacks capped effect. Juveniles are brown to chocolate, lacking warm tones. Head unicolored, not streaked; nape unstreaked. Primaries dark to tip. Tail shows short, broad, rounded extensions on central rectrices. All juveniles (even dark birds) show pale uppertail coverts. **IMMATURE:** Similar to winter adult, but underwings with variable amount of white barring.

Similar Species Most similar in size to Parasitic Jaeger, but shapes differ: Pomarine appears broader-winged, more potbellied, thicker-necked, and shorter-tailed (not including streamers). Breeding pale-morph adult Pomarines easily identified by spoon-tipped, twisted central tail feathers; generally more mottled and broader breast band; bicolored bill; extension of dark cap below gape; brighter yellow face coloration. Dark-morph adults separated by structure and tail spoons. In fall, Pomarine is the only jaeger expected to show wing molt while in our waters; the two other jaegers tend to molt south of N.A. and later in the season. In winter, adult Pomarine is separable from Parasitic by shape, more heavily barred body plumage, and thicker barring on rump. Growing broad-tipped tail streamers are not pointed as on Parasitic. Overall, Pomarine is less warmly colored than Parasitic, lacks head streaking, and consistently shows coarse black-and-white barring on rump; this is less obvious on Parasitics and is not present in

darker morphs. Pomarines have dark primary tips, while Parasitics' primary tips are fringed with pale. Long-tailed Jaegers are much smaller and slimmer, with narrower and longer wings as well as a longer tail. Adult Long-taileds (no dark morph) show long and pointed tail streamers as well as a small, crisply demarcated black cap. Pomarines at all ages show at least four white shafts on outer primaries; only two on Long-tailed Jaegers. Juvenile Long-taileds are paler and grayer than Pomarines: The palest extremes are contrastingly white-headed, whereas Pomarines are solidly dark-headed. Other than darker juveniles, Long-taileds show an unbarred white belly,

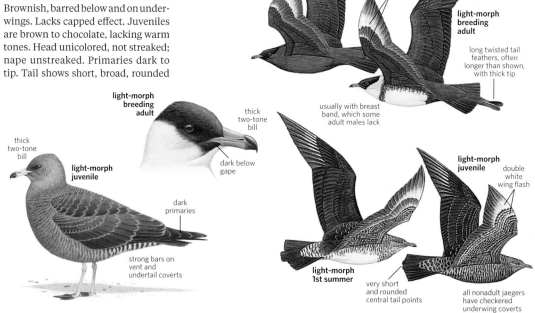

dark-morph breeding adult

double white wing flash

light-morph breeding adult

long twisted tail feathers, often longer than shown, with thick tip

light-morph breeding adult

thick two-tone bill

thick two-tone bill

dark below gape

light-morph juvenile

dark primaries

usually with breast band, which some adult males lack

light-morph juvenile

double white wing flash

strong bars on vent and undertail coverts

light-morph 1st summer

very short and rounded central tail points

all nonadult jaegers have checkered underwing coverts

but Pomarines are consistently barred on the belly.

Voice LONG CALL: A series of *yowk* notes, roughly two per second, given for several seconds. Not heard south of breeding areas.

Status & Distribution Numerically, more common, but less likely to be seen from land. **BREEDING:** Nests in wet tundra, often near coast. **MIGRA-TION:** Pelagic on both coasts. Adults move south before juveniles. First arrivals in July; peak movements Sept.–late Oct., with some wintering in N.A. waters. Juveniles typically arrive after Oct. Northbound birds peak late Apr.–late May, arriving in Arctic early to mid-June. Rare in autumn migration on Great Lakes (nearly all juveniles), where a few lin-ger into Dec. **WINTER:** Most common wintering jaeger, found as far north as northern CA, NC, and Gulf of Mexico. Main wintering areas farther north than either of the other two jaegers, with concentrations in Caribbean and northern S.A. **RARE STATUS:** Casual throughout interior of continent, mainly juveniles in late fall.

Population Stable; no data on trends.

PARASITIC JAEGER *Stercorarius parasiticus*   PAJA ▪ 1

Perhaps the most frequently seen jae-ger and the one most likely to be seen from land, often harassing terns. Mono-typic. L 19" (48 cm) WS 42" (107 cm)

Identification Polymorphic; medium size; intermediate between other spe-cies. Slim yet powerful. Adult central tail extensions are narrow and pointed, not long ribbon-like streamers. Narrow, long bill. Outer three to five primaries have white shafts. Upperwings unicol-ored; no contrast between wing coverts and secondaries. **SUMMER LIGHT-MORPH ADULT:** Dark brown above, pale below with dark vent. Black cap; paler forehead; contrasting pale yellow neck. Most adults have a breast band. **DARK MORPH:** Chocolate overall, blacker cap. **WINTER ADULT:** Shows dark cap; vari-ably barred breast and flanks; complete breast band. Tail streamers less well developed than in summer. **JUVENILE & FIRST-YEAR:** Variable. Brownish; barred below and on underwings. Lacks dark cap. Juveniles light brown to black-ish with cinnamon fringes, but usually cinnamon brown. Streaked pale nape patch. Densely barred throughout; pale primary tips. Paler juveniles show pale uppertail coverts. **IMMATURE:** Similar to winter adult, but underwings barred white.

Similar Species Compared to Long-tailed, not as slim and ternlike; has shorter tail, longer bill. Compared to Pomarine, slimmer, longer tailed, nar-rower winged. Breeding pale adults identified by short, pointed tail exten-sions; narrow, even (unmottled) breast band; black bill. In all ages, upper-wings are evenly dark above; coverts not paler than secondaries. Parasitics show three or more white primary shafts on outer wing. Juvenile Parasit-ics are warm colored, showing rusty or rufous tones, unlike more chocolate Pomarine or the colder gray-brown Long-tailed. Nape paler than head and shows streaking, unlike Pomarine. Above, paler individuals showing pale barred rump have narrower, wavier bars than Pomarine. Primaries pale fringed, unlike other species.

Voice LONG CALL: A series of three to four bisyllabic notes, roughly one per second, rising in pitch. Only in breeding areas.

Status & Distribution Fairly common. **BREEDING:** In Arctic, wet and moist tundra, and coastal wetlands. **MIGRA-TION:** Pelagic, but frequently seen from

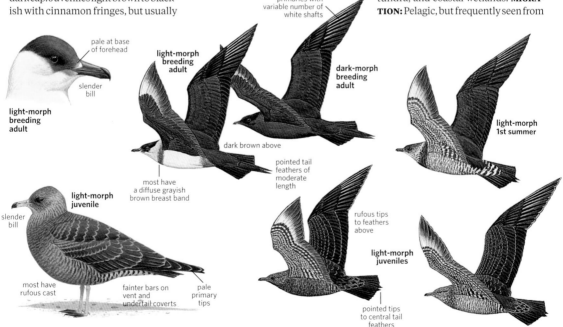

pale at base of forehead

light-morph breeding adult

slender bill

light-morph breeding adult

dark brown above

most have a diffuse grayish brown breast band

light-morph juvenile

slender bill

most have rufous cast

fainter bars on vent and undertail coverts

pale primary tips

primaries with variable number of white shafts

dark-morph breeding adult

pointed tail feathers of moderate length

rufous tips to feathers above

light-morph juveniles

pointed tips to central tail feathers

light-morph 1st summer

shore. Small numbers pass through Great Lakes region and Salton Sea in autumn. Adults move south before juveniles. First arrivals July; peak late Aug.–late Sept. Juveniles arrive late Aug., normally at least six weeks before the first juvenile Pomarines appear well south. Northbound birds peak late Apr.–late May; arrive in Arctic late May to mid-June. **WINTER:** Small numbers winter off Southern CA and FL; very rarely in the Gulf of Mexico; most winter in S.A. Found closer to shore than other jaegers, where they habitually harass tern flocks. **RARE STATUS:** Casual throughout interior of continent.

Population Apparently stable.

LONG-TAILED JAEGER *Stercorarius longicaudus* LTJA ■ 1

The smallest and most elegant jaeger, Long-tailed is locally common on its northern breeding grounds. Its flight is relatively buoyant and graceful, often hovers on breeding grounds. Unbroken tail streamers add 6–8 in. (15–20 cm) to length of breeding adult. Monotypic. L 14.5–16" (37–41 cm) WS 37–41" (94–104 cm)

Identification Wings relatively narrow based; tail projection behind wings usually longer than width of wing base, even without streamers. Bill relatively small but appears thick; legs have pale markings at all ages (dark on adults of other species). Wing molt occurs south of N.A. On upperwing of all ages, outer two to three white primary shafts most prominent (typically only outer two shafts). Adult lacks dark (or intermediate) morph. **BREEDING ADULT:** Neat black cap; no breast band; belly to vent dusky. On upperwing, blackish remiges (primaries and secondaries) contrast with medium brown-gray coverts; no white flash on underwing. Tail streamers long and finely pointed (often shed in fall). **WINTER ADULT:** Unlikely in N.A. Dusky chest band (partially shown by some fall migrants); barred tail coverts; short tail streamers. **JUVENILE & FIRST-SUMMER:** Polymorphic. Underwing has bold white flash lacking in older ages. Most show bold pale barring on underwings, bold whitish barring on uppertail coverts. Juvenile has bluntly pointed tail streamers; primary tips (at rest) with little or no whitish edg-

two to three white primary shafts in nearly all plumages

breeding adult

no white wing flash

gray upperparts contrast sharply with blackish flight feathers

very long, pointed central tail feathers

1st summer

light-morph juvenile

long slender wings

long tail has rounded central tail feathers with pale tips

pale tips to feathers above

some with pale belly and dark breast

dark-morph juvenile

stubby bill

clean blackish cap

color varies from grayish to chocolate brown

dark primaries

breeding adult

stubby bill

strong, straight bars on vent and undertail coverts

typical juvenile

ing; pale tips to upperwing coverts. First-summer is more capped and lacks pale tips to upperwing coverts; tail streamers have needle-like tips. Dark morph blackish brown overall, usually with bold whitish bars on tail coverts. **SECOND-SUMMER:** Resembles adult but tail streamers average shorter, underwing with variable pale barring; a few have dark body and messy whitish belly.

Geographic Variation Breeding adults from N.A. average less extensive dusky on belly than those in northern Europe. Previously considered polytypic.

Similar Species See other jaeger accounts. As a rule, Long-tailed may be likened to Mew Gull, Parasitic likened to Ring-billed Gull, and Pomarine likened to California Gull. Juvenile Long-

taileds lack warm reddish tones often shown by Parasitic.

Voice Shrill, mewing chippers *kyi-kyi-kyik*, etc., and high, clipped yelps. Mostly silent away from breeding grounds.

Status & Distribution Holarctic breeder, winters mainly off S.A. and South Africa. **BREEDING:** Common on dry tundra, late May–Aug. **MIGRATION:** Mostly well offshore and not likely to be seen from land. Mainly May and in fall when much more numerous from late July to early Oct., stragglers into Nov.; immatures off coasts June–July. Casual to rare inland (mainly fall) and the most likely jaeger inland away from large bodies of water; also casual off the Gulf Coast.

Population Most abundant and widespread jaeger in Arctic.

AUKS, MURRES, AND PUFFINS Family Alcidae

Atlantic Puffin, breeding (ME, July)

Spectacular and with many species easily observed at their remote breeding colonies, alcids pose a challenge to scrutinize and identify at sea, where they spend most of their time—much of it submerged. A typical birder's view is of distant birds in flight, often in rough sea condition. Identifications are best based on pattern of dark- and light-colored plumage, bill shape (if visible), and subtle differences in body shape and flight style.

Structure Heavy compact bodies, dense waterproof plumage, very short tails, and small, short wings are standard. Bills are highly variable in shape and coloration, from dagger-shaped in murres, guillemots, and some murrelets, to hatchet-shaped (puffins), to tiny (auklets, Dovekie, and some murrelets). The legs are short and the feet, which lack a hind toe, are fully webbed; they are brightly colored in some species.

Behavior Alcids normally fly close to the sea surface in a very rapid and direct manner with continuous rapid wingbeats; a few smaller species zigzag among wave tops. Most species (except *Cepphus* guillemots and *Brachyramphus* murrelets) usually remain well offshore except when attending breeding colonies. Solitary sick, weak, or oiled birds sometimes enter bays and harbors. Large numbers occasionally fly by headlands, especially just after dawn, driven close to land during onshore winds. "Wrecks," in which hundreds of birds occur far inland, sometimes take place after late fall and early

winter storms. Alcids forage by wing-propelled pursuit diving with three feeding preferences: schooling fish (murres, puffins), bottom fish (*Cepphus* guillemots), and zooplankton (murrelets, auklets, puffins). The clutch consists of one or two eggs. Alcids are extraordinarily variable in breeding habits: Chicks may depart the colony two days after hatching (*Synthliboramphus* murrelets), when half grown (murres), or remain in their burrows until full-size (puffins, auklets). Adults of some species may be seen at sea with small chicks that resemble miniature adults. Most species breed colonially (except *Brachyramphus* murrelets) on cliff ledges, rock crevices, earth burrows, surface scrapes (Kittlitz's Murrelet), and even mossy tree limbs (Marbled Murrelets, in parts of their range)—invariably in areas where terrestrial predators are absent or scarce. A few species have entirely nocturnal colony attendance. Dexterity on land varies, with some species clumsy and unable to stand upright (murrelets), and others agile (auklets and puffins).

Plumage Alcids have black-and-white or dull gray-and-brownish plumage. A few display spectacular nuptial plumes during the breeding season. Sexes look alike. Most undergo minor seasonal changes in coloration; some change drastically between winter and summer (*Cepphus* guillemots and *Brachyramphus* murrelets), and a few look the same year-round (auklets). Least Auklet is strikingly polymorphic.

Distribution Alcids are restricted to cold seas of the Northern Hemisphere, with the notable exception of three *Synthliboramphus* murrelet species that inhabit warm subtropical waters off Southern CA and Mexico. Most alcids come ashore only to breed on remote islands and exposed headlands. Long-billed Murrelet is a casual visitor to North America from Asia.

Taxonomy Worldwide there are 24 living species in 10 genera. Twenty species breed in North America, with diversity highest in the North Pacific. Two murrelet species have been split in recent years: Marbled Murrelet into Marbled and Long-billed Murrelets (1997) and Xantus's Murrelet into Scripps's and Guadalupe Murrelets (2012).

Conservation The greatest threat has been the introduction of predators (rats and foxes) onto breeding islands (now diminishing due to management). Oil pollution events affect all species negatively. This and gill netting result in downed birds and depletion of food stock. Food stocks are also reduced due to warming waters brought about by climate change. Four species have experienced severe declines and have threatened or endangered status: Marbled Murrelet (loss of old-growth nesting habitat, gill netting), Scripps's and Guadalupe Murrelets (introduced predators on breeding islands), and Kittlitz's Murrelet (oil spills, gill netting). The exceptionally large and flightless Great Auk was hunted to extinction by the middle of the 19th century. BirdLife International codes: 4 NT, 4 VU, 2 EN, and 1 EX.

DOVEKIES, MURRES, AND AUKS Genera *Alle*, *Uria*, *Alca*, and *Pinguinus*

Crisp black-and-white plumages characterize this group, which includes the largest alcids and a small planktivorous species. There were five species; one was extinct by the mid-19th century. All species in these genera occur in the Atlantic (murres also in the Pacific). Males leave the breeding colonies with their single, partly grown chick and provision it at sea.

DOVEKIE *Alle alle*   DOVE ■ 2

Dovekie is by far the smallest alcid in the North Atlantic—nearly half the size of Atlantic Puffin. A swimming bird typically adopts a distinctive neckless posture with its head held low to water. An active bird often "drags" its wings on the surface between dives. Monotypic. L 7.8" (20 cm)

Identification Entirely black and white. Upperparts are black, with distinctive scapular stripes, formed by narrow white margins on scapulars;

breeding
adult

winter

blackish
underwing

white tips to
secondaries

dark
neck band

black
head

breeding adult

stubby
bill

white-edged
scapulars in all
plumages

whitish extends
up into ear
coverts

winter

underparts are white. Bill is short, blending evenly with forehead to give a bull-headed look; at close range, bill shape is noticeably short, deep, and broad with a strongly curved culmen. Feet are blackish. **BREEDING ADULT:** Face, throat, neck, nape, and upper breast (bib) are entirely black, except for a small highly contrasting white spot "headlight" above eye. **WINTER ADULT:** White extends to throat, sides of nape, and neck, forming an incomplete white collar. **FLIGHT:** Football-shaped body, small size, and rapid "buzzing" wingbeats are distinctive. Secondaries are tipped with white; underwings are blackish. Flies low over sea, zigzagging among waves.
Similar Species Unlikely to be confused with any Atlantic alcid. Small

in size, it is closest in appearance to a juvenile Atlantic Puffin, which is 50 percent larger and has a different bill and head shape, and a dark face, throat, and neck. In the Bering Sea, where it occurs in summer with Least, Parakeet, and Crested Auklets, Dovekie is closest in size to latter two species, but differs in body shape and crisp black-and-white plumage. Parakeet Auklet has broad wings, a potbellied body shape, a large red bill, and a shallow fluttering wingbeat. Dovekie is more than twice the size of a Least Auklet.
Voice Highly vocal. A high-pitched *ha-keek*, frequently given at sea when flocks are present. At breeding colonies, a variety of maniacal laughlike chattering and screeching, given when perched and in flight.
Status & Distribution Millions breed in high arctic Greenland, Norway, and Russia. In N.A. a few hundred breed along the Canadian side of the Davis Strait and presumably (very small numbers) in AK near the Bering Strait (Little Diomede I. and St. Lawrence I.). Casual summer visitor to other Bering Sea islands. **BREEDING:** Colonially in crevices on scree slopes on high Arctic islands. **MIGRATION:** Moves away from breeding areas in Sept.; returns by late May–June. **WINTER:** Abundant off Atlantic Canada, especially on Grand Banks, arriving Nov. Occasionally large "wrecks" of hundreds of birds are blown inland (usually late fall). Uncommon and irregular off New England (occasionally in large numbers) to NC, Dec.–Mar. **RARE STATUS:** Casual south to FL, Great Lakes, and bodies of water near coast, exceptionally well inland.
Population In N.A. trend decreasing.

DOVEKIE AND ATLANTIC PUFFIN

CANADA

Lake Superior

Lake
Michigan

Lake
Huron

L. Ontario

L. Erie

Gulf of St. Lawrence

Gulf of Maine

COMMON WINTER RANGE

UNITED STATES

ATLANTIC

OCEAN

MEXICO

Gulf of Mexico

BAHAMAS

TROPIC OF CANCER

0   mi   500
0   km   500

DOVEKIE
--- Wandering area
● Isolated record

ATLANTIC PUFFIN
--- Wandering area
● Isolated record

COMMON MURRE *Uria aalge*   COMU   ■   1

Common Murre is the largest living alcid. Its flight is rapid and direct, usually staying close to the sea surface. Small flocks of Common Murres moving to and from feeding locations often fly in lines or "trains." On land, the Common Murre stands nearly vertically upright, but it rests its weight on the full length of its tarsi. It swims buoyantly, when not foraging, with head erect and tail cocked. Polytypic (5 sspp.; 3 in N.A.). L 17.5" (44 cm)

Identification Long dagger-shaped bill, slender neck (contracted in flight), and distinctive head profile of nearly a straight line from crown through culmen (the angle of the gonys is not prominent). Upperparts are nearly uniform dark sooty gray; underparts are white except for sparse dark streaking on flanks. Trailing edge of inner wing is white. Underwing is white and variably mottled with brown on tips of greater underwing coverts; axillaries are usually heavily marked with brown. Bill, legs, and feet are blackish. Eyes are dark brown. **BREEDING ADULT:** Head and neck are blackish brown. Shape of black-and-white border on neck forms a smooth, U-shaped curve. Bridled-morph birds, which vary in numbers at different Atlantic colonies, have a prominent white spectacle-like facial mark. **WINTER ADULT:** Similar to a breeding bird, except that its throat, sides to neck, sides of nape, and face are white, save for a dark streak extending from behind eye toward nape. First-winter bird has a shorter

bill and mottled, less contrasting, facial plumage.

Geographic Variation Nominate *aalge* is widespread in Atlantic. The two Pacific subspecies (*inornata* and *californica*) are generally larger including wings and bill, and lack the bridled morph. Two additional subspecies (*hyperborea* and *albionis*) occur in the western Atlantic; *albionis* of western Europe is smaller and browner than other subspecies.

Similar Species Common Murre differs strongly from sea ducks, loons, and grebes in its distinctive symmetrical artillery shell-like flight shape, with trailing feet. All plumages of Common Murre appear more brownish gray dorsally (in strong light), whereas Thick-billed Murre and Razorbill have blacker backs. Bill shape (long and pointed, lacking a prominent gonydeal angle in Common Murre) is useful in close views, but first-winter Thick-billed and Common Murres have more similar bills. The presence of obvious flank streaking—visible on swimming birds—is a good indicator of Common Murre; Razorbill and Thick-billed Murre have pure white flanks. In winter, distinguish Common Murre from a Thick-billed by its pale face with white extending onto sides of nape, crossed by a prominent dark line extending from behind eye. Molting and subadult Common Murres may have freckled throats and intermediate head patterns; the best way to separate them from Thick-billed Murres is by assessing structural differences.

Most populations of Common Murre attain breeding plumage earlier (often by late winter) than Thick-billed, and this difference can be helpful in picking out the unusual bird. In breeding plumage, Thick-billed has a prominent white gape stripe, which Common Murre lacks; Thick-billed's neck border of black-and-white is V-shaped, whereas Common's neck border is U-shaped. And lastly, Common Murre has a more upright stance when standing and a thinner head and neck than Thick-billed Murre.

Voice Highly vocal near breeding colonies, giving a variety of harsh grating *argggggh* calls. Not usually vocal at sea and in winter.

Status & Distribution Common on both coasts, almost never seen inland. **BREEDING:** Nests in dense colonies on bare cliff ledges and rocky shelves on islands. Half-grown young leave the nest site before fledging and are tended at sea by the male alone. **MIGRATION:** Moves offshore after departing from breeding colonies Aug.–Sept. **WINTER:** Occurs farther offshore than Thick-billed Murre. **RARE STATUS:** Rare but increasing in the last two decades on Atlantic coast south of Cape Cod, MA, to MD; on Pacific coast recorded rarely as far south as Baja California.

Population Various pressures—gill net fishing, oil spills, food shortages, and introduced predators (e.g., foxes) on breeding islands—continue to severely affect some Common Murre populations on the Pacific coast.

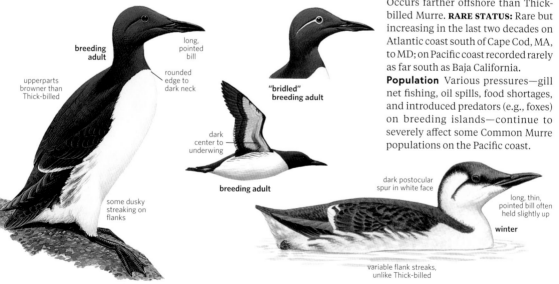

breeding adult

long, pointed bill

upperparts browner than Thick-billed

rounded edge to dark neck

"bridled" breeding adult

dark center to underwing

breeding adult

some dusky streaking on flanks

dark postocular spur in white face

long, thin, pointed bill often held slightly up

winter

variable flank streaks, unlike Thick-billed

THICK-BILLED MURRE *Uria lomvia*   TBMU ▪ 1

breeding adult
arra

distinct white line
on cutting edge of
upper mandible

white forms
inverted V into
blackish neck

breeding
adult
arra

bill slightly
less heavy
than adult

1st
winter
arra

short, thick bill
with curved
culmen

dark
face

black back

faint gape
stripe

winter
adult
arra

white flanks

Thick-billed Murre is a large black-and-white alcid restricted to cold seas. Its flight is rapid and direct. On land, it stands nearly upright but leans forward slightly, usually resting its weight against a cliff face (it breeds on narrow sea cliff ledges). Polytypic (4 sspp.; 2 in N.A.). L 18" (46 cm)

Identification Bill relatively thick with curved culmen and noticeable angle of gonys, thick neck, and head shape with relatively prominent forehead. Upperparts nearly blackish, head and neck very dark brown, underparts immaculate white.

arra

lomvia

Secondaries white tipped. **BREEDING ADULT:** Prominent white stripe along gape. Upward, V-shaped point at center of neck formed by black-and-white border. Underwing white, variably mottled with dark gray on underwing coverts, but overall whiter underwing than on Common Murre. Bill and feet blackish. Eyes dark brown. **WINTER ADULT:** Similar to breeding, except that throat and front of neck white with mottled border of entirely dark face. Gape stripe less prominent, but often still visible. First-winter birds have a shorter bill and mottled, less contrasting facial plumage.

Geographic Variation Pacific subspecies *arra* has a disproportionately longer bill with a less strongly curved culmen than nominate *lomvia* of the north Atlantic.

Similar Species Thick-billed has a distinctive, bulky, thick-necked artillery shell–like flight shape, with trailing feet, that differs strongly from sea ducks, loons, and grebes. It is similar to Common Murre but in winter is distinguished by its darker face, with dark mottling extending well below eye and onto sides of neck, and blacker back. In summer, Thick-billed has a prominent white gape stripe and a V-shaped white point in black center of neck, both lacking in Common Murre.

In flight, Thick-billed is noticeably blacker above than Common Murre. First-winter Thick-billed resembles a juvenile Razorbill, but it lacks that species' pointed tail and more extensive white on face. In flight at a distance, Thick-billed Murre is confusable at all ages and seasons with a Razorbill, which has a deeper bill (usually visible) and much longer, graduated tail (surprisingly hard to see).

Voice Highly vocal near breeding colonies, giving a variety of harsh grating *argggggh* calls. Not usually vocal at sea or in winter.

Status & Distribution BREEDING: Arctic colonies (numbering millions of individuals) south to Newfoundland (uncommon) and BC (rare). Nests colonially on narrow cliff ledges. **MIGRATION:** Moves mainly south after departing colonies Aug.–Sept., returning May–June. **WINTER:** Atlantic birds winter farther south than Pacific birds, occurring regularly to southern New England. **RARE STATUS:** In Atlantic, very rare south to mid-Atlantic, casual to FL and inland on Great Lakes. On Pacific coast, casual south of Canada to central CA (many records from Monterey Bay); once to Southern CA.

Population Both Atlantic and Pacific populations, although still large, have declined; large recent die-offs in AK.

RAZORBILL *Alca torda*   RAZO ▪ 1

Razorbill is a large Atlantic alcid with a massive head accentuated by a thick neck and an extraordinary, deep, laterally compressed bill with curved culmen and blunt tip. Its flight is rapid and direct, usually close to sea surface. On land, it stands nearly upright and

walks like a penguin. Compared to a murre, it breeds in a wider variety of habitats and regularly forages closer to shore in bays and estuaries. It swims with its head and long tail angled up. Polytypic (2 sspp.; nominate in N.A.). L 17" (43 cm)

Identification Long, graduated tail, extending beyond wing tips on swimming bird, unique among alcids. Upperparts jet black; underparts white. Trailing edge of secondaries white. White flanks extend onto sides of rump. Underwing mostly white.

Mouth interior yellow; bill and feet blackish; eyes dark brown. **BREEDING ADULT:** Face and throat black. Narrow white stripe from top of bill base to eye. Prominent vertical white stripe on grooved bill. **WINTER ADULT:** Similar to breeding, except that throat, front of neck, and ear coverts white, bill stripe less prominent, and face darker. First-winter with similar plumage, but all-blackish bill is shorter and more slender.

Similar Species Bulky flight shape distinguishes it strongly from sea ducks. (Beware Long-tailed Duck, which shows extensive white on flanks in flight like Razorbill.) Similar to Thick-billed Murre, but in winter it can be told by its large, blunt-tipped bill and paler face, with white extending well above and behind eye onto sides of nape. Razorbill's facial plumage resembles a Common Murre in win-

ter, but lacks the sharply defined dark postocular spur, and Razorbill always has a much deeper bill and is blacker above. In summer, Thick-billed has a prominent white gape stripe and Common Murre entirely lacks white marks on its bill. First-winter Razorbill has a small bill and resembles juvenile Thick-billed, but it has a pointed tail and more extensive white on its face. In flight at a distance, a Razorbill can be confused with a murre, but its bill is deeper (often hard to see), white extends farther onto sides of rump, more extensive white on underwing, and its long tail covers its feet, unlike the projecting feet of both murres.

Voice Highly vocal near breeding colonies, giving a variety of harsh growling calls.

Status & Distribution Locally common at colonies in Gulf of St. Lawrence, Newfoundland, and Labrador. **BREEDING:** Variety of habitats, including rock crevices and cliff ledges on coastal islands. **WINTER:** Most of N.A. population winters on Grand Banks and in Gulf of Maine. Occurs regularly south to VA. Winter movements unpredictable, and periodic irruptions send birds much farther south than normal (e.g., flock to off Miami Beach in the winter of 2012–13). **RARE STATUS:** Rare south to GA. Casual south to southern and western FL, and inland on Great Lakes; accidental AL, MS, and LA.

Population Near Threatened. Razorbill is the least common Atlantic alcid; populations are recovering.

GREAT AUK *Pinguinus impennis*   GRAU ▪ 6

The extinct Great Auk was the original "penguin"—a massive, flightless, black-and-white Atlantic alcid. It was hunted to extinction by the middle of the 19th century. It had a rapid, wing-propelled underwater flight. On land it was fearless, stood upright, and bred in large colonies on a few low islands. Its weight is estimated at about 10 pounds. Monotypic. L 30" (76 cm)

Identification Black and white. Resembled a giant Razorbill, with similar deep, laterally compressed bill with numerous concentric grooves. Wings tiny. **BREEDING ADULT:** Face and throat black, oval white face patch between eye and bill. Faint white stripes on grooved bill. **WINTER ADULT:** Similar to summer, but white face patch absent.

Status & Distribution Extinct; about 80 extant specimens of birds and eggs. Only three known breeding sites in N.A.: Funk I. (largest colony) and Penguin Is., NL, and Bird Rocks (Rochers aux Oiseaux), Gulf of St. Lawrence, QC. Five colonies in eastern Atlantic (two in Iceland, one in the Faroes, one on St. Kilda, and one on Orkney). Highly colonial, forming tightly packed colonies; a single egg laid on bare rock. **WINTER:** Atlantic Canada south to New England, mostly offshore, casual to SC. Remains have been found in prehistoric middens as far south as FL.

Population Extinct. Great Auk was slaughtered for food, oil, bait, and feathers. The last birds at Funk I. (where an estimated 100,000 pairs bred) were killed in about 1800. The last pair was collected at Eldey Stack, Iceland, on 3 June 1844.

GUILLEMOTS Genus *Cepphus*

Medium-size alcids, guillemots show a conspicuous white wing patch year-round. Thin necks and relatively small bills and heads make them look delicate. They fly low with shallow fluttering wingbeats, rather rounded wings, and a heavy-sterned appearance. They usually forage in shallow waters year-round, but some species move offshore in winter.

BLACK GUILLEMOT   *Cepphus grylle*   BLGU   ■   1

Black Guillemot's distinctive features include white, oval wing patches and gleaming white underwings. Agile on land, it stands upright and walks on tarsi and toes or on toes alone. Polytypic (5 sspp.; 2 in N.A.). L 13" (33 cm) **Identification** Black and white; underwings white. Legs and feet bright red (pink in winter). Eyes dark brown; bill black with mouth interior red. Unmistakable in Atlantic, where it is the only guillemot. **BREEDING ADULT:** Completely sooty black except for a broad, crisply defined white patch on greater and lesser coverts. White underwings, excepting tips of flight feathers and leading edge. **WINTER ADULT:** Very different. Underparts, uppertail coverts, rump, neck, and head white, with sparse dusky streaking and mottling around eye, on crown, and on back of neck. Mantle, scapulars, and uppertail coverts variably mottled. Individuals with patchy, mixed breeding and winter plumage occur late fall and early spring. **JUVENILE:** Similar to winter adult, but more extensively mottled with gray-brown overall; barred wing coverts (retained through first summer); dull-colored feet.

Geographic Variation Subspecies differences are most notable in winter plumage. In N.A., Arctic subspecies *mandtii* is much whiter overall than the East Coast breeding *arcticus*. In *mandtii*, an all-dark (in breeding plumage) color morph has been reported from western Greenland. There are three additional subspecies from Iceland, the Faroe Is., and the Baltic Sea. **Similar Species** In all plumages, Black Guillemot's distinctive upperwing patch separates it from all other alcids except Pigeon Guillemot (in northern Bering Sea, the small area of overlap); however, Black Guillemot lacks the dark bar (often obscured) seen in Pigeon Guillemot's wing patch. Black Guillemot's smaller size and faster wingbeats are useful identification cues to separate it from Pigeon Guillemot, especially from locations (e.g., St. Lawrence I.) where both species occur together. Pigeon Guillemot also appears plump. Black Guillemot's gleaming white underwings and axillaries remain the best field mark. **Voice** Variety of peeping and thin high-pitched screams and whistles, given near breeding colonies. **Status & Distribution** Common and

widespread. **BREEDING:** Small colonies among boulders and in rock crevices on low rocky islands and in coastal cliffs. **MIGRATION:** Most birds move only short distances from breeding areas. **WINTER:** Wherever ice-free water, even in high Arctic. Occurs regularly south to RI. **RARE STATUS:** Casual on Atlantic coast south of New England and in AK south of central Bering Sea. Also casual inland, mainly Great Lakes region; accidental to interior and southeast AK. Interior records all appear to be *mandtii*. **Population** Increasing in N.A.; global population ±1 million individuals.

winter plumages of *mandtii* paler than Pigeon Guillemot

immatures have barred coverts

1st winter

mandtii

winter adult

smaller and sleeker, with quicker wingbeats than Pigeon Guillemot

mandtii

breeding adult

solidly white, oval wing patch

winter adult

white underwing in all plumages

both guillemots have red legs

all guillemots have bright red mouth linings

winter adult and juvenile *arcticus* average darker than comparable plumages of *mandtii*

winter adult

arcticus

juvenile

white wing patch

breeding adult *arcticus*

PIGEON GUILLEMOT *Cepphus columba*   PIGU ■ 1

This Pacific species is similar to Black Guillemot, and like that species, it is often seen from shore. It breeds in small colonies on low rocky islands and in coastal cliffs, sometimes under wharves, and forages in shallow waters year-round. Polytypic (3–5 sspp.; 2–4 in N.A.). L 13.5" (34 cm)

Identification Black; conspicuous white wing patch has a black, wedge-shaped intrusion on its lower (or outer) edge. Some birds, particularly in western AK, have an additional dark wing bar. Underwings uniformly dusky. Bill is black; mouth interior bright red. Legs and feet bright red (pink in winter). **BREEDING ADULT:** Completely sooty black except for the wing patch. **WINTER ADULT:** Whitish head and underparts, with dusky markings on crown and back of neck. **JUVENILE:** Rather extensively mottled with gray-brown overall, including a dark cap; dull grayish pink feet.

Geographic Variation Nominate *columba* (Bering Sea, coast and islands) and *kaiurka* (central and outer Aleutians) have more extensive dark feathering on their wing patches. Some authorities merge *eureka* (CA to OR) and *adiantus* (WA to central Aleutians) with nominate *columba*. Extralimital *snowi* (Kuril Is.) is darkest, sometimes lacking white wing patches.

Similar Species In all plumages distinguished from Black Guillemot by darker underwings, and its oval wing patch always

has a dark bar, which Black Guillemot lacks. In the northern Bering Sea, where it overlaps in distribution with Black Guillemot, its larger size, plumper body, and slower wingbeats might also be apparent. Juvenile can be confused with the smaller Marbled Murrelet.

Voice Variety of thin high-pitched screams and whistles, normally given near breeding colonies.

Status & Distribution Common and widespread near rocky Pacific coastlines. **BREEDING:** Nests in cavities under rocks and driftwood, and in cliff crevices. **MIGRATION:** Mostly resident, but retreats from areas of heavy ice in Bering Sea. **WINTER:** CA and OR populations apparently move north. **RARE STATUS:** Casual to southernmost CA and northwestern Baja California (Islas Coronados).

Population Stable; global population ±230,000 individuals.

winter adult

dark underwing

dark covert bar obvious in flight

breeding adult

juvenile can have pale center to underwing, which causes confusion with Black Guillemot

compare to Marbled Murrelet

dark cap

mottled coverts

juvenile

winter adult

breeding adult

dark covert bar often obscured when swimming

MURRELETS Genus *Brachyramphus*

These murrelets are unique among alcids in having a cryptic summer plumage, which camouflages them at their open solitary nests on tree limbs and mountaintops. In fall they molt into black-and-white plumage. A streamlined body form and long, slender wings give them a very fast, direct flight style. They lack agility on land and cannot stand upright.

LONG-BILLED MURRELET *Brachyramphus perdix*   LBMU ■ 3

This enigmatic Asian species has occurred as a casual visitor in widespread, interior locations across N.A. and along Pacific coast. In 1997, it was split from the very similar Marbled Murrelet. Monotypic. L 11.5" (29 cm)

Identification Underwing coverts grayish. Eyes dark brown; bill, legs, and feet black. **BREEDING ADULT:** Extensively mottled dark grayish brown overall except for whitish throat, with mantle feathers and scapulars thinly edged with buff. **WINTER ADULT:** Black-and-white plumage similar to that of murres; however, conspicuous white scapular patches contrast strongly with otherwise dark upperparts. Blackish cap over most of face to well below eye.

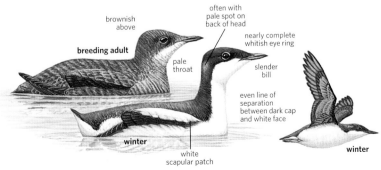

brownish above

breeding adult

often with pale spot on back of head

nearly complete whitish eye ring

pale throat

slender bill

even line of separation between dark cap and white face

winter

white scapular patch

winter

Nape, sides, and back of neck, mantle, rump, and upperwing coverts blackish, except for faint pale oval patch on side of nape. **JUVENILE:** Similar to winter

adult, but dusky brownish barring on breast and flanks.

Similar Species Along Pacific coast, carefully separate from Marbled. Long-billed is 20 percent larger and has a longer and thinner bill. In winter, it has more extensive dark plumage on lores, nape, and back of neck—entirely lacking pale collar of a winter Marbled. It also lacks dark barring on sides of breast that forms a projecting bar in Marbled and has grayish underwings (darker in Marbled). Long-billed's summer plumage is generally less rufous, more grayish brown overall, with a whiter throat. Beware of confusing Long-billeds with guillemots (especially juveniles), which

are larger and have longer bills and white wing patches.

Voice Needs study.

Status & Distribution Asian species. **BREEDING:** Northern Japan, Sea of Okhotsk, and Kamchatka Peninsula; nests in trees. **WINTER:** Normally near Japan. **RARE STATUS:** Casual in N.A. (late summer–early winter), with more than 50 records. Has strayed as far as NL, MA, NJ, PA, NC, SC, and FL. Accidental in Europe (UK and Switzerland). With increased observer awareness, a growing proportion of sightings are from the Pacific coast.

Population Near Threatened; likely declines due to logging.

MARBLED MURRELET *Brachyramphus marmoratus*   MAMU ■ 1

The small, slender-bodied, fast-flying Marbled has a dark, speckled breeding plumage. At sea, it keeps its head tilted up and its tail cocked nearly vertical, and its long-necked profile is accentuated by a shallow sloping forehead. It is frequently seen in "pairs" at sea. Monotypic. L 10" (25 cm)

Identification Bill longish, slender, and pointed. Wings narrow and pointed, seeming to blur with high wingbeat frequency; underwings blackish; tail entirely blackish year-round. Eyes dark brown; bill, legs, and feet black. **BREEDING ADULT:** Extensively mottled overall with dark chocolate brown. Mantle feathers thinly edged with rufous; scapulars fringed with white and rufous. Some birds paler with extensive white spotting—pairs often differing in plumage. **WINTER ADULT:** Black and white, recalling a miniature murre. Blackish cap over most of face to well below eye, except for a white loral spot. Neck and lower sides of nape white, giving a white-collared appearance, accentuated by dark barring on side of breast. Whitish scapulars contrast with dark upperparts. Some birds (presumably immatures) in winter-like plumage during summer. **JUVENILE:** Similar to winter adult, but dusky brownish barring on breast and flanks.

Similar Species Kittlitz's Murrelet in breeding plumage has paler, gray-brown mottling that does not extend onto lower belly, vent, or undertail coverts. In flight, note Kittlitz's contrasting dark wings. In all seasons Kittlitz's shows a pale face with a contrasting dark eye, tiny bill, and white outer tail

feathers, best seen in flight. (Beware of first-summer and transitional Marbleds that show white overlapping tail coverts.) Kittlitz's has a similar black-and-white winter plumage, but a much whiter face.

Voice Easily identifiable, loud penetrating *keer* calls. Noisy at sea year-round, with a complex vocal repertoire.

Status & Distribution Locally common in AK and BC, less numerous south of Canada. **BREEDING:** South of AK in large old-growth stands, sometimes far inland. In coastal southern and southwestern AK on ground on steep mountainsides. Often seen close to land, frequenting fjords, deep bays, saltwater lagoons, and even coastal lakes. **MIGRATION:** Some movement away from breeding areas; no evidence of long-distance migration. **WINTER:** Some birds remain near breeding areas; others more migratory. **RARE STATUS:** Very rare to casual in northern Bering Sea and south to northern Baja California. Apart from regular use of inland lakes near breeding areas, no confirmed records for interior N.A.

Population Endangered. Numbers are rapidly declining in southern part of breeding range due to loss of old-growth nesting habitat. The USFWS (in WA, OR, and CA) and Canada list the species as Threatened.

winter

juvenile

breeding adult

cap dips below eye

white collar

white scapular patch

winter

white lores

breeding adult

mottled brown chin and throat

overall chocolate brown

KITTLITZ'S MURRELET *Brachyramphus brevirostris* KIMU ▪ 2

Kittlitz's appears similar to a short-billed Marbled Murrelet—especially in wing and body shape and behavior—but it has a paler grayish buff breeding plumage and is more extensively white in winter. Often seen in pairs at sea. Monotypic. L 9.5" (24 cm)

Identification Outer tail feathers white year-round, best seen in flight. Underwing coverts dark. Bill very short, almost invisible; eyes dark brown; bill, legs, and feet black. **BREEDING ADULT:** Buffy upperparts and breast extensively mottled with grayish brown; mantle feathers and scapulars thinly edged buff. Face pale, with distinct dark eye. Belly, vent, and undertail coverts white. **WINTER ADULT:** Black and white. Blackish cap restricted to crown and forehead only. Face, neck, and nape white, giving a white-headed appearance. Blackish barring across breast, forming a nearly complete band. Back blackish, slaty gray appearance in good light, with contrasting whitish scapulars. Some birds (presumably immatures) in winter-like plumage during summer, others intermediate. **JUVENILE:** Similar to winter adult, but upperparts grayer, with dusky barring on breast and flanks for months after fledging.

Similar Species From Marbled, shorter billed and paler faced with contrasting dark eye; in breeding plumage, Kittlitz's has paler gray-brown mottling that extends only down to upper belly. White outer tail feathers are hard to see—beware of white overlapping upper and undertail coverts on Marbled. Dark underwings contrast more with pale underparts on Kittlitz's than on Marbled.

Voice Typical call is a nasal grunt. Apparently less vocal than Marbled Murrelet.

Status & Distribution Uncommon to fairly common, but sharply declining in N.A. **BREEDING:** Solitary ground-nester on high mountainsides among glaciers and snowbeds in southern part of range; frequents fjords, deep bays, and saltwater lagoons. **MIGRATION:** Recent studies indicate northward movements into the Bering Sea. **WINTER:** Poorly known; many birds apparently winter in Bering Sea near pack ice. **RARE STATUS:** Accidental to southwestern BC and Southern CA. The latter record, a juvenile, was picked up alive on a beach at La Jolla, San Diego Co., on 16 Aug. 1969. The specimen formed the basis for the description of the previously undescribed juvenal plumage.

Population Near Threatened. Kittlitz's numbers are rapidly declining in southeast and south-central AK, possibly due to climate change and competition with hatchery-raised salmonid populations. It is likely the bird species most seriously affected by the massive *Exxon Valdez* oil spill in 1989; many birds also drown in salmon gill nets.

winter
breeding adult
white outer tail feathers
white belly
underwing contrasts darker to body, more so than breeding adult Marbled

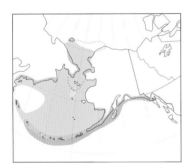

juvenile
extensive white face
short, stubby bill
breeding adult
winter
color overall paler than Marbled

MURRELETS Genus *Synthliboramphus*

Five species in the genus: three warm-water species off CA and Mexico, one widespread in the North Pacific, and one restricted to waters around central Japan. They lay a clutch of two eggs and take their precocial chicks to sea two days after hatching. "Xantus's Murrelet" was split into two species in 2012: Scripps's and Guadalupe.

SCRIPPS'S MURRELET *Synthliboramphus scrippsi* SCMU ▪ 2

Scripps's Murrelet spends most of its life at sea. When flushed, Scripps's, like its relatives, Craveri's and Guadalupe Murrelets, lifts directly off water, sometimes resettling a short distance away. Also like other *Synthliboramphus* murrelets, Scripps's lays just two eggs and takes its young to sea when they are two days old. The young are fed and reared far from land. At sea, all of these species are encountered singly or in pairs, often with flightless young chicks that dive to avoid boats. Monotypic. L 9.8" (25 cm)

identification A remarkably small seabird, resembling a tiny murre in plumage. Underparts bright white, including sides of breast. Upperparts slaty black, with very thin white crescents above and below eye; auricular often slightly paler and shows as a blended whitish splotch, diagnostic from Craveri's. Bill slim, short, and black. Legs and feet gray, with black webs. Iris dark brown. **JUVENILE:** Plumage like adult; chicks at sea have adults in attendance. **FLIGHT:** Wings narrow and pointed; flies with rapid wingbeats. Underwing coverts brilliant white; bases of remiges very pale, creating impression of a mostly white underwing.

Similar Species Craveri's Murrelet is very similar, but underwing appears mostly dark: Its underwing coverts are sooty, remiges blackish. Sometimes, when resting on water, *Synthliboramphus* murrelets raise their body vertically and flap their wings, revealing the underwings. Also note that black of face extends to lower mandible in Craveri's, whereas in Scripps's black extends to a point above the gape. Craveri's also has a spur of dark plumage on side of breast and a longer bill than Scripps's. Craveri's often cocks its tail up, unlike Scripps's and Guadalupe. In winter plumage, Marbled and Long-billed Murrelets both show conspicuous white patches on scapulars. See Guadalupe Murrelet.

Voice At breeding colonies, many birds vocalize simultaneously, giving high-pitched, rolling twitters. Very vocal at sea, unlike most alcids. **CALL:** A repeated piping whistle; also high-pitched chips.

Status & Distribution Uncommon. Never seen in large numbers except when gathering near colonies during breeding season; far more likely to be seen in nearshore waters than Guadalupe Murrelet. **BREEDING:** Colonially on arid islands in rock crevices and under dense shrubs; breeding begins in Feb. and concludes by Aug. **DISPERSAL:** Postbreeding dispersal apparently mostly to the north in summer and fall, with largest numbers recorded over outer continental shelf waters west of central CA, smaller numbers from northern CA to WA. Northernmost records are from Queen Charlotte Sound, north of Vancouver I., BC.

Population Vulnerable. Populations in CA have declined historically, mainly due to introduced predators (cats, rats, mice) on breeding islands; the species is listed as threatened there and is a can-

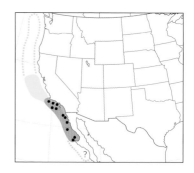

didate for the federal list as well. The Channel Is. are the species' stronghold in the US, with perhaps 2,000–4,000 breeding individuals, and most of the rest of the population is found off Baja California, Mexico, on Islas Los Coronados and Islas San Benito.

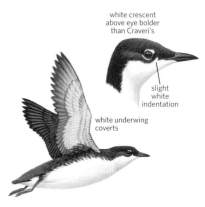

white crescent above eye bolder than Craveri's

slight white indentation

white underwing coverts

chick

chick accompanies adult to sea

GUADALUPE MURRELET *Synthliboramphus hypoleucus* GUMU ■ 3

Guadalupe Murrelet is an endemic Mexican species, nesting almost entirely on a few predator-free islets off Isla Guadalupe, about 150 mi from the Baja California coast. Guadalupe Murrelet was formerly combined with Scripps's Murrelet as a single species, "Xantus's Murrelet." Recent studies of genetics combined with different vocalizations led to their split into separate species. Both breed on Islas San Benito. Monotypic. L 9.8" (25 cm)

Identification Plumage almost identical to Scripps's Murrelet, its closest relative, but facial pattern quite different. Rather than having an even interface of black and white plumage from the bill to nape, as in Scripps's, the white in face of Guadalupe arcs both above the eye and well into auriculars. Below the eye, Guadalupe has a large amount of white that is very narrowly bordered in black, visible at close range. The white-cheeked impression of Guadalupe is discernible at several hundred

yards. A few individuals that appear intermediate in facial pattern between Guadalupe and Scripps's Murrelets occur, but their significance is unclear.

Voice Very unlike Scripps's; a cricket-like rattle, more like Craveri's, and also frequently heard at sea.

Status & Distribution Uncommon to rare away from colonies. **BREEDING:** Colonially on arid islands in rock crevices; breeding begins in Feb. and concludes by Aug. **DISPERSAL:** Postbreeding dispersal both south and north, and usually found well offshore. **RARE STATUS:** Casual to OR and WA. Two seen 17 mi west of the tip of Moresby I., Haida Gwaii, BC (2 Aug. 1994), represent the only record for Canada. A single bird found nesting at Santa Barbara I., CA, is the only confirmed US breeding record; birds netted in breeding season at night on San Clemente I. suggest breeding there.

winter range unknown

extensive white around eye is best field mark for Guadalupe

bill slightly longer than Scripps's

white underwing coverts like Scripps's

Population Endangered. Removal of cats and rodents from Isla Guadalupe, Mexico, would benefit the species immensely. Its very small breeding range makes Guadalupe Murrelet vulnerable to a variety of dangers, including light disorientation (and thus collision with vessels) and entanglement in nets.

CRAVERI'S MURRELET *Synthliboramphus craveri*   CRMU ■ 3

Craveri's Murrelet occurs as a rare postbreeding visitor to CA, where it is more regularly observed in warm-water years. Good looks are necessary to separate Craveri's from the very similar Scripps's and somewhat similar Guadalupe. In CA, it is normally seen singly or in twos at sea. Monotypic. L 8.5" (22 cm)
Identification Black, with slight brownish cast above and white below. Black cap extends on face below gape onto throat immediately underneath bill. Auricular area, nape, and back and sides of neck are black, extending onto sides of lower breast and giving a partial collared appearance. There are faint white crescents above and below eyes. Underparts, except sides of breast, are immaculate white. Underwing coverts and underside of flight feathers are dark brownish gray. Eyes are dark brown. Slender, pointed bill is black; legs and feet gray (with black webs). Wings are narrow and pointed; it flies with fast wingbeats. No seasonal variation. **JUVENILE:** Tiny, downy black-and-white chicks taken to sea by adults at two days of age. Fully grown juveniles resemble adults.

Similar Species Very similar to Scripps's. Craveri's has a more extensive black cap, black on sides of breast forming a partial collar, blackish to grayish underwings (sometimes with a whitish center, but never as white as the underwings of Scripps's or Guadalupe), and a relatively longer and thinner bill—all difficult to see. It is best identified at sea if it raises its wings. Potentially helpful is Craveri's rather long tail that is often cocked up to a point, unlike Scripps's and Guadalupe. Guadalupe has white extending up over eye, lacking in both Craveri's and Scripps's. All black-and-white murrelets off the West Coast require close scrutiny to confirm identification, especially in summer and fall. Marbled is differently patterned black-and-white in winter, with conspicuous white scapular patches; Ancient has a short pale bill, lacks white scapulars, and shows more extensive blue-gray upperparts.
Voice Vocal at sea. Typical call is a cicada-like rattle, rising to reedy trilling when agitated; also high-pitched chips.
Status & Distribution Rare to uncommon. **BREEDING:** Nests colonially on arid islands off both sides of Baja California. **DISPERSAL:** Some movement north after breeding season, thus

present irregularly in late summer (recently by mid-June off San Diego) and fall off southern and central CA, especially in El Niño years.
Population Vulnerable. Introduced predators (cats, rats) threaten Mexican breeding colonies.

often cocks tail when swimming, unlike Scripps's and Guadalupe

faint eye crescents

on shorter-billed, juvenile, eye crescents slightly bolder, dark does not extend as far under eye and may not extend under bill

adult

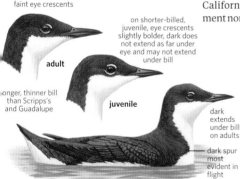

longer, thinner bill than Scripps's and Guadalupe

juvenile

dark extends under bill on adults

dark spur most evident in flight

underwing linings show extensive dark, unlike Scripps's and Guadalupe, but pattern variable; adults may have darker underwing

juvenile

adult

ANCIENT MURRELET *Synthliboramphus antiquus*   ANMU ■ 2

Ancient Murrelet is a strikingly marked, active, social murrelet with strictly nocturnal activity ashore at breeding colonies. Like other *Synthliboramphus* murrelets, it has narrow, pointed wings. In flight, it holds its head above the plane of its body during taking off and maneuvering. It flies with fast wingbeats. It normally stays in small groups offshore, sometimes coming close to land in bays and harbors. On water, it often appears neckless and flat-crowned. Monotypic. L 10" (25 cm)
Identification Ornate plumage. Entire back, rump, uppertail coverts, and upperwing coverts are blue-gray, contrasting with jet-black flanks and hood and immaculate white underparts. Distinctive short, laterally compressed bill has a black base and yellowish pink tip. Eyes are dark brown; legs and feet gray (with black webs). **BREEDING ADULT:** Black hood extends across upper breast, forming a distinctive bib that contrasts with white sides of neck. Variable, silvery white plumes

edge crown and nape (hence the name Ancient), with a second narrow band of similar plumes crossing upper back from sides of breast. Underwing coverts are white. **WINTER ADULT:** White plumes on crown and upper back are reduced, and black bib is smaller, less distinct with light flecking or barring. **JUVENILE:** Tiny, gray-and-white downy chicks are taken to sea by adults at two days of age. Fully grown juveniles resemble winter adults, but bib is less distinct.

Similar Species Ancient Murrelet is easily identified by its size, shape, and coloration. Marbled and Kittlitz's Murrelets have different bill shapes; show distinct black-and-white appearance in winter; have dark underwings and white scapular patches; and lack the black bib. Cassin's Auklet is evenly

colored above and lacks a black cap, but has extensive white on underparts that lead to frequent confusion.

Voice Vocal at sea, typically a loud *chirrup*. Varied repertoire of chattering, chips, and harsh calls given nocturnally at breeding colony.

Status & Distribution Common, especially when gathering near breeding colonies and at favored wintering areas. **BREEDING:** Colonially on predator-free islands in earth burrows (rarely rock crevices) on forested (southeast AK, BC, and perhaps northwest WA) and grassy (Aleutian Is.) slopes. **MIGRATION:** Some movement south after breeding season. **WINTER:** Regular from WA to northern CA; very rare and irregular to Southern CA. **RARE STATUS:** Casual inland on large lakes in late fall and winter, east to Gulf of Maine, VA, and as far south as LA, but most records are from the northern tier of states, many centered around the Great Lakes. Accidental off northwestern Baja California. Accidental UK.

Population Numbers are declining drastically due to introduced preda-

tors (rats, raccoons, foxes) on breeding islands. The population is recovering rapidly in the Aleutian Is. owing to the removal of foxes. Raccoons and rats seriously threaten the population on Haida Gwaii, BC. Canada lists it as a species of Special Concern.

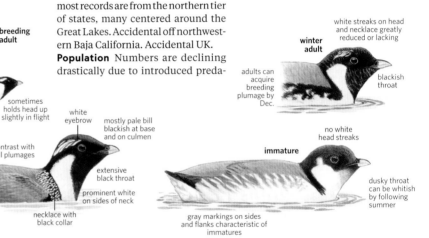

white underwing coverts

breeding adult

black stripe on sides and flanks

sometimes holds head up slightly in flight

white eyebrow

mostly pale bill blackish at base and on culmen

gray upperparts contrast with mostly black head in all plumages

extensive black throat

prominent white on sides of neck

breeding adult

necklace with black collar

white streaks on head and necklace greatly reduced or lacking

winter adult

blackish throat

adults can acquire breeding plumage by Dec.

no white head streaks

immature

dusky throat can be whitish by following summer

gray markings on sides and flanks characteristic of immatures

AUKLETS Genera *Ptychoramphus* and *Aethia*

The five small alcids in these genera have chunky bodies and short wings; they are not agile on land. Most species are restricted to the Bering Sea and attend colonies (some with millions of birds) in daylight; Cassin's Auklet is more widespread and nocturnal. All species are planktivorous and raise a single chick to full-size in a crevice or burrow.

CASSIN'S AUKLET *Ptychoramphus aleuticus* CAAU ▪ 1

Cassin's Auklet is a small, short-necked auklet with a dull plumage that remains fairly uniform at all ages and seasons. Its flight appears weak; it usually takes off from sea with difficulty at the approach of a vessel. At sea it is often seen in small to moderate aggregations year-round. Strictly nocturnal at breeding colonies. Monotypic. L 9" (23 cm)

Identification Overall brownish gray. Face and crown slightly darker than rest of upperparts. Small white crescents above (larger) and sometimes below (smaller) eye. Black bill

larger relative to other auklets' bills; triangular in profile with a straight or even slightly concave culmen, a sharp-pointed tip, and a broad base, with a pale base to lower mandible. White eyes; blue-gray feet with blackish webs. **JUVENILE:** Like adult, but brown or brownish gray eyes. **FLIGHT:** White undertail coverts and extensive whitish belly, and pale-centered underwing.

Similar Species Cassin's pale lower mandible, white iris, and white eye crescents are visible only at close range.

Ancient differs in shape (relatively long pointed wings and slender body) and coloration; Rhinoceros Auklet is much larger and has a larger, usually light-colored bill. Near the Aleutians, Cassin's may be seen with other similar auklets. Whiskered is smaller and darker (nearly black); its red bill is much shorter. Crested Auklet is larger and evenly dark colored; has longer, more pointed wings and a blunt, orange bill. Larger Parakeet Auklet has a different bill shape and color; in winter, underparts are white contrasting with dark underwings.

Voice Not vocal at sea, but sometimes makes a grating *krrrk* when alarmed.

Status & Distribution Common, the most widespread auklet. **BREEDING:** From the Aleutian Is., AK, to central Baja California. Nests colonially on predator-free islands and isolated coastal cliffs in earth burrows (also rock crevices). **DISPERSAL:** Moves offshore during nonbreeding season. **WINTER:** Mostly offshore, near breeding areas with some southward movement. **RARE STATUS:** No inland records. **Population** Near Threatened. Once decimated by introduced predators; now recovering in the Aleutians since removal of nonnative foxes. Rats and raccoons are still threats in BC.

wings appear rounder than murrelets

told from larger winter-plumaged Parakeet Auklet by pale underwing, especially in middle

side-to-side twisting flight distinctive from comparably sized murrelets and much larger Rhinoceros Auklet

overall gray and chunky, with whitish belly

adults

white crescent above pale eye visible at close range

pale base to lower mandible

PARAKEET AUKLET *Aethia psittacula*   PAAU ■ 2

One of the larger auklets, Parakeet Auklet has a chunky body shape, long neck, rounded wings, and relatively large feet. It breeds in small colonies. At sea it is normally seen singly or in small groups. Its wings appear broad and rounded compared to other auklets; its flight is strong and direct. Monotypic. L 10" (25 cm)

Identification Blackish above and white below. Bill bright red, with oval upper mandible and recurved lower mandible. White eyes. Underwing dark gray. Large bluish gray feet with a greenish tinge and black webbing. **BREEDING ADULT:** Upperparts blackish gray, with blackish barring variably extending onto throat, neck, and upper breast; appears dark overall on

sea. Slender white plumes extend from behind eye. **WINTER ADULT:** Underparts uniformly white, extending onto center of breast and throat. Bill dull-colored, white facial plumes reduced. **JUVENILE:** Similar to winter adult, but blackish bill and grayish blue eyes.

Similar Species Unmistakable when observed ashore; facial ornaments and bill shape unique. Parakeet Auklet could be mistaken for Rhinoceros Auklet, which has a longer, pointed bill, a less chunky body shape, and more blended (not black-and-white) plumage. It is closest in size to a Crested Auklet, but easily distinguished by its white underparts and rounded wings. In flight, note size, bill shape, rather slow, fluttery wingbeats, and broader wings. Much smaller Dovekie is sharply black-and-white. See Cassin's Auklet.

Voice Highly vocal. At colonies, birds give repetitive high-pitched whinnying calls.

Status & Distribution Locally common on Bering Sea and south-central AK islands. **BREEDING:** Colonially in

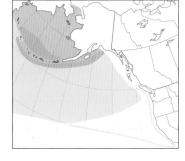

rock crevices, often along cliff-tops. **MIGRATION:** Moves south to avoid winter ice in Bering Sea, dispersing farther offshore than other auklets. **WINTER:** Most birds well offshore in North Pacific, irregularly to latitude of central CA. **RARE STATUS:** Casual in winter near coast from BC to CA. Accidental in HI and Europe (Sweden). **Population** Least Concern.

breeding adults

unique bill shape

orange-red bill

dark

juvenile

winter adult

all-dark underwings unlike Cassin's Auklet

dark throat and breast

white underparts

breeding adult

white belly

darker bill

white postocular line

winter adult

LEAST AUKLET *Aethia pusilla*   LEAU ■ 2

stubby dark red bill

breeding adults

light

white throat

dark

breeding adults

extensive whitish on underwing

winter adults

white underparts

juvenile

faint whitish scapular line

Least Auklet is a tiny, chubby alcid. Gregarious, active, and noisy, it breeds in a few large colonies. Flies in a characteristic side-to-side weaving style. On land, it stands erect and is very agile. Monotypic. L 6.3" (16 cm)

Identification Polymorphic. Short, rounded wings. Upperparts blackish with white patches on scapulars. Underwing coverts pale. Bill small, nearly black to bright red with a black base, usually with a straw-colored tip; small knob-shaped bill ornament on adult. White eyes. Bluish gray feet with black webbing. **BREEDING ADULT:** Underparts variably marked with irregular dark spotting, from almost black to nearly unmarked white; throat white with small black chin; scapulars variably white. Short, evenly scattered, white plumes on forehead, extending onto crown and lores and from behind eye. **WINTER ADULT:** Underparts uniformly white; no ornaments (bill black,

no bill knob, only a trace of white facial plumes); and white scapular patches more prominent. **JUVENILE:** Similar to winter adult, but eyes grayish and scapulars less white. **IMMATURE:** Brown forehead with sparse plumes. Dull-colored bill and smaller bill knob than breeding adult, bluish gray eyes.

Similar Species Tiny size easily eliminates all other alcids, except other auklets and Dovekie. Least Auklet usually shows white scapulars, which are lacking in all other auklets and Dovekie. It is closest in size to Whiskered Auklet, which is blacker, slightly larger, has somewhat longer wings, and shows pale feathering only on vent (only the darkest Leasts approach Whiskered's dark coloration). Parakeet Auklet is much larger and has slower wingbeats. Dovekie is slightly larger and has more contrasting black-and-white plumage.

Voice Highly vocal. At sea, birds taking flight give short high-pitched squeaks. At colonies, birds give a variety of high-pitched, high-frequency chattering.

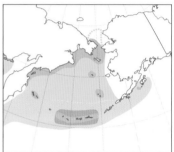

Status & Distribution Locally abundant in Bering Sea. **BREEDING:** Nests colonially in rock crevices on boulder beaches, talus slopes, boulder fields, lava flows, and cliffs. **MIGRATION:** Moves south to avoid winter ice in Bering Sea. **WINTER:** Poorly known, probably well offshore. **RARE STATUS:** Casual inland in western AK in late fall and winter after storms. Accidental in Arctic Canada and on West Coast south of AK (BC, once from central CA).

Population Introduced rats at Kiska I. threaten its largest colony.

WHISKERED AUKLET *Aethia pygmaea*   WHAU ■ 2

Whiskered Auklet is perhaps the most ornamented, living seabird while in breeding plumage. It is most often seen at sea, where it flocks together in tidal rips near colonies. It breeds at

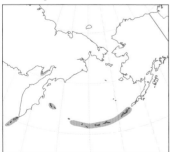

many small colonies throughout the Aleutian Is., where activity is mainly nocturnal. Monotypic (Asian sspp. *camtschatica* is sometimes recognized). L 7" (18 cm)

Identification Uniformly dark appearance on water. Upperparts blue-black; underparts blackish gray, except for light gray or whitish lower belly and vent. Underwing coverts dark gray. White eyes; bluish gray legs and toes, with black webbing. **BREEDING ADULT:** A slender, highly variable, forward-curving black forehead crest, and white facial plumes including a showy V-shaped, antenna-like ornament originating on lores, and long, white auricu-

lar plumes—all difficult to observe at sea. Bright red bill with a straw-colored tip. White vent is visible when bird takes flight. **WINTER ADULT:** Like breeding but ornaments much reduced and bill is dull red. **JUVENILE:** Similar to winter adult, but feather ornaments absent, bluish gray eyes, and blackish bill.

Similar Species Like the Crested Auklet, Whiskered Auklet is the only other small alcid that is similarly dark in overall coloration, but it is smaller with a smaller bill and a whitish vent. Least and Cassin's Auklets have whiter underparts and dumpier body shapes. Least also has white on scapulars.

Voice Highly vocal. At sea and at

breeding colonies, gives a thin kitten-like *mew*. At breeding colony, gives a variety of high-pitched mewing and trilling calls.

Status & Distribution Locally common on Aleutian Is. only; also locally Russian Far East. **BREEDING:** Colonially in rock crevices on coastal cliffs, beach boulders, and talus slopes on predator-free islands. **DISPERSAL:** Resident near breeding islands. **WINTER:** Mostly in large flocks in Aleutian inter-island passes. **RARE STATUS:** In AK, casual outside of breeding range. These records include several on the Pribilofs, St. Lawrence I. (Tatik Pt. and a specimen record from Gambell, 9 July 1931), and off Anak I., Bristol Bay.

Population Once decimated by introductions of exotic mammalian predators onto Aleutian Is.; now recovering since the removal of foxes from many islands.

CRESTED AUKLET *Aethia cristatella* CRAU ■ 2

The extremely gregarious Crested Auklet occurs in large flocks year-round; rarely seen alone. It breeds in a few large colonies on Bering Sea islands. Like Whiskered Auklet, it has strong, direct flight, is agile on land, and stands erect. It has a distinctive citrus-like plumage odor. Monotypic. L 9" (23 cm)

Identification Uniformly dark. White eyes; legs and toes bluish gray with black webbing. **BREEDING ADULT:** Spectacularly ornamented. Bright orange bill with peculiar, curved gape plates, white auricular plumes, and forward-curving, black forehead crest. Larger, more strongly hooked bills on males, thus sexes are distinguishable in the field, unlike other alcids. **WINTER ADULT:** Bill plates shed during breeding season, leaving a small dull orange bill by late summer. Black crest and white plumes reduced. **JUVENILE:** Similar to winter adult, but blackish bill, tiny crest, and bluish gray eyes.

Similar Species Closely resembles the smaller Whiskered Auklet, which shows pale feathering on belly and vent, as does Cassin's Auklet. Crested is closest in size to Parakeet Auklet, which has white underparts and less pointed wings. Often with much smaller Least Auklets that show white on scapulars and underparts. Compare first-summer birds, with reduced bill and feather ornaments, to Whiskered Auklet.

Voice Highly vocal. At sea, gives a sharp bark like the yap of a small dog. At breeding colony, gives a variety of trumpeting, cackling, hooting calls.

Status & Distribution Locally abundant in Bering Sea. **BREEDING:** Colonially in rock crevices and caves on talus slopes, lava flows, and cliffs on volcanic islands. **MIGRATION:** Like Least and Parakeet Auklets, shows some postbreeding movement into the Chukchi Sea; some reach Utqiagvik (Barrow), AK. Moves south to avoid heavy ice in Bering Sea. **WINTER:** Mostly in a few, food-rich, Aleutian inter-island passes. **RARE STATUS:** Rare in Gulf of Alaska east of Kodiak I.; accidental inland in western AK and along West Coast as far as northwestern Baja California, Mexico. Accidental to Iceland. **Population** Least Concern.

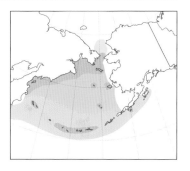

RHINOCEROS AUKLET AND PUFFINS Genera *Cerorhinca* and *Fratercula*

Puffins (three species) are medium-size alcids with big heads and thick, colored bills; elaborate facial ornamentation; and short, rounded wings. Rhinoceros Auklet has attributes intermediate between puffins and auklets. All are colonial and carry multiple fish externally in their bill to a single chick, which is raised to full-size in a burrow or rock crevice.

RHINOCEROS AUKLET *Cerorhinca monocerata*   RHAU ▪ 1

Although related to puffins, this alcid's body shape, different facial ornaments, and posture give it a more auklet-like appearance. Seen in moderate numbers near shore; singly or in small loose groups at sea. Nocturnal or crepuscular at colonies. Monotypic. L 15" (38 cm)
Identification Upperparts grayish brown. Underparts similar, except for whitish belly and undertail coverts. Yellowish eyes; dull yellowish legs and feet. **BREEDING ADULT:** Two sets of prominent white facial plumes, one extending from gape, other from above eye. Thick, dagger-shaped, orange bill, with decurved culmen, and single, whitish "rhino" horn at base. **WINTER ADULT:** Very similar, but facial plumes much reduced. Bill horn shed in fall, leaving birds with smaller, dull orange bill with blackish base. **JUVENILE:** Similar to winter adult, but no head plumes, darker eyes, and smaller, shallower bill, blackish initially, gradually changing to dull yellowish orange; much of plum-

age held into the following summer.
Similar Species Smaller head and more symmetrical body than a puffin. Larger Tufted Puffin lacks white underparts in all seasons, but compare with paler morph of a juvenile Tufted, which can show a pale belly. Horned Puffin always has pure white underparts. In flight, Rhinoceros Auklet resembles a grayish and white murre, but its head and bill are disproportionately larger. Cassin's is smaller and paler with a dumpier body shape, has a more wobbly flight style, and is less approachable. Parakeet Auklet is smaller and more pure white below, especially in winter. Also note markedly different bill shape.
Voice At and near breeding colony at night, gives sonorous *arr-aarrrgh* calls.
Status & Distribution Common and widespread along West Coast. **BREEDING:** Colonially in earth burrows on grassy and forested slopes of predator-free offshore islands. Center of breeding abundance is BC; also breeds in

Japan, Korea, and Russian Far East. **MIGRATION:** Moves south along West Coast Sept.–Oct.; returns to breeding colonies Mar.–Apr. **WINTER:** Along entire West Coast, regularly south to central Baja California. A few oversummer south of breeding range. **RARE STATUS:** Rare in summer north to Pribilofs; casual to northern Bering Sea.
Population Least Concern.

white plumes on face · "horn" · bright yellow-orange bill · breeding adult · immature · smaller yellowish bill · winter adult · plumes reduced and no horn · whitish belly · winter adult

ATLANTIC PUFFIN *Fratercula arctica*   ATPU ▪ 1

This medium-size, large-headed Atlantic alcid has a highly ornamented face and bill. Its flight is direct, and it is agile on land. It stands erect and on its toes, rather than on tarsi like most large auks. Its short neck and heavy, rounded body give it an overall chubby appearance. Polytypic (3 sspp.; nominate in N.A.). L 12.5" (32 cm)
Identification Upperparts blackish with lighter-colored facial disc, black neck collar; underparts immaculate white. Deep, laterally compressed bill with multiple concentric grooves (one to five, increasing with age). **BREEDING ADULT:** Colorful bill with dark bluish

gray crescent-shaped base outlined with pale yellow stripes, remainder bright reddish orange. Pale orange rictal rosettes at gape. Dark eye surrounded by thin red orbital ring, within a clownlike triangular dark gray wattle. Bright reddish orange legs and feet. White facial disc with silvery gray clouding; upperparts silky black fading to brownish black during breeding season. Dark face and winter-type bill on some (first-summer?) birds. Second- and third-summer birds have less brightly colored, more triangle-shaped bills with fewer grooves, and duller orange feet, compared to adults.

colorful tricolored bill · pale gray face · dark collar · breeding adult *arctica* · red-orange legs

Leucistic birds occur. **WINTER ADULT:** Very similar to breeding plumage, but bill plates, rosette, and eye ornaments shed in fall, leaving birds with smaller, dull reddish bill with constricted base and blackish face. Feet yellowish or orange-brown. **JUVENILE:** Similar to winter adult, but bill black, gradually changing to dull reddish, smaller, much shallower and daggerlike.

Geographic Variation Northward cline of increasing body size.

Similar Species Unmistakable in summer. In winter, its dirty blackish face, blackish underwings, short, triangular bill, and front-heavy flight shape, resulting from its large head and short neck, distinguish it from murres and guillemots. Superficially, it is similar to much smaller Dovekie, which has a weaving flight style and a very short, stubby bill.

Voice At breeding colony, gives a variety of low sonorous growling calls, mostly from within burrows.

Status & Distribution In N.A., locally abundant in Atlantic Canada and at a few breeding islands off central and northern ME. **BREEDING:** Colonially in burrow networks on grassy slopes of offshore islands. **MIGRATION:** Moves south out of Arctic Sept.–Oct.; returns to breeding areas in late Apr. **WINTER:** Most winter well off Atlantic Canada; some are regular south to VA, rarely NC (some remaining into May). **RARE STATUS:** Casual south to northern FL and inland on Great Lakes. Accidental to Talan I., Russian Far East.

Population Vulnerable.

orange legs and feet

breeding adult

dark underwing

winter adult

darker face

smaller bill

juvenile

HORNED PUFFIN *Fratercula corniculata* HOPU 1

Horned Puffin is a sister species to Atlantic Puffin, only it is larger, with an even more exaggerated bill and facial ornaments. Its flight is direct, typically high (50–150 ft) above sea surface, and its feet are prominent. Agile on land, it stands erect and on its toes, walking with a forward-hunched posture. Its short neck and heavy, rounded body give it an overall stocky appearance. Monotypic. L 15" (38 cm)

Identification Upperparts blackish; underparts immaculate white. Bill very large and deep, laterally compressed. **BREEDING ADULT:** Bill mostly bright lemon yellow with bright reddish orange tip. Pale orange rosettes at gape. Dark eye surrounded by thin red eye ring, within clownlike black wattle that includes a "horn" extending vertically from top of eye. Gleaming white facial discs; upperparts satiny black fading to brownish black during breeding season. Dark faces and winter-type bills on a few birds at colonies. Bright reddish orange legs and feet. Second- and third-summer birds have shallower, triangular bills, and duller orange feet, compared to older birds. **WINTER ADULT:** Bill plates, rosette, and eye ornaments shed in fall, leaving bill with smaller dull, greenish brown base, reddish tip, and grotesquely constricted base. Plumage relatively unchanged, but face becomes grayer, sooty black on lores. Yellowish or orange-brown legs and feet. **JUVENILE:** Similar to winter adult, but bill black, gradually changing to dull reddish, smaller, much shallower, and more daggerlike.

Similar Species Unmistakable in summer. In winter, its large head and short neck produce a front-heavy flight shape, which together with its dirty blackish face, blackish underwings, and triangular bill, identify it

"horn"

white face

breeding adult

much smaller and darker bill

white below

darker bill than breeding

breeding adult

immature

juveniles and winter adults darker faced

winter adult

as a puffin. Larger Tufted Puffin lacks pure white underparts in all seasons.

Voice At breeding colony, gives low sonorous growling calls, mostly from within breeding crevices.

Status & Distribution Abundant and widespread in western AK; locally common in southeastern and south-central AK; rare, perhaps extirpated, breeder in BC. **BREEDING:** Colonially in rock crevices (less commonly in earth burrows) on cliffs and rocky slopes of remote islands. **MIGRATION:** Moves out of northern Bering Sea Sept.–Oct., disperses widely offshore into North Pacific; returns May–June. **WINTER:** Almost entirely far offshore in North Pacific, irregularly south to latitude of CA. **RARE STATUS:** Casual, mainly in late spring, along West Coast to Southern CA (a few times in moderate numbers); very few have been seen in flight. Rare in summer east to Utqiagvik (Barrow), AK, where breeding has been attempted. Casual HI.

Population Apparent declines and some retraction at southern end of breeding range.

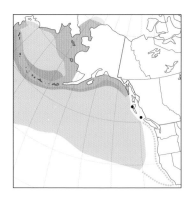

TUFTED PUFFIN *Fratercula cirrhata*   TUPU ■ 1

white face with long pale yellow head tufts

breeding adult

massive bill

immature

black underparts

breeding adult

adults with all-black underparts

immature

paler belly but variable

smaller yellowish orange bill

much darker faced in winter, loses tufts

thick red bill with black base

winter adult

The large, dark Tufted Puffin has a huge bill and spectacular golden-blond head tufts in summer. It takes flight from sea with difficulty, sometimes paddling across surface with wings, feet trailing; once airborne, its flight is direct, typically high (50–150 ft) above sea surface, and its feet are very prominent. It has a large head, short neck, and heavy body. Social on land, it breeds in large colonies. It stands erect on its toes and is quite agile on land, walking with a forward-hunched posture. Monotypic. L 16" (41 cm)

Identification Entirely sooty brownish black except for pale face. Deep, laterally compressed bill, with deep grooves (two to four, increasing with age). Yellowish white eyes; red eye ring.

BREEDING ADULT: Colorful bill, basal half olive green, remainder bright reddish orange, and orange gape rosettes. Face gleaming white, with long, blond, tuftlike plumes that curve backward. Plumage fades to brownish black during season. Second- and third-summer birds have shallower bills with fewer grooves, and duller orange feet. **WINTER ADULT:** Bill plates, rosette, and eye ornaments shed in fall, leaving bill with smaller blackish base, reddish tip, and grotesquely constricted base. Plumage similar to breeding, but head completely blackish, except for hint of pale buff behind eyes and along rear edges of crown. **JUVENILE:** Similar to winter adult, but underparts not as dark, with indistinct light brown areas on belly and throat (variable). Bill blackish, changing to dull orange, smaller than winter adult.

Similar Species Tufted Puffin is unmistakable in summer. In winter, its huge head and heavy body create a ponderous, front-heavy flight shape, which together with its dirty blackish face, blackish underwings, and triangular bill, identify it as a puffin. Smaller Horned Puffin has a crisp black-and-white body in all seasons. Rhinoceros Auklet also has whitish underparts, but it is paler and more grayish brown overall than juvenile Tufted; it also has a smaller head and a more symmetrical body shape in flight.

Voice At colony, gives low sonorous growling calls, mostly from burrows.

Status & Distribution Abundant and widespread breeder in western AK; locally common in south-central and southeast AK and BC; uncommon to rare and local south to northern CA; casual to Southern CA. **BREEDING:** Colonially in earth burrows (less commonly rock crevices) on grassy slopes and vegetated cliffs of remote islands. **MIGRATION:** Moves out of north Bering Sea in Oct., disperses offshore mainly into north Pacific; returns May–June. **WINTER:** Seldom seen; mostly offshore, but less pelagic than Horned Puffin. **RARE STATUS:** Accidental Machias Seal I., Gulf of Maine, and in HI. An older specimen claimed from ME, said to have been shot during winter of 1831–32 and used by John James Audubon to paint illustration in his *Birds of North America*. Accidental to the UK (Norfolk).

Population Decreasing in CA. Recent die-offs in Bering Sea are likely linked to climate change resulting in food shortages.

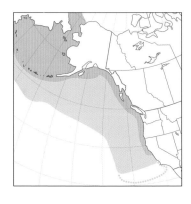

GULLS, TERNS, AND SKIMMERS Family Laridae

Black-headed Gull (center) with Bonaparte's Gulls, nonbreeding adults (NJ, Feb.)

Highly successful and cosmopolitan, gulls are very well known as a group. Even people with scant interest in birds can identify a "seagull." The other subfamilies (terns and skimmers) are less well known, but their relationship to the gulls is not difficult to ascertain. Easily identified to family, they are not so easily identified to species. Discussions of the identity of a gull or tern are classically lengthy. The difficulty in species-level identification is due in part to similarity among closely related species, but also to complex age-related changes and to hybridization in gulls. Looking at structure is an important consideration. Soft-part colors are also important. Immatures are usually more difficult to identify than adults, especially when worn, faded, or bleached. Exceptional care should be taken when identifying rare gulls and terns and photo-documentation is often essential.

Structure All show webbed feet, but they vary in wing structure and bill shape. Gulls are medium to large in size, with longish blunt-tipped bills. They have long pointed wings; short, square tails; and strong, medium-length legs. The narrow-winged terns are smaller, usually with notched or strongly forked tails, sharply pointed bills, and short legs. Skimmers are the most specialized of the group. They are very long-winged and short-legged, but they have a laterally compressed bill adapted to their unique feeding style, in which the lower mandible is much longer than the upper.

Behavior Flight is direct and strong, with much gliding in some species. Oceangoing gulls and some terns may arc up over the waves, especially under windy conditions, as tubenoses do. Gulls, terns, and skimmers roost in flocks, often mixed-species flocks. Gulls swim well, as do some species of tern, although terns prefer to perch on floating material when at sea. Skimmers can't take off from a swimming position, so they roost only on land. All members tend to breed in colonies, although some are solitary. Foraging varies, with the skimmer having the most impressive strategy: It flies over calm water, trailing its lower mandible in the water. When it senses a fish, it slams its bill closed, capturing the prey. Large gulls tend to be generalists, and a scavenging nature has allowed them to do well in urban areas. Smaller gulls specialize in seizing small prey on the surface. Most tern species prey on fish, usually by plunge-diving and capturing small fish with their bills.

Plumage Gulls and terns are usually gray above and white below, with notable exceptions. In summer, terns have black caps; large gulls have white heads, and small gulls have dark heads. Soft-part colors of most species will vary by age and season.

Distribution This is a cosmopolitan family. There are Arctic and Antarctic species, tropical and temperate species, and pelagic to montane species. Gulls are more diverse in temperate zones, while terns vary more in tropical and subtropical areas. Arctic Terns have record-long migrations between Arctic breeding sites and wintering locations off South Africa and Antarctica.

Taxonomy The 99 species (24 genera) are placed in three subfamilies, the gulls (Larinae), terns (Sterninae), and skimmers (Rynchopinae); two additional tern subfamilies are now often recognized, the noddies (Anoinae) and white-terns (Gyginae). Recent research has resulted in an increase in genera for both gulls and terns. Gull genera in North America increased from six to nine; changes include: placing Little Gull in the genus *Hydrocoloeus*; Bonaparte's, Black-headed, and Gray-hooded Gulls in *Chroicocephalus*; and Laughing and Franklin's Gulls in *Leucophaeus*. Tern genera in North America have increased from three to eight; changes include: placing Sooty, Bridled, and Aleutian Terns in the genus *Onychoprion*; Least Tern in *Sternula*; Gull-billed Tern in *Gelochelidon*; Caspian Tern in *Hydroprogne*; and Royal, Sandwich, and Elegant Terns in *Thalasseus*. The linear sequence of the genera has also been adjusted in both gulls and terns. Species-level taxonomy is equally complicated. Recently, Iceland and "Thayer's Gull" were lumped as a single species (Iceland Gull), and several gulls (for example, Herring and Mew) may be split in the future.

Conservation BirdLife International codes: 10 NT, 9 VU, 4 EN, and 1 CR (Chinese Crested Tern). The Critically Endangered Chinese Crested Tern was rediscovered, after no records in decades, from an islet off Taiwan, and scientists from Taiwan and China are working closely together to preserve this distinctive species.

Genera *Creagrus, Rissa, Pagophila, Xema, Chroicocephalus, Hydrocoloeus,* and *Rhodostethia*

The gulls in these genera are generally small and ternlike, except for the casual (to N.A.) Swallow-tailed Gull, an outlier species that is large and feeds nocturnally. The presence of a dark hood is not a taxonomically informative character. Males are larger overall and bigger-billed than females. An understanding of molt is helpful for identification. Most of the species in these genera are two-cycle (or two-year) species that attain adult plumage after their second postbreeding molt.

SWALLOW-TAILED GULL   *Creagrus furcatus*   STGU ▪ 5

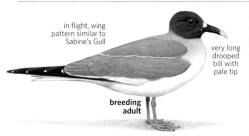

in flight, wing pattern similar to Sabine's Gull

very long drooped bill with pale tip

breeding adult

Among the most beautiful of the world's gull species, the Swallow-tailed Gull is equally remarkable for its nocturnal habits: It feeds only at night, on vertically migrant squid and fish; its large eyes are specially adapted for this strategy. No other gull, or species of seabird, is fully nocturnal. Monotypic. L 23" (58 cm) WS 52" (132 cm)

Identification Distinctive, large hooded gull. **ADULT:** Upperparts medium gray, underparts white shading to pale gray on breast in some. Long, forked tail is white. Long black, slightly decurved bill has whitish tip. **BREEDING:** Sooty black head with large white spot at base of maxilla (smaller white spot at base

of mandible); orbital ring vivid scarlet. **NONBREED-ING:** Dingy white head, dusky orbital ring. **FLIGHT:** Striking dorsal pattern: Blackish outer primaries contrast starkly with white outerwing coverts and secondaries, which contrast with gray mantle and inner wing coverts. Scapulars show neat, thin white edges. **IMMATURE:** Pattern similar to adults, but mantle and inner wing coverts mottled, rectrices black-tipped, head whitish, with blackish around eye and grayish brown ear spot.

Similar Species The striking dorsal pattern is superficially similar to Sabine's Gull and Large-billed Tern: The tern has a very large yellow bill and entirely black primaries; the gull, also found in pelagic habitats, has black, not white, outer primaries.

Voice Calls kittiwake-like, including a sharp, grating, rising *kweek.*

Status & Distribution BREEDING: Asynchronous breeding, throughout

the year, which reduces the effect of food shortages on the overall long-term population. Nests on rocky shores of Galápagos Is., mostly on Hood, Tower, and Wolf Is., with fewer on the other islands; also nests on Malpelo I. off Colombia in small numbers (50 pairs). **NONBREEDING:** Highly pelagic, found in deepwater areas off the coasts of Ecuador, Peru, and northern Chile often far from shore—most numerous from 3° north to 12° south, with smaller numbers south to central Chile, about 33° south. **RARE STATUS:** Accidental, four N.A. records (all adults): three (Mar., June, Sept.) records for central CA, and one (Sept.) record for WA. A few have been observed off the Pacific coasts of Panama, Costa Rica, and Nicaragua.

Population Global population estimated at 35,000 individuals in 2004; numbers at colonies fluctuate, probably because of prey availability, which is profoundly affected by changes in ocean temperatures, as during El Niño/Southern Oscillation cycles.

Parts of a Gull

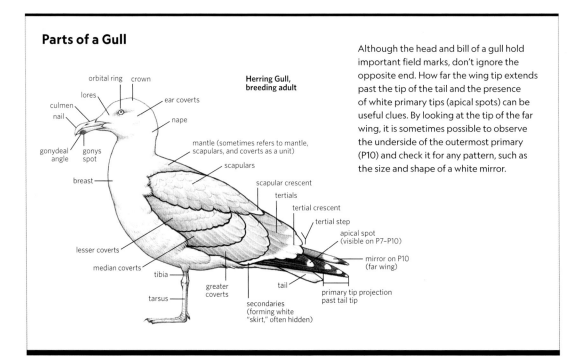

orbital ring   crown
lores
culmen
nail
ear coverts
nape

Herring Gull, breeding adult

gonydeal angle   gonys spot

breast

mantle (sometimes refers to mantle, scapulars, and coverts as a unit)

scapulars

scapular crescent
tertials
tertial crescent
tertial step
apical spot (visible on P7–P10)
mirror on P10 (far wing)

lesser coverts
median coverts
tibia
tarsus
greater coverts
tail
secondaries (forming white "skirt," often hidden)
primary tip projection past tail tip

Although the head and bill of a gull hold important field marks, don't ignore the opposite end. How far the wing tip extends past the tip of the tail and the presence of white primary tips (apical spots) can be useful clues. By looking at the tip of the far wing, it is sometimes possible to observe the underside of the outermost primary (P10) and check it for any pattern, such as the size and shape of a white mirror.

BLACK-LEGGED KITTIWAKE *Rissa tridactyla*   BLKI ▪ 1

This gull is buoyant and agile. It feeds mainly in flight by shallow plunge dives or by picking from surface. Polytypic (2 sspp.; both in N.A.). L 17–18" (43–46 cm) WS 37–41" (94–104 cm)
Identification A three-cycle gull. Medium-length bill; slightly cleft tail; short legs. **BREEDING ADULT:** Medium gray upperparts; upperwing's paler primaries contrast sharply with black wing tip (lack white mirrors); whitish underwing, contrasting black wing tip. Plain yellow bill; black legs. **WINTER ADULT:** Attains dark ear spot, dusky washed hind neck (sometimes on nonbreeding summer adults); primaries molted June–Jan. **JUVENILE & FIRST-SUMMER:** Juvenile has smoky-gray washed nape, blackish ear spot, blackish hind collar (often lost by midwinter on Atlantic birds). Note black M pattern on upperwings, whitish trailing edge; black tail band. By spring, dark bill mostly yellow; dark brownish legs blackish. Unlike most juvenile gulls, upperparts gray. Primaries molted late May–Oct. **SECOND-YEAR:** Resembles adult, but more extensive black on outer primaries' outer webs; usually some black on alula; bill sometimes dark tipped.
Geographic Variation Nominate subspecies breeds in North Atlantic; *pollicaris* in North Pacific. The latter averages larger, longer-billed, darker above, more extensive black wing tip, and later molt.
Similar Species Adult distinctive. Note medium gray upperparts, paler primary bases accentuating "ink-dipped" black wing tips, short black legs, plain yellow bill. In Bering Sea and Aleutians, see Red-legged. On

a distant first-year, upperwing pattern could suggest a Sabine's Gull, but note kittiwake's black ulnar bar, white underwings with small black wing tip. Juvenile Sabine's Gull has brownish hind neck and back; older plumages lack black tail band. Similar upperwing pattern on first-year Little Gull, but it is much smaller with shorter wings; slender bill; relatively longer, pinkish legs. Note Little Gull's dark cap, and by winter lack of black hind collar.
Voice Named for rhythmic *ketewehk ketewehk* call; flocks' laughing or honking calls can suggest small geese.
Status & Distribution Holarctic breeder, winters in northern seas. **BREEDING:** On sea cliffs. Returns Apr.–May, departs Aug.–Sept. **MIGRATION & WINTER:** Rarely seen from shore. Marked year-to-year variation in southward movements; winter wrecks periodically, with birds blown onshore and inland. Peak fall movements off BC in Sept.; off CA in mid-Nov. First arrivals to New England usually Oct., most Nov.–Dec. Northbound Pacific movement mainly mid-Mar.–Apr. In

Atlantic, move north from southern areas Feb.–Mar.; most depart New England mid-Apr. Very rare on Gulf Coast in winter. Nonbreeders oversummer irregularly south to CA and New England. **RARE STATUS:** Rare to casual (mainly late Oct.–Jan.) in interior.
Population Vulnerable.

white underside to primaries, except for black tips

breeding adult

base of primaries slightly paler than mantle

blackish M pattern across wings

juveniles

tail band

dusky postocular bar and wash on nape

unmarked yellow bill

winter adult

short blackish legs

blackish bill

black collar

RED-LEGGED KITTIWAKE *Rissa brevirostris*   RLKI ▪ 2

This attractive local gull of the Bering Sea region is best known (and readily seen) as a summer resident on the Pribilof Is. Monotypic. L 15.5–17" (39–43 cm) WS 33–35" (84–89 cm)
Identification A three-cycle gull with a relatively short, stubby bill; short legs; big dark eyes. Tail slightly cleft (most marked on juvenile). **BREEDING ADULT:** Slaty-gray upperparts; upperwing with narrow white trailing edge, black wing tip (lacking white mirrors); underwings with dusky-gray primaries, black wing tip. Plain yellow bill, bright red legs.
WINTER ADULT: Attains dusky ear spot,

hind-neck collar (sometimes on summer adults); primaries molted July–Feb. **JUVENILE & FIRST-SUMMER:** Juvenile has smoky-gray wash to nape, blackish ear spot, mottled slaty hind collar. Note upperwings' broad white trailing edge. Dark bill mostly yellow by spring; pinkish legs, orange-red by spring. Unlike most juvenile gulls, upperparts gray; lacks black on tail; primaries molted late May–Oct. **SECOND-YEAR:** Resembles adult, but black on outer webs of outer primaries more extensive; bill sometimes tipped dark.
Similar Species Black-legged is slightly

larger with more sloping forehead, longer bill, paler upperparts. Adult has black legs, whitish underwings; hence, black wing tips contrast more above and below. Head-on in flight, Black-legged shows white marginal coverts, unlike Red-legged's gray. First-year Black-legged has a black ulnar bar and tail band. Upperwing of first-year Red-legged suggests a Sabine's Gull, but white trailing triangle less extensive, bases of outer primaries gray.
Voice Squeakier than Black-legged.
Status & Distribution **BREEDING:** On sea cliffs in Bering Sea region; only five

known breeding locations (Pribilof Is., Buldir I., Bogoslof Is., and recently discovered from St. Matthew I.; also Commander Is.) Returns to colonies Apr.–May; departs Sept.–Oct. **WINTER:** Mainly found near pack ice in Bering Sea; a few probably to North Pacific. **RARE STATUS:** Casual (late Nov.–Mar., late June–mid-Aug.) south to WA and OR, also Japan; exceptionally to YT (Oct.), CA (three Feb. records), and near Las Vegas, NV (July).

Population Vulnerable. In the 1970s, estimated 232,000 birds worldwide. Since the 1970s, populations of Red-leggeds have declined on Pribilof Is. (no breeding in last few years) but increased on the Aleutian Is.

underside of primaries medium gray, except for black tips

base of primaries uniform with mantle, which is darker gray than Black-legged

breeding adult

breeding adult

shorter bill than Black-legged

no dark M-pattern

broad white trailing edge

no tail band

1st summer

very short coral red legs

pinkish legs

juvenile

IVORY GULL *Pagophila eburnea*   IVGU ■ 3

This sought-after icon of the high Arctic (*Pagophila* means "ice-loving") strays only casually into the lower 48. It can be confiding and usually associates only loosely with other gulls except when scavenging. Rarely, and only very briefly, it alights on water. Monotypic. L 17–19" (43–48 cm) WS 36–38" (91–97 cm)

Identification A two-cycle gull with a medium-size, slightly tapered bill; well-developed claws; fairly short legs. **ADULT:** All-white; no seasonal variation. Primaries molted Mar.–Sept., mostly before breeding (when molt suspended). Dark eyes; narrow orbital ring black to red in winter (rarely noticeable), red on breeding birds. Gray-green bill, orange to yellow-orange tip; black legs, feet. **JUVENILE & FIRST-WINTER:** Variable sooty blotching on face and throat; scattered dusky spots on rest of head, back variable blackish spots on upperwing coverts, flight feathers, tertials, and underparts. Variable blackish distal spots on upperwing coverts,

tertials, flight feathers. Grayish orbital ring; bill darker gray-green basally than adult, with yellow-orange tip and often dusky distal marks from late winter to spring. Attains all-white plumage by complete second postbreeding molt (mainly Apr.–Aug.).

Similar Species Unmistakable. Beware leucistic or albino individuals of any gull species. Some first-summer Glaucous and Iceland Gulls can bleach to almost all-white but are larger with longer pink legs, larger bills (pink-and-black on Glaucous, dark with a pinkish base on Iceland), and typical large-gull structure.

Voice Rarely heard away from breeding grounds. Calls include high, shrill to slightly grating whistles, unlike other gulls.

Status & Distribution Holarctic breeder on bare ground in high Arctic; mainly around pack ice in winter. **BREEDING:** Uncommon and local (June–Sept.); in N.A. (Canada) breeds on Ellesmere, Seymour, Devon, Baffin, and Perley Is. **MIGRATION & WINTER:**

Tied to pack ice. Main passage late Oct.–Dec. south into Bering Sea and Davis Strait, where most winter (Dec.–Mar./Apr.); irregularly NL. Moves north from Bering Sea late Mar.–Apr.; immatures linger through May in most years (when seen at St. Lawrence I.). **RARE STATUS:** Southern occurrences mainly Dec.–Feb. In West, casual BC, accidental Southern CA, western AZ, and CO. More frequent in East, where casual through Great Plains and Great Lakes regions, exceptionally TN and GA. Casual from Maritimes south to NJ. **Population** Near Threatened. Late-winter counts of over 35,000 birds in Davis Strait suggest that the estimate of 2,400 breeding birds in Canadian Arctic is low, although the origin of former numbers is not known (may include Greenland and Old World birds). Recent data suggest major declines in Canadian populations, perhaps linked to shrinking extent of pack ice due to climate change.

adult

greenish bill with yellow-orange tip

dusky face

pure white

long, narrow, rather pointed wings

short black legs

black markings on primary tips and near tail tip

1st winter

adult

SABINE'S GULL *Xema sabini*   SAGU ■ 1

This beautiful gull has a buoyant and ternlike flight. Polytypic (2–4 sspp.; 2 in N.A.). L 12.5–14" (32–36 cm) WS 33.5–35.5" (85–90 cm)

Identification Small, two-cycle gull; slender bill, forked tail. Striking upperwing pattern at all ages. **BREEDING ADULT:** Dark slaty hood; black neck ring; lacks white eye crescents; hood often spotted white in fall migrants. Bright white triangle on trailing edge of upperwings; whitish underwing; contrasting black outer primaries; medium-gray inner coverts. Underparts often flushed pinkish. Yellow-tipped black bill; blackish legs. **WINTER ADULT:** Unlikely in N.A. White head; slaty hind collar; ear spot; dark mottling on nape; neck sides washed smoky. Dull pinkish legs. **JUVENILE & FIRST-SUMMER:** Juvenile (plumage kept through fall migration) has dark gray-brown crown, hind neck, upperparts; scaly pale-edged back feathers with dark subterminal crescents. Wing pattern like adult, but with dark brown coverts; tail has black distal band. Black bill, dull pinkish legs. Complete post-juvenal molt in winter produces plumage like winter adult; initially gray back contrasts with brown upperwing coverts. After partial first pre-breeding molt, first-summer notably variable: Some have slaty-gray hind neck, mostly white head with dark ear spot; others have dark-slaty hood with whitish spotting. Black bill usually has variable yellow tip.

Geographic Variation Nominate (breeds northern AK, Arctic Canada, to Greenland; migrant off both coasts) is smallest, palest subspecies; *woznesenskii* (breeds western AK; migrates off Pacific coast) larger, upperparts average darkest. Two other described subspecies (breeding northern Russia; possibly migrating off Pacific coast) are intermediate. Now usually considered monotypic.

Similar Species Distinctive, but see first-year kittiwakes. Rarely seen juvenile Ross's have dark-brown upperparts and bold white trailing edge to upperwing similar to Sabine's, but tail strongly graduated with black-tipped central rectrices, bill smaller, white postocular area more extensive. Some heavily pigmented juvenile Franklin's Gulls (mainly July–Aug.) have dark brownish head and hind neck, but larger; stouter bill; blackish legs;

bolder white tips to primaries; different wing pattern in flight.

Voice Grating ternlike *kyeerr*, mainly given on summer grounds.

Status & Distribution Holarctic breeder, winters mainly off South Africa and western S.A. **BREEDING:** Fairly common (late May–Aug.) in N.A., on low-lying tundra and marshy areas, often near coast. **MIGRATION:** In fall, mainly late July–Oct.; stragglers into Nov., exceptionally Dec. Main passage off West Coast mid-Aug.–late Sept.; rare in interior (mainly Sept.); regular off Atlantic coast south to QC; rare to casual farther south of QC. In spring, mainly late Apr.–May (rarely by late Mar.) off West Coast; casual to interior; casual off Atlantic coast (late Apr.–mid-June). Nonbreeders oversummer locally off West Coast, rarely in northwest Atlantic. **WINTER:** Small numbers winter north to off southern Baja California. Accidental Lake Erie and FL.

Population Little data available; difficult to survey remote nesting populations.

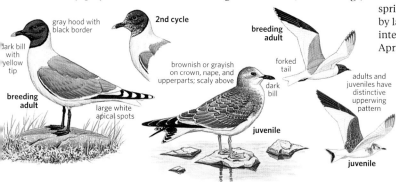

gray hood with black border

2nd cycle

dark bill with yellow tip

breeding adult

breeding adult

large white apical spots

brownish or grayish on crown, nape, and upperparts; scaly above

dark bill

juvenile

forked tail

adults and juveniles have distinctive upperwing pattern

juvenile

BONAPARTE'S GULL *Chroicocephalus philadelphia*   BOGU ■ 1

This gull is the only small gull that is common and widespread in N.A. In migration and winter, it occurs in flocks of up to a few thousand birds, feeding over tidal rips, power station outflows, and sewage ponds. It has a graceful, buoyant flight. Monotypic. L 13–13.5" (33–34 cm) WS 31.5–34" (80–86 cm)

Identification Small, three-cycle gull with slender black bill, white leading wedge on upperwing, and translucent underside to primaries. **BREEDING ADULT:** Slaty black hood. Bold white leading wedge on pale gray upperwing; underwing has translucent primaries with narrow black tips. Black bill, orange-red legs. **WINTER ADULT:** White head with dark ear spot; legs paler, pink to reddish pink. **JUVENILE & FIRST-WINTER:** Juvenile's hind neck and back are mottled pale cinnamon-brown, soon molting to pale gray by fall. Upperwing has dark brown ulnar bar, black trailing edge, and less prominent white leading wedge than adult, with mostly dark outer primary coverts. Pale pinkish legs; bill black or with dull pinkish base (not striking); legs pink

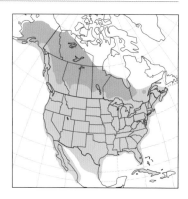

to orange. First-summer has variable head plumage: Some like first-winter, others with a partial blackish hood. **SECOND-WINTER:** Resembles winter adult, but with black marks on primary coverts, alula, and sometimes on tertials and tail.

Similar Species See rare Black-headed Gull and Little Gull, often found among flocks of Bonaparte's Gulls.

Voice Chatters and single *mew* calls all have relatively low, rasping, or buzzy tone; distinct from other gulls' calls.

Status & Distribution N.A.; winters to northern Mexico. **BREEDING:** Common (May–Aug.), in solitary pairs and loose colonies, around lakes and marshes in boreal forest zone; nests usually placed in conifers. Has nested locally

in eastern QC, probably ME. **MIGRATION:** In fall, first birds reach Great Lakes region and Bay of Fundy in July–Aug., but main passage in East is late Oct.–Nov., when rare north to NL. On Pacific coast, a few early migrants occur late July–Aug.; main migration late Oct.–Nov. In spring, mainly Mar.–May. Northbound influxes typically start late Mar. on Great Lakes, mid-May in Prairie Provinces. On Pacific coast, northbound influxes start late Mar.; peak movements Apr.–mid-May. Nonbreeders oversummer irregularly in Great Lakes region and Pacific states, more rarely on the East Coast south to mid-Atlantic. **WINTER:** Bulk of population spends winter period (Oct.–Mar.) on Lake Erie and Lake Ontario and along

Atlantic coast from MA south through mid-Atlantic states. During early winter most birds are on Great Lakes, but once lakes start to freeze, birds move toward Atlantic coast. Casual north to NL. In West, fairly common but local (Nov.–Mar.) in Pacific states, casual to southeastern AK. **RARE STATUS:** Casual (late May–June, Aug.) to Bering Sea islands. Also Europe, Africa, HI, Japan. **Population** Sharp declines in winter numbers on Atlantic and Pacific coasts. This species' largely inaccessible breeding range makes it difficult to survey.

blackish hood with gray cast | **breeding adult** | postocular spot | **winter adult** | **1st winter** | dark carpal bar | largely pale inner primaries | **1st winter** | **winter adult** | white primaries above and below with thin black tips | brownish above with pale fringes | **juvenile**

GRAY-HOODED GULL *Chroicocephalus cirrocephalus*　GHGU ▪ 5

This gull has occurred twice in N.A. The adult has a gray hood. Also known as "Gray-headed Gull." Polytypic (2 sspp.). L 16" (41 cm) WS 43" (109 cm)

Identification A slim, three-cycle gull. Large for a hooded gull, but much smaller than, say, Ring-billed. Longish bill, long legs. **SUMMER ADULT:** Hood distinctive: pale gray with darker border. Medium gray mantle; otherwise body white, tail white. Whitish eye, red orbital ring. Dark red bill, orange-red legs. Wing pattern distinctive: white wedge on outer primaries interrupted by large black wing tip; large mirrors on P9 and P10. Dark gray underwing; blacker on primaries with obvious mirrors showing. **WINTER ADULT:** Hood replaced by dark ear spot and smudgy area around eye, both connected by stripes on crown. Bicolored bill: red base, black tip. **FIRST-YEAR:** Dark ear spot, smudgy area around eye as in winter adult. Medium gray mantle; gray wings, dark tertials; brown centers to lesser and median coverts. Pinkish yellow bill with dark tip; brownish

orange legs. In flight, outer wing as in adult: white outer wedge with large black wing tip, but no mirrors. Inner wing gray, brown bar diagonally across inner wing, contrasting darker secondary bar. Tail white with black tail band. **SECOND-YEAR:** As adult, but smaller primary mirrors.

Geographic Variation Subspecies *cirrocephalus* in S.A. is slightly larger and paler than *poiocephalus* in Africa. The origin (and subspecies) of the two N.A. records is not known.

Similar Species Summer adult is easily identified by gray hood and wing pattern. First-year easily confused with Black-headed Gull of similar age. Gray-hooded is larger, with a longer bill and longer legs; shows slightly darker mantle. In flight, Gray-hooded has a solid black wing tip. Black-headed shows white on primaries, reaching slot-like toward tips. Gray-hooded's primaries are entirely dark on underside. Outer two primaries are white on Black-headed, producing long white underwing stripes.

Voice Similar to Black-headed Gull, but deeper and rougher.

Status & Distribution Accidental to N.A. Native to Africa and S.A. **YEAR-ROUND:** Estuaries and coastal concentrations of small gulls. **RARE STATUS:** Two records: Apalachicola, FL (26 Dec. 1998) and Coney Is., NY (24 July–early Aug. 2011). Also accidental to C.A. from Guatemala and Panama.

Population Combined African and S.A. population is estimated to be 50,000 pairs.

breeding adult *cirrocephalus* | distinctive black in wing-tip patte[rn] (all plumages) | grayish hood with dark border | red legs and bill

BLACK-HEADED GULL *Chroicocephalus ridibundus* BHGU ■ 3

This Old World counterpart of Bonaparte's Gull (with which it is often found) is generally rare in N.A. Polytypic (2 sspp.; both in N.A.). L 15–16" (38–41 cm) WS 35.5–39" (90–99 cm)
Identification Medium-small, two-cycle gull. Medium-long red to pink bill. **BREEDING ADULT:** Chocolate brown hood. Bold white leading wedge on pale gray upperwing; dark middle primaries on underwing, narrow white wedge on outer primaries. Deep red bill and legs. **WINTER ADULT:** White head, dark ear spot; paler red bill and legs. **JUVENILE & FIRST-WINTER:** Juvenile has mottled cinnamon-brown hind neck and back, molting to pale gray in fall. Upperwing has cinnamon-brown ulnar bar, dark trailing edge, less prominent white leading wedge than on adult. Pale peach to pinkish orange bill and legs; black-tipped bill. First-summer head plumage variable: some like winter adult; others with a solid, dark-brown hood, like breeding adult; legs and bill pink to reddish.
Geographic Variation Often considered monotypic. Breeding popula-

tions in N.A. presumably nominate (Europe). The subspecies *sibiricus* (eastern Asia) ranges to western N.A., averages larger (especially longer bill) and molts later.
Similar Species Only likely to be confused with smaller and much more common Bonaparte's Gull. A few key points (e.g., bill size, bill color, pattern on underside of primaries) should separate these species. Bonaparte's is smaller and daintier with relatively narrower wings; a more slender black bill; appreciably shorter legs; white underwings with translucent primaries; and slightly darker upperparts—thus white leading wedge on adult upperwing contrasts more. Breeding adult Bonaparte's has a blackish hood (attained a month or so later than Black-headed Gulls attain hoods in the East), orange-red legs; winter adult has a darker smoky-gray hind-neck wash, peach-pink legs. First-winter Bonaparte's averages a darker ulnar bar and has dark streaks on outer primary coverts (mostly white on Black-headed Gull).
Voice Varied screams and a grating

meeahr, higher and less buzzy than those of Bonaparte's Gull.
Status & Distribution Eurasia and eastern N.A.; winters to Africa. **BREEDING:** Rare (May–Aug.) in N.A., where first found breeding in 1977. Colonial or in pairs; on open or partly vegetated ground near water, usually in association with other waterbirds. Mainly NL; smaller numbers in Gulf of St. Lawrence region, QC; has nested, or attempted to, in NS, ME, MA; also possibly northwest IA and adjacent MN. **MIGRATION & DISPERSAL:** In western AK, rare in spring (mainly May–June, a few into summer) and casual in fall (late Aug.–Oct.), mainly to western Aleutians and Pribilofs, a few to St. Lawrence I., and mainland western AK. Casual in Great Lakes region and west to Great Plains. **WINTER:** Mainly Oct.–Apr. Locally fairly common in NL; uncommon to rare along St. Lawrence Seaway and Atlantic coast south to NJ; very rare to NC; casual farther south along Atlantic coast to FL and (mainly Nov.–Mar.) from TX east through Gulf Coast states. **RARE STATUS:** Casual (mainly Sept.–Apr.) along Pacific coast and inland from south-coastal AK to CA. Accidental elsewhere in interior West and to HI.
Population Marked increase in Europe in the 1900s, when spread across North Atlantic. First bred in Iceland in 1911 and in western Greenland since the 1960s.

dark inner primaries

dark slate underside to primaries, paler in immature

dark brown hood

breeding adult

1st winter

winter adult

dark brown hood

winter adult

long reddish bill

winter adult

1st winter

1st summer

orangish pink with dark tip

slightly paler above than Bonaparte's

LITTLE GULL *Hydrocoloeus minutus* LIGU ■ 3

This is the smallest gull, most often seen with flocks of Bonaparte's Gulls. In its fluttery and ternlike flight, it dips down to pick food from the water's surface. Monotypic. L 11–11.5" (28–29 cm) WS 27–29" (69–74 cm)

Identification Small, three-cycle gull; slender bill, fairly short legs. **BREEDING ADULT:** Black hood lacks white eye crescent. Pale gray upperwings, dark gray underwings; note white trailing edge out to wing tip. Red legs.

WINTER ADULT: White head with dark cap and ear spot. **JUVENILE & FIRST-WINTER:** Juvenile has blackish brown hind neck collar and back; pale-edged scapulars. Soon molts in pale gray back of first-winter. Upperwings have bold blackish M pattern. Tail white with black distal band. **SECOND-WINTER:** Resembles winter adult, but some have black wing-tip markings and paler wing linings.

Similar Species Bonaparte's is slightly larger and longer-billed; first-winter lacks dark cap, has upperwings with white leading wedge and black trailing edge. Ross's is similar, especially the first-winter, but has a shorter bill and longer, graduated tail; winter adult lacks dark cap; first-winter has broad white trailing edge on upperwing. First-winter Black-legged Kittiwake shares black M pattern on upper-wings, but is markedly larger, with bigger bill (pale based by winter), and dark legs. Note kittiwake's extensive whitish on trailing edge of wing and

lack of dark cap; often with blackish collar through winter.

Voice Nasal *kek* occasionally given in migration and winter.

Status & Distribution Eurasia. **BREEDING:** Rare and local; solitary pairs or small groups breed in marshes and around forested lakes/ponds from Great Lakes region to Hudson and James Bays. Breeding first detected in 1962 at Lake Ontario, Canada; most N.A. birds likely now breed in vicinity of Hudson and James Bays. **MIGRATION:** Mainly Aug.–Nov. and Mar.–May, primarily Great Lakes region, rare New England. **WINTER:** Nov.–Mar., mostly mid-Atlantic coast; movements closely tied to Bonaparte's Gull. Casual north to Maritimes and in southeast. **RARE STATUS:** Casual

west to Pacific coast, annual to CA. **Population** Small N.A. population; fewer noted on East Coast and elsewhere mirrors reduction of migrant and winter flocks of Bonaparte's Gulls.

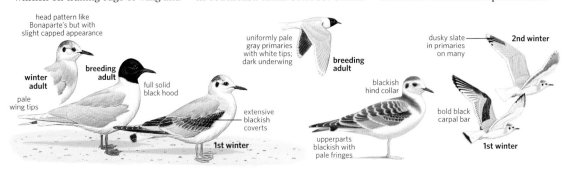

head pattern like Bonaparte's but with slight capped appearance

winter adult

breeding adult

pale wing tips

full solid black hood

extensive blackish coverts

1st winter

uniformly pale gray primaries with white tips; dark underwing

breeding adult

blackish hind collar

upperparts blackish with pale fringes

dusky slate in primaries on many

2nd winter

bold black carpal bar

1st winter

ROSS'S GULL *Rhodostethia rosea* ROGU ■ 3

This rare "pink gull" of the high Arctic only casually reaches the lower 48, usually with flocks of Bonaparte's Gulls. Its flight is buoyant and strong, with deep wingbeats; it dips down to pick food from water. Monotypic. L 12.5–14" (32–36 cm) WS 32–34" (81–86 cm)

Identification Small, two-cycle gull; long, graduated tail; small bill. All post-juvenal plumages can be strongly flushed pink on head and underparts. **BREEDING ADULT:** Narrow black neck ring. Pale-gray upperwings; smoky-gray underwings, broad white trailing edge to secondaries and inner primaries. Orange-red legs. **WINTER ADULT:** White head with dark ear spot; sometimes broken neck ring; duller legs. **JUVENILE & FIRST-WINTER:** Juvenile has sooty-brown crown, hind neck, and back; pale-edged scapulars. Upperwings have bold blackish M pattern, broad white trailing edge. White tail, black tips to elongated central rectrices. Pinkish legs. First-summer can attain black neck ring, all-white tail; orange-red legs.

Similar Species Distinctive, but might

be confused with Little Gull, which is similar in most plumages. Little Gull has slightly longer bill; shorter, blunter wings; and "normal" squared tail. Adult Little has narrower white trailing edge to wings extending to wing tip; winter adult has dark cap. First-winter has duller and less extensive whitish trailing triangle on upperwing (with dark secondary bar); black tail tip forms an even band. See juvenile Sabine's. Most small gulls can attain a pink flush on their head and underparts, but rarely as intense as most Ross's.

Voice Usually silent. In summer, mellow yapping *p-dew* and ternlike notes.

Status & Distribution Holarctic breeder, wintering near pack ice. **BREEDING:** Rare and local (June–

wedge-shaped tail

winter adult

dark gray underwing with bulging white trailing edge

1st winter

bold dark M

compare to Little Gull

1st winter

black collar

in all birds, underparts variably washed with pink

breeding adult

Aug.); solitary pairs and small colonies at marshes and small tundra lakes. **MIGRATION & DISPERSAL:** Postbreeding dispersal starts July–Aug. from breeding grounds in Russian Far East and Siberia; move east to western Beaufort Sea by Oct.; followed by late Oct.–Dec. passage south into Bering Sea. Best known in fall (Oct.) from Nuvuk (Pt. Barrow), AK. In spring, small numbers east irregularly (late May–June) to northeast Bering Sea. **WINTER:** Distribution probably linked to pack ice; main winter grounds (Dec.–Apr.) apparently Bering Sea south through the Sea of Okhotsk and Davis Strait/Labrador Sea. **RARE STATUS:** Casual central Canada (mainly spring) from southern MB east to ON and in lower 48 (mainly winter), mostly in Midwest and Northeast. Accidental in migration and winter south to CO, MO, DE, in Northwest (south to OR and eastern ID), and Southern CA (Salton Sea). **Population** Estimated 20,000–40,000 birds occur in northern AK waters in fall. No data on trends.

LARGER GULLS Genera *Leucophaeus* and *Larus*

The gulls in these two genera are the larger, "typical" gulls. Laughing and Franklin's Gulls are in the genus *Leucophaeus* and have dark hoods in breeding plumage. Most species are in the genus *Larus*, which includes the widespread Herring Gull and other large species that typically have white heads as breeding adults (the so-called "large white-headed gulls"). Adult plumages are patterned in gray, black, and white, often boldly; juveniles and younger immatures have much brownish in their plumage and are less distinctly marked. Males are larger overall and bigger billed than females; this can be striking in "large white-headed gulls" and is an important consideration in identification. In addition, hybrids are not uncommon, another factor to consider when identifying an atypical individual.

All of these species are either three- or four-cycle gulls with a variety of immature plumages that complicate identification. An understanding of molt is helpful for identification, as gulls can have up to four years of immature plumages. A four-cycle (or four-year) gull is one that attains adult plumage by its fourth postbreeding molt. Adults typically have a complete molt in fall and a partial molt in spring; this molt pattern starts with the second postbreeding molt at about one year of age and is repeated for life. However, molts in the first year are more variable: Most first-cycle large gulls have a single, protracted molt from fall through spring.

LAUGHING GULL *Leucophaeus atricilla*   LAGU ◼ 1

The dark-hooded Laughing Gull is the common and familiar "parking lot gull" of Atlantic and Gulf Coast beaches. Polytypic (2 sspp.; 1 in N.A.). L 15–17" (38–43 cm) WS 38–42" (97–107 cm)

Identification Medium-size, three-cycle gull with long, pointed wings. Relatively long bill often looks slightly drooptipped. **BREEDING ADULT:** Blackish hood, white eye crescents. Upperparts slaty gray with black wing tips; small white outer primary tips worn off by summer. **WINTER ADULT:** Whitish head, dusky auricular smudge. **JUVENILE & FIRST-WINTER:** Juvenile head, neck, chest, and back brown; upperparts with scaly buff edgings. Broad

long bill with slightly thicker tip

has dark "earmuffs," but no half hood like Franklin's

winter adult

dark underside to primaries

breeding adult

2nd winter

long wings

breeding adult

2nd winter

more extensively dark than Franklin's

1st winter

dark "earmuffs," but no hood

gray across breast and down sides and flanks

tail band includes outer tail feathers

overall brownish color

1st winter

broad, complete tail band

juvenile

black distal band on tail; underwings mottled. Back, neck, and chest molt to gray in fall; head to whitish with a dark mask. First-summer can attain partial hood. **SECOND-WINTER:** Resembles winter adult, but more black in wing tips, often some black on tail, neck sides grayish.

Geographic Variation N.A. populations are subspecies *megalopterus*, larger than subspecies *atricilla* of Caribbean.

Similar Species Franklin's Gull is slightly smaller, with a shorter, undrooping bill; all plumages have bold white eye-crescents. Adult Franklin's has a white band inside its black wing tip. In spring, adult Franklin's bright pink flush is often striking among white-breasted Laughing Gulls. Winter Franklin's has a blackish half-hood. First-winter has whitish underwings; its narrower black tail band does not reach tail sides. First-summer Franklin's Gull can suggest adult Laughing, but Franklin's smaller black underwing tip contrasts with pale gray primary bases.

Voice Laughing and crowing calls.

Status & Distribution N.A. to Caribbean; winters to northern S.A. **BREEDING:** Common (Apr.–Aug.), colonial on sandy and rocky islands, salt marshes. Breeding attempted in southern Great Lakes region, where it has hybridized with Ring-billed Gull. **MIGRATION & DISPERSAL:** In Northeast, peak movements late Aug.–Sept. (lingerers Dec.), and Apr.–early May (arrives Mar.). **WINTER:** Aug.–Apr. (some oversummer in much of winter range). Casual inland in Midwest. **RARE STATUS:** Rare inland in East and north to NL. Very rare to casual in Pacific coast states, except at south end of the Salton Sea and adjacent Imperial Valley, CA, where fairly common (mainly June–Oct.; has bred). Casual inland elsewhere in West. Casual Europe, Africa, Japan, and to Australia.

Population Stable.

FRANKLIN'S GULL   *Leucophaeus pipixcan*   FRGU   ■   1

Sometimes known as the "prairie dove," this dark-hooded gull is unusual in having two complete molts a year. Monotypic. L 14–15" (36–38 cm) WS 35–38" (89–97 cm)

Identification Medium-size, two-cycle gull. **BREEDING ADULT:** Blackish hood with thick white eye crescents; slaty gray upperparts. Black subterminal band on outer primaries is framed by large white primary tips and a white medial band. Red bill; dark reddish legs. **WINTER ADULT:** Head is whitish with a blackish half-hood and thick white eye crescents. Black bill has an orange tip. **JUVENILE & FIRST-WINTER:** Hood similar to winter adult. Juvenile has brown hind neck and back; back has pale edgings. Tail has black distal band not extending to sides of tail; underwings whitish overall with blackish wing tips. Back molts to gray in fall, hind neck becomes whitish. After complete molt in winter, first-summer resembles winter adult, but its wing tip lacks white medial band; but usually looks like winter adult after complete second postbreeding molt in fall.

Similar Species See Laughing Gull.

Voice Calls are higher pitched than Laughing Gull, often heard from migrant flocks.

Status & Distribution Breeds interior N.A., winters Pacific coast of S.A. **BREEDING:** Common (May–Aug.) in N.A., colonial in freshwater marshes. **MIGRATION:** In fall mainly Aug.–Nov., south through Great Plains to TX, lingerers into Dec. First spring arrivals Mar., peak movements late Apr.–early May. Uncommon to very rare to Pacific coast; casual north to Bering Sea. Uncommon to rare in Midwest and ON; casual to very rare along Atlantic coast from NL to FL. **WINTER:** Rare to casual north to TX. Occasional winter

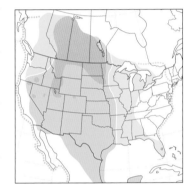

records elsewhere in US. **RARE STATUS:** To Europe, Africa, Japan, and Australia.

Population Breeding range shifted west and south through the 1900s, which—in combination with interannual colony fluctuations—makes population trends difficult to ascertain.

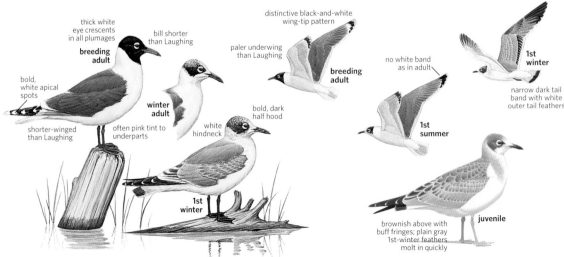

thick white eye crescents in all plumages

breeding adult

bill shorter than Laughing

bold, white apical spots

shorter-winged than Laughing

winter adult

often pink tint to underparts

white hindneck

1st winter

distinctive black-and-white wing-tip pattern

paler underwing than Laughing

breeding adult

bold, dark half hood

no white band as in adult

1st winter

narrow dark tail band with white outer tail feathers

1st summer

juvenile

brownish above with buff fringes; plain gray 1st-winter feathers molt in quickly

BELCHER'S GULL *Larus belcheri* BEGU ■ 5

This attractive gull was previously known as the Band-tailed Gull during the time it was lumped with the threatened Olrog's Gull of coastal Argentina. Monotypic. L 20" (51 cm) WS 49" (124 cm)
Identification A medium-size, stocky, four-cycle gull with long thick legs. Long, thick, parallel-sided bill lacks an expansion at the gonydeal angle. **SUMMER ADULT:** Black mantle; white head, underparts; pale gray neck; white tail with wide black tail band, narrow white tip. Tertial, scapular crescents narrow but obvious. Bright yellow bill; black subterminal bar on both mandibles; red tip, rarely missing or restricted to upper mandible. Dark eye, red orbital ring; bright yellow legs. In flight shows extensive black underwing tip. **WINTER ADULT:** Similar to summer adult, but

with dark hood and white eye crescents. **JUVENILE & FIRST-YEAR:** Brown with contrasting dark head, neck, breast; white eye crescents; white belly and undertail coverts. Juveniles with pale-fringed mantle feathers; older birds' mantle gray-brown. Uniform, dark brown wings. Blackish tail contrasts with whitish rump and uppertail coverts. Dull pink legs. Distinctive bill: pinkish or yellowish white with crisp black tip (outer third of bill); red nail. **SECOND-WINTER:** Blackish mantle, like adult. Browner wings, often with blackish inner median covert bar. Extensive brownish hood extends to lower breast. Tail band like adult. Bill bicolored; red tip now obvious. Bright yellow legs. **SECOND-SUMMER:** Similar, but white head and neck. **THIRD-YEAR:** Very like adult, but brownish wash to wings.
Similar Species Adults identified by crisp black tail band; brown hood distinctive in winter

plumage. Black-tailed shows pale eyes, dark gray mantle, thinner bill, slimmer body, longer wings. In comparison to Lesser Black-backed or Yellow-footed, Belcher's has a red-tipped bill, darker mantle, dark eye, no white primary tips. Strongly hooded look of first-year plumages distinctive. Second-year told from same-aged Kelp by brighter yellow legs, red bill tip, and structural differences.
Voice Undescribed.
Status & Distribution Casual in N.A., local resident along desert coasts of Peru and northern Chile. **YEAR-ROUND:** Strictly a coastal species, frequents rocky coastlines. **RARE STATUS:** One record from CA and four from FL. The FL records are unusual as this is a Pacific Gull, suggesting that a few crossed over C.A.
Population Estimated to be ±10,000 individuals; the population is currently increasing.

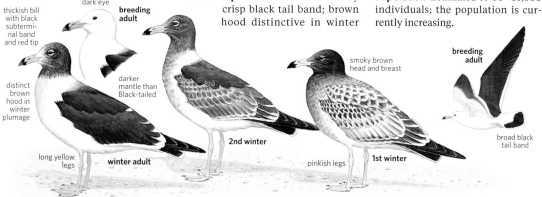

BLACK-TAILED GULL *Larus crassirostris* BTGU ■ 4

This Asian species has the potential to occur nearly anywhere. Adults are slim, with a black tail band and a characteristic red-tipped bill. Monotypic. L 18" (46 cm) WS 47" (119 cm)
Identification A medium-size, long-billed, four-cycle gull with long, narrow wings. Long bill is parallel-sided and lacks expansion at the gonydeal angle. **SUMMER ADULT:** Dark gray mantle. White tail has a wide black tail band and narrow white tip. Tertial and scapular crescents narrow but obvious. Bright yellow bill, black subterminal bar on both mandibles, red tip. Yellow eye, orange-red orbital ring. Bright yellow legs. In flight shows extensive black wing tip, to P5 or P4; no mirrors on outer primaries. **WINTER ADULT:** Like summer, but streaking on head and nape. **JUVENILE &**

FIRST-YEAR: Brown with paler facial area at bill base; contrasting white eye crescents. Juvenile with scaly mantle; first-winter birds more gray-brown above with darker feather centers.

Unicolored dark brown wings. Blackish tail contrasts with whitish rump and uppertail coverts. Bill distinctive: pale pink with crisp black tip (outer third). Pinkish legs. **SECOND-YEAR:** Dark gray mantle; browner wings. Brown head with white eye crescents and face. Black tail with crisp white tip. Bill greenish at base. **THIRD-YEAR:** As adult, but black on tail more extensive. **Similar Species** Adults are identified by their crisp black tail band and their pale eyes. Belcher's Gull (accidental in N.A.) has a tail band; but has dark eyes,

blacker mantle, thicker bill, and bulkier body. Told from adult California Gull by its red-tipped bill, darker mantle, pale eyes, and very small white primary tips. First-year Black-tailed resembles both California and Laughing Gulls. California shares bicolored bill, but Black-tailed is more uniformly colored and has a contrasting and white rump, well-defined white eye crescents, and a longer and slimmer shape. First-year Laughing shares long and slim shape, eye crescents, and uniform-looking plumage, but it has a dark bill and legs.

Voice LONG CALL: Mewing and deep. **Status & Distribution** Casual in N.A. Native to temperate Asian coast. **YEAR-ROUND:** Prefers rocky coastal habitats, usually associates with medium-size gulls. Most records in N.A. Mar.–Aug. **RARE STATUS:** Over 30 records, most in AK. Other records in BC, WA, CA, MB, IA, IL, TX, and FL; several sightings around Lake Michigan, OH, NL, NS, VT, MA, RI, NY, NJ, MD, and VA. Other extralimital records include Sonora, Mexico, and Belize. **Population** Stable.

HEERMANN'S GULL   *Larus heermanni*   HEEG ■ 1

A striking Pacific gull, entirely dark-bodied and white-headed when breeding—like a photo negative of our smaller dark-hooded and pale-bodied gulls. Monotypic. L 19" (48 cm) WS 51" (130 cm)
Identification A medium-size, four-cycle gull. Bill slightly drooped. **SUMMER ADULT:** Dark body, paler below. White head contrasts strongly. White head obtained in late fall to early winter, lost by midsummer. Narrow tertial and scapular crescents. Red bill with black tip; eyes dark with red orbital ring; black legs. In flight, all dark gray, slightly paler secondary coverts contrast with flight feathers; shows a narrow white trailing edge extending to inner primaries. No mirrors or white primary tips. Black tail has white tip and contrasts with pale gray rump and uppertail coverts. **WINTER ADULT:** Crisply streaked head; otherwise similar to summer adult. **JUVENILE & FIRST-YEAR:** Dark brown; scaly above due to pale feather edges; older birds gray-brown on mantle. Dull pink bill with dark tip. In flight, wings are brown with darker secondaries and primaries. **SECOND-YEAR:** Body dark gray; wings may be washed brown. Dark brown hood shows white eye crescent. Tertial crescents white; bill reddish at base. **THIRD-YEAR:** Much like adult, but head with more extensive brown wash.
Similar Species Young often mistaken for juvenile jaegers. Some Heermann's Gulls show a white patch on greater primary coverts, not primaries as on jaegers, but Heermann's separated by slower, less powerful flight; unbarred body and underwings; and lack of primary flash.
Voice CALL: A nasal *ahhh*. **LONG**

CALL: Short, deep, and nasal.
Status & Distribution Common. Over 90 percent breed on Isla Rasa in Gulf of California, so essentially it is a Mexican breeding species that migrates north to our region. **YEAR-ROUND:** Coastal, preferring sandy beaches. **BREEDING:** Very rare breeder in CA. **MIGRATION:** In central CA appears as early as late May in some years, late June in others; reaching Pacific Northwest in July, peaking late July–Sept. During warm-water years arrival is earlier and birds travel farther north. Moves south Sept.–Nov.; largely absent north of Monterey, CA, Jan.–May. **RARE STATUS:** Casual to southeastern AK. Very rare in lower Colorado River Valley, into AZ. Casual farther east; records from NV, NM, UT, WY, OK, TX, MI, OH, FL, VA, and ON.
Population Near Threatened. Very restricted breeding range. Population is approximately 300,000 breeding adults, but breeding success nearly collapses and adult mortality increases during strong El Niño years.

paler brown than 1st winter and scaly above

juvenile

streaked head

breeding adult

dark wings

broad black tail band

white head

red bill with small black tip

breeding adult

gray underparts

winter adult

pink-based bill with dark tip

1st winter

2nd winter

overall sooty brown color

MEW GULL *Larus canus*   MEGU   ■ 1

The smallest white-headed gull in N.A. Polytypic (4 sspp.; 3 in N.A.). L 16–17" (41–43 cm) WS 41–44" (104–112 cm) **Identification** Medium-small, three-cycle gull with rounded head, relatively small and slender bill, dark eyes. Adult has large white mirrors on P9–P10. N.A. Mew Gull described first; other subspecies ("Common Gull" and "Kamchatka Gull") follow. **BREEDING ADULT:** Upperparts medium gray; yellow legs and unmarked bill. **WINTER ADULT:** Head and neck with smudgy brown washing. **JUVENILE & FIRST-WINTER:** Overall brownish; back becomes gray in first-winter. Tail mostly dark brown (variable); uppertail coverts barred brown; underwing coverts brownish. **SECOND-WINTER:** Resembles winter adult, but more black on wing tips; tail usually with black markings. **"COMMON GULL":** Averages larger, longer billed. **ADULT:** Slightly paler gray upperparts, more black on wing tip than on Mew (variable); winter head markings more spotted; bill often has dusky ring. **JUVENILE & FIRST-WINTER:** Whiter on head, body, and underwing coverts than Mew; uppertail coverts and tail base mostly white with clean-cut black tail band. **"KAMCHATKA GULL":** Largest and bulkiest subspecies, with longer and stouter bill than

Mew. Adult upperparts average darker, wing tip with more black (variable), and eyes often pale (can be pale on Mew); coarser winter head markings. **JUVENILE & FIRST-WINTER:** Mostly white uppertail coverts and tail base; clean-cut, broad black tail band narrowest at sides.

Geographic Variation Population in N.A. is *brachyrhynchus* (Mew), which is smaller than nominate *canus* group ("Common" includes slightly larger *heinei* of Eurasia, and *kamtschatschensis* ("Kamchatka") of eastern Asia.

Similar Species Ring-billed is larger with flatter head and deeper, blunter bill; adult pale gray above with pale eyes, neat black bill ring. Juvenile and first-winter paler and more contrasty overall, with whitish underparts (variably spotted dusky) and underwings, and whiter tail base. "Common" and "Kamchatka" more similar to Ring-billed: On first-year Ring-billed, upperwing coverts have broader, pale-notched tips (often creating a checkered, rather than scaly, pattern); tertials tend to be darker with wider and more notched pale edgings; note tail patterns.

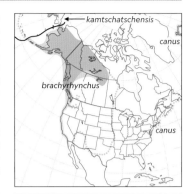

Breeding Range
heinei
kamtschatschensis
canus
kamtschatschensis
canus
brachyrhynchus

two white mirrors

2nd winter *brachyrhynchus*

1st winter *brachyrhynchus*

breeding adult *brachyrhynchus*

adult *kamtschatschensis*

adult *canus*

extensive gray on P8

large white mirrors

overall brownish compared to juvenile Ring-billed

black tail

brownish underwing

winter adult *brachyrhynchus*

brownish wash

2nd winter *brachyrhynchus*

most with dark eye

short, slender, yellow bill

breeding adult *brachyrhynchus*

darker above than Ring-billed

filled-in brown coverts with pale fringes

juvenile *brachyrhynchus*

more black on P8 than *brachyrhynchus*

more black on P8 than *brachyrhynchus*, like *kamtschatschensis*

long wings

kamtschatschensis

larger size

winter adult

pale base to tail

1st winter

1st winter intermediate between *brachyrhynchus* and *canus*

extensively dark tail

1st winter *brachyrhynchus*

head and breast pattern on *canus* more spotted, less washed with brown as in *brachyrhynchus*

often with darker ring on bill

winter adult "Common Gull" *canus*

black tail band contrasts sharply with white tail base

covert pattern like *brachyrhynchus*

1st winter *canus*

Voice Slightly shrill mewing calls in series; also clipped *kek* in winter.
Status & Distribution Following refers to Mew Gull unless noted. **BREEDING:** Common (May–Aug.), from tundra and marshes to sea cliffs, usually colonial. **WINTER:** Mainly coastal, also inland, commonly (mid-Oct./Nov.–Mar./Apr.).

Locally inland in BC and Pacific states. Casual (mainly Nov.–Mar.) in interior west, Great Plains, TX, and Great Lakes region. **RARE STATUS:** Mew casual on Bering Sea islands, accidental on Atlantic coast. "Common" rare in Atlantic Canada (mainly late Oct.–Apr.), most frequent to NL; casual south to mid-

Atlantic coast. "Kamchatka" rare spring and casual fall migrant in western Aleutians; casual on Bering Sea islands; no reports from the East are substantiated with specimens.
Population No data. Range of European subspecies increased in past 50 years—Iceland colonized in 1955.

RING-BILLED GULL *Larus delawarensis*   RBGU ■ 1

This is the common "seagull" across much of N.A., yet it is rare offshore. Adaptable and bold feeder, soars to catch insects, scavenges at dumps and parks, and even plucks berries from trees! Monotypic. L 17–20" (43–51 cm) WS 44.5–49" (113–124 cm)
Identification Medium-size, three-cycle gull. On adult, note yellow legs, pale eyes, and "ring bill." **BREEDING ADULT:** Pale gray upperparts; black wing tip with white mirrors on outer one to two primaries. Staring pale-yellow eyes set off by red orbital ring; yellow legs. **WINTER ADULT:** Head and neck with fine dusky streaking and spotting. **JUVENILE & FIRST-WINTER:** White head and underparts with dark spots and chevrons (gradually lost through winter). Fresh juvenile has neat, scaly upperparts; back usually molts in pale gray feathers by late fall. Underwings mostly white. Tail variable: base whitish; broad blackish distal band, sometimes broken. Dark brown eyes, pinkish bill with dark tip, pinkish legs. **SECOND-WINTER:** Like winter adult, but more black on wing tips (wing tip all black at rest); tail may have black markings; eyes pale.
Similar Species Larger California Gull is a four-cycle gull with a stouter bill and dark eyes in all ages. Adult California has darker medium gray upperparts (white scapular and tertial crescents contrast, unlike Ring-billed), a red gonys spot as well as a

black bill band. Legs often duller and more greenish in winter. California has a larger black wing-tip area and larger white mirrors; in flight from below, a dusky-gray subterminal secondary band (underwings more evenly white on Ring-billed). First-year California is brownish overall, unlike Ring-billed; but second-winter California is similar to first-winter Ring-billed: In addition to noting size and structure, note California's darker gray back; browner greater coverts; legs often have a greenish hue.
Voice Mewing calls and laughing series, higher pitched and less crowing than California Gull.
Status & Distribution Midlatitude N.A., winters to M.A. **BREEDING:** Usually colonial, on low, sparsely vegetated islands in lakes. Arrives

late Mar.–May, departs July–Aug. **DISPERSAL:** Postbreeding dispersal begins by late June, with juveniles recorded by mid-July as far south as Salton Sea, CA. **MIGRATION:** Nearly throughout N.A. where common to abundant in many habitats (parks, parking lots, etc.). Mainly Aug.–Oct. and Mar.–mid-May. Oversummering nonbreeders and immatures regular along Pacific (greatly outnumbered by California Gulls), Atlantic, and Gulf Coasts, local elsewhere. **WINTER:** Rarely north to NL and south-coastal AK. **RARE STATUS:** Casual north to central AK, Bering Sea, Arctic coast, and Japan. Very rare Europe, West Africa, and HI; accidental to Amazonian Brazil.
Population Recolonized Great Lakes region by the 1920s, and populations exploded during the 1960s and the 1970s. Now one of our most numerous gulls (most nest in Canada).

one small white mirror

paler underwing than Mew

2nd winter

1st winter

breeding adult

pale underside to secondaries

spots and streaks on head

outer wing pattern differs from Mew

black subterminal band

pale eye

smaller mirrors than Mew

winter adult

2nd winter

breeding adult

may have tail markings

pale gray above

yellow legs

dark eye

tail pattern variable

1st winter

juvenile

small dark centers to coverts

pinkish legs

WESTERN GULL *Larus occidentalis*   WEGU ■ 1

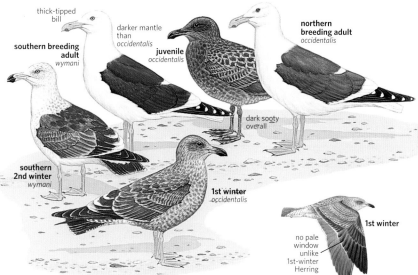

thick-tipped bill

southern breeding adult *wymani*

darker mantle than *occidentalis*

juvenile *occidentalis*

northern breeding adult *occidentalis*

southern 2nd winter *wymani*

dark sooty overall

1st winter *occidentalis*

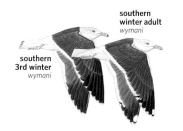

1st winter

no pale window unlike 1st-winter Herring

southern winter adult *wymani*

southern 3rd winter *wymani*

A bulky, dark-backed Pacific coast gull with a bright banana yellow bill. Polytypic (2 sspp.; both in N.A.). L 25" (64 cm) W 58" (147 cm)

Identification A stocky, four-cycle gull with moderately long legs; relatively short, broad wings; and a bill that expands noticeably at the gonydeal angle to give a "blob-ended" appearance. **ADULT:** White head, neck, and body contrast with a dark gray mantle. Consistently darker-mantled than commonly occurring large gulls on West Coast. Bright yellow bill with a reddish spot on the gonydeal angle. Eye varies from yellow to heavily speckled with dark; yellow orbital ring; and pink legs. In flight shows a broad white trailing edge to wing and one mirror on P10. **JUVENILE & FIRST-WINTER:** Dark and uniform-looking, with dark tail and underparts as well as a solid, unstreaked dark mask. Typically shows a paler area behind dark ear coverts, which extends forward to neck sides. Pale rump contrasts with darker tail and mantle. Dusky pink legs, with dark on front of tarsi. Dark inner primaries; no pale inner wing panel; secondaries neatly tipped white, creating a noticeable but narrow whitish trailing edge to wings. **FIRST-SUMMER:** Similar to first-winter, but more uniform and gray-brown on upperparts; slightly paler on head, neck, and underparts. **SECOND-YEAR:** Whitish head and body with streaking on nape and breast sides. Dark gray mantle, contrasting with browner wings, although median coverts usually gray at this age. Tertials have wide white tertial crescent. Black tail contrasts with white rump

and uppertail coverts. Usually pink-based bill color, with darker subterminal band and white tip. Dull pink legs. **THIRD-YEAR:** Much as adult, but retains some dark on tertials, greater primary coverts, and tail. Primaries tend to have small or no obvious white tips; head streaking is noticeable. Dark subterminal band on bill.

Geographic Variation The subspecies *occidentalis* breeds north from Monterey Co., CA, and *wymani* breeds south of Monterey Co. The more southern *wymani* differs in having darker upperparts and a tendency to show pale eyes; other ages are not separable.

Similar Species Adult Yellow-footed Gull has yellow legs; a thicker, even more blob-ended bill; and darker upperparts than *occidentalis*. First-year Yellow-footed resembles second-year Western, but its head and underparts are whiter, its bill shows a yellow base to lower mandible, and its legs begin showing yellow tones. Adult Slaty-backed is darker above and more slate-gray than *occidentalis*, lacking the blue-gray of Western Gull. Winter head streaking on Slaty-backed Gull is extensive, particularly around its bright yellow eye, and Slaty-backed shows a complex wing pattern with a "string of pearls" on primaries to P8, dividing gray primary bases from black wing tip. Western Gull also has a darker underwing than Slaty-backed. First-year Slaty-backed has pale inner primaries and pale underwings with narrow dark trailing edge to wing tips, whereas Western Gull is an even brown.

Hybrids with Glaucous-winged Gull are common and intermediate in appearance. Individuals that look like Western Gulls with odd features may be hybrids. Hybrid features include pinkish orbital rings, a paler mantle, and markings on head in winter.

Voice LONG CALL: Typical of a large *Larus*, but faster paced and higher pitched than Glaucous-winged.

Status & Distribution Abundant. **YEAR-ROUND:** Strongly coastal, venturing inland in San Francisco Bay Area, but otherwise usually on or near the coast or offshore. In recent decades, has spread well inland on coastal slope of Southern CA. **BREEDING:** Colonial, usually on rocky islands. **DISPERSAL:** Western Gulls of all ages disperse north after breeding season, as far as BC. Movements begin in July, peak Aug.–Sept., with some remaining into winter. Southbound movements peak in Mar. **RARE STATUS:** Very small numbers are now found at the Salton Sea. Casual to AZ and north to AK; accidental to CO, TX, IL, and NY.

Population Increasing; global population numbers ±117,000 individuals.

occidentalis

wymani

YELLOW-FOOTED GULL   *Larus livens*   YFGU   ■   2

A marine gull endemic to the Gulf of California, Yellow-footed Gull has a very thick bill. Monotypic. L 27" (69 cm) WS 60" (152 cm)

Identification A large, stocky, dark-backed gull with a thick and extremely blob-ended bill shape. A three-cycle gull, unlike other gulls of its size. **ADULT:** White head, neck, and body contrasting with a dark gray mantle. Bright yellow bill with a reddish gonys spot. Yellow eye with yellow to orange-yellow orbital ring; yellow legs. In flight it shows a broad white trailing edge to wing and one mirror on P10. **JUVENILE:** A gray-brown gull, brownish above with contrasting white belly and vent. White head and neck with fine but blurry streaking throughout. Dark bill with yellow patch at base of lower mandible; dull pink legs. Dark tail contrasting with whitish rump and uppertail coverts. Wings brownish and uniform, lacking pale inner primary panel; secondaries are neatly tipped white, creating a noticeable but narrow whitish trailing edge. **FIRST-WINTER & FIRST-SUMMER:** Wings and tail as in juvenile, but head and body showing more white, less streaking. Obtains a largely

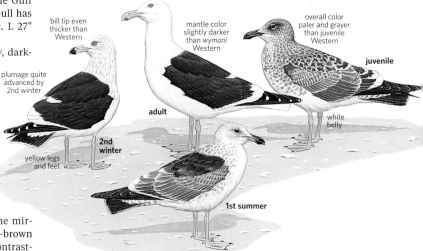

bill tip even thicker than Western

plumage quite advanced by 2nd winter

mantle color slightly darker than *wymani* Western

overall color paler and grayer than juvenile Western

juvenile

adult

white belly

2nd winter

yellow legs and feet

1st summer

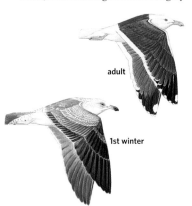

adult

1st winter

solid, dark-gray mantle starting in first winter, unlike other large gulls; mantle is well developed by summer. First-summer birds may begin showing yellow tones to legs. **SECOND-WINTER:** Similar to adult, with solid slate gray mantle and wings; dark on tail; often a darker secondary bar. Head is variably streaked; bill is pale pinkish yellow with a dark terminal third. Legs are yellowish by this stage. **SECOND-SUMMER:** Similar, but body, head, and neck average whiter.

Similar Species Most likely to be confused with Western Gull, particularly the darker subspecies *wymani* (found south of Monterey, CA). Both species (Western rare, but increasing) can be found together at the Salton Sea. Adult Yellow-footed Gull differs from Western Gull in its generally thicker bill and yellow legs. But beware, some adults have yellowish legs and have been mistakenly called Yellow-footed Gulls on the Southern CA coast. Once first-year Yellow-footeds obtain a gray back, they closely resemble second-year Westerns. These are perhaps best separated by looking for any yellow tones that may be present on legs and for Yellow-footed's generally thicker bill. In addition, Yellow-footed's dark bill tends to show a distinctive yellow patch at base of lower mandible. Second-year Western shows a strongly bicolored bill, with an extensive pinkish yellow base and a dark terminal (half to third). The Kelp Gull from S.A. is similar. It can be differentiated by its blackish mantle, reddish orbital ring, greener (often olive) legs, and a

strong bill not so obviously blob-ended as on Yellow-footed Gull. First-year Kelp Gull is not so contrastingly white below and has a thinner bill, and any adult-type mantle feathers are blackish, not slate gray.

Voice LONG CALL: A series of *keow* notes, starting with a slower and longer note and speeding up somewhat toward the end of the series, typical of a large *Larus*. Slower tempo and lower frequency than Western Gull's corresponding call.

Status & Distribution Uncommon. Breeds in Mexico, restricted to Gulf of California. In the US only as a nonbreeding visitor. **BREEDING:** Breeds on various rocks and small islands in Gulf of California (from 31° N to 24°30' N). **DISPERSAL:** Largely resident, but disperses north to Salton Sea, CA, mainly June–Aug., fewer in Sept. and only individuals by mid-Oct. Rare to very rare in winter and spring. Disperses south as far as Oaxaca during this period as well. Flocks at Salton Sea numbered in the hundreds, but many fewer in recent years. **RARE STATUS:** Casual to CA coast in Orange, Los Angeles, and San Diego (several records) counties. Otherwise, there are several records from AZ, NV, UT, and Mono Co. and San Bernardino Co., CA.

Population Current global population is ±60,000 individuals and is likely stable, although little data is available. Yellow-footed Gull was first recorded from Salton Sea in the mid-1960s. The numbers built up to about 50 by 1970, hundreds by the 1990s, but significant decline afterward.

CALIFORNIA GULL *Larus californicus*   CAGU ■ 1

A midsize gull of the Western interior, the California Gull is closely tied to saline lakes when breeding. Polytypic (2 sspp.; both in N.A.). L 21" (53 cm) WS 54" (137 cm)

Identification Long-billed and long-winged four-cycle gull. Bill characterized by being nicely parallel sided, lacking an expansion at the gonydeal angle. **SUMMER ADULT:** Medium gray mantle gull, darker than Ring-billed Gull. White head, neck, underparts, and tail. Bright yellow bill with a red gonys spot and a black subterminal band. During breeding period black bill band is reduced in size; nearly absent on some. Eye dark, with bright red orbital ring, carmine gape. Greenish yellow legs. Distinctive wing pattern shows extensive black on primaries, particularly so on P8 and P7, giving black wing tip a nearly square-cut shape. Mirrors on P9 and P10 are large. **WINTER ADULT:** Similar to breeding adult, but head and neck streaked brown, concentrated on back of neck, nape, and lower neck sides. Throat and front of neck are unstreaked. **JUVENILE:** Variable; usually dark, grayish brown, although some cinnamon brown, and often whitish on center of breast and belly. Bill all-dark. Wings lack paler inner primaries and have dark-based greater coverts. Tail is largely dark; legs pink. **FIRST-YEAR:** Like juvenile, but bill bicolored with pink base and black tip. Juvenal scapulars replaced by variable patterned feathers, but tend to show a solid gray-brown center and shaft streak and a large buffy gray tip with a narrow blackish terminal fringe. Summer bird has whiter head and worn and faded wings, contrasting with newer mantle.

SECOND-YEAR: Medium gray mantle; browner wings with marbled pattern on coverts and tertials; some gray inner median coverts often present. White head, neck, and underparts with streaks concentrated on ear coverts, nape, and breast sides. Bicolored bill, with grayish to gray-green base; legs similarly greenish gray. Wing pattern at this age shows blackish outer primaries clearly contrasting with paler gray inner primaries; dark secondary bar. Tail remains blackish, but now contrasts with white rump and uppertail coverts. In second-summer, head and body much more whitish; may obtain adultlike soft-part colors. **THIRD-YEAR:** Like adult, but retains dark on greater primary coverts and tail. White mirrors on primaries not as well developed.

Geographic Variation Nominate subspecies breeds east to CO, UT, and ID. Subspecies *albertaensis* farther east and north in NT, AB, SK, NB, ND, and SD; intermediates in MT. It is larger, larger billed, and paler on mantle than *californicus* and has a tendency to show less black on primaries and larger mirrors, with the tip of P10 often with entirely white.

Similar Species Distinguished from adult Herring Gull by darker mantle, dark eyes, greenish legs, and black and red on bill tip. Second-year birds similar, but note California Gull's darker mantle, dark eyes, grayish bill base and legs, as well as structural differences. See sidebar (p. 243) for separating immatures; also see Ring-billed Gull.

Voice LONG CALL: A series of *kyow* notes; first two are longer and more

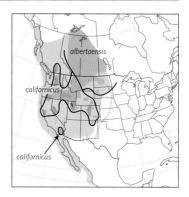

drawn out. Call is higher pitched than corresponding call of Herring Gull.

Status & Distribution Abundant. **BREEDING:** Colonies on flat islands, some on saline lakes. **MIGRATION:** Moves to coast after breeding. In summer and early fall, shows a generally northward movement. Southward movements begin in fall and winter, reaching southernmost winter range in midwinter, before moving north again. Interior birds, *albertaensis*, appear to move farther south than *californicus* and return slightly later in spring. **WINTER:** Shifts to Pacific coast with large numbers also offshore; also very common in interior near coast during winter. **RARE STATUS:** Casual throughout interior to East Coast, appears to be increasingly regular as a very rare to casual visitor to the East.

Population Stable. Global population ±621,000 individuals (2018). US population estimated to have doubled since 1930. For example, in San Francisco Bay, breeders increased from 400 to over 21,000 in last two decades.

HERRING GULL *Larus argentatus*   HERG ■ 1

The most widespread pink-legged gull in N.A., the Herring is common in the East and mainly a winter visitor in the West. Hybrids can be locally common (mainly in the West). Polytypic (4 sspp.; 3, possibly 4, in N.A.). L 22–27" (56–69 cm) WS 54–60" (137–152 cm)

Identification Large gull with four plumage cycles; sloping head and fairly stout bill with distinct (but not bulbous) gonydeal expansion. Pink legs at all ages. Subspecies *smithsonianus* described unless otherwise noted. **BREEDING ADULT:** Pale gray upperparts; black wing tip with white mirrors on outer one or two primaries. (In West, most have mirror only on outermost primary.) Pale yellow eyes; yellow-orange orbital ring; yellow bill, reddish gonys spot. **WINTER ADULT:** Dusky streaking and smudging on head and neck. Duller bill and legs; bill often develops black subterminal mark; orbital ring can be dark to pinkish. **JUVENILE & FIRST-WINTER:** Fresh juvenile sooty brown overall; neat, scaly upperparts; strong dark barring on tail coverts. Inner primaries form pale panel on upperwing. Tail is mostly blackish (but some with extensive whitish hue at base). **VARIABLE FIRST-WINTER MOLT:** Some (especially East Coast breeders) soon attain new barred and mottled back feathers; others (especially West Coast winter populations) retain most or all juvenal plumage through winter. Head often bleaches to whitish. Blackish bill soon develops dull pinkish color on base; rarely pink with clean-cut black

tip by midwinter. **SECOND-WINTER:** Resembles first-winter, but back usually with pale gray feathers (from none to a solid, pale gray saddle); tail mostly black. Pink bill with black distal third (more adultlike in second summer); eyes often pale. In second-summer, head and body whiter; bill rarely like adult, but usually with black distal mark. **THIRD-WINTER:** Highly variable. Some resemble second-winter; others resemble winter adult, but have more black on wing tips, some black on tail. Best aged by adultlike pattern of inner primaries. (Second-winter's inner primaries resemble first-winter's.) Bill usually pink with black distal band; eyes pale. In third-summer, head and body mostly white like breeding adult; bill like adult or with some black marks. **ADULT VEGAE:** Upperparts medium gray; eyes dark (mostly) to pale (rarely); reddish orbital ring; rich pink legs. **JUVENILE & FIRST-WINTER VEGAE:** Head and body paler overall than *smithsonianus*, with sparser dark barring on tail coverts; tail whitish based with broad blackish distal band. Older ages best distinguished by medium gray tone of upperparts. **ADULT ARGENTATUS & ARGENTEUS:** Not safely told from *smithsonianus*. **FIRST-WINTER ARGENTATUS & ARGENTEUS:** Paler and more checkered above and whiter on rump and tail base than first-winter *smithsonianus*. Resembles *vegae*, from which perhaps not safely told except by (presumed) distribution.

Geographic Variation Recent work

indicates that full-species status may be warranted for several taxa in the traditional Herring Gull complex, including "American" Herring Gull (*smithsonianus*), "European" Herring Gull (*argentatus* of Scandinavia and *argenteus* of Iceland, UK, and northwest Europe), and "Vega Gull" (*vegae* of east Asia and Bering Sea region).

Similar Species Adult fairly distinctive, but see "Thayer's" subspecies of Iceland Gull, which is smaller with more slender bill, shorter legs; eyes often dark; wing tips slaty blackish at rest and mostly pale from below, showing much more white on outer primaries than do Herrings in the West. Beware hybrids (of all ages) between Herring and Glaucous-winged or Glaucous. Juvenile and first-winter Herrings are extremely variable, but not like any other regular large gull in the East. Main problem in the West is separation of small female Herring from dark "Thayer's" subspecies of Iceland. Besides different structure, "Thayer's" outer primaries are more extensively pale on inner webs, creating venetian-blind pattern on spread outer primaries; unlike Herring, more

striking pale eye

extensive streaking

pale gray upperparts

juvenile *smithsonianus*

winter adult *smithsonianus*

pink legs

pale eye by late 2nd winter

often pale headed

2nd winter *smithsonianus*

often more adultlike

3rd winter *smithsonianus*

apical spots small or absent

some dark in tail

obvious pale window; compare to California Gull

1st winter *smithsonianus*

dark tail

1st winter *smithsonianus*

1st winter *smithsonianus*

pale underside to secondaries

breeding adult *smithsonianus*

Separating First-winter California Gulls from First-winter Herring Gulls

In first winter these two species look similar, especially Herrings that show a crisply bicolored bill. Both are brown with pinkish legs. Though they overlap in size, these gulls do consistently differ in structure. The long-winged California is slimmer, with a long, even-sided tubular bill. The Herring is bulkier, showing a deeper belly, thicker neck, and disproportionately shorter, broader wings; the bill tends to expand a bit at the gonydeal angle. Most first-winter Herrings, especially by midwinter, have a dark bill or a variable, but ill-defined pinkish base to the bill; very few show a crisply set-off black tip on a pink bill, like California's. Any black "bleeding" back toward the gape along the cutting edge suggests a Herring.

In flight, wing patterns differ. Californias do not show strikingly paler inner primaries, thus they look more evenly patterned. They also show dark greater coverts with paler tips, creating a second dark bar on the wing—the other dark bar being the secondaries. Herrings have strikingly paler inner primaries, and due to more banding on the greater coverts, they lack the second

dark bar. Many first-winter Herrings show a whitish head that contrasts with the dark body. Californias that have a whitish head also show a similarly pale neck and breast, with contrasting dark streaks on the breast sides and nape. At rest, California's dark greater coverts, with their whitish markings concentrated at the tip, create a whitish "skirt" pattern. The white markings also concentrate on the inner greater coverts, forming a "covert crescent" inside of the tertial crescent. These patterns are not shown by most Herrings. Finally, mantle molt begins early on Californias, and by the first winter they show a complex, messy mosaic of feather patterns. Herrings often look less messy above.

California Gull, juvenile (CA, Sept.)

Herring Gull, 1st winter (CA, Nov.)

solidly dark wing tip. See sidebar above for separation from California Gull. On older immatures, note overall shape, bill structure, and upperwing pattern. Some Herring × Glaucous-winged hybrids look like adult "Vega Gull" but are often bulkier, with slightly paler upperparts and wing tips. First-year "Vega Gull" is told from bulkier Slaty-backed by more solidly blackish outer primaries; narrower, blacker tail band. Bleached first-summer birds are not always identifiable.

Voice Varied. Long call has slightly honking or laughing quality. Call of "Vega Gull" is lower pitched and harsher.

Status & Distribution The following refers to *smithsonianus* unless otherwise noted. **BREEDING:** Breeds N.A.

Common (May–Aug.); colonial or in scattered pairs on coastal islands, islands in lakes, on buildings. Since the late 1980s has bred in LA, TX. Subspecies *vegae* is fairly common on St. Lawrence I. (May–Sept.). **MIGRATION:** Nearly throughout N.A., where common to abundant in many regions. Mainly Oct.–Nov. and Feb.–Apr. Subspecies *vegae* rare or casual in western and central Aleutians. **DISPERSAL:** Subspecies *vegae* is rare to uncommon to Arctic coast of AK and Bering Sea islands (mainly June–Sept.) and uncommon (Aug.–Sept.) to Seward Peninsula. **WINTER:** Coastal, inland, and offshore. In West, uncommon along much of immediate coast in range of more numerous Western and Glaucous-winged Gulls. Wintering

birds arrive continent-wide by Oct., with marked increase during mid- to late Oct. Departs most wintering areas by May, but some nonbreeders oversummer along Pacific, Atlantic, and Gulf Coasts; locally elsewhere. **RARE STATUS:** Casual *(smithsonianus)* to Europe and HI. Subspecies *vegae* possibly accidental in TX and CA, possibly elsewhere; none substantiated with specimens. Subspecies *argentatus* and/or *argenteus* is casual (mainly Nov.–Apr.) in the East, most records from NL. **Population** In N.A., recovered after egging and feather hunting in the late 1800s. (US population only 8,000 pairs in 1900.) East Coast population greater than 100,000 pairs in mid-1980s and spreading south. Numbers in Northeast and Atlantic Canada now declining.

white tail base

1st winter
argenteus

1st winter
vegae

more checkered
upperparts

paler than
smithsonianus

darker mantle
color than
smithsonianus

winter
adult
vegae

white tail
base

1st winter
vegae

breeding
adult
vegae

Large White-headed Gull Basics

The best way to learn these gulls is to learn the common species and study how they change throughout the year. Consider the following topics.

Structure

Structure is shorthand for shape and size. Relative size can be assessed by comparing the gull to well-known and wide-ranging standard species such as medium-size Ring-billed Gull and large Herring Gull. Shape is also best learned by comparison. For example, the Lesser Black-backed Gull is a long-winged species. You could quantify this by noting how far the primaries extend past the tail, and comparing that to a nearby, similar species.

Adult Features

Studying the features of the easier-to-identify adults simplifies identification of the more variable immatures. As immatures age, they become more and more like adults. In addition to noting structure when looking at adult gulls, one should concentrate on mantle shade, eye color, orbital ring color, leg color, streaking pattern (on winter birds), and wing-tip pattern (with particular reference to the extent of black and the number and size of mirrors). On a perched adult gull, the mirror on the outermost primary (P10) is often visible on the wing tip on the opposite side of the bird, where the underside of the feather is visible.

Herring Gull, juvenile (NJ, Jan.)

Ageing

Another topic to master—and it is the most difficult—is how gulls change with age. To understand these changes, consider the beginning point (the juvenile) and the

Herring Gull, winter adult (NJ, Feb.)

end point (the adult). All other plumages are intermediate between these two points. Juveniles are generally brown, with paler markings and edgings on the mantle and wings. On juveniles and first-year birds, concentrate on bill pattern, tail pattern. Note that while leg colors vary from pinks to yellow in adults, juveniles of even the yellow-legged species show pink legs. The juvenal is the first plumage obtained after the down, and it all grows in at the same time. This gives the juvenile an even appearance, with all feathers the same age. In older age classes, newer and fresher feathers always contrast with older and more worn feathers.

Molt

Plumage maturation is caused by molts, which can be complete (all feathers replaced) or partial (largely body feathers with no wing and tail feathers). Adult gulls have a complete molt after breeding and a partial molt before breeding. In larger gulls, the molt out of the juvenal plumage lasts a long time—often into the first spring—and is partial. Most noticeable is the change in the upperparts, with new back and scapular feathers. Note that early-molted feathers are more juvenal-like (brown) and late-molted feathers are more adultlike (gray-brown). By spring, the second molt starts, and this is a complete molt. After the first year, the immature bird's molts match up to the adult's. There is a complete molt in summer/fall and a partial molt in late winter.

Remember, first-year gulls are generally brownish but retain juvenal wings and tail. Second-year birds have gone through a complete molt, so they have a new wing and tail. In the three-year gulls (which obtain adult plumage in the third year), this second-year plumage is adultlike with some immature features. In four-year gulls, second-year immatures have an adultlike mantle contrasting with a browner wing. Third-year plumage is obtained through a complete molt, and it looks often very much like an adult except for minor features, such as showing some black on the tail, smaller primary mirrors, more extensive black on primaries, and smaller white primary tips. Finally, hybridization, feather wear, and bleaching can confound these issues, and not all immature gulls will be identifiable.

YELLOW-LEGGED GULL   *Larus michahellis*   YLGU   ■   4

The status in N.A. of this very rare visitor from Europe is masked by identification difficulties. Polytypic (2 sspp.; both may have occurred in N.A.). L 21–26" (53–66 cm) WS 52–58" (132–147 cm)
Identification Large gull with four plumage cycles. Fairly stout bill, slight to distinct (but not bulbous) gonydeal expansion. **BREEDING ADULT:** Medium gray upperparts; black wing tip with white mirrors on outer one to two primaries. Yellow legs; pale lemon eyes; reddish orbital ring. Orange-red gonydeal spot often extends to upper mandible. **WINTER ADULT:** Dusky head and neck streaking concentrated in half-hood and on lower hind neck. Streaking most distinct in late fall; by midwinter most birds are white headed. Bill quite bright yellow, rarely with black subterminal marks. **FIRST-WINTER & SUMMER:** Brownish overall; whitish ground color to head and underparts.

Head, neck, and chest often bleach to mostly whitish with variable dusky brown streaking and spotting by spring. White tail coverts sparsely barred dark brown. Bright white base to tail, with sparse blackish markings and broad blackish distal band—striking in flight. Slightly paler inner webs to inner primaries form an indistinct paler panel on spread upperwing. Blackish bill can show dull pinkish base by spring; legs pinkish. **SECOND-YEAR:** Second-winter resembles first-winter, but back usually has some medium gray feathers; tail ranges from extensively black to variably mixed with white; inner primaries average slightly paler, but do not form an obvious pale panel. Some second-winters (likely *atlantis*) have a distinctly dark-hooded appearance. In second-summer, head and body whiter; usually a solid medium gray saddle. Brownish to pale lemon eyes; reddish orbital ring in summer. Through winter, bill typically pinkish basally, blackish distally, with a pale tip and sometimes a blush of red at gonys. In summer, bill typically yellow with reddish gonydeal smudge and black distal band. Legs dull pinkish, yellowish by second summer. **THIRD-YEAR:** Resembles winter adult, but more black (less white) on wing tips; some have black on tail.

Geographic Variation Taxonomy complex. Two widely recognized taxa in western Europe: smaller and shorter-legged *atlantis* of Azores and larger, longer-legged *michahellis* of Mediterranean region; both may reach N.A., but most, if not all, appear to be *atlantis*. Other populations in western Europe may deserve recognition as subspecies, but more study is needed.

Adult *atlantis* is darker backed, and immatures are more brownish overall, suggesting Lesser Black-backed. Adult *michahellis* is paler backed, and immatures are more whitish on head and body, suggesting Great Black-backed. **Similar Species** Note adult's medium gray upperparts (intermediate between Herring and Lesser Black-backed) and yellowish legs. Be aware that Herrings can have yellow legs (mainly in spring) and that Herring × Lesser Black-backed hybrids may closely resemble Yellow-legged. (Hybrid adults typically have extensive dusky head streaking in winter, unlike Yellow-legged.) Lesser Black-backed averages a more slender bill, has narrower and relatively longer wings, and is slightly to distinctly darker above; it has less contrasting and slightly less extensive black wing tips (longer gray basal tongues on P8 and P9) and heavier dusky head and neck streaking in winter. Herring has less extensive black wing tips that often have more white (at least in Northeast, where Yellow-leggeds are most frequent), pinkish legs, and yellow-orange orbital ring; in winter, Herring has heavier dusky head and neck streaking and often a duller and more pinkish bill, with more distinct dark distal marks, but no red on upper mandible or gape (shown by many Yellow-leggeds). California is on average smaller and lighter in build with more slender bill, dark eyes; winter adult typically with heavier dark hind neck markings, often more greenish legs. First-winter Yellow-legged is most similar to Lesser Black-backed and perhaps not always separable. Subspecies *michahellis* is larger and bulkier than Lesser Black-backed, with blockier head, stouter bill, and longer legs; tail base averages more extensively white (and less barred) at sides; and head and underparts often whiter overall. Also see Great Black-backed

Gull. Subspecies *atlantis* is structurally more similar to Lesser Black-backed, but is on average bulkier and broader winged; outer rectrices usually unbarred at base (usually barred on Lesser Black-backed). Herring Gull is generally darker and browner overall with heavy dark barring on tail coverts, mostly black tail, pale inner primary panel, and often paler-based bill. On older immatures, note overall shape, bill structure, medium gray tone of upperparts, and upperwing pattern. **Voice** Harsher and more grating than Herring Gull; more similar to voice of Lesser Black-backed.

Status & Distribution Western Europe to North Africa. **RARE STATUS:** Casual (mainly Oct.–Apr.; also June and Aug. records from QC) to eastern N.A., from Atlantic Canada (mainly Newfoundland) south to mid-Atlantic coast. Accidental to TX. Specimen from QC referred to *atlantis*; provenance of other birds uncertain. **Population** In western Europe, *michahellis* has increased since the 1970s; no trends reported for *atlantis* (more than 8,000 pairs in the 1990s).

Breeding Range

michahellis

atlantis

1st winter
michahellis

no pale window

white tail base

breeding adult
michahellis

1st winter
michahellis

stouter bill than Herring

yellow legs

winter adult
michahellis

darker gray above than Herring

heavy streaking on head gives a hooded appearance

2nd winter
atlantis

large red spot on lower mandible

winter adult
atlantis

most if not all N.A. records appear to be darker-mantled *atlantis*

ICELAND GULL *Larus glaucoides* ICGU ■ 2

This polytypic species now includes the "Thayer's Gull," a taxon with a complicated taxonomic history extending back over a century. First described as a species by W. S. Brooks in 1915, it was soon merged into Herring Gull by J. Dwight. Work done by F. Salomonsen and especially A. H. Macpherson in the mid-20th century suggested that "Thayer's" was best placed with Iceland, not Herring. Published "research" by N. G. Smith in 1966 indicated that "Thayer's," Iceland, and Herring should all be treated as separate species, a recommendation followed by AOS (formerly AOU) in 1973. Studies by R. R. Snell in the mid-1980s suggested that "Thayer's" should be merged into Iceland; failures to replicate Smith's results raised serious accusations of fraud with Smith's studies. AOS followed Snell (and others) and lumped "Thayer's" with Iceland in 2017. Additional "actual" studies may further resolve this complex taxonomic issue, but accessing, let alone studying, the nesting colonies presents extreme difficulties (both groups nest on Arctic sea cliffs). The species is currently divided into two subspecies groups, Iceland (*glaucoides* and *kumlieni*) and "Thayer's" (*thayeri*). The Iceland group in N.A. are birds of Atlantic Canada and the Northeast in winter; it is an extremely variable-looking gull; most eastern N.A. records are of the subspecies *kumlieni*, commonly known as the "Kumlien's Gull." "Thayer's" breeds to the north and west of Iceland and winters primarily on the West Coast. Polytypic (3 sspp.; all in N.A.). L 20–24"

(51–61 cm) WS 51–56" (130–142 cm)

Identification A medium-large four-cycle gull with a relatively small and slender bill; all ages have dark pink legs and have extensive pale in wing tips. **BREEDING ADULT *KUMLIENI* & *GLAUCOIDES*:** Clean white head and neck. "Kumlien's" wing-tip pattern highly variable: Most birds have dark gray markings on outer four or five primaries and a large white tip to P10. Birds with reduced gray markings are not rare, though, and a few birds even appear to have all-white wing tips. (Darker gray restricted to outer webs of P9–P10 often not discernible in the field.) Birds with darkest and most extensive wing-tip markings approach "Thayer's" in pattern, and some may be intergrades with "Thayer's." Nominate *glaucoides* has white wing tips that lack gray markings, and its upperparts are slightly paler than "Kumlien's" (more like Glaucous). Eyes usually pale but can be dark; orbital ring reddish pink to purplish. Yellow bill with orange-red gonys spot. **WINTER ADULT:** Head and neck variably mottled and streaked dusky; duller bill often greenish based, rarely with dark subterminal marks. **FIRST-WINTER:** Retains full juvenal plumage through most of the winter. Pale overall; wing tips vary from medium brown to whitish; upperwing lacks contrasting dark secondary bar of "Thayer's." Bill black or with variable dull pinkish hue basally. Upperparts, including tertials, finely patterned. **SECOND-WINTER:** Resembles first-winter, but back often has some pale gray; bill usually pinkish with broad black distal band; eyes can be pale. **THIRD-WINTER:** Resembles winter

adult, but upperwings and tail usually washed brownish; bill duller, dull pink with a black subterminal band. **"THAYER'S GULL" (*THAYERI*):** In all ages shows a largely pale underwing, with dark trailing edge of outer primaries. "Thayer's Gull" averages larger than both subspecies of the Iceland Gull group. **BREEDING ADULT:** Medium gray mantle; white head, neck, underparts, and tail. Dull yellow bill, sometimes with greener base; red gonys spot. Eye color is variable: most are dark eyed, yet a few appear pale eyed, particularly in sunny conditions; dark red to purplish orbital ring. Bright pink legs. Distinctive variable wing pattern: Outer five to six primaries are black (or blackish) with extensive white tongues on inner and sometimes outer vanes, which breaks up black into thin strips—a venetian-blind effect. The mirror on P9 is confluent with white tongue. **WINTER ADULT:** Like summer adult, but head and neck streaked. **FIRST-WINTER:** Retains full juvenal plumage through most of winter. Variable, ranging from dark, Herring Gull–like birds to lighter extremes; exacerbated by bleaching and wear by late winter and early spring. Classic juvenile evenly warm brown, with mantle and wings

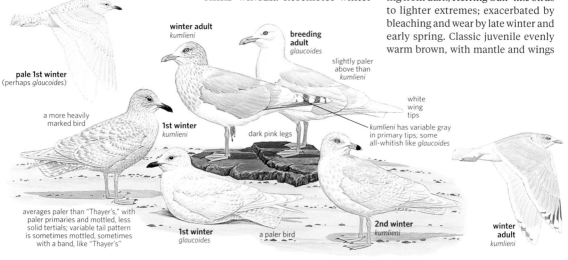

pale 1st winter
(perhaps *glaucoides*)

a more heavily marked bird

winter adult
kumlieni

breeding adult
glaucoides

slightly paler above than *kumlieni*

white wing tips

1st winter
kumlieni

dark pink legs

kumlieni has variable gray in primary tips; some all-whitish like *glaucoides*

averages paler than "Thayer's," with paler primaries and mottled, less solid tertials; variable tail pattern is sometimes mottled, sometimes with a band, like "Thayer's"

1st winter
glaucoides

a paler bird

2nd winter
kumlieni

winter adult
kumlieni

checkered with pale buff or whitish markings. Primaries dark brown but not blackish; show crisp pale fringes. Dark-centered tertials with variable pale markings on tips. Typically, body is palest, tertials darker, and wings and tail darkest. In flight, there is a noticeable dark secondary bar as well as paler inner primaries. Dark bill and bright pink legs. **FIRST-SUMMER:** Similar to juvenal, but has a tendency to bleach out and become very pale, contrasting with newer mantle. **SECOND-WINTER:** Medium-gray mantle; browner wings; greater coverts and tertials marbled or vermiculated. Extent of blackish primaries maximal at this age, contrasting with paler inner primaries. Tail mostly dark but contrasts with whitish rump and uppertail coverts. Bill usually bicolored with pinkish base and dark tip; some retain darker bill. **THIRD-WINTER:** Very much as adult, but some dark on greater primary coverts and tail.

Geographic Variation See above for detailed descriptions of all three subspecies. Separating palest-winged *kumlieni* from nominate *glaucoides* (rare in N.A.) is extremely difficult. Given the plumage variation, recognition of *kumlieni* as a subspecies is open to question—it does not fit the standard definition for a subspecies. Separating "Thayer's" from both subspecies of Iceland is more straightforward, but some (perhaps many?) should be left unidentified. On immatures wear and bleaching exacerbate the identification challenges.

Similar Species Glaucous is larger, with bigger and deeper bill and relatively short wing projection beyond tail; wing tips always white or with faint dusky subterminal marks in first-winter. First-winter Glaucous has brightly bicolored pink-and-black bill. Adult Glaucous has yellow to orange orbital ring (which can have pinkish tint in winter). "Thayer's Gull" is most often confused with Herring Gulls of corresponding ages. Structural differences are useful: "Thayer's" shows a smaller bill, steeper forehead, shorter legs, and more potbellied appearance. Adult "Thayer's" identified by reduced black on upperwings, extensive white tongues, extensively pale underwings, darker eye (when present), reddish orbital ring, greener bill, pinker legs, and often slightly darker mantle.

First-year "Thayer's" are paler than Herrings and they have paler underwings, more uniform upperwing pattern, and noticeably paler tertials than primaries that are best observed on perched birds.

Voice Shriller than Herring Gull; rarely vocal in winter.

Status & Distribution *KUMLIENI* & *GLAUCOIDES*: Subspecies *kumlieni* breeds in eastern Arctic Canada, winters northeast N.A. Nominate *glaucoides* breeds Greenland, winters to northwest Europe. Following applies to *kumlieni* unless stated. **BREEDING:** Fairly common, but local (June–Aug.) on sea cliffs. **WINTER:** The great majority of birds recorded south of New England are immatures. Fairly common Atlantic Canada (arriving late Oct.–Nov., departing Apr.–May; very rare in summer), smaller numbers south to NJ, inland to eastern Great Lakes. Rare (mainly Dec.–Mar.) south to NC; casual to FL, Gulf Coast, Great Plains. **RARE STATUS:** Casual (Sept.–Oct.) along Arctic coast of AK and south (mainly Dec.–Mar.) into the Northwest and south to CA (annual in recent years). Nominate *glaucoides* probably rare but regular (Nov.–Mar.) in Atlantic Canada, accidental (Oct.–Jan.) in the West (AK, YT, CA). **"THAYER'S"** (*THAYERI*): Common Pacific coast and some adjacent valleys in winter; rare to continental interior. **BREEDING:** More northerly breeding distribution than *glaucoides* and *kumlieni*. Colonial, on steep coastal cliffs from northwest Greenland to central Canadian Arctic. **MIGRATION:** Fall migrants reach coastal AK by early Sept., central CA by late Oct., larger numbers present Dec. and remain

until early Apr. **WINTER:** Variety of habitats, from coastal beaches and estuaries to inland garbage dumps and lakes. **RARE STATUS:** Casual to eastern seaboard.

Population Estimates of 5,000 breeding pairs of *kumlieni*, 50,000–100,000 breeding pairs of nominate *glaucoides* (±40,000 pairs in southwestern and eastern Greenland), and 6,300 breeding pairs of *thayeri*. There is limited data due to the remoteness and difficult terrain of breeding sites, but apparent increases in N.A. breeders and a stable European population have been reported. The species is extensively hunted in Greenland.

"Thayer's Gull"

1st winter

dark secondary bar

brownish on outer webs from above

pale underwing

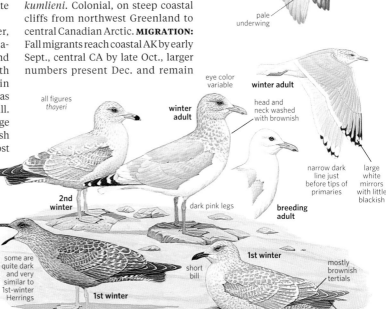

all figures *thayeri*

eye color variable

winter adult

winter adult

head and neck washed with brownish

2nd winter

dark pink legs

narrow dark line just before tips of primaries

large white mirrors with little blackish

breeding adult

short bill

1st winter

mostly brownish tertials

some are quite dark and very similar to 1st-winter Herrings

1st winter

brownish primaries with pale fringes; darker than *kumlieni*

LESSER BLACK-BACKED GULL   *Larus fuscus*   LBBG   ■   2

This European species, often seen with Herring Gulls, has "colonized" N.A. in the last 60 years, but breeding birds have yet to be found here. Polytypic (3–5 sspp.; 2 in N.A.). L 21–25" (54–64 cm) WS 52–58" (132–147 cm)

Identification Large four-cycle gull slimmer in build than Herring, with relatively longer and narrower wings and more slender bill. Subspecies *graellsii* described. **BREEDING ADULT:** Slaty gray upperparts; black wing tip with white mirrors on outer one to two primaries. Yellow to orange-yellow legs; pale lemon eyes; reddish orbital ring. Bright yellow bill with large orange-red gonydeal spot. **WINTER ADULT:** Dusky head and neck, streaking often concentrated around eyes. **FIRST-WINTER & SUMMER:** Brownish overall; whitish ground color to head and underparts. Many with whiter head and underparts that contrast with dark upperparts. White tail coverts sparsely barred dark brown. Tail has bright white base with blackish barring and variable, clean-cut blackish distal band. Inner primaries on upperwing not appreciably paler than outers; dark-based greater coverts often form a dark band. Blackish bill can show pale base by spring; pinkish legs. **SECOND-WINTER & SUMMER:** Like first-winter, but some feathers on back are medium gray. By summer, head and body whiter; usually a solid slaty gray saddle. Brownish to pale lemon eyes. Legs yellowish pink, yellowish by second summer. **THIRD-YEAR:** Resembles winter adult, but more black (and less white) on wing tips; some have black on tail; bill blacker in winter.

Geographic Variation Records in N.A. refer mainly to the paler-backed *graellsii* of western Europe, with fewer records of darker-backed southwest Scandinavian breeding *intermedius*. Many birds intermediate in appearance, however, and not safely assigned to subspecies. No substantiated records of Baltic-breeding *fuscus* for N.A.

Similar Species Adult and subadult Lesser Black-backed Gulls are fairly distinctive. First-winter Lesser Black-backed can be separated from Herring by smaller size, relatively longer and narrower wings, more slender black bill, and whiter ground color to head and underparts (especially tail coverts, which have sparser dark barring); in flight, upperwing pattern and tail/uppertail-covert pattern distinctive. Primary molt in first summer is later than Herring. Main problems arise with separation from rarer species and hybrids. Kelp is larger and bulkier with broader wings and a stouter bill (deeper at gonys than base, the reverse of a typical Lesser Black-backed). Adult Kelps are blackish above; legs often more greenish yellow. First-winter Kelp perhaps not distinguishable from Lesser Black-backed by plumage, but tail often has broader blackish distal band; note also bill shape, broader wings. Kelp x Herring Gull hybrids can resemble Lesser Black-backed closely, but are bulkier in build with a stouter bill (deepest at gonys); black underwing tip of hybrid adult contrasts more. Yellow-footeds larger and bulkier with much deeper, bulbous-tipped bill and broad wings; adults lack distinct dusky head markings in winter. Adult California is paler above, but structure similar to a small Lesser Black-backed; note California's dark eyes, often more greenish legs, wing-tip pattern.

Voice Deeper, hoarser than Herring.

Status & Distribution Northwestern Europe, wintering to Africa. Has increased dramatically in N.A., especially since the 1980s and continues to do so; first N.A. record in 1934 in NJ. **WINTER:** Rare to locally fairly common in East (mainly Sept.–Apr., some immatures oversummer); max. counts are from mid-Atlantic region south to FL. **RARE STATUS:** Casual to rare (mainly Sept.–Apr.) in interior and to West Coast, north to AK. A specimen from Shemya Is., AK, on 15 Sept. 2005 was determined to be *heuglini* from northwest Russia; this subspecies is variously treated as a subspecies of Lesser Black-backed, or its own species.

Population Large increase in *graellsii* and *intermedius* populations since the mid-1900s.

heavily spotted head and neck, dark around striking pale eye

slender bill with large reddish spot on gonys

breeding adult
graellsii

darker than *graellsii*

even longer winged

dark slaty gray above

winter adult
graellsii

winter adult
graellsii

thick dark tail band with white base

winter adult
intermedius

long winged

yellow legs

2nd winter
graellsii

rather small white mirror on P10

dark double bar

1st winter
graellsii

no pale window

extensive dark underwing

darker greater coverts

1st winter
graellsii

SLATY-BACKED GULL   *Larus schistisagus*   SBGU ■ 3

This marine gull of the Asian North Pacific regularly visits AK. Monotypic. L 25" (64 cm) WS 58" (147 cm)

Identification A large, long-necked, short-legged, potbellied, four-cycle gull. Strong bill is even in thickness, not showing a bulge toward the tip. Paradoxically, immature plumages show some resemblance to pale-winged gulls such as "Thayer's" and Glaucous-winged Gulls, although adults have dark backs and wings. **SUMMER ADULT:** Slate gray mantle, appearing blackish in some lights; darker mantle than Western Gull. White head, neck, underparts, and tail. Yellow bill, red gonys spot. Striking yellow eye with a red orbital ring. Legs are pink, often bright bubble-gum pink. Wing pattern distinctive: Slate gray of primary bases separated from black primary tips by a series of white tongues or spots, which begin at P4 or P5 and extend to P8; this line of white spots is referred to as a "string of pearls." Outer primaries are black to the base; mirrors on both or missing on P9. Underwing is similarly distinctive, appearing tricolored. White wing linings contrast with dark gray (or in some lights pale gray) underside of secondaries and primaries and a restricted black area at tip of primaries separated from gray underwing by the "string of pearls." White trailing edge on secondaries is broad, with inner secondaries often looking entirely white. **WINTER ADULT:** Head and neck are densely but crisply streaked, concentrating around eye and also lower nape. **JUVENILE & FIRST-WINTER:** Variable and difficult to characterize at this age. Tend to look uniform, lacking contrast; in this way they resemble young Glaucous-winged Gulls, although primaries, tail, and secondaries are darker than rest of plumage. Pale underwings show a dark trailing edge to primaries. Upperwings show a large pale inner primary patch, which extends out at least to P8, unlike on other similar large gulls. Other features include uniform or poorly marked greater coverts, dark centered tertials without much pale patterning, a dark tail, black bill, and bright pink legs. **FIRST-SUMMER:** Similar to first-winter, but has a tendency to bleach out and become very pale, with darker tertials, tail, and primaries. **SECOND-WINTER:** At this age, dark slate-gray of mantle is obvious, contrasting with pale brownish wings. Coverts are uniform and pale brown; folded wing shows very little contrast or pattern whatsoever. Tertials are dark centered, with a well-developed white tertial crescent; folded primaries are blackish, contrasting with paler coverts. In flight, underwings are still pale, with a contrasting dark trailing edge to primaries. Dark tail contrasts with white uppertail coverts and rump. Head streaking is as on adults, concentrated into a dark patch around eye and highlighting the now obviously pale eye. Bill shows a pale base, but it is still largely dark. **SECOND-SUMMER:** Similar, but body, head, and neck average whiter; wing coverts are often incredibly bleached, whitish, and contrast strongly with dark mantle. **THIRD-YEAR:** Looks much as adult, but retains some dark on tertials, greater primary coverts, and on tail.

Similar Species Adults likely to be confused with Western Gull, but Slaty-backed has a darker, cold gray mantle and a strongly streaked head in winter, especially around gleam-

ing yellow eye. Wing pattern of Slaty-backed—particularly the tricolored nature of the underwings—is not found on Western Gull, which is more evenly dark below. Slaty-backed is also shorter legged, potbellied, and longer necked and lacks a blob-ended bill. Identification of first-year Slaty-backed is still uncertain, but useful features are structure, uniform greater coverts, and extensive pale on primaries and underwings. Immature Herring × Glaucous-winged hybrids may show some or all of these features. After the second year, dark mantle, pale eye, and streaking pattern make identification more straightforward.

Voice LONG CALL: Slower and deeper than Western, resembling Glaucous-winged.

Status & Distribution Breeds in northeast Asia; uncommon in western AK. **BREEDING:** One record from Cape Romanzof, AK; also records of mixed pairings. **RARE STATUS:** Annual in winter to Pacific Northwest and south to central CA; casual elsewhere on continent, recorded east to NL and south to FL.

Population Little studied, but believed to be declining.

heavily spotted head and neck, dark around striking pale eye

bill of even thickness

2nd summer

winter adult

blackish slate above

broad white tertial crescent

deep pink legs

dark mantle contrasts with white coverts

broad white edge to secondaries

breeding adult

white spots form "string of pearls"

1st summer

1st summer

rather plain greater coverts

primaries like "Thayer's Gull"

GLAUCOUS-WINGED GULL   *Larus glaucescens*   GWGU   ▪ 1

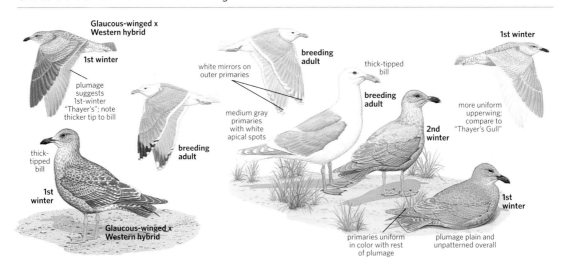

Glaucous-winged x Western hybrid

1st winter

plumage suggests 1st-winter "Thayer's"; note thicker tip to bill

white mirrors on outer primaries

breeding adult

thick-tipped bill

1st winter

breeding adult

medium gray primaries with white apical spots

breeding adult

thick-tipped bill

2nd winter

more uniform upperwing; compare to "Thayer's Gull"

breeding adult

thick-tipped bill

1st winter

Glaucous-winged x Western hybrid

1st winter

primaries uniform in color with rest of plumage

plumage plain and unpatterned overall

This thick-billed Pacific gull is the only large gull that shows primaries similar in darkness to the body. Monotypic. L 26" (66 cm) WS 58" (147 cm)

Identification A large, stocky, pale-winged four-cycle gull with broad wings, longish legs, and a thick, blob-ended bill. **SUMMER ADULT:** Gray primaries nearly unicolored with gray upperparts. White head, neck, body, and tail. Dark eye, pinkish orbital ring, pink legs, and a yellow bill with red spot at the gonys. In flight, wings look gray above; primaries slightly darker, showing white tongues, a "string of pearls," and a mirror on P10. **WINTER ADULT:** Heavily marked head, most typically finely barred or vermiculated with dark. Often bill becomes a duller yellow. **JUVENILE & FIRST-WINTER:** Pale and uniform, with grayish brown primaries similar in darkness to upperparts. Markings on coverts and tertials reduced, often looking vermiculated. Black bill, dusky pink legs with dark anterior tarsus. In flight looks uniform, lacks darker secondary bar or contrasting pale rump; dark tail. Underwings pale, primaries translucent. **FIRST-SUMMER:** Much paler, bleached, worn than first-winter. Primaries may appear whitish at this age. New scapular and upperparts feathers gray. **SECOND-WINTER:** Gray mantle contrasts with browner wings. Coverts and tertials often very uniform dull brownish gray, lacking obvious pale patterning. Gray-brown tail now contrasts with whiter rump and uppertail coverts. Bill begins to show a pink base at this time, but many retain largely dark bill into older age classes. **SECOND-SUMMER:** Similar, but body, head, and neck average whiter. **THIRD-WINTER & THIRD-SUMMER:** Much as adult, but retains some dark on tertials, wing coverts, greater primary coverts, and tail. Primaries tend to have small or no white tips. Bill has at least a dark subterminal band; sometimes terminal half is dark. **HYBRIDS:** Hybridizes commonly with Western (in WA) and also with Herring (in southwest and south-central AK).

Similar Species No other large gray-mantled gull shows primaries that are uniform in darkness with upperparts. Some "Kumlien's" Iceland Gulls may show similar grayish primaries, but petite-billed Iceland is much smaller and longer-winged than Glaucous-winged. In addition, older Iceland Gulls tend to show pale eyes. Hybrids with Western Gull, informally known as "Olympic Gulls," range from nearly identical to parental types to intermediate in appearance. A Glaucous-winged that shows a yellow orbital ring, noticeably darker primaries than upperparts, a darker than average mantle color or, in younger ages, a dark secondary bar, and contrasting paler rump or well-marked coverts is likely a hybrid. Hybrids with Herring Gull may resemble "Thayer's" in plumage features, but note that they are larger billed, bulkier, and longer legged, and often show barred or vermiculated head markings in winter. Juvenile hybrids begin molting their upperparts early in winter; "Thayer's" retains full juvenal plumage into late winter or early spring.

Voice LONG CALL: A typical loud series of evenly spaced *haaaw* notes. Slower pace and lower pitch than Western.

Status & Distribution Abundant. Also breeds in Asian Pacific, south to Japan. **YEAR-ROUND:** Rocky coasts preferred. **BREEDING:** Colonial, often uses small rocky islands, or rooftops in Seattle and Vancouver, BC. **MIGRATION:** Migrates south after breeding; first adults arriving in central CA as early as late Aug. Numbers peak in midwinter; most are gone from central CA by early Apr.; some, mostly immatures, oversummer. First-year birds migrate longer distances than adults. **WINTER:** Some winter near breeding areas; others move south as far as Baja California Sur, Mexico. Some winter inland within 50–80 mi or so of coast. It is noted annually at Salton Sea. **RARE STATUS:** Casual to interior with records east to IL; accidental in NL, Canary Is., Europe, and Morocco. **Population** Estimated 200,000 breeders in N.A.; increase in last 50 years likely due to garbage dumps and waste from industrial fishing.

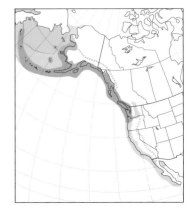

GLAUCOUS GULL   *Larus hyperboreus*   GLGU ■ 1

This northern gull basically looks like a larger white-winged Herring Gull. Even more aggressive than that species, it scavenges at dumps and harbors, often with gatherings of other large gulls. Polytypic (4 sspp.; 3 in N.A.). L 22–29" (56–74 cm) WS 56–63" (142–160 cm)

Identification A large four-cycle gull. Stout but not bulbous-tipped bill often expands slightly at culmen base; can be relatively parallel-edged and "slender" on immature females. All ages have pinkish legs and lack black in wing tips. **BREEDING ADULT:** Clean white head and neck. Pale gray upperparts; pure white wing tips. Pale yellow eyes (dusky flecking on some may indicate hybridization); orange-yellow orbital ring can be orange-red on *pallidissimus* subspecies. Yellow bill; orange-red gonys spot. **WINTER ADULT:** Light to moderate dusky mottling and streaking on head and neck. Duller bill often has pinkish base and dark subterminal marks; orbital ring often fades to dull pinkish. **FIRST-WINTER:** Whitish overall; variable pale brownish patterning. Wing tips white to creamy, often with neat dusky subterminal spots or chevrons on outer primaries. Pink bill with clean-cut black distal third; dark eyes. Often retains most or all of juvenal plumage through winter; it can be bleached white overall by spring, when newly molted body feathers look contrastingly dark. **SECOND-WINTER:** Resembles first-winter, but back usually with some pale gray by spring; bill usually has distinct pale tip; eyes can be pale. **THIRD-WINTER:** Resembles winter adult, but upperwings and tail usually washed brownish. Bill duller, yellowish with a black subterminal band; often some reddish on gonys.

Geographic Variation Smallest and darkest-backed *barrovianus* (AK); larger and paler-backed *leuceretes* (Canada, Greenland, and probably Iceland); nominate *hyperboreus* (northwestern Eurasia) is smaller and darker than *leuceretes*; largest and palest-backed *pallidissimus* (northeast Asia). The subspecies *pallidissimus* occurs in northwestern AK and breeds on St. Lawrence and St. Matthew Is.; at the latter location it can be seen with *barrovianus*. Immatures not identifiable to subspecies in the field.

Similar Species Smaller females may be confused with Iceland Gull. Iceland's bill is smaller, shorter, and more slender (often greenish based on winter adult); relatively longer wing projection beyond tail: At rest, Glaucous's tail tip usually falls between tips of P7 and P8, Iceland's between tips of P6 and P7. Adult Iceland has a reddish to purplish pink orbital ring. First-winter Iceland lacks Glaucous's clean-cut, pink-and-black bill, but second-winter Iceland's bill can be similar. Glaucous-winged Gull has a more bulbous-tipped bill and darker wing tips. Note blackish bill on Glaucous-winged (second-winter Glaucous-winged can have a Glaucous-like bill, but its wing tips are grayish). Adult Glaucous-winged is slightly darker above, with gray wing-tip markings. Southern sightings of Glaucous Gull should be double-checked for hybrids with Herring or Glaucous-winged; good views often needed for this. Herring Gull hybrids can look very like a Glaucous, but their outer primaries have dusky markings (a ghosting of Herring's pattern). Glaucous-winged hybrids can look even more like a Glaucous;

some not safely identifiable in field. On first-winters, a bulbous-tipped bill and messy bill pattern are clues; on adults, check eye color and orbital ring color. (Faint gray wing-tip markings may be visible only in the hand.) Bleached and leucistic/albinistic birds (especially Glaucous-winged) have caused confusion—check bill structure and pattern. **Voice** Similar to Herring, but slightly hoarser and lower pitched.

Status & Distribution Holarctic breeder, winters to midlatitudes. **BREEDING:** Common on coastal islands, sea cliffs, tundra lakes. Arriving late Apr.–June; mostly departing through Sept. Hybridizes with Herring Gull in AK and Canada and with Glaucous-winged in western AK. **MIGRATION:** Mainly late Oct.–mid-Dec., and in spring Feb.–Apr.; some linger through summer in Canada south of breeding range. **WINTER:** Nov.–Apr. Rare (mainly late Dec.–Mar.) south of northern tier of states. **WINTER:** Very rare to Southern CA, Gulf states, and FL. Rare to casual in interior N.A., away from Great Lakes. **Population** Poorly known (AK population perhaps more than 100,000 individuals); may be fairly stable in N.A.

winter adult

larger and chunkier than Iceland

2nd winter

pale eye and bill tip in 2nd winter

winter adult

very pale gray above

shorter wing tip projection past tail than Iceland

pinkish legs

white wing tips

breeding adult

1st winter

long pink bill with blackish tip

1st winter

1st winter

color of 1st-winter birds varies; some by mid- to late winter are chalky white

GREAT BLACK-BACKED GULL *Larus marinus*   GBBG   ■   1

The largest gull in the world, the Great Black-backed is an efficient scavenger and predator. It commonly preys on ducklings and can snatch adult puffins in flight! Monotypic. L 25–31" (64–79 cm) WS 60–65" (152–165 cm)

Identification A huge four-cycle gull with very long, broad wings and a low-sloping forehead. Its very stout bill (which is notably smaller on females) has a swollen gonys. **BREEDING ADULT:** Slaty blackish upperparts (browner when worn, in summer) blend into black wing tips; large white tip to P10 merges with a large white mirror on P9. Pale pinkish legs; olive to dull, pale yellow eyes; a reddish orbital ring. Yellow bill with an orange-red gonydeal spot. **WINTER ADULT:** Inconspicuous, fine dusky head and neck streaking concentrated mainly on hind neck; birds look white headed at any distance. Bill is duller than in summer, often pinkish at base and with blackish subterminal marks; orbital ring can be pinkish. **FIRST-WINTER & SUMMER:** Head and underparts whitish overall (often bright white by spring) with relatively sparse brownish mottling and streaking. Upperparts contrasty and checkered. Often retains most juvenile plumage through winter. White uppertail coverts and tail; broad black distal tail band typically broken up by internal, narrow white barring; tail base and uppertail coverts with fairly sparse dark bars. Poorly to moderately contrasting pale panel on inner primaries on upperwing. Dark eyes; dull pinkish legs. By spring, black bill usually develops a pale base. **SECOND-WINTER & SUMMER:** Second-winter often looks very similar to first-winter, but primary tips more rounded; greater coverts plainer and browner, typically with fairly fine markings (unlike boldly barred first-cycle coverts); and bill has large pale tip. A few slaty back feathers appear on some birds. By spring, back has some to many slaty feathers; eyes paler brown. In summer, orbital ring can brighten to orange; some birds develop yellow bill with reddish gonydeal smudge and black subterminal band. **THIRD-WINTER & SUMMER:** Some third-winters resemble second-winters (but note bill pattern and adultlike inner primaries); others much

more adultlike, but with black on tail, less white in wing tips, some brownish on wing coverts. In winter, bill typically pinkish with broad black subterminal band and creamy tip. In summer, bill usually brightens and may be indistinguishable from adult, but typically has dark distal marks.

Similar Species Note very large size and stout bill with swollen gonydeal expansion. First-winter and second-winter separated from smaller first-winter Herring by much whiter head and underparts and more boldly checkered upperparts; in flight, note white rump and tail base with broken black tail band. On older immatures and adults, note slaty blackish upperparts and paler pink legs. Adult Kelp has blacker upperparts, limited white in wing tip, and yellowish legs. Adult Great Black-backed × Herring hybrid has been mistaken for Western, but that species is relatively bulkier and broader winged, with a more rounded head and a more bulbous-tipped bill.

Voice Very deep calls, much lower pitched than Herring Gull.

Status & Distribution Northeast N.A., northwestern Europe. **BREEDING:** Common (Apr.–Aug.); in small colonies or scattered pairs on rocky islands, beach barriers, locally on rooftops; fairly common on eastern Great Lakes, rare on western. **MIGRATION:** Mainly late Aug.–Nov. and Mar.–Apr.; small numbers of nonbreeders over-summer in winter range. **WINTER:** Mainly Oct.–Mar., at beaches, fishing harbors, dumps, etc. **RARE STATUS:** Casual to very rare (mainly Nov.–Mar.) on Great Plains, and north (May–July) to Hudson Bay and NU. Accidental in AK, the Northwest, and Southern CA.

Population Dramatic increase in N.A. since the early 1900s, with range expansion both south along Atlantic coast and inland through Great Lakes. Increased in US by about 17 percent per year from 1926 to 1965.

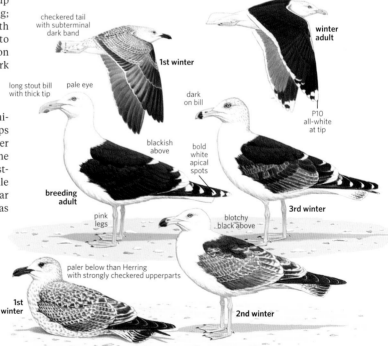

checkered tail with subterminal dark band

1st winter

winter adult

long stout bill with thick tip

pale eye

dark on bill

P10 all-white at tip

blackish above

bold white apical spots

breeding adult

pink legs

blotchy black above

3rd winter

paler below than Herring with strongly checkered upperparts

1st winter

2nd winter

KELP GULL *Larus dominicanus*   KEGU   4

one white mirror

This stocky, longish-legged, black-backed gull is widespread in the Southern Hemisphere, including western and southeastern S.A. It has bred in Louisiana and has hybridized with Herring Gull, complicating identification. Polytypic (2 sspp.; nominate in N.A.). L 23" (58 cm) WS 53" (135 cm)

Identification A four-cycle gull that shows an advanced maturation schedule: Blackish upperparts come in as early as first summer; becomes entirely blackish above by second summer. Bulky, large bill, and long greenish legs. Kelps have thick bills that lack an obvious expansion at the gonydeal angle. Adults have a black mantle, showing little difference in darkness of primaries and mantle; only Great Black-backed approaches this darkness. **ADULT:** Adults show a mix of blackish upperparts; bright yellow bill with red gonys spot; reddish orbital ring; variable but usually pale eye; greenish to yellow-green legs; and one mirror on P10. White head, neck,

underparts, and tail. Legs vary in color, being yellow-green during breeding season but becoming olive to olive-gray when not breeding. **FIRST-YEAR:** Brownish, with well-streaked whiter head and underparts. Juvenile mantle brown with neat pale fringes, many of which are replaced in first-winter by dark gray-brown feathers with darker shaft streaks; extent of gray-brown back and lightening of head and underparts continues into first-summer. Tertials are dark with narrow paler fringes; greater coverts dark at bases. Tail is entirely dark. In flight, wings lack paler inner primaries and have dark secondaries and greater coverts. **SECOND-YEAR:** Upperparts blackish by this age, contrast with browner wings. White head and body; moderate streaking on face, breast sides, flanks. Dark post-ocular line retained in even the whitest individuals. Pale pink-based bill by this age; dark eyes; grayish pink to grayish green legs. Tail is largely blackish or white with a broad black terminal band,

contrasting with white rump. **THIRD-YEAR:** Much like adult, but it retains some dark markings on greater primary coverts and sometimes on tail.

breeding adult

Similar Species The adult Great Black-backed is larger and has dull pink legs, large white tip to P10, and a second mirror on P9. First-year Kelp Gulls differ from Great Black-backeds of similar age by their largely dark tail, more uniformly patterned upperparts, and dark-based greater coverts. Lesser Black-backed Gulls (*graellsi* and *intermedius*) are smaller, slimmer, thinner billed, and much longer winged; adults are paler above than Kelp, which retains its unstreaked head all year long. First-year birds separated on structure and Kelp's darker tail. Kelp x Herring hybrids are similar in mantle color to Lesser Black-backed Gull, but they differ in stockier structure and larger bill.

Voice LONG CALL: Similar to Herring Gull's, although faster in tempo.

Status & Distribution Casual in N.A., widespread in Southern Hemisphere. **BREEDING:** Colonial; formerly bred and hybridized with Herrings on the Chandeleur Is. of LA. **RARE STATUS:** Casual in eastern N.A. (including breeding records off LA coast). Otherwise, records have been accepted from TX, FL, IN, OH, MD, and CA.

Population No estimates exist from S.A., but recent range expansion northward suggests population increasing.

plumage suggests Lesser Black-backed but chunkier with bigger bill

heavy bill

Kelp x Herring hybrid winter adult

slaty gray above

molting juvenile

winter adult

advanced 2nd cycle

greenish legs

blackish upperparts as dark or darker than Great Black-backed

NODDIES Genus *Anous*

Noddies are tropical terns—three species worldwide—with dark plumage, white caps, and long, wedge-shaped tails. Two species occur in N.A., but only the Brown Noddy breeds here (Dry Tortugas, FL).

BROWN NODDY *Anous stolidus*   BRNO   2

In N.A., this distinctive all-dark tern nests only on the Dry Tortugas, FL, where it builds a nest in low bushes. Its flight is steady and graceful, typically low over the water. Here flocks often feed with other tropical seabirds, milling and swooping to pick food from the surface. Brown Noddy does not usually dive for food and rarely alights on the sea. Polytypic (5 sspp.; nominate in N.A.). L 14–16" (36–41 cm) WS 31–35" (79–89 cm)

Identification Medium-size with

medium-long bill and long, graduated tail. No seasonal variation in appearance (unlike most terns). **ADULT:** Plumage dark brown overall (including underwings) with contrasting ashy-white forecrown offset by black lores and narrow white subocular crescent; ashy-gray hind crown and nape. On resting birds, primaries contrastingly blacker than upperparts, which, in fresh plumage, have gray cast often lost by summer. Black bill; dark gray legs. Molt begins Apr. to June, then

suspends during breeding, and is completed at sea by Mar., in time for return to colonies. **IMMATURE:** Juveniles dark brown overall with a neat white supraloral line that continues over bill base; forecrown is dull ashy gray, and brownish hind crown does not contrast with hind neck. Fresh fall birds have subtly paler brown tips to upperwing coverts and scapulars. Protracted complete post-juvenal molt starts in fall and continues into following summer (primaries molted mainly

long, pointed tail

overall chocolate brown color

adult
stolidus

blended white crown

very limited white on forehead

adult
stolidus

immature
stolidus

Feb.–Aug.). **FIRST-SUMMER:** Not illustrated. Forecrown mostly ashy white, often with some brownish smudging. Some birds' crown mostly ashy white; others still show a contrasting white supraloral line and have little ashy white on crown. Brownish to gray-brown hind crown merges with nape. Many spring birds still have some contrastingly pale, bleached (juvenile) upperwing coverts.

Geographic Variation Birds in N.A. are of nominate subspecies *stolidus*. Subspecies *ridgwayi* of west Mexico (not separable in field from *stolidus*) is a potential hurricane-assisted visitor to the West Coast or the Southwest.

Similar Species Unlikely to be confused when in the Tortugas, but casual visitors elsewhere may be puzzling. Juvenile and first-year Sooty Terns are slightly bulkier overall with a forked

tail, fine pale spotting on upperparts in fresh plumage, contrasting white underwing coverts, and a white central belly and undertail coverts. Black Noddy is smaller and blacker overall (primaries not contrastingly blacker at rest) with a more extensive and contrasting white cap on adults and immatures; its bill is distinctly thinner (obvious in direct comparison), and its white subocular crescent is shorter (lying mostly behind eye's midpoint). At sea, Bulwer's Petrel (accidental in N.A.) can suggest a small noddy, with its dark brown plumage, long wings and tail, and low flight over the water; it lacks any pale head markings, but has a pale-brown ulnar bar on upperwing. Its wingbeats are stiffer, and its flight is more wheeling, with frequent

weaving glides. Also compare dark juvenile jaegers, which have white primary flashes.

Voice Brown Noddy mostly silent except around colonies, where it gives varied guttural barks and a braying *keh-eh-eh-ehr*. First-year birds give higher, whistled calls.

Status & Distribution Pantropical. **BREEDING:** Fairly common but very local, arriving (at first nocturnally) at colony mid-Jan.–Feb. and departing through Oct., with breeding season mainly May–July. **DISPERSAL & RARE STATUS:** Casual in summer along Gulf Coast and north to Outer Banks, NC; accidental north to New England. **WINTER:** FL population presumably winters at sea in tropical Atlantic.

Population FL population is believed to be stable, fluctuating between 1,000 and 2,000 pairs.

BLACK NODDY *Anous minutus*   BLNO ▓ 3

In N.A., this species is most frequently recorded on the Dry Tortugas, where it associates with Brown Noddies. Its flight is typically low over water, but is often quicker than Brown Noddy, with more fluttery wingbeats. Polytypic (7 sspp.; 1 in N.A.). L 12–13.5" (30–34 cm) WS 26–29" (66–74 cm)

Identification Medium-small bird with a disproportionately long and distinctly slender bill and a long, graduated tail. No seasonal variation in appearance (like Brown Noddy).

adult
americanus

bill more slender than Brown Noddy

immature
americanus

smaller and blacker than Brown Noddy

white crown sharply delineated at rear

ADULT: Plumage blackish overall (including underwings) with a contrasting white crown offset by black lores, and a narrow white subocular crescent; hind neck is ashy gray. On resting birds, primaries not contrastingly blacker than upperparts. Molts poorly known, but likely similar to Brown Noddy. **IMMATURE:** Resembles adult, but worn first-summer birds often slightly browner overall, sometimes with bleached (juvenal) upperwing coverts. Rear border to white cap contrasts more sharply with blackish hind neck.

Geographic Variation N.A. birds are presumably of the subspecies *americanus*, which breeds on islands off the north coast of Venezuela.

Similar Species Brown Noddy is larger and bulkier with a thicker (and thus often slightly shorter-looking) bill and dark brown plumage overall.

Brown's primaries are also contrastingly blacker at rest. However, plumage tones can be difficult to evaluate in bright sunlight: On adults, look for Black Noddy's smaller size and, especially, its long, slender bill; also note more extensive white cap (beware that bright light can cause Brown to show a large white cap), shorter white subocular crescent, and grayish tail. Immature Black is more distinctive, with a large and contrasting white cap unlike any plumage of Brown.

Voice Rarely vocal away from colonies; likely to be mostly silent in N.A. Calls include guttural growls, *ahrrr* and *garrr*; first-year birds have a high, piping *swee*.

Status & Distribution Tropical Atlantic and Pacific Oceans. **RARE STATUS:** Rare spring-summer visitor to Dry Tortugas, FL. Accidental TX coast (mid-Apr.–July).

Population Stable with large global range.

| Genus *Onychoprion*

SOOTY TERN   *Onychoprion fuscatus*   SOTE ■ 2

This handsome, colonial-nesting tropical tern is largely pelagic and comes ashore only to breed. Adults from the FL population spend most of their nonbreeding time in the Gulf of Mexico and the Caribbean; juveniles migrate to waters off West Africa, where they remain at sea for two to five years before returning to breed. Sooty Terns have a graceful, buoyant flight and swoop down to pick food from the surface; they do not plunge-dive. They rarely alight on the water, and then only briefly; some birds may stay aloft for extended periods (perhaps for years as immatures). Formerly classified as a *Sterna* tern, but recently found to be only distantly related to that genus; it now occupies genus *Onychoprion* along with Aleutian Tern, Bridled Tern, and the extralimital Gray-backed Tern (*O. lunatus*). Polytypic (6 sspp.; 2 in N.A.). L 14–15.5" (36–39 cm) WS 34–36.5" (86–93 cm)

Identification Medium size with fairly broad wings and a deeply forked tail. **BREEDING ADULT:** Above black, with a broad, triangular white forehead patch; white below. Tail black with a white outer edge. In flight, note dark underside to remiges contrasting with white underwing coverts. Black bill and legs. **WINTER ADULT:** Not illustrated, but may be encountered in fall (especially storm-blown birds). Resembles breeding adult, but hind neck mottled pale gray, back feathers often have broad whitish tips, and forehead patch less neatly defined. **JUVENILE:** Sooty brownish black with white spotting above; white vent to undertail coverts. In flight, note dark body and whitish underwing coverts. **FIRST-SUMMER:** Lacks juvenile's whitish spotting, but back feathers often tipped whitish; underparts have variable dusky mottling.
Geographic Variation Atlantic breed-

ing populations are the nominate subspecies *fuscatus*, the breeding adults of which have white underparts with little or no smoky clouding on belly and vent. East Pacific populations (casual in Southern CA) are the subspecies *crissalis*; adults average smaller and have pale gray clouding on their underparts.
Similar Species Bridled Tern is slightly smaller and lighter in build, with narrower wings, a longer tail, and a more buoyant and lighter flight; it often stands on driftwood, weed patches, and other floating objects but rarely alights on the sea (which Sooty does). Only adultlike plumages of Bridled Tern are likely to be confused with Sooty's. Breeding adult Bridled Tern is dark brownish gray above, with remiges often appearing darkest part of upperparts (opposite of Sooty Tern) and with long tail looking mostly white (striking even at long range). Also note that underside of primaries is white based on Bridled (all-dark on Sooty); Bridled's paler hind neck often appears as a hind collar, and its white forehead patch is narrower and more chevron shaped, projecting past eye. Nonbreeding plumages differ in much the same ways, but Bridled has a shorter tail.
Voice Adult has a nasal barking or laughing *ka-waké* or *ke wéh-de-wek* (sounds like "wideawake") given year-round. Juveniles give high reedy whistles into their first winter: e.g., *wheeir*.
Status & Distribution Pantropical. **BREEDING:** Nest in large, very dense colonies. The colony on Bush Key, Dry Tortugas, FL, numbers about 40,000 pairs; birds arrive by Mar., depart late June–mid-Aug.; also nests on islets off LA and TX. **DISPERSAL:** Regular in

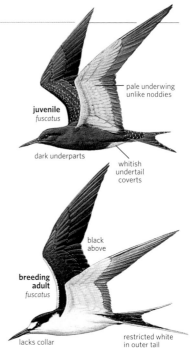

juvenile *fuscatus*

pale underwing unlike noddies

dark underparts

whitish undertail coverts

black above

breeding adult *fuscatus*

lacks collar

restricted white in outer tail

white restricted to forehead

white does not extend behind eye

adult *fuscatus*

summer (especially July–Sept.) north in Gulf Stream to NC. **RARE STATUS:** Casual to Southern CA. Storm-blown birds (mainly late summer and fall) north to Great Lakes and the Maritimes; accidental CO and Attu I., AK. **Population** Stable with a global population of 50–80 million birds.

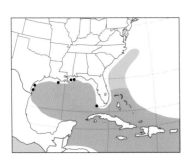

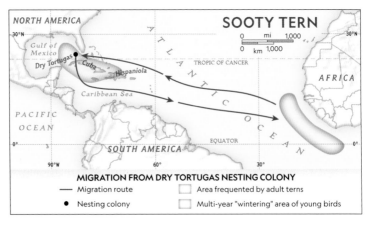

SOOTY TERN

NORTH AMERICA
Gulf of Mexico
Dry Tortugas
Cuba
Hispaniola
Caribbean Sea
PACIFIC OCEAN
SOUTH AMERICA
ATLANTIC OCEAN
TROPIC OF CANCER
AFRICA
EQUATOR

MIGRATION FROM DRY TORTUGAS NESTING COLONY
—— Migration route
● Nesting colony
☐ Area frequented by adult terns
☐ Multi-year "wintering" area of young birds

BRIDLED TERN   *Onychoprion anaethetus*   BRTE   ■ 2

This handsome tropical tern occurs in N.A. mainly as a nonbreeding pelagic summer visitor to warm offshore waters. Bridled Terns have a graceful, buoyant flight and swoop down to pick food from the surface; they do not plunge-dive. Often found resting on driftwood, sargassum weed mats, and even the backs of turtles, they rarely if ever alight on the sea. Polytypic (4 sspp.; 2 in N.A.). L 12.5–14" (32–36 cm) WS 31–33.5" (79–85 cm)

Identification Medium size; long, deeply forked tail. **BREEDING ADULT:** Black crown; chevron-shaped white forehead patch; gray-brown upperparts; whitish hind collar; outer rectrices mostly white. **WINTER ADULT:** May be encountered in fall. Resembles breeding adult, but back feathers tipped pale gray; forehead patch less neatly defined; whitish streaking in forecrown. **JUVENILE:** Pattern overall resembles adult, but much less well

defined. White forehead and short supercilium are framed by black auricular mask and dark-streaked crown; upperparts boldly barred pale buff to whitish. **FIRST-SUMMER:** Commonly seen in waters of N.A. Variable appearance intermediate between juvenile and winter adult. Upperparts have little or no whitish barring; head whitish overall, with an indistinct blackish auricular mask.

Geographic Variation East Coast birds are *melanopterus*, breeding adults of which have white underparts. East Pacific breeding populations (casual in Southern CA) are of subspecies *nelsoni*, breeding adults of which average larger than *melanopterus* and have pale gray clouding on their underparts.

Similar Species See Sooty Tern.

Voice Not especially vocal except around the nest, where adults give a mellow *kowk-kowk* or *kwawk kwawk*, and a harder *kahrrr*.

Status & Distribution Pantropical. **BREEDING:** Very local in FL Keys (Apr.– Aug.). **WINTER:** Some winter in Gulf Stream off FL. **DISPERSAL:** Regular offshore visitor (mainly May–Oct.) in Gulf of Mexico and shoreward edge of Gulf Stream north to NC, very rarely to off southeast MA. **RARE STATUS:** Storm-blown birds (mainly late summer) north to Atlantic Canada. Casual to Southern CA and AR.

Population Large global population, trends unknown.

grayish brown cast above

extensive whitish underside to primaries

breeding adult
melanopterus

extensive white in outer tail when spread

pale collar

white extends beyond eye

thinner bill than Sooty

white collar often less apparent on sitting bird

adult
melanopt

pale head

grayish brown above

1st summer
melanopterus

head pattern diluted

pale-edged upper parts

juvenile
melanopterus

ALEUTIAN TERN   *Onychoprion aleuticus*   ALTE   ■ 2

This handsome tern breeds on western AK coasts. Its calls are unique. Monotypic. L 11–12" (28–30 cm) WS 29.5–32" (75–81 cm)

Identification Deeply forked tail. On all plumages, note dark secondary bar on underwing. **BREEDING ADULT:** Black cap with white forehead, medium gray above and below. Black legs and bill. **JUVENILE & FIRST-SUMMER:** Fresh juveniles are relatively dark above, with cinnamon edging on upperparts; cinnamon wash bleeds onto chest and sides (soon fades to whitish). Bill pinkish with dark culmen and tip; legs pinkish. Presumably undergoes complete molt in first winter (similar to Common Tern), and first-summers are unrecorded in AK and presumably remain well south of AK waters.

Similar Species Arctic Tern is smaller

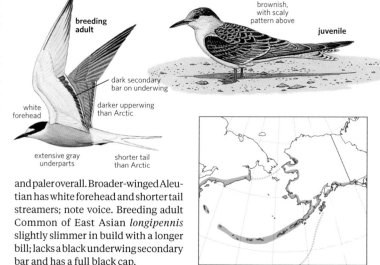

breeding adult

white forehead

dark secondary bar on underwing

darker upperwing than Arctic

extensive gray underparts

shorter tail than Arctic

brownish, with scaly pattern above

juvenile

and paler overall. Broader-winged Aleutian has white forehead and shorter tail streamers; note voice. Breeding adult Common of East Asian *longipennis* slightly slimmer in build with a longer bill; lacks a black underwing secondary bar and has a full black cap.

Voice Adult's piping whistled calls have a mellow to slightly rolled quality: *piiu* and *piirr-i-u* or *piiu-pi-pip*, and so on.
Status & Distribution BREEDING: North Pacific. Fairly common, but local; arriving southern AK (including Kodiak I.) and in Aleutians and Bering Sea mid-May–early June.

Departs colonies mid-Aug.–late Aug. **MIGRATION:** Casual in AK away from breeding grounds. Also casual Haida Gwaii, BC (mid-May–early June). **WINTER:** Winter grounds not well known, believed to lie off Indonesia and Malaysia; regular migrant off Hong Kong and Singapore. **RARE STATUS:** Accidental to UK.

Population Vulnerable. Ironically, not known to breed in the Aleutians before 1962, with subsequent increases perhaps due to removal of non-native foxes. Dramatic declines of over 90 percent in AK breeding colonies since 1960; trends in Russian Far East, where 80 percent of the population breeds, unknown.

TYPICAL TERNS Genera *Sternula*, *Phaetusa*, *Gelochelidon*, and *Hydroprogne*

Terns resemble small, angular gulls with pointed bills, shorter legs, and forked tails. Typical terns are pale gray to black above and white below, often with a black cap. First-summer birds resemble winter adults. Second-summers are variable—some look like first-summers, others like breeding adults. Many species hover and dive after fish.

LEAST TERN *Sternula antillarum*   LETE ■ 1

Colonies of this tiny tern "compete" with humans for beach space. Consequently, the species is declining in much of its range. It flies with hurried wingbeats and frequent hovering before plunge-diving for fish. Polytypic (4–5 sspp.; 3 in N.A.). L 8–9" (20–23 cm) WS 19–21" (48–53 cm)
Identification The smallest tern, with a relatively long bill and forked tail. All plumages have a pale smoky-gray rump and tail. **BREEDING ADULT:** Black cap has clean-cut white forehead chevron; outer one to three primaries form a contrastingly blackish upperwing wedge against fresher, pale gray middle primaries. Yellow legs; yellow bill tipped black. **WINTER ADULT:** Seen (Aug.–Sept.) before birds leave US. Lores and crown become white, with variable dark streaking on crown; bill becomes black. Dark-mottled lesser coverts form patagial bar on upperwing. Best told from first-summer (and juvenile) by strong contrast of old outer one to three primaries on upperwing. (Adult postbreeding primary molt mainly June–Dec., followed by pre-breeding primary molt includ-

ing seven to nine inner primaries.) **JUVENILE & FIRST-SUMMER:** Resembles winter adult, but fresh-plumaged juvenile has variable brownish barring and tipping on back and upperwing coverts; bill base often pale pinkish to yellowish. Most frequent first-summer plumage resembles winter adult, but with darker, blackish patagial bar and blackish outer four to five primaries, forming a contrasting upperwing wedge suggesting Sabine's Gull; bill usually black. Some birds resemble breeding adults, but less clean-cut chevron on forehead, forecrown speckled white, and yellow bill more extensively black distally.
Geographic Variation Nominate *antillarum* (breeds on East Coast and Gulf Coast) averages larger and paler than both *browni* (of CA) and, especially, *athalassos* (of the interior).
Similar Species Unmistakable by its small size. Little Tern (*S. albifrons*), its Old World counterpart, is very similar to Least Tern and should be sought on East Coast or in the western Aleutians, AK; all Little Tern plumages have contrasting white rump and tail, unlike

pale smoky gray of Least Tern and have distinctly different vocalizations.
Voice High reedy chippering; chatters with frequent disyllabic calls: *chi-rit* and *k-rrik*; a reedier *kree-it* or *kreet*; a clipped *k'rit*; and longer series, *kik kirvee*, etc.
Status & Distribution US south to Mexico and Caribbean. **BREEDING:** Fairly common but local on sandy beaches, also sandy islets on interior rivers. Arriving at southern breeding areas late Mar.–Apr., northern sites through May; fall movements start June–Aug. **MIGRATION & DISPERSAL:** Most birds depart US during Aug., a few linger into early Sept. and casually into Oct. Casual north to southern Canada. **WINTER:** Poorly known, presumably off coasts from Mexico and Caribbean to northern S.A. Accidental N.A. **RARE STATUS:** Accidental HI. Either a Least Tern, or more likely a Little Tern (on distribution), was photographed on Buldir I. (4–6 July 2005), the western Aleutians.
Population Pacific coast (*browni*) and interior (*athalassos*) populations are Endangered in US; main threats being habitat modification and human disturbance.

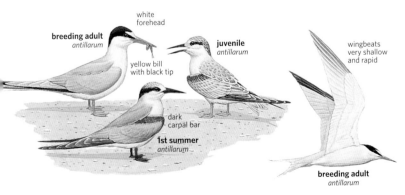

white forehead

breeding adult
antillarum

yellow bill with black tip

juvenile
antillarum

dark carpal bar

1st summer
antillarum

wingbeats very shallow and rapid

breeding adult
antillarum

LARGE-BILLED TERN *Phaetusa simplex*   LBTE ▪ 5

This striking, aptly named species nests in freshwater habitats of S.A. and moves to coastal regions there in the nonbreeding season. Its occurrence in N.A. is accidental. Its flight is graceful and it feeds by plunge-diving and hawking for insects in flight; often flies moderately high up and down rivers. Nests on river sandbars in S.A.

dark gray mantle and tail

black cap with narrow white forehead

breeding adult

wing pattern like Sabine's Gull

stout yellow bill

in mixed colonies with Black Skimmers and Yellow-billed Terns (*Sternula superciliaris*). Monotypic (some treat the species as polytypic, recognizing the more southerly S.A. subspecies *chloropoda*). L 14–15" (36–38 cm) WS 36–38" (91–97 cm)

Identification Disproportionately long bill is stout and yellow; legs are medium length and olive to greenish yellow; relatively short tail is shallowly forked. **BREEDING ADULT:** Black cap separated from bill base by a white forehead band. At rest, upperparts are dusky gray, with black primaries and a broad white band along bottom edge of wing. In flight, upperparts display a striking pattern recalling Sabine's Gull: bold white triangular panels on wings set off by dusky gray lesser upperwing coverts and black primaries and greater primary coverts (inner primaries are whitish). Underwings are mostly white, with dark wing tips. **NONBREEDING ADULT:** Resembles breeding adult, but forecrown mottled white.

JUVENILE & FIRST-YEAR: Resembles adult, but back and upperwing coverts mottled brownish, cap mottled whitish, and bill is duller yellowish.
Similar Species Should be unmistakable. More likely to be confused (at a distance) with Sabine's Gull than with any regularly occurring N.A. tern.
Voice CALL: A loud gull-like *kaay-rak*. Strays to N.A. are likely to be silent.
Status & Distribution S.A. freshwater species. Erratic visitor to coastal mangroves, beaches, and estuaries, primarily during May to Oct. Casual (mainly spring) north to Nicaragua.
RARE STATUS: Accidental, three records in late spring and summer from IL (Lake Calumet, Chicago, 15 July 1949), OH (near Youngstown, 29 May 1954), and NJ (Kearny Marsh, Hudson Co., 30 May 1988). Also accidental to Bermuda, Cuba (twice), and Grenada.
Population Stable; common throughout most of its range. Main threats are habitat disturbance and egg collecting.

GULL-BILLED TERN *Gelochelidon nilotica*   GBTE ▪ 1

This distinctive, medium-size tern of salt marshes and beaches has a smooth flight as it sweeps over open areas such as marshes or salt flats, picking its prey (insects, small crabs, etc.) from the ground, air, or even bushes! It does not hover like many other tern species or dive into the water to obtain prey. Polytypic (6 sspp.; 2 in N.A., weakly defined). L 13–14" (33–36 cm) WS 35–38" (89–97 cm)
Identification All plumages have ghostly pale upperparts. Stout bill and legs are black. **BREEDING ADULT:** Black cap becomes spotted white in late summer. **WINTER ADULT:** Note distinctive white-headed appearance with dark postocular mask or smudge; primaries in molt July–Feb.

JUVENILE & FIRST-YEAR: Resembles winter adult, but juvenile in fall is fresh plumaged, with variable buff wash and brownish subterminal marks on back; mask often less distinct; bill can have pinkish base into fall. Protracted complete molt (with primaries molted Jan.–Aug.) produces first-summer plumage, much like winter adult.
SECOND-YEAR: Second postbreeding molt averages later than in adults, so outer primaries fresher and paler in second summer than in adults.
Geographic Variation Minor within N.A. West Mexican (and Southern CA) *vanrossemi* is larger billed than eastern N.A. *aranea*. Of the four other subspecies (three Old World, one S.A.), the most distinct taxon is largest and palest

subspecies, *macrotarsa* from Australia. This nomadic breeding subspecies is treated by some as a distinct species.
Similar Species Gull-billed Tern's

vanrossemi

aranea

short, thick black bill

breeding adult

juvenile

rather long black legs

stocky body with "heavy flight"

very whitish overall, including head

winter adult

short tail

behavior distinct. Winter and juvenile Forster's Terns have a bolder black mask, slimmer in build, and dive in the water for food; at rest, note Forster's orange-red legs.
Voice Nasal but mellow laughing and barking calls, usually two to three syllables: *keh-wek* or *ku-wek*, and *keh-w-wek* or *kit-u-wek*; sharper than calls of Black Skimmer. Juvenile has high piping whistles given into first winter.
Status & Distribution Warmer climates worldwide. **BREEDING:** Fairly common but local in N.A., present Mar./Apr.–Aug. In the US, all colonies are coastal, except the one at the south end of the Salton Sea, CA. **MIGRATION:** Mainly Mar.–Apr. and Aug.–Sept. **WINTER:** N.A. birds winter along the Gulf Coast, Pacific coast of Mexico, and south to S.A. Casual north on coast to NC and at Salton Sea. **RARE STATUS:** Casual in interior N.A. except Salton Sea, CA, where it breeds. Casual to central CA and on Atlantic coast north to NS; accidental NL.
Population Locally erratic, but likely declining; population of *aranea* some 3,500–4,000 pairs; western *vanrossemi* perhaps fewer than 800 pairs (including Mexico), fewer than 200 in CA. In San Diego, CA, preys on chicks of the endangered Snowy Plover and the Least Tern, hence Gull-billed Terns have been shot in attempts to adjust the "balance" of nature. In addition, the San Diego Bay population declined suddenly by 50 percent a decade ago, apparently from a parasite that was acquired from the terns' mole crab prey. Populations there remain reduced.

CASPIAN TERN *Hydroprogne caspia* CATE ▪ 1

This, the largest tern in the world, is widespread in interior and coastal habitats but is rarely seen offshore. Its flight is steady and powerful, with fairly shallow wingbeats. Begging juveniles often follow adults on southbound migration; their loud plaintive whistles indicate their presence, which continues through much of the fall. Monotypic. L 20–22" (51–56 cm) WS 46–51" (117–130 cm)
Identification The heavy-bodied Caspian lacks tail streamers in all plumages. Stout red bill, with a black distal mark and pale tip. Blackish panel on underside of outer primaries is diagnostic of all plumages. **BREEDING ADULT:** Black cap becomes streaked white in fall. **WINTER ADULT:** Note densely flecked black-and-white crown, which blends into a broad black auricular mask; primaries molted Aug.–Feb. **JUVENILE & FIRST-YEAR:** Resembles winter adult, but juvenile fresh-plumaged in fall, with variable blackish subterminal marks on back and tail. Paler bill, orange to orange-red, with a dark subterminal mark; legs often yellowish, becoming dark by winter. A protracted complete molt (primaries molted Jan.–Aug.) produces first-summer plumage much like winter adult. **SECOND-YEAR:** Second postbreeding molt averages later than in adults, so outer primaries fresher and paler in second-summers than in adults; forecrown often streaked whitish.
Similar Species The smaller Royal Tern is a coastal species; its more slender (but still stout) bill is orange, without a dark tip (rarely orange-red in spring). In flight, note Royal's narrower wings; its deeper wingbeats; a mostly whitish underside to primaries; and a longer and more deeply forked tail (with streamers in breeding plumage). Winter and immature Royal Terns have an extensively white forehead and forecrown; these areas on Caspian are heavily streaked and appear dark.
Voice Adult has distinctive, loud rasping *rraah* or *ahhrr* and a drawn-out, up-slurred *rrah-ah-ahr* in chases and when diving at colony intruders. Begging young (into first-winter) give loud, plaintive, two-part whistled *ssíťuuh.*

Status & Distribution Worldwide except S.A. **BREEDING:** Common in N.A.; arrives at southern nesting sites Mar.–Apr. and northern sites (Canada, Great Lakes) mainly mid-Apr.–May, departs mainly Aug.–Sept. **MIGRATION:** Mainly Mar.–May and late July–Oct.; stragglers occur through Nov. **WINTER:** Winters south through C.A. and Caribbean, a few to northern S.A. **Population** Stable or increasing. Major increases on Pacific coast, in interior West, and in Great Lakes region.

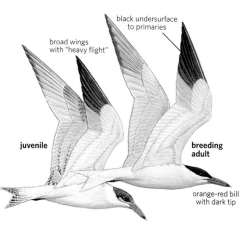

broad wings with "heavy flight"

black undersurface to primaries

juvenile

breeding adult

orange-red bill with dark tip

follows and begs from adult well into fall

juvenile

extensive dark markings above

orange-based bill

streaked crown

thick red bill with dark tip

winter adult

MARSH TERNS Genus *Chlidonias*

These terns are compact with dark underparts in breeding plumage. They usually feed by dipping down to the water's surface for insects or chasing insects in the air (hawking). Although all three species in the genus have occurred in N.A., only the Black Tern is a common breeder; the other two are casual or accidental visitors.

BLACK TERN *Chlidonias niger*   BLTE   ■   1

This handsome small tern occurs widely as a migrant and summer visitor at lakes and marshes across N.A., where it breeds in freshwater habitats. Its winter range is pelagic. Black Terns have an easy, slightly floppy flight as they swoop down to pick food from the surface; they also soar and catch insects in flight, but they do not plunge-dive. Migrants occur singly or, locally, in flocks of hundreds. Black Terns perch readily on posts and wires, and their nests are mats of floating vegetation. Polytypic (2 sspp.; 1 in N.A.). L 9–9.7" (23–25 cm) WS 23.5–25.5" (60–65 cm)

Identification Small with a fairly short, slender black bill and medium-length legs; tail with shallow cleft. **BREEDING ADULT:** Distinctive, with a black head and body, slaty-gray upperparts, and white undertail coverts. Underwing coverts are smoky gray.

Bill is blackish, legs dark reddish. Postbreeding molt starts in midsummer, when white spots appear on head and body; underparts of fall adults are mostly white or blotched black and white. **WINTER ADULT:** Head and underparts are white with a dark-streaked cap connected to a blackish ear spot and forming "headphones"; note dark bar on chest sides and slaty-gray mottling on flanks. **JUVENILE:** Resembles winter adult, but wings are uniformly fresh in fall; back and tertials have variable dark-brown distal markings that create a subtly mottled aspect. **FIRST-SUMMER:** Resembles winter adult, but underparts of some birds have scattered black spots.

Geographic Variation N.A. birds comprise New World subspecies *surinamensis*, which differs from nominate *niger* of the Old World (possible visitor to the East) in deeper black head and body of breeding adults and gray

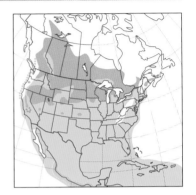

flanks of winter adults and juveniles.
Similar Species Distinctive; often strikingly small when seen perched among other terns, being barely larger than a Least Tern. See White-winged and Whiskered Terns.
Voice A piping, slightly reedy *peep* or *pseeh* and a quiet *kriih* given by birds in feeding flocks; a quacking *kek* in alarm; and a shrill, slightly grating *kehk* given by birds scolding in a breeding colony.
Status & Distribution N.A. and western Asia. **BREEDING:** Arrives on nesting grounds late Apr.–May, departing late July–Aug. **MIGRATION:** Mainly mid-Apr.–May and late July–Sept., can be numerous in Gulf of Mexico. Rare along and off Pacific coast. Casual visitor north to AK and to much of Atlantic Canada away from breeding sites. **WINTER:** Winters at sea; in Pacific waters from southern Mexico to Peru. Casual to CA. **RARE STATUS:** In Europe, *surinamensis* is casual.
Population Declining in most areas, mainly because of habitat degradation and loss. N.A. population in the early 1990s was a third of what it was in the late 1960s.

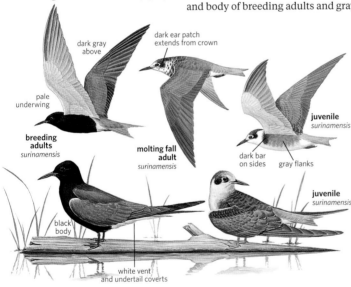

dark gray above

dark ear patch extends from crown

pale underwing

breeding adults *surinamensis*

molting fall adult *surinamensis*

dark bar on sides   gray flanks

juvenile *surinamensis*

black body

juvenile *surinamensis*

white vent and undertail coverts

WHITE-WINGED TERN *Chlidonias leucopterus*   WWTE   ■   4

This small tern of Eurasia (where it is more evocatively called "White-winged Black Tern") is a casual visitor to N.A., and breeding-plumaged adults are among the most handsome of birds. White-winged Terns might be found anywhere Black Terns occur and readily associate with them. (The

two species have hybridized in QC.) Whited-winged's flight is similar to Black Tern, swooping down to pick food from the water as well as soaring to catch insects in flight. They perch readily on posts and wires. Monotypic. L 9–9.5" (23–24 cm) WS 22–23" (56–59 cm)

Identification Small with a fairly short, slender bill and medium-length legs. Wings are slightly more rounded than Black Tern, and tail has only a shallow cleft. **BREEDING ADULT:** Unmistakable: Black head, body, and underwing coverts contrast with white "wings," tail coverts, and tail. Outer two to three primaries are dark (winter) feathers that contrast with newer middle primaries attained by the prebreeding molt. Bill is blackish to dark red; legs and feet are orange-red. Postbreeding molt starts in midsummer, when white spots appear on head and body. Molting adults in fall often lack most or all of the black, except for some underwing coverts. **WINTER ADULT:** Head, underparts, and underwings are white with a dark-streaked cap and a blackish ear spot often separated from cap by a white supraorbital area; back and tail are pale gray. Note dusky secondaries contrasting with whitish upperwing coverts. **JUVENILE:** Not recorded in N.A. Resembles winter adult, but wings are uniformly fresh in fall; its dark brown saddle contrasts with pale upperwings and uppertail coverts, and it has duller, pinkish legs. A protracted complete molt in first winter (primaries molted Jan.–Aug.) produces first-summer plumage. **FIRST-SUMMER:** Not illustrated, but resembles winter adult. **SECOND-SUMMER:** Not illustrated, but resembles breeding adult. Some have white spots on underwing coverts, and others have strongly piebald head and body in spring.

Similar Species Black Tern is slightly larger with more pointed wings, a slightly more cleft tail, a slightly longer bill, and slightly shorter legs. All

plumages of Black Tern have smoky-gray underwing coverts; its upperwings, back, and tail are essentially concolorous dusky gray, but marginal coverts of breeding adults can be paler and look whitish in some lights. Confusion is most likely in winter plumages, but note White-winged's paler upperwings with contrasting dark secondaries, its pale gray rump and tail, its whitish underwings, its lack of a dark mark on sides of neck, and its lack of dark mottling on flanks. On perched birds, note Black Tern's dark mark on sides of neck and its smokier gray upperparts; in direct comparison, Black Tern's slightly longer (and thus finer-looking) bill and shorter legs may be appreciated. White-winged's overall pale gray winter plumage recalls

many typical terns, but note its small size, short bill, marsh-tern flight, and shallowly cleft tail. If a suspected White-winged does not show the classic suite of features, the possibility of a hybrid with Black Tern should be considered. Also see Whiskered Tern.

Voice Mostly silent away from the breeding grounds, White-winged Tern may give a hoarse *kesch*, slightly deeper and harsher than Black Tern.

Status & Distribution RARE STATUS: Casual visitor (mainly late May–Aug.) to widely scattered locales in the eastern half of N.A., especially along the Northeast and mid-Atlantic coast, but only a few N.A. records over the last two decades. Accidental in the West (AK and CA).

Population Stable.

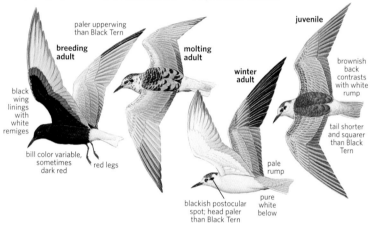

paler upperwing than Black Tern

breeding adult

molting adult

winter adult

juvenile

brownish back contrasts with white rump

black wing linings with white remiges

bill color variable, sometimes dark red

red legs

tail shorter and squarer than Black Tern

pale rump

pure white below

blackish postocular spot; head paler than Black Tern

WHISKERED TERN   *Chlidonias hybrida*   WHST ■ 5

This Old World marsh tern is an accidental visitor to eastern N.A. It is somewhat intermediate in appearance between typical terns (such as Common) and marsh terns. Its flight is similar to Black Tern but somewhat heavier, with less floppy wingbeats. Like Black Tern, Whiskered swoops down to pick food from the water, rather than diving. Polytypic (6 sspp.; presumed nominate in N.A.). L 9.5–10" (24–25 cm) WS 26.5–28.5" (67–72 cm)

breeding adult *hybrida*

stout, dark red bill

dark gray underparts contrast with white cheeks

Identification Medium-size tern with a medium-short bill. Wings are fairly broad and bluntly pointed; tail has shallow cleft. **BREEDING ADULT:** Black cap and smoky-gray underparts strongly set off white cheeks (or "whiskers"). Underparts are similar in tone to upperparts, except undertail coverts are whitish and rump and tail are smoky gray. In flight, smoky-gray underbody contrasts with whitish underwings and undertail coverts. Bill and legs are deep red. Postbreeding molt usually starts in fall, later than most other marsh terns. **WINTER ADULT:** Head and underparts are white with a narrow black postocular patch that merges into blackish streaking on hind crown; pale gray rump and tail; bill and legs blackish. **JUVENILE:** Resembles winter adult, but wings are uniformly fresh in fall and it has a dark brown saddle with broad cinnamon bars and pinkish legs. A protracted complete molt in first

winter produces first-summer plumage. **FIRST-SUMMER:** Resembles winter adult, but primaries in molt mainly Jan. through Aug.

Similar Species Separated from Common and Arctic Terns by typical marsh-tern flight; smaller size; and more compact shape, with a pale gray rump and shallowly cleft tail. Also, breeding adult Whiskered Tern has darker body plumage and more contrasting white cheeks. Winter adult White-winged Tern is smaller and more lightly built, with narrower wings and a shorter, finer bill; its black ear spot is more distinct and separate from dark cap.

Voice Flight call a rasping *krehk*.

Status & Distribution RARE STATUS: Accidental (July–late Sept.), one in 1993 from NJ and DE and NJ in 1998, both breeding plumaged adults; also accidental to Barbados and Great Inagua, Bahamas.

Population Stable.

STERNA TERNS Genus *Sterna*

This genus was recently reconfigured with fewer species. These medium-size terns are variably whitish with medium to long forked tails. They dive into the water to procure prey. Adults have black caps in breeding plumage. Overall structure and the timing of molt are important for identification.

ROSEATE TERN *Sterna dougallii*   ROST ■ 2

This tern feeds offshore; flies with a flight style distinct from other *Sterna* terns. Unbroken tail streamers add 2 in (5 cm) to breeding adult's length. Polytypic (5 sspp.; nominate in N.A.). L 12–13" (30–33 cm) WS 26–28" (66–71 cm)

Identification Relatively short wings; long tail. Breeding tail streamers project well beyond wing tips at rest. Bill black most of year. **BREEDING ADULT:** Solid black cap; bill develops red basally in summer, can be red with black tip by Aug.; rosy underparts flush usually subdued. Dark primary wedge visible at rest. **JUVENILE & FIRST-SUMMER:** Juvenile has variable blackish subterminal marks on back and tail (recalls juvenile Sandwich); forehead often dusky. Protracted complete molt produces first-summer plumage with white forehead, pale gray upperparts, dark patagial bar.

Similar Species Note shallow, rapid wingbeats unlike other *Sterna* terns. Note Roseate's dark upperwing wedge on outer few primaries, lack of dark trailing edge to primaries, whitish underparts. Juvenile patterned like Sandwich. First-summer Roseate told from first-summer Common by lack of dark secondary bar, finer bill.

Voice A scratchy *krrízzik* or *kír-rik*, often doubled, suggesting Sandwich; a rasping *rrahk* or *ahrrr*; a mellow *ch-dik* or *ch-weet*; and a chippering *cheut cheut*.

Status & Distribution Atlantic, Indian, western Pacific Oceans. **BREEDING:** Uncommon and local in N.A., arriving colonies Apr.–May. Departs in Sept. **MIGRATION:** Rare records Atlantic coast south of NJ, mostly off NC (late May, late Aug.–Sept.). **WINTER:** Mainly northeast coast of S.A.

Population Listed by USFWS as Endangered in US (about 4,500 breeding pairs in 2018). Threatened in Canada (about 50 breeding pairs).

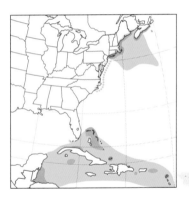

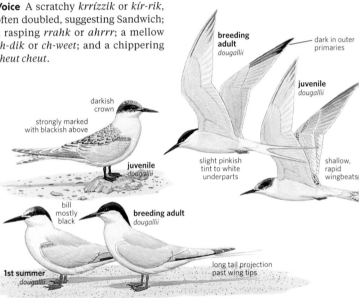

breeding adult *dougallii*

dark in outer primaries

juvenile *dougallii*

darkish crown

strongly marked with blackish above

juvenile *dougallii*

slight pinkish tint to white underparts

shallow, rapid wingbeats

bill mostly black

breeding adult *dougallii*

long tail projection past wing tips

1st summer *dougallii*

COMMON TERN *Sterna hirundo*   COTE ■ 1

Fairly common, along with Forster's our most widespread tern. Unbroken tail streamers add 1.5 in (4 cm) to breeding adult's length. Polytypic (3 sspp.; 2 in N.A.). L 11.5–12.5" (29–32 cm) WS 29.5–32.5" (75–83 cm)

Identification Medium size; medium-long bill. Breeding tail streamers fall about even with wing tips at rest. **BREEDING ADULT:** Solid black cap held through southbound fall migration; black-tipped red bill (mostly dark on early spring migrants, rarely all-red in late summer); red legs. Pale gray underparts; white rump and tail with dark outer web to outer rectrices. Pale gray upperwings usually have distinct dark wedge on primaries. **JUVENILE & FIRST-FALL:** Fresh juveniles have gingery wash on back and upperwing coverts (often fading by fall); pinkish red legs, bill base. Upperwing has contrasting blackish patagial bar, dusky gray secondary bar. **FIRST-SUMMER:** Forecrown, underparts white; black bill. Note dark patagial bar, dark secondaries. **SECOND-SUMMER:** Some resemble first-summer, but lack worn juvenal outer primaries; bill usually reddish basally. Others more closely resemble breeding adult.

Geographic Variation Nominate subspecies *hirundo* found in N.A., with red bill and legs. Breeding *longipennis* adults (East Asia) have darker plumage

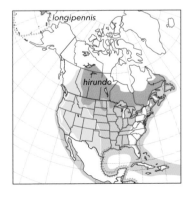

longipennis

hirundo

overall; bill and legs are blackish.

Similar Species See Forster's and Roseate Terns; compare *longipennis* subspecies with Aleutian Tern (see p. 256). Arctic Tern has shorter neck, narrower wings, shorter bill, and much shorter legs. Primaries are evenly translucent when backlit; outer primaries have narrower blackish tips; lacks dark wedge of Common. In flight, Arctic Tern has narrower, longer wings; shorter head-and-neck projection forward of the wings. Breeding adult Arctic has all-red bill, duskier gray underparts, tail streamers that project beyond wings at rest. Juvenile Arctic Tern has a slightly more distended black cap, a less contrasting dark pata-gial bar, which on upperwing, blends into gray coverts and whitish second-aries. First-summer Arctic best told by structure.

Voice Calls include a sharp *kik* and *kik-kik*; a grating *krrrih*; slightly drawn-out, disyllabic *eeeahrr*; also a rapidly repeated, grating *kehrr kehrr kehrr* in breeding-season chases.

Status & Distribution Holarctic breeder. **BREEDING:** Common in N.A., arrives at colonies Apr.–May, departs Aug.–Sept. **MIGRATION:** Mainly Apr.–May, July–Oct. Uncommon to rare in interior N.A. away from large bodies of water; also frequent offshore. Subspecies *longipennis* rare (mainly spring) on islands of western AK. **WINTER:** Off coasts from M.A. to S.A. Casual by midwinter to Gulf Coast and Southern CA. The subspecies *longipennis* winters in western Pacific and east Indian Ocean south to Australia.

Population Declining off West Coast, especially in spring, and on East Coast due to beach disturbance.

breeding adult *hirundo*
red bill with blackish tip
dark mask extends around nape
dark carpal bar
1st summer *hirundo*
grayish below
red legs
no tail projection past wing tips
juvenile *hirundo*
2nd summer *hirundo*
ckish ill
ensive k gray below
ackish legs
breeding adult *longipennis*
dark gray secondaries
dark in outer primaries, often present as a wedge, particularly evident in late summer and early fall
1st fall *hirundo*
breeding adult *hirundo*

ARCTIC TERN *Sterna paradisaea*   ARTE ■ 1

This northern-breeding tern winters in Antarctic waters and migrates mainly well offshore. The Arctic's flight is quick and graceful, with snappy wingbeats. It hovers briefly before plunge-diving for fish. Unbroken tail streamers add 1.5–2 in (4–5 cm) to the length of breeding adults. Monotypic. L 11.8–13" (30–33 cm) WS 30–33" (76–84 cm)

Identification Medium-small tern with a fairly short and slender bill, deeply forked tail, and notably short legs. Breeding adult's tail streamers project well beyond wing tips at rest. **BREEDING ADULT:** Solid black cap held through southbound fall migration; red bill and legs. Smoky gray underparts; has white rump and tail with dark outer webs to outer rectrices. Pale gray upperwings with translucent primaries, narrow dark trailing edge to outer primaries. **JUVENILE:** When fresh juveniles have variable brownish wash and barring on the back and upperwing coverts (often fading by fall). Upperwing has contrasting dark gray patagial bar and whitish secondaries and inner

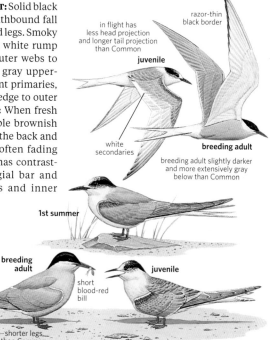

in flight has less head projection and longer tail projection than Common
razor-thin black border
juvenile
white secondaries
breeding adult
breeding adult slightly darker and more extensively gray below than Common
1st summer
breeding adult
short blood-red bill
juvenile
shorter legs than Common

primaries. **FIRST-SUMMER:** Forecrown and underparts white, or the underparts have gray smudging; upperwing has dark gray patagial bar; bill black to dull reddish.

Similar Species See Common Tern; see also Aleutian, Roseate, and Forster's Terns. At rest, note Arctic's short bill and very short legs. In flight, note Arctic's short neck, its long and narrow wings, its evenly translucent

primaries with narrow blackish tips.

Voice Calls include a high, clipped *kiip*; a shrill grating *keeahr*, higher and drier than Common; and a shrill, rapid-paced *ki-ki-kehrr* by scolding birds at colony.

Status & Distribution Holarctic breeder. **BREEDING:** Common in N.A., arriving at colonies from mainly early May (New England, southern AK) to June (high Arctic), departing late

July–Aug. **MIGRATION:** Mainly late Apr.–early June, late July–mid-Oct. Mainly casual inland. **WINTER:** Antarctic pack ice and adjacent waters, where rich food enables rapid complete molt. **RARE STATUS:** Rarely seen from shore. Very rare to casual from late spring to Nov. in interior N.A.

Population Recent declines in many areas, from AK to New England, likely due to food shortages.

FORSTER'S TERN *Sterna forsteri* FOTE ■ 1

This tern is rare offshore, but a familiar sight in interior and coastal habitats. Its flight is steady and graceful with strong, smooth wingbeats, hovering briefly before plunge-diving for fish. Unbroken tail streamers add 2.5 in (6 cm) to breeding adult's length. Monotypic. L 12.5–14" (32–36 cm) WS 30–33" (76–84 cm)

Identification Medium size; slightly bulkier, heavier billed, and longer legged than Common. It is the only medium-size "white tern" likely seen molting its outer primaries in N.A. Breeding tail streamers project well beyond wing tips at rest (but often broken in midsummer). **BREEDING ADULT:** Solid black cap (lost by late Aug.), orange-red bill with large black tip, orange-red legs. White underparts, contrasting pale-gray upperparts; pale-gray tail has white outer edges. **WINTER ADULT:** Note diagnostic bold black auricular mask with limited dusky wash on nape; blackish bill. Primaries molted July–Nov. **JUVENILE & FIRST-WINTER:** Resembles winter adult, but fresh juvenile has gingery wash and barring on head, back; first-winter has dark tertials, duskier uppersides to primaries, duskier legs than adult. In first-summer, legs brighter orange-red; bill can be orange-red with blackish culmen and tip. First-summer's protracted complete molt (primaries Apr.–Sept.) produces second-winter plumage. **SECOND-YEAR:** Resembles adult, but some second-summers have white-spotted forecrown; outer primaries darker and more worn than on adults, often forming an upperwing wedge recalling Common.

Similar Species Common Tern is slightly smaller with more slender bill and appreciably shorter legs; all Common Tern plumages have white tail with contrasting dark outer webs to outermost rec-

trices. Breeding adult Common is pale gray below; tail streamer tips fall about equal with, or shorter than, wing tips at rest; deeper red bill with smaller black tip; upperwing often shows dark wedge or contrast on trailing edge of primaries. Juvenile and first-summer Commons have dark patagial bar on upperwing; black partial cap extends solidly around nape. Arctic Tern is smaller with shorter neck, smaller bill, short legs, relatively longer and narrower wings, and quicker, more clipped wingbeats; all Arctic plumages have white tail, contrasting dark outer webs to outermost rectrices. Breeding adult Arctic is smoky gray below, all-red bill, primaries translucent but not silvery above. Juvenile Arctic Tern has dark patagial bar on upperwing; black partial cap extends solidly around nape. First-summer Arctic Tern has fresh primaries (replaced in winter); cap extends solidly around nape. Also see Roseate Tern.

Voice Hard, clipped dry *kik* or *krik*; slightly grating, shrill *krrih* or *kyiih* and *kyerr kyerr*; in breeding-season

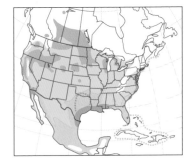

chases, gruffer, slightly shrill rasping series of *zzhi-zzhi-zzhi*.

Status & Distribution N.A. south to Mexico. **BREEDING:** Common, arrives at southern breeding sites Apr., northern sites (Canada, Great Lakes) mainly late Apr.–May. Departs northern breeding areas through Sept. **MIGRATION & DISPERSAL:** Fall migration mainly Aug.–Oct., ranging north regularly to Maritimes. **WINTER:** Southern US south to M.A. and Caribbean. **RARE STATUS:** Increasingly detected in Europe since the 1980s.

Population Apparently fairly stable.

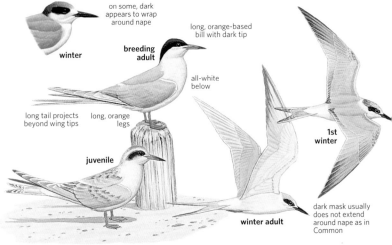

winter

on some, dark appears to wrap around nape

breeding adult

long, orange-based bill with dark tip

all-white below

long tail projects beyond wing tips

long, orange legs

juvenile

1st winter

winter adult

dark mask usually does not extend around nape as in Common

Genus *Thalasseus*

Seven species (three in N.A.) of large terns, including the critically endangered Chinese Crested Tern (*T. bernsteini*). These terns have a strong direct flight, and most have orange to yellow bills.

ELEGANT TERN   *Thalasseus elegans*   ELTE ▪ 1

This Pacific-coast tern nests in dense colonies and disperses north in late summer after breeding, with largest movements in warm-water years. Its flight is strong and graceful. Monotypic. L 14.5–16" (37–41 cm) WS 37–39.5" (94–100 cm)

Identification Medium-large size; note slender bill, which sometimes looks almost droop-tipped, and long, shaggy crest. All plumages can be flushed strongly pink on underparts. **BREEDING ADULT:** Black cap solid Jan.–June; spotted white in summer. Orange bill varies from bright orange-red with paler tip to uniformly mustard-yellow. Black legs rarely mottled or solidly orange. **WINTER ADULT:** Extensive white forecrown rarely has thin white postocular crescent separating eye from black crest; primaries molted July–Jan. **JUVENILE & FIRST-YEAR:** Resembles winter adult, but juvenile fresh plumaged in early fall, with dark gray centers to greater coverts and tertials, dark subterminal tail marks,

and dark secondary bar. Legs are often yellowish, usually becoming dark by winter. Protracted complete molt (primaries Oct.–July) produces first-summer plumage much like winter adult. **SECOND-YEAR:** Second postbreeding molt averages later than adult, so outer primaries fresher and paler in second-summer than in adult. Some have white spotting on forehead in breeding plumage.

Similar Species Closely resembles larger Royal Tern. Best feature is the disproportionately longer (males longer) and slimmer bill, often with a slight droop. Breeding plumaged Elegants sometimes have a pink flush on the underparts, unlike Royal in winter plumage (acquired earlier in summer by Royal). There is usually more white eye around the eye on Royal isolating the dark eye. When both are together (frequent in Southern CA) size and structure are apparent. Calls differ.

Voice A slightly screechy *rreeah* or *rreahk*, and *krreéh* or *krreíhr*; and a rough, grating *ehrrk*. Calls scratchy and grating; given incessantly by flocks of hundreds. Similar to Sandwich, but higher and screechier than Royal. Begging young (into first-winter) give high, piping *sii, siip-siip*.

Status & Distribution Breeds northwestern Mexico and Southern CA. Winters Pacific coast of S.A. **BREEDING:** Local Southern CA, Mar.–July; recently attempted nesting in Bay Area. **MIGRATION & DISPERSAL:** Common postbreeding visitor (mainly July–Oct.)

pale underside to primaries

breeding adult

juvenile

eye included in dark face

long crest

very long, slender bill

winter adult

bill of juvenile shorter, often yellowish

juvenile

to central CA, irregularly to southernmost BC. **WINTER:** Casual Southern CA. **RARE STATUS:** Casual inland in Southwest (annual at Salton Sea). Accidental on Atlantic and Gulf Coasts (has hybridized with Sandwich in FL), and Niagara River, NY/ON (found 20 Nov. 2013). Casual to Europe.

Population Increase on US Pacific coast since the 1970s may reflect changing ocean conditions. Colonized Southern CA as a nester in the 1950s.

ROYAL TERN   *Thalasseus maximus*   ROYT ▪ 1

A tern of southern coasts, the Royal nests in dense colonies. Its flight is strong and graceful. Monotypic (former ssp. *albididorsalis* from West Africa now a full sp.). L 17–19" (43–48 cm) WS 41–45" (104–114 cm)

Identification Large size; note stout orange bill at all ages. **BREEDING ADULT:** Black cap solid in Jan.–May, becoming spotted white by early summer. Black legs. **WINTER ADULT:** Extensive white forecrown often has a

white postocular crescent separating eye from black crest; primaries molted July–Feb. **JUVENILE & FIRST-YEAR:** Resembles winter adult, but juvenile is fresh-plumaged in early fall, with dark gray centers to greater coverts and tertials, dark subterminal tail marks, dark secondary bar. Bill paler orange; legs often yellowish, becoming dark by winter. Protracted complete molt (primaries molted Nov.–Aug.) produces first-summer plumage much

like winter adult. **SECOND-YEAR:** Second postbreeding molt averages later than in adults, so in second-summer plumage, outer primaries are fresher and paler than adults. White spotting on forecrown of breeding-plumaged spring birds may occur in all ages.

Geographic Variation CA populations average larger; they breed and molt earlier, and they are mostly resident or short-distance migrants. Atlantic coast populations average smaller, breed and molt later, and are medium- to long-distance migrants.

Similar Species See Elegant Tern. Caspian Tern is larger and heavier with broader wings and with a stout red, dark-tipped bill; in flight, note Caspian's diagnostic dark underside to outer primaries. Winter adult and juvenile Caspians have dark-streaked crowns.

Voice A fairly deep grating *ehrreh* and *rreh'k* or *rreh-eh*, lower and less shrieky than calls of Elegant and Sandwich Terns; a higher and shriekier *krriéh* or *rriehk*; a yelping or clucking *krehk* and *kehk*; and a more laughing *kweh-eh-eh*. Begging young (into first-winter) give a high whistled *see-ip*.

Status & Distribution Found along coasts in N.A. to S.A. **BREEDING:** Common in N.A. **MIGRATION & DISPERSAL:** Spring migration mainly late Feb.–Apr.; postbreeding northward movement and fall migration on East Coast mainly late June–Nov., regularly to southern New England, casually Atlantic Canada. **RARE STATUS:** Casual to northern CA (formerly regular in winter) and interior N.A.

Population Generally holding constant. Formerly common into the early 1900s as nonbreeding visitor (mainly Sept.–Mar.) north to San Francisco Bay (from Mexican colonies). Conversely, has colonized Southern CA as a breeding bird since 1959.

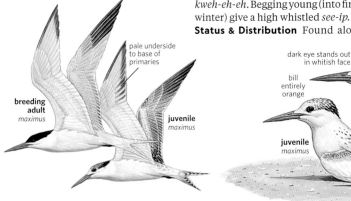

pale underside to base of primaries

breeding adult
maximus

juvenile
maximus

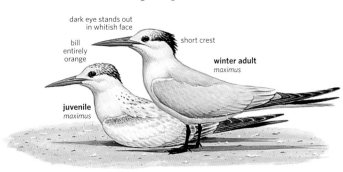

dark eye stands out in whitish face

bill entirely orange

short crest

winter adult
maximus

juvenile
maximus

SANDWICH TERN   *Thalasseus sandvicensis*   SATE ■ 1

This medium-size tern nests in dense colonies. Its flight is strong and graceful, with wings held slightly crooked. Polytypic (3 sspp.; 2 in N.A.). L 13.5–14.5" (34–37 cm) WS 34–36" (86–91 cm)

Identification Slender, yellow-tipped black bill, black legs, shaggy crest. **BREEDING ADULT:** Black cap solid mainly Mar.–June, spotted white late summer. **WINTER ADULT:** White forecrown, primaries molted mainly Aug.–Mar. **JUVENILE & FIRST-YEAR:** Like winter adult, but in fall juvenile has variable blackish subterminal marks on back, dark secondary bar. Black bill may have yellowish side patches (soon lost); develops yellow tip by spring. Complete molt (primaries Dec.–Aug.) produces first-summer plumage like winter adult. **SECOND-YEAR:** Second postbreeding molt averages later than in adult, so second-summer's outer primaries fresher and paler. White-spotted forecrown of breeding spring birds may occur in all ages.

Geographic Variation "Cayenne Tern" (*eurygnathus*) of Caribbean and S.A. has a yellow to yellow-orange bill; birds with partially yellow bills are common in Caribbean. The subspecies *acuflavida* breeds in N.A. and similar nominate *sandvicensis* in the Old World. Genetically, *acuflavidus* is closer to *eurygnathus* than *sandvicensis*, and some have recommended separate species status for these two very similar appearing subspecies. The two New World subspecies are warm water species, while Old World *sandvicensis* breeds north to throughout the British Isles and southern Scandinavia, thus suggesting different ecologies.

Similar Species In the West and on the Gulf Coast,

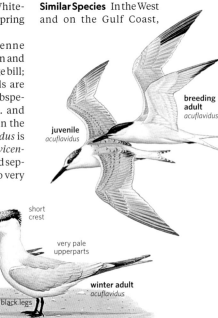

breeding adult
acuflavidus

juvenile
acuflavidus

long, slender black bill with yellow tip

short crest

juvenile
acuflavidus

very pale upperparts

winter adult
acuflavidus

black legs

beware hybrids with Elegant Tern. **Voice** Scratchy, penetrating *kree-ik* or *krrík*; slightly sharp, yelping *kehk*; and slightly reedy *ki-i wii-wii.* Begging young (into first-winter) has high, whistled *sree* or *sri-sree.* **Status & Distribution** Europe, eastern N.A., eastern S.A. **BREEDING:** Common in N.A., arriving colonies Apr.–early May, departing Aug.–Sept. **MIGRATION:** Spring mainly Apr.–May. Fall and postbreeding movement north on East Coast mainly Aug.–Oct. **RARE STATUS:** Casual Southern CA (where hybridized with Elegant and a variety of intermediate birds have been seen in the field), north to Atlantic Canada and north to Great Lakes, often after tropical storms. Accidental north TX and CO. "Cayenne Tern" accidental in NC. **Population** Stable.

SKIMMERS Genus *Rynchops*

This pantropical genus (three species) is sometimes treated as a separate family, Rynchopidae. Resembling large, angular terns, they are distinguished by long, laterally compressed bills with projecting lower mandibles. They feed in flight, mostly at night, by skimming with an open bill, then snapping it shut on contact with fish and crustaceans.

BLACK SKIMMER *Rynchops niger*   BLSK ▪ 1

This species favors coasts with sandy beaches and lagoons. Its flight is strong and buoyant, with smooth wingbeats. Flocks often fly in tight formation, at times wheeling like shorebirds. Males are larger than females, often noticeable in the field (male bills average 10–15 percent longer). Polytypic (3 sspp.; nominate in N.A.). L 17–18" (43–46 cm) WS 45–49" (114–124 cm) **Identification** Striking. Long, laterally compressed bill; fairly short legs; very long wings; fairly short, forked tail. Males larger. **BREEDING ADULT:** Crown, hind neck, and upperparts solidly black, contrasting sharply with white forehead, white trailing edge to secondaries and inner primaries; tail is mostly white with a black central stripe. Whitish underwings grade to dusky on remiges, with blacker wing tips. Black-tipped, bright red bill; red legs. Postbreeding molt starts in late summer on the breeding grounds (when a few inner primaries may be replaced before suspending molt for migration) or on the winter grounds. Primary molt completes Mar.–May, overlapping with pre-breeding molt of hind neck feathers. **WINTER ADULT:** Resembles breeding adult, but white hind neck, sometimes with a little dusky mottling, and bleached upperparts can look browner in early winter. **JUVENILE & FIRST-YEAR:** Juvenile crown and hind neck heavily streaked buffy and whitish; back and upper- wing coverts blackish brown; broad buff to whitish edgings create a bold, scaly pattern. Bill duller, less extensive red basally; legs paler, pinkish to pale orange-red. Protracted, complete molt starts Sept.–Dec. (primaries molted mainly Jan.–Aug.); produces plumage resembling winter adult by end of the first summer. Crown is dark sooty-brown to blackish with paler feather edgings; hind neck is white or mottled blackish; bill and legs average paler than adults.
Geographic Variation Birds in N.A. are of the nominate subspecies, with white underwings and an extensively white tail. The nomadic Amazonian-breeding *cinerascens*, a potential stray to N.A., has light sooty-gray underwings, reduced white on trailing edge of wings, and a dark tail.
Similar Species Black Skimmer is unique and unmistakable.
Voice A nasal, slightly hollow laughing *kyuh* or *kwuh* and a disyllabic *k'nuk* or *k'wuk*, with calling flocks producing chuckling choruses at times; alarm call a more drawn-out *aaawh*. Juvenile call higher pitched and squawkier than adult.
Status & Distribution Warmer regions of the Americas. **BREEDING:** Fairly common to common but local, arriving at northern colonies late Apr.,

departing Aug.–Sept. **MIGRATION:** Mainly Mar.–Apr. and Sept.–Nov. **WINTER:** Southern US to S.A. **DISPERSAL & RARE STATUS:** Casual inland in coastal states, also casual (mainly summer) in the Southwest, Great Plains, Great Lakes region, and (mainly after tropical storms) elsewhere interior East and Atlantic Canada.
Population Atlantic, Gulf Coast colonies fluctuate greatly: was considered declining in the 1970s; some evidence of recent stabilization. West Coast population has increased markedly in last 40 years, leveling off in last decade; Salton Sea numbers declining (recent breeding unsuccessful).

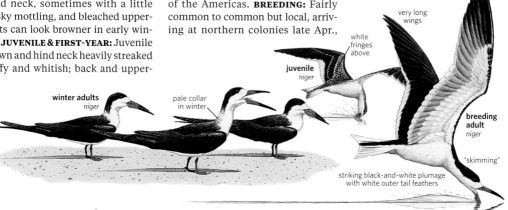

very long wings

white fringes above

juvenile *niger*

winter adults *niger*

pale collar in winter

breeding adult *niger*

"skimming"

striking black-and-white plumage with white outer tail feathers

TROPICBIRDS Family Phaethontidae

Red-billed Tropicbird (Galápagos Is., Jan.)

T ropicbirds are pelagic seabirds with long central tail streamers and direct flight styles. For field identification, focus on the distribution of black on the wings (especially on the primaries and primary coverts, which are visible only from above) and flight style; the bird's size, bill color, and tail streamers, if present, can be difficult to assess from a distance.

Structure Tropicbirds have heavy, pointed red to yellowish bills and long central rectrices forming spectacular streamers in adults. The pointed wings angle back and the body appears front-heavy in flight. Small, black fully

webbed feet, set well back on the body, make tropicbirds walk with clumsy lunges.

Behavior Flight is direct with species-unique wingbeats; glides are usually brief. On oceangoing trips, you may see them making a brief pass over the boat, or less frequently resting on the water in the distance, with tail streamers, if present, held in a high arc; white body stands out at great distance. Usually found singly. They plunge-dive for prey, preferring flying fish and squid.

Plumage Tropicbirds are mainly bright, satiny white with some black markings; the sexes look similar but juveniles are more heavily barred with black than adults. Molts are prolonged and poorly studied; adult tail streamers are often broken.

Distribution Found over most tropical and subtropical oceans, tropicbirds nest on oceanic islands or offshore islets or stacks. None nest in our area. They are uncommon to casual visitors to our southern waters, and strays (often storm-related) occur well north of their regular range, onshore and inland.

Taxonomy The world's three tropicbird species form a uniform group without close relatives; formerly classified within the order of pelicans and allies, they were recently classified as comprising a separate and unique order, Phaethontiformes. Geographic variation is minor.

Conservation Tropicbirds face threats from introduced predators on nesting islands. BirdLife International codes: Least Concern (all three species).

Genus *Phaethon*

WHITE-TAILED TROPICBIRD *Phaethon lepturus*   WTTR ■ 3

The White-tailed, the smallest and most graceful tropicbird, has the most buoyant flight of the tropicbirds. Its long, white tail streamers are especially prominent in flight, undulating and ribbonlike. The distinctive upperwing pattern is only hinted at from underneath, and can be difficult to judge from above in harsh light. Polytypic (6 sspp.; 2 in N.A.). L 15" (38 cm) Tail streamers 12–25" (30–64 cm) WS 37" (94 cm)

Identification ADULT: The large, black diagonal upperwing bar of adults is unique; it extends from the

longest scapulars and tertial centers out the wing to near the wrist joint. The outer wing is white, with black on outer four or five visible primaries; the black appears cut off by the white primary coverts halfway along the outer wing. The remainder of the upperparts and underparts are white

with no barring; the few black flank markings are not generally visible in the field. There is a small black patch through the eye. The long to very long white tail streamers are ribbonlike (they may have a pink or golden tinge) with black shafts. **JUVENILE:** Lacks tail streamers and is heavily barred with black on upperparts and upperwing coverts, but still shows sharp contrast between black outer primaries and white primary coverts. Black around eye does not extend back around nape. Bill is a dull yellow. Older immatures

slower, more ternlike flight

blackish diagonal bar on upperwing coverts

orange bill

adult *catesbyi*

greenish yellow bill

black cut off by white primary coverts

coarse bars on back and wing coverts

juvenile *catesbyi*

Pacific adult *dorotheae*

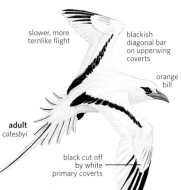

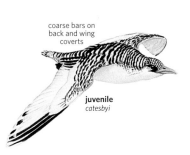

have lost most dorsal barring and have brighter yellow bills. **FLIGHT:** It is more graceful and buoyant than Red-billed Tropicbird; it appears slim-bodied and flies with ternlike wingbeats.

Geographic Variation There is geographic variation in size, bill color, and extent of black in the primaries. Caribbean and Atlantic records refer to *catesbyi*, with a deep orange bill (can approach red in color) and slightly more black on wing tips (showing black on five visible primaries). On Pacific *dorotheae* note greenish yellow bill and black on only four primaries.

Similar Species Separating Whitetailed from Red-billed Tropicbird in the Atlantic is the main problem. Differences in flight styles can be very helpful (see above). Bill color is unreliable, and the diagnostic upperwing pattern may be hard to see. Adult Redbilleds have barred (not white) upperparts and more black in the outer wing, extending through the primary coverts

closer to the bend of the wing. Juvenile Red-billeds show an upperwing pattern like adults, but species differences are subdued by extensive dorsal barring in both species. Juvenile Redbilled also has a longer black eye patch that extends to or across the nape. Redbilled and Red-tailed Tropicbirds are much larger, with broader wings and larger bills than White-tailed.

Voice Generally silent at sea. Grating *keek keek* series and shrill whistled notes given mainly on breeding grounds.

Status & Distribution Rare in our area. **BREEDING:** Pantropical; nearest colonies on Bermuda, Bahamas, and Greater Antilles. **NONBREEDING:** Regular visitor (mostly adults) mainly May–Sept., to Gulf Stream from off northern FL to NC; also formerly around Dry Tortugas, FL. Very rarely onshore after hurricanes. **RARE STATUS:** Casual in the Gulf of Mexico and off Atlantic coast from VA to MA; accidental to

NS. One onshore record from Southern CA (only certain record of *dorotheae*, which breeds as near as HI), and inland records for AZ (likely of Caribbean origin), PA, and NY.

Population Stable. The declines on Bermuda and in the West Indies have been partly offset by conservation measures (predator control, artificial nest sites); there are about 2,500 pairs on Bermuda and 3,500 pairs in the West Indies.

RED-BILLED TROPICBIRD *Phaethon aethereus* RBTR ■ 3

Red-billed Tropicbirds range off both of N.A.'s southern coasts, but are most numerous off CA, where most often seen on long-range day trips beyond the Channel Is. (Aug.– Sept.). The combination of barred upperparts, red bill, and long white tail streamers is diagnostic for adults. Polytypic (3 sspp.; *mesonauta* in N.A.). L 18" (45 cm) Tail streamers 12–20" (30–50 cm) WS 44" (112 cm)

Identification Large tropicbird with obvious black in primaries and barred upperparts in all plumages. **ADULT:** Mainly white with fine black barring above from nape to uppertail coverts; extensive black in the primaries and black in the longer primary coverts of upperwing. A long black mark extends from in front of the eye to the nape. Very long, thin white tail streamers lack a black shaft. Bill bright red. **JUVENILE:**

adult

red bill

extensive black in primaries and primary coverts

fine bars on back and wing coverts

black collar

flies with rapid, shallow wingflaps, almost like Peregrine Falcon

juvenile

extensive black

Finely barred with black above. Black mark through eye wraps around nape, unlike juveniles of other tropicbirds. Tail shows black tip; no tail streamers. Bill dull yellowish to orange. Older immatures have only sparse spots on crown and short white tail streamers. **FLIGHT:** Flies with rapid, stiff wingbeats, different from the more buoyant flight of White-tailed and the slower more powerful strokes of Red-tailed.

Similar Species White-tailed Tropicbird poses an identification problem in the Atlantic and Gulf of Mexico; see that species for separation. In the Pacific, a casual Red-tailed Tropicbird also has a deep red bill, but whiter

plumage; all ages of Red-tailed look completely white-winged at any distance. Red-billed superficially resembles large Royal (often seen well at sea) and Caspian Terns; but note the tropicbird's more rapid wingbeats, elongated central tail feathers (adults), black barring on the upperparts, dark tertials, and contrasting white secondaries and inner primaries.

Voice Generally silent at sea. Grating ternlike notes are given mainly around the nesting colonies.

Status & Distribution Rare in our area, but double figures have been seen on long day trips off Southern CA. **BREEDING:** Tropical and subtropical

eastern Pacific, southern Atlantic, and northwestern Indian Oceans. Nearest colonies in northern Gulf of California and Puerto Rico/Virgin Is. **NONBREEDING:** Regular at sea off southern (casually northern) CA, mainly May–Sept.; especially frequent in waters off Channel Is., CA. Rare off Southeast and casual in Gulf of Mexico (May–Sept.). **RARE STATUS:** North to WA (one record in 1945), New England (one returning bird to the Gulf of Maine, 2005–2019), along Gulf Coast (including several storm-wrecked birds), and interior in Southern CA and AZ. **Population** Worldwide, the Red-billed is the least numerous tropicbird. Nearby, some 1,800–2,500 pairs breed in the West Indies (mainly Lesser Antilles) and 500–1,000 pairs breed in the Gulf of California, Mexico.

RED-TAILED TROPICBIRD   *Phaethon rubricauda*   RTTR ■ 4

The large, broad-winged, robust Red-tailed Tropicbird flies with slower and more powerful wingbeats than the other tropicbirds. Casual in N.A., most records are from far off the CA coast. Polytypic (4 sspp.; likely *melanorhynchos* in N.A.). L 18" (45 cm) Tail streamers 12–15" (30–38 cm) WS 44" (112 cm)

Identification ADULT: Entirely white (adults may show a faint rosy tinge below) except for a black mark through the eye, black centers to the tertials (a hint of which may show from below), and black shafts to the primaries (not visible from below); the black centers to some flank feathers are not usually visible in the field. The central tail feathers are very thin, stiff, and red (with black shafts); though diagnostic, they tend to disappear against a blue sky or water background. The tail streamers frequently break or are lost. Bill is bright red. **JUVENILE:** It is heavily barred with black above, spotted black on the crown and nape; black in the primaries extends to vanes adjacent to shafts, thus slightly more extensive than in adults. The tail appears mainly white, wedge-shaped; lacks streamers. The bill is black, changing after fledging to yellow. Older immatures retain limited spotting on head and some black in the primary vanes adjacent to shaft, and show short, whitish or pinkish tail streamers. The bill becomes orange to orange-red in the second year. **FLIGHT:** Strikingly white in flight, it appears large, heavy, and broad winged.

Geographic Variation There is minor variation in size between subspecies; *melanorhynchos* breeds in the north-central Pacific Ocean, including on the Hawaiian Is.

Similar Species At-sea distribution in eastern Pacific overlaps with Red-billed Tropicbird; note Red-billed's extensive black in the primaries, long white tail streamers, and more rapid wingbeats.

Voice Generally silent at sea. Grating *ack*, often repeated, around nesting islands.

Status & Distribution Very rare (but perhaps annual) in our area. **BREEDING:** Tropical Indian and Pacific Oceans; nearest colonies are on HI, but a few may breed on islands far off southwestern Mexico; absent from Atlantic Ocean. **NONBREEDING:** Ranges widely over tropical and sub-

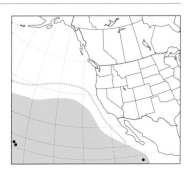

tropical Pacific and Indian Oceans. Probably regular more than 100 mi off southern and central CA coast (west of the cold California Current) where pelagic coverage is poor; most records Aug.–Jan. Recorded on the Farallones. **RARE STATUS:** Two onshore records from coastal CA (Orange Co. and San Mateo Co.); remains of one found at Buttle Lake Park on central Vancouver I., BC (June 1992). **Population** The Pacific Ocean breeding population consists of fewer than 15,000 pairs. Exploitation for human food threatens some important central Pacific colonies. Introduced mammalian predators also impact many breeding populations.

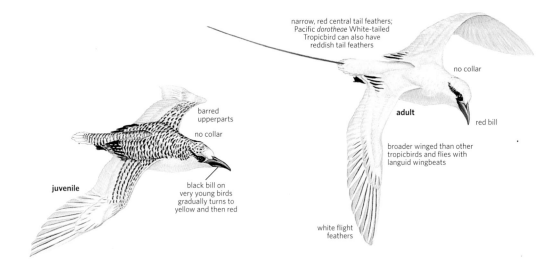

narrow, red central tail feathers; Pacific *dorotheae* White-tailed Tropicbird can also have reddish tail feathers

no collar

adult

red bill

barred upperparts

no collar

broader winged than other tropicbirds and flies with languid wingbeats

juvenile

black bill on very young birds gradually turns to yellow and then red

white flight feathers

LOONS Family Gaviidae

Pacific Loon, breeding (MB, June)

L oons, usually called divers in the Old World, present major identification challenges in both breeding and winter plumages: Swimming birds dive frequently. On swimming birds note the bill shape (stout or slender, straight or with any uptilt) and color, head shape, and posture. In breeding plumages, note the overall patterns, especially of the head, neck, and back. On winter and juvenal plumages, details of the dark/light patterning on the head and neck are important.

Structure Bodies are long and heavy, with large webbed feet set well back; bills are strong and dagger-shaped. Wings are relatively small with 10 primaries and 22 to 24 secondaries; tails are small and short, with 16 to 20 rectrices. Males are up to 10 percent larger than females.

Behavior Loons are aquatic birds that swim and dive with proficiency. Birds trying to avoid predators can swim very low in the water, whereas resting birds sometimes ride quite high and buoyantly, and preening birds often roll to the side, exposing their flanks and even belly. Determining posture is essential in assessing the extent and pattern of white along a bird's sides. Flight is fairly heavy and direct, with steady, measured wingbeats; the neck held outstretched and often slightly below the body plane; the large feet are visible beyond the short tail, especially in Common and Yellow-billed.

Plumage Plumages are mostly monochromatic, with breeding adults more boldly patterned than winter adults and juveniles; sexes look similar. Juveniles and first-summer birds resemble winter adults. Second-summer birds more closely resemble breeding adults but with messier, less complete breeding patterns, especially on the head and neck. Adult plumage is attained in the third winter.

Distribution High-latitude breeders throughout the Northern Hemisphere, loons are mid- to long-distance migrants that winter south in the New World to the southern US and northern Mexico. Nonbreeding immatures often remain south in winter range in summer.

Taxonomy Five species in one genus, *Gavia*, make up the modern-day loon family; all occur in North America. Loons are only very distantly related to the superficially somewhat similar grebes.

Conservation Oil spills present a major threat to nonbreeding populations in marine waters. Industrial poisons (such as mercury) may affect both breeding and wintering populations. Human disturbance and acid rain affect nesting Common Loons at more southerly lakes. BirdLife International codes: 1 NT.

Genus *Gavia*

RED-THROATED LOON   *Gavia stellata*   RTLO ■ 1

This smallest and most lightly built loon differs from other loons in its relatively subdued breeding plumage lacking white on upperparts, and in having a complete molt in late fall. Monotypic. L 25" (64 cm)

Identification Note slender, slightly upturned bill, often held slightly raised. Swimming birds often lack a chest bulge at the waterline, shown by other loons. **BREEDING ADULT:** Head and neck gray with black-and-white hind neck lines; brownish upperparts; dark red throat patch. Bill black; eyes red. **WINTER ADULT:** White face and neck; bill gray. Small, mainly subterminal, white spots and ovals on scapulars. **JUVENILE/FIRST-**

SPRING: Resembles winter adult, but face and neck variably smudged dusky, scapulars have narrower, chevronlike, subterminal whitish marks, and eyes brownish red. First-summer birds range in appearance from like winter adult to having patches of adultlike gray and dark red on neck. Complete molt in late summer produces adultlike winter plumage.

Similar Species Pacific Loon is slightly larger and bulkier with thicker neck, straight bill, white back patches, and has a fuller-chested profile when

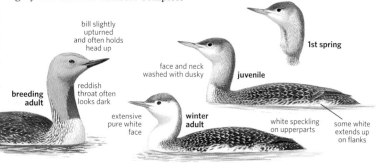

bill slightly upturned and often holds head up

breeding adult

reddish throat often looks dark

extensive pure white face

face and neck washed with dusky

juvenile

1st spring

winter adult

white speckling on upperparts

some white extends up on flanks

swimming. Winter/juvenile Pacific has dark cap extending down through eye, and hind neck more extensively dark (beware juvenile Red-throated, but its dark areas are messy, not well defined). **Voice CALL:** Flight call in summer a low, grunting or barking *ahrr* or *ahrk*, usually in series. **SONG:** A male-female duet of loud, wailing, drawn-out screams, *wheeahhr, heeahh*, etc. **Status & Distribution** Holarctic

breeder. **BREEDING:** Fairly common, mainly on lakes and ponds. **MIGRATION:** Mar.–May and late Oct.–mid-Dec., mainly coastal but also overland in East, especially Great Lakes. **WINTER:** Coastal waters; a few oversummer. Casual over much of interior N.A. and along Gulf Coast. **Population** AK population on western tundra declined about 50 percent between 1977 and 1993.

ARCTIC LOON *Gavia arctica*   ARLO ■ 2

This Old World species' breeding range extends into western AK, where it should be distinguished with care from the much commoner Pacific Loon. Polytypic (2 sspp.; *viridigularis* in N.A.). L 28" (73 cm)
Identification Resembles the Pacific Loon, but averages larger, with a longer and stouter bill, and often a more angular forehead. The Arctic's single best field mark in all plumages is the white flank patch on swimming birds or a white flank bulge on flying birds. The white patch extends up onto the sides of the rump. Otherwise, all age-related variation and molts are like Pacific Loon, except breeding. **BREEDING ADULT:** Hind neck is a slightly duskier, smoky gray. Throat patch is glossed green (very hard to see). **WINTER ADULT AND JUVENILE:** A dusky chinstrap (common on Pacific) is usually, perhaps always, lacking on Arctic. **Geographic Variation** N.A. records are of the eastern Eurasian subspecies, *viridigularis*, distinguished from nominate *arctica* of western Eurasia by its slightly larger size and a more greenish sheen (difficult to discern)

to the breeding adult's throat patch. **Similar Species** Pacific Loon averages smaller, with a rounder head that accentuates the smaller bill, and it lacks well-defined white femoral patches, but a swimming bird can show white along the waterline. Usually this is an untidy patch at about mid-body. Arctic Loon has a generally neater patch that flares up the sides of the rump. Breeding Pacific Loon has a paler crown and nape bordered by less contrasting white stripes that are separated from the chest striping by a narrow black bar (stripes typically merge on Arctic). Throat gloss color is unreliable. The great majority of winter adult and many juvenile Pacifics show a distinct dark chinstrap, unlike Arctic. Juvenile Pacific is paler naped than Arctic. A winter/juvenile Common Loon is larger and bulkier with a stouter bill, a broken whitish eye ring, and a more irregular dark/white border to the neck sides. **Voice** Similar to Pacific Loon, but lower pitched. **Status & Distribution** Eurasia and western AK. **BREEDING:** Uncommon to northwestern AK on small

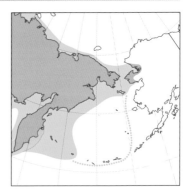

to large tundra lakes. **MIGRATION:** In N.A. best known at Gambell, St. Lawrence I., AK, where small numbers are regularly noted flying northeast in late spring a few in the fall. Rare in western and central Aleutians and elsewhere in Bering Sea. **WINTER:** Casual south along or near Pacific coast to Baja California in winter (a few summer records of immatures). Accidental to CO (fall) and OH (spring). **Population** Decreasing, especially European population.

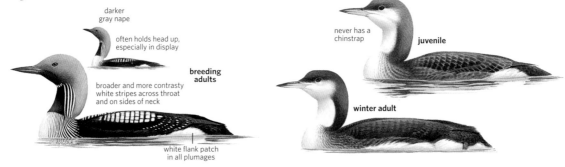

darker gray nape

often holds head up, especially in display

broader and more contrasty white stripes across throat and on sides of neck

breeding adults

never has a chinstrap

juvenile

winter adult

white flank patch in all plumages

SWIMMING JUVENILES FOR COMPARISON

Common   Yellow-billed   Arctic   Pacific   Red-throated

PACIFIC LOON *Gavia pacifica*   PALO ■ 1

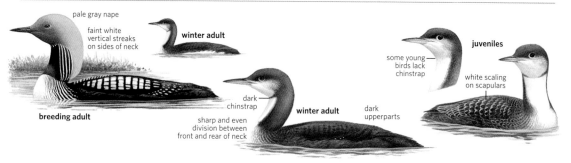

pale gray nape

faint white vertical streaks on sides of neck

winter adult

juveniles

some young birds lack chinstrap

white scaling on scapulars

breeding adult

dark chinstrap

sharp and even division between front and rear of neck

winter adult

dark upperparts

Pacific Loon is best known as a migrant and winter visitor along the Pacific coast, where it often occurs in flocks. Monotypic. L 26" (66 cm)

Identification Pacific Loon has a straight bill and smoothly rounded head. In winter/juvenile plumages, note the clearly defined and even dark/light border on the head and neck sides, and contrast between a paler hind neck and darker back. **BREEDING ADULT:** Head and hind neck are ashy gray with a purple-glossed (sometimes green) throat patch bordered by white stripes; upperparts are boldly patterned black-and-white. Bill is black, eyes red. **WINTER ADULT:** Throat and fore-neck are white with clean and even line of separation to dark sides of neck; about 90 percent of birds have a dusky to dark chinstrap, and the blackish upperparts have very muted paler checkering. Bill is pale gray with a dark culmen. **JUVENILE/FIRST-YEAR:** Juvenile resembles winter adult but paler feather tips to its upperparts create a scaly patterning; its upperwing coverts lack white spots; eyes brown-

ish red. About half of juveniles lack the dusky chinstrap. Many first-summers look like faded winter adults. **SECOND-WINTER:** Produced by complete molt in late summer, resembles winter adult but upperwing coverts lack distinct white spots. In spring of the third year, resembles breeding adult but the head and neck can be mottled white, the white back markings average smaller, and the bill may be piebald.

Similar Species See very similar Arctic Loon. Winter adult/juvenile Common Loon is distinctly larger and bulkier, and its stouter bill has a pronounced gonydeal angle. Note Common's broken whitish eye ring, the more irregular dark/white border to its neck sides (suggesting the breeding pattern), and that its hind neck is darker than the back, the opposite of Pacific Loon. Red-throated Loon is lighter in build, often has a paler (winter adult), or messier (juvenile) face, and has an uptilted bill. Winter adults and juveniles are usually paler overall with fine white flecks on the upperparts and either a more extensive white face or (juveniles) a messier and more

extensively dark face and neck than the Pacific.

Voice Rarely vocal in winter. **CALL:** Includes a low, grunting *awhrr.* **SONG:** Loud, slightly trumpeting, melodic wailing, *hooo-AHHr-uh,* and variations.

Status & Distribution N.A. **BREEDING:** Fairly common on small to large lakes. **MIGRATION:** Mainly Apr.–early June and Sept.–early Dec. over inshore waters, often in flocks. **WINTER:** Coastal and offshore waters, some oversummer. **RARE STATUS:** Rare inland throughout West, very rare in Midwest, and along East Coast in migration and winter. **Population** Stable or increasing.

LOONS IN FLIGHT

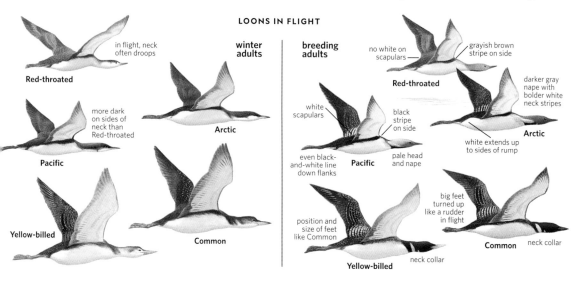

in flight, neck often droops

winter adults

breeding adults

no white on scapulars

grayish brown stripe on side

Red-throated

Red-throated

darker gray nape with bolder white neck stripes

more dark on sides of neck than Red-throated

Arctic

white scapulars

black stripe on side

Arctic

white extends up to sides of rump

Pacific

even black-and-white line down flanks

Pacific

pale head and nape

Yellow-billed

position and size of feet like Common

big feet turned up like a rudder in flight

Common

neck collar

Common

Yellow-billed

neck collar

COMMON LOON *Gavia immer*   COLO ■ 1

This large and widespread species is the most familiar loon throughout N.A. Monotypic. L 32" (81 cm)

Identification Distinctive stout, dark bill. Head shape fairly angular, often with a distinct forehead bump. In winter/juvenile plumages, white extends over eye and jagged dark/light borders on the neck sides. **BREEDING ADULT:** Head and neck oily green-black with two horizontal white neck bands striped dark; back boldly patterned black-and-white. Bill black, eyes red. **WINTER ADULT:** Throat and foreneck white, upperparts blackish brown. Bill pale gray with dark culmen. **JUVENILE/FIRST-YEAR**: Like winter adult, but paler feather tips on upperparts create scaly effect and eyes duller. Some first-summer birds retain almost all juvenal plumage, which can become faded and pale. Other first-summers attain new head, neck, and back feathers, so their upperparts have scattered darker feathers and sometimes small white spots.

Similar Species Yellow-billed Loon is slightly larger overall; longer bill has straight culmen, often held slightly raised, accentuating upturned shape; bill pale yellow (breeding) to creamy or pale ivory with a duskier base, but culmen always pale, at least near tip (dark on Common). Winter/juvenile Yellow-billed paler above, especially face, which usually has dark auricular mark; pale upperpart scaling on juvenile Yellow-billed usually more conspicuous.

Voice CALL: Includes a slightly manic, yelping tremolo. **SONG:** Loud, mournful, and eerie wailing cries are far-carrying, *whoo'oo ooh-ooh.*

Status & Distribution Breeds N.A. and on Greenland and Iceland. **BREEDING:** Fairly common, on large lakes. **MIGRATION:** Mar.–May and late Sept.–

Nov. **WINTER:** Mainly coastal, also on large, inland water bodies, some oversummer. Uncommon in Aleutians and mostly rare in Bering Sea region.

Population Bulk of world population nests in Canada.

heavy silver-colored bill with dark, slightly decurved culmen

winter adult

uneven line of separation of white and dark on neck

half dark collar

juvenile

white collar

breeding adult

YELLOW-BILLED LOON *Gavia adamsii*   YBLO ■ 2

This far-northern loon is closely related to the Common Loon. Monotypic. L 34" (86 cm)

Identification Readily identified, but on problem birds always check the bill shape and color: The bill is long and stout with straight culmen (that is pale, at least distally) and a distinct gonydeal angle; hence the lower mandible looks slightly recurved, an effect enhanced by the bill usually being held slightly raised. Winter adults and juveniles typically have a pale-faced aspect with a dark auricular mark and appear paler and browner above. **BREEDING ADULT:** Bill is straw yellow. **WINTER ADULT AND JUVENILE:** Bill is pale creamy yellow to ivory, sometimes with a grayish base.

Similar Species Yellow-billed Loon resembles Common Loon in all plumages, but winter Yellow-billed adults and juveniles are usually paler above. Common Loon is slightly smaller overall with a smaller bill that has a decurved culmen and less exaggerated wedge shape to the lower mandible; Common's bill is black (breeding adult) to pale gray with a dark culmen. The eye of a Yellow-billed appears smaller.

Voice Yellow-billed's yodel is similar to Common Loon but lower pitched

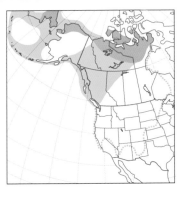

and hollow; tremolo is slower paced.

Status & Distribution Holarctic breeder. **BREEDING:** Uncommon to fairly common on tundra lakes and rivers. **MIGRATION:** Apr.–early June and late Sept.–Nov. **WINTER:** Mainly coastal AK waters, a few oversummer. **RARE STATUS:** Rare on Pacific coast south to northwest WA; very rare south to northern CA; casual to Baja California Norte; casual to very rare inland in West, and casual in East.

Population Near Threatened.

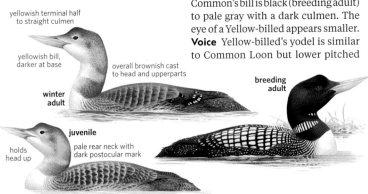

yellowish terminal half to straight culmen

yellowish bill, darker at base

overall brownish cast to head and upperparts

breeding adult

winter adult

juvenile

holds head up

pale rear neck with dark postocular mark

ALBATROSSES Family Diomedeidae

Black-footed Albatross (CA, Nov.)

Albatrosses, open ocean birds, spend the majority of their lives at sea, some wandering thousands of miles in search of food. They are characterized by their large size and long, narrow wings; the Wandering and Royal Albatrosses possess the largest wingspans of any living bird, measuring just over 11 feet! While the North American species are smaller, all are strikingly large when compared to other seabirds. Separation of similar species relies on observation of subtle plumage differences and bill color. Albatrosses live long lives, typically tens of years.

Structure Albatrosses are characterized by their long, narrow wings and thick, heavy bodies. Unlike the procellariids, which have a single tube on the culmen, albatrosses have very large bills with single tubes on either side of the culmen. The bills, made up of individual plates, are sharply hooked. Bill size and coloration can be critical field marks, and bill color often changes with maturation. Albatrosses have large, webbed feet, which sometimes extend past the tail in flight.

Behavior Flight style varies with wind speed and sea condition, but almost always incorporates a large amount of energy-efficient arcing and gliding, especially in stiff oceanic gales. Wingbeats are heavy and labored, and in light winds albatrosses work hard to stay aloft, often sitting on the water in calm conditions. They typically feed by surface dipping, alighting on the water and picking prey—including squid, fish, flotsam, offal, and carrion—from the surface. Their acute sense of smell helps in detecting food; many are attracted to fishing boats and chum. Although vocal on breeding grounds, most albatrosses are silent at sea. Most species nest in large colonies on remote oceanic islands; pairs mate for life.

Plumage Albatross plumages are largely variations of black and white, with some having gray or brown tones. The contrasting patterns of black and white are important identification field marks. Most species' plumages change with maturation; others exhibit less variation. Albatrosses show little sexual dimorphism; however, more work is needed. Molt can be complex and protracted, particularly in the larger species; it is not possible to replace all the feathers between breeding cycles, hence the long maturation.

Distribution Albatrosses reach their greatest diversity in the Southern Hemisphere; the doldrums of the equatorial waters presumably represent a barrier to their northward dispersal. In the North Pacific there are three breeding albatrosses: Laysan, Black-footed, and Short-tailed. The Short-tailed Albatross was historically common and is slowly recovering from near extinction. All others occur as casual to accidental visitors in North America. Except when nesting, albatrosses are entirely pelagic.

Taxonomy Albatross species-level taxonomy continues to undergo scientific debate. Worldwide 13 to 24 species in two to four genera are recognized. The AOS lists 10 species in four genera as occurring in North America: five mollymawks or southern albatrosses, one sooty albatross, one great albatross, and three North Pacific albatrosses.

Conservation Threats include feral predators (especially rats, pigs, and cats) at nesting sites; longline fishing at sea; ingestion of floating plastics by adults and the subsequent regurgitation of this material to chicks; and the human harvesting of birds and eggs for food. BirdLife International codes: 6 NT, 6 VU, 7 EN, and 2 CR; of all the world's albatrosses, only the Black-browed is considered Least Concern.

MOLLYMAWKS Genus *Thalassarche*

Five of the seven mollymawk species, medium-size albatrosses of southern oceans, occur as strays to N.A. waters. They are largely white below with blackish brown backs and upperwings. Important distinctions are the extent of dark margins on the underwings and bill coloration. All suffer drowning from longline and trawl fisheries.

YELLOW-NOSED ALBATROSS *Thalassarche chlororhynchos* YNAL ▪ 4

The mollymawk most often recorded in N.A.'s Atlantic waters. Polytypic (2 sspp.; nominate in N.A.). L 28–30" (71–76 cm) WS 74–85" (188–216 cm) **Identification** Small and slim. **ADULT:**

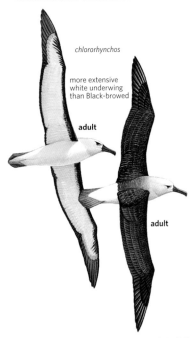

chlororhynchos

more extensive white underwing than Black-browed

adult

adult

Gray head in fresh plumage, otherwise strikingly black-and-white. Upperwings blackish with white primary shafts; back grayer; underwings white, with thin black margins, widest at the leading edge. Gray tail. At a distance, bill appears black; at close range, yellow ridge along the length of the culmen and orange-reddish tip visible. **JUVENILE:** Dark bill, white head; can show wider dark margins on the leading edge of the underwing.

Geographic Variation Similar *carteri* of the Indian Ocean, often treated as a full species ("Indian Yellow-nosed Albatross"), is unrecorded in N.A. Adult has whiter head and smaller, less triangular eye patch; yellow of culmen ridge has a sharply tapered,

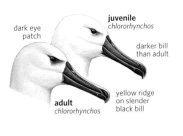

dark eye patch

juvenile
chlororhynchos

darker bill than adult

adult
chlororhynchos

yellow ridge on slender black bill

V-shaped base (rather than broader and U shaped).

Similar Species Adult Black-browed Albatross is more heavily built, with a broader dark leading edge on the underwings, and a thicker, yellowish bill. Immature's bill is dusky yellowish with a dark tip. Juvenile and immature Black-browed appear gray-headed, often with a grayish collar, turning whiter with age. Juvenile and immature Yellow-nosed appear white-headed, becoming grayer with age; gray hood pattern can overlap with Black-browed at some ages.

Status & Distribution BREEDING: Nominate *chlororhynchos* breeds on Tristan da Cunha and Gough Is. in South Atlantic; *carteri* on isolated island groups in southern Indian Ocean. **NONBREEDING:** Wanders north to subtropical and subantarctic waters of the Atlantic and Indian Oceans. **RARE STATUS:** Casual offshore from the Gulf of Mexico to the Canadian Maritimes, with a handful of records from onshore or inland where recorded in the interior as far as ON. **Population** Endangered.

WHITE-CAPPED ALBATROSS *Thalassarche cauta* WCAL ▪ 4

The species formerly known as Shy Albatross was recently split into three species—White-capped, Chatham, and Salvin's. Polytypic (2 sspp.; both perhaps recorded in N.A.). L 34–39" (86–99 cm) WS 90–104" (229–264 cm) **Identification** A large black-and-white albatross with pale gray mantle. **ADULT:** White crown contrasts with the pale gray face and dark brow. Large olive-gray bill has a yellow tip. Wings dark above and white below with very narrow dark border and a telltale "thumb mark"—a small dark spot on the leading edge of the underwing where it meets the body. Importantly, white on the underwing extends onto the primaries. Note grayish upper back, white rump, and gray tail. **JUVENILE:** Bill is grayish blue with a black tip; face grayer. **SUBADULT:** Head becomes whiter with age, starting with nape; bill often retains a dark tip.

Geographic Variation Sight records from CA likely pertain to almost identical, slightly larger *cauta*. Some adult *cauta* have yellow at base of upper bill ridge (culminicorn), unlike *steadi*. **Similar Species** Adult Chatham has dark gray head and bright orange-yellow bill with a dark tip; subadult has paler gray head and duskier yellowish (not grayish) bill. Adult Salvin's has a medium gray head and a grayish

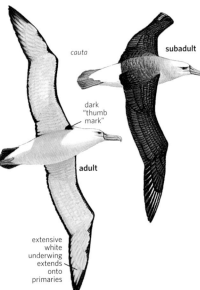

cauta

subadult

dark "thumb mark"

adult

extensive white underwing extends onto primaries

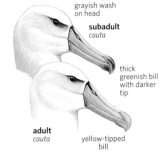

grayish wash on head

subadult
cauta

thick greenish bill with darker tip

adult
cauta

yellow-tipped bill

bill with a yellowish ridge and a less contrasting blackish tip; subadult Salvin's has a paler head. Diagnostically, Chatham and Salvin's have all-dark undersides to the primaries. Compared with Laysan Albatross, White-capped is much larger, with a paler gray mantle, larger gray bill, dark "thumb mark" on much whiter underwing, and has a longer, gray (not blackish) tail. **Status & Distribution BREEDING:** Nominate *cauta* breeds on islands off Tasmania; *steadi* breeds on Auckland, Antipodes, and Chatham Is., New Zealand. **NONBREEDING:** Disperses east and west from breeding areas across southern oceans, ranging to waters off western S.A. and southern Africa. **RARE STATUS:** Casual. One specimen of *steadi* off WA (1 Sept. 1951). Other photographed birds off northern CA, OR, and WA—all adults—were thought to be *cauta*: one subadult off OR was not identified to subspecies. **Population** Near Threatened. Population perhaps around 200,000+ (*steadi*) and 60,000 (*cauta*).

CHATHAM ALBATROSS   *Thalassarche eremita*   CHAL ▪ 5

This accidental visitor from New Zealand has been recorded once in the Northern Hemisphere. Monotypic. L 34–38" (86–97 cm) WS 91–99" (231–251 cm)
Identification Size and structure like White-capped and Salvin's Albatrosses; all three have a dark "thumb mark" on the underwing (can be absent in juveniles). Like Salvin's,

dark gray head

orange bill

adult

orange tint to bill; 1st year very similar to Salvin's

grayish head

subadult

white on underwing does *not* extend onto dark primaries. **ADULT:** Complete dark gray hood. Bill is bright yellow-orange with a dark tip restricted to the lower mandible. **JUVENILE:** Head paler gray than adult; bill is dusky with a hint of yellow and a fully dark tip; "thumb mark" less distinct or lacking. **SUBADULT:** Similar to juvenile, head becomes progressively darker with age; bill more obviously tinted yellow-orange, but still dusky on sides, with dark tip; older subadults with black subterminal band and small pale tip. **Similar Species** Most similar to Salvin's Albatross, which also has fully dark primaries. Adult Salvin's has a paler gray head; bill is dusky (not bright yellow) with a pale yellow upper ridge (culminicorn) and a more extensive dark tip. Immature Salvin's has a darker bill that lacks the yellow tones of similar-age Chatham; juveniles not separable on bill color. On White-capped, white underwing extends onto primaries. Adult has a mainly white head with a yellow-tipped grayish bill; head of immature is washed with pale gray, but is much whiter than similar-age Chatham and has a gray bill with a dark tip.
Status & Distribution BREEDING: Only breeding site is a tiny, isolated volcanic stack in the Chatham Is. known as The Pyramid. **NONBREEDING:** Ranges west across the Pacific to S.A. and up the coast from Chile to Peru in the Humboldt Current, returning via a more northerly route. Recently recorded off South Africa. **RARE STATUS:** Accidental: subadult, 27 July 2001 at Bodega Canyon off Pt. Reyes, CA. **Population** Vulnerable. Single breeding site in Chatham Is. with small (though stable) breeding population of about 5,250 pairs.

SALVIN'S ALBATROSS   *Thalassarche salvini*   SAAL ▪ 5

Salvin's is accidental in the North Pacific and easily confused with White-capped Albatross and immature Chatham Albatross. Monotypic. L 35–38" (89–97 cm) WS 93–101" (236–257 cm)
Identification Similar in size and structure to White-capped and Salvin's Albatrosses; all three have the dark "thumb mark" on the underwing (can be absent in juveniles). Like Chatham, underside of primaries completely dark. **ADULT:** Head and neck gray, forming a gray hood. Bill olive gray with pale yellow upper ridge (culminicorn) and black tip most prominent on the lower mandible. **JUVENILE:** Similar to adult except bill darkish gray with dark tip, eyebrow less contrasting, and "thumb mark" reduced or lacking. **SUBADULT:** Bill becomes paler and begins to develop a pale yellow upper ridge, dark tip reduced on upper mandible. **Similar Species** White-capped can have a fairly extensive pale gray hood, but note bill colors, more extensive white on the underwings, and completely white top of head. All ages of Chatham, except younger juveniles, have bills with distinct yellow or orange tones. Gray hood of subadult Chatham is darker than Salvin's.
Status & Distribution BREEDING: Most breed in the Bounty Is., the remainder in The Snares group, New Zealand; recent colonization of the Crozet Is. in the southern Indian Ocean. **NONBREEDING:** Ranges across the Pacific to western S.A. as far north as Peru; and east across the Indian Ocean to waters off South Africa. **RARE STATUS:** Accidental to N.A.; a subadult near Kasatochi I., western Aleutians, AK (4 Aug. 2003) and an immature off Half Moon Bay, central CA (26 July 2014); also HI (subadult on land, Apr. 2003). **Population** Vulnerable. Recent estimate of ±110,000 individuals; trend unknown.

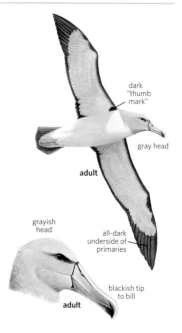

dark "thumb mark"

gray head

adult

grayish head

all-dark underside of primaries

blackish tip to bill

adult

BLACK-BROWED ALBATROSS *Thalassarche melanophris*   BBAL ■ 5

This visitor from the subantarctic region is one of the most numerous and widespread of all albatrosses. Even so, the endangered Yellow-nosed Albatross—which favors somewhat warmer waters than Black-browed Albatross—is more likely to be encountered in waters off eastern N.A. Polytypic (2 sspp.; nominate in N.A.). L 31–34" (79–86 cm) WS 81–91" (206–231 cm)

Identification Heavily built with thick neck. **ADULT:** Unique combination of yellow-orange bill with bright orange tip and whitish head with dark brow. Broad dark margins on underwings, broadest on leading edge. **JUVENILE:** Grayish to dark horn-colored bill, typically with dark blackish tip. Grayish head, turning whiter with age, with darkest gray on the hind neck often connecting across the breast to form a partial collar. Underwing often darker than adult; can appear all-dark at a distance. **SUBADULT:** Bill changing from dark to mostly pale with dark tip. Underwing becoming mostly adultlike and head mostly white with dark brow.

Geographic Variation Subspecies *impavida* (Vulnerable), often treated as a full species ("Campbell Albatross"), is unrecorded in N.A. It has a

striking pale iris, more extensive dark eye patch, and less extensive white on underwings.

Similar Species See Yellow-nosed Albatross—the major identification problem in the western North Atlantic. Black-browed is larger and bulkier than Yellow-nosed, with different underwing pattern (note the mostly dark underwings of immature Black-browed), and head and bill coloration. Distant Northern Gannets and even adult Great Black-backed Gulls have been confused with albatrosses.

Status & Distribution Circumpolar in southern oceans. **BREEDING:** Nominate *melanophris* nests around tip of S.A. on subantarctic oceanic islands;

impavida breeds only on Campbell I., in subantarctic New Zealand. **NONBREEDING:** North to waters off Peru and off Brazil. **RARE STATUS:** Casual in North Atlantic Ocean; most records are from European waters. Accidental to Greenland (twice) and Martinique. A number of reports along the East Coast, but only a few substantiated. Documented records from NL to VA.

Population Serious declines in the 20th century, but recent increases noted. Worldwide population estimated at 1.4 million.

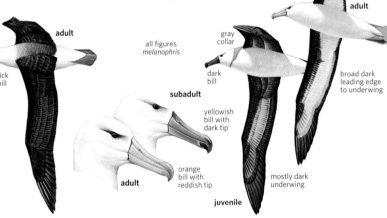

adult

all figures
melanophris

gray
collar

rather thick
orange bill

dark
bill

broad dark
leading edge
to underwing

subadult

yellowish
bill with
dark tip

orange
bill with
reddish tip

mostly dark
underwing

adult

juvenile

SOOTY ALBATROSSES Genus *Phoebetria*

LIGHT-MANTLED ALBATROSS *Phoebetria palpebrata*   LMAL ■ 5

This elegant, rather small albatross has long, narrow wings and a long tapered tail. Monotypic. L 31–35" (79–89 cm) WS 72–86" (183–218 cm)

Identification ADULT: Dark head with prominent white eye crescents. Pale mantle contrasts with dark head, wings, and long, tapered tail. Bluish purple cutting edge to the upper mandible (sulcus). **JUVENILE:** Like adult, but with less prominent eye crescents and all-dark bill.

Similar Species From Black-footed Albatross, Light-mantled has a longer tail, eye crescents that are prominent and white. Related Sooty Albatross (*P. fusca*, unrecorded in N.A.) has an all-dark body and a yellow sulcus.

Status & Distribution Circumpolar in southern oceans. **BREEDING:** Subantarctic islands. **NONBREEDING:** Widespread in southern oceans; ranges north to subtropical latitudes. **RARE STATUS:** Accidental, one photo-

documented record off northern CA (Cordell Bank, 17 July 1994), seen with Black-footed Albatrosses.

Population Near Threatened. Losses occur due to drowning from longline fishing.

adult

white eye
crescents

long, dar
wedge-
shaped t

dark head
contrasts
with pale
gray mantle

GREAT ALBATROSSES Genus *Diomedea*

WANDERING ALBATROSS *Diomedea exulans*   WAAL ■ 5

This huge seabird's wingspan reaches over 11 ft, but it often looks surprisingly compact when on the water—as do all albatrosses—with the wings

folded into three approximately equal sections. Many authorities split the Wandering group into five species: "Snowy" (*exulans*), "Gibson's"

(*gibsoni*), "Antipodes" (*antipodensis*), "Amsterdam" (*amsterdamensis*), and "Tristan" or "Gough" (*dabbenena*). Polytypic (5 sspp.). L 42–53"

(107–135 cm) WS 100–138" (254–351 cm) **Identification** Massive pinkish bill and white underwings with narrow dark trailing edge and dark primaries. **ADULT:** Extensive white back on most subspecies; females average less white. Some adult female taxa (e.g., *antipodensis*) retain mostly dark plumage with a white face; all ages of *antipodensis* resemble juveniles of other taxa. **JUVENILE:** Dark brown with conspicuous white face (all subspecies similar). **SUBADULT:** Complex plumage maturation takes 15 years or more. In general, becomes white first on the mantle, body, and head, eventually spreading across the upperwing coverts on most taxa.

Geographic Variation Tentatively, *antipodensis* and *gibsoni* are thought to have occurred in N.A., but complex plumage maturation and sex differences make it impossible to be certain. See other seabird references for more detailed accounts of subspecies.

Similar Species Superficially similar to Short-tailed Albatross (see that species for details). Very similar to Royal Albatross (*D. epomophora*, unrecorded in N.A.) of the Southern Hemisphere, which in turn is often split into "Southern Royal" and "Northern Royal" Albatrosses. "Northern Royal" has all-dark wings as an adult, unlike "Southern Royal" that has white patches on the upperwing coverts.

Status & Distribution Circumpolar in southern oceans. **BREEDING:** Biennially on subantarctic islands. **NONBREEDING:** Disperses north after breeding, but typically stays south of Tropic of Capricorn. **RARE STATUS:** Accidental. Two records in N.A.: one perched on a sea cliff, Sea Ranch, Sonoma Co., CA, 11–12 July 1967; one at Perpetua Bank off central OR, 13 Sept. 2008. Five European records.

Population Vulnerable; "Tristan" and "Amsterdam" taxa are Critically Endangered. Global population declining; highly susceptible to drowning from longline fishing.

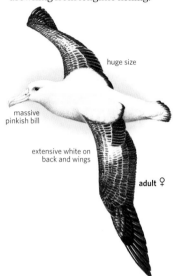

huge size

massive pinkish bill

extensive white on back and wings

adult ♀

NORTH PACIFIC ALBATROSSES Genus *Phoebastria*

Of the four species in this genus, three occur in the North Pacific. Laysan and Black-footed are small for albatrosses, while Short-tailed is much larger. Waved Albatross breeds on the Galápagos Is. and has not been recorded in N.A. They often follow ships and suffer mortality from commercial fishing operations.

LAYSAN ALBATROSS *Phoebastria immutabilis*   LAAL ■ 2

The only regularly encountered white-bodied albatross in the eastern North Pacific. Monotypic. L 28–31" (71–79 cm) WS 77–85" (196–216 cm) **Identification** All plumages similar. Blackish mantle and wings with a white flash in the primaries from white feather shafts; white below. Distinctive blackish patch above and in front of eye; cheeks washed gray. Bill variably pinkish, often orangish at base, with dark tip. Variably dark underwings with central white area (mostly white underwings exceptional) and dark underwing margins. **JUVENILE:** Like adult, but plumage fresh overall; cheeks usually white; bill lacks orange tones.

Similar Species Resembles larger and much rarer to N.A. "Shy Albatross" group (White-capped, Salvin's, and Chatham), all of which have whiter underwings with a dark "thumb mark," grayish mantle, longer gray tail, and heavier bill. Laysan lacks white patches on the upperwings and mantle and yellowish wash on head of adult or older subadult Short-tailed. See Black-footed Albatross for rare hybrids between the two species.

Status & Distribution North and central Pacific. **BREEDING:** On Hawaiian Is., Wake I., Bonin Is., and since the 1980s on islands off western Mexico. **NONBREEDING:** Moves mostly north and east into the Northern Pacific Gyre; found widely scattered in the eastern Pacific. Most commonly observed from CA to WA in

mainly dark wings and back

mostly white underwing with variable amount of dark

white rump

dark underwing primary coverts

white primary flash

darker underwing

rather thin pinkish bill with dark tip

dusky face

deep water Sept.–May; regular to cold AK waters in summer. **RARE STATUS:** Casual in spring in desert Southwest (southeastern CA and southwestern AZ), where (presumably) individuals traveling north up the Gulf of California continued overland. Accidental elsewhere inland (all seasons) in CA. **Population** Near Threatened. Stable to increasing, current population ±800,000 pairs. Threats include drowning from fishing operations, introduced predators at breeding colonies, and ingestion of plastic that is regurgitated to nestlings.

BLACK-FOOTED ALBATROSS *Phoebastria nigripes*   BFAL ■ 1

The Black-footed is the only regularly occurring dark albatross in the eastern North Pacific. Monotypic. L 28–32" (71–81 cm) WS 79–87" (201–221 cm)
Identification Typically appears overall dark brown at all ages and seasons; worn individuals can have paler brown wing coverts. Variable amounts of white at bill base, around eye ("teardrop"), and on uppertail and undertail coverts, occasionally onto the lower belly. White wing flash due to white primary feather shafts. Black tail. Grayish black bill, sometimes paler with yellow or pink tones on adults.
ADULT: Older adult often has a whiter face, sometimes even appearing white headed due to wear and sun bleaching. Amount of white on the upper- and undertail coverts appears to be age-related; uppertail coverts develop white only after the undertail coverts are mostly white. Birds with extensive white in both areas can be assumed to be at least six years old; although some birds 10 and older (females?) still have dark under- and uppertail coverts.
JUVENILE: Like adult, but generally browner with fresh unworn plumage; dark upper- and undertail coverts;

dark underwing

variable white in uppertail coverts

some with pale head and undertail

white primary flash

darkish bill

older adult

bill typically darker than adult; and "teardrop" below eye grayish and less defined. **HYBRID:** Rare hybrids with a Black-footed Albatross usually appear as darker versions of the Laysan's plumage pattern; often washed gray on the body and head. On darker hybrids, the white facial markings of a typical Black-footed are exaggerated, giving the face a prominent white appearance.
Similar Species A mostly dark-plumaged juvenile or young subadult Short-tailed Albatross differs from a Black-footed by its larger size and large pink bill. An exceptionally pale, adult Black-footed has been confused with a subadult Short-tailed, but note the different distribution of white plumage areas and the Short-tailed's pink bill.
Status & Distribution Uncommon to locally common year-round across the northeast Pacific Ocean; generally found in warmer waters and closer to shore than Laysan. **BREEDING:** Atolls in the northwestern HI islands, on islands off Japan, and since 2000 on a few islands off northwest Mexico. **NON-**

BREEDING: Disperses widely during summer. Moves largely north and east in the Northern Pacific Gyre, reaching the Aleutians. Rare in Bering Sea. Most numerous in deep water off CA, OR, WA, and BC from Mar.–Sept. During Nov.–Feb., most of population moves closer to breeding islands but immatures and nonbreeders uncommon to rare from off Baja California to BC, very rare to Gulf of Alaska.
Population Near Threatened. Still numerous (±55,000 pairs, almost all on Midway Atoll and Laysan I.), but has shown an alarming and steady decline. Threats include human disturbance, habitat loss, introduced predators, longline fishing, and oil pollution.

SHORT-TAILED ALBATROSS *Phoebastria albatrus*   STAL ■ 3

This large albatross with its striking "bubblegum pink" bill was brought to the brink of extinction in the early to mid-20th century; recently, it is becoming more regular in N.A. waters. Our most variably plumaged species, the Short-tailed Albatross—sometimes known as Steller's Albatross—has a succession of age-related plumages similar to the great albatrosses found in the southern oceans. It is obviously larger and more robust than Black-footed, with which it often appears in association off the West Coast. Most recent records off the West Coast are of chocolate brown juveniles, especially away from the Aleutians, AK. Monotypic. L 31–35"

(79–89 cm) WS 87–94" (221–239 cm)
Identification Where it occurs with other North Pacific species, the Short-tailed's large size, massive pink bill with pale bluish tip, and dark humeral patches are distinctive in all plumages, except juvenal. **ADULT:** Plumage pattern similar to that of great albatrosses of Southern Hemisphere. Largely white both ventrally and dorsally with golden wash on head. White scapulars and upperwing coverts contrast with black humerals, primary coverts, and outer greater upperwing coverts creating a somewhat pied pattern on the upperwing. Underwings white with narrow dark margin. **JUVENILE:** Recently fledged juvenile (most fledge in June) has dark bill with traces of pink. Bill soon becomes pink, timing uncertain (possibly within first two months at sea). Fresh plumage wholly dark brown with small variable white around eye and below bill, but often appearing cleanly dark-headed. Worn, older juveniles more pale-headed overall. Underwings dark. **SUBADULT:** Slow change (10 years or more) from the mostly brown juvenile-like plumage of younger subadult, to the whiter adultlike plumage of older subadult. Scapulars, upperwing coverts, and back change from brown to white as bird ages. Underwing changes from all-dark to white. Older subadult can appear nearly adultlike but often retains dark nape patch on the otherwise unmarked yellowish head.
Similar Species Adult Short-tailed is not likely to be confused with any other North Pacific albatross. Note the disproportionately large, bright pink bill visible at long range. The combination of dark underwings in juvenal and early subadult plumages and smaller size separate it from accidental Wandering Albatross. Recently fledged juveniles with dark bills are very similar to Black-footed Albatross, and older juveniles and younger subadults are superficially similar. On all ages, note much larger size of Short-tailed. Black-footed always lacks white on the upperwing unlike the pied upperwing of a subadult and adult Short-tailed. Hybrids between Black-footed and Laysan Albatrosses can resemble a Short-tailed; however, the hybrids are smaller and lack white on the upperwing.
Status & Distribution Rare but increasing. **BREEDING:** Historically on multiple islands off Japan; currently restricted to Torishima and Minami-Kojima (Senkaku Is.). Single pairs have occurred in albatross colonies on Midway and have bred there. **NONBREEDING:** Disperses to surrounding waters and follows the trade winds to the northeast Pacific and southern and central Bering Sea, casually farther north. Rare off West Coast, south to central CA. Records off the West Coast are becoming more regular.
Population Vulnerable. Listed as

Endangered by the USFWS. Hunting by Japanese plume hunters and several volcanic eruptions decimated this species by 1900; by 1949, it was believed to be extinct, but it was rediscovered in 1951. Historically (prior to 1900), large numbers of Short-taileds occurred during the nonbreeding season from the Aleutians to CA. Now fully protected on their breeding islands—although eggs and chicks still suffer predation by introduced rats, and future volcanic eruptions threaten nesting habitat—the Short-tailed population is slowly recovering; thus the higher percentage of juveniles and subadults than in a stable population. The population has rebounded to around 4,200 individuals in 2014 (3,540 on Torishima). Today, the main threat is from drowning associated with longline fisheries.

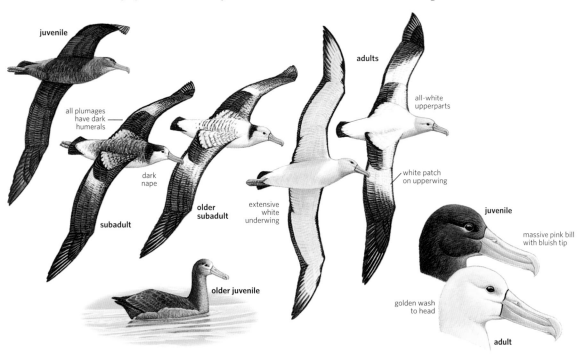

juvenile

all plumages have dark humerals

adults

all-white upperparts

dark nape

subadult

older subadult

extensive white underwing

white patch on upperwing

older juvenile

juvenile

massive pink bill with bluish tip

golden wash to head

adult

SOUTHERN STORM-PETRELS Family Oceanitidae

Wilson's Storm-Petrel (NC, Aug.)

The small size, quick flight low over the water, and monochromatic, generally dark plumages of storm-petrels challenge the abilities of many birders. Almost all observations of storm-petrels are made during organized pelagic birding trips into offshore waters. Fortunately, many storm-petrel species will closely approach boats when attracted by chum. Storm-petrels have well developed olfactory systems and use their sense of smell to locate food sources—and even individual nest burrows and crevices.

Structure The recent split of the storm-petrels into two families (Southern and Northern Storm-Petrels) is mirrored by structural differences. Members of the Southern Storm-Petrel family have shorter, rounded wings and long legs; members of the Northern Storm-Petrel family generally have longer wings (especially the inner arm) and shorter legs; there is some overlap in the tropics. Southern storm-petrels range from small to large; the Polynesian Storm-Petrel (*Nesofregatta fuliginosa*) of the tropical Pacific is the largest storm-petrel in the world, and Elliot's Storm-Petrel (*Oceanites gracilis*) of the Humboldt Current the smallest. Tails range from short to quite long; the nostril tubes on top of the bill are often prominent.

Behavior Flight style varies, but most stay low to the water due to their small size and foraging techniques. The three North American species in this family all have unique flight styles (see accounts), but that changes significantly with activity and wind conditions. Prey items are generally phytoplankton or other small invertebrates. Most prey is taken in flight by dipping the bill into the water's surface while foot pattering. Although vocal at night around colonies, most are silent at sea.

Plumage Of the nine species in this family, most have distinctive plumage patterns in black, white, and gray. Only Wilson's Storm-Petrel has the combination of white rump, pale carpal bar, and otherwise blackish brown plumage that closely resembles some of the northern storm-petrels. Ages and sexes generally look alike, but plumages fade considerably with wear. Most birds complete a single annual molt on the nonbreeding grounds at sea.

Distribution As the family name implies, most (seven of nine) species in this family breed only in the Southern Hemisphere. The exceptions are White-faced Storm-Petrel, which breeds on islands off West Africa and in southern oceans, and Polynesian Storm-Petrel, which breeds on some tropical Pacific islands north of the Equator. In North American waters, just one species (Wilson's Storm-Petrel) is a common nonbreeding visitor from its subantarctic breeding areas; two others (White-faced and Black-bellied Storm-Petrels) are rare visitors.

Taxonomy A deep evolutionary division between northern and southern storm-petrels has led to the recent recognition of the two separate families of storm-petrels. Worldwide, there are nine species in five genera, including the newly discovered Pincoya Storm-Petrel (*Oceanites pincoyae*) from Chile and the rediscovered New Zealand Storm-Petrel (*Fregetta maoriana*). The taxonomy of some species is still a matter of debate.

Conservation Storm-petrels are subject to avian (especially skuas) and mammalian predation. The introduction of non-native predators to breeding islands has resulted in the decimation of some breeding colonies. BirdLife International codes: 1 EN and 1 CR.

Genus *Oceanites*

WILSON'S STORM-PETREL   *Oceanites oceanicus*   WISP ■ 1

The small, dark, white-rumped Wilson's Storm-Petrel is abundantly observed in the Atlantic. Its distinctive feeding style includes pattering on the surface with long, dangling legs and feet. Polytypic (3 sspp.; nominate in N.A.). L 7.3" (18 cm) WS 16" (41 cm)
Identification Overall dark plumage, with bold white rump extending well down onto the undertail coverts. Well-defined pale carpal bar, frosty gray when fresh, fading to pale tan when worn, on the upperwing. Tail shape variable, often appearing squared in direct flight, but can appear slightly notched or even rounded during feeding. Long dark legs with yellow-webbed feet distinctive, often visible during feeding behavior. Ages and sexes look similar. Adults undergo molt during spring and summer in N.A. waters. **FLIGHT:** Flies low to the sea surface, occasionally arcing up into the air over wave tops. Direct flight with fluttery wingbeats interspersed with short glides. Feet extend past the

tip of the tail. When feeding, holds its wings above the horizontal, fluttering while pattering the surface with its feet; often appears to be "walking on water."
Geographic Variation Nominate *oceanicus* is disproportionately smaller than the extralimital *exasperatus* and *chilensis*, but the three are inseparable under field conditions.
Similar Species Wilson's has similar plumage to other North Atlantic storm-petrels, but Leach's and Band-rumped Storm-Petrels have distinctly longer wings and different flight styles. Leach's flight is nighthawk-like; Band-rumped's is shearwater-like. European Storm-Petrel is very similar, but is smaller and lacks both the foot projection and prominent pale carpal bar of the Wilson's. See the European Storm-Petrel account for more.
Status & Distribution Common. **BREEDING:** Nests colonially on oceanic islands

in the southern oceans. **NONBREEDING:** Common offshore in the Atlantic May–Sept.; regularly seen from shore at favored locations. Rare in the Gulf of Mexico and rare off the CA coast late summer and fall. **RARE STATUS:** Casual elsewhere on West Coast. Accidental inland in East after hurricanes; surprisingly few records.
Population The global population of this species, considered one of the most abundant seabirds, is likely stable.

short "arm"

feet project beyond tail

white extends well below tail

long legs

feet with yellow webs

European Storm-Petrel

molt gap

dropped greater coverts

old (faded) outer primaries

new inner primaries

Wilson's different stages of wing molt from spring to summer

typical foot-pattering behavior

extensive white on sides of undertail

Genus *Pelagodroma*

WHITE-FACED STORM-PETREL   *Pelagodroma marina*   WFSP ■ 3

This much sought-after species occurs rarely in the western Atlantic, where it is now seen almost annually in late summer. Polytypic (6 sspp.; 2 in N.A.). L 7.5" (19 cm) WS 17" (43 cm)
Identification Where it occurs in the western Atlantic with other dark storm-petrels, this small, dynamic bird is distinctive in having primarily whitish underparts and a bold white

facial pattern with dark auriculars and dark half-collar. Broad, paddle-shaped wings; clean white underwing coverts contrast with the dark flight feathers. The brownish mantle and darker upperwing coverts and remiges create a pale-saddled look; contrasting pale gray rump. Long tarsi dangle below the body during feeding and extend past the tail in direct flight.

Ages and sexes look alike. **FLIGHT:** Its unique foraging technique, where it appears to hop across the water in pogo-stick fashion, its long legs dipping into the water as it moves across the sea surface, is unlike any other storm-petrel. It appears to spring off the surface of the water while rapidly changing direction. It flaps with stiff, short wingbeats, and skips across the water with short glides; in direct flight, utilizes longer glides on outstretched wings held slightly bowed down and feet projecting beyond the tip of the tail.

Geographic Variation The majority of N.A. records are of the Cape Verde Is. *eadesi*. It has a whiter forehead and pale collar compared with the darker faced *hypoleuca* of the Salvages (Ilhas Salvagens). More southern taxa (*marina* and *maoriana*) have shorter legs and longer, more strongly forked tails.

Similar Species White-faced is unlikely to be confused with any other storm-petrel in the western Atlantic due to combination of distinctive plumage features and feeding behavior. Potentially it could be confused with a winter-plumaged phalarope while swimming, but note the White-

faced's different posture, wing shape, and flight style when flushed.

Status & Distribution Rare in N.A. **BREEDING:** Widely distributed on oceanic islands across the southern oceans and, north of the Equator, in the eastern Atlantic on the Cape Verde Is. and Salvages. **NONBREEDING:** Disperses to waters north and west of breeding grounds, reaching the continental shelf of N.A. in late summer and fall. In the Atlantic, found sparingly from waters off southern New England south to NC (early Jul.–Sept.), mainly inshore of the Gulf Stream and over submarine canyons; a few onshore records after storms. In the Pacific, the nearest record to N.A. is from waters off Costa Rica (*maoriana*).

Population Declining.

Large Southern

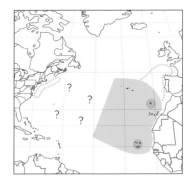

Hemisphere population (±4 million), but two North Atlantic subspecies (together) number ±200,000.

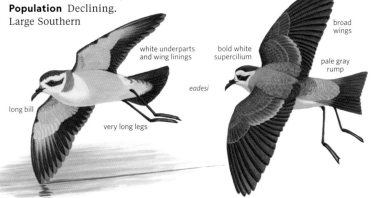

pushes off water and changes direction

long bill

white underparts and wing linings

very long legs

bold white supercilium

eadesi

broad wings

pale gray rump

Genus *Fregatta*

BLACK-BELLIED STORM-PETREL *Fregetta tropica*   BBSP ■ 5

This widespread southern ocean species has a distinctive flight style and plumage, but its black belly stripe can be hard to see, especially from the side. Polytypic (2 sspp.; N.A. records presumably of nominate). L 8" (20 cm) WS 18" (46 cm)

Identification Black-and-white pattern. Mostly white below with a variable dark stripe (absent on some) that connects the dark chest with dark undertail coverts. The upperparts are dark, with pale-tipped greater upperwing coverts; the rump is white. White underwing coverts contrast with dark flight feathers. Long legs and feet project past the tail. Ages and sexes are similar. **FLIGHT:** Foraging technique of splashing its breast into the water and then springing off forward with its long legs is unique. In direct flight it flies low across the water, rarely flapping.

Geographic Variation The taxonomy of *melanoleuca* (breeds on remote

Tristan da Cunha and Gough I. in the South Atlantic) is poorly understood; it lacks the black belly stripe of nominate *tropica*.

Similar Species When seen well, unlikely to be confused with other storm-petrels in the western Atlantic. The combination of white underparts, dark stripe down belly, and distinctive foraging action are unique. It is separated from the similar-looking White-bellied Storm-Petrel (*F. grallaria*, not recorded in N.A.) by more extensive dark hood, darker upperparts, and black belly stripe (except *melanoleuca*). Its flight style is similar to White-faced Storm-Petrel; however, it splashes into the water with its breast and does not pogo-hop with both feet.

Status & Distribution BREEDING: Nests on subantarctic islands. **NONBREEDING:** Disperses to tropical seas north of breeding areas, reaching the Equator in the Atlantic. **RARE STATUS:**

Four N.A. records (late May–mid-Aug.) from Gulf Stream waters off NC, all since 2004.

Population Likely declining; global population ±500,000.

striking black-and-white plumage

black line through white belly hard to see

long legs

tropica

NORTHERN STORM-PETRELS Family Hydrobatidae

European Storm-Petrel (Ireland, Jun.)

The small size, quick flight low over the water, and monochromatic, generally dark plumages of storm-petrels challenge birders' abilities.

Structure Most Northern Storm-Petrels are small, lightly built birds with long, narrow wings. They have large heads, with slim bodies and tails that range from short to quite long. Most species have short legs that do not extend past the tail in direct flight.

Behavior Flight style varies, but most species stay low to the water. They often appear swallowlike when searching for small prey items on the surface, but some larger species act more like small shearwaters, gliding and arcing on stiff wings. Several species have exaggerated, languid wingbeats like those of a nighthawk. Flight style changes significantly with activity and wind conditions. Prey items are generally phytoplankton or other small invertebrates. Most species make only nocturnal visits to their breeding colonies and are rarely observed in the immediate vicinity. Although vocal at night around colonies, most are silent at sea.

Plumage For the most part, Northern Storm-Petrel plumages are dark brown to blackish, but the Fork-tailed Storm-Petrel is largely grayish and pale. Many have extensively white rumps; other species are all-dark. Most dark storm-petrels show a pale diagonal bar across the upperwing; its extent and prominence can aid identification. Ages and sexes generally look alike, but plumages fade considerably with wear. Most birds complete a single annual molt on the nonbreeding grounds at sea.

Distribution Northern Storm-Petrels occur in both the Pacific and Atlantic Oceans. In North America at least 11 species have occurred in the eastern Pacific; in the western Atlantic two white-rumped species commonly occur (Leach's and Band-rumped), and two other species (European and Swinhoe's) have been recorded as casual visitors. Adding to the mix in the Atlantic are the three species of Southern Storm-Petrels, one of which (Wilson's) is common and also occurs in the Pacific. Migratory routes are poorly understood, but Black and Least Storm-Petrels regularly migrate into the Southern Hemisphere during their nonbreeding season.

Taxonomy The 17 species (one presumed extinct) are placed in two genera, *Hydrobates* and *Oceanodroma*. Recent research supports placing all Northern Storm-Petrels in a single genus, which by priority would be *Hydrobates*. The taxonomy of some species is still a matter of debate; the Leach's Storm-Petrel was recently split into three species, and the Band-rumped Storm-Petrel likely comprises two or more species. The AOS (2018) lists 12 species in two genera as occurring in North America.

Conservation Storm-petrels are subject to avian and non-native mammalian predation. The Guadalupe Storm-Petrel (*Oceanodroma macrodactyla*) from Guadalupe I. off Baja California is presumed extinct. BirdLife International codes: 2 NT, 4 VU, 2 EN, and 1 CR (PE).

Genus *Hydrobates*

EUROPEAN STORM-PETREL *Hydrobates pelagicus*   EUSP ■ 4

This very rare nonbreeding visitor from the eastern Atlantic usually associates with Wilson's Storm-Petrels and is easy to overlook. Polytypic (2 sspp.; N.A. records presumably of nominate). L 5.5–6.5" (14–17 cm) WS 14–15.5" (36–39 cm)

Identification Tiny size and blackish overall. Bold, clean, white rump patch extends marginally onto rump sides. The white greater underwing coverts are best looked for when it feeds with wings held high, otherwise they are difficult to see in flight. Upperwing typically dark, lacking the pale carpal bar of other Atlantic storm-petrels; juveniles can have pale-tipped greater upperwing coverts. Head and bill are small and delicate. Feet and legs are dark. Ages and sexes look alike. **FLIGHT:** Flies with flexing bursts recalling a large swift or Spotted Sandpiper; changes direction frequently. Wings rarely appear to break the horizontal in direct flight.

Similar Species Most likely to be confused with the Wilson's Storm-Petrel. In a flock of Wilson's, it can be picked

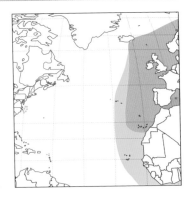

out by its smaller size and darker upperwings lacking a pale carpal bar, lack of foot projection, longer narrower wings with rounded tip, different flight style, and white greater underwing coverts.

Status & Distribution BREEDING: Nests on oceanic islands in the northeastern Atlantic and Mediterranean. **NONBREEDING:** Transequatorial migrant; wintering largely in southwest African waters. **RARE STATUS:** More than a dozen N.A. records. Single Aug. 1970 specimen record from NS; all

recent records (mid-May to early June) from off Cape Hatteras, NC.

Population Stable.

smaller and darker overall than Wilson's

lacks obvious carpal bar

long "arm"

white underwing bar

short legs

Genus *Oceanodroma*

FORK-TAILED STORM-PETREL *Oceanodroma furcata* FTSP ▪ 2

The Fork-tailed Storm-Petrel is found year-round in the northern Pacific, but only rarely south of central CA. Polytypic (2 sspp.; both in N.A.). L 8.5" (22 cm) WS 18" (46 cm)

Identification Only grayish storm-petrel in the northeastern Pacific. Underparts pale gray with darker grayish black underwing coverts. Head grayish, with dark blackish mark through eye. Upperparts gray, with pronounced M pattern across wings. **FLIGHT:** Wingbeats shallow; flight relatively direct.

Geographic Variation West Coast subspecies *plumbea* is smaller and darker than nominate *furcata* of Asia and the Aleutians.

Similar Species Other storm-petrels in the northeastern Pacific are largely blackish brown. Fork-tailed has been confused with juvenal and winter-plumaged Red-necked and Red Phalaropes, but they have white underparts and underwings; note differences

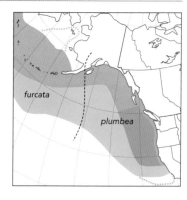

furcata

plumbea

in posture, structure, and flight style.

Status & Distribution Uncommon to fairly common. **BREEDING:** Nests on islets from AK south to northern CA. **NONBREEDING:** Disperses south and west after breeding. Rarely seen from shore and off Southern CA; accidental inland in CA (Yolo Co.).

Population Losses from native and introduced predators on some breeding islands.

rounded wing tip

prominent pale carpal bar whitens near body

distinctive fork to tail

pearly gray with blackish wing lining

dark mask can suggest a phalarope

RINGED STORM-PETREL *Oceanodroma hornbyi* RISP ▪ 5

This most distinctive of the world's storm-petrels has been recorded with certainty once in N.A. It is widely known by its alternate English name, Hornby's Storm-Petrel, which was given in honor of Admiral Sir Phipps Hornby of the British Royal Navy.

Monotypic. L 8.3–9" (18.9–21 cm)

Identification Very strikingly patterned fairly large storm-petrel with rather long bill; flight swooping and erratic, with rapid wingbeats. Underparts white except for the broad band of dark gray across the breast. Upper-

parts grayish, palest on mantle and uppertail coverts, browner in marginal, lesser, and humeral coverts and scapulars, with a starkly contrasting cream-colored carpal bar (median and greater coverts). A thin band of white across the nape, extending to the throat, separates the gray back from the blackish cap. Remiges and rectrices dark gray; tail strongly forked.

Similar Species Essentially unmistakable. Polynesian Storm-Petrel (*Nesofregetta fuliginosa*) is similar in pattern but has a completely dark face, brown rather than gray upperparts, white underwing, dark undertail coverts, and less fork in the tail.

Status & Distribution Forages in the cold Humboldt Current off northern Chile and Peru; a few to Ecuador. The location of its breeding colonies was a long-standing mystery. Finally, in Apr. 2017, a colony was discovered breeding in natural cavities in the desolate Atacama Desert (Chile), 47 mi inland. **RARE STATUS:** In N.A. waters one certain record: one well-photographed bird about 25 mi southwest of Santa Rosa I., CA (2 Aug. 2005); another about 50 mi off Coos Co., OR (3 May 2007), was seen only briefly, but accepted by the state records committee, a decision questioned by many (thought likely to be a phalarope). One specimen record at Isla Gorgona, Colombia (July 1979), is the only other record north of the Equator.

deeply forked tail

striking pied plumage with black cap and breast band

Population Insufficient data, but population potentially quite large.

SWINHOE'S STORM-PETREL *Oceanodroma monorhis* SSTP ■ 5

A rather mysterious species in the Atlantic Ocean, single Swinhoe's Storm-Petrels were discovered on islands around Madeira and England in the late 20th century and have since been observed on other islands and at sea elsewhere in the North Atlantic. Monotypic. L 8" (20 cm) WS 18" (46 cm)

Identification The only completely dark storm-petrel in the North Atlantic. Has stout bill, moderate pale buff bar across upperwing, and forked tail. Wings somewhat broad, like Band-rumped. On the upperwing, the outer four to five primaries show white shafts at the base, visible at close range. Flight remarkably different from Leach's, typically loping and unhurried, with relatively shallow wingbeats.

Similar Species In the Atlantic, no similar species, but beware Leach's Storm-Petrels with extensive dark brown plumage in central uppertail coverts, showing whitish uppertail coverts only on the sides; in all plumages, Leach's lacks white primary shafts. All-dark Bulwer's Petrel is larger, with larger bill, longer wings, and very long tail, also lacks white primary shafts. In the eastern North Pacific, where Swinhoe's is a plausible stray (unconfirmed report off AK, Aug. 2003), dark-rumped Leach's are common, and many other dark-rumped species must also be considered. All of these differ in structure, flight style, and plumage from Swinhoe's, but extensive study would be required to confirm identification. Because viewing conditions at sea can be challenging, many dark-rumped storm-petrels observed in the Atlantic and elsewhere are commendably left unidentified.

Status & Distribution BREEDING: Islands in the western North Pacific, mostly off Russia, Japan, Korea, and Taiwan; postbreeding dispersal into the northern Indian Ocean. Since the 1990s, small numbers have been detected at breeding locations of other species of storm-petrels on islands off Europe and Africa, from Scandinavia to the British Isles and south to Great Salvage I. (near Madeira) and the Azores, as well as in the Mediterranean (Italy, Israel). Breeding at any of these locations has yet to be confirmed.

only all-dark Atlantic storm-petrel

pale carpal bar

white base to primary shafts visible at close range

RARE STATUS: Off Hatteras, NC, single birds have been documented four times (20 Aug. 1993; 8 Aug. 1998; 2 June 2008; and 6 June 2009). One was documented 192 mi southeast of Cape Sable I., NS (9 Aug. 2010), just inside N.A. waters.

Population Near Threatened. Island breeding colonies imperiled by introduced predators, mining, and tourism.

LEACH'S STORM-PETREL *Oceanodroma leucorhoa* LESP ■ 1

Essentially a Northern Hemisphere species, Leach's Storm-Petrel occurs well offshore in both oceans from the Equator to the Arctic. Two subspecies of Leach's in the Pacific were recently elevated to full species: Townsend's Storm-Petrel (*O. socorroensis*, next species) and Ainley's Storm-Petrel (*O. cheimomnestes*, unrecorded in US). Polytypic (2 sspp.; both in N.A.). L 7–8" (18–20 cm) WS 17–19" (43–48 cm)

Identification Dark brownish black below. Wings dark brown except for pale carpal bar on the upperwing extending to the leading edge. Rump pattern variable, typically lacks a clean white appearance, often smudged gray through the middle, resulting in a "split-rumped" appearance in all Atlantic and most North Pacific populations; rump can range from all-white to all-dark in more southerly Pacific breeders. Entirely dark-rumped birds, often the subspecies *chapmani*, regularly occur in Southern CA waters.

FLIGHT: Very active. Flies with deep, rowing wingbeats incorporating frequent direction changes, bounding from side to side and up and down. Arcs and glides over wave tops on bent wings. In strong winds it may fly more shearwater-like.

Geographic Variation Nominate *leucorhoa* breeds in the North Atlantic and Pacific Oceans. It is larger and less variable in terms of plumage. Individuals

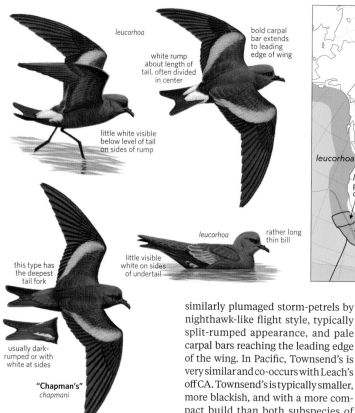

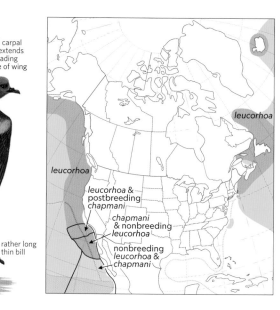

in the southern part of the range begin to show intermediate rump characteristics; totally dark-rumped individuals are rare. The subspecies *chapmani* ("Chapman's Storm-Petrel") breeds off Baja California. It has a high proportion of dark-rumped individuals (some have white at sides of the rump), slightly rounder wing tips, and a longer and more deeply notched tail.

Similar Species Separated from other similarly plumaged storm-petrels by nighthawk-like flight style, typically split-rumped appearance, and pale carpal bars reaching the leading edge of the wing. In Pacific, Townsend's is very similar and co-occurs with Leach's off CA. Townsend's is typically smaller, more blackish, and with a more compact build than both subspecies of Leach's; it also has a shorter tail and a shallow tail fork; white-rumped Townsend's lacks dark dividing stripe on rump. Dark-rumped "Chapman's" is easily confused with Ashy, Black, and Least Storm-Petrels. It differs from Ashy in having darker underwings, darker overall plumage, and flies with deeper wing beats; it differs from the Black by smaller size and less forked tail that often appears shorter; Least is much smaller and darker with a short, wedge-shaped (not forked) tail,

fainter carpal bar, and flies like a Black Storm-Petrel. Some dark-rumped individuals may be unidentifiable under field conditions. Ainley's Storm-Petrel (*O. cheimomnestes*) breeds in winter on Guadalupe I., has a white rump often with a dark center line, and is unrecorded in US. In the Atlantic and Gulf of Mexico, see Band-rumped.

Status & Distribution Usually scarce. Prefers deep water; rarely seen from shore. **BREEDING:** In burrows on offshore islets, where it is a nocturnal visitor. **NONBREEDING:** Generally disperses to surrounding waters, but some populations undertake long migratory movements across the Equator to tropical oceans. **RARE STATUS:** Casual from shore and inland during hurricanes.

Population Vulnerable, but widely distributed and still numerous.

TOWNSEND'S STORM-PETREL *Oceanodroma socorroensis* TOSP ■ 3

Formerly classified as a subspecies of Leach's Storm-Petrel, Townsend's was recently elevated to full species status. In summer and fall it disperses into Southern CA waters, most usually near or beyond the continental shelf. Monotypic. L 6.3–7" (16–18 cm) WS 16–18" (41–46 cm)

Identification Townsend's averages smaller, darker, and more compact than Leach's. It has a fainter carpal bar, more bluntly pointed wings, and a relatively short, forked tail. Most Townsend's have white rumps, and many lack the dividing line of Leach's,

appearing solidly white-rumped in the field. Note how the length of the tail is about equal to or less than the length of the white rump. However, the white rump patch is variable, and dark-rumped individuals occur and are easily confused with other species. **FLIGHT:** More direct and less erratic than Leach's.

Similar Species Positive separation of Townsend's from Leach's is complex and sometimes impossible. White-rumped Townsend's can be confused with white-rumped Leach's (usually shows a dark line up the center of the

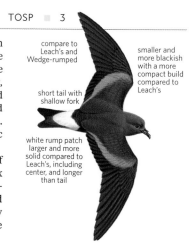

compare to Leach's and Wedge-rumped

smaller and more blackish with a more compact build compared to Leach's

short tail with shallow fork

white rump patch larger and more solid compared to Leach's, including center, and longer than tail

white rump, longer wings, bold carpal bar, bounding flight), Wilson's (small, short triangular wings and long legs), and Wedge-rumped (much smaller with disproportionately large white rump, shallow tail fork). Dark-rumped Townsend's are not field-identifiable from dark-rumped Leach's, except by structural differences that can be accurately judged only by in-hand measurements. Larger Ainley's Storm-Petrel (*O. cheimomnestes*)

(unrecorded N.A.) is very similar but rump is less pure white.
Status & Distribution Uncommon. **BREEDING:** In summer, only on islets near Guadalupe I. off the coast of Baja California Norte, Mexico. Recorded well offshore north to Santa Barbara Co. (June–early Nov.). **NONBREEDING:** Disperses south to off southern Mexico (Oct.–Apr.).
Population Endangered. Population about 7,000.

ASHY STORM-PETREL *Oceanodroma homochroa* ASSP ■ 2

Endemic to CA and waters off northwestern Baja California, the Ashy Storm-Petrel is best identified by its direct, but fluttery flight style, gray-brown (not blackish) plumage, and pale silvery underwing coverts.

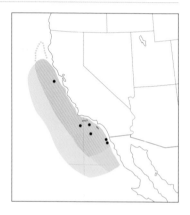

molting tail looks more like Least

molting fall adult

distinctly long tailed with deep tail fork

slightly paler gray on uppertail coverts

some pale visible in center of underwing, but can be hard to see

paler than Black Storm-Petrel with ashy-gray plumage

Monotypic. L 8" (20 cm) WS 17" (43 cm)
Identification Midsize, all-dark storm-petrel with medium brown plumage washed gray when fresh. The unique pale underwing coverts can appear silvery at sea, but they are often difficult to see under field conditions. Overall plumage is paler than other dark storm-petrels, but difference is subtle and subject to wear and fading and lighting conditions at sea can affect color perception. Ages and sexes alike. **FLIGHT:** Steady, measured wingbeats, rarely raising wings above the horizontal, and rarely gliding.
Similar Species From dark-rumped Leach's (*chapmani*) by Ashy's partly silvery wing linings and steady, but fluttery, wingbeats and direct flight style. Compared with Black and Least Storm-Petrels, Ashy has shallower wingbeats and paler plumage; also note its disproportionately longer tail length, especially compared with the short-tailed Least. But beware that an Ashy in molt can also appear short-tailed.
Status & Distribution Rare to locally common. **BREEDING:** Nests colonially

on islets off the coast of CA north to Marin Co. and south to the Coronados Is. off Tijuana, Mexico. **NONBREEDING:** In early fall, it congregates in large numbers in Monterey Bay and Cordell Bank, where much of the world population stages. Main wintering area(s) unknown; rarely recorded in winter off CA or from shore during storms.
Population Endangered. Small population (±5,000 pairs) and restricted range make it highly susceptible to accidental oil spills and non-native predators on breeding islands.

BAND-RUMPED STORM-PETREL *Oceanodroma castro* BSTP ■ 2

This blackish, long-winged storm-petrel with a clean-cut white rump patch has a complicated and still unsettled taxonomic history. Monotypic (but see Geographic Variation). L 9" (23 cm) WS 17" (43 cm)
Identification Indistinct carpal bar typically does not reach leading edge of wing; worn birds in late summer can show a more prominent bar. Tail square to shallowly forked. White rump band extends onto rump sides. **FLIGHT:** Shallow, measured wingbeats; often glides like a small shearwater.

Geographic Variation Different populations are perhaps best treated as subspecies or full species. Four populations breed on islands in the northeastern Atlantic (tentatively named "Grant's," "Cape Verde," "Madeiran," and "Monteiro's"). Some breed in winter, some in summer. "Monteiro's" (*monteiroi*, summer breeder on Azores, unrecorded N.A.) now a full species. Two or more populations breed in the South Atlantic and four in the Pacific. Identification criteria are yet to be determined. Most N.A. birds believed to be

widespread, winter-breeding "Grant's."
Similar Species Leach's in the Atlantic has a more extensive white rump, usually with a dark centerline. Adult winter-breeding "Grant's" undergoes primary molt (Feb.–early Aug.) after breeding, unlike summer-breeding Leach's in the Atlantic that begins primary molt in late summer; immatures and nonbreeders molt somewhat earlier. Smaller Wilson's has feet projecting past the tail, shorter rounded wings, and a fluttery flight style.
Status & Distribution BREEDING: Nests on the Azores, Berlengas, Canary Is., Madeira Is., Salvages, Cape Verde Is., and farther south in the Atlantic; also in Pacific, including HI, Galápagos, and Japan. NONBREEDING: Disperses west from eastern Atlantic breeding colonies to the Gulf Stream (FL to VA) late May–early Sept., some as far north as MA. Also regular over deep water in Gulf of Mexico. RARE STATUS: Casual inland after hurricanes in East. Accidental

CA (one banded on Farallones, 10 Nov. 2017, breeding population unknown).
Population Breeding population in North Atlantic ±10,000 pairs, all types. Listed as Endangered in HI by USFWS.

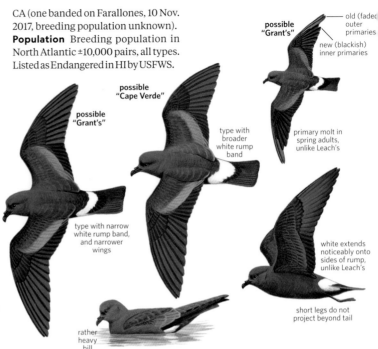

possible "Grant's"

old (faded) outer primaries

new (blackish) inner primaries

possible "Cape Verde"

possible "Grant's"

type with broader white rump band

primary molt in spring adults, unlike Leach's

type with narrow white rump band, and narrower wings

white extends noticeably onto sides of rump, unlike Leach's

short legs do not project beyond tail

rather heavy bill

WEDGE-RUMPED STORM-PETREL   *Oceanodroma tethys*   WRSP ▪ 4

This diminutive, casual species has reached CA and AZ. Polytypic (2 sspp.; one recorded in N.A.). L 5.9–6.7" (15–17 cm) WS 13–15.7" (33–40 cm)
Identification Very small. White rump patch extends well down uppertail coverts, giving the appearance of being largely white-tailed, but does not extend down the sides of the rump. On water, folded wings mostly hide the white rump. Pale carpal bar restricted primarily to the wing's inner half. FLIGHT: Deep, strong wingbeats.
Geographic Variation Nominate *tethys* breeds on the Galápagos Is.; smaller *kelsalli* on islets off Peru.
Similar Species Wedge-rumped's tiny size (especially *kelsalli*) extensive triangular white rump patch and lack

of foot projection separate from Wilson's; on Wilson's white extends down sides of rump, unlike Wedge-rumped. Larger Townsend's has different shape to its solid white rump.
Status & Distribution Casual to waters off CA. BREEDING: Visits colonies during the day, unlike other storm-petrels. NONBREEDING: Disperses to tropical eastern Pacific waters off Colombia and Ecuador and north to Panama. RARE STATUS: Seven records from central and Southern CA (July–Oct.); one *kelsalli* specimen found at Carmel (Jan.). More than 15 were found inland in Pima Co. and Santa Cruz Co., AZ, after passage of Hurricane Newton (Sept. 2016).
Population Least Concern.

triangular white rump almost reaches end of tail

kelsalli

small size and relatively dark plumage

kelsalli

white on sides of undertail often hidden by wings

BLACK STORM-PETREL   *Oceanodroma melania*   BLSP ▪ 2

The Black Storm-Petrel is the largest regularly occurring dark-rumped storm-petrel in the eastern Pacific. Monotypic. L 9" (23 cm) WS 19" (48 cm)
Identification Appears more substantial at sea than other dark, eastern Pacific storm-petrels. Entirely blackish brown, with extensive pale carpal bar formed by tawny upperwing coverts. Tail deeply forked. FLIGHT: Wingbeats deep and deliberate, Black

Tern-like, rising high above the horizontal before each downstroke. Glides frequently and arcs over wave tops like a small shearwater.
Similar Species Separation from other dark-rumped Pacific storm-petrels is difficult, especially from a dark-rumped Leach's. Note the distinctive flight style of Black Storm-Petrel, which lacks the nighthawk-like, bounding quality of Leach's. Black

Storm-Petrel is noticeably larger than Least Storm-Petrel and appears to have a longer tail and wings. It is darker overall than Ashy Storm-Petrel, with longer wings and more deliberate flight style. It is similar to the paler Tristram's (three N.A. records), which has a more deeply forked tail. Also similar to Markham's and Matsudaira's Storm-Petrels (unrecorded in N.A.).
Status & Distribution Fairly com-

mon to common from May (mid-Apr. off San Diego) to Oct. off Southern CA coast; regularly seen from shore. **BREEDING:** Colonially on islands off western Mexico and in the Gulf of California. A very few nest on rocks off Santa Barbara I., CA. **NONBREED-ING:** Regular postbreeding visitor farther north along the CA coast in late summer (large numbers in Monterey Bay), more during warm-water years; casual north of Sonoma Co. to northern OR. Most winter off C.A. and south to Peru. **RARE STATUS:** Casual to northern OR; also inland at the Salton Sea and along the lower Colorado River, particularly after strong cyclonic storms in early fall. **Population** Global population more than 500,000. Predation by invasive species at breeding colonies.

large and dark

flies with languid wingbeats like Black Tern

rather long neck and small head

dark wing linings

long winged

TRISTRAM'S STORM-PETREL *Oceanodroma tristrami*   TRSP ▪ 5

A species of the subtropical western Pacific Ocean, Tristram's Storm-Petrel breeds as close to N.A. as the northwestern Hawaiian Is. The species name commemorates 19th-century English naturalist Henry Tristram. Monotypic. L 10" (25 cm) WS 22" (56 cm)
Identification Very large, bulky, long-winged storm-petrel with deeply forked tail. Plumage rich brown overall, with prominent buffy carpal bars in upperwing and subtle gray sheen to head and rump, often visible in favorable light at close range. In moderate winds, flight tends to be direct rather than erratic, with shallow flaps.
Similar Species Easily confused with the comparably large Black Storm-Petrel; Tristram's is even larger and paler (especially head and rump) and also has a more deeply forked tail and shallower, less-decisive wing-beats. Markham's and Matsudaira's storm-petrels, unrecorded in N.A., should also be considered. Markham's is smaller and more delicately proportioned and lacks gray in the rump; Matsudaira's is larger than Tristram's but rather lightly built, with shallower tail fork, generally more buoyant flight style. Both typically show white shafts in outer primaries dorsally, not present in Tristram's.
Status & Distribution BREED-ING: Breeds in the Hawaiian Archipelago on Nihoa, Necker, French Frigate Shoals, Laysan, and Pearl and Hermes Reef, as well as on islets of the Bonin and Izu Is. off Japan. The first adults return to court in late Oct., and the last fledglings depart in June. Postbreeding dispersal is little known, but recorded east of Japan in summer. **RARE STATUS:** Three records: one banded at Southeast Farallon I., CA (22 Apr. 2006), and another found dead there on 18 Mar. 2015; another photographed about 22 mi north of San Nicolas Is. on 15 July 2018.
Population The world population is ±10,000 pairs.

larger and grayer than similar Black Storm-Petrel with paler carpal bar and more deeply forked tail

LEAST STORM-PETREL *Oceanodroma microsoma*   LSTP ▪ 3

The diminutive, all-dark Least Storm-Petrel is endemic to Mexican waters and disperses northward to near-shore CA waters during the late summer and early fall. Its overall appearance has been described as batlike due to its unusual shape, and its flight is quick and erratic on deep wingbeats. Monotypic. L 5.1–5.9" (13–15 cm) WS 12.6–14.2" (32–36 cm)
Identification Very small and dark storm-petrel that often appears almost tail-less in flight due to short wedge-shaped (tapered) tail. Dark upperwing, with a tawny carpal bar primarily on the inner wing. **FLIGHT:** Wingbeats unusually deep, strong, and direct for its small size; little or no gliding. Wings held high above body while feeding on the surface. Flight style strongly

tiny size and dark coloration

wedge-shaped tail sometimes apparent

short tail

slow wingflaps suggest Black Storm-Petrel

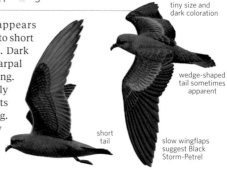

suggestive of the much larger Black Storm-Petrel with which it often associates off Southern CA.

Similar Species Least Storm-Petrel differs from other all-dark storm-petrels in its small size, flight style, and shorter wings and tail. Superficially resembles Black Storm-Petrel, but much smaller with much shorter wings and tail. Except for Least, other all-dark storm-petrels have notched tails. Molting Ashy Storm-Petrels with shorter tails are often confused with Least in Monterey Bay. Ashy is paler with a grayish cast and flies with a more fluttery flight, and in most years is absent from Monterey Bay.

Status & Distribution Uncommon in N.A. most years. **BREEDING:** Colonially on islands off west coast of Baja California and on islands in the northern Gulf of California. **NONBREEDING:** Disperses north to Southern CA waters during warm-water years (sometimes many), late Jul.–Oct.; rare to Monterey Bay, exceptionally in large numbers. Casual in late spring off San Diego. **RARE STATUS:** Casual inland at the Salton Sea (hundreds in 1976 after tropical storm Kathleen) and elsewhere in the desert Southwest, particularly after strong cyclonic storms.

Population Threats include limited breeding habitat on offshore islands and losses from introduced mammalian predators at breeding colonies.

SHEARWATERS AND PETRELS Family Procellariidae

Pink-footed Shearwater (CA, Oct.)

The identification of shearwaters and petrels—the procellariids—is challenging. Focus on the light-dark plumage patterns, body proportions, and flight styles.

Structure Heavy bodies, short tails, and long, narrow wings are the standard for procellariidae. Their bills are made up of individual plates and are sharply hooked. The nostrils are situated in tubes on the upper surface (culmen) of the bill—hence the term *tubenoses*.

Behavior Shearwaters and petrels can be found singly or in large groups, with the shearwaters being more prone to flocking behavior than the others. Their flight is usually a series of wingbeats interspersed with stiff-winged glides. The speed, duration, and depth of the wingbeats and the frequency and duration of the glides define a species' flight style. Wind speed and a bird's activity will alter its flight style. Feeding styles vary from surface dipping to deep diving. Prey items range from small copepods to squid and fish, and many species are attracted to fishing boats and chum. Their acute sense of smell helps them to detect food. Procellariids are generally silent at sea.

Plumage Procellariids appear monochromatic. Their plumages are generally variations on black, gray, brown, or white; some are tinged with blue. Male, female, and immature plumages look very similar. Many species have dark upperparts and light underparts. Some species have distinct color morphs (e.g., light or dark), and several species have individuals that show intermediate characteristics between light and dark, further complicating identification. All undergo a single annual molt after breeding. In our spring, most austral breeders will be noticeably in molt; fresh-plumaged birds of the same species are likely to be juveniles. Subadults and nonbreeders may molt earlier than breeders, complicating the issue. When worn and bleached, dark colors become paler and browner. Light conditions also affect color perception.

Distribution Shearwaters and petrels reach their greatest diversity in the Southern Hemisphere, but they inhabit all oceans. With few exceptions they come ashore only to breed; most nest in colonies on remote oceanic islands. Many nest in burrows underground. Only two procellariids—the Northern Fulmar and the Manx Shearwater—breed in North America; the others only migrate through or spend their nonbreeding season in North American waters. A number of species are casual or accidental.

Taxonomy Taxonomy is unsettled. Worldwide, 84 species in 16 genera are now recognized. In North America, 35 species in eight genera have occurred; informally, these are often subdivided into three groups: fulmars (one species), gadfly petrels (19 species), and shearwaters (15 species).

Conservation Many species are imperiled; many nesting sites are threatened by feral predators, especially rats, pigs, and cats. Longline fishing drowns birds at sea. Harvesting of birds and eggs threatens some species. BirdLife International codes: 11 NT, 19 VU, 14 EN, 10 CR, 1 CR (PE), and 2 EX.

FULMARS Genus *Fulmarus*

NORTHERN FULMAR *Fulmarus glacialis* NOFU ▪ 1

The Northern Fulmar is the stocky seabird often seen on winter pelagic trips off both coasts. In the Atlantic and in the high Arctic, light-morph birds predominate and appear gull-like in plumage. Along the West Coast, dark-morph birds are common and are often confused with dark shearwaters and petrels. Polytypic (2–3 sspp.; all in N.A.). L 16–18" (41–46 cm) WS 37–45" (94–114 cm)

Identification Polymorphic: light, dark, and intermediate plumages. Key features are its stocky build and acrobatic flight style in strong winds. Ages and sexes similar. **LIGHT MORPH:** White below and gray dorsally; white head with dark eye smudge. Pale primary bases appear as wing patch on upperwing. Rump white or gray. Tail pale gray (Atlantic) to dark gray (Pacific). Some Pacific birds are dramatically pale above. **DARK MORPH:** Plumage wholly chocolate brown to brownish gray, often with darker smudge around eye. Underwing coverts concolorous with body, but primaries appear paler. Upperwing may or may not have pale primary patch. **INTERMEDIATE:** Any shade of color between dark brown and white. Most often appears light, pearly gray with darker upperwing. Blotchy birds often seen on West Coast. **FLIGHT:** Flies on stiff, outstretched wings with little bend at the wrist, arcing high off the water in strong wings. In light winds, it flies direct and low to water with stiff, short wingbeats alternating with glides.

Geographic Variation Polymorphic *rodgersii* of the Pacific has a darker tail that contrasts with the rump and a more slender bill (on average). Nominate *glacialis* describes Atlantic high-arctic breeders, predominately light morph; polymorphic *auduboni* (not recognized by some authorities) describes

intermediate morph
rodgersii

light morph
rodgersii

white flash on inner primaries

Pacific birds have dark tails that contrast with paler rump

variable, some individuals even paler

light morph
rodgersii

Pacific birds' plumage more variable than Atlantic birds

heavier bill than Pacific birds

Atlantic dark morph
glacialis

dark morph
rodgersii

uniformly dark

Atlantic light morph
glacialis

tail blends with rump in both Atlantic morphs

heavy yellowish bill

light morph
rodgersii

Atlantic populations breeding in low Arctic and boreal areas. Both Atlantic subspecies have tails similar in color to the rump and the uppertail coverts.

Similar Species Light morphs are similar to large gulls. Note the fulmar's stocky overall shape; relatively short, narrower wings; and complicated bill structure with multiple plates and nostril tube on the culmen. Dark morphs can be confused with dark shearwaters, which have thinner, darker bills and longer, narrower wings. Confusion with dark *Pterodroma* petrels is a problem, but note the small, stocky, dark bills of those species and their more dynamic flight on longer, crooked wings.

Voice Generally silent at sea when alone; feeding flock can be noisy. At breeding colonies, it makes a variety of cackling and croaking sounds.

Status & Distribution Common.

BREEDING: Breeds on sea cliffs along the coasts of the north Pacific and north Atlantic as well as on high-arctic islands. **NONBREEDING:** Moves south in varying numbers to winter off the coast of New England and off the West Coast south to Mexico. Numbers fluctuate annually, and in some invasion years hundreds are visible from shore along the West Coast. Found year-round off the West Coast, but typically few are present during the summer months.

Population Increasing in the Atlantic and stable in the Pacific. Breeding colonies are threatened by introduced predators, pesticides, and oil pollution.

rodgersii

glacialis

GADFLY PETRELS Genus *Pterodroma*

This wide-ranging genus includes 31 species worldwide, of which at least 14 have occurred in N.A. In many species, an M pattern appears across the upperwing. Their flight style in strong winds is striking, rapidly covering large tracts of ocean in a series of dynamic, high arcs. For identification, note overall size, head and neck pattern, underwing pattern, bill size, and differing flight styles. Most nest underground in burrows on oceanic islands, spending the majority of their lives at sea.

GREAT-WINGED PETREL *Pterodroma macroptera*   GWPE ▪ 5

This gadfly petrel from the southern oceans has been recorded casually off CA. Subspecies *gouldi* ("Gray-faced Petrel"), which the CA birds are thought to be, will likely be split from nominate *macroptera* in the near future. Polytypic (2 sspp.). L 17–18" (43–46 cm) WS 39–42" (99–107 cm)

Identification Large and dark except pale grayish white around bill and on chin and throat. Chocolate brown plumage can appear darker on head and breast; thick, stout, black bill. Nominate *macroptera* has less heavy bill and less whitish around the bill. **FLIGHT:** Languid wingbeats for a *Pterodroma*, but still performs dynamic arcs in strong winds.

Similar Species Compare to Murphy's Petrel, which is smaller overall and has prominent pale-based flight feathers and grayer plumage. Pale feathering around Murphy's bill base is more restricted to area below bill. Murphy's legs and feet are pink, not black. Providence Petrel is also very similar but has a more prominent double white flash on underwing and a subtle M pattern across the upperwing.

Status & Distribution BREEDING: Subspecies *gouldi* breeds

on islands off New Zealand; *macroptera* on islands in southern Indian and Atlantic Oceans. **NONBREEDING:** Subspecies *gouldi* ranges into the southwest Pacific. **RARE STATUS:** Casual off central and Southern CA from late summer to early winter (about five records, one photographed from shore at La Jolla, San Diego Co., on 18 Dec. 2012).

Population Decreasing, but with large global population. Threatened at breeding colonies by introduced cats and rats; still legally harvested on some New Zealand islands.

primary flash fainter than Murphy's

uniform dark underwing

white evenly distributed around bill

gouldi

overall color browner, less gray, than Murphy's

PROVIDENCE PETREL *Pterodroma solandri*   PRPE ▪ 4

This gadfly petrel (also known as Solander's Petrel) performs an annual transequatorial migration from breeding areas in Australia to the North Pacific during the boreal summer. Monotypic. L 17–18" (43–46 cm) WS 39–42" (99–107 cm)

Identification A large, mostly brownish, stocky gadfly petrel. Relatively heavy bill and broad wings. Upperparts dark gray-brown with a silvery sheen when fresh, browner when worn; head usually appears darker; whitish feathering encircles bill. Underwing dark brown except for silvery white bases of primaries (except most of P10) with distinct dark tips to the pale-based primary coverts creating a double white flash (the best field mark).

Similar Species Murphy's Petrel is slightly smaller, with shorter, narrower wings, smaller bill, and more subdued, single white underwing flash; in Murphy's, the white in the face tends to be most extensive in the throat, less so above the bill, the reverse of Providence Petrel. See Kermadec Petrel.

Status & Distribution
BREEDING: Lord Howe I. and Phillip I., off Norfolk I. Adults begin to arrive in late Feb.; last fledglings depart in late Nov. **NONBREEDING:** Dispersal northward into the central North Pacific, where mostly recorded Sept.–Apr. In N.A. waters, ranges north into western Bering Sea:

dark crescent formed by tips of greater primary coverts

white around bill most prominent on lores

heavy bill

11 were studied during research cruises in the Aleutians 4–25 Aug. 1994, and 25 were documented 15 Sept. 2011, close to Russian waters, where previous records (e.g., 18 on 23 Sept. 2006) indicate that the species is likely regular there. Accepted record off WA on 11 Sept. 1983; one photographed off Vancouver, BC, likely this species.

Population Vulnerable. About 32,000 pairs remain; extirpated from Norfolk I. by 1800 (one million birds were harvested for food between 1790 and 1793).

KERMADEC PETREL *Pterodroma neglecta*   KEPE ▪ 5

This accidental gadfly petrel recorded just once off central CA. Polytypic (2 sspp.: likely *juana* in N.A.). L 15–16" (37–40 cm) WS 38–42" (97–106 cm)

Identification Medium-size petrel occurring in two color morphs (dark and light), but light-morph plumages are highly variable. All morphs have grayish brown upperparts with a white

upperwing flash (white primary feather shafts), an extensive white primary patch on the underwings (that extends broadly onto the outermost primary), and a short tail; taken together, they impart a remarkably skua- or jaeger-like appearance. **DARK MORPH:** Dark brown overall; some have paler, medium brown heads and underparts.

LIGHT MORPH: Variable. Some have mostly white head; others have dark head and chest with white underparts; still others have mottled brown underparts and a variable chest band. Underwing pattern often has more white on the median and greater coverts.

Geographic Variation Widespread nominate averages smaller than *juana*,

which breeds on islands off Chile.
Similar Species All plumage patterns resemble those of jaegers (particularly immatures without long tails) and skuas; Kermadecs are also known to pirate other seabirds, as do jaegers and skuas. Jaegers have a gull-like flight with more flapping and less soaring and banking. Jaegers and skuas lack raised nostril tubes (visible at close range). Compared with dark-morph Kermadec, Murphy's and Providence Petrels have a somewhat different distribution of white around bill and head. Providence is most similar, but note how the white on its underwings is less extensive on the outer two primaries; it also is larger and has a heavier bill. Light-morph Kermadec with a pale head and underparts

resembles a light-morph South Polar Skua, especially when on the water.
Status & Distribution Pacific Ocean; a few reports from eastern Atlantic Ocean. **BREEDING:** South Pacific islands from Lord Howe I. east to Easter I., and on a few islands off Chile; possibly on Ilha da Trinidade in the tropical Atlantic Ocean. **NONBREEDING:** Disperses over much of tropical and subtropical Pacific and into the central North Pacific as far as 40° N. **RARE STATUS:** One N.A. record, a dark morph photographed from the Farallones, off central CA (8 Sept. 2017).
Population Least Concern, but population greatly reduced from historic highs by introduced mammalian predators and human exploitation.

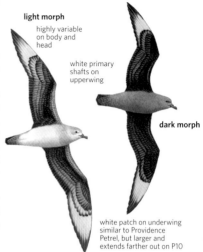

light morph
highly variable on body and head

white primary shafts on upperwing

dark morph

white patch on underwing similar to Providence Petrel, but larger and extends farther out on P10

TRINDADE PETREL *Pterodroma arminjoniana*   TRPE ▪ 3

This gadfly petrel is rare but regular in Gulf Stream waters off the East Coast; it was formerly lumped with Herald Petrel (*P. heraldica*) of the South Pacific. Most encountered in N.A. are dark morphs. Monotypic. L 14.2–15.4" (36–39 cm) WS 37–40.2" (94–102 cm)
Identification A medium-size and polymorphic petrel. Its upperparts are generally a smooth velvety brown, often with subtle, darker brown M pattern (usually absent on dark morph). The three morphs vary mainly in body color. Underwing variably pale with dark trailing edge, coloration generally matches morph type (e.g., dark morph has darker underwings overall). **LIGHT MORPH:** White below and brownish gray above; brownish head and neck sides often give dark-hooded look. Flanks dusky brownish; belly and undertail coverts bright

white. **DARK MORPH:** Plumage wholly chocolate brown to brownish gray, often with darker head and upper breast. Underwing coverts concolorous with body, but primary and greater primary covert bases flash white. **INTERMEDIATE:** Ventrally the body can be a mix of dark brown and white, generally palest on belly. **FLIGHT:** Acrobatic and agile, arcing high above the water on stiff set wings, angled back at the wrist. Direct flight low with stiff, measured flapping.
Similar Species Light-morph Trindade Petrel similar to Fea's Petrel, but it lacks Fea's bold upperwing pattern and pale tail. Dark-morph Trindade told from Sooty Shearwater by more restricted white on underwing primary bases and lacks pale underwing secondary coverts. Dark-morph immature jaegers (without tail stream-

ers) have broader wings with white flashes on the upper primaries and a more gull-like flight style.
Status & Distribution Rare visitor to Gulf Stream. **BREEDING:** Nests on islands off Brazil: Trinidade and Martin Vaz Is., where two populations breed at different seasons (possibly two species?); also Round I. in the Indian Ocean. **NONBREEDING:** Rare but regular as far north as VA in deep offshore waters from May–Sept. **RARE STATUS:** Casual north to NY; accidental ME, and inland after hurricanes as far north as PA and NY. PA record (documented with a movie camera) at Hawk Mtn. on 3 Oct. 1959 had been treated by many as a Kermadec Petrel. After careful review of the evidence by several authorities, the record is now treated as an aberrantly plumaged (likely leucistic) Trindade Petrel.
Population Vulnerable. Small population size (1,130 breeding pairs, 2008) and limited breeding range.

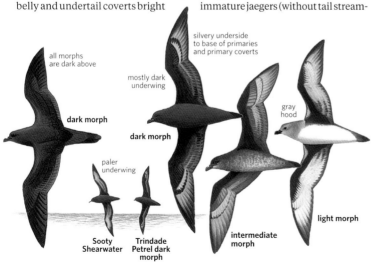

all morphs are dark above

silvery underside to base of primaries and primary coverts

mostly dark underwing

dark morph

gray hood

dark morph

dark morph

paler underwing

light morph

Sooty Shearwater

Trindade Petrel dark morph

intermediate morph

MURPHY'S PETREL   *Pterodroma ultima*   MUPE ■ 3

Murphy's Petrel is the dark *Ptero-droma* that is most likely to be observed off the West Coast. Although it superficially resembles dark shearwaters and dark-morph *rodgersii* Northern Fulmars, the Murphy's Petrel flies on stiff, angled wings with dynamic agility. Monotypic. L 14–15"

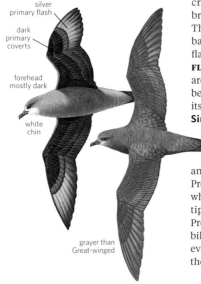

silver
primary flash

dark
primary
coverts

forehead
mostly dark

white
chin

grayer than
Great-winged

(36–38 cm) WS 35–38" (89–97 cm)
Identification Small, dark, thick-necked, and large-headed. Bill is all-dark, slender for a *Pterodroma*, and short. Plumage is overall gray-brown, with a subtle dark M pattern across upperwing. It appears white-throated, with a few white feathers on the fore-crown. The upperwing is grayish brown, concolorous with upperparts. The underwing is dark, with white bases to primaries showing a white flash below. Ages and sexes are alike.
FLIGHT: Murphy's flies with dynamic arcs above the horizon with stiff wings bent back at the wrist. In light winds, its flight is low with stiff wingbeats.
Similar Species Similar to Great-winged Petrel; see that species account for details. Providence Petrel is also very similar and should be considered. The larger Providence shows a prominent double white flash in the underwings and dark tips to the greater primary coverts. Providence has a bigger, more bulbous bill, and the white on its face is more evenly distributed above and below the bill (mostly below in Murphy's).

Status & Distribution Uncommon to fairly common visitor to West Coast. **BREEDING:** Breeds on low islands in the central South Pacific. **NONBREEDING:** Disperses to tropical Pacific waters into the central North Pacific. Regular from waters off central CA to southern WA, rarely to southern BC; mainly seen over deep water Mar. (a few late Feb.)–May; also a few later into summer.
Population Near Threatened. Presumably declining, but global population still large (±1 million birds).

MOTTLED PETREL   *Pterodroma inexpectata*   MOPE ■ 2

A gadfly petrel of the North Pacific that favors cooler waters. This species is most easily identified by its striking wing pattern and its diagnostic gray belly. Monotypic. L 13–14" (33–36 cm) WS 33–36" (84–91 cm)
Identification Medium-size and heavy-bodied with diagnostic grayish

extensive
gray belly

thick
black
bar

dark
M-pattern
above

black belly patch that contrasts with the white chest and undertail coverts. Wear and molt can make dark belly somewhat paler. Underwing bright white with very bold black ulnar bar across the underwing coverts that extends up to the axillaries; dark trailing edge. Upperwing gray, with black M pattern. Short tail gray and white undertail coverts. Mostly gray head; dark eye smudge; stout black bill. Juvenile has pale-tipped upperparts, imparting a more scaled appearance.
FLIGHT: Fast and strong; arcs high over the water on stiff, crooked wings.
Similar Species Dark belly patch contrasting with white chest and undertail is unique among *Pterodroma* species. Smaller Cook's Petrel lacks the contrasting pattern below, appears largely white on the underwings, and has a bolder dark M pattern across the upperwing. Mottled Petrel can appear dark in poor light and suggest a Murphy's Petrel.
Status & Distribution BREEDING: Islands off New Zealand. **NONBREEDING:** Transequatorial migrant. Migrates to the North Pacific; fairly

common off the Aleutians, the southern Bering Sea, and in the Gulf of Alaska (mostly May–Oct.). Rare to uncommon farther south far off the West Coast from BC to northern CA (mostly mid-Nov.–mid-Dec.); some off Northwest as late as May. Many have been found onshore, dead or exhausted. **RARE STATUS:** Remarkable specimen record from upstate NY (Apr. 1880).
Population Near Threatened. Population estimated at ±1.5 million birds and thought to be declining. Apparently vulnerable to changing ocean temperatures and El Niño events.

BERMUDA PETREL *Pterodroma cahow* BEPE ■ 3

Once thought to be extinct, this endangered species (also known as the Cahow) was rediscovered breeding on islets off Bermuda in the mid-20th century. In N.A., nearly all records of the Bermuda Petrel are from Gulf Stream waters off NC. In recent years, records have increased (first recorded in N.A. in 1996). Its rebounding population may account for this, but increased observer awareness and coverage are also factors. Monotypic. L 14–15"

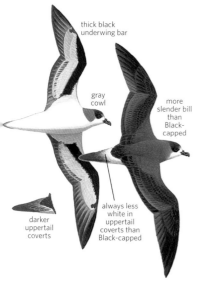

thick black underwing bar

gray cowl

more slender bill than Black-capped

darker uppertail coverts

always less white in uppertail coverts than Black-capped

(36–38 cm) WS 34–36" (86–91 cm)
Identification Distinctive small, slim structure, and dark-hooded appearance (typically lacking pale collar). Dark back, sometimes blackish with faint M pattern when fresh. Pale uppertail coverts usually restricted. Dark blackish gray head, typically lacking white collar; pale throat and dark neck sides. Underparts white. Underwing white, with dark flight feathers and bold dark ulnar bar diagonally across the underwing coverts. Ages and sexes similar. **FLIGHT:** Arcing flight typically lower than Black-capped Petrel.
Similar Species Most similar to Black-capped Petrel, but overall smaller and slimmer, with a smaller bill. Black-capped has a distinct white collar (lacking on some) and bright white uppertail coverts that impart a white-tailed appearance. Head of Bermuda is darker than Black-capped, with a cowled appearance. Fea's and Zino's Petrels have darker underwings and gray rumps.
Status & Distribution Atlantic Ocean. **BREEDING:** Breeds only on islets off Bermuda. **NONBREEDING:** Disperses to surrounding waters and to the north. **RARE STATUS:** Casual in warm Gulf Stream waters off NC mainly May–

Aug.; accidental MA (28 June 2009), but satellite tagging indicates it regularly ranges as far north as waters off Atlantic Canada and even Ireland.
Population Endangered; also listed as Endangered by the USFWS. Slaughtered for food during the 17th century, leading to its near extinction. Rediscovered—after an absence of more than 300 years—breeding on islets off Bermuda. In spring 2013, 105 breeding pairs were counted (overall population estimated at 500). Protection of colony on Nonsuch I., Bermuda, largely responsible for this species' recovery. Predators (primarily rats, hogs, cats, and dogs, even a Snowy Owl) have been the primary threat.

BLACK-CAPPED PETREL *Pterodroma hasitata* BCPE ■ 2

Birders can expect to encounter numbers of Black-capped Petrels on a typical pelagic trip off Cape Hatteras from May to Oct. Its highly acrobatic flight style is distinctive, and its bright, white plumage details make it identifiable at long range. Polytypic (2 sspp.; nominate in N.A.). L 15–18" (38–46 cm) WS 39–41" (99–104 cm)
Identification Large and stocky; with

dark blackish upperwings, a bold white collar (sometimes subdued or absent), and a very prominent and broad white rump band. Upperwings dark blackish with indistinct, dark M pattern. Large, stocky, dark bill. White below, with white underwings and variable dark ulnar bar across the underwing coverts. Two plumage types occur that may represent different subspecies, or even distinct species. **WHITE-FACED:** Mostly white head with a smaller black cap, broad white collar, and a thin, white supercilium that makes the dark eye stand out on the face. This type predominates in late spring off NC and molts about a month earlier than dark-faced birds. **DARK-FACED:** More extensive cap surrounds eye, dark nape, thicker dark neck spur. Predominates in July–Aug. off NC. Intermediate types occur. Ages and sexes similar; molt can cause unusual whitish areas on dark upperwing. **FLIGHT:** Dynamic,

arcing high above the water; rarely flapping on stiff wings angled back at wrist. Arcs higher above the water than shearwaters.
Geographic Variation Nominate *hasitata* (dark-faced and light-faced birds) breeds on Hispaniola—this is the subspecies seen off the East Coast.

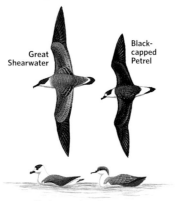

Great Shearwater

Black-capped Petrel

Black-capped Petrel

Great Shearwater

The all-dark *caribbaea* ("Jamaican Petrel," possibly a separate species), bred only on Jamaica; it was last seen in the 1800s and is now feared extinct. **Similar Species** Most similar to Bermuda Petrel, but larger and stockier with a heavier bill, bold white collar (often absent on dark-faced birds), and broad white band across rump and uppertail coverts. **Status & Distribution** Common in restricted Atlantic range. **BREEDING:** Mainly on Hispaniola at high elevations. Most confirmed nest sites have been in Haiti; recent information suggests possible breeding in Cuba and other small Caribbean islands. **NONBREEDING:** Frequent over deep Gulf Stream waters off NC, May–Oct.; rare in winter. Perhaps regularly ranges to VA and FL. Disperses to surrounding waters north of the Gulf Stream, rarely to the Canadian Maritimes, and south to Brazil. **RARE STATUS:** Casual from shore and inland to the Great Lakes and other large bodies of water after hurricanes. **Population** Endangered. Global

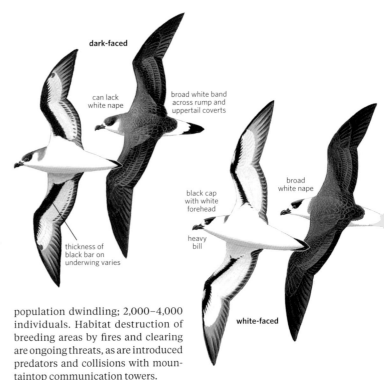

dark-faced

can lack white nape

broad white band across rump and uppertail coverts

black cap with white forehead

broad white nape

thickness of black bar on underwing varies

heavy bill

white-faced

population dwindling; 2,000–4,000 individuals. Habitat destruction of breeding areas by fires and clearing are ongoing threats, as are introduced predators and collisions with mountaintop communication towers.

JUAN FERNANDEZ PETREL *Pterodroma externa*   JFPE ▪ 5

dark comma

gray hindneck (can be paler)

mostly white underwings

There is one extraordinary record of this large tropical petrel for the US. It was not recorded over the ocean, but in the desert suburbs of Tucson, AZ! A tropical storm that worked its way up from the Sea of Cortez to AZ deposited numerous tubenoses in reservoirs and other locations, including this species as it flew over a birder's driveway. He alertly took some photos. Monotypic. L 16.5–17.7" (42–45 cm) WS 40.6–44.9" (103–114 cm)
Identification Long wings, long tail, and dark M pattern across its gray upperparts. Head with variable dark cap and eye surround; gray hindneck (sometimes paler). Underparts pure white. Underwings mostly white with dark primary tips and narrow dark trailing edge; importantly, the leading edge shows a small dark comma mark at the bend of the wing.
Similar Species Cook's Petrel is very similar in pattern but is much smaller,

shorter winged, and has a gray crown. Hawaiian Petrel has an extensive black crown and much more black on the leading edge of its underwings. The closely related White-necked Petrel (*P. cervicalis*) breeds in Kermadec Is. and ranges into the central Pacific; unrecorded in N.A. Usually has a prominent white hindneck collar, darker back, broader dark underwing borders, and a prominent dark chest spur.
Status & Distribution Pacific Ocean. **BREEDING:** Only at mid-levels of mountains on Alejandro Selkirk I. in the oceanic Juan Fernandez Is. group, Chile. **NONBREEDING:** Ranges north in small numbers to off western Mexico (Apr.–Oct., fewer Nov.–Mar.). **RARE STATUS:** Accidental Tucson, AZ (7 Sept. 2016), one storm-related bird associated with Hurricane Newton.
Population Vulnerable. Large population (about three million), but very restricted breeding range and a host

of threats (introduced predators, fires, flooding, light pollution, etc.) imperil the species.

HAWAIIAN PETREL *Pterodroma sandwichensis*   HAPE ▪ 4

This rarely encountered gadfly petrel was split from the very similar Galápagos Petrel (*P. phaeopygia*). Both taxa were previously classified as a

single species, the Dark-rumped Petrel. Monotypic. L 15–16" (38–41 cm) WS 38–41" (97–104 cm)
Identification Medium-size; sleek;

elegant pattern of clean white below, black above. Blackish gray, uniform upperparts. Dark tail and rump; occasionally flecked white on uppertail coverts. Head largely dark; white forehead; dark neck sides form partial collar. Underwing largely white; darker flight feathers; restricted black ulnar bar. Axillaries white; dark patch visible when arcing high into the wind. Ages and sexes alike. **FLIGHT:** Strong flight on long wings bent back at wrist; quick deep flaps followed by high arcs above the sea.

Similar Species Very similar in the field to Galápagos Petrel. Note Hawaiian Petrel's lighter build and cleaner white underparts; white on the face extends up in a point behind the ear coverts. There is no evidence that the Galápagos Petrel has ever occurred anywhere near N.A.

Status & Distribution Pacific Ocean. **BREEDING:** Once nested on all the Hawaiian Is. Now primarily restricted to small colonies on islands of Maui and Hawaii; small numbers possibly remain on three other islands. **NONBREEDING:** Disperses north and east, where small numbers are seen annually off CA and OR; has occurred as far north as BC. Most recorded are in spring (mostly late Apr.–May), but also in late summer. Usually well offshore, but some are closer and one was even recorded from a coastal headland (Pt. Dume, Los Angeles Co., 12 Aug. 2006). **RARE STATUS:** Accidental in the interior Southwest; one (specimen) was obtained at Yuma, AZ, on 24 Aug. 2013.

Population Endangered. Also listed as Endangered by the USFWS. Small global population estimated at 5,000 pairs. Decimated by habitat loss and predation by introduced predators (including humans) and feared extinct until rediscovery in 1948.

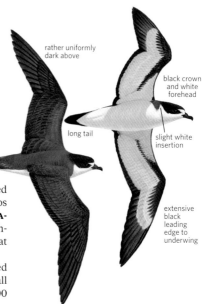

rather uniformly dark above

black crown and white forehead

long tail

slight white insertion

extensive black leading edge to underwing

FEA'S PETREL *Pterodroma feae*   FEPE ■ 3

Part of a confusing species complex (including Zino's Petrel), this infrequent visitor to Gulf Stream waters is one of the more strikingly plumaged tubenoses in the region. Polytypic (2 sspp.; both in N.A.). L 14–15" (36–38 cm) WS 34–38" (86–97 cm)

Identification Small gadfly with bold dorsal M pattern; blackish underwings; pale uppertail and rump. Upperwing grayish, often with distinct darker gray M pattern; some have brownish tinge above. Head generally dark, imparting a grayish hooded appearance; throat pale. Tail distinctive: grayish to occasionally whitish above; pale rump typically paler than mantle. White below; dark gray sides of neck typically do not connect in breast band. Ages, sexes similar. **FLIGHT:** Typical gadfly style, arcing high over the water in strong winds; direct flight on stiff, clipping wingbeats.

Geographic Variation Subspecies taxonomy unclear; recent information suggests that the two currently recognized taxa might be separate species. Nominate *feae* nests on the Cape Verde Is.; the slightly larger-billed *deserta* nests on Bugio I. in the Desertas Is. (part of the Madeira Archipelago). Based on current knowledge, field separation of *deserta* from *feae* is not possible. Fea's Petrel (likely *deserta*) may also nest in the Azores.

Similar Species Unlike any other species in the western Atlantic, but similar to the accidental Zino's Petrel (*P. madeira*). Zino's Petrel is smaller overall, notably so in the field; but perhaps most importantly, it is smaller billed. Zino's usually also shows extensive white on the underwing, unlike Fea's. **Status & Distribution** Rare. **BREEDING:** Colonies on islands in the eastern Atlantic. **NONBREEDING:** Disperses to surrounding waters. Rare from May–Sept. in warm Gulf Stream waters off Hatteras, NC (100+ records). **RARE STATUS:** Casual north in the Gulf Stream to the Canadian Maritimes; accidental inland after hurricanes.

Population Near Threatened. Global population is fewer than 1,000 pairs. Suffers predation from introduced predators, direct persecution from humans, and indirect habitat loss due to feral grazers.

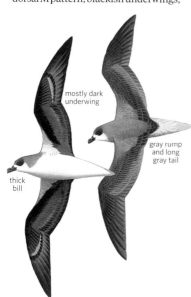

mostly dark underwing

gray rump and long gray tail

thick bill

? ? ? ?

desertae

? ?

? ? ?

feae

ZINO'S PETREL *Pterodroma madeira*   ZIPE ■ 5

The endangered Zino's Petrel, breeding only on Madeira I. in the eastern Atlantic, has been recorded once in N.A. waters. Its English name honors Paul Zino and his son Frank, who were largely responsible for rediscovering the species in 1969 and for adopting conservation measures to save the species. It is also known as Freira and Madeira Petrel. Monotypic. L 13–14" (33–36 cm) WS 33–35" (84–89 cm)

Identification A small *Pterodroma* that closely resembles Fea's, but is smaller with a slighter build, including a thinner and more delicate bill. Diagnostic from Fea's, if present, is a white underwing bar of variable width on the primary and secondary coverts. Some Zino's lack the white bar, but many have it. If such a bar is present on Fea's, it is more indistinct, but there is some overlap. Zino's commences primary molt in Sept. (believed to finish by Feb.), while in Fea's, nominate *feae* molts from Mar.–

Sept. and *deserta* from Nov.–May.

Similar Species From the more likely (but still rare) Fea's Petrel, see above. The slightly larger Bermuda Petrel is darker above with white at the base of the uppertail coverts and has much more white on the underwing.

Status & Distribution BREEDING: Nests only on Madeira on the highest part of the island. Subfossil remains indicate that prior to human settlement, it nested on the entire island. NONBREEDING: Geolocators on adults have shown summer movements west to the Azores and as far north as well off Ireland. In winter, most were off Mauritania and Senegal, northeast Brazil, and in tropical waters along the south mid-Atlantic Ridge to St. Helena. **RARE STATUS:** Accidental off Hatteras, NC: one commencing primary molt was photographed on 16 Sept. 1995. It was identified as Zino's years later.

Population Endangered. Also listed as Endangered by the USFWS. Total

diagnostic (from Fea's) white underwing bar of variable width present on many Zino's

smaller, has a more petite bill than Fea's

population is about 200 birds. Introduced predators are a problem, and in Aug. 2010 a forest fire enveloped its nesting area and killed 38 chicks and four incubating adults.

COOK'S PETREL *Pterodroma cookii*   COPE ■ 3

Cook's Petrel is a small, sharply patterned gadfly petrel of the Pacific. It is often grouped with Stejneger's and several other very similar extralimital species that are referred to collectively as Cookilaria petrels. Monotypic. L 12–13" (30–33 cm) WS 30–32" (76–81 cm)

Identification Combination of small size, pale gray upperparts with striking dorsal M pattern, and clean white underparts is unlike most other *Pterodromas*. Tail gray above with dark, central tip; white outer tail feathers

difficult to see under field conditions. Head largely grayish; dark eye patch; white throat. Clean white underparts, including underwing; thin, dark mark at wrist. Ages and sexes alike. **FLIGHT:** Dynamic but erratic compared to other *Pterodromas*, with quick direction changes.

Similar Species In N.A., Stejneger's is the most similar, but it is darker above with a less distinct M pattern and a darker head, imparting a capped or hooded appearance.

Status & Distribution BREEDING: Nests on three islands off New Zealand. NONBREEDING: Moves northeast to eastern Pacific. It is rare to fairly common in deep water well off the West Coast from spring through fall from Baja California to southern OR. **RARE STATUS:** Very rare to northern OR; casual to WA; accidental in AK. Has been found several times at the Salton Sea in mid-summer (presumably birds traveling north across the desert from north end of the Gulf of California).

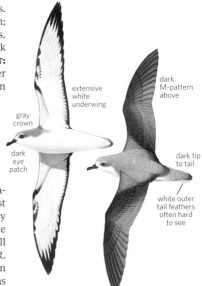

extensive white underwing

gray crown

dark eye patch

dark M-pattern above

dark tip to tail

white outer tail feathers often hard to see

Population Vulnerable. Threatened from predation of introduced mammals and from a native rail, the Weka.

STEJNEGER'S PETREL *Pterodroma longirostris*   STPE ■ 4

Similar to Cook's Petrel in its overall appearance, the dark-capped Stejneger's Petrel has been recorded just

a handful of times off the West Coast. The first N.A. record was in 1979, off Monterey Co., CA. Monotypic. L 11.4–

12.6" (29–32 cm) WS 28–30" (71–76 cm)

Identification A small, slim *Pterodroma* showing a dark M pattern on

its gray wings and upperparts, with clean white underparts. It has a notably black-capped (or half-hooded) head; the black half-hood encompasses the eye and contrasts strongly with the gray mantle and pale forehead. Normally the dark half-hood is very noticeable, but it can be difficult to see on birds flying away. The white of the throat hooks up slightly behind the eye, accentuating an intrusion of dark feathers just behind it. The tail is long and appears uniformly dark; the white in the outer tail feathers is restricted and difficult to see under field conditions. Underwings are clean white, with a thin, dark mark at wrist (similar to the Cook's Petrel). Ages and sexes are alike. **FLIGHT:** Dramatic and bounding, Stejneger's Petrel arcs high off the water in strong winds. It is prone to erratic maneuvers, though somewhat less so than Cook's Petrel.
Similar Species Similar to Cook's and to other unrecorded Cookilaria petrels.

When comparing it to Cook's Petrel, note Stejneger's black cap, its darker gray upperparts, its blacker outer wing, and its darker tail. Structurally, Stejneger's Petrel is slightly smaller and shorter-winged.
Status & Distribution Casual (mainly fall) to well off the CA coast. **BREEDING:** Nests on the Juan Fernandez Is. off Chile. **NONBREEDING:** Disperses to surrounding waters with movements north of the Equator. It has been suggested that Stejneger's Petrel makes a loop migration through the northern Pacific, passing through waters east of Japan as it moves north, and through the eastern Pacific (closer to our West Coast) as it moves south in Oct. and Nov. **RARE STATUS:** Casual off CA; fewer than 10 records from mid-summer through fall (mostly Nov.) in deep offshore waters. Likely more regular much farther offshore, beyond the 200-mi limit, but confusion with similar species confounds status. Also a remarkable specimen

record from coastal TX: one salvaged at Port Aransas 15 Sept. 1995.
Population Vulnerable. Threatened at breeding colonies, mostly by feral cats.

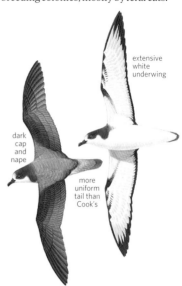

extensive white underwing

dark cap and nape

more uniform tail than Cook's

Genus *Pseudobulweria*

TAHITI PETREL *Pseudobulweria rostrata* TAPE ■ 5

The only N.A. record—off Hatteras, NC—was a completely unexpected occurrence for a bird that is known to breed only in the South Pacific. Polytypic (2 sspp.; unknown in N.A.). L 14.6–16.5" (37–42 cm) WS 40–42.5" (101–108 cm)
Identification Mostly dark brown, including the underwings, which can show a white central stripe due to variably pale bases of underwing primary and secondary coverts. Head and chest also dark brown, sharply contrasting with the white underparts. Tail relatively long; usually held closed and mostly covered below by long white undertail coverts that are tipped dark at the sides; uppertail coverts and tail base often show whitish areas. Wings long and narrow. Very heavy bill and disproportionately small head and long neck. Bill black, legs and feet bicolored pink and dark gray. **FLIGHT:** In moderate to strong winds, usually flies with wings held straight out from the body (albatross-like) with little flapping, unlike the crooked-wing, acrobatic flight of most *Pterodroma* petrels in similar wind conditions. In light winds, flies lower with languid flaps and long glides.
Geographic Variation Not separable

at sea. Nominate *rostrata* breeds in the central tropical Pacific and is slightly smaller with less heavy bill than *trouessarti* from the western tropical Pacific.
Similar Species Tahiti Petrel is a distinctive species. Another petrel of the South Pacific, in the genus *Pseudobulwaria*, Beck's Petrel (*P. becki*), is very similar but much smaller; Beck's has a tiny world population of perhaps fewer than 250 individuals and is poorly known. In the genus *Pterodroma*, Phoenix Petrel (*P. alba*, unrecorded in N.A.) and light-morph Kermadec Petrel (see that account) with a dark head and chest have similar plumage patterns but show much more white on the underwings, have smaller bills, and fly like a typical *Pterodroma*. Atlantic Petrel (*P. incerta*, unrecorded in N.A.) of the South Atlantic has dark undertail coverts, paler throat, and broader wings with paler flight feathers.
Status & Distribution Pacific Ocean. **BREEDING:** In burrows on islands in the South Pacific (mainly Sept.–Feb.). **NONBREEDING:** Ranges north and east to waters off Mexico and C.A. (May–Nov.) and into the eastern Indian Ocean. **RARE STATUS:** One remarkable record of a bird well photographed in Gulf Stream waters off Hatteras,

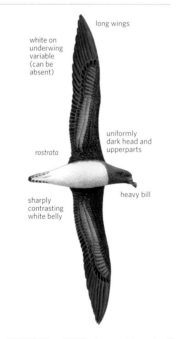

long wings

white on underwing variable (can be absent)

uniformly dark head and upperparts

rostrata

sharply contrasting white belly

heavy bill

NC (29 May 2018); also accidental off Durban, South Africa (11 Nov. 2018).
Population Near Threatened. World population estimated at fewer than 30,000 individuals; threats include introduced mammalian predators and mining operations.

BULWER'S PETREL *Bulweria bulwerii*   BUPE ■ 5

This unusual petrel—which recalls an oversize, dark storm-petrel with very long wings and an exceptionally long, pointed tail—is accidental in N.A. waters. The genus *Bulweria* contains only one other species, the larger Jouanin's Petrel (see next account) of the Indian Ocean. Monotypic. L 11–12" (28–30 cm) WS 25–27" (64–69 cm)
Identification A medium-size and all-dark petrel with a long, pointed

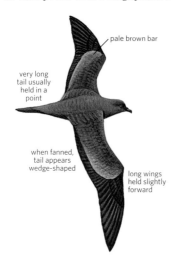

pale brown bar

very long tail usually held in a point

when fanned, tail appears wedge-shaped

long wings held slightly forward

tail, recalling a large storm-petrel. The long, pointed wings are dark brown above and below, with a bold tawny carpal bar extending forward to the leading edge of the wing. The tail is long, dark, and pointed; its wedge shape is concealed except during twisting aerial maneuvers. Head and underparts are dark brown; the black bill is rather heavy. Bulwer's Petrel appears dove-headed, lacking the thick neck and heavy build of most *Pterodroma* petrels. Ages and sexes are alike. **FLIGHT:** Buoyant, erratic, and zigzagging with long wings slightly bowed and held well forward; usually close to the water's surface; in calm conditions, sometimes a few wingbeats then a short glide. Bulwer's has the lowest wing-loading of any tubenose. Unlike *Pterodroma* petrels, it does not normally arc high over the water, nor does it flutter like storm-petrels.
Similar Species Bulwer's Petrel is most similar to Black Storm-Petrel, but note its larger size and medium brown coloration as well as its long, wedge-shaped tail; Black Storm-Petrel has a deeply forked tail. Dark-morph Wedge-tailed Shearwater has a similar

tail, but it is much larger and has a different flight style: slow, languid wingbeats, as well as prolonged soaring with bowed wings angled forward to the wrist, then swept back. See Jouanin's Petrel.
Status & Distribution Accidental to N.A. **BREEDING:** Breeds on oceanic islands in the three major oceans, including the islands of Macaronesia in the eastern Atlantic and the Hawaiian Is. (The majority of the world's population nests on Nihoa, in the northwest chain of HI.) These are the two likely sources for the N.A. records. **NONBREEDING:** At-sea distribution is poorly understood, but it disperses offshore to surrounding tropical waters. **RARE STATUS:** Two photographed records: off Monterey Bay, CA (26 July 1998), and off Oregon Inlet, NC (8 Aug. 1998); also an accepted sight record from off Cape Hatteras, NC, 1 July 1992. There are other sight records from VA to FL, and there are a least a dozen sight reports from the West Indies.
Population Stable, but populations are experiencing predation at breeding colonies by introduced mammalian predators.

JOUANIN'S PETREL *Bulweria fallax*   JOPE ■ 5

This poorly known species of the Indian Ocean has been recorded once in N.A. It was first described in 1955 by the French ornithologist Christian Jouanin. Monotypic. L 13–14" (33–36 cm) WS 30–31" (76–79 cm)
Identification Plumage is dark overall with very little variation, although some birds show an indistinct carpal bar on the upperwings, especially in strong light and on birds in worn plumage. Long, broad wings. Bill is black and noticeably thick and heavy (more so in males). **FLIGHT:** Effortless and buoyant. Flies low over the sea in a zigzag or meandering flight path with short glides and arcs, similar to flight style of Bulwer's Petrel but with slower, less hurried wingbeats.
Geographic Variation A population from around the Comoros Is. has recently been reported as being smaller overall, with a narrower bill, and more noticeable ulnar bar—features rendering it closer to Bulwer's Petrel.

Similar Species Bulwer's Petrel is very similar but substantially smaller overall (closer in size to Black Storm-Petrel; Jouanin's is closer in size to Black-vented Shearwater). Bulwer's has a much less heavy bill, pale brown carpal bar (often, but not always, absent on Jouanin's), a very long, more tapered tail, and narrower wings.
Status & Distribution BREEDING: Socotra I. in the northwest Indian Ocean (and perhaps other locations in the region). **NONBREEDING:** Disperses regularly to waters off southern India and in small numbers to Pacific waters off northwestern Australia. **RARE STATUS:** Accidental or casual to the Pacific. One caught and photographed (not measured) at night on Santa Barbara I., CA (1 June 2016). Other photographed CA sightings may pertain to this species or Bulwer's Petrel. Single specimen record for Lisianski I., HI (4 Sept. 1967), and three birds off Italy in the Mediterranean (specimen).

Population Near Threatened. Small world population (probably under 10,000) is threatened by introduced mammalian predators at its relatively few known breeding locations.

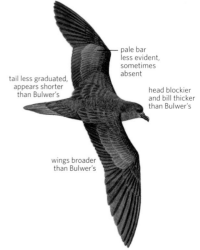

pale bar less evident, sometimes absent

tail less graduated, appears shorter than Bulwer's

head blockier and bill thicker than Bulwer's

wings broader than Bulwer's

Genus *Procellaria*

The five species of *Procellaria* petrels inhabit southern oceans, mostly at higher latitudes—two species are strays to N.A. waters. They are robust tubenoses with rather broad wings and short, thick bills. *Procellaria* petrels are larger than shearwaters but smaller than giant-petrels or albatrosses and occupy an intermediate ecological niche between them.

WHITE-CHINNED PETREL   *Procellaria aequinoctialis*   WCPE ■ 5

Among petrels, White-chinned is a bruiser, built for foraging in some of the planet's roughest seas. Heavy-billed and amply proportioned, this species is not likely to be overlooked among the more delicate shearwaters and gadfly petrels observed on N.A. pelagic trips. Monotypic. L 20–23" (51–58 cm) WS 52–57" (132–145 cm)
Identification A very stout-bodied, wide-winged *Procellaria* petrel, entirely chocolate brown in plumage except for a small amount of white in the chin (interramal space at the base of the mandible) that can be difficult to see with confidence or even impossible to see on flying birds. White on chin averages more extensive in Atlantic population. Heavy bill is yellowish with each section (or plate) of the bill bordered narrowly in black.
Similar Species Both the comparably built Westland Petrel (*P. westlandica*; unrecorded in the northern hemisphere) and the more delicately built Parkinson's Petrel lack the white chin and have dusky-tipped bills, but close

studies are required to discern these details. Spectacled Petrel (*P. conspicillata*, breeding only on Inaccessible I. in the South Atlantic) was formerly considered a subspecies of White-chinned. It has diagnostic white arcs (sometimes patchy) around eyes that often join across the forehead and a variably dark bill tip. Flesh-footed Shearwater is smaller overall with a less heavy, pinkish (not yellowish) bill and pinkish (not black) feet and legs.
Status & Distribution BREEDING: On South Georgia, Prince Edward Is., Crozet Is., Kerguelen Is., Auckland I., Campbell I., Antipodes Is., and Falklands; when breeding and afterward, ranges widely at sea, from Antarctic pack ice to subtropical latitudes. **RARE STATUS:** In N.A. waters, five records of single birds: near Rollover Pass, TX (27 Apr. 1986); off Bar Harbor, ME (24 Aug. 2010); Half Moon Bay, CA (18 Oct. 2009); near San Miguel I., CA (6 Sept. 2011); and Cordell Bank, off Marin Co., CA (16 Oct. 2011).
Population Vulnerable. Documented

decreases at various breeding sites, but large global population (about 3 million individuals). Inveterate ship follower and incidental mortality in longline fisheries is exceptionally high in this species.

PARKINSON'S PETREL   *Procellaria parkinsoni*   PAPE ■ 5

Simply called Black Petrel in its native New Zealand, Parkinson's Petrel is named for botanical illustrator Sydney Parkinson, who traveled on Captain James Cook's first voyage. These petrels are frequent patrons at fishing vessels, where they feed on by-catch and offal. Monotypic. L 16–17" (41–43 cm) WS 45–49" (114–124 cm)
Identification A sturdy, blackish brown petrel, with black legs and feet. Comparatively elegant for a *Procellaria* petrel, in flight recalling a large shearwater. Stout bill is greenish yellow, with each part of the bill bordered narrowly in black, which has a dusky tip.
Similar Species Compared to its beefier, more bull-headed congeners, White-chinned and Westland Petrels (the latter not recorded from N.A.), Parkinson's has disproportionately longer, lankier wings, noticeable in flying birds; on the water, the wings project well past the tail, giving a different profile than White-chinned or

Westland. White-chinned Petrel has a diagnostic small patch of white feathering under its bill; however, this can be exceedingly difficult to see, and its bill lacks the dark tip of Parkinson's bill. Westland Petrel is essentially identical in plumage and bill coloration, but is larger with a more massive head, heavier bill, and shorter wing tip projection past the tail. Parkinson's might be passed over for a smaller Flesh-footed Shearwater, which has pink legs and feet and a more slender, pink bill that lacks the neat black plate outlines of *Procellaria* bills.
Status & Distribution BREEDING: Only on Great Barrier and Little Barrier Is., New Zealand. Adults arrive in Oct.; fledglings depart by June. **WINTER:** Ranges widely over Pacific waters from southern Mexico to northern Peru and west to Australia. **RARE STATUS:** Only one valid N.A. record: 20 mi northwest of Pt. Reyes, CA (1 Oct. 2005, well-photographed).

large and heavily built

usually all-pale bill tip

variable white chin usually difficult to see

smaller and more lightly built than White-chinned

thick yellowish bill with blackish tip

dark tip to bill

Population Vulnerable. Small global population of about 5,500 individuals appears to be stable. Longline fisheries and introduced mammalian predators on breeding islands are the main threats.

SHEARWATERS Genus *Calonectris*

These shearwaters are typically large and long-winged; their slow, languid wingbeats are quite unlike the shorter, shallow wingbeats of the *Puffinus* shearwaters. They are able to arc high off the water, when they appear uncharacteristically buoyant for their size. Three species are now recognized: Cory's and Cape Verde (Atlantic) and Streaked (Pacific).

STREAKED SHEARWATER *Calonectris leucomelas*   STRS ■ 4

This casual wanderer from the western Pacific is a much sought after specialty of West Coast fall pelagic trips. Monotypic. L 19" (48 cm) WS 48" (122 cm)
Identification A large pale shearwater giving a dirty white-headed appearance; overall dark above and white below. Variable head pattern, but almost always appears white at a distance; at close range streaked with brown, heaviest on nape and palest around bill and eye. The bill is bluish or yellowish gray (rarely pinkish) with a darker tip. Dark brownish upperparts, often appearing scaly due to pale fringes on the upperwing coverts and mantle feathers; the upperparts become paler and browner when worn and bleached. Long, dark tail. The uppertail coverts are variable: they can be all-dark or white, forming a pale U shape on the rump. The underwings are largely white with a broad rear border and wing tip formed by the dark flight feathers. Just out from the wrist, the mostly dark, median primary coverts form a conspicuous dark patch on the underwing. The legs and feet are pinkish. Ages and sexes alike.
FLIGHT: Agile, with long, broad-based wings. Flies with slow, languid wingbeats unlike most shearwaters; arcs high above the sea surface when traveling.
Similar Species Most likely to be confused with Pink-footed Shearwater. Pink-footed appears dusky headed with a pink bill; has dark undertail coverts and darker (not scaly) upperparts; and usually has darker underwings than those found on Streaked.

languid, soaring flight
dark median primary coverts
white uppertail coverts on some birds form a pale "horseshoe"
white underwing
variably streaked white head
scaly upperparts
bill color varies from pale gray to pale pink
rather long tail

Voice Generally silent at sea, but gives excited cackling in feeding groups.
Status & Distribution Casual off the West Coast. **BREEDING:** Nests on islands off Japan and Asia. **NONBREEDING:** Migrates south to waters off the Philippines, Indonesia, and eastern Australia. **RARE STATUS:** Casual in the eastern Pacific. Most records from off central CA (Sept.–Oct.); one Sept. record from off OR. Two interior records: one captured at Red Bluff, CA (Aug. 1993), and one (found dead) near Medicine Bow, WY (June 2006).
Population Near Threatened. Large global population, but it is threatened at breeding colonies by introduced mammalian predators.

CORY'S SHEARWATER *Calonectris diomedea*   CORS ■ 1

Cory's Shearwaters, known among fishermen as "tuna ducks" for their habit of following and foraging above schools of tuna, are the largest of the world's shearwater species. Often found in mixed flocks with Great Shearwater in summer and fall off the Atlantic coast, Cory's dive deeply in pursuit of baitfish and squid, using wings and tail to propel and steer them with remarkable grace underwater. Polytypic (2 sspp.; both in N.A.). L 18" (46 cm) WS 46" (117 cm)
Identification A large shearwater, sometimes appearing gull-like in shape, with long, lanky wings and rather heavy yellow bill with dusky mark near the tip. Upperparts medium to pale brownish, often palest on the mantle and slightly grayish in the head. Upperwing sometimes shows carpal-ulnar M pattern. Underparts and underwing coverts bright white. Pinkish legs and feet. Molting birds, frequently seen in summer months, show bluish gray, fresh flight feathers contrasting with worn brownish ones; may show whitish wing bars due to dropped coverts. **FLIGHT:** In low winds, can appear somewhat ungainly, with deep, laborious flaps and rather loping flight style. In higher winds, far more agile, sweeping up and down in sinusoidal arcs, sometimes steeply.
Geographic Variation With a few exceptions, colonies of the smaller nominate subspecies *diomedea* (called "Scopoli's Shearwater") are located on islands within the Mediterranean Sea, whereas colonies of the larger subspecies *borealis* (called Cory's Shearwater) are on oceanic islands from the Azores south to the Canary Is.

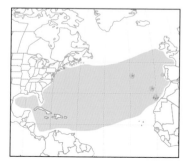

Distinguishing these subspecies, both of which occur commonly off the East Coast, is possible for many individuals, though it is helpful to take photographs to confirm identification: In the underwing, *borealis* shows a clean break between the white of the primary coverts and the dark primaries, whereas in nominate *diomedea*, the bases of the primaries show some white, especially the outer primaries, producing a different look in the "hand" from below. "Scopoli's" also averages smaller, with slimmer bill and narrower wings.

Similar Species Distant Cory's might be confused with Great Shearwater, which usually appears darker above and cleanly dark-capped; Great flies more boldly and snappily, on outstretched wings that look more pointed than those of Cory's. See also Cape

Verde Shearwater. From Pink-footed in Pacific, Cory's (*borealis* casual in the Pacific) is white below, much more extensively white on its underwings, has a thicker, yellow-based (not pink-based) bill, and adults molt in the fall.

Status & Distribution BREEDING: Nests on islands in the eastern Atlantic and Mediterranean. Adults return to nesting areas in Apr.; the last fledglings depart by Oct. **NONBREEDING:** Largely in the South Atlantic off Africa and S.A. Most of the birds observed in N.A. waters (May–Oct., in this time frame averages later to the north) are

believed to be immatures. Most common to MA, some to southern NS; very rare farther north. **RARE STATUS:** Casual inland after hurricanes. Accidental off central and Southern CA (three records; one bird spent two summers on the Coronodos Is., off Baja California Norte).

Population Declining, probably from a combination of factors that threaten many pelagic seabirds: introduced predators, fishing operations, and overexploitation of marine resources. Global population about 500,000.

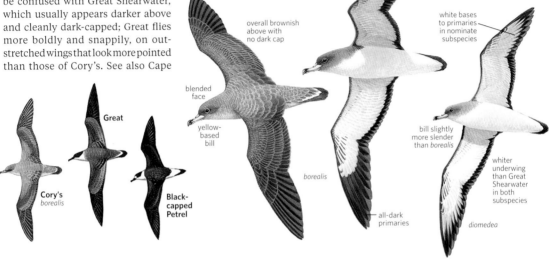

overall brownish above with no dark cap

white bases to primaries in nominate subspecies

blended face

Great

yellow-based bill

bill slightly more slender than *borealis*

Cory's
borealis

borealis

whiter underwing than Great Shearwater in both subspecies

Black-capped Petrel

all-dark primaries

diomedea

CAPE VERDE SHEARWATER *Calonectris edwardsii* CVSH ◼ 5

This large shearwater resembles a smaller version of Cory's Shearwater, with which it was considered conspecific for most of the 20th century. Endemic to the Cape Verde Is. off Africa, it is still a little-known species, particularly in the nonbreeding season. Monotypic. L 15.5" (39 cm) WS 39.5" (101 cm)

Identification Large shearwater, nearly as large as Cory's, brown above and white below; longest uppertail coverts usually show some white. Bill slim, olive-gray with dusky tip, some showing dull pinkish or olive-yellow tones at close range.

Similar Species Cory's Shearwater is very similar in plumage but averages bulkier and larger in all measurements; the slightly smaller "Scopoli's Shearwater" (currently classified as the nominate subspecies of Cory's) is closer to Cape Verde in proportions but still has broader wings and greater bulk. In direct comparison to Cory's

and "Scopoli's," Cape Verde has a notably smaller head and slimmer bill showing little or no yellow. The head often looks darker than in Cory's, with the dark of the crown contrasting more with the white underparts, unlike in Cory's, in which these colors are less contrasting, more diffuse where they meet. Capped appearance is never so strong as seen in smaller Great Shearwater, which is also darker above and has dark markings in underwing coverts and a dark patch on the belly.

Status & Distribution BREEDING: Nests in summer only in the Cape Verde Is. First adults arrive in late Feb.; last fledglings depart in Nov. **WINTER:** Ranges at sea off western Africa near breeding grounds to waters off Brazil and possibly Argentina, but very little known about range in nonbreeding season. **RARE STATUS:** The first N.A. record was of one found 30 mi southeast of Hatteras Inlet, NC (15 Aug.

2004); the only subsequent record is off Worcester Co., MD (12 Oct. 2006).
Population Near Threatened. About 10,000 pairs nest on five or six islands.

smaller than Cory's and has a slightly longer tail

slim olive-gray bill

more contrast on face than Cory's

SHEARWATERS Genus *Ardenna*

All the shearwaters are now placed in three genera representing three distinct lineages. The seven *Ardenna* shearwaters were formerly placed in genus *Puffinus*. In general, they are mid- to large-size shearwaters with plumages ranging from all-dark to variably patterned with white, gray, and black. Some species occur in different color morphs.

WEDGE-TAILED SHEARWATER *Ardenna pacifica*   WTSH ■ 4

This rare visitor, always a surprise species, is casual to the West Coast. The provenance of individuals recorded in N.A. is unclear, as breeding populations exist on islands off Mexico and in HI. Its graceful shape and flight style immediately separate it as something unique among Pacific shearwaters. Monotypic. L 18" (46 cm) WS 40" (102 cm)

Identification Polymorphic, occurring in both light and dark morphs. A fairly large, lanky shearwater char-

acterized by a slim build and a long, wedge-shaped tail. Head appears small; the long, thin, blue-gray bill has a dark tip. Pinkish legs and feet. **LIGHT MORPH:** White below and brownish above with pale tips on the upperwing coverts and scapulars giving a scaly appearance. White throat, chest, and belly, but head can appear largely dark at a distance. Underwing variable, but generally light; some with darker leading edge and dark line across the underwing coverts connecting with the axillaries. Dark flight feathers. **DARK MORPH:** Upperparts as in light morph, but often slightly darker. Underwing and ventral body wholly dark, with dark flight feathers. Some intermediate birds occur: dark morphs with grayer underparts or paler underwings, and light morphs with a grayish breast band or gray below with a white throat. **FLIGHT:** Flies with slow, languid wingbeats; usually stays low to the water. Holds broad-based wings slightly forward, bent back at the wrist and bowed down. The long tail is held in a point, so the wedge shape is visible only when it is spread.

Similar Species Flight similar to Buller's Shearwater. Light morph told from Pink-footed by tail shape, dark bill, and whiter underwings. The dark morph is similar to other dark shearwaters in plumage; note unique shape and flight style; most similar to Flesh-footed Shearwater, but note

languid flight with much soaring

dark morph

white underparts and wing linings

slender grayish bill

long wedge-shaped tail usually held in a point

dark morph

light morph

Wedge-tailed Shearwater's longer tail, lighter build, thinner black bill, and darker flight feathers.

Status & Distribution Casual to N.A. waters. **BREEDING:** Nests colonially on oceanic islands in the tropical Pacific (including HI and Isla San Benedicto, Revillagigedo Is., off western Mexico) and Indian Oceans; dark morphs predominate off Mexico; pale morphs from HI and Japan. **NONBREEDING:** Disperses to surrounding tropical waters; migration poorly known. **RARE STATUS:** Casual; about a dozen records for the Pacific coast, most from CA; one interior record at the Salton Sea (31 July 1988) and another at Amado, AZ, on 7 Sept. 2016, after Hurricane Newton. Both color morphs have been recorded.

Population Large global population. Historically impacted by human exploitation, predation pressure from introduced mammalian predators, and habitat loss.

BULLER'S SHEARWATER *Ardenna bulleri*   BULS ■ 2

This strikingly plumaged Pacific species can be mistaken for a rare *Pterodroma* petrel due to the bold M pattern across the surface of its wings. Widely considered the most elegant shearwater in N.A., its plumage, structure, and flight style are unusually graceful. Monotypic. L 16" (41 cm) WS 40" (102 cm)

Identification Distinctly black-capped head, with a clean demarcation between the dark cap and light sides of face and throat. Dove gray upperparts set off by a bold,

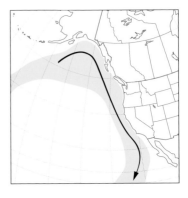

blackish M pattern, creating a black-on-gray pattern unique among shearwaters. Outer wing (primaries and primary coverts) is mostly dark above. Gleaming white underwings and underparts; the underwings are neatly outlined in black. Some fresh-plumaged birds have a very subdued M pattern; worn birds are browner above. Ages and sexes alike. **FLIGHT:** Elegant, with smooth, unhurried wingbeats and elongated appearance. In direct flight uses slow, measured wingbeats; in strong winds

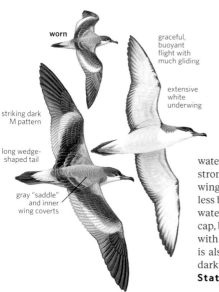

worn

graceful, buoyant flight with much gliding

extensive white underwing

striking dark M pattern

long wedge-shaped tail

gray "saddle" and inner wing coverts

arcs high like a *Pterodroma*, flapping little, if at all.

Similar Species Most similar to *Pterodromas*, especially Cook's Petrel, but note the larger size of Buller's, its slower wingbeats (if flapping), and its long, thin bill. Distinctive, unlikely to be confused with other shearwaters. A light-bodied Pink-footed Shearwater can show a subdued M pattern in strong sunlight, but has darker underwings and is larger and heavier with a less buoyant flight style. Great Shearwater (casual to Pacific) shares dark cap, but is more uniformly dark above with U-shaped white rump band, and is also darker on underwings with a dark belly patch (hard to see).

Status & Distribution Uncommon to fairly common off the West Coast. **BREEDING:** Nests colonially (late Nov.–Mar.) only on the Poor Knights Is. off New Zealand's North I. **NONBREEDING:** Moves north and east of breeding islands to feeding areas in the North Pacific from late July (casual June) through Nov. (peak in Sept.–mid-Oct., sometimes in large flocks). Occurs northward to the Gulf of Alaska; rare south of Pt. Conception. **RARE STATUS:** Accidental inland at the Salton Sea (6 Aug. 1966, specimen) and off NJ (28 Oct. 1984, photo). **Population** Vulnerable. Historically impacted by human exploitation and predation by introduced predators; breeding islands now a protected nature reserve and predator-free with about 200,000 breeding pairs.

SHORT-TAILED SHEARWATER *Ardenna tenuirostris*   STTS ▪ 2

This dark shearwater visits the West Coast primarily during late fall and winter, where separation from the more common Sooty Shearwater is problematic. Concentrations of up to a million birds have been seen off parts of AK, where this species congregates in late summer. Monotypic. L 17" (43 cm) WS 39" (99 cm)

Identification Dark with largely dark underwings and usually a pale chin. Bill structure and head shape are important identification features. Upperparts wholly chocolate brown, lacking strong contrasts. Underwings variable, typically dark brownish, but many have a pale silvery underwing panel (narrower and of even width, unlike similar panel on Sooty Shearwater). Typically round headed, imparting a dove-like appearance. Bill short, thin, and all-dark. Head often appears dark-capped due to the pale feathers around the bill, on the throat, and often on the lower face. **FLIGHT:** Flaps in short bursts, stiffly

from the shoulder. Does not appear long winged; rather its wings often look short and narrow (stick-like) for its relatively bulky body. Arcs high above water in strong winds.

Similar Species Notoriously similar to Sooty Shearwater and difficult to tell with certainty as characters are subjectively interpreted in the field. Short-tailed is smaller overall. It is shorter- and thinner-billed with a thinner tip (nail), and often shows a steeper forehead and slightly rounder head with a darker cap and a paler throat. Its underwings usually appear darker than Sooty's, but some have a pale central panel and are more similar to Sooty. On average, the pale on Short-tailed's underwings forms a narrower, less wide pale panel and is more restricted to the inner wing, whereas on Sooty the pale panel is broader and extends fully out onto the primary coverts. Differing light conditions can make assessment of this feature impossible. Short-tailed has a more buoyant flight style than Sooty due to is lighter weight and lower wing loading. Told from other dark shearwaters by overall shape, size, and plumage pattern and from dark-morph Northern Fulmar by smaller size, narrower wings, and thinner black bill.

Status & Distribution Common off AK in summer, less common farther south in late fall and winter. **BREEDING:** Nests colonially on islands of southern Australia. **NON-**

BREEDING: Moves north during the austral winter (our summer) to waters off AK. Most abundant off the AK Peninsula, the Aleutians, and in the Bering Sea where often very large numbers present off St. Lawrence I. in late summer and fall (a few sometimes by late spring). Small numbers move farther north into the Chukchi Sea, a few reaching Nuvuk (Pt. Barrow); also some move farther south along the West Coast in late fall and winter, where they can be found regularly off BC to central CA. Scarcer in Southern CA and Baja California waters. Accidental FL (specimen off southwest FL, 7 July 2000); sight record off VA.

Population Still common but recent sharp substantial die-offs.

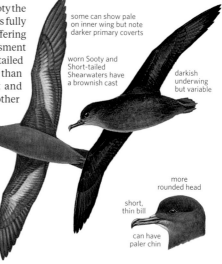

some can show pale on inner wing but note darker primary coverts

worn Sooty and Short-tailed Shearwaters have a brownish cast

darkish underwing but variable

more rounded head

short, thin bill

can have paler chin

SOOTY SHEARWATER *Ardenna grisea*　SOSH　▪ 1

The commonest dark shearwater in both oceans, but much more abundant in the Pacific, this transequatorial migrant reaches our waters during the austral winter. Spectacular concen-

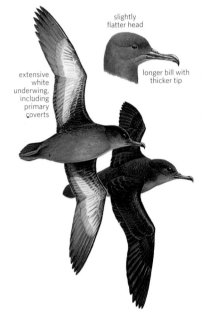

slightly flatter head

extensive white underwing, including primary coverts

longer bill with thicker tip

trations occur during mid-summer off West Coast, where hundreds of thousands were formerly seen from shore, fewer such concentrations in recent years. Monotypic. L 18" (46 cm) WS 40" (102 cm)

Identification Medium-size, dark shearwater with silvery underwings. Bill long and black. Underwing pattern distinctive, with pale silvery coverts and dark flight feathers. Lighting can affect the visibility of this character: In strong morning light, underwings can appear very white. **FLIGHT:** Varies with wind speed like all seabirds, but generally intersperses glides with bursts of clipped, stiff-winged flaps. Flapping often appears hurried. Arcs high over the water in strong winds when it can be confused with dark *Pterodroma* petrels.

Similar Species Most similar to the Short-tailed Shearwater; see that account. Told from other dark shearwaters and dark-morph Northern Fulmar by silvery underwing coverts and dark bill.

Status & Distribution Abundant in nearshore waters off West Coast; fairly common off East Coast. **BREEDING:**

Nests colonially on oceanic islands off southern S.A., Australia, and New Zealand. **NONBREEDING:** Undertakes long, trans-equatorial migration in both oceans moving north to spend the austral winter there. In North Pacific (Mar.–Oct., peak May–Aug.); rare winter. Far fewer in North Atlantic waters when most numerous in spring and early summer; numbers off ME and Atlantic Canada into early fall. **RARE STATUS:** Recorded eight times at the Salton Sea during summer; accidental southwestern AZ.

Population Near Threatened. Large global population (±20 million) is declining.

GREAT SHEARWATER *Ardenna gravis*　GRSH　▪ 1

This strikingly plumaged large shearwater of the Atlantic often attends boats, where it forages for refuse as well as for handouts. Monotypic. L 18" (46 cm) WS 44" (112 cm)

Identification Large and dark above; more neatly patterned than other Atlantic shearwaters. Brownish black upperwings; pale tips on the upperwing coverts and mantle create a scaled appearance. Secondaries and greater upperwing coverts washed pale gray when fresh. Dark tail, with

white uppertail coverts often forming a bold, U-shaped rump patch. Distinctive dark cap is set off by variable white collar; brownish mark on sides of neck frames prominent white on the side of the head; often prominent on distant birds, this is completely lacking on the Cory's Shearwater. Dark bill is long and thin. Clean white underparts, with gray belly patch (often hidden, lost in shadow, or occasionally absent; but diagnostic if seen). The pale-centered underwings are broadly framed in dark with variable dark markings on the axillaries and coverts. Does not molt in N.A. waters. **FLIGHT:** Although a large shearwater, the Great Shearwater flies with stiff wingbeats like the smaller *Puffinus* shearwaters and unlike the slow, languid wingbeats of Cory's Shearwater. Great Shearwater arcs and glides on stiff wings that are not

mottled underwing

distinct dark cap

variable white collar

dark belly smudge easily overlooked

white U-shaped band on upper-tail coverts

crooked back at the wrist, but held straight out, generally staying close to the water.

Similar Species Most similar to Black-capped Petrel, but it lacks the clean white underparts and dark ulnar bar as well as the white-tailed appearance

(when viewed dorsally) of that species. The black cap of Black-capped Petrel is variable, but usually more restricted than on Great Shearwater; and most Black-cappeds show a prominent white forehead and a broader white collar. At a distance, Great Shearwater often stays lower to the water than Black-capped Petrel and, noticeably, holds its wings more straight out. Great Shearwater is smaller and more boldly patterned than Cory's Shearwater, with a dark cap and pale collar. **Status & Distribution** Common off the Atlantic coast from May through Nov., a few into Dec. **BREEDING:** Nests on oceanic islands in the South Atlantic Ocean. **NONBREEDING:** Moves north to spend the austral winter in the North Atlantic, where large numbers congregate in late summer. **RARE STATUS:** Casual inland after hurricanes. Casual off West Coast north to AK at all seasons, mostly in fall. Casual to Gulf of Mexico.
Population Large global population estimated at 10 million.

PINK-FOOTED SHEARWATER *Ardenna creatopus*   PFSH ▪ 1

Pink-footed Shearwater is a large, pale-bellied species of offshore Pacific waters and is easily seen on pelagic trips from late spring through fall. Monotypic. L 19" (48 cm) WS 43" (109 cm)
Identification Large; lacks strong plumage contrasts; appears mostly grayish above and pale below. Heavy pink bill with dusky tip; legs and feet also pink, often visible against the dark undertail coverts in flight. Upperwing coloration ranges from gray to brownish, with odd white patches visible when molting. Underparts and underwings are quite variable: some authors consider it polymorphic, with light, dark, and intermediate morphs. Head generally dusky gray-brown, occasionally wholly dark on the darkest individuals; typically has a whitish throat. Whitish underparts, often dusky tinged, typically not appearing bright white below (beware strong sunlight); palest on belly. Underwing variable, but consistently shows dark flight feathers and pale underwing coverts, with variable dark markings on the axillaries. Dark individuals can have extensive grayish brown flanks and undertail coverts. **FLIGHT:** Generally flaps with slower, more labored wingbeats than other shearwaters, arcing and gliding high off the water in

strong winds. Wings are usually angled back slightly at the wrist, showing a long-handed appearance unlike most other *Ardenna* shearwaters.
Similar Species Black-vented is similarly colored and patterned, but is much smaller and flies with much faster wingbeats and shorter glides, and has a narrower dark bill. At a distance, darkest individuals can be confused with Flesh-footed, but dark Pink-footeds are grayer and show some pale on throat and belly.
Status & Distribution Fairly common in offshore Pacific waters from spring through fall (a few to southeastern AK), rare in winter. **BREEDING:** Nest colonially on oceanic islands off Chile. **NONBREEDING:** Moves north from breeding areas to spend the austral winter in N.A. Pacific waters.
Population Vulnerable. Small population of about 40,000 is threatened at breeding sites by introduced mammalian predators.

slow, lumbering flight

dark

light

uniform dark upperparts

typical

darker underwing than Buller's but variable

in molt

dark undertail coverts

dark

pink base to bill

FLESH-FOOTED SHEARWATER *Ardenna carneipes*   FFSH ▪ 3

This large, all-dark shearwater is rare in fall where it is often encountered with other shearwaters. Monotypic. L 17" (43 cm) WS 41" (104 cm)
Identification All-dark plumage, slow-flapping, languid wingbeats, and pale bill are distinctive. Upperparts chocolate brown, lacking contrast. Underparts and underwing coverts dark brown; primaries flash silvery in strong sunlight (compare to the Sooty Shearwater). Stout pink to pinkish horn bill with dark tip. Pink legs and feet, visible against the dark undertail coverts in flight. **FLIGHT:** Flaps slowly and deliberately, lacking the hurried wingbeats of most other shearwaters. Arcs and glides high over sea surface in strong winds, with wings bent back slightly at the wrist.
Similar Species Size and flight style similar to Pink-footed Shearwater. Larger than Sooty and Short-tailed, lacking pale underwings, and with a

thick, pinkish-based bill. Dark-morph Northern Fulmar is grayer-bodied has thicker all-yellowish bill. Dark-morph Wedge-tailed is more delicately built and longer tailed, with a thinner dark bill, a wedge-shaped tail, and dark flight feathers. Also compare carefully to Parkinson's and White-chinned Petrels and to first-winter Heermann's Gulls.

Status & Distribution Rare (sometimes uncommon) off the West Coast. **BREEDING:** Nests on oceanic islands off Australia and New Zealand and in the Indian Ocean. **NONBREEDING:** In the Pacific, migrates north across the Equator. Small numbers move east to the eastern Pacific, occurring from Gulf of Alaska to Baja California (Aug.–Nov., very rarely through spring and into summer); rare south of Pt. Conception, CA.

Population Near Threatened. Impacted at breeding sites by habitat loss and introduced mammalian predators.

Heermann's Gull
1st winter

bill color somewhat variable; very pale pink on some

silvery flight feathers contrast with blackish wing linings

pink bill with dark tip

pink feet often hard to see

SHEARWATERS Genus *Puffinus*

This genus now comprises 15 smaller species; seven former members are now in genus *Ardenna*. Five *Puffinus* occur in N.A. waters; Manx is the only breeding representative. The other four species visit offshore waters during migration or dispersal. For identification, note head and neck pattern, tail length, color of undertail coverts, underwing pattern, and bill size.

MANX SHEARWATER *Puffinus puffinus*   MASH ■ 2

This is one of a group of small black-and-white shearwaters that are confusing to identify and were once considered subspecies. Monotypic. L 13.5" (34 cm) WS 33" (84 cm)

Identification Small, stocky build; sharply demarcated black-and-white plumage. Upperwings typically blackish, occasionally fade browner. Distinctive head pattern: eye encompassed by dark cap, contrasting sharply with white throat and a pale area that wraps up behind the ear coverts (the pale "ear surround"). Clean white underparts. Underwing variable, often pure white coverts contrast with blackish flight feathers; occasionally a dark intrusion across coverts connects with axillaries. **FLIGHT:** Wingbeats stiff, hurried; in strong winds arcs high above water. Wings held straight out with little or no bend at the wrist.

Similar Species In the western Atlantic, most similar to Audubon's Shearwater. Note Audubon's quicker wingbeats, longer tail, and whiter face; most have dark undertail coverts. Manx in the eastern Pacific (where rare) resembles Black-vented Shearwater (see that species).

Status & Distribution Regular off the Atlantic coast (sometimes seen from shore off northern New England and Atlantic Canada). Most regular from NL to MA. Rare off the West Coast. **BREEDING:** Oceanic islands in the North Atlantic; in N.A. only on islands off NL and MA. **NONBREEDING:** In winter (Sept.–May) most move south to waters off the east coast of S.A. **RARE STATUS:** Rare from mid-Atlantic to NC; very rare to FL, mostly in late fall and winter. Rare off the West Coast (e.g., 100+ records for CA); breeding not yet confirmed in Pacific. Casual inland to the Great Lakes; accidental western MT (June 2004).

Population Large global population of ±740,000.

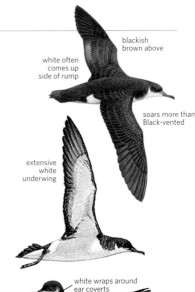
blackish brown above

white often comes up side of rump

soars more than Black-vented

extensive white underwing

white wraps around ear coverts

white undertail coverts

BLACK-VENTED SHEARWATER *Puffinus opisthomelas* BVSH ■ 2

This small shearwater occurs during the fall and winter months in the nearshore waters off Southern CA. Large flocks are routinely seen from shore. Monotypic. L 14" (36 cm) WS 34" (86 cm)

Identification Dingy; stocky build. Dark undertail coverts; individuals with white undertail coverts have been recorded. Upperwings brownish; fade substantially with wear. Head a dusky grayish brown, lacks strong contrast between dark cap and paler throat. Underparts variable: whitish, variably washed on flanks with dusky gray-brown. Ages and sexes alike. **FLIGHT:** Flaps with stiff hurried wingbeats, wings held straight out with little bend at the wrist. In strong winds, arcs high over the water.

Similar Species Distant birds resemble Pink-footed Shearwater, but fly with much faster wingbeats and shorter glides. Cautiously separated from much rarer Manx: Note Manx's cleaner white appearance below, more contrasting black-and-white plumage, white undertail coverts, and pale "ear-surround"; flight style slightly more languid with more soaring. Beware that harsh lighting can make Black-vented appear more black-and-white. The accidental Newell's Shearwater is very similar, but has much more contrasting, black-and-white plumage.

Status & Distribution Common within limited West Coast range. **BREEDING:** Oceanic islands off Baja California with 95 percent on Isla Natividad. **NONBREEDING:** Disperses primarily north after breeding; large numbers spend the winter (Aug.–Mar.) off Southern CA; present year-round off San Diego; in occasional years, some to large numbers to northern CA, even southern OR; casual north to BC.

Population Near Threatened. Declining global population (±75,000) with restricted breeding range.

labels: brownish above; flies with rapid flaps and brief glides; underwing darker than Manx; typical; dark undertail coverts; face pattern more blended than Manx; light; extent of dark on underwing variable; dark; dark bill; dark; dark undertail coverts

NEWELL'S SHEARWATER *Puffinus newelli* NESH ■ 5

The endangered Newell's Shearwater is a small tropical species that nests in HI. Monotypic. L 13.8–15" (35–38 cm) WS 30.3–33.5" (77–85 cm)

Identification A small shearwater, blackish above, white below, with white partly wrapping up toward the rump at the flank, just past the trailing edge of the wing. A small amount of white also extends from the throat to the rear of the auriculars.

Similar Species Newell's Shearwater was until recently treated as a subspecies of Townsend's Shearwater (*P. auricularis*). Townsend's (as now defined is unrecorded in N.A.) nests only on Socorro I. in the Revillagigedo Is., 370 mi off western Mexico and disperses to adjacent waters after breeding; it is Critically Endangered with about 100 pairs remaining (2008). Newell's averages larger (13.8–15" versus 12.6–13.8"). In Townsend's, the

undertail coverts are blackish, whereas Newell's has white undertail coverts with blackish tips. Newell's also shows a cleaner, more distinct division of black from white plumage at the neck and breast. Manx Shearwater is larger and more heavyset, with short tail and long white undertail coverts that nearly extend to the tail tip. The longer-billed Black-vented shows diffuse border of dark/light plumage in the head and neck, is browner above, and has extensively dark undertail coverts.

Status & Distribution **BREEDING:** Breeds only in the main Hawaiian Is., with about 90 percent nesting in the mountains of Kauai (May–Oct.). Post-breeding dispersal at sea mostly south and east of the nesting areas, possibly as far as Clipperton I., Mexico. **RARE STATUS:** Accidental coastal Southern CA: One wrecked on land at Del Mar, San Diego Co., CA (1 Aug. 2007).

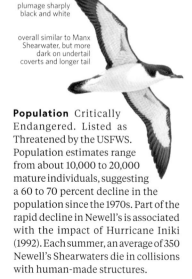

labels: plumage sharply black and white; overall similar to Manx Shearwater, but more dark on undertail coverts and longer tail

Population Critically Endangered. Listed as Threatened by the USFWS. Population estimates range from about 10,000 to 20,000 mature individuals, suggesting a 60 to 70 percent decline in the population since the 1970s. Part of the rapid decline in Newell's is associated with the impact of Hurricane Iniki (1992). Each summer, an average of 350 Newell's Shearwaters die in collisions with human-made structures.

AUDUBON'S SHEARWATER   *Puffinus lherminieri*   AUSH ■ 1

This small shearwater is found in the Caribbean and Gulf Stream. Polytypic (2 sspp.; both in N.A.). L 12" (30 cm) WS 27" (69 cm)

Identification Small and lightly built with a long tail. Blackish brown upperparts; worn plumage is much browner. Head variable, but typically dark capped; sharply demarcated from the white cheek and throat, bordered behind with a dark half-collar; some show white completely encircling the eye. Underparts bright white, including underwings; occasionally with variable dark bar across underwing coverts to axillaries. Long tail is dark. Undertail coverts vary from solidly blackish to extensively white with black tips. Medium-length bill is

thin and dark. Legs and feet typically pink, but can be blue or a mix of both. **FLIGHT:** Flight is low and fluttering with buoyant glides on slightly arched wings.

Geographic Variation Taxonomy is in a state of flux. The nominate subspecies from the Caribbean is the most likely to be encountered based on proximity. The status of the smaller *loyemilleri*, which nests in the Caribbean off Panama, is uncertain. Subspecific identification at sea is impossible due to lack of identification criteria.

Similar Species Most similar to Manx. Note Audubon's brownish cast to upperparts, longer tail, dark undertail coverts, and lack of pale tips to secondary coverts. Also similar to Barolo Shearwater, but larger, with darker face, larger bill, longer tail, and dark undertail coverts. In the Pacific, Galápagos Shearwater (*P. subalaris*)—a potential stray—is smaller and darker, and lacks the half-collar (recorded north to off southwestern Mexico).

Status & Distribution Common in warm waters (mainly in the Gulf Stream) typically along temperature breaks and around weed mats. **BREEDING:** Nearest populations nest on islands in the Caribbean. **NONBREED-**

all figures
lherminieri

short wings

dark undertail coverts

longer tail than Manx

pinkish feet

often feeds around sargassum

at rest, tail extends beyond folded wing tips

ING: Caribbean breeders disperse north into the Gulf Stream (mid-Apr.–Sept.) north to VA; small numbers in late summer, north to MA. Uncommon in Gulf of Mexico. Small numbers regular off eastern FL in winter. **RARE STATUS:** Casual inland after hurricanes (exceptionally to western KY).

Population Numbers off the East Coast have declined.

BAROLO SHEARWATER   *Puffinus assimilis*   BASH ■ 5

Until recently Barolo Shearwater, a casual visitor off the East Coast, was considered a subspecies of Little Shearwater (*P. assimilis*), but the current taxonomy may undergo further revision. Monotypic. L 11" (28 cm) WS 25" (64 cm)

Identification Very small shearwater with relatively short and slightly rounded wings. Upperparts are largely blackish, but whitish tips to upperwing secondary coverts (barely visible when

worn). Dark-capped head; bold white face, eye completely encircled with white. Tiny, all-dark bill. White underparts, including underwing coverts and undertail coverts. Grayish blue legs and feet. **FLIGHT:** Flies low to the water on hurried wingbeats and short glides; sometimes raises head upward at the end of glides.

Similar Species Most similar to Audubon's Shearwater but Barolo is smaller, more compact, with shorter, more rounded wings, and, importantly, has a white face and white undertail coverts. The treatment of the taxon *boydi* (breeds on Cape Verde Is.; unrecorded N.A.) is unresolved. Some authorities treat it as a subspecies of Audubon's; others treat it as a subspecies of Barolo; and some treat it as a full species ("Boyd's Shearwater"). Birds of the taxon *boydi* have a dark face and lack the white eye-surround of Barolo and are unlikely to be separable from the larger Audubon's in the field.

Status & Distribution Casual or very

flies with stiff, shallow wingbeats

white extends up to sides of rump

white extends prominently over eye

short bill

bluish feet; white undertail coverts

white tips to greater and median coverts (when fresh)

white extends onto primaries

wing tips fall about even with tail tip

rare to N.A. **BREEDING:** Breeds on islands in the eastern North Atlantic, from the Azores east to the Canary Is. **NONBREEDING:** Poorly known; some likely sedentary, others dispersing to surrounding waters. **RARE STATUS:**

Casual. Recorded more than a dozen times in N.A. (several other unsubstantiated reports); most recently from off Sable I., NS, and well off MA. Likely will prove to be annual in late summer and fall in the Gulf Stream, from MA to well off the Maritimes.

Population Small global population (±6,000) is declining. Probably best considered as Endangered, but lumped with Little Shearwater by BirdLife International.

STORKS Family Ciconiidae

Wood Stork (FL, May)

S torks superficially resemble other wading birds (especially ibises), but are larger and bulkier.
Structure Storks are large to huge birds, with long necks and legs and heavy bills that may be straight, slightly upturned, or decurved. The tails are short and the wings long and rounded.

Plumage Nearly all Ciconiidae species have a combination of white and black feathering. Many have featherless faces or heads. Soft parts are reddish or orange in most species.

Behavior Most species are colonial; the Jabiru and a few other species nest solitarily or in small groups. Tactile foragers, storks stand or walk slowly in marshes or lakes with their bills open; the mandibles close instantly upon contact with prey. They feed on a wide variety of aquatic animals, primarily fish; some Old World species also scavenge carrion. Unlike wading birds, but similar to cranes, storks fly with the head and neck fully extended. They soar to great heights on thermals, allowing them to efficiently travel great distances from nesting colonies.

Distribution Found on all continents except Antarctica. Most are in the Old World, especially Africa and Asia. Only the Wood Stork and Jabiru occur in North America.

Taxonomy Once thought to be closely related to New World vultures (Cathartidae) as well as other long-legged waders such as herons, storks are now placed in the revamped order Ciconiiformes with a new position following the tube-nosed swimmers (Procellariiformes). Of the 19 species of storks (in six genera), only three occur in the New World and only two are found in North America, including the Jabiru, a casual visitor.

Conservation BirdLife International codes: 2 NT, 2 VU, and 4 EN.

Genus *Jabiru*

JABIRU *Jabiru mycteria*   JABI   ● 4

This huge stork, found in tropical freshwater wetlands from southern Mexico to S.A., is a casual visitor to the southern US. Monotypic. L 52" (132 cm) WS 90" (229 cm)

Identification Unmistakable, with a massive, upturned bill. **ADULT:** Sexes similar. Plumage entirely white, including wings and tail. Unfeathered head and neck black, with conspicuous red band at base of neck. **JUVENILE:** Upperparts whitish, edged grayish brown, heaviest on wing coverts. Inner webs of primaries pale brown. **IMMATURE:** Scattered brownish feathers on upperparts, develops red patch on lower neck.

Similar Species Smaller Wood Stork has much smaller, slightly decurved bill, entirely black neck.

Voice Generally silent.

Status & Distribution Uncommon. Native from southeastern Mexico to southern S.A. **BREEDING:** Solitary but sometimes in loose colonies; nest usually in crown of tall palm or in mangroves. **DISPERSAL:** Nonmigratory, but prone to summer-fall postbreeding dispersal. **RARE STATUS:** Casual stray to northern Mexico (Veracruz) and TX (±8 reports); accidental to OK, LA, MS.

Population Near threatened in C.A. but more numerous in S.A.

large, upturned black bill

juvenile

red lower throat patch

breeding adult

all-white feathers

Genus *Mycteria*

WOOD STORK *Mycteria americana*   WOST ▪ 1

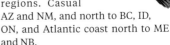
extensive black on flight feathers

unfeathered head and neck

adult

The Wood Stork is the largest wading bird regularly encountered in N.A. Flocks forage in freshwater with the bill held open to obtain prey, mainly fish; also frequents garbage dumps. Monotypic. L 40" (102 cm) WS 61" (155 cm)

Identification ADULT: Large white wading bird with black flight feathers. The head and upper neck are featherless, scaly, and dark gray. Dark bill is wide at the base and slightly decurved. Legs are dark; feet pale pinkish. **BREEDING ADULT:** The feet turn pink. Develops plume-like undertail coverts. **IMMATURE:** Adult plumage attained in fourth year. Neck and most of head covered with grayish feathering, lost by the second year. Head and upper neck become scaly. Straw-colored bill becomes blackish. **FLIGHT:** Black flight feathers contrast strikingly with white underparts and wing linings; neck extended.

Similar Species Unmistakable. In flight, wing pattern similar to American White Pelican but note head and neck coloration and long projecting legs.

Voice Silent except at nest.

Status & Distribution Locally common. **BREEDING:** Colonially, typically nesting in tall trees. Locally from coastal SC and GA throughout the FL Peninsula. Also in the American tropics and West Indies. **DISPERSAL:** Partly resident, but significant postbreeding dispersal north to southeast OK and southeast NC. **RARE STATUS:** Formerly moderate numbers to south end of Salton Sea until 2010 and coastal Southern CA until about 1960; now casual both regions. Casual AZ and NM, and north to BC, ID, ON, and Atlantic coast north to ME and NB.

Population Listed as Threatened by the USFWS. The US population of about 8,000 pairs is slowly recovering from a low of about 5,000 pairs in the late 1970s. Colonies extirpated from TX, LA, and AL, and much reduced in western Mexico.

grayish brown wash to neck feathers

adult

yellow feet

yellowish bill

unfeathered blackish gray head and neck

long, heavy decurved bill

juvenile

FRIGATEBIRDS Family Fregatidae

Magnificent Frigatebird, male (FL, Apr.)

Consummate aerialists, frigatebirds are lightly built tropical seabirds with long, angular wings and distinctive, long, deeply forked tails. Only Magnificent Frigatebird is found regularly in North America; two others occur as accidentals. Identification is fraught with difficulties.

Structure Light wing loading allows for long periods of effortless flight. The tail is deeply forked, and the small legs and feet are useful only for perching. The long bill is deeply hooked at the tip; the male's gular pouch is bare and red, and can be greatly inflated during courtship.

Behavior Soars great distances over warm waters, occasionally flapping with deep, slow wingbeats; does not alight on the water, but selects a perch such as rocks, shrubs, rigging, and guy wires. Frigatebirds pluck fish, squid, or offal from near the water's surface, often attending seabird feeding-flocks or concentrating around fish cleaning operations. They frequently pirate prey from other seabirds. Nests colonially in shrubs or mangroves on oceanic islands, or coastal bays.

Plumage Adult males are mostly black; females and juveniles show more white. Plumage transitions to adult, taking three or more years, are complex and variable.

Distribution Frigatebirds are found throughout tropical waters, but show great potential for long-distance movements well beyond their normal ranges, especially after tropical disturbances.

Taxonomy A uniform group of five species in one genus. The family is now placed in the recently created order Suliformes, which also includes the boobies and gannets (Sulidae), cormorants (Phalacrocoracidae), and darters (Anhingidae).

Conservation Introduced exotic species and human disturbance to nesting colonies have caused strong declines in many populations. BirdLife International codes: 1 VU and 1 CR.

Genus *Fregata*

MAGNIFICENT FRIGATEBIRD *Fregata magnificens*   MAFR   1

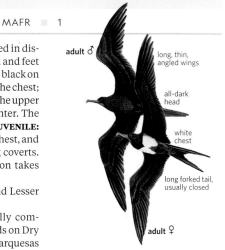

adult ♂

long, thin, angled wings

all-dark head

white chest

long forked tail, usually closed

adult ♀

Our only expected frigatebird, the Magnificent is numerous in southern and central FL, but occurs widely along the Gulf Coast and more rarely on the Atlantic and Pacific coasts and in the interior. Monotypic. L 40" (102 cm) WS 90" (229 cm)
Identification ADULT MALE: Entirely blackish overall, with a slight purple

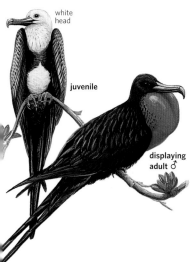

white head

juvenile

displaying adult ♂

gloss; the gular pouch (inflated in display) is orange to red. The bill and feet are gray. **ADULT FEMALE:** The black on the head comes to a point on the chest; the lower breast is white and the upper belly is constricted in the center. The bill is gray, the feet are pink. **JUVENILE:** Blackish, with a white head, chest, and belly patch, and brown wing coverts. Complex plumage maturation takes four to six years.
Similar Species See Great and Lesser Frigatebirds.
Status & Distribution Locally common. **BREEDING:** In US, breeds on Dry Tortugas and formerly on Marquesas Keys, FL. **DISPERSAL:** Small numbers along the entire US Gulf of Mexico coast. **RARE STATUS:** Rare but regular, mainly June–Sept., on Southern CA coast and inland at Salton Sea; fewer in recent years. Very rare on Atlantic coast north to NC. Casual or accidental north on coasts to southern AK, NL, and through interior of continent.
Population About 200 pairs on the Dry Tortugas; large colonies exist in western Mexico.

GREAT FRIGATEBIRD *Fregata minor*   GREF   5

Great Frigatebird has been recorded four times in N.A. Polytypic (5 sspp.; N.A. sspp. unknown). L 37" (95 cm) WS 85" (216 cm)
Identification ADULT MALE: From Magnificent Frigatebird, note rusty brown bar on the upperwing coverts, pinkish feet, and (often) whitish scallops on axillaries. **ADULT FEMALE:** Pale throat is gray, not black; has red orbital ring and a rounder (less tapered) black belly patch. **JUVENILE:** Rust wash on head and chest, at least

in fresh plumage, and pink feet.
Status & Distribution Accidental. **BREEDING:** Extensive range in Pacific and Indian Oceans; large colonies in northwestern Hawaiian Is. and in east Pacific from off western Mexico to Ecuador. West Atlantic subspecies (*nicolli*) on Trinidade and Martín Vaz Is. is nearly extinct. **RARE STATUS:** One specimen from OK (3 Nov.

adult female most distinctive with dark head, white throat, and red orbital ring

all plumages similar to Magnificent Frigatebird

adult ♀

1975); adults have been photographed in CA in Mar. (Farallones) and Oct. (Monterey Bay) and a sub-adult was photographed (Monterey Bay) in early Nov.
Population Nearest source colonies to N.A. include up to 10,000 pairs in Hawaiian Archipelago and about 165 pairs on Revillagigedo Is. off western Mexico.

LESSER FRIGATEBIRD *Fregata ariel*   LEFR   5

The smallest frigatebird, Lesser Frigatebird has been recorded four times in N.A. Polytypic (3 sspp.; N.A. sspp. unknown). L 30" (76 cm) WS 73" (185 cm)
Identification In all plumages, white extends from the sides into the axillars. **ADULT MALE:** Differs from the Magnificent Frigatebird by its much smaller size and the white flank patch that extends into the axillaries. **ADULT FEMALE:** Black head and chest (like

Magnificent Frigatebird), but the white breast sides extend into the axillars. **JUVENILE:** Pale rusty head; a white axillar spur is present.
Status & Distribution
BREEDING: Nearest breeding colonies are in South Atlantic at Trinidade I. and Martín Vaz I., east of Brazil (*trinitatis*). **NONBREEDING:**

white spur extends from flanks into axillaries

adult ♂

smallest frigatebird

Widespread in southwest and central Pacific and Indian Oceans. Regular north to western Pacific. **RARE STATUS:** Accidental: four widely scattered records, three in July (ME, WY, CA) and one in Sept. (MI). Casual north to southern coastal Russian Far East.
Population Stable.

BOOBIES AND GANNETS Family Sulidae

Northern Gannet (NL, July)

Boobies and gannets are supreme plunge-divers, entering the water from great heights to catch prey up to 30 feet underwater. Flight silhouette is distinctive, with long pointed wings and a body pointed both fore and aft. All are generally pelagic in our area, but gannets are readily seen from shore, as is the occasional booby. For field identification note especially the distribution of dark and light in the plumage, foot and bill color, and overall size. At least some plumages of each species can resemble one or more other species.

Structure Sulids are very large (gannets) to medium-size seabirds with long, pointed wings set well back on the body, wedge-shaped tails, and moderately long, thick necks; the pointed bills lack a hook on the tip and have serrated cutting edges. All sulids have thick, short legs and fully webbed feet. The bare gular pouch, orbital ring and loral area, and the forward-directed eyes give the birds a distinctive countenance.

Behavior The steady flap-and-glide flight may be low over water or (especially gannets) well above the water surface. The plunge-diving foraging behavior is directed at fish and squid. Sulids ride buoyantly on the water and swim well. Masked and Blue-footed Boobies generally perch on beaches, rocks, and gravel bars; Brown and especially Red-footed Boobies perch on bare tree limbs, buoys, and ship's rigging (often riding long distances). Gannets usually roost at sea. In most boobies sexes differ in voice, with males (except Red-footed) giving wheezy whistles, and females giving quacking calls; usually silent at sea.

Plumage Most species as adults are largely white below and variably all-dark to white above. The primaries are dark; other flight feathers are variably dark to white. Juveniles are generally darker than adults; plumage maturation takes three years (up to six for gannets). Red-footed Booby has distinct color morphs. Bare parts are often brightly colored, especially in breeding birds, and may differ between sexes and ages. Gannets show more extensive feathering on face.

Distribution Boobies are found in tropical and subtropical waters worldwide; gannets occupy colder waters in the North Atlantic, southern Africa, and Australasia. Nonbreeders disperse to coastal or pelagic waters. Gannets are migratory.

Taxonomy The Sulidae is now placed in the recently created order Suliformes, which also includes frigatebirds (Fregatidae), cormorants (Phalacrocoracidae), and darters (Anhingadae). The family comprises three closely related genera, *Papasula* (the endangered Abbott's Booby of the Indian Ocean), *Sula* (seven species of typical boobies, five in North America), and *Morus* (three gannets, one in North America). Boobies show weak to moderate geographic variation; a subspecies of Masked Booby—the Nazca Booby—was accorded full-species status in 2000.

Conservation Nesting sulids have long been exploited for human food, and introduced predators may also decimate colonies. Gannets are largely protected and well recovered from past declines (except the endangered Cape Gannet of Africa), but few booby colonies enjoy effective protection. BirdLife International codes: 2 EN.

BOOBIES Genus *Sula*

MASKED BOOBY *Sula dactylatra*   MABO   ■   3

This large booby is the most highly pelagic, staying well at sea except when breeding. Polytypic (4–7 sspp.; 2 in N.A.). L 32" (81 cm) WS 62" (158 cm) **Identification** Large with a heavy bill and relatively short tail. Facial skin is bluish black; legs olive-yellow to blackish. **ADULT:** White with black trailing edge to wing, mostly black tail, black tips to scapulars; yellow bill. **JUVENILE:** Dark brown hood with white collar, back and upperwing coverts brown with pale edges, and underparts and most of wing linings white; bill gray to dull yellowish. **SUB-ADULT:** Replaces brown body feathering with white during its second and third years; bill shows yellow by second year.
Geographic Variation Subspecies differ slightly in size and bare part colors. Nominate *dactylatra*, occurring in the Caribbean, is smaller than the Pacific subspecies. Birds recorded in CA are likely *californica* (although central Pacific *personata* is very similar).
Similar Species Adults can suggest the much smaller, white-morph Red-footed Booby (see species) and larger Northern Gannet, which lacks black on its secondaries and tail, has a blue-gray bill, and shows a yellow wash on its head and neck. Near-adult gannets with all-black secondaries have been reported, resembling the pattern of an adult Masked, but they have had all-white scapulars and white in the tail. Juvenile Masked

can be confused with juvenile or first-winter Northern Gannet; however, note the gannet's darker underwing (only the axillars are white), lack of white collar, and feathered gular area. See Nazca Booby.
Status & Distribution Pantropical. Very rare breeder and uncommon to rare nonbreeding visitor in N.A. **BREEDING:** Small numbers have nested in Dry Tortugas Is., FL, since the mid-1980s; closest large colonies in the Caribbean are off the Yucatán Peninsula, with small colonies in the northern Caribbean. A small Pacific colony is located at Alijos Rocks west of Baja California; larger colonies are on the Revillagigedo Archipelago and especially Clipperton I. off southwestern Mexico. **NONBREEDING:** Regular at Dry Tortugas Is., with up to double figures in a day; rare visitor elsewhere off the Atlantic

coast north to Outer Banks, NC, and in the Gulf of Mexico west to TX. **RARE STATUS:** Over 50 records for CA, plus over 20 juvenile Masked/Nazca; accidental to OR. Casual on Atlantic coast north to MA.
Population Least Concern.

all-white wing linings

adult
dactylatra

mostly white underwing

adult
dactylatra

juvenile
dactylatra

black mask

subadult
dactylatra

dark extends to body

dull yellow-green bill

juvenile
dactylatra

subadult
dactylatra

white collar on most young birds

adult
dactylatra

NAZCA BOOBY *Sula granti*   NABO   ■   4

Formerly a subspecies of Masked Booby, this eastern Pacific endemic ranges north to Mexico. All Masked-type boobies along the West Coast

should be scrutinized for this species. Monotypic. L 32" (81 cm) WS 62" (158 cm)
Identification Resembles Masked Booby in all plumages. Best character is bill color. **ADULT:** Orange-pink bill and more orange (not yellow) iris. **JUVENILE:** Extent of plumage variation unknown. Juvenile Nazca averages paler dorsally than Masked (gray-brown, not dark chocolate brown) and is believed to mostly lack—or to have an inconspicuous—white collar (most Masked juveniles show a broad white neck collar). Identification features of juveniles and subadults need clarification.

most juveniles lack a pale collar

adult with orange-pink (not yellow) bill and more orange eye

juvenile

all plumages very similar to Masked Booby

adult

Similar Species Nazca Booby averages a shorter and thinner bill, shorter legs, and longer wings and tail; it often shows more extensive white at base of tail feathers. Adult Masked shows a yellow to yellow-green bill (not orange-pink) and a yellow (not orangish) iris. Younger immatures and many sub-adults cannot be identified to species. **Status & Distribution** Common in its limited Pacific range. First N.A. record on coast of Southern CA in 2013, now over 80 records reflecting invasions north in recent years, many of which were full adults. **BREEDING:** Main colonies are on the Galápagos and off western Colombia; about 200 breed on Clipperton I. far off southwestern Mexico, and a few breed on the Revillagigedo Archipelago and possibly Alijos Rock southwest of Baja. **DISPERSAL:** Generally inshore, north to west coast of Mexico (a few into the Gulf of California). **RARE STATUS:** In addition to the recent cavalcade of records off and along the coast of CA; single records from OR, BC, and south coastal AK. **Population** Global population of ±30,000 suspected to be in decline due to food shortages.

BLUE-FOOTED BOOBY *Sula nebouxii* BFBO ■ 4

This booby of the eastern tropical Pacific with sky blue feet is an inshore species. Polytypic (2 sspp.; nominate in N.A.). L 32" (81 cm) WS 62" (158 cm)
Identification Large, very long tailed, long-billed booby with dirty brown and white patterning and blue (adults) to blue-gray legs and feet. Central tail feathers mostly whitish. **ADULT:** Streaked head and neck diagnostic. Brownish upperparts with white barring; white patches on upper back and rump. Underwing with white linings divided by dark bar. Pale yellow eyes; dark gray bill; blackish facial skin. **JUVENILE:** Head and neck more solidly and extensively brown than in adult, whitish dorsal patches more limited. Grayish blue feet; dull grayish bill; gray-brown eyes.
Similar Species Compare juveniles: Juvenile Brown Booby is smaller and solidly dark above; dark brown breast normally contrasts sharply with slightly paler underparts; underwings lack dark median covert bar, tail all-dark. Juvenile Masked Booby usually shows a complete white collar, shorter all-dark tail, and more extensively white underwing. Juvenile Nazca has whiter underwing with white on primary coverts.
Status & Distribution Irregular visitor to N.A. **BREEDING:** Common on islets in Gulf of California south to the Galápagos Is. and Peru. **NONBREEDING:** Rare and irregular (mainly mid-July–mid-Oct.) to Salton Sea, with occasional larger irruptions, but absent many years. Casual to coastal and central CA, southern NV, and western AZ usually during irruption years. Two major flight years since 1977: a small irruption in 1990 (several hundred) and the largest ever recorded in 2013. **RARE STATUS:** Records north to OR, WA, and BC; and single records from NM and a long-staying bird in central TX.
Population Least Concern.

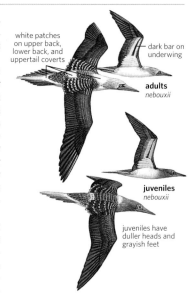

white patches on upper back, lower back, and uppertail coverts

dark bar on underwing

adults
nebouxii

juveniles
nebouxii

juveniles have duller heads and grayish feet

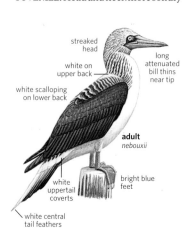

streaked head

white on upper back

white scalloping on lower back

long attenuated bill thins near tip

adult
nebouxii

white uppertail coverts

bright blue feet

white central tail feathers

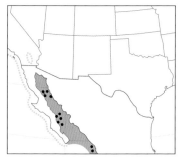

BROWN BOOBY *Sula leucogaster* BRBO ■ 3

The Brown Booby is found in tropical and subtropical waters and now occurs regularly in southern N.A. waters, mainly in southern FL and CA. Polytypic (4 sspp.; 2 in N.A.). L 30" (76 cm) WS 57" (145 cm)
Identification Bill and foot colors vary with population, age, and season (brighter at onset of breeding). The following refers to nominate *leucogaster*.
ADULT: Uniformly dark brown above and on neck, meeting bright white underparts in a sharp line across the lower breast. Axillars and underwing secondary coverts white. Legs and feet pale green to yellowish. On female, yellow to pinkish-colored bill, yellow facial skin, and a dark spot in front of the eye. On male, dusky yellowish bill, blue facial skin. **JUVENILE:** Dusky brown head, breast, upperparts; underparts dull whitish with dusky scalloping, but can appear all-dark from head to tail. Axillars and underwing secondary

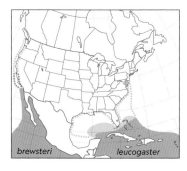

brewsteri

leucogaster

coverts dull whitish to mottled pale gray, but showing some contrast with remaining underwing. Bill bluish gray. **SUBADULT:** Increasingly white on belly, wing linings.

Geographic Variation Nominate *leucogaster* (Caribbean and tropical Atlantic) occurs in our southeastern waters. In *brewsteri* (Gulf of California to islands well off southwestern Mexico), adult males are distinctive with whitish frosting on the head and upper neck; juveniles are often very dark on the underparts.

Similar Species Compare juveniles: Juvenile Brown usually shows some contrast between the breast and slightly paler lower breast and belly, but it can be very subtle. A paler Red-footed Booby juvenile is mostly tan below with a dark band across the chest; a darker juvenile is more uniform, but it still shows a chest band. Red-footed's underwings are darker overall and more uniformly dark. Legs and feet can appear pinkish in both species, but head shape differs subtly; Brown has a more continuous contour from the culmen to the forehead, Red-footed has a slightly more concave culmen and more rounded forehead. Juvenile Masked and Nazca Boobies are larger than a Brown, have a dark hood limited to the head and neck, and show white on underwing primary coverts. Juvenile Northern Gannet is much larger, spotted with white, and has a dark underwing with only the axillars white.

Status & Distribution N.A. status has changed radically over the past decade, especially on the West Coast. **BREEDING:** Common in most tropical and subtropical seas including Gulf of California and parts of Caribbean. Recently established as a breeder on the Los Coronados Is. off northern Baja California and moderate numbers now resident on and around Santa Barbara I. off Southern CA. **NONBREEDING:** Found in small numbers off southern and central CA coast and on Dry Tortugas, FL. **RARE STATUS:** Casual along Pacific coast from northern CA to AK; also casual to Southwest. Very rare along Atlantic and Gulf Coasts (mostly July–Sept.) north to MA; casual to Canadian Maritimes. Also casual to Great Plains and the interior East.

Population Least Concern.

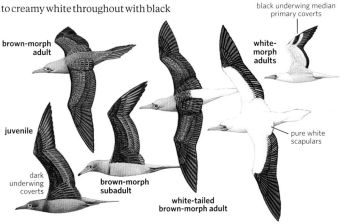

flatter head shape than Red-footed

blue facial skin on male

whitish head and neck

yellow facial skin on female

adult
leucogaster ♀

adult ♂
brewsteri

adult
leucogaster ♂

yellow legs and feet

sharply contrasting white belly and underwing

subadult
leucogaster ♀

under-wing pattern muted

juvenile
leucogaster

contrast between neck and belly can be slight to almost invisible

RED-FOOTED BOOBY *Sula sula*   RFBO   ■   4

This smallest booby is widespread in tropical oceans, but occurs only rarely in N.A. Adults show an array of color morphs but always have striking red feet. Polytypic (3 sspp.; 2 in N.A.). L 28" (71 cm) WS 60" (152 cm)

Identification Relatively slender with long, narrow wings; forehead more rounded than in other boobies; small bill is slightly concave along culmen. **ADULT:** Plumage variable. In all morphs, bill is pale blue, facial skin pink, orbital ring pale blue, and brilliant red feet. **WHITE MORPH:** White to creamy white throughout with black primaries and secondaries (innermost secondaries white), black median primary coverts on underwing, pure white scapulars; tail white or black (tipped white). **BROWN MORPH:** Brown throughout, but lighter tan on head and underparts, and darker dusky on flight feathers; underwings entirely dark. **WHITE-TAILED BROWN MORPH:** Similar, but rump, vent, tail coverts, and tail white. **JUVENILE:** Brownish throughout, but usually paler (even creamy tan) on head, neck, and underparts,

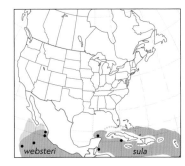

webeteri   sula

some bluish in lores

many with banded effect on breast

some pink in bill

gentle, round head

juvenile

dull pinkish feet

black-tailed white-morph adult

red feet

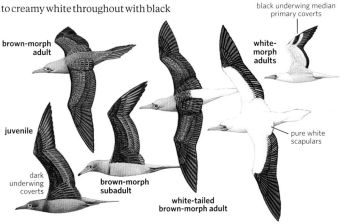

black underwing median primary coverts

white-morph adults

pure white scapulars

brown-morph adult

juvenile

dark underwing coverts

brown-morph subadult

white-tailed brown-morph adult

with dark chest band. Bill blackish, feet dull pinkish. **SUBADULT:** Approaches respective adult color morph; white-morph subadult often shows brown mantle (retained in some adults). Bill pink with dark tip, gradually changing to blue-gray; feet gradually brighten; adult plumage attained in third year. **Geographic Variation** Caribbean birds (nominate *sula*) and eastern Pacific birds (*websteri*) differ little; all-brown morph is rare in the Caribbean. **Similar Species** White-morph adult

with black tail resembles the much larger Masked and Nazca adults, but note Red-footed's blue-gray (not yellow or orange) bill, pink and pale blue facial skin, and bright red feet, and in flight, the black on trailing edge of wing of white-morph Red-footed does not reach the body; note black "comma" mark on the underwing primary coverts. See Brown Booby for juvenile differences.
Status & Distribution Pantropical distribution. Casual in our area. **BREEDING:**

Well-vegetated, tropical islets; nearest colonies off western Mexico and in the eastern Caribbean (off Yucatán). **RARE STATUS:** Casual but most regular on Dry Tortugas Is. and Gulf Stream waters off FL; recorded north to SC on Atlantic coast; one record off upper TX coast. Casual in CA (over 75 records), along the coast and well offshore; most recent records mainly Aug.–Nov., north to San Francisco area. Accidental in Sept. off Kenai Peninsula, AK.
Population Least Concern.

GANNETS Genus *Morus*

NORTHERN GANNET *Morus bassanus*   NOGA ■ 1

This is the only common sulid along the Atlantic coast. Readily seen from shore in winter and on migration. Flies in low lines or may soar high above the ocean and makes spectacular aerial feeding dives, wings folded back just before hitting the surface. Monotypic. L 37" (94 cm) WS 72" (180 cm)
Identification Larger and with longer wings than a booby. Adult plumage attained in about four years. **ADULT:** Mainly white with black primaries; golden buff wash on head. Pale gray bill, black facial skin, black feet with greenish lines. **JUVENILE:** Slate brown, spotted with white, with white upper-tail coverts; underwings dark with white axillars. **SUBADULT:** Dark body, wing coverts, secondaries, and rectrices gradually replaced by white, with much variation within age classes. Third-year birds often have a check-

ered or "piano key" look to the wings and a yellowish head. Central tail feathers and secondaries are usually the last dark feathers replaced.
Similar Species A distant immature flying low over the water can suggest a large shearwater or even an albatross. See Masked Booby for differences between immatures.
Status & Distribution Common. **BREEDING:** Colonially on islets and rocky cliffs, mainly Gulf of St. Lawrence and off NL. **MIGRATION:** Atlantic coast, mainly Mar.–May, Oct.–Dec. **WINTER:** Atlantic coast from NS to FL, and Gulf of Mexico, mainly Nov.–Apr. **RARE STATUS:** Casual on Great Lakes; accidental elsewhere in interior Northeast, Midwest, once (Sept.) on Victoria I., NT. Recent sight records off northern AK. One long-staying (from Apr. 2012

to 2020) adult on the Farallones and around San Francisco Bay region CA.
Population Least Concern.

juvenile
no pale collar
1st summer
2nd year
3rd year
juvenile
tawny head
dark bill
fine white speckles
adult
black primaries
adult
adult courtship display

CORMORANTS Family Phalacrocoracidae

Double-crested Cormorant, immatures (NJ, Jan.)

Cormorants (including shags) are long-necked diving birds represented in North America by the widespread Double-crested Cormorant and five more localized species. For identification, concentrate on shape and color of bill and gular region, overall size, tail length, neck thickness, flight silhouette, and distinctive markings of breeding adults.

Structure Cormorants have long necks, long, heavy bodies, stiff tails, and long bills strongly hooked at the tip. Their feet are fully webbed for underwater propulsion but serve also for perching on branches, cables, and cliff faces. **Behavior** Cormorants chase fish in shallows or dive for them in deeper water. The dense, wettable body plumage reduces buoyancy; after a diving bout cormorants perch with wings spread to dry. Flies with shallow wingbeats; briefly glides but rarely soars. Cormorants nest in small to very large colonies on sandy or rocky islands, cliff faces, or mangroves and other shrubs or trees. Grunting calls are usually heard only around the colonies.

Plumage North American species are mainly black as adults, often with a green or purple gloss. Some species show scaly patterning above; many extralimital species are white below. Cormorants variously show wispy plumes, crests, bright gular pouches, and white flank patches early in the breeding season. Sexes similar, juveniles and immatures are duller and browner than adults and lack ornaments. Adult plumage is attained in two to four years.

Distribution Cormorants frequent coastlines and interior lakes and wetlands worldwide. Some species are sedentary, others migratory. Four North American species are marine; two others also occur inland.

Taxonomy Divergent opinions exist on the recognition of species and genera of cormorants; the 31 to 39 species of cormorants worldwide (six in North America) are all placed in the genus *Phalacrocorax* by many current authors, but several additional genera are sometimes recognized.

Conservation Many cormorant species have declined, but North American populations are generally stable or expanding. BirdLife International codes: 3 NT, 7 VU, 3 EN, 1 CR, and 1 EX.

Genus *Phalacrocorax*

BRANDT'S CORMORANT *Phalacrocorax penicillatus*   BRAC ▪ 1

Common along the Pacific coast, this species is also the most numerous cormorant well offshore. Monotypic. L 35" (89 cm) WS 48" (122 cm)
Identification Fairly large with a short tail. Always shows pale buff on throat. **BREEDING ADULT:** Black throughout (with slight gloss) except for buff throat feathering; gular pouch bright blue; thin white plumes on neck, back. **NONBREEDING ADULT:** Similar to breeding, but no gloss or plumes and pouch dull gray. **JUVENILE:** Dark brown above; lighter brown below with band of pale buff on throat. Adult plumage attained third year. **FLIGHT:** Straight neck; low over water, often in flocks (see p. 323). **Similar Species** Double-crested Cormorant has a yellow to orange gular pouch and facial skin, deeper bill, and scaly upperparts; in flight, often well above water, shows thick neck with elevated head and longer, more pointed wings. Slimmer Pelagic has longer tail, smaller head, and thinner bill; juveniles are darker and all plumages lack buff throat patch.
Status & Distribution Common. **BREEDING:** Colonially on coastal cliff and offshore islands. **NONBREEDING:** General northward postbreeding movement July–Oct., and southward movement in fall and winter. **RARE STATUS:** Rare, mostly in winter to southeast AK. Accidental in spring from interior CA (Fresno Co. and Imperial Co.).
Population Overall numbers stable.

buffy band under bill in all plumages

juvenile

blue skin

winter adult

breeding adult

appears shorter tailed than Pelagic

NEOTROPIC CORMORANT *Phalacrocorax brasilianus*   NECO   ◼ 1

Resembling a small, long-tailed version of the Double-crested Cormorant, this aptly named, primarily Neotropical species is fairly common in N.A. along the TX and western LA coasts. Polytypic (2 sspp.; *mexicanus* in N.A.). L 26" (66 cm) WS 40" (102 cm)

Identification The combination of the bird's characteristic field marks—small size, relatively long tail, short bill, black feathered lores, and dull yellow gular pouch bordered by white feathering and distinctly pointed at the rear—identify Neotropic from Double-crested Cormorants, though the two species are most easily told when together. Easily separated in flight by proportional differences described above. **BREEDING ADULT:**

Blackish throughout with a slight bluish gloss; the back feathers and wing coverts are olive-gray with black borders. The gular pouch varies from brownish to yellow, but it is always bordered at the rear by a distinct band of white feathering coming to a point behind the gape. Short, thin white plumes are concentrated on the sides of the head and neck. **NONBREEDING ADULT:** Resembles the breeding bird, but lacks the white plumes, body plumage is less glossy, and gular pouch is a paler yellow with a slightly less distinct white border. **JUVENILE:** Brown throughout, often considerably paler on the head, neck, and breast. Gular pouch is a pale yellow; the white border is indistinct or absent. Older immatures gradually attain the dorsal patterning, blacker body plumage, and white throat border of adults.

Geographic Variation N.A. birds (*mexicanus*) are considerably smaller than nominate birds found ranging from Panama through S.A.

Similar Species To distinguish immature Neotropic from the closely similar Double-crested Cormorant, see sidebar below. The small and relatively long-tailed Pelagic Cormorant

is unlikely to be confused with Neotropic Cormorant, especially given the habitat and range differences.

wispy white plumes on face

long tail

adult
mexicanus

breeding adult
mexicanus

smaller bill than Double-crested

yellowish bill

dark orange gular with white border

juvenile
mexicanus

slender build

feathers more pointed

nonbreeding adult
mexicanus

long tail

Identification of Immature Double-crested and Neotropic Cormorants

The Double-crested Cormorant occurs widely, especially in winter, within the more limited US range of the Neotropic, and stray Neotropics are often seen with Double-cresteds—the two species have hybridized—so close comparison of the two species is often available.

Some characters that readily distinguish adults, such as the shape of the gular pouch, the distinct white throat border of Neotropic, the absence of bare yellow or orange supraloral skin in Neotropic, and the twin curled crests of Double-crested can be absent, obscure, or difficult to assess in juveniles and younger immatures. Thus close scrutiny of facial features in combination with careful attention to (and ideally comparison of) size and shape characters are important.

The yellowish gular pouch of Neotropic comes to a point behind the gape of the bill, with the throat feathering coming forward to a point under the pouch. In Double-crested the gular pouch is squared or rounded at the rear, with less feathering protruding forward underneath. This shape difference holds for all post-natal plumages. In all Double-cresteds there is a patch of yellow-orange facial skin in front of the eye and above the dark lores; this area is dark and feathered in the

Neotropic. (But beware: Some individuals show a thin yellowish streak above and in front of the eye.)

Double-crested Cormorants average larger than our *mexicanus* Neotropics in all standard measurements except tail length, but the smallest Double-crested can overlap Neotropic in bill length. Neotropics have slightly slimmer bodies and smaller heads. The most distinctive shape difference is the relatively much longer tail of Neotropic (the tail is 60 percent of the wing length in Neotropic, but less than 50 percent in Double-crested); in flight the tail length of Neotropic roughly equals the length of the neck (with head), whereas the tail is shorter than the neck in the Double-crested.

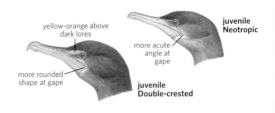

yellow-orange above dark lores

more acute angle at gape

juvenile Neotropic

more rounded shape at gape

juvenile Double-crested

Status & Distribution Common and widespread from Mexico through S.A. **BREEDING:** Fairly common resident in most of US range; found in coastal bays, inlets, and freshwater lakes and ponds. More local inland on reservoirs, lakes, and rivers; breeds in southern AZ, NM, southern OK, recently south FL (mostly mixed pairs with Double-crested), and southeast CA. Increasing, especially in Southwest. **RARE STATUS:** Rare to casual in Southern CA, away from the Colorado River and the Salton Sea, southern NV, UT, CO, Great Plains, the upper Midwest, southern ON, and on East Coast north to NJ, recently once to NH. Stray records are increasing rapidly.

Population After suffering severe declines in the 1960s, the populations in the US have largely rebounded, but there remains strong yearly variation in colony site use and population size.

DOUBLE-CRESTED CORMORANT *Phalacrocorax auritus*   DCCO ■ 1

The Double-crested is N.A.'s most widespread and familiar cormorant, found in both marine and freshwater habitats. It often flies higher than other cormorants, often in goose-like V formations. Polytypic (5 sspp.; 4 in N.A.). L 32" (81 cm) WS 52" (132 cm)

Identification Stocky, with medium-length tail, thick neck, and moderately thick bill; yellow to orange gular pouch, rounded or square at rear; bare supraloral skin yellow to orange; lores dark. **BREEDING ADULT:** Black; coverts and back feathers with grayish centers, black fringes. Facial skin orange; mouth lining bright blue. Nuptial crests black, white, or peppered. Bill blackish above, spotted with pale below. **NONBREEDING ADULT:** Similar to breeding adult, but crests lacking and facial skin often more yellow. **JUVENILE:** Dark gray-brown above with paler centers to back feathers and wing coverts; underparts color varies with individual and feather wear, but breast always paler than belly. Throat and breast dark brownish gray, becoming paler, even nearly white, with wear and fading (flanks and lower belly darker). Bill is extensively yellow to pale orange. Older immatures resemble nonbreeding adults but plumage paler, duller, and browner.

Geographic Variation Subspecies differ in overall size and in color and shape of crest in breeding adults. Nominate *auritus* (widespread from the Great Basin and Rocky Mtn. region east through central and eastern N.A.) is moderately large with black crests; *floridanus* (FL) is our smallest subspecies, with dark crests; *cincinnatus* (coastal AK to northern BC) is largest, with straight white crests; and *albociliatus* (Pacific coast from southern BC to western Mexico, and presumably inland to the Great Basin) is large and most show white to partially white crests. Subspecies range limits in interior western N.A. are not clear.

Similar Species Neotropic Cormorant is similar (see sidebar opposite) as are Great and Brandt's Cormorants (see those species). Very pale-breasted immatures might be mistaken at a distance for loons or, when perched, boobies.

Status & Distribution Common to abundant. **BREEDING:** A wide range of lakes, marshes, coastal estuaries, and offshore islands; colonially on rocky

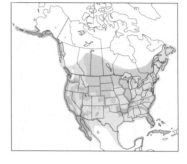

or sandy islands, marsh vegetation, waterside trees, or artificial structures. **MIGRATION:** Northern breeders move north mainly Mar.–Apr., return south mainly Aug.–Nov. **WINTER:** Mainly coastal habitats, but also on inland lakes, reservoirs, and fish farms. **RARE STATUS:** Some summer wandering north of regular range (e.g., Great Slave Lake). Very rare to central and northern Bering Sea region, including Chukotka, Russian Far East.

Population Currently there are about 350,000 breeding pairs in N.A. Most populations declined greatly from the 1800s through the early 1900s because of persecution and introduced predators. Major declines occurred again from the 1950s to about 1970 because of pesticide impacts. Numbers have strongly rebounded since the 1970s in the interior and Atlantic coast regions. Some populations are now controlled through management (especially in the Great Lakes region).

straight neck

Brandt's for comparison

juvenile

pale breast

yellow-orange stripe above dark loral stripe

orangish gular pouch

winter adult

thick kinked neck

shorter tail than Neotropic

Double-crested

dark belly

dark belly

wispy white crest

western breeding adult

some 1st years have white bordering gular pouch

mostly dark below

some 1st years have paler bellies

wispy black crest

breeding adult

2nd year

1st year

1st year

GREAT CORMORANT · *Phalacrocorax carbo* · GRCO ■ 1

This species, N.A.'s largest cormorant, has a huge worldwide range, but in N.A. it (all subspecies *carbo*) breeds only in easternmost Canada and ME; nonbreeders occur southward along and near the Atlantic coast in winter. Polytypic (7 sspp.; *carbo* in N.A.). L 36" (94 cm) WS 63" (160 cm)

Identification Large size, large blocky head, short tail, and limited yellow gular pouch bordered by white feathering. **BREEDING ADULT:** Black with scaly upperparts; broad white throat patch bordering small lemon yellow gular pouch and facial skin; thin white plumes on sides of head and bold white patch on flanks. **NONBREEDING ADULT:** Similar to breeding, but white throat patch more limited and white head plumes and flank patches

absent. **JUVENILE:** Gray-brown head, neck and breast contrast with white belly; flanks, wing linings, and upperparts dusky; pale feathered border to small yellow facial skin area. Older immatures progressively darker on body; definitive plumage attained in third year.

Similar Species Double-crested Cormorant (now wintering north into Great Cormorant's winter range), the only other cormorant in Great Cormorant's N.A. range, is smaller with a slightly slimmer neck and bill. Breeding adult Double-crested lacks Great Cormorant's white feathering on the face and flanks and has deeper, more orange, facial skin. A nonbreeding adult is more similar, but note that the Great Cormorant's limited pale yellow facial skin is pointed, not rounded at rear, and its broad pale

feathering bordering the facial skin points forward on throat toward bill. Most immature Great Cormorants are readily told from immature Double-crested Cormorants by their two-toned underparts: dark on the breast and flanks, whitish on the belly; a few Greats are more uniformly dark below and best told by facial characters (above) and grayish (not extensively orangish) bill. Also compare with much smaller and longer-tailed Neotropic Cormorant.

Status & Distribution Fairly common. **BREEDING:** Maritime habitats; nests colonially on ledges, plateaus on rocky islands, coastal points. **NONBREEDING:** Rare (immatures) south in summer to NJ. Seacoasts, bays, and lower portions of major rivers, and larger inland water bodies; south to NC. **RARE STATUS:** Very rare SC to FL. Casual well inland in Northeast as far as Lake Ontario. Casual along Gulf Coast (FL, MS).

Population In N.A. (*carbo*), only about 5,500 pairs breed, 80 percent of these in QC and NS. Declining and now much scarcer in southern part of winter range in N.A.

large blocky head

yellowish throat pouch

white

olive-bronze sheen on back

nonbreeding adult
carbo

long, heavy bill

white on throat more diffuse

dark neck

white neck feathers soon lost

dark throat pouch

breeding
carbo

white belly

juvenile
carbo

white flank patch

breeding adult
carbo

RED-FACED CORMORANT · *Phalacrocorax urile* · RFCO ■ 2

A close relative of the smaller Pelagic Cormorant, the Red-faced Cormorant is found only from southwest AK west through the Pribilofs and Aleutians to the northern Sea of Japan. It is not especially gregarious and nests in small colonies on cliff faces and steep rocky islands. Monotypic. L 31" (79 cm) WS 46" (117 cm)

Identification Extensive yellow to pale yellowish gray on bill in all plumages. A bit stockier in build

than Pelagic, with a slightly thicker bill. **BREEDING ADULT:** Glossy black with browner wings, large white flank patch; bill yellowish with blue base; bright red pouch and facial skin surrounding eyes and extending across forehead; thin white neck plumes. **NONBREEDING ADULT:** Similar to breeding, but body less glossy, red face and blue bill base duller, and white neck plumes and flank patch lacking. **JUVENILE:** Dark brown, slightly

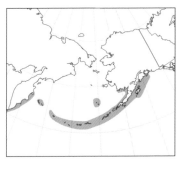

browner wings

all have thicker two-tone bill than Pelagics

winter adult

narrow pinkish eye ring

juvenile

extensive bright red on face

breeding adult

large white lower flank patch

paler on breast; facial skin dull yellow to pinkish (but forehead feathered); some yellow on bill.

Similar Species Closely resembles Pelagic Cormorant, but Red-faced is heavier with a stocky build, including a thicker neck and blockier head, apparent at rest and in flight; also slightly thicker bill. Male Pelagics in northern part of range can appear quite large. The Red-faced Cormorant's bill is extensively pale (usually yellow), whereas the bill of Pelagic Cormorant is slate gray or blackish. In adult Red-faced, the dark brown wings contrast with black body (Pelagic is more uniformly black). The facial skin of adult Red-faced Cormorant is a bright red and extends across forehead and broadly surrounds the eyes; in Pelagic, the red is deeper in color and limited to the lower face (fore-

head is feathered black). Juveniles are best told apart by the pale yellowish bill and face of the Red-faced (Pelagic is dark in these areas).

Status & Distribution Uncommon to fairly common year-round. **BREEDING:** Resident within range. Nests on sea cliffs with kittiwakes and alcids, commonly on the Pribilofs and Aleutians; some fall and winter dispersal within the general breeding range. It is less common and more local in the remainder of the AK range. **RARE STATUS:** Casual in AK north to Norton Sound and southeast to Sitka. Recorded twice (Apr., June) on Haida Gwaii, BC. Sightings in spring from St. Lawrence I. are not substantiated; some were clearly of Pelagic Cormorants, and all were likely that species. Well out-of-range claims of this species must be substantiated with photos.

Population About 50,000 birds breed in AK; populations are apparently stable, but little trend data exists.

PELAGIC CORMORANT *Phalacrocorax pelagicus* PECO ▪ 1

Despite its name, this small, slender cormorant frequents rocky coastlines and is rarely seen far at sea. Polytypic (2 sspp., both in N.A.). L 26" (66 cm) WS 39" (99 cm)

Identification Slender neck (held straight in flight), small head, very thin blackish bill, and relatively long tail. Plumage entirely dark, except for white flank patches of adults early in the breeding season. **BREEDING ADULT:** Shiny black, glossed purple and green, except for large white patch on flank and scattered thin white plumes on the neck. Short crests on crown, nape. Deep red facial skin limited to area around chin and eyes. **NONBREEDING ADULT:** Similar to breeding, but no flank patch, neck plumes, or crests. Body plumage slightly less

glossy; red facial skin less evident. **JUVENILE:** Blackish brown above, only slightly paler on underparts; our darkest juvenile cormorant. Adult plumage attained second year.

Geographic Variation Differences are slight. Nominate *pelagicus* breeds from southern BC through AK and northeast Asian range; *resplendens*, from southern BC to extreme northern Baja California, Mexico, is smaller, with a more slender bill.

Similar Species See Redfaced Cormorant. Along the Pacific coast, Pelagic Cormorant is confused with Brandt's; the latter is larger, with a heavier body and neck, larger head, thicker and longer bill, and shorter tail. Brandt's also shows a buff throat (dark and uniform with neck in all Pelagic plumages); Pelagic is not usually found at sea!

Status & Distribution Common, but rarely in large groups. Regularly forages up rivers and in bays well away from the immediate coast. **BREEDING:** Colonially on steep rocky slopes or cliff faces on coastlines and islands. **NONBREEDING:** Generally resident, but most of northernmost population withdraws south. Winters south to Pacific coast of Baja California. **RARE STATUS:** Casual on northern coast of

thin neck

longish tail

all have slender, dark bill

winter adult

juvenile

uniformly dark brown, including chin

limited dark red facial skin

breeding adult

white lower flank patch

AK and on leeward Hawaiian Is. Accidental east of Sierra Nevada at Silver Lake, Mono Co., CA (8 Dec. 1976).

Population The N.A. population of about 130,000 birds (mostly in AK) is generally stable.

DARTERS Family Anhingidae

Anhinga, breeding male (FL, Apr.)

Darters resemble slender, long-tailed cormorants. They are often seen loafing or sunning on water-side perches.

Structure Darters have a thin neck and small head with a pointed bill and small gular pouch; the heronlike neck is kinked (serving as a hinge for quick strikes at prey). The long broad wings, thin-based tail, and straight neck give flying darters a crosslike appearance. The short legs have large, fully webbed feet.

Behavior Highly aquatic, they swim with the body submerged; wettable plumage reduces buoyancy but requires long periods of drying and sunning with wings and tail spread. Flight consists of several flaps and short glides; darters often soar on thermals. They use their sharply pointed bills to spear fish and other prey. They nest in waterside trees and mangroves, often alongside herons, ibis, cormorants, and storks.

Plumage Adult males are largely black with silver-white markings on the upperwing coverts, back, and scapulars; in Old World taxa the head and neck have brown and white markings. Females and immature males are extensively gray-brown to buff on the head, neck, and breast.

Distribution Darters occur in warmer regions of the world from the southeastern US through tropical South America, sub-Saharan Africa, South Asia, and Australia/New Guinea. Some populations are migratory.

Taxonomy The darter family is now placed in the recently created order Suliformes, which also includes frigatebirds (Fregatidae), boobies and gannets (Sulidae), and their sister group, the cormorants (Phalacrocoracidae). The entire darter family consists of the Anhinga in the Americas and three more species in the Old World.

Conservation Populations are generally stable. BirdLife International codes: 1 NT.

Genus *Anhinga*

ANHINGA *Anhinga anhinga*   ANHI ■ 1

Anhingas inhabit warm southern wetlands. Polytypic (2 sspp.; *leucogaster* in N.A.). L 35" (89 cm) WS 45" (114 cm)

Identification Unique. Long, snake-like neck, small head, sharply pointed yellow bill, and long fan-shaped tail with pale tip. **ADULT MALE:** Black throughout with extensive silver-white panel on upperwing coverts and streaks on lance-like back and scapular feathers. Breeding birds show wispy white plumes on sides of head and neck, green lores, and orange gular pouch. **ADULT FEMALE:** Like male, but with grayish tan head, neck, and breast (richer cinnamon on lower breast). **JUVENILE:** Similar to adult female but duller, with reduced white on scapulars, coverts.

Similar Species A soaring Double-crested Cormorant is frequently confused with Anhinga; note soaring Anhinga's longer tail, straight neck, and different bill shape. Escaped African Darters have been seen in Southern CA; they lack a pale tail tip and have brown greater secondary coverts; males have rufous throat and neck and white stripe below cheek (females hint at this pattern).

Voice Generally silent away from nesting colony. Clicking and rattling calls sometimes given by perched birds.

Status & Distribution Fairly common. **BREEDING:** Wooded swamps, lakes, slow-moving rivers, and mangrove estuaries, with highest densities in FL, LA, southeast GA; formerly bred to OK, IL, MO, and KY. **NONBREEDING:** Northernmost populations withdraw south in Oct; return in Mar. Some northward postbreeding movement to southeast OK, central AR, and southeast VA. **RARE STATUS:** Casual to CA, AZ, NM, CO, Great Plains, and north to Great Lakes states, southern ON, and New England.

Population Least Concern.

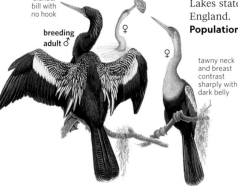

slender bill with no hook

breeding adult ♂

♀

♀

tawny neck and breast contrast sharply with dark belly

often soars

very long, square-ended tail

♀

PELICANS Family Pelecanidae

Brown Pelican (CA, Feb.)

Pelicans are instantly recognizable very large fish-eating birds with long, hooked bills and extensive gular pouches. In North America, the Brown Pelican inhabits subtropical and temperate coastlines and the American White Pelican breeds in interior western and central parts of the continent and winters along southern coastlines. Identification is straightforward in North America, less so in other parts of the world.

Structure Pelicans are huge waterbirds with long, broad wings, short tails, long necks, and long, hook-tipped bills. The gular pouch can greatly expand as the mandibles bow outward to form a large fishing "net" or scoop. The long bill rests on the foreneck in flight and often when at rest. The legs are short and stout, but the feet are fully webbed (totipalmate). The skeleton is lightweight, and pelicans ride buoyantly on the water. Despite such adaptations for weight reduction, American White Pelicans may weigh up to 20 pounds; the smallest Brown Pelicans (West Indian populations) only weigh about eight pounds.

Behavior Pelicans spend a lot of their time loafing on beaches, lakeshores, low islets or (Brown Pelican) piers, barges, rocky islets, and mangroves; more than 21 hours a day may be spent resting. They are highly gregarious, often feeding or loafing in groups of hundreds or even thousands; American White Pelicans breed in colonies on low islets on large lakes, while Brown Pelicans, a marine species, breed on rocky coastal islands, vegetated barrier islands, or mangroves. American White Pelicans feed by swimming, often in coordinated groups, and dipping for fish with the pouched bill. Brown Pelicans rarely feed this way; instead they plunge-dive from as high as 50 feet, hitting the water with a slight leftward rotation, and scoop up as much as 2 to 2.5 gallons of water, filling the pouch underwater. Mergansers, cormorants, and other fish-eating birds often accompany feeding flocks of White Pelicans, and Heermann's and Laughing Gulls frequently attend diving Brown Pelicans to steal captured fish. Pelicans have a distinctive flap-and-glide flight (a few deep flaps followed by a long glide is typical), often in lines or V's; they can soar for long periods, and migrating flocks of American White Pelicans are often seen high, soaring within thermals. Pelicans are generally silent, but they do hiss and grunt in breeding colonies.

Plumage Worldwide, six pelican species are predominantly white, with variably dark flight feathers; Brown and Peruvian Pelicans are darker bodied. Molts are prolonged and complex; definitive adult plumage is reached in about three years. Bare parts can be brightly colored, especially during courtship season.

Distribution Pelicans are found on interior lakes and coastlines over much of North America, southern Eurasia, Africa, Australasia, and the northern and western coasts of South America. The North American species differ in habitat; Brown Pelican is almost exclusively marine (it is the only pelican species normally seen at sea, though generally absent from waters more than about 100 mi from land). All species undergo postbreeding dispersal, and some are truly migratory.

Taxonomy The Pelecanidae family is placed in the recently reconfigured order Pelecaniformes; herons and ibises and spoonbills are now also included in this order, along with the curious African single-species families Balaenicipitidae (Shoebill) and Scopidae (Hamerkop). Worldwide the Pelecanidae family comprises a single genus (*Pelecanus*) with eight species. The Peruvian Pelican (*P. thagus*) has been split from the widespread smaller Brown Pelican.

Conservation The fish diet magnifies problems of organochlorine contamination of aquatic environments. Evidence of Brown Pelican reproductive failures owing to eggshell thinning mediated by the DDT metabolite DDE, providing a clear cause and effect, was a key motivation in the banning of DDT and related pesticides. Human disturbance of nesting colonies is also an important and persistent threat. BirdLife International codes: 3 NT.

Genus *Pelecanus*

AMERICAN WHITE PELICAN   *Pelecanus erythrorhynchos*   AWPE ■ 1

The huge American White Pelican has black primaries and outer secondaries and a yellow to orange bill. Breeding on interior lakes, this pelican is the only one likely to be seen away from seacoasts in most of N.A. Monotypic. L 62" (158 cm) WS 108" (274 cm)

Identification Unmistakable in all plumages. **ADULT:** Entirely white except for black primaries and outer secondaries. Bill, gular pouch, and bare facial skin orange-yellow, brightening to deep orange-pink in breeding birds; legs and feet yellowish, becoming orange in breeding birds. Early breeding season adults have pale yellowish plumes on crest and center of breast, and a fibrous knob up to two inches high two-thirds of the way out the culmen; after egg-laying a supplemental molt results in dusky gray feathering on crown and nape. Sexes similar. **JUVENILE:** Similar to adult but head, neck, and especially upperwing coverts marked with dusky gray; bill and feet duller yellow. Older immatures resemble adults, but lack yellowish plumes and supplemental crown markings.

Similar Species A distant flying bird could be confused with the smaller Wood Stork, also largely white with black on the flight feathers. Other white-plumaged pelicans could occur as escapes; of these only the Great White Pelican (*P. onocrotalus*) shares

the all-white body plumage, sharp black and white wing pattern, and yellow pouch. Great White Pelican is even larger, has black bill sides, and always lacks the bill knob.

Status & Distribution Common. **BREEDING:** Lakes in interior west and prairie regions; largest colonies in western and southern Canada. Small resident population in coastal TX. **MIGRATION:** Arrive breeding colonies Mar.–May; fall movements protracted. **WINTER:** Mainly from central CA and Gulf states and very locally along Atlantic Coast from NC south to southern Mexico on coastal bays and estuaries, but also on large inland lakes. Nonbreeders may summer well outside breeding range. **RARE STATUS:** Very rare in migration along Atlantic Coast from VA to Canadian Maritimes. Accidental in southeastern AK and far northern Canada.

Population The population is stable

or increasing since the 1960s. More than 20,000 pairs nest in the US, more than 50,000 in Canada. Now, much more frequent in East.

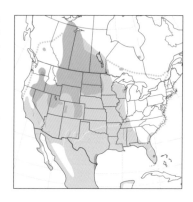

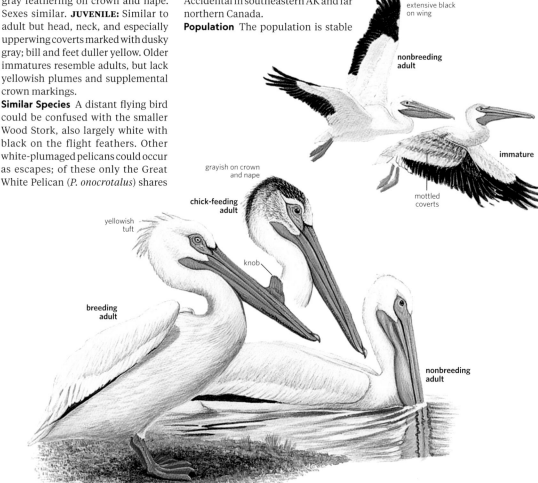

extensive black on wing

nonbreeding adult

immature

mottled coverts

grayish on crown and nape

chick-feeding adult

knob

yellowish tuft

breeding adult

nonbreeding adult

BROWN PELICAN *Pelecanus occidentalis* BRPE ■ 1

The small, dark Brown Pelican of marine coasts feeds mainly by plunge-diving. Although still considered endangered in parts of its range, it has generally rebounded well from severe declines caused by pesticide residues. Polytypic (5 sspp.; 3 in N.A.). L 48" (122 cm) WS 84" (213 cm)

Identification Diagnostic typical pelican form and extensively dark plumage. Sexes look similar, but age classes differ in plumage and body feather shape (narrower and more pointed in adults). Adults have white heads and dark bellies; juveniles the opposite. **BREEDING ADULT:** Upperparts appear gray (feathers slaty with silvery center streaks); underparts dark brown, streaked with silver on sides, flanks. Head and neck mainly white, with dark brown to chestnut hindneck stripe and foreneck (encircling yellow or whitish chest patch) and yellowish crown. Eyes white. Gular pouch dark greenish gray (bright red at base in courting *occidentalis*); bill with extensive pink or orange toward tip, and whitish area near base during chick-feeding stage. **NONBREEDING ADULT:** Similar, but entire neck whitish, eyes dark by late summer, bright colors on pouch and bill less evident. **JUVENILE:** Mainly brown on head, chest, and upperparts, but white from breast through remaining underparts; greater underwing coverts pale grayish white, forming a long stripe below. Eyes dark. **IMMATURE:** Year-old birds have some pale feathering on head and neck and some dark flecking on underparts; third-year birds resemble adults but show some whitish mottling on belly and hint of pale underwing stripe.

Geographic Variation The larger Pacific coast subspecies *californicus* is distinguished from most (maybe all?) Atlantic and Gulf Coast *carolinensis* by its bright red, rather than blackish green, gular pouch (adults early in breeding season). The small, darker-bellied nominate *occidentalis* (from the West Indies) has been collected once in coastal northwest FL.

Similar Species Unmistakable. Adults can look pale above when seen in strong lighting and momentarily may be mistaken for American White Pelicans. Even a heavily marked juvenile White Pelican is still mainly white and thus not likely to be confused.

Status & Distribution Common. **BREEDING:** Atlantic and Gulf Coast colonies nest mainly on low vegetated islands (mangroves in FL). Northernmost colonies are in MD and on northern Channel Is., CA. **DISPERSAL:**

Postbreeding dispersal May–Oct. north to DE, NJ, WA; a few north to NY and southernmost BC. Returns south to breeding regions by late fall. Regular postbreeding visitor to Salton Sea, where common; in the 1990s, some breeding attempts occurred. Elsewhere, irregular in Southwest, but in some years widespread. **RARE STATUS:** Rare (mainly late summer, fall) along Atlantic coast to NS; recorded rarely but widely in the interior to ID, ND, and Great Lakes region. Casual to southeast AK.

Population The species has recovered since the 1970s and is no longer listed as federally endangered. The West Coast population was the last to be removed from the US Endangered Species List.

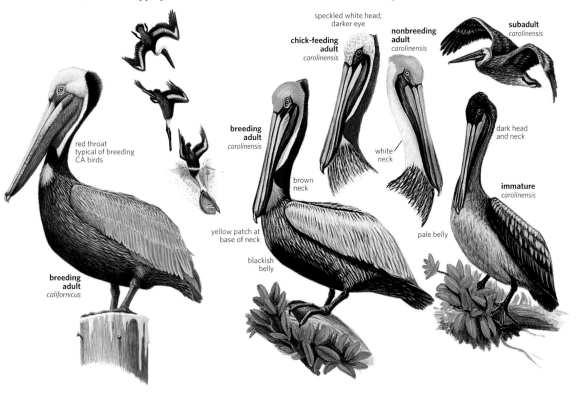

speckled white head; darker eye

chick-feeding adult
carolinensis

nonbreeding adult
carolinensis

subadult
carolinensis

breeding adult
carolinensis

red throat typical of breeding CA birds

dark head and neck

white neck

brown neck

immature
carolinensis

yellow patch at base of neck

pale belly

blackish belly

breeding adult
californicus

HERONS, BITTERNS, AND ALLIES Family Ardeidae

Yellow-crowned Night-Heron (NJ, Apr.)

One of the most distinctive and graceful bird families, the Ardeidae played a major role in the formation of the American conservation movement. Wading bird populations were decimated in the late 19th and early 20th centuries, when it was fashionable for women to wear hats adorned with feathers, wings, or even entire stuffed birds. Hundreds of thousands of wading birds were slaughtered for the fine plumes—aigrettes—grown by some species for courtship displays. The loss of the adults during the breeding season meant that their nestlings starved, causing the populations of most species to plummet within a few decades. Ardeids are called *wading birds* in North America, a term that should not be confused with the British term waders, which refers to shorebirds.

Structure Some ardeid species are small and chunky, but most species have moderate to large, slender bodies with long necks and legs—the legs of most species extend beyond the tail when in flight. The wings are generally long and rounded, the tail short. The long, dagger-shaped bill aids in capturing aquatic prey. The neck, which is held in an S curve while birds are resting, may be extended fully when foraging, and is drawn toward the body during flight.

Plumage Herons generally have a dull blue or gray plumage, while egrets typically show a plumage that is entirely white. Bitterns are plumaged cryptically, with brown or buffy streaking. Many species grow elongated plumes from the head, breast, or back that are used in elaborate courtship displays. Also associated with courtship (high breeding) is the intensifying of soft-part colors of the lores, bill, and legs of many species.

Behavior Most species nest colonially; nightly roosts or breeding colonies may contain thousands of individuals of several species. Wading birds use several foraging strategies that vary greatly by species. Many species stand in or near water and wait for prey to pass by—usually fish and small crustaceans, but snakes and other vertebrates are taken by larger species. Others walk slowly, perhaps also stirring up the water with their feet to flush prey. The Reddish Egret has a particularly active foraging behavior.

Distribution Cosmopolitan. The largest numbers of wading birds occur in areas with abundant wetlands and mild climate. The Everglades is perhaps the area most associated with wading birds in North America, but large populations occur along the entire Gulf Coast and in parts of CA.

Taxonomy Recent higher-level taxonomic revisions now place the family Ardeidae in the order Pelecaniformes with pelicans (Pelecanidae) and ibises and spoonbills (Threskiornithidae). Species limits remain uncertain for some taxa. Worldwide, 60 to 65 species in 16 to 21 genera are recognized. According to the AOS, 20 species in 10 genera occur in North America; these are often divided into three groups: bitterns (three species), herons and egrets (15 species), and night-herons (two species).

Conservation In North America, the end of the millinery trade in the early 1900s allowed many wading bird populations to recover—a recovery that continues for especially sensitive species such as the Reddish Egret. Wetland loss or degradation, pollution, colony disturbance, and overharvesting of prey, however, now threaten some populations. BirdLife International codes: 3 NT, 3 VU, 5 EN, 5 EN, 1 CR, and 5 EX (four are island-based night-herons).

BITTERNS Genera *Botaurus* and *Ixobrychus*

In contrast to other wading birds, which are showy and feed in the open, bitterns are secretive marsh dwellers with cryptic, streaked plumage that allows them to blend in with their surroundings. Bitterns are solitary species, but they can be common in an ideal habitat. Three species occur in N.A., one as an accidental.

AMERICAN BITTERN *Botaurus lentiginosus*   AMBI ■ 1

Cryptic in plumage and mostly cre- puscular in habits, the American Bit- tern is a denizen of marshes with thick vegetation. It attempts concealment by pointing its bill upward; it may sway its head and neck back and forth to mimic wind-rustled vegetation. It is detected primarily in flight and by hearing its eerie "pumping" vocaliza- tions. Most active during crepuscular hours of the day. Monotypic. L 28" (71 cm) WS 42" (107 cm)
Identification A large brown heron with a disproportionately long neck and short legs. **ADULT:** Sexes similar. Brown upperparts finely flecked with black. Dark cap, white supercilium, chin, and throat, and wide blackish malar streak. Yellow eyes and mostly yellow bill; legs and feet pale greenish or yellowish. White or pale underparts with bold rufous streaking. **JUVENILE:** Blackish malar stripe absent. **FLIGHT:** Strong and direct; may appear hunch- backed. Wings somewhat pointed. On upperwings, blackish flight feathers contrast with brown coverts; under- wings sooty.
Similar Species Immature night- herons are smaller, lack the dark malar stripe, have rounded wings and reddish eyes, and are often found in open habitats. The contrasting upperwing surface of an immature Yellow-crowned Night-Heron is more similar to American Bittern, but its bill is short and black.
Voice SONG: Distinctive, resonant "pumping" *oonk-a-lunk* is given on breeding grounds and late in winter and spring migration. **CALL:** A hoarse

wok or *wok-wok-wok* when flushed.
Status & Distribution Uncommon. **BREEDING:** Non-colonially in freshwa- ter wetlands with tall emergent vegeta- tion; also found in saline and brackish marshes in migration and winter. Throughout southern Canada and northern US; rarely in southern US. **MIGRATION:** Migratory, but some birds probably resident in Pacific region and in the southern portion of breeding range. **WINTER:** Southern US and in Pacific region as far north as Puget Sound, and Mexico. **RARE STATUS:** Rare in summer north to southeast AK. Rare in C.A. south to Costa Rica. Casual to central NT, northern QC, Bermuda, Lesser Antilles, Greenland, Canary I., and Europe.
Population Significant declines in parts of its range due to wetland loss and degradation.

often seen
motionless
with head up

juvenile

blackish
malar stripe

adult

dark brown
streaks

adult

wings slimmer
and more pointed
than night-herons

dark primary coverts,
primaries, and
secondaries contrast
with paler upperwing
coverts

YELLOW BITTERN *Ixobrychus sinensis*   YEBI ■ 5

This Asian relative of the Least Bittern has occurred once in N.A. Monotypic. L 15" (38 cm) WS 21" (53 cm)
Identification ADULT: Sexes similar. Head and neck buffy with black cap; brown back with black tail. Under- parts white with buffy streaking on neck and breast, extending to flanks. **JUVENILE:** Plumage browner, boldly streaked with black on upperparts (except rump); neck also streaked. **FLIGHT:** Black primary coverts and flight feathers. Adults with buffy wing patch on upperwing. Juveniles with brown secondary coverts heavily streaked with black.
Similar Species The highly migratory and uncommon Schrenk's Bittern (*I. eurhythumus*), breeding north to the Russian Far East, is distinguished by its slightly larger size, dark rufous upperparts (solid in male, distinctly black-and-white spotted in female, and similar in juvenile), and less contrast- ing upperwing pattern.
Voice Guttural grunt.
Status & Distribution Locally com- mon. **BREEDING:** Wetlands, even tall trees. From India to Japan through Malay Peninsula. **MIGRATION:** Par- tially migratory. **WINTER:** Much of breeding range (withdraws from Japan and China) to New Guinea and Guam. **RARE STATUS:** One specimen record in N.A. (Attu I., AK, 17–22 May 1989). Also accidental to Australia and Christ- mas I. (Indian Ocean).
Population Not threatened.

buffy plumage
overall

small size

adult

LEAST BITTERN *Ixobrychus exilis*   LEBI ■ 1

The Least Bittern, the world's smallest heron, is easily overlooked. It is partial to cattails, bulrush, or saw grass in freshwater marshes. It is very adept at clambering up reed stems. Like the American Bittern, it often freezes in place, with head and bill pointed skyward. Knowledge of its vocalizations and behavior will aid in finding this secretive, solitary species. It can often be spotted by scanning reed bases (hunting perches). Polytypic (5 sspp.; nominate in N.A.). L 13" (33 cm) WS 17" (43 cm)

Identification Tiny with bright buff wing patches, a dark crown, and two white or buff scapular stripes that contrast with a dark mantle. Whitish underparts are variably streaked on the foreneck with buff. Eyes, loral skin, and bill are mostly yellow; legs are greenish in front and yellow in back. **ADULT MALE:** Crown, back, rump, and tail are a glossy black. The loral skin can turn reddish pink during courtship. **ADULT FEMALE:** Similar, but upperparts are dark brown rather than black. **JUVENILE:** Similar to female, but colors are more muted.

Streaking on foreneck is browner and bolder, extending onto sides of neck; buffy wing coverts have dark centers. **DARK MORPH:** Very rare; known as "Cory's Least Bittern." It was best known from Ashbridge Bay, Toronto, ON, in the early 20th century, but these marshes were filled decades ago. It also bred at Lake Okeechobee, FL. It is perhaps unrecorded since the mid-20th century. In all ages, chestnut replaces pale areas on upperparts and wing patches, as well as on upperwing coverts. In adults, the white scapular lines are lacking. **FLIGHT:** Flies with quick wingbeats low over reed beds, dropping abruptly into them. All plumages (except "Cory's") show a large buffy patch on the upperwing that contrasts strongly with the dark flight feathers and deep rufous greater coverts.

Similar Species A juvenile Green Heron is also small with streaked underparts, but it is significantly larger and its wings are always entirely dark.

Voice Quite vocal with a varied repertoire. **ALARM:** Several calls, including *quoh* and *hah*. **CALL:** A rail-like *tut-tut-tut* or *kek-kek-kek-kek* heard year-round is the most frequent vocalization (often mistaken for a King, Clapper, or Ridgway's Rail). **SONG:** Male in breeding season only gives three or four short, low *coo* notes in rapid succession.

Status & Distribution Uncommon to locally common. **BREEDING:** Dense marshland vegetation, especially cat-

tails; rare in mangrove swamps. Local in much of N.A. **MIGRATION:** Partially migratory. Arrives in breeding areas Apr.–May; departs Aug.–Sept. **WINTER:** Resident in southernmost N.A. breeding areas. Majority of N.A. birds winter in C.A. and the Caribbean. **RARE STATUS:** Casual north to southwest Canada and NL. Accidental to the Azores, Bermuda, Clipperton I., and Iceland.

Population Not threatened.

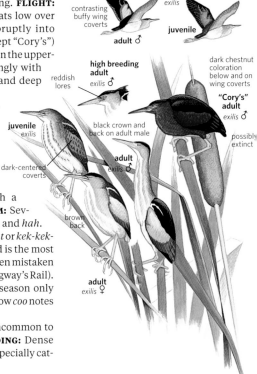

contrasting buffy wing coverts

exilis

adult ♂

juvenile

high breeding adult

exilis ♂

reddish lores

dark chestnut coloration below and on wing coverts

"Cory's" adult

exilis ♂

possibly extinct

juvenile *exilis*

black crown and back on adult male

adult *exilis* ♂

dark-centered coverts

brown back

adult *exilis* ♀

Genus *Tigrisoma*

BARE-THROATED TIGER-HERON *Tigrisoma mexicanum*   BTTH ■ 5

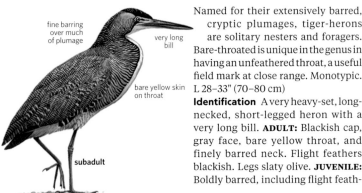

fine barring over much of plumage

very long bill

bare yellow skin on throat

subadult

Named for their extensively barred, cryptic plumages, tiger-herons are solitary nesters and foragers. Bare-throated is unique in the genus in having an unfeathered throat, a useful field mark at close range. Monotypic. L 28–33" (70–80 cm)

Identification A very heavy-set, long-necked, short-legged heron with a very long bill. **ADULT:** Blackish cap, gray face, bare yellow throat, and finely barred neck. Flight feathers blackish. Legs slaty olive. **JUVENILE:** Boldly barred, including flight feath-

ers, with cinnamon-buff and brown.

Similar Species Superficially similar to American Bittern, juvenile Black-crowned and Yellow-crowned Night-Herons, but these species have much shorter bills, yellowish legs, and lack gray neck barring. Two other *Tigrisoma* are found in C.A. and S.A.; juveniles of all three species are similar. **Voice CALL:** Gives a heronlike *wah* or *hauk*, often in rapid succession in flight.

Status & Distribution Resident of coastal habitats from Mexico (found along coastal slopes north to southern Tamaulipas and southern Sonora) to northwestern Colombia.

Two TX records: a single subadult at Bentsen-Rio Grande Valley SP, Hidalgo Co., TX (12 Dec. 2009–20 Jan. 2010) and

an adult in Uvalde Co., early Feb. 2017 through late May 2018.
Population Trends unknown.

Genus *Ardea*

GREAT BLUE HERON *Ardea herodias*   GBHE ■ 1

The largest and most widely distributed wading bird in N.A., the Great Blue Heron is found in a range of wetland and drier habitats. The subspecies *occidentalis*, "Great White Heron," is perhaps better treated as a distinct species. Polytypic (4 sspp.; 3 in N.A.). L 46" (117 cm) WS 72" (183 cm)
Identification Blue-gray upperparts; grayish neck; large, daggerlike bill. **NONBREEDING ADULT:** Sexes similar. Bluish gray back and wings with scapular plumes and a black shoulder patch; chestnut thighs, and blackish tail. Crown white. Legs and feet brown, brownish green, or greenish black. **BREEDING ADULT:** Soft-part coloration intensifies. Lime green or blue lores; yellow, orange, or reddish bill; reddish or greenish yellow legs. **JUVENILE:** Darker overall without plumes; black crown; blackish lores and upper mandible. **FLIGHT:** Strong, slow. Grayish coverts contrast with dark flight feathers. **"GREAT WHITE HERON":** White in all plumages; legs and feet dull yellow in breeding adult; darker in nonbreeding and especially in younger birds. Bill coloration as in Great Blue in all ages. **"WURDEMANN'S HERON":** Intergrade between blue and white subspecies occurs in southern FL; similar to blue subspecies but head variably white and neck pale with inconspicuous streaking below.
Geographic Variation Nominate subspecies is found over most of N.A. Pacific Northwest *fannini* has darker plumage and a shorter bill; white *occidentalis* and intermediates are restricted to southern FL.
Similar Species See the closely related Gray Heron from the Old World. The "Great White Heron" is distinguished

"Great White Heron"

short white head plumes

long and heavy yellowish bill

dark gray remiges

adult
herodias

white crown with black lateral crown stripe

breeding adult
herodias

"Great White Heron" breeding adult
occidentalis

dark cap

pale legs

immature has darker legs and bill

white head

"Wurdemann's Heron" breeding adult

rufous thighs

juvenile
herodias

from Great Egret by its larger size and longer, thicker bill; legs and feet are seldom all-dark, though color is variable based on age and season. Breeding adult has head plumes unlike any Great Egret.
Voice CALL: Loud, hoarse *kraaank*.
Status & Distribution Common. **BREEDING:** Colonially (rarely solitary) in all lower 48 states, AK, all but northern Canada, most of West Indies, and C.A. **MIGRATION:** Partially migratory; withdraws from northern portion of breeding range, except along coasts. **WINTER:** Along coasts and in southern portion of breeding range. **RARE STATUS:** Wanders widely; casual to southwestern AK and the Arctic coast, Azores, Clipperton I., Greenland, HI, and Spain. **"GREAT WHITE HERON":** Uncommon. Restricted in US to southern FL (±850 pairs resident) and also resident on Cuban cays; strongly partial to salt water areas, especially around mangroves. Rare or casual along the coasts north to TX and north to Maritimes; accidental inland to PA, KY, and ON.
Population Stable.

GRAY HERON *Ardea cinerea*   GRAH ■ 5

adult
cinerea

paler overall
than Great Blue

white thighs
(rufous in Great
Blue Heron)

shorter legs
than Great Blue

A familiar species of town and countryside across Eurasia and much of Africa, Gray Heron is the Old World counterpart to Great Blue Heron of N.A., which it closely resembles. Polytypic (4 sspp.). L 33–40" (84–102 cm) **Identification** A large heron, medium slaty gray above, white crown, whitish below, with two rows of neat, dark stippling down most of neck to breast; mostly grayish bill and legs. **ADULT:** Breeding birds have bright yellow bill, with single black plumes extending behind each eye; legs also partly yellow. **JUVENILE:** Sides of neck and head grayish brown; dark crown. **Geographic Variation** Nominate *cinerea* is found through most of the Palearctic, Africa, India, and Sri Lanka. Other subspecies include East Asian *jouyi*, increasingly threatened *firasa* on Madagascar, and *monicae*,

restricted to the Cape Verde Is. off Mauritania, which some authorities consider a full species. **Similar Species** Only likely to be confused with larger, longer-necked, and longer-legged Great Blue Heron. In all plumages Gray Heron lacks chestnut thighs of Great Blue; in flight, leading edge of wing shows prominent white area, rather than chestnut. Gray Heron is paler above and on the neck than Great Blue. **Voice CALL:** An abrupt, resonant *ka-ark!* and raspy *yeehr!* often given in flight. Also a screeching *rraank!* similar to Great Blue Heron but higher in pitch. **Status & Distribution BREEDING:** Vast Old World range. **WINTER:** Mostly in breeding range; northern breeders fully migratory, moving southward in autumn. **RARE STATUS:** Casual to western AK (western Aleutians and

Pribilofs) and Newfoundland. Also over 20 records from Greenland and casual to the Lesser Antilles, except Barbados where now annual. **Population** In most areas, populations appear to be stable.

GREAT EGRET *Ardea alba*   GREG ■ 1

This tall, stately egret is found in a variety of freshwater, brackish, or saline habitats over much of the US. It feeds primarily on fish and aquatic invertebrates, but is also found locally in fields and pastures, stalking small mammals. The plumes of Great Egrets were among the most sought after for the millinery trade, resulting in severe population declines of this once again common species. Polytypic (4 sspp.; *egretta* in N.A.; Asian *modesta* and possibly Eurasian *alba* are casual visitors). L 39" (99 cm) WS 51" (130 cm) **Identification** Large, with a long, slender neck. **NONBREEDING ADULT:** Entirely white. Eyes, lores, and bill are yellow; legs and feet are black. **BREEDING ADULT:** Long, graceful plumes (aigrettes) on the back that extend beyond the tail. Unlike many other herons and egrets, there are no plumes

on the head or neck. The lores flush lime green and the bill turns orangish with a dark culmen. **IMMATURE:** Similar to a nonbreeding adult. **Geographic Variation** In breeding plumage, Eurasian *alba* and slightly smaller Asian *modesta* have black bills. **Similar Species** Most similar to the white morph of Great Blue Heron ("Great White Heron") but somewhat smaller and less bulky, with thinner bill, and black legs and feet; lacks the short head plumes of adult "Great White Heron." The bill of the white

morph Reddish Egret is either all-dark (immature and nonbreeding adult) or bicolored with a pink base (breeding adult). Other egrets are much smaller and lack the combination of yellow bill and black legs and feet. **Voice** Generally silent except when nesting or disturbed, when it may utter *kraak* or *cuk-cuk-cuk* notes. **Status & Distribution** Found on all continents except Antarctica. **BREEDING:** In shrubs or trees in colonies with other wading birds, along the Atlantic, Gulf, and Pacific coasts, the Mississippi River floodplain, and locally elsewhere in the interior. Also in extreme southern Canada, the West Indies, C.A., and S.A. **MIGRATION:** Partially migratory; birds mostly resident along coasts and in southern US. Postbreeding dispersal may carry birds north of regular summer range. **WINTER:** Locally in the West

lime green
facial skin

more blackish bill
than nonbreeding

very long
neck

high breeding adult
egretta

yellow facial
skin

long
yellow bill

breeding
modesta

Asian *modesta* is slightly
smaller than Eurasian
alba (not shown); both
have black bills in
breeding plumage

long
plumes

nonbreeding
egretta

dark legs
and feet

and throughout the Southeast. **RARE STATUS:** Casual to southeastern, central, and northern AK and Atlantic Canada. Asian *modesta* casual (including two specimens) to Aleutians, AK. A black-billed bird from VA may have been *alba*, on distribution probability.

Population Although decimated by the plume trade in the late 1800s and early 1900s, the species has recovered; some populations are increasing.

INTERMEDIATE EGRET *Ardea intermedia* INEG ▪ 5

The serviceably, if not poetically, named Intermediate Egret is indeed intermediate in size between the smaller and larger white egrets, for which it is easily mistaken. It is accidental to N.A., with just two records from the Aleutian I., AK. Polytypic (3 sspp.). L 22–28" (56–72 cm)

Identification An all-white egret with dark legs and feet and a yellowish bill with a dark tip. In N.A. context, Intermediate is a bit larger than a Snowy Egret and well smaller than a Great Egret. **ADULT:** In high breeding condition soft parts change color variably among the different subspecies: The bill becomes reddish, red and yellow, or black; loral skin becomes green or yellow; iris color changes in some subspecies to red; and legs in some subspecies take on yellow or pinkish tones.

Geographic Variation Medium-size, nominate *intermedia* is found from southeastern Asia and western Indonesia to Japan, northern populations are migratory; largest, *brachyrhyncha*, inhabits Africa south of the Sahara and is not migratory; and smallest, *plumifera*, is found from east Indonesia to New Guinea and Australia and is also not migratory.

Similar Species Smaller Snowy and Little Egrets have yellow-soled feet and always have longer, slenderer black bills. Great Egret has a much longer neck, usually with a kink in the middle, and a longer bill in which the culmen aligns with the flat top of the head. (Intermediate shows a "bump" at the forehead.) When foraging, Snowy Egret tends to chase prey actively, while Intermediate Egret forages more like Little Blue Heron, with neck extended forward, watching downward, moving slowly or scarcely at all before seizing prey.

Voice A soft buzzing call is said to be unique among herons, but more typical egret-like calls *grak, glok, kroo, kraa* are heard from flying and nesting birds.

Status & Distribution MIGRATION: Mostly resident in its fragmented Old World range. Japanese breeders are migratory and depart in mid-autumn, wintering in Borneo and the Philippines. **RARE STATUS:** Two specimen records of nominate subspecies, from AK: Buldir I. (30 May 2006) and Shemya I. (28 Sept. 2010).

Population Significant declines have been documented in urbanized Japan.

dark-tipped yellow bill

breeding adult
intermedia

Genus *Egretta*

This genus contains 13 to 14 species. Four species (all called *egrets*) are white, six (mostly called *herons*) are blue or blackish, and three or four others contain both white and dark morphs. The species inhabit a variety of freshwater and saline environments.

CHINESE EGRET *Egretta eulophotes* CHEG ▪ 5

This elegantly plumed egret, a rather rare and highly migratory coastal Asian species, has strayed to N.A. just once. Monotypic. L 27" (65 cm) WS 41.5" (105 cm)

Identification Similar to Snowy and Little Egrets in size and plumage but shorter legged (especially tibia) and thicker billed (especially distally). Ages similar; skin on lores dips down, more straight on Little Egret. **BREEDING ADULT:** Develops a shaggy crest. Soft-part colors intensify: all-yellow bill, turquoise lores, black legs, and yellow feet. **NONBREEDING:** Largely yellow-green legs and feet. Lores usually appear darkish. Bill brownish; yellow on bill base extends up to upper mandible. Often has traces of the shaggy crest.

Similar Species Snowy and Little Egrets are similar. (Little would be the confusion species on Aleutians.) On Chinese Egret, note shorter legs (especially tibia) and the shape and color of the lores and bill. Chinese also has a steeper forehead, giving it a more dome-headed appearance. In breeding plumage, Chinese much more obvious with yellow bill, blue lores, and shaggy crest. In nonbreeding plumage, separation best done using multiple features.

Voice Undescribed.

Status & Distribution Highly migratory. **BREEDING:** Colonially on islands off Korea, China, and the Russian Far East. **WINTER:** Mainly from the Malay Peninsula to the Philippines, northern Borneo, and Sulawesi. **RARE STATUS:** One specimen record in N.A., on Agattu I., AK (16 June 1974).

Population Vulnerable. Declining due to the destruction of tidal wetlands; population likely under 15,000 individuals.

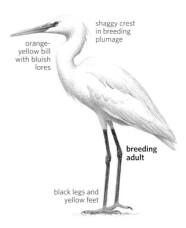

shaggy crest in breeding plumage

orange-yellow bill with bluish lores

breeding adult

black legs and yellow feet

LITTLE EGRET *Egretta garzetta*   LIEG ▪ 4

This Old World counterpart to the Snowy Egret inhabits a variety of aquatic environments. It feeds primarily on small fish. It has two color morphs, a typical white morph (all N.A. records) and a rare dark morph. Some authorities consider the Little Egret conspecific with the Western Reef-Heron (*E. gularis*). Polytypic (3 sspp.; nominate in N.A.). L 24" (60 cm) WS 36" (91 cm)

Identification Medium-size egret, very similar to Snowy Egret. Sexes similar. White morph described. **NON-BREEDING ADULT:** Plumage entirely white. Yellow eyes, black legs, and yellow feet. The lores are usually bluish gray, but can be greenish, pale yellow, or whitish. **BREEDING ADULT:** Two white occipital plumes, as well as numerous plumes on lower neck and back. Lores and feet flush reddish, but can be yellow. **IMMATURE:** Pale base to lower mandible, gray lores, brownish green legs.

Similar Species Little Egret has a flatter crown, a somewhat thicker and longer bill, duller yellow feet, and a more upright, longer necked posture (often appearing larger than Snowy Egret). Snowy Egret forages in a hunched manner and tends to move around more frenetically. Little Egret is best told from the Snowy Egret in nonbreeding plumage by its blue-gray rather than yellow lores; in breeding plumage, it has only two or three occipital plumes. Mostly white, immature Little Blue Heron often has dark tips to the primaries, and greenish yellow legs and feet. AK records of Little need to be carefully separated from Chinese Egret, especially in nonbreeding plumage (see that species account).

Voice Usually silent. **CALL:** A low guttural *kraak*.

Status & Distribution Common. Widespread in Old World. First recorded in the West Indies in 1954 (originating likely from West Africa) from Barbados and breeding there by 1994 (now ±20 pairs breed at Barbados); widely recorded elsewhere in Lesser Antilles and eastern Greater Antilles. **RARE STATUS:** Casual along the Atlantic coast north to Maritime Provinces and QC (most frequently recorded in spring and summer) and also to Puerto Rico and Lesser Antilles. Accidental to the western Aleutians, AK (male salvaged on Buldir I., 27 May 2000).

Population Stable, and range is spreading north in parts of Old World. West Indian populations have been increasing since the 1950s. Little Egrets will presumably be found with greater frequency in the US as the West Indies population increases.

Illustration labels: lores often greenish gray, unlike Snowy; nonbreeding adult *garzetta*; two long plumes; bill appears longer than Snowy; breeding adult *garzetta*; always all-black legs; yellow feet

WESTERN REEF-HERON *Egretta gularis*   WERH ▪ 5

This Old World species has recently colonized the West Indies and has strayed several times to the US and Atlantic Canada. It has two color morphs, white and dark, with rare intermediates; only the dark morph has been recorded from the New World. The Western Reef-Heron is sometimes considered conspecific with the Little Egret. Polytypic (2 sspp.; *gularis* in N.A.). L 23.5" (60 cm) WS 37.5" (95 cm)

Identification Dark morph described. **NONBREEDING ADULT:** Entirely slate gray overall, with a white chin and throat. Eyes yellow; lores and bill dusky yellow. Legs black and feet yellow. **BREEDING ADULT:** Develops two long, wispy, slate occipital plumes. Soft-part colors intensify: Lores flush red and feet turn black. **IMMATURE:** Overall brown or dark gray-brown plumage with a white chin and throat.

Similar Species Adult Little Blue Heron is entirely dark; in breeding plumage it has a shaggy head and neck plumes and entirely dark legs and feet. Rare dark morph of Little Egret (no N.A. records) is very similar, but it often has a dark throat and is shorter billed than *gularis* Western Reef. White-morph Little Egret is best distinguished by bill structure. Western Reef-Heron's bill has a subtle decurvature toward the tip.

Voice Usually silent. **CALL:** A throaty squawk.

Status & Distribution Occurs coastally from West Africa east to India. **RARE STATUS:** Recent arrival in the New World, beginning in mid-1980s. About 15 records from the West Indies, primarily Barbados. Six summer records in N.A. (1983 and 2005–2007) from Atlantic Canada and the Northeast, perhaps the more recent records could have involved only one individual visiting a number of locations.

Illustration labels: overall slaty-gray with white chin and throat; dark-morph adult *gularis*; yellow feet

Population Old World population is stable; records in the West Indies are increasing.

SNOWY EGRET   *Egretta thula*   SNEG   ■   1

The graceful Snowy Egret occurs throughout the New World and is found in a wide range of freshwater and saline habitats. It feeds on a variety of prey such as fish, aquatic invertebrates, and even snakes and lizards. Monotypic. L 24" (61 cm) WS 41" (104 cm)

Identification Medium-size white heron with a slender build and a thin black bill. **NONBREEDING ADULT:** Sexes similar. Plumage all-white with yellow eyes and lores yellow; legs black, feet distinctly yellow ("golden slippers"). **BREEDING ADULT:** Numerous elegant, shaggy plumes on its crown, nape, foreneck, and back. Some soft-part colors intensify: Lores flush red, and feet flush orange or reddish. **JUVENILE:** Bill has a pale gray base; lores are grayish green; legs often extensively greenish yellow, but most show varying degrees of black on the foreleg.

Similar Species Little Egret closely resembles Snowy Egret. The Snowy is distinguished by its yellow lores, more rounded crown, and shorter bill, narrower at the base. It also has numerous breeding plumes, compared to only the two or three long occipital plumes found on Little Egret. When foraging, the Little's posture is more upright, in a manner similar to a Great Egret. Mostly white, immature Little Blue Heron has a thicker, more bicolored bill, greenish yellow legs and feet, often has dark gray tips to the primaries, and feeds

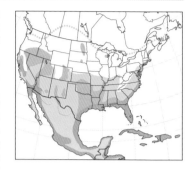

much more deliberately and slowly in a fixed hunched position.

Voice Generally silent except away from breeding colony. **CALL:** A harsh, raspy *aah-raarrh.*

Status & Distribution Common. **BREEDING:** Colonially with other wading birds. Locally from southeastern OR, the interior West, and north on the Great Plains to the Dakotas. Also in the West Indies and south to S.A. **MIGRATION:** Partially migratory. **WINTER:** Coastally from OR and NJ south to S.A.; also all of FL, along the entire Gulf Coast, and the West Indies. **DISPERSAL:** Wanders irregularly north to south Canada from BC to NL; also to HI, Bahamas, Lesser Antilles, Bermuda, and the Galápagos. Accidental to Clipperton I., Iceland, UK, and South Georgia I.

Population Its breeding plumes made it among the most sought after species for the millinery trade in the late 1800s and early 1900s, resulting in severe declines, but numbers have now recovered. The species' adaptability to a range of environmental conditions has allowed it to expand its range beyond its original distribution in the past several decades.

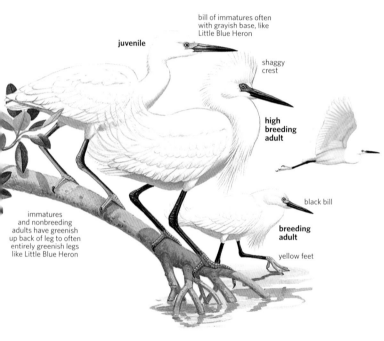

juvenile

bill of immatures often with grayish base, like Little Blue Heron

shaggy crest

high breeding adult

black bill

breeding adult

yellow feet

immatures and nonbreeding adults have greenish up back of leg to often entirely greenish legs like Little Blue Heron

Identification of White Egrets

Separating various white egret species is fairly straightforward as long as a few features are considered. Great Egret is the largest species; Cattle Egret, the smallest, has disproportionately short legs. Young juvenile Cattle Egrets can cause confusion due to their dark bills. With their sharply separated pink-based bill and shaggy head plumes during the breeding season, white-morph Reddish Egrets are easily identified. In contrast, immatures and winter adults have entirely dark bills. The "dancing" feeding behavior is an excellent differentiating characteristic for the species at all ages and times of year. Immature Little Blue Herons

during their first fall and winter closely resemble Snowy Egrets, but they have duller lores and a somewhat thicker and more obviously bicolored bill. Most, but not all, have some dusky in the wing tips, often only visible in flight. They also have more uniformly greenish yellow legs and feed much more deliberately than Snowy Egrets.

In flight, Great Egret is very long-necked and flies with slow wingbeats. Much smaller Snowy Egret has more rapid wingbeats. The small, short-legged Cattle Egret appears short and round-winged and flies with the most rapid wingbeats.

LITTLE BLUE HERON *Egretta caerulea*   LBHE ■ 1

Unique among herons, the Little Blue has a nearly all-white juvenal plumage and an all-dark adult plumage. Found in freshwater and saline habitats, it forages for aquatic invertebrates, fish, and amphibians by walking slowly or by standing still, with neck bent forward at a 45-degree angle. Monotypic. L 24" (61 cm) WS 40" (102 cm)

Identification Medium-size heron. **ADULT:** Sexes similar. Dark slate blue body with purple head and neck. Yellow eyes, dull greenish lores, blue-gray bill with black tip; gray to greenish gray legs and feet. **BREEDING ADULT:** Bright reddish purple head and neck with long lanceolate plumes on the crest and back. Lores and base of bill turquoise; legs and feet black. **JUVENILE:** Entirely white, but often with dark gray tips to most of the primaries. **FIRST-SPRING:** Adult plumage acquired gradually, resulting in splotchy white and blue (calico) plumage; like the adult by second fall.

Similar Species Dark-morph Reddish Egret is larger and paler overall, with shaggy head and neck plumes, pale eye, and a pink-based, black-tipped bill in the breeding adult. White-morph Reddish Egret is larger with pale lores and dark legs and feet; it also has a different foraging behavior. Tricolored Heron has white underparts in all plumages. Immature Little Blue is distinguished from Snowy Egret by its stouter, more bicolored bill, gray-green lores, greenish yellow legs and feet, and usually dark tips to the primaries. Smaller and shorter-legged Cattle Egret has a stout yellow bill.

Voice Generally silent away from breeding colony. **CALL:** A harsh, croaking *aarr-aarrh*.

Status & Distribution Common. **BREEDING:** Colonial. Along Atlantic coast from NY (local north to southern ME) through FL, along entire Gulf Coast to northern S.A., up the Mississippi River drainage to southern IL and IN, and in southern Great Plains. Along Pacific coast from Sonora and Baja California south, and locally at San Diego, CA; rare and sporadic breeder to ND and SD. **MIGRATION:** Partially migratory; withdraws from northern portion of breeding range. **WINTER:** Along entire Gulf Coast, along Atlantic coast to VA, and from extreme southwestern CA south along Pacific coast. **DISPERSAL:** Widely after breeding; casual to southern Canada, HI, MT, NL, WA. **RARE STATUS:** Accidental to the Azores, Chile, and western Greenland.

Population Threatened by wetland degradation or loss and colony disturbance. Recent declines in northern Atlantic coast range.

juvenile

dusky on wing tips visible on some juveniles

adult

mottled plumage seen on one-year-old birds in spring and summer

bicolored bill with grayish lores

molting immature

maroon head and neck

blue base to bill

juvenile

breeding adult

greenish yellow legs

nonbreeding adult

very deliberate feeding behavior

some juveniles show dusky wing tips; best viewed in flight

thick-based, slightly decurved bill

TRICOLORED HERON *Egretta tricolor*   TRHE ■ 1

This colorful, slender-necked heron of southern affinity is a graceful and active hunter. When feeding, it often dashes about in the shallows pursuing prey with wings slightly raised, and may change direction swiftly and with precision; it often wades in deeper water than most other herons. Polytypic (2–3 sspp.; *ruficollis* in N.A.). L 26" (66 cm) WS 36" (91 cm)

Identification Medium-size heron with a long, slender neck and long,

thin bill. **NONBREEDING ADULT:** Slate-gray head, neck, back, wings, and tail; back covered with elongated purplish maroon scapular feathers. White underparts, throat, and neck stripe. Reddish brown eyes, yellow lores, mostly yellow bill, and grayish yellow legs and feet. **BREEDING ADULT:** A few long, white occipital plumes and elongated purplish plumes on the lower neck. Soft-part coloration intensifies: red eyes; turquoise lores and base of

bill; and maroon, orange, or pink legs and feet. **IMMATURE:** Rich chestnut head and neck; scapulars and wing coverts edged with chestnut. **FLIGHT:** Swift and direct. White underparts and wing linings contrast with dark flight feathers.

Similar Species Distinctive.
Voice Relatively silent away from breeding colony. **CALL:** A raspy *aaah*.
Status & Distribution Uncommon to common. **BREEDING:** Colonial, usually on islands. Along the entire Atlantic and Gulf Coasts from NY (locally to southern ME) to northern Mexico, and throughout FL and most of LA, and occasionally inland to KS and AR; along the Pacific coast from Baja California south to northern S.A. Irregular breeder inland in SC, ND, and SD. **MIGRATION:** Withdraws from northern portions of breeding range; some US breeders winter in the West Indies. **WINTER:** Along the Atlantic coast from VA south, peninsular FL, and the Gulf Coast; now casual to Southern CA (formerly more regular). **RARE STATUS:** Casual to northern CA, OR, the Southwest, and southern Canada from MB to NL. Accidental in the Azores.
Population Declining in northern portion of breeding range along Atlantic coast. FL populations are showing a recent rapid decline, perhaps due to wetland loss or degradation.

bright blue lores and bill base in breeding plumage

rufous on neck and wing coverts

white underwing coverts and white belly in all plumages

breeding adult
ruficollis

juvenile
ruficollis

long bill

long, slim neck

white belly

nonbreeding adult
ruficollis

REDDISH EGRET *Egretta rufescens*   REEG ▪ 1

Scarcer and more range-restricted than other regular N.A. herons, the Reddish Egret is confined as a breeder to the Gulf Coast. It has two distinct color morphs; the dark morph predominates in N.A. It is also notable for its curious, active foraging behavior, quickly pursuing fish in shallows—"dancing" around with wings held out and flapping. It feeds on fish and aquatic invertebrates. Polytypic (2 sspp.; both in N.A.). L 30" (76 cm) WS 46" (117 cm)

Identification Large wading bird restricted to shallow estuaries. **NONBREEDING DARK MORPH:** Entirely rufous head and neck with short, shaggy plumes. Remainder of body dark gray; some dark morphs have a few scattered white feathers on wings or body. Pale eyes, bill entirely dark, blackish legs and feet. **BREEDING ADULT DARK MORPH:** Elongated plumes on head, neck, and back. Basal portion of bill pink, lores violet-blue, legs and feet bluish gray. **JUVENILE DARK MORPH:** Plumage brownish gray; wing coverts edged with rufous; worn birds often appear uniformly grayish. Pale eyes; black bill. **WHITE MORPH:** Plumage entirely white. Soft part colors as in dark morph.

Geographic Variation Two weakly

differentiated subspecies. Nominate *rufescens* is found in the US Southeast and the West Indies; *dickeyi*, with browner head and neck and darker plumage, is found from Baja California to C.A. (It wanders north to CA and the Southwest.) Birds from northwest Mexico consist of mainly or entirely dark morphs; all birds recorded for western states are dark morphs.

Similar Species Dark morph is distinctly larger than the Little Blue Heron, and it has black legs and feet; note differences in feeding behavior. Compared to a nonbreeding white-morph Reddish Egret, an immature Little Blue has greenish legs and feet and many have dusky wing tips. The nonbreeding white-morph

dark-morph breeding adult
rufescens

shaggy rufous feathers on head and neck

overall gray with some pale cinnamon

dark-morph breeding adult
rufescens

dark-morph juvenile
rufescens

when feeding, rushes about with wings open

gray body

only birds in breeding plumage have pink-based bills; bill otherwise blackish in all birds in fall and winter

distinctly larger than Little Blue Heron

dark legs and feet

white-morph winter adult
rufescens

Reddish Egret is distinguished from Snowy Egret by its larger size, all-black legs and feet, and lack of yellow lores. **Voice** Generally silent. Low *raaaah* and other notes when foraging or disturbed and flushed.
Status & Distribution Uncommon to locally common. **BREEDING:** Colonially with other wading birds; typically islands in coastal estuaries. Currently ±2,000 pairs in the US, most

in TX. **DISPERSAL:** Nonmigratory, but disperses along Gulf Coast and southern Atlantic coast to NC during spring and summer; some move inland. **WINTER:** Perhaps some withdrawal from northern parts of Gulf Coast; a few regularly move north from Baja California to Southern CA coast. **RARE STATUS:** Casual to the Southwest and north to central CA, NV, WY, CO, MI, IL, NY, and MA. Records in the interior of the

West come mainly from late summer and fall and have increased. Records of Little Blue Heron in the interior West are mainly from late spring and early summer.
Population Near Threatened. Continues to reclaim historic breeding range in FL, although some colonies are threatened from disturbance by boaters; coastal development has greatly reduced foraging areas in FL.

Genus *Bubulcus*

CATTLE EGRET   *Bubulcus ibis*   CAEG   ▦   1

This gregarious species often feeds in fields, where it follows cattle and grazing animals (even farm machinery) to forage on insects and other invertebrates that are flushed. It walks with an exaggerated, head-pumping strut. Often found well away from water, especially when feeding. Polytypic (2 sspp.; both in N.A.). L 20" (51 cm) WS 36" (91 cm)
Identification Nominate subspecies described. Small, stocky, white egret, with a short, thick neck and relatively short and stocky bill. **ADULT:** White with yellow bill; black legs and feet. **BREEDING ADULT:** Buff plumes develop on the crown, nape, lower back, and foreneck. Bill with a reddish base and yellow tip; lores flush purplish pink; legs and feet dark red. **JUVENILE:** Like nonbreeding adult, but with black bill. **FLIGHT:** Wingflaps much more rapid than other egrets.
Geographic Variation Accidental (AK) subspecies *coromandus*, sometimes treated as a separate species, from Asia is larger and longer-necked and has a rich cinnamon wash on head and neck in breeding plumage.
Similar Species Larger Snowy Egret is slimmer with black bill and black legs with yellow feet. Immature Little Blue Heron has a gray bill with a black tip, dark-tipped primaries, and greenish gray legs and feet. Much larger Great Egret has a longer neck, longer legs, longer bill, and black legs; it flies with slow wingflaps.

Voice Emits a coarse nasal *rick-rack* around the nest or roost.
Status & Distribution Common to abundant. **BREEDING:** Colonially with other wading birds in shrubs or small trees, often on islands. **MIGRATION:** Partially migratory, withdrawing from the northern part of breeding range. **WINTER:** In the southern US, especially in southeastern CA, along the entire Gulf Coast, and throughout the FL Peninsula; also in the West Indies and American tropics. **DISPERSAL:** Wanders well north. **RARE STATUS:** Casual to southeast AK and the northern Canadian provinces. One specimen record to AK of the Asian subspecies *coromandus* at Agattu, Aleutians (19 June 1988).
Population First documented in the New World in the West Indies in the 1930s, the species colonized FL by

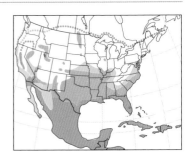

1941. Its arrival in the New World probably originated with birds from Africa crossing the Atlantic to S.A. By the early 1970s, had colonized most of the continental US. Most expansion has slowed or stopped, and major population declines have been seen in the Northeast and elsewhere since the peak in the 1970s and early 1980s.

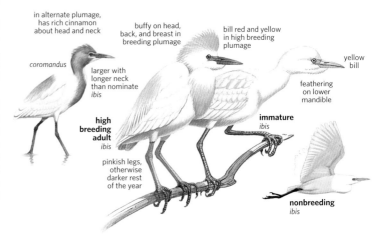

in alternate plumage, has rich cinnamon about head and neck

coromandus

buffy on head, back, and breast in breeding plumage

larger with longer neck than nominate *ibis*

high breeding adult *ibis*

pinkish legs, otherwise darker rest of the year

bill red and yellow in high breeding plumage

yellow bill

feathering on lower mandible

immature *ibis*

nonbreeding *ibis*

Genus *Ardeola*

CHINESE POND-HERON   *Ardeola bacchus*   CHPH   ▦   5

There are six small heron species in the genus *Ardeola*, all found in the Old World. These include the familiar, to

Europeans, Squacco Heron (*A. ralloides*), and the three Asian pond-herons: Indian Pond-Heron (*A. grayii*), Javan

Pond-Heron (*A. speciosa*), and the most migratory, the Chinese Pond-Heron, the only one occurring in northeast

Asia. Two other members of the genus occur in Africa. Most *Ardeola* herons have white wings that are particularly striking in flight. Monotypic. L 18" (46 cm) WS 34" (86 cm)

Identification NONBREEDING ADULT: Short and stocky with entirely white wings, rump, and tail. Head, neck, breast, and back dark. Yellow bill with black tip; yellow eyes, legs, and feet. **BREEDING ADULT MALE:** Head, neck, and upper breast bright chestnut. Base of bill bluish. Lower breast slaty colored. Long plumes. **BREEDING ADULT FEMALE:** Plumes less showy, foreneck pale; no slaty patch on breast. **WINTER ADULT AND IMMATURE:** Head and breast buffy, streaked with dark brown; mantle brown; white is usually hidden

on folded wing. **FLIGHT:** White wings, rump, and tail contrast vividly with dark body.

Similar Species Breeding adults are distinctive, but nonbreeding birds are almost identical to Indian Pond Heron (*A. grayii*) and Javan Pond Heron (*A. speciosa*), residents from southern Asia, and Squacco Heron (*A. ralloides*) from Europe, western Asia, and Africa.

Voice Undescribed; other pond-herons squawk.

Status & Distribution Some populations migratory. **BREEDING:** Common in China and Indochina. **WINTER:** Throughout much of Southeast Asia. Has strayed to Japan (where increasing) and Korea. **RARE STATUS:** Casual. Three summer records of

breeding-plumaged birds from western AK islands: Attu I., St. Paul I., and St. Lawrence I.

Population Common and increasing in its native range.

chestnut head and neck in breeding plumage

white wings in all plumages; striking in flight

breeding adult ♂

Genus *Butorides*

GREEN HERON   *Butorides virescens*   GRHE   ■   1

The Green Heron (found throughout N.A. and C.A.) was once taxonomically combined with the Striated Heron (*B. striata*, found throughout S.A. and much of the Old World, and recorded once on Bermuda) as the Green-backed Heron, based on reported widespread hybridization where their ranges overlap in the southern West Indies, southern C.A., and northern S.A. Their status as separate species was restored after it was learned that hybridization was not as widespread as was previously thought. The Green Heron is generally found alone or in family groups, occurring in many wetland habitats ranging from freshwater to saline, and open to wooded. It feeds primarily on small fish, but also takes a variety of aquatic invertebrates. The Green

Heron is one of the few N.A. birds to use tools to forage: It places a leaf, feather, piece of bread, or other object on the surface of the water. When a fish swims in to investigate, the heron grabs it. Polytypic (4 sspp.; 2 in N.A.). L 18" (46 cm) WS 26" (66 cm)

Identification Small and stocky, with short legs, a thick neck, and a small crest. **NONBREEDING ADULT:** Sexes similar. Crown and short crest blackish, sides of head and neck rufous. White chin, throat, underside of neck, and breast. Yellow-orange eyes; dark lores. Blackish yellow bill with greenish yellow base. Yellow legs and feet. Greenish gray back, wings, rump, and tail; upperwing feathers edged with buff. Gray belly and undertail coverts. **BREEDING ADULT:** Soft-part colors intensify: The lores flush bluish black; the bill turns largely black; the legs and feet flush orange. **JUVENILE:** Sides of face, neck, and breast heavily streaked with brown. Upperparts browner; upperwing feathers marked with buff edges and whitish spots at the tips. **FLIGHT:** Underwings of adult are uniformly dark gray; those of immature are paler with contrasting pale feather edges.

Geographic Variation The larger, paler subspecies *anthonyi* is found along the Pacific coast of Mexico and in the US Southwest; subspecies *virescens* occurs in east and central N.A., Mexico, the West Indies, and all of C.A.

Similar Species The only other small wading bird regularly encountered in N.A. is the smaller Least Bittern, which shows white scapular lines and bright buffy wing coverts in all typical plumages.

Voice Sharp *skeow* given when migrating at night or when flushed (often accompanied in flight by defecating). **ALARM CALL:** A series of *kuk* notes; raises crest and flicks tail.

Status & Distribution Uncommon to fairly common. **BREEDING:** From ND east to southern NS south to TX through FL. Also, mostly coastally from Puget Sound south through Baja California east to western AZ. Rare and sporadic through much of the interior. **MIGRATION:** May be seen casually through much of US. **WINTER:** Many birds resident along coasts and in the Deep South; distribution in C.A. more widespread. **RARE STATUS:** Casual to HI, Greenland, Iceland, British Isles, and Azores.

Population Numbers are stable.

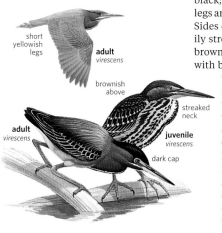

short yellowish legs

adult *virescens*

brownish above

streaked neck

adult *virescens*

juvenile *virescens*

dark cap

NIGHT-HERONS Genera *Nycticorax* and *Nyctanassa*

BLACK-CROWNED NIGHT-HERON *Nycticorax nycticorax* BCNH ▪ 1

This stocky heron feeds day and night on fish and aquatic invertebrates and even upland prey. Polytypic (4 sspp.; 2 in N.A.). L 25" (64 cm) WS 44" (112 cm)
Identification **NONBREEDING ADULT:** Black back; pale gray upperwings, rump, and tail. Black crown and nape; white forehead, face, neck. Two or three white occipital plumes. Red eyes; black bill; yellow legs and feet. **BREEDING ADULT:** Legs and feet

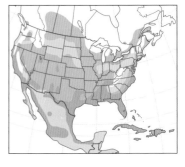

bright pink or red. Occipital plumes lengthen. **JUVENILE:** Brown above and paler below, with upperpart feathers tipped with large white spots, and underparts heavily streaked. Orange eyes; yellow bill with a dark culmen and tip. **FIRST-SUMMER:** Dark brown crown, back, and upperwings. Face and neck pale with diffuse light brown streaking; mostly pale below. **SECOND-SUMMER:** Like adult, but white forehead, hind neck, and underparts washed with brown or gray. **FLIGHT:** Wings short and rounded; only the feet project beyond tail.
Geographic Variation Subspecies *hoactli* is widespread in N.A.; nominate *nycticorax* from Eurasia is smaller; supercilium all-white.

Similar Species Larger, American Bittern has a blackish malar stripe; in flight, it shows more

pointed wings, and contrasting dark flight feathers. Juvenile Black-crowned is told from juvenile Yellow-crowned by overall browner plumage, paler crown color, shorter legs, and extensive greenish yellow on slightly slimmer bill. By its first summer, an immature begins to resemble an adult, and is readily told from a similarly aged Yellow-crowned.
Voice A harsh *wock*, lower pitched than the Yellow-crowned.
Status & Distribution Locally fairly common. **BREEDING:** Colonial. Also in Eurasia and Africa. **MIGRATION:** Partially migratory; withdraws from most of breeding range. **WINTER:** Along Pacific coast, entire Gulf Coast, along Atlantic coast from MA south. **RARE STATUS:** Casual to AK, Clipperton I., Greenland, HI. Nominate Eurasian subspecies recorded from western Aleutians and Pribilofs, AK.
Population Declining in northern part of range.

Illustration labels: pale wings · *hoactli* · juvenile · feet barely project past tail · adult · overall buffy brown and conspicuously marked with whitish spots · juvenile *hoactli* · extensive yellow on bill · brownish crown · 1st spring *hoactli* · 2nd spring *hoactli* · brownish tint · black crown and back · adult *hoactli* · white body · pink legs · high breeding adult *hoactli*

YELLOW-CROWNED NIGHT-HERON *Nyctanassa violacea* YCNH ▪ 1

Ranging from coastal beaches to inland cypress swamps, this species is more likely to be found nesting alone or in scattered pairs compared to other colonial heron species. It forages for aquatic invertebrates. Polytypic (5 sspp.; 2 in N.A.). L 24" (61 cm) WS 42" (107 cm)
Identification Longer neck and more upright posture than Black-crowned Night-Heron. **NONBREEDING ADULT:** Head black with bold buffy white cheek patch, pale yellow crown, and a few wispy occipital plumes. Largely blue-gray; feathers on back and upperwings black with wide blue-gray borders. Orange eyes; short, thick,

black bill; yellowish green legs and feet. **BREEDING ADULT:** Eyes scarlet; lores dark green; legs scarlet or bright orange. **JUVENILE:** Mostly grayish brown above with small white tips to feathers on upperparts; head and neck streaked. Dark bill shows some pale near base. **FIRST-SUMMER:** Like juvenile, but darker crown and mostly brown upperparts. **FLIGHT:** Slower wingbeats than Black-crowned; feet and a portion of the legs extend beyond tail.
Geographic Variation Nominate subspecies occurs in central and eastern US and Atlantic coast of C.A. Subspecies *bancrofti* has paler plumage,

narrower dorsal streaking, and a larger bill, and is found from Southern CA down the Pacific coast to C.A.; also West Indies.

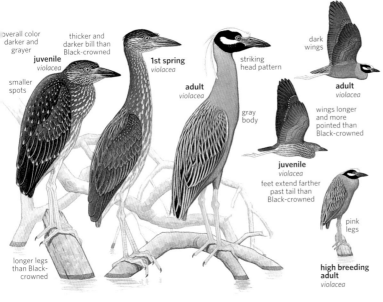

overall color darker and grayer

thicker and darker bill than Black-crowned

juvenile
violacea

smaller spots

1st spring
violacea

adult
violacea

striking head pattern

dark wings

adult
violacea

gray body

wings longer and more pointed than Black-crowned

juvenile
violacea

feet extend farther past tail than Black-crowned

pink legs

high breeding adult
violacea

longer legs than Black-crowned

Similar Species Juvenile is told from juvenile Black-crowned by somewhat darker and grayer plumage with less conspicuous white markings on upperparts; shorter, slightly thicker, and darker bill; and longer legs.

Voice Harsh *wock*, higher pitched than Black-crowned Night-Heron.

Status & Distribution Uncommon to fairly common, and local. **BREEDING:** Primarily coastal, from MA through FL to TX; a few inland north to MN, OH, and WI; now established coastal Southern CA. **MIGRATION:** Rarely seen in migration. **WINTER:** In the US, primarily central and south FL, but north to NC and west to TX. **RARE STATUS:** Casual to central and southeast CA, AZ, NM, ND, and southern Canada from SK to NS; also Clipperton I.

Population Thought to be stable.

IBISES AND SPOONBILLS Family Threskiornithidae

Roseate Spoonbill (TX, Mar.)

Behavior Colonial breeders, ibises and spoonbills nest and roost with other wading birds. They tend to forage in flocks away from other species. Tactile feeders, they feed on fish and aquatic invertebrates (especially crabs) in shallow fresh or salt water. Spoonbills feed most often by sweeping their bills back and forth, while ibises tend to probe to locate prey. Ibises also forage visually in uplands, feeding on terrestrial invertebrates such as grasshoppers.

Distribution The family is represented on all continents except Antarctica, reaching their greatest diversity in the tropics.

Taxonomy Worldwide, there are about 34 species in 13 genera (28 ibises in 12 genera; six spoonbills in one genus). Eleven ibises and one spoonbill occur in the New World. Recent higher-level taxonomic revisions now place the family Threskiornithidae in the order Pelecaniformes with the pelicans (Pelecanidae) and herons (Ardeidae). The species-level taxonomic status of ibises in North America has been debated for centuries. Glossy and White-faced Ibises have often been considered conspecific, but hybridization has only recently been documented. White and Scarlet Ibises are also often considered conspecific, based on hybridization in South America, where the two species occur sympatrically, as well as in FL, where Scarlet Ibises were unsuccessfully introduced in the 1960s.

Conservation The *Plegadis* ibises are expanding beyond previous range limits, but all threskiornithids remain at risk from colony disturbance and wetland loss or degradation. BirdLife International codes: 3 NT, 1 VU, 4 EN, 3 CR, and 1 EX.

Although easily distinguished from each other by bill shape, plumage coloration, and foraging behavior, ibises and spoonbills are closely related; hybridization between the two groups has been documented in the Old World.

Structure Threskiornithids are medium to large birds with relatively long necks and legs. Bill shape defines this family: long, decurved, and slender in ibises, and long, straight, and flattened, with a broad, spoon-shaped distal end in spoonbills. Flight is strong and direct on moderately long, rounded wings; flocks of ibises often soar.

Plumage Species are mostly unicolored, typically white in the spoonbills (but pink in Roseate Spoonbill) and white or blackish in the ibises. Immature plumage is dusky or mottled with brown in ibises, and whitish in spoonbills.

Genus *Eudocimus*

SCARLET IBIS   *Eudocimus ruber*   SCIB ■ 5

This wading bird graces wetlands in the New World tropics. Monotypic. L 23" (58.5 cm) WS 36" (91 cm) **Identification** Unmistakable. **BREEDING ADULT:** Entirely brilliant red with black tips to outer four primaries (most visible in flight). Red legs and bill with blackish tip. **NONBREEDING ADULT:** Pinkish bill. **IMMATURE:** Head and neck margined brown and gray. Brown upper back and wings; white lower back and rump tinged with pink. Underparts white tinged with

pinkish buff; dark gray legs and feet. **Similar Species** Roseate Spoonbill is larger with a spoon-shaped bill and pink and white plumage, rather than red. Juveniles indistinguishable from juvenile White Ibis until pink feathering develops. **Voice** Usually silent. **FLIGHT CALL:** Low, grunting *hunk-hunk-hunk*. **Status & Distribution** Common. Resident in northern S.A. and Trinidad. **BREEDING:** Colonial. **DISPERSAL:** Nonmigratory but prone to dispersal. **RARE**

STATUS: Perhaps to FL but valid occurrence clouded by issue of a few (mostly lost) 19th-century

adult

specimens and, more recently, by possibility of escapes. Observations elsewhere in US presumed to be escapes. **Population** Declining overall, but still common to abundant in places.

WHITE IBIS   *Eudocimus albus*   WHIB ■ 1

The White Ibis is found in virtually all wetland types and is often seen foraging in grassy fields. It feeds primarily on aquatic crustaceans in freshwater marshes, or on insects in fields. Monotypic. L 25" (64 cm) WS 38" (97 cm) **Identification ADULT:** White except for the black tips of the four outermost primaries (often hidden). Soft parts orange or orange-red; decurved bill

with a blackish tip. Eyes pale blue. **BREEDING ADULT:** Soft part coloration intensifies: face, bill, legs, and feet bright red or red-orange. **JUVENILE:** Head and neck streaked with dark brown. Brown mantle, upperwings, and tail; white rump. Underwings white with blackish band on trailing half of flight feathers. **FIRST-SUMMER:** Juvenile-like head and neck pattern with increasing white feathering on the back, wings, and finally head and neck as adult plumage acquired gradually. **FLIGHT:** Rapid wingbeats alternating with glides. Flies in V formation or tight flocks. **Similar Species** *Plegadis* ibises have dark rumps and underparts.

juvenile

black wing tips

breeding adult

dark eye

red face and white eye

pinkish red bill

white belly

juvenile

1st spring

breeding adult

red legs

Voice Often silent. **CALL:** Identical to Scarlet Ibis. **Status & Distribution** Locally common. Resident south through Bahamas (rare) and most of Greater Antilles and M.A. south to northern S.A. **BREEDING:** Colonially, with other wading birds or in large single-species colonies. Along Atlantic coast from VA south; along the entire Gulf Coast. **MIGRATION:** Partially migratory in northern parts of breeding range. Uncommon to northern TX, OK, AR. **RARE STATUS:** Rare and increasing on Atlantic coast from mid-Atlantic to Northeast. Casual to CA and the Southwest, north to SD, MI, and southern Canada; also to the Bahamas, Puerto Rico, Dominica, Bermuda, and Clipperton I. **Population** Increasing over the past 40 years.

DARK IBISES   Genus *Plegadis*

GLOSSY IBIS   *Plegadis falcinellus*   GLIB ■ 1

The predominantly eastern Glossy Ibis inhabits a variety of wetlands, from freshwater to saline. Its range now

overlaps with its western counterpart, the White-faced Ibis. Monotypic. L 23" (58 cm) WS 36" (91 cm)

Identification BREEDING ADULT: Rich chestnut-red on head, neck, upper back, and underparts. Lower back and tertials, glossed mostly with purple, wing coverts mostly with green. Uniformly grayish bill tinged horn-colored or dull red, thin and decurved. Dark gray facial skin bordered by pale bluish skin above and below the eye; brown eyes; dark gray legs and feet with dark maroon ankle joints. **NON-BREEDING ADULT:** Dark brown head and neck feathers edged with white, appearing finely streaked. **JUVENILE:** Sooty brown on the head, neck, and underparts with variable amounts of white splotching on the throat and chin. By late fall, many juveniles have

acquired the pale blue stripes bordering the facial skin. **FLIGHT:** Flocks fly in a weak V formation; birds appear all-dark at a distance.

Similar Species White-faced Ibis is very similar. See sidebar, p. 346.

Voice CALL: A sheeplike *huu-huu-huu-huu*. Flocks sometimes emit a subdued chattering or grunting, particularly on taking flight.

Status & Distribution
Fairly common but somewhat local. **BREEDING:** Coastally from southern ME south to VA and south along the Atlantic coast to coastal and inland locations in southern FL and locally along the Gulf Coast west to LA; also uncommon on Cuba. Widespread in the Old World. **WINTER:** Mainly along the Gulf Coast, throughout FL, and along Atlantic coast north to SC. **RARE STATUS:** Rare in the interior East and north to southern

Canada. Casual to western states.
Population Rare in N.A. through the 1930s, but its numbers have increased since then.

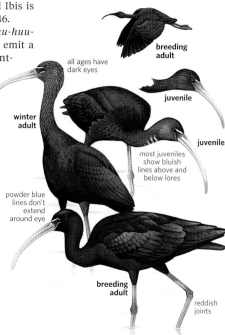

breeding adult

all ages have dark eyes

juvenile

winter adult

juvenile

most juveniles show bluish lines above and below lores

powder blue lines don't extend around eye

breeding adult

reddish joints

WHITE-FACED IBIS *Plegadis chihi*   WFIB ■ 1

The White-faced Ibis is the quite similar, western counterpart to the eastern Glossy Ibis, mostly replacing it west of the Mississippi River. Monotypic. L 23" (58 cm) WS 36" (91 cm)

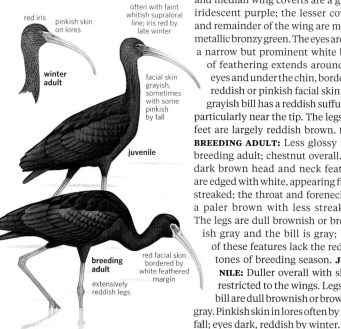

red iris
pinkish skin on lores

often with faint whitish supraloral line; iris red by late winter

winter adult

facial skin grayish, sometimes with some pinkish by fall

juvenile

breeding adult

red facial skin bordered by white feathered margin

extensively reddish legs

Identification Tall bird with long legs and a long decurved bill. **BREEDING ADULT:** Rich chestnut maroon plumage on the head, neck, upper back, and underparts. The lower back, tertials, and median wing coverts are a glossy iridescent purple; the lesser coverts and remainder of the wing are mostly metallic bronzy green. The eyes are red; a narrow but prominent white band of feathering extends around the eyes and under the chin, bordering reddish or pinkish facial skin. The grayish bill has a reddish suffusion, particularly near the tip. The legs and feet are largely reddish brown. **NON-BREEDING ADULT:** Less glossy than breeding adult; chestnut overall. The dark brown head and neck feathers are edged with white, appearing finely streaked; the throat and foreneck are a paler brown with less streaking. The legs are dull brownish or brownish gray and the bill is gray; both of these features lack the reddish tones of breeding season. **JUVE-NILE:** Duller overall with sheen restricted to the wings. Legs and bill are dull brownish or brownish gray. Pinkish skin in lores often by mid-fall; eyes dark, reddish by winter.
Similar Species The Glossy Ibis is very

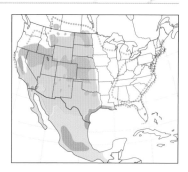

similar in appearance (see p. 346).
Voice Sound identical to Glossy Ibis. **CALL:** A sheeplike *huu-huu-huu-huu*. Flocks sometimes emit a subdued chattering or grunting.

Status & Distribution Locally common; south to S.A. **BREEDING:** Colonially, usually in dense marsh vegetation in low shrubs or trees, often on islands. **MIGRATION:** Withdraws from northern part of breeding range; migrates primarily south through the Great Basin and the Colorado River Valley; migrants in the desert southwest usually seen singly or in small groups. **WINTER:** Southern CA, southwestern AZ, and coastal TX and LA. The largest concentrations are found mostly in late summer through winter near the Salton

Sea, with a peak population estimated in the tens of thousands. **RARE STATUS:** Casual to the Midwest and the East Coast north to ON, QC, and NB; and south to FL (has bred). Accidental to southeast AK, Clipperton I., and HI. **Population** White-faced Ibis populations declined during the 1970s, but, in recent years, the breeding range and populations have expanded in the West. Populations in TX and LA have shown declines.

Identification of *Plegadis* Ibises

As both *Plegadis* ibises have expanded their ranges, their identification has become a continent-wide issue. White-faced Ibises are now annual along the Atlantic and eastern Gulf Coasts, and in the Midwest, while Glossy Ibises now occur regularly to the Great Plains, and some have reached the Pacific coast. The breeding ranges of both species overlap in LA and perhaps elsewhere. Hybridization in the wild has been documented and noted widely through the range of both species, including strays to the West Coast. A hybrid can be very difficult to recognize and is often misidentified as a Glossy Ibis.

Distinguishing *Plegadis* ibises in breeding plumage is easier than in other plumages. Breeding adult Glossy Ibises have brown eyes, dark lores with narrow pale blue borders that do not extend behind the eyes or under the chin, and gray legs and feet with contrasting reddish ankle joints. In contrast, breeding adult White-faced Ibises have red eyes, pink or red loral skin with wide white borders that extend behind the eyes and under the chin, and more extensive pinkish legs and feet.

The identification of *Plegadis* ibises in nonbreeding and immature plumages requires close views in good light, and even then, some immatures cannot be identified with certainty. Nonbreeding adult Glossy Ibises retain from breeding plumage the dark eyes and dark gray facial skin with a narrow but well-defined pale blue border above and below the eye. The legs are entirely grayish. Adult White-faced Ibises retain their red eyes and pale pinkish facial skin, but lose the white outline around it. Juvenile *Plegadis* ibises cannot always be identified conclusively in the field. However, by their first fall, even late summer, many juvenile Glossy Ibises have acquired the pale blue facial stripes, and many juvenile White-faced Ibises have acquired a reddish cast to their eyes and a touch of pink facial skin by mid-fall. Leg color and bill color between the two species are too variable to be used as useful field marks.

SPOONBILLS Genus *Platalea*

ROSEATE SPOONBILL *Platalea ajaja* ROSP ■ 1

The Roseate Spoonbill's pink and red plumage makes it one of the most flamboyant wading birds. Monotypic. L 32" (71 cm) WS 50" (127 cm)

juvenile

breeding adult

breeding adult

crimson marginal coverts

pale pink wings

juvenile

spoon-shaped bill

Identification Unmistakable. All ages have a large, flat, spoon-shaped bill and pink wings and back. **ADULT:** Head featherless, yellowish or greenish gray with black collar. Bill grayish with black markings; eyes, legs, and feet reddish. Red patch on white breast; lower back pink; rump red. Tail orange. Wings pink with wide red carpal bars. **FIRST-WINTER:** Head fully feathered. Plumage white except for wings and tail, which are entirely pale pink. Yellowish bill. **SECOND-WINTER:** Like adult, but plumage paler; wings lack red carpal bars.
Similar Species Flamingos share the spoonbill's pink plumage, but they have much longer and thinner necks and longer legs, as well as very different bill shape and foraging posture.
Voice Silent except at breeding colonies; various low, guttural grunts and clucking.
Status & Distribution Found from southern N.A. to southern S.A. In N.A., locally common. **BREEDING:** Colonially on estu-arine islands, rarely inland. ±5,000 pairs breed in FL, LA, and TX. **DISPERSAL:** Nonmigratory, but prone to summer-fall postbreeding dispersal along Gulf and south Atlantic coasts; uncommon inland to northeast TX, southeast OK, and AR. Northern dispersals usually made by immatures. **WINTER:** Resident in coastal southern FL, LA, and TX. **RARE STATUS:** Casual or accidental north to OH, PA, QC, and ME; and west to CO, UT, and NV. Casual to Southern CA (once to central CA) and the Southwest from colonies in northwestern Mexico, but numbers reduced in recent decades. **Population** Increasing in places.

NEW WORLD VULTURES Family Cathartidae

Turkey Vulture (FL, Nov.)

New World vultures share their naked heads and the practice of feeding on carrion with Old World vultures. Usually found in warm climates, they have a unique habit of urinating down their legs, letting the evaporative process cool their bodies in the process. Banding and tagging studies must therefore use wing tags instead of the standard metal or plastic leg band, which would interfere with that cooling process, and in some cases harm the bird.

Behavior Solitary nesters, New World vultures are otherwise gregarious in roosting, searching for food, and gang-feeding at carcasses. Bills are hooked for tearing into flesh, but their feet are relatively weak, not adapted for killing or grasping prey. A single bird spiraling down out of the sky toward food will attract the attention of other birds from a great distance. At the roost site in the morning, they often will perch spread-winged in the sun before leaving to forage. Eggs are laid on bare surfaces, such as cliff ledges, caves, hollow logs, abandoned buildings, and the ground.

Plumage All species are primarily black on the wings, body, and tail, with some pattern of white or silver on the underwings. All share the naked head, in order to feed on carcasses without fouling head feathers. Immatures usually have dark skin on the head, with the adult's skin lightening or becoming more colorful. The King Vulture (*Sarcoramphus papa*), distributed from central Mexico to northern Argentina, is unique in having extensively white plumage as an adult.

Distribution Limited to the Americas. Primarily warm-weather species, New World vultures are colonizing more northerly areas, perhaps as a consequence of climate changes, or perhaps due to increased availability of roadkills.

Taxonomy Recent genetic data strongly refute this family's close relationship to the storks (Ciconiidae). Some authorities place the New World vultures in their own order, Cathartiformes, although the AOS retains them in Accipitriformes (Hawks, Kites, Eagles, and Allies). Worldwide there are seven species, three of which occur in North America.

Conservation The California Condor is on the US Endangered Species List, and is the focal point of federal and state programs where captive-bred individuals are released into prospective habitats to re-establish it as a viable wild species. Farmers occasionally persecute the Black Vulture for aggressive behavior toward newborn farm animals. BirdLife International codes: 1 NT and 1 CR.

Genus *Coragyps*

BLACK VULTURE   *Coragyps atratus*   BLVU ■ 1

The gregarious and widespread New World species (except West Indies) roosts, feeds, and soars in groups, often mixed with Turkey Vultures. A carrion feeder that will bully a Turkey Vulture away from a carcass, it occasionally kills smaller live prey. Polytypic (3 sspp.; nominate in N.A.). L 25" (64 cm) WS 57" (145cm)

Identification ADULT: Glossy black.

soars on flat wings; wingflaps are rapid and shallow

extensive white base to outer primaries

adult
atratus

pale grayish legs

short tail

atratus

Whitish inner primaries often hard to see on the folded wing. Whitish legs. Skin of head wrinkled; bill dark at base and tipped ivory or yellowish. **JUVENILE:** Black body and wing feathers usually duller, less iridescent. Skin of head smooth, darker black than an adult. **FLIGHT:** Conspicuous white or silvery patches at base of primaries that contrast with black wings, body, and tail. Whitish legs extend almost to tip of relatively short tail. Soars and glides with wings held fairly flat. If seen at a distance, the quick, shallow, choppy wingbeats interspersed with glides are usually enough for identification.

Similar Species The Turkey Vulture shows silvery inner secondaries and a pronounced dihedral while in flight, along with deeper, slower wingbeats.

Voice Hisses when threatened.

Status & Distribution Abundant in the Southeast, expanding up the East Coast into southern New England. Less common in southern Great Plains, local in southern AZ. **BREED-**

ING: Nests in a sheltered area on the ground, including abandoned buildings. **MIGRATION:** Sedentary, northern breeders may migrate with Turkey Vultures to warmer winter territory. **RARE STATUS:** Very rare to CA, northern New England, and southern Canada; accidental to southwestern NM, and north to BC, southern YT, central MB, central ON, central QC, and Newfoundland. Casual West Indies.

Population The species adapts well to human presence, feeding on roadkills and at garbage dumps.

Genus *Cathartes*

TURKEY VULTURE   *Cathartes aura*   TUVU ■ 1

The most widespread vulture in the New World, the Turkey Vulture is locally called "buzzard" in many areas. It is unique among our vultures in that it finds carrion by smell as well as by sight. When threatened, it defends itself by vomiting powerful stomach acids. Polytypic (4 sspp.; 3 in N.A.). L 27" (69 cm) WS 69" (175 cm)

Identification Overall, it is black with brownish tones, especially on the feather edges. Legs are dark to pinkish in color, the head unfeathered. **ADULT:** The red skin color of the head contrasts

with the ivory bill and dark feather ruff on the neck. **JUVENILE:** The skin of the head is dark, the bill dark with a pale base. **FLIGHT:** A large dark bird that flies with its wings held in a noticeable dihedral. Usually rocks side-to-side, especially in strong winds. Underneath, the silvery secondaries and primaries contrast with the black wing coverts, giving a two-toned look to the wing. The tail is relatively long.

Geographic Variation Three subspecies are found in N.A. (*aura, meridionalis, septentrionalis*). Only a few minor differences in size and overall tone separate them.

Similar Species The Black Vulture has a quicker, shallow flap, shorter tail, and obvious white patches at the base of the primaries on its shorter, broader wings. The Zone-tailed Hawk and Golden Eagle will mimic the wing dihedral when hunting, and dark-morph Swainson's Hawks and Rough-legged Hawks also can fly in a dihedral, but at closer range all have feathered heads and different wing

long, narrow wings, unlike Black Vulture

adults

long tail

two-toned underwing

soars on slight dihedral; wingflaps are slow and deep

reddish head

grayish head

adult

juvenile

grayish legs on juvenile

shapes. Zone-tailed also shows banded tail and yellow cere and legs.
Voice Hisses when threatened.
Status & Distribution Year-round in southern US, migrates into northern US and Canada. **BREEDING:** Nests on the ground, using a shallow cave, hollow log, or thick vegetation. **MIGRATION:** Northern populations migratory, some heading to C.A. Also resident in S.A. and West Indies. **WINTER:** Increasing numbers in snowbound states. **RARE STATUS:** Casual to AK, YT, NT, northern MB, and NL. **Population** Stable.

Genus *Gymnogyps*

CALIFORNIA CONDOR *Gymnogyps californianus*   CACO ■ 6

The largest raptor in N.A., the California Condor is unmistakable when seen flying with any other bird. It will soar for hours on thermals in search of carrion, rarely flapping, and can cover vast amounts of territory. The wingbeat is slow and deep, the wing tips appearing almost to touch at the bottom of the stroke. Monotypic. L 47" (119 cm) WS 108" (274 cm)
Identification ADULT: The body is black; the head shows bare red-orange to yellow skin. Large areas of white are visible on the underwings. Legs are whitish. **JUVENILE:** Body and legs are similar to adult, but the head is dark. Mottled white underwings. **FLIGHT:** Wings held flat or with the tips slightly higher, with large splayed primaries ("fingers") evident. The dark tail is relatively short. Seen from above, the wings have a thin white diagonal line at the greater coverts.
Similar Species Smaller Turkey Vulture lacks white on underwings, and flies with a more pronounced dihedral. Smaller Golden and Bald eagles have feathered heads and longer tails.
Voice Hisses and croaks.
Status & Distribution California Condor fossil evidence has been found through much of western N.A. and even southern Canada. It was found north to the Columbia River to about 1900. After the early 1900s the species was restricted to a relatively small area in Southern CA (casual to central CA into the early 1970s). Many condors died by feeding on poisoned carcasses set out by farmers to control coyotes and by ingesting lead bullets from deer carcasses. The US listed the species as endangered in 1967. As the population dwindled, the remaining 22 wild birds were live-trapped (the last on 19 Apr. 1987) and placed into a captive breeding program. After successfully building up the captive breeding population, two captive-bred condors were released into the wild in 1992. Today, free-flying birds (over 312 in Dec. 2018) are gaining a toehold in their native range in CA and northern Baja California; also increasing where introduced in their non-native range in the Grand Canyon area of AZ and UT. Some released condors have produced young in the wild (starting in 2002), but about 200 birds remain in the captive breeding program. **YEAR-ROUND:** Nonmigratory. **BREEDING:** Solitary nesters, usually on a cliff face in a shallow cave; easily disturbed by man. Fledged chicks are looked after for up to eight months.

Population Critically Endangered. Listed as Endangered by the USFWS. Even though the Condor Recovery Plan has achieved a large measure of success, the California Condor is still considered critically endangered, with a total population (captive and free flying) in 2019 of about 1,000 birds.

juvenile

whitish wing linings

wings are long, broad, and of rather uniform width

unlike Turkey Vulture soars with no distinct dihedral

orange head

adult

white wing linings contrast with black flight feathers

adult

short tail

huge size

adult

white tips to greater coverts and white on edges of secondaries visible from above

OSPREY Family Pandionidae

Osprey (Scotland, July)

T his widespread fish-catching specialist, once part of the diverse Accipitridae family, is now placed in its own family.

Structure A large, long-winged, hawklike bird, it has a narrow, rounded head that lacks the bony "brows" of most other large raptors. The long legs have strong talons and remarkable spikelike pads on the toes that help grip slippery fish; the outer toe can be reversed to better hold prey.

Behavior Consummate fish-catchers. With strong, gull-like flapping flight, Ospreys are less reliant on rising air currents than most other large raptors and more readily cross large water gaps.

Plumage See species account.

Distribution Ospreys breed widely through North America, Eurasia, and northern Africa, as well as the Australasian region. They are nearly or entirely absent from most oceanic island groups, except West Indies.

Taxonomy The family consists of one genus and one species with four subspecies; some authorities have considered the subspecies *cristatus* of the Australasian region and the pale-headed *ridgwayi* of the Caribbean to be distinct species. Recognition of the family Pandionidae as distinct from the other diurnal birds of prey is based on studies of DNA, chromosomes, and morphology.

Conservation Least Concern.

| Genus *Pandion*

OSPREY   *Pandion haliaetus*   OSPR ■ 1

Known to many as the "Fish Hawk," the Osprey is almost exclusively a fish-eater, capturing its prey by plunging into the water feet-first. Polytypic (4 sspp.; widespread *carolinensis* and Caribbean *ridgwayi* in N.A.). L 22–25" (56–64 cm) WS 58–72" (147–183 cm)

Identification Large raptor; dark above and white below with a white head and a prominent dark eye stripe. **ADULT:** Dark above, white below; most Osprey have a necklace of breast streaks; females are larger and usually more heavily streaked on breast, but there is overlap in this feature. **JUVENILE:** Upperparts fringed with pale buff, giving it a scaly appearance. **FLIGHT:** Long, narrow wings held back at the "wrist," dark carpal patches on underwing.

Geographic Variation Caribbean *ridgwayi*, rare or casual to FL Keys, has broader wings and larger feet; it is largely white headed and is paler above and on underwing.

Similar Species Fairly unmistakable.
Voice Series of loud, whistled *kyew*'s.
Status & Distribution YEAR-ROUND: Present year-round in southern US. BREEDING: Large stick nest on dead treetops or human-made structures. MIGRATION: Late Mar.–Apr. in spring; mostly late Aug.–Sept. in fall. RARE STATUS: Casual to Aleutians, Bering Sea region, and northern AK.
Population Controls on pesticides (chiefly DDE) have resulted in rebounding populations.

dark "wrist"
pale wing linings
barred flight feathers
gull-like flight
carolinensis
long angled wings

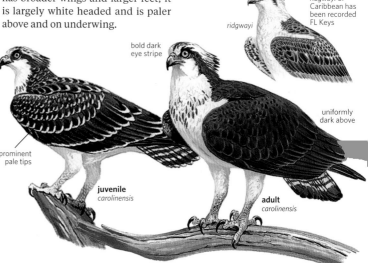

ridgwayi

prominent pale tips
bold dark eye stripe

whiter-headed *ridgwayi* of Caribbean has been recorded FL Keys
ridgwayi
uniformly dark above

juvenile *carolinensis*
adult *carolinensis*

HAWKS, KITES, EAGLES, AND ALLIES Family Accipitridae

Bald Eagle, adult (left) and subadult (AK, Mar.)

The North American members of family Accipitridae include a wide range of body sizes, structures, flight and hunting styles, prey selection, and habitat preference.

Structure Common features of this family are a hooked beak, strong feet with curved talons. The buteos have long, wide, rounded wings for soaring, as do the eagles. The kites and harrier have slimmer wings with pointed or rounded wing tips and long tails to help with hovering or maneuvering. Short, wide wings and long tails give the accipiters the ability to make quick turns in close quarters. Most juveniles have shorter wing feathers and longer tails than adults. Females are usually larger than males, and among species that inhabit a wide range of latitudes, northern breeders are usually larger than southern birds.

Behavior Not surprisingly, birds with different body and wing types employ different hunting strategies. The soaring buteos and eagles make long glides or dives from above on small mammals, reptiles, and ground birds. Species called *kites*—now in three subfamilies—not surprisingly use a variety of hunting techniques: White-tailed Kite hovers and then drops on prey; Swallow-tailed and Mississippi Kites obtain insect prey in flight; Hook-billed and Snail Kites feed largely on snails (Snail Kite routinely hovers). The accipiters are bird-catchers, speeding through the trees and brush to capture their prey on the wing. All species except the harrier build a stick nest,

usually in a tree, but western species may use a rock ledge. Harriers use grasses and sedges in their ground nest, which is usually concealed in longer grasses or reeds. Most breeders are site-faithful, and the larger birds will reuse a nest site for many years, adding to the nest structure. Most species are migratory, following well-known migratory paths and are observed at well-known hawk-watch sites. The southwestern species are more sedentary, but may wander outside of the breeding season. Mississippi and Swallow-tailed Kites are becoming regular wanderers in late spring up to the northeastern states, upper Midwest, and southern Ontario, and it is theorized that climate change may be contributing to possible range expansions.

Plumage Immatures are distinct from adults, with most species acquiring adult plumage in one year; the eagles take four years. Body feathers are replaced annually, as are flight feathers on the smaller hawks. The eagles and many buteos typically do not replace all their flight feathers in one year, creating visible patterns of old and new feathers. All show significant feather wear by spring, bleaching out colors on the upperparts.

Distribution The species in this family inhabit virtually every habitat in North America, from tundra to desert, mountain forests to coastal marshes. A species' relative abundance is usually tied to its habitat and prey preference. All are regular breeders in North America (though not necessarily common) except two of the sea-eagles, Double-toothed Kite, Crane Hawk, and Roadside Hawk, which are accidental or casual strays.

Taxonomy Worldwide the family comprises about 240 species in 69 genera; 30 species in 16 genera in North America. The genus *Buteo* is best represented in North America, with nine species. Recent genetic work has shown that traditional groupings (e.g., kites and Old World vultures) are not monophyletic. In North America there are now three subfamilies recognized: Elaninae (White-tailed Kite in North America), Gypaetinae (Hook-billed and Swallow-tailed Kite in North America), and Accipitrinae (all other North American species).

Conservation Traditionally, raptors have suffered most from indiscriminate shooting, although in recent years that practice has diminished. Fish-eating raptors have suffered population declines in the past due to pesticide-laden food sources. After banning most of the offending substances for use in North America, increases in breeding populations have been encouraging. Species that winter in Central America and South America, where some of these compounds remain in use, are still considered at risk, but to a lesser degree than previously. Old World vultures in Asia and Africa have suffered catastrophic losses in recent years from ingesting meat of domestic livestock that were given drugs fatal to vultures—eight species are Critically Endangered. Birdlife International codes: 33 NT, 24 VU, 17 EN, 13 CR, and 1 EX.

KITES Genera *Elanus, Chondrohierax, Elanoides,* and *Harpagus*

WHITE-TAILED KITE *Elanus leucurus* WTKI ■ 1

A striking, mostly white species of the Pacific states, TX, and FL. Often seen hovering or perched on telephone wires or dead snags; when hunting they slowly descend on small rodents with their wings held up in a deep V. Gregarious in fall and winter. Polytypic (2 sspp.; *majusculus* in N.A.). L 16" (41 cm) WS 42" (107 cm)

Identification ADULT: The white head, tail, and undersides contrast with black upperwing coverts (the "shoulders," formerly called Black-shouldered Kite), and gray crown, back, and upper flight

feathers. Below, the black carpal patch contrasts with gray primaries and white secondaries and coverts. Deep red eyes are framed with dark feathers. **JUVENILE:** Best told by brownish feathers on the back and crown, rufous wash across the breast, brown to orange eye. The tail has faint gray subterminal band, flight feathers with white tips. The darker body feathers are molted within a few months, giving the bird a more adult appearance, but still with the retained flight feathers. **FLIGHT:** Fast, shallow wingbeats interspersed with glides, pausing to hover tail-down, wings held high when hunting. Glides with a slight dihedral.

Similar Species Mississippi Kites never hover. They have a slate gray body with a black tail. Juveniles are heavily streaked. An adult male Northern Harrier has white rump above darker tail, no carpal patches on underwing, and a low, swooping flight with no hovering.

Voice Various whistled calls.

Status & Distribution YEAR-ROUND: Resident in coastal CA to the Sierra Nevada, grasslands of southeastern AZ and into TX, uncommon in FL,

rarely in southern NM. **BREEDING:** Builds a small nest in a solitary tree or brushline. **RARE STATUS:** Casual north to BC and MN, east to the East Coast.

Population Recent declines in Pacific states; FL population increasing.

habitually hovers when foraging

dark spot near "wrist"

adults *majusculus*

long whitish tail

buffy on chest

juvenile *majusculus*

black shoulders

adult *majusculus*

HOOK-BILLED KITE *Chondrohierax uncinatus* HBKI ■ 3

A tropical hawk that is a rare resident in the Rio Grande Valley area in TX. Most sightings are of birds taking flight in the morning; they rarely venture up in the air later in the day. In south TX feeds largely on tree snails (*Rabdotes alternatus*). Polytypic (3 sspp.; nominate in N.A.). L 18" (46 cm) WS 36" (91 cm)

Identification Featuring a long, slim body with long, square tail, the wing silhouette is unique: Wide, rounded primaries and bulging secondaries taper down to a narrow wing base, giving a diagnostic paddle-shaped wing.

Dark morph is rare in the US. **ADULT MALE:** Dark gray with light barring across the belly, tail with two wide bands, white eye color. **ADULT FEMALE:** Brown with heavy rufous barring on underside, rufous on underwings, two tail bands, white eye. **BLACK MORPH:** Rarely seen in US; adult is all black except for a single white tail band and tail tip. **JUVENILE:** Like adult female, but with thinner barring underneath, no rufous in wings, three narrow bands on tail, brownish eyes. **FLIGHT:** Slow, deep wingbeats for a bird of its size; will occasionally soar on thermals.

Similar Species Wing shape is unmistakable in flight. Red-shouldered Hawk shows crescents at base of primaries, narrow white tail bands.

Voice Loud rattling notes given near nest when disturbed or during courtship.

Status & Distribution Largely resident, with some migratory movement. Found from Mexico

all figures *uncinatus*

extensive white face

variable dark barring below

whitish underparts

juvenile

large pale eye

hooked bill

barre grayis unde

adult ♀

rufous collar

adult ♂

barred rufous underparts

to S.A. Rare and local along Rio Grande (Hidalgo Co. and Starr Co.). **RARE STATUS:** Casual in Zapata Co. and Cameron Co. Accidental to Smith Pt., Chambers Co., TX, on 29 Oct. 2011. **Population** The US population is small and is likely declining, possibly as a result of droughts and declining tree snail populations.

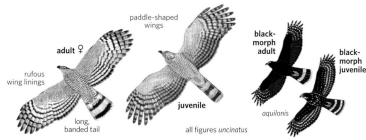

paddle-shaped wings

adult ♀

rufous wing linings

long, banded tail

juvenile

all figures *uncinatus*

black-morph adult

black-morph juvenile

aquilonis

SWALLOW-TAILED KITE *Elanoides forficatus*   STKI   1

usually seen in flight; white body and wing linings contrast with black flight feathers

long forked black tail

adults
forficatus

Elegant and graceful in flight, this species readily soars but does not hover. It feeds mostly on insects, lizards, snakes, and frogs, usually taken on the wing. Polytypic (2 sspp.; *forficatus* in N.A.). L 23" (58 cm) WS 48" (122 cm) **Identification** Head, underparts, and wing linings pure white; otherwise blackish. **ADULT:** Has red eyes. **JUVENILE:** Brownish eyes, the shorter tail is less deeply forked, flight feathers have lighter tips. **FLIGHT:** Unmistakable. **Voice** Mostly silent. **Status & Distribution** Uncommon. Found from southern US to S.A. **BREEDING:** Builds a flimsy stick nest in a treetop. **MIGRATION:** Most pass through the FL Keys, increasing numbers through southern TX hawk-watch sites. **WINTER:** Radio-tagged individuals from the Orlando, FL, area have been satellite-tracked to S.A. and back.

RARE STATUS: Increasing frequency of records into the upper Midwest and northeastern US, mostly in spring and summer. Casual Southwest and across Canada from MB to NS. Accidental CA. Rare migrant in West Indies. **Population** Small (about 1,000 pairs in US), but stable. Historical range much larger (north to MN, IL, OH).

DOUBLE-TOOTHED KITE *Harpagus bidentatus*   DTKI   5

The Double-toothed Kite's scientific name, *bidentatus* (two teeth), derives from the notches on the cutting edge of the maxilla. This feature, as in shrikes and falcons, is used to dispatch prey quickly with a bite to the neck. Partial to forests, but also seen in flight often circling high overhead. Polytypic (2 sspp.; likely northern *fasciatus* in N.A.). L 13.5" (34 cm) WS 27" (69 cm) **Identification** A small compact raptor with distinct, dark central throat stripe; recalls an *Accipiter* in shape. Long tail with three white bars. **ADULT:** Underparts barred vivid rufous on breast (more solidly rufous on adult female), gradating to gray and white barring on belly (rufous and white on adult female) and contrasting with fluffy white feathering of undertail coverts that in flight extend up on the sides of the rump. Underwing coverts white; primaries and secondaries barred blackish. Upperparts soft

gray. **IMMATURE:** Head and upperparts brownish; underparts buffy white with brown streaks and bars. **Geographic Variation** Larger nominate subspecies *bidentatus* (S.A. east of the Andes) has more intensely rufous and more densely barred (solidly rufous on female) underparts than *fasciatus* (central Mexico to western Ecuador). **Similar Species** A soaring Double-toothed Kite may bear an uncanny resemblance to a small *Accipiter*. The dark line down the middle of the throat and fluffy white undertail coverts distinguish Double-toothed from all other New World raptors. **Voice** Sharp, high-pitched whistled *tswee!* in series or singly, sometimes ending with a short, rising flourish, *yip!* Some calls sound uncannily like those of pewees (*Contopus*). **Status & Distribution BREEDING:** Lowland and mid-elevation forests. **MIGRANT:** Not known as a

checkered primaries

adult
fasciatus

rufous barring on underparts

two notches on bill

puffy white undertail coverts

shape and structure recall an *Accipiter*

underparts with bars and streaks on subadult

subadult
fasciatus

banded tail

long-distance migrant, but Trinidad records likely involve over-water crossings from S.A. **RARE STATUS:** Subadult at High I., TX (4 May 2011), and an adult in Hernando Co., FL (15 Oct. 2018). **Population** Least Concern.

EAGLES Genus *Aquila*

Worldwide, there are 11 species in this genus, only one of which occurs in N.A.: the Golden Eagle. Large raptors with wide wings, all are fierce hunters of mammals and birds, and many are prized for falconry. The human descriptive term "aquiline nose" refers to the hooked beak of *Aquila* eagles.

GOLDEN EAGLE *Aquila chrysaetos*   GOEA ■ 1

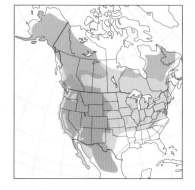

This large, dark eagle is a fierce hunter, preying on small to medium-size mammals but also taking ground birds, reptiles, insects, and even carrion in winter. This species is much more common in western N.A. Females are larger than males. Polytypic (6 sspp., *canadensis* in N.A.). L 30–40" (76–102 cm) WS 80–88" (203–224 cm)

Identification All ages a uniform dark brown, with varying amounts of golden feathers on nape, often appearing quite bright. Feathered tarsus. Bill appears tricolored: dark tip, lighter base, and yellow cere. **JUVENILE:** Has noticeable white base to tail, cleanly separated from dark terminal band, inner primaries with white patch visible in flight, sometimes even from above. Amount of white in the wing varies by individual, not simply age. Wings are wider and shorter than those of older birds. **SUBADULT:** Second-, third-, and fourth-year plumages are distinctive. Second-year birds will have an irregular trailing edge to the wing as new, shorter feathers mix with retained juvenal feathers. New tail feathers have a less clear-cut boundary between white tip and dark base. Upperwing coverts become tawny, giving a diagonal bar across the wing. Third- and fourth-year birds lose the white base of tail and white on primaries also disappears. **ADULT:** All-dark plumage with no white in wings or tail. Adult male has two or three dark gray bands on brown tail, while adult females have a single gray region in center of tail. **FLIGHT:** Powerful flaps can accelerate it surprisingly fast; it often power-dives after prey. Soars with wings flat or in a dihedral, and sometimes flies with and mimics Turkey Vultures.

Geographic Variation Six subspecies, five of which are found in the Old World; *canadensis* occurs in N.A. Size similar to European nominate subspecies but blacker in coloration, including on crown, with a sharply contrasting golden nape. Acquires adult plumage slightly earlier than nominate in about its sixth year. North Asian *kamtschatica*, found east of the River Yenisey east to western Chukotka and Kamchatka, similar but much larger.

Similar Species Immature Bald Eagle always has white by the body at the base of the wing, a longer head and shorter tail projection in flight, and a larger, bicolored bill. Turkey Vulture has silvery flight feathers underneath and a naked head.

Voice A rather faint and thin *kee-yep* or *yep*, sometimes in a series; usually silent away from nesting area.

Status & Distribution Found Eurasia and N.A. Subspecies *canadensis* found south to northern Mexico. **BREEDING:** Builds large stick nest on a high ledge or cliff, occasionally in a large tree. **MIGRATION:** Northern nesters move south late (late Oct.–Dec.) and return north mostly in late Mar. and Apr. with some into May. Southern birds (mostly adults) are sedentary. **WINTER:** Very local in the East; casual to the Gulf Region.

Population Formerly persecuted by farmers and ranchers. Breeding numbers are thought to be slowly declining in the western US but stable in Canada and AK.

golden nape

uniform dark underwing

adult
canadensis

longer tail projection than Bald Eagle

adult
canadensis

dark or faintly barred tail

juvenile
canadensis

short head projection

whitish wing patch

whitish tail base of variable width

HARRIERS Genus *Circus*

There are 14 species in this genus; one or more species occurs on every continent except Antarctica. Only the long-winged, slim-bodied Northern Harrier is found in N.A. It has recently been split from the Hen Harrier of the Old World. Harriers have facial discs that help focus the sounds made by hidden rodents.

NORTHERN HARRIER *Circus cyaneus*   NOHA ■ 1

This long-winged, long-tailed hawk flies low over grassy fields or marshes. Females are larger than males. Adult males are distinctive; females and juveniles are similar in coloration. Monotypic. L 17–23" (43–58 cm) WS 38–48" (97–122 cm)

Identification Long, slim, rounded wings, and a long tail with white on the rump. Small head, with pale eyes (except juvenile female) and well-defined facial disc like an owl. **ADULT MALE:** Pale gray above with darker markings, white wings and body below, and black wing tips. **ADULT FEMALE:** Brown above, underparts are lighter brown with distinct streaking on breast and belly. **JUVENILE:** Similar to adult female, with darker brown head, underparts less streaked and belly is rich cinnamon in color. Undersides of secondaries usually darker than primaries. Juvenile male has yellowish to gray eyes; female's eyes dark brown. **FLIGHT:** Wings held in a dihedral, rocks side-to-side in quartering flight over marshes. Sometimes soars high.

Similar Species Rough-legged Hawk has dark carpal patches on under-wing, white at base of tail. Swainson's Hawk has pointed wings, shorter tail. A wing (salvaged and retained) found on Attu I., AK (June 1990), has been tentatively identified as the smaller Eurasian Hen Harrier (*C. cyaneus*) on measurements and is awaiting genetic testing. This recently split (from Northern Harrier), highly migratory species breeds mostly in northern Eurasia east to Anadyrland and Kamchatka. Adult male Hen Harrier is uniformly pale gray above, has no chestnut spotting below, has black on six primaries (not five), and lacks barring on tail. Female and juvenile have a less distinct face pattern and are less cinnamon in coloration; juvenile is streaked below.

Voice Usually silent, except around nest or during courtship when both sexes give a fast series of chattering *kek* notes.

Status & Distribution Fairly common. **BREEDING:** Nests on ground in tall reeds or grasses. Nesting location and success are dependent on local rodent populations. **MIGRATION:**

Prefers to follow coastlines, but will ride thermals along ridges. East Coast migration is composed mostly of juveniles. Unafraid to cross open water. **WINTER:** Ranges from coastal and southern US into Mexico. Often associates and competes with Short-eared Owls at dusk. **RARE STATUS:** Casual to offshore western AK and HI.

Population Widely dispersed. Breeding has declined in many regions as agricultural areas have disappeared, especially in the eastern states.

roundish owl-like head

unstreaked belly

juvenile *hudsonius*

gray above

adult *hudsonius* ♂

white underwing with dark wing tip and trailing edge

adult *hudsonius* ♂

long, thin wings with rounded tips

adult *hudsonius* ♀

juvenile *hudsonius*

darker secondaries

barred remiges

streaked belly

whitish rump

juvenile *hudsonius*

adult *hudsonius* ♂

long tail

long, thin wings with rounded tips

flies with slight dihedral; acrobatic flight with quick turns and dives

ACCIPITERS Genus *Accipiter*

Three species of these short-winged, long-tailed raptors occur in N.A., 46 worldwide. Primarily bird-catchers, they hunt by ambush, flying rapidly around and even through brush to grab their prey with long legs and slender toes.

SHARP-SHINNED HAWK *Accipiter striatus*   SSHA ▪ 1

Our smallest *Accipiter*, often frequents bird feeders in winter. Females are larger than males. Polytypic (10 sspp.; 3 in N.A.). L 10–14" (25–36 cm) WS 20–28" (51–71 cm)

Identification A small, round-winged,

curved leading edge

short head projection

juvenile
velox

shorter square tail

juvenile ♀
velox

adult ♂
velox

thin legs

long-tailed hawk. Head rounded; eye appears more centered in head. **ADULT:** Blue-gray crown and nape are same color as back, creating a hooded effect against buffy cheeks. Eyes are red-orange (females) to deep red (males). Underparts are barred rufous. **JUVENILE:** Brown back feathers have rufous tips, white spots on wing coverts. Eye is pale yellow. White below with blurry, reddish brown lines and spots through the belly. **FLIGHT:** Quick, choppy wingbeats interspersed with short glides. When gliding or soaring, wings are held forward with "wrists" bent, head barely projecting in front of wings.

Geographic Variation Subspecies *velox* is common throughout most of N.A.; *perobscurus*, found on islands of BC, is darker and more heavily barred; *suttoni* of Mexico, into AZ and NM, is lighter, with fainter barring.

Similar Species Cooper's Hawk noticeably larger, although males approach female Sharp-shinned in length, but have a longer, more graduated tail with a rounded tip. Flatter crown; adult Cooper's has a darker

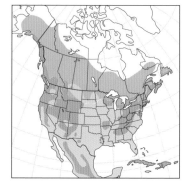

cap. Legs of Cooper's are thicker and wingbeats are a little slower.

Voice A high, chattering *kew-kew-kew* is heard around the nest; otherwise it is mostly silent.

Status & Distribution Widespread in northern and western forests. **BREEDING:** Prefers coniferous and mixed forests; small stick nest usually high and close to the trunk. **MIGRATION:** Commonest *Accipiter* at most hawk-watches; adult birds prefer to follow mountain ridges, while many juveniles end up following the coasts. Juveniles migrate first, followed by adults. **WINTER:** Throughout much of N.A. and into Mexico.

Population Major declines at eastern hawk-watch sites since the 1980s.

Juvenile Sharp-shinned versus Juvenile Cooper's

Hawk-watchers in the 1960s and 1970s would get into heated arguments over whether a passing juvenile *Accipiter* was a Sharp-shinned or Cooper's Hawk (henceforth, Sharpie and Cooper's). Since then, watchers and hawk banders have gradually worked out many of these field-identification problems.

Physically, Sharpies are grackle-size, noticeably smaller than Cooper's, which are about the length of a crow. This is handy when a bird is seen in a backyard or perched on an object of known size. Sharpies have a rounded head, with a noticeable notch in the profile where the forehead meets the bill. Cooper's has a flatter crown, especially if the hackles

Sharp-shinned Hawk, juvenile (NJ, Oct.)

are raised, and the profile from crown to bill is smooth and continuous. The eye of a Cooper's appears larger in the head and nearer to the bill. Both can show a pale supercilium, both have pale yellowish eyes. Back and wing coloration are similar—brown with rufous feather tips that slowly wear down through the winter and spring. The scapulars often show large white spots. The tail is brown with dark, evenly spaced bands. The undersides are white, with Sharpie showing blurry, reddish brown streaking and spotting down through the belly, while Cooper's has thinner, darker breast streaks that thin out or stop at the belly. Both have a white undertail,

COOPER'S HAWK *Accipiter cooperii* COHA ■ 1

The "chicken hawk" of colonial America, this medium-size *Accipiter* is a common sight at home bird feeders across the country. Females are larger and bulkier than males, juveniles differ from adults. The Gundlach's Hawk (*A. gundlachi*), a rare endemic of Cuba, is very closely related to Cooper's Hawk. Monotypic. L 14–20" (36–51 cm) WS 29–37" (74–94cm)

Identification The long tail is rounded at the tip; note also the relatively short wings and flat-topped head with eye close to beak. **ADULT:** Blue-gray upperparts, crown is darker and contrasts with lighter nape and buffy cheeks, giving the look of wearing a "beret." Eye color is orange to red. Undersides with rufous barring, undertail is white. **JUVENILE:** Brown above, with rufous edges and white spots on upperwing coverts. Tail long, with straight bands and wide, white tip that wears down by spring. Head usually buffy, eyes pale yellow. Undersides are white with thin brown streaks, white undertail. **FLIGHT:** Wings typically held straight out from body, head and neck projecting forward; this and tail length create a "flying cross" appearance. Shallow, quick wingbeats alternate with short glides.

Geographic Variation Hunting in more open country, western populations are smaller, with longer wings and shorter legs than eastern birds.

Similar Species Frequently confused with Northern Goshawk, the latter is usually larger, heavier appearing, and has relatively longer wings but a disproportionately smaller head. Smaller Sharp-shinned has a square tail.

Voice A low *keh-keh-keh* uttered around nest, occasionally mimicked by jays. Juvenile gives a squeaky whistle.

Status & Distribution Widespread through US and southern Canada, now commonly seen in forested suburbs. **BREEDING:** Nests in a variety of forest types, preying on small- to medium-size birds and small mammals, hunting from perches under the canopy. **MIGRATION:** Increasing numbers seen at eastern hawkwatches. **WINTER:** Winters south to Isthmus, Mexico; uncommon to rare in C.A. Juveniles winter farther north than adults.

Population Fairly common and increasing. Now widely nesting in suburban areas and seen more regularly there than Sharp-shinned, the opposite of what occurred in the 1900s.

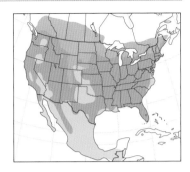

larger head, longer neck, and tawny nape

juvenile ♀

adult ♂

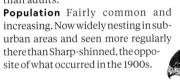

juvenile

longer head projection than Sharp-shinned

long rounded tail

straight leading edge

unstreaked undertail coverts

---

although some western Cooper's show thin streaks there.

When viewing a perched bird from the front, the tail can be diagnostic—Cooper's has a disproportionately longer, rounded tail, created by the feathers decreasing in length from inner to outer. Sharpie's tail feathers are almost the same length, creating a square tail tip. Seen from behind, Cooper's tail has a broad white terminal band, while Sharpie's tail only has a thin white edge. By spring, much of the white has worn off a Cooper's tail tips, especially the longest (central) feathers. In flight, they have different characteristics, or "jizz." The shorter wings of a Sharpie allow it to flap more

Cooper's Hawk, juvenile (NJ, Sept.)

quickly in a rapid, choppy motion with almost no flexing. Cooper's wings are much longer, and thus its wingbeats have a more fluid motion, almost like a wave traveling out the wing. Both fly in bursts of flapping and gliding. When gliding, the Sharpie pushes its wings forward, cocking the wrists. As a result, the head barely protrudes in front of the leading edge of the wing. Cooper's holds its wings almost straight out from its body, giving it a noticeable head and neck projection. This, along with the longer, rounded tail, gives a Cooper's the appearance of a "flying cross." As in all aspects of bird identification, the more birds you see in the field, the easier it is to name them.

NORTHERN GOSHAWK *Accipiter gentilis*   NOGO ▪ 1

Our largest *Accipiter*, the Northern Goshawk is found in northern forests and in montane West. Aggressive around the nest; noisily attacks humans that approach too closely. It preys on birds and small mammals. Females larger; adults distinct from juveniles. Polytypic (10 sspp.; 4 in N.A.). L 21–26" (53–66 cm) WS 40–46" (102–117 cm)

Identification A large, thick-bodied, long-winged, but small-headed hawk and with a long tail with graduated feathers. **ADULT:** Dark gray above, pale supercilium, lighter below with fine barring, fluffy white crissum. Red eye. **JUVENILE:** Brown above with heavy mottling, often with a checkerboard effect. Undersides light, with thick dark streaks, including crissum. Dark tail bands are wavy, with white borders between bands. **FLIGHT:** From above, adult has darker flight feathers than coverts, from below looks more pointy-winged than other accipiters. Prominent, pale supercilium easily visible. Juvenile from above has a pale diagonal band across wing coverts, irregular tail bands.

Geographic Variation Subspecies *laingi*, found from coastal southeastern AK to Vancouver I., BC, is heavily barred and darker than widespread *atricapillus*. Subspecies *apache*, from mountains in Southwest, is larger and darker than *atricapillus*. In adults of Old World subspecies, upperparts are browner (less blue), head pattern

weaker, and underparts more heavily marked with blackish bars and shaft streaks. Two northern Old World subspecies are dimorphic with pale and dark morphs. In northeastern *albidus*, the largest subspecies, about 50 percent of adults are white with faint brown markings.

Similar Species A juvenile female Cooper's causes frequent confusion. Cooper's is smaller, has shorter wings, and a larger-appearing head. Juvenile Cooper's is more finely streaked underneath with an all-white undertail; tail pattern differs; in flight, lacks pale wing covert band. Immature Red-shouldered Hawk in flight has buffy crescents at base of primaries.

Voice Around the nest, a loud, accelerating *kek-kek-kek-kek* and a wailing *kee-ah*. Young birds remain in the nesting area for weeks after fledging and are very vocal—an excellent way to locate this secretive species in summer.

Status & Distribution Rare to uncommon in mature northern forests and mountains. Also rare in Sierra Madre Occidental of western Mexico from northern Sonora and Chihuahua south to Jalisco; also Guerrero. **BREEDING:** Stick nest placed high, usually in main fork of a deciduous tree. **MIGRATION:** Late-fall migrant; numbers cyclical. **WINTER:** Usually

only juveniles winter in the Northeast. Fall passage adults usually work their way back north before winter's end. Mountain residents sometimes head to lower elevations. **RARE STATUS:** Casual in Southeast, but most reports likely represent misidentifications. Two records of subspecies *albidus* from Shemya I., western Aleutians (one white morph with photos on 17 Sept. 2001).

Population In eastern US, trend toward reforestation has created more habitat, with migration counts indicating a population increase. Although, large fall and winter irruptions in the East occur less frequently. Subspecies *laingi* is listed as Threatened by the USFWS.

prominent pale supercilium

juvenile ♀
atricapillus

long wings
for an *Accipiter*

long, pale gray supercilium contrasts sharply with dark auriculars

appears small-headed

pale greater covert bar, also visible in flight

juvenile
atricapillus

adult
atricapillus ♂

gray barred underparts

streaked undertail coverts

rather pointed wing tips

thin, pale, wavy bands border dark bands

Genera *Milvus* and *Ictinia*

BLACK KITE   *Milvus migrans*   BLAK ■ 5

This highly gregarious Old World species is accidental in N.A. Polytypic (6 sspp.; likely *lineatus* in N.A.). L 18–26" (46–66 cm) WS 47–60" (120–153 cm) **Identification ADULT:** A dark, medium-size raptor, with a slightly forked tail when not fully spread. Overall dark brown with long narrow wings with paler bases to primaries and large pale underwing primary window. **JUVENILE:** Like adult but with more contrasting dark mask and more streaked below with pale tipped wing coverts. **Geographic Variation** Moderate within the six named subspecies; *lineatus*, found over much of Asia, is the largest subspecies, with darker color-

ation on crown and underbody with creamy-buff belly and undertail and bold white patches on underwings. Adults of the two small African subspecies have yellow, not dark, bills. **Similar Species** Unmistakable, but compare to various dark-morph buteos. **Voice** Whining and mewing calls. **Status & Distribution** Widespread in the Old World, northern populations migratory. **RARE STATUS:** Accidental to western AK, one on St. Paul I., Pribilofs, 2–3 Jan. 2017. Also two winter records from Midway Atoll, HI. Casual to Bahamas (Inagua), Virgin Is., and Lesser Antilles. Accidental Chukchi Peninsula, Russian Far East.

Population Still very numerous, but declines in Europe and parts of Asia.

long, narrow wings

pale window

dark mask

tail fork (can disappear when spread)

streaked underparts

juvenile *lineatus*

MISSISSIPPI KITE   *Ictinia mississippiensis*   MIKI ■ 1

This small, pointed-wing kite looks more like a falcon than any other of our kites. A buoyant flier, it soars on flat wings, often high up in the air on thermals, catching and eating insects on the wing. Monotypic. L 14.5" (37 cm) WS 35" (89 cm) **Identification ADULT MALE:** Dark gray overall, lighter head with red eyes, dark primaries and tail. Seen from above, light secondaries form a bar across the wings. **ADULT FEMALE:** Like male, but darker head. **JUVENILE:** Dark eyes with wide, creamy supercilium line and gray cheeks. Back and wings are dark brown

with buffy edges; scapulars have white spots. Underparts heavily streaked, and dark tail has multiple thin white bands. **SUBADULT:** Body plumage similar to adult, but with a blend of juvenal and adult feathers, especially on tail and flight feathers in late summer/first spring. **FLIGHT:** Does not hover. The pointed wings are notable in that the outer primary is much shorter than the next one. Tail is square-tipped, usu-

white secondaries

black tail

adult ♂

gray body and underwing with darker wing tips and tail

♀

female has mostly gray outer tail feathers (all-black in male) and mottled undertail coverts (solid gray in male)

pale gray head

adult ♂

mostly dark gray body

falconlike wing shape

juvenile

banded tail held through 1st summer

long banded tail

1st summer

ally flared in flight. Underwing coverts are gray in adults, mottled in juvenile. **Similar Species** Most other kites are whiter. Adult Peregrine Falcon is larger, but with similar flight silhouette; its moustache mark and more powerful flight are diagnostic. **Voice** Downward whistle, given mainly on breeding grounds. **Status & Distribution BREEDING:** Central and southern Great Plains states, the South, and up the Atlantic coast to VA. Isolated colonies in NM and AZ. A stick nest built in a tall tree may be part of a loose colony of up to 20 pairs. **MIGRATION:** Most migrate Aug.–early Sept., stragglers into Oct., often in large groups through coastal TX. Return late Mar.–late Apr. **WINTER:** Well down in S.A. **RARE STATUS:** Casual in spring and summer (has nested) from New England, the mid-Atlantic, ON, and the upper Midwest. Also casual to SK, UT, NV, and CA. Accidental to NL. Rare in fall to western Cuba. **Population** Slowly spreading north.

SEA-EAGLES Genus *Haliaeetus*

These birds are very large raptors with a preference for fish; there are eight species worldwide, one breeding in N.A., and two casual visitors. Large, wide wings allow them to soar effortlessly as they search for prey that can also include waterfowl and carrion. They are skilled at kleptoparasitism, stealing prey from Ospreys, hawks, and falcons.

BALD EAGLE *Haliaeetus leucocephalus*   BAEA ■ 1

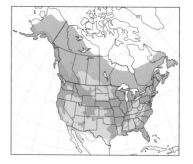

The official symbol of the US, the Bald Eagle, feeds primarily on fish and carrion (mainly in winter). It also takes waterfowl, mammals, and prey from other raptors. Monotypic. L 31–37" (79–94 cm) WS 70–90" (178–229 cm)
Identification A large, dark raptor with big head and bulky body. Females larger than males. Flight feathers take two to three years to replace. Head and neck project about one-half the length of the tail. **ADULT:** The pure white head and tail contrast with the dark brown body, wings, and legs. The bill and eyes are yellow. **IMMATURE:** There are four distinct stages, corresponding to each year of life. **JUVENILE:** Uniformly dark brown, with white on underwing coverts and axillaries; trailing edge of the secondaries is even. Beak and eye are dark. **SECOND-YEAR:** The eye turns gray-brown or whitish; the head develops a whitish supercilium. The back and belly become speckled with varying amounts of white. Trailing edge of wings is irregular due to new (shorter) and retained older (longer) secondaries. **THIRD-YEAR:** The cheeks and head become whiter, with a contrasting dark eye stripe; bill becomes yellow; eye color turns yellow. Back and belly still with white mottling, underwing coverts not as white as before. Secondary feathers are even in length. **FOURTH-YEAR:** Body and wing coverts mostly dark, but with white spots throughout. Head white with dark eye line, bill yellow. Tail is usually white, but often has a black terminal band. **FLIGHT:** Soars on flat wings; wingbeats are heavy, slow, and powerful.
Similar Species Turkey Vulture glides in a dihedral and rocks side-to-side when turning. Its two-tone underwings are visible in flight. Golden Eagle at all ages has dark axillaries, dark wing linings; head and neck project less than half the tail length. Immature Golden has white at the base of the primaries, white base of tail.
Voice It has a remarkably weak twittering or whistled call, often delivered in a staggered rhythm.
Status & Distribution Throughout most of N.A., always near water. Southern birds are nonmigratory. **BREEDING:** Long-lived, they build impressive stick nests in the crotch of a tall tree, adding to the structure every year. **MIGRATION:** They follow coastlines, rivers, and mountain ridges in fall. **DISPERSAL:** Southern juveniles wander north to the Great Lakes and Northeast in early summer. **WINTER:** Northern birds keep to the coastlines or south of the freeze line to find open water. Hydroelectric dams are a favorite location. **RARE STATUS:** Casual to Bering Sea islands and northern AK; also Ireland, UK, Russian Far East, and Japan.
Population The species formerly was common throughout the continent; pesticides in the food chain dramatically reduced numbers in the 1950s and 1960s. Conservation efforts were highly successful; nesting reported in most states. In 2007, the Bald Eagle was removed from the US endangered species list.

2nd year

juvenile

whitish underwing coverts and axillaries

note Osprey-like face pattern

3rd year

tail shorter than Golden

whitish in underwing and on belly

longer head projection than Golden

larger bill than Golden

juvenile

white head

adults

white tail

WHITE-TAILED EAGLE *Haliaeetus albicilla* WTEA ▪ 4

This large fish-eating eagle from Eurasia most closely resembles the Bald Eagle. Immatures take four to five years to attain adult plumage. Monotypic. L 26–35" (66–89 cm) WS 72–94" (183–239 cm)

Identification At rest, wing tips reach tip of tail. **ADULT:** Dark brown body and wings. Wedge-shaped white tail and creamy brown head contrast with dark body. Bill yellow. Undertail coverts are dark brown. **IMMATURE:** Variable, shows some white on underwing and axillaries; some pale mottling above. Tail longer and less wedge-shaped. Wings shorter and wider than on adults. **FLIGHT:** Shows seven emarginated (notched) primaries or "fingers" in flight.

Similar Species Adult Bald Eagle has white head and undertail coverts. Immature Bald has white axillaries ("wingpit"), more white on underwing coverts, dark band on tip of tail. Bald has six emarginated primaries.

Voice Gives a series of shrill calls, particularly in vicinity of the nest.

Status & Distribution Widespread across Eurasia. Also resident on Greenland. **RARE STATUS:** Most N.A. records are from western AK, predominantly from the western Aleutians, where it nested on Attu I. (at least in 1982 and 1983). One record from central Aleutians (Kiska I.). Casual to Pribilof Is. and in spring to St. Lawrence I., AK. Accidental to Kodiak I., AK, and MA.

Population Historic major declines in Europe, but conservation measures have led to increases in recent decades. Re-established in UK.

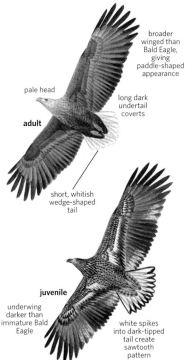

broader winged than Bald Eagle, giving paddle-shaped appearance

pale head

long dark undertail coverts

adult

short, whitish wedge-shaped tail

paler head

overall scaly appearance

short tail

adult

juvenile

underwing darker than immature Bald Eagle

white spikes into dark-tipped tail create sawtooth pattern

STELLER'S SEA-EAGLE *Haliaeetus pelagicus* STSE ▪ 4

A sea-eagle of eastern Asia, Steller's is a casual visitor to Alaska. Sexes are similar; females are larger than males. Monotypic. L 33–41" (84–104 cm) WS 87–96" (221–244 cm)

Identification A large eagle with a very long, wedge-shaped tail, and a massive bill. Wing tips on perched bird fall well short of tail tip. Legs are densely feathered. **ADULT:** Striking white upper and lower wing coverts, tail coverts, tail, and leg feathering, otherwise mostly dark brown but with

white forehead; bill, legs, and feet are bright orange-yellow. **IMMATURE:** Bill, legs, and feet slightly duller than adult. Plumage variable. Head brown with pale streaking on neck and breast. Underwing coverts brown streaked with white, white axillaries, flight feathers dark with

paddle-shaped appearance

massive size

white leading edge to wing

long white acutely wedge-shaped tail

huge orange bill

adult

huge bright orange bill

juvenile with mostly orange bill

white axillaries

white shoulders

juvenile

adult

white primary patch

long wedge-shaped tail is white in adult, dark tipped in juvenile

white leggings

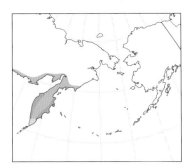

light bases to primaries. Leg feathers and undertail dark; white tail is less wedge-shaped than adult and tipped with dark. **FLIGHT:** Paddle-shaped wings; soars in a strong dihedral, secondaries wider than most eagles. A rare localized morph (or *niger* if recognized) from the Korean peninsula (one specimen Ussuriland, Russia, unrecorded for 70 years) was all black in adults apart from the white tail. **Similar Species** Adults are unique;

immatures are like Bald Eagle but with longer, wedge-shaped tails and a huge yellow bill.

Voice Loud gull-like calls described during aerial flight display; also barking notes like White-tailed Hawk, but lower pitched.

Status & Distribution Breeds coastal northeast Asia around Kamchatka, Koryakland, and Sea of Okhtosk and northern Sakhalin; winters south to Korean Peninsula and Japan (most

present eastern Hokkaido). **RARE STATUS:** About a dozen widely distributed records from AK, as far southeast as Juneau. A probable Bald Eagle x Steller's Sea-Eagle hybrid was photographed in BC during the winter of 2004–2005.

Population Vulnerable. Declining for a variety of reasons, including habitat destruction, persecution, and declining fish stocks. Total population now likely well under 10,000.

Genera *Geranospiza* and *Rostrhamus*

CRANE HAWK   *Geranospiza caerulescens*   CRHA ▪ 5

This rather large tropical hawk has a lanky body, small head, and long tail. The Crane Hawk is unusual for having very long, double-jointed legs that

white crescent on primaries

long orange legs

adult *nigra*

long banded tail

allow it to reach into tree crevices and nest holes in search of prey. Polytypic (6 sspp.; likely *nigra* in N.A.). L 18–21" (46–53 cm) WS 36–41" (91–104 cm)

Identification Dark gray body; broad and rounded wings; long, slightly rounded tail with two bright-white tail bands; long orange legs. When perched, the wing tips barely reach the base of the tail. **ADULT:** Slate gray, often with fine barring on leg feathers and belly. **JUVENILE:** Browner and slightly paler than adult, white on forehead and cheeks. **FLIGHT:** Long winged with distinct white crescents on primaries; flight feathers are darker than underwing coverts. Wingbeats are loose and floppy; usually soars only for brief periods.

Geographic Variation Subspecies *nigra* occurs in eastern Mexico north

to southern Tamaulipas and on Mexico's west coast from southern Sinaloa south; paler *livens* is found in northern Sinaloa and southern Sonora.

Similar Species Common Black-Hawk is stockier, has broader wings, shorter tail, and yellow legs. Also compare to rare black-morph Hook-billed Kite.

Voice A loud, plaintive whistle *wheeeooo*.

Status & Distribution YEAR-ROUND: Rare to uncommon resident from Mexico to Peru and southern Argentina. Prefers tropical lowlands, usually near water. **RARE STATUS:** Accidental; a Crane Hawk was present and well-photographed from 20 Dec. 1987 to 17 Mar. 1988 at Santa Ana NWR, Hidalgo Co., TX.

Population Little data on this low-density species.

SNAIL KITE   *Rostrhamus sociabilis*   SNKI ▪ 2

This specialist feeds on apple snails using its hooked beak to extract its prey. It flies low over marshes. and is often seen perched on bushes. Polytypic (3 sspp.; *plumbeus* in N.A.). L 17" (43 cm) WS 46" (117 cm)

paddle-shaped wings

adult *plumbeus* ♂

white tail base

adult *plumbeus* ♂

white undertail

gray tail tip

adult *plumbeus* ♀

adult *plumbeus* ♀

whitish supercilium and chin

thin, hooked bill

strong eye line

heavy streaks on under parts

juvenile *plumbeus*

white tail base

Identification Paddle-shaped wings and dark tail with white base and gray tip. **ADULT MALE:** Dark gray overall; flight feathers darker than back and wing coverts. **ADULT FEMALE:** Dark brown above, thin white streaking on underparts, white markings around red eye. **JUVENILE:** Brown eye, fringed tawny above and on crown, undersides pale with dark streaks. **FLIGHT:** Slow, graceful wingbeats.

Similar Species Female and juvenile Northern Harrier have white rumps, and fly with wings in a dihedral.
Voice Mostly silent.
Status & Distribution YEAR-ROUND: Resident in southern FL up to the Orlando area. Also resident from central Veracruz, Mexico, to S.A. and on Cuba. **BREEDING:** Nests in low trees or bushes, often surrounded by water. Will nest colonially. **RARE STATUS:**

Casual to northern FL. Casual or accidental to SC, NC, and TX; four late spring and summer coastal records from TX likely birds from Mexico (*major*, largest subspecies); accidental Presque Isle SP, Erie, PA (22 Oct. 2019).
Population Endangered (subspecies *plumbeus*, USFWS, intensely managed, approximately 500 pairs in FL). Stable; dependent on freshwater marshes for snails.

NEAR-BUTEOS Genera *Buteogallus, Rupornis, Parabuteo,* and *Geranoaetus*

COMMON BLACK HAWK *Buteogallus anthracinus* COBH ■ 2

This raptor found in the Southwest often near water, usually along a year-round watercourse, hunts from a perch. Juveniles are strikingly different from adults. Polytypic (5 sspp.; nominate in N.A.). L 21" (53 cm) WS 50" (127 cm)
Identification Large with wide wings and a short tail. **ADULT:** Body and wings black, cere and legs are orange-yellow. Tail has a broad white band and narrow white tip. **JUVENILE:** Dark brown overall, with rufous markings above and heavy dark streaks on buffy breast and underparts. The head is patterned with dark crown, eye line, and malar mark contrasting with buffy supercilium and throat. The tail is white with numerous dark, wavy bands, and is longer than in an adult. **FLIGHT:** Wide wings make the tail appear very short in comparison to other dark buteos. Adult has small, whitish crescent at base of primaries;

tail band is very evident. Immature has large buffy patch at base of primaries; tail appears lighter with dark terminal band. Soars on flat wings.
Similar Species Zone-tailed Hawk has a grayish cast and soars with a Turkey Vulture–like flight with a dihedral, and has a much slimmer wing shape and a longer tail. With close views, Zone-tailed has a white forehead, gray lores. Dark-morph buteos will show an underwing contrast between flight feathers and coverts, and have thinner wings and longer-appearing tails.
Voice Call is a series of loud whistles, falling in intensity near the end.
Status & Distribution A tropical hawk of Mexico, C.A., and northern S.A. **BREEDING:** Uncommon to rare in West TX, southwest NM, AZ, southwestern UT, and southeastern NV. **MIGRATION:** Very rare except for numbers in early spring along Santa Cruz

River (Tubac), north of Nogales, AZ. **RARE STATUS:** Casual north to CA, southern UT, southern NV, CO, and southern TX.
Population Probably stable. About 250 pairs breed in the US, mostly in AZ. The species is threatened by habitat destruction and human disturbance.

extensive pale yellow to whitish in lores

all figures *anthracinus*

white supercilium

broad, heavy blackish streaks on sides of neck

juvenile

heavily streaked underparts

solid dark patch on lower flank

adult

stocky body shape with short tail

longer tail than adult, banded with wavy black bars and much thicker terminal band

finely banded remiges

juvenile

banded tail

whitish patch

adult

short tail with single, broad, white band

short broad wings

GREAT BLACK HAWK *Buteogallus urubitinga* GBHA ◼ 5

A relative of the Common Black Hawk, the tropical Great Black Hawk is accidental in N.A. Polytypic (2 sspp.; identity of US record unknown). L 22–26" (55–65 cm) WS 45.5–51" (115–130 cm)

Identification ADULT: Northern subspecies (*ridgwayi*) described. Overall blackish with fine whitish barring on tibia feathers and two white bands on tail, the upper one narrow and often concealed. Facial skin (lores) gray and pale does not extend under eye. Cere and long legs yellow. In flight from below evenly black, lacks whitish areas at base of primaries. Long yellow legs extend well beyond white tail band. **JUVENILE:** Brown above and buffy below with streaks. Distinct head pattern with broad pale supercilium, but lacks broad dark malar streak.

Geographic Variation Two distinct subspecies, northern *ridgwayi* (described above) found in Mexico and C.A. south to central Panama. Larger S.A. nominate *urubitinga* (range extends into eastern Panama) differs in adult by lacking white bars on tibia feathering and especially by all-white tail base; lores yellow. Juveniles not separable in field.

Similar Species Common Black Hawk is shorter legged and shorter tailed. Adult has faint white patch at base of outer two primaries and lacks narrow white band at base of tail; legs extend to, not beyond, white tail band. Lores yellow. Juvenile Common has broad dark malar streak, throat and face more streaked; primary projection longer, almost reaching tail tip; tail has fewer and coarser dark bars and a broad darker terminal band.

Voice A long, harsh *keeeeeee*, most often given in flight.

Status & Distribution Lowlands and foothills from northwest and northeast Mexico south to Uruguay. Less tied to water than Common. **RARE STATUS:** Accidental US, one juvenile seen South Padre I., south TX, on 24 Apr. 2018, then remarkably again near Biddeford and Portland, ME, from 29 Oct. 2018 until 1 Feb. 2019, when injured bird was euthanized (specimen retained as a mount). Adults seen and photographed intermittently on Virginia Key, FL (1972–2015), of nominate S.A. subspecies were not accepted to FL list, presumably on origin.

Population Stable.

adult
ridgwayi

two white tail bands, upper often concealed (entire base of tail white in nominate)

fine whitish barring on thigh feathers (lacking in nominate)

juvenile
ridgwayi

broad white supercilium

finely barred tail and underwings

lacks heavy dark malar of juvenile Common Black Hawk

long tail lacks dark terminal band of juvenile Common Black Hawk

juvenile
ridgwayi

long yellow legs

wing tip falls well short of tail tip

both subspecies essentially identical in juvenal plumage

ROADSIDE HAWK *Rupornis magnirostris* ROHA ◼ 4

A small tropical hawk, the Roadside Hawk is common in much of Mexico, but only casual to south TX. Polytypic (12 sspp.; *griseocauda* in N.A.). L 14" (36 cm) WS 30" (75 cm)

Identification Small, long-tailed, long-legged. Frequently perches and hunts from fence posts, telephone poles, or trees alongside fields and roads in its native range. TX birds have been less cooperative. Wing tips reach halfway down the long tail, which has a buffy or whitish U along the upper-tail coverts. Legs are long and slim. **ADULT:** Gray to brownish on head and upperparts; pale yellow eyes. Long tail has three to four dark bands. Below, note brown bib and barred belly. **JUVENILE:** Browner than adults, with a wide supercilium and darker eyes. Chest has vertical streaks, the belly horizontal barring. Long tail has four to six dark bands. **FLIGHT:** Wings wide and rounded, with barred flight feathers, rufous on inner primaries. Stiff, rapid wingbeats and long tail make it look very like an *Accipiter*.

Geographic Variation Populations farther south show more rufous in the wings and grayer plumage.

Similar Species See Broad-winged Hawk and Gray Hawk. Long, evenly banded tail and rounded wings are

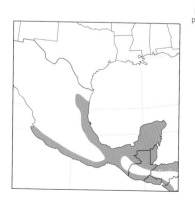

barred rufous patch on inner primaries

short rounded wings

adult
griseocauda

pale eye and bright yellow cere

brown bib

adult
griseocauda

short, pale supercilium

long yellow legs

wing tips extend about halfway down tail

streaked breast

juvenile
griseocauda

long, evenly banded tail

distinctive. Barred belly and dark, unstreaked bib on adult; vertical streaks on breast of juvenile.
Voice Birds in TX have been silent.

Roadside's call is a drawn-out complaining scream.
Status & Distribution Range stretches from Mexico to Ecuador and northern

Argentina. **RARE STATUS:** Casual to the Lower Rio Grande Valley, TX, with about a dozen accepted records.
Population Apparently stable.

HARRIS'S HAWK *Parabuteo unicinctus*   HASH ▪ 1

This large, dark raptor of the Southwest is noted for its habit of cooperative hunting. Polytypic (2 sspp.; *harrisi* in N.A.). L 21" (53 cm) WS 46" (117 cm)
Identification A long-tailed, long-legged, dark raptor with broad, rounded wings. It is often seen on open perches. The dark tail has a broad white terminal band and a white base with white undertail coverts.

The wing tips reach halfway down the tail. **ADULT:** Uniformly dark head, body, and wings, with chestnut wing coverts and leg feathers. The facial skin, cere, and legs are yellow. **JUVENILE:** Dark brown body with lighter streaks on head and neck, whiter streaking on belly than on breast. The tail has fine dark barring, as do the flight feathers. The white at the base of the tail is not as extensive as in an adult. **FLIGHT:** Very active hunter, flying with quick, shallow wingbeats. Its short, broad wings and long tail allow it to be very maneuverable; it can hover for short periods. From below, adult has chestnut coverts and dark flight feathers; juvenile shows barred gray secondaries, barred primaries with lighter bases and dark tips. Soars with flat wings; rounded wing tips.
Geographic Variation Some authorities recognize two N.A. subspecies—*harrisi* (TX) and *superior* (AZ)—but they are virtually identical; the nominate occurs in S.A.
Similar Species Dark-morph buteos have two-tone underwings with silvery-tone flight feathers. Juvenile Northern Harrier at a distance, when compared to a juvenile Harris's Hawk, has slimmer, longer wings, a white rump (not the base of the tail), and glides with a dihedral.
Voice Call is a long, harsh, grating *eeaarr*.

Status & Distribution Common, permanent resident in southern TX, into southern NM and southern AZ; recent records (including nesting) in Southern CA (now normally absent) and southern NV. **BREEDING:** Their complex social structure including multiple adults and pairings is still being investigated by researchers. Breeding groups consist of a dominant pair, related helpers that are offspring of that pair, unrelated helpers, and, sometimes, additional females with an intermediate position of dominance. **MIGRATION:** Mostly sedentary. **WINTER:** Often gathers in large groups to roost and hunt. **RARE STATUS:** A popular falconer's bird; any records substantially outside of normal range should be treated with caution.
Population Apparently stable, but under pressure from habitat (especially mesquite) destruction, falconry, and electrocution from power lines.

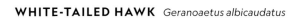

hort rounded wings

juvenile *harrisi*

adult *harrisi*

heavily streaked below with blackish brown

white undertail base and tip

chestnut wing linings

chestnut wing coverts

adult *harrisi*

chestnut thighs

white undertail coverts

long tail with white tip

WHITE-TAILED HAWK *Geranoaetus albicaudatus*   WTHA ▪ 2

This large, striking raptor found in the coastal region of southern TX. It feeds on small mammals, insects, reptiles, and even birds. It will gather at a prairie fire as well as burning sugarcane fields to pursue displaced prey. Polytypic (3 sspp.; *hypsopodius* in N.A.). L 20" (51 cm) WS 51" (130 cm)
Identification When perched, the wing tips reach past the tip of the tail, even in juveniles. Females are larger than males. It has an adult plumage and three immature plumages: juvenile, second-year, and third-year.

ADULT: Uniform gray on back, head, and wings, with a rufous patch on the lesser coverts forming an easily visible shoulder patch. Throat is white, as are the underparts, though some individuals will show fine barring underneath. The white lower back and uppertail coverts lead to the distinctive white tail with a black subterminal band and multiple faint bands. **JUVENILE:** Above, dark brown to blackish on head, back, and wings, with rufous feather edges on upper coverts. Head is dark with whitish

areas on cheek, and dark throat; most show a white patch on the breast.

Amount of dark on belly and white on chest varies greatly among individuals, with some heavily mottled to completely dark, but with lighter undertail coverts. Tail is light gray with many fine dark bands, contrasting with a white U on the uppertail coverts. Wings are narrower, and tail is longer, than adult. **SECOND-YEAR:** A blend of juvenile and adult features, including wing and tail proportions of an adult and coloration of a juvenile. Head is more uniformly dark, the back and wings are dark with a rufous shoulder patch. Chest is whitish with variable streaking, undersides barred with black or rufous. Uppertail coverts are white, rump is black, and tail is gray with fine bands and a dark subterminal band. Below, undertail coverts are white with variable mottling. **THIRD-YEAR:** Similar to adult plumage overall, but with blackish head, throat, neck, and back; tail is a mix of gray and white feathers, rump is a mix of black and white. Underneath, whitish barred with rufous. **FLIGHT:** Long, pointed wings pinch

in at the body. In flight, often soars in a dihedral; will hover while hunting. The wingbeats of White-tailed are heavier than those of Swainson's Hawk. On adults, the white underwing coverts contrast with darker flight feathers, secondaries being lighter than primaries. Its namesake tail with wide black subterminal band is unique among buteos. All immature plumages show variable dark streaking on the underwing coverts, lighter at bases of flight feathers than at tips. A white chest spot is usually visible.

Geographic Variation Subspecies *hypsopodius* found TX south through C.A. to western Colombia. Smallest *colonus* in northern S.A. and Trinidad, includes gray-morph birds; nominate *albicaudatus* from S.A. south of Amazonia is largest and includes dark-morph birds.

Similar Species In flight, adult light-morph Swainson's Hawk has two-tone underwings, but primaries and secondaries are the same color; uniform dark trailing edges to wing and dark bib on chest. Wing does not pinch in

at the body like White-tailed. Dark-morph Swainson's Hawk is similar to juvenile White-tailed, but has a more noticeably barred tail with a wide, dark terminal band, and does not show the white chest patch or white uppertail coverts. Ferruginous Hawk has all-white undersides, but legs are dark, underwings are uniformly white, tail lacks dark subterminal band. All other dark buteos are distinguished from juvenile White-tailed by the silvery flight feathers and lack of white U on the uppertail coverts.

Voice Rarely heard except when disturbed at nest site.

Status & Distribution Resident along Gulf Coast region of TX to southeast of Houston (just west of Anahuac). Widespread in similar habitat throughout C.A., into S.A. **RARE STATUS:** Casual to southwest LA and well inland in TX.

Population The species is listed as threatened in TX, with a small population of 200 to 400 pairs that appears to be stable.

grayish head and back

rufous scapulars and marginal coverts

smudgy white cheek patches

dark wing linings contrast with pale flight feathers

size of white breast patch and undertail coverts color variable

long pointed wings

white wing linings and underparts

juvenile *hypsopodius*

adult *hypsopodius*

palish secondaries

short white tail with black subterminal tail band

dark juvenile *hypsopodius*

some juveniles all-dark below except for undertail coverts

bold white chest patch

adult *hypsopodius*

juvenile *hypsopodius*

whitish tail base

pale tail

2nd year *hypsopodius*

whitish tail with smudgy band

BUTEOS Genus *Buteo*

Called *buzzards* in many parts of the world except N.A., there are 26 species in this genus—nine in N.A.; all are regular breeders except Roadside Hawk. They soar on wide, rounded wings, hunting small mammals, lizards, snakes, and the occasional unwary bird. They also perch-hunt along tree lines or under the canopy. Sexes are similar in appearance; females are slightly larger than males. Juveniles and, sometimes, older immatures have different plumage than adults. During fall migration, buteos travel in groups (called *kettles*), riding thermal updrafts along mountain ridges or coastlines. In late Sept., nearly one million Broad-winged Hawks pass over TX on their way south.

GRAY HAWK *Buteo plagiatus* GRHA ■ 2

This bird was called the "Mexican Goshawk" in older literature due to its overall gray coloration, longish tail, and barred undersides. Sexes look alike; females are larger and juveniles are different from adults. Formerly placed in the genus *Asturina* and now split from the polytypic Gray-lined Hawk (*B. nitidus*) from southwest

longer tail than Broad-winged

juveniles

whitish uppertail coverts

adult

banded black-and-white tail

distinct head pattern with dark eye line, dark malar, and pale cheek

juvenile

adult

fine gray bars on underparts

Costa Rica to S.A. Monotypic. L 17" (43 cm) WS 35" (89 cm)

Identification A small raptor with wide, rounded wings and long tail; all ages have uppertail coverts that form a distinct white U. **ADULT:** Overall gray, with fine white barring on breast and belly. White wing linings have fine gray barring, appearing whitish at a distance. Undertail coverts white; tail has two distinct white bands. **JUVENILE:** Dark brown body; distinct head pattern with light supercilium and cheek contrasting with dark eye line and malar stripes. Undersides heavily streaked; tail has multiple stripes that grow progressively wider toward the tip. Tail is longer than adult. **FLIGHT:** Accipiter-like; short, choppy wing-

beats are interspersed with flat glides. From underneath, adult shows black tips to primaries.

Similar Species Juvenile Broad-winged Hawk lacks the face pattern and uppertail U mark and has more pointed wings, a shorter tail, and flies with slower wingbeats. Juvenile Red-shouldered Hawk has buffy crescent at base of primaries.

Voice Calls include a loud, descending whistle.

Status & Distribution Resident in the Lower Rio Grande Valley; breeds in Big Bend NP and into southeastern AZ, rarely to southwestern NM. Very rare in winter in southern AZ. Nests in large trees, usually near water. **RARE STATUS:** Accidental to Southern CA.

Population Population stable, possibly expanding. About 80 pairs breed in the US.

RED-SHOULDERED HAWK *Buteo lineatus*   RSHA ■ 1

A common hawk of wet deciduous woodlands, it is the noisiest of the buteos, especially during spring courtship. A perch-hunter of the forest understory, it feeds on frogs, snakes, lizards, and small mammals. Females are larger than males, sexes are similar in appearance, and juveniles differ from adults. Polytypic (5 sspp., all in N.A.). L 15–19" (38–48 cm) WS 37–42" (94–107 cm)

Identification Medium-size *Buteo* with rounded wing tips that do not reach the tip of the tail. **ADULT:** Brown above with lighter feather edges and some streaking on head. Rufous on the upperwing coverts gives the "red shoulders." The primaries are barred or checkered black and white; the dark tail has three white bands. Underparts are rufous with white barring. **JUVENILE:** Mostly brown above, with less rufous on shoulders. Undersides are buffy with variable dark streaks; the brown tail has multiple narrow bands. **FLIGHT:** All ages show a distinct light crescent at the base of the primaries. Soars on flat wings held forward,

glides with wings cupped, giving a hunched appearance. Wingbeats are rather quick and shallow—especially in *elegans* which is like an *Accipiter*—interspersed with quick glides.

Geographic Variation Eastern subspecies (*lineatus*) descibed above, with adult showing dark streaks on breast. Juvenile tail from below is light with dark bands. Southeastern subspecies (*alleni*) and TX subspecies (*texanus*) are very similar to each other, with adults showing no dark streaks on breast; *texanus* is slightly brighter rufous. Juveniles have darker underside of tail, heavier markings on underparts. South FL subspecies (*extimus*) is the palest subspecies. Adults are pale gray above, pale rufous underneath, with the head appearing light at a distance. Juveniles are less rufous, have thinner streaking underneath. CA subspecies (*ele-*

gans) is the brightest subspecies, with juveniles appearing more like adults. Adults have unbarred rufous on the breast, wider tail bands. Juveniles have similar upperwings to the adult, with the most rufous on the shoulders of any subspecies, and a white crescent on the primaries. Underparts are barred rufous and buffy. The dark tail shows whitish bands.

Similar Species Broad-winged Hawk

elegans

lineatus

alleni & wintering *lineatus*

texanus

extimus

adult has no barring on primaries and only one or two tail bands; in flight shows more pointed wings. Juveniles are similar, but Red-shouldered shows three bands on the folded secondaries; Broad-winged lacks the pale crescent at the base of the primaries. Juvenile Northern Harrier is rufous underneath, but long tail, thin wings, and flight style are very different. A perched juvenile Northern Goshawk shows heavy streaking underneath and a long tail, but dark-banded tail pattern and shorter wings help the identification process.

Voice A loud, repeated *kee-ahh*.

Status & Distribution Widespread breeder throughout East and into southern Canada. Found throughout the South, into eastern and southern TX. CA subspecies is found into OR, southern WA, and AZ. **BREEDING:** Nests in deciduous woodlands, usually river bottoms, near lakes or swamps. Coexists in similar habitat with the Barred Owl. **MIGRATION:** In East, northern birds migrate to southern states and into Mexico. Southern subspecies are nonmigratory. **RARE STATUS:** There are records of eastern birds west to CO; *elegans* is rare NV, southwest UT, and western AZ and casual to ID and NM.

Population Stable.

juvenile elegans

adult elegans

rufous underparts

juvenile elegans — pale crescent — rufous wing lining — heavily marked underparts — banded tail

adult elegans — rufous wing lining — pale crescent

long banded tail — *juvenile elegans* — reddish shoulders — white crescent

juvenile lineatus

pale gray head

FL adult extimus

Eastern adult lineatus

pale crescent — banded tail — *adult lineatus* — rufous wing lining — *juvenile lineatus*

longer tail — *juvenile lineatus* — pale crescent

streaked underparts — pale crescent — *juvenile lineatus*

BROAD-WINGED HAWK   *Buteo platypterus*   BWHA   ▪   1

This bird made hawk-watching famous—thousands of birders gather to watch the annual fall migration of Broad-winged Hawks. The hawks start migrating in Sept. in New England, traveling down Appalachian ridges on their way to wintering grounds in S.A. They occur in light and dark (rare) morphs; juveniles different from adults. Polytypic (6 sspp.; nominate in N.A., others resident in West Indies). L 16" (41 cm) WS 34" (86 cm)

Identification A small *Buteo*. **LIGHT-MORPH ADULT:** Head, back, and wings are brown, throat is white, wing tips dark, dark tail with one wide white band. A second, thinner band may be visible on the fanned tail. Undersides are white with brown or rufous barring across breast, less on the belly. Some individuals may have a solid-colored dark breast, giving the bird a dark bib. **JUVENILE:** Brown above like adult, but with pale supercilium on head, dark malar stripe; brown tail has multiple darker bands, widest band at tip. Underparts are white with dark streaking on breast and belly, but

juvenile
platypterus

black wing tips,
more pointed than
Red-shouldered

juvenile
platypterus

black
moustachial
streak

variable
whitish
underwing

juveniles
platypterus

rufous-
brown
barred
underparts

variably
marked
underwings
and underparts

adult
platypterus

black
trailing
edge

adult
platypterus

one broad
white band
shows on
blackish tail

dark-morph
adult (rare)
platypterus

amount of streaking is highly variable, sometimes almost absent. **DARK MORPH:** All-dark body with dark wing coverts and silvery flight feathers; the dark tail has a wide white band. The juvenile is dark with variable light streaking on body and wing coverts. **FLIGHT:** Wing tips more pointed than those of the other common buteos, and trailing edge is almost straight. Adult has pale wing linings and flight feathers contrasting with dark primary tips, a wide dark band along the trailing edge of the wing. Juveniles have slightly longer tails, but the same wing silhouette

with the trailing edges not as dark. Backlit wings show a light rectangle at the base of the primaries. Underwing coverts are variably streaked, as is the belly. Soars on flat wings.

Similar Species In flight, juvenile Red-shouldered Hawk shows light crescents at base of primaries and

longer tail; wingbeats more rapid, especially *elegans*. Perched, it has a brown tail with dark bands; three bands on folded secondaries. Juvenile Cooper's Hawk can show the same overall markings, but no malar stripe; shorter, barred wings and much longer tail give a different shape.

Voice Thin, whistled *kee-eee*, often mimicked by Blue Jays, rarely given in migration.

Status & Distribution Common. **BREEDING:** Nests in eastern woodlands to eastern TX and MN, in Canada west to AB and BC. Rare dark morph nests in western Canada. **MIGRATION:** Famed for migrating in groups (called *kettles*), generally utilizing updrafts along mountain ridges. Reluctant to cross open water. During the last six days in Sept., typically more than 700,000 Broad-wings pass over Corpus Christi, TX. Small numbers are seen in fall (fewer in spring) at favored hawk-watching sites in the West. **WINTER:** Small numbers, usually juveniles, in southern FL and southern TX; casual in CA. **RARE STATUS:** Dark-morph birds casually seen in the East and West, in migration.

Population Stable, as far as is known.

SHORT-TAILED HAWK *Buteo brachyurus*   STHA ■ 2

This small and long-winged *Buteo* of the tropics occurs in two color morphs—light and dark. Usually catches its prey below the forest canopy, but most often seen soaring high in the sky or holding steady while facing into the wind. Rarely seen perched. Polytypic (2 sspp.; *fuliginosus* in N.A. and south through

Panama. L 15" (39 cm) WS 35" (89 cm) **Identification** Perched, the wing tips reach the tail tip. **LIGHT MORPH:** Adult dark above, including head and cheeks, giving a helmeted appearance. Light below, usually white or creamy-white on throat, body, and undertail coverts; usually shows small

rufous area on sides of upper breast. Tail from below has wide, dark terminal band and lighter, thin bands. Juvenile has more patterning on face, brownish secondary coverts, variable streaking on sides of breast; tail from below is light with many thin bands and narrow terminal band. Both show white underwing coverts with darker, barred flight feathers with black tips, and paler base to primaries. **DARK MORPH:** Adult uniformly dark brown, flight feathers lighter than coverts,

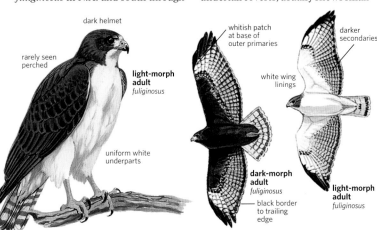

dark helmet

rarely seen
perched

light-morph
adult
fuliginosus

uniform white
underparts

whitish patch
at base of
outer primaries

white wing
linings

dark-morph
adult
fuliginosus

black border
to trailing
edge

darker
secondaries

light-morph
adult
fuliginosus

but with dark trailing edge, lighter bases to primaries. Tail is lighter than body with dark terminal band. Juveniles are mottled with white on underparts and underwing coverts; darker throat forms bib; tail has fine bands and dark tip. **FLIGHT:** A highly aerial species, Short-tailed Hawk shows a pointed-wing silhouette, especially when gliding. It typically soars at high altitudes when hunting, and dives or glides down when prey is sighted. When wind conditions allow, it often kites for considerable periods of time. White body and wing coverts stand out from darker flight feathers. White spot on forehead and lores is often visible.

Similar Species Broad-winged Hawk is similar size, but has lighter flight feathers and shorter wings when perched. Juveniles are spotted or streaked underneath. Light-morph Swainson's Hawk is larger, has two-tone underwings, but flight feathers are more uniform in color.

Voice Usually silent, especially in nonbreeding season. Has a high-pitched, drawn-out, slightly descending, two-syllable call.

Status & Distribution Widespread from northern Mexico to northern Argentina. **BREEDING:** Uncommon resident in central and southern FL; most birds in central FL move to southern FL in winter. **RARE STATUS:** Rare in southeastern AZ; most records from summer in the mountains (only proven breeding from Chiricahua Mts.). Casual to central AZ and southwestern NM. Casual to southern TX, including a failed nesting attempt with a Swainson's Hawk; also casual to central (Edwards Plateau), and West TX. Accidental to northeast TX and Upper Peninsula, MI.

Population Stable in FL.

ZONE-TAILED HAWK *Buteo albonotatus* ZTHA ■ 2

This dark raptor of the Southwest closely mimics a Turkey Vulture both in appearance and flight, and as a result is easily overlooked. Monotypic. L 20" (51 cm) WS 51" (130 cm)

Identification Appears blackish at a distance, has barred flight feathers, and the tail shows a distinct pattern. When perched, wing tips equal or exceed tail length. **ADULT:** Uniformly black with a grayish cast (Turkey Vulture has a brownish cast); yellow cere and legs. The tail of the adult male has one broad, white, mid-tail band and narrower, white inner band; the adult female's tail also has a mid-tail band, with at least one other pale band; from above paler bands are gray. **JUVENILE:** Variably spotted with white on underparts, dull yellow legs and cere (brighter by fall); tail is dark from above, light below with many thin dark bands and a dark tip. **FLIGHT:** Long, dark wings are two-tone, with heavily barred flight feathers. Adults show a broad dark trailing edge to the wing, thinner, more indistinct, on juvenile. Wingbeats are deep, but not as floppy as a vulture. Head and banded tail project farther than a vulture. Usually glides and tips side-to-side in a dihedral, just like a Turkey Vulture, but can also glide on flat wings; drops from low glide onto small birds, rodents, lizards, and fish.

Similar Species Similarly colored adult Common Black Hawk is utterly different in shape, particularly in flight where the shorter and much broader wings and shorter tail are apparent; when perched, adults are blacker in coloration and have yellow lores. Common Black Hawks have thicker and longer legs, and a bigger bill; wing tips do not reach tip of tail. Juvenile dark-morph Red-tailed Hawk has light panels at base of primaries; juvenile White-tailed Hawk usually has a white chest patch and has pale undertail coverts and a pale crescent on the uppertail coverts. Most often confused with the more numerous and larger Turkey Vulture and flight style and flight shape are remarkably similar to that species. Often joins Turkey Vultures, often flying below the main flock. On Zone-tailed note grayish cast to plumage and feathered, not bare, head. At close range, note yellow cere and legs and faint barring on underside of flight feathers; on adults white bars on tail. Behavioral clues, such as sudden dives for Zone-tailed to chase prey (often birds) are important.

Voice A loud screaming call, similar to a Red-tailed, but more of a whistle and less harsh, especially at the end.

Status & Distribution Found from Mexico south to Paraguay; also on Trinidad. Northern birds are migratory. **BREEDING:** Uncommon and local in wooded hills, mountains, and mesas often near watercourses from central and western TX through AZ, casual in Southern CA (has nested in the Santa Rosa Mts. and the east Mojave). **MIGRATION:** South into Mexico and beyond. **WINTER:** Rare in the Rio Grande Valley, TX, Southern CA, and southern AZ; found in lower elevations, not in mountains as in summer. **RARE STATUS:** Casual to NV, UT, and CO. Accidental or casual to LA and East Coast from MA to VA.

Population Expanding north in Southwest.

Turkey Vulture for comparison

long, narrow wings

flies with dihedral very similar to Turkey Vulture

juvenile

barred remiges differ from Turkey Vulture

finely banded tail

adult ♂

smaller bill than Common Black-Hawk

long tail with one broad white mid-tail band and one white band at tail base

adult

SWANSON'S HAWK *Buteo swainsoni*   SWHA ▪ 1

smaller bill than Red-tailed

light-morph juvenile

often with darker lateral breast patches

light-morph adult

brown breast band

broad dark subterminal tail band

light-morph adult

paler wing linings

juvenile

light-morph adult

long, narrow, pointed wings

whitish wing linings

A large hawk with long, narrow pointed wings, it is found in open grasslands and agricultural areas. Widely distributed across the West. Highly variable in plumage. Females are larger, and juveniles and subadults differ from adults. Feeds mostly on small mammals and insects, primarily grasshoppers and caterpillars. Monotypic. L 21" (53 cm) WS 52" (132 cm)

Identification Individuals grade evenly from light (most numerous) to rufous to dark morphs. Dark morphs predominate in Central Valley of CA. When perched, adult's long wings extend past the wing tips; a juvenile's wings almost reach the tail tip. Year-old birds are similar to juveniles, but have wide subterminal bands on tail, flight feathers. **LIGHT-MORPH ADULT:** Dark above and dark-headed with white on the forehead, lores, and

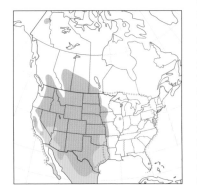

throat. Dark breast contrasts with white or lightly barred underparts, forming a bib. Tail is gray with dark subterminal band and many thinner dark bands, and a light U on the uppertail coverts. Underwings are distinctly two-tone with white coverts and darker gray flight feathers and dark gray trailing edges. **LIGHT-MORPH JUVENILE:** Dark back and wings with buffy feather edges, buffy head with a pale supercilium, buffy cheek, and dark malar line. Underparts are light-colored with variable streaking, usually with dark patches on the sides of the breast. Tail is brown with narrow bands, darker tip, and white undertail coverts. **INTERMEDIATE (RUFOUS) MORPH:** Above, similar to light-morph adult, but underparts are rufous, either evenly colored or with a darker chest than the rufous underparts. Throat is white, as are the undertail coverts. Juveniles are similar to light morph except for buffy underparts with heavier streaks. **DARK MORPH:** Adults are overall dark above and below, varying from black to dark brown. Undertail coverts always lighter, and underwing coverts can feature whitish or rufous mottling, paler flight feathers with dark trailing edge. Juveniles from above are similar to other morphs, but have darker wing coverts and heavy, dark mottling on body. **FLIGHT:** Best told by the combination of long, slim, pointed wings with the two-tone underwing coloration, relatively long tail, and dark bib on light-morph adults. Often soars with wings in a dihedral; wingbeats relatively quick and light for a large

Buteo; often hovers when hunting.
Similar Species LIGHT MORPH: Red-tailed Hawk has shorter wings, a dark patagial mark on the leading edge of the evenly colored underwing; most have a dark belly band. Perched, it appears larger billed and headed and its wing tips do not reach the end of the tail. Juvenile White-tailed Hawk has dark underwing coverts and usually has a white chest patch. **RUFOUS & DARK MORPHS:** All other dark adult buteos have dark undertail coverts and silvery, not gray, flight feathers.
Voice A drawn-out scream, usually heard near nest site.
Status & Distribution BREEDING: From the Great Plains westward to central CA, north to Canada and locally into AK and YT, and south into Mexico. **MIGRATION:** In fall, migrates in huge flocks through the western states, often descending into fields to feed on insect swarms. They pass through southern TX in late Sept. and early to mid-Oct. Rare fall migrant in the East, almost always juveniles. **WINTER:** Primarily in Argentina. Rare in southern FL and in the Rio Grande Valley, TX—especially immatures around the sugarcane fields—and a few in the Sacramento Valley, CA. Some also winter in west Mexico. **RARE STATUS:** Very rare fall and casual spring migrant in the East, almost always juveniles. Rare AK. Accidental Europe.
Population Declining in CA and likely elsewhere. Periodic large die-offs on winter grounds in Argentina.

wing linings on most slightly paler than flight feathers

dark-morph adult

prominent yellow cere

pale undertail coverts

intermediate-morph adult

RED-TAILED HAWK *Buteo jamaicensis*   RTHA   ▮ 1

The Red-tailed Hawk's widespread breeding range and large population make it the default raptor in most of the US and Canada. It utilizes a wide range of habitats, from wooded to open areas, farmland to urban settings. Red-tails come in a variety of color morphs, from pale to rufous to dark; average size varies from north (largest) to south (smallest). Prey species include rodents and small mammals, snakes, occasionally birds, even carrion. They are prone to albinism, occasionally appearing totally white. Polytypic (14 sspp.; 6 in N.A.). L 22" (56 cm) WS 50" (127 cm)

Identification A large, chunky, short-tailed raptor, with a large head and bill. All show a diagnostic dark patagial mark on the leading edge of the underwing, more easily seen in light-morph birds. The bulging secondaries give the wing a sinuous trailing edge. Juveniles have more slender wings and longer tails than adults, but when seen perched the wing tips usually fall short of the tail. Most adults have a reddish tail, varying in intensity by color morph and subspecies. **LIGHT-MORPH ADULT:** Occurs in all subspecies. Brown head is offset by darker malar stripe, white throat (except in some western populations and FL). Back and wings are brown, with white spots on scapulars, giving the appearance of a whitish V on the back. Chest and underparts white, crossed by a belly band of spots or streaks (darkest in Western and FL subspecies, missing in *fuertesi*). Tail is orange to brick-red with a dark terminal band and often with multiple thinner bands (Western). **LIGHT-MORPH JUVENILE:** Brown head with a dark malar stripe, often with a lighter supercilium. Throat is white (Eastern) to streaked darker (Western). Dark back and wings are mottled with white on the scapulars, forming a light V. Primaries are lighter than the secondaries, giving a two-tone look to the wing. Below, white undersides are separated by a belly band of heavy dark streaks. Tail is brownish with multiple thin, dark bands, often with a slightly wider terminal band. Underwings are light, coverts occasionally washed with rufous (Western), flight feathers tipped dark and lightly barred, and a pale rectangle at the base of the primaries. **DARK ADULTS (WESTERN BIRDS):** Darker brown above, usually without white spots on wing coverts. Dark belly

is offset by rufous to black chest. Tail is rufous, with thin dark bands and wider subterminal band. On rufous morph, undertail coverts are unbanded rufous, uppertail coverts barred brown. Under-

wing coverts are rufous with variable dark barring, patagial mark is visible. Darkest birds are uniformly dark; the dark undertail coverts are barred rufous. Tail is dark rufous with thin

white wing linings with sparse, black markings

adult light-morph "Harlan's"

mostly white below

white head with variable gray

tail variable with smudgy dark terminal band

adult "Krider's"

light rufous tail often with white base

pale underparts and wing linings

moderate to faint patagial bar

pale head

black wing linings with variable spotting

adult dark-morph "Harlan's"

variable white on forehead and around eye

slightly paler scapulars

adult dark-morph "Harlan's"

white marks on black underparts

extensive pale above

variably pale head

adult "Krider's"

pale underparts

juvenile *calurus*

heavy bill

adult *calurus*

white chest with heavily mottled belly

belly band

warm buffy wash to underparts

rufous tail

dark-morph
adult
calurus

pale breast
contrasts with
darkish head and
variably marked
belly band

adult
borealis

pale rufous tail with
narrow or no dark
subterminal band

adult
fuertesi

dark
patagial bar

uniform pale
underparts

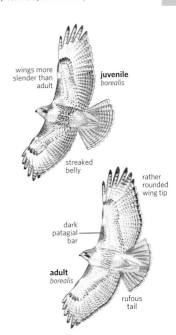

wings more
slender than
adult

juvenile
borealis

streaked
belly

rather
rounded
wing tip

dark
patagial
bar

adult
borealis

rufous
tail

barring and a wider dark subterminal band. **DARK JUVENILES:** Above, similar to light morph with white speckles on secondaries, often darker on the head and throat. Variable below, they are heavily streaked with rufous across chest; dark belly band with white or rufous streaks. Underwing coverts are mottled rufous or dark, with patagial mark often hard to discern. Tail is brown with many darker bands, like light-morph Western. **"KRIDER'S" RED-TAIL:** A pale, Great Plains color morph of *borealis*, treated by some as a separate subspeces, *kriderii*. Head whitish often with little or no malar stripe. Back and upperwings heavily mottled with white, underparts almost pure white with reduced patagial mark on underwing. Adult has orangish tail

fading to white at the base; juvenile has light tail banded with dark bars. It is considered a morph rather than a subspecies as the majority of individuals within the described breeding range do not resemble "Krider's." Definitions of what constitute a valid subspecies vary, but one widely accepted rule is that 75 percent of one population of subspecies must differ from 100 percent of an adjacent subspecies population. **"HARLAN'S" RED-TAIL:** In *harlani*, the dark morph is by far the most numerous. The adult is blackish, similar to other dark-morph birds, but with variable white mottling or streaking on chest and belly. Tail is gray with wide, dark subterminal band. Rare light-morph adult "Harlan's" similar to "Krider's," with light head but dark malar stripe, darker wings and flight feathers. Tail has a gray subterminal band and gray mottling fading to white at tail base. **FLIGHT:** Wingbeats are heavy, usually slow. Red-tails glide with wings level, occasionally soar in a slight dihedral. Immatures show a light rectangle at the base of the primaries.

Geographic Variation Eastern *borealis* is found west through the Great Plains, Western *calurus* west from the Great Plains, north into

Canada, and south to the range of "Fuertes's" *fuertesi* in AZ, NM, southern TX. AK *alascensis* along the AK coast, to the Pacific Northwest, "Harlan's" *harlani* from northwestern Canada into southern AK, and *umbrinus* in peninsular FL.

Similar Species Red-shouldered has light crescents at base of primaries in all plumages. Adult Ferruginous has mostly white underparts with dark legs, slimmer wing silhouette. Juveniles lack patagial mark, have dark crescent at the end of underwing coverts and pale primaries. Dark Ferruginous has white comma at end of underwing coverts, unbanded tail. Unlike Ferruginous and especially Rough-legged, which habitually hover while hunting in flight, Red-tailed seldom hovers unless there is a stiff breeze.

Voice A husky scream, rising then dropping in pitch *shee-eeee-arrr*.

Status & Distribution Found across N.A., into Mexico, and across the Caribbean. **MIGRATION:** Late fall along mountain ridges, the Great Lakes, fewer along the East Coast. Reluctant to cross open water; late July–Aug. movement along the south shore of Lake Ontario is primarily juveniles in postbreeding dispersal. Spring peak is Mar.–early Apr. along the south shores of the Great Lakes. **WINTER:** Many northern breeders winter in the southern US, subspecies mixing together. **RARE STATUS:** Casual to Bermuda and Newfoundland.

Population Stable.

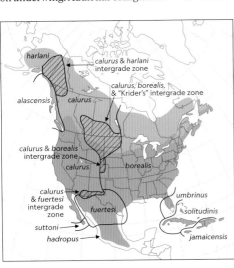

harlani

calurus & harlani
intergrade zone

calurus, borealis,
& "Krider's" intergrade zone

alascensis   calurus

calurus & borealis
intergrade zone

calurus   borealis

calurus
& fuertesi
intergrade
zone

umbrinus

solitudinis

fuertesi

suttoni

hadropus

jamaicensis

FERRUGINOUS HAWK *Buteo regalis* FEHA ■ 1

The largest of our buteos, the Ferruginous Hawk is sometimes called the Ferruginous Rough-leg for its feathered tarsi. Sexes are similar, although the females are larger. Variations include light, rufous, and dark morphs, light birds being far more commonly found. Ferruginous Hawks often perch on the ground. Generally feeds on small mammals (prairie dogs are among their favorite prey). The species is notable for gathering in winter roosts. Monotypic. L 23" (58 cm) WS 56" (142 cm)

Identification Large-headed, big-chested raptor with leg feathering that reaches to the toes. Wings are long and tapered with pale flight feathers. The light tail is of moderate length. Perched birds show the wing tips almost reaching the tip of the tail, shorter on juveniles. Head features a large bill and an enormous yellow gape reaching past the centerline of the eye. **LIGHT-MORPH ADULT:** Head is variably colored whitish or gray with rufous streaks and dark eye line. Back and upperwings are chestnut with dark markings and primary coverts, and darker wing tips. Primaries are white at the base. Underparts are

long gape extends back under rear of eye
paler head with whitish supercilium and dark postocular
brown above
yellow cere and gape
juvenile
pale head
extended gape
rufous
wings long, broad, and rather pointed
juveniles whiter below than adults
juvenile
dark-morph adult
mostly white below
feathered legs
adult
whitish tail lacks bands
rufous leggings
basal third of tail is white

variable rufous feathering in underwing coverts
rufous leg feathering
adults
soars with a shallow dihedral
all show broad whitish crescents

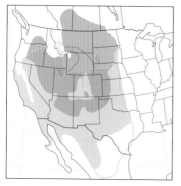

white with variable rufous barring on belly and flanks; legs barred rufous. Underwing coverts are variable, white to heavily barred, with light flight feathers almost unbarred, and dark crescent ("comma") at "wrist." Tail is plain, varying from white to gray, often with a rufous wash. **LIGHT-MORPH JUVENILE:** Dark brown back and upperwings are similar to many Red-tailed juveniles, but the head shows a large gape, dark eye line, and lighter cheeks. Primaries have light bases that make a noticeable white spot on the upperwings in flight. White undersides are usually sparsely marked, and leg feathers are white. Underwings are mostly white with scattered dark markings, and with a dark comma at wrist. Light tail has light bands on outer half and whitish base. **RUFOUS & DARK-MORPH ADULTS:** Very uncommon with variations from rufous to very dark. Dark brown or gray head, back, and wings, outer primaries show gray. Unbanded tail is gray above, silvery on underside. Undersides are dark, with variable amounts of white or rufous streaking. Underwings have silvery flight feathers and dark brown or rufous coverts, and white comma at wrist. Legs are dark, barred brown, or rufous. **DARK JUVENILES:** Dark brown above, head and breast slightly lighter and more rufous than belly; underwing coverts dark with white comma at wrist, dark primary tips; silvery flight feathers with light barring. Note

dark forehead. Tail is gray with darker bands. **FLIGHT:** From above, white at base of primaries is very visible, as is the light tail. Long, pointed wings are often held in a dihedral. Wingbeats are slow and powerful, interspersed with gliding. Hunts by soaring at high altitudes followed by long dives or low-level pursuit-flights near the ground to flush and ambush prey.

Similar Species Light-morph Red-tailed Hawks have a dark patagial mark on the underwing that Ferruginous Hawks lack, and they are less rufous or chestnut above. Red-tailed and other dark-morph buteos have different wing shapes, lack the pale area on the upperwing at the base of the primaries, and have more patterned tails. Dark-morph Rough-legged Hawk has a light forehead.

Voice Quite vocal during the breeding season, especially when near its nesting site or during confrontations with other raptors. Gives harsh alarm calls, *kree-a* or *kaah*. May call to signal alarm or location, to beg for food, or while engaged in territorial defense.

Status & Distribution Uncommon. **BREEDING:** Breeds in the western US into Canada and south to northwestern TX, NM, and AZ in open, dry country, often hilly. The large stick nest is built on a lone tree or on a rocky outcrop, and usually reused over many years. **MIGRATION:** Northern birds migrate into southwestern US, TX, and northern Mexico, and west to CA. Moves south mid-Sept.–early

Nov.; in spring, adults move north in Feb.–Mar., juveniles tend to migrate later, into Apr. Southern breeders are more or less sedentary. **RARE STATUS:** Casual east to KY, IN, OH, VA, NJ, and FL, most regularly in MN and WI.

Population Declining, fewer recorded in parts of eastern winter range (Great Plains) and in Southern CA.

ROUGH-LEGGED HAWK *Buteo lagopus* RLHA ■ 1

The most northerly of our buteos, the Rough-legged Hawk is a winter visitor to the lower 48 states from its Arctic breeding grounds. The numbers of visiting birds vary geographically and from year to year, most likely in response to the abundance or lack of prey within the birds' northern range. Rough-leggeds have small bills and have the smallest feet and toes of the large buteos. They often perch on thin and delicate branches, which would not support Red-tailed Hawks and other large species of buteos. Rough-legs feed heavily on voles, lemmings, and other small rodents and occasionally on small birds captured on the ground, including Snow Buntings, shorebirds, and others. Females are larger than males and have different plumage characteristics; juveniles are different from adults; all occur in light and dark morphs. Polytypic (4 sspp.; *sanctijohannis* breeds in N.A.; Asian *kamtschatkensis*, which is often paler and lacks a dark morph, is a rare migrant on the western Aleutians). L 22" (56 cm) WS 56" (142 cm)

Identification All have their legs feathered down to the toes. When perched, the wing tips reach past the tip of the tail in adults, but just reach it in juveniles. Most show a white forehead. **ADULT:** All morphs have a dark trailing edge to the underwings and a subterminal band on the tail. Light morph shows dark carpal patches on underwing. Light-morph male has breast more heavily marked than its belly (can be almost white), dark back and wings, light tail with multiple dark bands and a wide, dark subterminal band. Light-morph female has a browner back and upperwings and a darker belly than the male, usually with a lighter band between breast and belly, and the undertail has a large, dark subterminal band. Dark-morph adults have a black carpal patch against dark brown coverts and silvery flight feathers with a dark trailing edge. Males are uniformly dark, with three or four light bands on a dark tail. Females show a single dark band on the light undertail. **JUVENILE:** Light morph has markings similar to light-morph adult female, but with fewer dark markings on the underwing coverts, breast, and legs, including a narrower dark band on the trailing edge of the wing. The tail has a single broad, diffuse dark terminal band and a light base. Head often appears whitish. Seen from above, primaries have a light patch at the base; coverts are darker. Uppertail coverts are light. Dark morphs are similar to dark adult females, with narrower dark trailing edges to the underwings and tail, sometimes with lighter heads. **FLIGHT:** Long wings flap slowly and frequently are held in a dihedral. Juveniles show a large white patch at the base of the prima-ries, especially from above. Wing tip is rather blunt and shows five "fingers."

Similar Species From Rough-legged, larger Ferruginous Hawk lacks conspicuous blackish tail bands; note gape line; light morph lacks dark carpal patch on the underwing and breast streaking. Dark-morph juveniles more similar to Rough-legged but have a white comma at the wrist and dark foreheads. Dark "Harlan's" Red-tail usually has white markings on the breast or head, a blurry terminal band on the undertail, and gray on the uppertail.

Voice Most vocal during breeding season when male gives a soft, plaintive courting whistle. Alarm call is a loud screech or squeal.

Status & Distribution Fairly common. **BREEDING:** Throughout the Holarctic region, nesting on tundra and less commonly in subalpine forests. Nest is built on a rocky outcropping, on the ground, or in a lone tree, if present. **MIGRATION:** The entire breeding population leaves the tundra to winter in southern Canada and northern and central US, but the timing and abundance of the migration is determined by the abundance of their food source. Few arrive in northern states as early as late Sept.; most arrive in late Oct. and Nov. Sept. reports away from northern states are likely in error. Spring migration sees adults moving north in Mar., followed by juveniles in Apr. and well into May. **WINTER:** Juveniles often winter farther south than adults; adult males winter farther south than females. A few winter as far south as north-central Mexico. Now accidental to FL and Gulf Coast.

Population Insufficient data. Winter range has shifted significantly north.

thin, black ubterminal band

pale head with broad blackish belly band

black patch

light-morph adult
sanctijohannis ♀

adult
sanctijohannis ♂

adult males with multiple blackish bands at tail base

feathered tarsus

small bill

long and rather slender wings

dark-morph adult
sanctijohannis ♂

tail with multiple black bands, terminal band broadest and darkest

juvenile with pale head and blackish belly

white tail base with broad, black subterminal band

juvenile
sanctijohannis

BARN OWLS Family Tytonidae

Barn Owl (CO, May)

Tytonidae owls share many characteristics with owls of the larger Strigidae family (see p. 377).

Structure Tytonids differ from the strigids by having a heart-shaped facial disc; relatively small eyes; a short, squared tail; and serrated central claws. Their legs are long and feathered, and their toes are bare. Females generally larger than males.

Behavior These owls are essentially nocturnal and sedentary. Their habitats vary, from grasslands and open areas to closed forests. They hunt small mammals and other vertebrates from a perch or in low, quartering flights over open areas; sometimes they hover in place. Voices include shrieks, whistles, hisses, and screeches, unlike the resonant hooting of strigids.

Plumage Plumage is similar across ages and sexes, but females are generally darker. Upperparts range from pale gray and rust through orange-red to sooty blackish, often with sparse contrasting spots. Underparts are often paler, and the facial disc is paler still.

Distribution Tytonidae is a mainly tropical family reaching its greatest species diversity in Australasia, with about 13 species present. North America's sole representative, the Barn Owl, is the most widely distributed nocturnal raptor, resident on every continent except Antarctica.

Taxonomy The Tytonidae family comprises two genera (*Tyto* and *Philodus*), and about 19 species are recognized. Vocal differences between New and Old World Barn Owls (*T. alba*) indicate they might represent separate species.

Conservation Species inhabiting tropical forests are threatened by the logging of large trees. The restricted range of island-dwelling species renders them particularly vulnerable to habitat destruction. BirdLife International codes: 4 VU and 1 EN.

Genus *Tyto*

BARN OWL *Tyto alba*   BANO   ■ 1

By night, an unearthly shriek may alert a birder to a ghostly white Barn Owl flying overhead. The Barn Owl nests and

underparts vary from whitish to cinnamon-buff; males average paler

looks pale in flight with rounded wings and no dark carpal patches as in Short-eared Owl

dark eyes with whitish heart-shaped face

pratincola ♀

pratincola ♂

long legs

roosts in dark cavities in city and farm buildings, cliffs, and trees. Polytypic; Old World subspecies may be separate species (24 sspp.; *pratincola* in N.A.). L 16" (41 cm) WS 42" (107 cm)

Identification The Barn Owl's flight features shallow, slow wingbeats, often with long legs dangling. May hover in place while hunting. **ADULT:** A very pale nocturnal owl with a white, heart-shaped facial disc. Dark brown eyes; horn-colored bill. Head and upperparts mottled rusty brown and silvery gray, with fine, sparse black streaks and dots. Underparts vary from cinnamon to white. **MALE:** Palest birds are males. **FEMALE:** Darkest birds are always females; average larger than males.

Similar Species Other than Snowy Owl, this is the palest N.A. owl. Its flight and wing shape suggest slightly darker Short-eared Owl, but paleness of Barn Owl's plumage is usually evident; also note dark underwing crescents and buffy upperwing crescents on Short-eared.

Voice Male's territorial song is long, raspy, hissing shriek, *shrrreeee!* Often

given many times in sequence, usually in flight and near the nest. Female's vocalization similar.

Status & Distribution Rare to fairly common. **YEAR-ROUND:** Occupies low-elevation, open habitats, urban and rural. Density very low in more northerly areas. **WINTER:** Withdraws in winter from colder areas.

Population Has suffered sharp declines in eastern N.A. Canada lists it as endangered in the east and of special concern in the west. More than a dozen states in the Midwest and Northeast list it as endangered, threatened, or of special concern.

TYPICAL OWLS Family Strigidae

Great Gray Owl (MN, Jan.)

Owls are chiefly nocturnal predators, ecological counterparts of the diurnal birds of prey. Being chiefly nocturnal, they are difficult to detect, identify, and study. In the daytime, a collection of regurgitated pellets on the ground or a noisy mob of songbirds will often point a birder to a roosting or nesting owl. At night, voice is the best means to detect an owl and usually the best way to identify it.

Structure All owls have a relatively large head with immobile eyes that face forward to provide binocular vision and good depth perception. Acute hearing complements owls' keen eyesight. Most have a prominent facial disc formed of stiff feathers whose shape can be altered to help focus sounds. In some genera, ear openings are asymmetrically located in the skull, improving their ability to distinguish the direction and distance to a sound source. Asymmetric external ear structures on many genera provide further discrimination based on the frequency of the emitted sound. Some owl species can detect and capture prey in total darkness, by sound alone. Feather structure contributes to quiet flight. Soft body feathering absorbs sound and comb-like fringes on the leading edge of the outermost primaries reduce the sound of wings cutting the air. Some largely diurnal owls lack these fringes; their flight is not as quiet as that of nocturnal owls. Owl bills are strongly curved, with a sharp point for tearing prey. They can position their toes two forward and two backward for seizing prey, and their long claws are needle-sharp for clutching and quickly

dispatching it. Unlike tytonids, strigids generally have circular facial discs, relatively large eyes, and rounded tails; in addition, their central toe's claw is not serrated. Females are generally larger than males.

Behavior Most owl species forage by perching and watching for prey. They nest in natural or human-made cavities or in old woodpecker holes; some use old tree nests of raptors, crows, or squirrels. The larger owls prey mainly on mammals and the smallest on insects, with fish and small birds taken by some owls. Owls consume small prey whole but tear apart larger prey before consumption. They periodically regurgitate a dense pellet containing indigestible parts such as bones, chitin, hair, and feathers.

Voice The most commonly heard vocalization is a species-specific territorial song (both sexes vocalize in most species), delivered mainly in the month or two during which male owls establish a territory, locate potential nest sites, and attract a mate. This period of active singing varies among species: It begins about Oct. for the early nesting resident Great Horned Owl and about Apr. for the migrant Flammulated Owl. Owls also utter a variety of hoots, whistles, screams, whines, screeches, barks, or rasps when threatened, alarmed, or begging, for example. Sometimes these cannot be pinned down to a particular species; indeed one of the charms of nighttime owling is hearing a new, unidentifiable sound.

Plumage Owl plumage is similar between sexes, but females are sometimes darker than males. Downy young of most species molt directly into adultlike plumage; only a few hold a distinctively colored juvenile plumage for an appreciable period. Owls lack a distinctive breeding plumage, so after first attaining adultlike plumage, an owl's appearance changes little during its lifetime. Most adult strigids are cryptically patterned in shades of brown, though their facial disc may be quite distinctive. Many have erectile feather tufts at the sides of their crown, called ear tufts, although they are unrelated to the bird's ears or their hearing; their function is not known for certain.

Distribution Owls occur worldwide in virtually all terrestrial habitats, from the tropics to the Arctic, from below sea level to elevations above 14,000 feet. Most species are mainly resident and sedentary.

Taxonomy Worldwide, about 194 strigid species in about 28 genera are currently recognized. New species are still being described—about 10 in the last 20 years—as their voices (which are innate, rather than learned) and other attributes (including genetics) become better known. Twenty-two species in 13 genera occur in North America (four species as strays).

Conservation Many island-dwelling strigids are imperiled worldwide, four species are known to have gone extinct since 1600, and many more are at some level of risk. BirdLife International codes: 26 NT, 23 VU, 14 EN, 3 CR, and 5 EX.

SCOPS-OWLS Genus *Otus*

These are small to medium-size, mainly nocturnal Old World owls. Most have ear tufts and short, rounded wings. They inhabit forests and semi-open areas with scattered bushes or groups of trees. All have only one song type. About 41 species are currently recognized (one accidental visitor to N.A.), but more than half are confined to one or a few islands.

ORIENTAL SCOPS-OWL *Otus sunia*   ORSO ■ 5

The Oriental Scops-Owl is a small, nocturnal, primarily insectivorous owl of Asia's southeast rim. Its geographically variable vocalizations suggest that perhaps more than one species is involved. Polytypic (9 sspp.; *japonicus* in N.A.). L 7.5" (19 cm) WS 21" (53 cm)
Identification ADULT: Fine dark streaks on head; dark shaft streaks and thin horizontal pencil lines on breast; short ear tufts. Gray-brown, reddish gray, and rufous morphs exist. Yellow eyes; horn green bill with blackish tip. Legs only partially feathered.
Similar Species No other small owl is likely to occur on islands far offshore mainland AK.

Voice On the breeding grounds, territorial song of *japonicus* (the subspecies found in N.A.) is a rather low, whistled *tu-tu-tu*, repeated monotonously at short intervals. Probably silent on migration.
Status & Distribution Uncommon on breeding grounds in Japan. **BREEDING:** Favors deciduous and mixed forest, from Pakistan and India through China to eastern Russian Far East and

small size with short ear tufts

three color morphs

rufous morph *japonicus*

the main islands of Japan. **MIGRATION:** Four northerly subspecies, including *japonicus*, are largely migratory, wintering mostly in southern China, Thailand, and Malaysia and returning about May. Southerly subspecies are resident. **RARE STATUS:** Two N.A. specimen records of rufous-morph *japonicus* from the Aleutian Is., AK: Buldir I. (5 June 1977; a desiccated wing) and Amchitka I. (20 June 1979). **Population** Not globally threatened.

Genus *Psiloscops*

FLAMMULATED OWL   *Psiloscops flammeolus*   FLOW ■ 2

Once considered rare, the strictly nocturnal, insectivorous Flammulated Owl may actually be the most common owl in its breeding habitat. Its cryptic plumage provides excellent daytime camouflage, but a persistent flashlight-bearing birder tracking a singing owl at night may eventually be rewarded with deep red eyeshine returned from a pinecone-size shape perched close to a tree trunk. It nests and roosts in old woodpecker holes or natural tree cavities, sometimes in loose colonies. This New World species has recently been moved to its own monotypic genus. Monotypic. L 6.7" (17 cm) WS 20" (51 cm)

Identification Flight is nervous, darting, sometimes jerky, interspersed with occasional hovering as it pauses to check for prey. **ADULT:** Small; dark brown eyes; small and often indistinct rounded ear tufts; and variegated rufous and gray plumage. Pale grayish facial disc with variable rufous wash strongest around the eyes. Rufoustinged creamy spots form bold line on scapulars. Grayish brown bill; feathered legs, bare toes.
Geographic Variation Plumage and size vary clinally. Birds in the northwestern part of the range are the most finely marked; those in the Great Basin mountains are grayish and have the coarsest markings; those in the southeast are reddish. Wingspan and weight increase from southeast to northwest, presumably correlated with migration distance.
Similar Species Resembles screech-owls, but smaller; also dark eyes, shorter ear tufts, and relatively longer, pointier wings.

Voice Male's advertising song is a series of single or paired soft, deep, hollow, short hoots repeated every two to three seconds, lower pitched than that of a screech-owl (which has a more varied repertoire). Single hoots are often preceded by two grace notes at a lower pitch. Hoots resemble a distant Long-eared Owl. Female's hoots are higher pitched and more quavering. Ventriloquial; difficult to localize and usually closer than the listener estimates.
Status & Distribution Uncommon to common during breeding season, but overlooked. Silent and little known the rest of the year. **BREEDING:** Inhabits primarily open montane coniferous

reddish type

birds often more reddish in southeastern part of range

short ears are often indistinct

dark eyes

grayish type

forest, especially with ponderosa pine or firs, mixed with oaks or aspen. **MIGRATION:** Highly migratory; winters in C.A. Breeders in N.A. depart late Sept.–Oct., return beginning late Mar.–late May. Rarely detected during migration. **WINTER:** Range, diet, and habits are little known. **RARE STATUS:** Accidental in FL, AL, LA, east TX, and 75 mi offshore in the Gulf of Mexico. **Population** Canada considers it a species of special concern.

SCREECH-OWLS Genus *Megascops*

Most of these small to medium-size nocturnal owls of the New World have ear tufts and short, rounded wings. They inhabit forests, open woodlands, parks, and arid open and semi-open areas. All have more than one song type and are best identified by voice, combined with habitat and range. This genus (25 species; three in N.A.) split in 2003 from the genus *Otus*.

EASTERN SCREECH-OWL   *Megascops asio*   EASO ■ 1

In eastern wooded suburbs, this small owl is often the most common avian predator. It nests and roosts in old woodpecker holes or natural tree cavities and readily uses nest boxes. The range separation from Western Screech-Owl is not yet fully known. Polytypic (5–6 sspp.; all in N.A.). L 8.5" (22 cm) WS 21" (53 cm)

Identification ADULT: Small; yellow eyes; bill yellow-green at base with a paler tip. Ear tufts are prominent if raised; when flattened the bird has a round-headed look. Facial disc is prominently rimmed dark, especially on lower half. Underparts are marked by vertical streaks crossed by widely spaced dark bars that are nearly as wide as streaks. Scapulars have blackish edged white outer webs, forming a line of white spots across shoulder. Feet are disproportionately large. Occurs in rufous and gray morphs as well as intermediate brownish plumages; plumages are alike, but female is larger. Markings on underparts are less distinct on rufous-morph birds. **JUVENILE:** Similar to adult in coloration, but indistinctly barred light and dark on head, mantle, and underparts; ear tufts not yet fully developed.

rufous morph more numerous in Southeast, essentially unknown in West and south TX

rufous morph

gray-morph juvenile

ear tufts

yellow eyes

pale greenish bill

strong vertical and horizontal bars on underparts

gray morph

overall pale

western Great Plains *maxwelliae*

Geographic Variation Body size and intensity of markings vary clinally: smaller and darker in the south and east, larger and lighter in the north and west. Color morph distribution is more complex. In most areas intermediate brownish birds compose less than 10 percent of the population; but in FL, gray, rufous, and brownish birds are evidently about equally common. Rufous morph becomes more common in the Southeast and outnumbers the gray morph in some areas. Normally only gray-morph birds are found on the Great Plains (*maxwelliae*, the palest and most faintly marked) and in southernmost TX (*mccallii*).

Similar Species Western Screech-Owl's bill is blackish or dark gray at the base, but gray-plumaged individuals are otherwise nearly identical in appearance and habits to gray-morph Eastern Screech-Owls. Where their ranges overlap, the two species are best identified by voice.

Voice This species' territorial defense song is a strongly descending and quavering trill up to three seconds long, reminiscent of a horse's whinny (apparently lacking in *mccallii*). Contact song (three to six seconds) is a single low-pitched quavering trill; it may rise or fall slightly at the end. Female's voice is slightly higher pitched.

Status & Distribution Common. Range overlaps that of the Western Screech-Owl in eastern CO, along the Cimarron River, in extreme southwest KS, and in TX east of the Pecos River to near San Angelo. Hybrids with Western Screech-Owl are known from Big Bend NP and from eastern CO. **YEAR-ROUND:** Resident in a variety of tree-dominated habitats: woodlots, forests, river valleys, parks, even suburban gardens below about 4,500 ft. **RARE STATUS:** Accidental eastern NM. **Population** Stable.

maxwelliae

asio

hasbroucki

floridanus

mccallii

WESTERN SCREECH-OWL   *Megascops kennicottii*   WESO ▪ 1

This small, widespread, eared owl of the West inhabits a broad range of semi-open, low-elevation habitats. At dusk it may be seen as it emerges from its roost cavity to perch and look for prey; or it may be heard at the onset of breeding season giving its familiar "bouncing ball" song. Courtship begins in Jan. and Feb., with male singing near nest. Eggs are laid between late Mar. and late Apr. It preys on a wide variety of small animals, especially mammals, birds, and invertebrates, and nests in old woodpecker holes or natural tree cavities. It is nocturnal and somewhat crepuscular. Polytypic (8 sspp.; 6 in N.A.). L 8.5" (22 cm) WS 21" (53 cm)

Identification ADULT: Small, with yellow eyes; bill blackish or dark gray at base with pale tip. Ear tufts prominent if raised; when flattened, bird looks round headed. Underparts marked by vertical streaks crossed by much narrower and more closely spaced dark

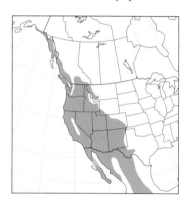

bars. Feet are disproportionately large. Sexes alike in plumage; females average slightly larger than males.

Geographic Variation Plumage generally monomorphic in a given area: brown or gray-brown in the Northwest, gray in southern deserts. Northwest coastal *kennicottii* more variable; generally dark brownish gray, but a small percentage is reddish. Size increases from south to north and from lowland to higher elevations. Toes feathered in northern populations, bristled in southern deserts.

Similar Species Whiskered Screech-Owl is slightly smaller, with disproportionately smaller feet, different colored bill, and usually bolder cross barring below. Where ranges overlap, Western is generally found at lower elevations and in less-dense woodland. Eastern's bill is greenish. Best separated from other screech-owls by voice. Flammulated Owl, in its own distinct genus, has dark eyes, is much smaller, and generally occupies higher elevations in areas where the species overlap.

Voice Two songs; both are common. A two-part tremulous whistled trill, first part short and second long, dropping slightly near the end: *dddd-ddddddr*; also a short sequence of five to 10 hesitating notes, accelerating in "bouncing ball" rhythm, ending in a trill: *pwep pwep pwep pwep pwepwepwepepepep*. Both sexes sing each song; the female's voice higher pitched. Agi-

dark bill

weaker crossbars than Eastern and Whiskered

Northwest coast *kennicottii*

tated barking notes are also given, often in advance of the song.

Status & Distribution Uncommon to fairly common. **YEAR-ROUND:** Inhabits a wide variety of semi-open woodlands, including streamside groves, pinyon-juniper deserts, suburban areas, and parks. Range has expanded eastward along the Arkansas River in CO, along the Cimarron River into extreme southwestern KS, and across the Pecos River in TX, increasing its area of sympatry with Eastern Screech-Owl. Hybrids with Eastern are known from CO and historically from Big Bend NP, TX.

Population Significant declines in Pacific Northwest, perhaps due to Barred Owl predation. Also declining in Central Valley, CA.

WHISKERED SCREECH-OWL   *Megascops trichopsis*   WHSO ▪ 2

This nocturnal and mainly insectivorous species reveals its presence with pair's duets at the onset of the breeding season. They nest in natural tree cavities or abandoned woodpecker holes. Polytypic (3 sspp.; *aspersus* in N.A.). L 7.2" (18 cm) WS 18" (46 cm)

Identification ADULT: Small, gray overall with yellow eyes; bill yellowish green at base with pale tip. Facial disc feathers at base of bill have whisker-like extensions. Ear tufts prominent if raised; when flattened, bird looks round headed. Small feet.

Similar Species Very similar to Western Screech-Owl (see above).

Voice Territorial song a series of four to eight equally spaced notes, *po po po*

po po po, mostly with emphasis on the third note, falling slightly in pitch at the end. Courtship song a syncopated series of hoots, like Morse code: *pidu po po, pidu po po, pidu po po* (pitch same as

Western Screech-Owl), often in a duet with slightly higher-pitched female.

Status & Distribution Fairly common.

dark bill

strong crossbars like Eastern Screech-Owl

aspersus

small feet

YEAR-ROUND: Inhabits dense oak and oak-conifer woodlands in the mountains of southeastern AZ and adjacent NM, in less open and generally at higher elevations than the Western Screech-Owl. Found at 4,000–6,000 ft, but mostly around 5,000 ft. **WINTER:** May move to lower elevations. **Population** Trends not known.

EAGLE-OWLS Genus *Bubo*

These nocturnal owls, most of which possess prominent ear tufts and powerful talons, are found in Eurasia, Indonesia, Africa, and the Americas. They occur in virtually all habitats except the densest forests and the largest deserts. About 16 species are currently recognized (two in N.A.). Some authorities include the fish-owls of Asia (genus *Ketupa*) in *Bubo*.

GREAT HORNED OWL *Bubo virginianus*   GHOW ■ 1

Many birders first meet this formidable owl in late winter by finding a female sitting in a large stick nest in a leafless tree. Others are introduced to it as a hulking, eared shape atop a power pole at dusk—perhaps a male bending nearly horizontally as it sings. Primarily a nocturnal perch hunter, Great Horned Owl is a fierce predator that takes a wide variety of prey, but most commonly mammals, up to the size of a large hare. It favors disused tree nests of other large species, such as Red-tailed Hawk, for nesting but also uses cavities in trees or cliffs, deserted buildings, and artificial platforms. It breeds early, with first eggs laid by Jan. in Ohio, later farther north. Young climb onto nearby branches at six to seven weeks and fly well from approximately 10 weeks. It often spends its daylight hours dozing in a tree, where the raucous cawing of a chorus of crows may lead one to the bird. Polytypic (approximately 16 sspp.; 9 in N.A.). L 22" (56 cm) WS 54" (137 cm)
Identification In flight, wings are broad and long, pointed toward tip; ear tufts are usually flattened and head is tucked in, producing a blunt profile. Flight is rapid and direct; wingbeats are stiff, steady, and mostly below the horizontal. **ADULT:** A very powerful, bulky owl with a broad body. Females larger than males and generally darker. Large head, stout ear tufts, and staring yellow eyes create a catlike appearance. Broad facial disc rimmed with black; whitish "eyebrows." White foreneck often conspicuous and ruff-like, especially when vocalizing. Upper chest coarsely mottled; rest of underparts crossbarred. Gray bill; densely feathered legs and toes. Plumage and size vary geographically. **JUVENILE:** Downy plumage grayish to buff, with dusky barring. By about Oct., young birds acquire complete set of flight feathers and tail feathers that differ in pattern, shape, and wear from those of adults: they have broader and more numerous crossbars, wings with uniformly fresh-looking rather than variably worn flight feathers, and tail feathers tapered rather than blunt ended.
Geographic Variation In general, birds of the eastern subspecies (including nominate) are medium size and brownish, with medium pale feet. Birds of the "Pacific Coast" subspecies group are smaller and dark, with dusky feet. Birds of the "Interior West" subspecies group are large and variably pale, with whitish feet. The palest subspecies, *subarcticus*, is resident across the far north-central portion of the range and has wandered southeast in winter as far as NJ.
Similar Species Long-eared Owl is smaller and more slender and weighs much less; it has a dark vertical stripe through its eye and longer, more closely set ear tufts; it also lacks the white throat. Its flight is floppier, and its voice is different. All other large N.A. owls lack ear tufts. Female and young Snowy Owls somewhat resemble *subarcticus*, but they have a white face and lack ear tufts.
Voice SONG: Territorial song is a series of three to eight loud, deep hoots in a rhythmic series; the second and third hoots are often short and rapid: commonly *hoo hoo-hoo hoooo hoo*; often longer, *hoo huhuhoo*

hooooo hoo. Also gives single hoots. Mostly heard near dusk and dawn. Male territorial singing begins about Nov. Duetting commonly begins one to two months before the first egg is laid. Female's voice is higher pitched. Juvenile gives a loud raspy screech.
Status & Distribution Common and widespread. **YEAR-ROUND:** Resident, sedentary, and territorial within the varied habitats of its breeding range, from forest to suburbs to open desert. **Population** Stable.

prominent ear tufts; color of facial disc varies geographically and individually

pale overall

bulky body shape

barred below

subarcticus

SNOWY OWL *Bubo scandiacus* SNOW ■ 2

At all times, this charismatic owl of the north is highly nomadic with most showing no site fidelity, even during the breeding season. During winter, it might appear far to the south. Primarily nocturnal, but will forage at all hours during continuous summer light, mainly from a perch. Lemmings and voles are its main prey, but a wide variety of prey items are taken including waterfowl and gulls, especially in winter. Monotypic. L 23" (58 cm) WS 60" (152 cm)

Identification Flight strong, steady, direct, jerky. Often glides on horizontal wings. **ADULT:** White, large; rounded head; small yellow eyes; blackish bill; heavily feathered legs, feet. Females larger than males. Male's plumage usually broken with narrow, sparse dark bars or spots. Female's markings larger, darker. Markings less intense as birds age; old males may be pure white. **IMMATURE:** More heavily marked than adult females.

Voice Fairly vocal when breeding, otherwise largely silent. Male's song a far-carrying series of two to six deep, low hoots, last often the loudest; higher-pitched female rarely hoots. When disturbed, either sex may utter a repeated *kre* call. High-pitched, drawn-out scream protests intrusion into winter territory.

Status & Distribution Uncommon to fairly common. **BREEDING:** Open tundra; nests on ground. Nomadic; breeds only where and when prey is abundant. **WINTER:** Winters in open areas; also in urban areas where it roosts on buildings. Some withdrawal from northernmost part of its range; others move north onto Arctic pack ice, often wintering at polynyas and feeding on waterfowl. Irruptive; in major southward flights birds may reach as far south as central CA and northern TX, LA, MS, AL, and northern FL. Fall migrants appear south by late Oct., more generally in Nov. and Dec. Northbound movement is in late Feb. and Mar.; a few remain well into May. **RARE STATUS:** Accidental Oahu, HI, and Bermuda, the latter (now a specimen) preyed on Bermuda Petrels. **Population** Vulnerable. Global populations now lower than originally thought. Various factors, especially climate change, represent threats.

round head

immature

color varies from all-white to heavily barred depending on age and sex

HAWK OWLS Genus *Surnia*

NORTHERN HAWK OWL *Surnia ulula* NHOW ■ 2

This largely diurnal Holarctic owl is often seen perched atop a conifer in partly open areas. Its flight is low and fast (with quick, stiff wingbeats) and highly maneuverable; it occasionally hovers. It stoops onto nearby prey with a smooth, gliding dive off its perch. It may pump its tail, when perched, especially just before it flies. Nests are found in cavities atop broken trunks, natural tree holes, and old holes of large woodpeckers. Polytypic (3 sspp.; *caparoch* in N.A.). L 16" (41 cm) WS 32" (81 cm)

Identification ADULT: Has a long, graduated tail, a falconlike profile, and a black-bordered facial disc. Its underparts are barred brown.

Geographic Variation Eurasian *ulula*, found from Scandinavia to Kamchatka, averages smaller and is paler overall than N.A. *caparoch*; two possible AK records (19th-century

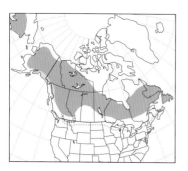

black border to facial disc

caparoch

dark barring on underparts

long tail

specimens) of Old World *ulula*, from St. Matthew and Bethel, may instead be pale *caparoch*.

Similar Species Unmistakable.

Voice Male's song is a trilling, rolling whistle, *ululululululululul*...±12 notes per second, lasting up to 14 seconds, then repeated. The female's version is shorter. Heard mainly at night, but also in morning; reminiscent of Boreal, but longer, higher pitched, sharper. A variety of alarm calls are given including flicker-like notes and screeches.

Status & Distribution Uncommon.

BREEDING: Across taiga belt, from the edge of the forested steppe north to timberline. Utilizes broken boreal forests with bogs, old burns, and semi-open areas. Nomadic; breeds only where and when prey is abundant. Rare, but increasing breeder in northwest MT; has bred in northern ID and northern WA. **WINTER:** Retreats slightly from northernmost part of mapped range. Irruptions probably tied to vole population cycles, which occasionally send some Northern Hawk Owls farther south; movements are mostly in Nov.–Dec. **RARE STATUS:** Casual to OR, NE, OH, PA, and NJ in winter; accidental Bermuda.

Population No studies in N.A., but studies in Finland showed a significant northern shift in the breeding range. The suspected cause is climate change.

PYGMY-OWLS Genus *Glaucidium*

These small but aggressive long-tailed owls (approximately 26 species; two in N.A.) lack ear tufts; many are mainly diurnal. Species in this genus are best identified by voice where ranges overlap, they inhabit deserts, deciduous bottomlands, wooded foothills, and high-elevation coniferous forests; they occur on all temperate continents except Australia. Many species prey primarily on other birds.

NORTHERN PYGMY-OWL   *Glaucidium gnoma*   NOPO   ■   2

The Northern Pygmy-Owl is an aggressive diurnal predator that sometimes catches birds or mammals larger than itself. Mobbing songbirds may lead a birder to the owl. Chiefly diurnal, it is most active at dawn and dusk. It makes its nests in natural tree cavities and old woodpecker holes. Polytypic (7 sspp.; 5 in N.A.). L 6.7" (17 cm) WS 15" (38 cm)

Identification ADULT: Long tail, dark brown with pale bars; rusty brown or gray-brown upperparts; spotted crown; underparts white with dark streaks; prominent "false eyes" on nape. Females average redder or browner, males grayer. **JUVENILE:** Crown spots indistinct or lacking; other whitish markings indistinct. Flight is undulating, with bursts of quick wingbeats; not as quiet as nocturnal owls.

Geographic Variation Birds of interior *pinicola* group are grayest; Pacific *californicum* group is browner. Much more distinctive than plumage are vocalizations, which are geograph-

ically variable and likely indicate that the complex is composed of multiple species: two, three, perhaps up to five. Three of these groups are in N.A.: the *californicum* group of three subspecies from southeast AK to CA; gray *pinicola* of the Rockies; and slightly smaller nominate *gnoma* from the sky islands of southeast AZ west to the Pajarito Mts., and from Hidalgo Co., NM, then south to Oaxaca. While rare in the Santa Catalina Mts., AZ, both subspecies might occur, *gnoma* being found at lower elevations. The *hoskinsii* group from the Cape District of Baja California and the polymorphic *cabanense* group south of the Isthmus from Chiapas to central Honduras may represent additional distinct species.

Similar Species Other small owls within Northern Pygmy-Owl's geographic and elevation range have short tails, lack nape marks, and have different plumage patterns.

Voice SONG: Calls of *californicum* group mellow, well-spaced, whistled *took* notes. Rocky Mtn. *pinicola* similar but lower pitched and with shorter pauses between notes. Nominate *gnoma* gives doubled notes, a *took-took*, repeated in a spaced series. Southern Baja *hoskinsii* also gives double notes, but they are delivered more

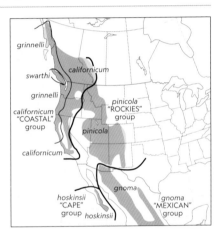

slowly than *gnoma*. C.A. *cabanense* gives a series of rapid single or doubled notes. At least the *californicum* group also gives a rapid series of *hoo* or *took* notes followed by a single *took*. These are often the first notes given to greet the predawn.

Status & Distribution Uncommon and mostly resident with local wandering and some altitudinal migration. Found through much of western N.A. and south in the mountains to central Honduras. **YEAR-ROUND:** Woodlands in foothills and mountains. In colder areas of its range, descends downslope in winter.

Population The subspecies *swarthi* of *californicum* group, an endemic subspecies on Vancouver I., BC, is on the provincial Blue List of vulnerable species due to limited range, small population, logging, and predation from Barred Owls.

californicum
group

"false eye"

spotted
crown

gnoma and *pinicola* groups
average grayer than
californicum group

gnoma
group

ng tail with
pale bars

FERRUGINOUS PYGMY-OWL *Glaucidium brasilianum*   FEPO ■ 3

This small, mainly diurnal owl is rare to uncommon in the US. It roosts in crevices and cavities and nests chiefly in old woodpecker holes or natural tree cavities. Polytypic (approximately 12 sspp.; 2 in N.A.). L 6.7" (17 cm) WS 15" (38 cm)

Identification Not as quiet in flight as nocturnal owls. Flight over longer distances is rather straight, with several rapid wingbeats alternating with glides; like a woodpecker. **ADULT:** Long tail, reddish with dark or dusky bars. Gray-brown upperparts; faintly streaked crown. White throat puffed

streaked crown
cactorum
"false eye"
long tail with rusty bars

out and more apparent when bird sings. Yellow eyes; prominent "false eyes" on nape. White underparts streaked reddish brown. **JUVENILE:** Like adults, but crown streaking very indistinct or lacking.

Geographic Variation AZ's *cactorum* is slightly grayer than TX's *ridgwayi*. Some authorities treat these two subspecies as constituting a separate species, "Ridgway's Pygmy-Owl" (*G. ridgwayi*), and S.A. birds as another species (*G. brasilianum*).

Similar Species The much smaller, nocturnal Elf Owl is short tailed and lacks false eyes. Northern Pygmy-Owl's range overlaps in AZ, but it inhabits higher elevations, its crown and breast are spotted, and it has fewer, more-whitish tail bands.

Voice The male's territorial and courtship song is a series of up-slurred, high-pitched whistled *pwip!* notes, monotonously delivered at about 2.5 notes per second. Phrases of about 10 to more than 100 notes are repeated for minutes at a time, spaced five to 10 seconds apart. Female's voice is higher and wheezier.

Status & Distribution Rare sedentary

resident in southeast AZ; uncommon sedentary resident in south TX. Common throughout most of its range south of the US. Found through C.A. and S.A. to Uruguay. **YEAR-ROUND:** Inhabits live oak-honey mesquite woodlands, mesquite brush, and riparian areas in TX, where insects are its main prey. In AZ, inhabits riparian woodlands and Sonoran desert scrub (including saguaro cactus), hunting mainly reptiles, birds, and small mammals.

Population Continues to decline, probably because of ongoing loss and fragmentation of habitat. The subspecies *cactorum* was federally listed in AZ until recently but remains very rare.

Genus *Micrathene*

ELF OWL *Micrathene whitneyi*   ELOW ■ 2

This is N.A.'s smallest owl. It nests and roosts in old woodpecker holes or other cavities in saguaros and trees. Polytypic (4 sspp.; 2 in N.A.). L 5.8" (15 cm) WS 14.7" (37 cm)

Identification ADULT: Tiny; round head; yellow eyes; lacks ear tufts. Wings fairly long, tail very short. Upperparts grayish brown to brown with buff to

yellow eyes
rounded head with no ear tufts
whitneyi
tiny size
very short tail

cinnamon spots on widespread southwest *whitneyi*; face and underparts with substantial cinnamon. **JUVENILE:** Like adult, but head and back markings slightly less distinct. Flight composed of rapid wingbeats in straight-line hunting strikes; less quiet than other nocturnal owls. Occasionally glides and hovers while feeding.

Geographic Variation Weak. Upperparts of *idonea* (south TX) are grayish, with little or no brown; face and underparts have little or no cinnamon.

Similar Species Pygmy-owls are chiefly diurnal, with long tails and "false eyes." The larger screech-owls have ear tufts and streaks below.

Voice The male's territorial and courtship song is an irregular series of usually five to seven high-pitched *churp* notes, delivered at about five notes per second. Both sexes utter a short, soft, whistle-like contact call, *peeu*; sometimes precedes male's song.

Status & Distribution Fairly common to common. **BREEDING:** Inhabits des-

ert lowlands, foothills, and canyons, especially among oaks and sycamores. **MIGRATION:** Northern populations migrate to southern Mexico by Oct. and return by Mar., males first. Casual in winter in southernmost TX. **RARE STATUS:** Accidental in fall to Los Angeles Co., CA.

Population Numbers reduced where habitat has been destroyed or degraded. Extirpated in southern NV; state-listed as endangered in CA, where nearly extirpated.

Genus *Athene*

BURROWING OWL *Athene cunicularia* BUOW ▮ 1

During daylight, this owl often perches at the entrance to its burrow or on a low post. It bobs its head when agitated. Primarily nocturnal, it hunts insects, small mammals, and birds from a perch or in flights, where it often hovers. It nests singly or in small colonies; western *hypugaea* is often associated with burrowing mammal colonies, where it usually occupies a disused mammal burrow; *floridana* usually excavates its own burrow. Polytypic (±20 sspp.; 2 in N.A.). L 9.5" (24 cm) WS 23" (58 cm)
Identification Ground dweller; long legs, unlike all other small N.A. owls. **ADULT:** Round head lacking ear tufts. **JUVENILE:** Plain brown upperparts

with few or no distinct markings; dark brown chest; pale buff underparts.
Geographic Variation Western *hypugaea* pale brown with buff mottling and spotting. Slightly smaller *floridana* (FL Threatened species) is darker brown, with whitish mottling and spotting, and paler below.
Voice Male's primary song is a soft, repeated *coo-cooooo*; also gives a chattering series of *chack* notes.
Status & Distribution Species' range extends to southern S.A. **BREEDING:** Open country: agricultural areas, grasslands, golf courses, airports.

MIGRATION: Northerly populations migrate. **RARE STATUS:** Casual in eastern N.A., north to western and southern ON, southern QC, and NS. Both subspecies have been found along the Atlantic coast.
Population Greatly reduced in much of the northern Great Plains by extermination of prairie dogs, conversion of prairies to cultivation, pesticide use, and habitat destruction. Declines continue. Many states list it as endangered, threatened, or of special concern. Endangered in Canada, with fewer than 1,000 pairs thought to remain.

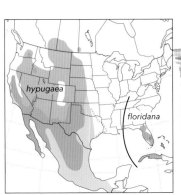

western
hypugaea

heavily spotted with buffy white above

round head

juvenile

no dark barring below

western
hypugaea

WOOD OWLS Genera *Ciccaba* and *Strix*

Medium-large to large nocturnal owls with large eyes and large, rounded heads, lacking ear tufts. Several species have different color morphs. They occupy wooded habitats in much of the world and prey mainly on small mammals and birds. *Ciccaba* is often merged in *Strix*; approximately 14 species in *Strix* (three resident), four in *Ciccaba* (one accidental in N.A.).

MOTTLED OWL *Ciccaba virgata* MOOW ▮ 5

This medium-size, nocturnal owl inhabits wooded habitats in C.A. and S.A. Polytypic (7 sspp.; likely *tamaulipensis* from northeastern Mexico in N.A.). L 14" (36 cm) WS 33" (84 cm)
Identification ADULT: Round head, no ear tufts; dark brown eyes. Light morph in drier areas (e.g., northern Mexico) has brown facial disc with bold white brows and whiskers; dark brown above with faint brownish barring; mottled dark brown on chest, rest of underparts streaked dark brown.
Geographic Variation Northerly subspecies are smaller, with fine distinct barring above and thin striping below.
Similar Species Closely related, larger Barred Owl has prominent barring

across upper chest and paler facial disc.
Voice Territorial song deep hoots: single *wh-owh* and *wooh*, and longer series, three to 10 hoots, often accelerating and becoming stronger before fading.
Status & Distribution RESIDENT: Subspecies *squamulata* from southern Sonora south to Guerrero; *tamaulipensis* from central Nuevo León and northern Tamaulipas (within 75 mi of Rio Grande); five other subspecies south to northeastern Argentina. **RARE STATUS:** One certain record: a likely roadkill found near Bentsen–Rio Grande SP (23 Feb. 1983). An accepted record from Weslaco, TX (5–11 July 2006), could well represent a misidentification.
Population Believed stable.

brown facial disc bordered by white

round head with dark eyes

streaked underparts

SPOTTED OWL   *Strix occidentalis*   SPOW ■ 2

This gentle-looking owl can be quite confiding when found dozing on a shaded limb during the day, perhaps alongside its mate. Imitating its call at dusk in a mature western forest may summon a mellow, echoing response from a far-away owl, but sometimes a Spotted Owl will fly in silently and unseen and then announce its presence with hair-raising barks. From a perch, it seeks wood rats, flying squirrels, and other small mammals. It nests mainly in tree cavities but also in debris on tree limbs or in old nests of other species. In the southwest, it also uses cliff ledges and caves. Polytypic (3–5 sspp.; 3 in N.A.). L 18" (46 cm) WS 41" (104 cm)
Identification Strictly nocturnal. Flight is slow and direct, with methodical wingbeats interspersed with glid-

color varies from darkest (Northwest) to palest (Southwest)

dark eyes

spotted below, including on breast

ing. **ADULT:** Brown overall with round, elliptical, or irregular white spots on its head, back, and underparts; lacks ear tufts. Indistinct concentric circles set in a rounded facial disc emphasize its dark eyes. Sexes alike in plumage; female larger.
Geographic Variation Three subspecies generally recognized in N.A. vary in color (dark to light) and spots (small to large) from north to south. "Northern Spotted Owl" (*caurina*), which resides from Marin Co., CA, to BC, is the largest. "California Spotted Owl" (nominate) is resident in CA in the Sierra Nevada and from Monterey Co. southward. "Mexican Spotted Owl" (*lucida*, or *huachucae*, according to some authorities) is resident in the Southwest and in Mexico. Some authorities recognize up to two additional subspecies in Mexico.
Similar Species The slightly larger Barred Owl is barred and streaked below, not spotted. Hybridization has occurred. See Barred Owl.
Voice A series of usually four nearly monotonic doglike barks, *hoo! hu-hu hooooh*, is used by either sex to proclaim and defend territory. Female's is higher pitched. Variants include ending with a sharper, louder bark, *hoo! hu-hu ow!*, and renditions of seven to 15 notes in a series. Contact call, given mainly by females, is a hollow, up-slurred whistle, *cooweeeeip!* Females use a rapid series of three to seven loud barking notes, *ow!-ow!-ow!-ow!-ow!*, during territorial disputes.

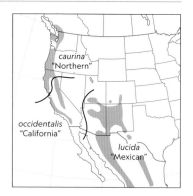

caurina "Northern"

occidentalis "California"

lucida "Mexican"

Status & Distribution Uncommon resident; decreasing in number due to habitat destruction. Further threatened by predation and hybridization as Barred Owl expands into its range in the Northwest and CA. **YEAR-ROUND:** Inhabits mature coniferous and mixed forests and wooded canyons, usually with multileveled, closed canopies and uneven-aged trees. Some individuals descend to lower elevations in winter.
Population Near Threatened. Subspecies *caurina* and *lucida* are listed as Threatened by USFWS; subspecies *caurina* is Endangered in Canada, where only a few known pairs remain in BC. Relatively widespread within ranges that are presumed to have changed little overall, but numbers probably have declined dramatically within specific habitats because of clear-cutting and even-aged forest management.

BARRED OWL   *Strix varia*   BADO ■ 1

This widespread woodland owl dozes by day on a well-hidden perch but seldom relies on its good camouflage to avoid harm. Instead, it flies away at the least disturbance, seldom tolerating close approach. But where foot traffic is heavy, such as along boardwalks in southern swamps, Barred Owl may sit tight and provide good views at close range. It hunts mainly from a perch but will also hunt on the wing, preying on small mammals, birds, amphibians, reptiles, and invertebrates. It prefers to nest in a natural tree hollow, but it will also use an abandoned stick nest of another species. Polytypic (4 sspp.; 3 in N.A.). L 21" (53 cm) WS 43" (109 cm)
Identification Large and chunky. Flight is heavy and direct, with slow,

methodical wingbeats; occasionally makes long, direct glides. **ADULT:** Dark brown barring on its ruff-like upper breast; rest of underparts are whitish with bold, elongated dark brown streaks; lacks ear tufts. Central tail feathers expose three to five pale bars between their tips and tips of uppertail coverts. Sexes alike in plumage; female larger. Chiefly nocturnal. **JUVENILE:** By about Sept., young birds acquire a complete set of fresh flight feathers; may be variably worn on adults. Central tail feathers expose four to six pale bars.
Geographic Variation Weakly differentiated. Southeastern *georgica* is darker brown than the widespread nominate *varia*. Subspecies *helveola* of southeast TX is paler.

Similar Species The underparts of the slightly smaller Spotted Owl are spotted overall, not barred and streaked. Hybridization has occurred where

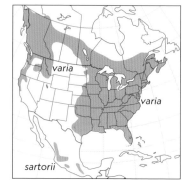

varia

varia

sartorii

ranges overlap; the hybrids, sometimes called "Sparred Owls," have plumages, voices, and sizes that are intermediate between the two species. **Voice** Highly vocal, with a wide range of calls. Much more likely than other owls to be heard in the daytime. Its most common vocalization is a rhythmic series of loud *hoot* or *whoo* notes: *who-cooks-for-you, who-cooks-for-you-all.* Also often heard is a loud, drawn-out *hoo-waaah* that gradually fades away; it is sometimes preceded by an ascending agitated barking. Often a chorus of two or more owls will call back and forth with these and other calls; the female's voice is higher pitched.

Status & Distribution Common in eastern N.A. Has expanded its range westward in the 20th century along Great Plains river corridors and through Canada's boreal forest to southeastern AK, WA, OR, and northern CA. **YEAR-ROUND:** Resident in mature mixed deciduous and uniform coniferous forests, often in river bottomlands and swamps; also in upland forests. **MIGRATION:** More northerly populations may drift south during late autumn if prey are scarce. **RARE STATUS:** Casual to central Sierra and Owens Valley, CA, and western NV. Accidental on Bermuda.
Population Increasing in N.A.

dark eyes

barred breast

vertical streaks

GREAT GRAY OWL *Strix nebulosa*   GGOW ▪ 2

Our largest owl is not our heaviest: Great Gray Owl's uncommonly dense, fluffy feathering creates an illusion of great bulk. This Holarctic owl is most often found perched low on the edge of a clearing, listening and looking intently for the small rodents on which it preys. It hunts from dusk to just before dawn from a low perch overlooking an open area. It is able to take prey moving unseen beneath snow by plunge-diving feet first into shallow snow or head-first into deeper snow, before thrusting its legs to grasp its target. It favors the abandoned nests of other birds of prey, but it also uses broken tops of large trees, preferring

huge size

dark rings on facial disc

black and white at bottom of facial disc gives "bow tie" effect

nebulosa

shaded locations. Polytypic (2 sspp.; nominate in N.A.). L 27" (69 cm) WS 55" (140 cm)
Identification It flies with slow and deep wingbeats and very little gliding. It often hovers above a suspected prey location before pouncing. **ADULT:** Relatively long tailed, with a disproportionately large head; lacks ear tufts. Pale gray facial disc with (usually) five distinct dark gray concentric circles around each yellow eye, making the eyes seem particularly small. Black-and-white "bow tie" pattern beneath the bill. Plumage largely gray, subtly patterned with combinations of whites, grays, and browns. Upperparts marked with dark and light; underparts boldly streaked over fine barring. Yellowish bill; legs and toes heavily feathered. Sexes alike in plumage; female distinctly larger. **JUVENILE:** Downy plumage is cryptic gray and white. By about Aug., young birds acquire a complete set of uniformly fresh flight feathers, not variably worn as on adults; terminal bands on flight feathers and tail feathers are whitish when present.
Similar Species Great Gray Owl is unlikely to be mistaken for any other species. The smaller Barred and Spotted Owls have dark eyes, browner plumage overall, and different patterns on their underparts. The similar-size Great Horned Owl has prominent ear tufts and larger yellow eyes; the northern subspecies of Great Horned (*subarcticus*) is grayest and palest.
Voice The male's courtship song is a series of five to 10 deep, res-

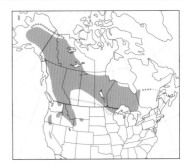

onant, muffled *whoo* notes that are longer and higher pitched than the Sooty Grouse; they gradually drop in frequency and decelerate toward the end of the series. The song may be repeated as many as 10 times. The female may answer with a soft, mellow *whoop.*
Status & Distribution Uncommon. Inhabits boreal forests and wooded bogs in the far north, coniferous forests with meadows in mountains farther south. **BREEDING:** Nomadic; breeds where it finds abundant prey and may not breed in years that prey is scarce. **WINTER:** Withdraws downslope or southward. Extreme prey shortages in the north may cause tens or hundreds of owls to winter in areas where normally few are seen. An unprecedented 1,700 Great Gray Owls were recorded in northern MN during the 2004–2005 irruption. **RARE STATUS:** Rare and irregular to dashed line on the map. Casual or accidental to northwest CA, SD, IA, OH, and PA.
Population Seems to be stable; genetically distinct population in Sierra Nevada, CA, is of great conservation concern.

EARED OWLS Genus *Asio*

Most of these medium-size owls have prominent ear tufts, long wings, well-developed facial discs, bold streaking below, and cryptic plumage. Some are open-country specialists; others occupy wider habitat ranges (e.g., forests, forest edges, shrubby growth, and open country). They occur on all temperate continents except Australia (seven species; three in N.A.).

STYGIAN OWL *Asio stygius*   STOW ■ 5

close-set ear tufts

dark facial disc

robustus

This is a medium-size, strictly nocturnal forest-dwelling owl. Polytypic (6 sspp.; 2 N.A.). L 17" (43 cm) WS 42" (107 cm)
Identification ADULT: Chocolate-brown overall, with long, close-set ear tufts. Whitish forehead contrasts with blackish facial disc. Underparts dirty buff, with bold dark brown streaks and crossbars. Legs feathered; toes bristled.
Similar Species The closely related Long-eared Owl lacks white forehead; has a rusty facial disc; underparts less boldly patterned; legs and toes fully feathered.

Voice Male's song is a series of deep, emphatic *woof* or *wupf* notes spaced six to 10 seconds apart.
Status & Distribution Little known. Resident in Mexico (within 200 mi. of US in Sierra Madre Occidental) and south to northern Argentina; also on Cuba and Hispaniola. **RARE STATUS:** Accidental at Bentsen-Rio Grande SP, southern TX (9 Dec. 1994, and 26 Dec. 1996), subspecies likely *robustus* (eastern Mexico to C.A.). Recent record at Key West, FL, on 1 June 2018, subspecies likely *siguapa* (Cuba).
Population Considered endangered by the Mexican government.

LONG-EARED OWL *Asio otus*   LEOW ■ 2

long ear tufts that are close together

rufous facial disc

blackish around eyes

overall slender body

heavily streaked and barred below

finer, more numerous wing markings than Short-eared

The Holarctic Long-eared Owl roosts in a dense tree or thicket. It may flush, or attempt to hide, stretching into a slim, erect profile with its ear tufts raised like little twigs. It hunts small rodents nocturnally, coursing low over the ground. It makes a nest in another bird's abandoned stick nest, preferably in a clump of trees. Polytypic (4 sspp.; 3 in N.A.). L 15" (38 cm) WS 37" (94 cm)
Identification ADULT: Slender; medium size; long, close-set ear tufts. Rusty facial disc; yellow eyes. Upperparts a mix of black, brown, gray, buff, and white. Bold streaks and bars on breast and belly. Central tail feathers with five to seven dark bars. Females average darker, richer buff, more heavily streaked. **JUVENILE:** Tail feathers have more exposed dark bars.
Geographic Variation Weak and clinal in N.A. where widespread *wilsonianus* meets the smaller and paler

tuftsi of western Canada south to northwest Mexico. Old World *Otus*, accidental AK, has reddish eyes, paler facial disc with more poorly defined outline, and more faintly marked underparts. Calls of Old World birds are lower pitched. They may represent a separate species.
Similar Species The larger, much heavier Great Horned Owl has a white throat. Short-eared's ear tufts are tiny; its underparts are boldly streaked.
Voice Generally silent except in breeding season. Male's advertising song is a series of 10 to more than 200 low, soft *hooo* notes, evenly spaced about two to four seconds apart. The female's answering call is a softer, higher-pitched *whoof-whoof-whoof*. The alarm call is a harsh, barking *ooack ooack ooack*. Many reports of calling birds in fall/winter are likely Great

Horneds giving single notes.
Status & Distribution Rare in Southeast; uncommon elsewhere in N.A. **BREEDING:** Wooded areas in lowlands and mountains with nearby open foraging areas having fertile, boggy, or arid sandy soil. **MIGRATION:** Those breeding in northern parts of the range with heavy winter snow cover migrate south to areas with more favorable climates. **WINTER:** Day-roosting groups of 10–100 or more cluster close together in one or a few neighboring trees. **RARE STATUS:** Casual southeastern

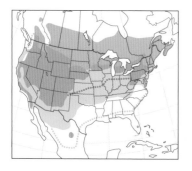

reddish eyes

paler facial disc

Eurasian *otus*

crossbars much fainter on Old World birds

AK, south TX, and FL; accidental to Bermuda and Cuba. Old World *otus* recorded (photos) twice from western AK, from Buldir I. and the Bering Sea. **Population** Numbers much reduced in some areas. State-listed as endangered in Illinois, threatened in Iowa, and a species of special concern in many other states.

SHORT-EARED OWL *Asio flammeus* SEOW ▪ 1

Memorable is the male Short-eared Owl giving courtship song high overhead, then dropping toward the ground with wing claps. It hunts voles and lemmings at any time of the day but is usually crepuscular, coursing low over open areas. Its nest is a shallow ground scrape under cover. Polytypic (11 sspp.; 2 in N.A.). L 15" (38 cm) WS 39" (99 cm) **Identification MALE:** Holarctic *flammeus* has a boldly streaked tawny breast and a pale, lightly streaked belly. Round head with closely spaced ear tufts barely visible; yellow eyes in blackish patches. Upperparts mottled brown and buff; tawny uppertail coverts. **FEMALE:** Larger, generally darker: upperparts browner; underparts rustier, heavier streaking.

Geographic Variation Little variation within the Holarctic (*flammeus*), including *sandwichensis* from HI. Three Greater Antillean subspecies are smaller with buffier underparts, darker upperparts, and a dark area around facial disc; also very fine or no streaking on belly and upperparts, and uppertail coverts dark brown. Calls differ. These are part of a group of seven subspecies (*domingensis* group) also found in S.A., including the Galápagos, and are probably best treated as a separate species.

Similar Species Closely resembles Long-eared Owl in flight (see sidebar). Northern Harriers have longer tails and white uppertail coverts. They hunt and compete with Short-eared Owls for prey, especially during twilight hours and sometimes rob them.

Voice Male's courtship song is a series of 13–16 rapid, deep hoots. Both sexes give hoarse *cheeaw* calls when disturbed. The *domingensis* group gives moaning, tomcat-like calls.

Status & Distribution Nominate still fairly common in much of the northern part of its N.A. range. Rare visitor, likely *cubensis* on probability of *domingensis* group, occurs on the FL Keys and Dry Tortugas in spring and summer. Most appear to be juveniles, probably dispersing from Cuba, where an uncommon and local resident. **BREEDING:** Tundra, prairie, coastal grasslands, and marshes. **MIGRATION:** Southward to escape areas with complete snow cover. Numerous records far at sea and colonization of distant islands attest to the species' propensity to wander; very rare on Bermuda. **WINTER:** More gregarious; flocks may roost together, rarely seen, in trees.

Population Has declined in many areas. Canada lists it as a species of special concern because of loss of habitat. Listed as endangered, threatened, or special concern species by seven northeastern states, where it is all but eliminated as a breeder.

domingensis

West Indian birds are darker above and around edge of facial disc than *flammeus*

upperwings more coarsely barred

slow, floppy wingbeats

dark primary covert patch on underwing

buffy and blackish wing patches

very short ear tufts

flammeus

blackish around eyes

flammeus

streaked below

flammeus

flammeus

domingensis group

Flight Identification of Long-eared and Short-eared Owls

The Long-eared Owl and the Short-eared Owl appear very similar in flight. They hunt by coursing low over open ground in slow, moth-like flight. Short-eared generally hunts during crepuscular hours, and Long-eared at night. Their upperwings have a broad tawny patch at the base of the outer primaries. Their underwings show a distinct carpal mark ("wrist mark"), formed by dark tips on the outermost under primary coverts. Both marks are usually more prominent on Short-eared, but differences are not diagnostic. The long ear tufts of Long-eared are flattened and not visible. Identification can be made, however, by noting underwing details in flight.

Short-eared Owl's outer four or five primaries are broadly tipped dark brown. In from that, a single dark brown band cuts across the outer primaries, separated from the dark tips by a wide buffy band. Two or three wide dark bands cut across the secondaries and the inner primaries near their tips. The overall impression is of a pale underwing that is coarsely barred on the outer half and has a dark wrist mark.

Long-eared Owl's primaries and secondaries are crossed by numerous dark and light bands of varying width. The overall impression is of a mainly pale, but finely barred underwing with a dark wrist mark.

FOREST OWLS Genus *Aegolius*

Small and nocturnal, with large, rounded heads lacking ear tufts, forest owls have well-developed, rounded or square facial discs. They have yellow or orange-yellow eyes and relatively long wings, and they reside in extensive forests, preying mainly on small rodents and shrews. One species is Holarctic; the others reside in the Americas (four species; two in N.A.).

BOREAL OWL *Aegolius funereus*　BOOW ■ 2

Males of this nocturnal species (named Tengmalm's Owl in Old World) give their far-carrying song in late winter. It is most often seen around nest cavities (old woodpecker holes and nest boxes). Polytypic (7 sspp.; 2 in N.A.). L 10" (25 cm) WS 23" (58 cm)

Identification ADULT: N.A. *richardsoni* has white underparts broadly streaked chocolate brown; whitish facial disc with distinct black border; umber-brown crown densely spotted white; pale bill. Female noticeably larger. JUVENILE: Chocolate-brown plumage held about June to Sept.

Geographic Variation N.A. *richard-*

soni is larger, darker, and more boldly patterned than nominate *funereus* of Europe. Within Eurasia there is a cline from west to east, from darker birds in Europe to palest and grayest (largest subspecies) *magnus* of Russian Far East. One Old World *magnus* reached St. Paul I., AK (Jan. 1911, specimen). New and Old World birds show strong genetic differences and males' territorial song slightly differs.

Similar Species Northern Saw-whet is smaller; adult has streaked crown, buffy facial disc with no dark border, dark bill, and cinnamon-brown breast streaks; juvenile tawny rust below.

Voice Call given by both sexes is a high, raspy *skiew*; also an ascending whistle. Male's territorial song is about a two-second-long rapid series of low, whistled *toot* notes; like a winnowing Wilson's Snipe.

Status & Distribution Uncommon to rare. YEAR-ROUND: Inhabits boreal forest belt, also subalpine mixed conifer-deciduous forests in high Rockies and mountains of Northwest. Has bred in ME and NH. MIGRATION: Adult males usually sedentary, females

nomadic; juveniles disperse widely. Prey shortages periodically drive large numbers south, to winter in areas where normally few are seen. Unprecedented irruption of 2004–2005 saw more than 400 in upper Midwest. RARE STATUS: Casual south to northern IL, northern OH, NJ, and on Bering Sea islands.

Population Stable.

prominent black border to facial disc

pale bill

richardsoni

juvenile *richardsoni*

darker body coloration than juvenile Northern Saw-whet

NORTHERN SAW-WHET OWL *Aegolius acadicus*　NSWO ■ 2

This widespread owl usually tolerates close approach without flushing. Strictly nocturnal, it hunts rodents from a perch and roosts by day in thick vegetation, often near the end of a lower branch. In some regions, often on day roost. Nests in a woodpecker hole or a nest box. Polytypic (2 sspp.; both

brooksi, endemic to Haida Gwaii

acadicus

no dark border to facial disc

dark bill

white areas replaced by golden buff

rufous streaking below

juvenile *acadicus*

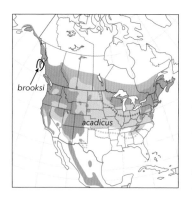

brooksi

acadicus

in N.A.). L 8" (20 cm) WS 20" (51 cm)

Identification ADULT: Reddish brown above with crown streaked white; white below with broad reddish streaks; dark bill; reddish facial disc, without dark border. JUVENILE: Plumage held about May to Sept. Dark reddish brown above, tawny rust below.

Geographic Variation Distinctive subspecies *brooksi*, with different

ecology and perhaps song, endemic to Haida Gwaii, BC, is much buffier on facial disc and belly than nominate; white only on eyebrows. Juveniles are darker on head, almost sooty black-brown, more richly colored below. It perhaps is as distinct as the Unspotted Saw-whet Owl (*A. ridgwayi*) that sings like Northern Saw-whet.

Similar Species Boreal Owl is larger;

adult has a spotted crown, whitish facial disc with black border, pale bill, chocolate-brown breast streaks; juvenile chocolate brown below.

Voice Male's territorial song is a monotonous series of single, low whistles, about two per second, on a constant pitch, said to be higher pitched in *brooksi*. A two- to three-second-long rising and screeching nasal whine is also frequently given.

Status & Distribution Fairly common in breeding range; uncommon to rare over much of winter range. **BREEDING:** Dense coniferous or mixed forests, wooded swamps, tamarack bogs. **MIGRATION:** Some remain year-round on breeding range, but considerable numbers migrate south or downslope in autumn. Casual on Bermuda. **WINTER:** Wide range of habitats; dense vegetation for roosting.

Population Stable, but no data on *acadicus*; small population (estimate of 1,900) of *brooksi* declining since 1970 due to logging; listed as threatened by Canadian government.

Genus *Ninox*

NORTHERN BOOBOOK *Ninox japonica* NOBB ▪ 5

round head with yellow eyes

japonica

long banded tail

The *Ninox* hawk-owls have indistinct facial discs compared with other owls. Northern Boobook was recently split from Brown Boobook (*N. scutulata*) of southern Asia based on strikingly different calls. Polytypic (3 sspp.; likely *japonica* in N.A. based on probability, but no specimen). L 12.3" (31 cm)

Identification The Northern Boobook is medium-size, hawklike brownish owl with a round head, no ear tufts, golden yellow eyes, and a long, banded tail; underparts heavily streaked with reddish brown. A nocturnal owl, but sometimes active at dusk and dawn.

Similar Species None; in flight suggests a small falcon or *Accipiter* hawk.

Voice Unlikely to be heard in N.A.

Status & Distribution Migratory Asian species. **BREEDS:** Northeast Asia to central China. **WINTER:** Some to southeast Asia; a stray reached Ashmore Reef, Australia. **RARE STATUS:** N.A.'s two records are from AK: St. Paul I. (27 Aug.–3 Sept. 2007) and Kiska I. (found dead on 1 Aug. 2008; not retained as a specimen!).

Population Apparently stable.

TROGONS Family Trogonidae

Elegant Trogon, male (AZ, Mar.)

Trogons are stunning tropical birds with bright colors. Two species barely reach the Southwest. Their plumages and rarity in the US place them among the most sought-after birds.

Structure They have a long, relatively broad tail; a large, rounded head with large eyes; a small, broad, notched bill; a short neck; and a compact body.

Behavior Trogons sit upright and motionless for extended periods, making them hard to find. Their calls are often the easiest method of locating them, but the calls can carry some distance and have a ventriloquial quality. Flights are usually short; the flash of color gives their presence away. They primarily eat large insects and fruit. Food is taken on the wing, either flycatching or, more typically, via sallies where the bird plucks an item from the end of a branch without landing. They are usually seen singly or in small groups, and nest in tree cavities.

Plumage Trogons have soft, dense plumage. Males are brilliantly colored, females more subdued. Males of the two species seen in the US are green and red. The metallic green upperparts occasionally appear greenish blue. Juveniles slowly molt into adultlike plumage over their first year. Adults typically molt in late summer.

Distribution A pantropical family of 44 species, trogons reach their greatest diversity in the Neotropics. Nine species occur in Mexico; two barely reach the US.

Taxonomy The 29 Neotropical trogon species are placed in four genera, from the large quetzals (*Pharomachrus*) to the most speciose genus, *Trogon*, which includes all the similar, smaller-size birds. Another 15 species in three genera occur in the Old World tropics.

Conservation Habitat destruction and human development threaten some species. Birder disturbance at well-known Trogon breeding locations in AZ may result in nest failures. BirdLife International codes: 8 NT and 1 VU.

Genus *Trogon*

ELEGANT TROGON *Trogon elegans* ELTR ■ 2

This colorful species is regular only in southeastern AZ. Polytypic (4 sspp.; 2 in N.A.). L 12.5" (32 cm)

Identification Both sexes have white breast band, yellow bill, and red orbital ring. **MALE:** Bright green head, chest, and back. Very fine barring on under-side of tail, each feather with a broad white tip; above, tail gold to green-ish copper, with two central rectrices tipped broadly with black. **FEMALE:** Gray-brown. A broad, white teardrop below and behind eye. **JUVENILE:** Like female, but lacking red. Large whitish buffy spots above.

Geographic Variation Subspecies *ambiguus* has reached TX; the male has deeper and more extensive red underparts; subspecies *canescens* occurs in AZ.

Similar Species Larger Eared Quetzal is stockier; its bill is black or gray; and it lacks a white breast band. Mountain Trogon (*T. mexicanus*), resident as far north as Chihuahua, Mexico, is a possible stray.

Voice CALL: Varying croaking or *churr* notes. **SONG:** A series of croaking *co-ah* notes.

Status & Distribution Found from US Southwest to Oaxaca; then in C.A. from Guatemala to northwest Costa Rica. **BREEDING:** In US, found in pine-oak woodlands, in association with streamside woodlands, primarily sycamores, mostly at elevations of 4,000–6,500 ft. Uncommon, but local, in southeastern AZ. The small population in Peloncillo Mts. of southwest NM and southeast AZ, now believed extirpated. **MIGRATION:** Routes, duration, and distances largely unknown. In spring, arrives Apr.–early May. Departs in fall by early Nov. **WINTER:** Withdraws from most of AZ, although still annual at that season. **RARE STATUS:** Casual to central AZ, eastern NM, western (Big Bend NP) and southern TX.

Population AZ breeding population small; possibly declining in Chiricahuas. AZ lists the bird as a candidate species on its Threatened Native Wildlife list. Disturbance at nest sites in AZ is a concern.

extensive white spotting on wing coverts

juvenile

broad white stripe extends down from behind eye

no red below

white spot on ear coverts

brownish on head and upperparts

banded tail

bright green head and upperparts

bronze tail

blackish termi-nal tail band

all figures *canescens*

yellow bill

white breast band

bright geranium red belly

delicately barred underside of tail

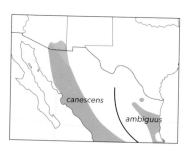

canescens

ambiguus

Genus *Euptilotis*

EARED QUETZAL *Euptilotis neoxenus* EAQU ■ 4

This large and showy (but shy) species, an endemic to the pine and pine-oak mountains of western Mexico, is casual to our area. Formerly called the Eared Trogon. Monotypic. L 14" (36 cm)

Identification Name refers to its wispy, postocular plumes. **MALE:** Dark head with bright green lower throat, upper breast, and most of upperparts; bright red belly. Tail steely blue above, mostly white from below. **FEMALE:** Like male but head, throat, and most of breast gray; belly paler red. **JUVENILE:** Like female, but duller; more black at base of underside of tail.

Similar Species Only superficially similar to Elegant Trogon (see that species).

Voice CALL: Long up-slurred squeal ending in a *chuck* note. And a loud, hard cackling *ka-kak*, sometimes given in flight. **SONG:** A long, quavering series of whistled notes that increase in volume.

slaty gray head and breast on female

dark bill

no breast band

all have a hunchbacked appearance and a disproportionately small head

unlike Elegant, tail feathers extensively pure white with no fine barring

blue-black tail when viewed from above

both sexes lack pale marking on head; "ears" visible only at close range

Status & Distribution Casual in mountain woodlands of southeastern AZ; two records in central AZ (Mogollon Rim and Superstition Mts.). Most records from late summer and fall, but some recorded year-round and some represent winter-only records. It is best known from the Huachuca Mts., where nesting was attempted in upper Ramsey Canyon. Its preference for remote canyons might result in underreporting. **Population** Widespread in west Mexican range; at risk from forest destruction and with loss of nest cavities.

HOOPOE Family Upupidae

Eurasian Hoopoe (Spain, May)

This unique family is composed of a single, utterly distinctive and very flashy species—the Eurasian Hoopoe.

Structure It has a long, curved bill, a head crest, and rounded wings.

Behavior Hoopoes feed primarily on the ground. Nests in cavities, either on the ground, in rocks, or in trees. The slow and undulating flight style is distinctive.

Distribution Widespread in Eurasia and Africa.

Taxonomy Hoopoes and the related wood-hoopoes of Africa (Phoeniculidae) are placed in their own order, Upupiformes, which is in turn most closely related to the hornbills. Most authorities recognize one species with eight subspecies. Some authors elevate the Madagascar and African subspecies to one or more separate species; notably, *marginata* from Madagascar has distinctly different vocalizations.

Conservation The island-dwelling St. Helena Hoopoe (*U. antaios*) was likely a flightless species. It is extinct and known only from subfossil skeletal remains.

Genus *Upupa*

EURASIAN HOOPOE *Upupa epops*   EHOO ■ 5

The Eurasian Hoopoe, also known as the Common Hoopoe, is an Old World stray to N.A. It has a distinctive behavior with walking gait on ground in open situations. Polytypic (8 sspp.; nominate in N.A.). L 10.5" (27 cm).

Identification Sexes look similar. Overall buffy pink with paler belly; rounded black-and-white striped wings. Crest feathers with black tips; normally flattened; raised most often after alighting and when excited. Bill long, narrow, and slightly decurved. **FLIGHT:** Floppy, with short undulations. Secondaries and secondary coverts are boldly striped black-and-white, while black primaries have a broad white band near tip and black tail has a white band near base.

Similar Species Unmistakable.

Voice CALL: A high-pitched *scheer*. **SONG:** A low resonant *poo-poo-poo*, from which the name hoopoe derives.

Status & Distribution Widespread Old World species found in Eurasia and Africa; northerly populations are migratory; most European breeders winter in sub-Saharan Africa; Asian breeders winter in southern and southeast Asia. Breeds east to Korea and Ussuriland, Russian Far East. **RARE STATUS:** Accidental in fall to western AK, one specimen record of the northern nominate subspecies from the Yukon Delta, western AK (2–3 Sept. 1975), and one photographed from a boat on the Chukchi Sea, 24 Sept. 2016.

Population Most populations stable.

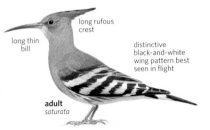

long rufous crest

long thin bill

distinctive black-and-white wing pattern best seen in flight

adult
saturata

KINGFISHERS Family Alcedinidae

Green Kingfisher, female (TX, Nov.)

In general, kingfishers sit on low perches watching for prey below them. Their flight is direct and strong, with rapid wingbeats.

Structure Tremendous variation in size exists within the family. The majority of species are small to medium size. All kingfishers have large heads with long, strong beaks and short legs. Tail length varies from very short and stubby to long with streamers.

Behavior Most species are solitary except during the breeding season. Exceptional are the large Kookaburras of Australia and New Guinea, which are more social. Kingfishers nest in burrows. Kingfishers in North America hunt from low perches over water. Many forest species elsewhere hunt primarily for terrestrial vertebrates. Species on the smaller end of the spectrum feed primarily on invertebrates.

Plumage Almost all are dark above, most commonly blue, green, or brown. Underparts are normally white or rufous, with a few species exhibiting banding or barring. For most species the sexes are similar, with minor differences between the color of underparts or tail color.

Distribution Kingfishers are found on all continents except Antarctica. They are also absent from the northernmost reaches of North America and most of Russia and central Asia. The family reaches its greatest diversity in an area encompassing Southeast Asia to Australia. There are three regularly occurring species in North America and one casual species; two more occur from Mexico to South America, for a total of six species in the New World.

Taxonomy The family is divided into three subfamilies, only one of which occurs in the New World. Worldwide, there are 91 species in 18 genera recognized. Two North American species (Ringed and Belted Kingfishers) formerly in the genus *Ceryle* now reside in the genus *Megaceryle*. Closely related to the kingfishers are two other New World families, the todies (Todidae) and motmots (Momotidae).

Conservation Many species of kingfisher live in primary forest and are therefore subject to pressures from deforestation. Most kingfisher species considered at risk are found on islands. BirdLife International codes: 25 NT, 10 VU, 2 EN, 4 CR, and 1 EW (Guam Kingfisher, extinct in the wild; 140 individuals survive in a captive breeding program).

Genera *Megaceryle* and *Chloroceryle*

RINGED KINGFISHER Megaceryle torquata   RIKI   ■   2

The largest of New World kingfishers, the Ringed normally frequents larger rivers but can also be found around ponds and lakes. It typically perches higher than the other, smaller kingfishers found in N.A. and often flies at surprisingly high altitudes with slow, deep wingbeats. Polytypic (3 sspp.; nominate in N.A.). L 16" (41 cm)

Identification Large with a big-headed appearance, short crest, and long, heavy bill. **ADULT MALE:** Slate blue above with a prominent white collar and rufous underparts. White underwing coverts easily visible in flight. Massive bill is gray at base, becoming black on distal half. **ADULT FEMALE:** Similar to adult male, but with broad slate blue breast band, bordered by a white band, and rufous underwing coverts. **JUVENILE:** Similar to adults, but male has a narrow slaty band across upper breast that is mixed with cinnamon brown. Slaty breast band of juvenile female is also mixed with cinnamon brown.

Similar Species Distinctive. Smaller Belted Kingfisher has predominantly white underparts.

Voice CALL: A harsh rattle that is slower and lower than Belted Kingfisher. Also a loud single *chack*, somewhat suggestive of Great-tailed Grackle; primarily given in flight.

Status & Distribution Locally common in the Lower Rio Grande Valley, becoming uncommon and local farther north. **BREEDING:** In burrows close to or along rivers. **WINTER:** Wanders from nesting areas during winter, including farther up the Rio Grande and its tributaries. **RARE STATUS:** Some regularity to the TX Hill Country east of current breeding distribution. Accidental to LA, OK, northern TX, and southeast AZ.

Population Range expansion in US has led to an increased population.

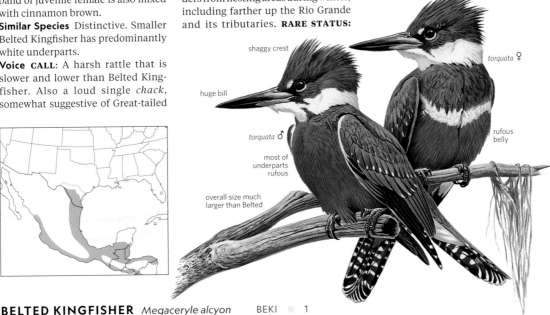

shaggy crest

huge bill

torquata ♂

most of underparts rufous

overall size much larger than Belted

torquata ♀

rufous belly

BELTED KINGFISHER *Megaceryle alcyon* BEKI ■ 1

The Belted Kingfisher is the most numerous and widespread kingfisher in N.A. These birds need clear, still water for fishing, with elevated perches from which to hunt. Monotypic. L 13" (33 cm)

Identification Medium size with a big-headed appearance, prominent shaggy crest, and long, heavy bill. **ADULT MALE:** Slate blue above with a prominent white collar. Underparts white with a single blue breast band. Large bill has a gray base with a black outer half. **ADULT FEMALE:** Similar to adult male, but with an obvious rufous band across upper belly. Rufous also extends down flanks. **JUVENILE:** Very similar to adults in both sexes. Juvenile male has a tawny breast band that is mottled. Juvenile female also has a much reduced rufous belly band.

Similar Species Distinctive. For most of its distribution there is nothing that can cause confusion. In southern TX, the much larger Ringed Kingfisher has rufous underparts.

Voice CALL: A loud, dry rattle. Belted Kingfishers also make harsh *caar* notes while perched and in flight.

Status & Distribution Fairly common. **BREEDING:** In burrows close to or along water. **MIGRATION:** A partially migratory species, with the northernmost populations leaving the breeding grounds. **WINTER:** Uncommon through the northern and central US, becoming more numerous in the southern third. Rare along northern Pacific coast from southeastern AK through BC. Winters south through C.A. to northern S.A. and in the Caribbean. **RARE STATUS:** Casual Pribilofs; also to Greenland, the Azores, and western Europe.

Population Stable.

♀

shaggy crest

♀

blue and rufous breast bands

♂

single bluish breast band

♂

AMAZON KINGFISHER *Chloroceryle amazona* AMKI ■ 5

Amazon Kingfisher is a widespread, adaptable species found in many aquatic habitats in the Neotropics. They prey mostly on crustaceans and fish, patiently watching from a perch over water until prey is spotted, then diving. Amazon Kingfishers tend to hover much less than Belted or Ringed Kingfishers, but they are much more conspicuous than Green Kingfisher. Monotypic. L 11–12" (28–30 cm)

Identification The only medium-size kingfisher with vivid green upperparts in Mexico and most of C.A. Long, heavy blackish bill. **ADULT:** Upperparts and head a deep forest green with an almost oily-looking metallic sheen, broken by broad white collar. Underparts white with a few dark streaks in flanks; male has broad rufous breast band, and female has extensive green on sides of chest. **JUVENILE:** Similar to female, with breast band sometimes broken around the midpoint, spotted with buff.

Similar Species Green Kingfisher is much smaller, with a smaller bill and smaller crest and usually forages from a concealed perch. Green-and-rufous Kingfisher (*C. inda*), a retiring species also smaller than Amazon and found north only to Nicaragua, is easily distinguished by its extensively rufous underparts.

Voice Often gives a harsh *teck* call. The rarely heard song, given from a treetop, is a whistled *see see see see*.

Status & Distribution Fairly common resident of lakes, ponds, oxbows, and rivers from northern Mexico south to central Argentina and Uruguay, mostly east of the Andes. The northernmost populations are found in Mexico, north to southern Sinaloa and southern Tamaulipas. **RARE STATUS:** Three well-documented records of females in winter along or near Rio Grande from Webb Co. and Cameron Co., TX.

Population Stable.

crested

much larger size than Green Kingfisher with more massive bill

no white on wings

♀

GREEN KINGFISHER *Chloroceryle americana* GKIN ■ 2

Smallest of the N.A. kingfishers, the Green Kingfisher frequents clear streams and ponds, often perching very close to the surface. The presence of a Green Kingfisher is often betrayed by its nervous calling when approached. Polytypic (5 sspp.; 2 in N.A.). L 8.75" (22 cm)

Identification Small with a big-headed appearance, an inconspicuous crest, and long, fairly heavy bill. **ADULT MALE:** Green above with a prominent white collar. Underparts white with a wide rufous band across chest. Wings are heavily spotted with white. Tail is green with white outer rectrices that are spotted with green. **ADULT FEMALE:** Similar to adult male, but with a green breast band that is mottled with white. There is also green mottling along flanks and across upper belly. Some individuals have a buffy wash across throat and chest. **JUVENILE:** Very similar to adult female but with buffy spotting on the crown and upperparts. **FLIGHT:** Direct and very fast, the bird's white-based outer tail feathers are conspicuous in flight.

Geographic Variation Differentiation of the two subspecies that occur in the US is weak. Birds occurring in TX belong to *hachisukai* and are more heavily spotted with white on lesser wing coverts. Individuals occurring in southeastern AZ are placed in *septentrionalis* and lack white spotting on lesser coverts.

hachisukai

septentrionalis

Similar Species Unmistakable. Both of the other N.A. kingfishers are much larger and slate blue in color.

Voice CALL: Recalls two pebbles knocked together. Another call is a long series of rapid, but subdued tick notes. A squeaky *cheep* is given in flight.

Status & Distribution Found through Mexico and south through C.A. and in S.A. to northern Chile and Argentina. In US, uncommon and often inconspicuous. **BREEDING:** In burrows close to or along rivers. **WINTER:** Susceptible to very cold weather, withdraws from northern portion of its range during severe winters. Rare in AZ, primarily in the San Cruz and San Pedro River drainages. **RARE STATUS:** Casual in TX outside normal range north to the panhandle, and in southwestern NM along the Gila River.

Population Stable, but there are possible declines in AZ.

white at base of outer tail feathers visible in flight

extensive white spotting on wings

♂

green upperparts

rufous breast

♀

very long bill

green breast band

WOODPECKERS AND ALLIES Family Picidae

Pileated Woodpecker, nestlings (MO, June)

Field identification of woodpeckers is not too complex except within a few close species pairs or groups such as the Yellow-bellied Sapsucker complex. These cases are compounded by hybridization, and not all individuals can be identified. Twenty-five species occur in North America, two as casual or accidental strays and one almost certainly extinct.

Structure Woodpecker structure and posture render them instantly recognizable (but see creepers and nuthatches). Our species vary from large sparrow to crow size. Adaptations for trunk foraging and excavation include a chisel-like bill, skull and neck muscle adaptations to reduce brain impacts from blows with the bill, stiffened rectrices, and strong claws. Woodpeckers have short legs; their feet have two toes forward and two (rarely one) backward. The extensible tongues are housed in a sling that wraps over the skull to (or nearly to) the nostril.

Behavior Woodpeckers perch along trunks and limbs, moving up in jerky "hitching" movements, using the stiff tail feathers as a prop. The flight of most species is strongly undulating. Most hop when on the ground. Foraging often involves drilling into or flaking bark and dead wood for grubs, but the diverse feeding repertoire includes gleaning, lapping ants from the ground, consuming and storing acorns and other mast, excavating seeds from cones, drilling into living plant tissue for sap, and aerial sallies for insects. Most species are quite vocal, but song is replaced by drumming. Woodpeckers excavate nest cavities in trunks or branches, usually in dead wood, that are later used by a variety of birds and other animals.

Plumage Most of our woodpeckers are black and white in various patterns. Red adorns the plumage of nearly all our species, but is usually limited to the head (and frequently present only in males).

Distribution Woodpeckers occur on all continents except Australia and Antarctica. They are most diverse in wooded regions, but some species have adapted to arid scrub, grassland, and alpine tundra. Most are sedentary; a few species (notably sapsuckers and Northern Flickers) can be strongly migratory, and other species move short distances or are nomadic depending on food availability.

Taxonomy There are 217 species of woodpeckers in 33 genera worldwide. These include the two wrynecks of the subfamily Jynginae, the 29 small piculets (subfamily Picumninae), and roughly 186 true woodpeckers (subfamily Picinae). Within the true woodpecker subfamily, several smaller groupings (tribes) are recognized.

Conservation Some species are among our most familiar. Most require standing snags for nest sites. Species requiring old-growth forests have generally declined, most strikingly the endangered Red-cockaded and the almost certainly extinct Ivory-billed. BirdLife International codes: 25 NT, 10 VU, 5 EN, 2 CR, 2 CR (LE), and 1 EX.

Genus *Jynx*

EURASIAN WRYNECK   *Jynx torquilla*   EUWR ▪ 5

The Eurasian Wryneck, an accidental stray from the Old World, hardly resembles a woodpecker. It perches horizontally and has a long tail without stiffened rectrices. Polytypic (4 sspp.; *chinensis* recorded from N.A.). L 6" (17 cm).

Identification Small with cryptic patterning, broad gray mantle stripes, buff throat, dark line through eye, another dark line along center of crown and back, sharply pointed bill, and long tail.

Similar Species At a glance it can suggest a songbird such as a thrush or sparrow.

Status & Distribution Breeds widely across temperate and boreal Eurasia east to northeast Asia; highly migratory, wintering in sub-Saharan Africa and from India through southeastern Asia. **RARE STATUS:** Three records for western AK: Cape Prince of Wales (8 Sept. 1945; specimen) and Gambell, St. Lawrence I. (2–5 Sept. 2003 and 9–16 Sept. 2019); San Clemente I., CA (25 Sept. 2017); one found dead in Feb. 2000 in southern IN was probably artificially transported.

Population Declines and extirpations in portions of European range.

dark mask and line on side of back

adult

cryptic pattern in browns and grays

Genus *Melanerpes*

The varied diets of these generalized New World woodpeckers include seeds and fruit; many take flying insects on the wing, and some store acorns and other nuts. Their tails are flat and only moderately stiffened, and the bills are medium to long and very slightly curved. Some species are highly social.

LEWIS'S WOODPECKER *Melanerpes lewis* LEWO ■ 1

This distinctive large, glossy black woodpecker flies with slow, steady wingbeats, recalling a crow in flight. It usually perches in the open and makes long and often acrobatic sallies for insects. Most populations are migratory; large irruptions south sometimes occur. Monotypic. L 10.5" (27 cm)

Identification Unmistakable. Sexes similar. **ADULT:** Glossy greenish black on upperparts and a broad pale gray breast band extending around hind-neck; belly is pink and face is deep red. **JUVENILE:** Head, face, and foreparts dusky, with no gray collar. Adultlike plumage is attained during winter.

Similar Species Unmistakable. Acorn and Red-headed Woodpeckers show conspicuous white patches on wings and much white on underparts. Distant flying Lewis's strongly suggest small crows.

Voice Generally silent for a woodpecker. **CALL:** Soft calls, including a series of short, harsh *churr* notes and clicking, squeaky *yick* notes. **DRUM:** Infrequent; a weak roll followed by a few individual taps.

Status & Distribution Uncommon to fairly common; often gregarious. **BREEDING:** Open conifer, oak, and riparian woodlands in interior West; rare in coastal areas. **MIGRATION:** Large diurnal flights are sometimes noted. Arrives in northern interior breeding areas during first half of May; most depart these by mid-Sept. **WINTER:** Oak savannas, orchards, shade trees in towns. Fall and winter movements are irregular, depending upon availability of acorns, conifer seeds; large flights sometimes occur into southern portions of CA, AZ, and NM. Winters irregularly north to WA, BC; and south to northern tier of Mexican states. **RARE STATUS:** Casual to the upper Midwest and Great Plains to southwestern TX; accidental east to NL, New England, mid-Atlantic states, and to southeast AK.

Population Largely eliminated from coastal Northwest due to degradation of pine, oak, and riparian woodlands.

dark red face

browner head and no collar

juvenile

gray collar

pink belly

adults

slow crowlike wingbeats

oily green upperparts, no red on face, and little pink on belly

dark wings

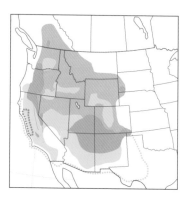

RED-HEADED WOODPECKER *Melanerpes erythrocephalus* RHWO ■ 1

This flashy woodpecker is a familiar sight over much of central and southeastern N.A. Partial to woodlands, it often perches openly and sallies out for insects; at other times it is surprisingly inconspicuous and quiet, especially in winter. Now regarded as monotypic. L 9.3" (24 cm)

Identification All ages show white secondaries and a white rump, contrasting with dark remaining upperparts. **ADULT:** Bright red head contrasts with black back and pure white underparts; a narrow ring of black borders red throat. Sexes similar. **JUVENILE:** Brownish head and upperparts, with blackish bars through white secondary patch and some brown streaking and scaling below; adultlike plumage is attained gradually over first winter; by spring most retain some black in secondaries.

Geographic Variation Although considered monotypic, birds west of the Mississippi River Valley ("*caurinus*") average larger and sometimes show a tinge of red on the belly.

Similar Species See Red-breasted Sapsucker (ranges do not overlap), which shares the all-red head, but with red extending through the breast, white on the wing coverts (not secondaries); note retiring sapsucker habits. Red-bellied Woodpecker commonly co-occurs and has somewhat similar calls, but has a whitish face and throat, barred upperparts, and very different wing pattern.

Voice **CALL:** A loud *queark* or *queeah*, given in breeding season, is harsher and sharper than rolling *churr* of the Red-bellied Woodpecker; also a dry, guttural rattle and, in flight, a harsh *chug*. **DRUM:** A simple or two-part roll, lasting about a second and consisting of 20–25 beats.

Status & Distribution Uncommon to fairly common. **BREEDING:** Occupies a variety of open woodlands, orchards, and open country with scattered trees. Summers rarely in northeastern UT. **MIGRATION:** Small parties of migrants noted in early fall and late spring. **WINTER:** Withdraws southward from most of the breeding range in the Great Plains and Great Lakes regions. Unrecorded in Mexico. **RARE STATUS:** Now rare to New England, mainly in fall. Casual to Maritimes, and west to BC, ID, CA, NV, AZ. Accidental NL.

Population Overall declining. New England breeding populations are essentially gone, and strong declines have been noted in the mid-Atlantic states, some Great Lakes states, FL, and elsewhere.

red head

brownish head

juvenile

barring on secondaries

pure white secondaries

adult

adult

ACORN WOODPECKER *Melanerpes formicivorus*   ACWO   ￭ 1

white primary patches
♂

female has black forecrown bar
♀

clown head pattern
♂

This conspicuous clown-faced woodpecker of western oak woodlands—the inspiration for the cartoon character Woody Woodpecker—is remarkable for its social habits, living over much of its range in communal groups of up to four or more breeding males and as many as three breeding females. These groups maintain and protect impressive granaries in which thousands of acorns are stored in holes drilled in tree trunks or utility poles for future consumption; in a study, a single tree contained more than 50,000 acorn-storage holes. They also feed by sallying for flying insects and gleaning trunks, and they often eat ants (as reflected in the species' scientific name). Polytypic (7 sspp.; 2 in N.A.). L 9" (23 cm)

Identification A boldly patterned black-and-white woodpecker with a white patch at base of primaries, a white rump, black chest, streaked black lower breast, and white belly. Striking head pattern, with a ring of black around base of bill, a red crown patch, a white forecrown narrowly connected to yellow-tinged white throat, and black sides of head setting off a staring white eye. **ADULT:** Iris white. Male has white forehead meeting red crown; female similar, but white forehead is separated from red crown by a black band. **JUVENILE:** Resembles adult but black areas are duller and iris is dark; juveniles of both sexes have a solid red crown like that of adult male.

Geographic Variation Pacific coast birds, *bairdi*, have slightly longer and stouter bills than nominate birds of the interior West. There is considerable additional variation in the remaining range south to Colombia, with five additional subspecies.

Similar Species Unmistakable with its plumage and loud calls. Larger Lewis's also sallies for prey and is gregarious, but it lacks white areas in plumage, including having no wing patch.

Voice Acorn is noisy and conspicuous in communal groups, with raucous "Woody Woodpecker" calls. **CALL:** Loud *wack-a, wack-a* or *ja-cob, ja-cob* series. Also, a scratchy, drawn-out *krrrrit* or *krrrit-kut*, and a high, cawing *urrrk*. **DRUM:** A simple, slow roll of about 10–20 beats.

Status & Distribution Common. **YEAR-ROUND:** Oak woodlands and mixed oak-conifer or oak-riparian woodlands. Most abundant where several species of oaks occur together. Isolated breeding populations are found on the east side of the Sierra Nevada, CA; on the central Edwards Plateau, TX (virtually extirpated); and possibly in extreme southwestern CO. **RARE STATUS:** Rare or casual, primarily in fall and winter, away from woodland habitats along the immediate Pacific coast and north to BC, and in western deserts. Accidental to the Great Plains states from ND and MN south to coastal TX.

Population Stable, apart from some local declines resulting from degradation of oak woodlands.

bairdi

formicivorus

augustifrons

GILA WOODPECKER *Melanerpes uropygialis* GIWO ■ 1

The Gila is a zebra-backed woodpecker of southwestern desert woodlands, where it can be noisy and conspicuous in tall cacti such as saguaros. Its US range does not overlap that of the similar Golden-fronted and Red-bellied Woodpeckers. Polytypic (4 sspp.; nominate in N.A.). L 9.3" (24 cm)

Identification Within its range, Gila Woodpecker's barred black-and-white upperparts, pale, grayish tan head and underparts (with a touch of pale yellow on belly), and broken white patch at base of primaries are diagnostic. Rump and uppertail coverts are barred with

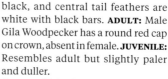

red crown on male — grayish brown forehead

grayish brown nape

uropygialis ♀

uropygialis ♂

barred central tail feathers

black, and central tail feathers are white with black bars. **ADULT:** Male Gila Woodpecker has a round red cap on crown, absent in female. **JUVENILE:** Resembles adult but slightly paler and duller.

Geographic Variation There are two additional subspecies in Baja California and one other in mainland western Mexico (Sonora); birds from farther south tend to be more heavily barred and darker overall.

Similar Species Gila's range does not overlap with that of Golden-fronted (except at southern end of range in Jalisco, Mexico), but potential strays of either species (e.g., in southern NM) would need to be carefully documented. Golden-fronted has extensive solid white on the rump, solid black central rectrices, and yellow-orange nape patch and nasal tufts. The immature female Williamson's Sapsucker is superficially similar to Gila, but pale bars on the back are tan rather than white; chest and sides are barred with blackish; head is darker gray-brown; and bill is shorter. Note also the sapsucker's more retiring behavior. See also Northern and Gilded Flickers, which are larger, show solid white rumps, yellow or red color in wings and tail, and a black crescent on breast.

Voice CALL: A loud, rolling *churrr* or *whirrrr*, often doubled. Also an insis-

tent, laughing *yip*, *yip* series. **DRUM:** Infrequent drum is a steady, loud roll.

Status & Distribution Fairly common. **YEAR-ROUND:** Desert woodlands, including cactus country, mesquite woods, riparian corridors, and lower canyon woodlands; often common in residential areas, date palm groves. Range extends west in CA to the Imperial Valley, east to southwestern NM, and at least formerly barely into southern NV. **DISPERSAL:** There is some movement into wooded foothills in southeastern AZ in fall and winter. **RARE STATUS:** Accidental west to eastern San Diego Co. and on coastal slope in San Bernardino Co. and Los Angeles Co., and in the Bay Area, CA.

Population Some declines have occurred with clearing of cottonwood-willow riparian associations, as along the lower Colorado River; colonization of the Imperial Valley occurred in the 1930s.

GOLDEN-FRONTED WOODPECKER *Melanerpes aurifrons* GFWO ■ 1

Closely related to Red-bellied Woodpecker, it largely replaces that species from central and southern TX south into C.A. Polytypic (12 sspp.; nominate in N.A.). L 9.8" (25 cm)

Identification All show a yellow area on nasal tufts just above bill, as well as a golden yellow to orange nape and hindneck. Rump is extensively pure white, and central rectrices are solidly black. Underparts are largely pale grayish white, with a touch of yellow on lower belly. **ADULT MALE:** Orange-yellow nasal tufts, red crown patch, and mixed red and orange-yellow nape patch. **ADULT FEMALE:** Similar to male, but red crown patch is absent and nasal tufts and nape are purer yellow, less orange. **JUVENILE:** Duller than adults; yellow or orange nasal tufts and nape lacking, but males (and some females) show a small red crown patch.

Geographic Variation Nominate *aurifrons* in US, but considerable geographic variation, with 11 subspecies in remainder of range. Appearance and vocalizations differ in southern Mexico; birds from southeastern Mexico to Honduras have fine white barring on back, continuous red from crown through nape, and some red on belly. Given the vocal differences, it is perhaps best considered a distinct species.

Similar Species Red-bellied overlaps marginally with Golden-fronted from east-central TX to southwestern OK, with hybridization frequent in the latter area. Red-bellied differs in its extensive red rear crown and nape (extending to bill in males), white bars on central rectrices (solidly black in Golden-fronted), and pink or red tinge on belly (yellow in Golden-fronted).

red crown on male

gold feathering above bill

aurifrons ♂

aurifrons ♀

gold nape

pure white rump area

dark central tail feathers

Rare individual Red-bellieds can show yellow or orange on crown and nape (see Gila Woodpecker account). **Voice CALL:** A rolling *churrr* and cackling *kek-kek*, both slightly louder and raspier than Red-bellied; also a scolding *chuh-chuh-chuh*. **DRUM:** A simple roll, often preceded or followed by single taps. **Status & Distribution** Fairly common. **YEAR-ROUND:** Dry woodlands and brushlands such as oak-juniper savannas and mesquite thickets; also riparian corridors, pecan groves, sub-urban areas. **RARE STATUS:** Wanders casually to northeastern TX, eastern OK, and southeastern NM. Accidental in Cheboygan Co., MI (20 Nov.–2 Dec. 1974), and FL; the latter record, at least, may pertain to an aberrant Red-bellied, and SC reports of Golden-fronted have conclusively been shown to be abnormal Red-bellieds. **Population** Apparently stable. Colonized southwestern OK in the 1950s and now hybridizes there with Red-bellied. Fairly common in the Big Bend region of TX only since the 1970s.

RED-BELLIED WOODPECKER *Melanerpes carolinus* RBWO ▧ 1

The Red-bellied Woodpecker is the familiar zebra-backed woodpecker eastern woodlands and towns. The name is confusing as the belly is just barely washed with reddish. Now regarded as monotypic. L 9.3" (24 cm)
Identification All Red-bellied Woodpeckers show a black-and-white barred back, white uppertail coverts, grayish white underparts, black chevrons on lower flanks and undertail coverts, and barred central tail feathers. In flight a small white patch shows at base of primaries. **ADULT MALE:** Entire crown, from bill to nape, is red; there is a suffusion of pink or red on center of belly. **ADULT FEMALE:** Red on head is limited to nasal tufts (just above bill) and nape; wash of color on belly is paler, less extensive. In rare individual females, nape and nasal tufts can be yellow-orange instead of red. **JUVENILE:** Resembles adults but duller, with red nasal tuft and nape patches lacking; bill is brownish (black in adults).
Similar Species Compare with Golden-fronted Woodpecker, which has solid black central rectrices, lacks pink or red on belly, and has a different pattern of color on the head.
Voice In breeding season, Red-bellied Woodpecker gives a rolling *churrr*; *it also gives* also a conversational *chiv chiv*; softer than calls of Golden-fronted Woodpecker. **DRUM:** A simple roll of up to a second, with about 19 beats per second.
Status & Distribution Common in the Southeast, uncommon to fairly common in the Northeast, Midwest, and Great Plains. **YEAR-ROUND:** Pine and hardwood forests, open woodlands, suburbs and parks. Small populations exist west to southeastern ND, central SD, and northeastern CO.

DISPERSAL: Not migratory, but at least some individuals in northern range withdraw southward in fall. **RARE STATUS:** Wanders casually north to central ON, southern QC, and the Maritime Provinces of Canada and west to AB and eastern NM. Accidental in southeastern WY, ID, OR, and NV.
Population Red-bellied's range has expanded northward in the Great Lakes region and New England over the last century and is also expanding northwestward in the Great Plains.

solid red crown and nape on male

red nape

♀

♂

whitish primary patch in Red-bellied, Golden-fronted, and Gila

♂

speckled white rump

pink on lower belly

barred central tail feathers

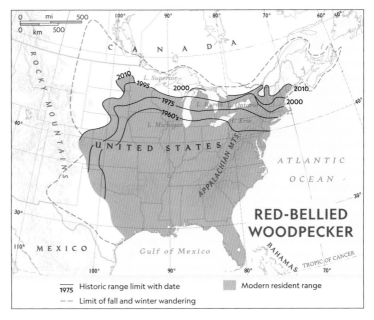

RED-BELLIED WOODPECKER

1975 — Historic range limit with date

– – Limit of fall and winter wandering

▧ Modern resident range

SAPSUCKERS Genus *Sphyrapicus*

The four N.A. sapsuckers are shy relatives of *Melanerpes* that feed on insects but most characteristically on living plant tissue and sap obtained from rows of drilled holes. All species give distinctive "Morse code" display taps; otherwise soft tapping. Sapsuckers are long winged, with strong undulating flight; all species are partially to highly migratory.

YELLOW-BELLIED SAPSUCKER *Sphyrapicus varius* YBSA ▪ 1

The only sapsucker normally found in the boreal and eastern parts of the continent, Yellow-bellied Sapsucker is our most highly migratory woodpecker. Monotypic (smaller, darker resident birds in southern Appalachians sometimes separated as *appalachiensis*). L 8.5" (22 cm)

Identification Shows less red on head than related Red-naped and Red-breasted, and back is more extensively scalloped with yellow-buff. **ADULT MALE:** Forecrown, chin, and throat red, outlined completely in black; red normally lacking on nape. **ADULT FEMALE:** Similar to male, but chin and throat are entirely white. **JUVENILE:** Head and underparts pale brownish barred with dusky black; upperparts extensively pale buff with dusky barring. Unlike Red-breasted and Red-

naped, this juvenal plumage is retained well into winter, with most not acquiring more adult-like plumage until late winter.

Similar Species See the very similar Red-naped Sapsucker (formerly, along with Red-breasted, considered conspecific with Yellow-bellied).

Voice This species and Red-breasted and Red-naped Sapsuckers are similar in calls and drums. **CALL:** A nasal *weeah* or *meeww*; on territory a more emphatic *quee-ark*. **DRUM:** A distinctive rhythm of a short roll of several beats, a pause, then two to several brief rolls of two to three beats each.

Status & Distribution Fairly common. **BREEDING:** Deciduous forests, mixed hardwoods and conifers of boreal regions and the Appalachians. **MIGRATION:** Main fall movement is Sept.–Oct.; spring migrants arrive in the Upper Midwest and Northeast during mid-Apr., and the northernmost breeding populations arrive late Apr., early May. **WINTER:** Widespread in the East south of New England and Great Lakes states, south to West Indies and Panama. **RARE STATUS:** Rare in summer AK (mostly eastern). Rare west to CA in fall and winter; a few

records north to WA. Accidental in Iceland, UK, and Ireland.
Population Generally stable.

pure white throat

adult ♀

red throat bordered by solid black frame

adu...

juvena pluma usuall well in winter

golden buff spots scattere liberally above

juvenil

RED-NAPED SAPSUCKER *Sphyrapicus nuchalis* RNSA ▪ 1

The Rocky Mts. and Great Basin representative of the Yellow-bellied Sapsucker complex, the Red-naped Sapsucker closely resembles the Yellow-bellied Sapsucker, and the two hybridize in southwestern AB. Red-naped hybridizes much more frequently with Red-breasted from BC south to eastern CA; in migration (especially) and winter, these hybrids are noted in small numbers in the Great Basin and southeastern BC. Monotypic. L 8.5" (22 cm)

Identification Very similar in all plumages to Yellow-bellied, but with slightly more red on head. **ADULT**

MALE: Red crown is bordered by black, with a small red patch below black nape bar; chin and throat are red, with red color partially invading black malar stripe that outlines throat; breast is black. **ADULT FEMALE:** Similar to male, but chin is usually white (red on rare individual females), the extent of white being variable, and red on crown and throat is slightly less extensive, more completely bordered by black on sides of throat; red on nape may be nearly or completely absent. **JUVENILE:** Closely resembles juvenile Yellow-bellied, but adultlike face pattern is attained by beginning

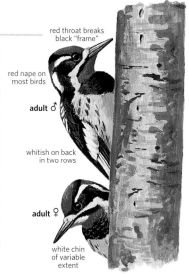

red throat breaks black "frame"

red nape on most birds

adult ♂

whitish on back in two rows

adult ♀

white chin of variable extent

of Oct. (brown retained on breast at least into midwinter).

Similar Species Male Yellow-bellied Sapsuckers can rarely show some red on nape. In Red-naped red throat invades or completely covers black border along malar, and pale markings on back are whiter and more restricted. Female Red-napeds with maximal red on chin and throat closely resemble male Yellow-bellied, but usually have white on uppermost chin and a hint of red on nape; note also back pattern differences. Male Red-naped Sapsucker, with maximal red invading the auricular and malar regions, may not be distinguishable

from Red-breasted × Red-naped hybrids; such hybrids usually have only limited black on the breast, auriculars, and sides of crown.

Status & Distribution Fairly common. **BREEDING:** Aspen parkland and deciduous groves within open coniferous woodlands or adjacent to montane forest. Breeding range narrowly overlaps that of the Yellow-bellied Sapsucker in AB, with some hybridization. **WINTER:** Riparian and pine-oak woodlands, orchards, and shade trees south to northwestern and north-central Mexico. Winters rarely north on the Pacific coast to WA, and BC. **RARE STATUS:** Casual east to KS, NE, OK,

southern TX, and southeastern LA. **Population** Generally stable.

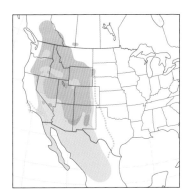

RED-BREASTED SAPSUCKER   *Sphyrapicus ruber*   RBSA ▪ 1

The Pacific coast representative of the Yellow-bellied Sapsucker complex, the Red-breasted Sapsucker differs from other sapsuckers in its almost entirely red head and breast. Identification is complicated by frequent hybridization with Red-naped from BC south to eastern CA and more limited hybridization with Yellow-bellied in BC. Polytypic (2 sspp.; both in N.A.). L 8.5" (22 cm).

Identification Head and breast are almost entirely red, and black breast patch is lacking. Back shows two rows of whitish or yellow-buff spots or bars, but this patterning is very limited in northern birds. **ADULT:** Red head and breast, pale yellow belly, back black with geographically variable yellow-buff to white barring on sides of back and lower back; white rump and white patch on wing coverts resemble other sapsuckers. Sexes are similar, but females (at least in southern populations) tend to show more pale markings on rectrices. Many birds, especially when worn, show a ghost of underlying black in auriculars and chest and black-and-white patterning in malar region;

red of worn birds may appear paler and more orange-red. **JUVENILE:** Brown head and extensive brown mottling below; darker than both juvenile Red-naped and Yellow-bellied, with less facial patterning. Adultlike plumage attained by Sept., though underparts are duller and mottled with brown.

Geographic Variation Nominate *ruber* breeds from southern OR northward; *daggetti* occupies the remainder of the range. Nominate birds show deeper and more extensive red on the breast, which is more sharply delineated from the pale belly; more limited white spotting on flight feathers; a deeper yellow wash on belly; and a more extensively black back with more limited cross bars that are yellowish buff.

Similar Species Compare with Red-naped Sapsucker; extensive hybridization renders many individuals with intermediate head patterns unidentifiable.

Status & Distribution Common. **BREEDING:** Moist coniferous forests, mixed oak-conifer riparian woodlands in coastal mountain ranges, usually in lower and wetter habitats than Williamson's. Subspecies *daggetti* is largely limited to montane habitats from about 4,000–8,000 ft. Red-breasted Sapsucker frequently hybridizes with Red-naped in the Cascades and eastern Sierra Nevada. **WINTER:** Some southward migration and higher-elevation breeders generally withdraw to lower elevations. Winters widely around deciduous trees, orchards, and parks in lowlands. Winters to northwestern Baja California. Northern nominate birds have been found as far south as San

ruber ♂

more extensive and solid red head than *daggetti*

daggetti ♂

ruber

daggetti

marks on back whitish and more extensive than *ruber*

small yellow spots on back

ruber

daggetti

Diego, CA, and southern AZ. **RARE STATUS:** Many claims may involve hybrids. Casual east to AZ, NM, CO, central TX, Sonora, and northwest to Kodiak I., AK. Accidental IA.

Population Generally stable.

WILLIAMSON'S SAPSUCKER *Sphyrapicus thyroideus* WISA ▪ 1

This sapsucker of western montane conifer and aspen forests is notable for its striking sexual dimorphism. Polytypic (2 sspp.; both in N.A.). L 9" (23 cm)

Identification Slightly larger than other sapsuckers. **ADULT MALE:** Largely black, with white rump, white postocular and moustachial stripes,

white head stripes

red throat

♂

white wing patch

both sexes have black on breast and yellow on belly

brown head

♀

barred back

white rump in both sexes

large white wing patch, red chin and throat. Belly is bright yellow; flanks scalloped black and white. **ADULT FEMALE:** Head gray-brown; black back has fine pale grayish tan bars; rump white. No white wing patch. Chest black, belly yellow, sides and flanks tan, barred with black. **JUVENILE:** Like respective adults, but duller; male has white chin, female lacks black breast patch.

Geographic Variation Nominate *thyroideus* breeds from south-central BC to northern Baja California; more easterly subspecies *nataliae* from southeastern BC to AZ, NM, has smaller, narrower bill.

Similar Species Males unmistakable; females suggest flicker, but Williamson's is smaller, shorter billed, lacks red or yellow in flight feathers, has yellow belly. See Gila Woodpecker. Juvenile lacks large white wing covert patch of other juvenile sapsuckers.

Voice CALL: A strong, slightly harsh *cheeur* or *queeah*. **DRUM:** Short roll followed by several shorter rolls; slower and more regular than other sapsuckers.

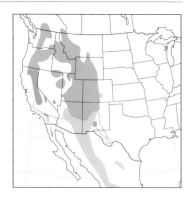

Status & Distribution Fairly common. **BREEDING:** Dry pine, spruce, or fir forests, often mixed with aspen groves. **MIGRATION:** Most move south in Sept.–Oct. and return Mar.–Apr.; in CA, AZ, NM more sedentary or make short altitudinal movements. **WINTER:** Dry conifer or pine-oak woodlands from CA and southwestern US to central Mexico. Casual north to OR, WA, CO; very rare in lowlands of CA, AZ in planted conifers. **RARE STATUS:** Rare in fall/winter to West TX, casually to the Edwards Plateau, central KS, and central NE. Accidental LA, NY.

Population Some declines, especially in Pacific Northwest.

Genus *Picoides*

These two medium-size woodpeckers have three toes, two forward-facing and one rear-pointing. They drill or flake away bark on dead and dying trees and are partial to burns, especially Black-backed. Two species are endemic to N.A.; a third is widespread across Eurasia. Males and juveniles have yellow crown patches, lacking in older females.

AMERICAN THREE-TOED WOODPECKER *Picoides dorsalis* ATTW ▪ 2

The male of this boreal and western montane species has a yellow crown patch. Forages on dead or dying trees, often in burned-over areas. Back pattern varies geographically, from solid white to heavily barred with black. It

fasciatus

bacatus

dorsalis

has recently been split from Old World Eurasian Three-toed Woodpecker (*P. tridactylus*). Polytypic (3 sspp.; all in N.A.). L 8.7" (22 cm)

Identification A three-toed woodpecker with white patch or barring on otherwise black upperparts; black barring on sides and flanks. Tail black with white outer rectrices. **ADULT:** Male has yellow crown, becoming spotted with black and white on forecrown; female lacks yellow. **JUVENILE:** Duller than adult; both sexes show some yellow on crown, reduced in female.

Geographic Variation Three subspecies differ in back pattern and size. Density of black barring on the back is greatest in eastern and smallest *bacatus* of eastern Canada south to

fairly solid white back, suggestive of Hairy, but note barred sides and flanks

Rocky Mts.
♀ *dorsalis*

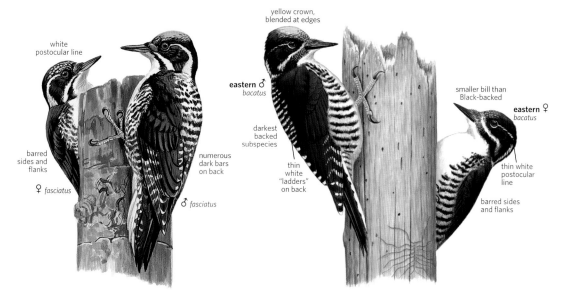

yellow crown, blended at edges

white postocular line

eastern ♂
bacatus

smaller bill than Black-backed

eastern ♀
bacatus

darkest backed subspecies

barred sides and flanks

numerous dark bars on back

thin white "ladders" on back

thin white postocular line

♀ *fasciatus*

♂ *fasciatus*

barred sides and flanks

New England and the Adirondacks. Dorsal barring and size are intermediate in *fasciatus* of boreal regions from AK to SK, and in the Cascades south to southern OR. Rocky Mts. *dorsalis*, the largest subspecies, has a pure white center of back with irregular barring on sides of back.

Similar Species See Black-backed Woodpecker. Can be mistaken for Hairy Woodpecker, especially white-backed Rockies *dorsalis*. Three-toed has heavily barred sides and flanks. Face pattern of Hairy shows more white on sides of neck and supercilium. Male Three-toed has a yellow crown patch. Beware some juvenile Hairies (e.g., in Newfoundland), which show barring on back.

Voice CALL: *Pik* or *kik*, higher than call of Black-backed Woodpecker, less sharp than Hairy; also a longer rattle. **DRUM:** Deep, resonant drums frequent and variable, range from 11–16 beats a second.

Status & Distribution Uncommon. **YEAR-ROUND:** Coniferous forests of spruce and fir, especially where there are large stands of dead trees in burned areas. Has bred south to western SD, MN, Upper Peninsula of MI, and MA. **DISPERSAL:** Small numbers move irruptively to areas adjacent to breeding range in winter, usually after early Oct.; irruptions generally more minor than in Black-backed. **RARE STATUS:** Accidental to KS (summer), NE, southern MN, IA, southern WI, RI, NJ, and DE.

Population Generally stable. Northerly and often remote range makes censusing difficult.

BLACK-BACKED WOODPECKER *Picoides arcticus* BBWO ■ 2

This is the larger relative of the similar American Three-toed Woodpecker. Both species feed in dead or dying conifers and can be especially prevalent in burned-over forests; they flake away sections of loose bark to obtain insects and their larvae. Worked-over trees are quite evident. Monotypic. L 9.5" (24 cm)

Identification Distinctive with solid black upperparts and heavy black barring on sides and flanks. Outer tail feathers white; head pattern shows long white submoustachial stripe; at best only a hint of a short white postocular mark. **ADULT MALE:** Roundish yellow patch on crown. **ADULT FEMALE:** Similar but crown all black. **JUVENILE:** Black areas duller than in adults with flight feathers more brownish; females have a few yellow feathers on crown; males have extensive yellow on crown.

Similar Species Most similar to darkest eastern subspecies *bacatus* of American Three-toed Woodpecker, which has extensive black barring on back. Black-backed is told by its solid black upperparts, lack of white postocular streak, and bolder white submoustachial stripe; it is also larger overall, with a longer, stouter bill.

Voice CALL: A single sharp *kuk*, lower and sharper than call of American Three-toed. **DRUM:** As in American

black face

sharply delineated yellow crown

solid black upperparts

long, stout bill

♂

barred sides and flanks

solid black crown

♀

Three-toed, the deep and resonant drum roll speeds up and trails off slightly toward the end. Black-backed's drum is slightly deeper, faster, and longer than that of American Three-toed.

Status & Distribution Uncommon, but

can be locally fairly common when responding to insect outbreaks in burnt or otherwise stressed forests. **YEAR-ROUND:** Conifer forests of spruce, fir, or pine. South of boreal regions and the northern Rockies, the Black-backed is found locally in the Adirondacks of NY, Black Hills of SD and adjacent WY, and the Cascades and Sierra Nevada from WA to central CA. **DISPERSAL:** Occasional irruptive movements south of regular range into New England, Great Lakes, and Canadian Maritimes, taking advantage of outbreaks of wood-boring beetles. **RARE STATUS:** Exceptional irruptions have brought birds as far south as IL, OH, PA, NJ; records at the southern periphery of range have decreased since the 1970s. Accidental to offshore islands in AK and MA. **Population** Forest practices of quickly removing dead trees after burns excludes feeding and nesting opportunities for this burn specialist.

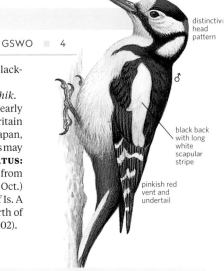

distinctive head pattern

♂

black back with long white scapular stripe

pinkish red vent and undertail

Genus *Dendrocopos*

GREAT SPOTTED WOODPECKER *Dendrocopos major* GSWO ▪ 4

This Eurasian woodpecker is a casual stray to AK. Polytypic (14–24 sspp. in Old World; AK specimen *kamtschaticus*). L 9" (23 cm)
Identification Medium size, white auriculars, black back bordered by large white scapular and covert patch, and red undertail coverts. Females lack red nape patch; juveniles have red in crown.
Similar Species Hairy and Downy Woodpeckers have white backs, black auriculars, and lack red undertail coverts and solid white wing patch.

See American Three-toed and Black-backed Woodpeckers.
Voice CALL: Sharp *kix, kick,* or *chik.*
Status & Distribution Resident nearly throughout Eurasia from Great Britain and northwestern Africa east to Japan, Kamchatka; northern populations may migrate irruptively. **RARE STATUS:** Casual in AK; at least six records from Aleutian Is. (mainly May, Sept.–Oct.) and one record in May on Pribilof Is. A male was observed wintering north of Anchorage, AK (fall 2001–Apr. 2002).
Population Stable.

Genus *Dryobates*

Small to medium-size woodpeckers that include the Downy Woodpecker, N.A.'s most numerous and familiar species. Twenty-five species are found in the New World, except for the Lesser Spotted and Crimson-breasted from Europe and southern Asia, respectively. Thirteen are restricted to S.A., and seven are found in N.A.

DOWNY WOODPECKER *Dryobates pubescens* DOWO ▪ 1

Our smallest woodpecker, the Downy is also among our most widespread and familiar species; it is confiding and often visits feeders. In all respects it suggests a small version of the Hairy Woodpecker, sharing the distinctive broad white stripe down the back. Polytypic (7 sspp.; all in N.A.). L 6.7" (17 cm)
Identification The small size and often

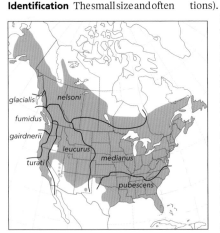

glacialis
nelsoni
fumidus
gairdnerii
leucurus
medianus
turati
pubescens

acrobatic foraging on small branches and twigs are distinctive. Has hybridized with Nuttall's. **ADULT:** Black crown, auricular, and malar; upper back, scapulars, and rump black, but a broad white stripe extends down center of back. Underparts unmarked white (to grayish buff in some populations). Outer tail feathers white with limited black spotting; variable white spotting on upperwing coverts and barring on remiges. Male has a small red nuchal patch, lacking in female. **JUVENILE:** As in other pied woodpeckers, both sexes have a pale red patch in center of crown, more extensive in male.
Geographic Variation The seven subspecies differ mainly in size (northern birds generally larger), underparts color (white to gray tinged), amount of black in rectrices, and amount of white spotting in wings. South-eastern birds are smaller and

male has red hindcrown spot

♂

short, stubby bill

♂

barred outer tail feathers

♀

Rockies
leucurus ♂

slightly grayer below than boreal and northeastern birds. Pacific coast birds have reduced white spotting on the wing coverts and secondaries; such white spotting is most highly developed in birds east of the Rockies. Birds of the Pacific Northwest are tinged gray on the back and gray-buff below.
Similar Species Nearly identical in patterning to Hairy. Downy is much smaller, with a short bill (much shorter than head); outer tail feathers usually show black spots (but these can be

lacking, and darkest Hairy subspecies may show a few spots). Pale nasal tufts of Downy are relatively larger than in Hairy. Hairy shows a larger wedge of black from the rear of the malar stripe onto the breast. Note call differences.
Voice CALL: *Pik* call is higher and much softer than Hairy's sharp, ringing *peek*. Commonly gives a distinctive high, slightly descending and accelerating whinny, *kee-kee-kee-kee*. DRUM: A soft roll, slightly slower than that of Hairy Woodpecker; about 17 beats a

second, with drum lasting 0.8–1.5 secs.
Status & Distribution Common; uncommon in northern boreal regions. **YEAR-ROUND:** Resident in a variety of deciduous woodlands and, more sparsely, in coniferous forests; also found in parks, gardens, and orchards, even in urban regions. Not migratory, but some individuals can disperse long distances. Absent from desert Southwest. **RARE STAUS:** Casual in southern AZ, Haida Gwaii, BC, and Pribilofs, AK.
Population Stable.

NUTTALL'S WOODPECKER *Dryobates nuttallii*   NUWO ▪ 1

Endemic to oak and mixed woodlands in CA and northwestern Baja California, Nuttall's is closely related to the Ladder-backed. Monotypic. L 7.5" (19 cm)
Identification ADULT: "Ladder-back"

on males, red restricted to rear crown; more extensive on juveniles of both sexes

black face with white borders

white nasal tufts

narrower white bars stop on upper back

white underparts

♂

♀

juvenile with orange-red crown

outer tail feathers with fewer bars than Ladder-backed

juvenile

pattern; spotted sides; barred flanks; auriculars almost wholly black; uppermost back black. Sexes similar but females lack red patch. **JUVENILE:** Similar but with extensive orange red on crown in both sexes.
Similar Species From Ladder-backed, shows more black on face; white bars on back are narrower; more extensive black on upper back; white outer tail feathers sparsely spotted rather than barred, purer white below, more cleanly spotted and barred with black; Ladder-backed's underparts washed with buffy, and markings are finer but often extend across breast as short streaks. Nuttall's nasal tufts are usually white (buffy to dusky in Ladder-backed). Red of crown male is restricted on Nuttall's. Calls differ markedly.
Voice CALL: Short, rolling *prrt* or *pitit*, may be followed by a longer trill, *prrt prrt prrrrrrrrrrrrr*; also a

loud *kweek kweek kweek* series. **DRUM:** A roll of about 20 taps.
Status & Distribution Common. **YEAR-ROUND:** Oak, mixed oak-conifer, and riparian woodlands; sea level to about 6,000 ft. Small populations extend onto deserts along riparian corridors. A few wander out onto the western Mojave Desert, CA. **RARE STATUS:** Casual or accidental to Imperial Valley; southwestern OR, and western NV.
Population Major increase in Southern CA into residential areas.

LADDER-BACKED WOODPECKER *Dryobates scalaris*   LBWO ▪ 1

This common desert woodpecker replaces the closely related Nuttall's in

arid regions; the two species are known to hybridize at a few localities in Southern CA. Polytypic (9 sspp.; *cactophilus* north of Mexico). L 7" (18 cm)
Identification Barred black-and-white back pattern of Ladder-backed Woodpecker extends up to hindneck, with very little solid black on upper back. Underparts are tinged creamy or buffy, with spots on sides, thin bars on flanks, and sparse, short streaks across breast. Outer tail feathers are barred with

buffy white face with narrow black frame

extensive reddish crown

buffy white nasal tufts

♂

white bars extend to nape

black crown

white bars extend to nape

buffy cast to underparts with spots on sides and flanks

♀

tail more barred than Nuttall's

black. There is as much white as black in face pattern, with lower auriculars being white. Sexes are similar, but male Ladder-backed Woodpecker has extensive red on crown, which is lacking in female.
Similar Species See closely similar Nuttall's Woodpecker.
Voice CALL: A fairly high, sharp *pik*; suggestive of Downy Woodpecker but louder, sharper, and slightly lower

in pitch. Ladder-backed also gives a slightly descending *jee jee jee* series and a louder, slower *kweek kweek kweek*. **DRUM:** A simple roll, like Nuttall's, but longer, averaging 1.5 secs.
Status & Distribution Common; extensive range south of US to Nicaragua, El Salvador. **YEAR-ROUND:** Dry desert woodlands with yuccas, agaves, cacti; pinyon-juniper foothills; mesquite woodlands; and riparian

corridors. Often common in southwestern towns. Overlaps (and sometimes hybridizes) with Nuttall's very locally on the western edge of the CA deserts from Inyo Co. and Kern Co. south to northwestern Baja California. **RARE STATUS:** Casual east to vicinity of Houston, TX, and on the Pacific coast near San Diego, CA.
Population Generally stable, though declines have been noted in TX.

RED-COCKADED WOODPECKER *Dryobates borealis* RCWO ▪ 2

This endangered species has a highly fragmented distribution in southeastern pine-woods, especially old-growth longleaf pine. Living in family groups, or clans, its territories can be identified by distinctive nest and roost cavity trees: large living pines with heartwood disease and with abundant resin flowing from holes drilled around the cavity entrance. This oozing pitch protects nests from snakes and other predators. Nesting trees are usually marked by forest rangers with colored rings around the bases. Monotypic (FL Peninsula birds slightly smaller). L 8.5" (22 cm)

Identification Unique face pattern, with black crown and nape, long black malar, and extensive white auriculars and sides of neck; sides and flanks with short streaks. Adult male's red "cockade" is essentially invisible in the field.
Similar Species White face is diagnostic from other N.A. woodpeckers.
Voice CALL: A raspy *sripp* or *churt* and high-pitched *tsick* or *sklit*, utterly different from other N.A. woodpeckers, are the most frequent and distinctive calls. **DRUM:** Quiet, infrequently heard.
Status & Distribution Rare and local. **YEAR-ROUND:** Mature lowland woods of long-leaf, loblolly, or other pines with open understory and suitable cavity trees afflicted with heartwood disease. Core remaining populations occur from eastern TX and adjacent OK, AR, LA, and from MS east to FL, the Carolinas, and southern VA (still a few). Populations in TN, KY, MD, and MO have been extirpated in recent decades. **RARE STATUS:** Accidental in northeastern IL, and twice in south-central OH.
Population Near Threatened, but listed as Endangered by the USFWS.

About 6,000 colonies estimated to exist in 2006, with two to five adults per colony. Most colonies are on federal lands, where management strategies have had mixed results, with declines continuing in many areas.

pure white cheek bordered by black bar

red spot seldom visible in field

♂

white bars on back

dark spots on sides and flanks

HAIRY WOODPECKER *Dryobates villosus* HAWO ▪ 1

Like a large, long-billed version of the Downy Woodpecker, the Hairy is a widespread generalist of a variety of forests and woodlands over most of the continent. Polytypic (about 17 sspp.; 11 north of Mexico). L 9.5" (24 cm)
Identification Plumage pattern is nearly identical to Downy Woodpecker, with long white patch down back, variable white spotting on wing coverts and flight feathers, and mostly unmarked underparts. Outer tail feathers are usually unmarked white. **ADULT:** Male shows red nuchal bar, often divided vertically by black (especially in some eastern populations); red is lacking in female.

Geographic Variation Variation is extensive but generally clinal. Nominate *villosus* is widespread in the East; southeastern birds (*audubonii*) are smaller, buffier (less pure white) below, and with less white on the back. Boreal *septentrionalis*, from interior AK east to QC, is the largest, whitest subspecies. Newfoundland *terraenovae* is distinctive, with white back reduced and barred (especially in immatures), some black spotting on outer rectrices, and often with fine black streaking on sides and flanks. In the West, *picoideus* from Haida Gwaii, BC, is most distinctive, with gray-brown underparts; black markings in the white

back stripe, sides, and flanks; and strong black bars on outer rectrices. Northwestern *harrisi* and *sitkensis* have gray-brown underparts and face and reduced white on wings. Subspecies *hyloscopus* of CA and northern Baja California is smaller, paler (light gray-buff) below, and whiter on head. Three additional subspecies of the Great Basin and Rocky Mts. regions (*orius, monticola, leucothorectis*) are moderate to large in size and white to very pale buff below; compared with eastern and boreal birds, they have reduced white on the back and greatly reduced white spotting on the wing coverts. Subspecies *icastus*, ranging

Rockies
orius ♂

Maritimes juvenile
terraenovae ♂

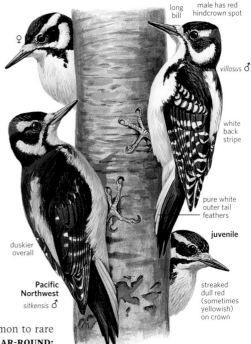

long bill

male has red hindcrown spot

villosus ♂

white back stripe

duskier overall

Pacific Northwest
sitkensis ♂

pure white outer tail feathers

juvenile

streaked dull red (sometimes yellowish) on crown

from southeastern AZ and southwestern NM south to central Mexico, is similar but smaller. Four M.A. subspecies from eastern and southern Mexico south to western Panama are smaller still (some nearly as small as Downy Woodpecker) and are variably buffy to deep buff-brown on the underparts. There are two additional subspecies on the Bahamas.

Similar Species Downy Woodpecker is similar in pattern but much smaller, with a small, short bill (much smaller than half the length of the head) and black bars on outer tail feathers. See American Three-toed Woodpecker. Note that some populations of Hairy Woodpecker (especially in Newfoundland) can show barred backs (especially as juveniles), and some Three-toed populations have nearly pure white backs.

Voice CALL: A piercing, sharp *peek*. The rattle call (whinny) is a fast, slightly descending series of these *peek* calls. **DRUM:** Rapid roll of about 25 beats in one second.

Status & Distribution Fairly common; uncommon to rare in the South and FL. **YEAR-ROUND:** Hairy Woodpecker occupies a wide range of coniferous and deciduous forests from sea level to tree line; such habitats are usually densely wooded, but in some areas are more open and parklike. Moves into burned areas, where most easily located in the Deep South and FL. **DISPERSAL:** Although generally nonmigratory, individuals can disperse long distances. Larger, more northerly birds from Canada

regularly occur in the Northeast in fall and winter. Recorded in fall and winter on the southern plains and Pacific coast lowlands well away from breeding habitats.

Population Declines that have been noted in many areas are thought to be due to fragmentation of forests, loss of old-growth trees, and nest site competition with European Starlings.

WHITE-HEADED WOODPECKER *Dryobates albolarvatus* WHWO ■ 1

white head

♂

white wing patch

♀

narrow patch folded wing

A striking bird of far western pine forests, the White-headed Woodpecker is unique among our species in its white head and solid black body. It often forages on pine cones, extracting seeds; otherwise it mainly flakes away bark on trunks or branches of conifers. Polytypic (2 sspp.; both in N.A.). L 9.3" (24 cm)

Identification A solid black body, mostly white head, and large white wing patches make this bird unmistakable. **ADULT:** Red nuchal patch, lacking in females. **JUVENILE:** Resembles adult, but both sexes have a pale red wash on crown, and white wing patch is often more interrupted with black spots.

Geographic Variation Birds of the Southern CA mountains, *gravirostris*, are slightly larger billed than the northern nominate subspecies.

Similar Species Acorn Woodpecker,

largely black above with white wing patches, may suggest White-headed dorsally, but Acorn's white rump and belly and black in the face simplify identification.

Voice CALL: Distinct call is a sharp, two- or three-syllable *pee-dink*, or *pee-de-dink*, dropping slightly in pitch. (Hairy gives a single note.) Also a longer series of these notes, a slower *kweek kweek kweek* series, and various softer calls. **DRUM:** A 1- to 1.5-second roll with about 20 beats a second. **Status & Distribution** Fairly common in CA; generally uncommon from OR

and western NV north. **YEAR-ROUND:** Montane coniferous and mixed forests usually dominated by ponderosa or Jeffrey pines, but also sugar pine, white fir, Douglas fir, and incense cedar. Also occupies burns. **RARE STATUS:** Rare in fall and winter at lower elevations adjacent to breeding mountains, and

casual in lowlands along CA coast and deserts. Casual south-central BC. Accidental western MT.
Population Generally stable in CA, but many populations, especially from OR north, are decreasing because of logging, even-age stand management, and long-term forest changes.

ARIZONA WOODPECKER *Dryobates arizonae*   ARWO ■ 2

This brown-backed species of pine-oak woodlands in the Southwest borderlands was formerly considered conspecific with Strickland's Woodpecker (*P. stricklandi*), a localized species of the high mountains surrounding Mexico City, which is smaller and has white on its back. Polytypic (2 sspp.; nominate in US). L 7.5" (19 cm)
Identification Solidly brown back

is unique among our woodpeckers. **ADULT:** Solid brown above, white below with heavy brown spotting on breast and bars on flanks, belly. Brown crown, auricular, and malar contrast with large white patch on sides of neck. Male has red nuchal patch, absent in female. **JUVENILE:** Limited red on crown. **Similar Species** Unmistakable. **Voice CALL:** Sharp, high *peeek* call is higher and hoarser than similar call of Hairy. **DRUM:** Rapid roll like that of Hairy, but longer.
Status & Distribution Fairly common, but usually wary, inconspicuous. **YEAR-ROUND:** Dry pine-oak woodlands and oak-riparian canyon woodlands from 4,000–7,000 ft. Found east to Peloncillo and Animas Mts., NM; and west to Santa Catalina and Pinaleño Mts., AZ; a few birds may move into lower foothill oak woodlands in winter.
Population Apparently stable.

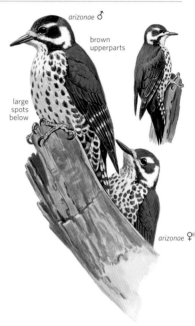

arizonae ♂

brown upperparts

large spots below

arizonae ♀

FLICKERS Genus *Colaptes*

Largely terrestrial, ant-eating woodpeckers of the New World, flickers have long, thin, slightly curved bills; large rounded wings; and flat tails that are only minimally stiffened. Plumage is barred olive to (in our area) brown on the back, spotted below, with colorful shafts on the wings and tail. Our species show a black chest crescent, and most show a bold white rump.

GILDED FLICKER *Colaptes chrysoides*   GIFL ■ 2

This "yellow-shafted" flicker of southwestern desert woodlands was re-split in 1995 from the Northern Flicker complex because of the very limited extent of interbreeding. Polytypic (3 sspp.; *mearnsi* in US and 2 more

in Baja California). L 11.5" (29 cm)
Identification A small flicker with yellow wing flash and yellow tail base; distal half of tail black from below; chest patch deeper and more rectangular than in other flickers, and black spots on lower underparts expand to bars or crescents. Crown and nape are rich brown, contrasting with gray face. Adult male has a red malar.
Similar Species Some hybrid "Yellow-shafted" x "Red-shafted" Northern Flickers approach Gilded in looks, having yellow in wings and tail but lacking a red nuchal crescent. Note Gilded Flicker's smaller size, more extensive black on undertail, paler more finely barred back, more

more cinnamon crown

larger black chest patch

mearnsi ♂

paler back with fainter bars

crescent-shaped black markings below

yellow underwing

♀

extensive chest patch, dark bars or crescents on lower underparts, and rich brown crown.
Voice CALL: High descending *klee-yer* and territorial *wick wick wick*; calls higher pitched than Northern Flicker.

DRUM: Like other flickers.
Status & Distribution Fairly common. **YEAR-ROUND:** Desert woodlands, especially where dominated by saguaro. Rare in limited range in CA and NV. Limited hybridization with

Northern Flicker in AZ. **RARE STATUS:** Casual to southwestern UT.
Population Noticeable and widespread declines from urbanization and destructive desert wildfires fueled by exotic grasses that kill saguaros.

NORTHERN FLICKER *Colaptes auratus* NOFL ■ 1

This familiar large woodpecker and the closely related Gilded Flicker show flashy color in the wings and a bold white rump in flight. Often feeds while on the ground, where it hops instead of walking, as do the two S.A. flicker species and Fernandina's Flicker (*C. fernandinae*) from Cuba. Polytypic (10 sspp.; 5 in N.A.). L 12.5" (32 cm)
Identification All Northern Flickers show a bold black chest crescent, a white rump, and bright color (salmon-red or yellow) in the shafts and much of the vanes of the flight feathers and on underwing coverts. All Northern Flickers are pale buffy white to rich buff below with black spotting and

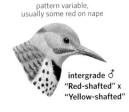

pattern variable, usually some red on nape

intergrade ♂
"Red-shafted" x "Yellow-shafted"

cafer

"Yellow-shafted" & "Red-shafted" intergrade zone

cafer

"Yellow-shafted Flicker"

"Red-shafted Flicker" & rare, wintering "Yellow-shafted Flicker"

luteus

collaris

auratus

nanus

chrysocaulosus

mexicanus

mexicanoides

gundlachi

have brown to gray-brown backs with black barring. **ADULT:** Sexes are similar, but males have a malar mark (red in "Red-shafted," black in "Yellow-shafted") that is lacking in females. "Yellow-shafted" has a gray crown with a red crescent on nape, a tan face and throat, and rich buff underparts with a relatively narrow chest crescent. Flight feathers and underwing linings are golden yellow. "Red-shafted" has a grayish head and throat with pale brown on forecrown and loral region, lacks red on nape, and has paler buff to creamy underparts with broader chest crescent; flight feathers and underwing linings are salmon pink. Introgressant individuals that combine characters of both groups are widely seen.
Geographic Variation The "Red-shafted" and "Yellow-shafted" groups, formerly considered separate species, are easily separated. In an extensive hybrid zone from northeastern NM and the TX Panhandle north to AB, BC, and southeastern AK, a large percentage of individuals encountered show traits of both groups. Intergrades are frequently seen in the area of overlap and throughout the West, but rarely noted in the East. Within the "Red-shafted" group, northwestern *cafer* (southern AK to northwestern CA) is darker than remaining *collaris* group,

which includes smaller *nanus* from southwestern TX. There is slight size variation within the "Yellow-shafted" group. Birds of the C.A. highlands (*mexicanoides* group, "Guatemalan Flicker") and the Cuba and Grand Cayman Is. (*chrysocaulosus* group) are distinctive representatives of the "Red-shafted" and "Yellow-shafted" groups, respectively. "Guatemalan Flicker" has different vocalizations and may well represent a distinct species. Birds of the *chrysocaulosus* group have barred and spotted rumps.
Similar Species Gilded Flicker closely resembles Northern Flicker and combines some features of "Yellow-shafted" (yellow wings and tail base) and the "Red-shafted" (head pattern). Gilded Flicker is smaller and shows black on the distal half of the undertail; Northern Flicker undertails are black on about the distal third (note that all flicker tails look mostly black from above). The crown of Gilded is more extensively brown than in "Red-shafted," and the back is paler, more gray-brown, with narrower and more widely spaced black bars (but note that interior western *collaris* "Red-shafted" are paler backed than northwestern birds). Black crescent on chest of Gilded is thicker and more truncated on sides. Spotting on underparts is broadened

all N.A. flickers have white rumps

"Yellow-shafted" ♀

yellow underwing

red nape and pale brown face

"Yellow-shafted" ♂

male has black "whisker"

brown nape and gray face

rounder spots below than Gilded

male has red "whisker"

"Red-shafted" ♂

pinkish red underwing

"Red-shafted" ♀

into short bars or crescents on the flanks of Gilded; Northern Flickers have round spots throughout underparts. Be careful of hybrids.

Voice SONG: On breeding grounds both sexes deliver a long *wick, wick, wick, wick* series in nesting season. **CALL:** A piercing, descending *klee-yer* or *keeew*; also a soft, rolling *wirrr* or *whurdle* in flight and a soft, slow *wick-a wick-a wicka* given by interacting birds. **DRUM:** A long, simple roll of about 25 beats.

Status & Distribution Common. **BREEDING:** Widespread in open woodlands, parkland, suburban areas, riparian and montane forests. **MIGRATION:** Northern populations of "Yellow-shafted Flickers" and northern interior "Red-shafted Flickers" are highly migratory. Small flocks of migrants and even large flights are evident from late Sept. through Oct. and in spring in late Mar. and Apr. **WINTER:** Uncommon to rare north to southern Canada (but common in southwestern BC). **RARE STATUS:** "Yellow-shafted Flicker" is a rare migrant and winter visitant west to the Pacific coast, though outnumbered by intergrades; also casual to Bering Sea islands. "Red-shafted" is casual east to MB, western MO, eastern TX, AR, LA; possibly also farther east, but most or all may not be pure "Red-shafted." "Yellow-shafted" is casual to western Europe.

Population Significant declines have occurred over much of the continent.

Genus *Dryocopus*

PILEATED WOODPECKER *Dryocopus pileatus*   PIWO ■ 1

Our largest woodpecker (other than the almost certainly extinct Ivory-billed Woodpecker), the Pileated Woodpecker is a crow-size, crested woodpecker of forested areas, and arboreous suburban neighborhoods, that feeds largely on carpenter ants and beetles extracted from fallen logs, stumps, and living trees. Foraging birds excavate large, rectangular holes in trunks and logs. Polytypic (2–4 sspp.; all in N.A.). L 16" (42 cm)

Identification Mostly black, with a slaty black bill, white chin, white stripe from bill down neck to sides of breast, a white patch at base of primaries, and extensive white on the underwing linings. Flight consists of deep, irregular crowlike wingbeats with little or no undulation. **ADULT MALE:** Red crown, crest, and malar mark. **ADULT FEMALE:** Forecrown mottled black, malar black.

Geographic Variation Differences between subspecies are minor; northern birds average larger.

Similar Species See Ivory-billed Woodpecker; all recent Ivory-billed reports have turned out to be Pileated Woodpecker or misidentified vocalizations of other species. Crows with aberrant white wing patches can momentarily suggest a Pileated.

Voice A long, flickerlike (but more maniacal) series *kee kee kee kee*, often slightly irregular in cadence, is given year-round. Also, single *wuk* or *cuk* notes. **DRUM:** Loud, deep, and resonant, lasting one to three seconds with about 15 beats a second; beat rate accelerates slightly, often trails off at very end.

Status & Distribution Common and widespread over much of East; uncommon in the Great Lakes region, boreal areas, and much of the mapped West, including the Sierra Nevada, and the Pacific coast. **YEAR-ROUND:** Dense coniferous and deciduous forests and woodlots, with suitable presence of large older trees, snags, and downed wood. **RARE STATUS:** Wanders casually slightly away from resident range, with documented records in CA from Malibu in coastal northern Los Angeles Co. and the San Joaquin Valley; there are also unsubstantiated reports for east-central AK, CO, UT, northwestern AZ, and southern NM.

Population Generally stable.

shorter, darker bill

♀

black above

♂

white wing patch

♀

extensive white underwing and black secondaries

♂

now N.A.'s largest woodpecker

Genus *Campephilus*

IVORY-BILLED WOODPECKER *Campephilus principalis* IBWO ▪ 6

This spectacular woodpecker of virgin southern bottomland forests is almost certainly extinct. The much publicized announcement in 2005 of its "rediscovery" in the White River–Cache River system and searches there and elsewhere (e.g., north FL), where reported sightings took place, yielded no credible evidence of their presence. All of these sightings are now questioned by most authorities. Polytypic (2 sspp.; nominate in US, likely extinct *bairdii* photographed in northeast Cuba in 1948, last undocumented sightings in the 1980s). L 19.5" (50 cm)

long black crest

almost certainly extinct

long ivory-colored bill

♀

black underwing bar

white secondaries

♂

♂

white scapular bar

large white wing patches visible on perched bird

Identification The largest US woodpecker. **ADULT:** Black with a massive creamy white bill, a large crest (male's is red and curves back, female's is black and curves forward), white stripes on sides of back. Eyes yellow. **FLIGHT:** Rapid and direct. Completely white secondaries and white tips to inner primaries. **JUVENILE:** Shorter crest, browner plumage, and brown eyes.

Geographic Variation Cuba (including Isla de Juventud) subspecies, *bairdi*, now also believed extinct, has a slightly shorter, narrower bill and averages more white on side of head.

Similar Species Obviously the smaller Pileated Woodpecker is fairly common over the entire Southeast and no doubt accounts for misidentifications. It has white, not black, secondaries, has more extensive white on underwing linings, lacks white back stripes, and has a white (not black) chin. Bill of Pileated is gray or blackish, not ivory.

Voice CALL: Tinny, toy trumpetlike *kent* calls, sometimes doubled (somewhat similar to White-breasted Nuthatch); also when birds are together, an up-slurred series of *kient* notes. **DRUM:** A strong single or double rap, typical of Ivory-billed's genus (*Campephilus*), often delivered when adult is disturbed or when one of pair is absent.

Status & Distribution Never numerous. Formerly resident in hardwood and cypress forests in the Mississippi River Valley, north to southern IL, Gulf Coast region, and southern Atlantic states from NC to FL. Logging of these forests throughout the southern US, particularly intensive from 1885 to 1915, along with localized scientific collecting (e.g., along the Suwannee River, FL, in 1892–93) extirpated populations, and after the mid-1920s the only well-documented remaining population was in the Singer Tract, Madison Parish, LA (discovered in 1932, last noted in Apr. 1944). See sidebar below.

Population Critically Endangered, but almost certainly extinct.

The Singer Tract and James T. Tanner

In 1937 James T. Tanner, a research associate at Cornell University, in coordination with the National Audubon Society, started a 21-month search and study of the Ivory-billed Woodpecker. Tanner scoured the Southeast, driving some 45,000 mi by car and uncounted miles by boat and horse and on foot. He found none except at the Singer Tract, LA, a vast 120 sq mi of mostly virgin forest. His detailed studies led to most of what we now know about the ecology of the species (*The Ivory-billed Woodpecker*, 1942). Tanner concluded that only in extensive virgin forests are there enough recently dead trees to extract their main food, large larval wood borers.

Set aside in 1920 as a state sanctuary, the Singer Tract was later sold to timber companies in 1937–38. Over the next six years or so, the entire forest was leveled. Tanner writes: "The story is much the same in all regions. Ivory-billed Woodpeckers have disappeared when the woods that they inhabited were cut over and the virgin timber removed. In many cases the disappearance of the birds almost coincided with the logging operations." He goes on: "There are no records of Ivory-bills inhabiting any areas for any length of time after those have been cut over." Any chance to save the species was forfeited when the Singer Tract was sold and logged.

CARACARAS AND FALCONS Family Falconidae

American Kestrel, immature male (NJ, Oct.)

Falcons are flashy, fast fliers and can reach high rates of speed on dives. For centuries captive birds have been trained by humans to hunt, leading to the term *falconry*. Historically, larger species, such as the Gyrfalcon, were reserved for royalty. The name *falcon* comes from the term *falcate*, meaning sickle-shaped, which describes the bird's wing silhouette.

Structure The typical falcon is a large-headed, dark-eyed raptor with pointed wings and a square-tipped tail. Most are strong fliers and take their prey in the air. Powerful flight muscles give them a thick-chested look. The beak has a notch on the cutting edge of the culmen, which severs the spinal cord of prey. The feet and toes are not designed for killing prey, but for grabbing and holding it immobile. Females are larger than males, dramatically so in larger species. Differing external features are found in the forest-falcons and caracaras, such as rounded or square-tipped wings.

Behavior These pointed-winged birds commonly take their prey in flight, either by diving powerfully from above, or by pursuing it from behind and below. Most falcons also hunt from an exposed perch, dropping down to pursue prey from behind. Small birds may be forced to high altitudes; then the falcon dives upon the exhausted prey. Many falcons hunt at dawn and dusk, even by city lights. Caracaras often feed on carrion and will rob food from other raptors. The *Micrastur* forest-falcons are ambush hunters in heavy foliage, much like North American accipiters. Caracaras build stick nests in trees, whereas falcons make scrape nests on a ledge or in a shallow cave. Kestrels favor a cavity and use human-made nest boxes. The nest of another hawk may be used. Northern-nesting species are usually migratory, although the Gyrfalcon does not follow any discernible pattern of passage. Eastern migrants prefer following coastlines, preying on shorebirds and migrant passer-ines. They often fly in wind conditions unfavorable to other migrating raptors and are not as reluctant to cross open water. Inland migrants soar on fixed wings along mountains' thermal updrafts, like other hawks, but take a direct flight line instead of the swirling kettles favored by the buteos. An owl decoy placed atop a long pole at a ridgeline hawk-watch site may attract a migrating falcon to swoop down and harass its historic enemy.

Plumage Except for the kestrels and the Merlin, adult falcons and caracaras of both sexes share similar plumage, while juvenile birds have different plumages than their adults, in some cases only by degree. All are darker above and generally light below. Most falcons have moustache marks on the face.

Distribution Falcons are found worldwide, caracaras and forest-falcons only in the New World. Falcons fly at all altitudes, in open areas throughout North America; caracaras are found in open areas with warm climates.

Taxonomy Recent genetic studies have shown that the Falconidae share a close affinity to the Psittacidae (parrots); both families are now placed just before the order Passeriformes (passerine birds). Within the family Falconidae, the subfamily Falconinae includes 37 species of *Falco* worldwide (nine in North America), plus seven Old World falconets and pygmy falcons. Eight forest-falcons and laughing falcons are placed in the Herpetotherinae family; the 11 living caracaras are now included within the subfamily Falconinae, though placed by some in a separate subfamily, Caracarinae.

Conservation Pesticides and other toxins in the environment are detrimental to nesting success. The prohibition of DDT and similar chemicals in North America has allowed populations to rebuild. Loss of grassland habitat in the East due to development and reforestation of farmland is a concern. BirdLife International codes: 8 NT, 5 VU, 1 EN, and 3 EX.

FOREST-FALCONS Genus *Micrastur*

COLLARED FOREST-FALCON *Micrastur semitorquatus* COFF ■ 5

The Collared Forest-Falcon looks more like an *Accipiter* than a falcon. Polytypic (2 sspp.; 1 in N.A., presumably *naso*). L 20" (52 cm) WS 31" (79 cm)
Identification Dark above and light below; short-winged, long-tailed, and long-legged. Black above, with white cheeks, neck collar, and undersides. Cheek has black crescent. Wings and back dark; tail dark with thin white bars, white tip. **ADULT:** Two morphs: a light morph with white or buffy, unmarked underparts and a rare dark morph. **JUVENILE:** Like adult above, but browner. Undersides heavily barred, ground color has buffy wash.
Similar Species Size similar to Cooper's Hawk. Completely unlike other N.A. raptors. Much smaller Sharp-shinned

Hawk subspecies *chinogaster* of southern Mexico is dark above, mostly white below, but lacks "sideburns."
Voice Quite vocal early and late in the day. Usual call a hollow *how* or *aow*, frequently repeated, plus rapid combinations getting louder and more slurred at the end. Also repeats a whiny *keer keer keer*, possibly a false alarm call to attract small birds.
Status & Distribution Resident from Mexico (north to southern Sinaloa and southern Tamulipas) to Peru and northern Argentina. **RARE STATUS:** Accidental, light-morph bird, Bentsen-Rio Grande SP, south TX (22 Jan.–24 Feb. 1994) is the only N.A. record.
Population Vast range, but like other forest-falcons, deforestation is a risk.

dark cap and dark point in white face; white collar

light-morph adult
naso

short wings

long graduated tail

CARACARAS Genus *Caracara*

Long-legged raptors of open areas throughout S.A. and C.A. Their structure and behavior are very different from that of a true falcon, appearing more like a vulture or *Buteo*. Eleven species are found in the New World (including the extinct Guadalupe Caracara, *C. lutosa*, from Guadalupe I., off Baja California). Only one caracara reaches N.A.

CRESTED CARACARA *Caracara cheriway* CRCA ■ 2

This large, long-legged raptor of parts of the southern US is often seen standing on the ground or perched on a pole. Its long, square-tipped wings are

unlike a falcon. It feeds on carrion and small prey. In flight the wingbeats are ravenlike; gliding, the wings are held slightly downward. Monotypic. L 23" (58 cm) WS 50" (127 cm)
Identification ADULT: Body, wings, and crown black, contrasting with white face and neck. Upper breast and back are white, finely barred with black. Tail white with dark barring and broad dark terminal band. Facial skin and cere are orange; legs yellow. **JUVENILE:** Similar pattern to adult, but with brownish and buffy coloration. Legs

blackish cap
slight crest
pink facial skin
adult
barred white chest
black belly
juvenile
streaked chest
conspicuous white wing patches
adults
long white-based tail barred with black
long legs

are gray, and facial skin is pinkish. In flight, white patches at base of primaries and tail pattern are conspicuous.
Voice Generally silent.
Status & Distribution Rare to common within range. **YEAR-ROUND:** Resident in central FL, along TX coastal prairies, and in a small range in southeast AZ. Range is expanding into central TX. **MIGRATION:** Generally nonmigratory. **RARE STATUS:** Many records over the last several decades have occurred over much of N.A., north to southern Canada.
Population Expanding in TX since the 1980s. Listed as Threatened ("*audubonii*") in FL by USFWS.

FALCONS Genus *Falco*

Also known as typical falcons or true falcons, there are 38 species described, nine of which appear in N.A. Six species are regular nesters here, including the restocked Aplomado Falcon into their former range. Known for their aerial acrobatics, and spectacular dives, falcons are often a symbol of sports teams and warplanes. All true falcons typically have pointed wings, a large head, and dark eyes. They occur worldwide, nesting on all continents except Antarctica.

AMERICAN KESTREL   *Falco sparverius*   AMKE ■ 1

Our smallest and most common falcon, it is usually found in close open areas, either perched on a snag or telephone wire or hovering in search of prey. Frequently bobs tail at rest. It hunts insects, small mammals, and reptiles from a perch or on the wing. Polytypic (17 sspp.; 2 in N.A.). L 10.5" (27 cm) WS 23" (58 cm)

Identification All plumages have two bold, dark moustache marks framing white cheeks on the face. **ADULT MALE:** Head has gray crown with rufous spot and two dark moustaches around white cheeks. Rufous back with black barring on lower back. Tail is patterned with variable amounts of black, white, or gray bands. Wings are blue-gray with dark primaries. Underparts are white, washed with cinnamon. **ADULT FEMALE:** Head similar pattern to male, but more brown on crown. Back, wings, and tail reddish brown with dark barring; subterminal tailband much wider than other bands. Underparts buffy white with reddish streaks. **JUVENILE MALE:** Head similar to adult, but less gray and with dark streaks on crown; back completely streaked, heavy streaks on breast. **JUVENILE FEMALE:** Very similar to adult female. **FLIGHT:** Typical falcon-shaped wings are slim and pointed, the tail long and square-tipped. Light, bouncy flight with wings swept back is usually not direct and purposeful—often with twitches or hesitation. Hovers above a field on rapidly beating wings, or soars in place in strong winds above a hillside. Light underwings and generally light body coloration. Males show a row of white dots ("string of pearls") on trailing edges of underwings. Fans tail when hovering.

Geographic Variation Widespread nominate *sparverius* is the typical migratory form. Subspecies *paulus*, from SC to FL, is smaller, the male with less barring on the back and fewer spots on its undersides; essentially nonmigratory. Four resident subspecies of *caribaearum* group from Greater, Lesser, and Netherlands Antilles are small and lack (or reduced) chestnut on crown; *sparveroides* of Bahamas, Cuba, and Jamaica with distinct light and rufous morphs. Behavior and ecology quite different: flight rapid and direct, more Merlin-like; does not hover or bob tail at rest. Might well be a distinct species.

Similar Species Merlin appears darker in flight due to dark underwings, shorter tail, which is not bobbed. When perched, looks darker, more heavy-bodied, facial pattern differs. Peregrine Falcon is larger, has wider wings, shorter tail, single heavy moustache. See larger Eurasian Kestrel.

Voice Loud, ringing *killy-killy-killy* or *klee-klee-klee* used all year round. Distinctive.

Status & Distribution Common in open areas, it ranges throughout N.A., including much of Canada and into AK. **BREEDING:** A cavity nester, it uses dead trees, cliffs, occasionally a dirt bank, and even a hollow giant cactus in the Southwest. It will also use human-made nest boxes. Up to five or six young per brood, depending on food availability. **MIGRATION:** Fall migration starts early, in Aug. Northern breeders winter in the southern US and northern Mexico. Eastern populations use the coastlines more than the inland corridors, and are not reluctant to cross water. Spring migrants are only concentrated along the Great Lakes watch sites. **WINTER:** The majority of birds winter in the southern US, often spaced out on every other telephone pole in agricultural areas. A small percentage winter in snow-covered states, the numbers depending on food sources.

Population Overall, numbers appear to be declining. Increases in the central US are being offset by declines in the Northeast and the West Coast (CA and OR). Eastern populations are thought to be affected by loss of open habitat due to two factors: human development and agricultural abandonment leading to reforestation, with a subsequent increase in Cooper's Hawk predation.

adult ♂

frequently hovers

adult ♀

lacks white primary marks of male

adult ♂

upperparts uniformly rufous

uniformly rufous-brown back and tail

adult ♀

all with two dark facial stripes

juvenile ♂

long rufous tail

male has bluish gray wings

adult ♂

1 sparveroides
2 dominicensis
3 caribaearum

sparverius

paulus

peninsularis

1 2 3

EURASIAN KESTREL *Falco tinnunculus* EUKE ■ 4

Also called Common Kestrel, this small falcon is from Eurasia. It shares many of the characteristics of the American Kestrel, but it is larger and has only one moustache mark on its face. Undersides on all ages are buffy and streaked. In flight the tail looks slightly wedge-shaped, and from above the wing gives a two-tone appearance. The bird feeds on small prey, including insects, small mammals, and birds, first hovering to spot its quarry, then descending in steps before stooping to the ground. Polytypic (11 sspp.; N.A. records likely of nominate). L 13.5" (34 cm) WS 29" (74 cm)

Identification ADULT MALE: Gray head and nape, with a single moustache and paler throat. Reddish back and wing coverts have black streaks, dark primaries. Gray rump and unmarked gray uppertail contrast with wide, dark subterminal tail band and a white tip. Undersides are whitish buffy, with spots on breast, streaked flanks, plain undertail coverts. ADULT FEMALE: Head with similar pattern to male, but more rufous-brown. Back is more heavily barred, rump gray with gray-brown barred tail and wide sub-terminal band. JUVENILE: Similar to female, but usually with buff tips to feathers on upperparts.

Similar Species American Kestrel is smaller, slimmer; has two moustache marks on face. Underparts are usually whiter, tail more rounded. Males have blue-gray wing coverts; red tail; females are less heavily marked.

Voice A repeated *kee-kee-kee*. Mostly quiet away from breeding grounds.

Status & Distribution Widespread in Old World, casual to western Aleutians and Bering Sea, accidental to East and West Coasts of US, most recently to WA, CA, FL, and MA. MIGRATION: Northern European and Asian birds are highly migratory, wintering in southern Europe, the Middle East, and southern Asia

Population Eurasian Kestrel seems to be stable across its range, with human persecution being the most common threat. No signs of range expansion to N.A.

two-tone upperwing

juvenile

adult ♂

wedge-shaped tail

pale gray head

adult ♀

single dark moustache

adult ♂

gray tail base

RED-FOOTED FALCON *Falco vespertinus* RFFA ■ 5

This medium-size, gregarious, and highly migratory Old World falcon of open country has been recorded once in N.A. It is larger than the Eurasian Kestrel and Merlin. It feeds mainly on insects, hawking them from midair, often hovering above a spot in search of prey. Monotypic. L 11" (27 cm) WS 29" (73 cm)

Identification ADULT MALE: Dark blue-gray body with dull red undertail coverts and tarsus feathering. Bright red legs. In flight note silvery primaries from above and dark slaty wing linings. ADULT FEMALE: Very different from male, with slate gray back and wings; brown head with black moustache; and white cheek and side of neck. Underparts are buffy with thin streaking. JUVENILE: Back and wings browner, with light feather edges. Head pattern similar to adult female. FIRST-SUMMER MALE: Dark gray head, mostly gray wings and body, and barred flight feathers and tail.

Similar Species Mississippi Kite is all gray with pointed wings, but does not hover when foraging.

Status & Distribution Highly migratory. Breeds Eastern Europe and western Asia; winters southern Africa. Small numbers occur annually to the British Isles and Scandinavia, particularly in late spring. RARE STATUS: Casual Iceland (at least four records). Accidental to N.A. One second-year male was at Katama Airfield, Martha's Vineyard, MA (8–24 Aug. 2004).

Population Near Threatened with declines in Eastern Europe.

wing is in molt, dark adult primaries are growing in

1st summer ♂

often hovers like a kestrel

male is slate-gray with rufous undertail region

MERLIN *Falco columbarius* MERL ■ 1

This small, dark falcon takes its prey in midair, often after spectacular chases. It primarily feeds on birds, sometimes ones larger than itself. Generally hunts on the wing, but it also hunts from a perch, ambushing passing prey. It occasionally soars on thermals on spread wings but never hovers. Moustache mark is less noticeable than on most falcons, often absent in *richardsoni*. Females differ from adult males. Polytypic (9 sspp.; 3 in N.A.). L 12" (31 cm) WS 25" (64 cm)

Identification Wings wider and tail much shorter than kestrel. Usually perches atop a snag. When perched, wing tips do not reach tip of tail. Descriptions below are of widespread N.A. *columbarius*, known as "Taiga Merlin." ADULT MALE: Blue gray above, undersides are whitish streaked with brown and rufous wash to sides and leg feathers. Tail dark with lighter gray banding. **ADULT FEMALE:** Slate-brown above. Tail bands are buffy on a dark brown tail. Below, buffy underparts are heavily streaked. **JUVENILE:** Apart from size, sexes very similar to each other; plumage like adult female; tail bands buffy. **FLIGHT:** Fast, flickering wingbeats and direct flight set it apart from American Kestrel.

Geographic Variation "Taiga Merlin" (*columbarius*) from Alaska to eastern Canada is smallest of the three; averages darker to the east of its range; and is highly migratory, wintering down to the Caribbean and C.A. "Prairie Merlin" (*richardsoni*) is largest of the three, much paler than "Taiga Merlin," especially adult females, which also have thinner streaks on body. Adult male is paler blue on back than "Taiga." All have wider and paler tail bands than "Taiga." Moustache mark can be very faint. "Black Merlin" (*suckleyi*) is a resident of the Pacific Northwest; generally sedentary, but does wander down to CA. About the size of "Taiga Merlin," but much darker overall, with almost no white on face, heavy and broad dark streaking on underparts with contrasting white spots on flanks, and pale tail bands are thinner and often incomplete.

Similar Species Slimmer American Kestrel has longer tail, less direct flight style. Much larger Peregrine Falcon has broader base of wings, large moustache mark on head. When perched, its wing tips reach or almost reach tip of tail. Much larger Prairie Falcon is similar to adult female and juvenile "Prairie Merlin" in coloration but has more a distinct moustache mark on face, and dark on axillaries and on underwing.

Voice Rapid *kee-kee-kee* or *klee-klee-klee* heard often around nest. Males are higher pitched than females. On migration, sometimes heard when harassing

other raptors, especially Peregrines.

Status & Distribution Widespread throughout the Holarctic. Relatively common on East Coast during fall migration, an uncommon winter bird. **BREEDING:** Uses a mix of habitats in conjunction with open spaces, anything from tundra to coastline, boreal forest to Northwestern rainforest, prairie edges, and parkland. In recent years they have been increasingly found within cities in prairie states and provinces; also spreading south in Northeast as far as WV. **MIGRATION:** Primary passage in fall, mid-Sept. through late Oct.; juveniles pass first, followed by adults. Majority move during morning and late afternoon. **WINTER:** Uncommon in northern

part of winter range, more likely to be found on coast. "Prairie Merlins" have recently been taking to wintering in Great Plains cities, feeding on small birds and rodents.

Population Increasing in N.A.

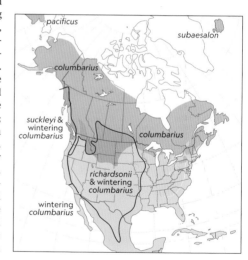

EURASIAN HOBBY *Falco subbuteo*   EHOB ■ 4

Also called Northern Hobby and "mini Peregrine" by raptor aficionados. A rapid, strong flier, this widespread highly migratory Eurasian species takes prey on the wing, usually small birds, also dragonflies and other large insects. Casual in the Bering Sea region and accidental elsewhere. Polytypic (2 sspp.; N.A. records likely nominate). L 12.3" (31 cm) WS 30.3" (77 cm)

Identification A medium-size falcon with long thin, pointed wings and a relatively short tail, dark above and whitish below with heavy streaking. **ADULT:** Dark gray head and upperparts, bold black moustache sharply contrasts with pure white cheek, short white streak above eye, whitish throat, with whitish underparts heavily streaked. Undertail coverts

and legs bright rufous, easier to see when perched. Cere and eye ring yellow. **JUVENILE:** Upperparts are more brownish, rufous tips to new feathers quickly wear off. Head pattern similar to adult, underside ground color may appear more buffy, tail more distinctly light-tipped, undertail coverts and legs buffy to dull rufous. **FLIGHT:** Long, slim wings and a relatively short tail, it

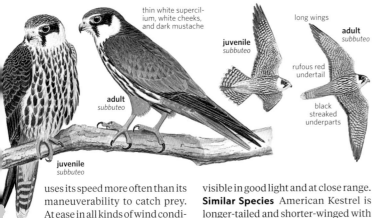

thin white supercilium, white cheeks, and dark mustache

long wings

adult
subbuteo

juvenile
subbuteo

rufous red undertail

adult
subbuteo

black streaked underparts

adult
subbuteo

juvenile
subbuteo

uses its speed more often than its maneuverability to catch prey. At ease in all kinds of wind conditions. Underwings are uniformly grayish with dark markings and darker wing tips. Rufous undertail of adult is visible in good light and at close range. **Similar Species** American Kestrel is longer-tailed and shorter-winged with a lighter flight; is lighter above, has two moustache marks on face. Merlin is stockier, darker, less heavily streaked underneath. Resembles far larger Peregrine Falcon in proportions, powerful flight, and dark over light plumage. **Voice** Silent away from breeding grounds. **Status & Distribution** Extensive breeding range in Eurasia; winters mainly in southern Africa, small numbers north to Kenya and in west Africa. Also Pakistan and the Indian subcontinent. **RARE STATUS:** Casual in AK, ±20 records (late spring, summer, and fall) mainly from the western Aleutians and Bering Sea region. Accidental to western WA (two Oct. records), MA (May), and NL (May). Also casual to Seychelles and accidental to Indonesia (Java and Timor). **Population** Stable.

APLOMADO FALCON *Falco femoralis* APFA ◾ 3

This medium-size falcon is a Neotropic species whose range formerly just reached into N.A. In recent decades captive birds have been released in southernmost TX. However, at this time, the South TX Aplomado Falcon is not yet treated as established. Often seen perched atop a single tree or cactus as it scans for prey. Polytypic (5 sspp.; *septentrionalis* in N.A.). L 15–16.5" (38–42 cm) WS 40–48" (102–122 cm)
Identification A slim, long-winged, long-tailed falcon, tail extending well past wing tips on a perched bird. Adults and juveniles have slightly different plumages. All show a distinctive head pattern of dark eye line and moustache mark, dark crown, and contrasting buffy cheeks and buffy supercilium that meets at nape. **ADULT MALE:** Head as described above. Dark gray back and upperwings contrast with buffy chest, black belly band, and rufous belly, undertail coverts, and legs. Flight feathers and tail are black-

juvenile
septentrionalis

juvenile ♀
septentrionalis

bold white supercilium and breast

long tail

buffy streaked breast

adult ♂
septentrionalis

blackish side patches and white-tipped secondaries

adult ♀
septentrionalis

dark belly band and underwing coverts

ish, with six or more narrow white bands on the tail. Black underparts are often finely barred in white. Eye ring and cere are yellow. **ADULT FEMALE:** Like the male but noticeably larger; buffy chest has black streaking, which is usually lacking in males. Females are noticeably larger than males. **JUVENILE:** Shows a similar pattern to adults, but with brownish back and wings, buffy tail bands, buffy underparts, legs, and undertail coverts. Buffy breast is more heavily streaked with black. Cere and eye ring are bluish gray. **FLIGHT:** Flies with deep, rapid wingbeats. It is highly maneuverable, able to turn and swoop in pursuit. Can also hover and buoyantly soar. Slim wings, long tail, plus light-dark combination of chest and underparts and white-tipped secondaries, make it fairly distinctive in flight. **Similar Species** Adult Peregrines have a dark moustache on face, dark back and wings, but are much larger, more stocky, shorter-tailed. Peregrine and Prairie Falcon lack black underparts. **Voice** Most typically a fast *ki-ki-ki-*

ki-ki-ki sound, males higher-pitched than females.
Status & Distribution Uncommon, local from Mexico to Argentina. Until the very early 20th century likely bred in open country from south TX to southeast AZ. Listed as endangered in 1986. A few wild pairs in southern NM appear to be hanging on, but additional captive-bred birds have been released there in recent years. A recent record from near the Davis Mts., TX, likely originated from a nearby wild population in Chihuahua, Mexico. Restocked into southern TX.
Population Very little data on Mexican populations, US numbers continue to grow under management. Listed as Endangered by the USFWS.

restocked

GYRFALCON *Falco rusticolus*   GYRF ■ 2

A visitor to the lower 48 states from its breeding grounds in arctic Alaska and Canada, the Gyrfalcon is the largest falcon in the world, capable of taking down flying geese and cranes. It is a powerful flier, with deceptively slow wingbeats and the longest tail of the large falcons. Gyrfalcons occur in three color morphs—white, gray, and dark (black). Juveniles are different from adults; sexes are similar, but females are larger. The Gyrfalcon is typically seen in the lower 48 states during fall migration and winter; an individual bird may take up winter residence and remain there for a period of months; may return subsequent winters. In medieval Europe, the Gyrfalcon was highly prized as a falconry bird, as it still is in some Arab countries. Monotypic. L 20–25" (51–64 cm) WS 50–64" (127–163 cm)

gray-morph adult

dark-morph juvenile

very dark brown above

darker wing coverts contrast with slightly paler remiges

gray-morph juvenile

long tail extends well beyond wing tips

white-morph adult

Identification A large, bulky falcon, with thick pointed wings wide at base, giving a broader wing silhouette than other large falcons. Females larger than males. Adult (all morphs) have yellow to orange cere, legs, and orbital ring; blue-gray to greenish gray on juveniles. Perched, they sit upright on cliffs, towers, buildings, and other human-made structures. Wings extend about two-thirds of the way down the tail. **ADULT LIGHT MORPH:** All-white with varying amounts of dark markings above, blackish primary tips; dark bands on tail often incomplete. Occasionally shows a faint moustache. **ADULT GRAY MORPH:** Dark gray to gray-brown ground color on back and wings is patterned with lighter gray to blue-gray feather tips, giving a scaled pattern. Tail strongly barred. Head can appear lighter than back, with a dark eye line and paler supercilium, moustache usually visible. Underparts with

barred flanks and spots on breast and belly. **ADULT DARK MORPH:** Blackish brown above with less obvious barring on back and tail; head usually with a hooded appearance; moustache usually lost in the overall pattern. Underneath, overall dark with some whitish streaks on breast and barring on flanks, belly, and undertail. **JUVENILE LIGHT MORPH:** Similar overall color to adult, but more heavily marked above with black or brown. Primaries are blackish with complete barring. Tail bands are complete, heavier than adults. **JUVENILE GRAY MORPH:** Darker gray-brown than adult, with back and wing coverts finely edged or spotted in lighter brown. Head color as dark as back, but still showing faint moustache and supercilium. Underparts are heavily streaked with brown, lightest on throat and undertail. **JUVENILE DARK MORPH:** Overall dark brown with heavy streaking underneath, barred undertail coverts. **FLIGHT:** A large falcon with heavy chest, wide pointed wings that are blunt on tips. Long tail is wide-based and tapers down to the tip when folded. Its wingbeats are slow and relatively stiff, yet it flies with great speed. Gray- and dark-morph birds show lighter flight feathers that contrast strongly with dark wing coverts, giving a strong two-tone look.

Similar Species White-morph Gyrfalcon is unmistakable. Albino or pale-morph Red-tailed Hawk may look similar, but wing shape and tail differences are easy to spot. Peregrine Falcon is slimmer, shorter-tailed, shows

heavier moustache, usually darker crown. Adult Peregrine is horizontally barred across breast; juveniles show uniformly dark underwings. Peale's Peregrine of the Pacific Northwest resembles a dark-morph Gyrfalcon. Prairie Falcon is lighter underneath, has dark axillaries contrasting with light underwing, and a pale cheek. Northern Goshawk has the same bulk, but differs with blunt rounded wings, wide secondaries, and two-tone upperwings.

Voice Usually silent.

Status & Distribution An uncommon species, it is a widespread breeder in the Holarctic. **BREEDING:** Nest scrape usually on a cliff, but also uses old Raven nests in a suitable location. **MIGRATION:** Gyrfalcon sightings along traditional hawk migration sites are usually late in the season—Nov.–Dec. There are enough late-Oct. records to make it a species to watch for then. **WINTER:** There seem to be no real patterns in wintering birds, but individuals have shown that they will return to the same general area in succeeding years. A favored roost site will be used for the night, with the bird leaving in the morning and arriving in the evening with sufficient fidelity to afford birders a viewing schedule. Most wintering birds are found above 40°N. **RARE STATUS:** Casual to CA, OK, northern TX, and east to the mid-Atlantic states.

Population Circumpolar and out of the range of most inhabited areas; populations are thought to be stable.

PEREGRINE FALCON *Falco peregrinus*   PEFA ■ 1

The Peregrine Falcon is world renowned for its speed, grace, and power in the air. It is distributed across the world, found on every inhabited continent, and on many islands. Formerly called the Duck Hawk in N.A. It can occur in almost any habitat in N.A., but is frequently seen around water—lakes, rivers, or coastal shorelines. Peregrines feed primarily on birds taken in flight, along with the occasional bat or rodent. Its large hind talon (the hallux) is anchored by heavy tendons, allowing it to use the force of its dive to kill its prey on contact. The true velocity of its high-speed dive has been the subject of much debate over the years, with some observers claiming speeds of up to 200 mph. Recently a falconer's Peregrine was outfitted with a skydiver's altimeter and data recorder and trained to chase down a lure released by the skydiver. The bird easily dove to a speed of 247 mph, and the feeling is that it could still go faster! Polytypic (at least 16 sspp.; 3 in N.A.). L 16–20" (41–51 cm) WS 36–44" (91–112 cm)

Identification Peregrine is a raptor with long pointed wings and a medium-length tail. All ages have a dark moustache mark on the face. When sitting wings extend almost to the tip of the tail. They favor an exposed perch, natural or human-made, to scan for prey. The majority of successful kills are made with high-speed dives that overtake from above or behind. Females are larger than males; sexes appear similar. **ADULT:** A dark head and nape, with dark moustache and white cheek. Cere and eye ring are yellow. "Tundra Peregrine" (*tundrius*) shows the thinnest moustache, whitest cheek, and often a light forehead. Some *anatum* have a dark cheek, blending with the moustache. Upperparts are dark gray, underparts are whitish, with clear breast and barring on belly and flanks. Subspecies *anatum* often shows a salmon wash on breast. "Peale's Peregrine" (*pealei*) is without salmon

color and has streaking on breast. **JUVENILE:** Browner above than adults, with brown streaking on buffy underparts. Cere and eye ring are blue-gray. "Tundra" has a light forehead, often a light crown (giving a blonde look), and thinnest moustache. Subspecies *anatum* shows a darker head, thicker moustache, thicker streaking on underparts. "Peale's" is overall very dark with a small cheek patch; underparts heavily streaked or almost uniformly dark. **FLIGHT:** Capable of swift level flight, power dives, and soaring on outstretched wings, Peregrine can fly in virtually any wind and weather condition. Typical flight is a smooth, quick, rhythmic, shallow wingbeat. With deeper wingbeats, it gains speed and dives, whereas it executes quick turns after fleeing prey with a fanned tail and choppy wingbeats. When soaring, wing silhouette tapers smoothly from body to tip. Underwings are barred on both flight feathers and coverts, giving a uniformly dark appearance.

Geographic Variation "Tundra" (*tundrius*) nests on the Alaskan and Canadian tundra, is overall the lightest N.A. subspecies, and is the longest-distance migrant. "Peale's" (*pealei*), the darkest and most sedentary, is found along the coast of the Pacific Northwest and presumably the Aleutians; some darker specimens may be darker *japonensis* of northeast Asia. Continental-breeding *anatum* is intermediate, but variable in features, while movement is on a smaller scale than "Tundra." The release of mixed-subspecies birds as part of the N.A. reintroduction program has resulted in many east-

adult
anatum

uniform
underwing

juvenile
tundrius

adult
pealei

juvenile
pealei

heavily
spotted
breast

heavily
streaked
underparts

ern birds showing characteristics of "Peale's," selected in hopes that the newly hacked birds will remain in the vicinity of their release points.

Similar Species Gray- and dark-morph Gyrfalcon is usually larger and darker, with smaller moustache, wider wing bases with two-toned underwing, and a longer tail that tapers from base to tip. Some small, dark male Gyrfalcons can appear similar to dark juvenile female "Peale's." Prairie Falcon is lighter overall, has a thin moustache and dark axillaries and underwing coverts contrasting with light underwing. When perched, Prairie's wing tips do not reach the tip of the tail.

Voice A rapid *kak-kak-kak*, not often heard away from the nesting area.

Status & Distribution The widespread continental form was decimated by the use of pesticides in the 1950s and 1960s, placing it on the Endangered Species List by 1970. A reintroduction program begun in the late 1970s has helped to restore Peregrine numbers

broad, dark
moustachial
stripe

long wing tips
extend nearly
to tail tip

adult
anatum

juvenile
anatum

tundrius

pealei

anatum

mixed stock

anatum

throughout. Wintering Peregrines may be found in large cities throughout the southern US. Recently taken off the Endangered Species List, the Peregrine is a showpiece for the ability of a species to rebound from the effects of pesticides in the environment. Most major cities in the eastern US now host Peregrine nests. However, some Peregrines have been introduced to locations where they did not formerly nest and have had a negative impact on native birds. Populations of "Tundra Peregrines" were never in real danger, but concern continues due to their wintering in countries where DDT use is still allowed. **BREEDING:** Historically nesting in a scrape atop cliffs, they have taken well to human-made structures like tall buildings, bridge beams, and towers, and even use old Osprey and Bald Eagle nests in coastal locations. **MIGRATION:** During the 1970s and early 1980s, a single Peregrine passing by a hawk-watch site was cause for celebration. A site in the FL Keys now records single days with more than 350 Peregrines passing by. Fewer numbers fly the inland ridges, often soaring in with kettles of other raptors to the delight of hawk-watchers. Prime fall dates in the East are early to mid-Oct., spring dates in late Apr. Recent surveys of offshore oil rigs in the Gulf of Mexico found Peregrines pick off exhausted migrants, with any one rig hosting up to a dozen falcons; frequent elsewhere offshore. **WINTER:** Usually found along the coast or around cities, wherever food is available.

Population Slowly increasing in N.A.

thinner dark moustache than *anatum*

juvenile *tundrius*

adult *tundrius*

PRAIRIE FALCON *Falco mexicanus*   PRFA ■ 1

The large, light-colored falcon of the arid western US, it inhabits foothills, grasslands, and other open country. While it primarily feeds on birds, especially flocking species in winter, it also takes ground squirrels when they are abundant. It takes birds on the wing. Monotypic. L 15.5–19.5" (39–50 cm) WS 35–43" (89–109 cm)

Identification A large raptor with pointed wings and a wide, square tail. When perched, wing tips fall noticeably short of tip of tail. Large head, a pale cheek, a pale supercilium, and a thin, dark moustache mark. Sexes appear similar, but females are larger. **ADULT:** Upperparts are brownish with pale barring on wing coverts and back; tail and flight feathers are lightly barred. Underparts are whitish with dark spots on sides and belly. Cere, legs, and eye ring are yellow. **JUVENILE:** Darker brown above than adults with less light barring on wing coverts and back, they appear darker at a distance. Underparts are more buffy and thinly streaked. Cere and eye ring are blue-gray. **FLIGHT:** Wing tips are not as sharply pointed as those of other large falcons, underwings feature light-colored flight feathers and outer wing coverts contrasting with dark axilliaries and longer underwing coverts, forming a dark "wingpit" that makes it look broader-winged. Females show more dark on longer underwing coverts than males.

Similar Species Peregrine Falcon shows more contrast between dark above and light below, a thicker, darker moustache, and uniformly darker underwings. Flight style is similar to Prairie, but hunts less often at low altitudes. A light-colored juvenile "Tundra Peregrine" shares the pale supercilium of Prairie Falcon, but lacks Prairie's pale cheek. A pale "Prairie Merlin" has a similar shape and coloration, but is much smaller with faster wingbeats, shows a faint moustache mark, and lacks the dark area of the underwing.

Voice A loud *kik-kik-kik* around the nest, usually silent at other times.

Status & Distribution Widespread throughout acceptable habitat from Canada to Mexico, but never abundant. **BREEDING:** Uses a small scrape on a cliff or ledge, high above the ground. Uses nests of other raptors or a raven. **MIGRATION:** Movements not well understood. Sightings well away from nesting locations begin by Aug. Prairie Falcons are noted for their postbreeding dispersal, wandering west to the Pacific and east to the Great Plains states. **WINTER:** Outside of mapped range, a few are found coastally from WA to CA and into Baja CA, Mexico. **RARE STATUS:** Records as far east as WI, IL, OH, PA, KY, and AL. The farther away from its traditional range, the greater the suspicion that an individual might be an escaped falconer's bird.

Population Declining as a breeder in some areas, especially where Peregrines are present.

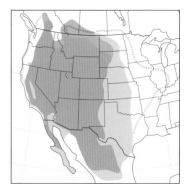

head pattern differs from Peregrine Falcon

pale brown above

adults

blackish axillaries and underwing coverts

PARAKEETS, MACAWS, AND PARROTS Family Psittacidae

Monk Parakeets (IL, Feb.)

The 167 species of parrots in this family are found throughout tropical and subtropical (rarely temperate) areas of the New World; 10 more live in Africa. As a result of escapes from the pet trade, numerous species are now at liberty (primarily in CA and FL).

Structure Psittacids have short necks and legs. Parakeets (many known as *conures* in the pet trade) and macaws are slim with long tails; parrots are chunky, with short tails. Psittacids are primarily arboreal. Their bills are short, thick, and curved, with a powerful, articulated tongue that aids in processing palm nuts and other plant food. In all species, two toes face forward and two backward. They show great size diversity, ranging from 4.7 to 39 inches (12–100 cm).

Behavior Most species are social, gathering in large mixed flocks at nighttime roosts or when foraging; some are communal breeders. Virtually all nest in cavities in palms, trees, or termite mounds; the Monk Parakeet is the sole exception. Typically noisy in flight, they often grow quiet and difficult to locate when feeding or roosting. Flocks can be identified to group fairly easily, based on shape and flight style, but multiple species are often present.

Plumage Psittacids are primarily green but often show some red, orange, or yellow on the head or in wings or tail. Sexes are usually similar.

Distribution Widespread. Greatest diversity is in South America; within US greatest exotic diversity is in southern FL and Southern CA.

Taxonomy Recent genetic studies have shown that the parrots (order Psittaciformes) share a close affinity to the Falconiformes (caracaras and falcons); both orders are now placed just before the order Passeriformes (passerine birds). Four groups of parrots are now recognized at the family level. The Kea and allies (Strigopidae), the cockatoos (Cacatuidae), the lories, lovebirds, and Australasian parrots (Psittaculidae, see p. 430), and this family, the Psitticidae, with 167 species in 36 genera. All psitticids are New World residents, apart from 10 African species: nine in the genus *Poicephalus* and one, the well-known Gray Parrot, in the genus *Psittacus*.

Conservation Capture for the pet trade during the late 1960s to early 1990s was vast. Unregulated capture and the massive degree of habitat destruction have endangered dozens of species. All psittacids in the US currently are exotic; our one native species, the Carolina Parakeet, went extinct a century ago. BirdLife International codes: 24 NT, 37 VU, 25 EN, 5 CR, 2 CR (PE), and 5 EX.

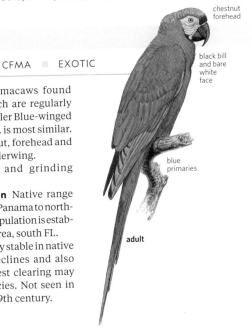

chestnut forehead

black bill and bare white face

blue primaries

adult

| Genus *Ara* |

CHESTNUT-FRONTED MACAW *Ara severus*   CFMA ■ EXOTIC

This small macaw, a native primarily from S.A., prefers openings in forest and forest edges. Monotypic. L 18" (46 cm)

Identification Overall green with chestnut forehead, lower auricular and chin; bare white face and black bill. There is extensive blue in the wings and tail and red on the leading edge of the wing. Some chestnut at base of tarsus feathering. Eye pale orange, legs blackish. Sexes and ages similar; juvenile may have dark eye.

FLIGHT: Extensive blue on primaries, inner secondaries, and tail; base of tail and most of underwing red.

Similar Species Several other similarly colored small macaws found in S.A., none of which are regularly seen in N.A. The smaller Blue-winged Macaw of eastern S.A. is most similar. It has red, not chestnut, forehead and mainly yellowish underwing.

Voice Gives harsh and grinding screeches.

Status & Distribution Native range extends from eastern Panama to northern Bolivia. A small population is established in the Miami area, south FL.

Population Generally stable in native range with some declines and also increases. Some forest clearing may actually benefit species. Not seen in Guyana since early 19th century.

NEW WORLD PARAKEETS Genera *Myiopsitta, Conuropsis, Thectocercus, Psittacara, Aratinga, Rhynchopsitta,* and *Brotogeris*

This diverse group of about 28 species ranges from northern Mexico to the southern tip of S.A. The Carolina Parakeet, sole member of the genus *Conuropsis*, is extinct; Thick-billed Parrot (genus *Rhynchopsitta*) was formerly a sporadic visitor to southeastern AZ and likely southwestern NM. Some species with allopatric native ranges occur sympatrically in CA and FL (e.g., the Red-masked and Mitred Parakeets) and may have hybridized. Other species not included in this guide may be encountered and likely represent recently escaped cage birds.

MONK PARAKEET *Myiopsitta monachus* MOPA ■ 2

This most numerous psittacid in N.A., the Monk Parakeet survives New England and IL winters by eating birdseed and roosting in its bulky stick

nest. Polytypic (4 sspp.). L 11.5" (29 cm) **Identification** Green, with forehead, throat, and breast grayish. Bill pinkish orange. Remiges and lesser primary coverts blue, remaining coverts green. Forehead barely tinged with green on immature.
Similar Species Unmistakable.
Voice Highly vocal, a variety of loud, harsh notes or quieter chattering.
Status & Distribution Exotic in US and many other countries. Native to lowlands of southern S.A. east of the Andes. **YEAR-ROUND:** Nonmigratory. Locally common in FL; smaller numbers in other states including IL,

CT, NY, and NJ; also OR, TX, and southernmost CA (Calexico); also Tijuana.
Population Expanding; many N.A. populations doubling every five to six years. Eradicated in some states (e.g., CA, GA); expanding in native S.A. range.

gray forehead

bluish remiges

gray throat and breast

CAROLINA PARAKEET *Conuropsis carolinensis* CAPA ■ 6

N.A.'s only native breeding psittacid became extinct about a century ago (in wild by 1904). Its stronghold was the great river basins of the Midwest and Deep South, where it favored cottonwoods, cypress swamps, and old-growth bottomlands and fed on a variety of fruits and seeds. Exact reasons for the bird's demise are not well understood. Polytypic (2 sspp.; nominate east of the Appalachians and *ludoviciana* to the west). L 13.5" (34 cm) **Identification** Sexes similar. **ADULT:**

Green body with yellow head and reddish orange face. Bill and orbital ring pale. Yellow and/or orange patches on shoulders, "thighs," and vent. Green coverts and long green tail, yellower below. **IMMATURE:** Entirely green except for orangish patch on forehead. **Similar Species** None in former range, early 20th-century reports from FL or GA possibly misidentified exotics.
Voice Vocal, especially in flight. **CALL:** Loud and harsh *qui* or *qui-i-i-i.*
Status & Distribution Formerly a

reddish orange face, remainder of head yellow

adult

locally common resident throughout much of eastern US—recorded north to NY, MI, and MN, and west to central CO.
Population Extinct. Last individual, "Incas," died on 21 Feb. 1918, at the Cincinnati Zoo.

BLUE-CROWNED PARAKEET *Thectocercus acuticaudatus* BCPA ■ EXOTIC

This parakeet is identified by its blue head and bicolored bill. Polytypic (5 sspp.; nominate in N.A.). L 14" (36 cm) **Identification ADULT:** Green. Most of head dull blue, which may appear green from a distance. Prominent white orbital ring, eyes orange, upper mandible pinkish orange, lower mandible dark. Flight feathers dull yellow-green. Reddish inner webs and dull yellow outer webs of tail feathers visible from below. **JUVENILE:** Blue on head restricted to forehead and forecrown.

Similar Species Other exotic parakeets found in N.A. do not have blue heads.
Voice A loud *cheeah-cheeah*, often repeated.
Status & Distribution Exotic in the US. Widely distributed and common in three separate regions of S.A. (north, east, and south-central). **YEAR-ROUND:** Nonmigratory. In US restricted to CA and FL; ±100 in CA at San Francisco, Los Angeles, and San Diego; ±125 in FL, at Ft. Lauderdale, Upper Keys, and St. Petersburg.
Population Increasing in CA and FL.

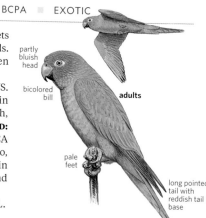

partly bluish head

bicolored bill

adults

pale feet

long pointed tail with reddish tail base

GREEN PARAKEET   *Psittacara holochlorus*   GREP ■ 2

This parakeet is established in the US in south TX, where colonies present in several cities along the Rio Grande. The subspecies *brevipes* on Socorro I., with different vocalizations, is treated as a different species by most. Polytypic (4 sspp.; nominate in south TX). L 13" (33 cm)

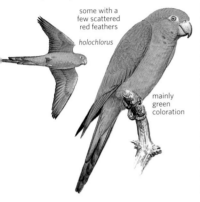

some with a few scattered red feathers

holochlorus

mainly green coloration

Identification Green overall with long tail. Sexes are similar. **ADULT:** Head often with scattered orange feathers. White or beige orbital ring; eyes orange, bill beige, legs and feet gray. May have scattered orange feathers on breast. Undersurface of flight feathers pale yellow; coverts yellow-green. **JUVENILE:** Similar to adult; eyes brown.
Similar Species Juveniles of other *Psittacara* such as White-eyed or Crimson-fronted Parakeet (*P. finschi*, not illustrated) may be encountered in FL. These have entirely green bodies but at least scattered red, orange, or yellow feathers on the underwing coverts.
Voice Various chattering calls. In flight, harsh screeches.
Status & Distribution Exotic in the US. Native to Mexico and C.A. **YEAR-ROUND:** Nonmigratory, but some movement of native birds in response to food supply. In US restricted to FL and TX; ±2,000 in TX, along the Lower Rio Grande Valley. Some also in FL at Ft. Lauderdale and Miami, where juveniles of other *Psittacara* species complicate identification.
Population Overall declining in native range, especially *brewsteri* of northwest Mexico. Increasing in TX and probably also in FL.

MITRED PARAKEET   *Psittacara mitratus*   MIPA ■ EXOTIC

Mitred Parakeet and the less numerous Red-masked Parakeet are found in CA and FL. Mitred Parakeets in FL roost—and apparently breed—in cavities and chimneys. Polytypic (2 sspp.; nominate in US). L 15" (38 cm)
Identification Large and mostly green; sexes similar. **ADULT:** Head has variable amount of red spotting. Wide, creamy white orbital ring; orange eyes; pale bill; pink legs and feet. Red spotting often on shoulders. Flight feathers yellowish below, green above—like other *Psittacara*. Underparts often with random red breast feathers. **JUVENILE:** Less red on cheeks and head; eyes brown.

Similar Species Smaller Red-masked Parakeet usually has more extensive, solid red hood, red on leading edge of wing, and gray legs and feet. Juvenile Mitred resembles other *Psittacara* not included here.
Voice Strident *scree-ah* and other shrieking calls.
Status & Distribution Exotic in the US. Native to Andes of Peru, Bolivia, and Argentina, where common. **YEAR-ROUND:** Nonmigratory. In US restricted to CA and FL.
Population Populations stable or increasing in CA (over 1,000, mostly in the Los Angeles area) and FL (500–1,000 in the Miami and Ft. Lauderdale areas); a few in New York City. Some local declines in native range; prior to 1990, exported in great numbers.

limited red on head and leading edge of wing

scattered red head feathers

adults
mitratus

long, pointed tail

RED-MASKED PARAKEET   *Psittacara erythrogenys*   RMPA ■ EXOTIC

Unlike the similar Mitred Parakeet, this species nests singly in natural cavities in oaks and palms. It flocks and roosts with other *Psittacara* species. Monotypic. L 13" (33 cm)
Identification A medium-size parakeet with a red hood and shoulders and a green body. Sexes similar. **ADULT:** Extent of bright red hood variable, may end near mid-bill or extend to throat. Wide white orbital ring; orange eyes; pale bill; gray legs and feet. Wings show red shoulders. Underwings are yellowish with red on lesser coverts.

Inner "thighs" red, but inconspicuous. **JUVENILE:** Red on head and underwing coverts much reduced; "thighs" green.
Similar Species See Mitred Parakeet.
VOICE: Strident s*cree-ah* or *skreet* calls, higher pitched than Mitred Parakeet.
Status & Distribution Exotic in the US. Native to western Ecuador and northwestern Peru **YEAR-ROUND:** Nonmigratory. In US restricted to CA, ±500 (San Francisco, Los Angeles, and San Diego areas), and FL, perhaps ±200 (Miami and Ft. Lauderdale).
Population Near Threatened.

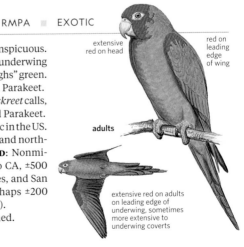

extensive red on head

red on leading edge of wing

adults

extensive red on adults on leading edge of underwing, sometimes more extensive to underwing coverts

WHITE-EYED PARAKEET *Psittacara leucophthalma* WEPA ■ EXOTIC

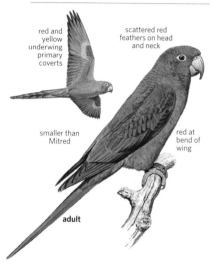

red and yellow underwing primary coverts

scattered red feathers on head and neck

smaller than Mitred

red at bend of wing

adult

Recently moved out of the genus *Aratinga* and placed in the genus *Psittacara*. Polytypic (2–4 sspp.). L 12.5" (32 cm)

Identification A medium-size parakeet, often seen in flocks. **ADULT:** Plumage mostly green with a white orbital ring and variable red flecking on head. Iris orange. **JUVENILE:** Tail shorter than adult, red flecking in head reduced or absent, underwing pattern muted. Iris brown. **FLIGHT:** Lesser and median primary coverts are red, and greater coverts are yellow in underwing.

Geographic Variation Most likely the nominate subspecies is found in N.A., but this has not been determined.

Similar Species Best told from other *Psittacara* parakeets, particularly Green, Mitred, and Red-masked, by distinct underwing pattern.

Voice A high, screechy chatter, similar to other *Psittacara* parakeets.

Status & Distribution The most widespread *Psittacara*; common over much of northern and central S.A. east of the Andes and south to northern Argentina; recently adapted to urban environments, where it nests in cavities in buildings and under eaves. In FL, where the species has been introduced, it is found in 12 southern counties and has become fairly common in the greater Miami area.

Population Stable.

NANDAY PARAKEET *Aratinga nenday* NAPA ■ 2

This large parakeet is one of the most successful psittacids in N.A. It has bred in sycamore woodlands in Southern CA, but prefers palm snags and telephone poles in urban areas in FL. Formerly known as Black-hooded Parakeet; recently placed in the genus *Aratinga* by S.A. researchers. Monotypic. L 13.8" (35 cm)

Identification Sexes similar. **ADULT:** Prominent blackish hood with an inconspicuous red or brown border on the hind crown; black bill and dark eyes. Light green body with powder blue breast patch. Bright red "thighs" distinctive. Lower half of uppertail bluish, undersurface of tail blackish. Orbital ring gray and inconspicuous. **IMMATURE:** Blue breast patch smaller; red "thighs" paler. **FLIGHT:** Primaries and secondaries dark blue above and blackish below, contrast with yellow-green wing linings.

Similar Species None in N.A.

Voice A harsh *kee-ah*, often doubled, or *chree, chree, chree*. Vocal, especially in flight.

Status & Distribution Exotic in the US. Native to interior of central S.A., where common. **YEAR-ROUND:** Nonmigratory. In US restricted to

black head

black remiges from below

black bill

bluish chest, reduced on immature

red thighs

blue-tinged primaries

lon poi blu tail

FL, where considered established (±1,000 individuals). Also found in Southern CA (±200 birds), mainly along the coast of southern Ventura Co. and Los Angeles Co., Huntington Beach, and the San Gabriel Valley.

Population Increasing in FL and CA.

THICK-BILLED PARROT *Rhynchopsitta pachyrhyncha* TBPA ■ 6

This majestic endangered parrot wandered infrequently from its western Mexican montane haunts to the southwestern US until about a century ago. The introduction of a flock into the Chiricahua Mts. of southeastern AZ in the 1980s was unsuccessful. Monotypic. L 16.2" (41 cm)

Identification A large green psittacid with a long, pointed tail and large, black bill. Sexes similar. **ADULT:** Red forehead extending in a broad line over the eyes; red shoulders. Underparts wholly green with red "thighs."

Orange eyes and dull yellow orbital ring. **IMMATURE:** Bill dusky; red on head limited to forehead; no red on wings. **FLIGHT:** Pattern unmistakable from below. A prominent yellow stripe (greater coverts) is set off by dark green wing linings and blackish flight feathers. Underside of tail also blackish; leading edge of the wing is red.

Similar Species No other psittacid within former N.A. range. Distinguished from *Amazona* by dark bill and long, pointed tail. Paler-billed

immature similar to smaller, white-billed Mitred or Red-masked Parakeet but orbital ring gray and inconspicuous. The Vulnerable congener, the Maroon-fronted Parrot (*R. terrisi*) of the Sierra Madre Oriental in northeast Mexico, is similar but head markings are maroon, not red, and it lacks the yellow underwing stripe. It nests in cavities on cliffs.
Voice Screeches, screams, and squawks.
Status & Distribution Uncommon to rare in old-growth pine forests in the Sierra Madre Occidental of northwestern Mexico north to northwest Chihuahua. Northern populations are migratory. Wanders in response to pine crop success. **RARE STATUS:** Formerly, irregular visitant (mainly in winter) north to southeast AZ where

recorded all major ranges north to the Pinaleno and Galiuro Mts. Most records were from the Chiricahua Mts. When present spent most of it time in pine stands, chiefly ponderosa pines. Most of what we know was recorded by ranchers in 19th and early 20th centuries; most recent invasions were in 1904 and 1917–18; sporadic records into the 1920s and 1930s. The last reliable report was in 1938, rumors until 1945. Sightings from Animas Mts., NM, until 1917. Record from near Truth or Consequences, NM (2003), was not accepted as a wild bird.
Population Endangered; still declining due to destruction of mature forest habitat in the high Sierra Madre Occidental of western Mexico.

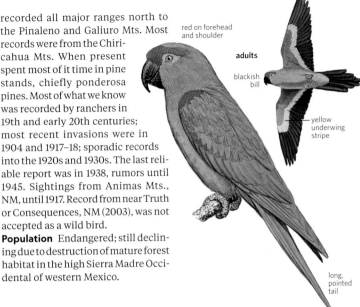

red on forehead and shoulder

adults

blackish bill

yellow underwing stripe

long, pointed tail

WHITE-WINGED PARAKEET *Brotogeris versicolurus* WWPA ▪ 2

Brotogeris parakeets are significantly smaller than *Psittacara* parakeets. They are mostly green with moderately long, pointed tails. White-winged and Yellow-chevroned Parakeets formerly

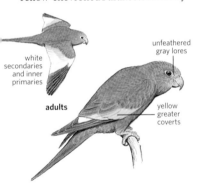

white secondaries and inner primaries

adults

unfeathered gray lores

yellow greater coverts

were considered conspecific under the name Canary-winged Parakeet. Monotypic. L 8.8" (22 cm)
Identification Sexes and ages similar. Perched birds often show little or no white in wings—only yellow of the greater coverts. An inconspicuous gray orbital ring merges with unfeathered gray lores. **FLIGHT:** Large white areas—formed by the white secondaries and inner primaries—conspicuous in flight. Outer primaries bluish.
Similar Species Yellow-chevroned Parakeet very similar. Best field mark on a perched White-winged is the unfeathered gray lores and orbital ring. In flight, white secondaries diagnostic, but note that hybrids with

Yellow-chevroned apparently occur at San Francisco and Ft. Lauderdale.
Voice Perched birds give various chattering calls. Flight call of *chree* or *chree-chree* richer and slightly lower-pitched than that of the Yellow-chevroned.
Status & Distribution Exotic in the US. Native to the northern Amazon Basin of northern S.A. (south to east-central Peru), where common. **YEAR-ROUND:** Nonmigratory. In US restricted to CA, ±50 at San Francisco and Los Angeles, and to FL, ±200 mainly in Ft. Lauderdale, smaller numbers at Miami.
Population US populations have declined considerably since the 1970s. Still common in S.A.

YELLOW-CHEVRONED PARAKEET *Brotogeris chiriri* YCPA ▪ 2

The Yellow-chevroned was formerly considered conspecific with the White-winged and known as the Canary-winged Parakeet (see account above). In FL both species nest in living date palms by burrowing into the insect debris surrounding the trunks. Polytypic (2 sspp.; apparently nominate only in N.A.). L 8.8" (22 cm)
Identification Sexes and ages similar. Perched birds show a yellow wing bar—similar to White-winged. Body and tail are entirely light green. Green lores are fully feathered, and there is a narrow white orbital ring. Dark eyes; pinkish yellow bill; pink legs and feet. **FLIGHT:** Yellow wing bar prominent in flight.

Primaries and secondaries are green.
Similar Species White-winged Parakeet is very similar. Fully feathered lores of Yellow-chevroned Parakeet best field mark when perched. In flight, shows no white in wings.
Voice Calls similar to White-winged Parakeet but higher and scratchier.
Status & Distribution Exotic in the US. Native to central S.A. from southern Amazon to northern Argentina, where common. **YEAR-ROUND:** Nonmigratory. In US restricted to CA, about 1,000 primarily in the greater Los Angeles area (established) with a few escapes in San Francisco (where hybrids with White-winged reported);

and FL, several hundred in the greater Miami area (Ft. Lauderdale to Homestead); hybrids also reported.
Population Increasing in US.

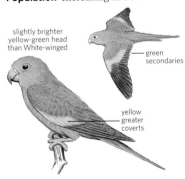

slightly brighter yellow-green head than White-winged

green secondaries

yellow greater coverts

AMAZONS Genus *Amazona*

Often called *amazons*, not *parrots*, the 30 species in this genus are large and chunky with short tails and are found from northern Mexico south through M.A. and S.A. to northern Argentina. Nine species are endemic to the West Indies.

RED-CROWNED PARROT *Amazona viridigenalis*   RCPA ■ 2

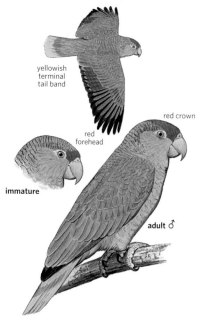

yellowish terminal tail band

red crown

red forehead

immature

adult ♂

The Red-crowned is the most widespread and common parrot in N.A. It roosts in large, noisy flocks with other species and nests singly in cavities in palms or other trees. Monotypic. L 13" (33 cm)

Identification Chunky green psittacid with variable amount of red on forehead and crown. Pale orbital ring, bill, and cere; yellow eyes. **ADULT MALE:** Extensive red forehead and crown; blue hind crown and nape. **ADULT FEMALE & IMMATURE:** Red on head restricted to forehead; crown bluish. **FLIGHT:** Upperwings have a red patch on secondaries and a dark blue trailing edge; yellowish band across tip of tail.

Similar Species Similar to Lilac-crowned, which has a burgundy forehead and purplish blue crown and nape, extending farther onto the auriculars. Longer tail of Lilac-crowned Parrot noticeable in flight.

Voice Loud grating and cawing calls, including a distinctive rolling, descending whistle.

Status & Distribution Exotic in the US. Native to northeastern Mexico, where endangered; ±3,000–6,500 remaining. **YEAR-ROUND:** Nonmigratory, but may wander in response to food availability. In US restricted to CA (3,000–5,000), southeast FL (±400), and southern TX (hundreds). Treated as established in TX and CA.

Population Endangered in native range in northeast Mexico. Increasing in CA.

LILAC-CROWNED PARROT *Amazona finschi*   LCPA ■ EXOTIC

lilac crown and nape

maroon forehead

green central tail feathers

longer tail than Red-crowned

A close relative of the Red-crowned Parrot, with which it flocks, roosts, and sometimes hybridizes in CA. Monotypic. L 13" (33 cm)

Identification ADULT: Sexes similar. Body green with thin dark scaling below, head green with burgundy forehead and purplish blue crown and nape that curves downward behind auriculars. Gray orbital ring, red-orange eyes, dark cere, and pale bill. **IMMATURE:** Eyes brown. **FLIGHT:** Upperwings green with blackish blue primaries and secondary tips; red outer secondaries. Underwings green. Yellowish band across tip of long tail.

Similar Species Very similar to Red-crowned Parrot. The primary differences are the lilac crown and nape versus the red crown (adult male) and bluish nape. Note also the dark cere and reddish eyes of Lilac-crowned Parrot. In flight, its longer tail is apparent.

Voice Various loud grating and cawing calls indistinguishable from those of Red-crowned Parrot, but also utters, especially in flight, a distinctive, squeaky, ascending whistled *ker-leek?*

Status & Distribution Exotic in the US. Native to western Mexico from southern Sonora and southwest Chihuahua to Oaxaca. **YEAR-ROUND:** Nonmigratory. In US restricted to Southern CA, where ±500 occur from Ventura south to San Diego Co., nesting in cavities in trees and telephone poles. Escapes (fewer than 200) seen in south TX and FL.

Population Endangered in native range in Mexico. Increasing in CA.

RED-LORED PARROT *Amazona autumnalis*   RLPA ■ EXOTIC

Also called Yellow-cheeked Parrot. Established for many years in several southern US states but perhaps overlooked, Red-lored Parrots have adapted readily to urban and suburban areas, where exotic plantings provide abundant food sources. Polytypic (4 sspp.; nominate presumably in N.A.). L 13" (33 cm)

Identification Nominate *autumna-* *lis* described. Bright green body with red frontlet and bluish crown. **ADULT:** Yellow below eye extends to auriculars. **JUVENILE:** Similar to adult, but little or no yellow in face, less blue in crown.

FLIGHT: Note yellow tips to tail feathers, dark blue wing tips, and brilliant red outer secondaries.

Geographic Variation Populations in the US are thought to be of the nominate subspecies, which has extensive yellow on face and whose range extends on the Atlantic slope from Mexico to northeast Nicaragua. Subspecies *salvini* ranges from Nicaragua to southwestern Colombia and northwestern Venezuela; *lilacina* is found in western Ecuador; and allopatric *diadema* is found only in northwestern Brazil along the lower Río Negro and adjacent northern bank of the Ama-

zon. It is possible that *salvini* (incorporating *lilacina*) and *diadema* could be recognized as distinct species.

Similar Species Adult Red-crowned Parrot has yellowish bill, red nape and crown, no yellow below the eye, and has blue limited to postocular area. Juvenile Red-crowned most similar to larger Red-lored but has yellowish bill and lacks blue in crown.

Voice Abrupt, very loud *kyeik*, *ack*, *eck*, *yoik* calls, similar to other *Amazona* parrots but more shrill than smaller Red-crowned.

Status & Distribution Fairly common and widespread in wooded lowlands

yellow on cheek

adult
autumnalis

from southern Tamaulipas, Mexico, to northern S.A. Present since the 1980s in southernmost TX, and smaller numbers have been present in southern FL and Southern CA for several decades.

Population Capture for the pet trade and habitat loss have caused some declines in its native range.

YELLOW-HEADED PARROT · *Amazona oratrix* · YHPA ▪ EXOTIC

This species native to the tropics is sometimes considered conspecific with the Yellow-naped (*A. auropalliata*) and Yellow-crowned (*A. ochrocephala*) Parrots under the combined name of Yellow-crowned Parrot. Polytypic (3 sspp.; *oratrix* in N.A.). L 14.5" (37 cm).

Identification Large size. **ADULT:** Head mostly or entirely yellow, depending on subspecies. White orbital ring, eyes orange, bill pale. Red and yellow markings on shoulder; yellow on "thighs." **IMMATURE:** Head mostly green with yellow crown and face, green wings; dark eyes. **FLIGHT:** Wings green with bluish tips to remiges. Red patch on outer secondaries. Underwings green. Tail green with yellow outer rectrices.

Geographic Variation Four subspecies: *oratrix* (head yellow, but nape and breast green) from southwest Mexico and very similar *magna* of east

Mexico north to eastern Nuevo León and central Tamaulipa presumably the subspecies in N.A.; *belizensis* from Belize to northwest Honduras (green nape and breast); *tresmariae* from Islas Marías (yellow head and breast, greater amount of yellow at the shoulders). Yellow-naped breeds in very small numbers in southeastern FL; Yellow-crowned is seen there occasionally.

Similar Species Pale-billed adults are unmistakable. Immature similar to adult Yellow-crowned Parrot (not illustrated) but has dark eyes.

Voice A variety of shrieks, squawks, and whistles; excellent mimic.

Status & Distribution Exotic in the US; very small numbers found in CA, TX, and FL.

Population Endangered in native range due to habitat destruction and capture for the pet trade. Populations declining in US.

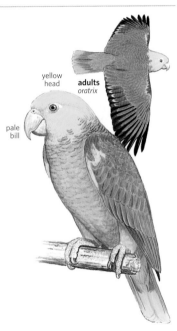

yellow head

adults
oratrix

pale bill

ORANGE-WINGED PARROT · *Amazona amazonica* · OWPA ▪ EXOTIC

The Orange-winged Parrot is one of several *Amazona* species breeding

bluish on head and nape

orange base to outer tail feathers

in small numbers in FL. It feeds on native and exotic fruits, nuts, and flowers. Monotypic. L 12.3" (31 cm)

Identification A small, green *Amazona*. Sexes similar. **ADULT:** Yellow face and crown divided by azure stripe above and through each eye, extending on to nape. Orange eyes with narrow, purplish orbital ring; pale bill with blackish edges; gray legs and feet. **IMMATURE:** Brown eyes. **FLIGHT:** Upperwings have a small orange-red patch on outer secondaries and a dark blue trailing edge. Underwings green. Tail has a yellowish band with orange stripes on outer tail feathers.

Similar Species None; yellow and azure head pattern distinctive.

Voice CALL: A shrill *kee-ik*, *kee-ik*, as well as various squawks, and whistled notes.

Status & Distribution Exotic in the US. Native to lowlands of northern S.A. east of the Andes from Colombia to southeastern Brazil, where common to abundant. **YEAR-ROUND:** Nonmigratory. In US, restricted to FL, where ±200 are found at Ft. Lauderdale and Miami; nests have been found in royal palm snags.

Population Stable or increasing in FL. Native birds are heavily trapped for the pet trade and shot for sport.

LORIES, LOVEBIRDS, AND AUSTRALASIAN PARROTS Family Psittaculidae

Peach-faced Lovebird (AZ, Apr.)

Behavior Highly varied with many adaptations. The nocturnal Night Parrot from the interior of Australia is little known but is believed to be critically endangered, while the Ground Parrot (also Australia, including Tasmania) is a skulker that is best located only by tracking down its calls at dawn and dusk during the calling period. The lories and lorikeets of Australasia to Polynesia have special structural adaptations that enable them to drink nectar.

Plumage Many species are green like New World Parrots (Psittacidae); many others have plumage with multiple colors. The lorikeets and rosellas are particularly colorful.

T his large Old World family of 45 genera and 184 species contains many colorful species. It has recently been split from the New World Parrots based on genetic distinctions. Along with New World Parrots they represent some of the most popular cage birds, and hundreds of thousands have been trapped for the pet trade. The small Australian Budgerigar is the most frequently kept species in captivity.

Structure Psittaculids vary greatly in size and structure, from the tiny pygmy parrots (six species in *Micropsitta*), as small as 3.1 in. through the slightly larger fig parrots (five species), lovebirds, and hanging-parrots and to the largest Greater Vasa Parrot of Madagascar at 19.7 in. and the Alexandrine Parakeet at 22.8 in. Some are square tailed; others (e.g., *Psittacula*) have long pointed tails.

Distribution Widespread from Africa (mostly central and southern) across southern Asia, and Indonesia to Australia and island groups in the South Pacific to Polynesia. New Guinea and Australia have the largest number of species.

Taxonomy Genetic studies show a strong divergence between the primarily New World Psittacidae and the Psittaculidae, which includes 184 Old World species in 45 genera. Psittaculids are most diverse in Australia and Southeast Asia; two groups—the Asian *Psittacula* parakeets and the African *Agapornis* lovebirds—are represented by naturalized populations in North America, along with the now extirpated Budgerigar

Conservation BirdLife International codes: 33 NT, 19 VU, 9 EN, 8 CR, and 10 EX.

Genus *Agapornis*

ROSY-FACED LOVEBIRD *Agapornis roseicollis*   RFLO ▪ 2

Also known as the Peach-faced Lovebird, this native of southwest Africa is a colonial nester. Polytypic (2 sspp., probably nominate in N.A.). L 6.2" (16 cm)

Identification Tiny, short-tailed parrot with pale bill. **ADULT:** Pale green overall with blue uppertail coverts. Rose-pink face and upper breast, reddish forehead; female duller. **JUVENILE:** Forecrown green, pink of face and throat pale buff-pink, base of upper mandible black. **FLIGHT:** Tiny size and erratic, twisting flight.

Similar Species Among the eight other species in the genus, Fischer's Lovebird (*A. fischeri*) and Yellow-collared Lovebird (*A. personatus*) and hybrids are sometimes encountered. These are escaped cage birds with no established populations.

Voice Calls include high-pitched, short *tweet!* and *squeer!* notes; also lower, throatier trills.

Status & Distribution First noted in 1987, several thousand now well-established around eastern portion of greater Phoenix area, AZ, where they have been found nesting under roof tiles and in fan palms and saguaros. Small numbers are also seen in FL and occasionally CA.

Population Severe declines in Angola due to trapping for the cage-bird trade.

red band across forehead

pink face and upper breast

adult ♂

also known as Peach-faced Lovebird

Genus *Melopsittacus*

This genus consists of a single species, the Budgerigar. One of the world's most abundant psittacids, it numbers perhaps five million individuals. These birds breed communally in cavities; formerly in FL in nest boxes, where it raised multiple broods annually. Well-established west FL populations of thousands now extirpated.

BUDGERIGAR *Melopsittacus undulatus*   BUDG ▪ EXOTIC

The Budgerigar is perhaps the most popular cage bird in the world. In captivity (bred since the mid-1800s), a wide variety of artificial color morphs exist, including white, yellow, or blue plumages. In native Australia (and former FL populations) birds are green. It is found in grasslands where it feeds on ground. Monotypic. L 7" (18 cm)
Identification This is a tiny parakeet. Sexes are similar. **ADULT:** Head is mostly yellow above and green below with black barring on auriculars, hind crown, and nape. Two black spots on each side of throat, with a small purplish patch on malar. Eyes are yellow; bill, legs, and feet grayish. Back and wing coverts are yellow, barred with black. White or yellow wing stripe in flight. Rump green. Tail equal to length of body, with blue central rectrices.

JUVENILE: Forehead barred, throat unspotted, eyes dark.
Similar Species None; Budgerigar is much smaller than any other psittacid found in N.A.
Voice A series of pleasant high-pitched chittering or chirping; some notes reminiscent of House Sparrows.
Status & Distribution Exotic in US. Common to abundant, but rather nomadic, in native interior Australia. **YEAR-ROUND:** Nonmigratory. In US was restricted to FL, where a breeding population along the central Gulf Coast since the early 1960s expanded to about 20,000 before crashing, the last ones recorded in Apr. 2014 around Hernando Beach and Bayonet Pt. Nesting competition with House Sparrows thought to be the primary cause of the decline and extirpation. Escapes of var-

ious color morphs possible anywhere.
Population Stable in extensive native Australian range.

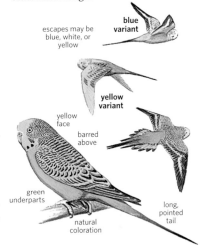

escapes may be blue, white, or yellow

blue variant

yellow variant

yellow face

barred above

green underparts

long, pointed tail

natural coloration

Genus *Psittacula*

ROSE-RINGED PARAKEET *Psittacula krameri*   RRPA ▪ EXOTIC

The Old World Rose-ringed Parakeet has a large native range across central Africa, south of the Sahara, and a separate population on the Indian Subcontinent. It is established in a few restricted areas of FL and particularly CA. The species is also called the Ring-necked Parakeet in the Old World. Polytypic (4 sspp.; 1–2 in N.A.). L 15.8" (40 cm)
Identification Overall, Rose-ringed Parakeet is yellowish green. It has a slender tail with very long central feathers. A narrow, red orbital surrounds pale yellow eye. Upper mandible is entirely bright red, but lower mandible is mostly or completely black. Remiges are darker green, contrasting with yellowish green coverts. **ADULT MALE:** Chin and throat are black. Black extends backward and encircles the head; back end of collar is tinged with rose pink. Nape is pale azure. **ADULT FEMALE & JUVENILE:** Chin and throat are yellow-green, collar is lacking, and nape is green.
Geographic Variation The introduced populations in N.A. are of subspecies *manillensis* and possibly

also *borealis*, both native to the Indian Subcontinent. The two African subspecies (*krameri* and *parvirostris*) have dark-tipped upper mandibles.
Similar Species Green Parakeets and immatures of other birds of the genus *Aratinga* may be mostly or entirely green, but these birds lack the red bill and extremely long central tail feathers of Rose-ringed Parakeet. Collar of the adult male Rose-ringed is diagnostic.
Voice Highly vocal at roosts. **CALL:** A loud flickerlike *kew* is common, along with various high, shrill notes uncharacteristic of most psittacids.
Status & Distribution The most widespread Old World psittacid. Common to abundant in its native range. **YEAR-ROUND:** Nonmigratory. In the US, a population at Bakersfield, CA, numbers about 1,000 birds, with fewer than 100 in the Los Angeles area. Fewer than 50 in FL, at Ft. Myers and Naples.
Population Stable in large Old World range. In US increasing in CA; FL trends unknown, but some populations now extirpated.

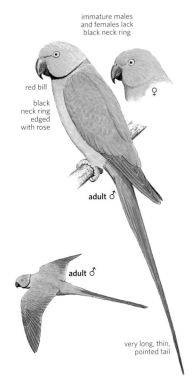

immature males and females lack black neck ring

red bill

black neck ring edged with rose

♀

adult ♂

adult ♂

very long, thin, pointed tail

BECARDS, TITYRAS, AND ALLIES Family Tityridae

This assemblage of 35 New World tropical and subtropical fruit- and insect-eating species has recently been given family rank now that the genera comprising it have been confirmed through molecular analysis to be each other's closest relatives.

Structure There are few outward characters that unite this family. They are small (purpletufts) to medium-size (tityras) passerines with short to medium somewhat stout bills that are slightly to strongly hooked at the tip. Becards have large, puffy heads and square to graduated tails; the tityras are especially strong billed and have relatively short tails.

Masked Tityra, female (Costa Rica, Dec.)

Behavior Becards and tityras are generally found singly or in pairs, feeding mainly on insects and fruit. Tityras are sluggish, perching openly high in the canopy; they sally for insects and hover for fruit. Becards feed sluggishly at mid-story and canopy levels, with some species joining mixed-species flocks. Songs range from musical whistles in many becards and schiffornis to grunting notes in the tityras. Nests vary greatly, from cavities (tityras) to cup nests (schiffornis, purpletufts) to large, globular structures with side entrances (becards).

Plumage Variable, from solidly black or rufous to strongly patterned. Most are sexually dimorphic; immature males resemble females. Tityras are striking pale gray, black, and white birds; two of the three species show bare red facial skin and bill bases.

Distribution A Neotropical family. Seven species range north to Mexico, and the northernmost species (Rose-throated Becard) barely reaches the US as a breeding bird. Two other species that reach northern Mexico have been recorded as strays near the US border. The family is most diverse in tropical South America. All are birds of woodlands and river edges.

Taxonomy It has long been known that the becards, tityras, and their allies were part of the vast radiation of tyrant flycatchers and other suboscine passerine families that are exclusively New World in distribution, but their exact placement has long been debated. Genera now placed in this family were until recently distributed among three families—cotingas, manakins, and tyrant-flycatchers. The family now comprises 35 species in seven genera. The becards, all but one in the genus *Pachyramphus*, make up about half the family, with the remainder including four tityras, two mourners (*Laniocera*), up to seven very similar species in the genus *Schiffornis* (formerly considered manakins), a cotinga-like bird in the genus *Laniisoma*, and three purpletufts (*Iodopleura*).

Conservation Populations of most species are stable. BirdLife International codes: 2 NT and 2 EN.

TITYRAS Genus *Tityra*

These four cavity-nesting species of the Neotropics are large, chunky, and boldly patterned; they have relatively short tails, heavy hooked bills, and bare reddish facial skin. There is some sexual dimorphism. Tityras are usually observed on exposed perches or in fruiting trees in the canopy of lowland forest.

MASKED TITYRA *Tityra semifasciata* MATI ■ 5

Common to our south, this unmistakable species has been found only once in the US. Polytypic (8 sspp.; likely *personata* in N.A.). L 9" (23 cm)
Identification MALE: Generally pale gray above (including inner secondaries, tertials, and upperwing coverts). Whitish gray below, including underwing coverts. Starkly contrasting black on face, most of wings, and thick subterminal tail band. Bare-skinned facial spectacles and base of heavy bill are pinkish red; tip of bill is black; eye is red. **FEMALE:** Darker and browner on head and back; darker gray tail contrasts less with black subterminal band; lacks black face. **JUVENILE:**

Like female but with subtle darker streaks on crown and back, and whitish edges to inner secondaries, tertials, and upperwing coverts.
Similar Species No other passerine with similar plumage pattern.
Voice CALL: A double, nasal grunt, *zzzr zzzrt*, given when perched or in flight; reminiscent of Dickcissel.
Status & Distribution Common. **YEAR-ROUND:** Lowland tropical forest and woodland from northern Mexico (southern Sonora and southern Tamaulipas) to S.A. **RARE STATUS:** Accidental, one record, Bentsen-Rio Grande Valley SP, TX (17 Feb.–10 Mar. 1990).
Population Stable.

pinkish red bare skin on face and pinkish red bill base

white with mostly black face

adult ♂
personata

BECARDS Genus *Pachyramphus*

Becards comprise 16 species of small- to medium-size birds; they are chunky, slightly crested, and disproportionately large headed and short tailed. Typically found in forest or woodland canopy, they are sluggish, almost vireo-like, and perch with an upright posture and sally short distances with fluttery flight for fruit or insects.

GRAY-COLLARED BECARD   *Pachyramphus major*   GCBE ◾ 5

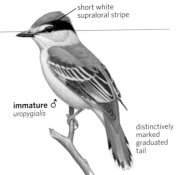

short white supraloral stripe

immature ♂
uropygialis

distinctively marked graduated tail

A close but very distinctive relative of the Rose-throated Becard, the Gray-collared Becard of Mexico and C.A. has been reported only once in the US, an immature male bird in AZ. Polytypic (5 sspp.). L 5.9" (15 cm)
Identification Disproportionately large headed, slightly crested, and thick billed. Note white supercilium extending from base of bill to back of eye, and graduated tail with black subterminal marks. **ADULT MALE** *uropygialis*: Black crown and back, pale gray hind collar and underparts; outer tail tipped with white. **ADULT FEMALE:** Cinnamon crown bordered black; brown back, buffy underparts, tail tipped with cinnamon. **IMMATURE MALE:** Intermediate, with cinnamon on back and rump but white markings on wings and tips of outer tail feathers.

In more easterly nominate subspecies, female has a black crown, and pale areas are more buffy cinnamon.
Similar Species Unlikely to be confused with any other species. Female Rose-throated Becard might be suspected but is larger, often bushy-headed, and has a gray crown with a buffy collar.
Voice CALL: A clear titmouse-like *peeu, peeu, peeu.* **SONG:** A repeated whistled *hoo hoo-dee* or *hoo wee-deet.*
Status & Distribution Resident in evergreen, pine-oak, and tropical deciduous forest as high as 8,200 ft, from southeastern Sonora and southern Nuevo León and Tamaulipas, south to north-central Nicaragua. Distinctive pale western Mexican subspecies *uropygialis* has bred at Sahuaripa, Sonora (60 mi from AZ),

indicating this subspecies' potential as most likely to reach the US. The single record north of Mexico, at Cave Creek Canyon, Cochise Co., AZ (5 June 2009), was identified as a second-calendar-year male *uropygialis* attaining adult plumage.
Population The west Mexican subspecies (*uropygialis*) has a limited range and is uncommon.

ROSE-THROATED BECARD   *Pachyramphus aglaiae*   RTBE ◾ 3

A widespread and numerous tropical lowland species that is very rare in the US. Polytypic (8 sspp.; 2 in N.A.). L 7.3" (19 cm)
Identification ADULT: Relatively small; short, thick, dark bill. **MALE:** Black crown; remainder of upperparts plain dark gray; light gray underparts with pink throat patch. **FEMALE:** Slate gray cap; pale rufous hindneck collar; brownish gray back; light buffy underparts. **JUVENILE:** Similar to female. First-year male more patchy gray and brown; has smaller pink throat patch.
Geographic Variation Subspecies *albiventris* (northwest Mexico; described above); *gravis* (eastern Mexico; casual to southernmost TX)

is larger. Male overall substantially darker, with more extensively black crown and more extensive, darker pink throat patch; female with rufous back, upperwings, and tail; more intensely rufescent below.
Voice CALL: Plaintive, descending *tseeeuuuu,* also a *pik* or *pidik* and a trill. **DAWN SONG:** A repeated, plaintive *see-cheew, wee-chew.*
Status & Distribution Very rare in US. **BREEDING:** Mature riparian forest in lowlands and lower mountain canyons. Where it breeds, this canopy dweller is sometimes easier to find by locating its large and peculiar, globular nest suspended from the tip of a high branch. Since about 2010, breeds only along Santa Cruz River, north of Nogales, AZ. Earliest arrivals mid-May, departs by mid-Sept. Formerly bred near Patagonia, AZ. **MIGRATION:** In AZ casual away from known breeding localities. **WINTER:** Northernmost populations move south into areas of year-round occurrence, south to Costa Rica; rare to west Panama and casual to central Panama. **RARE STATUS:** Casual in extreme southern TX,

mainly in winter (mostly Hidalgo Co.), but a few unsuccessful summer breeding attempts; single records north to Jeff Davis Co. (subspecies uncertain), Kenedy Co., and Aransas Co.
Population Always rare in US.

1st winter ♂
albiventris

blackish cap

thick, stubby bill

albiventris ♀

paler below than *gravis*

darker head than *albiventris*

dark cap

adult ♂
gravis

rose throat

adult ♂
albiventris

pale cinnamon underparts

♀
gravis

albiventris

gravis

insularis

five subspecies

TYRANT FLYCATCHERS Family Tyrannidae

Gray Kingbird (FL, Apr.)

With some flamboyant exceptions, flycatchers are characterized by shades of brown, yellow, and olive. Several, like many *Empidonax* and *Myiarchus*, present difficult identification challenges.

Structure Tyrant flycatchers perch upright and have a rather broad, flattened bill with a bit of a hooked tip. They usually have rictal bristles at the base of the bill; the one exception in North America is the Northern Beardless-Tyrannulet. There is considerable variation in structure, and understanding it is a critical part of identification. Overall shape, head shape, bill structure, primary projection, and wing formula are particularly important.

Behavior Tyrant flycatchers generally eat insects, but many at least occasionally eat small fruits, usually in migration or on the wintering grounds. Most sit on a perch and wait until they spot an insect. Some species, such as most species in the genus *Contopus*, sally out to grab a flying insect and return to the same or another prominent perch. Others, like *Empidonax* flycatchers, fly short distances within the canopy; after catching prey, they return and flick their tail up (most species) or bob their tail down

(Gray Flycatcher). Northern Beardless-Tyrannulets use their short warbler-like bill to glean stationary prey from foliage and branches. These differences in foraging style can be very helpful in identifying birds.

Plumage Flycatcher plumages are often studies in subtleties. Males and females are usually identical in plumage. Juveniles and immatures are similar to adults, with subtle differences in color and feather shape. Molt timing differs between some species and provides an excellent way to distinguish some difficult species in fall.

Distribution This is a large, exclusively New World family that reaches its greatest diversity and abundance in the New World tropics: 208 species have been recorded in Ecuador alone. Of some 420 species in about 100 genera, 47 in 15 genera are known from North America. Thirty-four species are native breeders; the other 13 are casual visitors. Most species in the US and Canada are highly migratory.

Taxonomy Passerines are classified into two suborders: oscines and suboscines. The tyrant flycatchers are the only suboscines that occur commonly north of Mexico. All have simpler syringeal morphology, which results in less impressive vocalizations (which are mainly innate rather than learned) than other North American passerines. Molecular work and analysis of vocalizations continue to clarify relationships among many difficult tyrant flycatcher groups. Ongoing genetic work suggests several deep divisions within the Tyrannidae, and as many as five or six additional families might ultimately be recognized; all North American species would remain in Tyrannidae.

Conservation The most threatened North American flycatcher is the southwestern subspecies of the Willow Flycatcher, which is a federally listed endangered subspecies. *Contopus* may also have a relatively high risk of decline, in part because it has the lowest reproductive rate of all North American passerine genera. Loss of habitat, habitat fragmentation, and decline in numbers of flying insects pose the biggest threats for most flycatchers. BirdLife International codes: 25 NT, 25 VU, 8 EN, and 2 CR.

TYRANNULETS AND ELAENIAS Genera *Camptostoma*, *Myiopagis*, and *Elaenia*

Birds of this large, diverse group of flycatchers generally feed on insects, but many at least supplement their diet with fruit. Only the Northern Beardless-Tyrannulet is regularly encountered in N.A.; two others have occurred as accidental visitors. They are best found by listening for their vocalizations and best identified by differences in facial pattern.

NORTHERN BEARDLESS-TYRANNULET *Camptostoma imberbe*   NOBT ■ 2

Otherwise easily overlooked, this small flycatcher is almost always first detected by its plaintive whistled vocalizations. Its foraging behavior is very different from any other flycatcher regularly found north of Mexico. It hops through foliage, frequently

flopping its tail up and down, like a vireo. Polytypic (3 sspp.; 2 in N.A.). L 4.5" (11cm)

Identification Grayish olive above and on breast, fading to dull white or pale yellow below. Short pale eyebrow contrasts with dark eye line. Darker

crown often raised, giving a bushy-crested appearance. Two indistinct pale buffy wing bars; very short primary projection. Short blunt-tipped bill has bright pinkish orange base and dark culmen. **JUVENILE:** Similar to adult but with cinnamon-colored wing bars and edging to secondaries. **Geographic Variation** Slight. Widespread nominate from southern TX averages smaller, with smaller bill, more grayish wash to olive upperparts, and less yellow on underparts than *ridgwayi* (southeastern AZ, southwestern NM, northwestern Mexico).

Similar Species Dull but relatively distinctive. Note small crest and tail-dipping behavior. **Voice CALL:** An innocuous, whistled *peeuuuu*; also a similar shorter call often followed by high trill: *peeut di-i-i-i-i*. **SONG:** A somewhat variable series of loud, clear, down-slurred notes, typically with one or two loudly stressed notes (e.g., *dee dee dee dee dee dee*). **Status & Distribution** Uncommon. **BREEDING:** Often found near streams in sycamore, mesquite, and cottonwood groves. Occurs from south TX and the Southwest to Costa Rica. **MIGRATION & WINTER:** Most depart northern breeding grounds by Oct., return by Mar.; but small numbers apparently winter in N.A., particularly in south TX. **RARE STATUS:** Casual to West TX (Presidio Co.) and outside

mapped range in southern TX north to Calhoun Co. and Goliad Co. **Population** Population in south TX appears to be increasing.

GREENISH ELAENIA *Myiopagis viridicata*   GREL   5

The Greenish Elaenia has occurred once in N.A. Polytypic (10 sspp., N.A. ssp. unknown). L 5.5" (14 cm)

Identification ADULT: Flat head; small, slender blackish bill (often at pinkish base below); long tail; very short primary projection. Mostly olive above, head more grayish; short white eyebrow; dark eye stripe. Wings and tail darker, noticeable olive edging, rather bright yellowish edging to secondaries. No wing bars. Olive breast; yellow belly, undertail coverts. **JUVENILE:** Similar but with brownish head and upperparts. **Similar Species** A distinctive elaenia

and unlikely to be confused with other elaenias or other flycatchers. **Voice CALL:** High and thin, somewhat burry, descending *seei-seeur* or *sleeryip*. **Status & Distribution** Accidental in US. Fairly common to common in resident range. **YEAR-ROUND:** Forest and woodlands, often in more open areas from northern Mexico to northern Argentina. **RARE STATUS:** Accidental at High I., Galveston Co., TX, 20–23 May 1984. **Population** Unknown.

WHITE-CRESTED ELAENIA *Elaenia albiceps*   WCEL   5

The White-crested Elaenia is a widespread S.A. flycatcher, one of two dozen tropical and subtropical elaenias that are difficult to identify. Some can only be told by their calls. Records should be substantiated by photos and recordings of its vocalizations. Polytypic (6 sspp.). L 6" (15 cm) **Identification** Often appears crested. Suggests a N.A. *Empidonax* species. Upper mandible blackish; lower mandible with pale base. **ADULT:** Grayish olive above and pale below. Broad white crown stripe sometimes conspicuous, sometimes partially concealed. Narrow whitish eye ring and lores. Bold white wing bars. **JUVENILE:** Lacks white base to crest. **Similar Species** Small-billed Elaenia (*E. parvirostris*) is most similar, but it is paler below with more contrast between olive face and white throat;

also has a third pale wing bar and slightly more prominent eye ring. A difference in wing shape may be helpful: The migratory White-crested subspecies *chilensis*—likely the source of the one confirmed record for N.A.— usually has a longer, more pointed wing tip than Small-billed. **Voice CALL:** In subspecies *chilensis* note is a *weeo* or *feeo*, different from short *prk* of Small-billed. **SONG:** On breeding grounds, *chilensis* has a short, raspy two-note or three-note phrase. **Status & Distribution** Common breeder in forest edge, second growth, and scrub in much of S.A. from southwestern Colombia through the Andes to southern Chile. The *chilensis* subspecies breeds in the southern two-thirds of Chile and Argentina and is a long-distance migrant northward to wintering grounds in the Andes of

Peru and Bolivia. **RARE STATUS:** Accidental to the US. The only accepted record was a calling *chilensis* on South Padre I., TX (9–10 Feb. 2008). An elaenia at Chicago, IL (Apr. 2012), was not accepted, but it may have been a Small-billed (*E. parvirostris*) from S.A. A bird at Santa Rosa I. in northwestern FL (Apr. 1984) was originally identified as a Caribbean Elaenia (*E. martinica*) but was subsequently considered to be an unknown elaenia species; it may have been a White-crested. **Population** Stable.

Genus *Myiarchus*

These rather big-headed, bushy-crested, relatively slim-bodied, long-tailed flycatchers (22 species; four breed in US; two more are rare visitors) are easy to recognize by virtue of upright posture, large all-dark bill, grayish brown upperparts, gray chest, yellowish belly, and often rufous-edged primary and tail feathers. Dark brown wings have two pale wing bars; rufous-edged primaries give the folded wing a prominent rusty wing panel. Moving from perch to perch, they sally within foliage (in wooded habitats, primarily sub-canopy) for insects or fruit. Distinctive vocalizations include many daytime calls, which can also be repeated or combined to form male's dawn song. *Myiarchus* are secondary cavity nesters.

Identification of *Myiarchus* Flycatchers

These species differ by size between species—in Brown-crested between subspecies. *Myiarchus* have only one complete molt per year, so by summer their plumage can become considerably duller and paler from wear and bleaching.

Juvenal Plumage

Young birds leave the nest in juvenal plumage. This first plumage is superficially similar to the adult but typically paler and of a softer, more delicate feather structure. It may include a more extensively rufous tail; secondaries have rusty or buffy edges (adults generally have white or yellow), making the rusty primary panel less well defined. In some species, coloration of the secondary edges is an important adult field mark, so individuals must first be correctly aged.

Great Crested Flycatcher (TX, Apr.)

Adult Plumage

Young birds look more like adults by mid-October. For identification use multiple characters; tail pattern and calls are particularly important. Determining the extent and pattern of dark brown versus rufous on the tail is vital, particularly for identifying silent individuals in areas of overlap. In adults, the central tail feathers and outer webs of the remaining pairs of feathers are dark brown, so that viewed from above, the closed tail above appears all brown, which is unhelpful for identification. Depending on the species, the inner webs of the outer five pairs have varying amounts of rufous and might have a dark shaft stripe of varying width along the feather shaft; the rufous is best studied on a perched bird from below, where the pattern of the outermost pair is visible even when the tail is closed. *Myiarchus* call frequently during the breeding season, and key vocalizations are diagnostic; migrants and win-

Great Crested Flycatcher (TX, June)

Ash-throated Flycatcher (CA, May)

tering birds call much less frequently. Overall size, bill size and proportions, and subtle differences in plumage coloration are important but much more subjective. Species that vary greatly in size or color will be easily distinguishable from each other, but there is extensive overlap among most species. Distance and lighting are also important with such subtle differences; distribution and breeding habitat can also be helpful in elimination.

Out-of-Range Birds

In general, strays are less likely in spring and summer, so identifications are usually safe based on probability during the breeding season in areas where only one species normally occurs. This approach works best with the Ash-throated and the Great Crested, which are the only species present across much of the west and east, respectively. All *Myiarchus* breeding in N.A. are migratory and normally winter south of the US border, except for small numbers of Ash-throateds in the extreme Southwest and the Great Cresteds in southern FL. Strays should be identified with extreme caution. A high percentage of rare or casual *Myiarchus* are found during fall and winter, usually in coastal regions. Species normally found in an area as breeders or migrants are not necessarily more likely to occur at the wrong season (e.g., the Great Crested Flycatcher is virtually unrecorded in winter in the East outside of FL). The vast majority of stray *Myiarchus* are Ash-throateds, the only rare species that has occurred in Canada, most of the interior US, and along the Atlantic coast north of FL. Thus, Ash-throated should be the first option to consider when encountering an out-of-season *Myiarchus* in the East.

DUSKY-CAPPED FLYCATCHER *Myiarchus tuberculifer* DCFL ■ 2

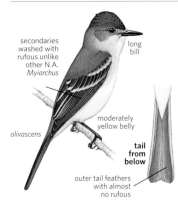

secondaries washed with rufous unlike other N.A. *Myiarchus*

long bill

moderately yellow belly

olivascens

tail from below

outer tail feathers with almost no rufous

Widespread in the Neotropics, our smallest *Myiarchus* barely enters the US. Its plaintive call reveals its presence in the leafy subcanopy of lower montane forests and woodlands. Range overlaps with Ash-throated and Brown-crested Flycatchers. Polytypic (13 sspp.; 2 in N.A.). L 6.4–7.3" (16–19 cm)

Identification ADULT: Grayish brown above, darker on head and back. Outer and middle secondaries edged rufous to rusty-yellow; inner secondaries whitish. Upperwing coverts tipped pale grayish brown, making wing bars less conspicuous than in other species. Gray throat and breast; yellow belly, with relatively sharp gray-yellow contrast; olive-suffused sides. Black, disproportionately long, slender bill; paler base of lower mandible. Orange mouth lining. **JUVENILE:** Similar but duller overall; tail broadly edged rufous on inner webs.

Geographic Variation Two subspecies reach the US: *olivascens*, described above (southeastern AZ, extreme southwestern NM, and northwestern Mexico), and *lawrenceii* (reaches south TX from eastern Mexico). The latter is slightly larger and darker above and has more extensive rufous inner edges of tail feathers.

Similar Species Other N.A. *Myiarchus* are larger, have different bill shapes, and more rufous in tail. Ash-throated and much larger billed Brown-crested have pinkish mouth lining; calls of both different.

Voice CALL: Mournful, descending whistled *peeeeuuuu*, sometimes preceded by *whit* note. **DAWN SONG:** A continuous series of mixed *whit* notes, whistles, and trills.

Status & Distribution Fairly common.

lawrenceii

olivascens

querulus

BREEDING: Lower- and middle-elevation montane riparian, oak, or pine-oak woodlands. **MIGRATION:** Migrants seldom detected in lowlands. In spring, exceptionally, arrives late Mar.–early Apr., more typically mid-Apr. In fall, most have departed US by mid-Aug., stragglers recorded to mid-Oct. **WINTER:** Mexico. **RARE STATUS:** Subspecies *olivascens* is a rare migrant and breeder in mountains of West TX; very rare but annual in late fall (Nov.) and winter west to coastal CA; accidental to OR and in spring to CO; *lawrenceii* is casual in winter in extreme southern TX. **Population** Stable.

ASH-THROATED FLYCATCHER *Myiarchus cinerascens* ATFL ■ 1

This is the default *Myiarchus* throughout much of the West. Monotypic. L 7.6–8.6" (19–22 cm)

Identification ADULT: Relatively small, slender; moderately long tailed. Whitish gray throat and pale yellow belly. Brown crown and brownish gray back. Blackish brown wings with two whitish wing bars; rufous edged primaries; secondaries edged white to pale yellow. Outer pairs of tail feathers extensively rufous on inner webs; dark shaft stripe flares at tip so that rufous does not extend to feather tips. Some lack typical tail pattern,

and pattern can vary among feathers. All-dark bill is relatively thin, short to medium length. Mouth lining pinkish. **JUVENILE:** Duller and paler; browner above; belly more whitish yellow; rufous edged secondaries; tail predominantly rufous with dark shaft stripes on outer webs. Some juvenile middle secondaries or tail feathers can be retained into first winter plumage.

Similar Species Very similar to Brown-crested Flycatcher, which averages darker gray and brighter yellow below, is larger in all aspects (especially *magister*), lacks typical Ash-throated tail pattern, and has different voice. Juvenile Ash-throated tail suggests Great Crested, but size, shape, and plumage should make identification easy. Nutting's best separated by voice, mouth color.

Voice CALL: Soft *prrrrt* (nonbreeders less vocal). **SONG:** *Ka-brick*. **DAWN SONG:** A repeated series of *ha-wheer* and other notes.

Status & Distribution Common.

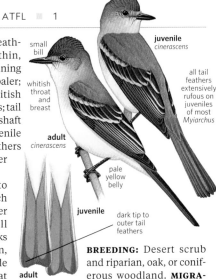

small bill

whitish throat and breast

adult *cinerascens*

juvenile cinerascens

all tail feathers extensively rufous on juveniles of most *Myiarchus*

pale yellow belly

juvenile

dark tip to outer tail feathers

adult

BREEDING: Desert scrub and riparian, oak, or coniferous woodland. **MIGRATION:** In spring, mid-Mar.–mid-May. In fall, late July–Aug., a few to mid-Sept.; stragglers Oct.–Nov. **WINTER:** Extreme southwestern US to Honduras. **RARE STATUS:** Rare/casual north to southeast AK, east to southeastern Canada and to Atlantic and Gulf Coasts in the US, mainly late fall and winter. **Population** Stable.

NUTTING'S FLYCATCHER *Myiarchus nuttingi* NUFL ■ 5

This casual visitor from Mexico closely resembles the Ash-throated Flycatcher in size, structure, and coloration. It also suggests the larger Brown-crested Flycatcher in overall coloration. Identifications are best confirmed by calls and, if possible, the orange-colored mouth lining. Polytypic (3 sspp.; *inquietus* in US). Based on vocal differences, it has recently been proposed that *flavidor* (Pacific slope from Chiapas to northwest Costa Rica) is a separate species. L 7.2" (18 cm)

Identification ADULT: Relatively small; small billed. Gray breast and yellow belly, with gray-yellow transition. Outermost secondary edge is rufous; remaining secondaries range (moving inward) from pale rufous or brownish white to white or grayish white. Tail pattern variable; typically, inner webs of outer five pairs of feathers extensively rufous with a dark shaft stripe of variable width (as in Ash-throated and Brown-crested), but rufous may or may not extend to the feather tip. **JUVENILE:** Essentially identical to Ash-throated; some rusty-edged juvenile middle secondaries can be retained in first winter plumage.

Similar Species Brown-crested (especially *magister*) and Great Crested Flycatchers are larger and have more extensively rufous tails. Brown-crested has a pinkish mouth lining, a dusky cap, and little or no rufous in its tail; and it is smaller and disproportionately longer billed. Ash-throated, which averages slightly larger, has a pinkish mouth lining; longer and more pointed wings; paler underparts with a whitish area between whitish gray breast and pale yellow belly; a subtle grayish collar across hindneck; grayer auriculars; and whitish or yellowish white secondary edges. Nutting's Flycatchers that share typical

Ash-throated pattern on outer (sixth) pair of tail feathers (e.g., feather tips dark, no rufous to tip) also have a dark shaft stripe on inner webs of second pair of feathers (lacking in Ash-throated).

Voice CALL: A sharp, whistled *wheep* (suggesting Great-crested) or *peer*; also a *pip*.

Status & Distribution Common in normal range. **YEAR-ROUND:** Thorn scrub and open tropical deciduous forest, from northwestern and central-eastern Mexico south to northwestern Costa Rica. **RARE STATUS:** Casual to AZ in all seasons (most records Bill Williams Delta, likely bred). Accidental to Southern CA and Big Bend, TX, in winter.

Population Presumably stable.

coloration very similar to slightly larger Ash-throated, but Nutting's has slightly yellower belly; best told by call

slightly more olive above than Ash-throated

inquietus

yellow tint to secondaries when in fresh plumage

dark tip of inner web of outer tail feather not as extensive as Ash-throated

tail from below

GREAT CRESTED FLYCATCHER *Myiarchus crinitus* GCFL ■ 1

The default *Myiarchus* of eastern N.A. is usually heard more than seen as it forages high in canopy or subcanopy. Monotypic. L 8.5" (21 cm)

Identification Key characters include a broad white stripe on innermost tertial, mostly rufous inner webs of all but central tail feathers, and overall darker plumage. Longer winged than other *Myiarchus*. **ADULT:** Dark gray face and breast (slightly paler throat) contrasts sharply with bright yellow belly; these colors blend at sides of breast to form olive-green patches. Olive-brown back. Bill rather thick, is black with pale at base of lower mandible (most). Mouth lining is bright orange-yellow. **JUVENILE:** Similar but has rusty secondary edges (but innermost two or three edged white) and wing bars; all tail feathers rufous. Immature brighter, like adult, but wing bars and most secondary edges still rusty.

Similar Species A pale or bleached Great Crested would be superficially similar to a Brown-crested or Ash-throated, but would still exhibit key field marks. See sidebar, p. 436.

Voice CALL: An ascending *whee-eep*; also *purr-it* and series of *whit* notes. **DAWN SONG:** A repeated series of modified *whee-eeps*.

Status & Distribution Common. **BREEDING:** Deciduous or mixed forest. **MIGRATION:** In spring, primarily western circum-Gulf, but some trans-Gulf, mid-Mar.–early June. In fall, both circum- and trans-Gulf, mid-July–mid-Oct. **WINTER:** Southeastern Mexico to Colombia and Venezuela; also southern FL. **RARE STATUS:** Very rare to casual/accidental, mainly in fall, to West Coast, AK, NT, NL, Bermuda, Bahamas, Cuba, and Puerto Rico.

Population Stable.

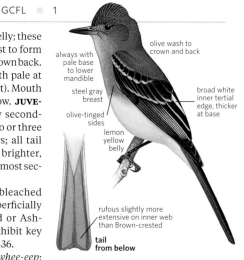

olive wash to crown and back

always with pale base to lower mandible

broad white inner tertial edge, thicker at base

steel gray breast

olive-tinged sides

lemon yellow belly

rufous slightly more extensive on inner web than Brown-crested

tail from below

BROWN-CRESTED FLYCATCHER *Myiarchus tyrannulus* BCFL ▪ 1

This is our largest *Myiarchus*. The western subspecies dwarfs the smaller Dusky-capped; the eastern subspecies is closer in size to the Great Crested and the Ash-throated. Brown-cresteds prefer more mature, undisturbed habitats. Polytypic

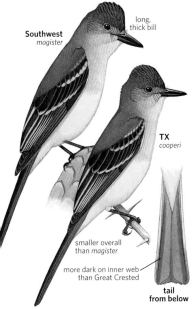

Southwest
magister

long, thick bill

TX
cooperi

smaller overall than *magister*

more dark on inner web than Great Crested

tail from below

(7 sspp.; 2 in N.A.). L 7.2–9.2" (18–23 cm) **Identification ADULT:** Rather bright yellow belly contrasts abruptly with gray breast. Outer pairs of tail feathers extensively rufous on inner webs; dark shaft stripes and rufous extend to feather tips. Bill disproportionately long, heavy, and black. Mouth lining typically pinkish. **JUVENILE:** Similar, but secondaries (except whitish inner 2) and wing bars are rusty edged and inner webs of tail feathers are more extensively rufous.
Geographic Variation Larger *magister* breeds in Southwest south through western Mexico; smaller *cooperi* breeds in southern TX and eastern Mexico south to Honduras.
Similar Species Great Crested has a darker gray face and breast; broad white edge to innermost tertial; all-rufous inner webs of tail feathers; and paler base to lower mandible. Smaller, paler Ash-throated has whitish transition between gray and yellow on underparts, and entire tail tip (from below) is dark not rufous. Smaller Nutting's and Dusky-capped have orange mouth lining; Dusky-capped has much less rufous in tail.
Voice CALL: A sharp *whit*. Breeding,

a rough, descending *burrrk* (or rasp) or *whay-burg*. **DAWN SONG:** Repeated *whit* notes, vibrato whistles, *burrrk* notes, and other complex phrases.
Status & Distribution Common. **BREEDING:** Riparian forest, thorn woodland, columnar cactus desert. **MIGRATION:** In spring, arrives in TX exceptionally by mid- to late Mar., more typically early to mid-Apr.; arrives in Southwest late Apr.–early May. In fall, generally departs Aug., rare after mid-Sept. **WINTER:** Mexico to Honduras. **RARE STATUS:** Very rare to coastal TX, LA, and FL (*cooperi*, but one LA specimen of *magister*). Casual, mainly fall-winter, to coastal CA (*magister*). **Population** Stable.

cooperi

magister

cozumelae

LA SAGRA'S FLYCATCHER *Myiarchus sagrae* LSFL ▪ 3

A rare visitor to our area, primarily from the Bahamas, the La Sagra's was formerly considered the same species as the Stolid Flycatcher (Jamaica and Hispaniola). It prefers to forage in dense second growth. At first glance, its coloration and more hunched posture may suggest an Eastern Phoebe. Polytypic (2 sspp.; both in N.A.). L 7.5–8.5" (19–22 cm)
Identification ADULT: A relatively small *Myiarchus* with a disproportionately long, thin, black bill. Pale gray on throat and breast; whitish on belly and undertail coverts, sometimes tinged yellow. Rufous edges to primaries reduced or absent; secondaries thinly edged white. Tail mostly blackish brown; extent of rufous on inner web

lucaysiensis

sagrae

of outer pairs of feathers ranges from a very narrow fringe to a fairly broad stripe along inner edge. Pale yellow mouth lining. **JUVENILE:** Secondaries and tail feathers edged with rufous.
Geographic Variation At least some (and presumably most) FL records pertain to *lucaysiensis* of the Bahamas, which is larger, with more rufous in tail. Subspecies *sagrae* (Cuba and Cayman Is.) is slightly smaller, with less rufous in tail, and has occurred once in Alabama.
Similar Species Other *Myiarchus* are yellower on belly or have more rufous in tail. Size, structure, and tail pattern suggest Dusky-capped Flycatcher (e.g., relatively small overall, disproportionately long bill, and reduced amount of rufous in tail).
Voice CALL: A distinctive high-pitched *wheep*, *whit*, or *wink*, sometimes doubled; also loud *teer teer*, softer *quip quip quip*.
Status & Distribution Common in normal range. **YEAR-ROUND:** Pine or mixed woodland, dense scrub and sec-

ond growth, mangroves; in FL almost always on or near immediate coast. **RARE STATUS:** Very rare in southern FL, mainly along extreme southeastern coast and Keys. First recorded in FL in 1982; records between late Oct. and mid-May. Accidental AL (specimen of *sagre*; 14 Sept. 1963, Orrville, Dallas Co.).
Population Stable.

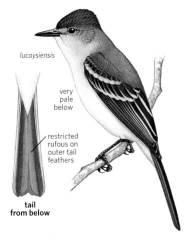

lucaysiensis

very pale below

restricted rufous on outer tail feathers

tail from below

Genera *Pitangus, Myiozetetes, Myiodynastes, Legatus,* and *Empidonomus*

GREAT KISKADEE   *Pitangus sulphuratus*   GKIS   ■ 2

dark mask

bold white supercilium

bright rufous wings and tail

white throat

texanus

bright yellow belly

texanus

This widespread, tropical species is common in south TX. Its large size, chunky shape, large black bill, loud voice, and striking coloration make it unmistakable. Noisy and conspicuous, it flycatches, but also plunge-dives for aquatic prey; it builds a bulky domed nest with a side entrance, usually situated in the fork of a tree or on a utility pole. Polytypic (10 sspp.; 2 in N.A.). L 9.8" (25 cm)

Identification ADULT: Black crown surrounded by white; central orange-yellow crown patch (larger in males; not often visible in the field). Black face contrasts with white supercilium and throat. Rest of underparts bright yellow. Brown back; wings and tail

extensively edged rufous. **JUVENILE:** Similar but duller; lacks central crown patch; has rustier brown back and more extensive rufous tail and wing edgings.

Geographic Variation Subspecies *texanus* breeds in TX; strays to AZ and NM reported as *derbianus*, which has a more grayish brown back. A specimen record of S.A. nominate subspecies (CA, 1926), lacking rufous in wings, considered unacceptable on origin.

Similar Species Social Flycatcher superficially similar but much smaller, with a much smaller bill and less contrasting head pattern; also lacks rufous on tail and wings.

Voice CALL: Year-round, a loud *crear*; named for breeding season's *kis-ka-dee*; also a loud, somewhat rising *reeee, chick-weer.* **DAWN SONG:** A repeated series of typical calls, uttered during twilight.

Status & Distribution Common. **YEAR-ROUND:** Lowland thorn forest, riparian forest, woodland, even suburban or urban situations; usually near water. Occurs from southern TX and northwestern Mexico to central Argentina. Introduced in Bermuda. **RARE STA-**

TUS: Rare beyond mapped range in TX. Casual/accidental in southeastern AZ, NM, CO, OK, KS, and southern LA. Some of these records involve breeding attempts or single birds that remain in an area for months or years and construct a nest. Accidental SD, IN, and ON in late fall/early winter.

Population Stable; slowly expanding breeding range northward in TX.

texanus

guatimalensis

derbianus

SOCIAL FLYCATCHER   *Myiozetetes similis*   SOFL   ■ 5

small bill

always lacks prominent rufous on wings and tail

adult *primulus*

This widespread Neotropical species has reached southern TX on two occasions. Polytypic (7 sspp.; *primulus* in N.A.). L 6.7–7.2" (17–18 cm)

Identification ADULT: Medium size; short black bill. Gray crown; white forehead, eyebrows connect indistinctly across nape; concealed reddish orange central patch; rest of face dark gray. Brownish olive back with darker wings. White throat; rest of underparts bright yellow. **JUVENILE:** Similar but duller; no crown patch; rusty-edged wings and tail.

Similar Species See Great Kiskadee.

Voice CALL: Strident *chee cheechee cheechee cheechee*; also harsh *cree-yooo.*

Status & Distribution Common, Mexico to northeast Argentina. **YEAR-ROUND:** Open woodland, second growth. **RARE STATUS:** Accidental in south TX; specimen 15 Feb. 1885 from Cameron Co. and one well-documented record in Bentsen-Rio Grande Valley SP (7–14 Jan. 2005). An accepted record at Anzalduas Co. Park, TX (17 Mar.–5 Apr. 1990), was likely misidentified.

Population Presumably stable.

SULPHUR-BELLIED FLYCATCHER   *Myiodynastes luteiventris*   SBFL   ■ 2

This handsome flycatcher reaches its northern limit in the mountain canyons (with flowing water) of southeastern AZ. A cavity nester that mainly forages in the canopy, the Sulphur-bellied Flycatcher is best located by its squeaky call. It usually stays within the canopy, but will engage in vigorous

aerial pursuits of prey. Monotypic. L 8.5" (22 cm)

Identification ADULT: Sexes similar. Moderately large overall with heavy, long, mostly black bill and disproportionately short tail. Strikingly patterned head: Dark mask and malar stripes contrast with whitish super-

cilium, moustachial stripe, and throat; grayish brown crown and back have blackish streaks; concealed yellow central crown patch. Uppertail coverts and tail extensively rufous. Dark brown wings with whitish edgings. Blackish chin, connecting malar stripes. Throat is lightly streaked blackish.

Rest of underparts pale yellow, heavily streaked blackish on breast and sides. **JUVENILE:** Similar to adult but duller and paler; wing feathers edged with cinnamon-buff; yellow crown patch small or absent.

Similar Species Piratic and Variegated Flycatchers are similar but smaller, with much smaller bills, less distinctly streaked and duller yellow underparts, and less rufous in tail. The very similar Streaked Flycatcher, with migratory northernmost (southern Mexico to Honduras) and southernmost (Argentina and

southern Bolivia) populations, is a potential stray to N.A. Streaked Flycatcher lacks dark chin connection between malar stripes, has a slightly heavier bill with more extensively pale base to lower mandible, and is whiter on belly.

Voice CALL: Loud *squeez-za* recalls a squeaky toy; also a rasping screech or *weel-yum*. Migrants usually silent. **DAWN SONG:** Repetitive, warbled *tre-le-re-re* or combinations of other notes.

Status & Distribution Fairly common. **BREEDING:** Mature sycamore-walnut dominated riparian forest at lower elevations in mountain canyons, usually with permanent water. Breeds south to Costa Rica. **MIGRATION:** Seldom detected in AZ lowlands. Arrives in AZ exceptionally by early–mid-May, more typically late May–early June; typically about one month earlier in northeastern Mexico than in northwestern Mexico and AZ. In fall, breeders depart AZ by early to mid-Sept., casually into Oct. and early Nov. **WINTER:** S.A., along eastern base of Andes

large bill

dark mask bordered by broad white lines

rufous tail

dark streaks below with yellowish background color

from Ecuador to northern Bolivia. **RARE STATUS:** Casual in spring and summer to TX (in spring as early as 5 Apr.; including breeding records), southeastern CO, southern NV, coastal LA, and the Gulf of Mexico off LA. Casual in fall, west to coastal CA and north and east to ON, NB, NL, MA, and along the Gulf Coast in TX, LA, AL, and FL.

Population Stable.

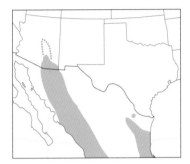

PIRATIC FLYCATCHER *Legatus leucophaius* PIRF ▪ 4

This migratory Neotropical species often chooses a high, conspicuous perch. It bears a superficial resemblance to a miniature Sulphur-bellied. Polytypic (2 sspp.). L 6" (15 cm)

Identification ADULT: Sexes similar. Dark brown crown has semi-concealed yellow central crown patch (larger in male) and is bordered by white eyebrows thinly connecting across nape. Blackish brown face; whitish throat with dark malar streaks. Rest of underparts progressively yellower toward lower belly and undertail coverts; breast and sides have extensive smudgy brown streaks. Dark brown wings and tail; slightly paler back. Wing coverts, secondaries, and tail thinly edged white to pale yellow. **JUVENILE:** Similar, but

lacks crown patch; edges of wing and uppertail coverts tinged rusty.

Geographic Variation Both subspecies at least partially migratory. Most US records probably pertain to northern *variegatus* (described above), which breeds from southeastern Mexico to Honduras; nominate, breeding from Nicaragua into S.A., is smaller and whiter on central underparts.

Similar Species Variegated is larger, with a disproportionately longer, narrower bill; pale-based lower mandible; somewhat more well-defined breast streaks; and conspicuously rufous-edged uppertail coverts and tail. The much larger, heavier-billed Sulphur-bellied has an extensively rufous tail. **Voice CALL:** Trilled, often repeated, *pi*

short all-dark bill

variegatus

dark malar streak

darker wings and tail than Variegated

ri ri ri ri ri.

Status & Distribution
BREEDING: Open woodland, forest edge, clearings. Breeds from southern Veracruz in Mexico south to Ecuador and northern Argentina. Early spring migrant with arrivals in Panama by mid-Jan. **WINTER:** S.A. **RARE STATUS:** Casual in spring and fall to FL (1), TX (5), KS (1), and NM (3).

Population Stable.

CROWNED SLATY FLYCATCHER *Empidonomus aurantioatrocristatus* CSFL ▪ 5

The Crowned Slaty Flycatcher, a S.A. species, has been recorded only once in the US. The more southerly breeding populations migrate north for the austral winter. Likely this one far overshot its normal distance, as Fork-tailed Flycatchers presumably do when they wander from S.A. far into the Northern Hemisphere. One other S.A. species

of this bird's genus *Empidonomus*, the Variegated Flycatcher, has several records in the US. Polytypic (2 sspp.). L 7" (18 cm)

Identification ADULT: Smoky or brownish grayish overall; crown black with a semiconcealed yellow patch down center, and only slightly crested. Gray supercilium extends

from bill to nape; lores and auricular area darker gray. Upperparts may be tinged brownish gray. Wings smoky brown, with narrow pale edges on inner flight feathers and coverts. Tail dull brownish gray. Underparts are paler brownish gray, slightly paler on belly. Eyes brown, bill, tarsi, and toes black. **JUVENILE:** Similar to adult but

crown dusky (not black) and lacking yellow patch; supercilium paler; tail with narrow rufous margins.
Similar Species Adult's combination of black crown, median yellow stripe, and dark gray face distinguish it from other flycatchers and elaenias. Slaty Elaenia (*Elaenia strepera*) is gray but crown is brown, not black, and a small crown patch is white, not yellow. Juvenile Crowned Slaty Flycatcher might be confused with Piratic Flycatcher or Variegated Flycatcher but lacks those species' whitish submoustachial stripe.
Voice Usually quiet in nonbreeding season. **CALL:** High *tzeer*. **SONG:** On breeding grounds rising *be-bee-beee-beeez*, low whistling *pree-ee-ee-er*, and

a series of squeaky notes, or a two-part *tsi-tsitsewt-tsi-tsébidit*.
Status & Distribution Widely distributed and generally common throughout its range. **BREEDING:** Deciduous forest, forest edges, and scrub with scattered trees, in interior Brazil, Paraguay, Uruguay, and south to central Argentina. **WINTER:** Some populations migrate to spend the austral winter (June–Aug.) as far north as western Brazil, Ecuador, and southeastern Colombia, where they prefer the canopy of humid forests, forest edges, and clearings. **RARE STATUS:** Strays have been reported farther north in S.A., once to Panama (Dec. 2007). The sole US record was an adult male (likely *pallidiventris*)

pale gray supercilium

dark crown with partially concealed yellow center

broad dark eye stripe

collected near Johnsons Bayou, Cameron Parish, LA (3 June 2008).
Population Stable.

VARIEGATED FLYCATCHER *Empidonomus varius*   VAFL ■ 5

This austral migrant from S.A. occasionally strays to N.A., where it might be confused with the Sulphur-bellied Flycatcher or the Piratic Flycatcher.

varius

larger bill than Piratic with pale base

streaking more distinct than Piratic

rufous on uppertail coverts, tail, and wings

Polytypic (2 sspp.; nominate in N.A.). L 7.3" (19 cm)
Identification ADULT: Relatively small, disproportionately long tailed. Black crown and mask; crown encircled by white and with concealed yellow central patch. Dark grayish brown back with subtle pale mottling. Blackish brown uppertail coverts; tail conspicuously edged rufous. Blackish brown wings; coverts and secondaries edged white; primaries faintly edged rufous. Dingy white throat, faintly mottled with gray and bordered by dark malar stripes. Rest of underparts gradually blend to pale yellow on belly and undertail coverts; breast and sides heavily streaked with blackish brown. **JUVENILE:** Similar, but duller below, browner above; wings extensively edged rufous; lacks crown patch.

Similar Species From Variegated, Piratic is smaller overall, with a shorter, broader, all-black bill; a shorter tail; more uniform brown back; little rufous in wings or tail; and blurrier breast streaks. Sulphur-bellied and Streaked are larger overall, heavier billed, more distinctly streaked below, and more extensively rufous on tail. Sulphur-bellied are also more yellow below.
Voice CALL: High, thin *pseee*.
Status & Distribution S.A. nominate subspecies breeds from central Bolivia to Argentina; migrates north for austral winter to northern S.A. **RARE STATUS:** Casual in N.A., single fall records from ME, ON, and WA, and two fall records for FL; single spring records from TN. Unrecorded Mexico or C.A.
Population Unknown.

KINGBIRDS Genus *Tyrannus*

These large, open-area birds (13 species) are conspicuous and noisy. Often seen on exposed perches or sallying to capture flying insects, they will also hover-glean fruit during migration or winter. They may be solitary or in loose flocks. Divided into three basic groups (white-bellied, yellow-bellied, and long-tailed), most are easily identifiable. Only the male sings a dawn song, and only adults have a colorful, semiconcealed, central crown patch and notched outer primaries. Juveniles have duller, paler, more delicately structured body plumage; and rusty-edged upperwing and uppertail coverts, secondaries, and tail. They molt to more adultlike body plumage by fall but retain juvenile wing and tail feathers.

TROPICAL KINGBIRD *Tyrannus melancholicus*   TRKI ■ 2

This wide-ranging tropical species reaches northern limits in southeastern AZ and in the lower Rio Grande Valley in TX; has nested along Rio Grande in West TX since 1997. Its deeply notched tail, and relatively heavy bill distinguish it from most other yellow-bellied kingbirds. The

Couch's (southern TX and eastern Mexico) is virtually identical to the Tropical; voice is the best way to distinguish between them. Extralimital records of silent birds (especially in the East, where Couch's has also occurred) are difficult to resolve. Polytypic (3 sspp.; *satrapa* in N.A.). L 9.3" (24 cm).

Identification ADULT: Gray head, dark mask; orange-red central crown patch usually not visible. Olive back; brownish wings and tail; tail conspicuously notched. Pale throat blends to olive-yellow on upper chest and bright yellow on belly.
Geographic Variation Note the slight

differences in coloration and wing, tail, and bill length; not safely separable in the field. US breeders and, presumably, most nonbreeding records involve northernmost *satrapa*.

Similar Species Couch's has a broader bill that appears shorter, different call, slightly shallower tail notch. On folded wing, Tropical's primary tips visible beyond secondaries appear unevenly spaced (adults), are evenly spaced on Couch's. Other yellow-bellied kingbirds differ in smaller bill, less extensively yellow below and differently colored tails.

Voice CALL: High-pitched rapid trill

or twitter. **DAWN SONG:** A repeated combination of one or two short *pip* notes followed by a trill.

Status & Distribution Uncommon in US, common elsewhere. **YEAR-ROUND:** Open areas with scattered trees, riparian woodland, forest edge and clearings, even open suburban situations; often near water. **MIGRATION:** Northwesternmost (AZ, Sonora) and southernmost (southern Bolivia and southeastern Brazil to Argentina) populations migratory. In spring, arrives AZ mid-May. In fall, rare in AZ after mid-Sept. **WINTER:** West-central and east Mexico south; southern TX breeding population present year-round. **RARE STATUS:** Rare mainly fall (after mid-Sept.) and winter, to West Coast (mainly CA but north to southwestern BC, casually to southeastern AK); less frequently inland in Southern CA, western AZ, and TX away from areas of local breeding. Casual NM. Casual in eastern N.A. north to southern Canada and NL. Accidental Bermuda (May record of nominate).

Population Stable or increasing. Slowly expanding breeding range northward; first bred in AZ in 1938, southern TX beginning in 1991, and West TX in 1997.

best told from Couch's by call

adults
satrapa

longer, thinner-based bill than Couch's

slightly narrower bill base than Couch's

spring adult ♂
satrapa

notched tail

adult ♂
primary tips
satrapa

uneven spacing of primary tips

COUCH'S KINGBIRD *Tyrannus couchii*　　COKI ▪ 2

This southern TX species is virtually identical to Tropical Kingbird and is best identified by voice. Tropical-type kingbirds found in southern TX, especially away from cities, are mostly Couch's, whereas breeding birds in AZ and most western strays will very likely be Tropicals. Monotypic. L 9.3" (24 cm)

Identification ADULT: Gray head, with blackish mask and concealed reddish orange central crown patch. Olive-gray back. Brownish wings

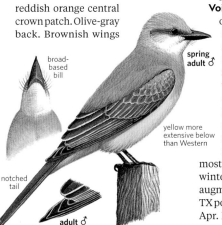

broad-based bill

spring adult ♂

yellow more extensive below than Western

notched tail

even spacing of primary tips

adult ♂
primary tips

and tail; tail prominently notched. Whitish throat blends to olive-yellow breast and bright yellow remainder to underparts. **JUVENILE:** Similar but with buff tips to upperwing and uppertail coverts.

Similar Species Combination of distinctly notched brownish tail and brighter underparts separate Couch's from other yellow-bellied kingbirds, except Tropical (see that species).

Voice CALL: Single or slow series of *kip* notes, suggestive of Western Kingbird; also *queer* or slurred *chi-queer*. **DAWN SONG:** A repeated *tuwit, tuwit, tuwitchew.*

Status & Distribution Common. **YEAR-ROUND:** Riparian forest, clearings, forest edge, woodland, overgrown fields with scattered trees, even suburban situations. **MIGRATION:** Partial migrant, with part of northernmost population moving south during winter. In spring, influx of migrants augments the overwintering southern TX population during mid-Mar.–early Apr. In fall, some of the southern TX population retreats south from late

Aug.–mid-Oct., with migratory flocks occasionally seen. **RARE STATUS:** Mainly fall and winter but occasionally spring and summer. Rare in TX outside breeding areas. Casual east to southern LA, including breeding record southwest; accidental, mostly in winter, west to NM, AZ, NV, and CA, and in eastern N.A. in AR, MI, PA, NY, MA, AL, and FL.

Population Stable or increasing. Since the mid-1900s, Couch's has expanded range in TX from immediate vicinity of extreme Lower Rio Grande north to Big Bend (a few), to near southern Edwards Plateau, and to Calhoun Co. on coast.

CASSIN'S KINGBIRD *Tyrannus vociferans* CAKI ▪ 1

Cassin's Kingbird is fairly widespread at middle elevations in the southwestern and west-central US, where it overlaps extensively with Western Kingbird. Monotypic. L 9.1" (23 cm)
Identification ADULT: Dark gray head and nape (mask less obvious); semiconcealed orange-red central crown patch; dark grayish olive back, brownish wings. Dark gray chest contrasts with white chin, blends to yellow belly. Tail squared, brownish black with an indistinct pale gray terminal

dark gray head, back, and breast
juvenile
all figures *vociferans*
adults
contrasting white chin
brownish gray wings usually paler than back
spring adult ♂
worn fall adult
usually pale tip

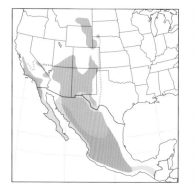

band. **JUVENILE:** Wings are edged pale cinnamon.
Similar Species Western has paler upperparts and chest; less chin-breast contrast; and darker, more blackish, less brownish wings and black tail with white outer web on outer feather (but beware Westerns with worn-off outer web). Thick-billed, Tropical, and Couch's have paler or yellower breasts, uniformly brown tails, and heavier bills; Tropical and Couch's also have paler backs and deeply notched tails.
Voice CALL: Single or repeated, stri-

dent *kabeer*; rapidly repeated *ki-dih* or *ki-dear*. **DAWN SONG:** A repeated *rruh rruh rruh-rruh rreahr, rruh ree reeuhr* (possibly confused with Buff-collared Nightjar).
Status & Distribution Common. **BREEDING:** Open, mature woodlands, including riparian, oak, and pinyon-juniper. **MIGRATION:** Relatively infrequently detected away from breeding sites. In spring, mid-Mar.–early June. In fall, departure late July–Oct., most in Sept. **WINTER:** Western to central-southern Mexico. Locally resident in coastal Southern CA. **RARE STATUS:** Casual/accidental to OR, MN, ON, NS, MA, NY, VA, AR, LA, and FL.
Population Stable.

THICK-BILLED KINGBIRD *Tyrannus crassirostris* TBKI ▪ 2

First recorded in the US in 1958, this West Mexican species now breeds in southeastern AZ and extreme southwestern NM. Polytypic (2 sspp., *pompalis* in N.A.). L 9.5" (24 cm)
Identification ADULT: Long, thick-based bill. Blackish brown head; slightly darker mask; concealed yellow central crown patch, paler back. Dark brown tail. Pale below with yellowish

(brighter in fresh fall plumage) belly.
Similar Species Massive bill, white throat, contrasting dark head, pale yellowish belly make identification easy.
Voice CALL: Loud, whistled *kiter-reer*; also a rapid trill similar to Eastern.
Status & Distribution Uncommon. **BREEDING:** Mature sycamore-cottonwood riparian forest. **MIGRATION:** In spring, arrives late May–early June.

In fall, departs in Sept., rare by Oct. **WINTER:** Western Mexico. **RARE STATUS:** Casual in summer to West TX (including breeding records). Casual to accidental, mainly in fall and winter, to western AZ, NV, CA (nearly annual), southwestern BC, northern TX, CO, ND, southern ON, and Baja California.
Population Range has contracted in southeastern AZ in last two decades.

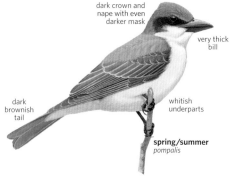

dark crown and nape with even darker mask
very thick bill
dark brownish tail
whitish underparts
spring/summer *pompalis*

1st fall *pompalis*
fresh fall birds quite yellowish on belly

WESTERN KINGBIRD *Tyrannus verticalis* WEKI ▪ 1

The Western is the default breeding yellow-bellied kingbird across vast areas of the West (especially arid low-

lands); but its distribution and habitat overlap with other yellow-bellied kingbirds (the Cassin's, Thick-billed,

Tropical, and Couch's) as well as with the Eastern and Scissor-tailed Flycatchers. From exposed perches on

juvenile

pale gray head and back

darker wings contrast with gray back

adults

worn fall adult

black tail with white edge

spring adult ♂

trees, shrubs, or wires, it chases flying insects; it will also take fruit during fall and winter. Monotypic. L 8.8" (22 cm) **Identification ADULT:** Pale gray head; darker mask; concealed orange-red central crown patch. Pale grayish olive back. Blackish wings, contrasts with paler back. Square-tipped, black tail; white outer web to outer feather. White throat blends to pearly gray chest and yellow belly. Relatively small, black bill. **JUVENILE:** Duller, paler. **Similar Species** Typical tail pattern unmistakable, but individuals with worn or missing white outer tail feathers might be mistaken for other yellow-bellied species. Cassin's is overall much darker on chest, upper-

parts; has contrasting white chin, paler wings, gray tail tip. Tropical and Couch's have heavier bills; olive-yellow chests; and dark brown, deeply notched tails. Cassin's, Tropical, and Couch's also have pale-edged upperwing coverts with a more scalloped appearance. Thick-billed has much heavier bill, darker upperparts, yellow central crown patch, paler underparts. (Juvenile's underparts yellower, more Western-like.) Superficially similar to juvenile or immature Scissor-tailed, immature white-bellied kingbirds with tinge of yellow on underparts, *Myiarchus*, or Say's Phoebe. **Voice CALL:** Single or repeated sharp *kip* notes. **DAWN SONG:** A repeated series of *kip* or *ker-kip* notes that is followed by another explosive, rapid

series of *kip* notes mostly on one pitch. **Status & Distribution** Common. **BREEDING:** Open country with scattered trees or shrubs; will nest on human-made structures (e.g., utility poles). **MIGRATION:** Mid-Mar.–early June; late July–mid-Sept.; scarce after early Oct., stragglers into Nov. **WINTER:** Central-western Mexico to Costa Rica; also southern and central FL. **RARE STATUS:** Rare migrant in East, mostly in fall on Gulf and Atlantic coasts; a few linger into early winter. Casual to AK, central and northern Canada, Bermuda, West Indies, and south to Panama. Accidental Azores. **Population** Stable or increasing.

EASTERN KINGBIRD *Tyrannus tyrannus* EAKI ■ 1

This species has the largest distribution of any N.A. kingbird. Monotypic. L 7.8–9.2" (20–23 cm) **Identification ADULT:** Black head blends to slate gray back; central crown patch varies from red to yellow. Dark gray wings; narrow white edgings to upperwing coverts and secondaries. Black tail with conspicuous white terminal band. White underparts; gray patches on sides of breast, paler gray wash across middle of breast. Extensively gray underwing coverts. **JUVENILE:** Generally similar but paler

grayish brown upperparts contrast with blackish mask; white tail tips narrower. **Similar Species** Distinctive. Immatures and worn adults can have somewhat paler or browner upperparts and reduced white tail tips; larger and bigger-billed Gray and Thick-billed Kingbirds can be superficially similar. Immature Fork-taileds are also superficially similar but have more head-back contrast, a longer tail, and a whiter center of breast. **Voice CALL:** Single or variety of *zeer*, *dzeet*, or trilled notes. **DAWN SONG:** A series of complex notes and trills, which are repeated over and over, *t't'tzeer, t't'tzeer, t'tzeetzeetzee*. **Status & Distribution** Common. **BREEDING:** Open areas in a variety of habitats that have trees or shrubs for nest sites. **MIGRATION:** Diurnal migrant, often observed in loose flocks; at least some trans-Gulf movement. In spring, mid-Mar.–mid-June, peaks mid-Apr.–mid-May; in West mid-May–June. In fall, late July–mid-Oct., peaks mid-Aug.–early Sept.; mostly gone by end of Sept., very rare after early Oct. **WINTER:** S.A., mainly western

blackish head

small bill

white tail tip

Amazonia (eastern Ecuador and Peru, western Brazil), but also casually as far south and east as northern Chile, Argentina, Paraguay, eastern Brazil, and Guyana. **RARE STATUS:** Rare during migration to Pacific coast, southwestern states, Bermuda, Bahamas, Cuba; casual to AK, southern YT, Hudson Bay, central QC, NL, Greenland. Accidental Europe. **Population** Generally stable.

GRAY KINGBIRD *Tyrannus dominicensis* GRAK ■ 2

This mainly West Indian species is a conspicuous summer resident of coastal FL and, locally, along the eastern Gulf Coast. Monotypic. L 9" (23 cm)

Identification **ADULT:** Relatively large billed. Gray crown and back with a dark mask and a reddish orange central crown patch. Tail distinctly notched. Underparts white with grayish wash across chest. **JUVENILE:** Similar but more brownish gray on back, with rusty-edged upperwing and uppertail coverts and tail feathers.

Similar Species Eastern is overall smaller and darker above, has obvious white tail tip, and lacks tail notch. Yellow-tinged immature Gray is superficially similar to Thick-billed, but the latter would be heavier billed, yellower below, and darker above, without notched tail. Loggerhead Shrike has a shorter, hooked bill and white on wings and tail.

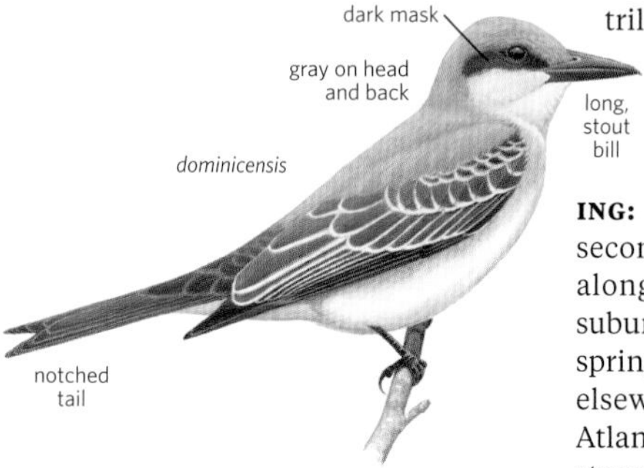
dark mask
gray on head and back
long, stout bill
dominicensis
notched tail

Voice **CALL:** *Pe-cheer-ry;* also a rapid trill similar to Eastern. **DAWN SONG:** Poorly known; described as "complex chatter."

Status & Distribution Rare to fairly common. **BREEDING:** Mangroves, open woodland, second growth, forest edge, mainly along immediate coast, including suburban situations. **MIGRATION:** In spring, arrives southern FL mid-Mar., elsewhere along Gulf and southern Atlantic coasts mid-Apr.–early May; stragglers into June. In fall, departs mid-Sept.–Oct. **WINTER:** Central and southern West Indies, central and eastern Panama, and extreme northern S.A.; rare southern FL, northern West Indies. **RARE STATUS:** Casual, mainly in spring, west to LA (recent breeding records) and TX. Also casual in eastern N.A., mostly in fall, north to ON, QC, and the Maritimes. Accidental to southwestern BC.

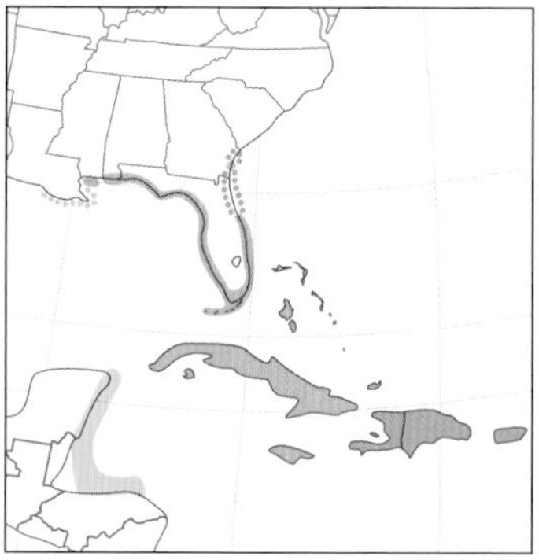

Population Stable.

LOGGERHEAD KINGBIRD *Tyrannus caudifasciatus* LOKI ■ 5

The Loggerhead Kingbird, a widespread West Indian species, has a disputed history of occurrence in FL. Only three of at least nine reports since 1971 are now considered valid, the first in 2007. Polytypic (6 sspp., likely nominate from Cuba). L 9" (23 cm)

Identification **ADULT:** Blackish crown and face, slight rear crest; rest of upperparts dark gray; white edges on wing coverts; white underparts; short primary projection compared with similar N.A. species.

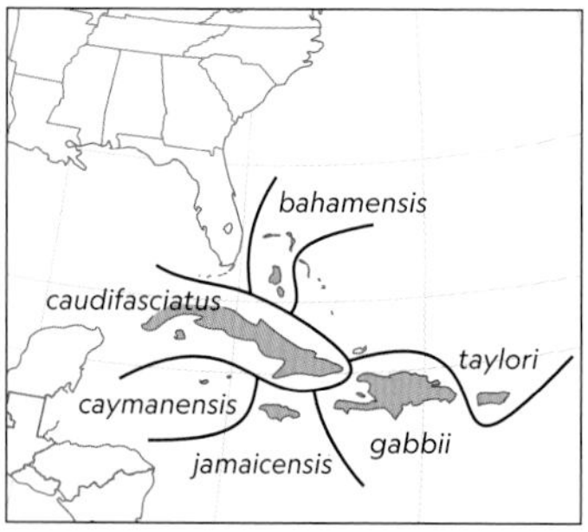
bahamensis
caudifasciatus
taylori
caymanensis
gabbii
jamaicensis

Geographic Variation Underparts yellowish in Bahamas (*bahamensis*); square tail with narrow white tip; white tip lacking in Puerto Rican (*taylori*) and Hispaniolan (*gabbii*) populations, which might be separate species.

Similar Species Eastern is smaller and much smaller-billed. Giant Kingbird (*T. cubensis*), an endangered Cuban endemic, is rounder headed and has a much thicker bill.

Voice **CALL:** Variable, but usually a loud buzzy *tireet,* or *teerrrp* often repeated; mostly silent outside the breeding season.

Status & Distribution Endemic to the Caribbean, year-round resident of tropical lowland evergreen forest and pine forests up to 6,000-ft elevation. The range includes the northern Bahamas, Cuba, the Greater Antilles, and at least formerly the Cayman Is. **RARE STATUS:** The three (all photographed) accepted US records are from southern FL. Previous reports represented misidentifications; although one photographed may have been a Giant Kingbird (*T. cubensis*), which has a half dozen late 19th-century specimen records in winter from Great Inagua in the southern Bahamas and from the Caicos Is.

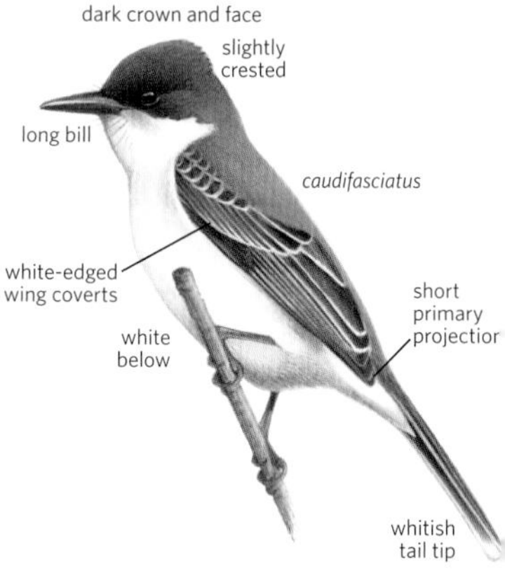
dark crown and face
slightly crested
long bill
caudifasciatus
white-edged wing coverts
white below
short primary projection
whitish tail tip

Population Stable.

SCISSOR-TAILED FLYCATCHER *Tyrannus forficatus* STFL ■ 1

The striking Scissor-tailed Flycatcher is our only regular long-tailed Tyrannus. It is not only graceful and beautiful, but also common and conspicuous. Monotypic. L 10–14.8" (25–38 cm)

Identification **ADULT:** Long, forked tail of male longer than female. Pale gray head and back; blackish wings; extensive white in outer tail. Whitish underparts; pinkish wash on belly. Salmon sides, flanks, underwing coverts; bright red axillaries. **JUVE-NILE:** Duller yellowish pink on underparts; tail much shorter. **Similar Species** Adult unmistakable. Immature is superficially like the

Western Kingbird, but it lacks pure yellow tones on belly and its tail is disproportionately longer, narrower, forked, more extensively white. **Voice CALL:** Sharp *bik* or *pup*; also a chatter. **DAWN SONG:** A series of *bik* notes interspersed with *perleep* notes.

Status & Distribution Common. **BREEDING:** Open country with scattered trees and shrubs. **MIGRATION:** In spring, arrives mid-Mar.–early Apr. Fall peak Oct., some into Nov. **WINTER:** Southern FL, southern Mexico to central Costa Rica, a few to Panama. Rare in TX, LA; locally regular central south TX and extreme southeastern LA. **RARE STATUS:** Rare to casual, to Pacific coast, southeastern AK, southern Canada, Atlantic coast. **Population** Range spreading east.

juvenile

juvenile

reddish pink "armpit"

adult ♂

long forked tail with extensive white; female has shorter tail

orange-buff belly

adult ♂

pale gray head and back

FORK-TAILED FLYCATCHER *Tyrannus savana* FTFL ■ 3

This distinctive, handsome, common Neotropical counterpart of the Scissor-tailed Flycatcher is an annual stray to N.A. Polytypic (4 sspp.; 2 in N.A.). L 14.5" (37 cm)
Identification ADULT: Unmistakable with a disproportionately very long, narrow, forked, mostly black tail. Black head contrasts with pale gray back and white underparts. Yellow central crown patch. **JUVENILE & IMMATURE:** Similar but much shorter tail, duller black head.
Geographic Variation Most of the more than 100 records in N.A. pertain to nominate (adult illustrated and described above), which has on average a darker back and less head-back contrast. Presumably, these records represent overshooting northbound austral migrants from southernmost S.A. breeding populations (occurring here May–July) or reverse spring austral migrants (originating from northern S.A. and occurring here late Aug.–early Dec.).

Northern *monachus*, occurring from southeastern Mexico to Panama, has been documented at least twice in early winter in southern TX; it has a somewhat paler gray back, and its more distinct white collar separates the crown and back. There are also subspecific differences in the notching pattern of outer primary tips (adults). Two other subspecies are found in S.A.
Similar Species The superficially similar Eastern Kingbird is larger and uniformly dark above and has a much shorter, broader, white-tipped tail.
Voice CALL: A sharp *bik* or *plick* and a chattering trill.
Status & Distribution Common. **YEAR-ROUND:** A variety of open and semi-open habitats. Generally present year-round from southeastern Mexico to central S.A. Breeds as far south as central Argentina during austral summer; generally absent as a breeder across upper Amazonia and north of the Amazon River, but widespread during austral winter. **RARE STATUS:** Casual/accidental, scattered records north to CA, WA, ID, AB, MN, WI, MI, ME, ON,

QC, NB, NS, NL; the great majority of records are from eastern N.A. (mostly fall, some spring, a few in winter).
Population Stable.

black cap

gray back

adult

very long, forked black tail

mostly savana

breeding *monachus sanctamartae* & *savana*

monachus & *savana*

monachus

immature

shorter tail

TUFTED FLYCATCHER   *Mitrephanes phaeocercus*   TUFL ■ 5

This distinctive montane tropical fly-catcher acts like a pewee, perching on conspicuous branches. After sally-ing, it may return to the same favored perches. Polytypic (4 sspp.; presum-ably 2 in N.A.). L 5" (13 cm)

Identification ADULT: Obvious tufted

brownish olive above

distinct pointed crest

cinnamon face and underparts

crest. Bright cinnamon face, under-parts; upperparts generally more brownish olive. Long brownish olive wings; dull cinnamon wing bars; pale edges to tertials and secondar-ies. **JUVENILE:** Similar but with pale cinnamon-tipped feathers on upper-parts; broader, brighter wing bars. Plumage held briefly.

Geographic Variation TX records pre-sumably *phaeocercus* (eastern Mexico to El Salvador); AZ record presumably *tenuirostris* (western Mexico).

Similar Species Buff-breasted Fly-catcher is paler; lacks tuft; has more-contrasting wing bars, blacker wings with shorter primary projection, and typical empid-like behavior, including flicking tail up.

Voice CALL: A whistled burry *tchur-ree-tchurree*, sometimes given singly. Also a soft *peek*, similar to Hammond's.

Status & Distribution Common to fairly common in resident range.

BREEDING: Favors somewhat open areas in pine-oak and evergreen woods from Mexico to Bolivia. **WINTER:** Some move to lower elevations; often found in a greater variety of habitats in win-ter. **RARE STATUS:** Casual or very rare in summer to southeastern AZ; has nested in Huachuca Mts.; acciden-tal western AZ. Casual in winter and spring to West TX.

Population Stable.

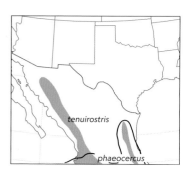

tenuirostris

phaeocercus

These rather plain, dark olive, medium-size flycatchers (14 species in the Americas; three breed north of the Mexico border; one casual to FL) have long wings. They sit motionless, with no tail movement, on fairly high, conspicuous perches, sallying forth to catch insects.

OLIVE-SIDED FLYCATCHER   *Contopus cooperi*   OSFL ■ 1

Olive-sided Flycatchers are usually easy to find, perching conspicuously on the top of a snag or on high dead branches or giving far-carrying vocal-izations. From their high and open perches, they launch out on long sal-lies to snag insects, and then often return to the same perch. Monotypic. L 7.5" (19 cm)

Identification ADULT: Large and dis-proportionately short tailed, with a rather large blocky head. Brownish gray-olive above. Distinctive whitish

tufts on side of rump, often hidden by wings but at other times is quite vis-ible. Dull white throat, center of breast, and belly contrast markedly with dark head. Brownish olive sides and flanks appear streaked, giving a heavily vested appearance. Mostly black bill; center and sometimes base of lower mandible are dull orange. All molting occurs on wintering grounds. **JUVENILE:** During fall separated from worn adults by fresh plumage, buff-brown wing bars, and brownish wash to upperparts.

Geographic Variation Populations differ subtly; breeders in Southern CA average slightly larger with a large bill. See below for geographic variation of vocalizations.

Similar Species Wood-pewees and Greater Pewees are similar in shape, but their underparts are less strongly patterned; they lack a sharply defined "vest" and are also relatively longer tailed, especially Greater Pewee; it also has an all-orange lower mandible. Wood-pewees are smaller and lack

a strong vested appearance. Eastern Phoebes are sometimes mistaken for this species, but they tend to perch lower to the ground and habitually dip tail downward.

Voice CALL: A repeated *pep*, often

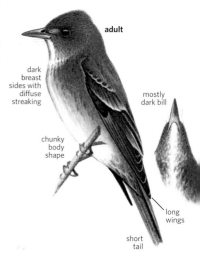

adult

dark breast sides with diffuse streaking

chunky body shape

mostly dark bill

long wings

short tail

repeated in groups of three or two: *pep-pep-pep pep-pep*; sometimes given singly. **SONG:** In the West far-carrying, distinctive, clear, whistled *quick-THREE-beer*; second note higher pitched; in the boreal north and East the accent is on the final note. Additional studies are needed.

Status & Distribution Uncommon to fairly common. **BREEDING:** Boreal forest, subalpine forest, spruce bogs, and mixed conifer or mixed conifer-deciduous forest; prefers relatively open areas, particularly burned areas. More numerous in the West. **MIGRATION:** In Southwest mostly mid-Apr.–mid-May; in Great Lakes May 10–mid-June. Generally rare migrant in East. **WINTER:** Mostly in Andes of western S.A.; smaller numbers in southern Mexico and C.A. Casual in winter on coastal slope of Southern CA (and outside Andes in S.A.). **RARE STATUS:** Casual on Bering Sea islands (Pribilofs and St. Lawrence I.).

Population Near Threatened. Widespread decline detected throughout most of range, including core range in Northwest. Once more widespread in WV, they are now scarce to absent as breeders; extirpated from southern New England, and locally in the

adults

white on belly extends up through center of breast

white flank patches show periodically from above

West. Threats include loss of habitat within breeding and wintering range, fire suppression, and a decline in flying insects.

GREATER PEWEE *Contopus pertinax* GRPE ■ 2

While restricted in the US to pine and pine-oak woodlands of the mountains of AZ and adjacent NM, this species is not hard to find. It perches conspicuously near the top of a tree, often larger trees than those frequented by Western Wood-Pewees. Its calls and distinctive song carries well. Birds outside of range are often identified by *pip* notes. Polytypic (3 sspp.; *pallidiventris* in N.A.). L 8" (20 cm)

Identification ADULT: Large; usually shows a short spiky, tufted crest. Large bill with a dark upper mandible and a distinctive entirely bright yellowish orange lower mandible. Generally dark olive above and on face with somewhat paler lores. Darker wings and tail, with indistinct wing bars. Paler underparts, especially on throat, which contrasts with grayish breast. Worn summer birds are paler overall, more grayish olive above. Adults molt on breeding grounds. **JUVENILE:** Similar to adult but with cinnamon wash to upperparts and buffy-cinnamon wing bars. More difficult to separate from adults than other N.A. *Contopus* due to adult postbreeding molt that occurs on breeding grounds.

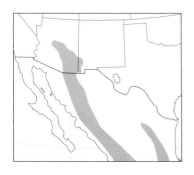

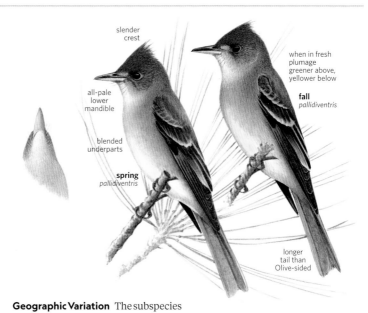

slender crest

when in fresh plumage greener above, yellower below

fall *pallidiventris*

all-pale lower mandible

blended underparts

spring *pallidiventris*

longer tail than Olive-sided

Geographic Variation The subspecies *pallidiventris* occurs in the Southwest; it is similar in size to nominate subspecies (central Mexico to Guatemala), but averages slightly paler and grayer. The subspecies minor of C.A. is smaller.

Similar Species Larger than woodpewees, with more noticeable tufted crest and larger bill that is bright orange below. (Even the brightest-billed Eastern Wood-Pewees do not have such a bright orange lower mandible.) Note vocalizations. Olive-sided Flycatcher shows a distinctive "vest," is shorter tailed, and lacks a crest and has a darker lower mandible. Eastern Phoebe is generally paler with a darker bill, tends to perch lower to the ground, and habitually dips tail downward.

Voice CALL: *Puip* or *beek* notes, often repeated in groups (e.g., *puip-puip-puip*); very similar to Olive-sided Flycatcher's call but averages somewhat higher pitched and softer. **SONG:** A far-carrying, distinctive, plaintive *ho-say ma-re-ah*. Song often begins with series of repeated introductory *whee-de* or *wee-de-ip* phrases.

Status & Distribution Uncommon to fairly common. **BREEDING:** Pine or pine-oak forest; in US often found in riparian areas in steep-sided canyons where pine-oak habitat borders canyon. Ranges south to Nicaragua and El Salvador. **MIGRATION:** Most return to AZ in mid-Apr. and leave by mid-Sept. Rarely noted in migration. **WINTER:** Most withdraw from northern portion of range and from higher elevations. **RARE STATUS:** Casual: TX, southern and central CA (mostly winter); also casual in winter in southern AZ.

Population Stable.

WESTERN WOOD-PEWEE   *Contopus sordidulus*   WEWP ■ 1

This species is extremely similar to the Eastern Wood-Pewee and is best identified by range and voice. Out-of-range birds should be identified with great care and preferably documented with photos and especially recordings of vocalizations. This identification is one of the most difficult ones in N.A. This is the only exclusively western-breeding passerine that winters almost entirely in S.A. Polytypic (5 sspp.; 2 in N.A.). L 6.3" (16 cm)

Identification ADULT: Extremely similar to Eastern Wood-Pewee, with long wings that extend one-third of the way down the tail. Average differences listed below, but plumage somewhat variable. Western Wood-Pewee tends to be slightly darker, browner, and less greenish than Eastern Wood-Pewee, with complete grayish breast band, darker centers to undertail coverts, less pale nape

contrast, and duller back. Wing bars, on average, are narrower and more grayish, contrasting less with wings. Base of lower mandible is primarily dark (usually darker than Eastern Wood-Pewee) but usually shows some dull orange. Adults molt on wintering grounds, and worn summer birds (and fall birds in N.A.) are essentially identical to Eastern Wood-Pewee in appearance. **JUVENILE:** During fall separated from worn adults by fresh plumage, buff-gray wing bars, and brownish wash to upperparts. Many have more extensive pale coloration on lower mandible than adults (i.e., more like Eastern Wood-Pewee). On average, wing bars contrast less than on juvenile Eastern Wood-Pewee; also note that lower wing bar is broader and more defined than upper: paler tips on greater coverts are broader and more defined than those on median coverts. Vocalizations may be the only diagnostic means of separating these look-alike species.

Geographic Variation Differences minor and clinal. Compared to widespread *veliei*, coastal breeders from southeastern AK through central OR (*saturatus*) usually have more of a yellow wash to flanks, a browner breast, and a duskier crown that contrasts more with brownish or olive back. Mexican and C.A. subspecies are darker and larger, except for southern Baja California's *peninsulae*, which are paler with a larger bill.

Similar Species Extremely similar to Eastern Wood-Pewee and best separated by range and voice. (Average visual differences compared above.) Most often confused with Willow Flycatcher; note Willow's (and Alder's) relatively short pri-

mary projection, smaller size, and as with other *Empidonax* the habitual upward tail flip. Compare with Greater Pewee, Olive-sided Flycatcher, and Eastern Phoebe.

Voice CALL: A harsh, slightly descending *peeer*; a short, even, rough *brrt*; and clear descending whistles similar to Eastern's up-slurred *pwee-yee*, but on one pitch or descending, not ascending. **SONG:** Has three-note *tswee-tee-teet*, usually mixed with peer notes; heard mostly on breeding grounds.

Status & Distribution Common. **BREEDING:** Open woodlands. **MIGRATION:** In spring, AZ and CA mid-Apr.–mid-June. From CO to OR, first individuals generally appear in early May; peak throughout the West in mid- to late May, but many into June, a few to late June. In fall, primarily Aug. and Sept. Most have left US by early Oct.; a few records into Dec. **WINTER:** Mostly northwestern S.A. No valid US winter records. **RARE STATUS:** Casual to eastern N.A. and north to western and northern AK and YT.

Population Breeding Bird Survey trends show widespread declines throughout most of range north of Mexico.

paler
adult

darker
adult

averages darker plumage
with darker lower man-
dible than Eastern, but
identification safely done
only by voice

variations
occasionally with
all-pale lower
mandible

EASTERN WOOD-PEWEE   *Contopus virens*   EAWP ■ 1

This species is extremely similar to the Western Wood-Pewee and is best identified by range and voice. Potential out-of-range birds should be identified with great care and preferably documented with photos and especially with recordings of vocalizations. Monotypic. L 6.3" (16 cm)

Identification ADULT: Plumage generally dark grayish olive above with dull white throat, darker breast; whitish or pale yellow underparts. Bill has

black upper mandible and dull orange lower mandible, usually with a limited black tip. Long wings extend one-third of the way down the tail. Very similar to Western Wood-Pewee, but spring and early summer adults are usually more olive with less extensive breast band (often produce vested appearance) and a pale smooth gray nape that contrasts slightly with a darker crown and face, the dark scooping under the eye; this overall effect creates a sub-

tly different aspect than in Western. Wing bars are often broader and more contrasty. Adults molt on wintering grounds, and worn summer birds (and fall birds in N.A.) are essentially identical to Western Wood-Pewee in appearance. **JUVENILE:** During fall separated from worn adults by fresh plumage, buff-gray wing bars, and brownish wash to upperparts. Many have more extensive dark coloration to lower mandible and appear more

usually adults have orange lower mandible

slightly paler nape than Western

juveniles can have all-dark bill

fresh spring adult

worn late summer adult

juvenile

long primary projection

variations

Status & Distribution Common. **BREEDING:** Variety of woodland habitats. **MIGRATION:** Primarily circum-Gulf migrant. Most return mid-Apr. (southern TX) to mid.-May (Great Lakes); in fall, remain later than the Western, regularly singing into Sept., and present into early Oct. **WINTER:** Mostly northern S.A. No valid US winter records. **RARE STATUS:** Casual to western N.A., mainly in migration, but also records of summering birds. Records in the West are best confirmed by vocalizations.

Population Breeding Bird Survey shows widespread declines, particularly in central N.A. Causes for decline unknown.

like Western Wood-Pewee. On average, wing bars stand out more than on Western Wood-Pewee, with upper and lower wing bars same color and prominence (unlike Western, which usually has a less noticeable upperwing bar).

Similar Species Extremely similar to Western Wood-Pewee and best separated by range and voice (see species). Most often confused with Willow and Alder Flycatchers. Note Willow's and Alder's relatively short primary projection (barely reaching beyond base of tail), smaller size, brighter wing bars, and habit of upward tail flipping. Wood-pewees also forage from higher prominent perches, to which they repeatedly return. Compare with Greater Pewee, Olive-sided Flycatcher, and Eastern Phoebe.

Voice CALL: A loud, dry *chip plit* and clear, whistled, rising *pawee* notes; often given together: *plit pawee*. **SONG:** A clear, slow plaintive *pee-a-wee*; second note is lower; often alternates with a down-slurred *pee-yuu*.

CUBAN PEWEE *Contopus caribaeus* CUPE ■ 5

This resident species from Cuba and the Bahamas is casual to south FL. It typically forages from low perches and is relatively tame and, usually, easily approached. Also called the Crescent-eyed Pewee, it was formerly treated as conspecific with the Hispaniola and Jamaican Pewees, as the Greater Antillean Pewee. Polytypic (4 sspp.; likely *bahamensis* in N.A.). L 6" (15 cm)

Identification Appears relatively small and long-tailed for a *Contopus*, almost *Empidonax*-like, with upright posture and erectile crest and large bill.

Best told by conspicuous and expanded white crescent behind eye, set off nicely by no trace of an eye ring on top and in front of eye. Olive above, darker on head. Noticeably paler underparts, especially belly and undertail coverts, which are yellowish when fresh. Wing bars and tertial edges are duller, especially with eastern *Empidonax*.

Geographic Variation US records likely *bahamensis* of the Bahamas.

Similar Species Wood-Pewees lack bold white crescent behind eye; Eastern and Western Wood-Pewees have longer primary projection and shorter tails. Most *Empidonax* have more uniform eye ring, prominent wing bars, smaller bill, and more regular upward tail flicking.

Voice CALL: Repeated *wheet* or *dee*; also a *vi-vi* similar to call of La Sagra's Flycatcher. **SONG:** A long thin descending whistle, sometimes likened to the sound of a bullet flying through the air: *wheeeooooo*.

Status & Distribution Common in resident range. **BREEDING:** Pine and

distinct crescent-shaped white eye ring

sometimes shows slight crest

short primary projection

longer tail than Eastern

other woodlands, forest edge, brushy scrub edges; Bahamas and Cuba. **RARE STATUS:** Casual in south FL (about six records), mostly in spring and fall; also one winter record. Several other FL observations either undocumented or not accepted by Florida Ornithological Records Committee.

Population Unknown.

EMPIDS Genus *Empidonax*

Eleven of the 15 species of empids breed in N.A. These small, drab birds are notoriously difficult to identify. A majority are various shades of greenish above and then paler below, and many have wing bars and eye rings. Most are long-distance migrants. Plumages become duller with wear. All eastern species but Acadian Flycatcher are circum-Gulf migrants.

YELLOW-BELLIED FLYCATCHER *Empidonax flaviventris*   YBFL ■ 1

Even in migration, these birds favor shady forest interiors. They are usually detected by their vocalizations or by

conspicuous circular yellow eye ring

short bill

1st fall

worn fall adult

yellow throat and belly

olive breast

spring

their active foraging with tail and wing flicking. Monotypic. L 5.5" (14 cm)

Identification Appears short tailed, with a large head and rounded crown that appears slightly peaked at back; moderate primary projection. Broad-based bill may appear somewhat large; entirely pale yellow-orange or pink lower mandible. **ADULT:** Mostly olive above; much more yellow below with a fairly extensive olive wash across sides of breast. Olive color from sides of face blends smoothly into more yellow throat. Conspicuous bold eye ring is relatively even, often yellowish, and often slightly thicker behind eye. Wings appear very black with bright white or pale yellow wing bars (less so on worn birds). Worn fall migrants are grayer above and paler below, occasionally nearly grayish white below. Molt occurs on wintering grounds. **FIRST-FALL:** Similar to spring adults but with bold yellow-buff wing bars.

Similar Species Compared to Acadian Flycatcher, note Yellow-bellied's smaller size, shorter primary projection, smaller bill, shorter and narrower tail, olive wash on breast, and molt timing. Very worn fall adults approach Least Flycatcher in appearance, but note Yellow-bellied's more blended throat, uniformly pale lower mandible, and call. See Pacific-slope Flycatcher

for differences with that species; see also Cordilleran Flycatcher.

Voice CALL: Includes a sharp whistled *chiu*, similar to Acadian. **SONG:** A hoarse *che-bunk* similar to Least, but softer, lower, not as snappy.

Status & Distribution Generally uncommon. **BREEDING:** Northern bogs, swamps, and other damp coniferous woods. **MIGRATION:** Circum-Gulf migrant. Late in spring, early in fall; e.g., south TX early May and southern Great Lakes mostly May 15–June 5, early Aug.–late Sept. **WINTER:** Eastern Mexico to Panama. Accidental in winter to southeastern AZ. **RARE STATUS:** Casual mostly in fall to western N.A.

Population Stable.

ACADIAN FLYCATCHER *Empidonax virescens*   ACFL ■ 1

The Acadian has a calm, even languorous look, with often-drooped wings and a minimum of wing and tail flicking. However, its vocalizations are

greenish above

spring

delivered with force and it frequently perches relatively high in trees. Monotypic. L 5.8" (15 cm)

Identification A large empid with long primary projection and a broad tail. There is usually a peak at back of head. Bill is largest of any empid: long and broad based with an almost entirely yellowish lower mandible. **ADULT:** Generally bright olive above and pale below. Usually with pale grayish throat, pale olive wash across upper breast, white lower breast, and faint yellowish belly and undertail coverts. Usually narrow, yellowish, and sharply defined eye ring (but may be faint on some birds). Quite dark wings with prominent wing bars.

Acadians typically appear very worn by mid-summer, almost white below. Molt occurs on breeding grounds. **FIRST-FALL:** Similar to adults but with bold buffy wing bars and more exten-

sive yellow coloration throughout. **Similar Species** Compare with similarly structured Alder and Willow Flycatchers and smaller Yellow-bellied and Least Flycatchers. **Voice CALL:** *Peek*, similar to first part of its song; louder and sharper than other empid call notes. On breeding grounds gives *pwi-pwi-pwi-pwi-pwi*, flicker like, but softer. **SONG:** An explosive *pee-tsup*.
Status & Distribution Common. **BREEDING:** Mature forest, includ-

ing swampy areas. **MIGRATION:** Mostly trans-Gulf migrant. In spring, arrives Gulf Coast in Apr.; southern Great Lakes, May 10–20. In fall, departs quickly after molting, July–late Sept.; rare in US after early Oct. **WINTER:** Mostly Panama to Ecuador. **RARE STATUS:** Casual in QC, NB, and NS. Least likely eastern empid in West. Accidental in NM, AZ, SK, and BC; also Iceland and UK.
Population Stable.

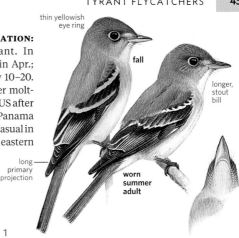

thin yellowish eye ring

fall

longer, stout bill

long primary projection

worn summer adult

ALDER FLYCATCHER *Empidonax alnorum* ALFL ■ 1

Until 1973, Alder Flycatcher was considered the same species as the Willow Flycatcher—the two collectively known as Traill's Flycatcher. Alders are difficult to separate from Willow, especially eastern birds, without hearing their distinctive vocalizations. Monotypic. L 5.8" (15 cm)
Identification A large empid with moderately long primary projection and a fairly broad tail. Moderately long and broad-based bill; almost entirely pinkish lower mandible, sometimes with a dusky area at tip. All molts occur on wintering grounds. **ADULT:** Most have strongly olive-green upperparts. Usually dark head, often with a grayish tone; contrasts with white throat. Paler underparts washed with olive across breast. Usually narrow and well-defined eye ring, but may be almost absent. Quite dark wings with prominent wing bars and tertial edges. Alders usually appear quite

worn by mid-summer and during fall migration, with paler upperparts and reduced wing bars and eye ring. **JUVENILE:** Similar to adults but with buffy wing bars. Post-juvenal molt occurs on wintering grounds.
Similar Species Extremely similar to Willow; should be identified by voice. On average, Alder has a slightly shorter bill, a more distinct eye ring, and a greener back. It usually has a darker head with a more contrasting white throat than nominate eastern Willow. The western subspecies of Willow Flycatcher appears less like Alder than do eastern Willows; note western Willow's duller and more blended tertial edges and wing bars, and shorter primary projection. Acadian is also very similar, but it has longer primary projection, and its white throat contrasts less with paler face. Compared with wood-pewees, note Alder's relatively short primary projection (barely reaching beyond base of tail), smaller size, brighter wing bars, and tendency to flick its tail.
Voice CALL: A loud *pip* similar to Hammond's Flycatcher but more robust. **SONG:** Harsh and burry *rrree-bee-ah*; last note is almost more of an inflection and may be inaudible at a distance. Sings occasionally on spring migration.
Status & Distribution Generally common on breeding grounds,

but surprisingly scarce as a migrant. **BREEDING:** Damp brushy habitats and wet woodland edges; bogs; birch and alder thickets. **MIGRATION:** Circum-Gulf migrant. Spring is late; rare in south TX before early May; southern Great Lakes peak in late May. Fall is primarily Aug. and Sept. Most have left US by early Oct. **WINTER:** S.A. **RARE STATUS:** Rare in Southeast in fall. Casual to CO, WA, CA, and Bering Sea islands. Accidental UK.
Population Unknown.

1st fall

slight greenish cast on back

moderate primary projection

thin eye ring

worn fall adult

very dark wings with sharply contrasting tertial edges

strong contrast between face and throat

spring

WILLOW FLYCATCHER *Empidonax traillii* WIFL ■ 1

Willow subspecies in eastern US and Canada (nominate *traillii* and *campestriss*, merged by some authorities) are more similar to an Alder Flycatcher than to northwestern *brewsteri* and "Southwestern Willow Flycatcher"

(*extimus*). Polytypic (5 sspp.; all in N.A.). L 5.8" (15 cm)
Identification Almost identical in structure to Alder. Bill averages slightly longer in western subspecies; primary projection in western sub-

species is shorter than Alder, but in eastern subspecies equal to Alder. All molts occur on wintering grounds.
ADULT: Eastern subspecies' features overlap almost completely with Alder, but upperparts average less strongly

olive; head usually paler and contrasts slightly less with white throat; eye ring less prominent. **JUVENILE:** Similar, but buffy wing bars.

Geographic Variation Subspecies fairly well defined but intergrade where ranges meet. Compared to eastern, all western subspecies have shorter primary projection; more blended tertial edges; and duller, buffy, wing bars. (They also differ from Alder in those characteristics.) Northwestern *brewsteri* has a darker brownish head (more like Alder) and upperparts and a brownish breast band. Rocky Mts. and Great Basin *adastus* is similar to *brewsteri* (farther west and north), but its upperparts and head average paler brownish olive to grayish olive; it also has a more extensive yellow wash to flanks and vent. Southwestern *extimus* (generally in lower elevations near range of overlap with *adastus*) has more grayish head, contrasting with olive-tinged grayish brown back and wing bars (averages paler than other western subspecies).

Similar Species See Alder Flycatcher. Also often confused with wood-pewees; note Willow's short primary projection,

smaller size, and habit of flicking tail upward.

Voice CALL: Thick, rich *whit*, like Least or Dusky but more resonant. **SONG:** Harsh, burry *fitz-bew* with variations. Sometimes only a husky, rough *rrrr-up*. Often sings on spring migration, sometimes early in fall migration. Songs of the "Southwestern" lower pitched, more drawn out (especially second phrase).

Status & Distribution Fairly common; "Southwestern" uncommon and local. **BREEDING:** Damp willow riparian habitats, edges of pastures, and mountain meadows. **MIGRATION:** Nominate *traillii* is a circum-Gulf migrant. Late in spring; rare in south TX before early May; arrives in southern Great Lakes mid-May. Fall primarily Aug., Sept. Most leave east by early Sept., west by early Oct. **WINTER:** Western Mexico to Ven-

ezuela. A few winter records for Southern CA. Nominate *traillii* believed to winter farther south (mostly in S.A.) than western subspecies. **RARE STATUS:** Casual migrant in Southeast and AK (has bred southeast).

Population Southwestern *extimus* is listed as Endangered by USFWS. Species faces widespread decline in N.A., largely due to loss of habitat.

worn
fall adult
traillii

1st
fall
traillii

extremely
faint eye ring

gray face with
less contrast
between face
and throat

spring
traillii

1st fall
brewsteri

long broad-based
bill with pale lower
mandible

darker
above
than
traillii

darker
face

extimus
similar to
but slightly
paler than
brewsteri

spring
extimus

spring
brewsteri

wing bars
and tertial
edges less
contrasty
than *traillii*

brewsteri
adastus

traillii
(includes campestris)

extimus

brewsteri
& adastus

all subspecies

LEAST FLYCATCHER *Empidonax minimus*   LEFL ▪ 1

This active and vocal species frequently flicks its wings and flips tail upward. Monotypic. L 5.3" (13 cm)

Identification Smallest eastern *Empidonax*. Primary projection usually short. Short, broad-based bill with a mostly pale lower mandible, usually

with a small dusky area at tip. Appears large headed with short tail. **ADULT:** Bold white eye ring. Grayish brown above with olive wash to back. Underparts whitish with gray wash across breast. Wings usually quite dark with contrasting wing bars. Adult's post-

breeding molt may begin on breeding grounds but occurs mostly on wintering grounds. A partial pre-breeding molt begins on wintering grounds but regularly continues into spring migration. **FIRST-FALL:** Similar but with buffy wing bars. Post-juvenal molt

short, broad-based bill

conspicuous white eye ring

short primary projection

1st fall

worn fall adult

grayish head, olive back

grayer wash confined to breast

pale flanks

spring

occurs primarily on breeding grounds. **Similar Species** May be confused with almost any other empid. In East, only Willow Flycatcher has a similar *whit* note. Compared to Willow, Alder, and Acadian, Least is smaller, a smaller bill, and shorter primary projection; it also usually has a conspicuous eye ring. It is similar to Dusky, Gray, Hammond's, and Willow Flycatchers in the West. Hammond's often appears similar because both have a relatively large head and a narrow tail; note Hammond's smaller, narrower bill; longer primary projection; and darker underparts, especially on flanks (always whitish on Least). Dusky and Gray Flycatchers usually appear longer tailed with duller wings; they usually have more extensive dark on underside of mandible. Gray is most easily separated by its tendency to dip its tail downward.

Voice CALL: A sharp *whit*, very similar to, but usually sharper and louder than, Dusky; not as thick as Willow. Given frequently even in migration. **SONG:** A snappy *che-beck, che-beck*, usually given in a rapid series. Similar to Yellow-bellied's *che-bunk* but snappier, higher, more accentuated, and usually quickly repeated. Sings regularly on spring migration.

Status & Distribution Common; local in West. **BREEDING:** Deciduous woods, orchards, parks, brushy understory important. **MIGRATION:** Circum-Gulf migrant. Spring to southern TX, mostly late Apr.–early May; southern Great Lakes early May–early June, peaking in mid-May. In fall, adults leave very early (e.g., peak at Long Pt., ON, mid- to late July). First-fall birds generally early Aug.–mid-Oct., depending on latitude. **WINTER:** Mexico to Costa Rica. Rare in winter in central and south FL, and very rare on Gulf Coast. **RARE STATUS:** Rare migrant, mostly in fall, through most of West; casual in winter to CA and southern AZ. Casual to AK. Accidental to Iceland. **Population** Stable.

HAMMOND'S FLYCATCHER *Empidonax hammondii*   HAFL ■ 1

The Hammond's is usually quite active, flicking its wings and tail upward at the same time. The *pip* call note is distinctive, but many Hammond's are relatively quiet in migration, especially in fall. Monotypic. L 5.5" (14 cm)

Identification This relatively small *Empidonax* appears large-headed and short-tailed, and it has long primary projection. Short, very narrow bill, with a mostly dark lower mandible (may be extensively pale on young birds). **ADULT:** Bold white eye ring, usually expands behind eye. Grayish head and throat; grayish olive back; gray or olive wash on breast and sides; yellow-tinged belly. Adult postbreeding molt occurs on breeding grounds. Fall birds are much brighter olive above and on sides of breast, much more yellow below. Pre-breeding molt is variable, so some spring birds appear much brighter than others. **FIRST-FALL:** Similar to adults but with buffy wing bars. Post-juvenal molt occurs primarily on breeding grounds.

Similar Species Overall, Hammond's is quite distinctive, with long wings, a tiny partly dark bill and, in fall, bright plumage. It is most often confused with Dusky and most easily separated by its *pip* call note if vocalizing. Experienced observers will note Dusky's relatively short primary projection and its bill, which is wider, longer, and usually with more pale coloration to base of lower mandible. Dusky has a slightly paler face with a pale supraloral area, yet fainter eye ring. Gray can be eliminated by these same characteristics and by its tendency to dip its tail, like a phoebe. Least has a larger, more triangular bill that is usually extensively pale; much

short, thin-based bill

conspicuous eye ring often almond-shaped

gray throat

olive breast and flanks

overall shape similar to Least

longer primary projection than Dusky

underparts, including flanks, duskier than Least

fall

spring

medium-length, slightly notched tail is edged with gray

paler underparts; shorter primary projection; more contrasting dark wings with prominent wing bars and tertial edges; and frequently given *whit* call note. Bright fall birds resemble "Western Flycatchers," but note gray throat and narrow dark-tipped bill.

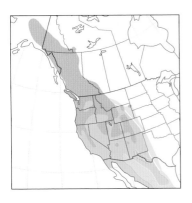

Voice CALL: A *pip*, similar to Pygmy Nuthatch and very different from calls of Dusky, Gray, and Least. **SONG:** First element of song suggests the *che-beck* of Least and particularly *che-bunk* of Yellow-bellied; may be given alone, particularly late in breeding season. Similar to Dusky but hoarser and lower pitched, particularly on second phrase. Phrases tend to be more two-parted. Also lacks the high, clear notes that are typical in Dusky's song. **Status & Distribution** Fairly common. **BREEDING:** Somewhat local in fairly solid mature coniferous forest usually without shrubby component; Duskies tend to be found in more open habitats at elevations both above and below Hammond's. **MIGRATION:** Spring is mostly early Apr.–early June. Fall is late Aug.–mid-Oct. Throughout most of the West, Hammond's is the most likely empid to be seen in Oct., especially inland; regular in CA into late Oct. **WINTER:** Southeastern AZ (very rare elsewhere in Southwest and CA) to western Nicaragua. **RARE STATUS:** Casual to Great Plains, mostly in fall (more frequent than Gray and slightly more frequent than Dusky). Casual in late fall and winter to eastern N.A. **Population** Likely declining with reduction of mature coniferous forest.

GRAY FLYCATCHER *Empidonax wrightii* GRFL ▪ 1

This empid has the distinctive habit of slowly dipping its tail downward, like a phoebe. Monotypic. L 6" (15 cm)
Identification Gray Flycatcher is a large empid with short primary projection and a fairly long tail. Head often appears disproportionately small and rounded. Straight-sided bill is narrow and averages slightly longer than Dusky Flycatcher. Most Gray's bills have a well-defined dark tip to lower mandible that contrasts with pinkish orange base, but a few have an entirely pinkish orange lower mandible. Molting generally occurs on wintering grounds. **ADULT:** Gray above, sometimes with a slight olive tinge in fresh fall plumage (in US, mostly seen in southeastern AZ, where Gray Flycatchers regularly winter). White eye ring is inconspicuous on pale head. Adult postbreeding molt occurs on wintering grounds. **JUVENILE:** Similar to adult but more brownish gray overall with buffy wing bars and somewhat fresher plumage in summer and early fall. Post-juvenal molt usually occurs on wintering grounds but may begin on breeding grounds.
Similar Species This species is very similar to Dusky Flycatcher in terms of plumage and structure. Gray Flycatcher does have a slightly longer, more two-toned bill, but is easily separated from Dusky Flycatcher and all other *Empidonax* by its distinctive relaxed phoebe-like tail dipping.
Voice CALL: A loud *wit*, very similar to Dusky Flycatcher; averaging somewhat stronger. Also very similar to calls of Least and Willow Flycatchers. **SONG:** A vigorous *chi-wip* or *chi-bit*, followed by a liquid *whilp*, trailing off in a gurgle.
Status & Distribution Fairly common. **BREEDING:** Dry habitat of the Great Basin and east of Cascades north to southern BC. Usually in yellow pine or pinyon-juniper. **MIGRATION:** An early spring migrant; Gray Flycatcher is the only western empid likely to be seen away from the Southwest or Pacific coast in Apr.; usually arrives on breeding grounds mid-Apr.–mid-May. Fall migration is primarily mid-Aug.–early Oct. Rare in fall to West Coast; more regular in spring. **WINTER:** Southern AZ to central Mexico. Rare in winter in Southern CA. **RARE STATUS:** Mostly in fall, but some in winter to eastern N.A. Casual to Great Plains. Accidental to AB, SK, ON, OH, ME, MA, CT, DE, and NC.
Population Breeding Bird Survey data indicate that this species is experiencing an overall increase.

head is small and rounded

eye ring inconspicuous on pale gray face

bill longer than Hammond's and has extensive pale base with dark tip

short primary projection

winter

spring

long tail, often bobbed down like a phoebe, unlike all other *Empidonax*

DUSKY FLYCATCHER *Empidonax oberholseri* DUFL ▪ 1

This is a relatively sedate species; it occasionally flicks its tail or wings upward. Dusky Flycatchers are generally intermediate between Hammond's and Gray Flycatchers, and many resemble the Least Flycatcher. Monotypic. L 5.8" (15 cm)
Identification A medium-size empid with short primary projection and a long tail. Straight-sided bill is intermediate in length between that of Hammond's and Gray Flycatchers, but closer to Gray; it is extensively pale at base, fading to a dark tip. Most molting occurs on wintering grounds. **ADULT:** Generally drab. Grayish olive upperparts, with little contrast between head and mantle. More yellowish below, with a pale gray or whitish throat and pale olive wash on upper breast. A moderately well-defined white eye ring may not stand out against head. Lores are often paler than on other empids. Adult postbreeding molt occurs on wintering grounds. During fall migration, these Dusky Flycatchers should not be confused with much brighter Hammond's Flycatchers. Fresh late-fall birds appear quite bright with

flanks and undertail coverts strongly washed yellow, but in US these birds are usually only seen in southeastern AZ, where they winter.

Similar Species Structure and plumage similar to Gray Flycatcher; but Grays have distinctive habit of dipping tail downward, like a phoebe. Dusky Flycatcher is also often confused with Hammond's Flycatcher. In addition to noting different molt timing, observe Hammond's longer wings and longer primary projection;

shorter, thinner, darker bill; and eye ring that expands behind eye. Also note vocalizations. Many Dusky Flycatchers appear similar to Least Flycatchers, particularly worn fall birds. Dusky Flycatcher has a slightly longer and narrower bill; the tail usually appears longer; wings are usually not as black and have low contrast wing bars and tertial edges; and eye ring is usually narrower and less pronounced. Least is also paler below, especially on the flanks, which are whitish, not dusky.

Voice CALL: A *whit* usually softer than Gray. Frequently gives a mournful *dew-hic* on breeding grounds, particularly early and late in day. **SONG:** Several phrases consisting of a quick, clear high *sillit*; a rough, up-slurred *ggrrreep*; another high *sillit* that may be omitted; and a clear, high *pweet*,

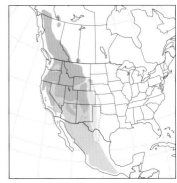

which is reminiscent of male Pacific-slope Flycatcher's contact note. This song is often confused with that of Hammond's Flycatcher, but Hammond never has high, clear notes characteristic of Dusky.

Status & Distribution Common on breeding grounds, but surprisingly very rare and often misidentified in migration west of the Sierra and southeastern CA. **BREEDING:** Open woodlands and brushy mountainsides. **MIGRATION:** Spring is mostly mid-Apr.–early June; most arrive on breeding grounds mid-May. Fall is mostly Aug.–late Sept. **WINTER:** Mostly Mexico (southeastern AZ south to Isthmus of Tehuantepec, Mexico). **RARE STATUS:** Casual to Great Plains and AK. Accidental to WI, ON, NS, DE, VA, and AL.

Population The Breeding Bird Survey shows a significant increase in N.A.

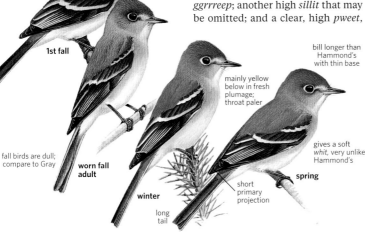

1st fall

fall birds are dull; compare to Gray

worn fall adult

winter

long tail

mainly yellow below in fresh plumage; throat paler

bill longer than Hammond's with thin base

short primary projection

gives a soft *whit,* very unlike Hammond's

spring

PINE FLYCATCHER *Empidonax affinis* PINF ■ 5

The Pine Flycatcher, a tropical species, has occurred once in N.A., from southeast AZ in summer. Similar to Hammond's and Dusky Flycatchers in coloration, but the bill coloration and eye ring shape are more suggestive of Pacific-slope or Cordilleran Flycatcher. Vocalizations are distinctive. Polytypic (5 sspp.; likely *pulverius* in N.A.). L 5.5" (14 cm)

Identification Dullest and grayest northwestern subspecies (*pulverius*) described. A moderate size and long-tailed *Empidonax*, olive-gray above, tinged yellow on pale belly and grayish throat; broad dusky-olive wash across breast. Wings dark with pale wing bars and secondary and tertial edges. Distinctive is the white eye ring that broadens behind the eye, much like Cordilleran and Pacific-slope, and the all orangish lower mandible, also

like those species. Moderately long primary projection.

Similar Species Hammond's and Dusky Flycatchers are similarly colored but have a more circular eye ring (especially Dusky) and have a dark tip to the lower mandible. Dusky and Gray Flycatchers have a shorter primary projection. Cordilleran and Pacific-slope Flycatchers greener above, yellower below; all vocalizations very different.

Voice Call a sharp *whip*, louder and lower than Dusky. Song of northern populations in Mexico consists of two to four hesitant phrases. Songs south of the Isthmus of Tehuantepec somewhat different.

Status & Distribution A resident montane species found in open woodland from central Chihuahua and southern Coahuila south to the mountains of

Guatemala. **RARE STATUS:** Accidental southeast AZ, one calling (not singing) well-documented bird at Aliso Springs, east side of the Santa Rita Mts. from 28 May–7 July 2016 (built and sat on a nest).

Population Stable.

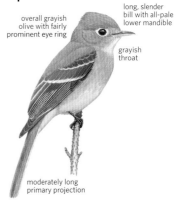

overall grayish olive with fairly prominent eye ring

long, slender bill with all-pale lower mandible

grayish throat

moderately long primary projection

PACIFIC-SLOPE FLYCATCHER *Empidonax difficilis*   PSFL   ■   1

This was formerly considered the same species as the Cordilleran Flycatcher, known as the Western Flycatcher. The two species are extremely similar, and silent birds are not separable away from the breeding grounds. Pacific-slope Flycatchers are usually found in shaded areas, even in migration. They tend to be quite active, often flicking the wings and tail. Polytypic (3 sspp.; all in N.A.). L 5.5" (14 cm)

Identification Medium-size empid; fairly short primary projection; tail may appear relatively long. Wide bill; entirely yellow-orange lower mandible. Most molting on wintering grounds. **ADULT:** Brownish green above, yellow below, brownish wash across breast. Broad pale eye ring often broken or very narrow above; expands behind eye. Adults usually appear quite worn by mid-summer and during fall migration. **JUVENILE:** Similar but with buffy wing bars, variably pale underparts. Some quite whitish; may be confused with Least. First post-juvenal molt on wintering grounds.

Geographic Variation Widespread nominate (south to Southern CA) has relatively small bill; relatively pale dull olive and yellow plumage with pale lemon wing bars (ochre-buff in juvenile and first-year). Subspecies *insulicola* (endemic to Channel Is.) has been suggested as a separate species: relatively dull; relatively long bill; whitish wing bars (buffy in juvenile and first-year); lower-pitched song; male's position note a rising *tsweep*, unlike more slurred mainland vocalizations. Longer-billed, smaller *cineritius* (accidental in winter to western AZ, breeds in southern Baja California) is dingy whitish in color with lemon tone.

Similar Species Virtually identical to Cordilleran Flycatcher. Both show extensive yellow throat and bright orange-yellow or pink lower mandible are different from those of other western *Empidonax*. Yellow-bellied is quite similar. Yellow-bellied usually has blacker wings; bolder wing bars and tertial edges; shorter tail; longer and more evenly spaced primary tips ("Western" types show much less even spacing); more uniform, round eye ring, rarely broken above; and more greenish upperparts. Its head tends to be rounder and not as peaked; calls are most reliable characteristic.

Voice Very similar to Cordilleran; some birds give intermediate vocalizations. **CALL:** Male's position note in mainland populations is a slurred *tseeweep*; male's position note on the Channel Is. is rising *tsweep*. Neither as two-parted as the typical male Cordilleran's position note. **SONG:** A sharp *tsip*; a thin, high slurred *klseeweee*; a loud *ptik!* Usually contains three separate phrases (usually repeated), all of which are higher pitched and thinner than for other *Empidonax* (except Cordilleran).

Status & Distribution Common. **BREEDING:** Moist woodlands, coniferous forests, shady canyons and residential areas. Often builds nests on human-made structures. **MIGRATION:** Moves more frequently in lowlands during migration than the Cordilleran.

Spring migration end of Feb.–mid-June. Fall is primarily Aug. and Sept., some through Oct. **WINTER:** West Mexico, usually at lower elevations than Cordilleran and closer to the coast. A few winter in coastal Southern CA. **RARE STATUS:** Out-of-range birds very similar to Cordilleran, so identification is extremely difficult and most should remain identified only to the species pair. Except under exceptional circumstances (e.g., a specimen or recordings of vocalizations), strays are best considered "Western Flycatchers." Accidental to CT, NY, PA, and LA. **Population** Little information.

rather long broad-based bill with all-orangish lower mandible

eye ring expands slightly behind eye

usually slight break to eye ring

pale 1st fall *difficilis*

1st fall *difficilis*

some immatures in fall are very dull; can be confused with Least Flycatcher

worn fall adult *difficilis*

olive sides to breast have slight brownish tint

longer tail projection than Yellow-bellied

spring *difficilis*

CORDILLERAN FLYCATCHER *Empidonax occidentalis*   COFL   ■   1

This and Pacific-slope Flycatcher are extremely similar and all but impossible to identify away from the breeding grounds. Recent studies indicate that over western and northwestern parts of range, vocalizations are intermediate and perhaps the two should be merged again into a single species. Polytypic (2 sspp.; *hellmayri* in N.A.). L 5.8" (15 cm)

Identification Cordilleran Flycatcher is only readily identifiable by range and by male's two-note *tee-seet* contact call. Average differences in song and other call notes are described below, but these may be matched by Pacific-slope Flycatcher. Some Cordilleran at least occasionally give contact notes that are indistinguishable from classic position notes of Pacific-slope Flycatcher. On average, the Cordilleran is slightly larger, darker, more green above, and more olive and yellow below, with slightly shorter bill. As with Pacific-slope, most molting takes

hellmayri

spring

almost identical
to Pacific-slope
in coloration

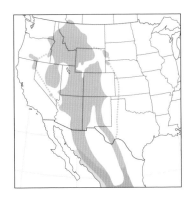

place on the wintering grounds. Post-juvenal and pre-breeding molts average more extensive in Cordilleran, but this not evident in the field. The only subspecies in N.A. is *hellmayri*.
Similar Species Essentially identical to the Pacific-slope Flycatcher. Average differences presented above, but individual variation makes it impossible to identify any one individual by anything other than in some parts of range or perhaps by vocalizations. See the Pacific-slope Flycatcher

for separation from other *Empidonax*.
Voice Extremely similar to Pacific-slope Flycatcher. **CALL:** Cordilleran usually gives a two-note *tee-seet*. Male's position note is the most distinctive difference between Pacific-slope and Cordilleran Flycatchers. A minority of Cordillerans give position notes, at least occasionally, that are extremely similar to those given by Pacific-slope Flycatchers. The whistled *seet* note often seems sharper in Cordilleran than in Pacific-slope. **SONG:** Very similar to Pacific-slope Flycatcher's song, but the first note of the first phrase is higher than the second note.
Status & Distribution Fairly common on breeding grounds, rare to uncommon in migration. **BREEDING:** Coniferous forests and canyons of the West. Often nests on cabins and other human-made structures. **MIGRATION:** Generally considered rare in lowlands during migration, even within core breeding range. Spring migration is late Apr.–early June, later than Pacific-slope; fall is primarily Aug. and Sept., some into Oct. **WINTER:** Mexico, usu-

ally in foothills to about 2,000 ft. **RARE STATUS:** Extreme similarity to Pacific-slope makes the identification of out-of-range birds extremely difficult, so true status of strays is unresolved. Thought to be casual on western Great Plains, mostly in fall; however many of these records could pertain to the Pacific-slope. Except under exceptional circumstances, strays are best considered just "Western Flycatchers." Accidental to LA.
Population Stable.

BUFF-BREASTED FLYCATCHER *Empidonax fulvifrons*   BBFL ■ 2

The Buff-breasted Flycatcher, the most distinctive *Empidonax*, reaches north of the Mexico border, where it is found locally in southeastern AZ and southwestern NM. Polytypic (6 sspp.; *pygmaeus* in N.A.). L 5" (13 cm)
Identification Smallest empid in N.A. Short, small bill; appears entirely pale orange from below; fairly long primary projection. Birds molt before leaving for wintering grounds. **ADULT:** Warm brown above with bright cinnamon buff wash to breast; throat paler, but blends into face. Whitish eye ring often somewhat almond shaped; pale wing bars. Worn summer birds paler, more grayish above, usually with at least a hint of buff across breast. **JUVENILE:** Similar but with duller upperparts,

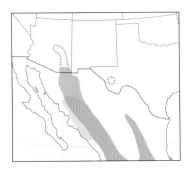

well-defined buffy wing bars. Post-juvenal and adult molts occur mostly on breeding grounds.
Geographic Variation Subspecies *pygmaeus* (AZ and NM) has relatively dark brownish upperparts, often tinged with gray.
Similar Species Distinctive. Even dullest birds usually have a light buff wash to breast. Tufted is much richer and deeper cinnamon in color, has a distinctive crest and longer wings, and does not flick tail.
Voice CALL: A soft to sharp *pwic* or *pwit*, sharper and higher than casual *whit* of Dusky. **SONG:** Jerky, with two phrases that are repeated one after the other: *chiky-whew, chee-lick*.
Status & Distribution Uncommon. **BREEDING:** Dry pine and oak woodlands, usually near openings with scattered shrubs. Very local in Huachuca and Chiricahua Mts. of AZ. Largely casual or accidental in US outside southeast AZ, but breeds in other mountain ranges as far north as central AZ and in Peloncillo Mts., NM. A small outpost also found at Davis Mts. Preserve, TX, where breeding was detected annually from 2000 to about 2010 until a devastating forest fire. Northern limit in eastern Mexico is

in mountains near Monterrey (within 100–120 mi of TX). **MIGRATION:** Casual in migration. Spring migration in AZ early Apr.–mid-May. Fall departs in Aug.–mid-Sept. **WINTER:** Mexico to central Honduras. **RARE STATUS:** Accidental to CO (accepted sight record from El Paso Co., May 1991) and Southern CA (photographed from Kern Co., May 2016).
Population Formerly bred north to central AZ and west-central NM. No information on trends south of the US.

conspicuous whitish eye ring

fresh
pygmaeus

cinnamon-buff wash across breast and down sides

tiny bill with orangish lower mandible

worn summer adult
pygmaeus

PHOEBES Genus *Sayornis*

These plump flycatchers have a distinctive trait of dipping their tails downward. Relatively short-distance migrants, they are found in open habitats, often near water or near human structures and often nest under building overhangs or under bridges. The Vermilion Flycatcher is very closely related and shares many traits with the phoebes.

BLACK PHOEBE *Sayornis nigricans*   BLPH ▦ 1

A distinctive black-and-white phoebe of the Southwest, this species is almost always found near water. Polytypic (6 sspp.; *semiatra* in N.A.). L 6.8" (17 cm)

Identification Black overall with white belly and undertail. **JUVENILE:** Plumage similar to adult, but browner,

semiatra

phoebes drop their tail down

mostly blackish except for sharply contrasting white belly

juvenile *semiatra*

with two cinnamon wing bars; some buffy tipping above.

Geographic Variation N.A. *semiatra* (south to western Mexico) has duller and duskier head; birds south of Isthmus of Panama have extensive white in wings.

Similar Species Distinctive. Has hybridized with Eastern Phoebe (CO); offspring appear intermediate.

Voice CALL: Includes a loud *tseew* and a sharp *tsip*, similar to Eastern Phoebe but sounding more plaintive and whistled. **SONG:** Thin whistled song consists of two different two-syllable phrases: a rising *sa-wee* followed by a falling *sa-sew*; usually strung together one after the other.

Status & Distribution Uncommon to common. **BREEDING:** Woodlands, parks, suburbs; often near water. **MIGRATION:** Resident over much of range. Breeders return to CO late

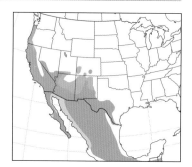

Mar.–mid-Apr.; depart early Sept. Fall migrants detected on the Farallones (CA) early Sept.–late Nov. **RARE STATUS:** Casually appears north and east to northern OR, WA, ID, northern UT, central TX, OK, and KS. Accidental to southwestern BC, south-central AK, FL, and NJ.

Population Increasing, with range slowly spreading north.

EASTERN PHOEBE *Sayornis phoebe*   EAPH ▦ 1

The Eastern Phoebe is a rather dull phoebe found in the East and across central Canada. It frequently nests under eaves, bridges, or other overhangs on human-made structures. This species is most easily separated from other dull flycatchers by its characteristic habit of dipping its tail in a circular motion. Monotypic. L 7" (18 cm)

Identification Eastern is brownish gray above; darkest on head, wings, and tail. Underparts mostly white, with pale olive wash on sides and breast. Fresh fall adult is washed with yellow, espe-

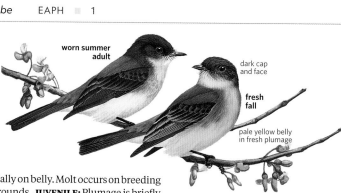

worn summer adult

dark cap and face

fresh fall

pale yellow belly in fresh plumage

cially on belly. Molt occurs on breeding grounds. **JUVENILE:** Plumage is briefly held and similar to adult but browner, with two cinnamon wing bars and cinnamon tips to feathers above.

Similar Species Pewees are darker and they have longer wings, but they are most easily separated from phoebes by phoebes' distinctive tail wagging. *Empidonax* flycatchers have eye rings and wing bars, which are absent in Eastern Phoebe. An *Empidonax* flycatcher flicks its tail upward; only Gray Flycatcher dips its tail downward.

Voice CALL: Typical call is a sharp *tsip*, similar to Black Phoebe but less strident. **SONG:** Distinctive, rough whistled song consists of two phrases:

schree-dip followed by a falling *schree-brrr*; sometimes strung together one after the other.

Status & Distribution Common. **BREEDING:** Woodlands, farmlands, and parks, often near water (less so in other seasons). **MIGRATION:** Breeders return to the Midwest mid-Mar.–late Apr. and depart late Sept.–early Nov. **RARE STATUS:** Rare in fall and winter to CA; otherwise casual west of the Rocky Mts. and northwestern Great Plains. Accidental to southern YT, northern and western (nested once near Nome) AK, and NL; sight record for England.

Population Apparently stable.

SAY'S PHOEBE *Sayornis saya* SAPH ▪ 1

This widespread western species is frequently seen perching on bushes, boulders, fences, and utility wires. Polytypic (4 sspp.; 3 in N.A.). L 7.5" (19 cm)
Identification Grayish brown above with contrasting black tail. Grayish breast contrasts with tawny-cinnamon belly and undertail coverts. **JUVENILE:** Like adult but with two cinnamon wing bars.

Geographic Variation Variation is complicated by individual variation and wear. Compared to more widespread nominate, northwestern breeding *yukonensis* (AK and NT to coastal OR) is smaller billed, with deeper orange underparts and deeper gray upperparts. Resident *quiescens* (deserts of southeastern CA and southwestern AZ) is paler overall.
Similar Species Female and immature Vermilion Flycatchers have a white throat and white chest with brown streaks; also has pale supercilium contrasting with dark auriculars.
Voice CALL: Typical call is a thin, plaintive, whistled, slightly downslurred *pee-ee*. **SONG:** A fast whistled *pit-tsear*, often given in flight.
Status & Distribution Common. **BREEDING:** A variety of open habitats from tundra to desert, usually with cliffs, canyons, rocky outcroppings, or human-made structures for nesting. **MIGRATION:** Early in spring, late in fall.

Bulk of migrants arrive from CO to east of the Sierra Nevada in CA late Mar.–mid-Apr.; southern BC late Mar.–mid-Apr.; and AK early to mid-May. In fall, most depart AB late Aug.–early Sept.; CO and OR late Aug.–Sept. **RARE STATUS:** Casual over most of the East (late fall–winter) with records for every state and province (or territory) except WV, NU, and PE; also Bering Sea islands.
Population Apparently stable.

grayish back

tawny belly

blackish tail

Genus *Pyrocephalus*

VERMILION FLYCATCHER *Pyrocephalus rubinus* VEFL ▪ 1

The stunning Vermilion Flycatcher frequently pumps its tail like a phoebe. Polytypic (13 sspp.; 2 in N.A.). L 6" (15 cm)
Identification Adult male red and brown. Adult female grayish brown above, darker tail; pale throat and supercilium; streaked pale breast; tawny belly and undertail. **JUVENILE:** Similar to adult female but spotted below; belly with yellowish tinge. Immature male acquires red feathers by late fall/early winter.
Geographic Variation Southwestern *flammeus* has paler brown upperparts with a grayish tinge. Head and breast of adult male average more orange-red in color, often with pale mottling. Adult

male *mexicanus* (TX) has deep, bright red underparts and darker brown upperparts without grayish tinge.
Similar Species Compare an immature with Say's and Eastern Phoebes, noting white throat and streaked breast.
Voice CALL: A sharp *pseep*. **SONG:** A soft tinkling repeated *pit-a-set*, *pit-a-see*, *pit-a-see*; often given in fluttery fight display. Also sings while

perched.
Status & Distribution Common. **BREEDING:** Open woodlands, parks; often near water. **MIGRATION:** Returns to northern breeding areas by late Apr.; most depart Sept.–Oct. **RARE STATUS:** Casual throughout most of lower 48 (fewest along Canadian border and in northeast) and to ON, NS, and QC. Has bred in CO and OK.
Population Apparently stable.

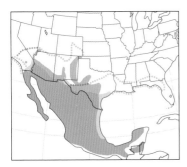

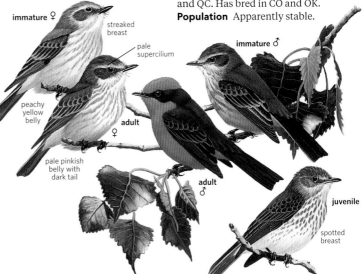

dark face

immature ♀

streaked breast

pale supercilium

immature ♂

peachy yellow belly

adult ♀

pale pinkish belly with dark tail

adult ♂

juvenile

spotted breast

SHRIKES Family Laniidae

Loggerhead Shrike (CA, Apr.)

W hat could be more incongruous than predators that attack mice, bats, birds, and insects, yet sing a soft catbird-like song? Add to this their habit of impaling prey and one gets extraordinarily fascinating birds.

Structure Shrikes' morphology is designed for carnivorous, raptor-like capturing and killing: broad wings and a long tail combine for speed and maneuverability; large jaws and powerful, hooked bills enable them to break the necks of their prey and carry them away. The legs and feet of shrikes are much weaker than those of true raptors and are of little use in hunting, although they occasionally do grab and carry small prey with their feet.

Behavior From an elevated perch, a shrike may watch attentively over long periods for prey. It strikes in a smooth movement, swooping down, attacking quickly, and carrying its quarry to another perch. Impaling prey on sharp objects may serve various purposes, including storage for a later meal, holding the food to tear it apart, and to mark territory and attract mates with a ring of kills.

Plumage The Northern and Loggerhead Shrikes (residents of North America) are patterned distinctively in black, gray, and white with black masks. Both are gray above and paler below, with black wings marked by white patches and long, black, white-edged tails.

Distribution The 34 species in four genera (30 in the genus *Lanius*) now comprising this family are distributed mainly in the Northern Hemisphere and Africa, although the ranges of some extend well south in Asia.

Taxonomy Shrikes are part of a primarily Old World radiation of crows and related families (Corvoidea), explaining their current placement near the corvids and vireos. A recent revision has split the Holarctic Northern Shrike from more southerly and western Eurasian populations, resulting in a change in the scientific name of our birds.

Conservation Many species around the globe are in trouble, some seriously. Suggested causes include loss of habitat, elimination of prey by pesticides, warming climate, and trapping for food in Southeast Asia. BirdLife International codes: 2 NT, 1 VU, and 1 CR.

BROWN SHRIKE *Lanius cristatus*   BROS ■ 4

This small Asian stray to N.A. usually perches on the side of a bush or tree, often less conspicuously than our N.A. shrikes. It feeds primarily on insects but will also take small birds. Polytypic (4 sspp.; likely nominate in N.A.). L 7.5" (19 cm)

Identification No white patch on wings. **ADULT MALE:** Warm brown upperparts; brighter rump and upper-tail coverts; buffy underparts; narrow black mask; white supercilium. **ADULT FEMALE:** Mask less distinct; sides, flanks barred pale brown. **JUVENILE:** Mask limited to brown ear coverts; upperparts, sides, flanks faintly barred.

Geographic Variation Nominate most likely to occur in N.A., although other subspecies could occur.

Similar Species Juvenile Northern Shrike (only brownish shrike in N.A.) is larger, longer billed, distinctly barred across underparts. See Red-backed Shrike. The Tiger Shrike (*L. tigrinus*), another highly migra-

female has duller head pattern and some barring on sides and flanks

juvenile
cristatus

overall brown above

white supercilium above dark mask

stubby bill

buffy-brown wash on sides and flanks

adult ♀
cristatus

adult ♂
cristatus

tail uniform with back

tory Asian species, could reach N.A. **Voice CALL:** Harsh *chacks* and *chur-ucks*. **SONG:** Soft warbles interspersed with sharp notes.

Status & Distribution Fairly common in Asia. **BREEDING:** Forest edges and open areas with bushes from central Siberia to Kamchatka and Anadyr (Russian Far East); also China and Japan. **WINTER:** Similar habitats in India, Southeast Asia, and East Indies. **RARE**

STATUS: Casual to AK, mostly in fall and a few spring records; also casual in fall and winter to coastal northern CA (four records). Accidental in NS. Most fall and winter records are of juveniles. Two wintering CA birds remained and molted into adult plumage.

Population Declining. Causes unclear, although habitat loss and trapping for food may account for declines in Southeast Asia.

RED-BACKED SHRIKE *Lanius collurio*   RBSH ■ 5

This Old World species has been tentatively recorded once from N.A. Adult males are distinctive; females and particularly juveniles, as well as hybrids with the central Asian Turkestan Shrike (*L. phoenicuroides*), create an abundance of identification issues. Monotypic. L 6.7" (17 cm)

Identification Rather small billed and long winged with a less graduated tail than Brown Shrike. **ADULT:** Male has bluish gray crown with black mask and chestnut back; below, throat white with rest of pale underparts with a light pink cast. Black tail with white base to outer tail feathers (conspicuous in flight). Female much duller with reddish brown back, sometimes barred, and mask brown; underparts with variable barring. Tail brown with white edges. **JUVENILE:** Similar to adult female but more barred above, the back being reddish brown with distinct subterminal dark bars. Most have brown tails, a few with warmer brown uppertail.

Similar Species Juvenile Brown Shrike lacks reddish cast to back. Juvenile Turkestan has warmer tail, colder and grayer back, and small pale patch at base of primaries.

Status & Distribution BREEDING: From western Europe to western Siberia. **WINTER:** Tropical Africa. **RARE STATUS:** One record, a juvenile at Gambell, AK, 3–22 Oct. 2017. One coastal Mendocino Co., CA, 5 Mar.–22 Apr. 2015, was determined to be a Red-backed x Turkestan after spring molt.

Population Declines in western Europe; extirpated as a breeder in UK.

frequently hybridizes with Turkestan Shrike in Central Asia

back with reddish cast

tail less graduated than Brown Shrike

longer primary projection than Brown Shrike

juvenile

LOGGERHEAD SHRIKE *Lanius ludovicianus*   LOSH ■ 1

This species is found in a variety of open habitats, where it perches on trees, shrubs, poles, fences, and utility wires. It captures small rodents and birds but favors large insects. Polytypic (7 sspp.; 6 in N.A.). L 9" (23 cm)

Identification ADULT: Medium-gray upperparts; stubby bill; wide black mask extends thinly across forehead; black wings. **JUVENILE:** Like adult, but faintly barred overall and with white tips to wing coverts.

Geographic Variation Subspecies differ slightly in bill shape and overall coloration; *mearnsi* (San Clemente Is., CA) has darkest gray upperparts.

Similar Species Northern Shrike is larger, paler, and larger billed; its mask does not extend across forehead; juveniles are brownish.

Voice CALL: Harsh *kee*, *kaak*, and *chek* sounds. **SONG:** Repeated chirps, squeaks, warbles, buzzes, and chips.

Status & Distribution YEAR-ROUND: Open country with scattered trees and shrubs, desert scrub, grasslands, farms, parks. **MIGRATION & WINTER:** Northernmost birds move into southerly portions of range; in West, fall migration starts by early July.

Population Near Threatened. Listed as Threatened (*mearnsi*) by USFWS; in Canada, *migrans* listed as Endangered and *excubitorides* as Threatened. Declining rapidly or extirpated in most regions, especially in the East, where *migrans* is now very rare to casual in eastern Midwest, Northeast, and southern Canada.

adult

black wings with white primary patch

black extends across forehead

darker gray back

stubby black bill

juvenile

adult

NORTHERN SHRIKE *Lanius borealis*   NSHR ■ 1

At a distance, this hunter may appear kestrel-like, but it perches more horizontally. It is best known during winter, when it perches on bare treetops or shrubs and frequently bobs tail. It is merciless in pursuing prey. The pursuits (sometimes largely aerial) at times go on for minutes. Polytypic (4 sspp.; 1 in N.A.). L 10" (25 cm)

Identification ADULT: Pale gray upperparts; long, heavy, sharply hooked bill, base paler in winter; black, narrow mask, tapering on lores; white forehead; black wings with white patch across base of primaries; grayish white underparts; long, black, white-edged tail. **FIRST-WINTER:** Slight brownish tint, grayish as season progresses; mask indistinct; underparts barred pale brown. **JUVENILE:** Brownish overall; dark brown patch behind eye; underparts conspicuously barred; wing patch indistinct.

Geographic Variation Recently split from other Old World subspecies west of Ob River, Russia; this arrangement has three east Asian subspecies and one (nominate *borealis*) in N.A. Northeast Asian *sibiricus*, slightly paler than *borealis*, has been recorded four times (two specimens) on Shemya I., AK, in Oct. **Similar Species** Loggerhead is smaller, smaller billed, darker on back; its mask extends thinly across forehead. **Voice CALL:** Loud *keek*, *shak*, other sharp sounds. **SONG:** Soft, catbird-like trills, twitters, whistles, warbles, mews, squeaks, and harsh notes. **Status & Distribution** Uncommon.

BREEDING: On taiga and edge of tundra. **MIGRATION:** Late fall (late Oct.–Nov.) and early spring migrant. **WINTER:** Regular to northern tier of US states; irregular farther south. **RARE STATUS:** Rare but regular in northern AZ, NM, and TX. Recorded south

to Southern CA, TN, and NC. Casual to Bermuda. **Population** Breeding range remote and little data; fewer noted on winter range in recent decades.

gray forehead

faint mask with white eye ring

brownish on head and upperparts

longer bill with distinct hook

paler gray back

adult
borealis

barred underparts

juvenile
borealis

immature
borealis

VIREOS Family Vireonidae

Black-capped Vireo, male (TX, Apr.)

Vireos are known more for their vocal repertoire than for their colorful plumage. Most are a rather uniform greenish or grayish, a few are more patterned, but all blend with their environs—thickets, dense brush, and trees—or have a preference for skulking. They often sing from one position for minutes. In sum, vireos are more often heard than seen.

Structure Vireos are rather stocky. Their bills are rather thick and blunt, with a hook at the end of the upper mandible. Their legs are thick and usually bluish gray.

Behavior Vireos are slow moving and often remain still for a few seconds or much longer, especially when they are singing. They frequently twist their head; some cock or move their tail. Food is mostly insects on the summering grounds; in winter many species eat primarily fruits and seeds. Vireos glean prey from the underside of leaves and limbs, often while hovering. Some are good mimics (e.g., White-eyed Vireo); since their songs are

learned, many develop notes from nearby breeding birds. **Plumage** The sexes look generally alike with differences readily apparent only in Black-capped Vireo. Some species have more subtle differences between the sexes. Plumages are usually similar all year, although most species are brighter in fresh fall feathers after molting (which occurs on the breeding grounds). Vireos are typically gray or green above, white or yellowish below. Keys to identification are presence or absence of wing bars, distinctiveness of wing bars; presence of an eye ring and spectacles, presence of a supercilium, and presence and boldness of adjacent eye lines or lateral crown stripes.

Distribution Widely distributed across the US and Canada; nearly all vireos are migratory. Some species undertake short migrations; others move from Canada to the Amazon Basin.

Taxonomy Fifty-two species of New World vireos make up the core of this family, but vireo taxonomy is evolving. Work is ongoing to determine if some species might involve multiple species (e.g., Hutton's and particularly Warbling). The 17 North American species (three occurring only as strays) are all currently placed in the genus *Vireo*. While superficially similar to warblers, vireos are most closely related to shrikes (Laniidae). Surprisingly, nine southeast Asian shrike-babblers (*Pteruthius*) and one *Erpornis* (once thought to have been part of a group of babblers known as yuhinas) appear to belong to the vireo family, resulting in a family of 62 species in six genera.

Conservation Many North American vireo species are susceptible to Brown-headed Cowbird parasitism; local control programs have resulted in strong population rebounds in *pusillus* Bell's and Black-capped Vireos. Habitat changes or losses impact several species. Bird-Life International codes: 4 NT, 2 VU, and 1 EN.

Genus *Vireo*

BLACK-CAPPED VIREO   *Vireo atricapilla*   BCVI ▪ 2

Small and very active, the males sing persistently throughout the breeding season. Their penchant for skulking in thick scrub makes viewing difficult. Monotypic. L 4.5" (11 cm)

Identification Prominent white spectacles on which eye ring is broken at top. Also note two yellowish wing bars, a greenish back, and reddish eyes. **MALE:** Glossy black cap. **FEMALE:** Slaty gray cap. **IMMATURE:** Cap gray, with males molting in some black during first year. Eyes brown. Females are more buffy below.

Similar Species Unmistakable, more likely to be confused with a Ruby-crowned Kinglet than any other vireo.

slaty cap

black cap, white spectacles

adult ♀

adult ♂

immature ♀

Voice CALL: *Tsidik*, recalling Ruby-crowned Kinglet, and *zhree*, like Bewick's Wren. **SONG:** A crisp, emphatic, and hurried, restless *which-er-chee*, *chur-ee*; two- or three-note phrases, repeated with variations. Whisper songs are softer, less warbled. Persistent vocalizer, singing through the day late Mar.–Aug.

Status & Distribution Uncommon and local. **BREEDING:** Scrubby deciduous vegetation, usually oaks, in rocky hill country. **MIGRATION:** Spring migration occurs late Mar. in TX (earliest 13 Mar.), mid- to late Apr. in OK. Fall migration occurs Aug.–Sept.; most birds depart by early Sept. **WINTER:** Pacific slope of Mexico. No US records.

RARE STATUS: Accidental eastern NE, east-central NM, and southern ON.

Population Vulnerable. Delisted in 2017 by USFWS due to active conservation measures leading to recovery; current US population estimated at 14,000. The species was eliminated from KS in the 1930s and now is nearly gone from much of former range in OK. Major factors include habitat destruction or deterioration and brood parasitism by Brown-headed Cowbird. Overgrazing removes key vegetation and attracts cowbirds; fire suppression adversely impacts the species.

WHITE-EYED VIREO   *Vireo griseus*   WEVI ▪ 1

A loud song from a thicket announces the White-eyed. Polytypic (6 sspp.; 3 in N.A.). L 5" (13 cm)

Identification ADULT: Bold face pattern with yellow spectacles and white iris. Olive above, gray nape, yellow forehead, and whitish wing bars. Whitish below, pale yellow sides and flanks. **IMMATURE:** Gray or brown iris.

Geographic Variation Compared to widespread nominate, *maynardi* (FL Keys) is grayer above, with less yellow below and a larger bill. Subspecies

micrus (southern TX) is smaller than *maynardi* but similar in color.

Similar Species Compare to eastern subspecies of Bell's, which has a dark eye, fainter wing bars, and different face pattern. See also Thick-billed.

Voice CALL: A raspy *sheh-sheh*, often repeated, suggestive of a House Wren. **SONG:** Loud, often explosive, five- to seven-note phrase; usually begins and ends with a sharp *chick*. Great mimic.

Status & Distribution Common. **BREEDING:** Secondary deciduous scrub, wood margins. **MIGRATION:**

Northern breeders migratory; southern populations mostly resident. Arrives southern Great Lakes late Apr.–mid-May. In fall most gone by early Oct. **WINTER:** Southeastern US, Mexico, and Caribbean. Recorded as far north as southern ON in Dec. **RARE STATUS:** Rare to the Maritimes and NL, primarily in fall; very rare but annual to CA (mostly spring), otherwise casual to West. Casual to Azores (fall).

Population Stable.

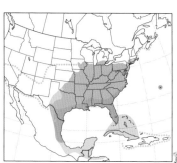

pale gray nape

white iris, yellow spectacles

griseus

FL Keys
maynardi

dark eyes

yellow sides and flanks

thin white wing bars

immature
griseus

THICK-BILLED VIREO *Vireo crassirostris* TBVI ▮ 4

This Caribbean species is a casual visitor to coastal southeastern FL from the Bahamas. Polytypic (5 sspp.; nominate in N.A.). L 5.5" (14 cm)
Identification Larger, stouter billed and more uniformly olive-brown above (lacks gray nape) than White-eyed; breast also tinged brownish. White eye ring is broken above eye, and yellow

broken eye ring above eye

uniform brownish olive upperparts

larger bill than White-eyed

crassirostris

brownish tint to sides of breast

supraloral mark is thick, but does not extend across forehead, like White-eyed. Two bold whitish wing bars.
Similar Species From White-eyed note Thick-billed Vireo's broken white, not yellow, eye ring, pale olive-brown underparts, and particularly lack of gray on nape or contrasting clear yellow on flanks. Iris is darker than on adult White-eyed, but similar to darker-eyed, young White-eyed Vireos; no doubt some reports of Thick-billed represent misidentifications.
Voice CALL: Slow and harsh *sheh*, or *chit* notes. **SONG:** Similar to that of White-eyed, but harsher.
Status & Distribution Common resident in the Bahamas in mangroves and more commonly thickets. **RARE STATUS:** Very rare Aug.–May visitor to

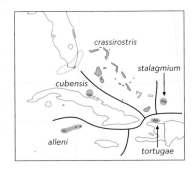

crassirostris

stalagmium

cubensis

alleni

tortugae

coastal southeastern FL and accidental Pinellas Co.
Population Localized and more yellowish Cuban *cubensis* known only from just a few north Cuban cays (mainly Cayo Peredón Grande), is nearing extinction.

CUBAN VIREO *Vireo gundlachii* CUVI ▮ 5

This Cuban endemic has occurred twice on Key West. In Cuba found singly, or in pairs; often joins mixed species flocks, often staying low in the vegetation. Polytypic (4 sspp.; likely nominate in N.A. on probability). L 5.3" (13 cm)
Identification A moderate-size vireo with large eye in a blank-appearing face, diffuse, but broad pale supraloral line and pale area behind eye; two rather thin wing bars. Olive to more grayish above and variably colored below from pale yellow to more whitish. Disproportionately large and extensively pale bill.

Similar Species Distinct. Most closely resembles a Hutton's Vireo from western N.A. Smaller than Ruby-crowned Kinglet with thicker bill and lacks black panel below lower wing bar.
Voice CALL: Includes a rapid descending series of *chi* notes and a scolding *kik*, along with a variety of other notes. **SONG:** Loud whistled and variable *chuee-chuee* (local Cuban name is "Juan Chivi").
Status & Distribution Endemic to Cuba, where widespread (including some large offshore cays and Isla de Juventud) in a variety of woodlands.

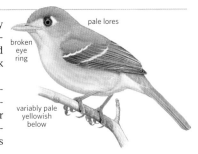

pale lores

broken eye ring

variably pale yellowish below

RARE STATUS: One at Key West, FL, on 19–24 Apr. 2016 and one at Key Largo on 29 Apr. 2017.
Population Stable.

BELL'S VIREO *Vireo bellii* BEVI ▮ 1

Bell's is usually located by its distinctive song, often given from inside dense low and mid-level thickets. Polytypic (4 sspp.; all in N.A.). L 4.7" (12 cm)
Identification Plumage coloration varies with subspecies. All have two ill-defined white wing bars (lower one more prominent), indistinct white spectacles that are broken in front and back, and dark lores.
Geographic Variation Four subspecies become progressively greener above and yellower below from west to east. Endangered West Coast *pusillus* ("Least Bell's Vireo") is dullest. Tail length increases from east to west. The two eastern subspecies bob their tails. Distinct genetic differences between

the eastern (*bellii* and *medius*) and western (*arizonae* and *pusillus*) groups, perhaps representing separate species.
Similar Species Larger Gray Vireo has a longer tail, poorly defined wing bars, and an eye ring without a hint of spectacles.
Voice CALL: A somewhat nasal, wrenlike *chee*. **SONG:** A questioning *cheedle-ee, cheedle-ew*; often delivered in couplets, ending with ascending or descending notes.
Status & Distribution Uncommon to fairly

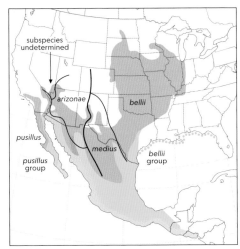

subspecies undetermined

arizonae

bellii

pusillus

pusillus group

medius

bellii group

common. **BREEDING:** Moist woodlands, bottomlands, mesquite, and, in Midwest, shrubby areas on prairies. **MIGRATION:** Seldom seen on migration. First spring arrivals to southern breeding range in late Mar., May to northern breeding locales. Fall migration occurs Aug.–Sept. **WINTER:** Not well known, but primarily Mexico; scattered records for southern tier of states. **RARE STATUS:** Casual north to OR and in Midwest to ON and along Gulf and Atlantic coasts north to ME. **Population** Declines in Southern CA (*pusillus*), AZ (*arizonae*), and eastern

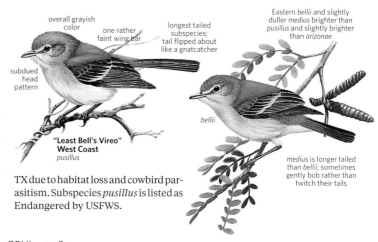

overall grayish color

one rather faint wing bar

longest tailed subspecies; tail flipped about like a gnatcatcher

Eastern *bellii* and slightly duller *medius* brighter than *pusillus* and slightly brighter than *arizonae*

subdued head pattern

bellii

"Least Bell's Vireo" West Coast *pusillus*

medius is longer tailed than *bellii*; sometimes gently bob rather than twitch their tails

TX due to habitat loss and cowbird parasitism. Subspecies *pusillus* is listed as Endangered by USFWS.

GRAY VIREO   *Vireo vicinior*   GRVI ◼ 2

The Gray Vireo is drab and long-tailed. It is an easy bird to miss due to its choice of rather warm, out-of-the-way habitats and its penchant for hiding in undergrowth. It is best located by its persistent song. Gray is an active forager and constantly flicks and whips its tail in a manner reminiscent of a gnatcatcher. Monotypic. L 5.5" (14 cm)
Identification Rather featureless with gray upperparts, whitish underparts, and a thin white eye ring. Wings are

thin whitish eye ring but no spectacled effect

thick, stubby bill

short primary projection

faint wing bar

long tail, which is flipped about

brownish gray, with faint wing bars when fresh, the lower one more prominent. Long tail is gray and edged white. Bill is short and thick, even for a vireo. In fresh fall plumage, Gray Vireo shows the slightest hint of green to rump and uppertail coverts and a slight yellowish wash along flanks.
Similar Species Plumbeous Vireo is similar, especially when worn, but is more boldly marked and has white spectacles; it has a longer primary projection and a shorter tail. Compared to the similarly grayish West Coast subspecies of Bell's (*pusillus*), the larger Gray has a heavier bill.
Voice CALL: A House Wren–like scolding note. The shrill, descending whistled notes, sometimes delivered in flight, are completely unlike any other vireo species. **SONG:** A series of *chu-wee chu-weet* notes, faster and sweeter than Plumbeous.
Status & Distribution Uncommon and local. **BREEDING:** Semiarid foothills and mountains with a variety of scrub habitats, including junipers. **MIGRA-**

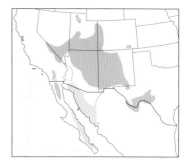

TION: Seldom seen. In spring, arrive CA and TX early Apr.; early May in CO. Some fall migrants depart by late Aug., most by early Sept. **WINTER:** Mostly northwestern Mexico, local in western TX, southern AZ, and a few have recently been discovered in Southern CA (Anza-Borrego SP). **RARE STATUS:** Casual along Southern CA coast and offshore islands, one remarkable specimen record for WI (Oct.).
Population TX population expanding over past 30 years; CA's fragmented and contracted.

HUTTON'S VIREO   *Vireo huttoni*   HUVI ◼ 1

This active vireo often flicks its wings in a kingletlike manner. Quite vocal, Hutton's is often heard before seen. Outside the breeding season, Hutton's vireos form mixed-species flocks. Polytypic (11 sspp.; 7 in N.A.). L 5" (13 cm)
Identification Similar to Ruby-crowned Kinglet and habitually confused with that species. Hutton's is greenish to olive gray above, with an eye ring broken above eye; pale lores. Two whitish wing bars; greenish yellow edges of secondaries and primaries connect to lower wing bar.

Geographic Variation Seven subspecies in N.A. are divided into "Pacific" group and "Interior" (or "Stephen's") group. "Pacific" group birds are smaller

thicker bill than Ruby-crowned Kinglet

eye ring broken above eye, unlike Cassin's Vireo

pale supraloral spot

overall dull

worn summer

fine bill

black bar

Ruby-crowned Kinglet for comparison

pale edges on primaries connect to lower wing bar

interior subspecies average more grayish than coastal subspecies and fresh plumaged fall birds are greener than worn summer ones

overall bright

fresh fall

and greener. "Interior" group birds are larger, paler, and grayer. More than one species might be involved. See map for separate (allopatric) ranges.

Similar Species Most likely to be confused with Ruby-crowned Kinglet; Hutton's has a thicker bill and paler lores and lacks a dark area below its lower wing bar. Immature female Cassin's Vireo can be greenish, but has prominent white spectacles and whitish throat.

Voice CALL: Low *chit* and raspy *rheee*, often followed by nasal, descending *rheee-ee-ee-ee*. **SONG:** "Pacific" group birds give a repeated or mixed rising *zu-wee* and descending *zoe zoo*. "Interior" group, a harsher *tchurr-ree*.

Status & Distribution Uncommon to fairly common. **BREEDING:** Mixed evergreens and riparian woodlands, partial to oaks. **YEAR-ROUND:** Largely resident, some seasonal dispersal to lower elevations, as early as July in CA. **RARE STATUS:** Casual in southeastern CA and southwestern AZ.

Population Recent expansion to the Edwards Plateau in TX.

huttoni group

stephensi group

mexicanus

vulcani

YELLOW-THROATED VIREO *Vireo flavifrons* YTVI ◼ 1

This is a large, colorful vireo and a strong, though slow-paced, singer. It moves sluggishly, which combined with its camouflaged coloration, can make it difficult to locate high in the leaves of the tall shade trees it favors. Yellow-throated often cocks its head as it surveys its surroundings or methodically searches for insects. Monotypic. L 5.5" (14 cm)

Identification Bright yellow spectacles, throat, and breast of this vireo are distinctive. Its wings are dark gray, with two bold, white wing bars. Crown and back are olive, rather bright, contrasting with a gray rump. Immature plumage is similar to that of adult but

paler yellow, sometimes with a slightly buffy throat.

Similar Species Unlike other vireos, but compare with Pine Warbler, with which it is confused, particularly in winter. Pine Warbler has a greenish yellow rump, streaked sides, thinner bill, and less complete and distinct spectacles. Vocalizations of the two are completely different—Pine Warblers often give a high, thin note when moving between branches.

Voice CALL: Includes a rapid series of harsh *cheh* notes, similar to those of "Solitary Vireo" complex. **SONG:** Slow repetition of *de-a-ree, three-eight*; burry, low-pitched two- or three-note phrases separated by long pauses: It often gives a whisper song, which is more warbled and less burry. The burry song is distinct from the clear song of Blue-headed Vireo, but similar to the burrier Plumbeous and Cassin's Vireos.

Status & Distribution Fairly common. **BREEDING:** Deciduous and mixed deciduous-coniferous habitats. **MIGRATION:** Long-distance, trans-Gulf migrant. Early spring migrant, with arrivals in southern states mid–

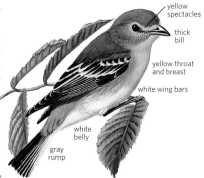

yellow spectacles

thick bill

yellow throat and breast

white wing bars

white belly

gray rump

late Mar., mid Apr. farther north, early May in Great Lakes. Fall migrations Aug.–Oct. (migrants noted as early as late July). Latest records are mid-Oct. in northern and middle latitudes, early Nov. in south. **WINTER:** Tropical lowlands of C.A., Bahamas, and Caribbean to northern S.A. Very rare in southernmost FL, casual in Southern CA, southern TX. Over-reported in the Southeast in winter; most reports are likely of Pine Warblers. **RARE STATUS:** Very rare in the West, more in spring than fall, but some have summered.

Population Apparently stable, with some local fluctuations.

CASSIN'S VIREO *Vireo cassinii* CAVI ◼ 1

Cassin's Vireo is the western counterpart of—and formerly considered conspecific with—the Blue-headed Vireo. Cassin's, very similar to Blue-headed in both plumage and behavior, is routinely heard singing on the breeding grounds, often throughout the day. The movements and behavior of the Cassin's are like those of the Blue-headed. Occasionally, the Cassin's flicks its wings like a kinglet. Polytypic (2 sspp.; nominate in N.A.). L 5.3" (13 cm)

Identification Olive-gray head of Cassin's contrasts slightly with its greenish back. There is a strong pattern of white spectacles and wing bars (the latter can be yellowish) contrasting with dark upperparts. Flanks are heavily washed with olive-yellow; throat and breast are whitish, typically dingy. Tail is dark blackish brown above, with olive or gray edges. From below, outer tail feathers are edged with white when fresh. Sexes

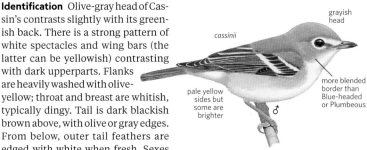

cassinii

grayish head

more blended border than Blue-headed or Plumbeous

pale yellow sides but some are brighter

generally look alike, although plumage of males is slightly brighter than that of females. Plumage of immature Cassin's vireos is duller, especially that

cassinii

lucasanus

of females, which can have entirely green heads and lack any white in tail. **Similar Species** Cassin's Vireo is similar to Blue-headed, but is duller with less face/throat contrast. From the smaller Hutton's Vireo, note the more grayish head on most and especially the prominent spectacles rather than broken eye ring, and the different vocalizations. Also see Plumbeous Vireo. **Voice CALL:** Call of Cassin's consists of scold notes similar to those of Blue-headed. **SONG:** Jerky two- to four-note song almost identical to the song of Plumbeous; much more burry than Blue-headed. Also very similar to the song of Yellow-throated Vireo. **Status & Distribution** Fairly common. **BREEDING:** Coniferous and mixed for-

est. **MIGRATION:** Early spring migrations typically occur in late Mar., peak in late Apr. west of the Sierra Nevada in CA, a little later east of the mountains; early May to BC, east of the Cascades. Fall migrations are more protracted: Aug. to Sept. is the bulk of passage; lingerers through Oct., particularly on the coast. Migrants are more likely throughout the western Great Plains in fall than in spring. **WINTER:** Primarily Mexico. Rare in coastal CA and southern AZ, casual in interior CA, NM, and western TX. **RARE STATUS:** Cassin's is casual north to AK; accidental in the East.

Population Unknown, but droughts and wildfires have likely reduced numbers.

BLUE-HEADED VIREO *Vireo solitarius* BHVI ■ 1

The Blue-headed Vireo is typically found in late spring or summer by its song. It forages at mid-level, yet can be difficult to find among the leaves. In summer it primarily eats insects and appears quite inquisitive, frequently cocking its head as it slowly forages in the branches of a tree. Polytypic (2 sspp.; both in N.A.). L 5.3" (13 cm) **Identification** Bright blue-gray to gray hood clearly contrasts with a bright olive back and white spectacles and throat. Wing bars and tertials are yellow-tinged, and dark secondaries have greenish yellow edges. Tail is dark above, with greenish edges, and white is easily seen in outer tail from below. Underparts are clean white with

bright yellow (sometimes mixed with green) on sides and flanks. Female slightly duller. **Geographic Variation** Larger Appalachian *alticola* has more slaty back; only flanks are yellow. **Similar Species** Told with difficulty from Cassin's Vireo. Note Blue-headed's overall brighter coloration and sharp contrast between blue gray hood and white throat, but some best left unidentified. Songs differ. **Voice CALL:** A nasal *cha-cha-cha-cha*, given as a single *cha* or repeated; a typical vireo scold. **SONG:** Short, clear notes with various intervals, similar to Red-eyed, but slower. Blue-headed Vireo can sing the song of a Yellow-throated Vireo (with which it has hybridized) and is a good mimic (e.g., singing a White-eyed Vireo song or giving a Yellow-bellied Flycatcher call). **Status & Distribution** Fairly common. **BREEDING:** Mixed woodlands. **MIGRATION:** Short-distance (*alticola*) to medium-distance migrant. Earliest vireo to move north in spring, to mid-Atlantic and southern Midwest by mid-Apr.; higher elevations or lati-

♀
solitarius

overall brighter than Cassin's with darker and more contrasting blue-gray head

average duller than male

wing bars often tinged yellow

sharp contrast between auricular and throat in both sexes

♂
solitarius

bright yellow sides with some olive

tudes not until May. Latest vireo to depart in fall, more protracted migration in fall. Most migration starts mid- to late Sept. away from Appalachians. Not expected in southern states until Oct. Midwest peak late Sept.–early Oct.; some remain into Nov. **WINTER:** Winters in southern states south to northern C.A. Rare north of mapped range. **RARE STATUS:** Annual to the West, primarily CA in fall (most records late Sept.–Oct.), casual there in winter and spring. Casual in other western states.

Population Stable.

solitarius

alticola

PLUMBEOUS VIREO *Vireo plumbeus* PLVI ■ 1

Formerly part of the "Solitary Vireo" complex—along with Blue-headed and Cassin's—Plumbeous is the grayer, Rocky Mts. species. Appearing large and bulky, the bird moves through trees in a deliberate man-

ner foraging for insects and sings frequently, even in nonbreeding seasons. Polytypic (4 sspp.; nominate in N.A.). L 5.5" (14 cm) **Identification** Nominate described. A rather colorless species, but with a

strong pattern composed of bold white spectacles and wing bars contrasting with lead gray upperparts. Generally white below, except on sides of breast, which are gray, sometimes tinged olive. Outer tail feathers are broadly edged white, visible from below; tail looks long on this species. In fresh fall plumage, yellow, if present on underparts, is restricted to flanks, and rump might show a greenish tint.

Similar Species Larger and bigger billed than Cassin's, with sharper head and throat contrast and gray upper parts. Avoid confusion with worn spring Cassin's, which can be dull gray above and seemingly lack green or yellow. Compare worn summer Plumbeous to Gray. Gray has a complete eye ring, but lacks supraloral mark, has a shorter bill, and waves its longer tail like a gnatcatcher. Grays also have short wings and a shorter primary projection.

Voice CALL: *Cheh* notes similar to Blue-headed. **SONG:** Burry notes almost indistinguishable from

Cassin's, but usually starts with clear notes; more burry sounding than Blue-headed.

Status & Distribution Uncommon to fairly common. **BREEDING:** Montane forests of pine, and oak-juniper, locally in deciduous woodlands. **MIGRATION:** Late spring migrant, primarily May. Late fall migrant, primarily late

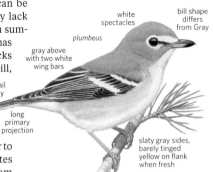

white spectacles
bill shape differs from Gray
plumbeus
gray above with two white wing bars
shorter tail than Gray
long primary projection
slaty gray sides, barely tinged yellow on flank when fresh

Sept.–mid-Oct., with stragglers to Nov. **WINTER:** Primarily Mexico. Rare in southern coastal CA, south-central AZ, casual southern NM, TX. **RARE STATUS:** Casual OR. Casual in eastern N.A.; spring records from ND, IL, MA, and ON; fall records from NJ, LA, AB, and NS.

Population Stable. Recently found breeding in northwest NV.

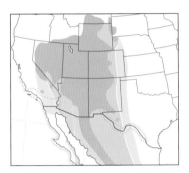

PHILADELPHIA VIREO *Vireo philadelphicus* PHVI ■ 1

This smallish green vireo is usually seen singly amidst tall, leafy trees. Its rather bright coloration may suggest a warbler. Monotypic. L 5.3" (13 cm)

Identification Face pattern diagnostic with dull white eyebrow and distinct dark eye line. Green above, with contrasting grayish cap and variably yellow below, the yellow of even intensity across the breast; palest on belly. No wing bars or spectacles. **FALL:** Fresh plumage, usually brighter green above and brighter yellow below.

Similar Species Warbling Vireo is very similar. Note Philadelphia's dark eye line extending through lores, darker cap, darker primary coverts, and yellow continuing to center of throat and breast. Some spring Philadelphias can be very dull with detectable yellow only with the best of views. Tennessee Warbler, superficially similar, has a thinner bill, white undertail coverts (usually), and usually utters high-pitched flight notes while it forages.

Voice CALL: An *ehhh* sometimes given though often silent. But may vocalize in response to "pishing." **SONG:** Very closely resembles that of Red-eyed Vireo but averages slightly slower, thinner, and higher-pitched.

Status & Distribution Uncommon. **BREEDING:** Open woodlands and riparian habitats. **MIGRATION:** In spring first arrivals usually mid-Apr. (TX),

dark eye line extends through lores
fall birds often quite bright yellow
fall
shorter tail
dark primary coverts
spring
yellow of equal intensity extends across breast

and early May (southern Great Lakes); peak late Apr.–early May, and mid-May, respectively. Broad fall migratory path; more regular on East Coast and in Southeast than in spring. Movement mid-Sept.–mid-Oct.; arrivals in southern states not expected until late Sept. Lingering birds casual to Nov. **WINTER:** Southern C.A. Accidental in Southern CA. **RARE STATUS:** Casual to rare (CA, >150 records) in the West. Casual to the Azores; two Oct. records for Europe.

Population Overall stable.

WARBLING VIREO *Vireo gilvus* WAVI ■ 1

The Warbling is most often located by its song. They tend to work mid and top parts of broad, leafy trees. Polytypic (5 sspp.; 3 in N.A.). L 5.5" (14 cm)

Identification Among the dullest of vireos, lacking wing bars and spectacles; gray with brownish or greenish tones to upper parts. Face pattern is ill-defined, with a dusky postocular stripe and pale lores; white eyebrow lacks a dark upper border. Underparts are typically whitish. **FALL:** In fresh plumage, greener above with yellow wash from sides to undertail coverts.

Geographic Variation Two western

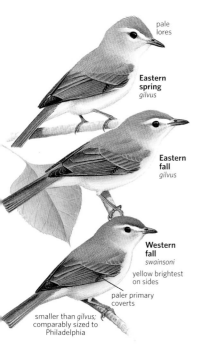

pale
lores

**Eastern
spring**
gilvus

**Eastern
fall**
gilvus

**Western
fall**
swainsoni

yellow brightest
on sides

paler primary
coverts

smaller than *gilvus*;
comparably sized to
Philadelphia

subspecies, especially *swainsoni*, are much smaller (by about 25 percent) than the nominate eastern subspecies, have a slighter bill, and tend to be more olive above. "Eastern" and "Western" groups likely represent two species; they breed almost sympatrically in a wide band from southern AB to CO Front Range. Two other subspecies in Mexico fit into the "Western" group.

Similar Species Fall birds greenish above, often with extensive yellow below, and can be confused with Philadelphia. On Warbling Vireo, yellow restricted to sides and flanks; it lacks dark lores of Philadelphia.

Voice CALL: A nasal *eahh* mobbing call; typical vireo. Note commonly uttered in flight. **SONG:** Eastern *gilvus* song is delivered in long, melodious, warbling phrases. Song of western *swainsoni* similar but less musical, higher tones.

Status & Distribution Fairly common to common. **BREEDING:** Deciduous woodlands, primarily riparian areas. **MIGRATION:** Western birds have prolonged spring migration (late Feb.–

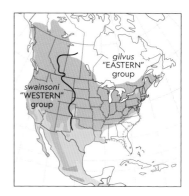

gilvus
"EASTERN"
group

swainsoni
"WESTERN"
group

early June). Eastern subspecies a circum-Gulf migrant, rare on eastern Gulf Coast; arrives mid-Apr. in TX, by early May to Great Lakes. Peak fall migration in northern US late Aug.–mid-Sept. Mostly gone from US by mid-Oct.; stragglers at southern areas to Nov., rarely Dec. **WINTER:** Mostly Mexico and C.A. Casual to Southern CA, southern AZ, and southern LA. **RARE STATUS:** Casual western AK. **Population** Overall stable.

RED-EYED VIREO *Vireo olivaceus*   REVI ▪ 1

A large vireo, the Red-eyed Vireo is one of the most common songbirds in eastern woodlands. It moves sluggishly through the canopy of broadleaf forests, making it hard to detect, and often picks food by hover-gleaning. It sings incessantly, often throughout the day. Monotypic (S.A. breeders a separate species). L 6" (15 cm)

Identification Bold face pattern with white eyebrow bordered above and below with black. Ruby red iris of adult (the origin of its name) visible at close range. Gray to blue-gray crown contrasts with olive back. Lacks wing bars. **FALL:** Flanks and undertail coverts usually washed pale yellow. **IMMATURE:** Brown iris; often extensive yel-

breeding

black lateral
crown stripe

dark line
through eye

1st fall

olive above, white below

pale yellow
lower flanks
and undertail
coverts

low wash on undertail coverts and flanks, which may extend up to the bend of the wing.

Similar Species Resembles Black-whiskered Vireo, but Red-eyed has a bold, black lateral crown stripe above its white eyebrow; more green above, less brown; red eye (adult); and will not show the diagnostic dark whisker. Yellow-green Vireo can be similar.

Voice CALL: A whining, down-slurred *myahh*. **SONG:** A deliberate *cheer-o-wit, cher-ee, chit-a wit, de-o*; a persistent singer of a variable series of robinlike short phrases

Status & Distribution Common. **BREEDING:** Woodlands. **MIGRATION:** Long-distance migrant. First spring arrivals on Gulf Coast by late Mar., late

Apr. in East/Midwest; peaks during Apr. on Gulf Coast, mid- to late May in East/Midwest. Migration continues into June farther north and west. Fall migration peaks late Aug. through Sept., most depart by early Oct., some linger to early Nov. Southern peak is early Sept.–early Oct. **WINTER:** Winters in northern S.A. No documented winter records for N.A.—reports at this season suspect. **RARE STATUS:** Rare but annual across Southwest; a few annually along CA coast. Casual to western AK. In Old World, very rare on the Azores and 75+ records for Europe, primarily late Sept.–mid-Oct.

Population Perhaps stable. Expanded into OR, UT, and then Newfoundland in the mid-20th century.

BLACK-WHISKERED VIREO *Vireo altiloquus* BWVI ■ 2

This Caribbean counterpart of the Red-eyed Vireo is a summer visitor along the FL coast. Territorial birds are vocal throughout the day, but are heard much more often than seen. Polytypic (6 sspp.; 2 in N.A.). L 6.3" (16 cm)

Identification Dark malar stripe can be quite evident at times, but often hard to see, owing to lighting, angle, and wear. White eyebrow is prominent, bordered below by black eye line and above by gray crown and, sometimes, a faint black border. Bill is large, and eye is amber. Lacks wing bars and is whitish below, with variable, pale yellow wash on sides and flanks. **IMMATURE:** Somewhat duller

barbatulus

altiloquus

than adult, with buffier underparts.
Geographic Variation Breeding subspecies is *barbatulus* (FL, Cuba, Cayman Is., Haiti); *altiloquus* (the rest of the Greater Antilles; casual to N.A.) has an indistinct, brownish gray eyebrow and even larger bill.
Similar Species Most similar to Redeyed Vireo. Compare also to Yellowgreen Vireo and Yucatan Vireo.
Voice CALL: A thin, unmusical *mew*.
SONG: Deliberate phrases with one to four notes (most typically two to three notes); notes are loud and clear, separated by a distinct pause, less varied and more emphatic than Red-eyed.
Status & Distribution Common.
BREEDING: Coastal mangroves and adjacent hardwoods. **MIGRATION:** Spring migrants arrive late Mar.–mid-Apr. Fall status more uncertain as males are less vocal. Most depart by early Sept. **WINTER:** Mostly Amazonia, but limits of subspecies' winter ranges poorly known. **RARE STATUS:** Subspecies *barbutalus* is casual along the Gulf Coast from coastal TX (about 20 records, mostly early Apr.–late May; twice in fall) to

northwestern FL. Accidental up Atlantic coast to VA. A few specimen records have been identified as nominate *altiloquus* (FL, LA).
Population Most FL populations are stable; the Tampa Bay population has been adversely affected by cowbird parasitism.

less evident dark lateral crown stripe than Red-eyed

longer bill than Red-eyed

FL
barbatulus

longer bill than *barbatulus*

blackish whisker

altiloquus

coloration similar to *barbatulus* but with buffier face; crown more brownish gray

YELLOW-GREEN VIREO *Vireo flavoviridis* YGVI ■ 3

This C.A. and Mexican species is sighted along the US border, particularly in southern TX. Unlike Red-eyed Vireo, the Yellow-green often cocks its tail. Yellow-green Vireo was re-split from Red-eyed Vireo in 1987. Polytypic (5 sspp.; at least nominate in N.A.). L 6" (15 cm)

Identification Typically yellowish green above, with rather bold, yellow edgings to flight feathers; lacks wing bars or spectacles. Head pattern is similar to that of Red-eyed but more blended—a gray crown blends into

back, and pale supercilium is bordered by rather indistinct lines. Dark line above supercilium is often very faint or absent. Underparts are whitish, with a yellow wash to sides and undertail coverts. Yellow continues up sides of breast and blends into ear coverts. Iris ruby red, and long bill, often with pale coloration on upper mandible as well as lower.
FALL: Brightest in fresh plumage, with more extensive yellow ventrally. **IMMATURE:** Similar to adult but iris is brown.
Geographic Variation Subspecies are weakly differentiated; nominate in northeast and west Mexico accounts for TX records. Records from AZ and CA could be nominate or *hypoleucos* from west Mexico; both breed in Sonora.
Similar Species Similar to Red-eyed, but bill larger. Strong yellowish wash on flanks extends to sides of face, while yellow of Red-eyed is usually

duller gray cap than Red-eyed with less-evident lateral crown stripe

yellowish on sides of neck and yellowish olive above

East Mexico
flavoviridis

large bill

immatures have brownish eyes

often co tail

hypoleucos averages duller, with smaller bill

immature west Mexico
hypoleucos

yellow sides and flanks

restricted to flanks. Yellow-green has a more diffuse face pattern; dark borders to supercilium reduced or absent. Bill has pale tones (pink, blue) versus black bill of Red-eyed. Upperparts are more yellowish green, with brighter edges to remiges. Black-whiskered is

much duller, with less green above, substantially less yellow below, and a longer, dark bill. Compare to Warbling and Philadelphia Vireos.
Voice CALL: A soft, dry *rieh*; chatter often repeated. **SONG:** A rapid but hesitant series of notes suggestive of a House Sparrow.
Status & Distribution Rare to casual.

BREEDING: Very rare, but probably annual breeder in southern TX (Rio Grande Valley). Main breeding range is from western and eastern Mexico south to Panama, primarily utilizing lowland forests and forest edges. Winters in S.A.
RARE STATUS: Casual in spring on the upper Gulf Coast (mid-Apr.–May) east to FL, and on Atlantic coast north to

MA. Very rare, but now recorded annually from coastal CA (late Sept.–Oct.; one July record); a few interior Southern CA records. Casual to southern AZ and southern NM in summer.
Population Stable, but current studies needed. Reliance on forest edge, in part, protects populations from effects of deforestation.

YUCATAN VIREO *Vireo magister*   YUVI ■ 5

A large brown vireo from the Yucatán Peninsula, it is accidental in the US. Monotypic. L 6" (15 cm)
Identification Perhaps the most noticeable features are brown plumage and large, heavy bill. A broad pale supercilium contrasts with a dark eye line. Throat and underparts are dull whitish, sometimes tinged buff; sides and flanks are washed with grayish brown. No spectacles or wing bars. Short primary projection, as one would expect with a generally sedentary species.
Similar Species Like a large-billed Red-eyed that lacks olive tones. Its

shorter wings and overall structure give it an ungainly look—very different from more streamlined Red-eyed. Yucatan's bill is much larger, and it lacks dark lateral crown stripe and has a brownish (not gray) crown.
Voice CALL: Sharp, nasal *benk* notes, often strung together in a series; also a softer, dry chatter.
SONG: Rich phrases, delivered hesitantly.
Status & Distribution Fairly common within home range. **YEAR-ROUND:** Resident on Yucatán Peninsula and some nearby islands; also Grand Cayman.
RARE STATUS: Accidental in TX—one

remarkable record from the upper TX coast, near Crystal Beach (28 Apr.– 27 May 1984).
Population Trend unknown; no indications of concern.

large heavy bill

short primary projection

grayish brown sides and flanks

CROWS AND JAYS Family Corvidae

Canada Jay, subspecies *canadensis* (ON, Jan.)

While most species are easy to identify, crows and ravens present some of the greatest identification challenges in North America. For these species, focus on bill size and structure, tail shape, and overall proportions. Vocalizations are also important. Much remains to be learned about regional variation in many species of corvids (e.g., American Crow, Canada Jay), and detailed observations and photographs will aid in our understanding.
Structure Corvids vary considerably in size, but all spe-

cies have strong legs and feet and a straight bill. Males are generally larger than females.
Behavior Corvids are omnivorous, and many species store seeds, nuts, and other foods for consumption during winter months. Some corvids gather in large flocks, particularly during the nonbreeding season. Several species are cooperative breeders, with helpers that assist with nest rearing. Corvids tend to have a diverse array of vocalizations. Most species give a few vocalizations most frequently, but do not be surprised to hear calls not described in this guide. Corvids can also mimic sounds. While jays are most known for hawk imitations, several species give very soft songs into which they incorporate a variety of mimicked sounds.
Plumage Coloration varies considerably, from the brightly colored Green Jay to the grayish coloration of the Clark's Nutcracker and the Canada Jay. Many North American jays are blue, and all crows and ravens are entirely or mostly black. Most corvids appear very similar throughout the year. Sexes are identical. Juvenile plumage is held only briefly and is generally duller but similar to that of the adult. Juveniles and immatures of some species have pale coloration to the bill that they may hold for one or more years. All corvids undergo a single annual (postbreeding) molt after breeding. The post-juvenal molt is usually partial; adult postbreeding molts are complete. In some species first-year birds may be aged by looking for differences in wear in the wing

coverts. Crows and ravens retain juvenal flight feathers for one year. By spring immatures are often quite worn, and retained brownish flight feathers can contrast markedly with fresher, replaced black wing coverts.

Distribution Corvids are found on all continents except Antarctica. In North America these birds range from the high Arctic to the Sonoran Desert and utilize nearly every habitat in between. Migration and dispersal is diurnal, but most corvids are largely nonmigratory. Some species stage occasional irruptions, during which time individuals can be far from their normal range.

Taxonomy Corvid taxonomy continues to undergo revision. Worldwide there are about 125 species in 22 genera in the family Corvidae. There are 21 species recognized in North America (including a casual visitor from Eurasia and two marginally-occurring east Mexican species), but a strong case could be made for placing the Northwestern Crow with the American Crow. In recent years the "Scrub-Jay" has been split into three, and more recently four species, and a further split in the Mexican portion of the range is possible. Corvids are believed to have arisen from an Australian ancestor that gave rise to a large group of some 650 species, which includes such diverse Old World groupings as the birds-of-paradise, Old World orioles, wood-swallows, and paradise flycatchers. The only other families from this assemblage that are found in North America are the vireos and the shrikes.

Conservation Some species are quite adaptable and do well near humans. Habitat destruction and fragmentation pose the most serious threats to this family. Populations of many corvids in North America declined dramatically with the spread of the West Nile virus. BirdLife International codes: 10 NT, 11 VU, 4 EN, 3 CR, and 1 EW (Hawaiian Crow or Alala, was extinct in the wild, but releases of captive birds began in 2016; 21 individuals are surviving in the wild as of Oct. 2018).

Genus *Perisoreus*

CANADA JAY *Perisoreus canadensis*   CAJA ▪ 1

Bold and cunning, Canada Jays (formerly named Gray Jay) usually travel in pairs or family groups, but they often stay in cover and remain quiet. While many corvids store food, the Canada Jay and its northern Eurasian counterpart, the Siberian Jay (*P. infaustus*), produce saliva that allows them to "glue" food together for storage. They sometimes visit feeders. Polytypic (8 sspp.; all in N.A.). L 11.5" (29 cm)

Identification This is a fluffy, long-tailed jay with a small bill. **ADULT:** Grayish above, paler below, with variable dark markings on back of head and pale tip to tail. **JUVENILE:** Sooty-gray overall, with a faint white moustachial streak. They have molted to an adult-like plumage by early fall. **FLIGHT:** Typically a burst of wingbeats, followed by slow unsteady glides.

Geographic Variation A highly variable species, but variation is somewhat clinal and many intermediate individuals occur where subspecies meet. Eight subspecies now generally recognized. Nominate *canadensis* (boreal forest from NT and northern AB to NL and PEI) is intermediate in overall coloration, with a white collar and forehead and medium-gray hindcrown and nape; upper parts only slightly paler than hindcrown nape, with prominent dark shaft streaks; wings are moderately edged with white. Subspecies *sanfordi* (NL and NS) is darker overall with more extensive black on crown. Subspecies *capitalis* (central and southern Rockies from eastern ID and WY to AZ and NM) has an extensive pale crown that makes head appear mostly white; upperparts considerably paler than *canadensis*; wings edged with frosty white. Subspecies *bicolor* (western MT to southeastern BC and northeastern OR) is intermediate between *capitalis* and *canadensis*,

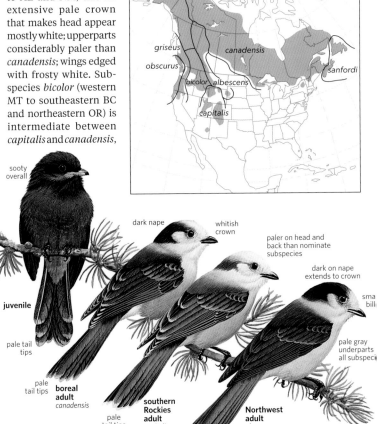

sooty overall

juvenile

pale tail tips

pale tail tips

boreal adult *canadensis*

dark nape

whitish crown

pale tail tips

southern Rockies adult *capitalis*

paler on head and back than nominate subspecies

dark on nape extends to crown

sma bill

pale gray underparts all subspeci

Northwest adult *obscurus*

pacificus

griseus

obscurus

bicolor albescens

capitalis

canadensis

sanfordi

with mostly whitish crown and frosty white wing edges of *capitalis* but with darker upperparts that have dark shaft streaks. Subspecies *albescens* (east side of Rockies from southeastern YT and east-central BC to the Black Hills east to central MB and northwestern MN) is also intermediate between *capitalis* and *canadensis*, but darker coloration on nape and hindcrown is more extensive than on *bicolor*. Subspecies *obscurus* (Pacific coast from southwestern BC to northwestern CA) has dark brownish gray upperparts (with white shaft streaks), with extensive brownish gray on crown and nape and paler underparts than other subspecies. Subspecies *griseus* (southwestern BC to northern CA) is slightly larger than *obscurus* and is paler gray. Subspecies *pacificus* (AK to YT and northwestern BC) similar to *obscurus* with brownish gray upperparts, but underparts heavily washed brownish gray.

Similar Species Clark's Nutcracker has a much longer bill, pale head, and black-and-white wings and tail. Juvenile Canada Jays are much darker; they usually are accompanying adults.

Voice Relatively quiet. **CALL:** A whistled two-part *wheeoo* and a low *chuck*.

Status & Distribution Fairly common. **BREEDING:** Northern and mountain coniferous forests; fewer in mixed forest. **DISPERSAL:** This species only infrequently strays short distances south in the East, and to lower elevations in the West; accidental to northeastern IA, MA, and southern NY.

Population Range in East may be retracting northward. Vulnerable to traps for smaller fur-bearing mammals.

TROPICAL JAYS Genera *Psilorhinus* and *Cyanocorax*

These two genera include 17 species of tropical jays, only two of which are found north of Mexico—the Green Jay and the Brown Jay. The Brown Jay was recently moved out of *Cyanocorax* into its own monotypic genus. Those species that have been studied have social breeding systems.

BROWN JAY Psilorhinus morio   BRJA ■ 4

The range of this Neotropical species barely reached southern TX. Brown Jays usually travel in boisterous small flocks. In TX it was most readily found at feeders. Polytypic (3 sspp.; *palliatus* in N.A.). L 16.5" (42 cm)

Identification A very large jay with long broad tail. Dark sooty-brown overall except for dirty whitish belly. Juvenile has a yellow bill, yellow legs, and narrow yellow eye ring, all turning black by second winter. Transitional birds have bills that intermediate. **FLIGHT:** Slow and unsteady, with heavy wingbeats interspersed with short glides.

Geographic Variation All birds north of Mexico belong to *palliatus*, characterized by uniformly brown tail and whitish belly with sooty wash. More southerly subspecies are polymorphic.

Similar Species Nearly unmistakable; crows and ravens uniformly black.

Voice Noisy; less varied than most other jays. **CALL:** A loud, displeasing *kaah kaah* or *kyeeah*, similar to Red-shouldered Hawk but higher and more offensive; usually given endlessly. Also produces a less voluminous popping or hiccuping sound.

Status & Distribution Discovered in TX in the early 1970s, it was uncommon and local in woodlands mainly below Falcon Dam. By 2000 numbers had markedly decreased and it disappeared about 2006; now casual.

Population Stable over most of range in tropics.

thin yellow orbital ring

yellow bill

heavy black bill

juvenile *palliatus*

pale belly

very large size

adult *palliatus*

GREEN JAY Cyanocorax yncas   GRJA ■ 2

The range of this Neotropical species extends into southern TX. While brightly colored, Green Jays blend in well with woodlands, and they are best found by listening for the characteristic vocalizations. The Green Jay is a regular visitor to feeding stations throughout the Lower Rio Grande Valley. Polytypic (11–14 sspp.; *glaucescens* in N.A.). L 10.5" (27 cm)

Identification Unmistakable with blue crown, complex black-and-blue face pattern, and black breast. Upperparts are bright green; underparts are paler and tinged with yellow; bright yellow on undertail coverts and outer tail feathers. Juveniles are duller with brownish olive head and throat, dull green upperparts, and paler yellow underparts. **FLIGHT:** Quick wingbeats interspersed with short glides; generally prefers to fly in forest interior.

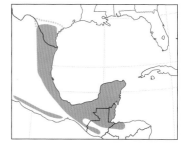

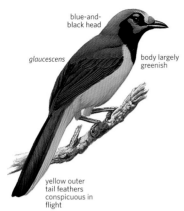

blue-and-black head

glaucescens

body largely greenish

yellow outer tail feathers conspicuous in flight

Geographic Variation Birds north of Mexico are northern *glaucescens*. Eleven to fourteen subspecies in the New World. The five Andean subspecies are sometimes considered a separate species, Inca Jay (*C. yncas*); if so, the scientific name of M.A. and N.A. birds would become *C. luxuosus*. Interestingly, the species is absent from Nicaragua to Panama.

Similar Species Unmistakable.

Voice An array of chatters, clicks, mews, rattles, squeaks, and raspy notes are frequently given, although generally quiet when feeding and less vocal during the breeding season. **CALL:** A series of four or five harsh electric calls, *jenk jenk jenk jenk*, and a dry scolding *cheh-chech*.

Status & Distribution Locally common resident in brushy areas and streamside growth of the Lower Rio Grande Valley as far west as Laredo and north to Live Oak Co. and sporadically north to San Antonio. Found in suburbs (e.g., McAllen, Brownsville), where habitat is suitable. **RARE STATUS:** Casual north to Brazos Co., Johnson Co., and Midland Co.

Population Trend unknown.

Genus *Gymnorhinus*

PINYON JAY *Gymnorhinus cyanocephalus* PIJA ■ 1

Finding this social species is usually feast or famine. Listen for the Pinyon Jay's far-carrying calls and scan for roving flocks of a few individuals to several hundred birds. Pinyon Jays nest in loose colonies beginning in winter. Considered monotypic, but three subspecies sometimes recognized. L 10.5" (27 cm)

Identification A short-tailed blue jay with a long spikelike bill. **ADULT:** Blue overall with white streaks on throat. **JUVENILE:** Ashy gray underparts without much blue coloration. **FLIGHT:** Direct with rapid wingbeats, unlike scrub-jay's undulating flight.

Geographic Variation Subspecies variation has been described as follows: Birds from northeastern portion of the range, *cyanocephalus*, have a shorter, more decurved bill, with pale plumage; western *cassini* (from north-central OR and southern ID to southern NV and central NM) average darker with straighter bill; and *rostratus*, in southwest portion of range, has a longer and wider bill and intermediate plumage coloration.

Similar Species Scrub-jays and Mexican Jays have paler underparts that contrast with bluer upperparts and do not occur in large flocks; shorter, thick bills; and much longer tails. The smaller Mountain Bluebird is often found in similarly large flocks but has a much shorter bill, thrush shape, and white undertail coverts.

Voice Suggestive of Gambel's or California Quail, or nasal like a nuthatch. **CALL:** Most commonly heard call, frequently given in flight, is a soft *hwaau* given repeatedly, or a single *hwauu'hau*.

Status & Distribution Fairly common but nomadic resident in pinyon-juniper wood-

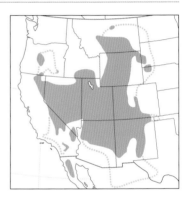

lands of interior mountains and high plateaus; also yellow pine woodlands. **DISPERSAL:** Generally nonmigratory. When cone crop fails, may irrupt outside of normal range. **RARE STATUS:** Casual: chiefly in fall to the Plains states (including West TX and north-central Mexico, where it may winter), southwestern deserts, and coastal CA. Accidental to southern WA, western IA, and southwestern SK.

Population Vulnerable. Major declines may have occurred with loss of pinyon-juniper woodlands in the mid-20th century.

overall blue color; duller on immatures

short tail

long, thin bill

CRESTED JAYS Genus *Cyanocitta*

STELLER'S JAY *Cyanocitta stelleri* STJA ■ 1

The Steller's Jay is a characteristic species of coniferous and mixed forests of western N.A. It is a bold and aggressive species frequently found scavenging in campgrounds, picnic areas, and feeding stations in the West. It is less gregarious than most other jays. The bird's flight is strong and steady, with wings rarely flexed above horizontal. Polytypic (13–16 sspp.; 8 in N.A.). L 11.5" (29 cm)

Identification A nearly unmistakable dark blue, black-crested jay with variable white or blue markings on head. Wings and tail are a vivid blue, with fine black barring. Head, including crest; back; and throat are blackish. Juveniles are washed with brownish or grayish to upperparts and are duller below.

Geographic Variation Extensive among the 13–16 subspecies from

AK to Nicaragua, but more limited and clinal among seven of eight subspecies north of Mexico, all of which have blue streaks on forehead. The most distinctive subspecies in N.A. is *macrolopha* of central and southern Rockies to northern Mexico, which has a long crest, paler back, and white streaks on forehead and over eye. It is also strongly genetically distinct from the other N.A. subspecies, which are shorter crested and differ primarily in size, head patterning, and overall coloration. Some, including nominate *stelleri* (Pacific coast from AK to southwestern BC) are darker backed.

The largest subspecies, *carlottae*, from Haida Gwaii off BC, is almost entirely black above. Another group of subspecies from central Mexico south to Nicaragua have short blue crests. **Similar Species** Nearly unmistakable. Crest, shorter tail, and lack of white in body separate Steller's Jays from scrubjays. Blue Jay, our other crested jay, is paler blue, has white in wings, tail, and face, as well as pale underparts. **Voice** Vocal with a diverse array of squawks, rattles, harsh screams. **CALL:** A piercing *sheck sheck sheck* and a descending harsh *shhhhhkk*. Steller's Jay frequently mimics other species,

particularly raptors, and also incorporates calls of squirrels and household animals, such as dogs and chickens. **Status & Distribution** Common. **BREEDING:** A variety of coniferous and mixed coniferous forests, including residential areas. **DISPERSAL:** Generally resident, but irruptions casually occur in fall and winter in both the blue-fronted subspecies group and in *macrolopha*, which is casual in southeastern CA. During such irruptions, individuals can appear to lower elevations of the Great Basin, the Great Plains, Southern CA, and southwestern deserts. Accidental east to northeastern IL, eastern NE, eastern KS, and central TX. **Population** Apparently stable.

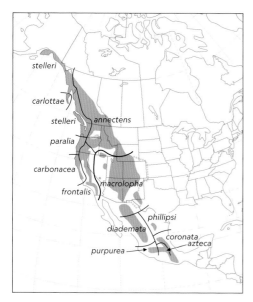

stelleri
carlottae
stelleri · annectens
paralia
carbonacea
frontalis
macrolopha
phillipsi
diademata
coronata
azteca
purpurea

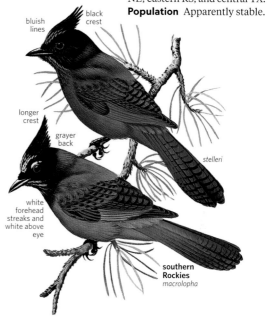

black crest
bluish lines
longer crest
grayer back
stelleri
white forehead streaks and white above eye
southern Rockies *macrolopha*

BLUE JAY *Cyanocitta cristata*   BLJA ■ 1

The Blue Jay is a familiar and widespread bird throughout the East. It is a frequent visitor to backyard feeding stations; also moves stealthily through the forest, plundering nests, searching for nuts, or quietly raising its young. Polytypic (4 sspp.; all in N.A.). L 11" (28 cm)

Identification Nearly unmistakable with black barring and white patches on blue wings and a rather long wedge-shaped tail; underparts pale gray with a dark necklace. Juveniles more grayish above with gray lores and more limited white on wings. **Geographic Variation** Minor, obscured by broad overlap. **Similar Species** Nearly unmistakable. Occasionally hybridizes with Steller's Jay (e.g., in the vicinity of the Front Range of CO). **Voice** Vocal with a diverse array of vocalizations. **CALL:** A piercing *jay jay jay*; a musical *yo-ghurt*; frequently imitates raptors, particularly Red-shouldered Hawk. Actively migrating birds are typically silent. **Status & Distribution** Common.

white tips to secondaries
bluish crest
blackish throat band
extensive white in wings
white tail tips

BREEDING: A variety of mixed forests, woodlands, suburbs, and parks.

MIGRATION: Diurnal migrant. Northern populations move south in varying numbers from year to year. During migration, large loose flocks occur, particularly along the shores of the Great Lakes and other northern locations known for raptor concentrations.

The spring peak is during May. In fall, first detected away from breeding grounds as early as July (typically earliest in big flight years); peaks in Great Lakes late Sept.–mid-Oct. **WINTER:** Distribution generally similar, but departs from northernmost breeding

locales. **RARE STATUS:** Casual chiefly in fall and winter west of the Rockies, recorded most frequently in the Northwest. Casual to northern CA and accidental Southern CA and AZ. Accidental to Bermuda.

Population Increasing in Northwest.

SCRUB-JAYS Genus *Aphelocoma*

Of the seven species of *Aphelocoma* jays, five are found north of Mexico. Generally medium in size, they have relatively long tails, are predominately blue above and gray below (or uniformly blue), and lack a crest. Species limits are poorly understood. Some species are social breeders.

FLORIDA SCRUB-JAY *Aphelocoma coerulescens*   FLSJ ■ 2

The only bird species endemic to FL. These birds are cooperative breeders and are usually encountered in small family groups. Many Florida Scrub-Jays are extremely tame. Monotypic. L 11" (28 cm)

Identification Similar to other scrub-jays but has a shorter and broader bill, a whitish forehead that blends into the eyebrow and blue auriculars;

back is grayer; and underparts faintly streaked. Juvenile is duller with a sooty grayish or brownish wash on head and back.

Similar Species Only scrub-jay found in FL. Blue Jay has black-and-white wings and tail, and is crested.

Voice CALL: Raspy and hoarse calls are vaguely reminiscent of other scrub-jays, but are lower and harsher.

Status & Distribution Fairly common, but local. Resident in scrub and scrubby flat woods of FL. Optimal habitat is produced by fire, consisting of scrub, mainly oak, about 10 ft high with small openings. Also found along roads and vacant lots near overgrown scrub habitat. **MIGRATION:** Nonmigratory. Most travel only a few miles during their entire life, making

it extremely unlikely for an individual to show up out of range.

Population Vulnerable. Listed as Threatened by USFWS (estimated

long tail

grayish back

whitish forehead

population 7,500 in 2018). Declined by some 90 percent during the 20th century due to habitat loss and fire suppression.

ISLAND SCRUB-JAY *Aphelocoma insularis*   ISSJ ■ 2

The Island Scrub-Jay is restricted to Santa Cruz I. in the northern Channel Is. off Southern CA. Its behavior is quite similar to that of the California Scrub-Jay, and like that species, pairs actively defend year-round territories. Island Scrub-Jays forage at all levels of

foliage and forage on the ground more regularly than the California Scrub-Jay. As with many island endemics, this species has a more varied diet than the similar mainland representative. It regularly caches acorns, lizards, and even deer mice. Monotypic. L 12.5" (31 cm)

Identification ADULT: Very similar to California Scrub-Jay but 15 percent larger, 40 percent heavier, and with a heavy bill that is up to 40 percent larger than largest-billed mainland scrub-jays. Island Scrub-Jay is darker blue above, with a darker brown back, blacker auriculars, and bluish undertail coverts. **JUVENILE:** Duller.

Similar Species This is the only scrub-jay found on Santa Cruz I.; no scrub-jays have been found on any of the other Channel Is.

large bill

larger overall

only jay on Santa Cruz I.

bluish undertail

Voice CALL: Similar to calls of California Scrub-Jay but slightly louder and harsher.

Status & Distribution Locally fairly common on Santa Cruz I. Most are

found in mid- to higher elevations in vegetation dominated by island scrub-oak, but also near sea level in canyons (e.g., Prisoners Cove). Approximately half of the adults on Santa Cruz I. are nonbreeding "floaters" that are generally found in marginal habitat, including pine and riparian scrub. Nonmigratory; unrecorded away from Santa Cruz I.

Population Recent estimates suggest a population of about 1,700 individuals (USFWS), much lower than a 1997 estimate of 12,500. This decline is perhaps due to drought or unknown causes.

CALIFORNIA SCRUB-JAY *Aphelocoma californica*   CASJ ■ 1

thick bill

blue breast band

darker blue than Woodhouse's

whitish flanks

Unlike Woodhouse's Scrub-Jay, the California Scrub-Jay is confiding and easily seen. Until recently, these two species were treated as conspecific. Their ranges are largely separate but do overlap in the Pine Nut Mts. (south of Carson City, NV) and some 25 mi south to Walker, CA. There, they hybridize, perhaps frequently; scrub-jays around Carson City and Reno, NV, and Alpine Co., CA, are California Scrub-Jays, though Woodhouse's are often and erroneously reported. Unlike Florida Scrub-Jay, but like Island and Woodhouse's, pairs hold territories year-round. Polytypic (5 sspp.; 4 in N.A.). L 11" (28 cm)

superciliosa

californica

obscura

hypoleuca

Identification Long-tailed jay with a stout bill and no crest. **ADULT:** Bluish above with a contrasting brownish back, whitish below, including the under-tail coverts, with a distinct bluish band on the sides of the breast. **JUVENILE:** Much grayer and duller overall, showing more brownish gray and little blue on head. **FLIGHT:** Usually undulating with quick, deep wingbeats.
Geographic Variation Weak. Five subspecies, four in N.A. Southwestern CA and northwest Baja California *obscura* is smaller and darker than *californica* and similar *superciliosa* to the north; it sometimes has bluish cast to undertail coverts. Isolated *cana* (sometimes not recognized) from Eagle Mt. (Riverside Co.) is lighter and grayer than *cana*. An additional subspecies (*hypoleuca*) is found in central and southern Baja California. It is overall paler than subspecies farther north.
Voice CALL: A variety of harsh calls, hoarser than Woodhouse's; most frequently a raspy *shreeep*, often repeated in a short series of shorter notes: *shuenk shuenk shuenk shuenk*.
Status & Distribution Common. **BREEDING:** Scrubby and brushy habitats, particularly with oak. Also found in gardens, orchards, and riparian woodlands and is a well-established urban species. In eastern part of range in northeast CA and western NV, also found in pinyon-juniper, the same habitat preferred by Woodhouse's Scrub-Jay. **DISPERSAL:** Largely resident. **RARE STATUS:** Rare to southwestern BC. Casual to southwestern ID and southeastern CA. Accidental Yuma, AZ.
Population This species is spreading northward in the Northwest.

WOODHOUSE'S SCRUB-JAY *Aphelocoma woodhouseii*   WOSJ ■ 1

Until recently this species was treated as conspecific with the California Scrub-Jay, known as the Western Scrub-Jay. Genetically, Woodhouse's differs more from California than California does from Island Scrub-Jay. Woodhouse's and California Scrub-Jays do not overlap, except in the Pine Nut Mts., NV, where they hybridize. This interior species is shyer than the California Scrub-Jay, and often only fleeting views are obtained. Polytypic (7 sspp.; 3 in N.A.). L 11" (28 cm)
Identification ADULT: The plumages resemble the California, but the blue is paler, closer to the color of the Pinyon Jay, and the back is more tinted with blue and is less contrasting; below, the breast band is more ill-defined and the underparts are dingier, less whitish, with a brownish gray cast to the sides and flanks, and the undertail coverts are pale blue, not white, as in most California Scrub-Jays. **JUVENILE:** Like California, the juvenile is duller than the adult.

Geographic Variation The more eastern *woodhouseii* is thicker billed than *nevadae*; *texana* from central TX is richer blue. Four other subspecies are found in Mexico; two more southerly subspecies (*sumichrasti* and *remota*) are placed in their own group and should perhaps be split as a separate

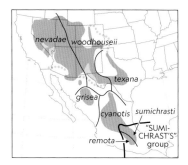

nevadae woodhouseii

texana

grisea

cyanotis sumichrasti

"SUMI-CHRAST'S" group

remota

nevadae and very similar *woodhouseii* paler blue above than California Scrub-Jay with duller and more blended breast band

grayish underparts

nevadae

thinner bill than California Scrub-Jay

species, "Sumichrast's Scrub-Jay." Vocalizations are distinctly different. **Similar Species** See the identification section (above) for separation from California Scrub-Jay. Beware that juvenile Californias much more closely resemble Woodhouse's in overall coloration. Woodhouse's, especially the western subspecies, *nevadae*, has a thinner bill than California Scrub-Jay. **Voice** Not quite as harsh as California Scrub-Jay and more two-syllabled— an up-slurred *jrr-eee*. **Status & Distribution** Resident in a variety of woodlands, especially

pinyon-juniper; found in suburban, but not usually urban, areas. Prone to occasional irruptions, particularly *woodhouseii* and *nevadae*. Irruptions often coincide with movements of Pinyon Jay, Steller's Jay, and Clark's Nutcracker. **RARE STATUS:** Rare and irregular, or casual, to the Sierra Nevada, southeast CA, and southwest AZ, and the western Great Plains in fall (some winter). Accidental to southern MB, northeastern IL. A record from northwestern IN was either this species or California Scrub-Jay. **Population** Stable.

deeper blue above than *woodhouseii* or *nevadae* with more contrasty back

heavier bill

underparts paler than *woodhouseii* or *nevadae*

TX Hill Country *texana*

MEXICAN JAY *Aphelocoma wollweberi*   MEJA ■ 2

This species travels in large, raucous flocks and has a cooperative breeding system like the Florida Scrub-Jay; it regularly visits bird feeders. Polytypic (4 sspp.; 2 in N.A.). L 11.5" (29 cm)
Identification A relatively thick-bodied, broad-tailed, broad-winged jay with a large bill. **ADULT:** Generally bluish on face and above, with a slight grayish cast to back and a brownish patch on center of back. Underparts pale gray. **JUVENILE:** Much grayer overall, very little blue on head. Juvenile *arizonae* has a yellowish bill that gradually turns black by third year. All ages of *couchii* have black bills.

Geographic Variation Two subspecies groups, both in N.A.: western *wollweberi* group includes *arizonae* (reaches AZ and NM), larger with dull pale grayish blue upperparts and a uniformly gray throat and breast; eastern *couchii* group (represented only by *couchii*, found north to Chisos Mts., TX) with blue head and whitish throat that contrasts with pale grayish underparts. The Mexican Jay was split in 2011 from Transvolcanic Jay (*A. ultramarina*) of central Mexico.
Similar Species From Woodhouse's Scrub-Jay, best distinguished by its slimmer body, thinner and disproportionately longer tail, contrasting dark cheeks, white eyebrow, whitish throat offset by at least a faint blue breast band, gray back, and voice. Woodhouse's usually does not gather in flocks.

arizonae

couchii

wollweberi

gracilis   potosina

Voice Most common call is a rising *week*, often repeated in a short series; calls generally less raucous than scrubjay. TX birds similar, but less musical. TX birds also give a mechanical rattle. **Status & Distribution** Common. **RESIDENT:** Pine-oak canyons of southwestern mountains. **DISPERSAL:** Largely resident; does not wander like Woodhouse's Scrub-Jay. **RARE STATUS:** Accidental two records from West TX, a specimen of *couchii* from Alpine on 24 Mar. 1935, and one (*arizonae*) from 24–25 Jan. 2001 at El Paso; a record from southwest KS, Clark Co., near Mt. Jesus (Mar. 1906), is based on lost specimen.
Population Not threatened.

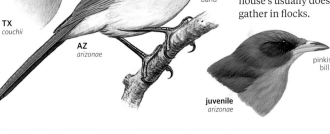

darker blue above

paler blue above

adult
no breast band

TX *couchii*

AZ *arizonae*

pinkish bill

juvenile *arizonae*

NUTCRACKERS Genus *Nucifraga*

CLARK'S NUTCRACKER *Nucifraga columbiana*   CLNU ■ 1

Clark's Nutcrackers are residents of the higher ranges in the West and are very adaptable and are commonly seen at scenic overlooks and picnic grounds. During summer and fall, nutcrackers store thousands of pinecone seeds, which they eat during winter and feed to their offspring during the very early

nesting season (most lay eggs by Mar. or early Apr.). Monotypic. L 12" (31 cm)
Identification A short-tailed, long-winged corvid with a long bill. Head and body pale grayish overall with boldly contrasting black-and-white wings and tail. **ADULT:** Paler whitish area on throat, forecrown, and around

eyes. **JUVENILE:** More uniform face and generally washed with brown. **FLIGHT:** Direct, with deep crowlike wingbeats. **Similar Species** Canada Jay has much shorter bill and more uniform grayish wings and tail, without boldly contrasting black-and-white wings. **Voice** Quite varied; most are nasal

and harsh, but some with clicks and cackles. **CALL:** Commonly gives a very nasal, grating, drawn-out *shra-a-a-a*; a nasal *whaah* similar to that of Pinyon Jay; and a slow rattle sometimes likened to the croaking of a frog.
Status & Distribution Locally common. **BREEDING:** Prefers forests dominated by at least one species of large-seeded pine. In summer, often found in higher coniferous forests near timberline. **MIGRATION & DISPERSAL:** Generally resident, but some populations regularly move to lower elevations in fall. In late spring, following breeding, most birds move upslope to higher elevations. Nutcrackers irrupt out of

overall gray with whiter face and black wings

long, slender bill

white outer tail feathers

adults

white secondaries

crowlike wingbeats

white undertail coverts

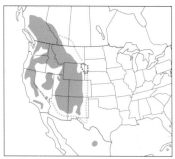

core range and into deserts and lowland areas of the West, often far from mountain where the species is resident; these irruptions likely follow major cone crop failures. Most individuals are seen in late Sept. and Oct., but may appear as early as late July or early Aug. and as late as Nov. Some may winter. Major irruptions generally occur every 10–20 years and are thought to follow two or more years of good cone production immediately followed by a failure of all seed sources. **RARE STATUS:** During irruptions casual to the Plains states, southeastern AK, southern YT, and coastal CA, including offshore islands. **ACCIDENTAL:** As far east as western ON, central MB, northeastern MN, northern IL, AR, and LA, and north to the central NT. Most extralimital records are Aug.–Nov.
Population Stable.

MAGPIES Genus *Pica*

BLACK-BILLED MAGPIE *Pica hudsonia* BBMA ■ 1

With their large size, bold pied plumage, and fondness for open areas, Black-billed Magpies are easily seen. Monotypic. L 19" (48 cm)
Identification An unusually long-tailed black-and-white corvid with a black bill. In good light, black on wings and tail shine with iridescent green, blue, and violet. Juvenile has milky grayish-colored iris and whitish gray gape flange. Upperparts are washed dull brownish, and belly is more cream-colored. **FLIGHT:** Relatively slow, with steady wingbeats. Magpies usually swoop up or down to perch.
Geographic Variation While monotypic, birds in the south average smaller and tend to show bare dark grayish skin below eye.
Similar Species Black bill (can appear pale in some light) and range distinguish this species from smaller Yellow-billed. Until recently, Black-billed Magpie was lumped with the widespread Magpie (*Pica pica*) of Eurasia and northwestern Africa, but all evidence suggests a much closer relationship with Yellow-billed.
Voice Quite varied, but most vocalizations are rather harsh. All are very similar to that of Yellow-billed Magpie. They are strikingly faster and lower pitched than Old World populations of magpie. **CALL:** Frequently gives a whining, rising *mea*; sometimes these are more drawn out and questioning: *meeaaah*. Also a quickly repeated *shek-shek-shek* with each phrase repeated three to five times.
Status & Distribution Common. Resident of open woodlands and thickets in rangeland and montane valleys; also suburban areas. Nests along watercourses and other areas

large white wing patch

black-and-white coloration

long tail

with trees and shrubs, but foraging birds use very open areas. **MIGRATION & DISPERSAL:** Generally considered to be nonmigratory, but some movement. Dispersing flocks form as early as July and typically consist of a few to a hundred birds; occasionally forms flocks of several hundred. Movements may be upslope, downslope, or in any direction. Banding recoveries have shown atypical movements of more than 300 mi. Most birds are thought to return near where they hatched to breed. **RARE STATUS:** Casual toward Pacific coast in Pacific Northwest (fewer records in recent decade); also east to western WI, IA, southwest ON, and northern TX, mostly in fall. In summer found north as far as northern AK, northeastern NT. Other sightings occur casually in the East and may pertain to escaped cage birds, particularly away from the Great Lakes.

Population Has declined over last 20 years in eastern part of its range on Great Plains. West Nile Virus has likely negatively impacted populations.

YELLOW-BILLED MAGPIE *Pica nuttalli*   YBMA ■ 2

This highly-prized CA endemic is gregarious; roosting and feeding in flocks; nests in small colonies. The range does not overlap with that of the larger Black-billed Magpie. Monotypic. L 16.5" (42 cm)

Identification Like Black-billed Magpie but with yellow bill and yellow skin below eye; extent of yellow is variable, sometimes fully encircling eye, sometimes confined to below eye. Differences may be related to age, state of molt, individual variation, or some combination thereof. **JUVENILE:**

Milky grayish iris; brownish wash, most noticeable on belly. **FLIGHT:** Wingbeats are somewhat faster than Black-billed.

Similar Species Yellow bill, yellowish skin around eye, and range should easily distinguish this species from Black-billed. Beware of Black-billed carrying pale objects or its bill appearing pale in some lighting—such sightings have resulted in erroneous reports of Yellow-billed Magpies.

Voice Very similar to that of Black-billed Magpie, perhaps higher pitched.

Status & Distribution Fairly common resident, but declining and extirpated from parts of CA range. Prefers oaks, particularly more open oak savanna; also orchards and parks. Found in rangelands and foothills of central and northern Central Valley, CA, and coast range valleys south to Santa Barbara Co. **RARE STATUS:** Casual north almost to OR. Not prone to wandering, and observations of strays may represent escapes.

Population Vulnerable. West Nile

variable yellow skin around eye

yellow bill

juvenile with extensive yellow bare skin around eye, but some adults show more yellow than illustrated

juvenile

virus has severely impacted this species. Loss of habitat poses another significant risk. Some populations have been locally extirpated (e.g., extirpated in 1981 from Pacific Grove, Monterey Co.; gone by 1900 from western Los Angeles Co. and Ventura Co.), and others have become fragmented due to habitat loss.

| CROWS AND RAVENS Genus *Corvus*

Crows and ravens form the largest group of corvids—40 species worldwide, seven in N.A. Crows and ravens are large, with mostly black plumages, sometimes with grayish or whitish markings on the head or neck. Ravens are larger than crows with a bigger bill and deeper vocalizations. For separation of similar species, focus on structural differences (particularly the bill and nasal bristles) and vocalizations, and beware of regional differences.

EURASIAN JACKDAW *Corvus monedula*   EUJA ■ 4

The Eurasian Jackdaw, a widespread Eurasian and North African species, is strictly a casual stray to N.A. During the 1980s, it staged a small invasion into the Northeast, with singles and small groups found in several northeastern states and provinces. Some (or all) of these birds may have been assisted by ships. Polytypic (4 sspp.). L 13" (33 cm)

Identification A small, mostly black crow with pale gray nape wrapping around sides of head. **ADULT:** Eerie pale gray iris. **JUVENILE:** Similar to adult but with dark iris and darker nape and side of head.

Geographic Variation No information on subspecies that appeared in N.A.; presumably *monedula* of north-central and central Europe or *spermologus* of western and southern Europe.

Similar Species Other crows and

gray nape and face with pale gray eye

small size

ravens are larger without pale area to nape.

Voice CALL: A sharp, abrupt *chjek*, typically repeated several times.

Status & Distribution Widespread Old World species from British Isles and northwestern Africa east to south-central Siberia and Kashmir; generally sedentary. Within native range found in a variety of open habitats. **RARE STATUS:** Casual to Iceland and aboard ships in the mid–North Atlantic; two

records for Greenland. Other records from Faroe Is., Canary Is., and Japan. First recorded in N.A. at Nantucket, MA, in late Nov. 1982; this bird was joined by another in July 1984; at least one remained until 1986. During the mid- and late 1980s found at several locations in the northeast, including a flock of 52 near Port-Cartier, QC; others at Miquelon, NS, ON, ME, RI, CT, and PA. A pair at the Federal Penitentiary at Lewisburg, PA, from May

1985 through at least 1991 nested or attempted to nest several times. The CT, RI, and PA records were ultimately rejected by state records committees, and it is generally thought that many (or all) N.A. records may pertain to ship-assisted birds.

Population Common and increasing throughout most of Old World range. Many that reached QC were poisoned or shot by the province's Fish and Game Department.

AMERICAN CROW   *Corvus brachyrhynchos*   AMCR ■ 1

juvenile crows (all N.A. species) are browner than adults

begging calls can resemble Fish Crow

juvenile

larger and heavier and bigger billed than Fish Crow

wing tip more rounded than Fish Crow

The American Crow is the default crow across most of N.A. It overlaps broadly with the Common Raven, and to a lesser extent with the Chihuahuan Raven, Fish Crow, and Northwestern Crow. Study of vocalizations, bill structure and size, tail shape, and overall structure of this species will greatly aid in the identification of other crows and ravens. Regional variation in size of the American Crow poses challenges, particularly in the Northwest. Polytypic (4 sspp.; all in N.A.). L 17.5" (45 cm)

Identification Largest crow in N.A., with uniformly black plumage and fan-shaped tail. Bill is larger than other American crows, but distinctly smaller than either raven. On rare occasions individuals show white patches in wings. **JUVENILE:** Brownish cast to feathers; grayish eye, and pinkish gape flange (quickly darkening after fledging). **IMMATURE:** Tends to show worn brownish wings that contrast with fresher black wing coverts. **FLIGHT:** Steady, with low rowing wingbeats. Does not soar.

Geographic Variation Four poorly defined subspecies generally recognized. While variation is largely clinal, differences between extremes

are apparent. Northern *brachyrhyn-chos* and eastern and southern *palus* are essentially inseparable. FL Peninsula *pascuus* has relatively long bill, long tarsus, and large feet. Also differs in behavior, never forming flocks; not found in urban areas and has more extensive vocal repertoire. It has been suggested that smaller western subspecies *hesperis* is more closely related to Northwestern Crow than to other subspecies of American Crow—the entire relationship between American and Northwestern Crow remains unclear.

Similar Species Compare with very similar Fish Crow and nearly identical Northwestern Crow (the former easily separated by voice); Common and Chihuahuan Ravens are much larger with much heavier bills and wedge-shaped or rounded tails.

Voice CALL: Adult's familiar *caw* generally well known. Voice of *hesperis* generally lower pitched than other subspecies. Juvenile's begging call is higher pitched, nasal, and resembles the call of Fish Crow but is not doubled.

Status & Distribution Common to abundant. **BREEDING:** A variety of habitats, particularly open areas with scattered trees. **MIGRATION &**

DISPERSAL: Diurnal migrant. In spring, arrive mid-Feb.–late Apr. Fall migration generally more protracted than in spring. Most depart north-central BC and AB by late Sept; peak in Great Lakes early Oct.–mid-Nov. Uncommon to rare migrant and winter visitor in deserts of the West. **WINTER:** Throughout much of the lower 48. **RARE STATUS:** Casual to southeast CA, southwestern AZ, southwestern TX, northwestern Sonora, Mexico.

Population Expanded with clearing of forests and planting of woodlots in prairies. Many populations experienced dramatic declines with the spread of West Nile virus early this century. Nevertheless long-term populations generally stable.

NORTHWESTERN CROW   *Corvus caurinus*   NOCR ■ 1

This small crow of the Pacific Northwest is virtually identical to *hesperis* American Crows. Field identification is essentially impossible. Most authorities consider the Northwestern Crow to be at best a subspecies of the American Crow; they will likely soon be lumped together. Monotypic. L 16" (41 cm)

Identification Virtually identical to American Crow. Northwestern averages smaller, with quicker wingflaps than most American Crows. Unfortunately, *hesperis* American Crows found in the Pacific Northwest are essentially identical. A tendency for nasal bristles to be placed along sides of bill on American Crow versus more on top of culmen in Northwestern Crow has been suggested, but there is considerable overlap, and this is unreliable. Voice often reported to be the most useful characteristic in separating this species from American Crows, but *hesperis* American Crows are so similar as to render this nearly impossible to use.

Similar Species Over most of range, likely to be confused only with Common Raven, which is much larger, with large heavy bill, and wedge-shaped tail. Separated from American Crow in most areas by range.

Voice CALL: Distinctly lower and more nasal than most populations of American Crow; *hesperis* American Crows also have lower and more nasal calls similar to Northwestern Crows.

Status & Distribution Resident of coastal areas and islands in Pacific Northwest. Common from south-coastal AK

nearly identical to American Crow

(Kodiak I.) to southern BC. Once a coastal resident in WA south through Puget Sound to Grays Harbor. Deforestation in the late 1800s allowed American Crows to invade southern portions of this region by the early 1900s. Puget Sound population and perhaps those around Vancouver, BC, are now generally considered to be small *hesperis* American Crows or hybrids. Within WA, birds thought most likely to be phenotypicly pure Northwestern Crows are limited to western Olympic Peninsula and San Juan Is. **MIGRATION & DISPERSAL:** Generally sedentary.

Population Stable or increasing.

TAMAULIPAS CROW   *Corvus imparatus*   TACR ■ 3

The range of this species barely extended into south TX near Brownsville; it is now extirpated. It does not overlap with any other crow. It was formerly considered the same species as the Sinaloa Crow (*C. sinaloae*) of northwestern Mexico with the English name of Mexican Crow. The two have very different vocalizations. Monotypic. L 14.5" (37 cm)

Identification Small, very glossy crow with a short, small bill. Recalls Fish Crow, but glossier; no range overlap. **JUVENILE:** Duller than adult, but still glossier than adults of other N.A. crows.

Similar Species Only crow regularly found in southern TX. Chihuahuan Raven is common in the same habitat, but is larger, stockier, and has a much larger and heavier bill. May be confused with male Great-tailed Grackle, particularly molting individuals that can have short tails. Still readily separated by crow's larger bill, stockier build, thicker legs, and dark eyes. Also note distinctive vocalizations.

Voice CALL: Rough, nasal, froglike croaking *ahrrr*, typically repeated several times and sometimes doubled, *rah-rahk*. Juvenile's begging call reportedly similar to that of Fish Crow.

Status & Distribution Common to fairly common in northeastern Mexico to northern Veracruz. Invaded the Brownsville area in 1968, and was common (low thousands) through the 1970s and 1980s. Found principally in winter but a few summered and nested there (through 2007). The species declined greatly in the 1990s and was rare by 2000. It is now strictly casual in south TX. It was formerly best known in TX from the Brownsville Sanitary Landfill.

Population Presumably stable in Mexico.

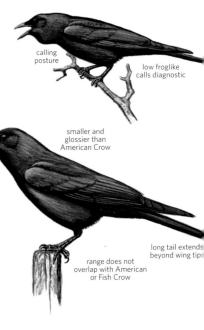

calling posture

low froglike calls diagnostic

smaller and glossier than American Crow

long tail extends beyond wing tips

range does not overlap with American or Fish Crow

FISH CROW *Corvus ossifragus*   FICR   ■   1

This glossy, gregarious corvid largely replaces the American Crow in coastal and tidewater regions of the Southeast. Its range is expanding up the Mississippi River Valley, its tributaries, and elsewhere away from the coast. While subtle structural differences exist, the most reliable way to identify this species is by vocalizations. Monotypic. L 15.5" (39 cm)

Identification Very similar to American Crow but with disproportionately longer tail, longer wings, shorter legs, and smaller feet. Fish Crows have a bluish violet or greenish gloss over most of wings and body (less so on lower underparts) that is more extensive than in American. Bill is usually more slender, but female crows have smaller bills and small-billed female Americans may have bill that appears very similar to Fish Crow. **JUVENILE:** Brownish cast to feathers; grayish eye, pinkish gape flange. **IMMATURE:** Tends to show worn brownish wings that contrast with fresher black wing coverts. **FLIGHT:** Similar to American Crow but with slightly stiffer and quicker wingbeats. When gliding, wings often appear swept back at tips, creating a less-even trailing edge to wing than American Crow. Small-headed and long-tailed appearance

is often most noticeable in flight. Fish Crow is apt to hover and soar, unlike American Crow.

Similar Species American Crow averages somewhat larger with disproportionately shorter wings and tail, but such differences difficult to use in the field; best identified by vocalizations. Note that some juvenile begging calls of American Crow resemble Fish Crow vocalizations. Closely related to Tamaulipas Crow, but does not overlap in range; Tamaulipas glossier, smaller, with disproportionately longer tail.

Voice Possibly more limited repertoire than American Crow. **CALL:** Typically a high nasal *ca-hah*, second note lower; has a similar inflection and emphasis to American's sometimes casual dismissal of something with an audible *uh-uhh*. Also gives low, short *awwr*. Begging calls of Fish Crow similar in quality to American's begging calls, but are shorter and cut off abruptly.

Status & Distribution Common. Resident usually near lakes, rivers, and marshes; also large numbers gather at landfills. **MIGRATION & DISPERSAL:** In many regions (e.g., MD, TN, central NY), spring migration/ return of breeders appears to peak in mid-Mar.–Apr. Fall migration more

protracted: AR late Aug.–early Nov.; departs central NY Sept.–Oct.; MD mid- or late Sept.–mid- to late Dec. (peaks 20 Oct.–10 Dec.). **RARE STATUS:** A few now found southwest MI and northeast OH. Casual in MI, southern ON, southern FL Peninsula and Keys, NS, and Bahamas. **WINTER:** Becomes more localized throughout range as species gather in large flocks. Because birds become less vocal in winter, the northern limit of wintering birds is somewhat uncertain. Withdrawal from northern portions of range, particularly in the interior.

Population Increasing in abundance, and populations spreading northward. First recorded OK in 1954; KY 1959; MO 1964; ME 1978; KS 1984; IN 1988; VT 1998; MI 2009; and OH 2011.

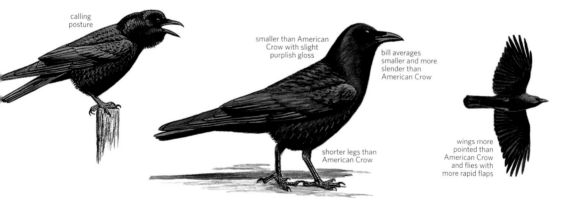

calling posture

smaller than American Crow with slight purplish gloss

bill averages smaller and more slender than American Crow

shorter legs than American Crow

wings more pointed than American Crow and flies with more rapid flaps

CHIHUAHUAN RAVEN *Corvus cryptoleucus*   CHRA   ■   1

Separating the smaller Chihuahuan Raven from the Common Raven is exceptionally difficult, if relying on subtle differences in size, bill and tail shape, and vocalizations. White bases to the neck feathers are the most reliable field mark, but these are usually hidden, so concentrate on bill shape and feathering on upper mandible.

Chihuahuan Ravens are very gregarious, particularly during the winter, when they often form large flocks (occasionally up to several thousand birds). Where trees are lacking, they will nest on utility poles, windmills, and abandoned buildings or under bridges. Monotypic. L 19.5" (50 cm)

Identification Very similar to Com-

mon Raven. Major difference is white (not gray) bases to neck feathers. Bill is slightly shorter; tail is slightly less wedge shaped and shorter; nasal bristles on top of bill extend farther out onto bill; throat feathers are less thick and shaggy. **JUVENILE:** Brownish cast to feathers; grayish eye, pinkish gape flange. **IMMATURE:** Tends to show worn brownish wings that contrast with fresher black wing coverts. **FLIGHT:** Very similar to Common Raven, but less wedge-shaped tail is often most easily seen on soaring birds.

Geographic Variation Birds of central and southern TX average smaller, but more study is required to determine if slight differences warrant subspecies status.

Similar Species Chihuahuan Raven is extremely similar to Common Raven. Most reliable difference is white-based feathers to neck, often visible only under windy conditions. Additional clues include shorter bill with longer nasal bristles; less shaggy throat feathering; smaller size; shorter, less wedge-shaped tail; and, on average, slightly higher-pitched vocalizations. All these characteristics are subjective, and where range overlaps some

white on neck shows when wind ruffles feathers

nasal bristles extend well over half the length of the culmen

shorter, more rounded tail than Common Raven

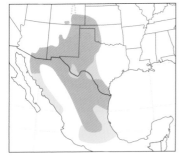

birds should probably be left unidentified, particularly given distant or brief views. Heavier bill and wedge-shaped tail distinguish Chihuahuan Raven from crows.

Voice Less varied vocalizations than Common Raven. **CALL:** Drawn-out croak is usually higher pitched and more crowlike than most Common Raven vocalizations. Common Ravens may give Chihuahuan Raven–like vocalizations. Listening to a bird (or a flock of birds) for a period of time is preferable in order to hear full variety of calls given.

Status & Distribution Uncommon to common. Resident of desert areas and dry open or shrubby grasslands, usually with scattered trees or shrubs. **MIGRATION & DISPERSAL:** Generally considered to be nonmigratory, but forms large roaming flocks in winter. Birds in northern portion of range are thought to withdraw, but this may vary between years. In some years, local movements through much of the range.

Population Declined in eastern CO and western KS during the late 1800s and the early 1900s; extirpated from southeastern WY. Now strictly casual in NE.

COMMON RAVEN *Corvus corax*   CORA ▪ 1

In northern or desert areas, this is one of the first species to greet the dawn as they patrol the roads for overnight roadkill. This large raven found throughout much of the Northern Hemisphere is more frequently found singly, in pairs, or in small groups, but in many regions is sometimes found in foraging or roosting flocks of several hundred. Polytypic (11 sspp.; 4 in N.A.). L 24" (61 cm)

Identification Largest corvid in the

Americas, with uniformly glossy black plumage, long, heavy bill, and long wedge-shaped tail. Bases of neck feathers are gray. Nasal bristles on top of bill cover basal third to half of

bill. Throat is covered by thick and shaggy feathers. **JUVENILE:** Brownish cast to feathers, grayish eye, and pinkish gape flange. **IMMATURE:** Tends to show worn brownish wings

larger than Chihuahuan but beware of smaller and smaller-billed Common Ravens from CA

longer bill than Chihuahuan and nasal bristles extend about half way out culmen

long, wedge-shaped tail

shaggy throat feathers

large size, larger than Red-tailed Hawk; frequently soars

that contrast with fresher black wing coverts. **FLIGHT:** Wingbeats shallower than crows. Frequently soars; pairs frequently engage in a variety of aerial acrobatics, sometimes even turning upside down. Glossy black plumage is often most apparent in flight, when ravens often appear "greasy," as if covered with oil.

Geographic Variation While subspecies variation is largely clinal, differences between extremes sometimes apparent. Northern and Eastern *principalis* is large with long bill of medium depth. Residents of southwestern AK to Russian Far East (*kamtschaticus*) are largest, with broadest and longest bill. Western *sinuatus* and particularly southwestern *clarionensis* are smaller, with smaller bill and shorter wings and tail.

Similar Species See Chihuahuan Raven, which can be extremely similar. Crows are much smaller with much smaller bills and fan-shaped tails.

Voice Extremely varied, with local dialects and individual specific calls

reported. **CALL:** Common call is a low, drawn-out croak *kraaah*; also a deep, nasal and hollow *brooonk*. Juvenile begging calls are relatively high-pitched, but there is much individual variation. Calls can be similar to Chihuahuan Raven.

Status & Distribution Generally common, but more local on southern periphery of range in East. **BREEDING:** Diverse array of habitats. Tends to prefer hilly or mountainous areas, but found on tundra, prairies, grasslands, towns, cities, isolated farmsteads, forests, even Arctic ice floes. **MIGRATION & DISPERSAL:** Generally considered sedentary, but poorly understood. Regular spring passage noted along Front Range of Colorado late Jan.–late Mar. **RARE STATUS:** Casual chiefly in winter to Great Plains, southern Great Lakes, and lower elevations of Atlantic coast states.

Population Declined greatly in the 19th and early 20th centuries due to loss of habitat, shooting, poisoning, and disappearance of bison on the

Great Plains; extirpated from AL, ND, SD, and the southern Great Lakes. Populations are now expanding into some of their former territory in parts of the East (e.g., a few are now regular again in eastern OH), Great Lakes, and northern Plains. In Northeast, a major increase in recent decades south to the coast of NJ, also south of there at lower elevations of Appalachians. Now present over much of CA Central Valley and to central CA coast.

LARKS Family Alaudidae

Horned Lark (UT, June)

These small, generally cryptically colored terrestrial passerines occur on all continents except Antarctica.

Structure Bill shape varies from long and decurved to thick and conical. The hind toe has an unusually long and straight nail. Most species have long tertials that cover the folded primaries.

Plumage Larks are generally dull-colored, in browns, rufous, buff, black, and white, with many species streaky or with nondescript uniform plumage; some species are more boldly marked. Juveniles are recognized by pale-fringed feathers on the back and wings.

Behavior These ground-dwelling birds walk or run rather than hop. Most give complex songs in display flight.

Distribution Larks reach their greatest diversity in arid regions of Africa, where 80 percent of species occur. South America is represented by one isolated breeding population of Horned Lark (*E. a. peregrina*) near Bogotá, Colombia. Two species occur in North America: the Horned Lark (native) and the Eurasian Skylark (introduced European *arvensis* to Vancouver I., BC, and Asian *pekinensis* as a natural stray to AK and the West Coast). The Horned Lark, known as the Shore Lark in the Old World, is the most widely distributed species in the family, with 42 subspecies occurring across North and South America, Eurasia, and North Africa.

Taxonomy Relationships to other families remain unclear. Historically the alaudids are thought to be the most primitive of the oscine passerines despite their complex songs. Some DNA evidence places the larks near the Old World sparrows, while recent genetic studies place them closer to the Old World warblers, bulbuls, and swallows; surprisingly, an especially close relationship to the Bearded Reedling (Panuridae) is suggested by molecular studies. The 93 species of larks are placed in 21 genera; species-level taxonomy is undergoing constant revision.

Conservation Lark habitat in general is threatened by human agricultural activities, as well as by urbanization. Many species have very restricted ranges and habitat requirements. BirdLife International codes: 2 NT, 1 VU, 4 EN, and 2 CR.

Genus *Alauda*

EURASIAN SKY LARK   *Alauda arvensis*   EUSK ■ 3

This Eurasian species raises a slight crest when agitated. Polytypic (11 sspp.; 2 in N.A.). L 7.2" (18 cm) **Identification** Upperparts heavily streaked; buffy white underparts with necklace of fine dark streaks on breast and throat. Narrow white trailing edge on secondaries and inner primaries and white outer tail feathers visible in flight. **ADULT:** Sexes similar. **JUVENILE:** Upperparts spotted and speckled, throat less streaked.

Geographic Variation Western European nominate *arvensis* is described above. The northeastern Asian *pekinensis* is more richly colored on breast, darker and more heavily streaked above, and has longer wings.
Similar Species Pipits have thinner bills than Eurasian Sky Lark; longspurs have thicker bills. They all lack white trailing edge on wing.
Voice FLIGHT CALL: A liquid buzzy *chirrup*. **SONG:** A very long (2–4+ min.) series of buzzy, trilling, churring notes, given usually in aerial display flight.
Status & Distribution Locally introduced population (*arvensis*) on south Vancouver I., BC; sedentary. Asian subspecies (*pekinensis*) rare in spring and fall to Aleutians and Bering Sea islands (casual in summer, has bred), casual into summer and has bred on Pribilofs and

likely on Attu I. **BREEDING:** Open habitats with short grasses and low herbs in BC; tundra in AK. **MIGRATION:** AK records primarily May–early June and Sept. **RARE STATUS:** Casual (records all likely *pekinensis*) from coastal areas from BC to northern CA. Nominate *arvensis* introduced on Hawaiian Is. Casual to Leeward Is. (*pekinensis*).
Population Introduced population in BC declined from about 1,000 (1965) to about 20 and faces extirpation. These birds reached San Juan Is., WA, where small numbers were present until 1998. Major declines in the Old World.

arvensis

white trailing edge to secondaries

short crest can be raised

pekinensis

darker, richer, and more heavily streaked than *arvensis*

fresh *arvensis*

worn *arvensis*

juvenile *arvensis*

Genus *Eremophila*

HORNED LARK   *Eremophila alpestris*   HOLA ■ 1

The head pattern with black "horns" and white or yellowish face and throat with broad black stripe under eye is distinctive. The Horned Lark usually

flava
arcticola
hoyti
merrilli
alpina
strigata
lamprochroma
sierrae
rubea   insularis
actia
utahensis
ammophila
leucansiptila
occidentalis
adusta
enertera
leucolaema & enthymia
praticola
giraudi
five subspecies
hoyti & alpestris intergrade zone
alpestris

nests on ground, including plowed fields, and lays two to five eggs (mid-Mar.–mid-July). Polytypic (42 sspp.; 21 in N.A.). L 6.8–7.8" (17–20 cm)
Identification Black bib. Tail black with two outer tail feathers edged light gray. Breast and belly yellow to white, breast patch black. Head boldly marked with black lores, cheek, and ear tufts; white to yellow eyebrow, ear coverts, and throat. Short, stout bill. Black legs. **ADULT:** Sexes similar, but males generally larger and darker. Female duller on head and face; breast band smaller and gray instead of black; "horns" smaller and rarely erected. **JUVENILE:** Lacks black facial and breast markings of adult, making dark eye

more conspicuous on brownish head. Back spotted with whitish to gray, giving a mottled appearance. Underparts generally creamy-buff with streaking on sides and breast. Tail as adult. Bill and legs paler yellowish to pinkish in younger birds.
Geographic Variation In N.A., 21 subspecies are recognized, with most intergrading. Migratory subspecies have longer wings, and birds in hotter environments have longer legs. Back color matches the color of the soil in local habitat; face and throat varies from yellow in Northeast to whitish or white in prairie and desert subspecies. Nominate *alpestris* of the Northeast is the largest and darkest of five eastern subspecies, and has a yellow throat. White-throated "Prairie" subspecies (*praticola*) is widespread in the upper

Lapland Longspur for comparison

casual to AK
♂ *flava*

northwestern Canada, AK
♂ *arcticola*

widespread in East in winter

winter ♂

♂

"Northern Horned Lark" *alpestris*

central Arctic coast breeder
♂ *hoyti*

CA Central Valley
♂ *rubea*

alpestris

northeastern CA
♂ *sierrae*

coastal Northwest
♂ *strigata*

black tail with white outer tail feathers

upperparts spotted with white

juvenile *ammophila*

alpestris

faintly streaked breast

small "horns"

praticola breeder over much of Midwest and East

"Prairie Horned Lark" *praticola*

juvenile often mistaken for Sprague's Pipit

winter ♂ *ammophila*

Gulf Coast *giraudi*

except juveniles, all have dark breast band

southwestern deserts
♂ *ammophila*

duller than male

♀ *ammophila*

Midwest. Ten subspecies, generally pale, are western in distribution; one of the palest is Great Plains subspecies *enthymia*. Another western group consists of six small subspecies that are largely rufous on upperparts. Restricted to northeastern CA, subspecies *sierrae* is very yellow below and russet above. Confined to the Channel Is., CA, *insularis* is rather streaked above and below. The ruddy-colored *rubea* is found in central CA. Eurasian *flava* has yellow throat (unlike AK breeding *arcticola*, which has a white throat).

Similar Species Juvenile Horned Lark frequently mistaken for Sprague's Pipit. Note thicker bill, longer tail and wings, and pale spotted upperparts of Horned, and paler buffy face of Sprague's.

Voice CALL: Includes a high-pitched *tsee-titi*, given from ground or in flight. Flight call resembles American Pipit's,

but softer. **SONG:** Typically, two or three thick introductory *chit* notes, followed by a high-pitched, rapid, jumbled series of tinkling notes rising slightly in pitch, given from ground or in display flight.
Status & Distribution Common over much of the West in open country; uncommon to fairly common in East. Movements complex; some subspecies resident and others highly migratory. **BREEDING:** Open, barren country, avoiding forests. Prefers barren ground and shorter grasses, deserts, brushy flats, and alpine habitat. **MIGRATION:** In spring, diurnal migrant in flocks.

Over much of Midwest, winter birds are largely nominate *alpestris*, but *praticola* returns to more northern breeding areas in the East by early Feb. in southern ON to early Apr. in MB. Timing is complex in much of the West, including altitudinal movements by alpine subspecies. Arrive in AK generally by mid-Apr. In fall, peak movements in the East late Oct.–Nov. Begin departing from AK in Aug. In Southwest, move as early as mid-June–Sept. **WINTER:** Occur in similar habitat to breeding grounds, including beaches, sand dunes, and airports.
RARE STATUS: Casual to southern FL; one Palearctic subspecies (*flava*) casual to west, southwest, and south-central AK.
Population Has declined in central Prairie Provinces and many western states. Subspecies *strigata* is listed as Threatened by USFWS.

SWALLOWS Family Hirundinidae

Cliff Swallows (CA, Apr.)

This family consists of a group of accomplished aerial-foraging songbirds. The terms *swallow* and *martin* are used fairly interchangeably, with the square-tailed species generally being referred to as martins and the fork-tailed species as swallows. In the Old World, the Bank Swallow is called the Sand Martin.

Structure Hirundinidae is a very homogeneous family of birds, all with very similar body structures, and generally unlike any other passerines. Swallows have long pointed wings with 10 primaries (the tenth extremely reduced); small compressed bills with a wide gape, sometimes with rictal bristles; short legs; and square to deeply forked tails. Feathers of the lores are directed forward, which shades the eyes, a useful adaptation for aerial insectivores. The structure of the syrinx (vocal apparatus) is well differentiated from other passerine families, though for the most part, swallows are unremarkable singers.

Plumage Swallows and martins have variable plumage on the whole, but with much consistency within genera.

The birds often are iridescent blue or green above, and dark, white, or rufous below. Many species show pale rumps or dark breast bands, and a few Old World species are striped below. Sexual dimorphism is weak, with the New World martins (nine species in genus *Progne*) a notable exception.

Behavior All species are insectivorous, though one species (Tree Swallow) is able to feed on wax myrtle berries in winter, allowing it to winter farther north than most species. Swallows feed on the wing, and all are very accomplished fliers. Many temperate species undergo long-distance migrations, some among the longest of any passerines. Many species, including all those breeding in North America, are social in the breeding season, some nesting in colonies and sometimes with other swallow species. Some species are less gregarious or solitary. Nests are built either of mud—sometimes reinforced with vegetation and attached to vertical faces of natural and human-made structures—or within cavities excavated by other species. The Bank Swallow is the only cavity-nesting species that excavates its own nest. The rare Blue Swallow (*Hirundo atrocaerulea*), from southern Africa, is the only species that nests belowground, using potholes or aardvark burrows in open grassland.

Distribution Eighty-four species in 20 genera are found worldwide, except in Antarctica and the high Arctic, with the greatest diversity in Africa (29 breeding species), and secondarily in Central and South America (19 species). Eight species breed in North America and seven additional species have been recorded as casual or accidental strays.

Taxonomy Swallows are well differentiated from other passerine families, and within the family there is little variation. Two species of river-martin, one occurring in West Africa and the other (likely extinct) known only from Thailand, form the subfamily Pseudochelidoninae. All other swallow species are placed in the subfamily Hirundininae. Although swallows were formerly considered to be most closely related to tyrant flycatchers (Tyrannidae), recent DNA evidence suggests that they are actually more closely related to Old World warblers (Sylviidae), babblers (Timaliidae), white-eyes (Zosteropidae), chickadees (Paridae), and long-tailed tits (Aegithalidae).

Conservation Many species, including all North American breeding species, have adapted well to the influence of humans on their environment. In fact, some species now rarely nest in natural situations, preferring to nest on human-made structures. BirdLife International codes: 5 VU, 2 EN, and 1 CR.

MARTINS Genus *Progne*

These largest swallows, with broad-based wings (adapted for sustained soaring), moderately forked tails, and strongly decurved upper mandibles, are the only swallows of N.A. showing obvious sexual dimorphism. Taxonomy of this genus is unsettled, with up to nine species described. All members of *Progne* are called martins but are not directly related to martins of the Old World.

PURPLE MARTIN *Progne subis* PUMA ■ 1

The Purple Martin is the largest swallow in N.A. Adult males appear all-dark, glossy blue-black; females and immatures duller above and grayish below. This is a popular and well-known bird in East due to its nesting in human-made structures, often around homes, and their melodic vocalizations. Polytypic (3 sspp.; all in N.A.). L 8" (20 cm)

Identification ADULT MALE: All glossy blue-black above and below (appears black in most lights), notched tail. In hand, small concealed white on sides of rump and sides of body. **ADULT FEMALE:** Dull above, with scattered patches of blue-black above. Grayish collar on hind neck. Underparts dusky brown, paler on center of belly. Undertail coverts with dusky centers. **IMMATURE:** Resembles adult female; some to many young males show some blue-black on head and underparts (fewer in subspecies *hesperia*); also with dark shaft streaks on ventral feathers, and sometimes a less distinct collar on hind neck; females, paler below and browner above, and lack dusky centers on undertail coverts. **FLIGHT:** Somewhat starling-like shape. Graceful wingbeats interspersed with gliding and soaring.

Geographic Variation Females of western *arboricola* and desert *hesperia* have whiter underparts and forehead. Immature males tend to look more like females of eastern nominate subspecies (*subis*).

Similar Species Like other martins except Brown-chested. Female Purple Martin is the only species with contrasting gray collar on hind neck and pale forehead. Mottled undertail coverts.

Voice CALL: Most frequently gives a *chur* call. When alarmed or excited, gives a *zwrack* or *zweet* call. **SONG:** Usually a series of chortles, gurgles, and slightly harsher croaking phrases. Also gives a churring, chortling dawn song around potential nest sites on arriving on the breeding grounds in early spring.

Status & Distribution Fairly common but a local and declining summer resident. **BREEDING:** In the East, colonially almost exclusively in artificial sites near human habitations. In the West, more often in natural cavities in forested areas, and in saguaro cactus in desert Southwest. **NEST:** In cavity excavated by another species, or in artificial structures; three to six eggs (late Mar.–late May). **MIGRATION:** In

purplish on forehead (often lacking on smaller southwestern *hesperia*)

1st-summer ♂
subis

western ♀
arboricola

purple variable and blotchy on underparts

paler below than nominate eastern *subis*

adult ♂
subis

♀
subis

broad wings

often soars; wingbeats are slow and deep

eastern ♀
subis

darker overall than western subspecies

dark purple overall but often appears black, especially in flight

adult ♂
subis

arboricola

subis

hesperia

spring, arrives as early as mid-Jan. in TX, FL, and Gulf Coast; early Mar. in VA and KS; mid-Apr. in southern Canada and AZ. During fall, in the East, very large aggregations of thousands of birds form locally in late summer. Passage peaks late July–Sept., beginning as early as late May, with stragglers until early Oct., exception-ally to early Nov. In Southwest, scarce Aug.–late Sept. **WINTER:** S.A. lowlands east of the Andes south to northern Argentina (rarely) and southern Brazil. **RARE STATUS:** Casual to NV, ID (has nested), and AK. Accidental in Bermuda and UK.

Population Causes of long-term declines unknown. Competes for nest cavities with the introduced European Starling and House Sparrow. In the West, logging has reduced availability of natural nest cavities. Increased availability of human-provided nest sites in eastern N.A. has had a positive effect on populations. Sharp declines in northern New England, Maritimes, and southern and central CA.

CUBAN MARTIN *Progne cryptoleuca*   CUMA ◼ 5

This stray from Cuba has been confirmed only once in N.A. Monotypic. L 7.5" (19 cm)

Identification ADULT MALE: Like Purple Martin but with disproportionately longer tail and some white (usually mostly veiled) in vent area. **ADULT FEMALE:** Similar to Purple Martin, with darker upperparts, duskier brown throat, breast, and flanks contrasting with pale belly and undertail; lacks paler grayish collar. **IMMATURE:** Similar to adult female.

Similar Species Male indistinguishable from Purple in field; in hand, shows some concealed white feathers on belly, longer more deeply forked tail. See above for female. Caribbean Martin (*P. dominicensis*) and Sinaloa Martin (*P. sinaloae*), which show clean, white belly and undertail coverts in adult.

Voice CALL: Gurgling; a high-pitched *twick-twick*. **SONG:** A strong, melodious warble close to that of Purple Martin.

Status & Distribution Breeding endemic on Cuba and Isle of Pines. Nonbreeding (Sept.–Feb.) range unknown, presumably S.A. **RARE STATUS:** Accidental south FL on Key West (imm. male specimen; 9 May 1895).

Population Locally fairly common on Cuba and presumed stable.

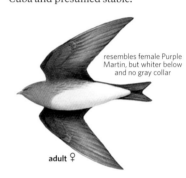

resembles female Purple Martin, but whiter below and no gray collar

adult ♀

GRAY-BREASTED MARTIN *Progne chalybea*   GYBM ◼ 5

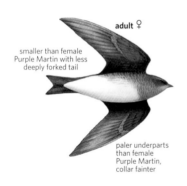

adult ♀

smaller than female Purple Martin with less deeply forked tail

paler underparts than female Purple Martin, collar fainter

This swallow has occurred in N.A. twice, both in TX. Polytypic (3 sspp.; nominate in N.A.). L 6.8" (17 cm)

Identification ADULT MALE: Purple-blue glossed upperparts. Face, breast, sides, and flanks gray-brown or sooty gray; chin and throat paler. Belly and undertail coverts pure white. **ADULT FEMALE:** Duller than adult male with paler chin and throat; whiter below, browner forehead. **IMMATURE:** Duller than adult female.

Similar Species Female Purple Martin has distinct gray collar; female South-ern Martin has mostly or entirely dark underparts; female Cuban Martin has more contrasting white belly.

Voice CALL: In contact, a *cheur*. Alarm and aggressive calls include *zwat*, *zurr*, or *krack*. **SONG:** A rich, liquid gurgling.

Status & Distribution Lowland resident from coastal northeast and west Mexico to northern Argentina. **RARE STATUS:** Two TX specimens, Starr Co., 25 Apr. 1880, and Hidalgo Co., 18 May 1889.

Population Possibly declining, but still abundant in parts of its native range; uses artificial nesting sites.

SOUTHERN MARTIN *Progne elegans*   SOMA ◼ 5

This species of S.A. has occurred once in N.A. Monotypic. L 7" (18 cm)

Identification ADULT MALE: Upperparts and underparts glossy dark, violet blue. Wings and tail black with slight bluish green gloss. Tail slightly forked. **ADULT FEMALE:** Smaller than male. Sooty black above, glossed blue on back; dusky brown below, often with pale edgings giving a scaly look. Most uniformly dark below. **IMMATURE:** Similar to adult female.

Similar Species Male slightly smaller than Purple Martin, with a slightly longer and more forked tail. In hand, lacks concealed white patch on sides and flanks. Female darker below than any other female martin.

Voice CALL: A harsh contact call and a high-pitched alarm call. **SONG:** A short warbling.

Status & Distribution Breeds in southern S.A.; migrates north in austral winter to western Amazon Basin. **RARE STATUS:** One specimen from Key West, FL (14 Aug. 1890).

Population Fairly common and presumed stable.

smaller than Purple Martin

darker below than female Purple Martin

adult ♀

BROWN-CHESTED MARTIN *Progne tapera* BCMA ▪ 5

adult
fusca

distinct breast band with
downward point like
smaller Bank Swallow

This species suggests an oversize Bank Swallow. Polytypic (2 sspp.; *fusca* and possibly nominate in N.A.). L 6.5" (16 cm)

Identification ADULT: Sexes similar. Subspecies *fusca* described. Upperparts brown. Dark brown band on upper breast; with pear-shaped dusky marks from center of breast to upper belly; sides and flanks brownish. The more northerly breeding subspecies *tapera* is slightly smaller and overall a paler brown and lacks dark extension of breast band.

Similar Species Larger female Purple Martin lacks breast band. Bank Swallow is much smaller.

Voice CALL: Gives a *chu-chu-chip* call when in flocks.

Status & Distribution S.A. from Colombia to northern Argentina, where locally known as the "River Martin." Southern breeding subspecies (*fusca*) is an austral migrant to northern S.A. **RARE STATUS:** Casual (mostly *fusca*); fewer than 10 records to eastern N.A., recorded MA to FL. One photographed at Patagonia Lake SP, AZ (3 Feb. 2006), is sole record for the West.

Population Presumed stable.

Genus *Tachycineta*

These swallows are mainly iridescent blue or green above and white below. The genus contains nine species, all occurring in the New World. Two species breed in N.A., and two species have been recorded as casual or accidental strays. The tails of *Tachycineta* are shallowly notched to deeply forked.

TREE SWALLOW *Tachycineta bicolor* TRES ▪ 1

Familiar denizens across N.A., Tree Swallows nest in abandoned cavities in dead trees or nest boxes provided by admiring humans. These birds are hardier than other swallows and can feed on seeds and berries during colder months. Monotypic. L 5.8" (15 cm)

Identification Small white area on

adult

1st spring ♀

dark
brownish
above

adult

white tertial tips

juvenile

fall adult

spring adult

dark blue
upperparts

sides of rump. **SPRING ADULT:** Upperparts iridescent greenish blue. Wings and tail dusky blackish, underparts entirely clean white. Lores black, ear coverts blue-black. Tail slightly forked. A few females retain brown back of immature into first spring, possibly longer. Tertials and secondaries edged pale grayish. **IMMATURE:** Upperparts gray-brown, with pale grayish edges on tertials and secondaries. Underparts clean white with indistinct dusky brown wash across breast, faintest in center. **FLIGHT:** Small but chunky swallow with a shallowly forked tail and broad-based triangular wings.

Similar Species Smaller Violet-green Swallow has white (male) to dusky (female) extending behind and above eye, and has more white on sides of rump. Tree Swallow flies with slower wingbeats and glides more than Violet-green. Bank Swallow similar to immature, but smaller and with distinct, clean-cut brown breast band, different face pattern (white behind ear), and paler brown rump.

Voice SONG: An extended series of variable chirping notes—*chrit, pleet, euree, cheet, chrit, pleet.*

Status & Distribution Common. **BREEDING:** Open areas, usually near water, including

marshes, fields, and swamps. **NEST:** Brings a few pieces of vegetation to nest cavity and often includes feathers; two to eight eggs (May–July). **MIGRATION:** In spring, arrives earlier in north than other swallows, usually mid-Mar.–early Apr. Departs later than other species, peaking late Sept. in MO and VA, lingering until late Nov. as far north as the Great Lakes. Very large aggregations of thousands of birds form locally from fall and into winter on East Coast. **WINTER:** Southern US, south through C.A. and the Caribbean. Open areas near water and nearby woodlands. **RARE STATUS:** Casual on Aleutians and Bering Sea islands and from Bermuda and the Azores; accidental in Guyana, near Trinidad, Greenland, and England.

Population The breeding range is expanding southward.

MANGROVE SWALLOW *Tachycineta albilinea* MANS ▪ 5

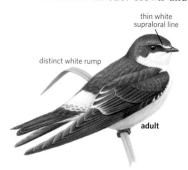

thin white supraloral line

distinct white rump

adult

This very small swallow is never found far from water, frequently in coastal areas. Monotypic. L 5.2" (13 cm)
Identification ADULT: Crown and back iridescent greenish, auriculars and lores black. Narrow white line above lores usually meets on forehead. Rump white, uppertail coverts dark. Wings and slightly forked tail dusky blackish, underwing coverts clean white. Tertial and inner secondary edges white. Underparts white, extending onto side of neck as a partial collar, sometimes with a dark diffuse wedge on sides of upper breast. **JUVENILE:** Gray-brown where adults are greenish blue. **FLIGHT:** Small size evident, with short, broad wings and short tail. Rapid, shallow wingbeats.
Similar Species Tree Swallow is noticeably larger, lacks white rump and supraloral. Common House-Martin is larger, bluish above, lacks white supraloral, and has a more deeply forked tail. White-rumped (*T. leucorrhoa*) from S.A. is unrecorded from N.A. It is larger, bluish above, and has narrower white edges on tertials.
Voice SONG: A series of *chrit or chriet* notes mixed with buzzier chirps.
Status & Distribution Common non-migratory resident of lowlands, especially near water, from northwestern and northeastern Mexico through Panama. **RARE STATUS:** One adult at Viera Wetlands, Viera, Brevard Co., FL (18–25 Nov. 2002).
Population Local declines.

VIOLET-GREEN SWALLOW *Tachycineta thalassina* VGSW ▪ 1

The adult male Violet-green Swallow is one of the world's most beautiful swallows. The western counterpart of the Tree Swallow, the smaller Violet-green Swallow has narrower wings with a different flight style and a shorter tail. This species frequently uses nest boxes. Polytypic (2 sspp., nominate in N.A.). L 5.3" (13 cm)
Identification ADULT MALE: Upperparts of male are dark velvet green, more bronze on crown, becoming purple on rump and uppertail coverts. Underparts are strikingly white, the white extending onto cheek, to behind and above eye. Lores are dusky. White of flanks extends onto sides of rump, forming two white patches from above in flight. Wings and short, slightly forked tail are black. Underwing coverts grayish. **ADULT FEMALE:** Somewhat duller than male, browner on head, face, and ear coverts. Throat is washed slightly with ashy brown. **JUVENILE:** Similar to adults, but green and purple are replaced with brownish. White on cheek and behind eye is much less extensive. Slight pale brownish wash on breast. Tertials and inner secondaries are very narrowly edged pale grayish. **FLIGHT:** Longer, narrower wings and shorter tail than Tree Swallow; Violet-green has a more fluttering flight style.
Similar Species Tree Swallow is larger, entirely greenish blue above, and lacks white or dusky around back of eye; has reduced, though some, white on sides of rump. The larger White-throated Swift has narrower pointed wings and black-and-white underparts. Strays reaching the Bering Sea region should be carefully compared to the all-white-rumped, barely smaller Common House-Martin.
Voice CALL: Various short notes, including a twitter, and *chee-chee* notes. Alarm call is a *zwrack* similar to that of Purple Martin.
Status & Distribution Common. **BREEDING:** Occurs in a variety of habitats, including open areas in montane coniferous and deciduous forests, coastal regions, and even in higher-elevation desert areas around cliffs. **MIGRATION:** In spring, arrives in southern AZ and Southern CA by early Feb., AK by early May. Departs AK by mid-Aug., lingering into mid-Oct. in BC and WA; elsewhere departs late Sept.–late Oct. Overall, a late fall migrant. Frequent early fall reports of migrants from CA's Central Valley, southeastern CA desert, and elsewhere in the desert Southwest are misidentified Tree Swallows. **WINTER:** Tidal flats to interior mountains. A few birds in CA, but most from Mexico south to Guatemala, El Salvador, and Honduras. **RARE STATUS:** Casual in western AK, including Aleutians and Bering Sea islands, and east of the Rockies, with most sightings in the East late Sept.–Nov.
Population Generally stable; the introduction of the House Sparrow and European Starling may have reduced Violet-green Swallow populations in urban areas of southern Canada and perhaps elsewhere.

adults

♂

small size with fluttery wingbeats

♂

juvenile

white up sides of rump

white around eye and lores

primary tips extend well past short tail

adult ♂

bright olive green upperparts, violet rump, and snowy white underparts

BAHAMA SWALLOW *Tachycineta cyaneoviridis* BAHS ▪ 4

Endemic to the northern Bahamas, the Bahama Swallow species has one of the most restricted ranges of any swallow in the world. Monotypic. L 5.8" (15 cm) **Identification** Underwing coverts strikingly white. **ADULT MALE:** Iridescent green upperparts, more bluish on wings and tail; largely white face with dusky lores. Underparts white. Wings and deeply forked tail dusky bluish black above. Margins of outer tail feathers sometimes whitish. **ADULT FEMALE:** Slightly duller

than male, with some slaty brown on ear coverts and breast washed with pale, sooty brown. Shorter tail. **JUVENILE:** Duller than adults, upperparts brownish with greenish mainly on mantle and wing coverts. Underparts white washed with sooty brown on breast. Shorter, less forked tail. **FLIGHT:** Resembles Barn Swallow in shape but Tree Swallow in plumage.
Similar Species Similar Tree Swallow has darker face, much shallower tail fork, and dusky underwing coverts.
Voice CALL: A metallic *chep* or *chi-chep*.
Status & Distribution Uncommon. **BREEDING:** Pine forests in the northern Bahamas (Grand Bahama, Abaco, and Andros; also a few on New Providence). Uses abandoned cavities in trees and sometimes on buildings; two to four eggs (Apr.–July). **WINTER:** Some winter on the breeding grounds,

with some moving south to the southern Bahamas and rarely or casually to eastern Cuba (Nov.–Mar.). **RARE STATUS:** Casual in southern FL (9+ records, but most are prior to 1990).
Population Endangered. Vulnerable due to small geographic range and hurricanes; world population estimated at approximately 2,500 pairs.

green above

adults

♂

♂

white wing linings

forked tail

Genus *Stelgidopteryx*

NORTHERN ROUGH-WINGED SWALLOW *Stelgidopteryx serripennis* NRWS ▪ 1

Named for the inconspicuous serrations on the first primary feather on the wing, unique among N.A. birds. Polytypic (6 sspp.; 2 in N.A.). L 5" (13 cm) **Identification ADULT:** Brown above; throat and breast pale brownish; rest of underparts dirty white. **JUVENILE:** Similar to adults, bright cinnamon-edged tertials and tips of wing coverts. **FLIGHT:** Long winged; flies with slow floppy wingbeats.

Geographic Variation Southwestern *psammochrous* is paler than nominate. Some treat *ridgwayi*, mainly from the Yucatán and with different vocalizations, as a separate species.
Similar Species Smaller Bank Swallow has white throat that wraps around auriculars, a distinct brown breast band, and a paler brown back and rump that contrasts with darker wings.
Voice CALL: Series of loose, low-pitched, upwardly inflected *brrt* notes, a buzzy *jrrr-jrrr-jrrr-jrrr*, or a higher pitched *brzzzzzt*.
Status & Distribution Fairly common. **BREEDING:** Open areas; often near rocky gorges, road cuts, gravel pits, or exposed sand banks. **NEST:** Singly or

ridgwayi

in small colonies in natural burrows or crevices, and on human-made structures; four to eight eggs (May–June). **MIGRATION:** In eastern N.A. arrives in early Mar. in FL to mid-Apr. in MI. In West, arrives in early Feb. in Southern CA to mid-Apr. in OR and WA. In eastern N.A. fall migration prolonged, late July–early Oct.; a few into early winter. In West, mainly in Aug. with some into Sept. **WINTER:** Mainly in lowlands and foothills from Mexico to C.A., a few in Caribbean. **RARE STATUS:** Rare in YT and southeastern AK; casual northern AK.
Population Stable.

long wings; flies with slow and floppy wingbeats

upperparts uniformly colored

juvenile

cinnamon wing bars

dusky wash on throat and breast

Genus *Riparia*

BANK SWALLOW *Riparia riparia* BANS ▪ 1

Our smallest swallow, the Bank is one of the few passerines with nearly worldwide distribution. Polytypic (2 sspp.; *nominate* in N.A.). L 4.8" (12 cm)
Identification ADULT: Pale grayish brown upperparts contrast with darker wings, underparts clean white with distinct brown breast band. Slightly forked tail. **JUVENILE:** Similar to adult. Wing coverts and tertials narrowly edged buff. **FLIGHT:** Sweeps wings backward with quick flicking wingbeats. Narrow, pointed wings, slim body. Flight feathers contrastingly dark from above.
Similar Species Northern Rough-winged Swallow is larger, and has slower floppier wingbeats and a dusky throat gradually blending into dirty white underparts. Immature Tree Swallow is larger, has a faint breast band and white sides to rump.
Voice CALL: Most frequently heard is a series of buzzy *trrrt* notes, not too different from the buzzing sound of high-tension lines.
Status & Distribution Locally common. **BREEDING:** Lowland areas along coastlines, rivers, lakes, and other wetlands; also in gravel quarries and road cuts. Small to large colonies in burrows in sand banks or cliffs; one to nine eggs (Apr.–June). **MIGRATION:** Departs S.A. in Feb., through US from early Mar. in South to late May in North. During fall migration, often in large flocks. Among the earliest to depart in the East, with most birds departing the Great Lakes by early Sept.; late July to early Nov. in FL; peak in late July and Aug. **WINTER:** Most often in wetlands areas, nearly throughout S.A. as far

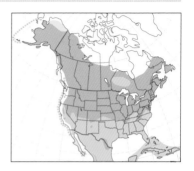

south as central Chile and northern Argentina, Pacific slope of southern Mexico, rarely in eastern Panama and in the Caribbean. Rare in Southern CA, southern TX, and FL.
Population Road building and quarries have changed breeding distribution in many areas.

paler and more scaly above than adult

juvenile

all figures *riparia*

rump and back contrast paler to darker wings

small size

pale coloring wraps around back of ear coverts

dark breast band with vertical extension forming point if viewed from front

flies with rapid shallow flaps

Genus *Petrochelidon*

The genus is composed of 11 species: eight Old World and three New World species, of which two are found in N.A. Characterized by square tails, some rufous on the head, orangish or chestnut rumps, and white-striped dark backs, all build mud nests. This genus is sometimes merged with the more widespread *Hirundo*.

CLIFF SWALLOW *Petrochelidon pyrrhonota* CLSW ▪ 1

Cliff Swallows nest colonially on cliffs and buildings and under bridges. Polytypic (4 sspp.; all in N.A.). L 5.5" (14 cm)
Identification ADULT: Square tail; orangish buff rump; dark cap extending below eye; chestnut cheeks, sides of neck, and throat; dark throat; whitish to buffy to chestnut forehead patch (feathering immediately above bill darkish). Underparts, including flanks, whitish. **JUVENILE:** Similar to adult, but entire head usually dark brownish black, sometimes with small pale grayish, whitish, or rusty forehead patch. Rump paler buff. Chin and upper throat pattern quite variable, some mixed with white, full black, gray, or cinnamon. **FLIGHT:** Short triangular wings and square tail. Whitish underparts and paler underwings contrast with dark head and throat. **HYBRID:** Very rare; hybrids with Barn Swallow and Tree Swallow (once) known. Cave Swallow hybrids possible but not confirmed.
Geographic Variation Taxonomic treatments differ, and there is considerable intergradation between the four recognized subspecies: Northern nominate (much of N.A.), *ganieri* (Southeast, west of Appalachians from central TN south), *tachina* (much of Southwest), and *melanogaster* (southeast AZ, southwest NM, and to southwest Mexico). They differ slightly in size
and overall darkness, including color of forehead and throat. Most distinctive is

melanogaster with chestnut forehead.
Similar Species Compare to Cave Swallow. (See sidebar, p. 498.)
Voice CALL: A single low-pitched *churr*, and a higher *keer* when disturbed. **SONG:** A subdued squeaky twittering given in flight and near or from within nest.
Status & Distribution Locally common. **BREEDING:** Various habitats, including grasslands, towns, open forest, and river edges wherever there are cliff faces or escarpments for nesting.

Small to large colonies, on cliff faces and on human-made structures (sometimes with Barn Swallows), rarely at cave entrances. A gourd-shaped structure built entirely of mud and saliva; one to six eggs (Apr.–June). **MIGRATION:** Always via C.A. Departs winter range in early Feb., a few by mid-Jan. to CA. Arrives in IL in early Apr., and AK in mid-May. Departs after nestlings fledge, sometimes as early as late June. Peak is Aug.–early Sept., earlier in Southwest (July–early Aug.). A few may linger to

early Nov. in East and exceptionally in West. **WINTER:** Grasslands, agricultural areas, near towns, and in marshes. S.A., from southern Brazil south to south-central Argentina. **RARE STATUS:** Rare in fall to Newfoundland. Casual to Bering Sea islands, AK (spring), and the Azores. Casual to Wrangel I., Russian Far East, southern Greenland, Europe, the Azores.
Population Has expanded its range into the Great Plains and eastern N.A. in the past 150 years.

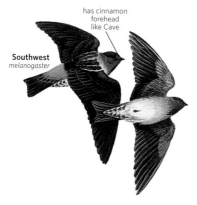

has cinnamon forehead like Cave

Southwest
melanogaster

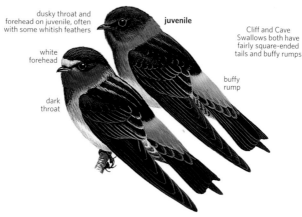

dusky throat and forehead on juvenile, often with some whitish feathers

white forehead

dark throat

juvenile

Cliff and Cave Swallows both have fairly square-ended tails and buffy rumps

buffy rump

CAVE SWALLOW *Petrochelidon fulva*   CASW ■ 1

This small, square-tailed, buff-rumped swallow has greatly expanded its breeding range by utilizing human-made structures. Polytypic (6 sspp.; 3 in N.A.). L 5.5" (14 cm)
Identification ADULT: Square tail and orangish buff to russet-chestnut rump. Has a small blackish cap extending to top of eye, which gives a somewhat masked look; broad pale wraparound buff collar (further isolates black cap), cheeks, and throat; chestnut forehead patch. Wings, tail, and uppertail coverts blackish; back is black with narrow white stripes. Underparts are whitish with pale buff flanks. **JUVENILE:** Similar to adult, but with less extensive forehead patch, variably paler chin and throat, browner back lacking distinct white stripes, and duller flanks. **FLIGHT:** Has a small black cap with pale buff collar and face visible at some distance. Glides frequently. **HYBRID:** Rarely with Barn Swallow, suspected with Cliff Swallow.
Geographic Variation Subtle. Subspecies breeding in FL and Cuba, *cavicola*, is slightly smaller, shows tawny-buff flanks, richer rufous-buff rump; subspecies breeding in southwestern US, *pallida*, is slightly larger, shows grayish buff flanks, variable orange-

Cliff and Cave Swallows both have fairly square-ended tails and buffy rumps

buffy rump

cinnamon forehead

pale collar contrasts with dark cap

adult
pallida

juvenile
pallida

rump slightly more cinnamon than Cliff

Southwest
pallida

buff rump. Slightly smaller *fulva* from Hispaniola is not separable in the field from *cavicola*.
Similar Species Cliff Swallow, especially southwestern subspecies. See sidebar, p. 498.
Voice CALL: Various chattering notes, including a short, clear *weet* or *cheweet*, ascending or descending in pitch. **SONG:** A complex squeaky twittering, slightly higher pitched and more prolonged than that of Cliff Swallow.
Status & Distribution Locally common, but patchy in distribution.

Migration little known, and wintering grounds of most US breeding birds not fully known. **BREEDING:** Natural or human-made structure often near water. Has only recently expanded northward into the US; first recorded nesting in TX in 1910, NM in 1930, and FL in 1987. Continues to expand in TX and NM, but still fairly restricted in southern FL. In colonies in twilight zone of caves, sinkholes, culverts, and under bridges (sometimes with Barn and Cliff Swallows), often high up and inaccessible in caves. A crescent-shaped half bowl built entirely out of mud and saliva, sometimes with bat guano; three to five eggs (Apr.–Aug.). **MIGRATION:** Movements and routes little known. In spring, arrives mid-

sides and flanks more cinnamon tinged than *pallida*

cavicola

darker cinnamon on rump than *pallida*

Cuban *cavicola*

Feb.–early Mar. in NM. Departs from NM late Oct.–mid-Nov. Routes largely unknown, but has been reported from western Mexico, Curaçao, Costa Rica, and Panama (sight records). **WINTER:** FL birds may winter in Caribbean. Winter range of Cave Swallows nesting in NM unknown. TX birds move south to unknown areas, but since the 1980s hundreds have overwintered in southern TX, often along streams, rivers, or lakes. **RARE STATUS:** Recent annual late fall invasions on eastern seaboard and in Great Lakes. (See sidebar below.) A few spring records there as well. Casual to accidental in spring or summer in AZ, CA (also in winter), central Great Plains, MS, AL, KY—all or nearly all records thought to refer to *pallida* from the Southwest. Two spring specimens from Sable I., NS, (others seen) have been ascribed by various authorities as either *cavicola* or *fulva* from the West Indies.

Population Cave Swallows are potentially vulnerable to disturbance at nesting colonies, although populations appear stable. The species is expanding its breeding range to north and east using human-made bridges and culverts, most dramatically in TX since the mid-1980s. Cave Swallows are vulnerable to temperature extremes and adverse weather conditions in NM and TX.

pallida

citata

fulva group

Cliff versus Cave Swallow in the East

Strong weather systems, characterized by prolonged strong west or southwest winds, sometimes bring Cave Swallows to the Atlantic and Great Lakes states late Oct. through Nov. Cliff Swallows have typically left for their S.A. winter areas by then, but occasionally lingerers are recorded, presenting an identification challenge. The subspecies of Cave Swallow most likely occurring in the East during and following these weather events is thought to be the southwestern *pallida*, not a bird from the geographically closer *fulva* West Indies group.

Adults of the eastern subspecies of Cliff Swallow (*pyrrhonota*) show a bright cream-colored forehead, unlike the Cave's chestnut forehead. Cliffs can be distinguished by their dark heads, including the throat, con-

trasting with whitish underparts. The Cave has a pale buff throat contrasting little with the breast.

Juvenile Cliffs typically have dark heads, but some have a few paler markings on the throat. Juvenile Caves are similar to adults; a small dark cap is relatively easy to see against pale buff cheeks and throat, even in flight.

Cave Swallow (NJ, Nov.)

Cliff Swallow (CA)

Genus *Hirundo*

BARN SWALLOW   *Hirundo rustica*   BARS ▪ 1

This most widely distributed and abundant swallow in the world is familiar to birders and nonbirders. Polytypic (6–8 sspp.; 3 in N.A.). L 6.8" (17 cm)

Identification ADULT MALE: Deep iridescent blue crown, back, rump, and

wing coverts; deeply forked tail with large white spots; rich buff to rufous forehead and underparts; iridescent blue patches on sides of breast, sometimes with very narrow connection in center. Wings and tail black. **ADULT FEMALE:** Similar to male but with paler

underparts and less deeply forked tail. **IMMATURE:** Duller above than adults, with shorter but still deeply forked tail, buffy throat and forehead, whitish underparts. **FLIGHT:** Very pointy wings, deeply forked tail, and low zigzagging flight are distinctive. **HYBRID:**

Rarely reported. Cliff and Cave Swallows in N.A., and Common House-Martin in Old World.

Geographic Variation Subspecies *erythrogaster* breeds in N.A. Eurasian subspecies (*rustica*) is accidental; it has clean white underparts with broader dark breast band. East Asian subspecies (*gutturalis*) is casual; similar to *rustica* but smaller, usually with incomplete blue breast band.

Similar Species No other swallow in N.A. shows such a deeply forked tail. Shorter-tailed juveniles in flight may suggest Tree Swallow but will always show partial dark breast band and buff throat, white spots on tail.

Voice CALL: In flight repeats a high-pitched, slightly squeaky *chee-jit*. **SONG:** A long series of squeaky warbling phrases, interspersed with a nasal grating rattle.

Status & Distribution Common. **BREEDING:** In various habitats in lowlands and foothills with nearby open

areas and water. Former natural nesting locales, including caves and cliff faces, have now mostly been abandoned in favor of a variety of human-made structures, including barns. A cup-shaped bowl made entirely of mud and the bird's saliva; three to seven eggs (May–June). **MIGRATION:** In spring, arrives in extreme southern US late Jan.–early Feb., peaking in mid-May in northeastern US. Early arrival in southeastern AK in mid-May. Departs northeastern US as early as mid-July, peaking late Aug.–early Sept.; late Sept.–early Oct. in southern CA. Main routes through C.A. and through Caribbean, but also a trans-Gulf migrant. **WINTER:** Often in fields and marshes mainly in lowlands; rarely in southern US; more frequently now in Southern CA, and southern AZ. Uncommon from Mexico south through C.A. Most common throughout S.A. as far as central Chile and northern Argentina; breeding noted in Buenos Aires. **RARE STATUS:** N.A. subspecies casual or accidental in western and northern AK, and HI,

southern Greenland, Tierra del Fuego, Falkland Is. Eurasian *rustica* accidental in northern AK (two specimens from Utqiagvik), NT, NU, and southern Labrador. East Asian subspecies *gutturalis* casual Aleutians and Bering Sea islands, AK; accidental mainland AK and Haida Gwaii, BC.

Population Still common worldwide, but substantial declines including throughout N.A.

erythrogaster

bluish above

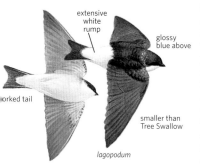

Genus *Delichon*

COMMON HOUSE-MARTIN *Delichon urbicum* COHM 4

This Old World species is casual in migration in western AK. Polytypic (3 sspp.; 1, possibly 2, in N.A.). L 5" (13 cm)

Identification ADULT MALE: Blue upperparts, white rump and underparts, forked tail, whitish underwing coverts. White feathered feet. **ADULT FEMALE:** Like male but grayer below. **JUVENILE:** Duller upperparts, less deeply forked tail.

Geographic Variation Mostly European subspecies, *urbicum*, shows dark uppertail coverts below white rump, forked tail. Asian subspecies, *lagopodum*, likely a separate species, shows white uppertail coverts continuous with white rump, slightly less deeply forked tail.

Similar Species Larger Tree and comparably sized Violet-green Swallows lack a solidly white rump. Asian House-Martin (*D. dasypus*), not recorded from N.A., has duller upperparts, gray-brown underwing coverts, pale gray wash on throat and undertail coverts, less deeply forked tail.

Voice CALL: A scratchy *prrit.* **SONG:** A soft twittering.

Status & Distribution BREEDING: Western Europe to Russian Far East. **RARE STATUS:** AK: one specimen of *lagopoda* from Nome, over 20 records from Bering Sea islands (likely *lagopoda* on probability). Accidental St.-Pierre I. (likely European *urbica*). Casual to Greenland (specimens of *urbica*).

Population Stable in most regions.

CHICKADEES AND TITMICE Family Paridae

Bridled Titmouse (AZ, May)

These lively sprites top many people's lists of favorite birds. The North American Paridae divides neatly into two groups: brightly patterned chickadees (seven species) and mostly drab titmice (five species). They inhabit tundra, high mountains, desert scrub, old-growth forests, small woodland patches, urban parks, and suburban backyards. Nearly all are sedentary. The northernmost species can endure frigid winters, migrating only in intermittent irruptions south or moving short distances down-elevation for the winter when food is scarce. Among the most intensively studied North American birds, parids have provided immense knowledge and understanding of physiological adaptation, hybridization, vocalization, genetic variations and relationships, socialization, and many other aspects of avian behavior.

Structure Parids are small birds, either round-headed (chickadees) or prominently crested (titmice). Their relatively short, rounded wings and long tails are aerodynamically suited for agility in crowded habitats. Their short, sharply pointed bills are designed to grasp tiny arthropods and crack small seeds.

Behavior Acrobatic feeders, parids glean high and low, flitting from leaf to leaf and branch to branch in noisy chatter, nearly always easy to see and hear. Their vocal repertoires are exceedingly complex. Most species sing in clear whistles, and all have sputtering variations of *tsick-a-dee* or *chirr-chirr*, plus a seemingly endless array of gargles, twitters, and *zeets*—each sound used in a particular behavioral context. Parids eagerly visit bird feeders. All are cavity-nesters, and many use nest boxes.

Plumage The chickadee's standard features include a dark cap, white or partly gray cheeks, and a black bib;

their body colors are various shades of gray or brown. Titmice are typically dull, unpatterned gray or brown, brightened only by rusty flanks in several species and by a brightly patterned face in one. The sexes in most species look alike, and juveniles differ little from adults.

Distribution The Paridae includes some 59 species around the globe, with 12 occurring in North America. At distributional extremes, Black-capped and Boreal Chickadees span the continent, while only the northernmost Mexican Chickadees and Bridled Titmice and the easternmost Gray-headed Chickadees reach North America. The family includes habitat generalists such as the Carolina Chickadee, at home in remote southeastern river bottomlands and crowded urban parks, as well as species with strong habitat preferences such as the Oak Titmouse, which relies heavily on the woodlands of its name.

Taxonomy The classification of North American chickadees has long been stable, but titmice have undergone considerable flux. The AOS divided the Plain Titmouse into Oak and Juniper species in 1997. Black-crested and Tufted Titmice, traditionally classified as separate species, were merged in 1976 and then separated again in 2002. The Black-capped Chickadee is the most variable of our parids, with at least nine subspecies recognized in three geographic groups. The Paridae includes the Asian Hume's Ground Jay, now rechristened as Ground Tit after molecular studies showed it not to be a corvid.

Conservation Loss of habitat may be a potential threat to a few species, especially for the Oak Titmouse in CA, where its favored oak woodlands are being destroyed for suburbs and agriculture. BirdLife International codes: 3 NT, 1 VU, and 1 EN.

CHICKADEES Genus *Poecile*

Worldwide this Northern Hemisphere genus consists of 15 species, named chickadees in the New World and tits in the Old World, all of them formerly classified in the genus *Parus*. Seven of these species occur in N.A., but the Mexican Chickadee and the Gray-headed Chickadee breed here only at the limits of their ranges.

CAROLINA CHICKADEE *Poecile carolinensis* CACH ▪ 1

The Carolina Chickadee is quite at home in cities and towns, readily using nest boxes and bird feeders. In fall and winter, the Carolina Chickadee forages in mixed flocks with nuthatches, woodpeckers, warblers, and other woodland species. The only chickadee in the Southeast, Carolina is smaller and duller than most other chickadee species. Polytypic (4 sspp.; all in N.A.). L 4.8" (12 cm)

Identification Cap and bib are black; cheeks are white, tinged at rear with pale gray when fresh; back and rump are gray, sometimes with an olive wash; greater coverts are gray without pale edgings; secondaries and tertials are indistinctly edged in dull white or pale gray; flanks are pale grayish, tinged buffy when fresh in fall.

Geographic Variation Four subspecies: largest and palest on upperparts in western portions of the range (*atricapilloides*); smallest and darkest in FL (*carolinensis*); the other two subspecies from south-central US (*agilis*) and east-central US (*extimis*) are intermediate. Differences are weak and clinal where ranges meet. Subspecies are not field-identifiable.

Similar Species Absence of white wing edgings on greater coverts and grayish wash on flanks, even when slightly tinged with buff, help to separate it from Black-capped. Carolina interbreeds extensively in some areas with Black-capped along a belt from KS to NJ; hybrid offspring are not easily identifiable. (See sidebar, p. 502.) Paler sides and flanks distinguish it from Mountain Chickadee when the usually conspicuous white supercilium of the latter species is not visible.

Voice CALL: A fast and high-pitched *chick-a-dee-dee-dee*. Compare to Black-capped's call. **SONG:** A four-note whistle, *fee-bee fee-bay*, the last note lowest in pitch. A Carolina may learn Black-capped's two- or three-note song in areas where their ranges overlap, so do not identify by song alone at the contact zone.

Status & Distribution Common. **YEAR-ROUND:** Open deciduous forests, woodland clearings and edges, suburbs, and urban parks; in the Appalachians it prefers lower elevations than Black-capped, but it has been recorded regularly as high as 6,000 ft. **RARE STATUS:** Wanders casually short distances north; single records in MI and ON are extraordinary.

Population Apparently declining in some regions in recent decades, particularly in the Gulf Coast states, but its range has expanded northward along much of the contact zone with Black-capped, particularly in OH, PA, and NJ.

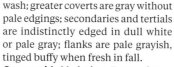

atricapilloides

slightly shorter tail than Black-capped

grayish tinge to rear cheek

fresh fall
extimis

slightly duller flanks than Black-capped

worn summer
extimis

BLACK-CAPPED CHICKADEE *Poecile atricapillus* BCCH ▪ 1

The most widespread, numerous, and geographically variable chickadee. It is curious, with little or no fear of humans, and it is famous for willingly, after a little training, taking seeds and nuts from the hand. During fall and winter, they are a main component of noisy feeding parties usually containing titmice, nuthatches, woodpeckers, and other species. A flock of Black-cappeds in fall often signal the presence of migrant warblers and vireos. Polytypic (9 sspp.; all in N.A.). L 5.3" (13 cm)

Identification Black cap; white cheek; black bib; greater coverts, secondaries, and tertials edged conspicuously white in fall and winter (less so by spring when worn); sides and flanks buffy or pinkish when fresh, fading to pale buff by summer when worn; outer tail feathers edged white.

Geographic Variation Northern and eastern birds include *turneri, fortuitus, septentrionalis, bartletti*, nominate *atricapillus*, and *practicus*. Northwestern *occidentalis* is small and dark-backed, with narrow white wing edgings and flanks heavily washed with buffy tan. Two subspecies from interior West, *garrinus* and especially *nevadensis*, are large and pale-backed with broad white wing edgings, and pale buffy sides and flanks.

Similar Species Carolina Chickadee is very similar,

turneri

bartletti

fortuitus

septentrionalis

atricapillus

occidentalis

nevadensis

garrinus

practicus

especially in late spring and summer when Black-capped's white wing edgings are worn. Where ranges overlap, the two species resemble each other and they also hybridize, intergrading characters. (See sidebar below.) Otherwise, the combination of bright white cheek, especially in fresh plumage, pure black cap and bib, gray upperparts, bright wing edgings, and pink-

darkest subspecies

occidentalis

nevadensis

palest subspecies

ish sides and flanks should distinguish Black-capped from other species.
Voice CALL: *Chick-a-dee-dee-dee*, lower and slower than Carolina. **SONG:** A clear, whistled two-noted *fee-bee* or three-noted *fee-bee-ee*, the first note higher in pitch. In the Pacific Northwest some birds sing *fee-fee-fee* with no change in pitch.
Status & Distribution Common. **YEAR-ROUND:** Deciduous and mixed woodlands, clearings, suburbs, and urban parks. Occurs in the Appalachians at higher elevations than Carolina. **FALL & WINTER:** Makes irregular irruptions south, usually not far into Carolina's range, but irregular to southeastern MO, northeast KY, eastern VA, and MD. **RARE STATUS:** Casual to northernmost AZ (recent breeding record) and OK Panhandle. Accidental West TX (specimen, El Paso Co.)
Population Range contracting northward in Midwest and mid-Atlantic.

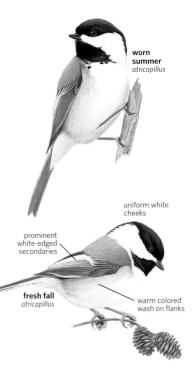

worn summer
atricapillus

uniform white cheeks

prominent white-edged secondaries

fresh fall
atricapillus

warm colored wash on flanks

Separating Black-capped and Carolina Chickadees

The differences between these two species are usually straightforward, except where the two species overlap and interbreed. Keep in mind the following:

Greater coverts. Brightly edged with white on the Black-capped, but inconspicuously bordered in dull whitish or gray on the Carolina.
Fringes of secondaries, tertials, and outer retrices. Prominently white on the Black-capped, but indistinctly dull whitish or grayish white on the Carolina.
Flanks. Pinkish tinged contrasting with pale gray breast and belly on the Black-capped, but dull grayish buff (and entirely grayish when worn) on the Carolina.
Wear of feather tips. The Black-capped's wing and tail edgings less bright in spring and summer, but still more conspicuous than those of the Carolina.
Cheeks. Black-capped has a purer white cheek; feathers on rear of cheek of Carolina are tipped with gray when fresh in fall and winter.
Hybrids. At the contact zone of the two species, a belt from KS to NJ, the hybrids or products of

Black-capped Chickadee, fresh

Carolina Chickadee, fresh

hybrid ancestry show a befuddling combination of Black-capped and Carolina characteristics. Intermediacy in plumage features is also evident in the Appalachian Mts., where the Black-capped and Carolina are found at higher and lower elevations respectively.

Vocalizations. Not particularly helpful. A bird that looks like one species may learn to sing the other's song or, bilingually, the songs of both species. Some sing aberrant songs that are not exactly representative of either species. The *chick-a-dee* calls (typically higher-pitched and faster for the Carolina) may be more reliable at the zone, but even these do not always exactly match the normal sounds or are confusingly intermediate.
Other features. Differences between the species frequently mentioned in field guides—for example, the extent of white on the cheek and relative sharpness of the bib's lower edge—are little or no help as well.

Where these two species occur together, it is probably best to leave many of them unidentified.

MOUNTAIN CHICKADEE *Poecile gambeli* MOCH ▪ 1

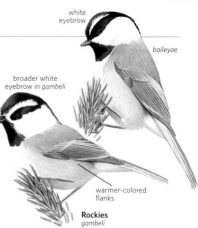

white eyebrow

baileyae

broader white eyebrow in *gambeli*

warmer-colored flanks

Rockies *gambeli*

This chickadee can be found above 12,000 ft, sometimes to timberline and beyond. It spends a lot of time in the postbreeding season caching conifer seeds for the winter. It forages higher in tall trees than most chickadee species, and it often spends the nonbreeding season foraging in mixed-species flocks. Polytypic (4 sspp.; 3 in N.A.). L 5.3" (13 cm)

Identification Black cap and bib, and white supercilium, which consists of white-tipped feathers, is unique among chickadees. White cheeks; grayish, brownish, or olive upperparts, depending upon the subspecies; greater wing coverts, secondaries, and tertials indistinctly edged with pale gray.

Geographic Variation Nominate *gambeli* in the Rocky Mts. is tinged with buff or brownish olive on back,

abbreviatus

gambeli

baileyae

atratus

inyoensis

has buffy sides and flanks, and has a prominent white supercilium. CA's *baileyae* is entirely grayish on back, sides, and flanks, sometimes with a slight olive tinge. It has a less distinct supercilium than *gambeli*. The other N.A. subspecies (*inyoensis*) has various features intermediate between these two. Recently, species recognition was proposed for the Rocky Mts. *gambeli* group (including *inyoensis*) and the Pacific *baileyae* group, based on genetic, morphological, and vocal differences. Somewhat intermediate *inyoensis* found east of the Sierra Nevada complicates the issue. Additional studies are needed.

Similar Species When supercilium is less conspicuous, as in worn *baileyae*, it could be confused with Black-capped. Wing edgings offer a helpful distinction between the two: pale gray and inconspicuous on Mountain, prominently white on Black-capped. Mountain's white eyebrow and smaller bib should distinguish it from Mexican Chickadee.

Voice CALL: A hoarse *chick-dzee-dzee-dzee*, as well as a variety of buzzes and chips. **SONG:** Typically a three- or four-note descending whistle, *fee-bee-bay*, or a *fee-ee-bee-bee* on one pitch, with many local dialects. The *fee-bee-bay* version has been compared to the melody of "Three Blind Mice"; another song (*baileyae* group) sounds like "cheese-bur-ger." Some of its vocal-

izations are very similar to those of Black-capped, and a good look may be necessary to ascertain the species.

Status & Distribution Common. **BREEDING:** Primarily found in montane coniferous forests and mixed woodlands, but locally near coastal areas, pinyon-juniper, and desert riparian woodlands. Sometimes use nest boxes. **WINTER:** Some descend to lower elevations in foothills, riparian woodlands in valleys, and suburban areas, where it sometimes visits feeders. **RARE STATUS:** Rare to southern AZ (away from Santa Catalina Mts.). Casual, mostly in winter, to southeast CA and east to SK, ND, SD, KS, NE, OK, and TX Panhandle.

Population Breeding bird surveys show a long-term decline in many parts of Mountain Chickadee's range, but the causes are unknown.

MEXICAN CHICKADEE *Poecile sclateri* MECH ▪ 2

This Mexican species barely reaches the US. It calls frequently, and in the postbreeding season it forages among mixed-species flocks of as many as

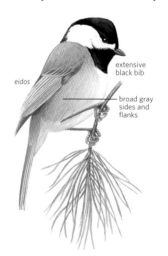

eidos

extensive black bib

broad gray sides and flanks

a hundred birds. Polytypic (4 sspp.; only *eidos* occurs north of Mexico). L 5" (13 cm)

Identification Black crown; black bib, largest of any chickadee species, extends well down onto breast; white cheeks; dark gray upperparts, sides, and flanks.

Similar Species Only chickadee within its limited US range.

Voice CALL: Husky, nasal buzzes, such as *zhree*; various *tsip* or *chit* calls. **SONG:** A rapid, slightly buzzy warble, *chee-lee, chee-lee, chee-lee*; no clear whistled song.

Status & Distribution Uncommon to fairly common, but local, at high elevations. **YEAR-ROUND:** Mainly in montane coniferous forests to at least 9,000 ft at its northern range limits—the sky islands in the Chiricahua Mts.

in southeast AZ and the Animas and Peloncillo (scarce) Mts. in southwestern NM. Some birds move to lower elevation (±6,000 ft) woodlands in the postbreeding season.

Population Declines in Mexico. Large forest fires in Chiricahua and Animas Mts. have reduced populations.

eidos

garzai

CHESTNUT-BACKED CHICKADEE   *Poecile rufescens*   CBCH ▪ 1

The Chestnut-backed Chickadee spends the postbreeding and winter seasons foraging noisily in mixed-species flocks. Unlike other parids (except for the Mountain Chickadee), it forages high in the canopies of tall conifers. This species is the smallest chickadee in N.A. One of the three brown-backed species (along with Boreal and Gray-headed), its nominate subspecies is the most richly colored parid. Polytypic (3 sspp.; all in N.A.). L 4.8" (12 cm)

rich chestnut back, sides, and flanks

pure white cheeks

rufescens

chestnut back

gray sides

coastal central CA
barlowi

Identification Chestnut color of back is unique, but it varies in intensity. Cap is dark brown, shading to black at its lower edge from bill through eye; cheek is white; bib is black; back and rump are rufous; greater wing coverts are edged white; breast and belly are whitish; sides and flanks are bright rufous or dull brown, depending on subspecies; tail is brownish gray.

Geographic Variation The widely distributed *rufescens* is the most colorful subspecies, with a rich chestnut back, bright white edgings on greater wing coverts, and extensively bright rufous flanks. The other two subspecies are restricted to coastal CA and around the San Francisco Bay Area; they differ notably on underparts. In Marin Co., *neglectus* has reduced, pale chestnut flanks that do not contrast conspicuously with whitish breast and belly; *barlowi*, found from the Bay Area to extreme northern Ventura Co., has grayish flanks, tinged with olive brown. The sparse and local populations at mid-elevations on the west slope of the Sierra from Butte south to northern Madera Co. (typically found in well-vegetated, moist canyons with tall, mature conifers, and other trees, notably madrones) are the nominate subspecies.

Similar Species Boreal Chickadee also has a dark brown cap and rich brown sides and flanks, but it has predominantly grayish, not white, cheek, lacks a rufous tint on back and rump, and shows no white wing edgings.

Voice CALL: All vocalizations are very different from other chickadees. A hoarse, high-pitched, rapid *sik-zee-zee* or just *zee-zee*; also a characteristic sharp *chek-chek*. SONG: No whistled song is known.

Status & Distribution Uncommon to common. YEAR-ROUND: Coniferous forests, especially of Douglas fir, and mixed and deciduous woodlands. DISPERSAL: Postbreeding movements have been noted to higher elevations in BC and to lower elevations in OR. RARE STATUS: Wanders irregularly inland as far as southwestern AB. Rare in Sacramento Valley, CA.

Population Its range has expanded southward and eastward in recent decades from humid coastal regions to the drier eastern San Francisco Bay Area (*barlowi*); it can now be found on the CA coast south to the northern Ventura Co. line. Numbers appear to be stable in the northern portion of the range, but recent declines in the interior expansion area have raised conservation concerns.

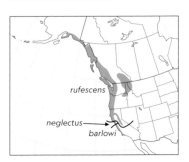

rufescens

neglectus

barlowi

BOREAL CHICKADEE   *Poecile hudsonicus*   BOCH ▪ 1

This species is one of N.A's largest chickadees and, like the Gray-headed Chickadee, it survives winters in the far north by hoarding large supplies of food. Unlike the Gray-headed, it sometimes makes long-distance irruptions southward (much less so in recent

decades), apparently in autumns when food is scarce, exceptionally to the central and mid-Atlantic states. Boreal was sometimes called the Brown-capped Chickadee in older literature—an apt name describing a diagnostic field mark over most of range, although the rest of the upperparts are brownish as well. Many observers comment on its tameness around humans in much of the year, but those who seek Boreal during the breeding season know that it is exceedingly quiet and inconspicuous. Polytypic (4–5 sspp.; all in N.A.). L 5.5" (14 cm)

Identification Largely gray cheek is unique among chickadee species. Cap is dull brown; bib is black; back and rump are brownish in eastern portions

of the range and grayer in western regions; wings are plain brownish with no pale edges on coverts or secondaries; sides and flanks are extensively washed with dull chestnut, brown with a reddish tinge, or tawny, depending upon subspecies.

Geographic Variation The four generally recognized subspecies are: *hudsonicus* (widespread from AK to eastern Canada; including *stoneyi* from northern AK and northwestern Canada); *columbianus* (southern AK, western Canada); *littoralis* (southeastern Canada, northeastern US); and *fanleyi* (south-central Canada). At their extremes the plumage differences are notable, but the characters vary clinally across the range, and

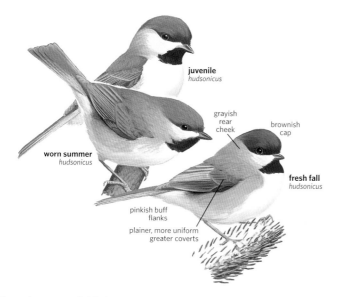

juvenile
hudsonicus

grayish
rear
cheek

brownish
cap

worn summer
hudsonicus

fresh fall
hudsonicus

pinkish buff
flanks

plainer, more uniform
greater coverts

subspecies are not field-identifiable. **Similar Species** Brown cap and dusky auriculars distinguish it easily from Black-capped and Carolina Chickadees; wings are plainer than Black-capped. Dull wings and darker face also separate it from Chestnut-backed and Gray-headed Chickadees. Relatively grayish northwestern Boreal could bring to mind Gray-headed, but wings and especially face differ enough to identify either species. **Voice CALL:** A wheezy, nasal, drawled *tseek-day-day*; also a characteristic single, high *see* or *dee*. **SONG:** A clear trill. No whistled song is known. It is usually less vocal than Black-capped with which it shares much of its range.

Status & Distribution Fairly common. **YEAR-ROUND:** Prefers spruce and balsam fir forests, ranging north to the tree line on the taiga; to a lesser extent mixed woodlands. Its range overlaps with the ranges of the Gray-headed in northern AK (formerly northern YT) and Chestnut-backed in BC, WA, ID, and MT, but no hybridization is known. **MIGRATION:** Regular short-distance movements southward have been reported within MN, SK, and MB. **WINTER:** Move slightly south of breeding range and have occurred casually as far south as IA, IL, IN, OH, WV, PA, NJ, DE, MD, and VA—but few records in recent years.
Population Trend data are unreliable because populations have not been surveyed in much of the range. Numbers increase greatly over short periods of years during spruce budworm outbreaks, then decline, making judgments about trends difficult. Numbers are considered stable in most regions but surveys suggest a gradual long-term decline in recent decades, particularly along the southern edge of the range in the northern tier of the US. Extensive logging in Canadian boreal forests poses a potentially serious threat.

GRAY-HEADED CHICKADEE *Poecile cinctus* GHCH ▪ 3

Known in Eurasia as Siberian Tit, this species is perhaps our hardest-to-see breeding bird and remains little known. It remains on its home territory through the Arctic winter, surviving on hoarded food supplies. The bird has a small head and long tail. Polytypic (4 sspp.; 1 in N.A.). L 5.5" (14 cm)
Identification Brownish gray cap; pure white cheek including auriculars; black bib, ragged at corners; brown back and rump with grayish tinge; greater coverts, secondaries, and ter-

pure white
cheeks

white edgings to
greater coverts and
secondaries

longish
tail

lathami

slightly paler
flanks than
Boreal

tials prominently edged white; buffy to cinnamon sides and flanks; whitish breast and belly.
Geographic Variation Three subspecies in northern Eurasia; *lathami*, in N.A. Differences are weak and clinal, although *lathami* averages paler and grayer overall, with slightly darker cinnamon flanks. Vocal and genetic studies of *lathami* would be useful to determine relationships to Old World subspecies.
Similar Species Brownish tint of its cap and back differentiate it from Black-capped Chickadee. Grayish northwestern Boreal Chickadee possibly confusing, but white auriculars and bright secondary edgings of Gray-headed are distinctive, as are the vocalizations.
Voice CALL: A loud *deer deer deer* or *chi-urr chi-urr.* Apparently no clear, whistled song.
Status & Distribution Rare in N.A. Resident in coniferous forests in Europe and across northern Asia; in N.A., mainly willows and stunted spruces along waterways in Brooks Range; formerly east through the

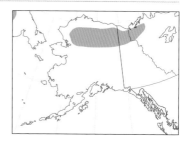

northern YT to Mackenzie (no recent records) in northwest Canada. Distribution in N.A. is not well known; for instance, there are only a half dozen records in YT since July–Aug. 1926, when 13 were collected on the Old Crow River. Traditional locations in AK include Kelly Bar at the confluence of the Kelly and Noatak Rivers north of Kotzebue and the Canning River area in Arctic NWR. **RARE STATUS:** A few winter records for Fairbanks several decades ago (at feeders) are the only valid records for central AK. No valid records for AK Range, including Denali NP.
Population The species is declining in much of its extensive Eurasian range.

TITMICE Genus *Baeolophus*

All five species in this genus are New World titmice formerly classified in the genus *Parus*. Three breed almost entirely in N.A., and two have substantial distributions in Mexico. Revisions to the taxonomy split the Plain Titmouse into Juniper and Oak species and separated the Black-crested Titmouse from the Tufted Titmouse.

BRIDLED TITMOUSE *Baeolophus wollweberi*   BRTI ■ 2

Slightly smaller than other titmice, the Bridled sometimes behaves chickadee-like and at other times forages more slowly, more deliberately, and less noisily than other parids. Bridleds commune in substantial flocks, first as family groups and then, through fall and winter, in larger numbers of two dozen or more of their own species. They also form the nucleus of mixed-species flocks, sometimes including

distinct crest with
black-and-white
head pattern

phillipsi

Mexican Chickadees and Juniper Titmice, which forage together. Polytypic (4 sspp.; 2 in N.A.). L 5.3" (13 cm)

Identification Bridled's strikingly patterned face is an unmistakable field mark. **ADULT:** A black and white (the "bridle") face; a tall, dusty crest; a black throat, appearing as a small bib. Gray upperparts sometimes have an olive tint; underparts are paler, washed with dull olive on flanks; belly sometimes has a yellow tint in fresh plumage. **JUVENILE:** Crest is shorter; face pattern is duller; bib is gray. **Geographic Variation** Weak variation. Subspecies *phillipsi* (AZ and southwestern NM, south of the Gila River) is slightly paler than *vandevenderi* (AZ and southwestern NM, north of the Gila River).

Similar Species Facial pattern and black throat separate it from all other titmice. Black bib might suggest a chickadee, when seen from below, but with a better look, crest and face will easily eliminate that possibility. **Voice CALL:** A rapid, high-pitched *tsicka-dee-dee* or a harsh *tzee-tzee-tzee-tzee*, similar to Juniper, but not the allopatric Oak Titmouse. **SONG:** A rapid series of clear, whistled, identical *peet-peet* notes. **Status & Distribution** Common.

YEAR-ROUND: Oak, juniper, and sycamore woodlands in mountains of southeastern AZ and southwestern NM, up to 7,000 ft. The main portion of its range is in Mexico. **DISPERSAL:** Flocks make local movements to riparian areas at lower elevations in winter. **RARE STATUS:** Accidental in winter to Bill Williams Delta, western AZ (17 Feb.–20 Mar. 1977). Reports from TX not well documented.

Population The trend is uncertain because of insufficient monitoring, particularly in Mexico, but Bridled's limited breeding range and the potential loss of its preferred habitats by logging and forest fires causes concern in limited range in Southwest US. The loss of oak woodlands has extirpated the species from some locations in Mexico.

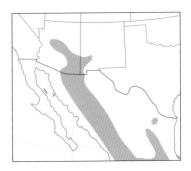

OAK TITMOUSE *Baeolophus inornatus*   OATI ■ 1

Once combined with the Juniper Titmouse and aptly named Plain Titmouse, the Oak Titmouse is easily separable from all other timice species except the Juniper. Its lack of field

marks is, ironically, an important field mark. Polytypic (4 sspp.; 3 in N.A.). L 5" (13 cm)

Identification The Oak Titmouse is a drab brownish or grayish brown color overall, but is paler below. It has a short crest.

Geographic Variation Four subspecies, one restricted to Baja California; variation is weak and clinal. The northernmost subspecies, *inornatus*, from OR to south-central CA, has medium brownish gray or olive-brown upperparts, pale gray underparts with pale brown tinge on flanks, and a short bill. Subspecies

worn
affabilis

short
grayish
crest

slight
brownish
tint above

fresh
affabilis

affabilis found in southwestern CA is notably larger, has darker gray-brown or olive-brown upperparts, flanks washed in dusky brown, and a longer bill; subspecies *mohavensis*, restricted to the Little San Bernardino Mts. in San Bernardino Co. and Riverside Co., is smaller, paler, and grayer than *affabilis*.

Similar Species The closely related Juniper Titmouse is slightly larger, paler, and grayer, but similarly plain. The two are best separated by range and vocalizations; they are sedentary and are sympatric only in a small area of northern CA (Modoc Plateau).

Voice CALL: Hoarse *tsick-a-deer* differs from Juniper Titmouse. **SONG:** A series of clear, whistled sets of alternating high and low notes, such as *peter peter peter* or *teedle-ee teedle-ee*, with many variations, but higher pitched and slower than Juniper.

Status & Distribution Common. **YEAR-ROUND:** Primarily oak and pine-oak woodlands, including suburban areas, on the Pacific slope. **RARE STATUS:** Casually reported east of its usual limits in CA, including once (specimen) near the south end of the Salton Sea, Imperial Co.

Population Declining in CA due to losses of its preferred oak habitat.

JUNIPER TITMOUSE *Baeolophus ridgwayi*   JUTI ▪ 1

The Juniper Titmouse travels through woodlands in small family parties. It will visit bird feeders. Formerly classified as a subspecies of Plain Titmouse, the Juniper has a much larger range but is more sparsely distributed than the Oak Titmouse, its close relative; its total population has been estimated at less than half that of the latter species. Polytypic (2 sspp.; both in N.A.). L 5.3" (13 cm)

Identification Sexes and ages similar. Drab medium-gray upperparts; grayish white underparts; short crest.

Geographic Variation Despite its wide distribution, Juniper Titmouse shows little geographic variation. Two different subspecies are recognized: nominate and the more westerly

almost identical to Oak Titmouse, but slightly grayer and narrower billed; best told by calls and song in narrow range of overlap

zaleptus—they are not identifiable in the field.

Similar Species Juniper's uniform gray coloration separates it from other titmice except for Oak, which is slightly smaller, darker, and brownish tinted. To the eye, differences appear slight, but genetic, morphological, ecological, and vocal characters set the species apart. They are best distinguished from each other by range and vocalizations; fortunately, they are mostly sedentary, coming into contact only at the periphery of their ranges in a small area on the Modoc Plateau in northern CA. Research at the contact zone indicates that there is little or no interbreeding and that physiological adaptations to temperature and precipitation differ significantly between the species.

Voice CALL: A hoarse, rapid *tsick-dee*, faster than Oak, more like Bridled Titmouse. **SONG:** A variety of long, rolling trills on a uniform pitch and three-part groups of *wheed-leah, wheed-leah, wheed-leah*. Male has a large repertoire of song types.

Status & Distribution Uncommon. **YEAR-ROUND:** Primarily juniper and pinyon-juniper woodlands in the intermountain region and south-

ward. It can be found at elevations up to 7,000–8,000 ft in AZ, NM, and eastern CA. **RARE STATUS:** Casual in CA's Mojave Desert south to northern Kern Co.

Population Its relatively small numbers, low density, limited habitat preferences, and recent declines reported in the southern Rockies and the CO Plateau region have raised conservation concerns.

TUFTED TITMOUSE *Baeolophus bicolor*   TUTI ▪ 1

The active and noisy Tufted Titmouse, N.A.'s most widespread titmouse, is remarkably uniform morphologically, genetically, vocally, and behaviorally throughout its range. Besides gleaning trees and shrubs for arthropods, it spends more time on the ground searching leaf litter than do chickadees. In the nonbreeding season Tufted Titmice are not as tied to joining mixed-species flocks as chickadees, though; after the breeding season they spend a lot of time in small foraging parties consisting of parents and their offspring. Tufted frequents well-vegetated urban and suburban areas, willingly uses nest boxes, and regularly visits bird feeders. Monotypic. L 6.3" (16 cm)

Identification ADULT: Gray crest; black forehead; gray upperparts; pale gray or whitish breast and belly; flanks prominently washed in rust or orange. **JUVENILE:** Gray forehead, rather than black; slightly paler crest than in adult. **HYBRID:** Tufted and Black-crested Titmice interbreed in a north-south belt 15 to 25 mi wide (hybrids predominate) extending from southwest OK to just east of the Balcones Escarpment and south to Refugio Co. on TX coast, producing offspring with variably intermediate dusky crests and chestnut or grayish foreheads that make identification near impossible.

Similar Species The only other titmice likely found in its range is Black-crested Titmouse, which has a pale forehead. Juniper Titmouse is much plainer, with a shorter crest. It also lacks a contrasting dark forehead and does not have brightly washed pinkish flanks.

Voice CALL: Calls are varied, but include a harsh, scolding *zhee zhee zhee*. **SONG:** A loud, whistled two-syllabled *peto peto peto* or *wheedle wheedle wheedle*, often repeated monotonously. It is delivered a little more rapidly and is more two-syllabled than Black-crested's similar song.

Status & Distribution Common. **YEAR-ROUND:** Deciduous woodlands, suburbs, urban parks, wherever else trees are large enough to provide nest holes. **RARE STATUS:** Wanders casually north of its usual limits in fall and winter; records from SD, MN, QC, and NB.

Population Expanded northward during much of the 20th century into New England and southern Canada. More recently, the expansion has stopped and there have been some range contractions.

juvenile

tall crest

dark forehead

adult

pinkish flank patch

BLACK-CRESTED TITMOUSE *Baeolophus atricristatus* BCTI ■ 2

Except for its black crest, which no other titmouse has, and white forehead, the Black-crested Titmouse could be a mirror image of the larger Tufted Titmouse in behavior and appearance. Throughout the summer, fall, and winter, this lively, noisy, and continually active bird forages in small family groups, searching trees and shrubs for insects, spiders, and their eggs. Confiding and curious, Black-crested shows little fear of humans and does not hesitate to use nest boxes. Polytypic (3 sspp.; all in N.A.). L 5.8" (15 cm)

Identification ADULT: Black crown and crest; dull white forehead; medium to dark gray upperparts, sometimes with a slight olive tinge; pale gray underparts with a cinnamon wash on the flanks. **JUVENILE:** Gray crest, sometimes washed with brown; dusky forehead. **HYBRID:** Black-crested and Tufted Titmice interbreed freely in

a north-south belt 15 to 25 mi wide extending from southwest OK southwards, just east of Balcones Escarpment to Refugio Co. on the TX coast. Here hybrids outnumber pure birds. Offspring show variably intermediate shades of dusky crests and gray or brownish foreheads, making it near impossible to assign them to a species. Most hybrids have a chestnut patch on forehead.

Geographic Variation Three subspecies in N.A.: *castaneifrons* (southwestern OK to south TX); *paloduro* (north and Trans-Pecos TX); and nominate (south TX and northeastern Mexico). They differ slightly in size and plumage coloration; they are not field-identifiable.

Similar Species Juveniles with gray crests may be difficult to distinguish from Tufted (where ranges overlap) or Juniper Titmouse. Look for juvenile Black-crested's whitish forehead and crown darker than upperparts.

Voice CALL: A harsh, scolding *jree jree jree*, sharper and more nasal than Tufted. **SONG:** A series of slurred *chew chew chew* or *pe-chee-chee-chee* phrases, faster and with each note tending more toward one syllable than the Tufted's similar vocalizations.

Status & Distribution Common resident. Black-crested inhabits oak and other deciduous woodlands along watercourses, open tracts with scat-

juvenile

black crest

white fore-head

adult

tered trees, arid scrubby habitats, and urban and suburban areas. Its range extends south in northeast Mexico south to Veracruz. **RARE STATUS:** Casual to Guadalupe Mts., West TX, and to High Plains in West TX Panhandle. It has not yet been recorded in NM, even though the northern part of the Guadalupe Mts. are in NM.

Population No trend data exist, but fragmentation of oak-woodland habitats has raised concern. The Black-crested's return to full species status in 2002 has brought renewed attention to monitoring its numbers and its potential conservation needs.

PENDULINE TITS AND VERDIN Family Remizidae

Verdin (NM, Nov.)

Penduline Tit of Eurasia are notably larger and longer tailed than other family members. The bill is short and sharply pointed. The legs are long and slender.

Behavior They are acrobatic, easily moving through dense foliage and even hanging upside down while foraging. They often hold larger prey items under a foot while tearing it apart with the bill.

Plumage Most of the species in this family have non-descript plumage. They primarily have gray, olive, or yellow upperparts with pale, or white, underparts. The multiple species of penduline tits have rather ornate plumage.

Distribution The family reaches its greatest diversity in Africa, with six species in the genus *Anthoscopus*. The penduline tits of Eurasia, genus *Remiz*, are now divided into three species (four by some authors). Only a single species, the Verdin, in its own monotypic genus *Auriparus*, is found in North America.

Taxonomy Worldwide, there are 10 species in three genera. The Remizidae family is sometimes classified as a subfamily of the Paridae; in any case, there is a sister group relationship between the two families.

Conservation No member of the family is considered of conservation concern, although two are considered rare, and many aspects of their biology are poorly known.

These are very small birds that generally resemble chickadees. All species have small, sharply pointed bills. Some species, like the Verdin, are rather solitary, while others are quite gregarious and nest in loose colonies. Nests of most species are a tightly woven bag or ball with a tubular side entrance.

Structure Most species are small-bodied, with a short tail. Although small, the Verdin of North America and the

Genus *Auriparus*

VERDIN *Auriparus flaviceps*   VERD ■ 1

An acrobatic species, the Verdin is generally solitary away from nest sites. It is most easily detected by its surprisingly loud calls. Nest is an intricately woven ball of spiderwebs and small twigs. The male often builds several structures during the nesting season; both sexes roost in these year-round. Polytypic (6 sspp.; 2 in N.A.). L 4.5" (11 cm)

Identification Very small body size, medium-length tail. Sexes similar. **ADULT:** Dull gray overall, darker on upperparts, with a yellow face and chestnut shoulder. Small, sharply pointed bill with a straight culmen. **JUVENILE:** Similar to adult, but paler

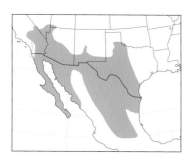

gray and lacking yellow face and chestnut shoulder making them appear very plain. Juveniles also have pale base to bill.

Geographic Variation Variation in two N.A. subspecies is weak and clinal. Western birds (*acaciarum*) tend to have more brownish upperparts than those from southern NM to central TX (*ornatus*).

Similar Species Yellow face and chestnut shoulder of adults distinguish them from other chickadee-like birds. Juvenile similar to Lucy's Warbler (particularly female) but can be distinguished by its slightly heavier bill with a pale base. Bushtit has a longer tail and smaller bill with a curved culmen. Gnatcatchers have longer, black tails and prominent eye rings.

Voice CALL: A clear *tschep* and rapid chip notes. **SONG:** A plaintive three-note whistle, *tee tyew too*, with the second note higher.

Status & Distribution Common; uncommon in UT and rare in OK and TX Panhandle. **YEAR-ROUND:** Resident in desert scrub and other brushy habi-

tats, including mesquite woodlands. Formerly resident in coastal southern San Diego Co., CA. **RARE STATUS:** Otherwise casual in southwestern CA north to coastal southern Santa Barbara Co. **Population** Stable, although very little data concerning this species.

LONG-TAILED TITS AND BUSHTIT Family Aegithalidae

This family is characterized by very small birds with short wings and long tails. The small body size combined with rather loose contour feathers gives these birds the look of tiny fluff balls.

Structure Small bodied; long tailed. All species have small black, conical bills; legs generally long and slender.

Behavior Sociable birds that live in small family groups during breeding season and congregate in larger flocks in nonbreeding season. Typically thought to be monogamous, but cooperative breeders and nest helpers recorded.

Plumage In general, species in this family are gray or brown above and white below. Most have a black or brown mask, often combined with a black bib. Others

Bushtit, male (WA, Apr.)

are fairly uniform in color, such as the Bushtit. Sexes are similar in plumage, although juveniles are often distinctive.

Distribution This family's greatest diversity is in the Himalaya and western China; single species in Europe and North America.

Taxonomy Sometimes included as a subfamily of Paridae; however, they differ in several important aspects, and DNA analysis confirms that this family is only distantly related to the parids. The Bushtit, the only New World representative, is in its own genus. Worldwide there are 10 species in four genera.

Conservation No member of the family is considered of conservation concern, but some Asian species are not well known.

Genus *Psaltriparus*

BUSHTIT *Psaltriparus minimus*   BUSH ■ 1

This acrobatic species is most often encountered, except when breeding, as medium-size vocal flocks moving through the various types of woodland. The nest is an intricately woven hanging structure. Polytypic (10 sspp.; 5 in N.A.). L 4.5" (11 cm)

Identification Very small body size and long tail distinguish this species from other chickadee-like birds. **ADULT:** Pale gray overall, darker on upperparts, with very small bill. Interior birds have brown ear patch with a gray crown; coastal birds have a brown cap. Adult females have pale eyes. **JUVENILE:** Similar to adult except in Chisos Mts., TX, where juvenile males can show a black ear patch, as can some adult males. Juvenile females develop

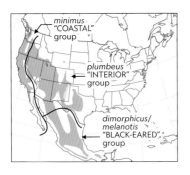

pale eyes a few weeks after fledging.

Geographic Variation Three well defined subspecies groups: "Coastal" *minimus* group with five subspecies, three in N.A.: *saturatus* (Pacific Northwest), *minimus* (Pacific coast), and *californicus* (inland from coast). "Interior" *plumbeus* group with one subspecies, *plumbeus* (east of the Sierra Nevada). "Black-eared" *melanotis* group with four subspecies; *dimorphicus* in the Chisos Mts., TX. The "Coastal" and "Interior" groups meet near Lone Pine, Inyo Co., CA.

Similar Species Immature Verdin more uniform in overall body plumage and has a shorter tail. Gnatcatchers have longer, black tails, prominent eye ring. **Voice** Sharp twittering *tsip* or *tseet*. "Interior" *plumbeus* group has lower-pitched and sharper calls and might be a separate species.

Status & Distribution Common. **YEAR-ROUND:** Resident in a variety of woodland, scrub, and residential habitats; south to Guatemala. **DISPERSAL:** Some movement to lower elevation in northern part of range. **RARE STATUS:** Very rare ("Interior" group) to western Great Plains and "Coastal" group to Salton Sea region, CA.

Population Stable in most of the US.

black ear coverts

"Black-eared Bushtit" juvenile ♂ *plumbeus*

gray crown

"Interior" *plumbeus*

pinkish buff face

female with pale eye, dark in male

"Interior" ♀ *plumbeus*

long tail

grayish face

"Coastal" ♂ *minimus*

brown crown

NUTHATCHES Family Sittidae

Pygmy Nuthatch (CO, Feb.)

Nuthatches are familiar acrobatic birds, with one or more species occurring throughout much of the US and Canada. All but the Brown-headed Nuthatch are commonly seen at bird feeding stations. When not nesting, nuthatches frequently gather in mixed-species flocks that include chickadees, creepers, kinglets, and warblers. Experienced observers listen for nuthatch calls to detect other species. Identification is straightforward, except when a Brown-headed or Pygmy Nuthatch wanders far from its normal range.

Structure Nuthatches are small and chunky, with a short tail. They are well adapted to creeping about at all angles on tree trunks and branches with legs that are relatively short and with strong toes and claws. Their bills are rather long, straight, and pointed; their wings are short.

Behavior Nuthatches are usually seen on the surfaces of tree trunks and branches, climbing in any direction in pursuit of invertebrates and seeds. Unlike woodpeckers and creepers, they do not use their tail for support and may go down trees headfirst. The smaller species often forage among the outermost needles on branches. They also drop to the ground to feed. To open a seed, a nuthatch wedges it into a crevice and pounds on the seed with its bill. This behavior, known as hacking, likely gave rise to the family name. During fall and winter, nuthatches store nuts, seeds, and invertebrates in caches. The Brown-headed Nuthatch is one of few bird species known to use a tool: Individuals have been seen holding bark in their bill and using it to flake off another piece of bark. All nuthatches nest in cavities, mostly tree holes and cavities. Some Brown-headed and Pygmy Nuthatches are cooperative breeders. Most are vocal year-round, with their loud whistles, trills, and calls often belying their presence even before they are revealed by their active foraging style.

Plumage Nuthatches are blue-gray above, typically with black or brown markings on the head. The underparts are pale, sometimes washed with buff or rufous coloration. The sexes are similar or identical in all nuthatches; with good views, White-breasted and Red-breasted Nuthatches can be sexed in the field. Plumages are similar throughout the year with a pre-breeding (early spring) molt that is limited or absent. All species have a partial post-juvenal (fall) molt. Under ideal field conditions, aging is possible by looking for worn, brownish colored wing coverts that contrast with the blue-gray mantle on first-year birds. By late spring the worn appearance of adults makes aging difficult, even in the hand.

Distribution Nuthatches are widely distributed throughout the Northern Hemisphere. They reach their greatest diversity in south Asia, where 15 species occur. Most nuthatches are nonmigratory, but in North America even the generally sedentary species are occasionally seen far from their normal range.

Taxonomy This is a relatively small family with roughly 28 species and three genera worldwide (four species in North America); all but one species belong to the genus *Sitta*. The exceptions are the Wallcreeper, in its own subfamily Tichodrominae, and the two African and Indian species in the genus *Salpornis* that had previously been placed in the creeper family Certhiidae, or in their own family. Species limits among the nuthatches continue to be debated. A recent study suggests that the critically endangered subspecies of the Brown-headed Nuthatch found on Grand Bahama I. may be a separate species: the Bahama Nuthatch (*Sitta insularis*). Work continues to determine whether the White-breasted Nuthatch may involve as many as three species.

Conservation Habitat loss and degradation largely caused by shifts in cultivation, grazing, and timber harvesting pose the biggest threats to this family. BirdLife International codes: 2 NT, 2 VU, 3 EN, and 1 CR.

RED-BREASTED NUTHATCH *Sitta canadensis* RBNU ■ 1

The distinct nasal calls usually announce the presence of the Red-breasted Nuthatch. During the breeding season, the Red-breasted is usually found in forests dominated by firs and spruces; during migration and winter, it is found in a variety of habitats. When not breeding, the Red-breasted can be seen in small flocks with other nuthatches, chickadees, kinglets, and Brown Creepers. Monotypic. L 4.5" (11 cm)

Identification Small, with a prominent supercilium that contrasts with darker crown and eye line. **ADULT MALE:** Bold face pattern: inky black crown and nape; prominent white supercilium extending from sides of forehead to sides of nape and separating crown from very broad, black eye line. Upperparts otherwise a deep blue-gray. Wings are edged blue-gray. Whitish chin and throat blend into buff breast and rich buff belly, flanks, and undertail coverts. **ADULT FEMALE:** Similar to male, but black on head is paler, more lead-colored, often contrasting with blacker nape. Wings are edged dull gray with a brownish or olive tinge. Upperparts are duller gray; underparts are less richly colored. **IMMATURE:** Usually duller than adults of the same sex. Wing coverts, primaries, and secondaries are uniformly brownish gray without edging visible in adults. Many individuals not safely aged under normal field conditions, particularly in late spring when even bright adults have become quite worn. **FLIGHT:** Undulating and resembling a short-tailed woodpecker; white diagonal subterminal band on tail sometimes visible in flight.

Similar Species This is the only N.A. nuthatch with a broad white supercilium and contrasting broad dark eye line. Chinese Nuthatch (*S. villosa*) is found as close as southern Ussuriland, Russia, but is unlikely to occur as a stray to N.A. It would be separated by its ill-defined eye stripe, duller underparts, and lack of a diagonal white subterminal band on tail.

Voice CALL: A nasal *yank* that is variable but typically repeated. Classically described as similar to a toy tin horn. Short versions are sometimes given in flight. Calls are higher and more nasal than White-breasted Nuthatch. **SONG:** A rapid, repeated series of *ehn ehn ehn* notes; reminiscent of calls.

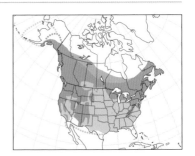

Status & Distribution Common to abundant. **BREEDING:** Northern and subalpine conifers, particularly spruces and firs. Occasionally breeds south of mapped breeding range, usually in conifer plantations or residential neighborhoods with conifers. **MIGRATION:** Irruptive; often moving in two- to three-year cycles but variable. Northernmost migrate annually; southernmost are generally resident. First detected away from breeding grounds as early as July (typically earliest in big flight years); peaks in Great Lakes Sept.–mid-Oct. Spring migration less pronounced, but migrants are seen through May in much of the lower 48 states. **WINTER:** Highest densities typically occur along the US-Canada border (e.g., Northeast; MI-WI border; south-central BC and northwestern WA). **RARE STATUS:** Casual in mainland north Mexico, north Baja California, and west AK; Bermuda (four records); Iceland (one record); England (one record).

Population Bird Breeding Survey shows significant increase throughout the breeding range. In the East, resident range is expanding southward.

grayer cap and eye line on female

duller below than male

deep cinnamon underparts

prominent white supercilium

♀

♂

WHITE-BREASTED NUTHATCH *Sitta carolinensis* WBNU ■ 1

Less gregarious than other nuthatches. Commonly seen at bird feeders. Polytypic (11 sspp.; 6 in N.A.). L 5.8" (15 cm)

Identification The species has a black crown and nape that contrast with a white face and breast. **ADULT MALE:** It has a uniformly black crown and nape. Upperparts are blue-gray. **FEMALE:** Paler crown often contrasts with a blacker nape; in the Southeast, head pattern is more similar to male.

Geographic Variation There are nine subspecies of the White-breasted Nuthatch in Mexico and N.A., five of which occur north of the US-Mexico border. They are placed into three groups on the basis of plumage, genetics, and voice. These groups likely represent separate species. The "Eastern" *carolinensis* group is monotypic. It is the palest and the centers of the tertials and wing coverts are sharply defined, black, and contrast distinctly. Crown stripe is broader, bill is thicker and shorter, and underparts are whiter than in either western group. Their nasal calls—*yank, yank*—are the classic, slow, low-pitched calls associated with most published descriptions of this species. The "Interior West" *lagunae* group consists of Rockies *nelsoni* and more westerly *tenuissima*, and five additional subspecies in Mexico, including *lagunae* from the Cape District of southern Baja California. It is the darkest group with darker upperparts and more extensive dark coloration to flanks. Their very short, rapid succession of high-pitched nasal calls—*nyeh-nyeh-nyeh-nyeh*—have a laughing quality about them and are very different from the calls of the *carolinensis* and *aculeata* groups. The

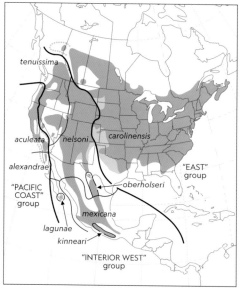

tenuissima

aculeata nelsoni carolinensis

alexandrae

"EAST"
group

"PACIFIC
COAST"
group

oberholseri

mexicana

lagunae

kinneari

"INTERIOR WEST"
group

well-defined
black center
to tertials

bluish
upperparts

black cap
on male

"Eastern" ♂
carolinensis

white face contrasts
sharply with black
cap and nape

"Eastern" ♀
carolinensis

grayer
cap

western subspecies
have grayer flanks
than carolinensis;
darkest in Pacific
aculeata

"Interior" ♂
tenuissima

longer, more
slender bill

in western subspecies,
dark tertial centers more
blended; upperparts less
bluish; white extends
farther above eye

bill like tenuissima
but slightly shorter

"Pacific" ♂
aculeata

"Pacific Coast" *aculeata* group consists of *aculeata* and one other subspecies from northern Baja California. It closely resembles the *lagunae* group but is slightly paler overall, particularly on mantle and flanks, and paler centers of tertials and wing coverts show little contrast with the rest of upperparts. Their calls—*wheer, wheer*—are higher-pitched and longer than the "Eastern" *carolinensis* group.

Similar Species The large size and white face with an isolated dark eye are distinctive.

Voice CALL: Varies between subspecies. **SONG:** "Eastern" group: series of repeated whistles on one pitch. "Pacific Coast" group: similar but higher pitched. "Interior West" group: Distinctly different, more staggered in delivery.

Status & Distribution Fairly common. **BREEDING:** A variety of deciduous and mixed-forest habitats, generally preferring relatively open woods. **DISPERSAL & MIGRATION:** Poorly known. Does not undertake large-scale irruptions (unlike Red-breasted), but some *lagunae* group birds move onto the Great Plains, a few *aculeata* move to coastal lowlands, and a few *carolinensis* group birds disperse west. Found in southeastern CO (where it does not breed) early Aug.–early May. **RARE STATUS:** Casual to deserts of southeastern CA, mostly fall (mostly *aculeata*). Accidental ("Interior West") to Ventura Co., CA, in winter. Also casual (subspecies undetermined) to Vancouver I., BC; and offshore CA (Santa Cruz Is. and the Farallones); Sable I., NS; and one remarkable fall record for Bermuda. **Population** Generally stable, possible declines in the Southeast.

PYGMY NUTHATCH *Sitta pygmaea*   PYNU ■ 1

The Pygmy Nuthatch is easily found, due to its unremitting vocalizations. It is one of only a few cooperative-breeding passerines: A third of breeding pairs are assisted by one to three male helpers, usually relatives. Pairs roost together, and juveniles roost with parents. In winter, congregations of a dozen or more birds may roost together: There are reports of more than 150 individuals roosting in a single tree. This species forages well out on branches but also regularly feeds on the ground. Polytypic (6 sspp.; 3 in N.A.). L 4.3" (11 cm)

Identification Tiny. **ADULT:** Dusky-olive crown; small pale nape spot. Lores, eye line contrast darker. Rest of upperparts blue-gray. Underparts pale, variably washed with buff on flanks. **IMMATURE:** Averages duller with brownish gray wing coverts, secondaries, and primaries. **FLIGHT:** Undulating.

Geographic Variation Three subspecies are found north of the US-Mexico border (three confined to Mexico). Widespread *melanotis* has blackish lores and eye line that contrast noticeably with olive crown. Nominate *pygmaea* (coastal central CA) is smaller, with dusky lores and eye line that contrast only slightly with crown. Flanks extensively washed buff. Subspecies *leuconucha* (montane San Diego Co.) is larger, with uniformly gray-olive

grayish brown cap

darker eye line

creamy buff below

crown, lores, and postocular; relatively large pale spot on nape; flanks only lightly washed buffy gray.

Similar Species Separated from Red-breasted by face pattern and calls; from

Brown-headed by calls and range.

Voice CALL: Rapid, clear, high-pitched notes usually given in a series of three or more notes: *bip-bip-bip* or *kit-kit-kit*. Calls vary in pitch and intensity, but are given almost incessantly. Calls of nominate *pygmaea* are very rapid. **FLIGHT CALL:** Similar but softer: *imp imp*. **SONG:** A rapid, high-pitched sequence of two-note phrases: *ki-dee, ki-dee, ki-dee*.

Status & Distribution Common resident in long-needled pine forests in the West. **RARE STATUS:** Casual to lowlands during nonbreeding season (mostly late July–mid.-Dec.). Rare to southeastern CO. Casual to southwestern BC, central MT, western KS,

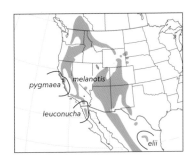

pygmaea melanotis

leuconucha

elii

western OK, and in TX from western Panhandle, Davis and Chisos Mts., and El Paso Co. Accidental in eastern ND and MN (same bird), IA, eastern KS, and eastern TX.

Population Species of special concern in ID, WY, and CO.

BROWN-HEADED NUTHATCH *Sitta pusilla*   BHNU ▪ 1

This energetic, vocal species spends most of its time foraging well out on branches and on treetops, frequently hanging upside down. This species is perhaps better known for using tools than any other bird in N.A.: An individual will take a bark scale, hold it in its bill, and use it as a wedge to pry off other bark scales as it searches for insects and spiders. Apparently, this behavior is more common in years of poor cone crops. Polytypic (2 sspp.; nominate in N.A.). L 4.3" (11 cm)

Identification Tiny; resident in pine forests in the Southeast. **ADULT:** Warm brown to chestnut crown and upper portion of auriculars; relatively large whitish nape spot. Underparts pale, washed with dull buff on flanks. Sexes identical. **IMMATURE:** On average

slightly duller. **FLIGHT:** Undulating.

Geographic Variation The longer billed subspecies *insularis* is found only on Grand Bahama I. and is Critically Endangered with possibly a few, if any, pairs remaining after two devastating hurricanes. Some authorities classify it as a full species ("Bahama Nuthatch").

Similar Species Easily separated from White-breasted and Red-breasted Nuthatches by face pattern and size; best separated from Pygmy by range. Also note Brown-headed's more brownish crown, larger pale nape spot, and less buff undertail coverts. Brown-headed has shorter primary projection, and its subterminal tail band is grayer and more limited, but this is difficult to see on these active birds. Note vocalizations, all of which tend to be lower pitched and less pure than Pygmy.

Voice CALL: A distinctive, repeated two-part note like the squeak of a rubber duck, *kew-deh*. At times, up to 10 or more notes may follow the first note. Feeding flocks also give a variety of twittering, chirping, and talky *bit bit bit* calls. Flight calls similar.

Status & Distribution Fairly common resident in pine forests of the

brown cap

Southeast, particularly the loblolly-shortleaf in the upper Coastal Plain and the longleaf-slash association in the lower Coastal Plain. Favors mature open woodlands with dead wood (e.g., burn areas) but regularly found in young and medium-age pine stands. **DISPERSAL:** Very limited postbreeding dispersal. **RARE STATUS:** Rare KY. Casual to Grimes Co. and Brazoria Co., TX, and to southern NJ. Accidental southeastern WI, northern IL, northern IN, eastern OH, and PA.

Population Declining throughout range; threatened by both clear cutting and fire suppression.

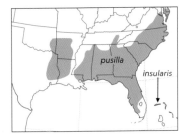

pusilla

insularis

CREEPERS Family Certhiidae

The inconspicuous Brown Creeper is the only certhiid found in the Americas. All certhiids are rather similar in size, plumage, and behavior. In parts of Europe, where two species occur (Eurasian and Short-toed Treecreepers), identification is extremely

difficult; vocalizations best separate these two species.

Structure Short legs, long stiff tail feathers, slender decurved bills, long curved claws and toes, relatively long and dense feathers, and short rounded wings.

Behavior Creepers are almost always found on trees,

using their stiff tail as a prop. When foraging, all species move jerkily up tree trunks and sometimes onto larger branches. In constant motion, they glean and probe for arthropods and insects. Creepers are vocal, but their high-pitched calls and songs may be difficult to hear. Their nests are placed behind loose bark. To maximize heat conservation in cold weather, at least three species (including Brown Creeper) may share a night roost of up to 20 birds, huddled together behind loose bark or in a tree cavity.

Plumage Intricately patterned and highly cryptic, their upperparts are a mixture of white, rufous, brown, black, and buff markings that suggests tree bark. The underparts are pale. Unlike other passerines, creepers retain the central rectrices until all others have been replaced. Post-juvenal molt is incomplete, but all tail feathers are replaced. There is no pre-breeding molt.

Distribution Found in forested regions of the Northern Hemisphere, creepers reach their greatest diversity in the Himalaya.

Taxonomy There are nine species in one genus, *Certhia*. Some DNA comparisons suggest that creepers are part of a much larger family that includes nuthatches,

Brown Creeper (OH, Feb.)

chickadees, wrens, gnatcatchers, and gnatwrens. The two species of spotted creepers (*Salpornis*), with a discontinuous range in Africa and India, have recently been transferred to the nuthatch family.

Conservation Mature and old-growth forests, the species' preferred habitat, face increasing threats from logging and other forms of habitat destruction. BirdLife International codes: 1 NT.

Genus *Certhia*

BROWN CREEPER *Certhia americana*   BRCR ■ 1

This cryptic solitary species is inconspicuous as it forages on tree trunks. It forages by starting close to the ground, spirals up a tree trunk probing bark crevices, and then flies to the bottom of a nearby tree and starts over. Often joins mixed-species flocks in migration and winter. Polytypic (15 sspp.; 10 in N.A.). L 5.3" (13 cm)

Identification Sexes alike. Upperparts streaked brown, buff, and black; underparts pale with buffy flanks; rump tawny. **FLIGHT:** Pale wing stripe at base of flight feathers prominent in flight.

Geographic Variation Subspecies are divided into three groups: "Western" group (*alascensis, occidentalis, stewarti, phillipsi, zelotes, montana, leucosticta*) small, dark, and long billed. "Eastern" group (*americana, nigrescens*) larger, generally paler, and shorter billed. "Mexican" group found to M.A. subspecies and darker with white spotting that contrasts more; range extends into southeast AZ and southwest NM. Three color morphs (reddish, brown, and gray) in several populations.

Voice Very high-pitched. **CALL:** A soft, sibilant *seee*, buzzier and doubled in western birds, *tseeesee*. **SONG:** Variable, but consists of several notes, *seee seeedsee sideeu*. "Eastern" songs more complex, quavering, and usually end on a high note; "Western" songs more rhythmic and often end on a low note; *albescens* ("Mexican" group) also differs.

Status & Distribution **BREEDING:** Coniferous, mixed, or swampy forests. Wider variety of habitats during migration and winter. **MIGRATION:** Peak late Mar.–mid-Apr. and Oct.–Nov. in Midwest/Northeast.

Population Declining due to reduction of old-growth forests.

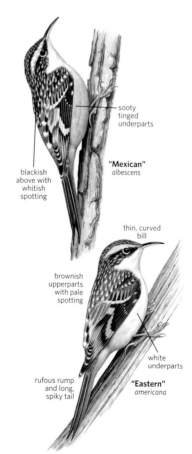

sooty tinged underparts

blackish above with whitish spotting

"Mexican"
albescens

thin, curved bill

brownish upperparts with pale spotting

white underparts

rufous rump and long, spiky tail

"Eastern"
americana

[Map labels:]
alascensis
stewarti
"WESTERN" group
occidentalis
phillipsi
zelotes
leucosticta
montana
montana
albescens
"MEXICAN" group
alticola
"EASTERN" group
americana
nigrescens

WRENS Family Troglodytidae

Rock Wren (CA, Apr.)

N orth America's wrens are diverse (11 species in North America; 82 worldwide) yet distinctive; members of Troglodytidae are rarely confused with species in other families. Our wrens are brown, and most have slender, decurved bills. Tail length and structure vary; in many postures, however, wrens exhibit a characteristic cocked tail. With good views, sight identification is straightforward. Wrens are often found in dense cover, though, and can be fidgety. Song and microhabitat preferences are important for identification.

Structure Wrens often strike a rotund, potbellied pose. When foraging or evading detection, they appear to flatten or prostrate themselves. The tail varies from very short to fairly long, but it is distinctive—typically cocked up and/or flipped about. Wrens have medium to very long, decurved, usually slender bills. Species in North America range in size from small to large.

Behavior Wrens are often observed creeping about substrates close to or on the ground. Specialized microhabitat preferences are cues to identification (e.g., Cactus Wrens within thorny shrubs, Rock Wrens on talus slopes). Songs are complex, variable, and distinctive, with several species (e.g., Winter and Canyon Wrens) among our finest vocalists. Birds of some species sing through the night (e.g., Marsh Wren) and/or during the winter (e.g., Carolina Wren). Several species (e.g., Marsh Wren) build conspicuous dummy nests that are abandoned after construction.

Plumage All wrens of North America are largely brownish, darker above and paler below. Some species tend toward grayish or reddish tones, but these differences are usually of limited relevance to species-level identification. Important marks to note include streaking and spotting on upperparts, barring and shading on underparts and tail, and the supercilium (strong, weak, or nearly absent). Coloration is important in assigning individuals to different subspecies.

Distribution The wrens' center of diversity is Middle America, with more than 30 species in Mexico. All of the lower 48 states host multiple wren species, but several do not range north to Canada, and only the Pacific Wren is regularly found in AK. The Eurasian Wren is the only troglodytid that occurs in the Old World (recent research, however, indicates that the Eurasian Wren comprises multiple species, one of which may include Pacific Wren). Migratory strategies range from sedentary (e.g., Cactus Wren), to intermediate-distance migration (e.g., Sedge Wren). Long-term range shifts appear to be under way for several species.

Taxonomy Eighty-two species in 19 genera. Relationships among and within lower-order taxa are not always clear-cut. For example, the genera *Salpinctes* and *Catherpes* are sometimes lumped, the taxonomic affinities of some *Troglodytes* are unclear, and the genus *Thryothorus* has been split into four genera. At the level of currently recognized species, future splits may be in the offing for House and Marsh Wrens.

Conservation Most wren species in North America appear to be stable or increasing. Below the species level, population-change patterns are complex, as with the Bewick's Wren (eastern populations are in severe decline; several western populations are increasing and expanding). Habitat loss threatens local populations of Marsh and Sedge Wrens. Climate change may be implicated in the Carolina Wren's ongoing range expansion. BirdLife International codes: 6 NT, 3 VU, 3 EN, and 2 CR.

Genus *Salpinctes*

ROCK WREN *Salpinctes obsoletus* ROWR ■ 1

The well-named Rock Wren inhabits rocky and pebbly habitats of all sorts: from scree fields above timberline, to talus slopes and bajadas, to washes and road cuts in open desert. The nest site is indicated by a peculiar trail of pebbles. Polytypic (10 sspp.; 2 in N.A.). L 6" (15 cm)

Identification Usually seen creeping about ground, but sometimes perches conspicuously and bobs up and down in exaggerated, jerky movements. **ADULT:** One molt a year; sexes similar. Brown and buff above. Tail pat-

tern in flight distinctive, with broad buffy terminal band, broken at center. Underparts paler; throat and breast gray-white; belly pale buff. Upperparts spotted, with white; underparts with faint dark streaks. Fresh birds (early fall) more contrastingly patterned than worn birds (midsummer). **JUVENILE:** Similar to adult, but less contrast.

Geographic Variation Nominate subspecies is widespread; *pulverius* is resident on San Nicolas and San Clemente Is., CA.

Similar Species Structure and behavior generally wrenlike, but rarely confused with other species. Canyon Wren's habitat superficially similar, but tends more toward sheer rock faces. **Voice** Varied, but most vocalizations diagnostic. **CALL:** An emphatic *ch'-pweee*, with overall buzzy quality; softer twittering audible from nearby. **SONG:** *Bweer bweer bweer, chiss chiss chiss, swee swee swee,* a repetitious complex of jangling series, suggesting a distant, weak-voiced Northern Mockingbird. **Status & Distribution** Fairly common.

Range extends south to Costa Rica. **BREEDING:** Arid country with sparse vegetation; usually in the immediate vicinity of loose, pebbly substrates. **MIGRATION:** Withdraws from northern portion of range; altitudinal withdrawal occurs throughout range. **WINTER:** Relative winter ranges of latitudinal versus altitudinal migrants poorly known. **RARE STATUS:** Casual in eastern N.A. in fall, winter, and spring. **Population** Stable.

long bill

frequently bobs body

faintly streaked breast

contrasting cinnamon rump

buffy tail tips

Genus *Catherpes*

CANYON WREN *Catherpes mexicanus* CANW ■ 1

The clear, sweet notes of the Canyon Wren are a characteristic sound of rimrocks and other mountainous country in western N.A. The species can be difficult to see, but a good view reveals distinctive behavior, morphology, and plumage. Polytypic (8 sspp.; 4 in N.A., differences weak and obscured by individual variation). L 5.8" (15 cm)

Identification Bill extremely long and decurved. Flattened profile perched and in flight. Rarely seen away from rock outcroppings or canyon walls; usually seen creeping or probing about crevices and rock faces. Behavior

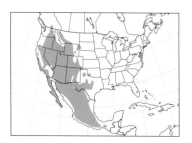

mouselike. **ADULT:** One molt a year; sexes similar. White breast contrasts with reddish brown belly. Tail, wings, and back mainly reddish brown; head grayish. With close view, note spotting and barring on tail, wings, back, and belly. **JUVENILE:** Similar to adult, but with less spotting and barring.

Similar Species Plumage, song, and habitat selection distinctive. Can overlap with Rock Wrens, but these typically found in more pebbly and sandy environments (e.g., talus slopes, quarries).

Voice Vocalizations distinctive. **CALL:** Shrill *beet*, given singly or repeatedly; year-round. **SONG:** Series of approximately 15 clear, whistled notes, descending in pitch and slowing somewhat; usually ends with fewer than five rasping notes, audible at close range. Audibility of song varies with terrain and orientation of observer; songs given fairly infrequently (interval less than 1 min). **Status & Distribution** Uncommon

to fairly common. Range extends to southern Mexico. **BREEDING:** Canyons, rock faces, outcroppings, often near water. Uses sheltered rock crevices, even stone buildings and rock walls, for nesting. **MIGRATION:** Mostly resident but some (perhaps many) that breed at higher elevations move to lower elevations in winter. **RARE STATUS:** Casual well-away from resident range; as far east as eastern NE. **Population** Stable, but little studied.

spotted grayish crown

very long bill

white throat

rufous tail

extensively chestnut belly

Genus *Troglodytes*

HOUSE WREN   *Troglodytes aedon*   HOWR ■ 1

Dull in appearance but notable for its effervescent song, the House Wren is a common summer inhabitant of scrublands and woodland edges throughout much of N.A. Variation in plumage and call notes is extensive and range extends south to S.A. Polytypic (31 sspp.; 2 in N.A.). L 4.7" (12 cm)

Identification A small wren with medium-length bill and tail, it responds readily to pishing. Overall jizz is of a plain, typical wren. **ADULT:** One molt a year; sexes similar. All populations brownish above, paler graybrown or gray below. Supercilium, often indistinct, is paler than crown and auriculars. **JUVENILE:** Variable, but many differ from adults in having warmer buff and rufous tones, along with indistinct scalloping on throat and breast.

Geographic Variation Two widespread, poorly differentiated subspecies in N.A.: Western *parkmanii* and eastern *aedon*. Eastern birds average more rufescent, western birds grayer. Breeding birds in the mountains of southeastern AZ, classified with *parkmanii*, have a slightly buffier throat

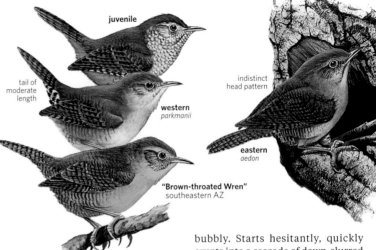

juvenile

tail of moderate length

western
parkmanii

indistinct
head pattern

eastern
aedon

"Brown-throated Wren"
southeastern AZ

"Southern
House Wren"

and breast and a bolder supercilium; *cahooni* and closely related *brunneicollis* of northern Mexico are more richly colored; treated as a separate species until 1973 ("Brown-throated Wren"). Some tropical subspecies likely represent separate species.

Similar Species Main points of confusion are Winter and Pacific Wrens; both are smaller, shorter-tailed, more heavily barred on flanks and *crissum*, and usually darker; Pacific Wren has the richest coloration. House Wren is further distinguished by differences in song, call, timing of migration, and microhabitat preferences.

Voice Most vocalizations have a dry quality. Vocalizations vary geographically within vast range, including (call notes) within N.A. **CALL:** Most given singly or in stuttering series. Another note, which is raspy, resembles that of Blue-gray Gnatcatcher. **SONG:** Loud,

bubbly. Starts hesitantly, quickly erupts into a cascade of down-slurred dry trills.

Status & Distribution Common. **BREEDING:** Most vegetated habitats, except for dense forests, open grassland, marshland, and desert. Favored microhabitats include clearings, edges, residential neighborhoods. **MIGRATION:** Most birds in N.A. migratory. House Wrens migrate later in the spring and earlier in the fall than Winter and Pacific Wrens. **WINTER:** Generally as breeding, but with greater tendency for dense cover. Not a hardy species; stragglers north of core wintering range uncommon. **RARE STATUS:** Sometimes noted to offshore or peninsular locales from which otherwise absent.

Population Subspecies *martinicensis* (Martinque I.) is extinct; *guadeloupensis* (Guadeloupe I.) may be extinct. Long-term population increases documented in many areas. Species is tolerant of humans and readily accepts nest boxes.

PACIFIC WREN   *Troglodytes pacificus*   PAWR ■ 1

Formerly classified as a single worldwide species, the Winter Wren; in 2010 the AOS separated western and eastern N.A. populations as distinct species, retaining Winter Wren for eastern birds and establishing Pacific Wren for western ones. Substantial genetic distance in mitochondrial DNA, distinct vocal differences, and an absence of interbreeding between eastern and western populations at their contact zone in AB, Canada, led to the split.

Polytypic (8 sspp.; all in N.A.). L 3.5–4.5" (9–11 cm)

Identification Sexes are similar in all plumages. Short tail and overall shape are very similar to Winter Wren. **ADULT:** Face dark brown; disproportionately long bill. Supercilium, throat, and breast are bright buff or grayish depending on subspecies. Rich buff and dark bars on wings, flanks, belly, and tail. **JUVENILE:** Similar to adult but with less distinct barring on belly;

darker and more richly
colored above than
Winter Wren

much
richer buff
below

pacificus

sometimes has faint breast-scalloping. **Geographic Variation** Differences among subspecies are complex and subtle, primarily based on size, plumage coloration, and pattern of dark markings. "Alaska Island" group is largest, with longest bill, and uniformly pale brownish to grayish. Within this group, *meligerus* of the western Aleutians is a dark subspecies, *kiskensis* of central Aleutians is paler, and *semidiensis* of the Semidi Is. and Chowiet I. is the palest subspecies. Populations in Southeast AK, Pacific Northwest and western Can-

ada are highly colored (*pacificus* is the most deeply rufescent N.A. subspecies); interior populations in western ID and WY are grayer. Some taxonomists would recognize up to four more subspecies, while others would recognize fewer. Subspecies pallescens, currently treated as Eurasian Wren (*T. troglodytes*) from Commander Is., Russia Far East, is genetically similar to Aleutian birds, so the range of Pacific Wren may actually extend to this island group and perhaps farther south in east Asia as far as Taiwan. **Similar Species** Winter Wren is paler, duller, with less richly rufous and buff coloration on underparts and supercilium. Variation in both species not fully known, so calls are always the best separating feature. House Wren is larger, longer tailed, paler, especially on throat and breast; barring does not extend forward of legs; supercilium is less distinct. **Voice** Vocalizations diagnostic between Pacific and Winter Wren. **SONG:** A long, loud, complex, and beautiful series of trills, similar to Winter Wren but faster, slightly

higher pitched, somewhat harsher, and less melodic. **CALL:** A *chimp*, often doubled, reminiscent of Wilson's Warbler. Winter Wren's call is more like a *kelp* than a *chimp*.
Status & Distribution Fairly common in its habitat. Nests primarily in old-growth coniferous and mixed forests with dense understory and often near water. **BREEDING:** Primarily Pacific coast from AK to central CA, and in scattered interior regions. **WINTER:** Within breeding range and south to Southern CA, AZ, rarely NM and CO. **RARE STATUS:** Accidental to northern AK and east to western KS.
Population Stable in much of range, but declining severely in the northwestern rainforest, particularly in BC.

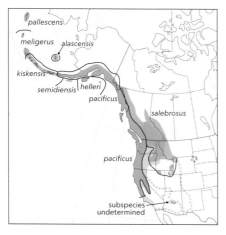

Map labels: pallescens, meligerus, alascensis, kiskensis, semidiensis, helleri, pacificus, salebrosus, pacificus, subspecies undetermined

larger, with much longer bill

central and eastern Aleutians
kiskensis

WINTER WREN *Troglodytes hiemalis* WIWR ■ 1

Tiny, nervous, and unforthcoming, the Winter Wren breeds mainly in cool, shady forests and tangles at northern latitudes and higher elevations. Its long, loud, complex, and striking song is frequently rated among the best among all N.A. birds. Polytypic (2 sspp.; both in N.A.). L 3.5" (9 cm)
Identification Diminutive in all respects. Small-bodied, stub-tailed; bill short for a wren. Hides among dense shrubbery, tangled roots, etc. Surprisingly hard to glimpse, even when close. Note constant twitching and jerking motions, with tiny tail held

overall coloration duller and paler than Pacific Wren; vocalizations diagnostic

hiemalis

straight up. Sexes similar. **ADULT:** All populations brown to dark brown, pale brownish to grayish underparts and prominent barring on belly, flanks, and crissum; whitish throat. **JUVENILE:** Similar to adult, but with less distinct barring on belly.
Geographic Variation Widespread nominate *hiemalis* has dark upperparts with rufous tinge, breast washed pale brownish; Appalachian *pullus* has less rufescent upperparts and moderate brownish wash on breast, decidedly darker and less rufescent, underparts lighter brown with vermiculations on abdomen and flanks heavier; bill smaller and more slender.
Similar Species Pacific Wren is brighter and darker overall, more richly rufescent, and with dark barring; throat and breast rich buff. House Wren is larger, longer tailed, paler, especially below; barring does not extend forward of legs; supercilium less distinct.
Voice Vocalizations are diagnostic between Pacific and Winter Wren.

SONG: A long, loud, complex, and beautiful series of trills, similar to Pacific Wren but slower, slightly lower pitched, and more melodic. **CALL:** A *kelp*, unlike Pacific's *chimp*.
Status & Distribution BREEDING: Fairly common in coniferous forest habitat across Canada, northern US, south in Appalachians to GA. **WINTER:** Winters south to Gulf Coast states and GA. **RARE STATUS:** Rare to northwestern FL; casual farther south in FL and to Southwest and CA.
Population Stable.

Genus *Cistothorus*

MARSH WREN *Cistothorus palustris*   MAWR ■ 1

The highly vocal Marsh Wren may be heard in cattail marshes throughout much of N.A. all through the summer. Like most wrens, it is reclusive, but its densities are often high in favored breeding areas. The ovoid dummy nests of the species are conspicuous. Vocal and plumage variation is extensive and sometimes visually discernible in the field. Given the vocal differences between groups (see below), they likely represent two separate species. Polytypic (15 sspp.; 14 in N.A.). L 5" (13 cm)

Identification Usually stays close to cover in dense cattails. **ADULT:** Most subspecies have a conspicuous supercilium sharply set off against dark brown crown and gray-brown auriculars; black back streaked with prominent white stripes; wings, tail, rump, and belly variably chestnut. **JUVENILE:** More smudgy-looking, less postocular marked than adult, with few if any white streaks on back.

Geographic Variation Extensive. Two distinct subspecies groups, "Eastern" and "Western" divide on the central Great Plains. These groups are separated mainly by song (see below). The "Eastern" *palustris* group is comprised of six subspecies, most of which are darker and more richly colored, typi-

fied by widespread *dissaeptus* (illustrated). The subspecies *griseus* of the south Atlantic coast is decidedly gray, the dullest Marsh Wren subspecies. The "Western" *paludicola* group, composed of eight subspecies in N.A. (one more in Mexico), is basically similar, but colors more muted and contrast somewhat weaker, especially on crown.

Similar Species See Sedge Wren.

Voice Geographic variation in songs extensive. **CALL:** Common call note a soft *chuck*, without the sharp, grating quality of the call notes of many other wren species. **SONG:** All songs have a liquid or gurgling sound, more so than the dry, stuttering song of Sedge Wren. Liquid quality is more pronounced in eastern birds than in western birds, which also have a much larger repertoire.

Status & Distribution Locally common. **BREEDING:** Dense and large cattail marshes. Western birds are more catholic for their selection of breeding habitat, nesting in small and large marshes. **MIGRATION:** Complex. Some populations highly migratory, others partially migratory, others nonmigratory. Migration in general poorly detected, owing to reclusive behavior of migrants. Migratory populations usually back on breeding grounds by mid-May. Withdrawal to wintering grounds

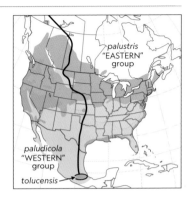

begins by midsummer; most migrants are detected by early fall in the West, but not until Oct. in the East. Even on migration, most individuals are found in cattails and similar vegetation. **WINTER:** Cattail marshes generally favored, but a broader array of habitats is accepted. Dense cover almost always required. **RARE STATUS:** Casual to NL. Accidental in fall to south-central AK (subspecies group unknown).

Population General pattern is of increasing numbers in western populations vs. decreasing numbers in the East, possibly reflecting western birds' tendency to be more generalized in habitat selection. Destruction or alteration of wetlands, especially in the East, has caused local declines and extinctions.

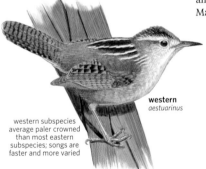

resident in coastal marshes from SC to east-central FL

western
aestuarinus

western subspecies average paler crowned than most eastern subspecies; songs are faster and more varied

a very gray subspecies

south Atlantic coast
griseus

more solid crown, more prominent white stripes on back, more prominent whitish supercilium, and longer bill than Sedge Wren

most eastern subspecies are, on average, a little darker and more richly colored than western subspecies

eastern
dissaeptus

SEDGE WREN *Cistothorus platensis*   SEWR ■ 1

Few N.A. birds show as much distributional complexity as the Sedge Wren. Nesting begins in the northern portion of the range in late spring and shifts southward by midsummer. Although a weak flier, it is the only wren of N.A. that withdraws completely from its breeding range in winter. Sedge Wrens are seldom detected on migration, but strays are recorded far from main

routes. Highly disjunct and largely nonmigratory populations occur south to islands off southern S.A. This species was formerly known as the Short-billed Marsh Wren but actually avoids marshes where Marsh Wrens are found. Polytypic (17 sspp.; only *stellaris* in N.A.). L 4.5" (11 cm)

Identification Small. Bill short for a wren. In winter and migration can be

difficult to glimpse, with most individuals staying under cover. Where locally common on breeding grounds easy to observe due to constant singing and frequent interactions. **ADULT:** Two molts a year (early spring, late summer); sexes similar. Buffy overall, subtly but extensively streaked and spotted. Note fine white streaks across brown crown, fairly coarse black-and-white steaks on buffy orange back. Supercilium faint but discernible; buff-tinged. Unmarked underparts warm buff, brighter on breast and throat. Tail and wings broad, rounded; extensively but weakly barred with brown and buff. **JUVENILE:** More muted overall than adult; streaking on upperparts more subdued, and buffy tones on underparts weaker.

Geographic Variation One subspecies, *stellaris*, in N.A.; variation discrete and extensive (17 additional subspecies) farther south to southern S.A., likely involving multiple separate species.

Similar Species Marsh Wren differs in many respects, including darker colors and more contrast overall, solid brown crown, conspicuous white supercilium, and black back streaked with white. The two species rarely overlap on breeding grounds, with Marsh in dense palustrine or estuarine cattail marshes, Sedge Wren in tall grasses in damp uplands.

Voice Most vocalizations sharp, chattering. **CALL:** Sharp *chap*, like introductory notes in song. Close-up, soft sputtering series, as given by other wrens. **SONG:** Several sharp notes followed by a short series of faster notes: *chip chip chip ch'ch'ch'ch'ch'ch'ch*, very different from the song of Marsh Wren. Frequently sings on its winter grounds in spring before departure.

Status & Distribution Complex. Locally common, but also scarce in or absent from many areas of seemingly suitable habitat. **BREEDING:** Loosely colonial in tall grasses; wet soil preferred, but extensive standing water a deterrent. Breeding in northern portion of range (upper Midwest, western Great Lakes) occurs earlier than breeding farther south (central Great

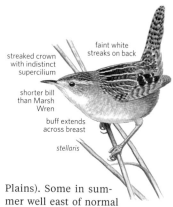

faint white streaks on back

streaked crown with indistinct supercilium

shorter bill than Marsh Wren

buff extends across breast

stellaris

Plains). Some in summer well east of normal range. **MIGRATION:** Begins northward movement by early May. Many individuals retreat from northern breeding grounds in early summer to staging or breeding grounds farther south. Fall migration rather late; bulk of records in Oct. **WINTER:** As breeding, but more general, with some individuals in wet marshes, others in drier scrub. **RARE STATUS:** Annually to hundreds of miles from area of regular occurrence; casual to West Coast. Most strays noted in fall.

Population Overall stable but hard to monitor given erratic and opportunistic movements.

Genus *Thryothorus*

CAROLINA WREN *Thryothorus ludovicianus* CARW ▪ 1

The adaptable and highly vocal Carolina Wren is a familiar inhabitant of gardens and woodlands in the Southeast. Climate-related range shifts in the species are well documented. Polytypic (10 sspp.; 6 in N.A.). L 5.5" (14 cm)

Identification Active, inquisitive. In most of range, the most brightly colored wren in its habitat. **ADULT:** One

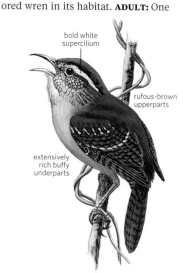

bold white supercilium

rufous-brown upperparts

extensively rich buffy underparts

molt a year; sexes similar. Upperparts bright reddish brown; breast and belly warm buffy orange; throat whitish. White supercilium; long, conspicuous. **JUVENILE:** Overall tones duller.

Geographic Variation Six subspecies north of Mexico; four others in Mexico and C.A. Populations fairly homogeneous north of about 32° N. More heterogeneous farther south, but field identification to subspecies is difficult.

Similar Species Where ranges overlap, Bewick's Wren may present confusion. Bewick's has colder colors, thinner bill, and longer, expressive white-cornered tail. Songs different.

Voice Loud and frequent. **CALL:** Varied. A hollow, liquid *dihlip*. Another call sounds like a stick being run across a wire-mesh fence. **SONG:** Rich and repetitious. Most songs consist of short, repeated phrases; song may start or end with single notes—*chip mediator mediator mediator meep*. Primitive antiphonal singing is sometimes heard: One bird begins with characteristic song; mate finishes with low rattle.

Status & Distribution Common.

BREEDING: Dense vegetation, frequently near human habitation. **MIGRATION:** Largely sedentary. **WINTER:** A few found north of resident breeding range. **RARE STATUS:** Very rare to casual to MB, MN, SD, CO, NM, and AZ. These individuals may be better thought of as vanguards in range expansion.

Population Range expanding northward; westward expansion is erratic. Range contractions follow harsh winters, but expansion then resumes; no doubt is a projected beneficiary of climate change.

Genus *Thryomanes*

BEWICK'S WREN *Thryomanes bewickii*   BEWR ■ 1

Vocal and plumage variation in the Bewick's Wren is extensive, but all adults have a very long tail, tipped in

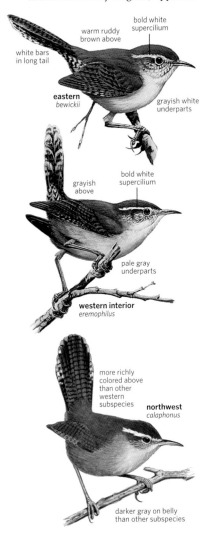

bold white supercilium
warm ruddy brown above
white bars in long tail
grayish white underparts
eastern *bewickii*

bold white supercilium
grayish above
pale gray underparts
western interior *eremophilus*

more richly colored above than other western subspecies
northwest *calaphonus*
darker gray on belly than other subspecies

white and flicked side to side. Eastern populations have disappeared from most of former range and are in continuing decline. Polytypic (14 sspp.; 10 in N.A.). L 5.1" (13 cm)

Identification ADULT: Sexes similar. Plumage variable, but all subspecies gray-brown to rufous-brown above, gray-white below, with long, pale supercilium. White-tipped tail has black bands.

Geographic Variation Complex, extensive. "Pacific Coastal" *spilurus* group comprises nine subspecies, six in N.A.; as a group brownish above, pale gray below; *calaphonus* from the Northwest is the most richly colored above and darkest below. Subspecies *leucophrys* from San Clemente I. and *brevicaudus* from Guadalupe I. are extinct. "Interior West" *eremophilus* group comprises five subspecies, four in N.A.; widespread *eremophilus* is dull and grayish above, subspecies farther east are more warmly colored above. "Eastern" *bewickii* group is rufescent brown above and whitish below.

Similar Species In most of range distinctive tail and supercilium prevent confusion with other wrens. See Carolina Wren.

Voice Varied. In many places in the West, Bewick's produces the local mystery song. **CALL:** Variable, but many notes with raspy or buzzy quality, some quite loud. **SONG:** Most songs combine one to five short, breathy, buzzy, notes (sometimes run together in a short warble) with a longer, often loose, trill. Single buzzy or nasal notes may be introduced. Overall rhythm and tone remind many observers of Song Sparrow, but elements in Bewick's song are usually thinner,

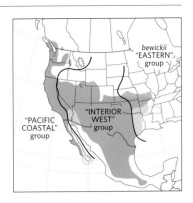

bewickii "EASTERN" group
"PACIFIC COASTAL" group
"INTERIOR WEST" group

buzzier. In the "Interior West" songs are simpler than those from the far West or *bewickii*.

Status & Distribution Western populations fairly common. Eastern *bewickii* has sharply declined in recent decades, reduced to remnant breeding populations centered in the Ozarks, a few up the Tennessee Valley. **BREEDING:** Shrubby vegetation always a requirement. Western individuals frequent arid juniper foothills, lush riparian corridors, residential districts, etc. Eastern *bewickii* inhabits open woodlands with thickets. **MIGRATION:** Western populations largely sedentary. Eastern populations migratory, with general pattern of dispersal south and west following breeding. **WINTER:** As breeding, with generally warmer and lower-elevation component. **RARE STATUS:** Except for possibly a few remnant sites, now casual east of the Mississippi River.

Population Subspecies *bewickii* once ranged northeast to NY and ON. Competition with House Wren thought a major factor in Bewick's decline.

Genus *Campylorhynchus*

CACTUS WREN *Campylorhynchus brunneicapillus*   CACW ■ 1

Calling to mind a small thrasher, the oversize Cactus Wren is a distinctive inhabitant of the thorn-scrub country of the desert Southwest. A good clue to this wren's presence is its characteristic dummy nest, placed conspicuously but inaccessibly amid chollas, acacias, etc. Polytypic (7 sspp.; 2 in N.A.). L 8.7" (22 cm)

Identification Large wren with bulky

build, broad tail, relatively thick bill. **ADULT:** One molt a year; sexes similar. Plumage spotted and streaked all over; overall tones brown, black, and white. Note variable dark spotting on undersides and conspicuous white supercilium. In flight rounded wings and tail are brown overall but heavily spotted and barred with white. **JUVENILE:** Similar to adult, but buffier overall,

with spotting and barring more muted. **Geographic Variation** Subspecies *anthonyi* (Southwest) and *couesi* and *guttatus* (TX) are densely spotted on breast. "San Diego" Cactus Wren (*sandiegensis*), resident in coastal Southern CA in southern Orange Co. and San Diego Co., has a different pattern on underparts—black spotting is finer and more uniform, and background color is more extensively white. **Similar Species** Other wrens smaller and usually in other habitats, so confusion unlikely. Similar in some respects to Sage and Bendire's Thrashers. **Voice** Most vocalizations low-pitched and grating. Species highly vocal, often singing through the hottest times of the day. **CALL:** Low, growling, clucking sounds, sometimes given in a loose series, *chut chut chut.* **SONG:** A distinctive, unmusical series of pulsing, chugging notes, reminiscent of an old car trying to start—*churr churr churr.* **Status & Distribution** Uncommon to locally common in southwestern US and northern Mexico. **YEAR-ROUND:**

Towns, washes, and open desert with thorn scrub. **RARE STATUS:** Very few records of wanderers away from known resident locations. **Population** "San Diego" subspecies uncommon, local, and declining; threatened by habitat loss. Other subspecies have declined elsewhere, especially southeast CA, where *anthonyi* now extirpated from parts of former range (e.g., Imperial Valley and parts of the Mojave Desert).

sandiegensis

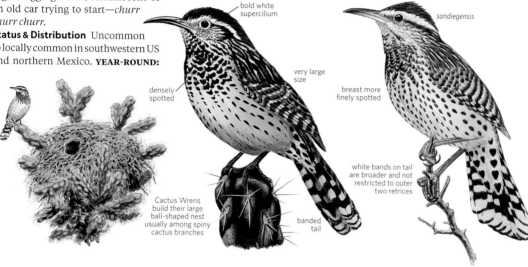

bold white supercilium

very large size

densely spotted

sandiegensis

breast more finely spotted

white bands on tail are broader and not restricted to outer two retrices

Cactus Wrens build their large ball-shaped nest usually among spiny cactus branches

banded tail

Genus *Thryophilus*

SINALOA WREN   *Thryophilus sinaloa*   SIWR ▪ 5

This nonmigratory Mexican species, resident as far north as northern Sonora and now casual to southeastern AZ. Polytypic (3 sspp.; likely the duller, northern *cinereus* in N.A.). L 5.3" (13 cm)
Identification A skulker. Shaped like a Carolina Wren, with a long, thin bill and a short, often upraised tail. Sexes similar. **ADULT:** White face with dark eye stripe and white supercilium; sides of neck streaked black and white. Upperparts brown, wings and tail brown with black bars. Underparts

with pale gray breast, pale buff lower breast and belly; undertail coverts barred black and white (hence alternate English name Bar-vented Wren). **JUVENILE:** Duller overall, with pale rusty underparts.
Similar Species Western interior Bewick's is gray above and pale grayish below. Carolina is much brighter, with extensively rich buffy underparts and deep rusty brown upperparts. Happy Wren (*T. felix*), another west Mexican endemic and potential stray into US, has black eye line, more boldly striped face, and entirely buff underparts including breast.
Voice SONG: Loud, rich phrases, often with rapid repetition of notes, such as *weet-weet—weet, chuee-chuee-chuee.* **CALL:** Wide variety of notes, rough, buzzy, sometimes chattering.
Status & Distribution West Mexican endemic, resident in tropical deciduous forest, gallery forest, mangrove forest, and thickets in thornscrub from northern Sonora south to west-

ern Oaxaca. **RARE STATUS:** Casual AZ: first recorded from near Patagonia in 2008 and since then records from year-round along Santa Cruz River at Tubac, north of Nogales, and north side of Huachuca Mts.
Population Stable; has expanded northward in Sinaloa, breeding at least to Santo Domingo, 70 mi south of AZ.

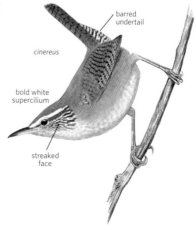

barred undertail

cinereus

bold white supercilium

streaked face

cinereus

sinaloa

russeus

GNATCATCHERS AND GNATWRENS Family Polioptilidae

Blue-gray Gnatcatcher, breeding male (ON, May)

Small and active, the gnatcatchers present an identification challenge in the southwestern border regions where two or more species may occur together, but only the familiar Blue-gray Gnatcatcher is found over most of North America. This small New World family also includes three Neotropical species known as gnatwrens.

Structure This is a generally uniform group of small (4–5 in.) insectivorous birds. The bills are very thin, moderately short in gnatcatchers but long in the gnatwrens. Gnatcatchers are all closely similar in structure, being slender with long, expressive tails; the two *Microbates* gnatwrens have shorter tails. All have wings that are relatively short and rounded.

Behavior Active insectivorous birds that flit through twigs and foliage, often cocking and waving their long tail. They forage by gleaning, but also by hover-gleaning or making short aerial sallies. The resident species are typically found in pairs year-round. They build small, soft deep cup-shaped nests. Both sexes incubate for about 14 days.

Plumage Generally gray in color, often with black markings on the crown or elsewhere on the head, and in most species white in the outer tail feathers. The sexes and age groups are generally similar, but breeding season males in temperate zone species acquire black head markings.

Distribution Exclusively New World; gnatcatchers are most diverse in middle and tropical South America, occurring mainly in lowland forest, thornscrub, and arid scrub; one species (Blue-gray Gnatcatcher) breeds north through much of the US and even southernmost Canada. The gnatwrens are found in tropical lowland forest understory and borders from southeastern Mexico to Brazil and Bolivia. Most species are sedentary, but the two North American Blue-gray Gnatcatcher subspecies are migratory.

Taxonomy Long linked taxonomically to the ill-defined "Old World Warbler" family Sylviidae, the 13 gnatcatchers (genus *Polioptila*) and three gnatwrens (*Microbates* and *Ramphocaenus*) are, in fact, more closely related to the predominantly New World wrens (Troglodytidae) and creepers (Certhiidae). There may well be more cryptic species within *Polioptila*, one of which involves a potential split within the two North American subspecies of Blue-gray Gnatcatcher. The Yucatán Gnatcatcher (*P. albiventris*) from the peninsula's north coast was split in 2019 from White-lored. Four species occur in North America north of Mexico; the Black-capped is rare and local along the AZ-Sonora border.

Conservation Some range-restricted species are considered rare and in some cases threatened. The California Gnatcatcher is listed as Threatened within its US range by USFWS. BirdLife International codes: 1 NT.

Genus Polioptila

BLUE-GRAY GNATCATCHER *Polioptila caerulea*   BGGN ▪ 1

This species is quite active. Polytypic (8 sspp.; 2 in N.A.). L 4.3" (11 cm)
Identification Long tailed, with outer tail feathers almost entirely white (tail from below looks white). Bill is thin and pale gray. **BREEDING MALE:** Blue-gray above, including most of head and back. Crown has a black line at forecrown that extends along sides of crown; white eye ring contrasts with gray face. Wings brownish gray; tertials blackish, edged white. Underparts entirely white. **NONBREEDING MALE:** Black on crown absent, resulting in grayish crown. **FEMALE:** Like

nonbreeding male but grayer above.
Geographic Variation Nominate *caerulea* more extensively white tail; western *obscura* has a black base to outer rectrices that extend beyond undertail coverts. Western male slightly less blue on back, with a thicker black forehead mark and a less contrasting supraloral line than nominate *caerulea*. Western females are dingier above. Vocalizations also differ, suggesting perhaps that the two represent separate species (more study needed).
Similar Species Easily confused with all other N.A. gnatcatchers, all of which

have different calls; California is also darker below. The best feature is tail pattern. Blue-gray is almost entirely white on outer rectrices; Black-tailed and California have mostly black outer rectrices with white tips or edges. Tail shape also differs. In Black-tailed in CA (and Black-capped) the outer tail feathers are short and progressively increase in length, giving a graduated tail shape with an intricate pattern of white tips. In Blue-gray, outer tail feathers are much longer, so on closed tail, only outers are visible. Be aware that in late summer gnatcatchers molt

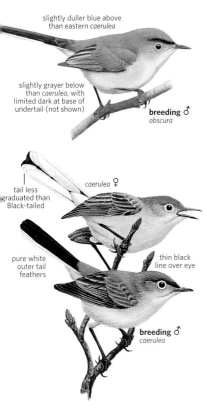

slightly duller blue above than eastern *caerulea*

slightly grayer below than *caerulea*, with limited dark at base of undertail (not shown)

breeding ♂
obscura

tail less graduated than Black-tailed

caerulea ♀

pure white outer tail feathers

thin black line over eye

breeding ♂
caerulea

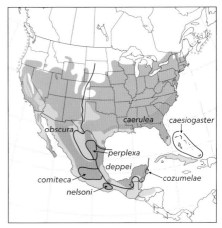

caerulea *caesiogaster*

obscura

perplexa

deppei

comiteca *cozumelae*

nelsoni

their tails. Blue-grays will look mostly dark from below when their outer rectrices are dropped. Female and winter male Black-capped Gnatcatcher best separated by tail shape and calls; winter male Black-capped has a dark line over the eye like winter male Black-tailed, but not like any winter Blue-gray.

Voice CALL: A querulous *pwee* or various mewing calls. Western birds have lower, slightly harsher notes; Easterns' common call slightly more wiry, thin. **SONG:** Thin, wiry notes; lower and harsher in western populations.

Status & Distribution Fairly common. **BREEDING:** Various woodlands. **MIGRATION:** Earliest migrants reach Great Lakes in late Mar., typically in early Apr. Peak late Apr. through early May, with stragglers to later in the month. Fall migration starts as early as late June or July in southern states. Farther north migration starts mid-Aug., with peak mid-Sept. Small numbers seen into Oct., rarely into Nov., even Dec. **RARE STATUS:** Annual in fall in small numbers to Atlantic Canada, Aug.–Nov. Casual in spring to Atlantic Canada; to BC and Ottawa in spring and fall. **WINTER:** Southern US, south to Honduras; casual to Nicaragua. Also winters regularly (*caerulea*) to Cuba and is rare in the Cayman Is. The status of *caerulea* in the Bahamas, where the distinctive and endemic *caesiogaster* is a common resident, is unknown.

Population Northward expansion in northeastern US and southeastern Canada occurred in the 20th century.

CALIFORNIA GNATCATCHER *Polioptila californica* CAGN ■ 2

A Southern CA and Baja California endemic, this gnatcatcher is restricted to coastal chaparral in our area. Formerly considered conspecific with the Black-tailed Gnatcatcher, it was split based on morphological and vocal differences. Polytypic (4 sspp.; darkest nominate in US, other 3 in Baja California; some treat the species as monotypic). L 4.3" (11 cm)

Identification A dark, rather dingy gnatcatcher with a dark tail. Black outer rectrices are edged white, including tip, but white does not extend noticeably inward from tip. From below, tail looks dark. Crown and upperparts dark gray. Wings decidedly brownish. Underparts grayish with buffy wash on flanks, particularly in females. **BREEDING MALE:** Crown is black to include slightly below eye. **NONBREEDING MALE:** Crown is gray, with black reduced to a streak over eye. **FEMALE:** Like nonbreeding male, but plumage browner; flanks more strongly washed with buff.

Similar Species Similar to Black-tailed Gnatcatcher, but with darker, dingier gray below. Tail of California is edged white, as opposed to white tips of outer tail feathers of Black-tailed, and California has a less distinct eye ring. Vocalizations also very distinctive and ranges do not overlap in CA.

Voice CALL: A rising and falling *zeeer*, rather kittenlike. **SONG:** A series of *jzer* or *zew* notes.

Status & Distribution Threatened. Local resident in sage scrub of southwest CA, north to southeast Ventura Co. **RARE STATUS:** Casual short distances away from known locations (e.g., nested in 2014 and 2015 near Castaic Lake, Los Angeles Co.).

more restricted white on outer tail feathers

californica ♀

black cap like Black-tailed

breeding ♂
californica

overall darker above and below than Black-tailed

Population Listed as Threatened (*californica*) by USFWS due to human development of its coastal sage-scrub habitat. The range of *californica* extends into northern Baja California to 30° N but only a few thousand pairs believed to remain in CA. Populations of the three subspecies in Baja California appear stable.

BLACK-TAILED GNATCATCHER Polioptila melanura BTGN ▪ 1

The Black-tailed is an inhabitant of desert thornscrub, partial to washes. Although small, this feisty songbird will aggressively defend its nest against larger predators. Formerly considered conspecific with California Gnatcatcher. Polytypic (3 sspp.; 2 in N.A.). L 4" (10 cm)

Identification Large white terminal spots on graduated tail feathers; short dark bill. **BREEDING MALE:** Glossy black cap extends past, but not under, eye, contrasting with thin, white eye ring. Tertial edges usually whitish. **NONBREEDING MALE:** Like breeding male but dark cap is replaced by thin dark line over eye. **FEMALE:** Like nonbreeding male, but lacks dark line over eye; back washed with brown.

Geographic Variation Three subspecies: One (*curtata*) is from Tiburon I. in the Gulf of California. In the US, *lucida*, which breeds in the Sonoran and Mojave deserts (CA and AZ), has less white in tail, a noticeably paler gray base to mandible, and, in female, a back with a rather brownish wash. Nominate *melanura*, which breeds in the Chihuahuan desert (NM and TX), has more white in tail, a rather uniform dark bill with only slight paling at base, and a lighter brown wash on back of female.

Similar Species Most likely confused with Black-capped Gnatcatcher as birders search for the latter species in southeastern AZ. Most easily distinguished by white tips to outer rectrices of Black-tailed, compared to nearly all-white outers of Black-capped; tail of Black-capped is also more graduated. Also note on breeding males the more extensive black cap coming below the eye and longer bill and a more faint eye ring, usually limited to a crescent below eye. In nonbreeding Black-capped, note pale face including auriculars, whereas in nonbreeding

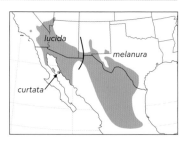

Black-tailed the face is rather more uniform with the crown. See Blue-gray Gnatcatcher.

Voice CALL: Rasping *cheeh* and hissing *ssheh*; in general, its vocalizations are rather wrenlike. **SONG:** A rapid series of *jee* notes.

Status & Distribution Uncommon to fairly common. **YEAR-ROUND:** Resident in variety of arid habitats in the Southwest US, as well as in northern and central Mexico. Rather rare western south TX. A few isolated populations exist in CA, notable in the Panamint Valley, Inyo Co., southwest of Mojave, Kern Co., and in the Antelope Valley in northeast Los Angeles Co. This species moves very little, if at all. A specimen record from San Antonio, TX, is one of few records that might come from outside its known range. Any suspected Black-tailed Gnatcatcher outside its known range should be thoroughly documented. **Population** Stable.

strongly graduated tail with white tips to outer tail feathers

♀

breeding ♂
black cap comes just to eye; in winter, male has gray cap with short black line above eye

BLACK-CAPPED GNATCATCHER Polioptila nigriceps BCGN ▪ 3

This western Mexican species discovered in the US in 1971 is rare in southeastern AZ. Polytypic (2 sspp.; *restricta* in N.A.). L 4.3" (11 cm)

Identification Long bill is usually dark. Tail is strongly graduated, with outer rectrices mostly white. **BREEDING MALE:** Black cap and lores; black extends below eye. Head and back blue-gray with indistinct white eye ring; face rather pale. Underparts whitish, with grayish wash on breast. **NONBREEDING MALE:** Black crown absent; dark line over eye, or dark flecks. **FEMALE:** Like nonbreeding, but no black on crown.

Similar Species Separated from Blue-gray Gnatcatcher by much more graduated outer tail feathers, longer bill; males in breeding plumage easily separated by extensive black cap. From Black-tailed Gnatcatcher, in all plumages, outer rectrices are mostly white; in breeding plumage, dark crown extends below eye, and eye ring, if present, is less distinct. In nonbreeding, note the different call and tail pattern of Black-capped, as well as its longer bill.

Voice CALL: A rising, then falling *mee-ur*, somewhat like California Gnatcatcher; also suggestive of Bewick Wren. **SONG:** A jerky warble.

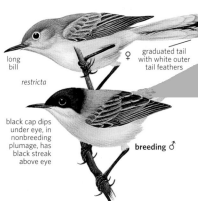

♀
graduated tail with white outer tail feathers

long bill

restricta

black cap dips under eye, in nonbreeding plumage, has black streak above eye

breeding ♂

Status & Distribution Western Mexican species very rare (primarily resident) in foothill canyons of southeastern AZ. A few pairs also present in Guadalupe Canyon in both AZ and NM. **Population** Stable.

restricta

nigriceps

DIPPERS Family Cinclidae

American Dipper (UT, June)

Few passerines are as specialized as the highly aquatic dippers. They spend their entire lives around, and often actually in, rushing streams and rivers. Accordingly, details of their physiology and behavior are unique. There is one North American species, which is widespread in the West. It is absent from the East.

Structure Portly overall, appearing bobtailed with strong legs and feet. The bill is short, compressed, and sturdy. Wings are short and rounded.

Behavior Highly distinctive foraging techniques. Often seen wading into rushing water, submerging head while looking for food—mostly aquatic larvae. Also swims on the surface for short distances, peering underwater, while paddling with feet. Using its wings, a dipper can submerge completely and maneuver near stream bottoms by grasping the substrate—remaining underwater for up to 30 seconds. When not in water, it stands on rocks and boulders, pumps its body up and down in highly stereotyped motion, and flashes white-feathered eyelids.

Plumage Dense and waterproof. All species exhibit significant brown or gray, some with white and/or chestnut. The American Dipper is one of the plainer species.

Distribution Found in large montane watershed complexes in the Americas, Europe, North Africa, and Asia.

Taxonomy Worldwide, five closely related species, all placed in a single genus. A close relationship to wrens (Troglodytidae) has been posited; others argue instead for a closer alliance with the mockingbirds and thrashers, thrushes, and Old World flycatchers.

Conservation Most dipper populations are naturally isolated and hence susceptible to local extinctions from drought, habitat alteration, and water pollution. BirdLife International codes: 1 VU.

Genus *Cinclus*

AMERICAN DIPPER *Cinclus mexicanus*   AMDI ▪ 1

One of the most charismatic species of the American West, the American Dipper is at home along rushing streams and rivers, from timberline to sea level. Although its attire is drab gray and brown, its behavior, physiology, and vocalizations are remarkable. It readily jumps into streams of rushing water and swims underwater, procuring its prey. Whitewashed rocks in streams are a likely sign of its presence. Polytypic (5 sspp.; *unicolor* in N.A., 4 others from Mexico to western Panama). L 7.5" (19 cm)

Identification Plump and monochromatic. Frequently bobs. Legs and feet pale; wings short; tail very short. **ADULT:** Slate gray with somewhat browner head; dark bill. One molt per year, in late summer. **JUVENILE:** May retain distinctive plumage into early winter; note paler tones overall, especially throat; distinctive white scalloping to contour and wing feathers; yellowish bill. **FLIGHT:** Alcid-like: low over the water, buzzy.

Similar Species This species is unmistakable.

Voice CALL: Loud, sharp *bzeet*, given frequently by flushing birds; easily heard. On landing, may give fast series of these notes. **SONG:** Amazingly long (exceptionally to 5 min) series of repeated, varied, thrasher-like phrases. Song often ventriloqual; nearby birds may sound faint, far away.

Status & Distribution Fairly common. AK to Panama. **BREEDING:** Near rushing streams. Year-round resident where streams remain partly ice-free. **MIGRATION:** High-elevation breeders move to nearby ice-free streams at lower elevations; some farther. Juveniles generally move to lower elevations than adults. **RARE STATUS:** Casual to TX; accidental MN.

Population Habitat naturally fragmented. Declines due to prolonged drought and water diversion projects.

unicolor

mexicanus

will swim underwater to obtain food

unicolor

overall paler than adult

juvenile

unmistakable— chunky and sooty gray overall

pinkish legs

BULBULS Family Pycnonotidae

Red-whiskered Bulbul

Bulbuls are native to tropical areas of the Old World. Popular as cage birds, several escaped species have established naturalized populations in many regions. In the US, bulbuls are associated with parks and neighborhoods that are landscaped with exotic plants and trees that provide fruit and nectar year-round.

Structure Bulbuls are slim with long tails; some have small crests.

Plumage Commonly brown, olive, or gray, with red, yellow, or black markings. Many species have bold head patterns and brightly colored undertail coverts. Juvenile plumage generally resembles that of adult.

Behavior Highly gregarious during the nonbreeding season, most bulbuls forage and roost in large flocks. Primarily frugivorous, but also feed on insects, which are gleaned from foliage or captured in midair by flycatching or hawking. Many pairs of Red-whiskered Bulbuls in FL are accompanied by a third adult, but cooperative breeding not documented.

Distribution Their native range extends from the western Palearctic to the Orient and the Philippines. The Red-whiskered Bulbul is found in North America as a result of escaped cage birds; another species is also naturalized in HI.

Taxonomy A large family with 130 species in 30 genera. Some genera have been transferred in and out of the bulbul family, testament to the complex relationships of the mainly Old World cluster of families jointly termed the Sylviida. The Bare-faced Bulbul (*Pycnonotus hualon*) was just described from Laos in 2009.

Conservation The US has banned further importation of Red-whiskered Bulbuls because of their potential threat to crops. BirdLife International codes: 20 NT, 9 VU, 3 EN, and 2 CR.

Genus *Pycnonotus*

RED-WHISKERED BULBUL *Pycnonotus jocosus*   RWBU ■ 2

The Red-whiskered Bulbul is the sole representative of this Old World family in N.A., although the Red-vented Bulbul (*P. cafer*) occurs in HI (Oahu) and small numbers are around Houston, TX. Bulbuls are popular cage birds in the US. Polytypic (9 sspp.; *emeria* in FL; CA sspp. unknown). L 7" (18 cm)

Identification A slender songbird with a conspicuous crest. Sexes similar. **ADULT:** Black forehead and crest. Small red "whiskers" just below and behind eyes. Narrow black line separates white cheek from white throat and malar. White breast and belly, with blackish spur on sides of breast forming incomplete band. Flanks and vent washed with buff; red undertail coverts. Tail broadly tipped with white on all but central rectrices. Black eyes, bill, legs, and feet. **JUVENILE:** Brown crown and crest, red "whiskers" lacking, buffy undertail coverts.

Similar Species Few other crested birds occur within narrow US range. Distinctive. May be confused with Phainopeplas in CA, which are wholly black or grayish, with white patches in outer primaries.

Voice CALL: *Kink-a-jou* or *chip-petti-grew*; *peet* given at roost or when alarmed. **SONG:** Rolling musical whistle *chee-purdee, chee-purdee-purdee*.

Status & Distribution Exotic in the US; importation now banned. Fairly common, but declining in native range from India to China and to the northern Malay Peninsula; subspecies found in FL native from India to southwestern Thailand. Resident, including US populations, restricted to Los Angeles Co. (hundreds, mostly in the San Gabriel Valley), where now considered established, and FL (perhaps

long black crest

small red auricular spot

black breast mark

red undertail coverts

juvenile

white tail tips

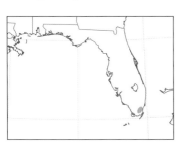

a few hundred individuals at Kendall and Pinecrest, southwest of Miami). FL birds nest Feb.–July in shrubs, palms, or small trees, usually those not native to the US.

Population In FL, the range is stable or expanding slightly; in CA population increasing and expanding.

KINGLETS Family Regulidae

Golden-crowned Kinglet, female (NY, Oct.)

The two North American kinglet species are tiny forest songbirds that are rather easily identified by size alone. Breeding in the upper canopy of boreal forest and in spruce bogs, kinglets are numerous in migration and in winter and often mix with other songbirds, including chickadeees and, during migration, warblers. **Structure** Kinglets have rounded bodies, medium-length tails, relatively short wings, and small, thin bills.

Plumage Grayish or olive with bright olive-edged flight feathers, two white wing bars, concealed colorful crown patches. Like the Golden-crowned Kinglet, most species worldwide have boldly marked heads, particularly lateral black crown stripes. The Ruby-crowned Kinglet is much less patterned and can be confused with vireos.

Behavior Frequently observed in feeding flocks in migration at all levels. Insectivorous and in winter feed on insects in buds and under bark, allowing them to winter farther north than other insectivores. When foraging, they tend to quiver or flick their wings, often hovering below leaves to glean insects.

Distribution Six species worldwide: two in North America, one in Eurasia, one in Europe to North Africa, one endemic to the Canary Is. (Madeira), and one endemic to Taiwan.

Taxonomy All species are in *Regulus*, formerly classified with the Old World warblers (Sylviidae, now split into multiple families); recent genetic studies suggest they are not closely related to sylviids, and their affinities remain unclear.

Conservation Generally stable; no species is considered at risk.

Genus *Regulus*

GOLDEN-CROWNED KINGLET *Regulus satrapa*   GCKI ■ 1

A tiny, thin-billed, wing-flicking insectivore, it has a conspicuously striped head. Polytypic (5 sspp.; 3 in N.A.). L 4" (10 cm)
Identification ADULT MALE: Dull grayish olive above, paler whitish below, head boldly marked with white supercilium, blackish lores and eye line. Yellow crown bordered broadly with black. Orange-red in center of yellow crown most visible during display or when bird is agitated. Black at base of secondaries, contrasting with white wing bar. **ADULT FEMALE:** Similar to male, no orange-red in crown. **IMMATURE:** More pointed tail feather tips than adult; a few males may lack orange in crown.
Geographic Variation Western sub-

species (*apache* and *olivaceus*) slightly smaller, somewhat brighter, with longer white supercilium, and longer bill than widespread eastern *satrapa*. In the West, *apache* is larger and paler than *olivaceus*.
Similar Species Ruby-crowned Kinglet, readily distinguished by head pattern and call.
Voice CALL: When flocking, a very high, sibilant jingling, *tsii tsii tsii*. Also, a quiet, high single note, *tsit*, and a thin sibilant *seee* similar to call of Brown Creeper. **SONG:** An extended version of the call, becoming louder and chattering toward the end: *tsii tsii tsii tsii tiii djit djit djit djit.*
Status & Distribution Common. **BREEDING:** Mainly boreal forests, a few in mixed or deciduous forests and conifer plantations. Also breeds in central Mexico and Guatemala. **SPRING MIGRATION:** Spring and fall migration difficult to detect in some areas with resident populations. Winter residents depart Gulf States before Apr. Peaks across continent late Mar.–late Apr. **FALL MIGRATION:** Begins late Sept. over much of range, peaking in East in Oct. and early Nov. **WINTER:** Many

orange-red median crown stripe, bordered by yellow

satrapa ♂

whitish underparts

yellow median crown stripe

satrapa ♀

bold whitish supercilium bordered by black on both sides

remain in breeding range through winter. Primarily south of Canada and north of Mexico in a wide variety of habitats. Numbers in Southern CA and the Southwest (always small) vary from year to year.
Population Breeding range increasing in the East and the Midwest due to plantings of spruce and pine. Adversely affected by logging, wildfire, and drought.

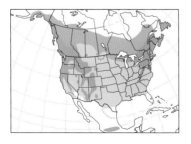

RUBY-CROWNED KINGLET   *Regulus calendula*   RCKI   ■ 1

This tiny, thin-billed, wing-flicking insectivore is grayish olive with a conspicuous broken white eye ring. Polytypic (3 sspp.; 2 in N.A.). 4.2" (11cm)

Identification ADULT MALE: Greenish above, paler below; broad, teardrop-shaped white eye ring, slightly broken at the top and bottom; lores olive. Ruby crown patch visible only during display, or when bird is agitated. Black at base of secondaries, contrasting with white lower wing bar. **ADULT FEMALE:** Lacks ruby crown patch. **IMMATURE:** A few males may have orange, yellow, or olive crown patch.

Geographic Variation Northwestern subspecies *grinnelli* (breeding coastal southeastern AK and BC) is slightly darker and more brownish than widespread nominate *calendula*. Subspecies *obscurus* on Guadalupe I., Mexico,

is extinct (nesting last confirmed 1953). **Similar Species** Compare with Hutton's Vireo.

Voice CALL: Most frequently heard is a husky two-syllable *ji-dit*. **SONG:** Often heard in migration; begins with two to three very high-pitched notes, abruptly changing to a rich, and surprisingly loud, warble: *tsii tsii tsii chew chew chew teedleet teedleet teedleet.*

Status & Distribution Common. **BREEDING:** Breeds in boreal spruce-fir forests, preferably near water and especially in black spruce bogs. Breeding begins immediately when females arrive in early May. **SPRING MIGRATION:** Mainly Mar. to early May, in southern US. Early Apr. to late May, peaking late Apr. to early May in central and northeastern US. **FALL MIGRATION:** Begins mid-Sept.; peaks late Sept. to mid-Oct.; continuing through mid-Nov. Arrives as early as late Sept. in FL and Mexico. **WINTER:** Not as hardy as Golden-crowned Kinglet, and winters farther south in a broad range of habitats. Primarily southern and western US through Mexico to Guatemala. **RARE STATUS:** Rare Bering Sea islands (mainly fall). Casual to Yucatán Peninsula, Bahamas, Cuba, Jamaica (sight record), Greenland (two

records). Accidental Aleutians, Ireland, and the Azores.

Population Breeding areas in the western US likely adversely affected by logging, wildfire, and drought. For instance, the species is now rare in the Sierra Nevada, CA, where formerly much more numerous in the early 20th century.

slightly broken white eye ring

black bar below lower wing bar

male's red crown patch usually concealed

calendula

calendula

pale olive underparts ♂

LEAF WARBLERS Family Phylloscopidae

Arctic Warbler (AK, June)

This is a uniform group of small insectivorous woodland species found across Eurasia as well as in Africa and Indonesia; eight species have occurred in North America. Seventy-three species are currently recognized in two genera, *Phylloscopus* (62) and *Seicercus* (11). Identification in both genera is very difficult between some species; vocalizations (song and contact notes) are often the best characters.

Structure These small insectivores have short, fine bills and slender legs. The wings are moderate in length to long and pointed in the more migratory species.

Behavior Leaf warblers are very active foliage gleaners and also take insects on aerial sallies; *Phylloscopus* warblers are mainly arboreal, while *Seicercus* warblers mainly inhabit undergrowth. Songs vary from simple to monotonous trills; rattles; or high, thin whistles.

Plumage *Phylloscopus* are dull, all species being olive-gray, greenish or brownish above and paler below, often with a yellow, buff, or pale olive wash; most show a dark transocular line and pale supercilium, and some species show one or two thin wing bars, lateral crown stripes and a pale central crown stripe, and/or a pale rump patch. The *Seicercus* warblers of Southeast Asia are mostly brighter yellow; most show a bold, pale eye ring; many show tawny or chestnut head markings; and some have bold wing bars.

Distribution Leaf warblers breed in Africa, Eurasia, and the Indonesian and Philippine Archipelagos, with one species occurring east to New Guinea; they are most diverse in the Palearctic, especially at higher elevations in eastern Asia and the Himalaya region. One Asian species (Arctic Warbler) breeds east to central AK. Most temperate-zone Palearctic species are strongly migratory, wintering in Africa, India, and Southeast Asia. Seven species (Willow, Common Chiffchaff, Wood, Dusky, Pallas's Leaf, Yellow-browed, and Kamchatka Leaf Warblers) have been recorded as strays to western and northern AK; Dusky and Arctic Warblers have also been recorded in fall south to Baja California. Yellow-browed has been recorded from BC, CA, and Baja California.

Taxonomy The leaf warblers were formerly considered part of the "Old World Warbler" family Sylviidae, but recent molecular data have led taxonomists to split that huge and ill-defined family into many smaller families.

Within the Phylloscopidae a number of new species have been recently described, particularly in Himalaya and eastern Asia; this follows intensive study by ornithologists incorporating vocal and molecular data. Some 77 species are now recognized; all are placed in the genera *Phylloscopus* and *Seicercus* (though 13 species in the former genus are sometimes placed in the genera *Rhadina* and *Abrornis*—including the Yellow-browed and Pallas's Leaf Warblers, which may belong in the genus *Abrornis*). The Arctic Warbler was recently split into three species, two of which occur in North America.

Conservation BirdLife International codes: 1 NT and 5 VU.

Genus *Phylloscopus*

WILLOW WARBLER *Phylloscopus trochilus*   WILW ■ 5

This casual visitor from Eurasia occasionally wags its tail. Polytypic (3 sspp.; presumably *yakutensis*, but no specimen). L 4.5" (11 cm)

Identification Subspecies *yakutensis*: uniform green to brownish olive above, without wing bars. A narrow, pale supercilium and a thin, dark eye line. Below, whitish (spring); fall birds have pale yellow wash on throat and breast. Legs are dusky; feet pale.

Geographic Variation Subspecies *yakutensis* (breeds eastern Siberia and Russian Far East) duller than other subspecies breeding to the west. Nominate *trochilus* and *acredula* from Europe are

greener above and more yellow below.

Similar Species The eastern subspecies of Common Chiffchaff, *tristis*, is grayish brown above; Willow Warbler is slightly greener above and has longer primary extension, which is at least 75 percent of the length of the exposed tertials, compared to shorter projection of Chiffchaff (approximately 50 percent). Otherwise, auriculars of Chiffchaff are uniformly dark, whereas they are slightly paler on Willow. Legs usually paler in Willow, but eastern *yakutensis* has dusky legs. Arctic Warbler is longer winged with a bolder supercilium, a pale wing bar, and a larger bill. Tennessee Warbler has a different shape, more pointed bill, and contrastingly white undertail coverts.

Voice CALL: A disyllabic *hoo-eet*. **SONG:** Beautiful, descending notes (not heard in N.A., suggestive of Canyon Wren).

Status & Distribution Breeds in Eurasia, winters in Africa. **RARE STATUS:** In recent years, recorded intermittently from St. Lawrence I., AK (some 20 records, late Aug.–early Sept.); also recorded from Pribilofs. Regular fall

migrant to Iceland. Accidental northeast Greenland (specimen, subspecies undetermined).

Population Declines in parts of Europe.

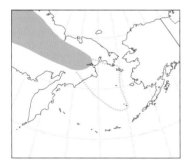

bill smaller than Arctic

fall
yakutensis

underparts mostly dull yellow

palish legs

small bill

overall plain with no wing bars

spring
yakutensis

long primary projection

COMMON CHIFFCHAFF *Phylloscopus collybita*   CCHI ■ 5

Common Chiffchaff is a widespread and common summer resident throughout much of Eurasia. It frequently bobs its tail. Polytypic (6 sspp.; *tristis* in N.A.). L 4.5" (11 cm)

Identification Active, medium-size leaf warbler with a fine dark bill, dark legs, and somewhat short primary projection (equal to about 50 percent of the length of the exposed tertials); four outer primaries (P5–P8) are emar-

ginated (three in Willow Warbler). Subspecies *tristis* has a gray-brown crown and upperparts (lacking green tinge). Moderately prominent supercilium is whitish, as is the breast (lacking strong yellowish tones). Ear coverts often tinged rusty. Wing and tail feathers often narrowly edged yellow-green. Fall immatures may show a short, pale wing bar.

Geographic Variation Easternmost

gray-brown crown and mantle

plain wings

"Siberian"
tristis

whitish buff underparts

dark legs

short primary projection past tertials

subspecies *tristis* ("Siberian Chiff-chaff") has white to pale buff under-parts and brownish upperparts that separate it from most subspecies found in Europe and Asia Minor, although

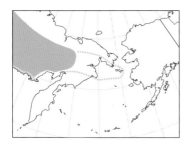

some northern European *albietinus* can be similarly dull. Plumage and especially different vocalizations may warrant separate species status for *tristis*.

Similar Species Eastern *tristis* is similar to *yakutensis* Willow Warbler; see that species for details on separation. Dusky Warbler is similar in coloration, but is browner, less grayish above, less whitish below, and has a longer supercilium. Dusky's behavior is much more skulking and its call note is very different.

Voice CALL: For *tristis*, typically a plaintive, flat *peep*; less often *swee-oo*

similar to European subspecies. The song of *tristis* (not heard in AK) is more varied than European subspecies.

Status & Distribution "Siberian Chiffchaff" breeds in coniferous taiga woodlands from the Urals east to the Kolyma River; winters chiefly from the Middle East to India; casual migrant to Japan. Regular migrant to Iceland. **RARE STATUS:** Casual, 10 records for N.A. all in AK, in spring and fall; all at Gambell, St. Lawrence I., except for one at St. Paul I., Pribilofs. This subspecies is also a regular stray, mainly in fall, to western Europe. **Population** Stable.

WOOD WARBLER   *Phylloscopus sibilatrix*   WOWA   ■   5

A casual stray from Eurasia to AK. Monotypic. L 5" (13 cm)

Identification Perhaps the most colorful of the leaf warblers, has bright green

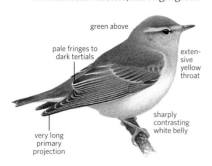

green above

pale fringes to dark tertials

extensive yellow throat

sharply contrasting white belly

very long primary projection

upperparts and plain wings and has pale fringes to the dark-centered tertials. Bold and long yellow supercilium is offset by dark eye line with a slight rear hook, and clean yellow ear coverts; back greenish. Below, throat and upper breast are a clear lemon yellow and sharply separated from remainder of pure white underparts. Note very long primary projection and short tail. Both bill and legs are pale, usually pinkish.

Similar Species Not similar to other potential *Phylloscopus*, as its throat and breast are distinctly yellow, contrasting with otherwise white underparts. Perhaps more likely to be

confused with a fall Tennessee Warbler (casual to Bering Sea islands), which is much more extensively yellow below.

Voice CALL: A Sharp *zip* and a plaintive *tew*.

Status & Distribution This long-distance migrant breeds in Eurasia, east to south-central Siberia; winters in tropical Africa south of Sahara. **RARE STATUS:** First recorded in 1978, now a dozen records (all in fall from AK), from the western and central Aleutians, Pribilofs, and St. Lawrence I., except for one at Middleton I.

Population Moderate declines in Europe.

DUSKY WARBLER   *Phylloscopus fuscatus*   DUWA   ■   4

A regular stray from Asia, this brown *Phylloscopus* spends more time close to, or on, the ground, but will feed in trees on occasion. It works its way through low bushes or on the ground, flicking its wings and calling regularly. Polytypic (3 sspp.; nominate in N.A.). L 5.5" (14 cm)

Identification Upperparts are a uniform dusky brown; no wing bars. Bill is short, thin, and dark; legs are dark. Dusky Warbler exhibits a subtle, but distinct face pattern: pale supercilium, dingier in front of eye with dark eye

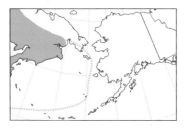

line; faint white eye ring. Underparts are dingy, darkest across breast, but otherwise creamy with buffy brown wash on flanks and undertail coverts. Unlike that of most *Phylloscopus*, tail is slightly rounded.

Similar Species Kamchatka and Arctic Warblers, the most likely *Phylloscopus* in our area, differ in several ways from Dusky. Both are greenish above, with a wing bar on greater coverts and a thicker bill that is usually pale on lower mandible, have pale mottling in auriculars; calls differ. Dusky is most similar to the larger Radde's Warbler (*P. schwarzi*), unrecorded in N.A. Radde's has a thicker bill that is noticeably pale on lower mandible. Radde's often has yellow or olive tones, which Dusky lacks, and undertail coverts of Radde's are particularly contrasting and yellowish buff. Supercilium in front of Radde's eye is more muted and yellowish.

Finally, the calls of the two birds differ.

Voice CALL: A hard *tschik*, recalling the chip note of Lincoln's Sparrow.

Status & Distribution Breeds Siberia, Russian Far East, Mongolia, and China; winters Southeast Asia. **RARE STATUS:** Casual (mostly fall) on islands off western AK, and in fall, off south-coastal AK and in CA; there are two fall records from Baja California.

Population Stable.

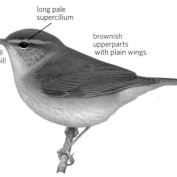

long pale supercilium

brownish upperparts with plain wings

small fine bill

PALLAS'S LEAF WARBLER *Phylloscopus proregulus* PLEW ■ 5

In plumage, structure, and behavior, this tiny denizen of taiga forest of eastern Asia bears a resemblance to kinglets (family Regulidae). Monotypic (formerly 5 sspp.). L 3.5" (9–10 cm)
Identification A tiny, kinglet-size bird, vaguely reminiscent of Golden-crowned Kinglet, with plump body and short tail. Overall, patterned contrastingly in green, brown, yellow, and white. Boldly striped head distinctive, with a bright yellow median crown

yellow median stripe on dark crown
tiny bill
bold wing bars
often hovers, showing yellow rump
1st fall

stripe, bold dark greenish brown lateral crown stripe, bold yellow supercilium (brightest in front), and blackish eye line; has whitish tertial tips and wing bars. Lemon yellow rump, especially conspicuous during flight, when the bird habitually flits and hovergleans around vegetation, much as kinglets do.
Geographic Variation Most authorities now treat Pallas's as monotypic. Current treatment of leaf-warblers in the complex that included Pallas's elevates many former subspecies to full species.
Similar Species Most likely to be confused with Chinese Leaf Warbler or other taxa recently split from the Pallas's complex (Gansu, Lemon-rumped, Sichuan; all unrecorded in northeast Asia and N.A.), but all have much less vivid tones overall, with the whitish supercilium. Yellow-browed Warbler

and Hume's Leaf Warbler (*P. humei*), the latter unrecorded in N.A., are also similar, but both lack the yellow rump and vivid yellow supercilium and distinct median crown stripe.
Voice CALL: A soft, nasal, rising *chuee*. **SONG:** A long, ringing series of trills and clear whistles.
Status & Distribution Coniferous and mixed taiga woodlands from northern Mongolia and southern Siberia east to Sakhalin I. and northeastern China. **WINTER:** Primarily in southeastern China and northern Vietnam, more rarely elsewhere in northern Southeast Asia. **RARE STATUS:** One photographed at Gambell, St. Lawrence I., AK (25–26 Sept. 2006), represents the only N.A. record. Small to moderate numbers are regular in fall to Europe, some winter there.
Population Considered widespread and common in its large range.

YELLOW-BROWED WARBLER *Phylloscopus inornatus* YBWA ■ 4

This bird is a casual stray from Asia. Monotypic (most authorities now recognize the polytypic Hume's Warbler, *P. humei*, from Inner Asia, as a separate species). L 4.5" (11 cm)
Identification When fresh, a small leaf warbler, greenish olive above, with an obvious yellowish supercilium and distinct whitish wing bars. Bill is small and

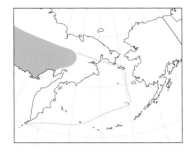

thin. Tertials are dark, with distinct pale edges. Dark patch at base of secondaries, like Ruby-crowned Kinglet.
Similar Species Distinguished by white wing bars and dark-centered tertial edges, unlike other expected *Phylloscopus*. Arctic and Kamchatka Leaf Warblers are larger, with a thicker bill, and have one thinner bar on greater wing coverts. Smaller Pallas's Leaf Warbler (one N.A. record) has a much bolder head pattern and a yellow rump, and it habitually hovers as it feeds. Hume's Leaf Warbler (unrecorded N.A. but annual stray to northwest Europe) is very similar, but slightly duller; calls differ.
Voice CALL: High, ascending *swee-eet*.
Status & Distribution BREEDING: Northern Asia, east of the Urals. **WINTER:** Southern Asia (mostly Southeast

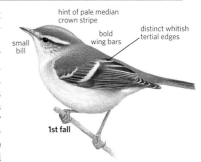

hint of pale median crown stripe
bold wing bars
distinct whitish tertial edges
small bill
1st fall

Asia), with a well-established pattern of reaching Europe, including Iceland, and the Middle East. **RARE STATUS:** Casual to western Aleutians and Bering Sea islands (about 10 records). Accidental in fall to Middleton I., AK; southwestern BC; eastern CA; and in winter to Baja California Sur.
Population Stable.

ARCTIC WARBLER *Phylloscopus borealis* ARWA ■ 2

The Arctic Warbler breeds across northern Eurasia and winters in the Philippines, southern Southeast Asia, and Indonesia. In N.A., it breeds in AK. Recent studies have revealed that the Arctic Warbler actually comprises three separate cryptic species, two of which occur in N.A.: the Arctic and Kamchatka Leaf Warbler. These studies are based on distinct molecular and

vocal differences (both songs and call notes). Based on present knowledge, silent birds are not field-separable. All specimens of migrants from the Aleutians belong to the Kamchatka Leaf Warbler. Monotypic. L 5" (13 cm)
Identification Long, yellowish white supercilium and a dark eye line. Auriculars are mottled. Upperparts are olive with thin lower whitish wing bar;

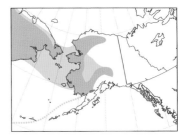

long, pale supercilium

faint pale wing bar

olive upperparts

spring

sometimes an indistinct lower wing bar; long primary extension. Underparts are whitish, with some olive on sides and flanks. Bill is rather large and extensively pale; legs and feet are straw-colored.

Similar Species Silent birds (most migrants) are not field-separable from the almost identical Kamchatka Leaf Warbler, which occurs as a migrant in the Aleutians. Kamchatka Leaf averages greener above and the bill and wing length average longer; in hand, suggestive, but not diagnostic from Arctic, is that P10 is longer than the greater primary coverts. Vocalizations, including call notes, differ. Another split from Arctic, the larger Japanese Leaf Warbler, breeds from Honshu to Kyushu, Japan; it is not recorded for N.A. It is larger and brighter, more yellowish, than either Arctic or Kam-

chatka Leaf and also differs vocally. Arctic is much more olive above (less brownish), and paler below than Dusky Warbler, with a larger bill and at least one wing bar. Willow Warbler lacks a wing bar and has a thinner bill.

Voice CALL: A buzzy *dzik*. **SONG:** A long, loud series of toneless, buzzy notes.

Status & Distribution Eurasian species. **BREEDING:** Fairly common to common breeder in shrubs of medium height (under three feet) east to the AK Range of central AK. **MIGRATION:** A Trans-Beringian migrant typically arriving in AK at the end of the first week of Jun.; fall migration is mostly mid- to late Aug.; regular migrant (rare spring, fairly common fall) on St. Lawrence I. Sight records elsewhere on Bering Sea islands are either this

species or Kamchatka Leaf. **RARE STATUS:** Accidental YT. There are six fall records for CA and additional fall records for NV and Baja California. These are regarded at present as either Arctic or Kamchatka Leaf, although some were handled and at least one was heard (thought to be Arctic), but not recorded.

Population Stable.

KAMCHATKA LEAF WARBLER *Phylloscopus examinandus* KLWA ▪ 5

This East Asian species was recently split from the very similar Arctic Warbler. In N.A., it is mainly known as a scarce migrant in the Aleutians, AK. Monotypic. L 5" (13 cm)

Identification Virtually identical to the Arctic Warbler, but slightly greener above and averages a longer bill and wing. Indicative, but not diagnostic, is that in hand, P10 is longer, not shorter, than greater primary coverts. Best told in field by vocaliza-

tions, but many migrants are silent.

Voice Often silent. **CALL:** Distinctive two-syllable *tzz-eet*. Arctic's call note is one syllable and higher pitched.

Status & Distribution BREEDING: Breeds in eastern Asia from Kamchatka and Sakhalin I. to the Kurile Is. to northern Hokkaido. **MIGRATION:** Coastal Asia, a rather late spring and fall migrant. **WINTER:** The winter range is not well known but is probably in Indonesia; perhaps some winter on the Malay Peninsula. **RARE STATUS:** Rare in spring (usually first two weeks of June, a few later) and rare to casual in fall from second week of Sept.–mid-Oct. to western Aleutians. Casual in fall (Oct.) to central Aleutians. All specimens taken so far in Aleutians are this species, not Arctic Warbler. Otherwise specimens have been taken on St. Matthew I., Old Chevak on AK mainland, and Prince Patrick I., NT. Casual north-

bill and wing average longer than Arctic; all vocalizations differ

averages brighter green than Arctic, but difference slight and individual, seasonal, and age variation make this unreliable as a field mark

spring

in hand suggestive, but not diagnostic, from Arctic is P10 longer than greater primary coverts

west Australia. The identity of stray migrants in N.A. is problematic. For instance, there are well-photographed records of Arctic Warbler or Kamchatka Leaf Warbler (formerly treated as Arctic) from the Pribilofs, and in fall from CA and NV. What we do know is based entirely on collected specimens.

Population Appears stable.

SYLVIID WARBLERS Family Sylviidae

The sylviids had long been a large and dominant assemblage of small insectivorous birds found throughout the Old World. The family, as formerly constituted, was the legacy of taxonomies based on superficial morphological characters, many of which have been proved to be less than reliable indicators of evolutionary relatedness. There is no current consensus on the composition of this family, but it is either dominated by or virtually restricted to the 25+ warblers of the genus *Sylvia*, one of which has been recorded as a stray

in western AK. Various other genera that had long been considered babblers in the family Timaliidae are now placed in the Sylviidae, including the enigmatic Wrentit of western North America (by the AOS).

Structure Sylviids are slender-billed and medium- to long-tailed warblers. The Asian parrotbills, sometimes placed within the Sylviidae, have remarkably thick bills with a strongly arched culmen, and long, graduated tails.

Behavior Foliage-gleaning; food is primarily arthropods, but most species take small fruits at least part

Wrentit (CA, Oct.)

in the Mediterranean region. Most winter in southern Europe or Africa. The easternmost Eurasian species, the Lesser Whitethroat, has been recorded once in western AK. Most species occupy scrub habitats and broadleaf woodlands. Other "babblers" currently placed in the Sylviidae by various authorities are primarily Asian.

Taxonomy The definition and taxonomic composition of the Sylviidae is definitely a "work in progress," and various genera are likely to be transferred into or out of the family in future years. The enigmatic Donacobius of tropical South America seems closely related to the Sylviidae, but for now, it is usually given family rank. As many as 70 species in some 17–19 genera are considered sylviids by recent authors; many of these were previously placed in with the babblers (Timaliidae) or in smaller families (e.g., Paradoxornithidae for the 20 species of parrotbills). The sole New World breeding species, the Wrentit, was formerly afforded unique family status (Chamaeidae) and then merged into the babbler family Timaliidae. The timaliids, ultimately, proved to be polyphyletic and some of its former members, including the Wrentit (which may be closest to the parrotbills), were transferred to the Sylviidae.

Conservation Birdlife International codes: 6 NT and 3 VU.

of the year and may even be important seed-dispersal agents. Songs range from rich warbling with fluty tones to scratchier jumbles in the genus *Sylvia*. Open cup nests.

Plumage *Sylvia* are dull above and whitish to gray or vinaceous below; many species have black crowns or other black markings. In many species the iris may be red, yellow, or white, and some have bright red orbital rings.

Distribution The genus *Sylvia* (including *Parisoma*) is found in Europe, Africa, and western Asia; most diverse

Genus *Sylvia*

LESSER WHITETHROAT   *Sylvia curruca*   LEWH   5

The Lesser Whitethroat, a Eurasian species, has been recorded once in N.A. from western AK. Polytypic (7–9 sspp.; subspecies of N.A. bird uncertain). L 5.5" (14 cm)

Identification Crown is gray with a dark mask from base of bill to auriculars. Wings and back are plain warm brown, with darker centers to tertials. Tail is dark from above, and square with some white in outer retricies. Throat and undertail coverts are white; flanks and sides have a tan-brown wash.

Geographic Variation Up to nine subspecies; some authorities have split the "Northern" group from the "Desert" group. The northern subspecies,

blythi and *curruca*, have more well-defined face masks and more contrasting brown wings compared to more southern subspecies.

Similar Species Unlike any N.A. species, but identification within the whitethroat complex is difficult, especially if Lesser Whitethroat is split into two or more species.

Voice CALL: A sharp *tik*, often repeated, and various chattering notes.

Status & Distribution BREEDING: Eurasia. **WINTER:** Africa to southwest Asia. **RARE STATUS:** Accidental. One record (photo) from St. Lawrence I. on 8–9 Sept. 2002

thought to pertain to the northern complex of subspecies. A few fall records from Japan and Korea suggest a weak pattern of dispersal to the northeast.

Population Stable.

gray crown

dark mask

1st fall
blythi

Genus *Chamaea*

WRENTIT   *Chamaea fasciata*   WREN   1

Wrentits are curiously long-tailed and skulking denizens of dense, non-forested chaparral on the West Coast. Their taxonomic placement has always been a conundrum; in the recent past they were placed with babblers (family Timaliidae), now with the sylviid warblers. Polytypic

(5 sspp.; all in N.A.). L 6.5" (17 cm)

Identification Recognized by distinctive round, fluffy body and long, round-tipped tail usually cocked at an angle. It is usually heard rather than seen but is curious and can approach very close. Whitish eye; short and stout bill; and lightly streaked buffy or cinna-

mon underparts are distinctive. Muted plumage is alike in all ages and sexes.

Geographic Variation Subspecies not clearly differentiated: Birds from northern, wetter areas are darker, more cinnamon below; subspecies from southern dry sites are grayish above, dull buffy grey below.

Similar Species Unmistakable.
Voice CALL: Wooden-sounding, rattled *churrrrrrr*, or longer *krrrrrrrrrrrr*.
SONG: Male song a *pit-pit-pit-pit-pit-pit-trrrrrrrr*, which has a bouncing-ball quality as notes speed up into the trill. Female song is similar, but lacks trill, and the pit notes are spaced evenly.
Status & Distribution Common. **YEAR-ROUND:** Found in dense chaparral or huckleberry-salal thickets from sea level to 6,000 ft. Resident within range. The Columbia River marks the northern end of its range; no records for WA.
Population Stable.

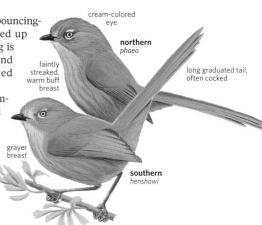

cream-colored eye

northern
phaea

faintly streaked, warm buff breast

long graduated tail, often cocked

grayer breast

southern
henshawi

REED WARBLERS Family Acrocephalidae

Sedge Warbler

Formerly in the Sylviidae, these are medium- to large-size, dull, and difficult to identify warblers; most inhabit reed beds and other marshes, with a few species in dry woodlands and brushlands.

Structure Moderately large warblers with rather long and strong bills, sloping foreheads, and rounded (*Acrocephalus*) or squared (*Iduna, Hippolais*) tails.

Behavior Most species skulk within reed beds and other dense low habitats, being most detectable when singing. Songs consist mostly of rapid jumbled chatters and churring notes, often incorporating mimicry and in many species given at night.

Plumage Generally brownish to grayish above and paler gray or whitish below. African "yellow warblers" (now in the genus *Iduna*) are bright yellow below and relatively short-winged and long-tailed.

Distribution Found widely in the Old World, from western Eurasia and Africa (plus Madagascar) east to the Australasian region and on some oceanic islands of the Indian and south and central Pacific Ocean. Eurasian species are mostly migratory, many wintering in Africa, but also to India, Southeast Asia, and the Philippines. Three species have been recorded as strays to western AK; another is resident in the leeward Hawaiian Is.

Taxonomy Molecular work has helped sort out the relationships among the "Old World Warblers," confirming the monophyly of the genera in this family. As currently constituted, this family includes 62 species in five genera, of which *Acrocephalus* (42 species) dominates. The African genus *Chloropeta* is now merged into *Iduna* (which itself was split off from the Eurasian genus *Hippolais*).

Conservation Many species are decreasing through habitat loss. Several oceanic island species are considered endangered, and several island endemic species or subspecies are now extinct. Europe's Aquatic Warbler has long been of conservation concern and is regarded as Europe's most endangered songbird. Birdlife International codes: 2 NT, 7 VU, 5 EN, 4 CR, and 6 EX.

Genus *Arundinax*

THICK-BILLED WARBLER *Arundinax aedon*　　TBWA　■　5

Although the reed warbler family consists of many species that are exceedingly difficult to identify, the Thick-billed Warbler from Asia is not one of them. The structure and face pattern are distinctive. This somewhat secretive species avoids reeds. Polytypic (2 sspp., weakly differentiated; some treat species as monotypic; AK bird more likely eastern *stegmani* on geographic probability). L 7.3" (19 cm)
Identification A large, long-tailed, and short- and rounded-winged reed warbler that lacks an eye stripe; large dark eye stands out in its blank face. The bill is thick and disproportionately short, and the culmen is curved. Uniform brownish above and pale below with a warm buff suffusion on the breast and flanks.
Similar Species Oriental Reed Warbler

(*A. orientalis*, unrecorded in N.A.) of eastern Asia is comparable in size but has a dark eye stripe and a more pointed bill.

Voice CALL: A harsh *chack* note, often delivered in a series.

Status & Distribution Migratory East Asian species that favors brushy areas, not wetlands. **BREEDING:** Breeds from south-central Siberia (from Ob River east) and northern Mongolia to southeast Russian Far East and northeast

China. **WINTER:** Winters in southern (also Sri Lanka) and eastern India east through Indochina and southern Yunan, rarely to southeast China. **RARE STATUS:** Accidental western AK, one at Gambell, St. Lawrence I., 8–13 Sept. 2017. Casual to Japan, Malaysia, and northwest Europe (mainly fall). Accidental Egypt. **Population** Stable.

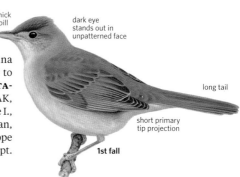

thick bill

dark eye stands out in unpatterned face

long tail

short primary tip projection

1st fall

Genus *Acrocephalus*

SEDGE WARBLER *Acrocephalus schoenobaenus* SEWA ■ 5

This warbler breeds primarily in Europe and is one of the more distinctive members of the *Acrocephalus* genus. It is somewhat secretive and getting good looks can be difficult. It has been recorded twice from St. Lawrence I., AK. Monotypic. L 4.5–5" (11.5–13 cm)

Identification Wedge-shaped tail and long primary projection are quite distinctive. Buffy olive brown above, with darker brown streaks in mantle; rump and uppertail coverts a contrasting rich rusty brown, most readily observed in flight. Bold supercilium; dark eye stripe; brownish auriculars; and dark-streaked crown (often showing indistinct brownish median stripe). **ADULT:** Whitish below, with buff tones in breast and flanks. **IMMATURE:** Shows light streaking or stippling on breast.

Similar Species Two European species

unrecorded in N.A. resemble Sedge. The Moustached Warbler (*A. melanopogon*) occurs east to Central Asia and is readily distinguished from Sedge by shorter primary projection, whiter and differently shaped supercilium, more richly colored upperwings, and darker crown. The Aquatic Warbler (*A. paludicla*), unknown from Asia, is similar but is more strikingly patterned. The very rare, little-known, highly migratory, and likely endangered larger Streaked Reed Warbler (*A. sorghophilus*) of eastern Asia (breeds northeast China, possibly north to Amurland, Russian Far East, winters Phillippines) is more diffusely patterned on the back, but is otherwise quite similar.

Voice CALL: Dry *chirr* or *errrrr*. **SONG:** A long, loud rollicking set of trills, whistles, and staccato notes, often includes mimicked calls of other species.

bold supercilium and dark eye line

faint streaks on back

1st fall

Status & Distribution BREEDING: Dense marsh vegetation from northwestern Europe to western Siberia and northwestern China. **WINTER:** Sub-Saharan Africa. **RARE STATUS:** Two fall N.A. records from Gambell, St. Lawrence I., AK (30 Sept. 2007 and 4 Oct. 2018).

Population Some population declines over much of this species' large range.

BLYTH'S REED WARBLER *Acrocephalus dumetorum* BREW ■ 5

This plain and brownish Eurasian species is atypical of most *Acrocephalus* as it usually avoids wetlands, preferring to move slowly through bushes, sometimes low trees, while it flicks its tail. It has been recorded twice from western AK. Monotypic. L 4.5" (11 cm)

Identification Overall an unmarked brownish warbler with a relatively long bill, a short supercilium, extending to just behind the eye, darkish legs, and relatively short round wings.

Similar Species Notoriously similar to other Eurasian *Acrocephalus*, notably Marsh Warbler (*A. palustris*) and the duller (less rufous brown) eastern *fuscus* subspecies of Reed Warbler (*A. scirpaceus*). Both are primarily European species; Marsh Warbler is unrecorded in East Asia and Reed

Warbler is accidental in eastern China. Blyth's Reed has two emarginated primaries and a short primary projection; in hand it has a distinct wing formula. The bill is rather long and the lower mandible is usually tipped dark. The very rare and poorly known Asian Large-billed Reed Warbler (*A. orinus*) is also quite similar but has a longer bill.

Voice CALL: A hard *tuck*.

Status & Distribution Breeds in overgrown clearings in deciduous forest and in riverine vegetation. In winter found in bushy areas, avoids wetlands. **BREEDING:** Breeds from northeastern Europe east to about Lake Baikal. **WINTER:** Indian subcontinent and coastal western Myanmar. **RARE STATUS:** Two

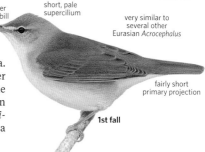

rather long bill

short, pale supercilium

very similar to several other Eurasian *Acrocephalus*

fairly short primary projection

1st fall

fall records for Gambell, St. Lawrence I. (9 Sept. 2010 and 18–21 Sept. 2015). Rare, but annual (mostly fall) to northwest Europe, even casual to Iceland. Also casual to Thailand, eastern China, and Japan.

Population Stable; increasing and spreading in northeastern Europe.

GRASSBIRDS Family Megaluridae

Middendorff's Grasshopper-Warbler (Japan, June)

This Old World family includes small to medium-size, mostly skulking birds that usually stay hidden near the ground in grassy thickets. The two species recorded in North America are casual migrants to the westernmost Aleutians and other islands in the Bering Sea region.

Structure Grassbirds are rather long- and stout-billed and long-legged as warblers go. They have strongly graduated tails and dense, long undertail coverts. They range from about 4.5 to 9 inches.

Behavior These are in many cases—especially for species within the genus *Locustella*—extremely secretive birds that skulk and walk well within vegetation and are reluctant to fly. The two Australasian songlarks, part of genus *Cincloramphus*, inhabit more open areas and are considered beautiful songsters (with elaborate flight songs). The songs of most *Locustella* are mechanical and repetitious, often insect-like in quality.

Plumage Nearly all species are nondescript, clad in shades of brown above and usually somewhat paler below; some species are streaked above, and a few have ventral streaking or diffuse spotting (often with strongly marked undertail coverts).

Distribution Grassbirds are found from Africa and Eurasia east to Australasia. They inhabit brushy and scrubby areas, forest interiors, swamps, marshes, and tall grasses. The two species recorded in AK have the most northeasterly Asian breeding ranges of any locustellid species; one of these (Lanceolated Warbler) has reached CA once.

Taxonomy The family includes 56 species in 10 genera. Like so many small insectivorous birds of the Old World, the grassbirds were formerly placed in the now-splintered "mega-family" Sylviidae. Asiatic species of *Bradypterus* are now in the genus *Locustella*, so the former genus (12 species) is now restricted to Africa. The most species-rich genus is *Locustella* (24 species); many of these species are migratory, some highly migratory. Some authorities consider the valid name of this family to be Megaluridae.

Conservation Populations of most species are stable; the subspecies of Fernbird (*Poodytes punctatus rufescens*) from the Chatham Is. off New Zealand is extinct. Birdlife International codes: 8 NT, 8 VU, 2 EN, and 1 EX.

Genus *Locustella*

MIDDENDORFF'S GRASSHOPPER-WARBLER *Locustella ochotensis*   MIGW ■ 4

If not singing from an exposed perch, this species, a visitor from Asia, skulks in grasses. Most treat the species as monotypic but some authorities recognize larger and grayer northeast breeding birds as *subcerthola*. L 6" (15 cm)

Identification A large, chunky, brown warbler with a relatively thick bill. Brown with a faintly streaked back; rump region warmer brown. A whitish supercilium is offset by dark eye line. Underparts are whitish, with buff on sides and flanks, and streaks at sides of breast. Graduated tail feathers are brown with dark subterminal marks and a whitish tip. **FALL:** Plumage more colorful when fresh, with much of underparts washed yellowish buff; supercilium buffy. **FLIGHT:** When flushed underfoot, note the large size, with an obviously warmer rump and tail; tail looks wedge-shaped and has whitish tips to retrices.

Similar Species The other congener, Lanceolated Warbler, is smaller, more obviously streaked above and below, and lacks pale tips to rectricies. A typical-appearing Pallas's Grasshopper-Warbler, *L. certhiola*,

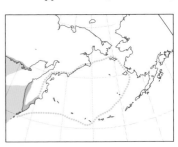

more distinct whitish supercilium than Lanceolated

faint back streaks

dull rufous rump

spring

skulks

wedge-shaped tail with pale tips

overall more richly colored than in spring; can be yellow-buff tinged below

fall

breeding to the west of Middendorff's, was recently photographed (9–12 Sept. 2019) at Gambell, St. Lawrence I., has recently been accepted (Dec. 2020). It is similar to Middendorff's. Its best feature is a strongly streaked back, but complications exist. Where the breeding ranges of the two species overlap in coastal regions of Russian Far East, hybridization is frequent; up to 70 percent are hybrids in some regions (lower Amur River and northern Sakhalin; morphological characteristics of these hybrids are not yet well studied).
Voice CALL: A soft *kit*. **SONG:** A few in AK have been singing. Harsh and *Acrocephalus*-like, less insectlike than European *Locustella*.

Status & Distribution Favors wet areas, including willows. **BREEDING:** Coastal northeast Asia. **WINTER:** Primarily in the Philippines and north Borneo. **RARE STATUS:** Casual to western Aleutians and Bering Sea islands, mainly in fall. Also casual Indonesia and northwest Australia.
Population Possibly declining.

LANCEOLATED WARBLER *Locustella lanceolata* LANW ◾ 5

The smallest *Locustella*, the Lanceolated Warbler is a casual visitor from Asia. Secretive, it runs and walks on the ground. Like other *Locustella*, very difficult to see when not singing. Its movements on the ground are certainly mouselike. One clue to its presence is agitated chiffing or chacking notes when alarmed. On its winter ground in southeast Asia, often found in quite dry situations. Singing birds perch up on grasses; otherwise unlikely to be seen much off the ground. (2 sspp.; *hendersoni* is coastal Asian breeder; often treated as monotypic; AK specimens not identified to sspp.). L 4.5" (11 cm)
Identification A midsize warbler with black streaks on a brown crown, man-tle, and rump. Tertials are dark, with paler fringes. Supercilium indistinct on an otherwise streaked face. Below, breast, flanks, and undertail coverts are streaked; ground color buff or whitish with wear. Throat is usually white and generally unmarked. Flicks its wings.
Similar Species Somewhat similar to the larger Middendorff's Grasshopper-Warbler, but that species lacks extensive streaking above, including on crown, and is streaked primarily on sides of breast. Middendorff's has a dark eye line, contrasting with a bolder, pale supercilium, and is larger. Finally, tail feathers of Middendorff's have pale tips and a dark subterminal patch.
Voice CALL: A distinctive, metallic *rink-tink-tink*, delivered infrequently;

also an explosive *pwit* and rapid series of *chack* or *chiff* notes when disturbed or agitated. **SONG:** A far-carrying but thin, insectlike reeling sound, like line moving through a fishing reel.
Status & Distribution **BREEDING:** From eastern Finland to western Kamchatka. **WINTER:** Southeast Asia, Sumatra, Philippines, and north Borneo. **RARE STATUS:** Up to 25 on Attu, western Aleutians, during the spring and summer of 1984. Two more were found at Attu on 2–6 June 2000; up to four recorded (with nesting) from Buldir I. in summer 2007. Accidental to St. Lawrence I. (Sept. 2013 and Oct. 2015) and to Farallones, CA (11–12 Sept. 1995). This is an annual fall stray to the UK.
Population Stable.

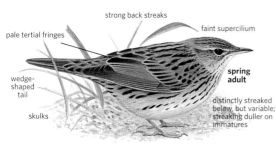

strong back streaks
faint supercilium
pale tertial fringes
wedge-shaped tail
skulks
spring adult
distinctly streaked below, but variable; streaking duller on immatures

RIVER WARBLER *Locustella fluviatilis* RIWA ◾ 5

The secretive River Warbler, recorded only once in N.A., is primarily an eastern European breeding species. For breeding, it favors low damp vegetation along streams and ditches and at other times of the year favors dense undergrowth. Monotypic. L 6" (15 cm)
Identification A dark brown rather elongate warbler with a broadly rounded tail. Above it is uniformly brown, has a short indistinct pale supercilium and an indistinct eye ring. Below, the underparts are dirty white with brownish flanks. Distinctive are the diffuse streaks across the breast and broad white tips to long dark undertail coverts. The tail is broad and rounded.

Similar Species Unlike other East Asian warblers. Most similar to Savi's Warbler (*L. luscinioides*, unrecorded in N.A.), which breeds from Europe to far western China. Savi's is more warmly colored and plain breasted and lacks white undertail marks.
Voice CALL: A soft *churr*, also a sharp *tschick*, often repeated. **SONG:** A remarkable throbbing insectlike song.
Status & Distribution **BREEDING:** Breeds from southern Sweden south to northern Romania and east to western Siberia to the Irtysh River and to western Kazakhstan. **WINTER:** East Africa from southeast Kenya to north-

short, indistinct supercilium
broad rounded tail
white tips to long undertail coverts
diffuse streaks across breast
skulks

east South Africa. **RARE STATUS:** One record from Gambell, St. Lawrence I. on 7 Oct. 2017. Very rare to UK (mostly spring) and three records for Iceland.
Population Some declines in Europe.

OLD WORLD FLYCATCHERS Family Muscicapidae

Bluethroat, male (AK, June)

The core of this family is a suite of upright perching birds. Of the 17 species recorded from North America, records of most are from the western Aleutians and Bering Sea islands, AK. The Northern Wheatear and Bluethroat breed in arctic North America. **Structure** Old World Flycatchers have rather flattened bills with a slight hook at the tip and well-developed rictal bristles. Their tails are short. The wings in most of the migratory species are long, and the legs and feet are short and relatively weak. Many of the chats are ground-inhabiting, with long legs; most have expressive tails (varying from short to long), which can be cocked,

shivered, or flicked open and shut. All have 10 primaries.
Behavior Resemble tyrant flycatchers but are not closely related. Old World flycatchers sally out after insects and return, often to the same perch. Most species are arboreal. Some use exposed snags at the tops of trees for hunting perches, while others forage lower in the undergrowth or from the ground. Many species engage in wing and tail flicking; others remain still. They are solitary overall. Call notes and especially songs are infrequently heard in migration. Wheatears and other chats are ground- and rock-loving species; many are conspicuous in open habitats while others skulk within underbrush. The Asiatic forktails forage along mountain streams.
Plumage In some genera the plumage is subdued, and all post-juvenal plumages look alike, while in others (e.g., *Ficedula*) there is strong sexual and sometimes seasonal and age variation. Chats vary from plain to strikingly colored; many show conspicuous tail markings.
Distribution Widespread in the Old World, throughout Eurasia and Africa. Most European and northern Asian species are strongly migratory.
Taxonomy Molecular studies show that many of the chats and relatives formerly placed in the thrush family (Turdidae) are closer to Old World flycatchers; the Muscicapidae has thus swelled to over 300 species in 57 genera.
Conservation BirdLife International codes: 32 NT, 20 VU, 13 EN, and 1 CR.

Genus *Muscicapa*

GRAY-STREAKED FLYCATCHER *Muscicapa griseisticta* GSFL ■ 4

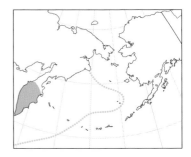

All *Muscicapa* (23 species) are dull, and their plumages, except juvenal and first fall, are alike. The Gray-streaked Flycatcher (formerly called Gray-spotted) is the most regularly occurring Old World flycatcher to N.A. (e.g.,

spots form distinct streaks on white underparts

spring

very long primary projection

1st fall

27 recorded from Attu I. on 2 June 1999). Monotypic. L 6" (15 cm)
Identification ADULT: Moderately large and very long winged with wing tips projecting nearly or actually to tail tip. Distinct stripes on sides of throat and well defined dark ventral streaks against a white background color; grayish brown above with somewhat ill-defined eye ring and darker streaks visible on forehead. Undertail coverts are always whitish and unmarked. **JUVENILE:** Palely spotted with dark bars above; most similar to adult by fall.
Similar Species Pattern of markings below distinctive from Dark-sided Flycatcher, though note that species is variable in this regard. Markings below are more distinctive than in Spotted Flycatcher and note unmarked white throat and much longer wings. Crown of Spotted more extensively streaked.
Voice CALL: A thin *seet* and a fairly loud *speet-teet-teet* have been described. **SONG:** Perhaps not described.
Status & Distribution Uncommon in

a variety of woodland habitats. **BREEDING:** Russian Far East east to Sakhalin and southern Kamchatka. **WINTER:** From Philippines and northern Borneo to western New Guinea. **MIGRANT:** Eastern China, rarely Vietnam. **RARE STATUS:** To western Aleutians and more recently small numbers to the Pribilof Is. Majority of records are for late spring (late May to mid-June), but also occurs in Sept.
Population Likely stable.

ASIAN BROWN FLYCATCHER *Muscicapa dauurica*   ABFL ■ 5

This casual stray to western AK is vaguely suggestive of a Least Flycatcher in appearance. Unlike Spotted and Dark-sided Flycatcher adults, Asian Browns have a complete post-breeding molt on the breeding grounds rather than after reaching the winter grounds. This bird tends to perch somewhat more within the canopy of woodlands than other flycatchers. Polytypic (3–4 sspp.; nominate in N.A.). L 5.3" (13 cm)

distinct whitish eye ring and supraloral

spring *dauurica*

unstreaked underparts

pinkish-based bill

1st fall *dauurica*

shorter primary projection than Dark-sided and Gray-streaked

Identification ADULT: Grayish brown above with distinct white eye ring and supraloral line. Whitish below with grayish wash on sides of breast and, more rarely, diffuse streaks. Primary projection is relatively short, with wing tips falling to only base of tail. Distinct and sharply separated pale base to lower mandible on bill. **JUVENILE:** Distinct white spots on upper parts, bolder white markings on wing, and some fine dark scaling on breast; most are similar to adult by fall.
Geographic Variation Complex. The one AK specimen belongs to the most migratory, northerly *dauurica*. Three to four other subspecies in Southeast Asia, one of which (*williamsoni*) is sometimes considered specifically distinct.
Similar Species This species is much less marked ventrally than the other *Muscicapa* flycatchers recorded from AK, especially Dark-sided or Gray-streaked. Also wing tips project far less down the tail in this species. Note, too, the distinctly pale-based lower mandible and distinct white eye ring and supraloral line. The other *Muscicapa* are darker billed, and only Dark-sided has an eye ring that approaches the

distinctness of Asian Brown.
Voice CALL: Includes a series of high, thin, sharp notes, and a short, thin, and high-pitched *tzi*. **SONG:** Similar to Dark-sided but shorter and less pleasant, consisting of short trills, squawky notes, and mixed with two- to three-note whistled phrases.
Status & Distribution Common and widespread in Asia. **BREEDING:** Nests in rather open mixed woodlands. Nominate subspecies breeds from southern Siberia and northern Mongolia to northeast China, Sakhalin, and southern Kuril Is. south to the Himalaya. **WINTER:** Found in a variety of open woodlands, including parks and gardens, from India, southern China, and throughout mainland Southeast Asia and the Greater and Lesser Sundaes. Other subspecies are less migratory and found in mainland Southeast Asia and northeast Borneo. **RARE STATUS:** Five AK records: a specimen from Attu I. (25 May 1985), Buldir I. (29 May 2005), St. Paul I. (6–9 Sept. 2013), and Gambell, St. Lawrence I. (9 June 1994 and 3 Sept. 2017).
Population Stable as far as is known.

SPOTTED FLYCATCHER *Muscicapa striata*   SPFL ■ 5

A familiar bird of Europe, this species has been recorded only once in N.A. It tends to perch at mid-levels in vegetation, usually on a prominent perch. Polytypic (7 sspp.). L 6" (15 cm)
Identification Often flicks its tail. **ADULT:** Grayish brown above including head and face, and indistinct whitish eye ring. Fine dark streaking on crown and forecrown. Below, lacks distinct dark malar and pale submustachial (faint dark markings are present) stripes on sides of throat, but does show faint blurry streaks against a pale buffy background on throat and across chest. Bill is mainly dark. Although appearing long winged, wing tips do not extend more than halfway down tail. **JUVENILE:** Has prominent buff spotting above and dark mottling below. Fall immature has more prominent edges to tertials and wing feathers and buff tips to inner greater coverts, which form an ill-defined wing bar.
Geographic Variation Moderate, but differences are largely clinal. Seven

subspecies are recognized, the closest to AK being the overall rather pale *neumanni*.
Similar Species Gray-streaked Flycatcher is much more prominently streaked below against a whiter background color, has streaking on crown restricted to forehead, and has wing tips that extend to about the tip of the tail. Asian Brown Flycatcher is cleaner underneath, has a more prominent whitish eye ring and supraloral line, and has a more extensive pale base to lower mandible. Dark-sided Flycatcher is overall darker; it has darker sides and flanks with inverted whitish stripe up the midsection, a more prominent whitish eye ring and supraloral line, and has longer primary tip projection.
Voice CALL: Includes a thin and squeaky *zeeee*, a sharp *chick*, and a doubled note. **SONG:** Weak, high-pitched, somewhat squeaky and disjointed, often with long intervals between phrases.
Status & Distribution BREEDING:

Nests in a variety of rather open woodland environments from western Europe north to Fennoscandia and south to northwest Africa, east to about Lake Baikal in eastern Siberia, and south to the Middle East and north and west Pakistan. **WINTER:** Africa south of the Sahara. **RARE STATUS:** A photo-documented record from Gambell, St. Lawrence I. (14 Sept. 2002), is the only record for N.A.
Population Declining in Europe.

fine streaking on crown

faint eye ring

indistinct streaks on breast

adult

DARK-SIDED FLYCATCHER *Muscicapa sibirica* DSFL ■ 4

This Asian species occurs more rarely to western AK than the Gray-streaked Flycatcher. In Asia it nearly always perches on conspicuous, exposed bare branches, often high in trees. When sallying out, usually returns to same perch. Polytypic (4 sspp.; nominate in N.A.). L 5.3" (13 cm)

Identification ADULT: Grayish brown above with dark head and rather conspicuous white eye ring; whitish below with brownish wash on sides and flanks and some diffuse to fairly

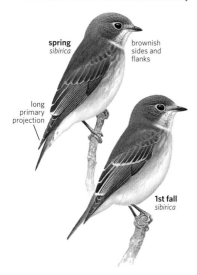

spring
sibirica

brownish
sides and
flanks

long
primary
projection

1st fall
sibirica

obvious streaks on center of breast; whitish half-collar and rather pale buffy brown supraloral spot; long primary projection that extends mostly or actually to tip of tail. **JUVENILE:** Darker above with pale buff spots/streaks. More streaking ventrally and with buff tips and edges to secondary coverts and tertials; most are similar to adult by fall.

Geographic Variation The AK specimens refer to the most migratory subspecies, nominate *sibirica*, which breeds in coniferous woodlands from central Siberia to Kamchatka, Japan, and northeast China and winters in Southeast Asia, including western Indonesia and the Philippines. Three other subspecies—all shorter-distance migrants—breed in the Himalayan region; some winter south to the Malay Peninsula.

Similar Species Like the slightly larger and bigger-billed Gray-streaked, Dark-sided is quite long winged, though usually is slightly shorter. It can be told from Gray-streaked by more diffuse breast streaking against a more brownish rather than white background color. Note also brownish cast to supraloral spot and white half-collar. Though diagnostic on Dark-sided, dark markings on undertail coverts are often

concealed. Asian Brown Flycatcher has more distinctly bicolored lower mandible, is much paler ventrally, has a bolder eye ring, and has much shorter wings.

Voice CALL: Very high, short and rapid metallic trills and a single high down-slurred note. **SONG:** A weak and subdued series of high thin notes followed by musical trills and whistles.

Status & Distribution BREEDING: Breeds mainly in Siberia and Russian Far East. **WINTER:** Southeast Asia. **RARE STATUS:** Casual to the western Aleutians (mainly late spring) and to the Pribilof Is., AK (late spring and fall). Accidental to Utqiagvik, AK, in spring. **Population** Stable as far as is known.

Genera *Erithacus, Larvilora, Cyanecula, Calliope,* and *Tarsiger.*

EUROPEAN ROBIN *Erithacus rubecula* EURO ■ 5

1st fall

orange-red
face and
breast

long, thin legs

The widespread European Robin is a beloved species and a familiar visitor to backyard gardens. It has occurred only once in N.A., in late winter from the Northeast. Polytypic (8 sspp.; N.A. record unknown). L 5.2" (13 cm)

Identification A small, upright perching species with long dark legs. It often droops its wings and cocks its short tail; remains motionless for long periods, then rapidly hops. Distinctive rusty red face, throat, and breast, bordered by gray; brownish on flanks; and brownish above, with a yellow buff wing bar. **JUVENILE:** Briefly held, spot-

ted over much of plumage and lacks red on face and underparts.

Similar Species Unmistakable.

Voice CALL: A *tic*, often repeated in a series. **SONG:** An elaborate series of short ethereal verses, with no two the same.

Status & Distribution Partial to open woodlands with undergrowth. **BREEDING:** Found in Europe, North Africa, some Atlantic islands, and east to western Siberia and Caucuses; eastern populations migratory. **WINTER:** To Middle East. **RARE STATUS:** One N.A. record from North Wales, Bucks

Co., PA (21 Feb.–7 Mar. 2015). Casual to north India and Japan and over 1,000 records for Iceland.

Population Stable and increasing.

SIBERIAN BLUE ROBIN *Larvilora cyane* SBRO ■ 5

The habitual, rapid tail-quivering behavior of this terrestrial species is distinctive. Polytypic (2 sspp.; likely *bochaiensis* in N.A.). L 5.5" (14 cm)

Identification ADULT MALE: Deep blue above; white underparts. **FEMALE:** Brownish olive above, buffy white throat, and buff breast and flanks with

darker scaling on throat and sides; some have blue on tail base. **IMMATURE:** Immature males resemble adult females, but have dull blue uppertail

coverts and usually have blue in the scapulars and some wing coverts. Immature females lack blue in tail.
Similar Species Female's olive-brown upperparts may suggest a small *Catharus* thrush, but note plainer underparts. Also see Red-flanked Bluetail.
Voice CALL: A hard *tuck* or *dak*. **FLIGHT NOTE:** Unknown. **SONG:** Where breeding, loud, rapid, explosive *try try try* and *tjuree-tiu-tiu-tiu-tiu*, usually

introduced by fine, spaced *sit* notes.
Status & Distribution BREEDING: Coniferous forests from central Asia east through Kamchatka. **WINTER:** Southern China through Southeast Asia to Borneo. **RARE STATUS:** Accidental on Attu I. AK (21 May 1985), and Gambell, St. Lawrence I. (2–5 Oct. 2012). One disputed report from YT. Also accidental to Europe.
Population Declining.

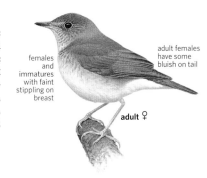

females and immatures with faint stippling on breast

adult females have some bluish on tail

adult ♀

RUFOUS-TAILED ROBIN *Larvilora sibilans*   RTRO ■ 5

To an observer in N.A., the Rufous-tailed Robin recalls a miniature Hermit Thrush. It is generally a secretive species, skulking near the ground in wet forested gullies in Asia. Also known as Swinhoe's Robin or Swinhoe's Pseudo-robin, the species was first described by 19th-century English naturalist Robert Swinhoe, who described many other Asian vertebrates. Monotypic. L 5.3" (13 cm)
Identification Male and female plumages similar. A small but rather plump thrushlike bird, typically found on the ground; pale pinkish legs appear disproportionately long for size. Has a habit of "shivering" rear of body and tail. Plain brown upperparts contrast with rufous uppertail coverts and tail and rusty wings. Underparts dingy cream color, brownish olive on flanks, with dusky scaling or scalloping

mostly on throat, breast, and flanks. Scaling pattern fainter in immature or worn plumages. Buff eye ring and supraloral area; white of throat inserts a bit below dusky auricular.
Similar Species Other species in genera *Larvilora* and *Cyanecula* should be considered. Female and young Siberian Blue Robins are less richly colored above, lack rufous tail, and often show bluish rump. Hermit Thrush has similar plumage but is much larger and has disproportionately shorter legs, with heavily spotted rather than scaly underparts; flicks, not shivers, tail.
Voice SONG: A loud, vibrato series of varied downward trills, given in phrases. **CALL:** A subtle *tuk-tuk*.
Status & Distribution BREEDING: In well-wooded forest with heavy undergrowth, often near streams, mostly in

the taiga zone, from Yenisey east to the Kamchatka Peninsula, south to Sakhalin I. and northeastern China. **WINTER:** From southeastern China through Indochina; a few west to Thailand. **RARE STATUS:** Six N.A. records, all from western AK and in spring (June) and fall (Sept.) from Attu I., St. Paul I., and St. Lawrence I.
Population Stable. Uncommon to fairly common within its range.

rufous tail

faint scaling on breast

BLUETHROAT *Cyanecula svecica*   BLUE ■ 2

This species is a skulker, though singing males at the start of the breeding season are conspicuous. Polytypic (11 sspp.; nominate in N.A.). L 5.5" (14 cm)
Identification Runs on ground, usually with its tail cocked. **MALE:** Distinctive blue throat with red spot, in winter reduced. **FEMALE:** Brown

above; distinct whitish supercilium bordered by blackish lateral crown stripe; white malar stripes; black lateral throat stripes; and white throat. Black necklace crosses otherwise whitish underparts. **JUVENILE:** Briefly held, pale spots above and dark bars below. **FLIGHT:** In all plumages, rufous

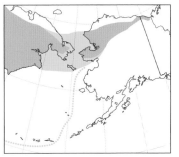

patches at base of tail are distinctive and easily seen in flight.
Similar Species Similar to Siberian Rubythroat in structure and in skulking, terrestrial habits.
Voice CALL: Common call a dry *tchak*; also a whistled *hiit* and a hoarse *bzru*.
SONG: Often given in circling flight with tail spread. Strong, clear, and

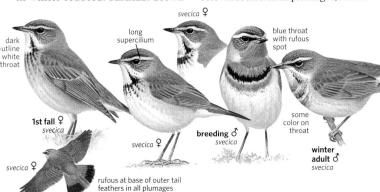

dark outline white throat

1st fall ♀ *svecica*

svecica ♀

long supercilium

svecica ♀

svecica ♀

breeding ♂ *svecica*

blue throat with rufous spot

some color on throat

winter adult ♂ *svecica*

rufous at base of outer tail feathers in all plumages conspicuous in flight

varied, usually beginning with metallic *ting* notes, speeding up and becoming a jumble of notes. Habitually includes imitations of other birds.
Status & Distribution Fairly common breeder from western Europe east through Russian Far East to north-ern AK and northern YT. **BREEDING:** Nests in willow riparian in US and Canada. **MIGRATION:** Medium- to long-distance migrant. Spring: Arrives in western AK ±25 May. In fall, most depart AK in Aug., but small numbers are still moving in Bering Strait area through early Sept. **WINTER:** Primarily from northeastern Africa east through southeastern Asia. **RARE STATUS:** Casual to Pribilofs and western Aleutian Is. Accidental in fall to San Clemente I., CA (14–18 Sept. 2008). **Population** Stable.

SIBERIAN RUBYTHROAT *Calliope calliope*   SIRU ■ 3

The male Siberian Rubythroat has an iridescent ruby throat. This Asian species is often quite difficult to see well, due to its skulking habits and its preference for dense vegetation. Monotypic. L 6" (15 cm)
Identification MALE: Crown and upperparts brownish. White supercilium; white malar stripe bordered by black; bright ruby throat with a black-spotted lower border. Gray breast; lower sides and flanks are washed cinnamon, and belly and vent area are white. **FEMALE:** Similar in pattern, but browner than male and lacks black on face, with streaked auriculars and a pale throat, some with varying amounts of red. Immature females have no color on throat.

Similar Species Smaller Bluethroat is distinguished by more complex head pattern, a distinct dark necklace, and a rufous tail base.
Voice CALL: A hard *chak*; a muffled, creaking *arrr*; and a whistling *ee-lyu*. **SONG:** A variable chatty and calm warbling; often mimics other bird species.
Status & Distribution Uncommon. **BREEDING:** In dense thickets with or without overstory, from Kamchatka west to the Ural Mts.; winters in southeast Asia. **RARE STATUS:** In western Aleutians, rare in spring; rare to casual in fall. Very rare in spring and fall elsewhere in western AK (nearly annual on Pribilof Is.). Accidental in winter to ON. Casual to Europe.
Population Stable.

ruby red throat bordered by black and white stripes

pink-throated adult ♀

shorter, fainter supercilium than Bluethroat

uniform-colored tail

adult ♂

1st fall ♂

♀

RED-FLANKED BLUETAIL *Tarsiger cyanurus*   RFBL ■ 4

The Red-flanked Bluetail, a primarily Asian species, has occurred in western AK and in a few western states. Some split the two more southerly subspecies as a separate species, Himalayan Bluetail (*L. rufilata*), the adult male of which has a pale blue forehead and supraloral stripe. Polytypic (3 sspp.; nominate in N.A.). L 5.5" (14 cm)
Identification Flicks tail down. Adult male (only a few records in N.A.) is distinc-tive. Female brownish above; faint supraloral stripe; distinct thin white eye ring. White chin and throat contrast strongly with brownish chest. Distinct orangish sides and flanks. Males from their first year not separable from females.
Similar Species Siberian Blue Robin somewhat similar, but lacks orange sides and flanks and white supraloral stripes. Females have brown tails.
Voice CALL: A whistled short *hweet* or *veet*, often repeated, and a hard, throaty *keck* or *trak*, usually doubled. **FLIGHT NOTE:** Unknown. **SONG:** Fast,

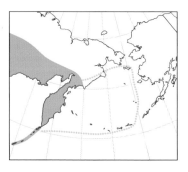

blue tail often flicked down

narrow whitish supraloral stripe

white throat

orange sides and flanks

adult ♂
cyanurus

cyanurus ♀

clear, melancholy verse, *itru-churr-tre-tre-tru-truur*.

Status & Distribution BREEDING: Nominate subspecies in taiga and other coniferous forests in northern Asia, west through northwestern Russia to Finland. WINTER: Korea and Japan south to northern Southeast Asia. RARE STATUS: Casual spring and fall to western Aleutians, Pribilofs, and St. Lawrence I. (fall); also casual or accidental in fall/winter to BC, OR, ID, WY, and CA. Rare in Europe west of Finland.
Population Stable.

Genus *Ficedula*

This is a large genus of about (taxonomy dependent) 31 species, three of which have been recorded in N.A. There is marked sexual, age, and in some cases seasonal dimorphism. Species in *Ficedula* often perch with their wings drooped. They nest in holes or cavities in trees or in nest boxes.

NARCISSUS FLYCATCHER   *Ficedula narcissina*   NAFL ■ 5

Adult male Narcissus Flycatchers are stunning. Females and immatures are more subdued. The species has occurred only twice in N.A., both second-year males on Attu I., AK. Polytypic (3 sspp.; nominate in N.A.). L 5.3" (13 cm)
Identification Nominate described. ADULT MALE: Unmistakable with black head and upperparts, except for orange supercilium and rump and white wing patch. Ventrally extensively orange. FEMALE: Brownish above with olive tinge on lower back and rump; whitish below with some mottling on breast and a wash on flanks; sometimes pale yellow wash on belly. IMMATURE MALE: Resembles female through winter but much more like adult males by spring, except for browner wings and patches of brown elsewhere on upperparts.
Geographic Variation Nominate *narcissina* described above. Subspecies *elisae* ("Green-backed Flycatcher") breeds in northeast China and winters primarily on the Malay Peninsula. Adult males have yellow supercilium and are green above; females duller and lack wing and rump patch, are olive above with a contrasting rufous rump and tail and yellowish below. Another green-backed subspecies, *owstoni* ("Ryukyu Flycatcher"), is resident on the southern Ryukyu Is., Japan. Both have distinctly different songs and are now usually split as separate species.
Similar Species Yellow-rumped Flycatcher (*F. zanthopygia*) is a possible stray to western AK; adult male resembles Narcissus but has an extended white spur from wing patch and white supercilium. Duller female has wing patch, spur, and a yellow rump.
Voice CALL: *Tink-tink*. SONG: A warble with repeated three-syllable notes.

Status & Distribution BREEDING: Nominate *narcissina* breeds south coastal Russian Far East, including Sakhalin, the southern Kuril Is., Korea, and Japan in deciduous, mixed, or coniferous woods with dense undergrowth. WINTER: Nominate in forests of the Philippines, sparingly to Borneo. RARE STATUS: Accidental, two specimen records, both from Attu I., western Aleutians (20–21 May 1989 and 21 May 1994).
Population Nominate *narcissina* stable. Subspecies *owstoni* has suffered severe declines; *elisae* no doubt vulnerable, but little data.

all figures *narcissina*

spring ♀

1st spring ♂

♀

brownish olive above

unmistakable black and yellow-orange with white wing patch

adult ♂

reddish-fringed tail

MUGIMAKI FLYCATCHER   *Ficedula mugimaki*   MUFL ■ 5

A long-winged flycatcher that feeds from mid- to upper levels in the canopy of forests in eastern Asia; recorded only once in N.A. Monotypic. L 5.3" (13 cm)
Identification ADULT MALE: Blackish head and upperparts with bold, but short supercilium, white wing patch. Below, extensively orange from chin to belly. FEMALE: Grayish brown above, orangish on throat and breast, two thin wing bars. IMMATURE MALE: Overall closer to female; usually shows some indication of a supercilium.
Voice CALL: A rattled harsh *trrrt*.

SONG: A fast, musical warble.
Similar Species No confusion with species in N.A. or northern Asia.
Status & Distribution Overall uncommon. BREEDING: From central Siberia to North Korea, northeastern China, and Russian Far East, including Sakhalin, in mature mixed forest. WINTER: In forests of Southeast Asia, Philippines, and east to Sulawasi. RARE STATUS: Accidental. A record from Shemya I., western Aleutians (24 May 1985), is supported by marginal photos (accepted by AOS and ABA but not by AK Checklist Committee).
Population Stable as far as is known.

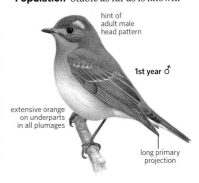

hint of adult male head pattern

1st year ♂

extensive orange on underparts in all plumages

long primary projection

TAIGA FLYCATCHER   *Ficedula albicilla*   TAFL ▪ 4

This Asian Flycatcher, also known as Red-throated Flycatcher, has a distinctive black tail, which is often flicked rapidly upward. It forages mostly from low to mid levels; often flicks tail up. Monotypic. L 5.3" (13 cm)

Identification SUMMER MALE: Distinctive red throat; brownish above, face and chest ash-grayish; whitish eye ring. Black uppertail coverts and tail with white lateral patches, not usually visible on folded tail; best noted in flight. **FEMALE & WINTER MALE:** Both lack red throat and face and have paler chests. Fall immatures show faint pale wing bars and tertial edges. The red throat of the breeding male is acquired as early as Feb.

Similar Species All plumages of Taiga have a black tail with white patches, and the red throat on the breeding adult male is distinctive. Taiga Flycatcher was formerly treated as a subspecies of Red-breasted Flycatcher (*F. parva*). On breeding male Red-breasted, the red throat extends to the chest. Female and immature male Red-breasteds are similar to Taiga, but in all Taiga the longest upper-tail coverts are black (brown in Red-breasted), breast is grayer, and bill darker. Red-breasted breeds farther west in central and eastern Europe and to southwest Siberia and Iran; it winters to Pakistan and northern India.

Voice CALL: Given year-round, a brief and fast series of dry rattled notes, *trrrrr*; a high-pitched *tzee.* SONG: An Old World bunting–like song.

Status & Distribution Common in Asia in open woodland. BREEDING: From eastern European Russia east to Mongolia and Kamchatka. WINTER: From eastern India and through mainland Southeast Asia. RARE STATUS: Very rare or casual to western AK (mainly spring); over 20 records from western Aleutians; multiple records from Pribilof Is. and St. Lawrence I. Accidental to northern CA in Solano Co. and Yolo Co. (25 Oct. 2006). Casual to northwest Europe.

Population Stable.

reddish throat

spring ♀

grayish wash on breast

breeding adult ♂

♀

1st fall

black uppertail coverts and tail with white oval patches, best viewed in flight, in all plumages

---

Genus *Phoenicurus*

COMMON REDSTART   *Phoenicurus phoenicurus*   CRET ▪ 5

On this familiar European species, the reddish tail is best observed in flight. Polytypic (2 sspp.; likely nominate in N.A.). L 5.5" (14 cm)

Identification A slim, thrushlike bird with alert demeanor and upright stance; frequently quivers tail, especially just after perching. ADULT MALE: In breeding plumage has soft gray upperparts; black face and throat set off from gray by white supraloral; breast and flanks a rich chestnut orange; tail rich rufous orange except for brownish gray central rectrices. Adult males in fall, in fresh plumage, have more muted coloration. FEMALE: Tail like male, but lacks head pattern and vivid colors; breast buffy, upperparts mouse brown. IMMATURE: Like female, but young males begin to show traces of adult colors and patterns.

Geographic Variation The range of nominate *phoenicurus* extends from Europe to central Siberia and northern Mongolia, while subspecies *samamisicus* with hint of a pale wing panel breeds from the southern Balkans and Greece eastward to Iran, Turkmenistan, and Uzbekistan.

Similar Species Any *Phoenicurus* redstart out of range, such as in N.A., should be studied carefully to rule out congeners such as Daurian (*P. aureorus*), which has white patch at base of secondaries and tertials, and the eastern subspecies of Black (*P. ochruros rufiventris*).

Voice CALL: A sweet, up-slurred, whistled *wheet* in nominate subspecies; less up-slurred in *samamisicus.*

Status & Distribution BREEDING: From Europe (and sparingly northwestern Africa) eastward to Siberia, south to Mongolia and Central Asia.

WINTER: Nominate subspecies winters in Africa, mostly south of the Sahara, while *samamisicus* winters in northeastern Africa and the Arabian Peninsula. **RARE STATUS:** The only N.A. record is of a first-fall male at St. Paul I., AK (8–10 Sept. 2013).

Population Stable.

diffuse supercilium

immature ♂
phoenicurus

rusty orange breast and flanks

rusty orange tail with dark central tail feathers

STONECHATS Genus *Saxicola*

STONECHAT *Saxicola torquatus* STON ■ 4

All N.A. records of Stonechat are attributable to the *maurus* or *stejnegeri* groups (specimens of latter) of subspecies, split by some as Eastern Stonechat. Recent molecular and vocal evidence suggests *stejnegeri* should be split too. Polytypic (24 sspp.; includes "African," "Asian," "East Asian," and "Canaries" groups, all sometimes split as a separate species). L 5.3" (13 cm)
Identification Small size, short tail, thin bill, small white patch on inner wing, contrastingly pale rump, and upright posture are distinctive. **MALE:** Black head, back, wings, and tail; wide white ear surround; white belly, sides, and rump; and orange wash on chest. **FEMALE:** Pale brown head with indistinct paler supercilium, buffy orange wash on breast. **IMMATURE:** Similar to adult female, but male has traces of black in face.
Voice CALL: Shrill, sharp whistle; a throaty, clicking *vist trak-trak*.

Status & Distribution Common. **BREEDING:** In Old World. Eastern Stonechat breeds in open scrubland in Asia and northeasternmost Europe. **WINTER:** From Japan south to southeastern Asia and west to northeast Africa. **RARE STATUS:** Casual to various AK locations (mostly at Gambell), but oddly not the Aleutians; one fall record on Grand Manan I., NB, and another on San Clemente I., CA. **Population** Stable.

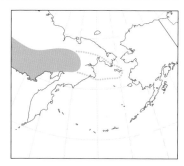

black face and throat, white nape

spring ♂
maurus group

faint supercilium

fall adult ♂
maurus group

orangish chest

whitish rump

small whitish wing patch

spring ♀
maurus group

1st fall
maurus group

WHEATEARS Genus *Oenanthe*

NORTHERN WHEATEAR *Oenanthe oenanthe* NOWH ■ 2

Northern Wheatears breeding in N.A. are among the longest-distance migrant passerines in the world. Polytypic (4 sspp.; 2 in N.A.). L 5.8" (15 cm)
Identification Male has upperparts gray with white rump; white underparts with buff tinge to chest. Wide black area across face. Wings blackish; tail pattern—black with large white basal corners—unique in N.A. Female similar in pattern, but less contrasty and with strong buffy brown cast.

Adult male in fall and winter similar to females. **JUVENILE:** Distinctive; upperparts with buff spotting.
Geographic Variation Nominate *oenanthe* is widespread breeder in Eurasia and breeds in AK and northwesternmost Canada; *leucorhoa* breeds in Greenland, NU, and northern Newfoundland. Both subspecies winter in sub-Saharan Africa. Males of *leucorhoa* similar, but a bit larger with richer buff on chest extending down

oenanthe

leucorhoa

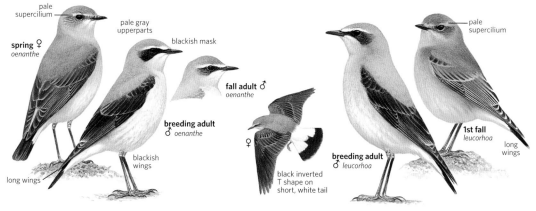

pale supercilium

pale gray upperparts

spring ♀
oenanthe

blackish mask

fall adult ♂
oenanthe

breeding adult
♂ *oenanthe*

♀

blackish wings

long wings

black inverted T shape on short, white tail

breeding adult
♂ *leucorhoa*

pale supercilium

1st fall
leucorhoa

long wings

sides to flanks; females on average, more richly colored, browner.
Similar Species Other wheatears are similar, particularly Isabelline Wheatear (*O. isabellina*), which is fairly widespread in Asia and casual east to Korea and Japan.
Voice CALL: A tongue-clicking *chack* and a short whistle, *wheet*. **SONG:** Scratchy warbling mixed with call notes and imitations of other birds; often given in flight.
Status & Distribution Common breeder across high-latitude Northern Hemisphere (absent north-central Canada), wintering in sub-Saharan Africa. **RARE STATUS:** Very rare along East Coast, primarily in fall; casual to accidental elsewhere south of N.A. breeding range; one wintered in LA and another in south TX. Determination to subspecies in fall is problematic; those in East likely *leucorhoa*.
Population Declining in Europe.

PIED WHEATEAR   *Oenanthe pleschanka*   PIWH ▪ 5

This European and west Asian species has occurred just once in N.A. Now monotypic with split of Cyprus Wheatear (*O. cypriaca*). L 5.5–6.3" (14–16 cm)
Identification All plumages with extensive white on rump and tail. **ADULT MALE:** Extensive black throat (some—morph "*vittata*"—have white throats), face, and upperparts. Black is heavily obscured in fall and winter; black limited in immature. **FEMALE:** Highly variable. Most are cold gray overall, many with diffuse breast streaks. Some darker birds (illustrated) are much darker, almost male-like.
Similar Species Females and immatures separated from Northern Wheatear, as well as from the more westerly Old World Eastern Black-eared Wheatear (*O. hispanica*), by darker and colder coloration.
Voice CALL: A scratchy *brsche* and clicking *tshak* notes.
Status & Distribution Prefers barren stony area on flatlands or in hills, cliffs, even settlements. Barren fields in migration and winter. **BREEDING:** From eastern Bulgaria and eastern Turkey east discontinuously to southern Urals, Transbaikalia, north China, and northwest Himalaya. **WINTER:** East Africa and to Yemen; some in Egypt. **RARE STATUS:** One N.A. record, a dark second-year female at Cape

females variable, often much paler than illustrated

2nd-year female ♂

Nome, western AK (4 July–4 Aug. 2017). Also casual to Korea and Japan, and rare to northwest Europe.
Population Stable.

THRUSHES Family Turdidae

Hermit Thrush (OH, Oct.)

The thrush family houses some very distinctive species but also includes some tough identification challenges. Chief among the latter group are the brown thrushes in the genus *Catharus*. Relatively close views are required to identify the thrushes, as is experience in variation, both individual and regional/subspecific, in plumages of the various species.
Structure Thrushes are generally plumper and more compact than other similar-size birds, but generally have quite long wings; in fact, thrushes average among the longest-winged passerines relative to body size. There is relatively little variation in bill shape among the family, with all North American species possessing short to medium-length and thin bills. Many species are terrestrial and thus have relatively long legs.
Plumage Although most thrushes are monomorphic in plumage (males and females appear similar), the family includes nearly the whole range of plumage variation from monomorphic to dimorphic. Some *Turdus* species exhibit sexual dimorphism, although in the American Robin this is confounded by individual and subspecific variation. The bluebirds are fairly dimorphic, with distinct male and female plumages. The forest-dwelling *Catharus* thrushes exhibit subtle to not-so-subtle subspecific differences in plumage coloration. All North American thrushes undergo a single annual molt after breeding; pre-breeding molts are absent or, at best, limited. The family also is characterized by the typically spotted plumage of juveniles, a trait that few other families match. Upon completing their post-juvenal molt, young birds achieve an essentially adult plumage.
Behavior Flight style is strong and fairly uniform across the family, with quick, flicking wingbeats followed by very short glides with the wings held closed. This style produces a fairly level flight, unlike the flight of most other passerines, which tends to undulate. Some *Turdus* (e.g., American Robin) forage in a very ploverlike manner, with short runs and stops when they search the ground for invertebrate prey. American Robins are purported to be able to hear earthworms (a staple for the species) under the surface of the ground, which may explain their common behavior of cocking their

heads sideways—to better listen (or watch) for prey.

Distribution The thrushes make up one of the most widespread bird families, with the largest radiations in the Old World. In fact, seven of the 27 species dealt with here are primarily birds of the Palearctic; another eight are Neotropical or West Indian in origin. Half of the North American thrushes have wide distributions across the continent: Four are primarily western species and two are eastern specialties. All species of thrush breeding in North America are at least partly migratory. Most North American species are medium-distance migrants; four species of brown thrush regularly occur south of Mexico, and two of these winter solely in South America.

Taxonomy The Old World chats (seven recorded from North America) were moved recently to the Old World Flycatcher family (Muscicapidae, see p. 540) as a result of molecular genetic studies. Some 159 species in 20 genera remain in the Turdidae.

Conservation Of the North American breeding species, Wood Thrush is threatened by habitat loss, forest fragmentation, and higher rates of nest predation and Brown-headed Cowbird parasitism; Bicknell's Thrush is threatened by climate change habitat loss on the winter grounds. Of the four HI thrushes/solitaires in the genus *Myadestes,* one is Critically Endangered (Puaiohi), one is Possibly Extinct (Olomao), and two are Extinct (Kamao and Amaui). BirdLife International codes: 28 NT, 11 VU, 2 EN, 3 CR, 1 CR (PE), and 4 EX.

BLUEBIRDS Genus *Sialia*

This is a strictly New World genus composed of three species. All are frugivorous (plucking small fruits from perches) and insectivorous (sallying to the ground for prey). The Mountain Bluebird frequently hovers over open areas while searching for food. All three species have hybridized with each other; regularly between Eastern and Mountain Bluebirds.

EASTERN BLUEBIRD *Sialia sialis*   EABL ▪ 1

sialis ♀

white belly and undertail coverts

The Eastern Bluebird is familiar to millions in eastern N.A. Polytypic (8 sspp.; 3 in N.A.). L 7" (18 cm)

Identification MALE: Bright blue above, with orange throat, ear surround, chest, sides, and flanks. **FEMALE:** Upperparts are less blue (often grayish); has partial whitish eye ring, and whitish throat bordered by brown lateral throat stripes. **JUVENILE:** Very similar to Western Bluebird, but with tertials fringed cinnamon; immatures discernable with duller upperparts and browner primary coverts. Flight as in other bluebirds; pale wing stripe less obvious than in Western Bluebird.

Geographic Variation Three subspecies in N.A.; compared with widespread and somewhat migratory, short-billed nominate *sialis,* resident *fulva* of southeastern AZ and likely southwestern NM (rare) slightly larger and paler; and southern FL *grata* with longer bill.

Similar Species Male Western Bluebird with all-blue head, including chin and throat, and is generally a deeper blue with gray vent and undertail coverts. Male Western usually has some rufous on upperparts, particularly on scapulars. (See female bluebird sidebar, p. 550.)

Voice CALL: Musical, typically two-noted *too-lee.* This call is also given in flight. **SONG:** Mellow series of warbled phrases; varied.

Status & Distribution Common in the eastern three-fifths of the lower 48 states and in southern Canada; uncommon and local in southeastern AZ. **BREEDING:** Nests in open woodland, second-growth habitats, and along the edges of fields and pastures, placing nest in cavity; readily accepts nest boxes. **MIGRATION:** Short- to medium-distance migrant. Spring: arrives Great Lakes ±25 Feb; southern SK ±1 Apr. Departs northernmost range during Oct. Usually migrates in flocks. **WINTER:** Almost always in flocks, often mixed with Yellow-rumped, Pine, and Palm Warblers and/or Dark-eyed Juncos; mainly central and southern US, but to south-central CO, central NM, and northeastern Mexico. **RARE STATUS:** Casual west to AB, northeast BC, and UT; accidental to ID and northwest OR.

Population Nest boxes have apparently helped reverse a decline.

juveniles of all bluebirds are spotted

rufous in both sexes wraps around sides of neck and includes throat

juvenile *sialis*

sialis ♂

overall paler than nominate *sialis*

southwestern ♂ *fulva*

sialis

grata

fulva

nidificans

guatemalae

meridionalis

WESTERN BLUEBIRD   *Sialia mexicana*   WEBL   ■   1

This species typically prefers more wooded breeding habitats than does the Mountain Bluebird, though co-occurs widely with it. Polytypic (6 sspp.; 3 in N.A.). L 7" (18 cm) **Identification** Shorter-winged and shorter-tailed than Mountain Bluebird. Undertail coverts grayish. **MALE:** Dark blue on head (including chin and throat), wings, and tail; variably rusty on scapulars and back. **FEMALE:** Grayer head and back; paler orange underneath; at least partial whitish eye ring. **JUVENILE:** Overall heavily spotted like other juvenile bluebirds. **FLIGHT:** Level. Migrates diurnally in

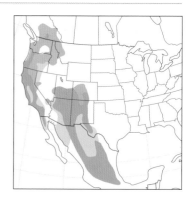

flocks, occasionally with Mountain Bluebirds, from which they can be distinguished by more obvious pale underwing stripe (created by darker wing linings and flight feathers) and, with experience, by shorter, rounder wings.

Geographic Variation The widespread subspecies *bairdi* (breeds west to central UT, southern AZ) is larger and with more extensive chestnut on upperparts than more westerly *occidentalis*; *jacoti* of southeastern NM and trans-Pecos TX is smaller, with extensive dark chestnut on upperparts.

Similar Species Unlike plumage of Eastern, male's head entirely blue; females similar to Eastern and Mountain Bluebirds. (See sidebar below.) **Voice CALL:** Similar to other bluebirds, but a harder *few*, though some calls are fairly strongly two-noted; this call also given in flight. **SONG:** Consists of a series of call notes and is primarily heard at dawn.

Status & Distribution Uncommon to fairly common at mid- and low elevations in western lower 48, extending north into Canada in BC. **BREEDING:** Nests in parklike, low-elevation pine and mixed forests, riparian bottom-

gray sides to neck

♀

all-blue head

some dark rufous scapulars

♂

gray undertail coverts

lands, and oak savanna. **MIGRATION:** Short-distance migrant that rarely strays far from nesting areas. Spring arrival central CO ±15 Mar.; eastern WA ±10 Apr. Fall: Most depart southern BC ±31 Oct.; northern CO ±30 Sept. **WINTER:** Primarily in southwestern US and northern Mexico, but range to southern WA on Pacific slope; northern extent of wintering variable, dependent on food (juniper or mistletoe berries). **RARE STATUS:** Casual east to ND, western KS, and eastern TX. **Population** Increasing in urban parks and suburban areas in some regions (e.g., CA).

Identifying Female Bluebirds

The Mountain Bluebird has a distinctive structure, being longer-winged, longer-tailed, longer-legged, and slimmer-billed than the other two bluebirds; and its penchant for hovering while foraging is another good clue. The female Mountain Bluebird's white eye ring is obvious on its otherwise bland face; both Eastern and Western Bluebirds exhibit partial whitish eye rings (behind the eye). Of the three, Mountain Bluebird appears most frequently well out of range. Eastern and Western Bluebirds have darkish lateral throat stripes, with the Eastern's being darker than those on Western (which can be quite vague) and

contrasting more with Eastern's white throat; Western's throat is gray. Eastern Bluebird has a distinctive orange ear surround.

Mountain Bluebird typically has gray underparts; however, it is more warmly colored on breast when fresh in fall. Its lower flanks are a cold gray-brown and contrast sharply with the white undertail coverts. Western Bluebird is usually orange on the chest, sides, and flanks. Eastern Bluebird is orange on the chest, sides, and flanks, typically brighter than Western. The belly and vent areas on both Mountain and Eastern Bluebirds are white, whereas on Western they are gray.

Eastern Bluebird, female (OH, Mar.)

Western Bluebird, female (CA, Apr.)

Mountain Bluebird, female (NM, Oct.)

MOUNTAIN BLUEBIRD *Sialia currucoides* MOBL ■ 1

The Mountain Bluebird is often seen hovering above open country in a wide variety of western habitats. Monotypic. L 7.3" (18.5 cm)
Identification Long, pointed wings. The very long primary projection of Mountain Bluebird sets this species apart in all plumages. Note thin bill and white undertail coverts. **ADULT:** Male's almost entirely sky-blue plumage (tipped with some brownish in fall when fresh) is unique. Females are extensively gray underneath, though some, in fresh plumage, have orangish cast to chest and upper sides. **JUVENILE:** Variably spotted white above, though nearly lacking in some, and vaguely spotted dark below. **FLIGHT:** Level and typically thrush-like. A flocking diurnal migrant, female has

translucent flight feathers.
SIMILAR SPECIES Females, particularly orange-chested individuals, can be mistaken for other bluebird species (see sidebar opposite), and grayer females for Townsend's Solitaire (see that species account).
Voice CALL: A somewhat soft bluebird whistle, less two-noted than that of Eastern and softer, less single-noted than that of Western. This call is also commonly given in flight. **SONG:** A series of notes similar to that of the call note, but more variable.
Status & Distribution Common and widespread in western N.A. **BREEDING:** Nests in huge elevational and latitudinal range (grasslands through forests to alpine tundra), requiring nesting cavities or niches in proximity to open country for foraging; readily takes to bluebird boxes. **MIGRATION:** Short- to long-distance migrant. Arrivals or migrants apparent in spring starting ±15 Feb. in southerly areas, northern extreme of breeding range occupied ±1 May. Fall: departs northernmost breeding areas ±30 Sept.; more southerly areas in Oct., though quite variable. **WINTER:**

thin bill
♂
♀
gray flanks have brownish cast when fresh
overall sky blue coloration
long primary projection

Mainly lower 48 states and northern Mexico, but extent and local occurrence of wintering highly variable and dependent on food resources (e.g., in CO, particularly juniper berries). Regularly found in flocks in agricultural fields, where hovering foraging behavior put to good use. **RARE STATUS:** Rare in spring to coastal Northwest. Casual to western and northern AK and to eastern N.A. in fall and winter; found alone and in flocks of Eastern Bluebirds.
Population Populations of Mountain Bluebird have increased concomitant with a great increase in the provision of nest boxes in the species' range.

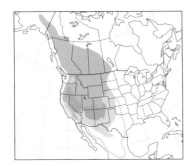

SOLITAIRES Genus *Myadestes*

TOWNSEND'S SOLITAIRE *Myadestes townsendi* TOSO ■ 1

The voice of the Townsend's Solitaire is often the first indication of its presence; it is a characteristic sound of western coniferous forests. Polytypic (2 sspp.; nominate in N.A.). L 8.5" (21.5 cm)
Identification A long and slender thrush, Townsend's Solitaire often perches conspicuously on the highest available perch. In some wintering situations, individuals hold territory with an extensive supply of small fruits (e.g., juniper berries) defended by call note and, rare in N.A. birds, song. **ADULT:** Monomorphic. Entire head and body medium gray, though underparts can be slightly paler. Thin, but crisp, white eye ring; short, black bill; and short, black legs are distinctive. Central third of long tail is bordered by black and with outermost rectrix on each side mostly white. Wings are distinctive, being primarily gray with whitish fringes to dark tertials and

greater coverts with striking buffy orange patches (which are duller in some individuals). **JUVENILE:** Juveniles are heavily spotted throughout body and wing coverts, with a less obvious wing pattern; sometimes seen in migration in this plumage. Older immatures are virtually identical to adults, except that occasional birds retain a few juvenal wing coverts, as is typical in thrushes; at least through early spring, most should have more worn flight feathers. **FLIGHT:** It is really in flight that Townsend's Solitaire is most distinctive, with floppier flight than is typical of thrushes, and a bold, buffy orange wing stripe, more obvious from below than from above. Tail is often flicked open and closed in flight.
Similar Species Sometimes confused with duller female Mountain Bluebird, but they exhibit blue at least in their tail and wings. They have shorter tails

and relatively longer wings and lack the solitaire's bold wing pattern. That wing pattern is surprisingly similar to that of Varied Thrush, but nothing else about the species is. Brown-backed Solitaire (next species account), a Mexican species of accidental occurrence in southeastern AZ, has a brown back

and wings, a white malar stripe, and a long, absolutely distinctive song.
Voice CALL: A single *thnn*, similar to toot of a pygmy-owl, usually repeated at intervals. **SONG:** A finch-like warble, at times quite long and disjointed, with occasional call notes (including those of Cassin's Finch) interspersed and with no distinct pattern.

Status & Distribution Common. **BREEDING:** Nests in a wide variety of open conifer-dominated habitats, often with rock outcrops and/or cliffs on which the birds place their nests. **MIGRATION:** Short- to long-distance migrant, with likely downslope movement. Spring: Most depart low-elevation winter areas by ±1 May, but some are still there in mid-May. Fall: Arrive at low elevations ±1 Sept., but general arrival there about ±25 Sept. **WINTER:** Throughout lower 48 breeding range, even at very high elevations (to 11,000 ft in CO), and in lowlands adjacent to foothills and mountains, partial to juniper woodlands with berries; winters to central Mexico. **RARE STATUS:** Very rare, but regular in fall and winter to the East, especially to the northern tier of states and southeastern Canada. Accidental St. Lawrence I., AK. **Population** Stable.

juvenile

boldly spotted plumage

overall gray color with prominent white eye ring

prominent buffy wing stripe

buffy wing patches

mostly white outer tail feathers

BROWN-BACKED SOLITAIRE *Myadestes occidentalis*   BBSO ■ 5

This montane Mexican and M.A. thrush has occurred at least twice in AZ. It is a close relative of Townsend's Solitaire. It is kept as a cage bird in Mexico and occurrences in the US are questioned by some. Polytypic (5 sspp.). L 8.3" (21 cm)
Identification ADULT: Gray and rufous overall. Head gray with darker lores and white eye crescents; upperparts brownish olive, wings slightly darker and rufous; underparts gray; tail dark grayish to blackish with mostly white outer rectrices. **JUVENILE:** Head and body whitish buff, scalloped dark brown.
Similar Species Townsend's is similar in shape but with bold and complete white eye ring, and dark gray wings with conspicuous buff wing patches. Compare also Slate-colored Solitaire (*M. unicolor*) of southern Mexico and M.A., which is slaty gray overall.

Voice CALL: Loud, up-slurred *wheeoo*. **SONG:** Amazing series of initial hesitant call notes, accelerating into a long rocking garble of flutelike notes, sung year-round.
Status & Distribution Resident in pine-oak and montane evergreen forests to at least 11,000 ft in the Sierra Madre Oriental and Occidental from northern Sonora and southern Nuevo León south to central Honduras. Range possibly advancing northward in Sonora; currently found north to Sierra Huachinera about 80 mi from AZ. **RARE STATUS:** Two accepted US records, both in southeastern AZ: Madera Canyon in the Santa Rita Mts. (4 Oct. 1996) was initially rejected based on doubt about its origin and then reassessed and accepted in 2011 as valid; Miller Canyon in the Huachuca Mts. (16 July 2009) and presumably the same individual 2.7 mi away at Ramsey Canyon (18 July–1 Aug. 2009). **Population** Declining due to habitat destruction.

white eye crescents nearly meet

rusty brown wings

occidentalis

long tail

BROWN THRUSHES Genus *Catharus*

This terrestrial New World genus consists of 12 species; the five breeding north of Mexico have similar plumage patterns and retiring behaviors. Identification is complicated by subspecific variation. Boreal species exhibit a wide, pale wing stripe in flight. Two species of the Neotropical radiation of this genus have occurred accidentally in N.A.

ORANGE-BILLED NIGHTINGALE-THRUSH *Catharus aurantiirostris*   OBNT ◼ 5

This typically shy denizen of moist, high-elevation forests in C.A. and S.A. has found its way north of Mexico at least four times, suggesting some migratory movements in this species. Polytypic (up to 14 poorly differentiated sspp. in the Neotropics; 1 in N.A., presumably *clarus*). L 6.5" (17 cm)

Identification Orange-brown above with brighter uppertail coverts, wings, and tail. Underparts generally pale gray and whitish, with chest and flanks darkest. Bright orange bill, legs, and eye rings provide obvious field marks. In flight, lacks pale wing stripe typical of boreal genus *Catharus*.

Similar Species Compared with Orange-billed Nightingale-Thrush,

Black-headed is darker, more olive, on upperparts with distinctly blackish crown and face. Russet Nightingale-Thrush (*C. occidentalis*, unrecorded in N.A.) is less bright above, and has vague darkish spotting below and a black bill.

Voice CALL: A nasal, Gray Catbird–like *meeer*. **SONG:** A jerky, short, scratchy though loud warble of varied phrases.

Status & Distribution Fairly common inhabitant of montane understory. **RARE STATUS:** This elevational migrant has occurred twice in spring in the Lower Rio Grande Valley of TX (8 Apr. 1996, at Laguna Atascosa NWR; 28 May 2004, at Edinburg). A territorial singing male was in the Black Hills, SD

(10 July–19 Aug. 2010), and another singing male was in the Zuni Mts., NM, on 18 July 2015.

Population Stable.

orange bill, orbital ring, and legs

orange-brown upperparts, pale gray below

BLACK-HEADED NIGHTINGALE-THRUSH *Catharus mexicanus*   BHNT ◼ 5

A 2004 record in south TX provided for this montane M.A. species' addition to the list of N.A. avifauna. Polytypic (4 sspp.; 1 in N.A., presumably nominate *mexicanus*). L 6.5" (17 cm)

Identification Orange eye ring, bill, and legs are only bright parts of otherwise dark bird. Crown and face blackish with nape a bit paler; rest of upperparts and wings dark olive. Chin, throat, and vent area whitish; rest of underparts gray. A skulker, stays low to the ground.

Similar Species Orange-billed Nightingale-Thrush has similar bright

orange soft parts, but Black-headed has blackish head and is grayer above.

Voice CALL: Buzzy or petulant *mew*, *rreahr*. **SONG:** Similar to Orange-billed Nightingale-Thrush, but more melodious; six to eight thin, high-pitched, flutey whistles and trills; some phrases repeated.

Status & Distribution Fairly common in humid montane forest from southern Tamaulipas and western Nuevo León to western Panama; some suggestion of altitudinal migration. **RARE STATUS:** Accidental, one record in lower Rio Grande Valley, at Pharr,

Hidalgo Co., TX (28 May–29 Oct. 2004).

Population Insufficient information to determine trends.

blackish crown and face

orange orbital ring, bill, and legs

VEERY *Catharus fuscescens*   VEER ◼ 1

Among a retiring genus, the Veery is shy and relatively poorly known. Polytypic (4–5 sspp.; all in N.A.). L 7.7" (18 cm)

Identification For a *Catharus*, generally rather distinctive; less spotted below, brighter reddish above, and lacking eye ring. **ADULT:** Depending on subspecies, upperparts entirely dull reddish brown to bright rufous-brown, a few western birds nearly gray-brown. Large, non-contrasting, pale grayish loral area and brown to reddish lateral throat stripes fairly weak. Chest washed buff, some pinkish buff, and vaguely to distinctly spotted with dull reddish brown on upper chest; any spotting on lower chest is grayish. Belly white, sides and flanks pale gray. Bill horn-colored with dark culmen. **JUVENILE:** Spotted whitish above, dark below; some older imma-

tures distinguishable by presence of retained juvenal wing coverts with buffy tips.

Geographic Variation Subspecies vary primarily in brightness and color of upperparts and distinctness of chest spotting; subspecific identification in field problematic due to slight differences and individual variation. More western *salicicola* has dullest upperparts (brown with reddish tinge) and distinct brownish spots below; central Canada and upper Midwest *levyi* (medium-dull reddish brown) and eastern Canada *fuliginosus* (bright reddish brown) with distinct reddish brown spots; northeastern US *fuscescens* medium-bright reddish brown with indistinct reddish spots; and southern Appalachian *pulichorum* dark reddish brown

with moderately distinct brownish red spots.

Similar Species Other *Catharus* more spotted underneath, less reddish above. See "Russet-backed" Swainson's Thrush.

Voice CALL: Abrupt, rough, descending *veer*; also a slow *wee-u* and a harsh

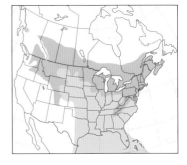

face pattern like
Gray-cheeked with
faint partial eye ring
around rear of eye

western
salicicola

faintly
spotted
chest

gray
flanks

eastern
fuscescens

chuckle; alarm call a sharp, low *wuck*. **FLIGHT NOTE:** Low *veer*. **SONG:** Flute-like, slow, somewhat mournful, downward-spiraling *veeerr veeerr veeerr*.

Status & Distribution Fairly common, but western birds restricted by limited suitable riparian habitat. **BREEDING:** Most nest in deciduous and/or mixed forest but *salicicola* is primarily a shrubby willow riparian inhabitant. **MIGRATION:** Long-distance migrant; eastern birds trans-Gulf migrants, though many

aspects of migration unknown. In spring, peaks on Gulf Coast ±25 Apr., southern Great Lakes ±15 May; casual-to-rare migrant in West, arrival in montane CO breeding areas ±25 May. In fall, Upper Midwest and Northeast peak ±10 Sept., though late migrants linger in northeastern US into Oct. **WINTER:** Southern S.A., a bit south of Amazon Basin. Accidental N.A., most winter reports erroneous. **RARE STATUS:** Casual in southeast AK, CA, and Southwest; accidental in Europe. **Population** Forest fragmentation may negatively impact populations.

GRAY-CHEEKED THRUSH   *Catharus minimus*   GCTH   ▪   1

faint partial eye ring
around rear of eye

grayish
lores

1st fall
aliciae

aliciae

grayish
brown
flanks

minimus

Gray-cheeked Thrushes that breed in the Russian Far East and winter in S.A. are among the longest-distance passerine migrants. Polytypic (2 sspp.; both in N.A.). L 7.3" (18 cm)
Identification A cold gray-brown, long-winged *Catharus*. **ADULT:** Upperparts brownish olive. Face quite plain with grayish loral area and partial eye ring behind eye. Thin, buff to gray malar stripe and white throat contrast strongly with black lateral throat stripes. Chest spotting black and rounded, extending above and below cream wash on breast. Sides and flanks washed brown (with gray or olive cast). Belly and vent unmarked white. Bill yellow, with dark culmen and tip; yellow usually extending beyond nostrils and contrasting with gray lores and

white throat. **JUVENILE:** Spotted buff above, blackish below; some older immatures distinguishable by presence of retained juvenal wing coverts with buffy tips.
Geographic Variation Subspecific identification in field not possible due to slight differences and individual variation. Widespread *aliciae* described. Newfoundland *minimus* upperparts and flanks slightly warmer colored, more like Bicknell's; breast lightly washed cream; and bill darker and duller, with yellow usually not extending beyond nostrils.
Similar Species Colder-colored than other *Catharus*; very similar to Bicknell's (see that entry).
Voice CALL: Variable; downward-slurred *wee-ah*, higher pitched, more nasal than Veery; lower pitched than Bicknell's; also a thin *pweep*. **FLIGHT NOTE:** *Che-errr*, which is similar to that of regular call note, but more nasal, stressing highest pitch in middle; ending more abruptly than that of Bicknell's Thrush. **SONG:** A Veery-like descending spiral, but thinner, higher, more nasal, and with a stutter.
Status & Distribution Uncommon to common. **BREEDING:** Nests in soggy

areas, usually conifer bogs or willow or alder thickets; also in stunted spruce at high elevation. **MIGRATION:** Long-distance, trans-Gulf migrant. Spring: Gulf Coast peak ±1 May; arrival in western AK ±1 June, when some still migrating through LA. Fall: peak in eastern US ±1 Oct. **WINTER:** At lower elevations in northwestern SA (south to eastern Ecuador). No documented winter reports for N.A. **RARE STATUS:** Casual to accidental in West; annual in spring in eastern CO; casual to northern Europe (>50 to UK), Azores; accidental to Surinam.
Population Stable.

BICKNELL'S THRUSH   *Catharus bicknelli*   BITH   ▪   2

Bicknell's Thrush is basically a slightly smaller version of the Gray-cheeked. Monotypic. L 6.3" (16 cm)
Identification Nearly indistinguishable from Gray-cheeked. **ADULT:** Upperparts brown, usually with a warmer cast; tail usually with reddish tinge. Compared to Gray-cheeked,

wings are color of upperparts, except primaries slightly warmer; wings shorter; primary projection about equal to length of longest tertial. Pale loral area not as contrasty; partial eye rings grayish; auriculars plain grayish brown. Black lateral throat stripes contrast with auriculars and white throat.

Chest spotting blackish, usually in a wash of buff. Mandible mostly dull to medium yellow with dark tip; maxilla all-dark. **JUVENILE:** Spotted buff above, blackish below; white throat. Some older immatures distinguishable by retained juvenal wing coverts with buffy tips.

Similar Species Gray-cheeked very similar, but larger, longer-winged, primaries edged with duller brown, and less rufous in tail. Unless in hand or diagnostic vocalizations are heard (and preferably recorded), migrants are best left unidentified, especially out of range.

Voice CALL: Similar to notes in song, single *wee-ooo*; also a sharper *shrip*. **FLIGHT NOTE:** Sharp, buzzy *peeez* or

cree-e-e is higher, less slurred than that of Gray-cheeked. **SONG:** Similar to Gray-cheeked, more even in pitch, beginning with low *chuck* notes; high-pitched phrases rise slightly toward end and terminate, after pause, with *shre-e-e*.

Status & Distribution Uncommon to rare and local. **BREEDING:** Mostly in dense stunted spruce and/or balsam fir forests at tops of northeastern mountains; lower in eastern QC. **MIGRATION:** Medium-distance migrant, generally along Atlantic coastal plain. Typically arrives on breeding grounds ±25 May, slightly earlier in Adirondacks, slightly later in QC. Departs New England mid–late Sept.; probably peaks southern NJ late Sept.–early Oct. **WINTER:** Mostly in montane forests on Hispaniola, primarily Dominican Republic where better habitat exists. Rare in mountains of eastern Cuba. **RARE STATUS:**

Accidental OH (Sept. specimen) and to Bermuda.

Population Vulnerable. Extirpated as a breeder in MA and on Cape Breton I., NS. Introduced exotic pests are killing trees in its breeding habitat. Deforestation a major concern on winter grounds on Hispaniola, especially Haiti.

almost identical to Gray-cheeked, but slightly smaller and averages warmer above, especially on tail

grayish brown flanks

SWAINSON'S THRUSH *Catharus ustulatus*   SWTH ■ 1

The most common migrant *Catharus* in much of N.A. Polytypic (6 sspp.: all in N.A.). L 7" (18 cm)
Identification Bold, buffy eye ring and supraloral patch distinctive. Two subspecies groups: "Olive-backed" *swainsoni* group and "Russet-backed" *ustulatus* group. "Olive-backed" group described. **ADULT:** Upperparts olive-brown. Round chest spots dark brown with underlying buffy wash. Sides and flanks olive-gray. Tail similar in

color to upperparts. **JUVENILE:** Spotted buff above.
Geographic Variation Subspecies identification possible in field only to group. "Olive-backed" group: Widespread *swainsoni* described above; *appalachiensis* upperparts slightly darker; *incanus* of western AK to northern BC and north-central AB has upperparts paler and grayer, underparts whiter, and chest spotting nearly black. "Russet-backed" group: Northwestern coastal *ustulatus* (southeastern AK to northwestern CA) and *phillipsi* (Haida Gwaii) upperparts brown, tinged reddish, with breast washed brownish with indistinct spots,

and flanks warm brown; and *oedicus* of rest of CA is duller above; eye rings of western subspecies usually thinner. The two groups have different breeding grounds (though they approach each other rather closely), migration routes, and winter grounds. Intergrade specimens are known from Hyder, southeast AK, the two groups might best be treated as separate species.
Similar Species "Russet-backed" subspecies is regularly confused with western *salicicola* Veery, but note differences in color of breast spotting, flank color, and facial pattern.
Voice CALL: "Russet-backed" gives a liquid *dwip*, "Olive-backed" a sharper

buffy eye ring and supraloral line

upperparts with a russet cast, like many Veeries

Pacific coast
ustulatus

buffy flanks

Alaska
incanus

olive-gray upperparts

back more olive, less grayish, than *incanus*

spots on breast blacker than *ustulatus*

swainsoni

brownish flanks

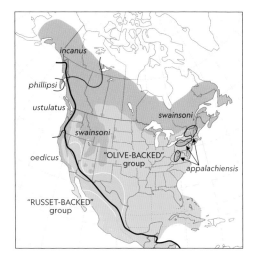

incanus

phillipsi

ustulatus

swainsoni

oedicus

swainsoni

"OLIVE-BACKED" group

appalachiensis

"RUSSET-BACKED" group

quirt; also rough, nasal chatter. **FLIGHT NOTE:** Clear, spring peeper-like *queep*. **SONG:** Flutelike, similar to Veery in spiraling pattern, but spiraling upward.
Status & Distribution Uncommon to common. **BREEDING:** Mostly in deciduous and/or mixed forest; birds in western US primarily nest in riparian habitat with combination of shrubby willow understory and deciduous or coniferous overstory. **MIGRATION:** "Olive-backed" birds are trans-Gulf migrants;

"Russet-backed" birds migrate through the CA deserts (spring) and along and off Pacific coast (fall). Peaks on Gulf Coast ±25 Apr., southern Great Lakes ±15 May, and CA deserts 10–25 May; arrival in montane WY breeding areas ±25 May; spring migration continues into early June as far south as KS and southeastern CA. In fall, most migrate Sept.–mid-Oct., arrival on CO plains ±25 Aug.; both groups are largely absent as migrants over much of the South-

west. **WINTER:** Numerous erroneous reports from N.A. "Russet-backed" birds winter from western Mexico to northwest Costa Rica; casual southwestern CA; "Olive-backed" birds winter in S.A. **RARE STATUS:** Casual to Bering Sea islands and northern AK, also NT; casual to Europe.
Population Declining over much of range, especially in West (e.g., extirpated as a breeder from much of Sierra Nevada, CA).

HERMIT THRUSH   *Catharus guttatus*   HETH ■ 1

The only *Catharus* expected in winter in the US. Polytypic (9–13 sspp.; all in N.A.). L 6.8" (17 cm)
Identification Slow tail lift is distinctive within the genus. **ADULT:** In eastern *faxoni*, upperparts brown with moderate rufous wash. Eye ring whitish, occasionally not complete in front. Loral area darker than on other *Catharus* and not contrasting. Tail distinctly reddish, contrasting with upperparts. Boldly spotted below. Sides and flanks washed buffy brown; brownish gray in other subspecies. Bill dark with pinkish base to mandible. **JUVENILE:** Spotted whitish to buff above, blackish below.
Geographic Variation Three subspecies groups. Subspecific identification to group is possible with good views. "Alaska and Northwest" *guttatus* group includes *guttatus* (coastal southern AK to western BC), *nanus* (southern AK islands), *verecundus* (Haida Gwaii), *vaccinius* (coastal southwestern BC and northwestern WA), *jewetti* (northwestern WA to northwestern CA), *slevini* (interior south-central WA to west-central CA), *munroi* (central BC and western AB to northern MT), and *oromelus* (interior southern BC east to northwestern MT and south to northeastern CA). All are small and most are dark brownish, with little to no reddish coloration on back, and with richly colored tails. "Interior Western Montane" *auduboni* group includes *sequoiensis* (Sierra Nevada of CA), *polionotus* (eastern CA east to northwestern UT and AZ), and *auduboni* (southeastern WA east to southern MT, south to southern AZ and NM, West TX, and into Mexico). All with pale, grayish brown backs and duller, less contrasting tails; the last subspecies having distinctive buff undertail coverts. "Eastern" *faxoni* group includes *faxoni* (southern NT south to southern AB, east to NL and

MD) and *euborius* (central AK and northern BC); birds in this group are richly colored above with bright reddish tails, and have tawny brown flanks (flank colored duller and variable in *euborius*).
Similar Species Some similarity to Gray-cheeked, Bicknell's, and Swainson's Thrushes (see those entries).
Voice CALL: Soft, low blackbird-like *chuck*; also a rising, whiny catbird-like *wheeee*. **FLIGHT NOTE:** Clear, somewhat complaining *peew*. **SONG:** Flutelike; begins with long, clear whistle followed by series of rather clear phrases; successive songs often alternate pitch direction of song, rising then falling.
Status & Distribution Common. **BREEDING:** Typically in conifer-dominated forests, usually in areas of relatively little undergrowth; also in deciduous forests. **MIGRATION:** Medium- to long-distance migrant, with western montane breeders wintering farthest south (to southern Guatemala). Early migrant in spring, with southern Great Lakes peak ±25 Apr.; arrival in farthest reaches of breeding range ±10 May, though with stragglers still in West in early June. Fall: Western montane birds begin migration in early Sept., but *faxoni* and *guttatus* groups peak ±15 Oct. **WINTER:** Western subspecies winter in coastal states from WA south to southern Guatemala; *faxoni* group winters in Southeast (from Long I., NY, south), a few to southern Midwest, south to northeastern Mexico. **RARE STATUS:** Rare to Bering Sea islands in migration and to Bermuda fall

through spring; casual north to northern AK and northern Canada, Greenland, Greater Antilles, and Europe.
Population Increasing.

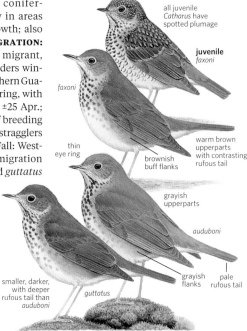

all juvenile *Catharus* have spotted plumage

juvenile *faxoni*

faxoni

thin eye ring

warm brown upperparts with contrasting rufous tail

brownish buff flanks

grayish upperparts

auduboni

smaller, darker, with deeper rufous tail than *auduboni*

guttatus

grayish flanks

pale rufous tail

[map labels:]
guttatus | munroi
"ALASKA and NORTHWEST" group
"EASTERN" group
nanus
jewetti
faxoni
oromelus
slevini
auduboni
sequoiensis
polionotus
"INTERIOR WESTERN MONTANE" group

Genus *Hylocichla*

WOOD THRUSH *Hylocichla mustelina*   WOTH ▪ 1

This is one of the finest songsters living in our eastern deciduous forests. It occupies a monotypic genus. Monotypic. L 7.8" (20 cm)
Identification A chunky, well-marked brown thrush of eastern N.A. **ADULT:** Orange-brown upperparts, much brighter on rear crown and nape. Obvious white eye ring barely broken by dark gray eye line. Distinctive black-and-white striped auriculars and spotting on throat forming streaks. Underparts white with large and distinct, oval black spots extending onto lower belly and flanks, the latter

nearly complete bold, white eye ring

rich rufous above, especially on crown and nape

speckled cheeks

boldly spotted with black below

washed warm brown. Bill grayish pink with darker culmen. **JUVENILE:** Spotted whitish above, dark below; some older immatures distinguishable by flanks possibly less warm, and through fall by thin cinnamon tips to inner coverts forming very slight, partial wing bar. **FLIGHT:** Wide, pale wing stripe as in boreal *Catharus*, but wing base wider with pronounced secondary bulge.
Similar Species *Catharus* thrushes similar, but are smaller, less potbellied, less heavily spotted, and most not as bright. Strong white eye ring not matched in *Catharus*, nor is auricular pattern (though Gray-cheeked has vaguely-streaked auriculars). Brown Thrasher has much longer tail and bill; yellowish eye.
Voice CALL: A rolling *popopopo* and a rapid, staccato *pit pit pit*. **FLIGHT NOTE:** Sharp, nasal *jeeen*. **SONG:** Beautiful, flutelike *eee-o-lay*, with last note accented and highest in pitch; this song usually introduced with quiet *po* notes, not audible at long distance, and ending with a buzzy or trilled whistle.
Status & Distribution Fairly common but declining in East. **BREEDING:** Nests in deciduous forest. **MIGRA-**

TION: Medium- to long-distance trans-Gulf migrant. Spring peaks Gulf Coast ±15 Apr.; southern Great Lakes ±15 May. In fall, some move early (late Aug.), but most still present near breeding grounds into Sept.; Gulf Coast peak in first half of Oct. Winters from eastern Mexico east and south to Panama. **WINTER:** Casual in the US. **RARE STATUS:** Casual to accidental migrant in West. Accidental in western AK, western Mexico, northern S.A., Iceland, UK, and Azores.
Population Near Threatened. Forest fragmentation and resultant cowbird parasitism has greatly reduced populations in parts of breeding ranges.

ROBINS Genus *Turdus*

Of the huge genus *Turdus* (as many as 80 species), N.A. north of southern TX supports just a single common breeding species, the American Robin. However, seven Old World species have occurred in N.A. Another three Mexican/Neotropical species have occurred in the US, with the Clay-colored Thrush now breeding in small numbers in south TX.

EURASIAN BLACKBIRD *Turdus merula*   EUBB ▪ 5

This all-dark *Turdus* is common and widespread in Europe, where it is one of the best known species. It commonly occurs in Europe's parks and backyard gardens. It is unlike other N.A. species, but similar to other blackish thrushes in M.A. and S.A. Polytypic (14–18 sspp., sometimes split into 2–3 species; presumably nominate *merula* in N.A.). L 10.5" (27 cm)
Identification Behavior and habitat similar to American Robin. **MALE:** All black; orange-yellow eye ring and bill (duller on immature); dark legs. **FEMALE:** Browner; pale throat and dark-streaked chest; eye ring and bill duller.
Similar Species Unmistakable but beware of smaller and short-tailed

European Starling, which has a slim pointed bill; Eurasian Blackbird is solitary.
Voice CALL: Many calls, one a deep *pok* and a hard *chack-ack-ack*. **FLIGHT NOTE:** Quiet, rolling *srrri*. **SONG:** Similar to that of American Robin.
Status & Distribution Found throughout Europe, south to North Africa and east to China; northern populations migratory. Introduced in Australia and New Zealand. **RARE STATUS:** One accepted record from NL (male specimen found dead 16 Nov. 1994). Two other records, from Montreal, QC, and Kent Co.,

ON, treated as possible escapes. About a dozen records for Greenland.
Population Overall increases, but some declines in the UK.

male all-black with orange-yellow orbital ring and bill

merula ♂

EYEBROWED THRUSH *Turdus obscurus* EYTH ■ 3

The Eyebrowed Thrush is a rare visitor to western AK, especially in spring. Monotypic. L 8.5" (22 cm)

Identification Somewhat similar to American Robin. Head and throat of male is largely slaty gray with strong white eyebrow, eye arcs, and chin; black lores. Female similar, but paler

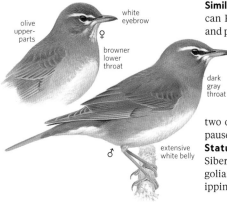

olive upper- parts

white eyebrow ♀

browner lower throat

dark gray throat

extensive white belly ♂

and browner with strong white malar stripe. Back, wings, and tail are olive- brown, the latter without white tail corners. Sides and flanks are orange- rufous; belly and undertail coverts unmarked white. Bill yellowish with variably dusky tip. Legs yellowish brown to pinkish.

Similar Species Duller female Ameri- can Robin told by overall coloration and pattern and precise head pattern.

Voice CALL: Variable; high, thin, penetrating *dzee*, a dou- bled *chack* or *tuck*, and a single *tchup*. **FLIGHT NOTE:** High *tseee*. **SONG:** Clear, whistled, mournful series of phrases of two or three syllables followed by a pause, then a lower-pitched twittering.

Status & Distribution BREEDING: Siberia, Russian Far East, and Mon- golia. **WINTER:** Southeast Asia, Phil- ippines, Sumatra, and north Borneo.

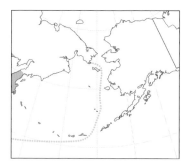

RARE STATUS: Rare spring migrant (irregularly numerous, e.g., 180 on Attu I. in May 1998) in western and cen- tral Aleutians and Pribilofs, very rare in fall; casual north to St. Lawrence I.; accidental elsewhere in western AK and north to Utqiaġvik. One in north- eastern Kern Co., CA (28 May 2001).

Population Likely declining due to trapping for food.

DUSKY THRUSH *Turdus naumanni* DUTH ■ 4

This striking Asian thrush occurs very rarely in AK. Polytypic (2 sspp.; both N.A.). L 9.5" (24 cm)

Identification Male blackish above; bright white supercilium. Underparts blackish with variable extent of white fringing; wings mostly rufous. Female similar to male, but upperparts brown rather than blackish; face less white; wings brown. Whitish chest crescent on both sexes variable in extent, but distinctive.

Geographic Variation Subspecies *eunomus* described above; "Nau- mann's Thrush" (*naumanni*, more southerly breeding) is rufous on head, where *eunomus* is white, and on belly, where *eunomus* is black; wings brown and reddish in tail. Inter- grades are apparently frequent. Most Old World authorities now split "Nau- mann's Thrush" as a separate species.

Similar Species Varied Thrush vaguely similar, but head and wing patterns distinctive.

Voice CALL: Variable; series of *shack* notes; a chattering *kwaawag*; *kwet-kwet*; European Starling–like *spir*. **FLIGHT NOTE:** Thin *shrree*, often repeated, or thin, high *huuit*. **SONG:** Flutey and melodious series of phrases, often ending in faint trill or *twitter, tryuuu-tvee—tryu*, with accent in first phrase.

Status & Distribution BREEDING:

broad whitish supercilium

black upper breast band bordered by white crescent

rusty rufous wings

eunomus ♂

rich buffy supercilium

eunomus ♀

extensive black markings below

reddish outer tail feathers obvious in flight

reddish underparts

naumanni ♂

Siberia and the Russian Far East. **WINTER:** Eastern China, also in Japan (common); a few in some winters as far south to northern Thailand. **RARE STATUS:** Casual to western and cen- tral Aleutians, Pribilofs, and St. Law- rence I.; also recorded mainland AK, YT, and BC. One likely (photos) record of "Naumann's Thrush" from Gambell, St. Lawrence I., on 5 June 2015; other sight records for Attu and St. Paul Is.

Population Stable.

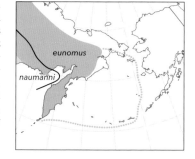

eunomus

naumanni

FIELDFARE *Turdus pilaris* FIEL ▪ 4

The Fieldfare is a distinctive Eurasian breeder of casual occurrence in northeastern N.A. Monotypic. L 10" (25 cm) **Identification** Head is primarily medium gray with a black loral area; buffy white to orange chin; throat with fine blackish streaking; and dark lateral throat stripe connected to black lower corner of auriculars. Bill varies from yellow to orange with a black tip of variable extent. Upper back, scapulars, and wing coverts are purplish brown; flight feathers are blackish. Lower back and rump

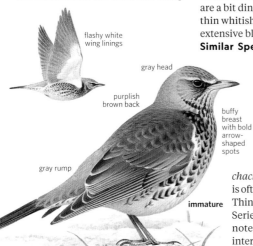

flashy white wing linings

gray head

purplish brown back

buffy breast with bold arrow-shaped spots

gray rump

immature

are pale gray and contrast strongly with rest of upperparts; this feature is particularly noticeable in flight. Chest and sides are buffy to orange with blackish feather centers forming arrowhead-shaped spots, lower ones typically larger. Belly and undertail coverts are white, the latter with dark centers or shaft streaks. Longish tail is black with vague gray corners. Legs are gray. Primary projection is about equal to length of longest tertial. In flight, contrastingly white wing linings are obvious. Birds in first-winter plumage are a bit dingier, less contrasty, with a thin whitish lower wing bar and more extensive black on bill.
Similar Species Vaguely similar to American Robin, but brown upper back and wings, pale gray lower back and rump, and white wing linings of Fieldfare are distinctive. Redwing browner, has rusty flanks and wing linings.
Voice CALL: Variable; *chack*, similar to Dusky Thrush, is often doubled. **FLIGHT NOTE:** Thin, nasal *tseee* or *weeet*. **SONG:** Series of chattery or warbling notes with squeaky chuckles interspersed; without flutelike

quality typical of genus. **FLIGHT SONG:** Chattering, more drawn out, and faster.
Status & Distribution BREEDING: Common west from western Russian Far East through northern Europe and locally in southern Greenland. **WINTER:** Primarily in southern Europe and southwestern Asia, but the species' fall migration has a strong facultative aspect, and many individuals continue only as far south as necessary to avoid snow. **RARE STATUS:** Casual in late fall and winter in northeastern N.A. from NL to DE; also casual in western and northern AK in late spring/early summer. Accidental to ON, MN, MT, BC.
Population Stable. Greenland population possibly extirpated.

REDWING *Turdus iliacus* REDW ▪ 4

A relatively small, short-tailed Old World *Turdus*, the Redwing has streaked underparts and reddish orange flanks and wing linings, the latter accounting for its English name. Polytypic (2 sspp.: both sspp. may have reached N.A., but no specimens). L 8.3" (21 cm)
Identification Upperparts are grayish brown, with a strong face pattern of long white supercilium, black eye line, white-streaked auriculars, white

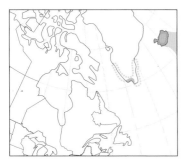

malar stripe, black lateral throat stripe, and white chin and throat. Wings are colored same as back; tail is blackish. Chest and upper belly are white with extensive dark brown to blackish streaking. Streaking of chest sides extends into reddish orange lower sides and flanks. Lower belly is unmarked white; outer undertail coverts have dark centers.
Geographic Variation Two poorly marked subspecies, with *coburni*, a breeder in Iceland, the Faroe Is., and, recently, southern Greenland, being slightly larger and slightly darker in all plumage aspects. Which subspecies accounts for northeastern N.A. records is uncertain (though quite possibly *coburni*), but recent records from north Pacific coast probably refer to nominate *iliacus*.
Similar Species Redwing is unlikely to

be confused with any other thrush in N.A. but a molting juvenile American Robin; however, distinctive face pattern and lack of other juvenile traits (such as white upperparts spotting) should easily rule out that option. Fieldfare is vaguely similar, but distribution of orange underneath is quite different, as is head pattern. Streaked

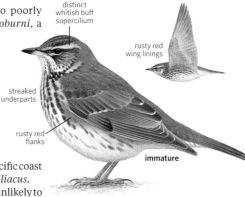

distinct whitish buff supercilium

rusty red wing linings

streaked underparts

rusty red flanks

immature

underparts and strong black-and-white head pattern might cause confusion with one of the tropical streaked flycatchers, but behavior and underparts pattern are quite different. In flight, Redwing appears short-tailed; note reddish orange wing linings. **Voice CALL:** Variable; a nasal *gack*; a harsh *zeeh*; an abrupt *chup*, sometimes extended to *chidik*. **FLIGHT NOTE:** A long, somewhat hoarse *stoooof* or a

thin, high *seeeh*. **SONG:** Quite variable, with tone and structure being more important than particular phrases; short phrases and squeaky twitters interspersed with long (up to 6 seconds) silent intervals. **Status & Distribution BREEDING:** From Russian Far East west to Iceland, where common; recent colonization of a small area in southern Greenland. **WINTER:** Primarily in

southern Europe, northwest Africa, and southwest Asia. **RARE STATUS:** The species is a casual visitor, primarily in late fall and winter, to the rest of Greenland and to NL, and is accidental farther south in northeastern N.A. (to Long I., NY, and eastern PA). In West, recent winter records from south-coastal AK, southwest BC, and western WA. **Population** Near Threatened.

MISTLE THRUSH *Turdus viscivorus* MITH ■ 5

The Mistle Thrush, a large Old World thrush, has turned up only once in N.A., in winter from NB. Polytypic (3 sspp.; likely nominate in N.A.). L 10.2–11.4" (26–29 cm)
Identification Flies with strong direct flight. **ADULT:** Large. Grayish brown above with paler rump and narrow whitish wing bars; whitish below, streaked on upper breast, otherwise conspicuously marked with round spots; often dark bar on sides of breast. Intricate head pattern with pale rear surround to ear coverts and two dark vertical facial bars. White tail corners and white underwing coverts. **IMMATURE:** Similar but with whitish panel on outer greater coverts.
Similar Species Much larger than all *Catharus*. Larger, paler, and grayer than Old World Song Thrush with different facial pattern, white (not rusty-buff)

underwing coverts, and round (not arrow-shaped) ventral spots. Calls very different.
Voice CALL: Most frequent vocalization an extended dry rattle. **SONG:** Suggestive of European Blackbird, loud with clear and varied phrases.
Status & Distribution Found in woods, orchards, woodland edges, and more open areas; in some parts of breeding range almost treeless areas. **BREEDING:** From British Isles and Europe (also Atlas Mts., North Africa) east to Siberia (Yenisey River and to near Lake Baikal), northwest China (rare), and western Himalaya. **WINTER:** Withdraws from northern and eastern parts of breeding range. Winters from southern part of breeding range to Middle East. **RARE STATUS:** Accidental to N.A., one record

intricate face pattern with two dark vertical bars

larger than American Robin; larger and grayer than Song Thrush

rounded black spots below

white tips to outer tail feathers

at Miramichi, NB, 9 Dec. 2017–21 Mar. 2018. In Old World rare to Iceland (has nested) and casual to Saudi Arabia, India, Azores, and Japan.
Population Moderate declines in Europe.

SONG THRUSH *Turdus philomelos* SOTH ■ 5

Among Europe's most cherished songbirds, Song Thrush has been an inspiration to musicians and writers throughout its range but has been a special muse for English poets, from Chaucer to Shakespeare, Wordsworth to Hughes. Its scientific name, *philomelos*, means "lover of song." Polytypic (3 sspp.). L 8–9.3" (20–23 cm)
Identification Larger and stockier than *Catharus* thrushes but smaller than American Robin. Upperparts warm brown. Underparts whitish with ochre-buff wash on breast and flanks, liberally spotted with dark arrowhead-shaped spots except on lower belly. Narrow eye ring, pale submoustachial bordered by dark malar stripe, and auriculars somewhat mottled, most strongly at edges. In flight, from N.A. *Catharus* by rusty buff underwing coverts. **JUVENILE:** Slightly paler above than adult, with buff-tipped

upperwing coverts in fresh plumage.
Geographic Variation Migratory nominate subspecies is found through most of the species' large Eurasian range. The dark subspecies *hebridensis*, restricted to Isle of Skye and Outer Hebrides, Scotland, is sedentary, while warm-toned *clarkei*, breeding in the rest of the British Isles and in western Europe (France, Belgium, and the Netherlands), is a partial migrant.
Similar Species See Mistle Thrush. In N.A. context, Song Thrush might suggest Swainson's Thrush but is larger and more heavily and extensively marked in face and underparts.
Voice CALL: A sharp *zip* and longer *seep*. **SONG:** Typically given at dusk, a bold, very long series of repeated notes in phrases, some ethereal sounding, others grating, incorporating imitations of other species and even human-made sounds.

Status & Distribution BREEDING: From the British Isles and most of northern and central Europe across Ukraine nearly to Lake Baikal. **WINTER:** Northernmost birds from Scandinavia and Russia migrate farthest south, into the Mediterranean and Middle East, whereas birds from the western portions of the range are

dark-bordered auricular

plain brown upperparts

extensively patterned below with arrow-shaped marks

fall immature

sedentary or partly migratory, some wintering near breeding grounds or on the Iberian Peninsula. **RARE STATUS:** The only N.A. record comes from Saint-Fulgence, eastern QC (11–17 Nov. 2006). Annual in small numbers to Iceland, chiefly in fall; one record from Greenland.

Population Song Thrush populations have declined by 50–70 percent in some areas of England and Europe since the 1970s.

CLAY-COLORED THRUSH *Turdus grayi* CCTH ▪ 3

Formerly just a winter resident in southernmost TX, this widespread tropical thrush, the national bird of Costa Rica, now breeds in TX in small numbers. It is somewhat shier than American Robin. Polytypic (8 sspp. recognized, differences are minor; *tamaulipensis* in N.A.). L 9.5" (24 cm)
Identification A typical *Turdus*, it is particularly common in Mexico and M.A. and essentially replacing American Robin there. **ADULT:** Lacks eye ring. Upperparts medium brown, underparts tan, with throat vaguely streaked brown and pale buff. Eyes are

reddish orange, bill is greenish yellow with a dark base, and variably colored legs range from dull pink through greenish brown to gray. **JUVENILE:** Similar to adult but with pale buff shaft streaks on back feathers, scapulars, and median coverts, and with buff tips to median and greater coverts.
Similar Species See White-throated Thrush.
Voice CALL: Variable; a distinctive cat-like and rising *jerereee* (similar to a call of Long-billed Thrasher) and some typical thrush notes, including doubled *tock*. **FLIGHT NOTE:** High, thin *siii*, weaker than American Robin. **SONG:** Clear, whistled phrases recalling that of American Robin, but with a much wider variety of phrases.
Status & Distribution Abundant in Neotropics; variably rare to uncommon in US. Most frequently seen at the various reserves in the lower Rio Grande north to Webb Co., but also seen in nearby well-wooded suburban yards, particularly in

winter. **BREEDING:** Nests in areas of dense thickets, streamside brush, and woodlands. **WINTER:** Small breeding population is supplemented in winter (numbers vary from year to year) with birds from northeast Mexico. Found from northeast and northwest Mexico south to northern Ecuador. **RARE STATUS:** Rare to casual to Edwards Plateau and northeast to Walker Co., TX. Accidental to Big Bend NP, northern NM, and southeastern AZ.
Population Stable.

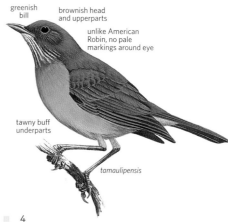

greenish bill

brownish head and upperparts

unlike American Robin, no pale markings around eye

tawny buff underparts

tamaulipensis

WHITE-THROATED THRUSH *Turdus assimilis* WTTH ▪ 4

The retiring White-throated Thrush is widespread in M.A. It is similar looking and closely related to the White-necked Thrush (*T. albicollis*), a widespread S.A. thrush found east of the Andes. In N.A., most records are from southernmost TX. Polytypic (14 sspp.; 1 in N.A., presumably *suttoni*). L 9.5" (24 cm)
Identification White-throated is a darkish *Turdus* of moist montane forests of middle and upper canopy levels. This species can be difficult to see well due to its shy nature. Upper-

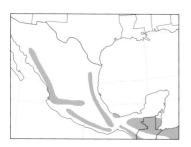

parts are dark brown; underparts dark tan, with throat strongly streaked blackish and white, and undertail coverts somewhat paler. A distinctive white band on upper chest requires a frontal view. Bold yellow-orange eye ring surrounds dark eyes; bill is greenish brown, and legs are dull pinkish brown.
Geographic Variation Two subspecies, *calliphthongus* and *lygrus*, are found in western Mexico; *calliphthongus* is paler, more of a grayish brown above, and has a grayish, not yellow, orbital eye ring. It occurs in southeast Sonora and might reach the Southwest (sspp. identification of two in AZ records is uncertain).
Similar Species From Clay-colored, note upperpart color and presence or absence of eye ring and bib color.
Voice CALL: Quite variable; short, guttural *ep* or *unk*; nasal *rreuh*; whistled *peeyuu*. **SONG:** Similar to that of Ameri-

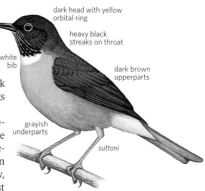

dark head with yellow orbital ring

heavy black streaks on throat

white bib

dark brown upperparts

grayish underparts

suttoni

can Robin, but with wider variety of phrases, and sometimes repeating phrases like a mimid.
Status & Distribution RARE STATUS: Casual to south TX (nearly all Rio Grande Valley) in winter. Accidental southeast AZ.
Population Declining due to habitat destruction.

RUFOUS-BACKED ROBIN *Turdus rufopalliatus* RBRO ■ 3

The Rufous-backed is a rare visitor from Mexico. Polytypic (2–3 sspp.; *rufopalliatus* in N.A.). L 9.3" (24 cm)
Identification Superficially similar to American Robin. Head and upper back are largely gray with variable extent (due to age?) of orangish brown on crown; lacks eye ring. Lower back and wing coverts bright orangy-rufous, contrasting with gray rump and tail. Black and white throat streaks extend onto upper chest; rest of underparts orange, except for white central belly and undertail coverts. Legs are orangish pink, eyes are orange, and bill is dull to bright orange. The overall paler "Grayson's Thrush" (sssp. *graysoni*, breeds on Tres Marias Is., off western Mexico) is sometimes regarded as a distinct species; it could possibly reach the US.
Similar Species Duller individuals might be mistaken for American Robin, but bill and leg colors, plumage pattern, and lack of eye-ring distinctive.
Voice CALL: Variable; a typical *Turdus* throaty trebled *chok*, a plaintive, drawn-out, descending whistle, *peeeuuuuu*. **FLIGHT**

NOTE: High *sseep*. **SONG:** A leisurely, clear, low-pitched, warbled series of phrases with a repetitive pattern, recalling that of American Robin.
Status & Distribution Rare but nearly annual fall and winter visitor to the southwestern US, primarily southern AZ (plus a few aseasonal records), but with records scattered from southern TX to coastal Southern CA.
Population Stable.

Figure labels: gray head with no white around eye; rufous back and wing coverts; long streaks on throat; white extends up to lower breast in a point; rufopalliatus

AMERICAN ROBIN *Turdus migratorius* AMRO ■ 1

This species' often-confiding nature, distinctive plumage, pleasing song, and acceptance of human-dominated habitats make it one of the most beloved of N.A. birds. Polytypic (7 sspp.; 5 in N.A.). L 10" (25 cm)
Identification A distinctive, potbellied bird. Forages on lawns and other areas of short vegetation for earthworms and other invertebrates in a run-and-stop pattern typical of terrestrial thrushes. **ADULT:** Depending on sex and subspecies, head with white eye arcs varies from jet black to gray, with white supercilia and throat, blackish lores and lateral throat stripe. Underparts vary, often in tandem with head color, from deep, rich reddish maroon to gray-scalloped, peachy orange. Males tend to be darker, females grayer, but overlap makes determining sex of many problematic. Throat streaked black and white; vent and undertail coverts white. Upperparts medium gray; tail blackish with white corners. Bill color yellow with variable, season-dependent, black tip. Legs dark. **JUVENILE:** Dark spots on underparts; whitish spots on upperparts and wing coverts. Older immatures not distinguishable from adults; small percentage retain a few juvenal wing coverts or other feathers. **FLIGHT:** Quick, flicking wingbeats are followed by short, closed-wing glides.

Figure labels: migratorius; white spot visible in flight

Figure labels: juvenile migratorius; heavily spotted underparts; bold white broken eye ring; yellow bill; brick red underparts; migratorius ♀; migratorius ♂

Wing linings are color of underparts; remiges blackish.

Geographic Variation Nominate *migratorius* is found in the north and much of the East; it is described above. North Pacific coastal *caurinus* and widespread western *propinquus* (larger, paler) with white tail corners small or lacking; Canadian maritime *nigrideus* dark brownish to blackish above, underparts deep rufous, medium-size tail corners; southeast US *achrusterus* smaller, upperparts browner, smaller white tail corners. The subspecies *confinis* from the Laguna Mts., in Baja California Sur's Cape District, is often recognized as a separate species, "San Lucas Robin."

Similar Species Duller females possibly mistaken for Eyebrowed Thrush. Juveniles possibly confused with spotted thrushes.

Voice CALL: Variable; low, mellow single *pup*; doubled or trebled *chok* or *tut*; shriller and sharper *kli ki ki ki ki*; high and descending, harsh *sheerr*. **FLIGHT NOTE:** Very high, trilled, descending *sreeel*. **SONG:** Clear, whistled phrases of two or three syllables *cheerily cheery cheerily cheery*, with pauses; lacks the burry quality of many tanagers; *Pheucticus* grosbeaks typically have different tempo.

Status & Distribution Common and widespread. **BREEDING:** Wide variety of wooded or shrubby habitats with open areas. **MIGRATION:** Short- to medium-distance migrant. Departs northerly winter-only areas by ±10 Apr.; arrival northern Great Lakes ±20 Mar.; central AK ±1 May. Strong facultative aspect (particularly in East) in fall, so variable timing; departs southern Canada ±20 Oct. **WINTER:** Mainly lower 48 and Mexico; also southernmost ON and BC, Bermuda, Bahamas (rare), northern Guatemala. **RARE STATUS:** Casual Bering Sea islands. Rare (fall and winter) to Europe and Azores; casual to Cuba, Jamaica, and Hispaniola. **Population** Stable.

RED-LEGGED THRUSH   *Turdus plumbeus*   RLTH ▪ 5

This resident Caribbean species has been recorded three times from FL. It is resident as close to FL as Grand Bahama. Polytypic (6 sspp.; *plumbeus* in N.A.). L 10.5" (27 cm)

Identification Sexes similar. Nominate from Bahamas described. Dark slate gray body with white chin and black throat; black bill; bright red orbital ring, legs, and feet. White in tail best seen in flight.

Geographic Variation Eastern subspecies *ardosiaceus* (Hispaniola and Puerto Rico) and *albiventris* (Dominica) have red bill, black-and-white streaked throat (may be a separate species). Widespread Cuba *rubripes* has rufous belly; Cayman Brac *coryi* similar. Eastern Cuba *schistaceus* more

like *plumbeus* but with whitish belly.
Similar Species Unmistakable.
Voice CALL: Shrill creaking and whistled notes, often in pairs. **SONG:** Usually two or three phrases repeated, such as *chirra, chirra, weeup, wheet*, interspersed with thin *wheet* or *weep* notes—much slower, lower pitched, and less musical than American Robin.

Status & Distribution Widespread resident of West Indies. Habitats include tropical deciduous and montane evergreen forests, scrub, thick undergrowth, yards—usually seen on the ground in leaf litter. **RARE STATUS:** Three FL records: one at Hammock Sanctuary, Melbourne Beach (31 May 2010); one at Lantana Nature Reserve,

Palm Beach Co. (25 Apr. 2019); and one seen gathering nesting material at Miami Beach (26–30 June 2019).
Population Stable.

mostly black throat with white chin

red orbital ring and legs

plumbeus

Genus *Ixoreus*

VARIED THRUSH   *Ixoreus naevius*   VATH ▪ 1

This species' ethereal song is a distinctive feature of wet northwestern forests. Polytypic (4 sspp.; *all* in N.A.). L 9.5" (24 cm)

Identification Orange legs and dark bill with yellow mandible base. Distinctively colored, although a rare pale variant has white rather than orange pigmentation (first for UK was in this pale plumage). **MALE:** Blue-gray above, orange below, with broad orange supercilium, black auriculars and chest band, complicated pattern of orange-on-black wings. **FEMALE:** Similar but upperparts brown; indistinct chest band. **JUVENILE:** Similar to adult females, but chest scalloped with dark gray-brown; central belly white.

juvenile
meruloides

orange wing bars and markings on wing

meruloides ♀

orange supercilium

meruloides ♂

black breast band

back color varies, gray to brown depending on subspecies

FLIGHT: Similar to American Robin, but orange-and-black wing linings and bold orange wing stripe.

Geographic Variation Subspecies differentiation based on female plumages. Northern *meruloides* paler above and below; north Pacific coastal *naevius* and *carlottae* upperparts darker, tawny tinged, underparts orange; southern interior *godfreii* paler, upperparts reddish tinged.

Similar Species The less-retiring American Robin has a longer tail and plain wings; also lacks orange throat, supercilium, and mottled flank pattern. In flight, Townsend's Solitaire's wing pattern is similar, but that species is more slender, with longer, thin, white-edged tail.

Voice CALL: A low *tschook* similar to Hermit Thrush, but harder; a high *kipf*; a thin, mournful whistle, *woooeee*.

FLIGHT NOTE: Short, humming whistle. **SONG:** Series of long, eerie whistles of one pitch, with successive notes at different pitch and long inter-note intervals.

Status & Distribution Common. **BREEDING:** Nests in moist, typically conifer-dominated, habitats in Northwest; in tall willow riparian north of tree line. **MIGRATION:** Short- to medium-distance migrant with some only moving altitudinally. Spring: Departs southern winter areas ±15 Mar., though some still there early May; arrives western AK ±30 Apr. Fall: Departs northern breeding areas ±15 Sept.; first arrivals in southern wintering areas ±10 Oct. **WINTER:** Coastal AK to Southern CA (numbers vary considerably from year to year in CA) and parts of northern Rockies; rare to casual south and east of northern

Rockies. **RARE STATUS:** Rare to Bering Sea islands, and very rare to East, particularly northern tier of states and southern Canada. Accidental to Iceland and UK.

Population Logging in breeding range negatively impacts this species.

AZTEC THRUSH *Ridgwayia pinicola*   AZTH ■ 4

The Aztec Thrush is a Mexican visitor to the US Southwest. Although boldly patterned, it is shy and inconspicuous, often perching motionless for extended periods. Monotypic. L 9.3" (24 cm)

Identification ADULT MALE: Blackish hood, browner on crown, face, and back; blackish tail broadly tipped white. White uppertail coverts form a U-shaped band. A dark brown vent strap separates white belly from white undertail coverts. Wings have an intricate pattern of black, white, and dark brown, with obvious white tertial, primary, and secondary tips. **ADULT FEMALE:** Similarly patterned to male, but all dark colors are paler and with a more obvious streaked aspect to head and chest; tips of tertials and secondaries are grayish. **JUVENILE:** Back, scapulars, and chest are streaked with buff; belly is creamy-colored and scaled with brown; the dark eye line

contrasts with a whitish, streaked supercilium. Some older immatures possibly distinguished by presence of retained juvenal wing coverts. **FLIGHT:** Bold pattern is distinctive; whitish wing stripe contrasts with blackish secondaries with white tips.

Similar Species Distinctive, but compare to juvenile Spotted Towhee.

Voice CALL: Harsh, slightly burry and whining *wheeerr* and a *whining*, slightly metallic *whein*. **SONG:** Probably a louder and steadily repeated variation of the call note.

Status & Distribution Uncommon to rare. **BREEDING:** Pine forests in western and southern Mexico. **MIGRATION:** Northernmost breeders may be partially migratory, perhaps accounting for pattern of occurrence in US Southwest. **WINTER:** Primarily in breeding habitat and range, often joining mixed-species flocks of other frugivores (*Turdus* thrushes, Gray Silky-flycatchers). **RARE STATUS:** Casual to southeastern AZ sky islands (mostly late summer and early fall, occasion-

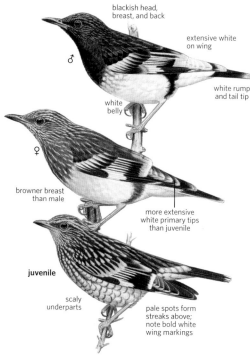

ally in small numbers) and Sierra Madre Oriental of Mexico (late fall, early winter). There are a handful of records from western and southern TX (fall, winter, and spring).

Population Likely declining due to habitat destruction.

MOCKINGBIRDS AND THRASHERS Family Mimidae

Long-billed Thrasher (TX, Apr.)

Thrashers and mockingbirds, along with the Gray Catbird, constitute the Mimidae found in North America. Most of the species are known for their long, varied songs of repeated phrases; some, the Northern Mockingbird in particular, are well-known mimics of other birds. The family reaches its highest diversity in the desert Southwest. They live in a variety of habitats, from wet thickets to desert washes and chaparral hillsides.

Structure Generally large for passerines, these birds have long tails and long, decurved bills used to probe for food items in leaf litter and holes in the ground. Their legs are rather long, with long toes used for running. Several species have distinctive yellow or orange eyes.

Behavior Many species are well adapted to a desert environment. They tend to be quite secretive, remaining well hidden during the heat of the day, but they often climb to the top of trees to sing. Members of the Mimidae family generally feed on insects and seeds, although several species take advantage of seasonal fruit production and feed on berries and cactus fruit, mainly in the fall and winter. Most are very territorial, defending both summer and winter territories. Some, particularly thrashers, form pairs that last multiple years. Most species are resident within their range. One species, the Gray Catbird, is a Neotropical migrant, whereas some others (Sage, Bendire's, and Brown Thrashers) move south seasonally and winter mainly in the southern US. Some species (e.g., Curve-billed Thrasher), although nonmigratory, disperse after the breeding season and have been detected well away from their breeding range.

Plumage Most thrasher species are brown, with varying amounts of streaking or spotting on the underparts. Others are gray and even blue, blue-and-white, or even black. Some are concolor and characterized by their rusty crissums (undertail coverts). The sexes are similarly plumaged in all species.

Distribution The Mimidae comprises about 34 species in 10 genera worldwide. Ten species in four genera breed in North America. Two additional species are casual or accidental visitors from the Neotropics: the Blue Mockingbird from Mexico and the Bahama Mockingbird from the West Indies.

Taxonomy The majority of the species are monotypic and show little, if any, geographical variation. Some, such as the LeConte's and the Long-billed Thrashers, have additional subspecies farther south in Mexico. The Curve-billed Thrasher can be divided into two distinct subspecies groups (that possibly warrant species status); the two groups may overlap in southeastern AZ. The Mimidae is the sister group to the starlings (Sturnidae).

Conservation Several North American thrashers, in particular those species with local ranges (e.g., LeConte's, California, Long-billed, and particularly Bendire's), are threatened by loss of natural habitat to urbanization and increased agriculture. Northern Mockingbird and Gray Catbird may benefit from an increase in disturbed habitats. Two island-restricted Mexican species (Socorro Mockingbird and Cozumel Thrasher) are Critically Endangered; the latter, not having been seen for several decades, is likely extinct. BirdLife International codes: 1 NT, 2 VU, 3 EN, and 2 CR.

Genera *Melanotis* and *Dumetella*

BLUE MOCKINGBIRD *Melanotis caerulescens*   BLMO ■ 5

This fancy Mexican endemic, a casual stray in the Southwest, is normally secretive, remaining hidden low or on the ground under dense thickets in thorn forest. It is usually detected first by hearing one of its loud, varied calls. Polytypic (2 sspp.; likely *caerulescens* in N.A.). L 10" (25 cm)

Identification Larger than Northern Mockingbird and not closely related, it is more thrasher-like in shape and behavior. Dark slaty blue overall (blackish in poor light); black mask. **ADULT:**

Appears to have paler blue streaking on head, throat, upper breast. Longish bill with a slight decurvature. Ruby red eye. **IMMATURE:** Grayer blue overall with brownish tinge to primary edges, less noticeable streaking; eye not as red.

Similar Species No other bird in N.A. is entirely blue with a black mask.

Voice CALL: Calls can be quite variable. Typical calls include a *chooo* or *chee-ooo*. Also a loud *wee-cheep* or *choo-leep*, a low *chuck*, and a sharp *pli-tick*. **SONG:** Thrasher-like varied

series of phrases that are often repeated.

Status & Distribution Endemic to Mexico. **WINTER:** Known to move altitudinally in fall and winter in Mexico. **RARE STATUS:** Casual in southeastern AZ and southernmost TX; origin questioned on records from coastal CA (e.g., a wintering adult in Long Beach) and NM. The AZ records all involve wintering birds.

Population Likely declining in Mexico.

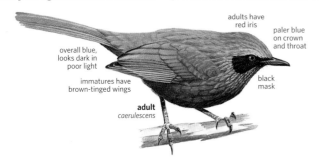

adults have red iris

paler blue on crown and throat

overall blue, looks dark in poor light

immatures have brown-tinged wings

black mask

adult
caerulescens

GRAY CATBIRD *Dumetella carolinensis*   GRCA ■ 1

This very distinctive mimid generally remains hidden in the understory of dense thickets in eastern woodlands and residential areas. It often cocks its longish, black tail, and it is usually detected by its harsh, down-slurred *mew* call, reminiscent of an agitated cat's *meow*. A chunky, medium-size bird, it is larger than *Catharus* thrushes yet smaller than other thrashers. No other N.A. bird has a uniform dark gray plumage. Monotypic. L 8.5" (22 cm)

Identification Sexes similar. Body entirely dark gray, with black cap, black tail, and chestnut undertail coverts. **JUVENILE:** Lacks black cap and chestnut on undertail much paler.

Similar Species Plumage unique. Its mimicking song vaguely resembles songs of other thrashers. *Mew* calls can be confused with calls of the Hermit Thrush or the Spotted and Greentailed Towhees.

Voice CALL: A nasal, catlike, downslurred *mew*. Also a *quirt* note and a rapid chatter alarm when startled. **SONG:** A variable mixture of melodious, nasal, and squeaky notes, interspersed with catlike *mew* notes. Some individuals are excellent mimics. Normally sings and calls from inside dense thickets.

Status & Distribution Common, but secretive. **BREEDING:** Nests in dense

thickets along edge of mixed woodland. In the West, where overall much less numerous, nests along willow- and alder-lined montane streams. **MIGRATION:** Nocturnal migrant. Trans-Gulf and Caribbean migrant. Sometimes abundant during migratory fallouts along the Gulf Coast of TX and LA and in southern FL. Spring peak in TX mid-Apr.–early May. **WINTER:** Mainly southeastern US (uncommon to rare elsewhere in East). Mexico, northern C.A., and Caribbean islands. **RARE STATUS:** Very rare during migration and winter in the Southwest and along Pacific coast. Casual to AK and northern Canada.

Population Western birds limited by loss of riparian habitats.

blackish cap

dark chestnut undertail coverts

short, slender dark bill

overall steel gray

THRASHERS Genus *Toxostoma*

CURVE-BILLED THRASHER *Toxostoma curvirostre*   CBTH   ■   1

The Curve-billed Thrasher is the common thrasher of the rich, cactus-laden Sonoran Desert and mesquite brushlands of the Chihuahuan Desert and southern TX. Polytypic (6–7 sspp.; 2 groups in N.A., thought by some to be two separate species). L 11" (28 cm)

Identification Sexes similar. A large thrasher with round, spotted underparts. Noticeable whitish wing bars and well-defined whitish tail tips on eastern birds; these features much fainter on western subspecies group. Long, black, decurved bill; eye distinctly orange-yellow to reddish (*curvirostre* group). **JUVENILE:** Recently fledged birds have less distinct spotting than do adults, and their bills are significantly shorter and less decurved.

Geographic Variation Two groups: The more easterly *curvirostre* group represented in US by *oberholseri* (includes *celsum*) is found from east side of the Chiricahuas, AZ, to southern TX. It has clearer spotting below against a whiter background, more distinct white wing bars, and more extensive

juvenile
palmeri

thick, decurved bill

striking orange-red iris

whitish wing bars

round spots below

oberholseri

iris averages more golden yellow than *oberholseri*

spots less obvious below than *oberholseri*

palmeri

distinct white tail tips

smaller, less contrasty tail tips and wing bars than *oberholseri*

celsum

palmeri

curvirostre group

maculatum

oberholseri

palmeri group occidentale

curvirostre

white tips to the tail feathers. The more westerly *palmeri* group (in US, only *palmeri* in southern AZ) has less distinct breast spots and less conspicuous white tips to the tail feathers. Calls differ. The degree of contact between the groups (west or south side of Chiricahuas) remains unstudied.

Similar Species Adults distinctive; note different habitat and calls compared with Bendire's. (See sidebar below.) Juvenile Curve-billed with shorter bill is easily confused with Bendire's.

Voice CALL: Very distinctive loud *whit-wheet* or *whit-wheet-whit* in *palmeri* group; in *curvirostre* group the call notes are all on one pitch. **SONG:** Long and elaborate, consisting of low trills and warbles, seldom repeating phrases.

Status & Distribution Common resident in desert habitats, particularly those rich in cholla and other cacti. Particularly common in suburban neighborhoods that retain natural desert vegetation. Also found in mesquite-dominated desert washes. **RARE STATUS:** Extralimital records mostly pertaining to *palmeri* from CA, NV, ID, AB, SK, MB, northern Plains states, various Midwestern states, LA, and FL Panhandle; *curvirostre* group once in winter to Yolo Co., CA. **Population** Stable.

Curve-billed and Bendire's Thrashers

In southern AZ and southwestern NM, Curve-billed and Bendire's Thrashers overlap in range and habitat. Although the Bendire's is very locally distributed, it can often be found with the more widespread Curve-billed. Adults in fresh plumage are more easily distinguished than worn adults. The larger Curve-billed has a longer, decurved bill; heavier, more blurry spotting on the underparts; and a frequent two-note *whit-wheet* call (*palmeri*) or *whit-whit* (*curvirostre* group). The slightly smaller Bendire's has a relatively short, mostly straight bill (sometimes with slight decurvature) and

Bendire's   Curve-billed

palmeri group   curvirostre group

finer, more distinct arrow-shaped spotting on the underparts. Its call is a seldom-heard low *chuck*. Juveniles and worn adults pose a greater identification challenge: The juvenile Curve-billed has a shorter, straighter bill and, typically, a pale area along the gape. Both birds when worn have virtually no obvious spotting on the underparts, making structural differences and call more important. White tips to tail are more reduced in Bendire's than in the Curve-billed *curvirostre* group, but they are quite similar to the tail pattern of the *palmeri* group.

BROWN THRASHER *Toxostoma rufum* BRTH ■ 1

The widespread thrasher of eastern N.A., this is a generally secretive bird of dense thickets and hedgerows that frequently sings from open exposed perches at the top of trees. Also visits suburban yards. Polytypic (2 sspp.; both in N.A., more western *longicauda* is paler). L 11.5" (29 cm)

Identification Upperparts entirely bright rufous; underparts white to buffy white, especially on flanks; extensive black streaking. Wing coverts with black subterminal bar and white tips, forming two wing bars. Bill long and slender with little decurvature. Yellow eye.

Geographic Variation Western *longicauda* larger, paler, with less extensive streaking than eastern nominate.

Similar Species Most similar to the Long-billed Thrasher of southern TX and often confused with that species. Brown Thrasher is always rare in the lower Rio Grande Valley. Long-billed is more grayish above and has a longer, more decurved bill, redder eye, and shorter primary projection; some calls are similar, but Long-billed gives diagnostic slurred whistles.

Voice CALL: A low *churr* and a loud, smacking *spuck*, somewhat resembling the call note of a "Red" Fox

Sparrow. **SONG:** A long series of varied melodic phrases, each phrase often repeated two or three times. Rarely mimics other bird species.

Status & Distribution BREEDING: Uncommon in dense thickets throughout the eastern US. **MIGRATION:** Birds from the northern portion of the breeding population migrate south in the fall, augmenting resident populations in the South. **RARE STATUS:** Rare in West to AZ and CA in migration and winter. Casual to AK, BC, YT, NT, NU, NL, and northern Mexico. Accidental UK.

Population Declines have been noted in the Northeast.

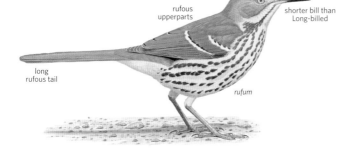

- rufous upperparts
- shorter bill than Long-billed
- long rufous tail
- *rufum*

LONG-BILLED THRASHER *Toxostoma longirostre* LBTH ■ 2

Superficially similar to the Brown Thrasher, the Long-billed Thrasher replaces Brown Thrasher in south TX and northeast Mexico where Brown Thrasher is rare. Long-billed is found in dense thickets and in the understory of remaining natural woodland habitat in south TX, particularly in the lower Rio Grande Valley. Habits similar to Brown Thrasher's, particularly in feeding behavior and singing from somewhat exposed perches. Away from the breeding season, Long-billed can be secretive and hard to locate, though it often gives call notes. Polytypic (2 sspp.; *sennetti* in N.A.). L 11.5" (29 cm)

Identification Sexes similar. Coloration of upperparts grayish brown; very black streaking below; face quite gray, contrasting with browner head and back; bill black, long, and distinctly decurved; orange eye.

Similar Species Most similar to Brown Thrasher, with which it barely overlaps in winter. (See account above.)

Voice CALL: Most common call is similar to the Brown Thrasher's loud smacking *spuck*; also known to give a mellow *kleak* and a loud whistle *cheeooep*. **SONG:** Similar to the Brown's

- *sennetti*
- *longirostre*

long series of melodious phrases, yet the individual phrases are not duplicated as often. Song differs from the overlapping Curve-billed Thrasher's song by being slower and more musical.

Status & Distribution Resident throughout breeding range in southern TX, north in very small numbers to the southern Edwards Plateau and on coast to Matagorda Co., and in northern Mexico. **RARE STATUS:** Very rare up the Rio Grande to Big Bend region, TX; casual north to Midland and to NM. Accidental CO.

Population Loss of native brushland to agriculture in the lower Rio Grande Valley has likely reduced populations.

- gray face
- grayish brown upperparts
- long decurved bill
- shorter primary projection than Brown
- *sennetti*

BENDIRE'S THRASHER *Toxostoma bendirei* BETH ■ 2

Although locally distributed in the desert Southwest, Bendire's Thrasher is easiest to find during the late winter and early spring, when singing activity is at its peak. It is normally somewhat secretive, with its pale sandy brown coloration blending in nicely with the sparse desert environment it prefers (Feb.–May). Males sit up on shrubs, fences, power poles, and roofs to sing, making them more conspicuous. Monotypic. L 9.8" (25 cm)

Identification Sexes similar. Easily confused with similar Curve-billed Thrasher, but slightly smaller. Overall plumage light brown above, slightly paler below; distinct arrow-shaped spots across the breast; buffy flanks and undertail coverts; relatively short, straight (or slightly decurved) bill, usually with pale at the base; orange eye. Tail with narrow pale tips to the feathers. **SUMMER ADULT:** Late spring and summer adults in worn plumage show little or no spotting.

Similar Species See sidebar, p. 567.

Voice CALL: A seldom heard low *chuck*. **SONG:** Different from Curve-billed Thrasher. A long series of warbling, melodic phrases, some with a harsh quality. Phrases often repeated two or three times each.

Status & Distribution Uncommon and local. **BREEDING:** Most numerous in the yucca-dominated Chihuahuan Desert and in the Sulphur Springs Valley in southeastern AZ. Rarer and more difficult to find at higher elevations across northern AZ, southernmost NV, and southern UT. A few nest in southeastern CA (southern Inyo Co. south to Riverside Co.). In CA and NV, found in Joshua tree woodlands. **WINTER:** Birds from the northern portion of the range migrate to low-elevation desert in southern AZ and northern Sonora, Mexico. Some birds move into suburban habitats. **RARE STATUS:** Casual in fall and winter mainly to coastal CA (formerly annual).

Population Vulnerable. Major declines throughout range.

spots on breast are triangular when fresh

shorter bill with pale base to lower mandible

fresh

worn

overall color a warmer brown than Curve-billed

CALIFORNIA THRASHER *Toxostoma redivivum* CATH ■ 2

The California Thrasher is found in chaparral foothills along the CA coastal slope and the west slope of the Sierra Nevada, and is usually the only thrasher found within its limited range. Most often detected by call or song. It often remains well hidden in dense vegetation on hillsides but will perch on exposed perches, particularly to call or sing. Especially in early spring, but throughout the year, resident males sing from exposed perches at the tops of shrubs. Polytypic (2 sspp.; both in N.A.). L 12" (30 cm)

Identification Dark overall, with long, black, decurved bill, heavier than Crissal and LeConte's. Tawny buff undertail coverts; paler than the Crissal's chestnut crissum. Pale throat and eyebrow contrast with dark, patterned face; dark eye.

Geographic Variation Northern *sonomae* and Southern CA nominate populations differ slightly.

Similar Species The California Thrasher is the only resident thrasher within its range. Similar in coloration to paler and grayer Crissal Thrasher, but note non-overlapping ranges. Stray Curve-billed and Bendire's Thrashers potentially overlap, but note California's more uniform coloration and heavy, blacker bill. California Thrasher does overlap with California Towhee, which is chunky and similarly colored, but note the thrasher's much larger size and strongly decurved bill.

Voice CALL: A low, flat *chuck* and *chur-erp*. **SONG:** Loud and sustained song made up of mostly guttural phrases, many repeated once or twice. Known to mimic other bird species and sounds.

Status & Distribution Fairly common resident on the chaparral hillsides and other dense brushy areas throughout most of coastal CA, the foothills of the western Sierra Nevada, extreme western edge of desert areas in Southern CA, and northwest Baja California. **RARE STATUS:** Very sedentary, but has wandered casually north to southern OR.

Population Losses in parts of its range due to development of preferred chaparral habitat and fires have no doubt reduced numbers.

white throat

long, strongly decurved bill

very long tail

tawny undertail coverts

overall brown coloration

LECONTE'S THRASHER *Toxostoma lecontei* LCTH ■ 2

This attractive, pale thrasher is generally found in sparser, more sandy desert than are other thrashers. The LeConte's behavior is similar to that of the Crissal, with which it overlaps: It is secretive but is often seen running between shrubs with its tail cocked in the air; it sings and calls from exposed perches but otherwise stays hidden. Although much paler than Crissal, both have unspotted underparts. Polytypic (3 sspp.; 2 in N.A.). L 11" (28 cm)
Identification Sexes similar. Very pale, creamy brown in coloration, with light tawny undertail coverts, thin black malar stripe, black lores, and a long,

thin, black decurved bill. Long, black tail contrasts with pale body. Dark eye.
Geographic Variation Nominate subspecies *lecontei* occupies much of range. Isolated *macmillanorum* from western San Joaquin Valley and adjacent Carrizo Plains, CA, is slightly darker. A third subspecies (*arenicola*), found in west coastal Baja California, has been proposed to be split as a separate species.
Similar Species LeConte's is paler and not as secretive as Crissal. Note its contrasting dark eye and lores; its tawny undertail coverts are lighter than Crissal. It overlaps with Curve-billed and Bendire's in southern AZ but is scarce and local, preferring more open areas, and differs strongly in plumage and behavior.
Voice CALL: Includes an ascending, whistled *tweeep* or *suuweep* and a shorter Crissal-like call. **SONG:** Very similar to Crissal Thrasher, being loud, melodious, and consisting of numerous phrases,

sometimes repeated. Little or no mimicry of other species. Heard mostly at dawn and dusk.
Status & Distribution Uncommon and local resident in sparse, sandy-soiled desert of southwestern AZ, Southern CA, southern NV, and extreme southwestern UT. Also found in northeastern Sonora and Baja California, Mexico. Very seldom wanders away from resident range.
Population Populations declining due to urbanization and conversion of habitat into agricultural land. Extirpated from Santa Barbara Co. and Fresno Co.

dark eye

paler overall than Crissal

thin, long, strongly decurved bill

tawny undertail coverts

very long tail

CRISSAL THRASHER *Toxostoma crissale* CRTH ■ 2

Very secretive and shy, the Crissal Thrasher prefers dense thickets. It sometimes sings quietly from exposed branches that stick up from the tops of shrubs. Found in a variety of habitats, from thickets in desert to juniper hillsides at higher elevation, it is one of the most difficult passerines to see well when not singing. Sometimes it can be seen running on the ground with tail cocked, similar to other desert thrashers. Crissal Thrasher feeds by probing its long bill into ground litter or holes. Along with California and the LeConte's Thrashers, it forms a

complex of uniformly colored thrashers with long (thin on Crissal and LeConte's), black, decurved bills and chestnut crissums. Polytypic (3 sspp.; 2 in N.A.). L 11.5" (29 cm)
Identification Sexes similar. Uniformly dark grayish brown; often shows a paler white throat, distinct black malar stripe, and chestnut undertail coverts. Worn individuals in late spring and early summer often lack the clear white throat, and the black malar stripe can appear quite faded. Small, dull yellowish eye.
Geographic Variation Differences

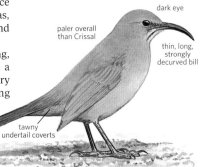

minor; eastern nominate is slightly darker than westerly *coloradense*.
Similar Species Crissal Thrasher is grayer and more uniform (not spotted below) than Curve-billed and Bendire's Thrashers, with which it overlaps in desert washes in southern AZ; all three can be found at the same locations. Range does not overlap with darker and browner California Thrasher, which has tawny, not chestnut, undertail. Overlaps with closely related LeConte's (sometimes found at the same location), but LeConte's is much paler overall and has a less-patterned malar and head, and lighter

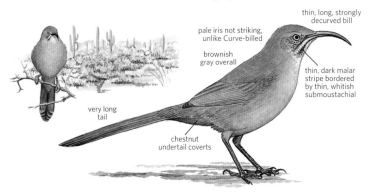

thin, long, strongly decurved bill

pale iris not striking, unlike Curve-billed

brownish gray overall

thin, dark malar stripe bordered by thin, whitish submoustachial

very long tail

chestnut undertail coverts

tawny undertail coverts; calls differ. **Voice CALL:** A loud repeated *chideery* or *churry-churry-churry* or *toit-toit-toit*. **SONG:** Varied long series of musical phrases, its cadence more leisurely and less harsh than the Curve-billed.

Status & Distribution Uncommon resident in the understory of dense mesquite and willows along streams; also ranging up to lower montane slopes at higher elevations. Very secretive, more so than other desert thrashers.

Not known to wander like the Curve-billed does.

Population Local population threats due to conversion of habitat into agriculture and urbanization. Not as adaptable to such threats as Curve-billed.

Genus *Oreoscoptes*

SAGE THRASHER *Oreoscoptes montanus* SATH ■ 1

This distinctive, small thrasher—characteristic of the open sagebrush plains of the montane West—is often seen singing atop sage bushes or along fence lines. It flies low from bush to bush and often runs on the ground with its tail cocked. On breeding grounds, Sage is the only thrasher with heavily streaked underparts and a short, relatively straight bill. Sage Thrasher,

which is smaller than other thrashers, is migratory, vacating breeding grounds during the winter. In migration and winter it favors open brushy areas, but can also be found feeding in fruiting trees, particularly junipers and Russian olives. More confiding than most other thrashers. Monotypic. L 8.5" (22 cm)

Identification ADULT: Generally gray-brown above, white below with heavy black streaking and a salmon-buff wash to the flanks. Wings with thin, distinct white wing bars. Tail long with white corners. Obvious black malar stripe, pale yellow iris, and short, relatively straight bill. **LATE SUMMER ADULT:** Can show virtually no streaking on the underparts when in very worn plumage. **JUVENILE:** Back and head streaked. Streaking on underparts reduced and not as black, especially on flanks. Eye darker.

Similar Species Brown and Long-tailed Thrashers also have heavily streaked underparts, but they differ by having more rufescent upperparts and longer, more decurved bills. Worn adults can lack streaking and look very similar to Bendire's Thrasher, but Sage Thrasher has a more contrasting head pattern and retains suggestion of white wing bars. Note difference in calls and song. (See juvenile Mockingbird.)

Voice CALL: Gives several calls, includ-

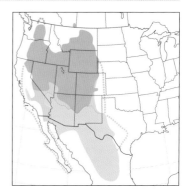

ing a *chuck* and a high-pitched *churrr*. **SONG:** Long series of warbled phrases.

Status & Distribution Fairly common **BREEDING:** Restricted to specialized sagebrush habitat, mainly within and west of the Rocky Mts. **MIGRATION:** Spring migration begins by late Jan.; in fall, most individuals leave breeding grounds by mid-Oct. Flocks occasionally detected during migration. **WINTER:** Mainly found in desert scrub habitat in the Southwest and northern Mexico. **RARE STATUS:** Disperses widely. Rare to coastal CA, casual east of breeding range, with numerous fall and winter records from many eastern states and provinces east to the Atlantic seaboard.

Population Likely decreasing due to habitat loss.

small size short bill streaked underparts

tish tips

worn

MOCKINGBIRDS Genus *Mimus*

Two members of this genus occur in N.A.; seven others are Neotropical. Generally medium-size with long tails and relatively straight bills, they have gray or brown coloration, with varying amounts of white in the wings and tail. *Mimus* mockingbirds, widely known as great songsters that mimic other species, are territorial and often sing at night.

BAHAMA MOCKINGBIRD *Mimus gundlachii* BAMO ■ 4

This Caribbean species occurs as a very rare stray to south FL, including the FL Keys and the Dry Tortugas. Generally more secretive than the Northern Mockingbird, the Bahama Mockingbird remains hidden in dense thickets. It behaves and looks more

like a thrasher than a mockingbird, often running between thickets. It is also browner above and more streaked above and below than the Northern Mockingbird. Polytypic (2 sspp.; nominate in N.A.). L 11" (28 cm)

Identification Sexes similar. **ADULT:**

Black streaks on a brown back extend up neck onto head; black streaking found on lower belly, flanks, and undertail coverts. Wings have two relatively narrow white wing bars but no white at base of primaries. Tail has white tips to outer feathers. Head has obvious black malar stripe and white eyebrow. Bill is longer and slightly more curved than bill of Northern Mockingbird. **IMMATURE:** Less patterned than adult, with less distinct streaking on back and flanks; dark malar less pronounced.

Similar Species Immature Northern has spotting or streaks below, but more confined to breast, compared with the Bahama's streaked flanks. All plumages of Bahama Mockingbird lack the white flash in primaries; all plumages also have white in tail confined to the tips of the outer feathers.

Voice CALL: A loud *chack* similar to Northern Mockingbird. **SONG:** More thrasher-like, with a complex mixture of phrases. Not known to mimic.

Status & Distribution The nominate subspecies is a rather uncommon resident in the Bahamas and is rare on cays off northern Cuba. The subspecies *hillii* is a local resident on Jamaica. **RARE STATUS:** Very rare visitor, primarily in spring, to parks with dense thickets along the coast of southern FL, with most records from between West Palm Beach and Key West. Also recorded from the Dry Tortugas in spring.

Population Potential loss of breeding habitat due to development, and hurricanes, may impact this species in the Bahamas and Cuba.

brownish tinge to back

gundlachii

distinct streaks on lower sides, flanks, and undertail coverts

lacks white wing patch of Northern Mockingbird

darker tail than Northern Mockingbird

whitish tail tips

gundlachii

hillii

NORTHERN MOCKINGBIRD *Mimus polyglottos*   NOMO ■ 1

This very common, well-known, and conspicuous mimid of the southern US is known for its loud, mimicking song, often heard during spring and summer nights in suburban neighborhoods. Both sexes aggressively defend nesting and feeding territories. They flash their white outer tail feathers and white wing patches conspicuously during courtship and territorial displays. Often seen on wires and fences in towns, Northern Mockingbirds frequently feed on berries during the winter. Monotypic. L 10" (25 cm)

Identification Sexes similar. **ADULT:** About the size of an American Robin, but thinner and longer tailed. Upperparts gray, unstreaked; underparts grayish white, unstreaked; long black tail has white outer tail feathers; conspicuous white wing bars; white patch at the base of primaries contrasts with blacker wings. Black line through yellow eye; bill relatively short and straight. **JUVENILE:** Underparts can be heavily spotted; upperparts with pale edging give back and head a streaked appearance; black line through eye less distinct; eye darker.

Similar Species The Loggerhead Shrike is easily separated by overall shape, face mask, and bill shape; Northern Mockingbird has large white wing patch, especially visible in flight Also see Sage Thrasher. A long-staying Tropical Mockingbird (*M. gilvus*) near Sabine Pass, TX (nested with a Northern Mockingbird, producing hybrid young), was not accepted on origin; it is resident as close as the Yucatán.

Voice CALL: A loud, sharp *check.* **SONG:** Long, complex song consisting of a mixture of original and imitative phrases, each repeated several times. Excellent mimic of other bird species. Often sings at night.

Status & Distribution Common and conspicuous. **BREEDING:** Nests in a variety of habitats, including suburban neighborhoods. **MIGRATION:** Northern birds are partly migratory, but winter well north. **RARE STATUS:** A frequent stray, birds are regularly found well north of mapped range, even to AK and northern Europe. Accidental to Europe.

Population Stable.

white outer tail feathers

extensive white wing patch

juvenile

spotted below

grayish above, whitish below, with white wing bars

STARLINGS Family Sturnidae

European Starling

With their glossy plumages and sociable ways, it is not surprising that starlings and mynas have been popular with bird fanciers for generations. The approximately 114 species of Sturnidae (in 34 genera) are native to the Old World, but several species have been introduced across the globe. The three species treated here have had varying degrees of success in North America: The widespread European Starling is the most successful exotic, the adaptable Common Myna is flourishing in south FL, and the Hill Myna is declining in FL. The Crested Myna (not covered here) was extirpated from Vancouver, BC, by 2003 and from FL.

Structure Most starlings are of medium size and medium build, with fairly thick, slightly decurved bills and sturdy feet. Mynas are on average larger, chunkier bodied, thicker billed, and shorter tailed than many starlings. Many species have ornate plumes or bare-part projections on the head and face.

Behavior Many species are gregarious, breeding colonially and gathering in immense numbers during the nonbreeding season. Social behavior is often advanced. Vocalizations are complex; many notes are coarse and unmusical, but several species are capable of a large array of sounds. Starlings and mynas adapt well to human-altered and human-inhabited landscapes, and introductions have succeeded in many regions.

Plumage Starlings vary in plumage, but many are largely black and iridescent, with patches of bright oranges, reds, and yellows on the feathers. Bare parts are frequently brightly colored. The species treated here are primarily black and iridescent with limited areas of white and bright yellow patches on some bare parts.

Distribution Ancestrally, starlings and mynas, which tend to be cavity nesters, were associated with woodlands in Africa, Southeast Asia, and their regional island complexes. For millennia, the family has been increasingly cosmopolitan in its range and generalist in its habitats. In North America, the European Starling is established continentwide; other established sturnids are local. Some species are resident; others highly migratory.

Taxonomy Recent evidence points to affiliations with the Turdidae, Muscicapidae, and especially the Mimidae. Within the Sturnidae, there may be a major break between the African and Asian taxa. The polyphyletic mynas belong to the Asian group, as does the European Starling. The two creeper-like *Rhabdornis* from the Philippines are now placed in the starling family.

Conservation Worldwide, the Sturnidae run the gamut from abundant nuisances (European Starling) to species in danger of extinction (Bali Myna). Habitat loss and capture for the cage-bird trade have been factors in the decline of several species. Six starlings with island-restricted ranges have gone extinct. Established populations of European Starling in North America are sometimes targeted for control. BirdLife International codes: 10 NT, 6 VU, 2 EN, 7 CR, and 6 EX.

Genus *Sturnus*

EUROPEAN STARLING *Sturnus vulgaris* EUST ■ 1

Widespread and abundant in much of N.A., the introduced European Starling is arguably and problematically the most successful bird on the continent. Often characterized as bold, this bird is actually fairly wary and can be difficult to approach. Polytypic (13 sspp.; nominate in N.A.). L 8.5" (22 cm)

Identification Stocky and short tailed, often seen strutting about lawns and parking lots. Flight profile distinctive: Wings look triangular in flight and can appear somewhat translucent. **ADULT:** One molt per year, but fresh fall adults look very different from summer birds. On freshly molted birds, black plumage has white spots all over; by winter, spots start to disappear from wear and then they actually break off; and by spring, the birds are glossy black all over, with strong suffusions of iridescent pinks, greens, and ambers. Slender and pointed bill usually gray in fall and yellow by winter. **MALE:** By spring has blue-based bill. **FEMALE:** By spring has pink-based bill; paler eyes. **JUVENILE:** Distinctive; dark gray-brown feathering all over. Birds begin a

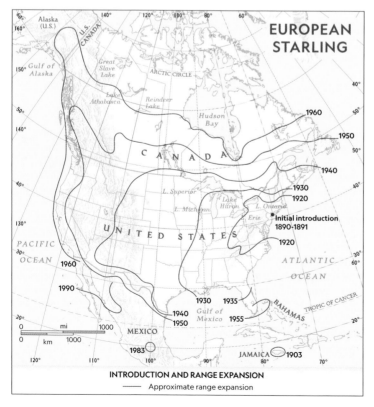

short tail

vulgaris

triangular wings

slender dark bill

plumage heavily spotted with whitish

fall *vulgaris*

short tail

overall grayish brown

yellow bill

glossy plumage

winter *vulgaris*

juvenile *vulgaris*

dull, blurred streaks on belly

breeding ♂ *vulgaris*

complete molt into adultlike plumage soon after fledging, and briefly exhibit a striking mosaic of juvenal and adult feathers, usually replacing the head feathers last.

Similar Species Structure is distinctive, but sometimes confused with unrelated blackbirds, which occur with starlings in large flocks. Blackbirds are more slender bodied, with longer tails and less pointed wings. Flight profile is more like a waxwing or a meadowlark than a blackbird.

Voice Highly varied. **CALL:** Commonly heard calls include drawn-out, hissing *sssssheeeer* and whistled *wheeeeoooo*. **SONG:** Elaborate, lengthy (>1 min. long), with complex rattling and whirring elements, and overall wheezy quality; call notes may be incorporated into song. Imitates other species, especially those with whistled notes.

Status & Distribution Old World species. Abundant, still expanding range in the Americas. **BREEDING:** Needs natural or artificial cavities. Often

EUROPEAN STARLING

1960
1950
1940
1930
1920
Initial introduction 1890-1891
1920
1930 1935
1940 1950
1955
1903
1983
1960
1990

INTRODUCTION AND RANGE EXPANSION
— Approximate range expansion

evicts native species from nest holes. **MIGRATION:** Withdraws in winter from northern portion of range. **WINTER:** Gregarious, with largest concentrations around cities, in nighttime roosts that can number into the tens of thousands or more. **RARE STATUS:** Casual northern Canada. Specimen from Shemya I., western Aleutians, AK (12 Sept. 2012), appears to be *vulgaris,* thus originating from N.A., rather than Asian *poltaratskyi.*

Population Successfully introduced in Central Park, NY, 1890–91; spread across continent by late 1940s. Population currently exceeds 200 million. Asian *poltaratskyi* is uncommon in winter to Japan.

Genus *Acridotheres*

COMMON MYNA   *Acridotheres tristis*   COMY ■ 2

This Asian native is well established in south FL. It is a popular cage bird (the "House Myna"), and escapes may be seen anywhere. Introduced in HI in 1866 and now common and widespread throughout the main islands. Polytypic (2 sspp.). L 9.8" (25 cm)
Identification Chunky; often in loose interspecific or intraspecific flocks. **ADULT:** One molt per year. Molt schedule variable; FL data lacking. Brown body; black head with yellow-orange

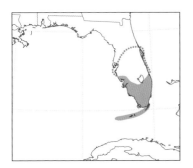

bare patch around eye; yellow-orange bill; dull yellow-orange feet; white patches at base of primaries (conspicuous in flight), tail tip, and undertail coverts. **JUVENILE:** Duller overall.
Geographic Variation Populations in N.A. presumably of widespread nominate subspecies.
Similar Species Resembles other *Acridotheres* mynas. Outside of N.A., hybridizes with the Crested and other *Acridotheres* mynas, especially in introduced settings.
Voice Varied, but most vocalizations raucous; accomplished mimic. **CALL:** Loud, slurred whistles; short, grating notes. **SONG:** Long series of short notes, given by both sexes, with liquid, rolling syllables. Call notes often incorporated into song.
Status & Distribution First detected in FL in 1983; now common in southern FL cities and suburbs. **YEAR-ROUND:** Warehouses, malls, fast-food restaurants, etc.; nests primarily in

blackish head

yellow bill

yellow skin around eye

tristis

white wing patch

brown above and below

white undertail coverts

white tail tip

artificial cavities, especially signage.
Population The FL population has increased considerably; perhaps thousands of pairs currently breeding.

Genus *Gracula*

HILL MYNA   *Gracula religiosa*   HIMY ■ EXOTIC

Even when judged by the overall standards of the myna clan's vocal excellence, Hill Myna is notable for its superior abilities as a songster and mimic, particularly for those in captivity. A population in south FL is waning. Polytypic (9 sspp.; *intermedia* in N.A. from northern India, Thailand, Indochina, and southern China; the Southern Hill Myna, *G. indica,* from southwest India and Sri Lanka, has been recently split). L 10.6" (27 cm)
Identification Large and chunky. **ADULT:** Sexes similar. Black overall; white primary bases, shows as a bold white patch in flight. Warty yellow nuchal and suborbital bare patches conspicuous. Dark eyes, orange bill, yellow-orange feet. **JUVENILE:** Plumage browner, duller; bare parts less colorful.

Similar Species Chunkier and larger than *Acridotheres* mynas. Wings are more rounded. Tail tip not white, as in *Acridotheres.* Partial to trees; *Acridotheres* mynas are found on the ground.
Voice Varied; many vocalizations, especially from birds in captivity, have human qualities. **CALL:** Tremendously varied, and with marked variations among subspecies in its native range in southern Asia. Most are of short duration, some extremely loud. **SONG:** Complex; short to long series of chuckles and whistles. Song elements on average mellower than those of *Acridotheres* mynas.
Status & Distribution A native of Asian forests from India to the Philippines. Uncommon around Miami, FL. **YEAR-ROUND:** Arboreal; nests primar-

ily in natural cavities in tall trees.
Population Remnant introduced population in Miami area currently numbers 25–100 breeding pairs.

thick orange bill

yellow wattle

intermedia

white wing patch

WAXWINGS Family Bombycillidae

Bohemian Waxwing (MN, Jan.)

Finding waxwings is often feast or famine. They may be abundant one year, gone the next. Flocks are often quite tame and allow observers to approach closely. Once waxwings are found, identifying them is straightforward: Differences in the wings, tail, and underparts between species are easily seen.

Structure Waxwings are relatively chunky birds with a pointed crest, conspicuous when perched. They have triangular wings and a short, square, or slightly rounded tail with long undertail coverts giving them an appearance similar to European Starlings. When perched, their silhouette is distinctive. The short, broad bill is well adapted for grasping and gulping down berries.

Behavior No other bird family in North America is more addicted to fruit than the Bombycillidae. From fall through early spring waxwings feed almost exclusively on sugary fruits; however, they will also consume developing fruits (flowers) and insects in springtime. Except when nesting, waxwings are usually seen in flocks. In winter flocks number in the hundreds, sometimes in the thousands. They frequently forage in fruiting trees, sometimes with American Robins, bluebirds, and Pine Grosbeaks. Even during the breeding season, nesting pairs are often found in close proximity to each other, near an abundant food source. Lacking defended territories, waxwings evolved an unusual characteristic among passerines—they do not sing.

Plumage Waxwing plumages are characterized by an elegant blend of soft and silky smooth browns and grays. Waxwings have dark throats and masks. The wings vary from the simple white edging on the inner edge of the tertials on the Cedar Waxwing, to the more elaborate patterns found in the Bohemian and Japanese Waxwings. Bombycillids have a partial post-juvenal molt. Differences between the sexes are subtle and may be challenging to see under most field conditions. Waxwings are named for the waxy "droplets" at the end of their secondaries (and, rarely, the tail feathers), which are unique to the family. Synthesized from carotenoid pigments, these droplets form as extensions of the rachis that project beyond the feather veins. These may be lacking, or at least reduced on young birds, and most developed in adult males. The link between diet and feather color has been well studied in Cedar Waxwings. Typically, Cedars only synthesize yellow carotenoids from their diets, resulting in the classic yellow tail tip. In much of eastern North America, however, the introduction of exotic honeysuckle—with the red carotenoid pigment rhodoxanthin—has led to Cedars molting in tail feathers with orangish tips; the exact color depends upon how much honeysuckle was consumed during the molt.

Distribution Waxwings are found only in the Northern Hemisphere. Bohemian Waxwing is Holarctic, while Japanese Waxwing inhabits eastern Asia, and the Cedar Waxwing resides primarily in North America.

Taxonomy There are three waxwing species, all in the genus *Bombycilla*. Waxwings are closely related to silky-flycatchers (Ptiliogonatidae). Waxwings differ from silky-flycatchers in having pointed wings, a minute tenth primary, and waxy "droplets" at the tips of the secondaries. They lack the well-developed rictal bristles found in silky-flycatchers. The Palmchat (*Dulus dominicus*) of Hispaniola is also closely related. The extinct Oo's of the Hawaiian Is. (family Mohoidae) are also related to this group of small families.

Conservation Bohemain Waxwing is likely stable in North America and the Old World. Cedar Waxwing is possibly increasing. Japanese Waxwing, thought to have a small population, is classified by BirdLife International as Near Threatened due to habitat loss and degradation.

Genus *Bombycilla*

BOHEMIAN WAXWING *Bombycilla garrulus* BOWA ▪ 2

Bohemian Waxwings often winter in flocks of several hundred birds. In summer, they travel in pairs or small flocks and frequently perch on the tops of black spruces. Stray individuals to the south are almost always found with Cedar Waxwings. Polytypic (2–3 sspp.; 2 in N.A.). L 8.3" (21 cm)

Identification Starling-size; sleek crest; grayish overall with face washed in chestnut; rufous undertail coverts; tip of tail yellow. White primary tips make wing appear notched with white. Sexes similar; male averages slightly larger throat patch, broader yellow tip to tail, and more extensive waxy

tips to secondaries. First-winter birds lack white tips to primaries and waxy tips reduced. **JUVENILE:** Streaky below with white throat (June–Oct.). **FLIGHT:** White bases to primary coverts appear as contrasting white band in flight.

Geographic Variation Subspecies *centralasiae* (often merged with nominate *garrulus*) is casual from Asia to western Aleutians (specimens) and Pribilof Is. Paler than *pallidiceps* of N.A. with little contrast between forehead and rest of head; often darker undertail coverts.

Similar Species See Cedar Waxwing. In flight, European Starlings appear similar in shape and flight style.

Voice CALL: Commonly a high, sharply trilled *zeeee*, lower and distinctly more trilled than Cedar Waxwing (more rattlelike).

Status & Distribution Uncommon to irregularly common. **BREEDING:** Open coniferous or mixed woodlands. **MIGRATION:** In fall departs interior AK in Sept. First arrivals in SK late

juvenile

streaked underparts

white tips of primary coverts conspicuous in flight

crest

gray belly

pallidiceps

cinnamon undertail coverts

centralasiae

darker cinnamon undertail than *pallidiceps* and paler overall

Sept.; huge flocks arrive Dec. First arrivals in northern US mid-Oct.–early Nov. In spring, most leave lower 48 by mid-Mar., but some remain into Apr. **WINTER:** Largest concentrations occur across southern BC and AB and south-central SK. Irregular to Northeast, usually in small numbers; almost annual in ME, Maritimes, NL. **RARE STATUS:** Rare and irregular to dashed line on map. Casual to Southern CA, northwestern AZ, TX (mainly Panhandle), northwestern AR, southern NJ, southern MD.

Population Stable.

CEDAR WAXWING *Bombycilla cedrorum*   CEDW ■ 1

Cedar Waxwings are easily found in open habitat where there are berries. This species times its nesting to coincide with summer berry production, putting it among the latest of N.A. birds to nest. It is highly gregarious; flocks of hundreds, occasionally thousands, are encountered during migration and winter. Now regarded as monotypic. L 7.3" (18 cm)

Identification Smaller than Bohemian Waxwing, with pale yellow belly and whitish undertail coverts. Tip of tail usually yellow, broadest in adult males, narrowest in immature females. Some birds (especially imma-

tures) have an orange tail tip, a result of consuming non-native honeysuckle fruit during molt. **JUVENILE:** Streaky below with white chin and bold malar stripe (July–Nov.).

Similar Species Bohemian Waxwing is similar but larger, grayer, has rufous undertail coverts, white bar on primary coverts, and chestnut wash on face. A juvenile Cedar can be separated from a Bohemian by its lack of white wing patches, and lack of any rufous on undertail coverts.

Voice CALL: Commonly a high trilled *zeeeee*, higher and less trilled than Bohemian. Also a long, high, pure *seeee*; and a shorter descending *sweeew*, longer than analogous call of Bohemian. Does not sing.

Status & Distribution Common. **BREEDING:** Open woodlands and old fields; rare but regular breeder in southeastern AK and southern YT. **MIGRATION:** Arrival in Great Lakes region is not usually until May and departure Sept.–Oct. **WINTER:** Southern US to C.A., some linger until early June. **RARE STATUS:** Rare to Panama and the West Indies, most frequent on Cuba. Casual to central and western

AK. Accidental to Iceland, UK, and Azores.

Population Increasing, likely due in part to spread of exotic fruiting plants.

waxwings have triangular wing shapes

juvenile

streaked underparts

crest

yellowish belly

yellow tail tip

white undertail coverts

SILKY-FLYCATCHERS Family Ptiliogonatidae

Phainopepla, male (AZ, May)

Members of this small Middle American family are often found in close proximity to fruit, although insects are just as important in their diet. The family's English name describes their soft, sleek plumage and their frequent flycatching. They are not at all closely related to flycatchers, which has led some people to call them by the simpler name "Silkies." Identification of all silky-flycatcher species is straightforward, with differences in coloration, structure, and patterning obvious.

Structure Silky-flycatchers are mostly slender thrush-size birds with long tails and small bills. Most are crested. Their rounded wings help them make quick sallies from branches to catch insects.

Behavior Silky-flycatchers occur in small flocks, pairs, or alone and typically perch near the tops or edges of shrubs and trees. Most call frequently, making them easy to find. During the breeding season, they may form loose colonies when there is an abundance of fruit; during these times adults defend nests from other members of the same species but usually do not defend food supplies. They construct shallow cup nests placed in crotches of trees and shrubs.

Plumage Ptiliogonatids have plumages characterized by black, gray, or pale brown, often with yellow or white highlights. The plumages are generally smooth and blended, without well-defined lines or distinct markings of color. Even juveniles lack spots, bars, streaks, or other distinctive markings. All species are sexually dimorphic, with males' plumage brighter or exhibiting more contrast. In Phainopepla there is a supplemental plumage (before the post-juvenal molt). The first post-juvenal molt varies from partial to complete. There is no prebreeding molt.

Distribution Silky-flycatchers are found from the southwestern US to western Panama. Aside from the desert-loving Phainopepla, they primarily inhabit mountain ranges. Movements in Middle America require more study, but since these species feed heavily on fruit, it is likely all species stage at least local movements.

Taxonomy Ptiliogonatidae breaks down into three genera with four species. The family is very closely related to waxwings and sometimes merged with that family. Silky-flycatchers differ from waxwings in having rounded wings and prominent rictal bristles; the tenth primary is also larger and the ninth primary is smaller. The monotypic genus *Phainoptila* (Black-and-yellow Silky-flycatcher found in the highlands of Costa Rica and western Panama) appears very different—much more thrushlike; however, unlike thrushes, the juvenal plumage is plain and unspeckled and the short tarsi are covered with broad transverse scales—both characteristics of silky-flycatchers.

Conservation Habitat destruction and degradation pose the biggest threat to silky-flycatchers. Habitat fragmentation may not pose as great a threat as it does for many families since all silky-flycatchers are found to some degree in edge habitats. Some species are caught and kept in captivity. All four species are listed as Least Concern by BirdLife International.

SILKY-FLYCATCHERS Genus *Ptiliogonys*

GRAY SILKY-FLYCATCHER *Ptiliogonys cinereus*   GRSF ■ 5

This M.A. species is an accidental visitor to TX and perhaps elsewhere. The Gray Silky-flycatcher feeds primarily on berries of mistletoe and other plants, as well as insects; it often perches conspicuously atop trees. Polytypic (2–4 sspp.). L 7.5" (19 cm)

Identification Structurally similar to Phainopepla, but with a bushy crest that is not as wispy. **ADULT MALE:** Generally blue-gray overall, with orangey-yellow flanks and bright yellow undertail coverts. Wings and tail dark; tail has a broad white base to outer tail feathers. Dark lores contrast with white eye ring and paler forecrown.

ADULT FEMALE: Similar to male, but coloration more subdued, with more brownish tones. Yellow often confined to undertail coverts. **JUVENILE:** Similar to adults, but duller overall.

Geographic Variation There is some discussion as to how much variation truly exists; differences in age and sex are subtle, and they have not been fully resolved. Records for the US are *otofuscus* from northwestern Mexico or nominate *cinereus* from eastern and central Mexico.

Similar Species This species is unmistakable; note its structure, upright posture, bushy crest, eye ring, yellow at least on undertail coverts, and white patches in the long tail.

Voice CALL: Gives varied, chattering notes; flight calls and some calls given when perched may suggest call notes of *Piranga* tanagers.

Status & Distribution BREEDING: Pine-oak and pine-evergreen forest in mountains of Mexico and Guatemala. Largely resident, but occasionally nomadic; moves in large flocks. May move locally downslope in winter; possibly withdraws from northwestern Mexico. Potential for extralimital records pertaining to escapes complicates understanding natural distribution. **RARE STATUS:** Accidental in TX: Laguna Atascosa NWR (21 Oct.–11 Nov. 1985) and El Paso (2 Jan.–5 Mar. 1995). Five seasonally scattered records from southwestern CA controversial; natural occurrence questioned, but could be wild strays.

Population Stable.

gray crest

yellow undertail coverts

white based tail

adult ♂

Genus *Phainopepla*

PHAINOPEPLA *Phainopepla nitens* PHAI ■ 1

Phainopepla—Greek for "shining robe"—is the only ptilogonatid regularly found north of the US-Mexico border. The Phainopepla is often solitary but may gather to feed on seasonally abundant crops. It feeds heavily on mistletoe berries in winter; its diet includes more insects at other seasons. It vigorously defends feeding territories, and excrement may pile several inches high under territorial perches. It perches upright, usually in the open on tops of trees. Polytypic (2 sspp.; both in N.A.). L 7.7" (20 cm)

Identification Slender thrush-size bird with a shaggy crest and long tail. Red eyes (both sexes). Males substantially larger than females. **MALE:** Shiny black; white wing patches conspicuous in flight. **FEMALE:** Grayish; white edging on all wing feathers (less so on primaries and secondaries). White wing patches absent. **JUVENILE:** Similar to adult female but with buffy wing bars and brownish upperparts; dark eye. Immature male typically begins to acquire black feathers in fall, producing black patches. **FLIGHT:** Fluttery but direct; bold wing patches revealed on adult male.

Similar Species Distinctive. Northern Mockingbird has white wing patches, but a very different shape and horizontal posture. Mockingbirds are paler gray with darker wings and tail (tail with white corners).

Geographic Variation Smaller *lepida* (US Southwest to Baja California and northwest Mexico) and larger *nitens* (southwest TX to central Mexico) differ in wing and tail length; they are not separable in the field.

Voice CALL: A distinctive, querulous, low-pitched, whistled *wurp?* May imitate other species (e.g., Red-tailed Hawk and Northern Flicker). **SONG:** A brief warble that includes a whistled *wheedle-ah*; infrequently heard.

Status & Distribution Uncommon to common. **BREEDING:** Late winter–early spring in mesquite brushlands; in summer moves into wetter habitats and raises a second brood. Highest densities occur in riparian woodlands and dense mesquite thickets. **MIGRATION:** Apparent short-distance migrant. Begins spring migration late Mar.–early Apr.; arrives northern CA early to mid-Apr. Fall return occurs Aug.–Nov. **WINTER:** Mostly lowlands, from Sonoran Desert to central Mexico. **RARE STATUS:** Casual in southern OR, CO, KS, south-central and extreme eastern TX. Accidental to eastern N.A. with records for ON, WI, RI, and MA.

Population Apparently stable; potential threats include continued destruction of mesquite forests and riparian woodlands in the Southwest.

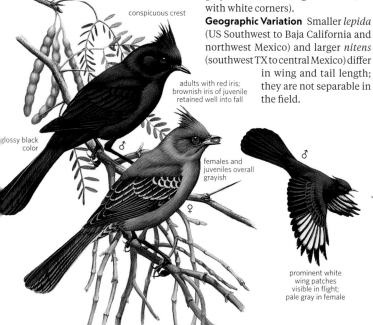

conspicuous crest

adults with red iris; brownish iris of juvenile retained well into fall

glossy black color

♂

females and juveniles overall grayish

♀

♂

prominent white wing patches visible in flight; pale gray in female

OLIVE WARBLER Family Peucedramidae

Olive Warbler, female or 1st-spring male (AZ, Apr.)

Long considered an aberrant wood-warbler, the Olive Warbler is strongly differentiated genetically and is now placed in its own family with uncertain relationships.

Structure Olive Warblers resemble *Setophaga* wood-warblers but have a long, very thin bill, a slightly decurved culmen, distinctly notched tail, and 10 primaries.

Plumage The tawny orange head coloration of the adult male is unlike any wood-warbler, and a white patch at the base of the primaries is shared with only the Black-throated Blue Warbler.

Behavior Vocalizations consist of loud whistled notes (the females also sing). They flick their wings like kinglets. Olive Warblers allow their nestlings to soil the nests with droppings, a behavior unknown in wood-warblers.

Distribution Found in montane coniferous and mixed pine-oak forest from southeastern AZ and southwestern NM to extreme western Nicaragua. The northern subspecies is partially migratory; the southern subspecies are sedentary.

Taxonomy On the basis of anatomy, Olive Warblers show some similarities to the vast array of Old World warblers, but some behavioral differences and recent genetic evidence contradict this and suggest a closer relationship to the finches and nine-primaried songbirds. More study will resolve the taxonomic position. Olive Warblers are generally considered to be more primitive than wood-warblers.

Conservation Populations likely affected by logging. Listed as Least Concern by BirdLife International.

Genus *Peucedramus*

OLIVE WARBLER *Peucedramus taeniatus*   OLWA ■ 2

A distinctive warblerlike resident of southwestern pine forests, with a loud song and kinglet-like wing flicking. Polytypic (5 sspp.; *arizonae* in N.A.). L 5.2" (13 cm)

Identification ADULT MALE: Plumage acquired by second fall. Tawny orange head and upper breast with distinct black ear patch. Gray back. Two broad white wing bars and a white spot at base of primaries. Tail distinctly notched, with large white spots. **ADULT FEMALE:** Crown, nape, and back grayish. Ear patch mottled gray and black, surrounded by yellowish. Throat and breast yellowish. **IMMATURE MALE:** Closer to adult male than adult female, showing more mottled black and gray cheek patch, more yellowish head, including crown and upper breast. **IMMATURE FEMALE:** Like adult female but head duller and back more brownish.

Geographic Variation Northern subspecies *arizonae* in N.A. is larger, paler, and greener-backed than the four southern subspecies.

Similar Species Immatures are similar to an immature female Hermit Warbler, which lacks white at base of primaries and has a yellower face, indistinctly notched tail, and different bill shape, calls, and songs.

Voice Very distinct. **CALL:** A soft, whistled *phew*, similar to Western Bluebird. Also a hard *pit*. **SONG:** A loud *peeta peeta peeta peeta*.

Status & Distribution Fairly common in mountains of southeastern AZ and southwestern NM. **BREEDING:** Higher elevation pine and fir forests. **NEST:** High in conifer, far from trunk, three to four eggs (May–June). **MIGRATION:** Arrive in AZ early Apr. Departure schedule unknown. **WINTER:** Most move south of AZ; very few remain in breeding range, sometimes at slightly lower elevations. **RARE STATUS:** West TX and portions of NM outside breeding range. **Population** Slight declines in N.A.

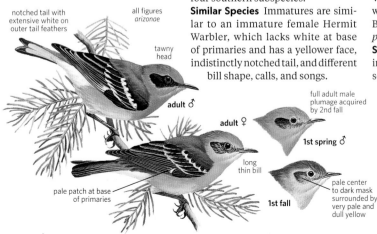

notched tail with extensive white on outer tail feathers

all figures *arizonae*

tawny head

adult ♂

full adult male plumage acquired by 2nd fall

adult ♀

1st spring ♂

long thin bill

pale patch at base of primaries

pale center to dark mask surrounded by very pale and dull yellow

1st fall

ACCENTORS Family Prunellidae

Siberian Accentor (ID, Feb.)

The accentors are an Old World family, found from Spain, across Eurasia, to the Russian Far East and Japan. The greatest diversity of species is found in the mountains of central Asia. Some species are migratory; others are resident.

Structure Accentors are small to medium-size passerines, similar in size and shape to our New World sparrows (in Passerillidae), although accentors have thin, warbler-like bills. Some species (e.g., Alpine Accentor) are stocky. The legs are short, and the tail is short (two species) to medium length and typically notched.

Plumage Accentors are brownish above, often with rufous tones, and streaked on the upperparts. Some species are largely grayish below while others are buffy or cinnamon below, some showing contrasting white or black throats. Many have a contrasting face pattern with a noticeable paler or warm-colored supercilium.

Behavior These ground foragers specialize on insects, but take seed in winter. Most are little known, except the familiar Dunnock of Europe and the largest member of the family, the Alpine Accentor, a widespread species across Eurasia.

Distribution Restricted to the Palearctic, from the British Isles to Japan, the family is confined to temperate latitudes as far south as North Africa, Yemen, and southern Asia. Several species are restricted to alpine or highland areas. One species has been introduced to South Africa and New Zealand.

Taxonomy There are 12 accentors, all in the genus *Prunella*. Thrushes were once thought to be closely related to accentors, but recent genetic studies place the accentors in the Passeroidea, a group of families that includes finches and Old World buntings, towhees and sparrows, tanagers, blackbirds, sunbirds, weavers, pipits, and other taxa. Therefore, accentors appear to be only very distantly related to the thrushes.

Conservation All 12 species are listed as Least Concern by BirdLife International.

Genus *Prunella*

SIBERIAN ACCENTOR *Prunella montanella*   SIAC ■ 4

The Siberian Accentor is a rare visitor from Asia to AK in fall. Polytypic (2 sspp.; *badia* in N.A.). L 5.5" (14 cm) **Identification** A sparrow-like bird that forages on the ground, with hunched posture. Unlike a sparrow, it shows a thin warbler-like bill. The structure is otherwise like a sparrow, with a medium-length, notched tail. **ADULT:** Rufous brown above with diffuse streaking, contrasting with marked face and tawny-to-cinnamon underparts. Lacks bold wing bars. Thin, buffy lower wing bar is well formed. Head is dark, with a bold and contrasting tawny supercilium, as well as a diffuse tawny spot on rear ear coverts, gray central crown stripe, and gray nape. Tawny throat contrasts with dark auriculars. Flanks are streaked rufous brown. Belly and vent are white. Legs are pink. Bill is dark, sometimes showing a paler horn base to lower mandible. **IMMATURE:** Much like adult, duller in plumage with less grey on nape; some retain dark juvenal speckling on breast.

Similar Species Bold, dark-and-tawny head pattern is distinctive for a small, ground-dwelling songbird. Rustic Bunting, another Asian visitor, shows a similar rusty brown plumage and bold face pattern. However, the bunting has a conical bill, whitish supercilium, and whitish throat.

Voice CALL: High-pitched *tsee-ree-seee*, with the quality of Brown Creeper calls. **SONG:** Not yet heard in N.A., is a high-pitched warble, like Pacific Wren, although slower and less complex.

Status & Distribution In eastern Asia,

this is an early spring and late fall migrant. **BREEDING:** Breeds in Asia. **WINTER:** In central and eastern China and the Korean Peninsula. **RARE STATUS:** Rare but annual in recent years in fall (mainly Sept.–early Oct.) from St. Lawrence I.; recorded elsewhere on Bering Sea islands and western Aleutians, AK. Winter and early spring records from interior (near Fairbanks) and southeastern AK, BC, AB, MT, ID, and WA indicate a few regularly winter in northwestern N.A. Most winter records have been at bird feeders. Casual to northern Europe. **Population** Stable.

badia

dark crown with gray center

gray nape

rufous streaks on back

bold, rich buff supercilium

thin bill

dark cheek with a few buffy spots

rich buff below

WEAVERS Family Ploceidae

Orange Bishop, breeding male (AZ, July)

K nown for their elaborate, suspended nests woven from grass and other plant fibers, the weavers are a large family of finchlike birds, primarily of the African tropics. Their colorful plumage makes them popular captives, resulting in naturalized populations in many regions of the world. One species of bishop (*Euplectes*) is established in some parts of North America, originating from the pet trade.

Structure They are small to medium size with conical bills; most have relatively short tails, short, rounded wings, and short but strong legs.

Plumage Breeding males are brightly colored—red, orange, or yellow bodies, often with contrasting black faces or masks—in many species. Nonbreeding males resemble females, which resemble sparrows.

Behavior Many species are highly colonial, foraging and roosting in large flocks; some are colonial nesters. Weavers are largely sedentary, making only local or seasonal movements, but some species are migratory. They feed primarily on seeds and grain, although some species also take insects, fruit, or nectar.

Distribution They are found primarily in Africa, but five species in genus *Ploceus* are found in southern Asia. Additionally some introduced species are also in Eurasia, Australia, and Oceania.

Taxonomy Authorities vary on taxonomy (about 115 species in 15 genera). The Estrildidae (waxbills) are the ploceids' closest kin.

Conservation Some are crop pests in their native ranges and are killed in large numbers; others have been reduced by capture for the pet trade. BirdLife International codes: 3 NT, 7 VU, 6 EN, 1 CR, and 1 EX.

BISHOPS Genus *Euplectes*

This genus comprises nine species known as bishops and eight as widowbirds. Breeding male bishops are brightly colored with short tails; breeding male widowbirds are less showy but grow spectacular tails that are replaced with shorter ones after breeding. All species are endemic to Africa, but many occur elsewhere as a result of escaped or released captives.

NORTHERN RED BISHOP   *Euplectes franciscanus*   NRBI ▪ EXOTIC

The Northern Red Bishop is the only weaver found in numbers in N.A. Polytypic (2 sspp.; presumably nominate in N.A.). L 4" (10 cm)

Identification Small finches with stout bills and short tails that are often flicked open. **BREEDING ADULT MALE:** Red-orange upperparts and throat with black crown and face. Black bill; pink legs and feet. Black wings and tail. Black breast and belly; orange undertail coverts. **FEMALE, IMMATURE &**

NONBREEDING MALE: Sparrowlike. Head buffy with pale supercilium and streaked crown and nape. Black eyes prominent on pale face; pink bill, legs, and feet. Brown upperparts heavily streaked with black. Breast and sides of neck buffy with dark streaking.

Similar Species Breeding male Southern Red Bishop (*E. orix*), an escape occasionally seen in N.A., is very similar but with black throat and chin. Duller plumages resemble sparrows (especially Grasshopper Sparrow); note stout, wholly pink bill, short tail, and pattern of black streaking above. **Voice CALL:** A sharp *tsip* and a *tsik-tsik-tsk*. **SONG:** High-pitched and buzzy.

Status & Distribution Exotic in US. Native to sub-Saharan Africa. **YEAR-ROUND:** Nonmigratory. In Southern CA (mostly Los Angeles Co. and Orange Co.), ±500 found in flood control basins and channeled rivers. A few are present in Phoenix, AZ.

Population Declines in CA.

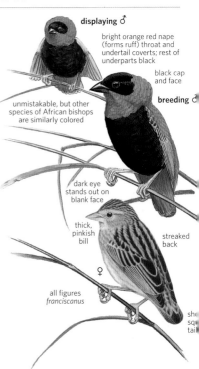

displaying ♂

bright orange red nape (forms ruff) throat and undertail coverts; rest of underparts black

black cap and face

breeding ♂

unmistakable, but other species of African bishops are similarly colored

dark eye stands out on blank face

thick, pinkish bill

streaked back

♀

all figures *franciscanus*

she squ tail

INDIGOBIRDS AND WHYDAHS Family Vinduidae

Pin-tailed Whydah, breeding male (CA, June)

One spectacular Vinduidae species, the Pin-tailed Whydah, is rapidly becoming established in Southern CA.

Structure Rather small with stubby bills. Male whydahs and paradise-whydahs have spectacularly long tails in breeding plumage.

Behavior Breeding males hold territories and are socially promiscuous. In the breeding season, male whydahs and paradise-whydahs have spectacular, often aerial, song displays. All are brood parasites and lay eggs in the nests of estrildid finches. Nearly all hosts raise both their own young and the foster nestlings. The female Cuckoo Finch (*Anomalospiza imberbis*) removes the eggs of the host before dropping her own single egg.

Plumage All are sexually dimorphic. Breeding males show extensive (whydahs and paradise-whydahs) to nearly entirely black (indigobirds) coloration. Females and nonbreeding males of all species are much duller, basically grayish brown with streaked back and most with strong head streaks. The male Cuckoo Finch is greenish yellow in color.

Distribution All 20 species are resident and confined to Africa, south of the Sahara.

Taxonomy Formerly placed in the Ploceidae, and later the Estrildidae, the whydahs and indigobirds now merit their own family, closest to the estrildids. The 19 species in the genus *Vidua* include short-tailed indigobirds and the whydahs (in which males can be spectacularly long-tailed). The unique Cuckoo Finch is placed in its own, aptly named genus, *Anomalospiza*.

Conservation All 20 species are listed as Least Concern by BirdLife International. A few have very limited ranges.

Genus *Vidua*

PIN-TAILED WHYDAH *Vidua macroura* PTWH ■ EXOTIC

red bill

black-and-white plumage with large white wing patch

nonbreeding male like nonbreeding female with red bill, but male whiter below, darker above, central tail feathers may extend slightly

blackish lateral crown stripe and postocular lines

dramatic long tail that is flopped about in a dramatic and buoyant circular flight display

black bill (red in nonbreeding female)

breeding ♂

breeding ♀

streaked with rufous-buff and black above

This widespread African species is established locally in coastal Southern CA. Breeding-plumaged males are striking with their long tails and coloration and have spectacular and buoyant circular flight displays with tail flopped about. Often in the company of Scaly-breasted Munias, which they parasitize. The host munias raise their young along with those of the whydah. Monotypic. L breeding ♂ 12" (30 cm), nonbreeding ♂ and ♀ 4" (10 cm)

Identification Males have a distinctive red bill in all plumages, while the red bill of females becomes dark in the breeding season. **BREEDING MALE:** Distinctive black cap and back with white collar and large white wing patch; upper chin and breast sides black, otherwise pure white below. Spectacularly long black tail, the extension involves the central two pairs of tail feathers. **NONBREEDING MALE:** Brownish gray overall with strongly streaked head; central tail feathers may extend slightly; bill red. **FEMALE:** Very similar to nonbreeding male but with dark bill when breeding. **JUVENILE:** Resembles juvenile Scaly-breasted Munia but is grayer overall, and young juveniles have a circular

white gape marking; older juveniles develop red bills.

Similar Species Breeding adult males have been confused with the Fork-tailed Flycatcher. Nonbreeding males and females similar to comparable plumages of Northern Red Bishop. Note the red (male) or black (female), not pinkish, bill of Pin-tailed and the much more strongly streaked head. Females of the other nine African whydahs and indigobirds are similar, but the breeding-plumaged male Pin-tailed is distinctive.

Voice CALL: A soft *chwit*, a low *peee*, and in flight double chip notes; also chattering notes in alarm or on taking flight; begging juveniles give a rapid series of chip notes. **SONG:** Repetitive, measured, and somewhat jerky *tseet tseet tsu-weet*.

Status & Distribution Native to sub-Saharan Africa, where resident in grassy and open shrubby habitats, marshes, cultivated lands, and gar-

dens. **RESIDENT:** Introduced and established in Southern CA, mainly from southeast Los Angeles Co. to central Orange Co. Often found in grassy and weedy areas in parks. Also introduced and fairly common in Puerto Rico, where they parasitize the also-introduced Orange-cheeked Waxbill.

Population Stable in Africa, greatly increasing in CA.

WAXBILLS Family Estrildidae

Nutmeg Mannikin (CA, Feb.)

A large family (±131 species in 34 genera), estrildids are found in tropical regions of the Old World, primarily in the African tropics. They are typically open-country birds. Several species are common in aviculture, and some have established exotic populations in many regions of the world, including North America.
Structure They range from tiny to small, and have pointed tails, short, rounded wings, and dispropor-

tionately large, conical bills; a few have slender bills.
Plumage Most estrildids are brightly colored, but in the genus *Lonchura*, most are brown, black, or white. They exhibit no seasonal variation. The bill is often brightly colored, azure in many *Lonchura*. Some species show marked sexual dimorphism; in others, the sexes look similar.
Behavior Many species are highly colonial outside the breeding season, but most breed solitarily. Estrildids feed chiefly on the ground or in low vegetation, primarily on seeds and grain with some fruit or insects taken.
Distribution They are found in Africa, Asia, Australia, and on South Pacific islands.
Taxonomy The taxonomy is confusing and undergoing extensive revision as a result of DNA studies. Two subfamilies are recognized: the waxbills (Estrildinae) of the African tropics (with two species in India and Southeast Asia) and the mannikins, munias, grassfinches, and allies (Lonchurinae) of Africa, Southeast Asia, Australasia, and the western Pacific. Estrildids are closely related to whydahs (Viduidae) and weavers (Ploceidae).
Conservation Most estrildids are common to abundant in their native ranges. BirdLife International codes: 7 NT, 4 VU, and 2 EN.

Genus *Spermestes*

BRONZE MANNIKIN *Spermestes cucullata*   BRMA ■ EXOTIC

One of the smallest of the munias, this common and widespread African species has been introduced and is present in small but growing numbers in parts of Southern CA. Somewhat gregarious, often in small family groups, but in Southern CA joins Scaly-breasted Munias and Pintailed Whydahs. The whydahs parasitize the Scaly-breasted Munias and perhaps Bronze Mannikins. Usually feeds on the ground. Within native African range, raises up to four broods per year. Polytypic (2 sspp.; nominate in N.A.). L 3.5" (9 cm).
Identification **ADULT:** Sexes similar. Nominate *cucullata* is grayish brown above with blackish head, throat, and scapulars; greenish gloss on shoulders and irregular barring on flanks, leg feathering, and undertail coverts. The sides of the breast have a solid patch of glossy green. **JUVENILE:** Dull brown above, paler brown bill with thick dark bill and blackish tail.
Geographic Variation Nominate subspecies is found in West Africa east to western Kenya, west of the Rift Valley. It has a glossy green patch on

the sides of the breast. East African *scutata* occurs from Ethiopia south and east of Rift Valley to central and southeast South Africa. Intergrades are found in the Rift Valley. Eastern *scutata* typically lacks greenish patch on sides of breast and has a more finely barred rump, uppertail and undertail coverts.
Similar Species In N.A., adults are distinctive, but juveniles very similar to Scaly-breasted Munia. Juvenile Bronze is much smaller and slightly colder brown with a slightly thicker-based bill and blackish tail.
Voice **CALL:** Buzzy, high-pitched *jik* or *jik jik*. **SONG:** Rather deep, slow, and measured and in two parts.
Status & Distribution Native to sub-Saharan Africa, where basically resident but somewhat nomadic. Range extends in East Africa up along the west side of the Red Sea. Also found on islands off Zanzibar, and the Comoros. Introduced during the slave trade to Puerto Rico, and now numerous and widespread. In Africa, found in farms, gardens, and light woodland. In CA, found in weedy areas, often in parks.

very thick bill

adult
cucullata

green highlights on sides and scapulars characteristic of West African *cucullata*

juvenile
cucullata

Now found in coastal Southern CA from southeast Los Angeles Co. south to central Orange Co.; locally fairly common and increasing.
Population Abundant in native range.

Genus *Lonchura*

The members (28 species) of this genus are known by several names: mannikin (generally those from Africa or New Guinea), munia (generally those from Asia), or sparrow. All are small, pointed-tailed granivorous finches, usually found in flocks. Many naturalized populations are found around the world, including two in N.A.

SCALY-BREASTED MUNIA *Lonchura punctulata*   SBMU ■ EXOTIC

The Scaly-breasted Munia, the only munia found in numbers in N.A., is largely restricted to Southern CA, where flocks feed on grass seeds in weedy river channels and residential areas. It is also known as the Nutmeg Mannikin and in the pet trade as "Spice Finch." Polytypic (9 sspp.; apparently only nominate in N.A.). L 4.5" (11 cm)

Identification ADULT: Sexes similar. Rufous face and breast; remainder of head and upperparts brown. Extensive black scaling on otherwise white lower breast to undertail. Yellow-orange uppertail coverts. **JUVENILE:** Upperparts wholly light brown. Auriculars and underparts peach, with paler throat and undertail coverts. Dark eyes; slate gray bill.

Similar Species Juvenile very difficult to separate from other *Lonchura* species that are occasionally seen in N.A., such as Chestnut Munia (*L. atricapilla*) or Tricolored Munia, although the latter has a paler belly and slightly smaller bill. Told from immature *Passerina* buntings by smaller size, thicker bill, and thinner tail; Indigo Bunting has blurry streaks below.

Voice CALL: Loud *beee* or *ki-bee*, repeated frequently. **SONG:** High-pitched *tiks* and whistles; nearly inaudible.

Status & Distribution Exotic in the continental US; established in HI. Native from India to China and much of the Orient. Birds in CA from Indian subcontinent; provenance of birds in FL unknown. **YEAR-ROUND:** Nonmigratory, but movements of more than 20 mi noted in CA are based on banding recoveries. In US, largely restricted to CA: primarily Los Angeles Co. and Orange Co., but now in good numbers north to Santa Barbara Co. and south to San Diego Co.; a few to northern Baja California. Found east to western San Bernardino Co. and western Riverside Co. Small numbers in San Francisco, TX (Houston area), and FL (Pensacola and Miami).

Population Increasing in CA.

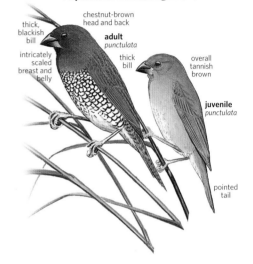

thick, blackish bill

intricately scaled breast and belly

chestnut-brown head and back

adult *punctulata*

thick bill

overall tannish brown

juvenile *punctulata*

pointed tail

TRICOLORED MUNIA *Lonchura malacca*   TRMU ■ EXOTIC

This small munia endemic to India is established in parts of the New World. Sometimes treated as conspecific with the Black-headed Munia (*L. atricapilla*) from Asia. The two hybridize in eastern India. Monotypic (taxonomy unsettled). L 4.8" (12 cm)

Identification ADULT: Sexes similar. Black head and center of belly to undertail coverts and thighs jet black. Upperparts warm chestnut, slightly darker on the flight feathers. Bill silvery gray. Up to four color morphs (occasionally seen in wild) have been described. **JUVENILE:** Brown head and upperparts, buff below with paler belly. Molts to more adultlike plumage once fledged. Bill initially dark, becomes paler with age.

Similar Species Adult Tricoloreds are distinctive. Younger juveniles very similar to Scaly-breasted Munia but are darker above; older juveniles have paler bills.

Voice CALL: Various notes, including a strong *peet*. **SONG:** A series of almost inaudible soft squeaks, followed by a long, thin descending whistle (*weeee*).

Status & Distribution Partial to wet areas, including marshes and irrigated crops (e.g., rice paddies). Resident and locally common in southern India and Sri Lanka. Introduced and locally established in HI (Oahu), C.A., S.A. (Colombia and Venezuela), and West Indies (Jamaica, Cuba, Hispaniola, Puerto Rico, and Martinique). Several records from south FL may have originated from Cuba, where locally fairly common.

Population Common in native range. Exotic populations spreading, especially in Caribbean and C.A., so additional N.A. sightings seem likely.

black head

bright chestnut upperparts and tail

adult

most of underparts white

thick bill initially with some dark but soon lightens to resemble adult bill

juvenile

OLD WORLD SPARROWS Family Passeridae

House Sparrow, female and male (IL, Oct.)

Old World sparrows are widespread in Africa and the Palearctic. Two species have been successfully introduced to North America. The House Sparrow is ubiquitous to about 60° N, whereas the Eurasian Tree Sparrow is more restricted in range.

Structure Most species have short legs, rather short wings, and short, thick bills.

Behavior Many species are gregarious, and some are common to abundant. Several passerid species have flourished in human-influenced environments, and many are tame and approachable. They feed mostly on the ground. The House Sparrow leaves feeding areas and forms noisy communal roosts well before dusk.

Plumage Old World sparrows are mainly clad in browns, grays, and russets. In many species, there are marked differences between juveniles and adults, as well as between the sexes. Seasonal variation can be pronounced, too.

Distribution About 38 species in eight genera. The genus *Passer* (to which both North American species belong) contains the most species and is the most widespread and most diverse in its habitat preferences. Two other genera, not introduced to North America, are more specialized: The widespread rock sparrows (*Petronia*) are restricted mainly to warm, arid climes; snowfinches (*Montifringilla*) are centered around the Tibetan Plateau and occur in barren habitats at high elevations.

Taxonomy Old World sparrows are not related to New World sparrows in the family Passerellidae. Instead, their closest alliance is with the family Ploceidae, in which they were formerly placed. Taxonomists differ in the division of species between the two families. The family includes 23 sparrows in the genus *Passer*, along with snowfinches, rock sparrows (petronias), and an odd Philippine endemic, the Cinnamon Ibon (formerly placed in the white-eye family Zosteropidae).

Conservation Especially in the genus *Passer*, many species enjoy commensal relationships with humans. The House Sparrow, however, long regarded as a textbook example of a beneficiary of human activity, is currently in sharp decline in parts of its native range. BirdLife International codes: 2 VU.

Genus *Passer*

HOUSE SPARROW *Passer domesticus*   HOSP ■ 1

The cheery and sociable House Sparrow is more closely associated with humans than any other widely established N.A. exotic. Introduced to New York City in 1851, the species today flourishes in both large cities and remote agricultural outposts—just so long as there is some trace of human influence. It aggressively defends nest cavities, possibly to the detriment of native species. It is gregarious in winter and near dusk gathers in communal roosts. Polytypic (12 sspp.; nominate in N.A.). L 6.3" (16 cm)

Identification Tame; gregarious. Flight more direct, often higher, than native sparrows. Bill thick, conical; legs short; stocky build. One molt per year, but seasonal variation pronounced. **ADULT MALE:** Worn (breeding) male contrastingly marked; throat and breast black, areas behind eye and on nape are chestnut. On a freshly molted bird, blackish and reddish regions are obscured by gray feather tips. Bill black in summer; yellowish base to lower mandible in winter. **ADULT FEMALE:** Mainly gray-tan. Buffy eye stripe; gray-brown crown and auriculars. Bill more yellowish than male; tip and culmen are dusky. **JUVENILE:** Variable. Plain overall. Resembles adult female.

Geographic Variation N.A. population (nominate *domesticus*) exhibits extensive geographic variation, with clinal variation: larger birds with shorter appendages in colder climes, and darker plumages in more humid environments.

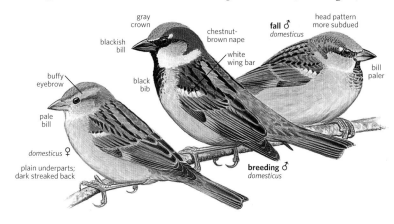

gray crown

blackish bill

chestnut-brown nape

fall ♂ *domesticus*

head pattern more subdued

white wing bar

buffy eyebrow

black bib

bill paler

pale bill

domesticus ♀

plain underparts; dark streaked back

breeding ♂ *domesticus*

Similar Species Males distinctive; plainer females and juveniles present a combination of structural and plumage characters that separate them from native sparrows. A female Northern Red Bishop may be confused with a female House Sparrow.

Voice All vocalizations simple. **CALL:** Varied, but three notes are prevalent: throaty *jigga*, usually given by agitated birds; soft *chirv*, often heard in flight; honest-to-goodness *chirp*, given in various settings. **SONG:** Short series of pleasant *chirp* notes.

Status & Distribution Fairly common to abundant. **YEAR-ROUND:** Cities, farms, and other human-transformed environments. **RARE STATUS:** Several records for northern Canada and AK,

including from Gambell, St. Lawrence I.; they probably originate from nearby Chukotka (Russian Far East) as a result of introductions.

Population Some Palearctic populations are in sharp decline, possibly due to changing land use patterns. Studies in N.A. may reveal similar declines.

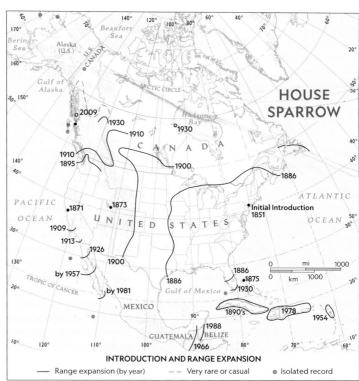

HOUSE SPARROW

INTRODUCTION AND RANGE EXPANSION
—— Range expansion (by year) – – Very rare or casual ● Isolated record

---

EURASIAN TREE SPARROW *Passer montanus* ETSP ■ 2

juvenile
montanus

chestnut-brown cap and bold blackish spot on white cheeks

black chin, whitish collar

adult
montanus

Like the House Sparrow, the Eurasian Tree Sparrow was deliberately introduced to N.A. (St. Louis, 1870). Unlike the House Sparrow, its range has not expanded greatly. Polytypic (9 sspp.; nominate in N.A.). L 5.9" (15 cm)

Identification Usually in small flocks. **ADULT:** One molt per year. Sexes similar. Key mark is white cheek with large black spot. Crown and nape chestnut; neck collar white; throat black. On worn plumage (spring, early summer), russet and chestnut regions

browner, duller. **JUVENILE:** Resembles adult; buffier, duller overall; bill straw yellow.

Geographic Variation None in N.A. Our subspecies is nominate *montanus*; there are eight other subspecies in the Old World.

Similar Species White neck collar, black cheek spot, and chestnut-brown across entire crown separate slightly smaller Eurasian Tree Sparrow from House Sparrow.

Voice Varied; all notes are simple. **CALL:** Monosyllabic utterances, sharper and more metallic than House Sparrow, typically without the pleasing, liquid qualities: *chet, kip,* etc. **SONG:** Series of sharp monosyllabic notes, interspersed with more liquid, disyllabic notes.

Status & Distribution Locally common. Species occupies an area of about

10,000 sq mi, mainly in west-central IL; range extends to west of St. Louis, MO, and north to southeastern IA. **YEAR-ROUND:** Near human habitation. **RARE STATUS:** Recorded north and east of main N.A. range. Casual to SK, MB, ND, SD, NE, MN, WI, MI, ON, IN, KY, and NJ (ship assisted?). Presumed escapes of ship-assisted birds from Asia, where common, have been observed on West Coast (including multiple records around Long Beach, CA, a major port).

Population N.A. range is expanding slowly, generally northward.

WAGTAILS AND PIPITS Family Motacillidae

Eastern Yellow Wagtail (AK, May)

Slender and highly migratory passerines from the Old World, wagtails and pipits have long tails with white outer tail feathers. Wagtails are black and white or yellow, while pipits are brown and streaked. Both share the distinctive behavior of walking on the ground and most wag their tails. They are generally found on or near the ground. Most species in North America are mainly found in the Arctic of western AK; five species (two wagtails and three pipits) breed in North America, while five Eurasian species occasionally occur as rare, casual, or accidental visitors.

Structure Pipits are mostly medium-size, slender birds with long tails. Their bills are also long and slender, and somewhat pointed. The pipits have long legs and toes. Wagtails are shaped similarly to pipits, but they have much longer tails.

Behavior Wagtails and pipits walk on the ground while searching for insects. Most species of pipits and wagtails forage at the edge of water. Some species of pipits (e.g., Sprague's) rarely if ever wag their tails and avoid aquatic habitats. During the spring and summer, wagtails sing from exposed perches on shrubs, small trees, or human-made structures. Pipits have elaborate and often lengthy flight displays, but also sing from the ground. Wagtails have a distinctive undulating flight, while pipits fly more directly, but both call often in flight; their species-specific calls aid in identification. Sprague's Pipit, when flushed, rises gradually and then circles around before it plunges back to Earth, breaks its fall immediately above the ground, and disappears into the grass. Mostly diurnal migrants, wagtails and pipits use rivers and coasts as navigation aids (except Sprague's, which migrates from the Great Plains to grasslands in the southern US and northern Mexico). American Pipits form large single-species flocks in winter.

Plumage Wagtails are sexually dimorphic with several age, seasonal, and sex-related plumages—generally combinations of black, white, and yellow, with long tails and white outer tail feathers. Telling identification features include the extent of white in the wing, the color of the back, and the pattern of black on the underparts. Pipits are not sexually dimorphic, but some do have different winter plumages. Most are brown with plain or streaked backs, varying amounts of streaking on the underparts, and white outer tail feathers. Important features to notice include the degree of streaking on the back and underparts, leg coloration, and the presence and extent of buff, or yellowish buff, on the underparts.

Distribution North American wagtails breed and migrate in coastal arctic AK, including the Aleutians and Bering Sea islands, and are casually seen along the West Coast, and accidentally inland, mostly in fall, but a few have wintered. Pipits generally restrict themselves to arctic regions as well, breeding on rocky tundra; however, the American Pipit also breeds above tree line in the Rocky Mts. and Pacific states, and the Sprague's Pipit only breeds on the short-grass prairies of the northern Great Plains.

Taxonomy Worldwide there are at least 68 species in six genera. Recent genetic studies have influenced the determination of species limits for some wagtails (including the White and Yellow Wagtail complexes) and pipits (American). Two enigmatic genera from other families have not only been shown to be motacillids, but are actually embedded within the two major genera. One is the Sao Tome Shortail from an island off West Africa, which was classified as a sylviid but is actually a *Motacilla* wagtail. The other is the Rufous-throated White-eye, an aboreal species from the Moluccas, classified as a white-eye but actually embedded with the pipit genus *Anthus*.

Conservation Most motacillids are not threatened in North America, with the exception of the Asian Eastern Yellow Wagtail with a limited breeding range in North America. Formerly common, AK populations have crashed over the last two decades. Overgrazing and the introduction of non-native grasses are limiting the breeding and wintering grounds for the Sprague's Pipit. BirdLife International codes: 6 NT, 5 VU, and 3 EN.

WAGTAILS Genus *Motacilla*

Motacilla is mainly an Old World genus, with two breeding (mainly arctic AK) and two as rare or accidental visitors to N.A. Highly migratory, they winter in Asia, Europe, and Africa. They have distinctive calls and an undulating flight pattern. The immatures are the species most difficult to identify.

EASTERN YELLOW WAGTAIL   *Motacilla tschutschensis*   EYWA ■ 2

Formerly considered a subspecies of the Yellow Wagtail, the Eastern Yellow Wagtail used to be one of the more characteristic arctic passerines of western AK. It sings from the tops of willows or feeding in the short, grassy tundra. It walks on the ground, where it searches for insects. Polytypic (3–4 sspp.; 1–2 in N.A.). L 6.5" (17 cm)

Identification BREEDING MALE: Quite distinctive. It is yellow underneath and olive green above with a gray crown and has a narrow white supercilium contrasting with dark gray cheeks. It has a relatively long black tail with white outer tail feathers; wing bars are indistinct. **FEMALE:** Slightly duller than male, not as yellow underneath, less dark on cheek, and less gray on crown. **JUVENILE:** Lacks yellow below and is a dull grayish brown above, with narrow white supercilium, and grayish white below, with varying amounts of speckling on sides of breast and buffy flanks. It also has narrow white wing bars and a white throat with variably dark malar. **FIRST WINTER:** Similar to juvenile, but it has less dark in malar region.

Geographic Variation Two subspecies

immatures are mainly grayish above and whitish below

conspicuous white or whitish supercilium in two subspecies occurring in Alaska

slight breast band

breeding ♂
tschutschensis

breeding ♀
tschutschensis

some have pale yellow on vent and undertail coverts

immature
tschutschensis

breeding ♂
tschutschensis

breeding ♂
simillima

averages brighter than nominate subspecies with less evident breast band

long tail is somewhat shorter than White Wagtail

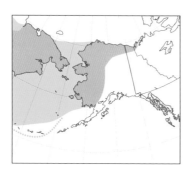

have been identified from N.A. The AK breeding nominate *tschutschensis* has a speckled breast band. The subspecies *simillima*, regular on the Aleutian and Pribilof Is., averages brighter yellow underneath and greener above, and typically lacks speckling on sides of breast. Some authorities do not recognize this subspecies and merge it into *tschutschensis*. Two other Asian subspecies are distinctive: *macronyx* with a dark face and *taivana* with a bold yellow eyebrow. These East Asian subspecies are genetically distinct and differ vocally from the Western Yellow Wagtail (*M. flava* with 10 subspecies; unrecorded in N.A.), which is found to the west in the Palearctic and now treated as a separate species by the AOS.

Similar Species Adults appear similar to the much rarer (in N.A.) Gray Wagtail female, but Gray Wagtail has blacker wings and a much longer tail. An immature can be confused with an immature White Wagtail, but Eastern Yellow Wagtail is browner and shorter tailed and usually lacks dark on sides of breast. See also Citrine Wagtail.

Voice CALL: A loud *tsweep*. **SONG:** A series of high-pitched *tszee tszee tszee* notes.

Status & Distribution Now uncommon with a restricted range in N.A. **BREEDING:** Nests in willows in tundra along the Bering Sea in western and northern AK. **MIGRATION:** Trans-Beringian migrant; regular on islands in Bering Sea. Formerly was seen migrating in large numbers in late summer along the Bering Sea. **WINTER:** Mainly Southeast Asia and Philippines. **RARE STATUS:** Casual in summer to northern YT. Casual in early fall along West Coast from BC to Baja California.

Population Severe declines, at least locally on AK breeding grounds, likely reflect trapping (for food) in Asia while on migration and on winter grounds.

CITRINE WAGTAIL   *Motacilla citreola*   CIWA ■ 5

An accidental stray to N.A. with three scattered records, the Citrine Wagtail is typically associated with water. Polytypic (2 sspp.; *citreola* in N.A.). L 6.5" (17 cm)

Identification BREEDING MALE *CITREOLA*: Bright yellow on head and underparts with a gray back and unique black nape. Bold white wing bar and tertial edges. Breeding male *calcarata* (central Asia) has a black back, a more solid white wing patch, and deeper yellow coloration. **FEMALE & WINTER ADULT MALE:** Hollow cheek completely surrounded by yellow; grayish back and crown; and white wing bars. **FIRST-WINTER:** Gray and white; resembles female, but lacks all yellow. Note diagnostic head pattern described above, but pale surround of auricular white, not yellow.

Similar Species Immature Citrine Wagtail resembles Eastern Yellow Wagtails, but note Citrine's broader and encircling supercilium, more prominent white wing bars, gray upperparts, and pale lores and forehead.

Voice CALL: Very similar to the loud *tsweep* call of Eastern Yellow Wagtail. **SONG:** Often given from atop a willow bush; resembles song of White Wagtail. **Status & Distribution** Locally common Eurasian species. Breeds from northeast Europe to Central Asia in wet areas with low vegetation, open bogs, soggy riverbanks, and meadows. **WINTER:** Mainly in southern Asia west to India, some to Middle East; accidental in N.A. **RARE STATUS:** Three N.A. records, all in winter: one adult at Starkville, MS (31 Jan.–1 Feb. 1992), an immature from Comox, Vancouver I., BC (14 Nov. 2012–25 Mar. 2013), and an adult near Davis, Yolo Co., CA (15–16 Dec. 2017). **Population** Breeding range extending west in eastern Europe; now to Poland.

completely lacks yellow but face pattern similar to adult

immature *citreola*

yellow-centered auriculars surrounded by yellow

bold white wing bars

winter adult ♂ *citreola*

GRAY WAGTAIL *Motacilla cinerea*   GRAW ■ 4

The Old World Gray Wagtail is a casual spring migrant to western AK. This long-tailed species is usually found near water. Like White Wagtail, often detected in flight by its distinct two-note call and undulating flight pattern. Polytypic (3–4 sspp.; eastern *robusta*, if recognized, in N.A.). L 7.8" (20 cm)
Identification BREEDING ADULT MALE: Yellow below, gray above, with greenish rump. Narrow white supercilium, extensive black throat. Black wings; tertials edged white; white base to secondaries forms a wing stripe that is visible only in flight. Pinkish legs. Similar to adult female in winter. **FEMALE:** Lacks black throat and is duller than male, lacking much of the yellow on underparts and brightest yellow on undertail coverts. Most distinctive field marks are its very long tail (longer than other wagtail species) and lack of wing bars. **JUVENILE:** Similar to female, but browner overall, buffy across breast, and has two buffyish wing bars. It already has bright yellow undertail coverts.
Geographic Variation In mainland Eurasia, geographic variation is clinal. Tail length decreases from west to east, thus East Asian *robusta* is shorter-tailed; it is also slightly deeper yellow below than *cinerea*. Some authorities merge *robusta* into *cinerea* and recognize only three subspecies. Two other subspecies are restricted to Madeira and the Azores.
Similar Species The combination of yellow underparts and black throat make the male unique. The female looks similar to Eastern Yellow Wagtail, but lacks wing bars and has grayer upperparts, a much longer tail, and duller underparts that contrast with brighter yellow undertail coverts. Gray Wagtail has pinkish legs; they are blackish on Eastern Wagtail.
Voice CALL: A two-note metallic *chink chink*, somewhat similar to White Wagtail, and very different from the single buzzy call of Eastern Yellow.
Status & Distribution Eurasia. **MIGRATION:** Casual spring migrant to the western Aleutians; fewer records from central Aleutians and north in Bering Sea to St. Lawrence I. Five scattered fall records for western AK islands. **RARE STATUS:** Accidental to a ship off northwestern Arctic Canada, BC, WA, and CA in fall.
Population Stable.

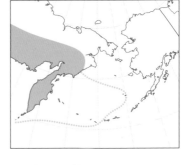

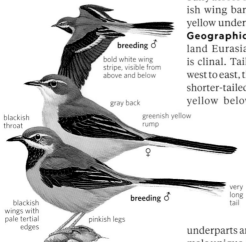

breeding ♂

bold white wing stripe, visible from above and below

gray back

blackish throat

greenish yellow rump

♀

blackish wings with pale tertial edges

pinkish legs

breeding ♂

very long tail

WHITE WAGTAIL *Motacilla alba*   WHWA ■ 3

Formerly considered two species (White and Black-backed Wagtails), the White Wagtail favors rocky breakwaters, rusting machinery, garbage dumps, and buildings. It sings from an exposed perch and walks on ground when feeding. Call notes and undulating flight are distinctive. Polytypic (9–11 sspp.; 3 in N.A.). L 7.3" (18 cm)
Identification Sexually dimorphic with complicated age differences. Adults distinct, overall gray, black throat, black on hind crown, white in wings; immatures with reduced black and less white in wings. Tail black with white outer feathers. Separation of two subspecies difficult in immatures; intergradation further complicates the issue. **BREEDING MALE:** In *ocularis*, a gray back, black hind crown and nape, white forehead, black throat extending to bill, and a thin black line through eye. Wing coverts white. In *lugens*, a black back, nape, and hind crown; black extending farther forward on crown. Black throat with white chin and black eye line. Wider black line through eye. Wing, including flight feathers, mostly white. **BREEDING FEMALE:** In *lugens*, some gray in back. **WINTER FEMALE:** In *ocularis*, a gray back and coverts have less white, forming more distinct wing bars. In *lugens*, some black on its scapulars and mostly white wing coverts (patch). **IMMATURE:** Variable. Gray above, whiter below, with varying amounts of black on bib. Separation to subspecies should be done very carefully or preferably not at all.

Geographic Variation Subspecies *ocularis* is the regular breeder on AK mainland and St. Lawrence I.; *lugens* is a casual summer visitor to the Aleutians and St. Lawrence I. These two subspecies are the only ones with a black postocular line; other subspecies are white-faced. They intergrade to an unknown extent where they come into contact in northern Kamchatka; no detailed study has been done from this region. Nine other subspecies occur across Europe and North Africa. Nominate European *alba*, breeding as close as Iceland and Greenland, has a white face.

Similar Species Juvenile looks similar to immature Eastern Yellow, but White is longer tailed and grayer above. Note different calls.
Voice CALL: A distinctive two-note *chizzik* given in flight; also a *chee-whee* given from the ground or an exposed perch. The vocalizations are identical between subspecies.
Status & Distribution Generally rare in western AK. **BREEDING:** Nests sparsely in villages on Seward Peninsula and St. Lawrence I. (uncommon); most breeding birds are *ocularis*, but mixed pairs with *lugens* or intergrades have been encountered; *lugens* nested once on Attu I. **WINTER:** Both *lugens* and *ocularis* winter in eastern Asia. **RARE STATUS:** Both *lugens* and *ocularis* are casual along the West Coast, mainly in fall and winter. Accidental in Sonora, Baja California, the Southwest, and the East. Records involve both *lugens* and *ocularis*, but some were not identified to subspecies. European *alba* is accidental on the East Coast.
Population Stable.

pearly gray back

breeding
ocularis

both subspecies have a black eye line

long tail

all wagtails fly with undulating flight and call often

black bib extends to chin

long tail

breeding adult ♂
lugens

adults show extensive white in wing

chin usually white

black back

breeding adult ♀
lugens

winter adult ♂
lugens

breeding adult ♂
lugens

identification of immatures to subspecies is problematic

immature
lugens

European subspecies has white face with no dark eye line

breeding ♂
alba

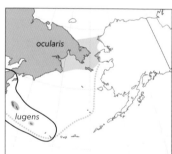

ocularis

lugens

lugens

PIPITS Genus *Anthus*

Three breeding and three rare to casual species occur in N.A. They exhibit little or no sexual dimorphism, except for Red-throated Pipit. The majority of N.A. species are gregarious, especially in migration and winter. All are highly migratory. They winter in grasslands and agricultural fields, where their cryptic plumage protects them.

TREE PIPIT *Anthus trivialis*   TRPI ■ 5

similar to Olive-backed Pipit, but browner above with more distinct back streaks

trivialis

buffy breast contrasts with white belly

This mainly western European species is a casual visitor to N.A. It will perch up in trees, but it also walks on the ground. Polytypic (2 sspp.; nominate in N.A.). L 6" (15 cm)
Identification Sexes similar. Brownish above, boldly streaked with black; heavily streaked across buffy breast with streaking down flanks. Face pattern diffuse with short, pale supercilium and faint eye line. Rump and uppertail coverts unstreaked; legs pinkish. In fall, fresher with rich buff wash across breast.

Similar Species Olive-backed has a more finely streaked olive back and a bolder face pattern with a split bicolored supercilium and a dark spot in the rear of the ear coverts. Pechora and non-adult Red-throateds have bold white streaks on back and bold black streaks down flanks; in Red-throated, bill is finer and background color of underparts is uniform throughout. Similar Meadow Pipit (*A. pratensis*), from Europe, unrecorded in N.A., breeds in Greenland.
Voice CALL: A thin, buzzy *teez*, very

different from American Pipit but similar to Olive-backed.

Status & Distribution BREEDING: Eurasia, in a variety of forested habitats. **WINTER:** Africa, south of the Sahara, and India. **RARE STATUS:** Seven N.A. records from Bering Sea region: five (four fall, one spring) records from Gambell, St. Lawrence I., one fall record for St. Paul I., and one speci-

men of *trivialis* from Wales at the west end of the Seward Peninsula (23 June 1972). The latter record was of a singing bird in display flight.

Population Declining in Europe.

OLIVE-BACKED PIPIT   *Anthus hodgsoni*   OBPI   ■   3

The Olive-backed Pipit from Asia is a rare migrant to N.A., mainly to western AK. It often pumps its tail when it walks on the ground. In Asia, partial to open woodlands. Polytypic (2 sspp.; *yunnanensis* in N.A.). L 6" (15 cm)

Identification Sexes similar. A boldly patterned pipit with a relatively plain, faintly streaked olive back; rich buff breast contrasts with white belly; boldly streaked black breast, sides, and flanks; and a broad, dark malar stripe. Uniquely patterned face has a supercilium that is buff in front of eye and white behind; blackish spot

at rear of auriculars is bordered above by a white spot. Bright pinkish legs. Freshly molted birds are buffier below, and more olive above.

Geographic Variation Subspecies *yunnanensis*, breeding in northern Asia and wintering in Southeast Asia, is only faintly streaked on back. The more sedentary subspecies *hodgsoni* from the Himalaya and most of Japan is more strongly streaked on its back. In Hokkaido, Japan, the two subspecies intergrade.

Similar Species No other pipit has a heavily black-streaked breast with underlying rich buff wash that contrasts with a white belly, and a relatively plain olive back; face pattern also distinctive.

Voice CALL: A high, thin, buzzy *tseee*, similar to Tree Pipit, but shorter, buzzier, and less descending than Red-throated Pipit call.

Status & Distribution BREEDING: Breeds across Russia to North Korea and Japan; also southern China and the Hima-

laya. **WINTER:** Philippines, Southeast Asia, and India. **RARE STATUS:** Rare (mainly spring) migrant to the western Aleutians (likely nested Attu I., 1998) and the Pribilofs. Very rare in spring and fall, north to St. Lawrence I. Casual to central Aleutians. Accidental in fall to Middleton I. (AK), CA, and Baja California, and in spring to NV (specimen, 16 May 1967, 10 mi south of Reno, established first N.A. record). Also rare (mainly fall) to western Europe.

Population Stable.

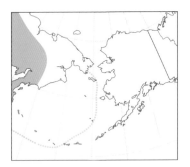

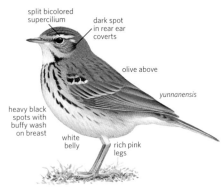

split bicolored supercilium

dark spot in rear ear coverts

olive above

yunnanensis

heavy black spots with buffy wash on breast

white belly

rich pink legs

PECHORA PIPIT   *Anthus gustavi*   PEPI   ■   4

back feathers with black centers and rufous edges

streaks through nape

white "braces"

distinct primary projection, unlike other pipits

buffy wash on breast

white belly

gustavi

The solitary and secretive Pechora Pipit is often silent when flushed, but its call is diagnostic. Polytypic (2–3 sspp.; *stejnegeri* in N.A., if recognized). L 5.5" (14 cm)

Identification Sexes similar. Primary tips project beyond tertials, unlike on other pipits. Rich rufous-brown above with heavy streaks, contrasting white "braces" on sides of back; streaked nape; rich buffy yellow wash across

breast; white belly; two distinct white wing bars; diffuse face pattern; rump streaked; pinkish legs.

Geographic Variation Nominate subspecies occupies most of range. Aleutian specimens were identified as *stejnegeri*, which breeds only on Commander Is. Recent authorities have merged that subspecies into nominate *gustavi*. Another, more southerly subspecies, *menzbieri,* breeding in northeast China and southeasternmost Russian Far East, is smaller and duller, and has a markedly different song; winter grounds unknown, but has been collected in Philippines.

Similar Species Immature Redthroated Pipit lacking red coloration routinely causes confusion.

Voice CALL: A hard *pwit* or *pit*. Very different from other pipits.

Status & Distribution BREEDING: Locally across much of subarctic

Russia, including Commander Is., where numerous. **WINTER:** Mainly in Philippines, also northern Borneo and parts of Indonesia. **RARE STATUS:** Casual migrant to western Aleutians (spring) and to the Pribilofs. Very rare to Gambell, St. Lawrence I., mostly in fall (over 20 records). Also casual to northwest Europe.

Population Stable.

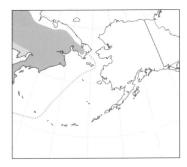

RED-THROATED PIPIT *Anthus cervinus*   RTPI ▪ 3

The male Red-throated Pipit is one of the more recognizable pipits in N.A. but can be secretive. It readily associates with American Pipits in migration. When flushed, its very distinctive call is often heard. In spring, the male sings in the tundra from atop boulders or in a display flight. Monotypic (some authorities recognize a second subspecies, *rufigularis*). L 6" (15 cm)

Identification Back is heavily streaked, with buffy or whitish edges. Crown is finely streaked, but nape is unpatterned. It also has noticeable whitish wing bars, little or no primary projection, and pale legs. **BREEDING MALE:** Variable, bright tawny-red head, throat, and breast, which varies in intensity; rest of underparts are buffy white with streaking along sides of breast and flanks. **ADULT FEMALE:** Red, if present, is mainly on throat and upper breast; heavier black streaking on breast and sides. **IMMATURE:** Lacks red coloration. Prominently streaked above and below with broad dark malar. **Similar Species** Adults in bright plumage are distinctive. Immatures and dull females easily confused with much rarer Pechora Pipit, but note Red-throated's uniformly colored underparts, plain nape, unbroken eye ring, lack of primary projection, and very different calls.

Voice CALL: Given in flight; a high, piercing *tseee* that descends at the end. **SONG:** A long, varied series of chirps, whistles, and buzzy notes, given from the ground or from a display flight.

Status & Distribution Common on arctic tundra from Fennoscandia through Russian Far East. **BREEDING:** Has bred in short, rocky tundra in western AK along Bering Sea. **MIGRATION:** Regular spring and fall migrant to western and central Aleutians and islands in Bering Sea (mainly St. Lawrence I.); spring peak in early June; fall peak in late Aug. **RARE STATUS:** Rare to casual fall migrant along West Coast (mostly CA), mainly in Oct., and casual inland to southeastern CA. Casual in spring to CA (also winter), AZ, and southwest BC. Casual in spring and fall to YT; also casual Mexico.

Population Stable.

breeding ♀

breeding adult females average more restricted red throat and more streaks on breast, but there is overlap

extensive brick red throat

pale nape

prominently streaked back

broad dark malar

breeding ♂

no primary extension past tertials

heavily spotted and streaked below with uniform underlying color

immature

AMERICAN PIPIT *Anthus rubescens*   AMPI ▪ 1

The American Pipit (known as Buff-bellied Pipit in Europe) breeds at high latitude from Siberia to western Greenland and on high-elevation, rocky alpine tundra in western N.A. Often common in migration and winter in wet agricultural fields and around sewage ponds in single species flocks. Forages out in the open, often in flocks, frequently bobbing its tail. Polytypic (3–4 sspp.; all occurring in N.A., can be divided into 3 sspp. groups). L 6.5" (17 cm)

Identification Sexes similar. Plumage highly variable, ranging from clear buffy underparts with no streaks, to whitish underparts with heavy streaks, depending upon subspecies and time of year. Upperparts generally gray to brown, unstreaked. Tail with white outer tail feathers. **BREEDING:** Subspecies *alticola* is pale gray above, unstreaked tawny-buff underparts, buffy auriculars with buffy supercilium, dark legs, and short hind claw. Subspecies *rubescens* (including *paci-* *ficus*, recognized by some) is darker gray above, buffy below, streaks across breast that extend faintly down flanks, gray auriculars with buffy supercilium, and dark legs. Subspecies *japonicus* is similar to *rubescens* but with blacker streaking, a browner, slightly streaked back, and often pale legs. **WINTER:** Subspecies *alticola* is not separable from *rubescens* in the field, perhaps not even in hand! Subspecies *rubescens* is grayish brown above with faint streaking, whitish below, buffier on flanks, heavy dark streaking across breast variable, decidedly white supercilium, and dark legs (usually). Subspecies *japonicus* is boldly patterned with more streaking on upperparts and thick black streaks against a white background on underparts, prominent white wing bars, and pink legs (sometimes pale on N.A. subspecies too).

Geographic Variation N.A. subspecies divide into three groups: the nominate *rubescens* group, breeding on N.A.

japonicus or pacificus?

rubescens group

alticola

tundra (*rubescens* and slightly paler *pacificus* not recognized by some); the "Rocky Mountains" group (*alticola*); and the "Asian" group (*japonicus*, a rare migrant, mainly in western AK). **Similar Species** All other N.A. pipits are browner and more streaked on upperparts, except Olive-backed. **Voice CALL:** Given in flight; a sharp *pip-it*. Call of *japonicus* is higher pitched, often single noted. **SONG:** A rapid series of *chee* or *cheedle* notes, typically given in flight display, lasting 15–20 seconds.

Status & Distribution Common. **BREEDING:** Widespread and scattered during the breeding season in rocky tundra across the Arctic (*rubescens*), high mountaintops in the Rocky Mts. and CA (*alticola*), and AK and the Pacific Northwest (*pacificus*). **WINTER:** Common and widespread in agricultural areas and open country from southern US south through Mexico. Rocky Mts. *alticola* believed to winter mainly in Mexico. Subspecies *rubescens* is much more widespread. **MIGRATION:** Found virtually anywhere in open country or by water across the US; casual to the Azores. **RARE STATUS:** Asian *japonicus* regular in fall (late Aug.–Sept.) and more rarely in spring to western Aleutians and Bering Sea islands. Very rare to casual (in fall) south to Southern CA; also casual Mexico. N.A. *rubescens* very rare to western Europe and the Azores.

Population Winter birds are susceptible to the overuse of pesticides in agricultural areas.

slender bill

dark malar stripe broadens at bottom

rather plain back

necklace

streaked underparts

leg color varies from dark to showing some pinkish tones

winter *rubescens*

breeding *rubescens*

white wing bars

more heavily and darkly streaked below

richly buff and unstreaked underparts

winter *japonicus*

pink legs

breeding *alticola*

SPRAGUE'S PIPIT *Anthus spragueii* SPPI ■ 2

In winter, the Sprague's Pipit is one of the more secretive grassland birds from the Southwest to eastern TX, yet in summer its song is a characteristic sound of the western prairies, when displaying males give continuous varied song from high altitude. Difficult to spot on the ground. When flushed, the bird calls alarmingly, rises to a great height while circling, plummets straight down with folded wings, opens its wings just before impact, and then disappears into the tall grass. However, it is usually reluctant to flush, prefer-

ring to walk away rather than fly. Unlike other pipits, Sprague's does not bob its tail. Monotypic. L 6.5" (17 cm)

Identification Sexes similar. It is stocky and rather short tailed for a pipit, heavily streaked on back and crown, rather buffy brown underneath, with fine streaking across breast, pale legs and base of mandible, and extensive white outer tail feathers. Very plain, blank face with large, dark eye that stands out. **JUVENILE:** It is more scaled on back and has bolder, white wing bars.

Similar Species Immature Redthroated Pipit looks somewhat similar, but note plainer face and lack of streaking on flanks of Sprague's; flight call and behavior when flushed also differ. Juvenile Horned Lark (see p. 488) is frequently confused with a Sprague's Pipit, but it has a thicker bill and different back and tail pattern, calls, and behavior.

Voice CALL: A loud squeaky *squeet*, often given two or three times in succession.

SONG: Given in flight; a series of descending *tzee* and *tzee-a* notes.

Status & Distribution Uncommon. **BREEDING:** Nests in grassy fields in open prairie. **WINTER:** Usually solitary. Found in tall patches of ungrazed grass in the southern US to southern Mexico. **RARE STATUS:** Very rare in fall and winter to CA. Casual in East to East Coast; accidental BC.

Population Vulnerable. Overgrazing and the introduction of non-native grasses are negatively affecting populations on both summer and winter grounds.

dark eye stands out in blank face

extensively pink at base of lower mandible

streaks on back give a scalier-looking pattern than other streak-backed pipits

faint necklace

juveniles very scaly above

pinkish legs

FRINGILLINE AND CARDUELINE FINCHES AND ALLIES Family Fringillidae

Pine Siskin (CA, Oct.)

Most fringillid songbirds have characteristic conical-shaped bills used for feeding on seeds. Nearly all have short, notched tails. Most species have undulating flights with distinctive flight call notes. Several species are widespread; others have more restricted ranges. Some fringillids are arboreal and prefer coniferous forests, while others live in tundra habitats or above tree line. Many visit seed feeders in the winter. The family also includes the euphonias of the Neotropics and the spectacular evolutionary radiation of Hawaiian honeycreepers.

Structure Our finches share the conical or finchlike bill, although its size varies considerably—from tiny in the Hoary Redpoll to massive in the Hawfinch and Evening Grosbeak. In addition, the crossbills have uniquely shaped bills with crossed tips, specialized for extracting seeds from pinecones. The Hawaiian honeycreepers show a great range of bill shapes. Fringillids' wings vary in shape. Many fringillids are long-distance migrants, and thus evolved elongated primaries, but some less migratory species (e.g., the House Finch and the Lesser Goldfinch) have shorter, rounder wings. The euphonias are specialized for feeding on mistletoe fruits.

Behavior While a few of our finches are fairly sedentary, most are at least partially migratory in winter, vacating breeding areas in search of a fluctuating food supply. Many fringillids are often irruptive—being present in areas in very large numbers some years, only to be absent in others; redpolls, crossbills, and Evening Grosbeaks are famous for this behavior. The rosy-finches and a few others vacate higher elevations during the winter. Each species can be recognized in flight by its species-specific flight call. In most species, the song is not as important as group contact calls;

the Evening Grosbeak, however, rarely, if ever, sings.

Plumage Most species in North America are sexually dimorphic—the adult males being brightly colored, as in goldfinches and *Haemorhous* finches, and females duller, browner, and often streaked. Some species, such as Purple Finch, Cassin's Finch, and Pine Grosbeak, take longer than one calendar year to achieve adult breeding plumage, usually molting into it by the second winter. Some species—American Goldfinch for instance—have a distinct winter plumage, while most have a distinct first-winter plumage duller than adult breeding birds. Many species have combinations of wing bars and white in the tail, both more conspicuous in flight.

Distribution The Fringillidae family consists of about 221 species in 57 genera worldwide. In North America, there are 24 species in 11 genera, of which 16 breed and eight occur as strays. Several species generally are found in coniferous forests across the boreal zone in Canada and northern US, as well as in pine and spruce-fir forests in the Appalachians and mountainous West. A number of species (e.g., goldfinches and Pine Siskin) are widespread during the breeding season, even more so in the winter when they form flocks. The Brown-capped Rosy-Finch has a very restricted breeding range and limited dispersal. The euphonias are found in Central and South America (three species ranging to northern Mexico), and the Hawaiian honeycreepers (39 species) are endemic to the Hawaiian Is.

Taxonomy Fringillidae members are most closely related to the Old World families of finchlike birds, including the weavers, estrildids, and Old World sparrows (as well as pipits and wagtails). Taxonomists divide the family into three subfamilies: the Fringillinae, which includes the Brambling and the Common Chaffinch; the Euphoniinae (Neotropical fruit-eating euphonias); and the Carduelinae, which includes all other fringillids, including the diverse Hawaiian species. Our knowledge of taxonomic relationships within the family is an evolving process. For instance, studies of the Red Crossbill suggest that as many as nine or more species in N.A. alone may be involved, based on differences in bill size, distribution, and flight call notes; one isolate in this group, the Cassia Crossbill, has recently been elevated to full species status.

Conservation Some species (e.g., House Finch) are expanding their range, while others (e.g., Purple Finch) have ranges that are contracting. Only two (Apapane and Hawaii Amakihi) of the 39 Hawaiian honeycreepers are *not* at risk; 16 are Extinct and six more are Possibly Extinct. BirdLife International codes: 7 NT, 15 VU, 11 EN, 7 CR, 6 CR (PE), and 17 EX.

OLD WORLD FINCHES Genus *Fringilla*

Of this highly migratory Old World genus, only the Brambling and the Common Chaffinch occur in N.A. as rare migrants or strays. Medium-size birds, the males are generally colorful, while the females appear drabber and browner. They usually feed on the ground.

COMMON CHAFFINCH *Fringilla coelebs*　CCHA　■　4

brown lateral crown stripes

blue-gray crown and nape

♀

♂

pinkish below

extensive white on wings in all plumages

The partially migratory Common Chaffinch, a very common songbird in Europe, is a casual visitor to N.A.'s Northeast. It often feeds on the ground. It has an undulating flight. The white markings on its wings and tail are conspicuous perched and in flight. Polytypic (18 sspp.; nominate *coelebs* from mainland Europe or *gengleri* from British Isles most likely to appear in N.A.). L 6" (15 cm)

Identification Sexes differ. **MALE:** Gray crown and nape surround a brown face. Brown back; pinkish below. White lesser coverts; each has a white wing bar and a white base to primaries; white outer feathers; rather long bill. **FEMALE & JUVENILE:** Much duller than the male—mostly brown, darker above and buffier below. A conspicuous gray area surrounds a brown cheek. Wing markings like male.

Similar Species Male distinctive. Winter American Goldfinch is somewhat similar to female Chaffinch, but Chaffinch is noticeably larger and uniformly buffier brown underneath and has a much larger bill and bolder wing pattern.

Voice CALL: A metallic *pink-pink* or a forceful up-slurred whistle, *hweet.* **SONG:** Introduced by several introductory notes and composed of trills of several pitches; endlessly repeated.

Status & Distribution BREEDING: Widespread in all types of woodland and in parks and gardens. Breeds Eurasia and North Africa east to western Siberia and northern Iran. Introduced to New Zealand and Cape Town, South Africa. **WINTER:** Found in southern portion of breeding range and western Europe; northern and eastern populations migratory. **RARE STATUS:** Casual or accidental stray from Eurasia to northeastern N.A., mainly in fall and winter. Accepted records from NL, NS, ME, and MA. Origin of records elsewhere, including in the West, considered problematic.

Population Large global population; increasing in Europe.

BRAMBLING *Fringilla montifringilla*　BRAM　■　3

The Brambling, a highly migratory Eurasian finch, is a rare migrant to western AK, sometimes occurring in small flocks. The male is strikingly patterned in black, white, and rufous. It has an undulating flight, and its characteristic white rump is best observed in flight. Monotypic. L 6.3" (16 cm)

Identification Sexually dimorphic. White rump in all plumages. **BREEDING MALE:** In spring, adult male has a black head and back; bright tawny-orange throat, breast, and shoulders; and white lower belly with dark spotting on flanks. Tail is black and deeply notched. Wings are black with white lesser wing coverts, a buffy lower wing bar, and white bases to primaries that form a small white patch on folded wing. Rump is white, but it is often covered by folded wings. Conical bill is relatively small. **WINTER MALE:** Similar to breeding male, but black plumage is overlaid by a brown edging that wears off throughout winter, revealing striking plumage underneath. **FEMALE:** Browner overall, head is grayish brown, with a grayer face and lateral black stripes on sides of nape. Below, a duller rufous.

Similar Species No other species has the same color combination and all-white rump.

Voice CALL: A nasal *check-check-check*; often given in flight. Also a nasal *zwee.* **SONG:** Short, buzzy.

Status & Distribution BREEDING: Southern Eurasia north and east to eastern Russian Far East and Japan. **WINTER:** Withdraws south from breeding range to Europe, Himalaya, China, and Japan. **RARE STATUS:** Rare to fairly common (spring) on western and central Aleutians and some to Bering Sea islands. Casual in fall and winter to Canada and northern US; recorded south to Southern CA, NV, and NC.

Population Declines in Europe.

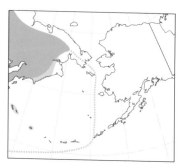

glossy black head and back

orange breast and shoulders

breeding ♂

pale-fringed feathers with blackish bases

white rump in all plumages

fall ♂

gray face bordered by black

pale nape spot

♀

EVENING GROSBEAK AND HAWFINCH Genus *Coccothraustes*

Both species in this finch genus are large with large bills, very short tails, and striking black-and-white wing patterns. The Evening Grosbeak breeds in N.A., whereas the Hawfinch breeds in Eurasia and occurs as a rare migrant to AK. Evening Grosbeaks form flocks during the nonbreeding season.

EVENING GROSBEAK *Coccothraustes vespertinus* EVGR ▪ 1

This noisy finch forms large, gregarious flocks during winter, when it frequents feeders. Found during summer mainly in coniferous forests across boreal Canada and in the Rockies; its winter movements are both erratic and irruptive. This species lacks a true song. Polytypic (3 sspp.; 2 in N.A.). L 8" (20 cm)

Identification A large, stocky, finch with very short tail and heavy bill. **MALE:** Rich golden brown, becoming darker olive-brown on head. Forehead and supercilium golden yellow. White secondaries and tertials are distinctive. **FEMALE:** Grayish brown above, buffier on underparts, collar, and rump. Throat whitish with dark malar stripe. Blackish wings with white patch at base of primaries (distinctive in flight); somewhat darker secondaries. White spots at end of tail. **JUVENILE:** Male overall duller than adult male.

Geographic Variation Of the three described subspecies in N.A., eastern *vespertinus* tends to have shorter bill and broader yellow eyebrows than southwestern *montanus*. Western *brooksi* is also long billed, and includes populations with three different flight calls.

Similar Species See Hawfinch.

Voice CALL: Flight calls vary geographically (five types from five regions have been described); individual birds give only one type of call. These vocalizations likely foster assortative mating

and future research may reveal that they are actually separate species with little or no interbreeding. Eastern birds (type 3) give a Loud, hoarse *clee-ip* or *peer*; there are four other call types in the West, all clearer than eastern *vespertinus*; best learned from audio recordings or by studying sonograms.

Status & Distribution BREEDING: Mature coniferous forest and mixed woods; mainly in mountains in West. **WINTER:** Sometimes common at lower elevations and south of breeding range; in the East southern invasions have been much more infrequent in recent decades for unknown reasons. Rare to casual in southern states. **RARE STATUS:** Casual in migration and winter north to AK. Casual to

yellow forehead and eyebrow
vespertinus ♂
large bill
vespertinus ♀
white primary patch
white patch on inner wing
breeding ♂ vespertinus
juvenile ♂ vespertinus
short tail with white tail spots
female with gray on head and back and greenish nape; buffy below
breeding ♀ vespertinus

Bermuda. Accidental to UK and Norway. **Population** Vulnerable. Significant declines in N.A., especially in East.

HAWFINCH *Coccothraustes coccothraustes* HAWF ▪ 4

This Eurasian species related to the Evening Grosbeak is a rare migrant to the outer Aleutians, and has been recorded from islands in the Bering Sea. Typically it is much more wary than Evening Grosbeak, with many reports of fly-by individuals. It sometimes occurs in small groups, but it tends to be elusive. When on the ground, it walks with a parrotlike waddle. Polytypic (6 sspp.; *japonicus* in N.A.). L 7" (18 cm)

Identification A large, stocky, short-tailed finch with a massive bill. **BREEDING MALE:** Body is yellowish brown above and pinkish brown

below, while head is a more golden brown. Throat and lores are black; nape and collar are gray. The black wings have white secondaries and bases to the primaries that form a conspicuous white band on the folded wing and make a striking black-and-white pattern in flight. Tail feathers have broad white tips. Bill is blue-black in the spring and turns yellowish in the fall. **FEMALE:** Overall, duller than the male and has grayer wing patches.

Similar Species Unmistakable.

Voice CALL: A loud *ptik*, usually given in undulating flight.

Status & Distribution BREEDING: Eurasia east to Kamchatka; found in deciduous and mixed woodlands,

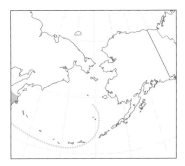

with a preference for mature forests. **RARE STATUS:** Rare spring and casual fall migrant in the western and cen- tral Aleutians, with dozens recorded in some years. Very rare spring and casual fall to Bering Sea islands; one mainland spring record west of Nome, AK. Very rare to Iceland. **Population** Stable.

all figures *japonicus*

yellowish bill in fall and winter

thick blue-black bill in breeding season

breeding ♂

unique club-shaped inner primaries

pale greater covert patch

blue on primaries and secondaries

fall/winter ♀

♂

similar to male but with grayish, not blue, secondaries

short tail

Genus *Carpodacus*

COMMON ROSEFINCH   *Carpodacus erythrinus*   CORO ■ 4

This Eurasian species is a very rare migrant in western AK. Virtually all reports are from the Aleutians and Bering Sea islands. Single individuals are typically found, but it has excep- tionally occurred in small groups. It is usually seen feeding on the ground. Polytypic (at least 5 sspp. recognized; *grebnitskii* breeding in eastern Siberia and Russian Far East has reached N.A.; it is darker than nominate sspp., and adult males have more red above, less red below). L 5.8" (15 cm)
Identification Note long primary pro-

jection past the tertials. **MALE:** Head, breast, back, and rump are a pinkish red. Back is marked with indistinct dark streaking, while flanks and lower underparts have little or no diffuse streaking. It lacks distinct eyebrows. Bill is finchlike, but it has a strongly curved culmen. Wings have two indis- tinct pinkish wing bars. **FEMALE & IMMATURE MALE:** Very drab brown with diffuse streaking above and below, except on paler throat. Immature male acquires bright plumage by second fall.
Similar Species This species is the more likely red finch to occur in the Aleutians and Bering Sea islands, although Purple Finch has been documented from St. Lawrence I. Pur- ple's culmen is decid- edly straight, making it appear more pointed. Female House Finch has shorter primary projec- tion, browner cheeks and back, and slightly more prominent streak- ing below.
Voice CALL: A soft, nasal *djuee*.

Status & Distribution BREEDING: Eastern Europe and Asia. **WINTER:** Indian Subcontinent and northern Southeast Asia. **RARE STATUS:** More likely to be found in spring, at least on the Aleutians; some have been colorful adult males. Casual spring migrant (late May–early June) on the western Aleutians and islands in the Bering Sea (e.g., St. Paul and St. Lawrence). Records north of the Aleutians are fairly equally balanced between spring and fall. Two spring records for central Aleutians. Accidental on Southeast Farallon I., CA (23 Sept. 2007, photo). Very rare to Iceland.
Population Decreasing trend in Europe.

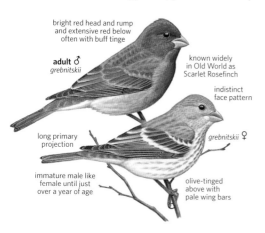

bright red head and rump and extensive red below often with buff tinge

adult ♂ *grebnitskii*

known widely in Old World as Scarlet Rosefinch

indistinct face pattern

long primary projection

grebnitskii ♀

immature male like female until just over a year of age

olive-tinged above with pale wing bars

PALLAS'S ROSEFINCH   *Carpodacus roseus*   PARO ■ 5

This partly migratory and somewhat nomadic northern Asian finch has been recorded just once in N.A., in fall from the Pribilof Is., AK. In migra- tion in winter found in small to mod- erate-size flocks. Polytypic (2 sspp.;

N.A. sspp. not known). L 6.5" (17 cm)
Identification A somewhat stocky finch with long notched tail and thick, short, dull blue-gray bill. All plum- ages other than adult male distinctly streaked overall; pale wing bars. **ADULT**

MALE: Overall raspberry pink over much of head, breast, and sides, with pale silver-pink on forehead and throat; nape and back distinctly streaked. **ADULT FEMALE:** Much duller with orange on forehead, pinkish on rump.

IMMATURE MALE: Reddish color intermediate between adult male and adult female. **IMMATURE FEMALE:** Orange tint limited to forehead and rump.
Similar Species Common Rosefinch does not show distinct streaking on back, and streaking below on immatures is fainter and more blended; wing bars less distinct.
Voice CALL: A variety of notes, including a short and subdued whistle.
Status & Distribution Breeds in mon-

tane taiga; winters forest edges and agricultural lands with trees. **BREEDING:** Yenisey Basin, south-central Siberia, southeast Altai, and northern Mongolia east to Magadan area and Sakhalin, Russian Far East. **WINTER:** Southern edge of breeding range south to northern China, Korea, and northern Japan. **RARE STATUS:** One at St. Paul I., Pribilofs, AK, 20–24 Sept. 2015. Accidental Europe.
Population Stable.

pale orange-red on forecrown, underparts, and rump

immature ♂
roseus

overall plumage finely streaked

Genus *Pinicola*

PINE GROSBEAK   *Pinicola enucleator*   PIGR ▪ 1

The large, plump, long-tailed and small-billed Pine Grosbeak is one of the more characteristic species of the boreal forest. The Pine Grosbeak frequently forms small groups during the nonbreeding season. Migratory. Polytypic (8 sspp.; 5 in N.A.). L 9" (23 cm)
Identification A large, and somewhat chunky, long-tailed finch with a short, stubby, curved bill. **ADULT MALE:** Mostly rich pinkish red body with dark wings and two white wing bars. Depending on subspecies, there are varying amounts of gray on underparts, particularly on flanks and belly, as well as varying amounts of dark centers to back feathers. **ADULT FEMALE & IMMATURE MALE:** They are mostly gray overall with two distinct white wing bars and a variably yellowish olive wash to head, back, and rump. **VARIANT:** Some females and immature males are quite russet on head and rump.
Geographic Variation The subspecies differ in size, tail length, bill shape, and coloration. The most widespread and largest is northern, boreal *leucura* (includes *alascensis* and NL *eschatosa*); coastal Pacific *flammula* (Unalaska, Aleutians, Kodiak I., and south to northwest BC) is slightly smaller and averages shorter tailed, but has a larger and more strongly hooked bill; *carlottae* (Haida Gwaii, and Vancouver I., BC) is the smallest and darkest subspecies; Rocky Mts. *montana* is large with a medium-size bill; *californica*, resident in the Sierra Nevada, CA, and the adjacent Carson Range, NV, is of medium size with a long tail and small bill; Asian *kamtschatkensis* (most of Siberia and Russian Far East) is smaller than *leucura* and *flammula*, with a narrow, blunt, and strongly hooked bill; it has been collected on the western Aleutians and the Pribi-

lofs, AK. Vocal studies done several decades ago revealed strong differences between the N.A. subspecies. Additional molecular and vocal studies might well reveal that multiple species may be involved in this complex.
Voice CALL: Geographically quite variable. Includes a whistled *pui pui pui* or *chii-vli*. The call of *californica* is loud and is strongly up-slurred. **SONG:** A short musical warble.
Status & Distribution Uncommon to fairly common in coniferous forest. **WINTER:** May move into deciduous woods and orchards. **RARE STATUS:** In the East, *leucura* is irruptive in fall and winter, and is casual south to IL, OH, and MD in fall and winter. Casual to Bermuda and Greenland. Accidental AR and SC. Very few have appeared well south of resident range in recent years. In the West, *montana* is irruptive and appears casu-

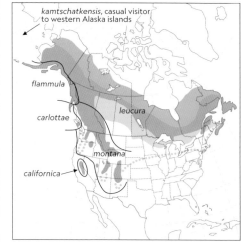

kamtschatkensis, casual visitor to western Alaska islands

flammula

carlottae

leucura

montana

californica

ally south to southeastern OR, eastern CA, AZ (formerly bred in White Mts.), West TX, and the Great Plains.
Population Perhaps stable, but far fewer *leucura* irruptions south in East, may reflect declines.

stubby black bill with curved culmen

yellow-olive head and back; body otherwise gray

leucura ♀

white wing bars

deep pinkish red

russet variant
leucura

adult ♂
leucura

Genus *Pyrrhula*

EURASIAN BULLFINCH *Pyrrhula pyrrhula*   EUBU   ■   4

This distinctive Eurasian species has been found in AK. Polytypic (10 sspp.; *cassinii* in N.A.). L 6.5" (17 cm)
Identification Chunky, with a thick neck and a very short, stubby bill. **MALE (cassini):** Intense reddish pink underparts and cheeks contrast with black crown and chin. Gray back contrasts with black wings (one prominent white wing bar) and tail and distinct white rump and undertail coverts. **FEMALE:** Similarly patterned, but brown where the male is pink. **JUVENILE:** Resembles female, but has a brown cap. Adultlike plumage acquired by fall.
Geographic Variation Extensive. Most geographic variation involves size and the presence, extent, and brightness of pink or pink-red on underparts. All AK

records are of *cassinii* from northeastern Asia; the type specimen was taken at Nulato, AK (10 Jan. 1867).
Similar Species No other N.A. bird has this color combination.
Voice CALL: A soft, piping *pheew.*
Status & Distribution BREEDING: Eurasia. **WINTER:** Withdraws from northern part of breeding range; winters slightly south of resident range.
RARE STATUS: Casual in spring (late May–early June) to the western Aleutians (one fall record) and the Bering Sea islands in spring and fall. Casual in

winter to AK mainland. Rare in winter to Iceland.
Population Moderate declines in Europe.

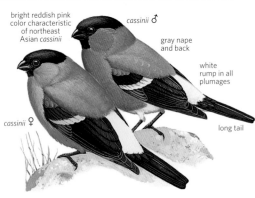

bright reddish pink color characteristic of northeast Asian *cassinii*

cassinii ♂

gray nape and back

white rump in all plumages

cassinii ♀

long tail

ROSY-FINCHES Genus *Leucosticte*

Characterized by their bright rose-pink bellies and wings, there are three N.A. members of this genus and one accidental species from Asia. They are usually found in the Arctic or above tree line in the West's high mountains, preferring the tundra often seen at the edge of snowfields. During winter they disperse to lower elevations in flocks and often gather at foothill feeders.

ASIAN ROSY-FINCH *Leucosticte arctoa*   ASRF   ■   5

The Asian Rosy-Finch inhabits some of the least hospitable, and accessible, parts of Central and northeastern Asia, nesting well above tree line in most of its remote range. Polytypic (5 sspp.; easternmost *brunneonucha* presumed in N.A.). L 6.3" (16 cm)
Identification ADULT: Male *brunneonucha* with blackish crown, face (no gray on head), throat, and upper breast, with warm caramel-colored nape and postocular stripe; rose spangling on underparts, wing, and uppertail coverts. Female similar but lacks pink. Immatures of both sexes slightly duller.
Geographic Variation There are five named subspecies, mostly allopatric as breeders, and poorly known. They differ substantially in plumage. Easternmost *brunneonucha* is believed to be the subspecies involved in the AK records, but neither record is supported by a specimen. Similar *gigliolii* to southwest (southwest Yakutia) is duller, largely lacking pink in plumage.
Similar Species Asian is only likely to appear as a stray in the range of *griseonucha* or *umbrina* Gray-crowned, which are notably larger, bulkier birds,

although smaller *tephrocotis* Gray-crowned has occurred once in June at Savoonga, St. Lawrence I. Adult Asian Rosy-Finches are fairly distinctive, appearing to N.A. birders perhaps like a Brown-capped Rosy-Finch above, a Black Rosy-Finch below in general terms, but Asian Rosy-Finch's colors and their distribution in the plumage are not shared by any other species: the contrast between blackish face and brown nape is distinctive.
Voice CALL: A brief *pyut* or soft *tyew*, similar to other rosy-finches. **SONG:** Similar to other rosy-finches, a series of twittering notes, delivered from the ground.
Status & Distribution BREEDING: On treeless montane plateaus, rocky tundra, moraines near snowfields and glaciers, scree slopes, cliffs, and rocky shorelines from southern Siberia and northern Mongolia east to Kamchatka; eastern *brunneonucha* breeds from Lena Delta south and east to Okhotsk coast and Kamchatka. Has bred in Hokkaido, Japan. **WINTER:** Movements may vary among subspecies, but most appear to

winter downslope from breeding areas. Eastern *brunneonucha* is the most migratory subspecies. Winters south to southeastern Russian Far East, Korea, northeast China, and in northern Japan, where often fairly common in coastal areas; readily observed at feeding stations. **RARE STATUS:** Accidental to AK with two photo-supported records: one female at Gambell, St. Lawrence I., on 25–26 Oct. 2008 was not identified until 2016; a male at Adak I., 30 Dec. 2011.
Population No data, perhaps stable.

blackish forecrown and face

warm brown nape

rose-pink on flanks and wings

winter ♂ *brunneonucha*

GRAY-CROWNED ROSY-FINCH *Leucosticte tephrocotis* GCRF ■ 1

The Gray-crowned Rosy-Finch is the most widespread of the N.A. rosy-finches. Found in alpine and rocky coastal areas (AK), it frequents steep rocky cliffs during the breeding season. Often seen at the edge of snowfields. Gray-crowneds frequently form mixed flocks with other rosy-finch species during winter at feeders at higher elevations in interior West. Polytypic (6 sspp.; all in N.A.). L 5.5–8.3" (14–21 cm)

Identification In general, Gray-crowned is a chunky ground-dwelling brown finch with bright rosy-pink tinting its lower belly, rump, lesser wing coverts, and edging to wing feathers. Both sexes have blackish bills in summer and yellowish bills in winter. All subspecies have a distinctive black forecrown and pale gray hindcrown. **MALE:** Belly and wings are a brighter rosy-pink. **FEMALE:** Overall, a paler brown than male; also less pink on underparts and a lighter pink edging to wing feathers. **JUVENILE:** It is entirely grayish brown with pale edging to wing coverts and flight feathers, and a darker forecrown. Bill is darker and not as yellow as on winter bird.

Geographic Variation Six subspecies are currently recognized. The migratory gray-headed "Hepburn's" (*littoralis*) is found in the near coastal West. Two largely resident subspecies—*umbrina* from the Pribilofs, St. Matthew and Hall Is., and slightly paler *griseonucha* (includes *maxima*) of Aleutians, western AK Peninsula, Shumagin and Semidi Is.; also Commander Is. (Russian Far East)—share the same extensive gray face of *littoralis* but are much larger. Migratory

interior West and Brooks Range, AK, *tephrocotis* and similar-appearing *wallowa* (northeast OR) and *dawsoni* (Sierra Nevada and White Mts., CA) have a narrower gray headband, the lower face being brown, not gray.

Similar Species Unmistakable gray-cheeked Gray-crowned does not overlap with other rosy-finches during breeding season. Head pattern of Gray-crowned subspecies with brown cheeks and Black look similar, but Black is always much blacker or a charcoal gray, lacking brownish tones. Female Brown-capped could also easily be mistaken for an immature female Gray-crowned.

Voice CALL: A high, chirping *chew*, sometimes given in repetition and often given in flight. **SONG:** A series of descending *chew* notes; sometimes given in flight display. It is seldom heard.

Status & Distribution Fairly common. **BREEDING:** Resident in the Aleutians and Commander Is. (*griseonucha*) and the Pribilofs, St. Matthew and Hall Is. (*umbrina*), where it frequents coastal cliffs and rocky tundra. Migratory *littoralis* mainly found in coastal mountains from AK south to WA, OR, and northern CA. Interior subspecies *tephrocotis* group mainly found above tree line from the Brooks Range of northern AK south through the northern Rocky Mts. to ID; also northeastern OR (*wallowa*) and in the Sierra Nevada and White Mts., CA (*dawsoni*). **WINTER:**

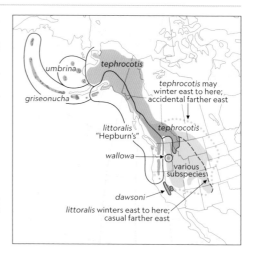

Both migratory subspecies (*littoralis* and *tephrocotis*) winter south of breeding range, mostly at higher elevations throughout the interior west, south to CO, UT, northern NM, and northern CA. The winter distribution is somewhat erratic, with species invading farther south some years in search of food. Storms drive birds to lower elevations. **RARE STATUS:** Casual in northern AZ (*tephrocotis* and *littoralis*) and various states and provinces east to MN and western ON. Accidental Savoonga, St. Lawrence I., 13 June 2008 (specimen of *tephrocotis*). Accidental CA Bay Area, PA, NY, QC, ME, and AR.

Population Perhaps stable.

extensive gray face

"Hepburn's Rosy-Finch"
winter ♂
littoralis

gray head band

winter ♂
tephrocotis

yellow bill on all fall and winter rosy-finches, becoming black by spring

extensive gray face, like *littroralis*

Pribilof, St. Matthew and Hall Is. *umbrina* and Aleutian and Commander Is. (Russian Far East) *griseonucha* are much larger than *littoralis*

winter ♂
umbrina

dark sooty overall with little pink

juvenile
tephrocotis group

brownish edges above, compare to female Black Rosy-Finch

winter ♀
tephrocotis

like male but with less pink

BLACK ROSY-FINCH *Leucosticte atrata* BLRF ■ 2

The Black Rosy-Finch is largely restricted to the central Rocky Mts. in the interior west. It frequents rocky tundra above tree line, often feeding at the edge of snowfields. During winter, it frequently joins with other rosy-finches to form large mixed-species flocks and habituates established seed feeders. Monotypic. L 6" (15 cm)

Identification Black Rosy-Finch is much darker than similarly patterned *tephrocotis* Gray-crowned Rosy-Finch. Both sexes have blackish bills in summer and yellowish bills in winter. **MALE:** Almost entirely black body has some silver edging to breast and back feathers, especially when fresh in fall and early winter. Pale silvery gray of hind crown extends down to level of the eye, forming a gray head band. A

pink blush extensively covers lower belly and uppertail coverts. Obvious pink coverts and on primaries and secondaries on otherwise blackish wings. **BREEDING FEMALE:** Patterned similar to male but more of a charcoal gray color than black. Whitish lower belly has little or no pink. Pink wing coverts are paler than male. **JUVENILE:** Body is uniformly gray with two buffy cinnamon wing bars and a pale eye ring. Bill is pale.

Similar Species The male Black Rosy-Finch is much blacker than any other rosy-finch. The female Black is also dark, but there are gray tones, too, though no brown, as in Gray-crowned Rosy-Finch.

Voice CALL: A high chirping *chew* is like other rosy-finch calls.

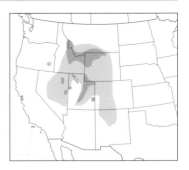

Status & Distribution Fairly common within its limited range. **BREEDING:** High-elevation rocky tundra, usually above tree line in the central Rocky Mts. from southern MT and ID to southeastern OR, northern NV, and northern UT, where it occurs on top of isolated high mountains. **WINTER:** It forms large flocks that move down to lower elevations and south to northern NM. **RARE STATUS:** Casual in winter to northern AZ and eastern CA (Inyo Co. and Mono Co.). **Population** Endangered (BirdLife International).

silver-gray edges above on fresh plumaged females and males

dark sooty gray overall with less and paler pink than male

winter ♀

stunning — black overall with contrasting rose and silver-gray head band

breeding ♂

BROWN-CAPPED ROSY-FINCH *Leucosticte australis* BCRF ■ 2

The breeding range of Brown-capped Rosy-Finch is restricted to CO and southern WY. Brown-capped can be told from other rosy-finches by the lack of gray on the hind crown or nape. It behaves much like other rosy-finches, breeding above tree line in rocky tundra and often feeding on the ground at the edge of snow patches. Like the interior Gray-crowned and Black Rosy-Finches, the Brown-capped migrates to lower elevations during the winter and frequents seed feeders. But unlike those species, it is completely restricted to the Rocky Mts. region, with no substantiated records of strays

elsewhere. Monotypic. L 6" (15 cm)

Identification Unlike other rosy-finches, most Brown-cappeds lack gray on head. Both sexes have blackish bills in summer and yellowish bills in winter. **MALE:** Body almost entirely rich brown with a darker, almost blackish crown. Lower belly is bright reddish pink, while wing coverts and rump are extensively edged in pink. **BREEDING FEMALE:** Much drabber than male, with a uniformly darker brown body and little or no pink on belly, rump, or wing coverts. **IMMATURE FEMALE:** Distinguished by a pale grayish brown body, more blended head pattern, light

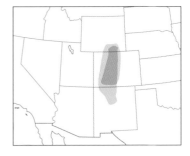

pink wing coverts, and very little pink on belly.

Similar Species Brown-capped is generally browner and paler than any plumage of Black Rosy-Finch. Some individuals have varying amounts of gray on hind crown, making field identification from an immature female *tephrocotis*-group Gray-crowned difficult or impossible.

Voice CALL: A high,

some worn 1st-spring and summer females from *tephrocotis* group of Gray-crowned look very similar

Brown-capped unrecorded away from Rockies region

rich brown back and breast with no or only slight gray head band

extensive rosy-pink on wings and below

winter ♀

dark sooty cap

breeding ♂

chirping *chew* is similar to other rosy-finch calls.

Status & Distribution Fairly common in proper habitat. **BREEDING:** Like other rosy-finches, it prefers high-elevation rocky tundra, but it has a more limited breeding range, which is restricted to above tree line in the eastern Rocky Mts. of CO and southern WY. **WINTER:** Moves to lower elevations and joins mixed-species flocks with other rosy-finches. In winter, regular south to the Sandia Mts., northern NM. Like the other two rosy-finch species, often visits seed feeders where all three species frequently occur together (CO and NM).

Population Endangered (BirdLife International).

Genus *Haemorhous*

In N.A., three breeding species (formerly in *Carpodacus*) make up this highly sexually dimorphic genus. The males are bright red or pink on the head and breast, to varying degrees. Females are brown and streaked with subtly different head patterns. Other important field marks are the bill size and shape, the degree of streaking on flanks and undertail coverts, and species-distinctive call notes.

HOUSE FINCH *Haemorhous mexicanus*   HOFI ■ 1

The attractive House Finch is one of the more common and recognizable species throughout the US. Originally a western species of semiarid environments, House Finch was introduced in the East in the 1940s; it has now expanded its range and spread to virtually every state, as well as a multitude of habitats. It has become very common in suburban areas and is easily attracted in large numbers to seed feeders. Polytypic (12 sspp.; 2 in N.A.). L 6" (15 cm).

Identification House Finch is a relatively small *Haemorhous* finch with a longish, slightly notched tail, short wings, and a distinctly small bill with a curved culmen. **MALE:** Breast, rump, and front of head are typically red, but color can vary to orange or occasionally yellow. Red breast is clearly demarcated from a whitish belly with dark streaks. Top of crown and auriculars are brown. Back is brown and noticeably streaked. Wings have two pale indistinct wing bars each. Immature male acquires red by first fall. **FEMALE:** Much drabber, lacking all-red coloration of male. Brown body has distinct, blurry streaking above and below; lacks distinct pale eyebrow found on male.

Geographic Variation There is clinal variation, as well as individual variation, and effects of diet on plumage coloration complicate the separation of different subspecies. Subspecies *frontalis*, found nearly throughout N.A., sports a generally more orange-red to yellow (more rarely) breast and has less distinct streaking on belly. The subspecies *clementis* from San Clemente I., CA, and the Islas Coronados, Mexico, is brighter red with bolder streaking on a whiter belly.

Similar Species The male House differs from the male Purple Finch not only by having a smaller, more curved bill, but also by lacking a distinct eyebrow, having a brown cap and auricular patch, and being heavily streaked on belly. Told from male Cassin's Finch by brown cap and eyebrow and curved bill. Other telltale differences between the species include the Cassin's pink cheek and pinkish tone on its back, and on female and immature male Cassin's, the much finer and crisp streaks on its belly. The adult male Common Rosefinch is more rose-pink overall and lacks distinct streaking on its belly. Females are more difficult. The female House Finch has a very plain face, unlike Purple and Cassin's, which both show distinct eyebrows and blurry streaks below. Also note House's smaller, more curved bill. The female-plumaged Common Rosefinch looks similar, but she is drabber, with less distinct streaking below.

Voice CALL: Most commonly a whistled *wheat*. **SONG:** Lively and high-pitched, consisting of varied three-note phrases that usually end in a nasal *wheeer*.

Status & Distribution A very common, often abundant resident throughout much of the US, extending north into much of extreme southern Canada and south into Mexico. Western population appears to be spreading. **MIGRATION:** Some northern populations appear to be migratory, moving south in winter. **RARE STATUS:** Casual southeast and south-central AK.

Population The human modification of natural habitats, particularly the increase of seed feeders throughout the East, greatly benefits the House Finch populations. Only natural island populations appear to be threatened.

males can be yellow to orange

variant ♂
frontalis

broad reddish eyebrow

red throat and breast

brownish streaks on flanks

typical ♂
frontalis

square-ended tail

decurved culmen

muted brown back

very indistinct face pattern

frontalis ♀

PURPLE FINCH *Haemorhous purpureus* PUFI ■ 1

short thick bill
with slightly
decurved culmen

adult ♂
purpureus

dark auricular
and distinct white
facial streaks

distinct
dark
streaks
above

white belly and
flanks with no
brownish, as in
californicus

**Pacific region
adult ♂**
californicus

distinct
and rather
broad streaks
below

This migratory rose red (not purple) finch is fairly common across much of N.A.'s boreal forest and south through the Pacific states. Polytypic (2 sspp.; both in N.A.). L 6" (15 cm)

Identification Short, strongly notched tail; eastern *purpureus* described. **ADULT MALE:** Body mostly rose red, brightest on head and rump. Pink eyebrow contrasts with darker cheek. Back with noticeable streaks and pinkish ground color. Below extensively streaked. Undertail coverts white. Bill rather large with a straight culmen. **FEMALE:** Whitish underparts heavily streaked; undertail coverts white. Head with whitish eyebrow and sub-moustachial stripe, contrasting dark brown cheek and malar stripe. Crown and back have distinct streaks. Immature males maintain female-type plumage for over a year.

unlike *purpureus*, has
a brownish tinge to
back, flanks, and belly

pale facial stripes
less obvious

overall a deep
rose-red color

♀ *purpureus*

Pacific region ♀
californicus

browner,
less contrasty
streaking below
against a buffier
background color

less obvious streaking
above against a more
olive background color

purpureus

californicus

Geographic Variation Two distinct subspecies. Nominate resides mainly in boreal forests of much of N.A.; male has longer wings and is brighter over-all, female has a bolder head pattern and whiter underparts. Pacific region's *californicus* adult male is less bright and has a brownish wash on its back and sides; female-plumaged birds are much buffier underneath with a less bold face pattern and more blurred streaking overall. They may represent two distinct species.

Similar Species Males and females most similar to Cassin's. Adult male Cassin's is lighter pink, particularly on underparts and eyebrow, but with brighter red crown. See sidebar below for females.

Voice CALL: A musical *chur-lee*, and a sharp *pit* (*californicus*; softer and more musical in nominate) often given in flight. **SONG:** A rich warbling; shorter than Cassin's, lower pitched and more strident than House. Songs of nominate subspecies longer and more complex, more like Cassin's.

Status & Distribution Fairly common. **BREEDING:** Open coniferous forests and mixed woodland in East and North, and coniferous forest and oak canyons in West. **WINTER:** Eastern birds migrate south to lower latitudes, sometimes irrupting with major invasions south to the southern US. Western birds primarily move to lower elevations. **RARE STATUS:** Rare throughout much of the interior West (both subspecies) and casual to southeast and western AK (both subspecies). Casual to Bermuda. **Population** Slight decreases.

Identification of Female Purple and Cassin's Finches

One of the more challenging identification problems in the West involves separating adult female-plumaged Purple and Cassin's Finches. The females (and immature males) of both species are basically brown with coarse dark streaking on whitish underparts. Both species also have similar face patterns with a pale eyebrow, a pale submoustachial streak, and a darkish malar stripe. Further complicating the identification, the Purple Finch has two distinct subspecies. Visible differences between the three are subtle, and perhaps best used in combination with each other; however, note that each has different call notes. The Purple Finch gives a quick, musical *chur-lee* when perched and a sharp *pit* in flight. The call of eastern *purpureus* is softer and higher-pitched. The Cassin's call when perched can be similar to the call of Purple Finch, but in flight it gives a

distinctive high-pitched *kee-up* or, the longer version, *tee-dee-yip*.

The female eastern Purple Finch (*purpureus*), overall the bolder patterned of the species, has a very bold face pattern, with a broad eyebrow, a whiter submous-tachial stripe, and a prominent brown malar streak. The underparts are white with very dark streaking and a slight buffy wash on the flanks. The white, unstreaked undertail coverts are a critical field mark. The bill is relatively long, with a slightly curved culmen.

The face pattern on the female western Purple Finch (*californicus*) resembles the face pattern on the Cassin's; the former, however, appears more washed-out. The underparts on *californicus* look more like those on a House Finch, with very blurry streaks; note the different face pattern with a pale supercilium and

CASSIN'S FINCH *Haemorhous cassinii*   CAFI   ■   1

A species of the montane West is slightly larger and longer winged than the similar Purple Finch, which it occasionally overlaps with during winter. It is often seen in small flocks, mainly in coniferous forest, but it often appears in the interior lowlands, particularly in spring when deciduous trees are budding. Monotypic. L 6.3" (16 cm)

Identification Note distinct streaked undertail coverts and rather prominent pale eye ring in all plumages. Bill is longer and more pointed than other *Haemorhous* finches. Sexually dimorphic with adult males pink and females and immature males brown. Lighter pink than other *Haemorhous* finches. **MALE:** Bright pinkish red crown contrasts sharply with a brown streaked nape. Back is heavily streaked and washed pink. Fairly wide eyebrow and submoustachial stripe are both pale pink. Light pink throat and breast blends into white on lower belly. Immature male acquires adult plumage by second fall. **FEMALE:** Upperparts are brown and streaked, while underparts are white with fine, crisp streaking, which is heaviest on breast and flanks. On head, a pale eyebrow and submoustachial stripe outline darker auricular.

Similar Species Male is most similar to male Purple Finch. Cassin's eyebrow and streaking on back tend to be wider and frostier; it usually has fine streaking on flanks and undertail coverts as well. Some Purples can show some streaks on undertail coverts so character is not diagnostic. Primary projec-

tion is noticeably longer in Cassin's, as is bill. Note different flight calls. See sidebar below for separating female-plumaged Cassin's and Purples. Cassin's and House overlap more, with Cassin's typically found in coniferous forest and House more in the lowlands, but they do overlap in winter when Cassin's populations irrupt to lowlands. Note Cassin's pink eyebrow, finer black streaking on flanks, frostier upperparts, longer primary projection, and longer bill with a straight culmen.

Voice CALL: In flight gives a dry *kee-up* or *tee-dee-yip*. **SONG:** A lively, varying warble, longer and more complex than Purple Finch or House Finch.

Status & Distribution Fairly common in montane coniferous forests. **BREEDING:** Found throughout much of the Rocky Mts., west into the Cascades in WA and OR, and the Sierra Nevada and southern mountain ranges in CA. **WINTER:** Unpredictable. Often stays in breeding range, but periodically drops to lower elevations and/or moves well south. Winters as far south as the mountains of central Mexico. Much more common in the lowlands of the interior West than Purple Finch, which is rare there. **RARE STATUS:** Casual to eastern CO, NE, KS, northern TX, AK, and West Coast. Accidental MN.

Population Near Threatened.

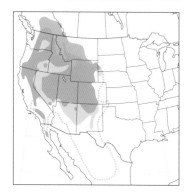

contrasting crimson-red crown

wide brown malar streak

adult ♂

pale pink throat and breast with white belly

pale eyebrow, but face pattern less distinct than Purple

larger bill than Purple with straight culmen

long primary projection, longer than Purple

cinnamon tinge to auriculars

adult ♀

notched tail as in Purple

all plumages have streaked undertail coverts, unlike most Purples

immature male plumage like female until about one year of age

purpureus Purple, female (NJ, Jan.)   *californicus* Purple, female (CA, July)   Cassin's, female (CA, Feb.)

a heavier bill with a slightly curved culmen on Purple.

   The Cassin's Finch is slightly larger than either Purple. It has very white underparts with finer black streaking, particularly on the flanks, and the streaking extends to the undertail coverts. The face pattern is similar to that of *californicus*, but it is more blended and often shows a pale eye ring. The long bill has a straight culmen. At close range, note the Cassin's longer and less evenly spaced primary extension and the more distinct white edging to its flight and tail feathers.

pale bill

dark grayish olive nape and crown

adult ♂
kawarahiba

ORIENTAL GREENFINCH *Chloris sinica*   ORGR ■ 4

A casual spring migrant in the western Aleutians, this Asian finch sometimes occurs in small groups. No other finch in N.A. has large yellow wing patches. Polytypic (5 sspp.; largest and migratory northeast Asian *kawarahiba* in N.A.). L 6" (15 cm)

Identification Subspecies *kawarahiba* described: Small, stocky brown finch with a deeply notched tail; short,

stubby bill; and a large yellow wing patch in primaries. **ADULT MALE:** Uniformly olive-brown, with greener head and rump; dark gray nape and crown; white edges to tertials and tips of primaries. Yellow undertail coverts and base of tail feathers. **ADULT FEMALE:** Duller, no gray crown or nape. **JUVENILE:** Similar to female but finely streaked below.

Similar Species Juvenile Oriental Greenfinch is vaguely similar to smaller Pine Siskin, but yellow in wing is much broader and back is unstreaked.

Voice CALL: Includes a twittering rattle and a nasal *zweee*.

Status & Distribution Asian species. **BREEDING:** Southeast Russian Far East, Korea, Japan, and eastern China. Subspecies *kawarahiba* breeds Kamchatka, northern Kurile Is., Sakhalin, and Hokkaido and winters central and southern Japan, south rarely to Taiwan. **RARE STATUS:** Casual (*kawara-*

extensive yellow on wings in all plumages

female duller than male

adult ♀
kawarahiba

juvenile streaked below

juvenile
kawarahiba

hiba) to western Aleutians in spring; a few summer and fall records. Accidental to southwest BC in early Nov. and northwest CA in winter.
Population Stable.

REDPOLLS Genus *Acanthis*

COMMON REDPOLL *Acanthis flammea*   CORE ■ 1

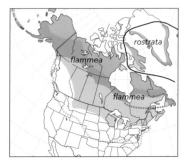

Common and Hoary Redpolls are two closely related finches of the boreal forest and Arctic tundra scrub. The two are very close genetically, and many believe that the two species should be considered conspecific. Despite the genetic evidence there has yet to be a comprehensive study of the two species on the breeding grounds where the two are conspecific in a broad band around the world. The anecdotal evidence compiled so far indicates that they act as separate species. The Common is the more widespread species of the two, usually inhabiting subarctic forest during the summer and frequenting seed feeders in southern Canada and northern US during the winter, when they form large flocks.

Adults have characteristic red cap or "poll." Polytypic (2 sspp.; both in N.A.). L 5.3" (13 cm)

Identification Common Redpoll is a relatively small, streaked finch with a small, pointed bill; short, deeply notched tail; two white wing bars; black chin; red cap; and varying amounts of red underneath. **BREEDING MALE:** Cap is bright red. Upperparts are brown with distinct streaking. Bright rosy red of throat and breast extends onto cheeks. White flanks and undertail coverts have fine black streaking; paler rump has distinctive streaking. **BREEDING FEMALE:** Lacks red breast of male and has variable amounts of streaking underneath, usually confined to sides. **WINTER MALE:** Duller. Buffy wash on sides and rump. **WINTER FEMALE:** Also buffier on sides. **IMMATURE:** First-year birds resemble adult female, but they tend to be buffier. **JUVENILE:** Brown and streaked, it acquires red cap in late summer molt.

Geographic Variation Two breeding subspecies in N.A. The small-billed and smaller *flammea* has less coarse streaking and is widespread across Canada to AK; the large-billed and larger *rostrata* has

coarser streaking underneath and is found on Baffin I. and Greenland. Both overlap during winter, but the clinal variation makes identification problematic.

Similar Species Great care is needed to separate Common Redpoll from the very similar-looking Hoary Redpoll. The breeding adult male Hoary is a very frosty white above, and white below with a very pale pink blush on breast. Females and immatures are much more difficult; rely on the differences in bill size and shape, the extent of streaking on rump, the ground color of the upperparts, the quality of the streaking on flanks and undertail coverts, and location.

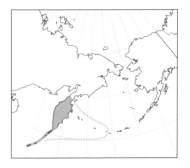

juvenile
flammea

breeding ♀
flammea

breeding ♂
flammea

extensive pinkish red breast

The juvenile Common can resemble a juvenile Pine Siskin, but it lacks yellow in the wing. The extent of interbreeding between Common and Hoary Redpolls is unknown.
Voice CALL: When perched, gives a *sweee-eet*; flight call a dry rattling *jid-jid-jid-jid*. **SONG:** A lengthy series of trills and twittering rattles.
Status & Distribution Common. **BREEDING:** Found in the subarctic forests and tundra across northern Canada and much of AK. Subspecies *rostrata* breeds in tundra scrub, where it overlaps with the Hoary. **WINTER:** Forms large flocks. Irruptive migrant south through much of Canada to

winter ♀
flammea

faint flank streaks

streaked undertail coverts

winter ♀
rostrata

larger and darker than *flammea* with more extensive streaking below

breeds in Iceland, Greenland, and Baffin I.; a few in winter in East in redpoll invasion years

winter ♂
flammea

northern US. Winters farther south than Hoary. **RARE STATUS:** Casual or accidental anywhere in southern US, and to Bermuda.
Population Major declines in N.A. in recent decades.

HOARY REDPOLL *Acanthis hornemanni*   HORE ▪ 2

winter ♀
exilipes

slightly smaller bill than Common

very pale pink

pale rump

faint flank streaks

winter ♂
exilipes

larger and overall paler than *exilipes*

paler upperparts than Common

winter ♂
hornemanni

The Hoary Redpoll is very similar to its sister species, the Common Redpoll, but it is generally found farther north in tundra habitats during the breeding season. The adults have the characteristic red cap or "poll" and are generally white or frosty above and below. During the nonbreeding season, Hoary forms large flocks with Common Redpolls and winters mainly in Canada, though it rarely reaches the northern US. Polytypic (2 sspp.; both in N.A.). L 5.5" (14 cm)
Identification Hoary Redpoll is a relatively small finch with a short, notched tail and a very short, stubby bill. Like Common Redpoll, it sports a distinctive red cap in all plumages except juvenile. **BREEDING MALE:** Very frosty white with contrasting red cap and pale pinkish blush on breast. Wings with rather bold wing bars and white edging on secondaries. Rump white, usually unstreaked. Flanks with fine to moderate black streaking, undertail coverts usually clean white, or with a hint of streaking. Bill very short. **BREEDING FEMALE:** Generally pale overall. Typically white below

with fine black streaking on sides and flanks. Rump white with little or no streaking. Back streaking quite white. **WINTER ADULT:** Buffier overall on sides. **IMMATURE:** Streaking more prominent on sides. **JUVENILE:** Lacks red cap. Pale brown overall with streaking above and below.
Geographic Variation Two subspecies breed in N.A. Nominate *hornemanni*, found in Baffin I. and Greenland, is larger, larger billed, and overall paler than the more widespread *exilipes*.
Similar Species Easily most confused with Common Redpoll. Separating adult males is fairly easy; the whiter (frostier) Hoary has a pale pinkish blush on its breast, while Common's breast is rose red. Females and immatures present much more of a problem. Main characteristics used to separate the two include size and shape of bill, presence or absence of streaking on rump, degree of streaking on flanks and undertail coverts, and location. The degree of hybridization between Hoary and Common is unknown, but those that have been identified as hybrids, could actually have been immature Hoary Redpolls. Juvenile Hoary has been confused with Pine

Siskin, but Pine Siskin has yellow in the wings.
Voice CALL & SONG: Very similar to Common Redpoll; more study needed.
Status & Distribution Common in tundra scrub. **BREEDING:** Found across the high Arctic, generally above tree line, or in willows and alders, across northern Canada and northern and western AK. Generally (at least for *exilipes*) found farther north in nonforested habitats as compared to Common Redpoll; in areas of overlap Hoary appears to arrive on the breeding grounds earlier than Common, and that alone may be a main causation for assortative mating between these two species. **WINTER:** Forms large flocks that migrate south across much of Canada. **RARE STATUS:** Rare into northern US, mostly the northern tier states; casual farther south; usually associated with Common Redpolls coming to seed feeders. An adult male (specimen) of *hornemanni* was taken at Fairbanks, AK, on 28 Mar. 1964.
Population Trends unclear. BirdLife International does not give species status to Hoary Redpoll.

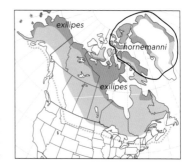

exilipes

hornemanni

exilipes

CROSSBILLS Genus *Loxia*

Found in coniferous forests, crossbills use their highly specialized bills with crossed tips to extract seeds from pinecones. Highly nomadic, the species disperse great distances in search of food. The taxonomy is complicated; in N.A., only three species are recognized, but variable populations with different bill sizes and calls suggest more.

RED CROSSBILL *Loxia curvirostra*   RECR ■ 1

Red Crossbills are found almost exclusively in coniferous forests across the boreal zone of Canada, the Appalachians, and in the coniferous forests of the West. Dependent upon the local pinecone crop, it moves around both seasonally and annually. An invasive and irruptive species, it often becomes numerous in an area, then vanishes when the cones are depleted. Polytypic (very tentatively 19 sspp.; 8 in N.A.). L 5.5–7.8" (14–20 cm)
Identification Sexually dimorphic. Chunky with a short notched tail, large head, and a stout, crossed bill. **MALE:** Brick red with darker, unmarked wings (some show narrow whitish wing bars). **FEMALE:** Entire body is a dull yellowish olive, while unmarked wings are dark and throat is gray. May show some patches of red on body. **JUVENILE:** Mostly brown, but paler below with heavy dark streaking. Varying narrow white wing bars (usually with upper bar thinner than lower). **FIRST-YEAR MALE:** Resembles female, but is more orange; patches of dull red often present.
Geographic Variation Very complicated. Red Crossbill subspecies are referred to as "Types," differentiated mainly by range, overall size, and bill

size, and most important, flight call notes, which are assigned to types. These types do not match the current taxonomy—part of the difficulty in determining the number of species involved. As many as nine different types have been identified so far in N.A. (another in C.A. and potentially another 13 in Europe and Asia). An indeterminate number of additional species may be involved. The identification of different types is very difficult in the field. Given that many of the types can turn up in virtually any part of Red's range and that different types sometimes flock together, a certain identification to type is problematic; the best way to document records is to record their sounds, which is easily done with a smartphone. Only broad extremes will be covered here (for more information refer to technical literature). The smallest and smallbilled crossbills have been put into *sitkensis* (Type 3); they generally are found farther north (into AK), but they are very widespread. The largest Red is *stricklandi* (Type 6), with a relatively huge bill; it is restricted to southeastern AZ and southern NM. Types 2, 5, and 8 (*percna*, Newfoundland only) are also large billed. The remaining Types 1, 4, and 7 are medium-size birds with medium-size

bills. Types 3 and 10 have small bills. Some are restricted to the West, others are more widespread.
Similar Species White-winged and Cassia Crossbills (see accounts) are the only other crossbills. White-winged has a distinctively patterned wing with all-white lesser coverts (forming a wider upper wing bar) and a prominent lower wing bar. An adult male White-winged is generally pinker than the brick-red male Red. Use wing patterns to separate all plumages. Some Reds have narrow wing bars, but note different width and shape. Flight notes are also different.
Voice CALL: Generally a *kip-kip* or *chip-chip* usually given in double notes, but sometimes difficult to discern individual calls as multiple individuals call simultaneously. Calls

variant ♂

some with
faint whitish
wing bars

juvenile

female typically has
olive body with darker
wings and gray throat

typical ♀

northern ♀
minor

overall size and bill size,
especially, vary significantly and geographically

southwestern ♂
stricklandi

all crossbills have
large heads and
short, notched tails

overall reddish
coloration

typical ♂

are given in flight and are different between types (see above). **SONG:** Begins with several two-note phrases, ending in a warbling trill; often given from the top of a conifer.

Status & Distribution Fairly common. **BREEDING:** Irregular breeder anywhere there are pine or spruce-fir forests, as far south as GA, and north to AK and NL. Different subspecies

appear to form single-species flocks and may have non-overlapping breeding ranges, but more study is needed. **MIGRATION:** Wander greatly in search of food. Known to "invade" the lowlands across much of N.A. well away from the breeding grounds; in those years often found in ornamental pines in parks and cemeteries. Casual to eastern Aleutians, Bering Sea islands,

and Greenland; in winter, to the Southeast and Bermuda.

Population Logging in old-growth coniferous forests may affect food supplies; trees generally produce cones once they reach 60 years old. In Newfoundland, *percna* is also threatened by introduced red squirrels, which outcompete the Red Crossbills for pinecones, and loss of breeding habitat.

CASSIA CROSSBILL *Loxia sinesciuris*   CACR ■ 2

The Cassia Crossbill was "discovered" by Craig Benkman in 1997 and recognized as a species by the AOS in 2017. Resident and restricted entirely to south-central ID. Here red squirrels are absent, so the evolution of lodgepole pinecones differs from elsewhere in the Rockies where squirrels are present. The specific epithet *sinesciuris* translates to "without squirrels." Monotypic. L 6.9–7.5" (18–19 cm)

Identification Thick bill, bill depth 0.3 mm deeper than Type 2 and 0.6 mm deeper than Type 5—the two types of Red Crossbill that can also occur in the

range of Cassia Crossbill. Plumages closely follow those of Red Crossbill.

Similar Species Separated with difficulty from most types of Red Crossbill by the deeper bill. Best separated by call.

Voice CALL: Low pitched and deep *dip-dip*. **SONG:** Differs slightly from Red Crossbill.

Status & Distribution Fairly common and totally resident in South Hills and Albion Mts., Twin Falls Co. and Cassia Co., ID.

Population Estimated population of 6,000.

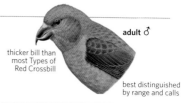

adult ♂

thicker bill than most Types of Red Crossbill

best distinguished by range and calls

WHITE-WINGED CROSSBILL *Loxia leucoptera*   WWCR ■ 2

The White-winged Crossbill is one of three species in N.A. with a distinctive finchlike bill that crosses at the tip; its cousins, the Red and Cassia Crossbills, are the others. White-winged inhabits the northern boreal forest from Newfoundland to AK; highly nomadic and irruptive, its populations move around both seasonally and annually because of fluctuating spruce cone crops. It forms large single-species flocks. Polytypic (2 sspp.; nominate in N.A.). L 6.5" (17 cm)

Identification Black wings, with contrasting white lesser coverts (often covered by scapulars),

bold white wing bar, and white tips to tertials. **ADULT MALE:** Pink body. Grayish flanks and undertail coverts. Black scapulars and lores. Dark on rear portion of auricular forms a crescent on side of head. **WINTER MALE:** Paler pink overall. **FEMALE:** Brownish olive body, with very indistinct streaking on underparts. Pale yellow rump. Dark lores, auricular patch more distinct. **IMMATURE MALE:** Like adult male, but yellow with patches of orange. **JUVENILE:** Brown with heavy streaks all over. Wings have thinner white wing bars than adult.

Geographic Variation Old World *fasciata*, breeding in

the boreal forest from Fennoscandia to Russian Far East, east to about Magadan, differs vocally from nominate *leucoptera* and likely represents a distinct species. A former subspecies of White-winged from the mountains of Hispaniola is now recognized as a full species, Hispaniolan Crossbill (*L. megaplaga*); it is closely tied to the West Indian Pine.

Similar Species Red and Cassia Crossbills are the only other N.A. species with a bill distinctly crossed at tip; however, White-winged male is pinker and all plumages have a bolder wing pattern. Know that some Reds

immature ♂
leucoptera

juvenile
leucoptera

leucoptera ♀

thinner crossed bill than Red Crossbill

pink body a little more reddish pink in summer

white tertial tips

black wings with thick white wing bars

winter adult ♂
leucoptera

show narrow white wing bars. Male Pine Grosbeak, which has white wing bars and is red overall, lacks crossed bill and is a third again larger.

Voice CALL: A rapid series of dry *chet* notes and a double or triple *chik-chik*; given in flight. Generally softer than call of Red. **SONG:** A series of mechanical trills and whistles, usually rising in pitch; given from a treetop or in flight display.

Status & Distribution Fairly common, but erratic. **BREEDING:** Frequents northern boreal forests, but irregular and nomadic. Present at locations one year, absent the next. Timing and success dependent upon spruce cone production. **WINTER:** Generally south of normal breeding areas, but inconsistent year to year. Regular in the northern states. **RARE STATUS:** Nominate *leucoptera* is casual to Bering Sea islands, northwestern CA, and northern NM, TX, and FL; also Greenland and Bermuda. Accidental Unalaska I., eastern Aleutians (sspp. uncertain).

Population Stable.

SISKINS AND GOLDFINCHES Genus *Spinus*

EURASIAN SISKIN *Spinus spinus*   EUSI ▪ 5

Eurasian counterpart of Pine Siskin. Monotypic. L 4.8" (12 cm)

Identification Structurally like a Pine Siskin but bill slightly longer. **MALE:**

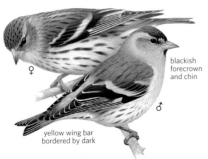

♀

blackish forecrown and chin

♂

yellow wing bar bordered by dark

Black forecrown and chin. Yellow throat and breast. Streaked olive upperparts and prominent black streaking on flanks. Olive auricular surrounded by yellow. Two bold yellow wing bars; extensive yellow edging on flight feathers and at base of tail. Yellow rump. **FEMALE:** Very similar to Pine Siskin with no black on crown or chin. Yellow restricted to sides of throat, wash on face, eyebrow, and rump. **JUVENILE:** Streaked above, whitish below, with dark streaks.

Similar Species Males unlike other carduelids in N.A. Female and especially juvenile very similar to female or juvenile Pine Siskin, but blacker wing coverts on Eurasian and slight yellow suffusion on head and breast and undertail.

Voice CALL: A two-note whistled *ti-lu* or a short rattle.

Status & Distribution Europe and eastern Asia. **RARE STATUS:** Two spring records from Attu I.; one winter record from Unalaska I. Casual in northeast N.A., with records from ON, ME, MA, NJ, and St. Pierre and Miquelon, but origin of some (most?) questioned.

Population Moderate declines in Europe.

PINE SISKIN *Spinus pinus*   PISI ▪ 1

A widespread and conspicuous breeding species of coniferous forest across the boreal zone of Canada and northern US, as well as in mountainous areas of the West, the Pine Siskin is an irregular and less predictable winter visitor virtually anywhere in the US. It forms large flocks during the nonbreeding season and is commonly attracted to seed feeders. Polytypic (3 sspp.; nominate in N.A.). L 5" (13 cm)

Identification It is entirely brown and streaked and has prominent yellow in the wing, a short deeply notched tail, long wings, and a longish, pointed bill. **MALE:** Brown and streaked above, below whitish with coarse dark streak-

ing. Two prominent wing bars, lower one extensively yellow. Distinct yellow edging to flight feathers and tail is conspicuous in flight and on folded wing. Some males very yellowish with reduced streaking. **FEMALE:** Similar to male, but yellow in wings and tail greatly reduced. **JUVENILE:** Quite buffy yellow, but fades by late summer.

Similar Species *Haemorhous* finches are considerably larger with thicker bills and lack yellow flash in the wing of siskins. See Eurasian Siskin.

Voice CALL: Most commonly a buzzy,

rising *zreeeeee*; also a harsh, descending *chee* in flight. **SONG:** A lengthy jumble of trills and whistles similar to American Goldfinch, but huskier.

Status & Distribution Common and gregarious. **BREEDING:** Found in coniferous forests of the north and mountainous west. **WINTER:** Range erratic from year to year, likely due to fluctuating food supply; can be found virtually anywhere. Often associates with goldfinches. **RARE STATUS:** Rare and irregular to Bermuda; very rare to Bering Sea islands, AK.

Population Stable.

prominent yellow wing stripe and yellow patches at base of tail

pinus

pinus

juvenile pinus

pinus

thin, sharply pointed bill

pale wing bars

green morph pinus

sometimes exhibits xanthochroism with a yellow suffusion; compare to female Eurasian Siskin (above)

streaked underparts

LESSER GOLDFINCH *Spinus psaltria* LEGO ■ 1

This very common carduelid finch of the Southwest breeds in a variety of habitats at different elevations. It is often seen in small flocks feeding along brushy roadsides, particularly where thistle grows. It is often detected by its distinctive flight calls.

Polytypic (5 sspp.; 2 in N.A.). L 4.5" (11 cm). **Identification** A small, sexually dimorphic finch. **BREEDING MALE:** Differs depending on subspecies. In green-backed *hesperophila*, bright yellow underparts and black cap contrast with an olive-green back. Note white at base of primaries forms a small white patch on folded wing. In male black-backed *psaltria* (*mexicanus* of some authors), bright yellow underparts contrast with entirely black upperparts. **FEMALE:** Yellow below and greener above. Note distinctive white at base of primaries. Some females very dull and drab, mostly gray with some yellowish green wash to body; wings duller, but still with white patch. **IMMATURE MALE:** It lacks full black cap of adult, but it still has black on forehead (*hesperophila*) or has some black intermixed with green on back and crown (*psaltria*).

Geographic Variation Green-backed *hesperophila* is more widespread in Southwest, with black-backed *psaltria* breeding from CO to southern TX. Intergradation occurs clinally between TX and CO.

Similar Species Males are very distinct from other goldfinch species. Female American is larger and has white undertail coverts, a wide buffy lower wing bar with very little white at base for flight feathers, and a pale pinkish bill. A very pale female Lesser can show whitish undertail coverts, but it has a different wing pattern than all American Goldfinches and is always greener than female Lawrence's.

Voice CALL: Includes a plaintive, kittenlike *tee-yee*. **SONG:** Very complex jumble of musical phrases, often mimicking other species.

Status & Distribution Very common. **BREEDING:** A variety of habitats at different elevations from arid lowlands to high pine forests, often found near water. **WINTER:** Northern and high elevation populations migrate to southern US and Mexico, augmenting resident populations there. **RARE STATUS:** Casual north and east of mapped range in Great Plains. Accidental in the East. **Population** Range has spread east and especially north in recent decades.

hesperophila ♀
smaller than American
pale yellow underparts

black-backed adult ♂ *psaltria*

green-backed adult ♂ *hesperophila*

pale ♀ *hesperophila*
blackish cap contrasts with green back

immature ♂ *hesperophila*
white patch at base of primaries

hesperophilus
psaltria
witti
jouyi
columbianus

LAWRENCE'S GOLDFINCH *Spinus lawrencei* LAGO ■ 2

The subtle combination of gray, yellow, and black makes the male Lawrence's Goldfinch one of the more striking carduelids in N.A. Largely restricted during the breeding season to the foothills and montane valleys in CA, it has a unique irregular dispersal pattern some winters to the southwestern US. Often seen in small groups, it frequents brushy areas, preferring those found along riparian corridors. Like other goldfinches, it favors thistle plants; particularly partial to fiddleneck in spring. During drier months, often most easily located at water sources. Forms flocks in nonbreeding season, sometimes with Lesser Goldfinches, but often in pure flocks. Monotypic. L 4.8" (12 cm). **Identification BREEDING MALE:** Very

sharp plumage, acquired through wear. Mostly gray body; lighter gray cheeks; black front of crown, face, and chin; and yellow breast. Black wings with extensive yellow forming two broad wing bars and yellow edging to flight feathers. Yellow rump. Distinctive circular white tail spots.

brownish wash on back in fresh fall plumage in both sexes, wears to more grayish by spring

winter ♀

winter ♂

black forehead, lores, and chin

yellow breast

distinctive white oval tail spots like a *Setophaga* wood-warbler

breeding ♂

prominent yellow on wings in all plumages

juvenile

BREEDING FEMALE: Much duller than male. Grayish body, browner above. Reduced amount of yellow on breast. Two prominent yellowish wing bars, with yellow edging to primaries. **WINTER MALE:** Like breeding male, but browner above and duller below. **WINTER FEMALE:** Like breeding female, but browner. **JUVENILE:** Streaked; unlike other goldfinches. Fresh adultlike plumage acquired by fall.

Similar Species At all seasons note extensive and diagnostic yellow in wings and circular white tail spots on outer tail feathers. Male unique. Female could be confused with very dull Lesser Goldfinch female, which is smaller and is still greenish; Lawrence's Goldfinch has whiter undertail coverts; note wing color and tail pattern. Call notes strongly differ among goldfinch species.

Voice CALL: A very distinctive, rather soft bell-like *tink-ul*. **SONG:** A series of jumbled musical twittering, often interjecting *tink* notes, and habitually mimicking other species like Lesser Goldfinch.

Status & Distribution Rather uncommon and local. **BREEDING:** Prefers drier interior foothills and montane valleys of CA and northern Baja, but seeks out water sources from streams and pools. Some populations resident, while others migratory. Spring migration begins by late Feb. and fall migrants noted into late Oct. Casual in central AZ. Breeding areas not consistent from year to year in many regions. **WINTER:** Irregular fall and winter movement to the southwestern US, occasionally wintering in moderate numbers in southeastern AZ and northern Mexico. AZ wintering birds begin arriving by mid-Oct. but there are occasional mid-summer records

too. Some years the wintering grounds remain largely unknown. **RARE STATUS:** Rare and has bred in northeast and northwest CA. Casual to southern UT, northern AZ, NM, western CO, West TX, and OR; accidental western WA and southwestern OK.

Population Stable but little studied.

AMERICAN GOLDFINCH *Spinus tristis*   AMGO   ■   1

The breeding male American Goldfinch is striking. It regularly visits seed feeders. Often gregarious, especially during the nonbreeding season, when it flocks to roadsides and brushy fields to feed on thistle and sunflowers, often far from woodlands. Polytypic (4 sspp.; all in N.A.). L 5" (13 cm).

Identification BREEDING MALE: Entirely bright lemon yellow in most subspecies with white undertail coverts. Jet black cap. Black wings with yellow lesser coverts; two white wing bars. Pink bill. **BREEDING FEMALE:** Underparts yellow with white undertail coverts, olive green above. **WINTER MALE:** Cinnamon brown above and on breast and flanks, with white lower belly and undertail coverts, yellowish wash on throat and face, and muted black on forehead. Wings boldly patterned; yellow lesser coverts; wide, whitish lower wing bar. Bill darker than in breeding season. **WINTER FEMALE:** Mostly drab gray body with black wings and two bold buffy wing bars. White undertail coverts. Dark bill. **IMMATURE MALE:** Black on forehead reduced or lacking. Lesser coverts duller. **JUVENILE:** Resembles adult female. Unstreaked.

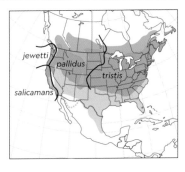

Geographic Variation Differences among subspecies slight, but breeding adult males *salicamans* (CA) do not achieve bright yellow coloration.

Similar Species Female told from smaller female Lesser by white in tail feathers, white undertail coverts, and more white in wings. All plumages of winter Americans are duller.

Voice CALL: A *per-chik-o-ree* or *ti-di-di-di*, sometimes rendered as "*potato chip*," given mainly in flight; distinctly different from other goldfinches. **SONG:** A long series of musical phrases. Not known to mimic other species.

Status & Distribution Common throughout much of US and southern Canada. **BREEDING:** A variety of habitats, from weedy fields to open second-growth woodland, and along riparian corridors, particularly in West. **RARE STATUS:** Casual to AK.

Population Some populations declining (e.g., CA).

breeding ♀
tristis

black wings and tail; white undertail coverts

breeding ♂
tristis

bright yellow body

black forehead and forecrown

yellow shoulder; brownish back

winter adult ♂
tristis

whitish undertail coverts

prominent wing bars

winter ♀
tristis

juvenile
tristis

LONGSPURS AND SNOW BUNTINGS Family Calcariidae

Snow Buntings (AB, Oct.)

Recent molecular work using mitochondrial and nuclear DNA has shown that this small group of sparrow-like tundra and grassland species is well differentiated genetically from the New World sparrows and Old World buntings of the family Emberizidae and best placed in its own family. Six species are recognized in three genera, all of which are found in North America.

Structure All are medium-small (6–7 in.) sparrow-like, conical-billed, seed-eating birds with relatively long wings and short legs.

Behavior Ground-dwelling seedeaters, longspurs and snow buntings walk in a shuffling gait; they are strong fliers, rising well into the air with an undulating flight and dropping steeply back to the ground. Territorial songs are given from the top of a shrub and also, in most species, in flight. They are gregarious in the nonbreeding season and prefer open country, often flocking with Horned Larks and often flushing in groups to avoid predators.

Plumage Snow Buntings are largely white with black markings in the breeding season, and washed with tan on head and upperparts in winter; they always flash much white in the wings and tail. Longspurs are generally brownish with conspicuous white markings in the tail; breeding males are marked with black on the head and variably on the underparts, and in most species show areas of rich buff or chestnut in the plumage. Females, winter males, and immatures are duller.

Distribution The family is limited to the Arctic and north temperate regions of North America and Eurasia, from tundra habitats to short-grass prairies and other open habitats. Two species (Snow Bunting and Lapland Longspur) occur across the northern Holarctic Region; McKay's Bunting is restricted in the breeding season to two Bering Sea islands, and the remaining three longspur species breed only in North America. All species are migratory but some winter within subarctic and north temperate regions; others move farther south.

Taxonomy Longspurs and snow buntings have traditionally been placed within the large family Emberizidae, and at least superficially seem to fit well within that family (some authorities had even recommended merging the longspur genus *Calcarius* into the Old World bunting genus *Emberiza*). Molecular data, however, indicate that the calcariids split off from the other groups of nine-primaried oscines quite early and deserve family rank of their own. Among the longspurs, McCown's has been treated with the other three species in the genus *Calcarius* but is now, once again, placed in its own genus, *Rhynchophanes*. McKay's Bunting is perhaps a well-marked subspecies of the Snow Bunting. Some authors use the name Plectrophenacidae for this family.

Conservation Populations are generally stable, although the two prairie longspurs (Chestnut-collared and especially McCown's) have fairly restricted ranges on the Great Plains and have suffered from habitat degradation of native grassland. The McKay's Bunting has a tiny breeding range and is therefore at some risk. BirdLife International lists Chestnut-collared Longspur as Vulnerable.

LONGSPURS Genera *Calcarius* and *Rhynchophanes*

Four longspur species (three endemic to N.A.; one widespread, circumpolar) occur in open areas and use trees or bushes only for song perches. In winter, they form large flocks, often with paler Horned Larks. They often call when flushed and while in their strong, undulating flight. Based on genetic evidence, McCown's was restored to its own monotypic genus.

LAPLAND LONGSPUR *Calcarius lapponicus*   LALO ◼ 1

In migration and winter often joins Horned Larks and Snow Buntings. Look for Lapland's darker overall coloring and smaller size. Polytypic (5 sspp.; 3 in N.A.). L 6.3" (16 cm)
Identification White on outer two tail feathers; reddish indented edges on greater coverts and on tertials. Very long primary projection past longest tertial. **BREEDING ADULT MALE:** Extensive black on head and under-

parts; a broad white or buffy stripe extends back from eye and down sides of breast; chestnut nape. **BREEDING ADULT FEMALE:** Duller. **WINTER:** Bold, dark triangle outlines plain buffy ear patch; dark streaks (female) or patch (male) on upper breast; dark streaks on sides; broad buffy eyebrow and buffier underparts; belly is usually white. **JUVENILE:** Buffy and heavily streaked above and on breast and sides.

Geographic Variation Five subspecies show weak variation in coloration. Eastern *subcalcaratus* breeds from north-central NT to NL and winters west to TX; the remainder of the N.A. range is occupied by *alascensis*, which is comparatively much paler. Commander Is.–breeding *coloratus* is blacker above and below and more richly colored than *alascensis*, has occurred casually in summer on Attu I.
Similar Species See sidebar (p. 616).
Voice CALL: A musical *tee-lee-o* or *tee-dle*. **FLIGHT NOTE:** A dry rattle with whistled *tew* notes; a trait shared with Snow and McKay's Buntings. **SONG:** Rapid warbling, frequently given in short display flights; only on breeding grounds.
Status & Distribution Common. Circumpolar distribution. **BREEDING:** High arctic tundra. **WINTER:** Fields and beaches. **MIGRATION:** Spring late Feb.–early Apr. (into May northern US and Canada); fall early Sept.–late Nov., peaking late Oct.–early Nov. in most of the US. **RARE STATUS:** Rare FL; casual Mexico.
Population Stable.

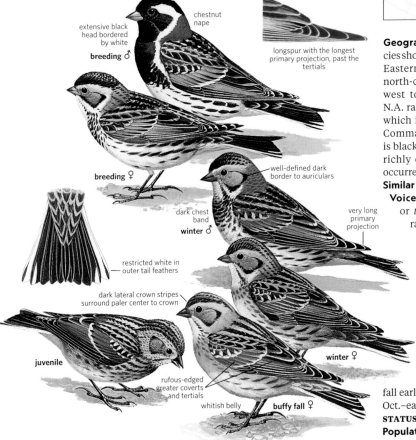

chestnut nape

longspur with the longest primary projection, past the tertials

extensive black head bordered by white
breeding ♂

breeding ♀

well-defined dark border to auriculars

dark chest band
winter ♂

very long primary projection

restricted white in outer tail feathers

dark lateral crown stripes surround paler center to crown

juvenile

rufous-edged greater coverts and tertials

whitish belly

winter ♀

buffy fall ♀

CHESTNUT-COLLARED LONGSPUR *Calcarius ornatus*   CCLO ◼ 1

The somewhat secretive Chestnut-collared Longspur favors denser grass in migration and in winter than does either the McCown's or Lapland Longspurs. Flocks of Chestnut-collared Longspurs flush from underfoot and

can vanish again into ankle-high grass. Chestnut-collared mixes less frequently with other longspurs. Except on breeding grounds, rather rarely seen in the open (unlike Lapland and McCown's Longspurs). Most

easily identified by distinctive call. Monotypic. L 6" (15 cm)
Identification White tail is marked with a blackish triangle. Primary projection past longest tertial is short and primary tips barely extend to base

of tail. Fall and winter birds have grayish, not pinkish, bills. **BREEDING ADULT MALE:** Black-and-white head, buffy face, and black underparts are distinctive; a few have chestnut on underparts. Whitish lower belly and undertail coverts. Upperparts are black, buff, and brown, with chestnut collar, whitish wing bars. **WINTER MALE:** Paler; feathers are edged in buff and brown, largely obscuring black underparts. Male has small white patch on shoulder, often hidden; compare to Smith's Longspur. **BREEDING ADULT FEMALE:** Usually shows some chestnut on nape. **WINTER FEMALE:** Like breeding female, but paler. **JUVENILE:** Pale feather fringes give upperparts a scaled look.

Similar Species Juvenile is best distinguished from juvenile McCown's Longspur by its tail pattern and bill shape. See sidebar, p. 616.

Voice FLIGHT NOTE: Distinctive two-syllable *kittle*, repeated one or more times. Also gives a soft, high-pitched rattle and a short *buzz* call. **SONG:** Pleasant rapid warble, given in song flight or from a low perch; heard only on breeding grounds.

Status & Distribution Fairly common. Winters south to northern Mexico. **BREEDING:** Nests in moist upland prairies and typically prefers moister areas with taller, lusher grass than does McCown's Longspur. **MIGRATION:** Spring migration mid-Mar.–mid-Apr.; fall migration late Sept.–mid-Nov. Migration is primarily through western and central Great Plains. Males may wander during midsummer. **RARE STATUS:** Rare in CA in migration and winter. Casual during migration to eastern N.A. and the Pacific Northwest. Accidental in winter and mid-summer to the East Coast.

Population Vulnerable. As with

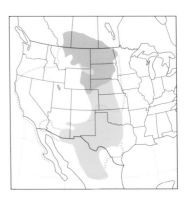

McCown's, Chestnut-collared has suffered due to the destruction of native prairies by overgrazing and the conversion to large-scale agriculture. Formerly, Chestnut-collared Longspur bred in Kansas and was more widespread throughout its current range.

chestnut collar

breeding males

primary tip projection shorter than illustrated (see wing tip illustration)

black breast and belly

winter ♂

veiled black breast and belly

winter ♀

small darkish bill

faint streaks on breast

longspur with the shortest primary projection, less than half an inch

dark triangle on white tail

SMITH'S LONGSPUR *Calcarius pictus*   SMLO   2

In winter Smith's Longspur is found in ankle-high grass (especially three-awn grasses) or alfalfa. Secretive like Chestnut-collared Longspur; Lapland and McCown's are less secretive. Smith's sometimes flocks with Laplands. Distinctive are the strong buff underparts and white shoulders (males) and the sharp call, given by all longspurs when flushed and in flight. This species has a polyandrous mating system. Monotypic. L 6.3" (16 cm)

Identification Outer two tail feathers are almost entirely white. Bill thinner and more pointed than on other longspurs. Note long primary projection past longest tertial, a bit shorter than on Lapland, but much longer than on Chestnut-collared or McCown's. **BREEDING ADULT MALE:** Bold black-

distinctive black-and-white head pattern

fine streaks on buffy underparts

breeding ♂

breeding ♀

long primary projection, a little shorter than Lapland

more white in outer tail feathers than Lapland

winter ♂

white lesser coverts often best noted in flight

and-white head pattern; rich buff nape and underparts; white patch on shoulder, often obscured. **BREEDING ADULT FEMALE:** Duller, crown streaked, chin paler. Dusky ear patch bordered by pale buff eyebrow; ear patch border often broken in rear. Underparts buff with thin reddish brown streaks on breast and sides. Much less white on lesser coverts than male. **WINTER & IMMATURE:** Like female.

Similar Species See sidebar below.

Voice CALL: Short, nasal *tseu.* **FLIGHT NOTE:** Dry, ticking rattle, harder and sharper than the call of Lapland and McCown's. **SONG:** Rapid, melodious warbles, ending with a vigorous *wee-chew*, delivered only from the ground or a perch; males sing during spring migration.

Status & Distribution Uncommon. **BREEDING:** Open tundra and damp, tussocky meadows. **WINTER:** Open, ankle-high grassy areas. **MIGRATION:** Spring late Mar.–late May in US and May in Canada; arrives on breeding grounds late May–early June; fall early Sept.–mid-Nov. Typically migrates earlier in fall and later in spring than Lapland. Regular spring migrant in the Midwest, east to western IN, rarely western OH (formerly more frequent). **RARE STATUS:** Rare migrant in western Great Lakes. Casual to both coasts, mainly in migration (mostly fall).

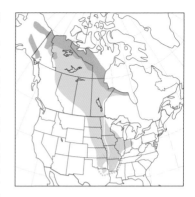

Population Perhaps declining; has disappeared from many breeding sites in AK and northwestern BC.

Winter Longspurs

Winter longspurs present one of the greater challenges among N.A. passerines, in part because the shifting flocks can be quite furtive on the ground, making viewing difficult. Details of plumage, shape, and tail pattern are useful, but identification is always aided by a consideration of call notes, habitat, and behavior. All species are gregarious. Lapland Longspur most resembles Smith's; Chestnut-collared is most like McCown's. The bird's tail pattern provides a good starting point and is best seen in flight, especially as the tail is flared upon landing. Lapland and Smith's have white outer tail feathers (whiter in the Smith's). Chestnut-collared and McCown's have largely white tails with differing patterns of black: a black triangle in Chestnut-collared and a black inverted T in McCown's. On the ground, Smith's and Lapland have a long primary extension, while McCown's and especially Chestnut-collared have a comparatively short primary extension past the longest tertial. When on the ground, Smith's is best told from Lapland by its wing panel, which is a contrasting chestnut color in Lapland and non-contrasting buffy in Smith's. Also note Smith's smaller, slimmer bill; finer, sparser streaking below; often broken rear border to the facial frame; and white (adult male) or white-edged median coverts. All Smith's are extensively buffy below, but this can be rarely approached in Lapland. McCown's is large and pale, with a large pinkish bill and distinctive dull face pattern (recalling female House Sparrow); and males have dark chestnut median coverts. Chestnut-collared is shaped more like McCown's but is not as plain and is darker with shorter primary projection and a smaller, darker bill. All four species' flight calls are distinctive when learned and invaluable to identification. Focus on Lapland's dry rattle, interspersed with *tew* or *jit* notes; Smith's slower, sharper, clicking rattle; McCown's softer, more abrupt rattle, interspersed with a distinctive *pink* note; and Chestnut-collared's unique *kittle* call. Habitat, while not diagnostic, can be an important clue. McCown's prefers the most open country (heavily grazed grasslands, plowed fields, flat dirt areas). Lapland similarly prefers open areas that are mostly lacking in vegetation, but it may be slightly more regular in short grass areas (e.g., airports) than McCown's. Both Chestnut-collared and Smith's are found in short grass areas, typically in ankle-high grass. Chestnut-collared is often found in areas that have some patches of bare ground interspersed; Smith's is usually in denser, more complete areas of grass. The two denser-grass species are also more furtive and often flush from almost underfoot; Lapland and McCown's are more likely to flush at a distance.

McCown's Longspur, winter

Lapland Longspur, winter

Smith's Longspur, winter

Chestnut-collared Longspur, winter

McCOWN'S LONGSPUR *Rhynchophanes mccownii* MCLO ■ 2

This species favors more barren country than do other longspurs and often flocks with Horned Larks. Look for McCown's chunkier, shorter-tailed shape and slightly darker plumage, its mostly white tail, its much thicker bill, and its undulating flight. On breeding grounds it is easily separated from Chestnut-collared by its unique display flight. Monotypic. L 6" (15 cm)

Identification McCown's white tail is marked by a dark inverted T shape. Note also stouter and thicker-based bill than found on other longspurs. Its primary projection is slightly longer than Chestnut-collared; in a perched bird, wing extends almost to tip of short tail. **BREEDING ADULT MALE:** Black crown, black malar stripe, black crescent on breast; gray sides. Upperparts streaked with buff and brown, with gray nape and rump; chestnut

median coverts form contrasting crescent. **BREEDING ADULT FEMALE:** Streaked crown; may lack black on breast and show less chestnut on wing. **WINTER ADULT:** Large pinkish bill with a dark tip; feathers edged with buff and brown. Winter adult female paler than female Chestnut-collared, with fewer streaks on underparts and a broader buffy eyebrow. Some winter males have gray on rump and variable blackish on breast; retain chestnut median coverts. **JUVENILE:** Streaked below; pale fringes on feathers give upperparts a scaled look; paler overall than juvenile Chestnut-collared.

Similar Species See Chestnut-collared Longspur for identification of fall and winter birds. See also sidebar opposite.

Voice FLIGHT NOTE: A dry rattle, a little softer and more abrupt than Lapland. A unique *pink* note is especially useful for identification; it may recall a soft Bobolink or Purple Finch flight note. **SONG:** Heard only on breeding grounds;

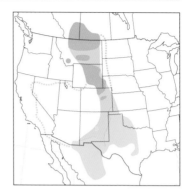

a series of exuberant warbles and twitters, generally given in a distinctive song flight, unlike that of Chestnut-collared. McCown's Longspur rises to a considerable height and then delivers its song as it floats slowly back to the ground with its wings held in a sharp dihedral. The appearance recalls a falling leaf or a floating butterfly.

Status & Distribution Uncommon to fairly common, but range has shrunk significantly since the 19th century. Winters south to northern Mexico. **BREEDING:** Dry shortgrass prairies. **WINTER:** Dry shortgrass prairies and fields, also plowed fields, airports, and dry lake beds. Very rare visitor to interior CA and NV. **MIGRATION:** Spring migration early Mar.–mid-May; fall migration mid-Oct.–early Nov. Migrates primarily through western Great Plains. **RARE STATUS:** Casual to coastal CA and OR; accidental to BC and to East Coast.

Population Breeding range drastically reduced since the 1800s; for example, the species formerly bred in Oklahoma, South Dakota, and western Minnesota. Contraction in range and overall decrease in population is due to land management practices that have greatly reduced shortgrass prairie. Conversion to large-scale agriculture is especially to blame.

short primary projection, but longer than Chestnut-collared

breeding ♂

black crown

black chest patch

breeding ♀

black inverted-T on white tail

plainer face than Chestnut-collared with buffy supercilium

unstreaked buffy breast

thick pinkish bill

winter ♀

veiled blackish chest patch

chestnut median coverts

winter ♂

short tail

juvenile

Genus *Plectrophenax*

SNOW BUNTING *Plectrophenax nivalis* SNBU ■ 1

The Snow Bunting's breeding and flocking behaviors are similar to the Lapland Longspur. Polytypic (4 sspp.; 2 in N.A.). L 6.8" (17 cm)

Identification Long black-and-white wings; breeding plumage acquired

from wear by end of spring. Bill black (summer) or orange-yellow (winter). Males usually show more white (especially wings). **JUVENILE:** Grayish, streaked; buffy eye ring. First-winter plumage browner than adult.

Geographic Variation Larger, whiter, bigger-billed *townsendi* is resident on Aleutians, Pribilofs, and Shumagin I. (also Commander Is.); nominate occurs throughout rest of N.A. range.

Similar Species See McKay's Bunting.

Sometimes confused with albinistic sparrows. Note plumage, behavior. **Voice CALL:** Often given in flight. Sharp, whistled *tew*; short buzz; musical rattle or twitter. Rattle and *tew* notes like Lapland (rattle softer, *tew* clearer). **SONG:** Loud, high-pitched musical warbling, only on breeding grounds. **Status & Distribution** Fairly common.

BREEDING: Tundra, rocky shores, talus slopes. **WINTER:** Shores, weedy fields, grain stubble, plowed fields, roadsides. **MIGRATION:** Fall late Oct.–early Dec.; spring early Feb.–late Mar. **RARE STATUS:** Very rare to northern CA and FL; casual to Southern CA, AZ, TX, and Bahamas. **Population** Stable.

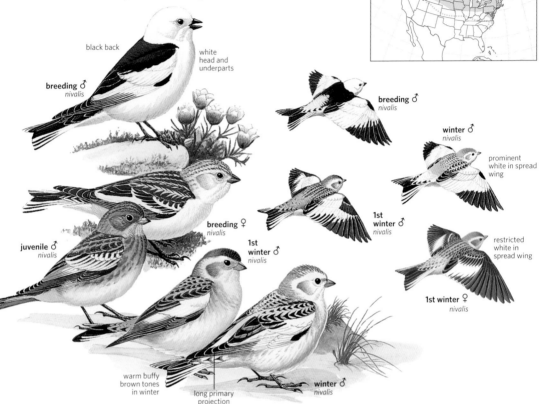

MCKAY'S BUNTING *Plectrophenax hyperboreus* MKBU ▪ 3

The McKay's Bunting has a tiny worldwide breeding range, confined to two Bering Sea islands. Monotypic. L 6.8" (17 cm)

Identification Resembles Snow Bunting, but has more white in the tail and primaries. **BREEDING:** Male mostly white, including back (black in Snow), with less black on wings and tail. Female like Snow, but white panel on greater coverts. **WINTER:** Edged with rust or tawny brown; male whiter

overall than Snow. **JUVENILE:** Like juvenile Snow.

Similar Species Snow Bunting is very similar; beware of hybrids, which appear regularly. Male McKay's more extensively white; black limited to feather tips. Female McKay's is whiter in tail, wing coverts, and primaries than male Snow. Juveniles very similar.

Voice Calls and song identical to Snow Bunting.

Status & Distribution Uncommon. **BREEDING:** Breeds regularly only on Hall and St. Matthew Is. in the Bering

Sea; arrives earlier in spring than the Snow Bunting, which is absent as a breeder. A few formerly present in summer on the Pribilofs. Uncommon migrant (has wintered) on St. Lawrence I., where a few have paired with Snow Buntings. **WINTER:** Uncommon in winter along Bering Sea coast of AK from Kotzebue south and to Cold Bay; casual in winter southward, in interior of AK and on Aleutians. **RARE STATUS:** Accidental BC, WA, and OR.

Population Stable, but vulnerable, given its tiny breeding range.

OLD WORLD BUNTINGS Family Emberizidae

Pallas's Bunting, female *pallasi* (Siberia, Russia, June)

This large and colorful Old World family is found throughout Eurasia and Africa. Nearly 10 species have reached North America.

Structure These are medium-size passerines with moderately long tails.

Behavior These species feed mostly on the ground, flying up into bushes or trees when alarmed. Best views are often obtained of singing males at first sunlight. Some species are highly gregarious in migration and winter.

Plumage The majority are colorful and show moderate to striking sexual dimorphism, the differences being most apparent in alternate plumage. Most show white outer tail feathers, although some are plain-tailed.

Distribution Many species are Eurasian; of these some two thirds are Asian and include species from Central and East Asia. Ten species are found mostly in Africa. Palearctic species are mostly migratory, some are highly migratory. The African species are resident or are somewhat nomadic, their movements tied to rainfall. Nine species have occurred in North America, mostly in Alaska as rare to accidental visitors. Two species (Little and Rustic Buntings) have occurred in North America away from AK. Not surprisingly, they occupy a wide variety of habitats from reed beds to brushy slopes to coniferous and even tundra edge.

Taxonomy With the New World sparrows recently placed in their own family, the Emberizidae now consists of 41 or 42 species. Nearly all were traditionally placed in the single large genus *Emberiza*, but many authorities now place the 10 mostly African species in *Fringillaria* and 14 of the Asian species (including eight of the nine species recorded in North America) in *Schoeniclus*.

Conservation Most species are not threatened. Socotra Bunting (*F. socotrana*) with a tiny range has an estimated population of 1,400. Yellow-breasted and Jankowski's (*E. jankowskii*) Buntings are critically endangered. Yellow-breasted is widely trapped (a million or more annually) in nets for food markets in eastern Asia and numbers have declined catastrophically. Jankowski's is now found only in very small numbers at a handful of sites. BirdLife International codes: 4 NT, 2 VU, 1 EN, and 2 CR.

| Genus *Emberiza*

PINE BUNTING *Emberiza leucocephalos*   PIBU ▪ 5

This Old World species has been recorded on just four occasions in N.A., all in fall from western AK. Polytypic (2 sspp.; nominate in N.A.). L 6.5" (17 cm)

Identification Rusty rump. **MALE:** Distinctive broad white stripe under eye is unique. In breeding plumage, male is rusty about face and throat, and it has

a white median crown stripe. In winter plumage, colors and pattern subdued, and median crown stripe is subdued. **FEMALE:** Rusty supercilium; underparts are whitish and finely streaked; white cheek is less prominent than on male; rusty rump. **IMMATURE:** Like female, but often duller and lacking white cheek patch entirely, though

it has a white spot (at least) at rear auriculars, similar to that found on Rustic Bunting.

Geographic Variation Pine Bunting comprises two subspecies, but only *leucocephalos* has been found in N.A.

Similar Species White cheek patch is distinctive on adults. Immatures could be confused with either Rustic or Little. Note, however, Pine's weak malar; its lack of both a strong eye ring and a crest; and its dull or whitish cheek, which is never chestnut like that found on Little. **FEMALE & IMMATURE:** Yellow-breasted Bunting is also similar to Pine; however, Yellow-breasted always shows some tint of yellow on face or underparts.

Voice CALL: Abrupt *spit* or *tic*, often doubled, *spi-tit*; also a hoarse *jeeit*. **SONG:** Long and variable, repetitious warbling phrases.

Status & Distribution BREEDING: Across northern Russia; winters south to central-south Asia. **RARE STATUS:** From Urals across Russia to Magadan region, Russian Far East, and south to northern China. **WINTER:** Mostly Central Asia to northern China. **RARE STATUS:** Four fall records (Oct.–Dec.) to western AK (recorded Attu, St. Paul, and St. Lawrence Is.). Rare to casual in the Middle East and western Europe. **Population** Possibly declining.

breeding ♂

pale spot in rear of auriculars

broad buffy eyebrow

head pattern more veiled in fall and winter

mostly chestnut head with distinct and broad white stripe, outlined in black, under eye

♀

fall/winter adult ♂

rusty cast to rump

all figures *leucocephalos*

YELLOW-BROWED BUNTING *Emberiza chrysophrys* YBWB ▪ 5

Among the Eurasian buntings, Yellow-browed has one of the smallest breeding ranges, limited to a rather small area in southeastern Siberia and the adjacent Russian Far East. Its scientific name, *chrysophrys*, is derived from the Greek words for "golden-browed," and in fact the golden portion of the supercilium, strongly recalling White-throated Sparrow, distinguishes this species from others in the genus. Monotypic. L 6" (15 cm)

Identification A relatively large-headed and crested bunting, similar in size to Rustic Bunting and slightly larger than Little Bunting. Stout bill with dusky maxilla, pink mandible. Legs pale pink. Head pattern quite striking in all plumages: narrow white median crown stripe, dark lateral crown stripe, supercilium yellow mostly (becoming whitish behind eye), dark auriculars with pale spot at rear corner, white malar, and blackish submalar line. Head pattern most striking in breeding adult male, in which auriculars are black. Underparts mostly whitish, with flanks and breast washed buff and with fine streaks. Rump and uppertail coverts tinged reddish brown in all plumages. White outer tail feathers. Crown feathers often raised to a crest.

Similar Species In female and immature plumages, care should be taken to exclude Tristram's Bunting (*E. tristrami*), unrecorded N.A., which lacks yellow in supercilium, and the much less strikingly patterned Yellow-breasted Bunting (*E. aureola*), which is washed with buffy yellow in face and underparts and lacks dorsal rufous tones of Yellow-browed.

Voice CALL: A short, sharp *tsick*. **SONG:** A short, variable series of sweet and more modulated notes; often starts with a soft, long note.

Status & Distribution BREEDING: Mixed forest edge in southeastern Siberia and Russian Far East. **WINTER:** Central and eastern China. **RARE STA-**

females and fall males duller but show similar head pattern

mostly black head, white median crown stripe

fall

bicolored supercilium; white rear auricular spot

spring ♂

blackish ventral streaking

TUS: One at Gambell, St. Lawrence I., AK (15 Sept. 2007), represents the only N.A. record. **Population** Uncommon but apparently stable in its limited range.

LITTLE BUNTING *Emberiza pusilla* LIBU ▪ 4

First recorded in AK in 1970, this species had only three records between 1970 and 1990. Since then it has been found repeatedly in fall at Gambell (St. Lawrence I.), and it may prove annual with continued coverage. Monotypic. L 5" (13 cm)

Identification Small bunting, with short pink legs, short tail, and a small triangular gray bill with straight culmen. Full, creamy white eye ring, chestnut ear patch, and two thin whitish wing bars. Whitish and heavily streaked underparts; white outer tail feathers. **BREEDING ADULT:** Chestnut crown stripe bordered by black stripes and chestnut auricular; some adult males have chestnut on chin. **IMMATURE & WINTER ADULT:** Similar, but duller often with a pale spot in rear of auricular.

Similar Species Rustic Bunting

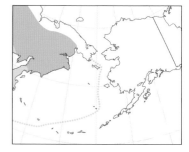

(female and fall birds) can be confused with Little Bunting. Note Rustic Bunting's larger size, heavier bill with pink lower mandible, diffuse rusty streaking below, and lack of eye ring. **Voice CALL:** Sharp *tsick*, similar to Rustic Bunting, but sharper. **SONG:** Short, variable series of rising and falling notes.
Status & Distribution Fairly common. **BREEDING:** Found in taiga and forest-tundra habitats from Fennocandia to Russian Far East northeast to Anadyr River Basin, and at least occasionally to northern Chukotka. **WINTER:** Open

woodlands and overgrown fields from Nepal and India to northern Southeast Asia. Rare in western Europe. **RARE STATUS:** Rare to St. Lawrence I. in fall (mid-Aug.–late Oct.) and once spring (2–4 June 2008); casual elsewhere in western AK (3+ records from western Aleutians and Pribilofs), WA (Oct.), OR (wintered), CA (four records, Sept.–Dec.), southeast AZ (Slaughter Ranch, 27 May 2017), and Baja California Sur (fall).
Population Stable.

chestnut median crown stripe and auricular with broad dark lateral crown stripes

fall birds, especially immatures, are duller

small size

prominent whitish eye ring

straight culmen

immature

breeding ♂

RUSTIC BUNTING   *Emberiza rustica*   RUBU   ▇   3

The Rustic Bunting is a rare migrant from Asia to the western Aleutians and Bering Sea islands. This and Little Bunting are the only Old World buntings to have occurred south of AK. Regarded by most as monotypic. L 5.8" (15 cm)
Identification Has a slight crest, whitish nape spot, pale spot in rear of auricular, and a prominent pale line extending back from eye; white outer tail feathers. All plumages have rusty streaking below. **BREEDING MALE:** Black head; prominent white supercilium; upperparts bright chestnut, with buff and blackish streaks on back; white underparts, with chestnut breast band and streaks on sides.

FEMALE: Brownish head pattern, with pale spot at rear of ear patch; streaks below a duller rust. **IMMATURE & WINTER MALE:** Similar to female.
Similar Species Lapland Longspur is similar to Rustic Bunting, but its behavior is different (e.g., longspur is unlikely to perch in a bush). Among other plumage differences, Lapland lacks pale ear spot, crested appearance, and rufous streaking below. Rustic differs from female and immature Reed Bunting and from Pallas's Bunting in its crested appearance; its stronger face pattern, pale median crown stripe, and small pale nape spot; extensive rufous streaking on breast and flanks; and its more

prominent wing bars. See also Little Bunting.
Voice CALL: Hard, sharp *jit* or *tsip*, lower and sharper than Little Bunting. **SONG:** A soft bubbling warble.
Status & Distribution Common. Eurasian species; breeds from northern Fennoscandia east to Kamchatka and northern Sakhalin I. (Russian Far East) and northeast to Anadyr River basin. It winters primarily in eastern China, Korea, and Japan. Rare in western Europe. **MIGRATION:** Rare spring and fall migrant on western Aleutians; very rare spring and casual in fall on central Aleutians; rare in spring and fall on Bering Sea islands. **RARE STATUS:** Casual in fall and winter in southern YT and southern AK south to CA. Accidental SK in winter.
Population Vulnerable. Declining.

pale spot at rear of crown and in ear coverts in all plumages

pinkish bill with straight culmen

chestnut markings below

♀

often raises slight crest

black-and-white head pattern

bright chestnut breast band and streaks down sides and flanks

breeding ♂

YELLOW-THROATED BUNTING   *Emberiza elegans*   YTBU   ▇   5

An East Asian species, Yellow-throated Bunting has been recorded only once from N.A., from Attu I., western Aleutians. Polytypic (2 sspp.; *elegans* in N.A.). L 6" (15 cm)
Identification Distinctive long, pointed crest. White outer tail feath-

ers. **MALE:** Black crest, broad black mask and breast patch, bordered above narrowly with white; supercilium white and narrow in front of eye, broadens and turns yellow behind eye and turns white again at rear; yellow throat; reddish brown, streaked back;

whitish breast with brownish streaking. **FEMALE & IMMATURE:** Duller version of male; auricular and crown brownish; no breast patch.
Geographic Variation Nominate subspecies *elegans* (includes *ticehursti*) breeds in southeastern Russian Far

East, northeast China, Korea, and Tsuhima I., Japan. Winters western Japan, Korea, and eastern China. Darker and more richly colored *elengantula* is largely resident in central China.

Voice CALL: Sharp *tzick*. **SONG:** Long series of sweet rollicking phrases.

Status & Distribution BREEDING: Eastern Asia. **WINTER:** Nominate subspecies winters Korea, Japan, and southeast China. **RARE STATUS:** One record: Attu I., AK (25 May 1998).

Population Stable.

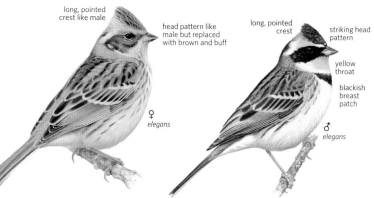

long, pointed crest like male

head pattern like male but replaced with brown and buff

♀
elegans

long, pointed crest

striking head pattern

yellow throat

blackish breast patch

♂
elegans

YELLOW-BREASTED BUNTING *Emberiza aureola* YBSB ■ 5

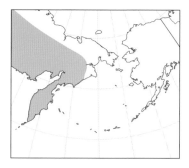

pale median crown stripe and pale spot in rear of auricular

♀
ornata

light yellow underparts

black face; east Asian *ornata* also with black on sides of breast and forehead

white in outer tail feathers, as with most *Emberiza* buntings

dark rufous-brown upperparts and white shoulder patch

bright yellow underparts

♂
ornata

This gregarious bunting has appeared a few times in western AK. Males are distinctive; females and immatures are subtle. Polytypic (2 sspp.; *ornata* in N.A.). L 6" (15 cm)

Identification Rather long bill with straight culmen and pinkish lower mandible. Bold streaks on back and white outer tail feathers. **BREEDING MALE:** Rufous-brown upperparts, black face, bright yellow underparts, white patch on lesser and median wing coverts. **WINTER ADULT MALE:** Overall pattern much more veiled. **FEMALE:** Distinctive head pattern: well-outlined median crown stripe; brown lateral crown stripes with internal black streaking; auricular with dark border and pale spot in rear; yellowish underparts with sparse streaking; unmarked belly. **IMMATURE:** Like female, often paler yellow.

Similar Species Pine, Reed, and Pallas's lack yellow wash, striking face.

Voice CALL: Sharp *tzip*, similar to Little Bunting. **FLIGHT NOTE:** Flat *stuck*.

Status & Distribution BREEDING: Fin-

land to Russian Far East; winters mainland Southeast Asia. Rare to western Europe. **RARE STATUS:** Casual in AK; St. Lawrence I. (late June, Sept.), Buldir I. (late June), and Attu I. (twice late May). Accidental Forteau Bay, NL (16–19 Oct. 2017).

Population Critically Endangered. Catastrophic declines due to illegal trapping for food at migration and, in particular, wintering sites in southern China and Southeast Asia.

GRAY BUNTING *Emberiza variabilis* GRBU ■ 5

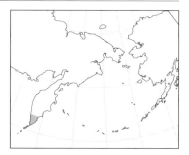

distinctive head pattern with gray on sides of nape

large size and mostly pale bill in all plumages

♀

rufous-brown rump, often best seen in flight

lacks white in tail in all plumages, unusual in *Emberiza* buntings

overall dark gray with black back streaks

♂

This is a large bunting with a heavy, pinkish bill and lacks white in its tail. Winters in Japan where it often joins flocks of Yellow-throated Buntings in dense woodlands. Regarded by most as monotypic. L 6.8" (17 cm)

Identification Large, mostly pinkish bill and no white in outer tail feathers. **ADULT MALE:** Gray overall, prominently streaked with blackish on back and wings. **IMMATURE MALE:** Intermediate to female, some gray but broadly edged with buff on upperparts, head, and underparts. Immature plumage is largely held through

first spring. **ADULT FEMALE:** Patterned like male but mostly brown instead of gray; gray on sides of nape; chestnut rump and tail is conspicuous in flight.

Similar Species Male is distinctive.

Female Gray is like other *Emberiza*, only large, with a thick bill and a distinctive rusty rump and tail, without white outer tail feathers.
Voice CALL: Sharp *zhii*. **SONG:** Warbling series of loud notes.
Status & Distribution Found in dense woodland with stands of dwarf bamboo. **BREEDING:** Southern Kamchatka, southern Sakhalin, Kuril Is., and northern Japan; winters primarily in Japan. **RARE STATUS:** Four spring records from western Aleutian Is., AK. Also rare in Korea; accidental east China and Taiwan.
Population Stable, but small population.

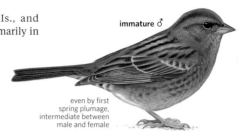

immature ♂

even by first spring plumage, intermediate between male and female

PALLAS'S BUNTING *Emberiza pallasi*   PALB ◼ 5

The Pallas's Bunting is a casual visitor from Asia. It often flicks its tail when perched. Polytypic (3 sspp.; northern breeding *polaris* in N.A.). L 5" (13 cm)
Identification White outer tail feathers. Streaked back with pale rump. Like Reed Bunting, but smaller; shorter tail; smaller, two-toned bill with pink base (black in breeding male) and straight culmen; grayish lesser coverts (diagnostic but often concealed). **BREEDING MALE:** Like breeding male Reed, but lacks all rust. **FEMALE & IMMATURE:** Like female Reed, but lacks median crown stripe, has less distinct eyebrow and lateral crown stripe, paler rump.
Voice CALL: Loud *cheeep*, recalling House Sparrow, very unlike Reed Bunting.
Status & Distribution BREEDING: North-central Siberia east to western Chukotka, Russian Far East; also Mongolia. Winters mostly in Korea and eastern China. **RARE STATUS:** Casual western AK, mostly at Gambell (six records, five in fall); a first-spring male specimen at Utqiaġvik (11 June 1968) was *polaris*. Accidental UK.
Population Stable.

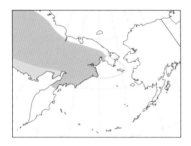

smaller and smaller billed than Reed with straight, not curved, culmen

pale lower mandible

no median crown stripe

♀

striking head pattern

pale wing bars

breeding ♂

gray (adult male) to grayish lesser wing coverts diagnostic from Reed, but often hidden

all figures *polaris*

immatures often have breast streaking in fall

juvenile

REED BUNTING *Emberiza schoeniclus*   REBU ◼ 4

The Reed Bunting is a casual visitor to western AK. Polytypic (10 sspp.; *pyrrhulina* in N.A.). L 6" (15cm)
Identification From Pallas's, larger Reed has chestnut lesser wing coverts (often concealed), large bill with curved culmen; cinnamon wing bars; dark lateral crown stripes, paler median crown stripe, darker rump, and different call. **BREEDING MALE:** Black head, throat; broad white submoustachial stripe; white nape; upperparts streaked black and rust; gray rump. White underparts with thin reddish streaks. **FEMALE:** Pale buffy rump, broad buffy white eyebrow.
Geographic Variation N.A. records are of pale East Asian *pyrrhulina*.
Similar Species See Pallas's and Rustic Buntings.
Voice CALL: A falling *seeoo*. **FLIGHT NOTE:** A hoarse *brzee*.
Status & Distribution Common. **BREEDING:** Eurasia east to Kamchatka. **WINTER:** Western Europe to eastern China and Japan. **RARE STATUS:** Casual western Aleutians in spring (about 10 records); accidental in fall to Gambell (28–30 Aug. 2002).
Population Declining in Europe.

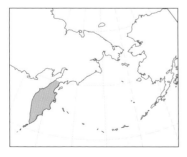

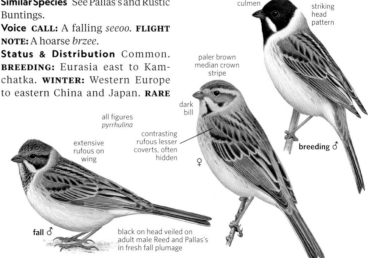

curved culmen

striking head pattern

paler brown median crown stripe

dark bill

♀

breeding ♂

all figures *pyrrhulina*

extensive rufous on wing

contrasting rufous lesser coverts, often hidden

fall ♂

black on head veiled on adult male Reed and Pallas's in fresh fall plumage

TOWHEES AND SPARROWS Family Passerellidae

Baird's Sparrow (ND, June)

The Passerellidae family is a New World family and is best represented by a bewildering array of sparrows, most of which are cryptically patterned with various shades of brown and are affectionately known to many birders as the quintessential LBJs: "little brown jobs." While sparrow identification is challenging for the beginning birder, familiarity with distinctive shapes and behavior of the genera, as well as with the habits and habitats of sparrows themselves, simplifies the process.

Structure Passerellids are fairly small perching birds, ranging in size from *Spizella* sparrows, the small *Ammospiza* sparrows, and the Grasshopper Sparrow (in *Ammodramus*) to the much larger towhees. Their bills are conical, and their wings are typically short and rounded.

Behavior Behavior is often quite skulking, particularly so in some species. The exception is during the breeding season, when males sing from more elevated perches or in a few species perform skylarking flight songs. The songs can be faint and unremarkable, but more often they are quite complex and beautiful. Most species have several calls, which may include a simple chip and a high, thin lisping note. The latter note is often given as the flight note, including from night migrants, but is used more frequently as a contact call when on the ground.

Plumage Typically, the passerellid plumage is a dull collection of brown, gray, black, and white. In many species the sexes are similar, but a few species (e.g., *Pipilo* towhees and Lark Bunting) show moderate to pronounced sexual dimorphism.

Distribution Passerellids are predominantly found in North and Central America, with a few species found in South America. Although diversity is centered in the tropics, there are 44 Old World species, most of which are in the genus *Emberiza*. Although New World diversity is centered in the tropics, some 60 species are known north of Mexico. Preferred habitats are often open country, including fields, deserts, gardens, marshes, riparian areas, and wood edges; a few species prefer the forest interior. Most North American species are short-distance migrants, but some species inhabiting the southern US are resident. Nests are typically an open cup that is placed on the ground or low in shrubbery. A typical clutch would include two to five eggs. The winter diet is primarily composed of seeds and other vegetable matter; in summer that diet is heavily supplemented with insects.

Taxonomy Passerellidae comprises some 127 species in 30 genera. Recent molecular work has resulted in the transfer of many sparrowlike genera into the tanager family (Thraupidae); these groups include the Sporophila seedeaters, the Tiaris grassquits, and a great many other West Indian and Neotropical genera (as well as the iconic "Darwin's Finches" of the Galápagos Is.). Furthermore, our New World sparrows are well-differentiated genetically from the Old World buntings (*Emberiza* and related genera), and this new family, the Passerellidae, was split off from the Emberizidae. Revisions of our sparrows at the generic level are ongoing; for example, the marsh and grassland sparrows formerly placed in the genus *Ammodramus* are now sorted into three different genera.

Conservation These small birds are often fairly catholic in their habitat requirements, so most species have stable populations. However, certain species are more specialized in their habitat requirements, and some populations are severely depressed, endangered, or extirpated; habitat loss is almost always to blame. This is particularly true of the grassland species (e.g., Baird's). BirdLife International codes: 7 NT, 4 VU, 8 EN, 1 CR, and 1 EX (Bermuda Towhee, *Pipilo naufragus*, described from bones).

Genus *Arremonops*

OLIVE SPARROW   *Arremonops rufivirgatus*   OLSP ■ 2

This fairly secretive sparrow stays close to dense cover, and retreats to undergrowth when it is startled. The loud distinctive song can be heard during the breeding season. Polytypic (8 sspp.; nominate in N.A.). L 6.3" (16 cm)

Identification Olive's plumage is a dull olive above; underparts are an unmarked pale gray. Distinctive head pattern with warm brown lateral crown and eye stripes. **JUVENILE:** Buffier than adult, with pale wing bars. Breast is finely streaked.

Geographic Variation Of about eight subspecies in M.A., only *rufivirgatus* occurs in the US.

Similar Species Green-tailed Towhee, rare in N.A. range of Olive Sparrow, is larger, and it has a distinct rufous cap. It also has a brighter greenish coloring on back, wings, and tail, as well as a white throat.

Voice CALL: A dry *chip*, given singly or repeated in a rapid series when agitated. Also a buzzy *speeee*. **SONG:** An accelerating series of dry *chip* notes.

Status & Distribution Common. Nonmigratory. Southern TX to Costa Rica. **YEAR-ROUND:** Dense undergrowth, brushy areas, mesquite thickets, live oak. **RARE STATUS:** Occurs casually slightly north of mapped range.

Population Although still common in its native habitat, Olive Sparrow has suffered significantly in the US as mesquite thorn forest has given way to agriculture and residential uses, especially in the Lower Rio Grande Valley.

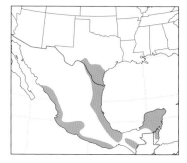

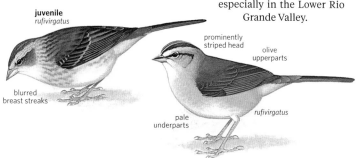

juvenile
rufivirgatus

blurred
breast streaks

prominently
striped head

olive
upperparts

pale
underparts

rufivirgatus

TOWHEES Genus *Pipilo*

All six species of N.A. towhees were all formerly placed in *Pipilo*. Three species—Canyon, California, and Abert's—were moved to *Melozone*, which includes four species of ground-sparrows from M.A. *Pipilo* towhees are large sparrows with long, rounded tails. They feed on the ground, scratching with both feet at once.

GREEN-TAILED TOWHEE   *Pipilo chlorurus*   GTTO ■ 1

The migratory ground-loving Green-tailed Towhee emerges infrequently in the open to feed. In breeding season, males sing from exposed perches. Monotypic. L 7.3" (18 cm)

Identification Olive upperparts and tail; gray head and underparts, fading to whitish belly. Distinct head pattern with obvious reddish crown, white loral spot, distinct white throat, dark moustachial stripe, and white malar stripe. **JUVENILE:** Two faint wing bars, streaked plumage overall, olive-tinged upperparts. Lacks reddish crown. Adultlike plumage acquired by fall migration.

Similar Species Distinctive. See Olive Sparrow.

Voice CALL: Catlike *mew*, like Spotted Towhee but clearer and more two-parted. Also a thin, high *tseeee* (possible flight note) and varied chips when excited. **SONG:** Whistled notes begin with *weet-chur* and end in a raspy trill. Quite similar to "Thick-billed" subspecies of Fox Sparrow, but has buzzier phrases.

Status & Distribution Fairly common. **BREEDING:** Dense brush and chaparral on mountainsides, high

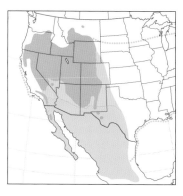

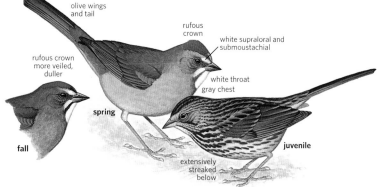

olive wings
and tail

rufous
crown

white supraloral and
submoustachial

rufous crown
more veiled,
duller

white throat
gray chest

spring

fall

extensively
streaked
below

juvenile

plateaus, and sage steppes. **MIGRA-TION:** Spring late Mar.–mid-May, most mid-Apr.–early May. Fall July–Oct., peaking Sept.–early Oct. Uncom-mon migrant generally, rare along West Coast. **WINTER:** South to central Mexico; favors brushy draws and des-ert thickets. **RARE STATUS:** Casual in fall and winter throughout the East. **Population** Has probably declined as sagebrush steppes have been converted to agricultural and grazing land.

SPOTTED TOWHEE *Pipilo maculatus* SPTO ■ 1

Spotted and Eastern Towhees (for-merly lumped as Rufous-sided Towhee) have a narrow hybrid zone in the central Great Plains. In general, a female Spotted differs less from a male than in Eastern. Polytypic (20 sspp.; 9 in N.A.). L 7.5" (19 cm)
Identification MALE: Plumage like Eastern Towhee, except for white tips on median, greater coverts forming two white wing bars, variable white spotting on back and scapulars, lack of rectangular white patch at primary bases. **FEMALE:** Similar to male Spot-ted, but with a slate-gray hood (variable by subspecies). **JUVENILE:** Like juvenile Eastern, but lacks white primary patch.
Geographic Variation Nine N.A. sub-species show weak to moderate varia-tion: Subspecies are *oregonus* (OR to BC), *falcifer* (coastal northwest CA), *megalonyx* (coastal central to Southern CA), *clementae* (certain Channel Is.); *arcticus* (Great Plains), *montanus* (Rocky Mts.), *falcinellus* (south-central CA to OR), *curtatus* (primarily in Sierra Nevada), and *gaigei* (resident in moun-tains of southeastern NM and West TX). "Interior" group birds have exten-sive white spotting above, prominent white tail corners; "Coastal" group birds dark overall with white back spotting, reduced tail corners (variable). Subspe-cies *oregonus* is darkest; white increases southward to *megalonyx*. Within the "Interior" group, *arcticus* shows the most white spotting and most extensive white cor-ners to tail; other subspe-cies have less white, but are similar to one another. Head color of females varies geo-graphically from brownish (*arcticus*) to blackish, thus more like males.
Similar Species Distin-guished from Eastern by white spotting on back (faint on *oregonus*), white wing bars, lack of white patch at bases of primaries, and call. Hybrids breed on the Great Plains, winter to south.
Voice Song and calls also show great geographical variation. **CALL:** A descend-ing and raspy mewing in

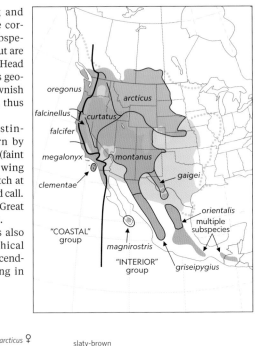

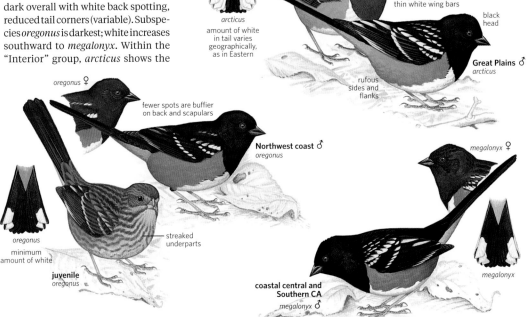

extensive white

arcticus ♀

slaty-brown head

red iris

white back and scapular spots and thin white wing bars

black head

arcticus
amount of white in tail varies geographically, as in Eastern

Great Plains ♂
arcticus

rufous sides and flanks

oregonus ♀

fewer spots are buffier on back and scapulars

Northwest coast ♂
oregonus

megalonyx ♀

oregonus
minimum amount of white

streaked underparts

juvenile
oregonus

coastal central and Southern CA
megalonyx ♂

megalonyx

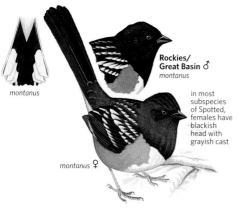

Rockies/
Great Basin ♂
montanus

montanus

in most
subspecies
of Spotted,
females have
blackish
head with
grayish cast

montanus ♀

montanus; an up-slurred, questioning *queee* in *arcticus* and coastal subspecies. All subspecies also give a high, thin lisping *szeeueet* that drops in middle (possible flight note), like call of Eastern, and various chips when agitated. **SONG:** "Interior" group gives introductory notes, then a trill. "Coastal" birds sing a simple trill of variable speed. **Status & Distribution** Common. Some populations largely resident; others are migratory.

The most migratory subspecies is *arcticus*. Resident south to Guatemala. **MIGRATION:** Fall primarily Sept.–Oct.; spring Mar.–early May. **RARE STATUS:** Casual (likely *curtatus*) to AK; subspecies *arcticus* is casual to East. **Population** Stable in most areas. Subspecies *clementae* is extirpated from San Clemente I., one of CA's Channel Is., due to overgrazing by introduced goats; persists on Santa Catalina and Santa Rosa Is. Another island subspecies (*consobrinus*) from Guadalupe I., off Baja California, is extinct.

EASTERN TOWHEE *Pipilo erythrophthalmus*   EATO ▪ 1

The Eastern behaves similarly to the Spotted Towhee. Polytypic (4 sspp.; all in N.A.). L 7.5" (19 cm)
Identification Conspicuous white corners on tail and white patch at base of primaries. **MALE:** Black upperparts, hood; rufous sides, white underparts. **FEMALE:** Black areas replaced by brown. **JUVENILE:** Brownish streaks below.
Geographic Variation Four subspecies show weak to moderate variation. Overall size and extent of white in its wings and tail decline from northern part of range to the Gulf Coast; bill, leg, and foot sizes increase. Large nominate subspecies (breeds in North) has red irides, most extensive white in tail. Smaller *alleni* of FL paler and duller, with straw-colored irides. Inter-

mediate southern subspecies *canaster* (west) and *rileyi* (east) have variably orange to straw-colored irides.
Similar Species See Spotted Towhee.
Voice CALL: Emphatic, up-slurred *che-wink*; in *alleni*, a clearer, even-pitched or up-slurred *swee*. Also a high-pitched *szeeueet*, dropping in middle (possible flight note). Various chips when agitated. **SONG:** Loud ringing *drink your tea*, sometimes with additional notes at beginning or shortened to *drink tea*.
Status & Distribution Fairly common.
BREEDING: Partial to second growth with dense shrubs and extensive leaf litter, coastal scrub or sand dune ridges, and mature southern pinelands.
MIGRATION: Resident, except for partially migratory nominate. Migration primarily Oct. and Mar. **RARE STATUS:**

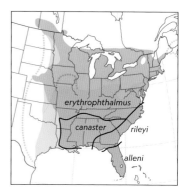

Casual to CO and NM. Accidental to AB, ID, OR, AZ, and Europe.
Population Recent declines, especially in North, due to urbanization. Southern populations more stable.

juvenile
erythrophthalmus

milk chocolate
upperparts and head

erythrophthalmus ♀

yellowish
white iris

FL ♂
alleni

black head
and back

limited
white

alleni

extensive
white

streaked
underparts

white base to
primaries in all
plumages

erythrophthalmus ♂

rufous
sides

erythrophthalmus

Genus *Aimophila*

RUFOUS-CROWNED SPARROW *Aimophila ruficeps*   RCSP ▪ 1

The resident Rufous-crowned Sparrow feeds on the ground and might flush away from an observer into a

bush or cactus, but it often sits up on rocks or bushes to survey the area. Habitat is one of the best clues for this

species: It is closely tied to dry, grassy and rocky slopes. Its distinctive song and calls aid detection. This species, like other *Aimophila*, is almost always found singly or in pairs and does not occur in flocks (though family groups may forage together in late spring and summer). Polytypic (17 sspp.; 6 in N.A.). L 6" (15 cm)

Identification Has gray head with dark reddish crown, distinct whitish eye ring, rufous line extending back from eye, and single black malar stripe on each side of face. Gray-brown above, with reddish streaks; gray-brown below; tail long, rounded. **JUVENILE:** Buffier above, with streaked breast and crown; may show two pale wing bars.

Geographic Variation Seventeen subspecies show moderate variation in size and coloration. Six US subspecies fall into two groups: the small, warm-toned, "Coastal" group and the large, pale "southwestern" group. Birds of the "Coastal" group are small, with reddish upperparts; they include the northern *ruficeps*, which is smaller and somewhat warmer in color than the *canescens* of Southern CA; the subspe-

cies *obscura* is similar to *canescens*, but is limited to Southern CA's Channel Is. The "Southwest" group includes the pale gray *eremoeca*, which is found over much of the range's eastern interior, the widespread southwestern subspecies *scottii*, which is pale and reddish, and the sometimes recognized, slightly darker *rupicola* from southwestern AZ.

Similar Species Somewhat similar to the smaller Rufous-winged Sparrow, but darker gray below and has a larger bill, just one whisker mark, a contrasting white malar, and a bolder eye ring; it also lacks Rufous-winged's rufous shoulder and pale mandible. Chipping Sparrow and other *Spizella* are superficially similar, but shape and behavior of Rufous-crowned are distinctive.

Voice Song and calls are extremely helpful for detecting and identifying this species. **CALL:** A distinctive, sharp, *dear*, often given in a rapid series; also drawn-out *seep* notes (possible flight note). **SONG:** A rolling, bubbly, series of rapid chip notes; can sound very similar to House Wren's song, but note

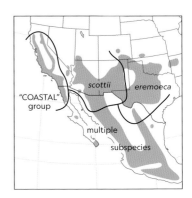

Rufous-crowned's shorter duration and more explosive quality.

Status & Distribution Fairly common. Occurs south to southern Mexico. **YEAR-ROUND:** Strongly prefers rocky hillsides and steep grassy slopes with areas of open ground or bare rock. Largely resident, with some very limited local movements downslope. **RARE STATUS:** Very rare or casual to central or southern Great Plains; accidental to northeastern Kern Co., CA, and, remarkably, to WI.

Population Fire suppression and development have degraded the scrub habitats that Rufous-crowned Sparrow prefers in TX and, especially, in Southern CA (*canescens*). Some Southern CA populations occur in highly imperiled coastal sage-scrub habitat. Populations in AR and eastern OK are small and isolated.

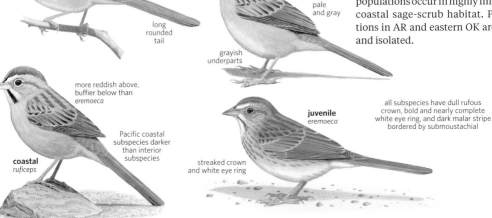

southwestern
scottii
more rufous above than *eremoeca*
dark malar stripe
long rounded tail

eremoeca
pale and gray
grayish underparts

more reddish above, buffier below than *eremoeca*
Pacific coastal subspecies darker than interior subspecies
coastal
ruficeps

juvenile
eremoeca
streaked crown and white eye ring

all subspecies have dull rufous crown, bold and nearly complete white eye ring, and dark malar stripe bordered by submoustachial

---

Genus *Melozone*

The three N.A. species in this genus were formerly in *Pipilo*. This reconfigured genus also includes four species of ground-sparrows from M.A. and an endemic Mexican species, White-throated Towhee (*M. albicollis*).

CANYON TOWHEE *Melozone fusca*   CANT ▪ 1

As its name implies, the Canyon Towhee is common on shallow, rocky canyon slopes and rimrock in the Southwest. It is similar to the Cali-

fornia Towhee, but is not genetically closely related and its calls and songs are completely different. The two were formerly considered the same spe-

cies—the Brown Towhee. The genetic affinities of Canyon Towhee are actually with White-throated Towhee (*P. albicollis*) from southern Mexico.

Polytypic (10 sspp.; 3 in N.A.). L 8" (20 cm)

Identification Plumage is pale gray-brown, fading to whitish on belly, with cinnamon-buff undertail coverts. Rufous-brown cap, buffy eye ring, buffy throat framed by necklace of black streaks typically forming black spot at base of throat. **JUVENILE:** Lacks rufous crown, has narrow buff wing bars, and is faintly streaked below.

Geographic Variation Three N.A. subspecies show weak and clinal variation in measurements, overall coloration, and prominence of rufous cap. Two

small, dark subspecies inhabit central and southwestern TX (*texanus*) and the Sonoran and Chihuahuan Deserts of AZ, NM, and West TX (*mesoleucus*); *mesoleucus* has a much stronger rufous cap. Northern *mesatus* is large and pale, with a brown cap that is tinged rufous.

Similar Species The resident California Towhee has never been known to overlap in range, even as a stray. Compared to California Towhee, Canyon is paler, grayish rather than brown; it has a shorter tail and more contrast in reddish crown, giving a capped appearance. Crown is sometimes raised as a short crest. Canyon has a larger whitish belly patch with a diffuse dark spot at its junction with breast, a paler throat bordered by finer streaks, lores same color as cheek, and a distinct buffy eye ring. Songs and calls are also very distinctive. Range of Abert's Towhee does overlap with Canyon Towhee, and while there are average habitat differences (Abert's more tied

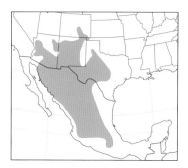

to riparian woodland along water courses), the two species do occur at the same locations.

Voice CALL: Shrill *chee-yep* or *chedup*. **SONG:** More musical, less metallic, than California; opens with a call note, followed by sweet slurred notes. Also gives duet of lisping and squealing notes, like California Towhee.

Status & Distribution Common. Resident; no regular movements. **YEAR-ROUND:** Arid, hilly country; desert canyons. **RARE STATUS:** Casual even a short distance out of range to southeastern UT and southwestern KS.

Population Stable.

pale rufous crown

buffy throat

mesoleucus

whitish belly with fairly conspicuous breast spot

CALIFORNIA TOWHEE *Melozone crissalis*   CALT ■ 1

The California Towhee is a common resident of chaparral through most of coastal CA. This species often occurs in pairs year-round, like the Canyon Towhee and the Abert's Towhee. The California Towhee and the similar Canyon Towhee have never been known to overlap in range, even as strays. Polytypic (8 sspp.; 6 in N.A.). L 9" (23 cm)

Identification Brownish overall. Buff throat bordered by a distinct broken ring of dark brown spots; no dark spot on breast, unlike Canyon. Lores same color as throat, contrast with cheek; warm cinnamon undertail coverts.

JUVENILE: Faint cinnamon wing bars; faint streaking below.

Geographic Variation Six subspecies in US show weak and clinal variation in size and overall coloration. Generally, size decreases from north to south, with the three more interior subspecies (*bullatus, carolae, eremophilus*) averaging larger than the three coastal subspecies (*petulans, crissalis, senicula*). Coloration is generally darker to the north and paler to the south, but is fairly dark in *senicula* of coastal Southern CA.

Similar Species See Canyon and Abert's Towhees.

Voice CALL: Sharp metallic *chink* notes; also gives some thin, lispy notes and an excited, squealing series of notes, often delivered as a duet by a pair. **SONG:** Accelerating *chink* notes with stutters in the middle.

Status & Distribution Common. Resident; no known movements. **YEAR-ROUND:** Chaparral, coastal scrub, riparian thickets, parks, and gardens.

Population Stable, except for the federally threatened subspecies *eremophilus*, which is limited to shrubby thickets in the Argus Mts. of southern Inyo Co., CA.

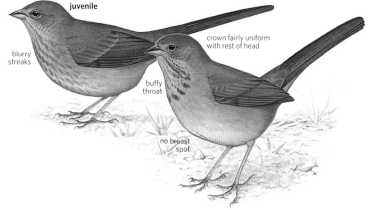

juvenile

blurry streaks

crown fairly uniform with rest of head

buffy throat

no breast spot

ABERT'S TOWHEE *Melozone aberti*   ABTO ▫ 1

This towhee of the Southwest's low riparian areas in deserts can be fairly secretive inside thickets but is locally common, especially along the Colorado River and at the Salton Sea. It is similar to the Canyon and California Towhees, but its range does not overlap with the California Towhee, and the Canyon uses somewhat different habitats where it overlaps with Abert's. Polytypic (2 sspp.; both in N.A.). L 9.5" (24 cm)

Identification Overall coloration is cinnamon brown with a darker tail. Has a distinctive black face (particularly lores).

Geographic Variation The two subspecies are poorly differentiated and the differences are slight. The nominate occurs in southeastern AZ, extending barely into NM. Subspecies *dumeticolus* occupies the rest of the range, including western AZ, southeastern CA (and northwestern Mexico) to southern NV and southwestern UT; it is slightly paler, with a faint reddish tinge to upperparts and underparts.

Similar Species Canyon and California Towhees are similar, but only the former overlaps in range. Prominent black face, lack of streaking on throat, and lack of a contrasting cap easily separate Abert's from both species. Note also habitat differences: Canyon is found on slopes and hills; Abert's prefers moister, lower-lying areas, including mesquite thickets and riparian scrub; also found in well-vegetated areas in Southwestern towns.

Voice CALL: Piping or shrill *eeek*. Also high *seeep*; various chips when agitated. **SONG:** An accelerating series of *peek* notes, often ending in a jumble; frequently sings in a duet.

Status & Distribution Common. Resident; no known movements. Occurs south only to northern Sonora and Baja California. Inhabits desert woodlands, mesquite thickets, riparian growth, orchards, and suburban yards. Found typically at lower elevations than the similar Canyon.

Population Abert's Towhee persists throughout its range, but it has declined rangewide due to the severe degradation of riparian habitat caused by the overuse of water in the Southwest, by grazing, and by the increase of invasive exotic plants. These impacts have been especially marked along the Colorado River.

pale bill
overall warm brown coloration
blackish around bill

Genus *Peucaea*

This genus is composed of eight species, four of which occur in N.A. During the breeding season, territorial males sing from exposed perches. Cassin's also frequently sings in flight. At other seasons most species secretive as they forage in dense cover. Most species do not flock. They fly low over the vegetation before diving back into cover; sometimes they will perch in a bush. Other than contact notes, they are silent outside the breeding season.

RUFOUS-WINGED SPARROW *Peucaea carpalis*   RWSP ▫ 2

This species is fairly common, but local, in southeastern AZ, being found primarily in the Santa Cruz and Avra Valleys. It sings from exposed perches, but even when it is not singing, it tends to be more conspicuous than other *Peucaea*. It often flies to the tops of bushes and fence lines when flushed. Polytypic (2 sspp.; nominate in N.A.). L 5.8" (15 cm)

Identification ADULT: Reddish eye line and cap with a faint gray median crown stripe. Distinctive double whisker with black moustachial and malar stripes on face. Two-toned bill, with pale mandible. Gray-brown back, streaked with black; whitish underparts; two whitish wing bars. Reddish lesser coverts distinctive but usually difficult to see as they are covered by the scapulars. Grayish white underparts, without streaking. **JUVENILE:** Less-distinct facial stripes; buffier wing bars; darker bill; lightly streaked, whitish breast and sides; plumage can be seen as late as Nov.

Geographic Variation Nominate *carpalis* in the US is larger than the Mexican subspecies *cohaerens*.

Similar Species Small bill and body, long tail, and rufous cap could recall Chipping Sparrow (or other *Spizella*), but note that tail is rounded rather than notched. Color of eye line also helps eliminate Chipping, as does double whisker mark. Rufous-winged is not found in flocks, unlike Chipping and the other *Spizella*. See also Rufous-crowned Sparrow, another non-flocking species.

Voice CALL: Distinctive, sharp, high *seep*. **SONG:** Two primary song types: an accelerating series of sweet *chip* notes lasting several seconds, and a

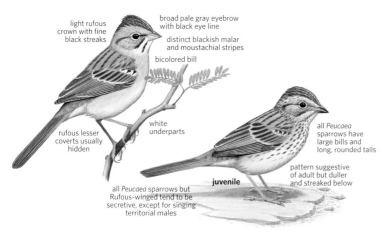

light rufous crown with fine black streaks

broad pale gray eyebrow with black eye line

distinct blackish malar and moustachial stripes

bicolored bill

rufous lesser coverts usually hidden

white underparts

all *Peucaea* sparrows have large bills and long, rounded tails

juvenile

pattern suggestive of adult but duller and streaked below

all *Peucaea* sparrows but Rufous-winged tend to be secretive, except for singing territorial males

shorter song consisting of two to four *chip* notes followed by fast trill, *tink tink tidleeeee.*

Status & Distribution Uncommon to locally fairly common. Almost no migration or vagrancy known, but a few recently discovered in Guadalupe Canyon, NM. Occurs south to central Sinaloa in western Mexico. **YEAR-ROUND:** Rather level areas of desert grassland mixed with brush, shrubs, and cactus; often occurs along washes.

Population Heavy grazing has impacted habitat of Rufous-winged Sparrow, and it is absent from many areas that it historically occupied. It is considered an indicator species for healthy grassland.

BOTTERI'S SPARROW *Peucaea botterii*   BOSP ■ 2

Behavior like Cassin's Sparrow, but does not skylark. Polytypic (9 sspp.; 2–3 in N.A.). L 6" (15 cm)

Identification Large and plain with large bill; fairly flat forehead; long, rounded, dusky-brown tail. Gray upperparts streaked with dull black, rust, or brown; underparts unstreaked; whitish throat and belly; grayish buff breast and sides. **JUVENILE:** Buffy belly, broadly streaked breast, narrowly streaked sides.

Geographic Variation Nine subspecies show marked variation in measurements and upperparts coloration. The subspecies *arizonae*, breeding in southeastern AZ (and extreme southwestern NM) and northern Mexico, is more reddish above and buff on breast;

texana, of extreme southern TX and northeastern Mexico, is slightly grayer.

Similar Species Best identified by voice. See Cassin's and Bachman's.

Voice CALL: Variable high *tsip* notes; also a high, piercing *seep* (possible flight call). **SONG:** Several high *tsip* or *che-lik* notes, followed by two short trills and a longer series of notes accelerating into a trill, like a bouncing ball. A secondary shorter song type just has introductory notes. May perform perch-to-perch or perch-to-ground song flights, but does not skylark like Cassin's does.

Status & Distribution Uncommon. Occurs south to Nicaragua. **BREEDING:** Fairly tall grasslands and prairies, often shrubby; also open grassy woodlands. Subspecies *arizonae* closely tied to tall, dense grasslands; *texana* found in coastal prairies. Returns late

arizonae

goldmani

botterii group

texana

petencia

vantynei

Mar. (*texana*) or early May (*arizonae*), depart by Oct. **WINTER:** Mainly Mexico, a few remain in AZ. **RARE STATUS:** Accidental (has bred; likely *mexicana* subspecies from nearby north-central Mexico on probability) in West TX.

Population Both subspecies found in the US have declined significantly (particularly *texana*) because of degradation of their grassland habitat as a result of grazing, development, and the conversion to agriculture.

head pattern more poorly defined than Bachman's

redder above than *texana*

all Botteri's have plain tails

arizonae

all *Peucaea* have long rounded tails

streaked upperparts

buffy, unstreaked rear flanks

texana

slightly larger bill than Cassin's

streaked below

juvenile *texana*

CASSIN'S SPARROW *Peucaea cassinii*   CASP ■ 1

Its melodic song and skylarking display flight are distinctive. When not singing, Cassin's Sparrow is extremely secretive and is difficult to flush and then find again. Its secretive behavior

is like Botteri's and Bachman's (also in *Peucaea*). Monotypic. L 6" (15 cm)

Identification A large, drab, grayish sparrow, with a large bill and fairly flat forehead. Long, rounded tail is dark

gray-brown; outer tail feathers have indistinct white tips, less apparent when worn. Gray above and streaked with dull black, brown, and variable amount of rust (a few are decidedly rusty above); underparts grayish white with a few short streaks on lower flanks. **JUVENILE:** Streaked below; paler overall than juvenile Botteri's.

Similar Species Best identified by voice. Botteri's Sparrow is browner, lacking dark flank streaks and barred tail with pale tail corners. Cassin's Sparrow has distinctive black crescents on its uppertail coverts and

overall appears like a large Brewer's

whitish eye ring

dark "anchors" on upperpart feathers

whitish tertial fringes

faint flank streaks

streaked below

juvenile

scapulars (where Botteri's Sparrow is streaked), its wings are more patterned, and its white-fringed tertials have black centers. Botteri's does not have a song flight.

Voice CALL: High *stit* given when excited; also *psyit* call (possible flight call). **SONG:** Often given in brief fluttery song flight, in which the bird rises to a height and then floats down. Song typically begins with a soft double whistle, a loud, sweet trill, a low whistle, and a final, slightly higher note; alternate versions include a series of *chip* notes ending in a trill or warbles. Also gives a trill of *pit* notes.

Status & Distribution Fairly common; variable abundance in response to rainfall. Occurs south to central Mexico. **BREEDING:** Breeds in grasslands with scattered shrubs, cactus, yucca, and mesquite. Returns to breeding grounds in early Apr. or later, departs Sept.–early Oct. **WINTER:** Border region of Mexico south to Guanajuato. **RARE STATUS:** Casual to eastern N.A. (mostly in

whitish tail corners

fall), CA (including in numbers in the East Mojave, San Bernardino Co., in spring 1978 and 2019 with breeding, after heavy winter rainfall), and southern NV. Accidental to OR and ID.

Population Loss of grasslands has caused declines.

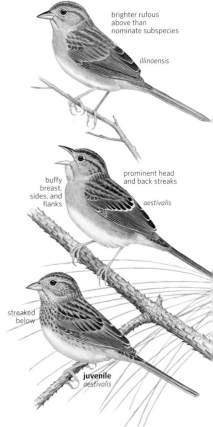

BACHMAN'S SPARROW *Peucaea aestivalis*   BACS ■ 2

This highly secretive species has been known to crawl down gopher holes to escape pursuit. Outside the breeding season, Bachman's and Cassin's Sparrows behave similarly, but in spring and summer Bachman's sings from high perches, often an open pine branch, and, as in Botteri's, it does not perform song flights. Polytypic (3 sspp.; all in N.A.). L 6" (15 cm)

Identification This large sparrow has a large bill, a fairly flat forehead, and a long, rounded tail. It is gray above, heavily streaked with chestnut or dark brown; sides of head are buff or gray; belly whitish. **JUVENILE:** Distinct eye ring; streaked throat, breast, and sides, often into first winter.

Geographic Variation Three subspecies show moderate color variation. Subspecies *bachmani* breeds in northeastern portions of the range and is very similar to the slightly richer reddish *illinoensis* from the western portions of breeding range. Southeastern *aestivalis* (SC to FL) is darkest and grayest.

Similar Species Botteri's Sparrow does not overlap in range and differs in voice. It also has richer coloration and more patterned wings. A stray Botteri's, and particularly a Cassin's, might overlap range of Bachman's; note Cassin's white tail corners and drabber plumage.

Voice CALL: High *tsit*. **SONG:** One clear, whistled introductory note, followed by a trill or warble on a different pitch. Pitch and speed of trill is highly variable. The song is ventriloquial, and singing bird can be hard to locate.

Status & Distribution Rare to uncommon. Apparently extended its range north in the early to mid-1900s (probably because farm abandonment and logging created vast areas of early successional forest), occurring north to PA, MD, OH, IN, IL, and south ON; subsequent declines by the 1950s led to

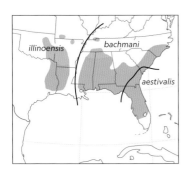

illinoensis

bachmani

aestivalis

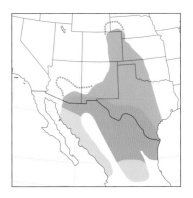

brighter rufous above than nominate subspecies

illinoensis

buffy breast, sides, and flanks

prominent head and back streaks

aestivalis

streaked below

juvenile aestivalis

extirpation from northern range during the 1960s (e.g., last record for OH in Scioto Co. was in 1978). **YEAR-ROUND:** Inhabits dry, open, grassy woods, especially pines, and scrub palmetto. Also occupies regenerating pine forests that are 7–20 ft high. **MIGRATION:** Northern populations are migratory; timing is poorly known, and it was only casually (now accidental) very rarely detected in migration, no doubt due to its secretive nature. Returns to northern breeding grounds by mid-Apr.; fall migration perhaps occurs early Aug.–mid-Oct., but little data. **RARE STATUS:** Accidental to KS, MI, NJ,

NY, and throughout its former range. **Population** Near Threatened. Formerly common in open, old-growth pine forests, but that habitat has been entirely lost to logging. Bachman's Sparrows currently persist in young pine stands (less than 15 years old) and what few older stands (more than 70 years old) remain in the Southeast. Fire suppression and cutting of pine stands favor shrubby undergrowth rather than the grass that Bachman's prefers. The disappearance from the former northern part of the range is puzzling as there appears to be much suitable habitat remaining.

BACHMAN'S SPARROW

RANGE REDUCTION
—— Approximate northern limit of range

Genus *Spizelloides*

AMERICAN TREE SPARROW *Spizelloides arborea* ATSP ■ 1

One of the hardiest sparrows, this is the only one likely to winter in much of the far northern US and southern Canada, where the Dark-eyed Junco can also be found. The American Tree Sparrow often occurs in flocks of up to 50 birds. In habitat and behavior, they are much like Field Sparrows, but American Tree Sparrows are more frequent at bird feeders. Polytypic (2 sspp.; both in N.A.). L 6.3" (16 cm)

Identification Rufous crown and postocular line; pale below with dark breast spot, rufous patch on breast sides, and buff flanks. Bicolored bill, dark upper and yellow lower mandible. Outer tail feathers edged in white on outer webs. **WINTER:** More buffy; rufous color on crown broken by a diffuse central crown stripe. **JUVENILE:** Streaked on head and underparts.

Geographic Variation Two subspecies are weakly differentiated (measurements and coloration). The smaller, dark nominate subspecies breeds eastward from the eastern NT and winters eastward from the central Great Plains. The western *ochracea* is larger and paler, and it winters from the central Great Plains west.

Similar Species See Field Sparrow. The smaller Chipping Sparrow rarely overlaps in range (except in certain areas in migration); it has a distinct

breeding *arborea*

bicolored bill

rufous patch at breast sides and buffy flanks

winter *ochracea*

breast spot

juvenile *ochracea*

dark eye line in all plumages.

Voice **CALL:** Sharp, high, bell-like *tink*; sometimes with a more lispy quality (possible flight note). Flocks also give a musical *teedle-eet*. **SONG:** Usually begins with several clear notes followed by a variable, rapid warble.

Status & Distribution Fairly common. Uncommon to rare west of Rockies. **BREEDING:** Breeds along edge of tundra, in open areas with scattered trees, brush. **WINTER:** Weedy fields, marshes, groves of small trees. **MIGRATION:** One of the late-fall and early-spring migrants. Fall migration in US typically mid- or late Oct.–late Nov.; spring mid-Mar.–early Apr.; a few later accidental in US south of Canada after

late May (late Apr. in midlatitudes). **RARE STATUS:** Casual to Southern CA, central TX, and to the Gulf Coast. **Population** Declining.

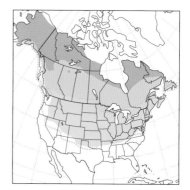

Genus *Spizella*

All six species in this genus occur north of Mexico, five of them regularly. They are fairly small, slender sparrows with small heads, small bills, and distinctly notched, disproportionately long tails. They typically feed in flocks in migration and winter on the ground, and they fly up to a tree or bush if disturbed. Their flight is light, buoyant, and undulating, with frequent delivery of their flight call. The American Tree Sparrow was recently placed in its own genus, *Spizelloides*.

CHIPPING SPARROW *Spizella passerina* CHSP ▪ 1

The Chipping Sparrow forms sizable flocks during the nonbreeding seasons, often mixing with juncos; Lark or Clay-colored Sparrows; Pine, Yellow-rumped, or Palm Warblers; or bluebirds. Their flight note is distinctive and often heard. When disturbed, they fly up to a tree or other elevated perch. Polytypic(5 sspp.; 2 in N.A.). L 5.5" (14 cm)

Identification Dark eye stripe extends through lores to base of bill, gray nape and cheek, gray unstreaked (except juvenal) rump, two white wing bars, and lack of a prominent malar stripe. **BREEDING ADULT:** Bright chestnut crown, distinct white eyebrow. **WINTER ADULT:** Browner cheek, streaked crown with some rufous color. **FIRST-WINTER:** Similar to winter adult, but brownish crown. **JUVENILE:** Underparts prominently streaked; crown lacks rufous; rump slightly streaked. Plumage often held well into Oct., especially in western subspecies.

Geographic Variation Five subspecies show slight variation in color and measurements. Nominate eastern subspecies is small and fairly dark,

with rich rufous upperparts. Western subspecies *arizonae* is larger and paler (breeds from Great Plains west).

Similar Species Clay-colored and Brewer's Sparrows differ from winter and immature Chippings by their pale lores, prominent malar and submoustachial stripes (particularly Clay-colored), and brownish rumps, though the rump is often concealed by the folded wings; lack chestnut on cap. Brewer's has a streaked crown, nape, and rump, and a duller face pattern with a more distinct eye ring; Clay-colored is typically warmer buff-brown on breast, especially in fall, and has a broader, pale supercilium, a darker and broader postocular line, and a more strongly contrasting gray nape.

Voice CALL: High *tsip*; sometimes a rapid twitter when excited. **FLIGHT NOTE:** High, sharp *tseet*; sharper at beginning than in Brewer's or Clay-colored. **SONG:** Rapid trill of dry chip notes, all one pitch; speed can vary considerably.

Status & Distribution Common.

Occurs south to Nicaragua. **BREEDING:** Lawns, parks, gardens, woodland edges, pine-oak forests. **MIGRATION:** Spring mid-Mar.–mid-May; fall late July–early Nov., peaking Sept.–late Oct. Rare in winter north of mapped range. **RARE STATUS:** Very rare to western AK, mostly in fall.

Population Stable. Has largely benefited from human activities, including the clearing of forests and creation of open, grassy parks.

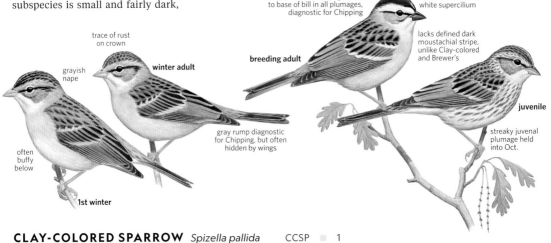

dark line runs through lores to base of bill in all plumages, diagnostic for Chipping

rufous crown with well-defined white supercilium

trace of rust on crown

lacks defined dark moustachial stripe, unlike Clay-colored and Brewer's

breeding adult

grayish nape

winter adult

gray rump diagnostic for Chipping, but often hidden by wings

juvenile

often buffy below

streaky juvenal plumage held into Oct.

1st winter

CLAY-COLORED SPARROW *Spizella pallida* CCSP ▪ 1

The Clay-colored's behavior is similar to that of Chipping. It may flock with that species or with Field Sparrows. Monotypic. L 5.5" (14 cm)

Identification Brown crown with black streaks and a distinct buffy white or whitish central stripe. Broad, whitish eyebrow; pale lores; brown cheek outlined by distinct dark postocular and moustachial stripes; conspicuous pale submoustachial stripe. Gray nape; buffy brown back and scapulars, with dark streaks; unstreaked brown rump does not contrast with back as

in Chipping. Adult in fall and winter is buffier overall. **IMMATURE:** Much buffier; gray nape and pale stripe on sides of throat stand out more. **JUVENILE:** Breast and sides streaked.

Similar Species Identification of fall *Spizella* can be quite challenging and should focus on details of head pattern. Brewer's Sparrow is most similar, but it is duller overall and less distinctly marked. Its face pattern is indistinct, lacking prominent white supercilium, whitish moustachial, and pale central crown stripe.

Unlike Clay-colored, nape of Brewer's is streaked and does not contrast with head and breast. Clay-colored Sparrow usually shows a buffy wash across breast, especially in fall, while Brewer's is paler on breast. Winter and immature Chipping are also similar but tend to be darker on back, with chestnut tones rather than buff and tan. Chipping Sparrow's face is much more strongly marked and set off by a prominent dark eye line that extends through lores. Chipping Sparrow typically does not have Clay-colored's prominent whitish supercilium and moustachial, and its usually grayish breast is unlike Clay-colored's buff breast.

Voice CALL: High *tsik*, given repeatedly when excited. **FLIGHT NOTE:** A thin *sip*, like Brewer's Sparrow. **SONG:** A series of three to four long, evenly spaced, insectlike buzzes; somewhat recalls the song of Golden-winged Warbler.

Status & Distribution Fairly common. Winters south to southern Mexico. **BREEDING:** Fairly common in brushy fields, groves, and thickets. **WINTER:** Primarily in Mexico, uncommonly in southern TX, rarely in southern FL and AZ. Very rare in winter to both coasts. **MIGRATION:** Primarily through interior mid-Apr.–mid-May and late Aug.–late Oct. Rare in fall, casual in spring to both coasts. **RARE STATUS:** Casual to NL, Arctic Canada, YT, and AK (fall). **Population** Stable.

bold head pattern, including strong median crown stripe and buffy cast to plumage

buffy auricular

breeding

all have pale lores, like Brewer's, but unlike Chipping

grayish nape contrasts with rest of buffy plumage

dark moustachial and broad, pale submoustachial stripes

streaking on underparts

juvenile

quite buffy below

fall

BREWER'S SPARROW *Spizella breweri*   BRSP   ■   1

This is a sparrow of the dry Great Basin desert, intermontane valleys, and mountain meadows. In winter and migration it is usually found in dry, sparse desert scrub; the similar Clay-colored Sparrow is more typically found in lusher, grassier habitats, though there is much overlap. Brewer's Sparrows may form flocks or mix with other sparrows in migration and winter. Polytypic (2 sspp.; both in N.A.). L 5.5" (14 cm)

Identification Brown crown with fine black streaks. Distinct whitish eye ring, grayish white eyebrow; overall indistinct face pattern. Pale brown ear patch with darker borders, pale lores, dark malar stripe. Buffy brown, streaked upperparts; buffy brown rump may be lightly streaked. Immature, fall adults, and winter adults are somewhat buffy below. **JUVENILE:** Lightly (*breweri*) to prominently (*taverneri*) streaked on breast and sides.

Geographic Variation Two subspecies show moderate variation in measurements and overall color and pattern. The widespread nominate subspecies occupies most of the breeding range, where it nests primarily in sage steppe habitat. The other subspecies, *taverneri*, known as "Timberline Sparrow," has a largely disjunct breeding range and is thought by some to represent a different species. It breeds in the subalpine zone of the Canadian Rockies from northwestern MT to east-central AK and is thought to winter in north-central Mexico. It has a slightly different song, a larger bill, heavier black streaking on nape and upperparts, a stronger face pattern, and a darker gray breast that contrasts more with the belly.

taverneri

breweri

Juveniles are heavily streaked with blackish below. Despite these average differences, reliable field identification, other than of juveniles, may not be possible (except by breeding range and habitat) due to apparent variation within *breweri*.

Similar Species Clay-colored Sparrow is most similar. Note Clay-colored's unmarked gray nape, which contrasts with face and breast, and more prominent facial pattern with a well-defined median crown stripe, whitish moustachial, and bold supercilium.

Voice CALL: High *tsik*, given repeatedly when excited. **FLIGHT NOTE:** A thin *sip*, like Clay-colored Sparrow. **SONG:** A series of varied, bubbling notes and buzzy trills at different pitches; entire song is often very long in duration.

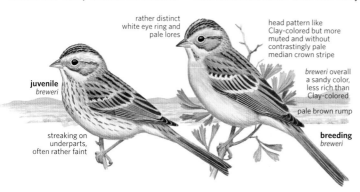

juvenile *breweri*

rather distinct white eye ring and pale lores

head pattern like Clay-colored but more muted and without contrastingly pale median crown stripe

breweri overall a sandy color, less rich than Clay-colored

pale brown rump

breeding *breweri*

streaking on underparts, often rather faint

Status & Distribution Common. Winters south to central Mexico. **BREEDING:** Sage steppe habitats across the West, especially in extensive stands of big sagebrush (*Artemisia tridentata*). **WINTER:** Found in dry desert scrub, including saltbush and creosote; also weedy roadsides. **MIGRATION:** Spring Mar.–mid-May, peaking mid-Apr.–early May; fall early Aug.–early Oct., peaking in Sept. **RARE STATUS:** Casual to Great Lakes. Casual to accidental in eastern N.A.

Population Breeding Bird Survey data show a gradual decline, presumably due to the loss of sage steppe habitats to grazing, agriculture, and the invasion of exotic plants.

larger than *breweri* with markings stronger and darker, more like Clay-colored

grayish auricular

"Timberline" *taverneri*

breeding

juvenile

streaks below heavier and darker than juvenile *breweri*

FIELD SPARROW *Spizella pusilla* FISP ■ 1

In migration and winter, the Field Sparrow forms small flocks, sometimes mixing with other sparrow species. It usually gives its flight call when flushed. Polytypic (2 sspp.; both in N.A.). L 5.8" (15 cm)

Identification Entirely pink bill is distinctive. Gray face with reddish crown, distinct white eye ring, and indistinct reddish eye line. Back is streaked except on gray-brown rump. Rich buffy orange unstreaked breast and sides; grayish white belly; pink legs. **JUVENILE:** Streaked below; buffy wing bars.

Geographic Variation Two subspecies show well-marked variation in measurements and overall coloration. Nominate breeds roughly east from eastern Dakotas and eastern TX; *arenacea* breeds to west, but occurs farther east in migration and winter.

The longer-tailed *arenacea* is larger, paler, and grayer; *pusilla* is especially rufous on auriculars, buffier below, and richer above.

Similar Species Larger American Tree has a two-toned bill and a black central breast spot. Breeding-plumaged Chipping shares rufous cap but has a dark eye line, is plain gray below without buff on breast or flanks, has a darker bill, and lacks bold eye ring. See also the endangered Worthen's Sparrow of northeastern Mexico (next account).

Voice CALL: A high, sharp *chip*, similar to call of Orange-crowned Warbler. **FLIGHT NOTE:** High, loud *tseees*. **SONG:** A series of clear, plaintive whistles accelerating into a trill.

Status & Distribution Fairly common. Prefers open fields with tall grass often near hedgerows, overgrown fields, power-line cuts. **MIGRATION:** Spring mid-Mar.–early May, peaking mid-Apr.; fall, notably late, mainly mid- to late Oct. Rare migrant in east CO (a few breed in southeastern CO) and Maritime Provinces. **WINTER:** Winters uncommonly to rarely south to northeast Mexico. **RARE STATUS:** Casual to NL and west to CA. Accidental to WA and BC.

Population Stable.

grayish median crown stripe and white eye ring

pink bill

buffy streaked breast

juvenile *pusilla*

some rufous color around edge of auricular

all plumages of both subspecies have a white eye ring, pink bill, and grayish median crown stripe

winter *pusilla*

fresh plumaged birds are very buffy below in *pusilla*, usually much grayer in fall and winter *arenacea*

pink legs

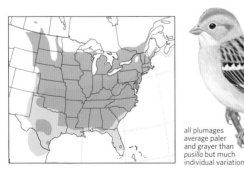

breeding *arenacea*

all plumages average paler and grayer than *pusilla* but much individual variation

WORTHEN'S SPARROW *Spizella wortheni* WOSP ■ 5

The type specimen of this poorly known Mexican species was taken in NM in 1884, but it has not been recorded in the US since. It closely resembles the Field Sparrow. Monotypic. L 5.5" (14 cm)

Identification Distinguished by bright pink bill, solid rufous cap, dark grayish face and underparts, and dark legs.

Similar Species Field is paler below and on face, has pinkish legs, less-solid but more extensive rufous cap (extends

solid rusty crown

blackish legs and feet

to forehead), and a different song.
Voice CALL: High, thin *tssip* (possible flight note); probably also high chips when excited. **SONG:** Dry, chipping trill suggesting Chipping Sparrow.
Status & Distribution Very rare and local. Breeds on the Central Mexican Plateau, primarily in Coahuila and particularly Nuevo León. Currently it is best known from one huge black-tailed prairie dog colony in Nuevo León, where it is often seen in flocks.

BREEDING: Breeds in shrubby deserts, overgrown fields, grassy woodland edge. **RARE STATUS:** One record from Silver City, NM (16 June 1884); could have been a member of an isolated population that was quickly extirpated by overgrazing.
Population Endangered. Its population may be as small as several hundred birds. Decline is presumably a result of overgrazing, which degrades its breeding habitat.

BLACK-CHINNED SPARROW   *Spizella atrogularis*   BCSP ■ 1

This attractive sparrow is typical of chaparral and brushy slopes, where usually detected by its distinctive song. In winter they may be found singly, in loose association with other sparrows, or in small flocks. Polytypic (4 sspp.; 3 in N.A.). L 5.8" (15 cm)
Identification Medium gray overall; rusty above with black streaks; bright pink bill. **BREEDING MALE:** Black lores and chin; long tail all-dark. **FEMALE:** Black chin reduced or absent. **WINTER:**

Lacks black chin. **JUVENILE:** Like winter, but faint streaks below.
Geographic Variation The four subspecies show slight variation in overall coloration and size. Subspecies *cana* breeds from south-central CA to northern Baja California; slightly darker *caurina*, breeds in central CA (Marin Co. to San Benito Co.); larger *evura* breeds east of the Sierra Nevada to AZ, NM, and West TX; and nominate *atrogularis* breeds in north-central Mexico.
Similar Species Gray head and underparts and rufous back are distinctive.
Voice CALL: A high *tsik* is given when agitated; very similar to other *Spizella*. **SONG:** Plaintive song begins slowly and accelerates rapidly, like a bouncing ball. Begins with slow *sweet . . . sweet . . . sweet*, continuing in a rapid trill; somewhat like the song of Field Sparrow, but accelerates much more rapidly and ends with a faster trill. **FLIGHT NOTE:** High, thin *seep*.
Status & Distribution Uncom-

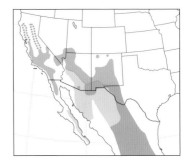

mon to fairly common. **BREEDING:** Inhabits brushy arid slopes in foothills and mountains. Moves into burn areas. **MIGRATION:** Very rarely detected in migration; spring arrival seems to peak in early Apr.; most birds seem to be gone from breeding grounds by late Aug., with a very few individuals noted in Sept. **WINTER:** Southern Southwest south to northern Oaxaca, Mexico. **RARE STATUS:** Rare to southern CO (probably nested). Casual to central TX and to southern OR.
Population Stable; possibly declining in some areas.

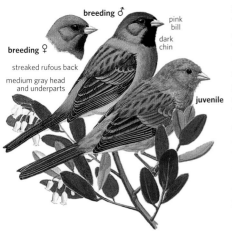

breeding ♂

pink bill

dark chin

breeding ♀

streaked rufous back

medium gray head and underparts

juvenile

Genus *Pooecetes*

VESPER SPARROW   *Pooecetes gramineus*   VESP ■ 1

The confiding Vesper Sparrow (monotypic genus) is a bird of open country often found feeding along roadsides or in open areas in grasslands and often perching on fence lines; avoids thick brush. It often associates with Savannah, Brewer's, or Lark Sparrows. Polytypic (3 sspp.; all in N.A.). L 6.3" (16 cm)
Identification Large; pale underparts marked with fine and distinct but short streaks; prominent white eye ring; ear patch with dark border in rear and outlined by white along lower and rear

edges; supercilium indistinct. Chestnut lesser coverts distinctive but often concealed. Distinctive white outer tail feathers visible on perched bird from below, but best seen in flight.
Geographic Variation Three subspecies show moderate differences in measurements and overall coloration. Eastern nominate subspecies *gramineus* is smaller and slightly darker overall than widespread *confinis*. Uncommon, small, and dark *affinis* with a buffy tinged breast breeds in coastal dunes of western WA and OR,

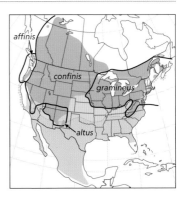

affinis

confinis

gramineus

altus

formerly Del Norte Co., northwest CA. **Similar Species** The shorter-tailed Savannah Sparrow is smaller, lacks eye ring and white outer tail feathers. **Voice CALL:** High chip notes when excited. **FLIGHT NOTE:** High, rising *pseeet*; calls less often in flight than does Savannah. **SONG:** Rich and melodious, two to three long, slurred notes followed by two higher notes, then a series of short, descending trills. **Status & Distribution** Fairly common (uncommon in East). **BREEDING:** Dry grasslands, farmlands, open forest clearings, sagebrush. **WINTER:** Dry grasslands, grassy margins of agricultural fields, roadsides. Prefers shorter grass with open areas interspersed.

Winters south to southern Mexico. **MIGRATION:** An uncommon migrant. Spring late Mar.–early May, peaking in mid-Apr.; fall early Sept.–mid-Nov., peaking late Sept.–late Oct; fall migration is later in the East. **RARE STATUS:** Casual to NL; accidental Arctic Canada and AK.

Population Moderate declines throughout its range are probably due to changes in farming practices and decline in quality of sage steppe and grassland habitats throughout range. Northwestern *affinis* has significantly declined in recent decades.

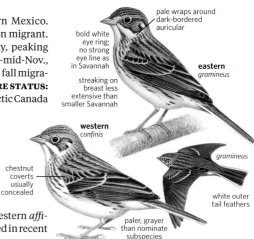

pale wraps around dark-bordered auricular

bold white eye ring; no strong eye line as in Savannah

streaking on breast less extensive than smaller Savannah

eastern *gramineus*

western *confinis*

chestnut coverts usually concealed

gramineus

white outer tail feathers

paler, grayer than nominate subspecies

Genus *Chondestes*

LARK SPARROW *Chondestes grammacus*   LASP ■ 1

The accommodating Lark Sparrow of the open-country is very gregarious in the nonbreeding season. Polytypic (2 sspp.; both in N.A.). L 6.5" (17 cm) **Identification** Long, rounded tail with prominent white corners. **ADULT:** Distinctive harlequin head. Dark central breast spot on white underparts. **JUVENILE:** Duller face; fine, black streaks on breast, sides, and crown. **Geographic Variation** The two subspecies are weakly differentiated. Eastern *grammacus* darker than western *strigatus*.

Similar Species None. **Voice CALL:** Sharp *tsik*, often a rapid series and frequently delivered in flight. High chip notes when excited. **SONG:** Very long, begins with two loud, clear notes, followed by a series of rich, melodious notes and trills and unmusical buzzes. **Status & Distribution** Fairly common. **BREEDING:** Primarily west of the Mississippi on prairies, roadsides, farms, open woodlands, mesas, oak savanna;

once bred as far east as NY and western MD. **MIGRATION:** In spring Apr.–May. An early fall migrant, as early as early July and peak in Aug.–mid-Sept. **RARE STATUS:** Rare migrant coastal Northwest and throughout the East, mainly in fall. Accidental to AK, northern Canada, and Europe. **Population** Stable, though eastern populations are declining and it is extirpated from some former breeding areas in the Midwest and East.

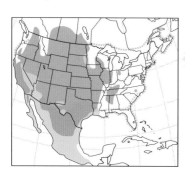

confiding behavior

distinctive head pattern and breast spot

whitish patch

juvenile
streaked below and more subdued head pattern than adult

extensive white in tail

Genus *Amphispiza*

Two species are currently recognized in this genus—the medium-size Black-throated Sparrow and the larger Five-striped Sparrow. Although they have striped faces and distinctive throat patterns, they strongly differ in behavior and all vocalizations.

FIVE-STRIPED SPARROW *Amphispiza quinquestriata*   FSSP ■ 3

The Five-striped Sparrow, a primarily Mexican species, is extremely local in southernmost AZ. The first US breeding was confirmed only in 1969. The small populations appear to be somewhat cyclical or irruptive; only a few locations are occupied consistently. During breeding season territorial males can be found singing from the tops of bushes in a few steep canyons; otherwise rather secretive. Previously placed in the genus *Aimophila*; it might best be placed in its own genus, *Amphispizopsus*. Polytypic (2 sspp.; *septentrionalis* in N.A.). L 6" (15 cm)

Identification Dark brown above; short white supercilium and five bold throat stripes; white belly with gray breast sides and flanks. Dark central

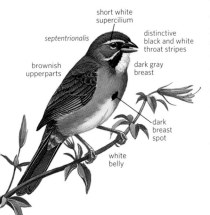

short white supercilium

septentrionalis

distinctive black and white throat stripes

brownish upperparts

dark gray breast

dark breast spot

white belly

breast spot. **JUVENILE:** Head pattern more indistinct, very diffuse dusky streaks throughout, underparts uniformly dusky with no breast spot and buffy wing bars.

Similar Species Distinctive. Juvenile Black-throated Sparrow has much bolder supercilium and stronger streaks on breast; they prefer sparser cover, on more-level slopes.

Voice Best located by song. **CALL:** Gruff *churp*; also higher *pip* and *seet* notes when excited. **SONG:** Slow series of short, high, varied phrases, with chipping and trilled qualities; typical phrase is a *churp* followed by a short, variable trill; fairly long pauses between phrases; some songs consist of just a few notes. Sings through the summer monsoon season.

Status & Distribution Uncommon. West Mexican species; locally west of the Sierra Madre from Jalisco north to southeastern AZ, with seasonal movements in certain areas; US range restricted to a few canyons in extreme

southeastern AZ (e.g., California Gulch, Sycamore Canyon; sometimes in canyons on west side of Santa Rita Mts.). **BREEDING:** Highly specialized habitat in the US: favors tall, dense shrubs on rocky, steep canyon slopes. **NONBREEDING:** Status outside of breeding season poorly known, perhaps present, but highly secretive or some (most?) may withdraw to Mexico. Most records late Apr.–late Sept.

Population The population in the US is small and therefore vulnerable, but it is not threatened by human activity.

BLACK-THROATED SPARROW *Amphispiza bilineata*   BTSP ■ 1

A beautiful and confiding sparrow. Not a gregarious species, but small family groups are seen foraging together by midsummer. Polytypic (9 sspp.; 3 in N.A.). L 5.5" (14 cm)

Identification Short-winged and long-tailed sparrow. Outer tail feathers white on outer web with white tip. **ADULT:** Black lores and triangular black patch on throat and breast with white eyebrow, submoustachial stripe, and underparts. Upperparts

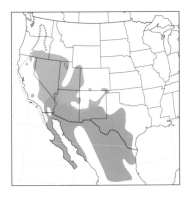

plain brownish gray. **JUVENILE:** Plumage often held well into Oct. No black on throat; breast and back finely streaked. Bold white supercilium as in adult.

Geographic Variation Nine subspecies (three in US) show moderate but clinal variation in overall color and tail pattern. Nominate, breeding in central and southern TX, is smallest, with darker back and more extensive white in tail. Larger *opuntia* (breeds west to southeastern NM and southeastern CO) and *deserticola* (breeds in remaining western portion of range) are paler with reduced white in tail.

Similar Species Juvenile similar to Sagebrush and Bell's Sparrows; note bold eyebrow.

Voice CALL: Faint, high, metallic *tink*; also gives series of soft, metallic tinkling notes. **SONG:** Two clear notes followed by a trill.

Status & Distribution Fairly common in a range of desert habitats. **BREEDING:** In arid areas over much of West, somewhat irruptive in northwest part

bold white supercilium and submoustachial stripe

deserticola

large black throat patch

bold white supercilium

white throat

juvenile *deserticola*

of range. Northern breeders arrive by early May. Fall movements by late Aug. through Sept. Resident to central Mexico. **RARE STATUS:** Rare to casual migrant to West Coast. Casual to eastern N.A. in fall and winter.

Population Stable.

Genus *Artemisiospiza*

Formerly these two species (previously treated as one, the Sage Sparrow) were placed in the genus *Amphispiza* with Black-throated Sparrow. However recent genetic studies indicated that they belong in their own genus, which is also supported by distinct behavioral differences.

SAGEBRUSH SPARROW *Artemisiospiza nevadensis* SABS ■ 1

The Sagebrush Sparrow is found in the sagelands in the intermountain West. It spends much of its time on the ground, running between shrubs with its tail raised in the air. When perching, it may seem agitated, delivering a tinkling call and twitching its tail nervously. Monotypic. L 6" (15 cm)
Identification Males larger than females. **ADULT:** Grayish brown head; white eye ring; white supraloral area, not extending past eye; broad white submoustachial stripe; weak malar stripe; pale brown, narrowly streaked back; whitish below with large dark brown central breast spot; buffy wash with distinct streaks on flanks; white edge outer web of outer tail feather and inner web at tip. **JUVENILE:** Duller,

more streaked, face pattern muted but still with pale supraloral spot.
Similar Species Smaller Bell's of coastal subspecies *belli* is much darker and largely unstreaked on back (streaks on scapulars); has duller pale edges to outer tail feathers. Interior Bell's (*canescens*) is paler, more like Sagebrush, but has very faintly streaked back and prominent dark malar stripe; tail pattern intermediate. See juvenile Black-throated Sparrow.
Voice CALL: Twittering series of thin, junco-like notes. SONG: Strongly defined, pulsating pattern of rising and falling phrases, seemingly in a minor key.

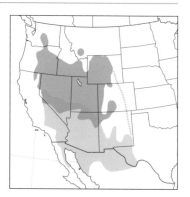

short whitish supercilium

streaked across back

runs on the ground with tail up

faint malar

juvenile

Status & Distribution BREEDING: Primarily in sagebrush and saltbush desert scrub in Great Basin; arrives late Mar. Some upslope movement by late summer; southward movement by mid-Sept. WINTER: Found across the Southwest in arid plains with sparse bushes. Greatly outnumbered by *canescens* Bell's in winter in southeast CA. **RARE STATUS:** Casual to Plains states (e.g., SD, KS, NE) and to West Coast; accidental to KY and perhaps NS (Sagebrush or Bell's).
Population Declining because of rangewide loss and degradation of habitat.

BELL'S SPARROW *Artemisiospiza belli* BESP ■ 1

Bell's Sparrow is a dark-headed, dark-backed resident in CA's coastal chaparral (favors chamise) and interior desert scrub. It is difficult to see (especially *belli*) when not singing atop a bush. For over a century, Bell's Sparrow and Sagebrush Sparrow were confusingly classified and reclassified as either one species or two. In 2013 the AOS separated them into two species based on differences in mtDNA, morphology, ecology, and limited gene flow at the contact zone in eastern CA (near Bishop, Inyo Co.).

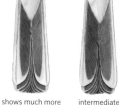

belli

canescens

clementae→

belli

cinerea

The taxonomic status of *canescens* is uncertain and may represent yet another species. Polytypic (3 sspp., 3 in N.A.). L 5.7" (15 cm)
Identification Males larger than females. **ADULT (*belli*):** Dark gray head, white eye ring, white supraloral spot, broad white submoustachial stripe; dark malar stripe; unstreaked brownish back. Large black central breast spot; streaked buffy flanks, dark tail with no white on outer feathers. **JUVENILE:** Similar to Sagebrush Sparrow but slightly darker.
Geographic Variation Nominate *belli* of coastal CA, with dark brown back and bold black malar; *canescens* of San Joaquin Valley and Mojave Desert similar but paler; *clementeae* of San Clemente I., CA, very similar to *belli* (not recognized by some authorities). Though *canescens* is intermediate between *belli* and Sagebrush, it is genetically allied to *belli*.
Similar Species Larger Sagebrush Sparrow is paler, streaked across back, weaker malar stripe, more white on

OUTER TAIL FEATHERS

Sagebrush

Bell's *canescens*

Bell's *belli*

shows much more white than *A. b. belli*, but *A. b. canescens* is intermediate

intermediate amount of white

shows almost no white on outer tail feather

outer tail feathers. See juvenile Black-throated Sparrow.
Voice CALL: High, metallic, thin junco-like notes. Vocalizations differ between Bell's and Sagebrush, but more study is needed. SONG: A jumbled series of rising and falling phrases. Song of *canescens* differs from *belli* and Sagebrush Sparrow but needs more study.
Status & Distribution BREEDING: Year-round resident in dense chaparral

and sage scrub in coastal CA; inland in saltbush desert scrub and along western slopes of the central Sierra Nevada, and south to central Baja California. **WINTER:** Resident within breeding range, but many *canescens* winter in southeast CA and western AZ. **Population** Declining because of habitat loss due to development and conversion to agriculture. Subspecies *clementeae* is listed as Threatened by the USFWS due to predation by introduced mammals.

plain back
streaks on flanks
central breast spot
coastal *belli*

faint back streaks
canescens
distinct malar
intermediate bewteen Sagebrush and *belli* Bell's

Genus *Calamospiza*

LARK BUNTING *Calamospiza melanocorys*   LARB ▪ 1

The Lark Bunting is a characteristic bird of fence lines and open prairie during summer. This species forms large, tight-knit flocks during the nonbreeding season. Flocks of other sparrows are more scattered: With the Lark Bunting, the entire flock seems

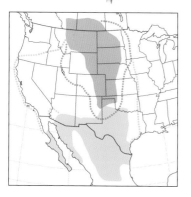

all with pale wing patches, buffy white in females

breeding ♂

to move as a cohesive unit. In flight, it looks short and round-winged. Monotypic. L 7" (18 cm)
Identification Stocky, thick-necked, and barrel chested for a sparrow, with a fairly short, squared-off tail and a thick, blue-gray bill. Males may migrate south in breeding plumage, which they often retain through Aug. **BREEDING MALE:** Entirely black with white wing patches; males in molt are a patchwork of brown and black. **FEMALE:** Brown above; pale below with regular sharp, brown streaking; buffy sides; brown primaries. **WINTER MALE:** Like female, but with black primaries, face, and throat. **IMMATURE:** Like female; males usually have some black on face, chin, or lores.
Similar Species Female and immature similar to some sparrows and *Haemorhous* finches. Diagnostic white or buffy white wing patch (sometimes concealed), thick blue-gray bill, and its shape and behavior. Most finches do not spend as much time on ground. **Voice FLIGHT NOTE:** Distinctive, soft *hoo-ee.* **SONG:** A varied series of rich whistles and trills.
Status & Distribution Common. Winters south to central Mexico. **BREED-**

early spring ♂
short tail with white terminal spots
all with thick, bluish bill
on winter males, black surrounds bill
streaking below darker than on females
winter ♂
indistinct face pattern

ING: Dry plains and prairies, especially in sagebrush. Annual shifts depending on precipitation. **WINTER:** Open grasslands, roadsides, dry, weedy fields. **MIGRATION:** Spring mid-Apr.–mid-May; fall early Aug.–mid-Oct. **RARE STATUS:** Rare to coastal CA. Casual to Pacific Northwest and to the East. **Population** Declining in northern part of breeding range.

Genus *Passerculus*

SAVANNAH SPARROW *Passerculus sandwichensis*   SAVS ▪ 1

The Savannah Sparrow is one of the most common sparrows of open country. When flushed they usually fly up and away from the observer. Occasionally, they fly low over the grass like an

Ammodramus or *Centronyx* sparrow. When flushed, they will often fly to an elevated perch; at other times they drop back to the grass. In the nonbreeding season, they often loosely

gather in small flocks. Their distinctive flight note is given frequently upon flushing. Polytypic (17 sspp.; 8 in N.A.). L 5.5" (14 cm)
Identification Savannah is a fairly

small sparrow with a short, square tail. Heavily streaked below; face pattern well marked with a supercilium (yellow or whitish), strong eye line, dark malar stripe, and pale moustachial; no eye ring. Plumage otherwise highly variable by subspecies.

Geographic Variation The eight subspecies found north of Mexico show marked variation in body and bill size and in overall color. They are divided into four distinct, field-identifiable groups. "Continental" group (8 sspp.) is fairly small and small billed. Eastern birds are generally more richly colored (darker in the north), while most western subspecies are generally paler and grayer; palest is the western interior breeding *nevadensis* and AK subspecies. Second, distinct *princeps*, known as "Ipswich Sparrow," breeds on Sable I., NS, and winters on East Coast beaches; note its larger size, larger bill, and even paler plumage. Third, "Belding's" group birds are like "Typical" Savannahs in size and bill shape, but they are quite dark in plumage; they are increasingly darker from north to south, with the darkest being *beldingi* (illustrated) of Southern CA salt marshes. Finally, "Large-billed Sparrow" is a very distinctive complex involving several subspecies that breed in Mexican salt marshes; only *rostratus* reaches the US. Its bill is very large compared to other Savannahs, being especially thick at base. It is pale brownish gray overall, and its underparts are marked with broad, blurry, reddish brown streaks.

Similar Species See Vesper Sparrow. Savannah Sparrow is similar to the secretive Baird's Sparrow, but has a solid eye line, and is not extensively orangish about head. The larger, browner Song Sparrow has a longer, rounded tail, lacks yellow above eye, and differs in behavior and habitat.

Voice All subspecies similar in voice, except for "Large-billed." **CALL:** High *tip*. **FLIGHT NOTE:** A thin, descending *tseew*; in "Large-billed" a more metallic *zink*. **SONG:** Two or three chip notes, followed by a long, buzzy trill and a final *tip* note. Compare to that of Grasshopper Sparrow. "Large-billed" has short, high introductory notes, followed by about three rich, buzzy *dzeee* notes; it usually does not sing in the US.

Status & Distribution Common. Occurs south to Guatemala. May be found in a variety of open habitats, usually with short grass, including grasslands, airports, agricultural fields, salt marshes, and beaches. "Large-billed" occurs uncommonly and exclusively on saline flats and rocky shorelines of the Salton Sea and also coastal Southern CA (formerly to central CA, where now casual); "Ipswich" breeds on Sable I., NS, and winters on East Coast

beaches to northeastern FL. **MIGRATION:** Timing varies, but generally in spring Mar.–mid-May, peaking in Apr.; and in fall mid-Aug.–early Nov., peaking in mid-Oct. "Large-billed" occurs in the US primarily from mid-July–Feb.; it has not yet been found breeding at the Salton Sea. **RARE STATUS:** Rare to high Arctic Canada; casual to central and western Aleutians, Bering Sea islands, Greenland, Europe, and northeast Asia.

Population Mostly stable. Pacific *beldingi* has declined drastically due to destruction of CA salt marshes. "Large-billed" has declined significantly due to destruction of marshes in the Colorado River Delta (Gulf of California). "Ipswich" is vulnerable due to its limited breeding area on a small island.

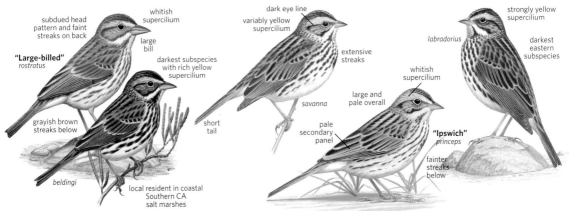

subdued head pattern and faint streaks on back

whitish supercilium

"Large-billed" *rostratus*

large bill

darkest subspecies with rich yellow supercilium

grayish brown streaks below

beldingi

local resident in coastal Southern CA salt marshes

dark eye line

variably yellow supercilium

extensive streaks

short tail

savanna

pale secondary panel

labradorius

strongly yellow supercilium

darkest eastern subspecies

whitish supercilium

large and pale overall

"Ipswich" *princeps*

fainter streaks below

Genera *Ammodramus*, *Centronyx*, and *Ammospiza*

The nine members (seven in N.A., two in S.A.) of these three genera were formerly all placed in one genus, *Ammodramus*, but genetic evidence indicates that they are better placed in three genera. *Ammodramus* is now restricted to Grasshopper and two S.A. species. Henslow's and Baird's are placed in *Centronyx* and the "marsh sparrows" (four species) are placed in *Ammospiza*. All are difficult to observe, except during breeding season, when males sing from more exposed perches. They are not gregarious. Their short, weak flights are low to the ground, ending with a sudden drop into the grass. All have comparatively flat heads; large bills; fairly short, sometimes spiky, tails. Several species migrate in juvenal plumage.

GRASSHOPPER SPARROW *Ammodramus savannarum*   GRSP ■ 1

Male Grasshopper Sparrows sing conspicuously from visible perches, but otherwise the Grasshopper is fairly secretive, except for juveniles, which often perch up on fence lines in late summer. Polytypic (12 sspp.; 4 in N.A.). L 5" (13 cm)

Identification Small and chunky; short tail, large bill, flat head. Buffy breast, sides; without obvious streaking. Pale, whitish central stripe on dark crown; white eye ring; yellow-orange spot in front of eye. Buffier below in fall, never as bright as LeConte's. **JUVENILE:** Brown streaks on breast, sides. Early fall migrants can be in this plumage.

Geographic Variation The four N.A. subspecies vary in dorsal color from FL's dark *floridanus* to reddish *ammolegus* of southeastern AZ and southwestern NM. Eastern *pratensis* is slightly richer colored than western *perpallidus*, ranging east through the Great Plains.

Similar Species LeConte's has a broad buffy orange eyebrow that extends well behind eye, a blue-gray ear patch, and heavy streaks on breast and sides. Beware of juvenile Grasshopper Sparrows with streaked breasts. Compare also with female Orange Bishop and Savannah Sparrow.

Voice CALL: High, sharp *tsik*, often doubled; sometimes more excited, higher *tik*, often in rapid succession, near nest. **FLIGHT NOTE:** High, lisping *tseee*, sometimes from perch or from inside cover. **SONG:** One to three high chip notes followed by a brief, grasshopper-like buzz; also sings a rambling series of squeaky and buzzy notes, often in flight.

Status & Distribution Uncommon to common. **BREEDING:** Pastures, grasslands, fields, airports, palmetto scrub. **MIGRATION:** Detected on breeding grounds (where some winter) in Southern CA late Mar.; midlatitudes of US mid-Apr.; fall migration hard to detect due to secretive behavior but mostly in Oct. **RARE STATUS:** Rare to Atlantic Provinces.

Population Subspecies *floridanus* is listed as Endangered by USFWS. Declines due to conversion of native grassland and suppression of grassland fires.

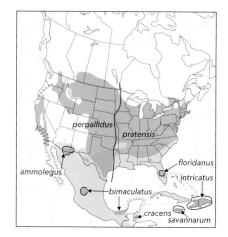

BAIRD'S SPARROW *Centronyx bairdii*   BAIS ■ 2

One of the rarer sparrows, this species is poorly known due to behavior and identification difficulties; migration and wintering areas incompletely known. The Baird's is secretive, especially away from breeding grounds. It looks most like Savannah, and a good view of the head is necessary to discern the different patterns. It typically flushes only at close range, flies short distances, and drops back into cover. This differs from the behavior of some Savannahs, which are more apt to flush at a distance and perch in the open on landing. Monotypic. L 5.5" (14 cm)

Identification Orange-tinted head (which is duller on worn summer birds); note especially two isolated dark spots behind ear patch, no postocular line. Double whisker (malar and moustachial) lines. Wide-spaced, short dark streaks on breast form a distinct necklace; chestnut on scapulars. **JUVENILE:** Head paler, creamier; central crown stripe finely streaked; white fringes give scaly appearance to upperparts; underparts more extensively streaked.

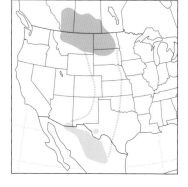

Similar Species See Savannah and Henslow's Sparrows.

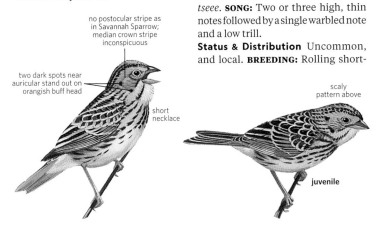

no postocular stripe as in Savannah Sparrow; median crown stripe inconspicuous

two dark spots near auricular stand out on orangish buff head

short necklace

scaly pattern above

juvenile

Voice CALL: Excited *tink* or *tsip* notes near nest. **FLIGHT NOTE:** High, thin *tseee*. **SONG:** Two or three high, thin notes followed by a single warbled note and a low trill.

Status & Distribution Uncommon, and local. **BREEDING:** Rolling short-grass prairies with scattered shrubs or weeds. Recently discovered in Larimer Co., CO. **WINTER:** Large open grasslands, with clumps of grass interspersed with bare areas. **MIGRATION:** Rarely seen, through Great Plains (west KS to west MN, also north NM). Spring migration primarily in early May; in fall primarily Sept., a very few to mid-Oct. No doubt overlooked on western Great Plains due to secretive behavior and access issues to appropriate grassland habitat. **RARE STATUS:** Casual breeder east to west MN; accidental in summer to WI. Accidental in fall to East (NY and MD) and CA (five records).

Population Significant declines throughout breeding range due to conversion of native prairie to agriculture.

HENSLOW'S SPARROW *Centronyx henslowii*   HESP ■ 2

On breeding grounds males sing from small shrubs or tall grass; at other times the Henslow's is very secretive. After being flushed several times it may perch in the open for a time before dropping back into cover. Polytypic (2 sspp.; both in N.A.). L 5" (13 cm)

Identification Large flat olive head; large gray bill; malar and moustachial stripes; streaks across breast, extend down flanks; dark chestnut wings and back. **JUVENILE:** Paler, yellower, less streaking below. Compare to adult Grasshopper Sparrow.

Geographic Variation Two weakly differentiated subspecies are not field

separable: *susurrans* in East; smaller-billed, paler nominate in Midwest.

Similar Species Most similar to Baird's, but greenish head, nape, and most of central crown stripe; wings extensively dark chestnut. Also confused with Grasshopper; note Henslow's greenish head and central crown stripe; malar and moustachial stripes; dark chestnut wings and back; breast streaking. Beware juvenile Grasshopper with streaked breast.

Voice CALL: Sharp *tsik*. **FLIGHT NOTE:** High *tseee*, like Grasshopper. **SONG:** Distinctive, short *tse-lick*.

Status & Distribution Uncommon and local. **BREEDING:** Wet shrubby fields, weedy meadows, reclaimed strip mines. **WINTER:** Weedy fields, wet second growth, and grassy understory of pine woods. Winters casually north to IL, IN; accidental to New England. **MIGRATION:** Poorly known; arrives at more southerly breeding areas by mid-Apr. **RARE STATUS:** Casual to New England, Maritimes, and coastal mid-Atlantic. Accidental west of mapped range to ND, CO (two sight records), and NM.

Population Formerly bred on mid-Atlantic coastal plain and in Northeast;

pea-soup green head with blackish lateral crown and postocular stripes

large bill

rich dark chestnut on upperparts

necklace

largely unstreaked breast

juvenile

now largely extirpated. Significant long-term declines due to loss of native grassland habitat. Recent management efforts have led to recovery in some areas.

LECONTE'S SPARROW *Ammospiza leconteii*   LCSP ■ 1

Fairly common but secretive, the pale buffy LeConte's Sparrow scurries through matted grasses like a mouse. Compared to Grasshopper, it prefers lower-lying, often wet fields and marsh edges. When flushed, it

flies away from the observer with a characteristically weak flight, flaring its narrow tail feathers upon landing. Monotypic. L 5" (13 cm)

Identification White central crown stripe, becoming orange on forehead;

chestnut streaks on nape; and broad buff streaks on back distinguish LeConte's from sharp-tailed sparrows. Bright, broad, buffy orange eyebrow; grayish ear patch; thinner bill; and orange-buff breast and sides distinguish it from Grasshopper Sparrow. Dark streaks on sides of breast and flanks. **JUVENILE:** Buffy; tawny crown stripe; breast heavily streaked. Some migrate in juvenal plumage, which can be retained into late Oct.

Similar Species Nelson's and Salt-marsh are most similar but have gray-ish central crown stripe; unstreaked

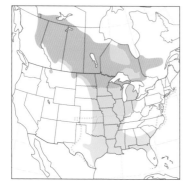

gray nape; distinct white lines on back; larger bill; and more extensive gray face. Juveniles more similar, but LeConte's is paler, sandier, and streaked on nape. See Grasshopper account.

Voice CALL: Sharp *tsit*. **FLIGHT NOTE:** High *tseeet*, like Grasshopper Sparrow. **SONG:** Short, high, insectlike buzz: *tsit-tshzzzzzzz*.

Status & Distribution Fairly common. **BREEDING:** Wet grassy fields, bogs, marsh edges. **WINTER:** High marsh and marsh edges, low-lying grassy fields, especially those with broom sedge or switchgrass components. **MIGRATION:** Commonly through Great Plains east to Mississippi River Valley; uncommon to rare east to Ohio River Valley. Mid-Sept.–mid-Nov.; spring Apr.–mid-May. **RARE STATUS:** Casual to CA and NM, and accidental to BC, WA, OR, and AZ. Occurs casually in migration along East Coast, primarily in fall; rare to

median crown stripe orangish on forehead, then white

orange-buff supercilium

purplish pink nape streaks

broad buff back streaks

necklace and fine streaking on sides and flanks

juvenile

some juveniles migrate south in juvenal plumage

casual in winter north to NJ and west to southeastern CO and western TX. **Population** Apparently stable.

SEASIDE SPARROW *Ammospiza maritima*   SESP ■ 1

This large *Ammospiza* species is found in salt marshes. Males sing from exposed perches or deliver flight songs (flying up from perch, then fluttering down as song is delivered). In winter, it is readily drawn into view by pishing and when they perch up during high tides, sometime loosely associating with Nelson's and Saltmarsh Sparrows. Polytypic (9 sspp.; all in N.A.; *nigrescens* extinct). L 6" (15 cm)

Identification Fairly chunky, it has a flat head, a long spikelike bill, and a short, pointy tail. Distinct yellow supraloral patch; dark malar stripe; distinct whitish throat; broad, pale submoustachial. Breast varies (gray, whitish, or buffy), some streaking. Upperparts grayish with variable brown or olive coloration; variable reddish cast to wings; back variably streaked. **JUVENILE:** Warmer, paler, buffier than adult, especially on face and breast; breast streaking more distinct; may lack distinct yellow supraloral.

Geographic Variation Seaside Sparrow varies widely in overall color. Most subspecies are dull except for the yellow supraloral patch. Subspecies form four distinct groups: "Atlantic Coast" *maritima* (breeds from VA north, winters from NJ to FL) and *macgillivraii* (breeds NC to northeast FL) are dull grayish olive above, smoky gray on the breast with diffuse

streaking. "Gulf Coast" (*sennetti* of southeast TX, *fisheri* from TX to north-western FL, *peninsulae* of western FL) has a buffier breast and face, more distinct breast and back streaking. "Cape Sable," the very localized *mirabilis*, inhabits a small area in the Ever-glades, with an olive cast to the back and nape, more distinct breast and back streaking. "Dusky" (*nigrescens*), by far the darkest, became extinct in June 1987. ("Cape Sable" and "Dusky" were recognized as separate species prior to 1973.)

Similar Species Juvenile regularly

confused with Saltmarsh Sparrow. Larger Seaside has longer bill; longer, less spiky tail; flimsy, fresh juvenal feathering.

Voice CALL: Loud *tsup*, sometimes

all with yellow supraloral spot

long bill

maritima

whitish throat and submoustachial

fisheri

buffy breast, especially in fresh plumage

blackish overall

slightly greenish above

juvenile *maritima*

nigrescens extinct

mirabilis

doubled. Rapid, high *tik* notes when excited on breeding grounds. **SONG:** Wheezy *tup zheee-errr*, like Red-winged Blackbird, but softer, buzzier. Secondary song more complex and stuttering.

Status & Distribution Uncommon to fairly common, but local. Grassy tidal marshes; favors taller grass (e.g., *spartina*). **MIGRATION:** Resident (excluding *maritima*); returns to breeding grounds between mid-Apr. and early May; most depart by mid-Oct.; stragglers often linger north of wintering areas into Nov., Dec. **RARE STATUS:** Rare postbreeding visitor to coastal ME (has bred). Casual to the Maritimes. Accidental inland in PA, NC, and TX.

Population Stable. Declines in New England and FL; habitat limited, populations small. Subspecies *mirabilis* listed as Endangered by USFWS; only 4,000 to 6,000 "Cape Sable" birds remain. The distinctive "Dusky" and a population along St. Johns River (sometimes recognized as *pelonota*) are extinct; caused by diking of salt marshes and spraying of insecticides.

NELSON'S SPARROW *Ammospiza nelsoni* NESP ■ 1

Secretive and hard to find away from the breeding grounds, Nelson's Sparrow (formerly Nelson's Sharp-tailed Sparrow) has warm, soft plumage colors that make it distinctive once it is found. Its behavior is like LeConte's Sparrow. Sometimes it responds to pishing, especially when forced to higher ground during high tide. It has a somewhat smaller bill and rounder head than Saltmarsh Sparrow, though this is less pronounced in the Atlantic coast subspecies (*subvirgata*). The two species come together and sometimes hybridize in coastal southern ME and NH. Polytypic (3 sspp.; all in N.A.). L 4.8" (12 cm)

Identification Orange-buff triangle on indistinctly marked face; gray median crown stripe and nape; streaked buffy breast contrasts sharply with white belly; back marked with black and white stripes; wings tinged with rich rufous. Narrow, sharply pointed rectrices. **JUVENILE:** Fainter median crown stripe; duller nape; variably thicker eye line; less contrast above; lacks streaking across breast.

Geographic Variation Three subspecies with disjunct breeding ranges. Interior *nelsoni* is brightest, with rich orange on face and breast; distinct streaking on breast and flank; distinct white streaks on darker back. Subspecies *subvirgata* of the Maritimes and coastal ME is much duller overall (recalling Seaside Sparrow); has diffuse streaking below; grayer underparts; longer bill. In *altera*, from James and Hudson Bays, brightness is intermediate, streaks blurred, still, it is doubtfully separable from *nelsoni* in the field.

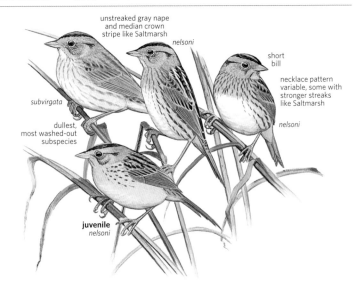

unstreaked gray nape and median crown stripe like Saltmarsh

nelsoni

short bill

necklace pattern variable, some with stronger streaks like Saltmarsh

nelsoni

subvirgata

dullest, most washed-out subspecies

juvenile *nelsoni*

Similar Species Saltmarsh has a longer bill and flatter head and a more sharply delineated auricular with more crisply declined breast streaks, differing especially from *subvirgata* Nelson's, the breeding subspecies that overlaps with Saltmarsh. LeConte's (occurs at the same interior marshes and other wet areas in migration) is a lighter buff color with broad buffy, not narrower white, back stripes. LeConte's has a white, not gray, median crown stripe and a pinkish stippled, not gray, nape. Note also that LeConte's winters across much of the southern US, unlike Nelson's, which in winter is confined to coastal marshes, and thus any orange-headed sparrow seen outside the narrow migration time frames for Nelson's is very likely a LeConte's.

Voice CALL: Sharp *tsik*. **FLIGHT NOTE:**

altera

nelsoni

subvirgata

nelsoni

all 3 subspecies

nelsoni & altera

A high, lisping *tsiis*, shorter and less strident than Grasshopper Sparrow; regularly delivered from ground in cover. **SONG:** A wheezy *p-tshhhhhhh-uk*, ending on a lower note; flight display unlike Saltmarsh, with song delivered at zenith of a short vertical flight, followed by a slow descent. In Saltmarsh, song flight is lower and more horizontal.

Status & Distribution Fairly common. **BREEDING:** Fresh marshes, wet meadows, bogs (*nelsoni*); salt marshes of James and Hudson Bays (*altera*); upper reaches of coastal salt marshes and coastal cranberry bogs (*subvir-* *gata*). **WINTER:** Coastal salt marshes, in same habitats as the Saltmarsh Sparrow, though may also occur in fresher upper reaches of marshes where Saltmarsh does not occur. Saltmarsh and all three subspecies of Nelson's can be found in migration and winter in the same coastal marshes in the mid- and south Atlantic states. Accidental in winter in the interior; very rare but annual at large tidal estuaries in winter in coastal CA. **MIGRATION:** Late-spring migration takes place during mid-May–late May; fall migration late Sept.–late Oct. Migration of *nelsoni* primarily through interior broadly through the Mississippi River Valley; rare inland migrant eastward and westward. No doubt overlooked during narrow time windows, but observers in recent decades are more regularly locating this species inland. **RARE STATUS:** Subspecies *nelsoni* is casual in migration (mainly fall) to CA and CO. Accidental to AZ, NM, WY, and WA.

Population The diking and draining of salt marshes in the Maritime Provinces and southern ME has reduced available breeding habitat and numbers of *subvirgata*. Interior populations are probably stable.

SALTMARSH SPARROW *Ammospiza caudacuta*   SALS ■ 1

Saltmarsh Sparrow and Nelson's Sparrow were formerly classified as a single species. More recently they had their English names shortened by dropping "sharp-tailed" from them. Saltmarsh Sparrow is restricted to coastal salt and brackish marshes and is virtually unknown in freshwater. It hybridizes with Nelson's Sparrow (*subvirgata*) where ranges overlap in southern coastal ME and NH. Its behavior is similar to Nelson's, though it may be less secretive. It has a flat head and a fairly long, pointed bill. Its narrow tail is composed of narrow, sharply pointed rectrices. Polytypic (2 sspp.; both in N.A.). L 5" (13 cm)

Identification Gray auricular surrounded by orange; dark postocular line; narrow dark malar stripe sets off white throat; gray nape and central crown stripe; reddish brown back with strong white stripes; indistinct buffy or whitish breast marked by extensive dark streaks across breast and along flanks. **JUVENILE:** Like juvenile Nelson's, but cheek, crown, nape, and back are darker; streaks below more extensive and distinct.

Geographic Variation Differences between the two subspecies are weak and clinal, and they are obscured by individual variation. The two differ subtly and are not safely field-separable. Nominate breeds from northern NJ to southern ME; *diversa* breeds from southern NJ to northern NC. Subspecies *diversa* is shorter-billed than nominate on average and has a slightly paler orange supercilium and darker upperparts. Winter ranges overlap.

Similar Species Similar to Nelson's, but longer bill and flatter head; eyebrow streaked with black behind eye; orange-buff face triangle contrasts strongly with paler, very distinctly streaked underparts; dark markings around eye and head more sharply defined. It is most different from the more washed-out and faintly marked *subvirgata* Nelson's, the subspecies that shares the same marshes in southern ME and NH. The smaller LeConte's Sparrow (would rarely associate with Saltmarsh Sparrow) is easily separated by the head and back patterns. See also Seaside (especially juveniles).

Voice CALL: Like Nelson's Sparrow. **FLIGHT NOTE:** Like Nelson's; regularly delivered from ground in cover. **SONG:** Softer, more complex than Nelson's. Flight display unlike Nelson's: song delivered in rapid, low flight over grass. Note Nelson's song flight rises to an apex with a descent back to the ground.

Status & Distribution Fairly common. **BREEDING:** Breeds in extensive coastal salt marshes (especially those with a large component of saltmeadow cordgrass, which forms

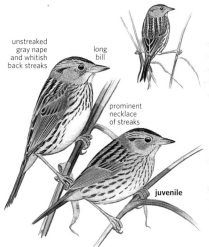

unstreaked gray nape and whitish back streaks

long bill

prominent necklace of streaks

juvenile

short-grass pastures within drier portions of the marsh). **WINTER:** Less tied to saltmeadow cordgrass; frequently forages along dikes, ditches, pool edges, often in stands of taller smooth cordgrass or even shrubs during high tides. Overlaps with all three subspecies of Nelson's Sparrow in migration and winter in marshes of mid- and south Atlantic coast. **MIGRATION:** Returns to breeding grounds between late Apr. and mid-May; most depart by early Nov.; stragglers often linger north of wintering areas into Dec. **RARE STATUS:** Rare on FL Gulf Coast where Nelson's is numerous. Unrecorded farther west on Gulf Coast, although secretive behavior and access to marshes would make detection difficult. Accidental to PA (two records), NS (one record).

Population Endangered. Diking and draining of salt marshes along Atlantic coast has depressed numbers.

FOX SPARROW *Passerella iliaca*   FOSP ■ 1

The Fox Sparrow feeds on the ground, usually in dense cover. It habitually scratches with both feet. Usually found in dense cover, but can be coaxed into view. Often visits feeders. Polytypic (16 sspp.; all in N.A.). L 7" (18 cm)

Identification This large, chunky sparrow has a moderately long tail. Bill shape and plumage are highly variable by subspecies group, but all have extensive triangular spots on the underparts, which may coalesce into a central breast spot as in most other streaked sparrows. Most subspecies have a reddish rump and tail and reddish coloring in wings.

Geographic Variation See sidebar opposite.

Similar Species "Red" Fox is occasionally confused with Hermit Thrush, but note "Red" Fox's bill shape, streaked flanks, streaked back, and different behavior. Lincoln's and Song Sparrows are both smaller, with thinner bills, sharper breast streaking and disproportionately longer tails; and they are as likely to be found in grassy or weedy

areas than in wooded thickets.

Voice CALL: "Thick-billed" Pacific subspecies give a sharp *chink* call, like California Towhee; "Red" and "Sooty" give a *tschup* note, similar to but louder than Lincoln's Sparrow, and similar to but softer than Brown Thrasher. "Slate-colored" is somewhat similar, a *tewk*, slightly down-slurred. **FLIGHT NOTE:** High, rising *seeep*, given commonly on ground and in thickets. **SONG:** Sweet, melodic, warble composed of seven or more phrases: for example, *too-weet-wiew too-weet tuck-soo-weet-wiew*. Sweeter in the northern reddish subspecies; includes harsher or buzzy trills in other subspecies. Frequently sings on the winter grounds.

Status & Distribution Fairly common. **BREEDING:** Dense willow and alder thickets ("Red"); montane willow and

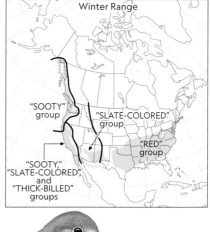

"Red" group
iliaca

reddish streaks on gray back

rufous in cheek

bright rufous tail

winter *iliaca*

very reddish *iliaca* from east of Hudson Bay illustrated here; *zaboria* from west of Hudson Bay is just slightly duller; both winter in Southeast

rufous streaking on underparts

on all "Sooty" subspecies, tail contrasts much less with back than other groups

winter *unalaschcensis*

palest and grayest "Sooty" Fox Sparrow

"Sooty" group
unalaschcensis

"Sooty" subspecies darken gradually in breeding range from north to south

like *schistacea* but slightly more rufous

"Slate-colored" and "Thick-billed" subspecies have rufous tails that contrast sharply with grayish backs

winter *altivagans*

one of the darkest subspecies

winter *townsendi*

"Thick-billed" group
megarhyncha

"Slate-colored" group
schistacea

broad dark, slightly blurry, streaking below

thickest bills on *brevicauda* and *stephensi*

all four subspecies groups have yellow bills for nonbreeding, blue-gray during breeding season

winter *schistacea*

sharp streaking below, including flanks

breeding *stephensi*

alder thickets ("Slate-colored") or coastal forests and thickets ("Sooty"); montane thickets and chaparral ("Thick-billed"). **WINTER:** Undergrowth and dense thickets in conifer-ous or mixed woodlands, chaparral. **MIGRATION:** Fall migration late Sept.– late Nov.; spring migration early Mar.– late Apr. "Red" group migrates several weeks earlier in the spring and later in the fall than western groups do. **RARE STATUS:** Rare to Bering Sea islands and Aleutians; casual Arctic Canada and Greenland; accidental Europe. **Population** Stable.

Fox Sparrow Subspecies Groups

Fox Sparrows comprise four distinct subspecies groups that differ by plumage, range, and voice. Some authorities consider these subspecies groups to be up to four separate species. Identification to exact subspecies is rarely possible, given clinal variation, extreme similarity of certain subspecies, and the occurrence of intergrade populations, but they can usually be identified to subspecies group. Key features are overall upperparts coloration; color and extent of spotting below; presence of back streaking; bill size; and call note, especially between the similarly plumaged "Slate-colored" and "Thick-billed" groups.

"Red" Group: Subspecies *iliaca* and the slightly duller, darker *zaboria* make up "Red" Fox Sparrow group. They are generally (including genetics) the most distinct group. They have bright reddish brown on tail, rump, wings, and face; back is gray with reddish brown streaks; underparts are white, except for a pale yellow-buff wash on throat and upper breast, and have contrasting, blurry reddish brown streaks; also have narrow white wing bars. Reddish breast spotting well defined on white underparts. The face pattern of "Red" Fox Sparrows is striking, with bright reddish brown crown and ear coverts prominent on otherwise pale gray face. Juveniles of these subspecies have less distinct breast spotting and a brownish gray crown and back with no streaking. These birds breed in the boreal forests and taiga of the far north from the Seward Peninsula, AK (*zaboria*), to NL (*iliaca*) and both subspecies winter in the southeastern US.

"Sooty" Group: Birds in "Sooty" group are similar in pattern to those in "Red" group but lack the wing bars, streaked back, and warm reddish brown tones of the "Red." Genetically, they are closest to the interior "Slate-colored" group. Color within the "Sooty" group varies from a cold blackish brown (*fuliginosa*, *townsendi*, including *chilcatensis* which is not recognized by many) in more southerly breeding subspecies to a somewhat paler dark brown (e.g., *sinuosa*, *annectens*, *insularis*, and the palest *unalaschensis*) in subspecies breeding farther north. Breast of "Sooty" group birds is so densely spotted that its central splotch extends across entire breast and its flanks are more mottled than streaked. Base coloration of underparts of a juvenile "Sooty" is a warm buff. Some in "Sooty" group show gray on head like "Slate-colored" birds, and are tinged with brown on wings, rump, and tail. Breast spotting is heavy and dense, often forming a checkerboard-like pattern. These birds breed along the Pacific coast from the AK Peninsula to northwestern WA and they winter south to Baja California Norte, typically in more mesic and denser chaparral than

"Slate-colored" and especially "Thick-billed" groups. The paler northern breeding birds migrate the farthest south, leapfrogging over the more southerly and darker subspecies. Migratory movements of western AK *unalaschensis* poorly known.

"Slate-colored" Group: "Slate-colored" has a medium gray head, nape, and unstreaked back, with contrasting dull red-brown wings, rump, and tail. Subspecies differ subtly, from *olivacea*, with a reddish cast to the head and back; to *schistacea* and *swarthi*, with more gray-brown on the head and back; to intermediate *canescens*, with less brown on the back and a larger bill (tending toward the "Thick-billed"). Markings below, while generally like those of "Red," are a much darker brown and are more sharply defined. "Slate-colored" and "Thick-billed" groups can sometimes show very faint whitish wing bars. A problematic subspecies of northwestern Canada, *altivagans*, is intermediate between the "Slate-colored" and "Red" groups but lacks the distinct rich reddish tones of "Red" group birds. It has only a very faintly streaked gray back. "Slate-colored" birds breed in many of the Great Basin ranges and the northern and central Rockies; and winter in CA, with a very few in AZ and NM.

"Thick-billed" Group: Birds in this group have consistently large, very thick-based bills; plumage is similar to "Slate-colored." This subspecies group typically has a gray mandible (as does "Slate-colored"), but its bill can be yellowish or orangish (as in "Red" or "Sooty" birds); bill color may vary seasonally. Bill size varies from the smaller-billed *megarhyncha* (includes *fulva* and *monoensis*, from OR to northern CA) to *brevicauda* (northwestern CA) and the southern *stephensi* (central and Southern CA mountains and northern Baja California), which have especially large bills. Placement of *canescens* in "Slate-colored" is uncertain in view that breeding populations in White Mts. (NV and CA) mostly chip like "Thick-billed." Breast spotting is most distinct and blackest in the "Thick-billed" group. The winter grounds of "Thick-billed" are not well known but are mainly in CA, especially Southern CA, in coast and transverse ranges, typically on drier (south-facing), more sunlit chaparral than "Sooty" group.

Voice: "Red" and "Slate-colored" subspecies give a *tchewp* call note; this is slightly less down-slurred in "Sooty," which has a smacking *tschup*; "Thick-billed" birds and some (White Mts.) *canescens* "Slate-coloreds" have a distinctive *chink*, similar to California Towhee. Song differences are subtle and complex, but "Red" Fox Sparrows tend to sing sweeter, clearer notes; others use more buzzes and trills.

Genus *Melospiza*

The three *Melospiza* have long, rounded tails, often pumped in flight. They favor brushy areas, often near water, and are also found in marshes. Secretive but inquisitive, they are responsive to pishing. All are quite vocal.

SONG SPARROW *Melospiza melodia* SOSP ▪ 1

The Song Sparrow is one of our most widespread sparrows. They are very responsive to pishing when they repeatedly utter their diagnostic *chimp* note. Polytypic (25 sspp.; 19 in N.A.; 6 restricted to Mexico; 14 other sspp. were recently merged). L 4.8–6.8" (13–17 cm)

Identification Long, rounded tail, often flipped in flight and when landing. Broad, grayish eyebrow; broad. Upperparts usually streaked. Underparts whitish; streaking on sides and breast often converge in central spot. **JUVENILE:** Buffier; fine streaking.
Geographic Variation Extensive, marked variation in measurements, overall color. Identification to individual subspecies is often not possible in the field, though birds are often separable to one of the five geographic groups. Perhaps most distinctive is "Alaska Islands" group: four subspecies resident on the Aleutian Is. chain and east to Kodiak I., all of which are very large (close to size of Fox Sparrow), large-billed, and generally sooty-colored. No other subspecies groups are as large in overall size; the subspecies from the western Aleutians (e.g., *maxima*) are largest. "Pacific Northwest" group (including *morphna*) is large and dark sooty or dark rusty colored; streaking on breast and flanks often has a dusky background. "California" group (including *heermanni* and those resident on the Channel Is.) are small and dark, with distinctly marked rich reddish wings, very gray faces; sharp blackish streaking below contrasts markedly with upperparts. "Desert Southwest" birds (*fallax*) are quite pale reddish brown with well-defined breast streaking, contrasting little with the back and wings. Typical of "Eastern" group is the nominate, which is medium-size and fairly brown-backed with moderate contrast in breast streaking.
Similar Species See Lincoln's, Swamp, Savannah, Fox, and Vesper Sparrows.
Voice CALL: A nasal, hollow *chimp*; also high chip notes

when excited. **FLIGHT NOTE:** A clear, rising *seeet*. **SONG:** Three or four short clear notes followed by a buzzy *tow-wee*, then a trill.
Status & Distribution Common. Winters south to northern Mexico; isolated resident population in central Mexico. **BREEDING:** Brushy areas, especially near water. **WINTER:** In East, found in a variety of brushy habitats and marshes, often with Swamp and Lincoln's Sparrows. In West, Song Sparrow is more closely tied to ponds or streams with lush growth. **MIGRATION:** Most populations are migratory; Pacific and southwestern populations generally resident. Spring early Mar.–late Apr., peaking late Mar.; fall mid-Sept.–mid-Nov., peaking late Oct. **RARE STATUS:** Casual to Pribilofs, AK, and to Europe.
Population Stable in most por-

tions of range. The subspecies resident on Santa Barbara I., *graminea*, became extinct due to overgrazing. Degradation of salt marsh habitats around the San Francisco Bay and desert riparian habitats in the Southwest has also negatively affected certain populations.

"ALASKA ISLANDS" group
maxima
sanaka
insignis
kenaiensis
caurina
rufina
"PACIFIC NORTHWEST" group
merrilli
morphna
cleonensis
samuelsis
maxillaris
pusillula
gouldii
heermanni
"CALIFORNIA" group
graminea
rivularis
"DESERT SOUTHWEST" group
"MEXICAN PLATEAU" group (four subspecies)
melodia
"INTERIOR WESTERN" group
montana
"EASTERN" group
melodia
atlantica
fallax
goldmani

many subspecies have blurry streaks below often with central spot on breast
melodia

juvenile *heermanni*
buffy wash with blurry streaks on breast and sides
compare carefully to Lincoln's Sparrow

thicker streaks on breast and sides than Lincoln's
CA subspecies are very dark
heermanni

northwestern subspecies are dark and ruddy
whitish submoustachial
morphna

AK subspecies, especially those on Aleutians, are very large
maxima

fallax
smallest and palest subspecies

all *Melospiza* have long, rounded tails

LINCOLN'S SPARROW *Melospiza lincolnii*   LISP ■ 1

This sparrow stays in or close to dense cover, flushing away from the observer with its short tail flipped on landing. It often raises its slight crest when disturbed. Polytypic (3 sspp.; all in N.A.). L 5.8" (15 cm)

Identification Buffy wash and fine streaks on breast and sides contrast with whitish, unstreaked belly. Note broad gray eyebrow, whitish chin, and eye ring. Briefly held juvenal plumage is paler overall than juvenile Swamp Sparrow, but closely resembles larger Song.

Geographic Variation The three subspecies show weak, clinal variation in measurements and general coloration. Nominate breeds in the northern portion of the breeding range (AK to the Maritimes), *alticola* breeds in the montane west, and *gracilis* breeds in the Pacific Northwest. Generally, *alticola* is largest and *gracilis* is smallest and darkest.

Similar Species Song is larger with a longer, rounded tail; it lacks an eye ring, and streaking on underparts is broader. Lincoln's has pencil-thin streaks, is grayer on face (especially supercilium), and has a buffy wash to breast that Song lacks, but beware of larger and longer-billed juvenile Song with a buffy breast and finer breast streaking. Similar-size Savannah is found more often in open fields and often flies high when flushed. (Lincoln's usually perches lower.) Savannah also lacks gray face, rich reddish brown back and wings, and buffy wash on breast. See Swamp Sparrow.

Voice CALL: A hardy flat *tschup*, repeated in a series as an alarm call. **FLIGHT NOTE:** Sharp, buzzy *zeee*, similar to that of Swamp Sparrow but sharper; given commonly on ground

often appears slightly crested when agitated

broad gray supercilium

finer streaking than Song against buffy breast

well-outlined buffy submoustachial stripe

juvenile

similar to adult but overall, pattern a little more muted and more streaked

and in thickets. **SONG:** A rich, loud, series of rapid bubbling notes, ending with a trill.

Status & Distribution Common in West, uncommon in East. **BREEDING:** Bogs, mountain meadows, wet thickets. **WINTER:** Thickets, overgrown fields, dense brush, cutover woods. Winters south to C.A. **MIGRATION:** Spring migration, earlier in the West, late Mar.–mid-May; fall migration early Sept.–early Nov. **RARE STATUS:** Casual to Bering Sea islands, Arctic Canada, Greenland; accidental Azores.

Population Stable.

SWAMP SPARROW *Melospiza georgiana*   SWSP ■ 1

Generally common, the Swamp Sparrow is found in swamps or marshes, but it is also found in rank brush or tall grass where it may associate with Song and Lincoln's. They can be brought into view by pishing. They deliver a loud, Eastern Phoebe–like metallic call note when excited and at dawn and dusk, when marshes and brushlands are filled with their chorus of *chip* notes. A Swamp often flies away, straight and low, when flushed, and usually flips its long tail (especially when landing) like the other two *Melospiza* species. It typically does not call in flight, but often calls upon landing. Polytypic (3 sspp.; all in N.A.). L 5.8" (15 cm)

Identification Gray face; white throat; rich rufous wings contrast with upperparts; reddish brown back streaked with black.

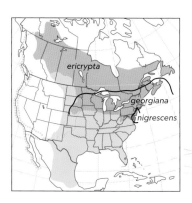

ericrypta

georgiana

nigrescens

variable reddish crown

ruddy rufous wings in all plumages

grayish breast

breeding

dark back streaks

juvenile

winter adult

dull, blurry streaks confined to sides

immature

BREEDING ADULT: Reddish crown, gray breast, whitish belly. **WINTER ADULT & IMMATURE:** Buffier overall, especially on flanks; duller crown streaked, divided by gray central stripe. **JUVENILE:** Briefly held plumage usually even buffier; darker overall than the juvenile Lincoln's or Song; redder wings and tail.

Geographic Variation Three subspecies show moderate variation. The nominate and *ericrypta* divide the breeding range roughly at the Canadian border (nominate south, *ericrypta* north). Both small-billed and warm; nominate slightly warmer flanks and cap, grayer back. The local *nigrescens* (breeds in brackish marshes from NJ to MD, winter range poorly known) has a longer bill, darker back, black streaking on crown, and no buff tones to flanks.

Similar Species The Song and Lincoln's are similar but heavily streaked below. Juvenile and first-winter Swamp can be streaked lightly, but never as prominently as the Lincoln's or Song. Unlike the Lincoln's, the Swamp lacks an eye ring; also has a somewhat longer tail and richer red-brown wings, with gray chest.

Voice CALL: Hard, metallic chip, somewhat similar to that of Eastern or Black Phoebe. **FLIGHT NOTE:** A prolonged, buzzy *zeee*, similar to Lincoln's, but softer. **SONG:** A slow, single-pitched musical trill. Some deliver a faster trill.

Status & Distribution Common. Winters south to central Mexico. **BREEDING:** Nests in dense, tall vegetation in marshes, bogs, alder swales, wet hay fields. **WINTER:** Tall grass marshes, brushy fields, wet open woodland, especially cutover forest. **MIGRATION:** Spring mid-Mar.–early May, peaking mid-Apr.; fall mid-Sept.–mid-Nov., peaking mid-Oct. **RARE STATUS:** Rare in migration and winter in West. Casual to AK; accidental Russian Far East.

Population Most populations stable; mid-Atlantic subspecies *nigrescens* sharply declined recently (reasons poorly known).

Genus *Zonotrichia*

This genus comprises five species (four breed in Canada and the northern US; one in the highlands of C.A., one throughout S.A.). Adults have strong face patterns; immatures are more indistinctly marked. Songs are loud whistled phrases, simple but melodic. They typically feed in flocks along dense cover, darting for safety at once when danger approaches.

WHITE-THROATED SPARROW   *Zonotrichia albicollis*   WTSP ■ 1

This species frequents feeders or woodland edges. Monotypic. L 6.8" (17 cm)

Identification Two distinct color morphs. White-striped morph has white eyebrow (yellow in front of eye) and black lateral crown stripes; eyebrow stripe is buffier in winter. Tan-striped morph has tan eyebrow (dull yellow in front of eye); dark brown lateral crown stripes; and usually with diffuse streaking below. Both morphs have an outlined white throat (with thin malar stripe in tan-striped morphs) and mostly dark bill. Upperparts rusty brown and underparts grayish. **FIRST-WINTER:** Duller than adult; poorly defined white throat patch bordered by dark malar stripe; some streaking on breast and sides. **JUVENILE:** Pale grayish eyebrow; breast and sides heavily streaked.

Similar Species See White-crowned.

Voice CALL: A sharp *pink*, like White-crowned, but higher; also a drawn-out, lisping *tseep*. **SONG:** Thin whistle, generally one to two single notes, followed by three to four long notes, sometimes tripled: *Pure sweet Canada Canada Canada*. Sings in winter.

Status & Distribution Common. **BREEDING:** Clearings in deciduous or evergreen forests, bogs. **WINTER:** Common in woodland undergrowth, brush, gardens. **MIGRATION:** Spring late Mar.–mid-May, peaking mid-Apr.; fall mid-Sept.–early Nov., peaking mid-Oct. Rare to uncommon in migration and winter in the West. **RARE STATUS:** Rare to southeastern AK; casual to northern AK. Rare to Arctic Canada, northern Mexico; casual to Caribbean region, Belize, Europe, and Azores.

Population Stable.

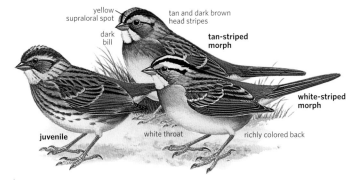

yellow supraloral spot

tan and dark brown head stripes

dark bill

tan-striped morph

white-striped morph

juvenile

white throat

richly colored back

HARRIS'S SPARROW   *Zonotrichia querula*   HASP ■ 1

The behavior of this sparrow is much like that of the White-throated. Regularly occurring in medium-size flocks, it mixes sometimes with the White-throated, and especially White-crowned (shares a more open habitat in migration and winter), or other sparrows. Monotypic. L 7.5" (19 cm)

Identification Large, larger than other *Zonotrichia*. Bright pink bill; black crown, face, and bib variable by age and plumage; all show postocular spot. **BREEDING ADULT:** Extensive black from crown to throat; gray face. **WINTER ADULT:** Buffy cheeks; throat all black or with white flecks or partial white band. **IMMATURE:** Resembles winter adult with less black; white throat bordered by dark malar stripe. **Similar Species** Male House Sparrow (not a true sparrow; see p. 586) has a black throat, but is smaller, has bolder wing bars and a chestnut crown, does not have bright pink bill, and behaves totally unlike any passerelid sparrow. **Voice CALL:** Loud *wink*. Flocking birds give a husky chatter. **FLIGHT NOTE:** A drawn-out *tseep*. **SONG:** Series of long, clear wavering whistles, often beginning with two notes on one pitch followed by two notes on another pitch. **Status & Distribution** Uncommon to fairly common. **BREEDING:** Stunted boreal forest. **WINTER:** Open woodlands, brushlands, hedgerows. **MIGRATION:** Spring late Mar.–late May; fall early Oct.–early Nov. **RARE STATUS:** Rare to very rare east and west of narrow regular range. **Population** Stable.

black crown and bib

large pink bill

breeding

dark postocular spot in all plumages

winter adults

black on chin and throat variable

dark malar stripe

dark chest patch

brownish flank streaking

immature

GOLDEN-CROWNED SPARROW *Zonotrichia atricapilla* GCSP ■ 1

The adult Golden-crowned Sparrow is distinctive, but the immature is the plainest-faced *Zonotrichia*. It prefers denser brush than White-crowned. Monotypic. L 7" (18 cm) **Identification** Brownish back streaked with dark brown; breast, sides, and flanks grayish brown. Bill

juvenile

dull yellow forehead; head pattern otherwise muted

yellow crown with broad black eyebrow

winter adult crown pattern much more muted

mostly dark bill

immature

winter adult

breeding

dusky above, pale below. **ADULT:** Yellow patch tops black crown; more obscured in winter. **IMMATURE:** Dull yellow restricted to forehead and overall plain-headed. **JUVENILE:** Dark streaks on breast and sides. Plumage briefly held. **Similar Species** White-crowned Sparrows have distinct head striping at all ages; immature Golden-crowned has a plainer head, darker underparts, and darker bill. Yellow crown is usually present as a trace of color on forecrown and above eyes. **Voice CALL:** A flat *tsick*. Flocking birds give a husky chatter. **FLIGHT NOTE:** A soft *tseep*. **SONG:** A series of three or more plaintive, whistled notes, often with each on a descending note: *oh dear me.*

Status & Distribution Fairly common to common. **BREEDING:** Stunted boreal bogs and in open country near tree line. **WINTER:** Dense woodlands, tangles, brush, chaparral. **MIGRATION:** Spring migration mid-Mar.–mid-May, peaking mid-Apr; fall migration mid-Sept.–early Nov., peaking mid-Oct.

RARE STATUS: Rare to Aleutians and Bering Sea islands; casual to Arctic Canada and eastern N.A. south to Gulf Coast.

Population Stable.

WHITE-CROWNED SPARROW *Zonotrichia leucophrys* WCSP ■ 1

The White-crowned Sparrow is very common in the West; except when breeding it often occurs in large numbers. Polytypic (5 sspp.; all in N.A.). L 7" (18 cm)

Identification For bill color, see sidebar opposite; whitish throat; brownish upperparts; mostly pale gray underparts. **ADULT:** Black-and-white striped crown. **IMMATURE:** Tan and brownish head stripes. **JUVENILE:** Brown and buff head; streaked underparts.

Geographic Variation See sidebar opposite.

Similar Species White-throated Sparrow is darker gray below and has a dark bill, prominent white throat, and yellow supraloral. Immature is similar to immature Golden-crowned Sparrow, but note the latter's dark gray bill, indistinct head pattern, and dingy underparts.

Voice CALL: Loud, metallic *pink*. Flocking birds give a husky chatter.

FLIGHT NOTE: Sharp *tseep*. **SONG:** One or more thin, whistled notes followed by a variable series of notes.

Status & Distribution Common in the West, uncommon in East, rare in the Southeast. **BREEDING:** Clumps of bushes or stunted trees on taiga and tundra; coastal scrub, chaparral for "Pacific" group birds. **WINTER:** Hedgerows, desert scrub, brushy areas, wood edges, and feeders. Winters south to central Mexico. **MIGRATION:** Spring mid-Mar.–late May, peaking early to mid-Apr., later in the Great Lakes Region (nominate) and in the West (*oriantha*); fall early Sept.–mid-Nov., peaking mid-Oct. **RARE STATUS:** Subspecies *gambelii* is rare on Aleutians and Bering Sea islands and in the East; *pugetensis* accidental southern

NV. Casual to Caribbean region, Greenland, Europe, Azores (*leucophrys*), and northeast Asia (*gambelii*).

Population Stable.

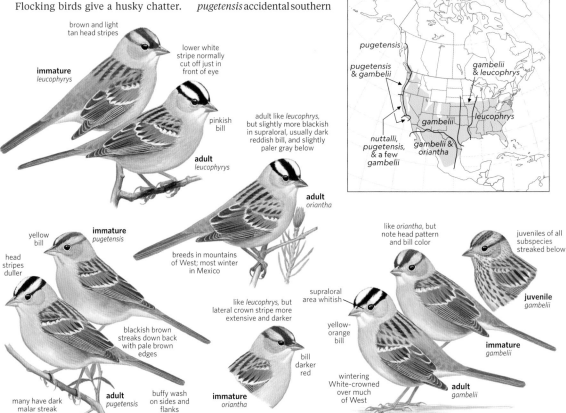

brown and light tan head stripes

immature
leucophryys

lower white stripe normally cut off just in front of eye

pinkish bill

adult
leucophryys

adult like *leucophrys*, but slightly more blackish in supraloral, usually dark reddish bill, and slightly paler gray below

adult
oriantha

like *oriantha*, but note head pattern and bill color

juveniles of all subspecies streaked below

yellow bill

immature
pugetensis

head stripes duller

breeds in mountains of West; most winter in Mexico

juvenile
gambelii

supraloral area whitish

blackish brown streaks down back with pale brown edges

like *leucophrys*, but lateral crown stripe more extensive and darker

yellow-orange bill

immature
gambelii

many have dark malar streak

adult
pugetensis

buffy wash on sides and flanks

immature
oriantha

bill darker red

wintering White-crowned over much of West

adult
gambelii

White-crowned Sparrow Subspecies

The White-crowned Sparrow is divided into three well-defined subspecies groups: "Taiga" (eastern dark "lored" *leucophrys* and pale "lored" western *gambelii*), "Pacific" (northern *pugetensis* and southern *nuttalli*), and "Western Montane" (dark "lored" *oriantha*).

Plumage Differences

The "lores"—actually the supraloral area above the lores—are the best first clue: dark in *leucophrys* and *oriantha*, pale in "Pacific" group and *gambelii*. Bill is yellow in "Pacific," orange in *gambelii*, and pink in *leucophrys* and *oriantha*. "Pacific" birds often show a malar stripe, which is rare in others. Primary extension is shortest in "Pacific," longest in "Taiga," and intermediate in *oriantha*. The "Pacific" group lacks gray edging on the back, showing only dark brown and tan; gray edging is fairly prominent on the back of the other subspecies. This group also has a buffier coloration to the breast and flanks. All ages show a distinctive difference that can be hard to spot in the field: "Pacific" birds have a small patch of yellow feathers at the bend of the wing; on the other subspecies the color is white.

Voice Differences

Songs differ somewhat: rising and falling *zuuuu zeee jeee jeee zee* for *leucophrys* and *gambelii*; clearer and more rapid with a different introductory note and more trills (especially the last two notes) for "Pacific"; song of *oriantha* is variable, more like "Pacific" in the Sierra Nevada, more like *gambelii* in the ranges of the Great Basin and the Rocky Mts. Within the resident subspecies *nuttalli*, there are many different song dialects, even from nearby locations. Call notes are flatter in "Pacific"; *oriantha* is sharp, almost like a Blue Grosbeak. Another behavioral distinction is that "Pacific" birds place their nests in low bushes, whereas other subspecies place them directly on leaf litter.

Geographical Ranges

Within the "Taiga" group, *leucophrys* breeds mainly in the Canadian tundra east of Hudson Bay, while *gambelii* breeds west of Hudson Bay. Hudson Bay birds comprise many intergrades. In winter, *leucophrys* winters in the East, *gambelii* in the West; *gambelii* is rare to casual in the East. "Pacific" *pugetensis* and *nuttalli* reside along the Pacific coast from extreme southeastern AK to CA. "Pacific" *pugetensis* is rare to uncommon in winter in the Central Valley and rare south of Los Angeles Co. on the coast; casual well-inland in Southern CA; accidental in southern NV. Subspecies *oriantha* breeds in the Rocky Mts. and in the higher elevations of the Sierra Nevada and the Cascades, and very locally in Southern CA. Dark "lored" birds, presumably *oriantha*, are casual in migration in coastal CA. Despite its very similar appearance to *leucophrys*, *oriantha* is perhaps more distantly related to *leucophrys* than *gambelii* is to *leucophrys*. Subspecies *pugetensis* has recently been found breeding in passes in the Cascades of WA alongside *gambelii*, where the two subspecies appear to be acting as separate species.

JUNCOS Genus *Junco*

This genus includes five species (or, according to some authors, just three), two of which occur north of Mexico. These birds are shades of gray above and white below, with pink bills and prominent white outer tail feathers. Their behavior—often feeding on the ground singly or in groups; flying up suddenly to a tree—may recall *Spizella* sparrows.

YELLOW-EYED JUNCO *Junco phaeonotus*   YEJU   2

This distinctive bird, with piercing yellow eyes, replaces the Dark-eyed Junco in certain mountain ranges adjacent to the Mexican border. It behaves much like Dark-eyed, but it rarely forms large flocks. Polytypic (4 sspp.; *palliatus* in N.A.). L 6.3" (16 cm) **Identification** Bright yellow eyes, black lores. Bicolored bill with dark maxilla, pale mandible. Pale gray above; bright rufous back; rufous-edged greater coverts and tertials; pale gray underparts. **JUVENILE:** Similar to juvenile "Gray-headed" Dark-eyed Junco, but with rufous wing panel like adult; eye is brown, becoming pale before changing to yellow of adult. **Geographic Variation** Moderate

striking yellow eye

bicolored bill

pale throat

rufous back

juvenile *palliatus*

palliatus

rufous greater coverts

variation among the four subspecies; only *palliatus* occurs north of Mexico. Baird's Junco (*J. bairdi*) from Baja California Sur, Mexico, was recently split as a separate species.

Similar Species "Gray-headed" group of Dark-eyed Junco is most similar. Yellow-eyed has a golden eye color and is rich rufous on greater coverts and tertials. Note that *dorsalis* ("Gray-headed" group) has bicolored bill.

Voice CALL: Sharp *dit*, softer than Dark-eyed Junco. **FLIGHT NOTE:** A high, thin *seep*, similar to call of Chipping Sparrow, lower than call of Dark-eyed Junco; also rapid twitter as in Dark-eyed. **SONG:** A variable series of clear, thin whistles and trills.

Status & Distribution Fairly common. Breeds in mountains south to Guatemala. In the US, found only in southeastern AZ and southwestern NM (Peloncillo and Animas Mts., Hidalgo Co.); recently a few discovered near Silver City. Coniferous and pine-oak slopes, generally above 6,000 feet. Generally resident, but some move to slightly lower elevation in winter. **RARE STATUS:** Casual to West TX. **Population** Stable.

DARK-EYED JUNCO *Junco hyemalis*   DEJU   ■   1

In winter, Dark-eyed Juncos typically form flocks and often associate with other species, including Chipping Sparrows, Pine and Palm Warblers (in the southeastern US), and bluebirds. The Guadalupe Junco (*J. insularis*) from Guadalupe I., Mexico, was recently split as a separate species. Polytypic (14–15 sspp.; 12 in N.A.). L 6.3" (16 cm)

Identification A sparrow with a long, notched tail and a small pinkish or horn-colored bill (bicolored in *dorsalis*). Two prominent white outer tail feathers in most subspecies; three in "White-winged." Most subspecies have a gray or brown head and breast sharply set off from a white belly, but are otherwise highly variable. (See Geographic Variation.) **MALE:** Typically darker with sharper markings. **FEMALE:** Typically browner with more indistinct markings. **JUVENILE:** Heavily streaked, often with a trace of adult pattern.

Geographic Variation The 12 subspecies in N.A. show marked variation and fall into four major groups: "Slate-colored," "Oregon," "White-winged," and "Gray-headed." The groups have at times been considered separate species. **"SLATE-COLORED" GROUP:** "Slate-colored" (two subspecies and *cismontanus*) is the most widespread and only subspecies group found regularly in the East. It breeds throughout the species' range east of the Rockies and in the northern region; it winters mainly in the East and is uncommon to rare over most of the West. The male "Slate-colored" has a white belly contrasting sharply with a dark gray hood and upperparts, usually with very little contrast between the hood and back; immatures can have some brown wash on the back and crown. In female, the amount of brown on head and at center of back varies; more extensive in immatures. "Slate-colored" comprises two subspecies: widespread nominate and larger, bluer-billed *carolinensis*, which is resident in the Appalachians from PA to northern GA. An additional subspecies, *cismontanus*, is often grouped with "Slate-colored." It breeds from the YT to central BC and AB and may winter throughout the West; it is casual to the East. In coloration, *cismontanus* is intermediate between "Slate-colored" and "Oregon," with males showing a blackish hood that contrasts with a usually grayish back (occasionally with some brown). **"OREGON" GROUP:** "Oregon" (six subspecies in N.A. and two in Baja California Norte) breeds in the West Coast states north to southern AK and east to central NV and western MT; winters throughout the West and Great Plains and is very rare to the East. Male "Oregon" has a slaty to blackish hood, contrasting sharply with its rufous-brown to buffy brown back and sides; female has duller hood color. Of the five "Oregon" subspecies, the more southerly subspecies are paler. Part of the "Oregon" group, "Pink-sided" (*mearnsi*) breeds in the northern Rockies, centered in Yellowstone NP and ranging from northern Utah to the Cypress Hills of southeastern AB and southwestern SK; it winters in the southern Rockies, Southwest, and western Great Plains, rarely to the West Coast, and is accidental to the East. "Pink-sided" has broad and extensive, bright pinkish cinnamon sides, a blue-gray hood, a poorly defined reddish brown back and wings that do not contrast markedly with flanks, and blackish lores. Females are duller

"Slate-colored" ♀
hyemalis

"Slate-colored" ♀
hyemalis

some have faint white tips to wing coverts, similar to "White-winged"

females browner above

all subspecies have white outer tail feathers

juvenile

all juvenile juncos are streaked

"Oregon" ♀
shufeldti

grayish hood

buffy rufous sides

blackish hood

"Oregon" ♂
shufeldti

but retain basic pattern; they are frequently confused with some "Oregon" females, mainly because of misrepresentation in many guides. **"WHITE-WINGED" GROUP:** "White-winged" (*aikeni*) is the most local, breeding exclusively in the Black Hills region and wintering along the Front Range of the Rockies; it is casual to accidental in western TX, AZ, and Southern CA. "White-winged" is mostly pale gray above, usually with two thin white wing bars; it is also larger and bigger-billed, with more white on its tail. It is most similar to "Slate-colored" (which can rarely have narrow wing bars) but is larger and paler, with contrasting blackish lores and more extensive white in tail. **"GRAY-HEADED" GROUP:** "Gray-headed" (*caniceps*) is the subspecies of the southern Rockies, breeding through much of NV, UT, and CO south to central AZ and western TX; it winters in the southwest and southern Rockies states and is rare to the West Coast and accidental to the East. Females and immatures are very similar to the "Oregon" Juncos, but are less distinctly hooded. In "Gray-headed," pale gray head and dark lores resemble head pattern of "Pink-sided," but flanks are gray rather than pinkish, and back is marked by a very well-defined patch of reddish hue that does not extend to wings and that contrasts sharply with rest of body. A distinctive subspecies, *dorsalis*, is sometimes known as "Red-backed Junco" and is largely resident from northwestern AZ through NM to the Guadalupe Mts. of western TX. It differs from the more widespread, migratory, northerly breeding *caniceps* in having an even paler throat and a larger, bicolored bill that is black above and bluish

below. **INTERGRADES:** Intergrades between some subspecies are frequent. Common intergrades are "Pink-sided" x "Oregon" and "Pink-sided" x "Gray-headed." Subspecies *cismontanus* may be a broad intergrade population of "Oregon" x "Slate-colored." Identification to subspecies group thus requires caution to eliminate the possibility of an intergrade; for intergrades, look for intermediate characteristics: For example, a darker, more contrasting hood on a "Pink-sided" indicates the influence of "Oregon" genes; reduced pink sides and a well-defined reddish back on a "Pink-sided" indicate "Gray-headed" parentage.

Similar Species See Yellow-eyed Junco.

Voice Songs and calls among subspecies generally similar, but songs and calls of "Gray-headed" *dorsalis* are more suggestive of Yellow-eyed, at least in parts of its range (e.g., Guadalupe Mts., TX and NM). **CALL:** Sharp *dit*. **FLIGHT NOTE:** A rapid twittering. **SONG:** A musical trill on one pitch; often heard in winter.

Status & Distribution Common. Breeds south to northern Baja California; winters south to northern Mexico. **BREEDING:** Breeds in coniferous or mixed woodlands. **WINTER:** Found in a wide variety of habitats but tends to avoid areas of denser brush; it especially favors feeders, parks, and open forest

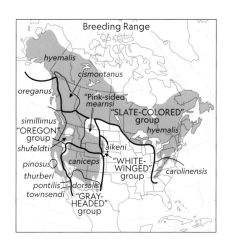

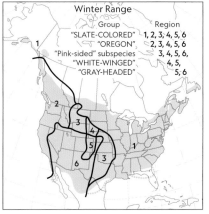

without an understory. Uncommon to Bermuda. **MIGRATION:** Withdraws from wintering areas during Apr., typically early–mid-Apr. Fall arrivals first appear in late Sept., peaking in late Oct. **RARE STATUS:** Very rare to Bering Sea islands and Aleutians. Casual to southern FL, northern Caribbean, Arctic Canada, Greenland, and Europe.

Population Stable.

SPINDALISES Family Spindalidae

These four species from the Caribbean, which were long treated as the Stripe-headed Tanager—a single species within the tanager family—were subsequently split as four species and recently placed in their own family. All give high-pitched soft notes, sometimes in a series. Only the Western Spindalis has occurred in North America.

Western Spindalis, male *pretrei* (Cuba, Dec.)

Structure Rather small to medium-size birds with short tails and somewhat stubby bills. The Jamaican Spindalis is the largest species.

Behavior Mainly arboreal and forages at all levels in the vegetation in search of fruit. Sometimes found in small flocks.

Plumage Shows strong sexual dimorphism: Males have a black-and-white-striped head and are brightly colored. Females are much more subdued and olive overall,

with a darker auricular and a faint supercilium. The Western Spindalis and Hispaniolan Spindalis show a whitish patch at the base of the primaries. The juveniles in all species look like the respective adults once they molt by fall.

Distribution Restricted to the Bahamas and the Greater Antilles in the Caribbean. (One subspecies of Western Spindalis is endemic to Cozumel I. off the Yucatán Peninsula.)

Taxonomy Molecular work has revealed a number of small groups of New World nine-primaried songbirds with unclear affinities, and many of these are now placed in their own families. The four closely related species of *Spindalis* (formerly with tanagers, Thraupidae) now constitute the family Spindalidae.

Conservation BirdLife International lists all four species as Least Concern.

WESTERN SPINDALIS   *Spindalis zena*   WESP   ■   3

The Western Spindalis is a rare visitor primarily from the Bahamas to coastal southeastern FL and the FL Keys. Typically found in fruiting trees; in its native range, often in small groups. Males are distinctive. Females are totally different and much duller. Polytypic (5 sspp.; 3 in N.A.). L 6.8" (17 cm)

Identification MALE: Black-and-white stripes on head; white wing patch on greater coverts. Back is black (or green in some subspecies), contrasting with tawny rump. White outer tail feathers. **FEMALE:** A drab grayish olive with a darker auricular, paler throat, superclium, and malar. Distinct

whitish patch at base of primaries.

Geographic Variation Nearly all males from FL are black-backed birds from central and southern Bahamas (*zena*). Green-backed birds from northern Bahamas (*townsendi*) reported. One record from Key West of a male Cuban *pretrei*. Two additional subspecies are found on Grand Cayman I. (*salvini*) and Cozumel I. (*benedicti*). Females not separable to subspecies.

Similar Species Males unmistakable. Females plainer, but distinctive head and wing pattern and short, thick bill.

Voice CALL: A thin, high *tseee*, which is given singly or in a short series.

SONG: A series of high thin notes ending in buzzy phrases.

Status & Distribution Resident in the Bahamas, Cuba, Cayman Is., and Cozumel I. **RARE STATUS:** Rare coastal southeast FL; casual on FL Gulf Coast and FL Keys. Most reports are from spring, fall, and winter. There is one breeding record.

Population Stable.

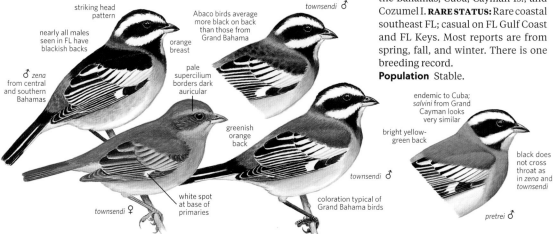

striking head pattern

nearly all males seen in FL have blackish backs

♂ *zena* from central and southern Bahamas

Abaco birds average more black on back than those from Grand Bahama

orange breast

pale supercilium borders dark auricular

townsendi ♂

greenish orange back

townsendi ♂

coloration typical of Grand Bahama birds

townsendi ♀

white spot at base of primaries

bright yellow-green back

endemic to Cuba; *salvini* from Grand Cayman looks very similar

black does not cross throat as in *zena* and *townsendi*

pretrei ♂

YELLOW-BREASTED CHAT Family Icteriidae

The Yellow-breasted Chat has recently been placed as a single species in its own family.

Structure This species is large with a long tail and very thick bill with a strongly curved culmen.

Behavior A skulking species, singing males on breeding territories can be conspicuous, especially in spectacular display flight with slow-motion fluttery wingbeats. A distinctive trait is holding food with its feet.

Plumage Largely green or grayish above with white spectacles on the face, a bright yellow throat and breast, and white belly. Sexual and age dimorphism is minor.

Distribution This migratory New World bird breeds from southern Canada to northern Mexico and winters in Mexico and Central America.

Taxonomy Endless argument and speculation over the affinities of *Icteria* (is it a wood-warbler, or blackbird, or mimid?) have been clarified by recent molecular studies, though current authors still disagree whether it should be considered an offshoot within the New World blackbird family Icteridae, or placed in its own family (with the confusingly similar name of Icteriidae). The latter view currently prevails.

Conservation Destruction of riparian woodland in the West and the maturing of forests in much of the East has caused declines.

Yellow-breasted Chat, male *virens* (OH, May)

YELLOW-BREASTED CHAT *Icteria virens*   YBCH ■ 1

This species is heard more often than it is seen: It is secretive in its brushy and often impenetrable habitat. Polytypic (2 sspp.; 2 in N.A.). L 7.5" (19 cm)

Identification Large songbird with a thick bill and a long tail. Nominate subspecies described. **ADULT MALE:** Bright yellow throat and breast sharply contrast with white belly and undertail coverts. Grayish olive head has white spectacles, black lores, and a very narrow white submoustachial stripe. Upperparts are all olive green. Bill is largely black. **ADULT FEMALE:** Very similar to male, but head more olive and lores duller. **IMMATURE:** Similar to, but slightly duller than, adult female.

Geographic Variation Subtle. Western *auricollis* is longer tailed (sometimes known as "Long-tailed Chat"), grayer above and has a broader white submoustachial stripe.

Similar Species Unmistakable if seen well, but usually skulks.

Voice CALL: A variety of calls, including a harsh *chough* and a nasal *air*. **SONG:** An extensive repertoire, consisting of a series of irregularly spaced scolds, chuckles, mews, rattles, and other unmusical sounds. It often incorporates harsh *sheh sheh sheh sheh* calls and a higher-pitched *tu-tu-tu* series. Sings in flight in spectacular display flight with tail tucked and deep, slow-motion wingbeats.

Status & Distribution Uncommon; a medium-distance migrant, nominate *virens* is both a trans-Gulf and circum-Gulf migrant. **BREEDING:** Low, dense vegetation with open canopy, including shrubby habitat in wetland areas and early second growth. In East (*virens*), favors second growth; in West (*auricollis*), favors riparian areas. **MIGRATION:** During spring migration in the East, it

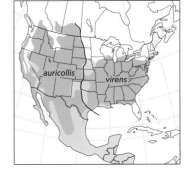

arrives on the Gulf Coast in mid-Apr., reaching the Midwest by early May. In the West, it arrives in CA and southern AZ in mid-Apr., reaching northernmost breeding areas in Canada by late May. Secretive behavior makes it difficult to detect fall migrants. It moves mainly late Aug.–late Sept., with some remaining into early winter along entire East Coast. **WINTER:** From northern Mexico south to western Panama. Rarely in southern US and casually farther north. **RARE STATUS:** Casual in spring migration north of breeding range; regular in fall north to Atlantic Canada; rare in fall to Bermuda; casual in fall and winter to Bahamas; very rare in Cuba and possibly Grand Cayman Is. Three specimens from Greenland. **Population** Declines throughout much of range and now virtually extirpated from southern New England.

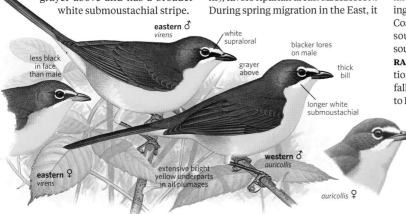

less black in face than male

eastern ♂
virens

white supraloral

blacker lores on male

grayer above

thick bill

longer white submoustachial

western ♂
auricollis

eastern ♀
virens

extensive bright yellow underparts in all plumages

auricollis ♀

BLACKBIRDS Family Icteridae

Altamira Oriole (TX, Feb.)

Contrary to popular thought, blackbirds are not all black. In fact, the family name, Icteridae, refers to the yellow color of many species.

Structure Blackbirds all have strong bills, ranging from short and finchlike, to sharp, to stout and heavy. Some blackbirds are stocky, but some orioles are slim, while the meadowlarks are rotund and stocky. Tail length in most species is medium, extremes being the short tails of the meadowlarks and the long and keel-shaped tails of the grackles. In the tropics, other groups, such as the caciques and oropendolas, are found, and the latter are among the largest Neotropical passerines.

Plumage Generally blackbirds show some black, yellow, orange, or red on the plumage. Females of some species are brownish and streaked. Many of the blackbirds and grackles have largely or fully black plumages. Orioles are the brightest blackbirds: Their plumages are a mix of yellow, orange, or chestnut with black. Female orioles vary depending on the species: Some are dull and yellowish, lacking the strong patterns of the adult male, while others are nearly identical to adult males in

plumage. Young males may look more similar to females.

Behavior All blackbirds share special musculature that allows them to open their bill with great force, allowing them to open objects and extract food. Other common behaviors include "bill tilting," in which an individual will point its bill up toward the sky. This aggressive signal often is given to a nearby individual during feeding and in territorial disputes. During territorial singing, blackbirds make the flash colors obvious, flaring red epaulets, exposing yellow breasts, or twisting yellow heads. In display, the grackles deeply keel the tail, ruffle the plumage, and drop the wings. Orioles tend to be largely monogamous. Females of more-migratory species exhibit duller plumages, and most territorial defense is performed by the male. Larger grackles and marsh-nesting blackbirds have polygamous (many mates) unions. Usually a male defends a large, high-quality habitat and several females settle in his territory and mate with him. Among the grackles, the females congregate in a breeding colony and the largest and most aggressive male defends the harem. Cowbirds lay their eggs in the nests of other species, doing away with parental care altogether. Many territorial species are highly social during the nonbreeding season, forming large mixed-species flocks. The Tricolored Blackbird is also highly social and colonial during the breeding season. It is the only strictly colonial land bird extant in North America.

Distribution Restricted to the New World, blackbirds are found from Alaska to Tierra del Fuego. The most migratory species breed in North America, with the Bobolink showing the longest and most impressive migration.

Taxonomy There are about 104 species of blackbirds in 30 genera, arranged in seven subfamilies. North American blackbirds fit into five of these subfamilies: the orioles (Icterinae), the blackbird-grackle group (Agelaiinae), the meadowlarks and allies (Sturnellinae), and two single species groups, the Yellow-headed Blackbird (Xanthocephalinae) and the Bobolink (Dolichonychinae). Extralimital are two Neotropical subfamilies of caciques.

Conservation The Tricolored Blackbird and Rusty Blackbird are at risk and their populations have drastically declined in the last few decades. BirdLife International codes: 3 NT, 7 VU, 7 EN, 1 CR, and 1 EX.

Genus Xanthocephalus

YELLOW-HEADED BLACKBIRD *Xanthocephalus xanthocephalus*   YHBL ▪ 1

This species is a beautiful marsh-dwelling blackbird of the prairies and the West. Monotypic. L 9.5" (24 cm)

Identification Males noticeably larger than females. **ADULT MALE:** Black with a bright yellow head (darker in winter), with triangular black mask. White primary coverts show as wing patch in

flight. **ADULT FEMALE:** Brownish, with yellow supercilium and breast. Whitish throat with dark malar stripes and yellow submoustachial above it. **IMMATURE MALE:** Like female, but substantially larger, more extensive yellow on head and neck, black lores; noticeable white on primary coverts, although not

a fully formed patch. **JUVENILE:** Rich buffy head, contrasting with whitish throat and brownish body.

Similar Species Adult male unmistakable; larger size is helpful in mixed blackbird flocks. Duller brown females have a contrasting yellow breast.

Voice CALL: A rich, liquid *check*. **SONG:**

Two song types, both with a mechanical, unpleasant, or at least unusual sound. The primary song, *kuk, koh-koh-koh…waaaaaaaa*; the final nasal scraping sound is separated from the introductory notes, accompanied by an asymmetrical display where neck and head are bent to one side. The second song is shorter, a croaking *kuuk-ku, whaaa-kaaaa*, lasts two seconds, accompanied by a symmetrical display. **Status & Distribution** Common. **BREEDING:** Wetlands of cattail, rushes, or *Phragmites*; feeds in adjacent grasslands or farmland. **MIGRATION:** A diurnal migrant, usually moving in single-sex flocks. In spring, males arrive one to two weeks before females, arriving mid-Apr.–mid-May. Fall

movement by July, but most move Aug.–mid-Oct. **WINTER:** A few birds in agricultural areas, especially feedlots, in border states from CA to southern TX; most in Mexico. Females winter farther south than males. **RARE STATUS:** Rare in East away from mapped range. Casual in AK, YT, NT, NU, and Bermuda. Accidental in Europe.
Population General increase during the 1970s. Local droughts can greatly alter their numbers.

♀

deep yellow throat and breast

deep yellow head

spring adult ♂

juvenile

white wing patch on primary coverts

spring adult ♂

immature ♂

Genus *Dolichonyx*

BOBOLINK *Dolichonyx oryzivorus*   BOBO ■ 1

The breeding-plumaged Bobolink is charismatic with attractive black-and-white plumage. Monotypic. L 7" (18 cm)
Identification Small with a sparrow-like bill, very long wings, and pointed tail feathers. **SUMMER MALE:** Black (feathers tipped buff when fresh), with buffy nape and extensive white above. **SUMMER FEMALE:** Streaked with pink-based bill and pinkish legs. Dark crown with crisp white median stripe, buffy face, a short but obvious postocular stripe; yellowish buff streaking above and streaked on sides and flanks. **FALL:** Similar to summer female, but warmer yellowish below.
Similar Species Breeding male is distinctive. Male Lark Bunting is black with white on wings, but not on nape or body. Females and fall plumage are superficially sparrow-like, but note Bobolink's rich buffy coloration, larger size, long wings, and bright pink legs. Dark crown shows a buff central stripe and a short postocular stripe. LeConte's Sparrow is much smaller, streaked on breast, white on belly, brighter orange-buff on face, and short winged.
Voice CALL & FLIGHT NOTE: A loud and sharp *pink*. **SONG:** A euphoric, complex bubbling and gurgling, often given in flight, gives the bird its name, *bob-o-link bob-o-link blink blank blink*.
Status & Distribution Locally fairly common. **BREEDING:** Old fields. **MIGRATION:** Arrive in FL in mid-Apr., and in the Northeast early May; after breeding congregates in marshes to molt, then heads south Aug.–late Sept. through Oct., rarely into Dec. In fall, regular in FL (rare farther west), but most appear to take an overwater route to S.A., hence it is very common on Bermuda in fall (exceptionally, after Hurricane Emily on 25 Sept. 1987, 10,000+ were counted there). **WINTER:** In south-central S.A. **RARE STATUS:**

Rare, primarily in fall on West Coast. Casual to NU, the Azores, Europe, and the Galápagos Is.; accidental AK.
Population Declining.

buffy nape

black face and underparts

buffy edges to feathers when fresh

early spring ♂

breeding ♂

dark pink bill

strong head pattern

rich yellow-buff overall

strong back streaks

fall

white rump

breeding ♀

very long primary projection

spiky tail tips

MEADOWLARKS Genus *Sturnella*

The meadowlarks are open-country blackbirds, with a distinctive rotund and short-tailed starling-like shape. They fly with an odd flight style, using shallow fluttery wingbeats. The two N.A. species are yellow below, with a black V on the breast; most of the seven species in this genus are red breasted.

WESTERN MEADOWLARK   *Sturnella neglecta*   WEME   ■   1

The song of the Western Meadowlark is emblematic of the West. Polytypic (2 sspp.; both in N.A.). L 9.5" (24 cm) **Identification** Rotund, stocky, medium-size blackbird with a long bill, short tail, strong legs, and pointed tail feathers. **BREEDING ADULT:** Yellow below with a black V on breast. Crown brown with white median crown stripe, dark postocular stripe, yellow supraloral area. Yellow on throat invades malar area. Gray-buff flanks streaked brown, vent and undertail coverts whitish. Fresh birds appear scaly due to complete pale fringing of feathers, but worn individuals look pale streaked as pale tip wears. Coverts pale brown with thin dark bars separated by the feather shaft. Similarly, central tail feathers are pale brown with discrete narrow dark brown bars. White on outer three tail feathers, a small brown strip remains on outer corner of outer two rectrices, but next one in (R4) largely dark with only a white wedge on inner vane. Bill gray with darker culmen and tip, legs dull pink, eyes dark. **WINTER ADULT:** Pale tips cloud black V on breast. Slightly more buffy yellow underparts; more scaly looking upperparts. **JUVENILE:** Similar to winter adult, but duller face pattern, paler yellow below, and breast streaked in a V, not solid.

Geographic Variation Subspecies *confluenta* of the Pacific Northwest is darker than the nominate, and it shows dark bars on tail feathers and coverts that widen at center of each feather and join up with adjacent dark bars, like the Eastern Meadowlark.

Similar Species Eastern Meadowlark is extremely similar. The southwestern subspecies of Eastern Meadowlark (known as "Lilian's") is even more similar to the Western than the more widespread eastern subspecies in its overall pale plumage. They can be separated by head and throat coloration, tail pattern, and voice. Some vocalizations are diagnostic, such as the blackbird-like call of Western. The two-parted song is longer and lower in frequency and lacks the ascending whistles of Eastern (including "Lilian's") song; however, the song is learned, and in rare cases the meadowlarks can learn each other's songs—this is not the case for the call. Western shows more yellow on throat; it extends to malar area and can be discerned best with a scope view. Western is generally paler than Eastern, but similar to "Lilian's," showing a pale gray-brown overall color, rather than the warmer, more saturated brown of Eastern. Western shows pale gray-buff flanks, like "Lilian's," and Eastern has darker, mid-tone buff flanks with stron-

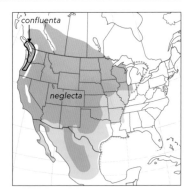

ger and more continuous streaks. Wing coverts are grayish brown with narrow dark bars on Western, while on Eastern they are warm brown to cinnamon brown, with wider dark bars. Western shows largely white outer two tail feathers, while Eastern shows largely white outer three tail feathers. Western prefers shorter grass than Eastern, even bare ground, and has shorter legs.

Voice CALL: A low *chupp* or *chuck*. Females give a dry rattle, males a slower rolling note. **FLIGHT NOTE:** A sweet whistled *weeet*. **SONG:** Males have melodious and flute-like song lasting approximately 1.5 secs. Two phrases, starting with several clear whistles, and a terminal phrase that is more gurgled, bubbling, and complex: *tuuu-weet-tooo-twleedlooo.*

Status & Distribution Common. **BREEDING:** Dry grasslands, agricultural areas. **MIGRATION:** Diurnal migrant; northern populations highly migratory, southern ones more resident. Spring arrival usually Mar.–Apr., fall movements late Sept.–Oct. **WINTER:** A variety of open habitats. **RARE STATUS:** Casual in AK, NT, and Hudson and James Bays. Rare in eastern Midwest. Casual to East Coast from QC and NS to GA. Accidental Russian Far East. **Population** Declining.

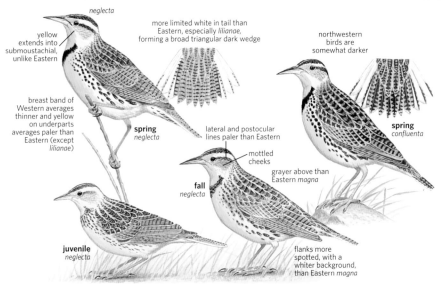

yellow extends into submoustachial, unlike Eastern

breast band of Western averages thinner and yellow on underparts averages paler than Eastern (except *lilianae*)

neglecta

more limited white in tail than Eastern, especially *lilianae*, forming a broad triangular dark wedge

spring *neglecta*

lateral and postocular lines paler than Eastern

mottled cheeks

fall *neglecta*

grayer above than Eastern *magna*

northwestern birds are somewhat darker

spring *confluenta*

juvenile *neglecta*

flanks more spotted, with a whiter background, than Eastern *magna*

EASTERN MEADOWLARK *Sturnella magna* EAME ■ 1

The whistled song betrays the presence of this grassland species. Polytypic (16 sspp.; 4 in N.A.). L 9.5" (24 cm)
Identification Longer legs. **SUMMER ADULT:** Nominate *magna* described. Intense yellow below with thick black V on breast. Warm buff flanks crisply streaked blackish. Upperparts more blackish than Western. Coverts and central tail feathers warm brown with dark bars that widen and meet adjacent dark bars at the feather shaft. Outer three tail feathers largely or entirely white. **WINTER ADULT:** Brighter when fresh, more rufous above, flanks more richly buff. **JUVENILE:** Paler yellow below and breast V streaked.
Geographic Variation Sixteen subspecies recognized, four in N.A. Nominate *magna* described above; southeastern *argutula* smaller and averages darker, especially those from FL. The most distinct is *lilianae*, "Lilian's." Found in the Southwest, it is smaller, has longer wings and legs, and is paler than other N.A. Easterns, more like Western Meadowlark, and has separate narrow bars on tail and greater coverts. It has extensive white on tail, with outer three rectrices entirely white, and next mostly white; vocalizes like Eastern. South TX *hoopesi* is intermediate. The subspecies *hippoprepis* from Cuba has very different vocalizations, more like Western (song), and likely

is a separate species.
Similar Species Western Meadowlark is very similar; see that account. Eastern's call is diagnostic; the higher-pitched chatter is unlike the lower-pitched rattle of a Western. Eastern lacks yellow on the malar, and most subspecies are both darker and richly colored. Eastern shows largely white outer three tail feathers; white more extensive on "Lilian's" Meadowlark. "Lilian's" shows the pale plumage and discrete, separate barring as in Western, but it lacks streaking on the pale cheek, thus showing a great deal of contrast with the dark eye line and crown, and whitish supercilium and cheeks.
Voice CALL: A buzzy *dzert*; also a chatter given by both sexes, higher pitched than rattle of Western Meadowlark. **FLIGHT NOTE:** A sweet whistled *weeet*. **SONG:** Three to five or more loud, sliding, descending whistles, *tsweee-tsweee-tsweeeooo*, lasting over a second.
Status & Distribution Fairly common. **BREEDING:** Grasslands and old field habitats; where sympatric with West-

ern, takes moister, taller grassland and shrubby edge habitats. "Lilian's" in desert grassland. **MIGRATION:** Diurnal migrant; northern birds (*magna*) migratory, southern ones (most *argutula* and *hoopesi* and *lilianae*) resident. Spring arrival usually Mar.–Apr., fall Oct. **WINTER:** Farmland, grasslands, and rangelands. **RARE STATUS:** Casual to Atlantic Canada, ND, CO, southwestern AZ, MB, and Bermuda. Accidental to WA and CA.
Population Near Threatened. Significant declines from the 1960s to the 1990s due to habitat loss.

ORIOLES Genus *Icterus*

The most colorful icterids, largely orange, yellow, or chestnut, orioles are generally slim and long tailed with sharply pointed bills. They weave a characteristic nest that looks like a hanging basket. Plumage patterns are plastic, and species with very similar patterns (e.g., Hooded and Altamira, Baltimore and Orchard) are quite distantly related. In addition, sexual dichromatism is heavily influenced by the role females have in territorial defense and migratory tendency. More migratory orioles show a greater degree of difference in plumage, while in tropical, resident species the female may be as brightly plumaged as the male. Immature male plumage is retained for at least a year.

BLACK-VENTED ORIOLE *Icterus wagleri*   BVOR ■ 4

The Black-vented Oriole is a casual visitor from Mexico. Polytypic (2 sspp.). L 8.7" (22 cm)
Identification Slim and long-tailed with narrow, long bill. **ADULT:** Black above with black hood. Bright orange-yellow below with narrow persimmon band below black breast. Black wings with bright orange-yellow shoulders; tail entirely black. **IMMATURE:** Olive

above and yellow-orange below with greenish shoulder patch. Black on face and breast variable from black chin and lores and scattered black feathers elsewhere to more extensive black feathering.
Geographic Variation Northwestern *castaneopectus* from northwest Mexico larger with narrower chestnut breast band than more easterly *wagleri*. TX records likely *wagleri*; AZ record perhaps *castaneopectus* on probability.
Similar Species Absence of wing bars, white edges to wing, and black vent separate it from all adult N.A. orioles. Peachy-yellow under-parts are distinct from other N.A. orioles. Immatures may be confused with Hooded Oriole; however, Black-vented

lacks wing bars, has a messy black bib (when present), and is streaked on back.
Voice CALL: A nasal *nyeh*; also a mechanical chatter. **SONG:** A series of nasal notes and squeaky whistles.
Status & Distribution Largely resident in semiarid areas from east-central Sonora and southern Nuevo Leon, Mexico, south to north-central Nicaragua. **RARE STATUS:** Casual to West (two records) and south TX (five records) north to Kleberg Co.; accidental near Patagonia, AZ (18 Apr. 1991).
Population Stable.

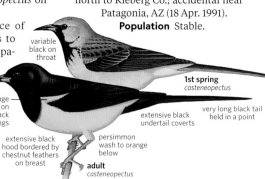

variable black on throat

1st spring
casteneopectus

very long black tail held in a point

orange bar on solid black wings

extensive black undertail coverts

extensive black hood bordered by chestnut feathers on breast

persimmon wash to orange below

adult
casteneopectus

ORCHARD ORIOLE *Icterus spurius*   OROR ■ 1

This smallest oriole is common in the East and Midwest. Polytypic (2 sspp.; both in N.A.). L 7.2" (18 cm)
Identification Small size. Bill slightly decurved. **ADULT MALE:** Overall black and chestnut with long black tail. **ADULT FEMALE:** Olive above; bright yellow below. Two crisp white wing bars and white edging to flight feathers. **IMMATURE MALE:** Similar to female, but by midwinter shows a neat

black bib and lores, and by spring often with some chestnut spotting on face or especially on breast.
Geographic Variation Nominate is breeding subspecies in N.A. In northeastern Mexico, the endemic breeder *fuertesi* (winters from Balsas Basin south on Pacific slope to Chiapas) has occurred twice in spring and summer in Cameron Co., TX. In adult males, ochre replaces standard chestnut coloration.

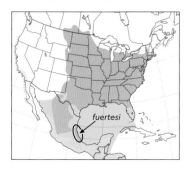

fuertesi

Female not separable from *spurius*. "Fuertes's Oriole" or "Ochre Oriole" is sometimes recognized as a full species.
Similar Species Immature and female Orchards yellow below, not orange or orange-yellow as in Baltimore. Female and immature Hoodeds similar to Orchard, although larger Hooded has longer, more graduated tail, with more decurved bill. Beware of short-billed juvenile Hooded, which are routinely

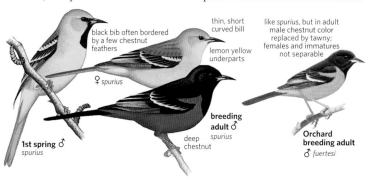

black bib often bordered by a few chestnut feathers

♀ *spurius*

thin, short curved bill

lemon yellow underparts

like *spurius*, but in adult male chestnut color replaced by tawny; females and immatures not separable

1st spring ♂
spurius

deep chestnut

breeding adult ♂
spurius

Orchard breeding adult
♂ *fuertesi*

seen in late summer and early fall and cause frequent confusion with Orchard. Female and immature eastern Hooded subspecies (*cuccullatus* and *sennetti*) are more orange than an Orchard; western Hooded similarly colored. Orchard's *chuck* call deeper and huskier than a similar call rarely given by young Hoodeds; *wheet* call of Hooded is not given by Orchard.

Voice CALL: A sharp *chuck*, often in a series. **SONG:** A musical, springy, and rapid warbled song interspersed with raspy notes.

Status & Distribution Fairly common. **BREEDING:** Open woodlands, urban parks, and riparian woodlands particularly in the west of range. **MIGRATION:** Trans-Gulf migrant in spring with arrival about Apr. 1, moves

south as early as mid-July, but most head south in Aug. Molts after winter grounds are reached. **WINTER:** From west Mexico to northern S.A.; rare to casual in N.A. **RARE STATUS:** Rare west to CA, AZ, and north to Maritimes. Very rare OR and WA. Casual to BC, NL, and Bermuda; accidental to southeastern AK.

Population Stable.

HOODED ORIOLE *Icterus cucullatus* HOOR ▪ 1

This slim oriole prefers palms. Polytypic (5 sspp.; 3 in N.A.). L 8" (20 cm) **Identification** Long, graduated tail. Thin, noticeably decurved bill. **ADULT MALE:** Black bib, face, and back contrasting with orange or yellow-orange head and underparts. Black wings, with black shoulders, two white wing bars, white fringes on flight feathers. **ADULT FEMALE:** Olive above; yellowish or dull orange below. **IMMATURE MALE:** Like female, but by spring shows a neat black bib and lores. **JUVENILE:** Like female, but bill short.

Geographic Variation Five subspecies in two groups: The *cucullatus* group, includes *sennetti*, from south TX; adult males are more orange, have a shorter bill and more black on forehead. Similar *cucullatus* from along the Rio Grande (Big Bend and a bit downriver) more deeply orange; adult males almost orange-red around bib. Females and immatures of *cucullatus* group are more orange, less yellow, ventrally. The *nelsoni* group (including *nelsoni* found from NM to CA) is more yellow, with a longer, more decurved bill; males have a more restricted bib with black not extending across forehead.

Similar Species Male is similar in pattern to Altamira, but slimmer, with a more slender bill, a black shoulder, and a white upperwing bar. Also see Orchard.

Voice CALL: A whistled *wheet* and a short chatter. Also a *chut*, from juveniles, like Orchard. **SONG:** A quick and abrupt series of springy, nasal, or whiny notes, lacking sweet whistled sounds of other orioles.

Status & Distribution BREEDING: Open areas with scattered trees, riparian areas, and suburban and park settings. **MIGRATION:** Arrives by late Feb., departs mostly in Aug.; immatures linger well into Sept. **WINTER:** Mainly in Mexico; very rare coastal CA. **RARE STATUS:** Casual to BC and WA, including in winter; annual in OR; casual to southern YT, southeastern AK, and eastern N.A.

Population Declines in southern TX likely due to the increasing population of Bronze Cowbirds with resulting nest parasitism.

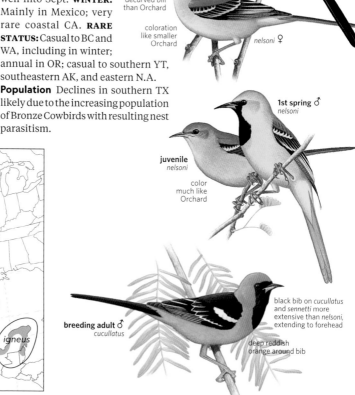

breeding adult ♂
sennetti

orange tint

♀ *sennetti*

black bib

scaly back

winter adult ♂
nelsoni

breeding adult ♂
nelsoni

longer, more decurved bill than Orchard

coloration like smaller Orchard

nelsoni ♀

1st spring ♂
nelsoni

juvenile
nelsoni

color much like Orchard

breeding adult ♂
cucullatus

black bib on *cucullatus* and *sennetti* more extensive than *nelsoni*, extending to forehead

deep reddish orange around bib

nelsoni

cucullatus

sennetti

trochiloides

cucullatus

igneus

STREAK-BACKED ORIOLE *Icterus pustulatus*   SBAO ▪ 4

This Mexican oriole has a stippled, streaked back. Polytypic (10 sspp.; nearly all *microstictus* with stippled back in N.A.). L 8.3" (21 cm)
Identification This oriole has a thick-based, straight, pointed bill. **ADULT MALE:** Head is reddish orange, otherwise largely bright orange. Restricted black is found on lores and narrow bib; entirely black tail. Back is stippled with blackish streaks, and black wings are densely edged with white, forming a nearly solid white panel on closed secondaries and primaries. There are bold white wing bars. **FEMALE & IMMATURE:**

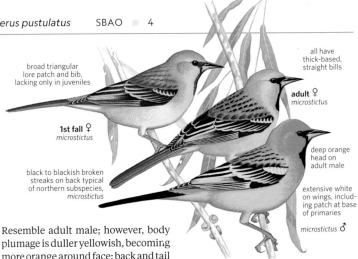

broad triangular lore patch and bib, lacking only in juveniles

1st fall ♀
microstictus

black to blackish broken streaks on back typical of northern subspecies, *microstictus*

all have thick-based, straight bills

adult ♀
microstictus

deep orange head on adult male

extensive white on wings, including patch at base of primaries

microstictus ♂

Resemble adult male; however, body plumage is duller yellowish, becoming more orange around face; back and tail are greenish. Wings have less extensive white edging than on adult male.
Similar Species Male's streaked back, extensive white on wings, and reddish orange head together are diagnostic. Female Streak-backed is similar to an immature male Bullock's, but note thick bill base and stippled back; also Bullock's shows a largely blue-gray bill with black culmen, while entire upper mandible of Streak-backed is black.
Voice CALL: A low *wrank*, also a sweet *chuwit* like a House Finch, a dry chat-

ter. **SONG:** Melodious whistled song similar to that of a Bullock's Oriole.
Status & Distribution Resident from northern Mexico to Costa Rica, on Pacific slope in open woodlands and forest edge. **YEAR-ROUND:** Open woodlands, forest edge. **RARE STATUS:** Casual mainly in fall and winter to southern AZ (has bred) and Southern CA; accidental to OR, UT, CO, NM, eastern TX, and WI.
Population Stable.

BULLOCK'S ORIOLE *Icterus bullockii*   BUOR ▪ 1

The widespread and common oriole of the West. Usually regarded as monotypic. L 8.7" (22 cm)
Identification Relatively short tail; straight, sharply pointed largely blue-gray bill. **ADULT MALE:** Black eye line, crown, nape, and back. Bright orange supercilium. Bright orange on underparts and rump. Very narrow black bib. Wings black with extensive white wing

patch on coverts. Black tail with an orange base to outer rectrices. **FEMALE:** Orange to orange-yellow on head and breast; shows hint of male's face pattern, with darker eye line and brighter yellowish supercilium. Back gray; pale whitish gray belly. **IMMATURE MALE:** Like female, but by spring shows black lores and bib as well as brighter orange breast. **JUVENILE:** Duller than female, bill pink or orange-pink at base. Wings duller, brownish black with buffier and less well-developed wing bars; throat and breast pale yellow.
Similar Species The adult male is

1st spring ♂
black lores and chin
gray rump

immature ♀

large white wing patch

orange supercilium and black eye line

pale yellow head and grayish back
dusky eye line

yellowish throat and breast

♀

breeding adult ♂

extensive white belly

dark "teeth" extend into white wing bar

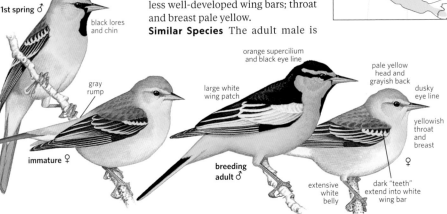

unmistakable. Female and immature easily confused with a dull Baltimore. Female and immature Hooded are entirely yellow below, slimmer, and have a longer and more graduated tail and also a thin, decurved bill. **Voice CALL:** A rapid series of *cha* notes on one pitch, given by both sexes; also a sweet but faint *kleek*, or *pheew*. **SONG:** A musical, lively series of whistles ending in a sweeter note: *kip, kit-tick, kit-tick, whew, wheet*. From Baltimore, songs are shorter and less variable. **Status & Distribution** Common. **BREEDING:** Mainly open woodlands and riparian areas, especially fond of cottonwoods. **MIGRATION:** Spring arrival in south Mar.–Apr., crosses into Canada by early to mid-May. Adult males southbound beginning early July, females and immatures late July–Aug., a few into Sept. **WINTER:** Most retreat to Mexico; rare in coastal Southern CA; casual northern CA and Southwest. **RARE STATUS:** Casual to the East, particularly in fall and winter, but overreported (many are dull Baltimores). Casual to AK, including from the Bering Sea islands in fall. **Population** Gradual declines.

SPOT-BREASTED ORIOLE *Icterus pectoralis*   SBOR ■ 2

The Spot-breasted Oriole from C.A. was introduced to the Miami, FL, area. Polytypic (4 sspp.; nominate in N.A.). L 9.5" (24 cm)
Identification A large oriole with a thick-based bill with a decurved culmen. **ADULT:** Bright orange body with a black back. A small face mask and narrow bib are black; black spots immediately below bib are distinctive. Black wings show an orange shoulder, a white spot at base of folded primaries, and a white wedge on folded tertials. Tail solid black. **IMMATURE:** Duller orange than adult, with olive-green back and tail, duller pattern on wings; breast spots often absent. **JUVENILE:** Duller than immature and lacking black lores and bib; bill may show pinkish base to lower mandible. **Similar Species** Spotted breast and white wedge on folded tertials are diagnostic; note bill shape. **Voice CALL:** A nasal *nyeh*, also a sharp *whip* and a short chatter. **SONG:** A lengthy, repetitive yet pleasing set of warbled whistles, some of which are delivered slowly. One song style more repetitive, another more variable. Female's song less complex than male. **Status & Distribution** In open woodlands from southwest Mexico to northwest Costa Rica. Introduced to south FL, where uncommon and local. **YEAR-ROUND:** Parks and urban habitats in FL; open shrubby woodlands in native range. **Population** Stable in native range. First found nesting in FL in 1949. The population has oscillated since the introduction but is now on a decline.

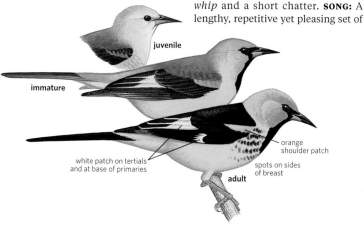

juvenile

immature

white patch on tertials and at base of primaries

orange shoulder patch

spots on sides of breast

adult

ALTAMIRA ORIOLE *Icterus gularis*   ALOR ■ 2

This Lower Rio Grande specialty builds a long hanging nest. Polytypic (3 sspp.; *mentalis* in N.A.). L 10" (25 cm)
Identification Largest and stockiest oriole in N.A. with very thick bill. **ADULT:** Bright orange body, with deeper orange face and bold black mask and narrow bib. Back black, contrasting with orange lower back, rump, and uppertail coverts. Wing black, with an orange shoulder, a well-marked white lower wing bar. (On primaries, most of white on primaries restricted to a bold white patch at their base.) Tail entirely black on male; a variable amount of olive on female tail. **IMMATURE:** Like adult, but duller orange and back and tail olive. Wings

olive back

immature *mentalis*

black lores

juvenile *mentalis*

very thick-based bill

orange shoulder bar

thin white wing bars

deep orange on head and underparts

adult *mentalis*

white primary patch

blackish, lacks orange shoulder, white edging greatly restricted. **JUVENILE:** Duller still, with buffy wing bars, no black on lores or bib.

Geographic Variation Subspecies vary in minor ways in size and saturation of color. Only *tamaulipensis* in TX.

Similar Species Hooded is similar to an adult Altamira in pattern, but Altamira is much thicker billed. Altamira also shows an orange shoulder patch. Immatures are also separable by size and bill, but note thinner bib on Altamira.

Voice CALL: Contact call is a nasal *ike*, or *yehnk*, often repeated. **SONG:** A series of loud musical whistles, often interspersed with harsher notes. Songs are repeated several times before switching to a different song type. Song is easily imitated by a human whistler.

Status & Distribution Resident south to northwest Nicaragua. Uncommon in riparian woodland in south TX.

Population Threatened in TX.

AUDUBON'S ORIOLE *Icterus graduacauda* AUOR ■ 2

The secretive Audubon's Oriole often stays low in the understory. Polytypic (4 sspp.; *audubonii* in TX). L 9.5" (24 cm)

Identification A large oriole with a moderately thick, straight bill. Basal half of lower mandible is blue-gray. **ADULT:** Black hood contrasts with greenish yellow back and lemon yellow underparts. Black wings have a yellow shoulder, a white lower wing bar, and white fringes on tertials and on secondaries, but fringes do not reach to base of secondaries, creating a dark bar there. Tail is black. Female shows a duller, more greenish back. **JUVENILE:** Lacks black hood of adult and is greener above; wings dull blackish, and tail greenish.

Similar Species Only oriole with an isolated black hood. Our other hooded orioles have black backs. A juvenile may be confused with a juvenile Altamira Oriole, but the Audubon's is yellowish, not orange, and shows a duller, grayish head color. A juvenile Hooded Oriole is also similar, but smaller, much slimmer, and slimmer billed, with clear buffy white to white wing bars.

Voice CALL: A nasal *nyyyee*; wrenlike calls and a high-frequency buzz. **SONG:** A long song of melancholy, tentative, and slow whistles. The whistles sound flat and in the same pitch. Both sexes sing.

Status & Distribution Resident from south TX to central Veracruz with separate and distinctive populations found in west (*nayaritensis*) and southwest (*dickeyae*) Mexico. Uncommon in TX in woodlands with dense understory.

RARE STATUS: Casual north of mapped range in TX to Bastrop Co., Guadalupe Co., and Midland Co. Accidental southern IN.

Population Range extending north in TX, now to southern Edwards Plateau.

juvenile audubonii
olive-yellow back
adult ♂ audubonii
black head
yellow

BALTIMORE ORIOLE *Icterus galbula* BAOR ■ 1

The common oriole throughout much of the East. Monotypic. L 8.7" (22 cm)

Identification Relatively short tail and straight, sharply pointed bill. **ADULT MALE:** Orange, with black hood and back. Orange shoulder with white lower wing bar. Tail black with black-based orange outer rectrices. **FEMALE:** Variable. Some like male, but usually lack solid black head and have greenish orange tail. Typical female orange below, with brownish orange face and back, spotted or blotched dark on back and crown. Wings blackish with two white wing bars (upper one wider). **IMMATURE MALE:** Variable. Like female, but no black on face, more extensively orange below, bill often with pinkish or orange tone. **IMMA-** **TURE FEMALE:** Much duller, often with a paler, grayish white belly. **JUVENILE:** Duller than immature, olive back. Pink tone to bill.

Similar Species Dull immature female Baltimores are very similar to Bullock's Oriole. These Baltimores have more open faces with just a trace of a postocular line, thus the dark eye stands out in a blank face. The wing bars are bolder and lack the "teeth" of Bullock's; this is most apparent on the tips of the median coverts. Below, these dull Baltimores show more of blend from breast to belly, while the contrast is sharper in Bullock's. Bullock's and Baltimore do rather frequently hybridize in a narrow band on the western Great Plains. A female Orchard Oriole is yellow below and greenish above, slimmer, and smaller; an immature male Orchard has a crisp black bib, often bordered with a few chestnut feathers by spring.

Voice CALL: A whistled *hew-li* and a

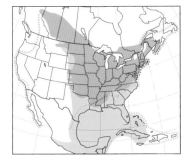

dry chatter, the latter qualitatively different from Bullock's. **SONG:** A series of musical, sweet whistles; quite variable. **Status & Distribution** Common. **BREEDING:** Deciduous forest, forest-edge parkland, riparian forest. **MIGRATION:** Some circum-Gulf, but many are trans-Gulf migrants. Arrive Gulf Coast by early Apr., and Canada by early May. Southbound late July–early Aug., peaking late Aug.–early Sept. in north, and mid-Sept.–mid-Oct. on Gulf Coast. Often a fairly common migrant on Bermuda (some winter there). **WINTER:** From southern Mexico to northern S.A. in moist forest and shade coffee plantations. A small number in the US South and FL, largely in urban settings. **RARE STATUS:** Rare in NL and CA, mostly in fall and winter. Very rare elsewhere in West. Casual to Pacific Northwest and western Europe. **Population** Slight declines.

all with thick wing bars and orangish below

yellow-ochre rump

fall immature females

maximum black spring adult ♀

black hood

fall immature ♂

strongly contrasting wing bars

upper bar lacks "teeth" found on Bullock's

1st spring ♀

breeding adult ♂

BLACK-BACKED ORIOLE *Icterus abeillei*   BBOR ■ 5

The Black-backed Oriole, a Mexican endemic, is accidental in N.A. The origin of all records (all adult males) is controversial; the PA record is accepted. Its plumages closely resemble Bullock's Oriole, with which it was once treated as conspecific; they hybridize in northern Durango. Monotypic. L 7.5" (19 cm) **Identification ADULT MALE:** Extensive black on head and upperparts with orange supraloral stripe and partial broken eye ring. Sides and flanks black. Extensive white on wing. **ADULT FEMALE:** Like female Bullock's (see p. 666). **IMMATURE MALE:** By winter shows blackish lores and throat stripe. **Similar Species** Adult male distinctive.

Female and immature like Bullock's. Note duskier face with faint supraloral and spectacles; sides duskier. **Voice** Vocalizations closely resemble Bullock's as far as known. **Status & Distribution** Native to central Mexico. **BREEDING:** From northern Durango and southern Nuevo León to Michoacán and central Veracruz. Northern populations migratory. **WINTERS:** In Mexico from the Trans-Mexican Volcanic Belt south to Oaxaca. **RARE STATUS:** Accidental N.A. The accepted record was in Reading, PA, 26 Jan.–10 Apr. 2017. However, the same individual in Sutton, MA, was

short orange supercilium

black back

broad black extends down sides and flanks

large white wing patch

winter adult ♂

not accepted by the state committee. Likely the same bird was at Stamford, CT, 14 May 2017. One from San Diego Co. from Apr. 2000 to Jan. 2002 was not accepted by the state committee. **Population** Stable.

SCOTT'S ORIOLE *Icterus parisorum*   SCOR ■ 1

Scott's Oriole is N.A.'s only desert specialist oriole. Monotypic. L 9" (23 cm) **Identification** Note long, straight, sharply pointed bill. **ADULT MALE:** Black head, back, and breast, sharply separated from lemon yellow underparts. Wings black with yellow shoulder. Tail black with yellow bases to outer rectrices. **ADULT FEMALE:** Olive above, with obscure streaking on back; few with black on throat. Below dull olive-yellow. **IMMATURE**

adult female and 1st-spring male can look quite similar

1st spring ♂

adult ♂

long, straight bill

stippled dusky black above

yellow shoulder

immature ♀

olive-yellow below, duller than female Hooded

bright lemon yellow underparts

MALE: Similar to female, and some not separable, but back more densely marked and underparts brighter yellow. Typically a black bib and face and variable black spotting on crown and sides of head. Immatures of both sexes hold plumage into winter and show no traces of black on throat or face.

Similar Species Audubon's and Scott's Orioles are the only yellow orioles in N.A. Its black back, two white wing bars, and yellow on tail identify Scott's. A female Scott's is much more olive and generally muddier and duller colored than other orioles.

Voice CALL: A harsh *chuck* or *shack* and a scolding *cheh-cheh*. **SONG:** A fluty warbled set of whistles; the low pitch and richness of notes resemble the song of a Western Meadowlark at times. Females sing a softer and weaker song than the males.

Status & Distribution Common. **BREEDING:** Primarily desert, particularly at interface of low desert to higher, more wooded elevations. Often in yucca or agave as well as juniper. **MIGRATION:** In spring arrive late Mar.–early Apr., leave in fall late July–mid-Sept. **WINTER:** Retreats mainly to Mexico; rare Southern CA. **RARE STATUS:** Very rare to CA coast, primarily in fall and winter. Casual to OR, northern Great Basin, MN, and LA. Accidental elsewhere in eastern N.A. **Population** Stable.

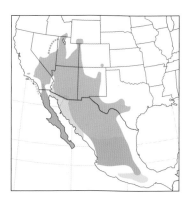

BLACKBIRDS Genus *Agelaius*

The five species belonging to this genus are restricted to N.A., C.A., and the Caribbean. A group of similar S.A. species were previously included. Largely black plumages; bright red, tawny, or yellow "epaulets" or shoulder patches; and unmusical and screechy songs characterize the males.

RED-WINGED BLACKBIRD *Agelaius phoeniceus*   RWBL   ▪ 1

This species is one of the most widely distributed, abundant, and well-named birds in N.A. Polytypic (22 sspp.; 14 in N.A.). L 8.7" (22 cm)

Identification Medium-size with a sharply pointed black bill. **SUMMER MALE:** Black with bright red shoulder patch or "epaulet," bordered by yellow in most subspecies. **WINTER MALE:** Feathers finely tipped with brown. **FEMALE:** Well streaked throughout, with whitish supercilium. Brown above with rusty feather edges, densely streaked below. Peachy wash on chin. Dull reddish edges to lesser coverts. **IMMATURE MALE:** As winter male, but feathers more broadly edged with brown. Trace of supercilium; red epaulet shows black spotting.

Geographic Variation Most subspecies poorly defined, but "Bicolored" group (*californicus* and *mailliardorum* from CA) and *gubernator* from Central Plateau, Mexico, characterized

"Bicolored Blackbird"

whitish throat

immature ♀

distinct pale supercilium

scapular feathers edged rusty, back feathers edged buff

many with pinkish throat

adult ♀

streaked belly against pale background color

red shoulders most visible when singing

adult ♂

1st year ♂

"Bicolored Blackbird" adult ♂

thick bill

yellowish border to red patch largely or completely absent in Bicolored

"Bicolored Blackbird"

back pattern similar to other subspecies of Red-winged

dark belly similar to Tricolored; also note bill shape and back pattern

adult ♀
californicus

central coastal *mailliardorum* is smaller billed than Central Valley *californicus*; female is darker and has a longer wing

by female with a dark belly; reduced supercilium. Males show black median coverts, so no yellow border.

Similar Species Male Tricolored has a shiny sheen to black plumage, a thinner, pointed bill, a white epaulet border, and colder gray-buff fringes in winter, and different vocalizations.

See sidebar below to separate Tricolored females. **Voice CALL:** *Chuk.* **SONG:** A hoarse, gurgling *konk-la-ree* (variable); female duets with male with a rattle-like song. **Status & Distribution** Abundant. Open or semi-open habitats. **BREEDING:** Cattail marshes, but also in moist

open, shrubby habitats. **MIGRATION:** Southern populations resident; northern ones migratory, arrive in Northeast and Upper Midwest by late Feb. and leave by Nov. **RARE STATUS:** Casual to Arctic AK, across Arctic Canada, and Greenland. Accidental to Europe. **Population** Stable.

Identification of Female Red-winged and Tricolored Blackbirds

Separating female Red-winged and Tricolored Blackbirds in CA, where their ranges overlap, is confusing and poorly understood. Birders tend to ignore the females and concentrate on the males, which is a valid way to identify the species, but really not that useful when one comes across a lone female or a group of females (sexes often forage separately). Key characters include the bill shape and tones and colors on the upperparts in fresh plumage. When feeding, note the cocked tail of Tricolored.

Compared with female Red-winged Blackbirds over much of N.A., which are strongly streaked throughout the underparts, the females of the two CA Red-winged "Bicolored" subspecies are darker and when worn can look largely solidly blackish above and unstreaked and solidly blackish on the belly and vent, with dark streaks on a pale background restricted to the throat and upperbreast. This plumage coloration is generally the same pattern shown by female Tricolored Blackbirds. When the plumage is fresh in fall and winter, the rusty edges, as opposed to gray (female Tricolored), are diagnostic.

Structurally, these species do differ somewhat. Tricolored Blackbirds show thinner-based and longer bills than most CA Red-winged Blackbirds, but note that the subspecies breeding in the Kern River Valley (*aciculatus*), Kern Co., is characterized by its long and slender bill. The Tricolored has a more pointed wing shape, which can be looked for on the perched bird.

In fresh plumage (during fall and early winter), a Red-winged Blackbird is edged with rufous, golden, and warm buffy edges on the upperparts, while these same areas are cold gray-buff on a Tricolored Blackbird. Often on a Red-winged there are two obviously paler, more yellowish lines of streaks on the back, like suspenders. The white tips to the median coverts (upperwing bar) are broader and more noticeable on a Tricolored. Finally, some adult female Red-wingeds show a peachy or pinkish wash to the throat, absent in Tricoloreds.

Worn summer females, which by this point lack the paler feather tips on the upperparts, may be unidentifiable if not studied closely, although a Tricolored usually shows a stronger white bar on the median coverts.

Tricolored, female (CA, May)

Red-winged, female (IL)

Red-winged, *aciculatus*, female (CA)

TRICOLORED BLACKBIRD *Agelaius tricolor*   TRBL   ▪   2

The Tricolored is similar to the Red-winged Blackbird. Feeding behavior with tail cocked is distinctive. Around 99 percent live in CA. Monotypic. L 8.7" (22 cm)

Identification Note long, sharply pointed bill. Seldom alone, breeding in colonies and wintering in medium- to large-size flocks. **SUMMER MALE:** Black with slight pale iridescence and dark

red shoulder patch or epaulet bordered broadly with white. **WINTER MALE:** As summer male, but shows gray-brown feather tips throughout body; epaulet border is creamy white. **FEMALE:** Dark. Streaks on throat and breast contrast with solid dark belly; gray edges form streaks above when fresh, and dull reddish edges to lesser coverts. **IMMATURE MALE:** As winter male,

when fresh, back and scapular feathers edged gray

thin, pointed bill

dark belly

fall/winter ♀

but more heavily edged and fringed. **Similar Species** See Red-winged Blackbird and sidebar (p. 671) for females. **Voice CALL:** *Kuk,* lower pitched than similar Red-winged call. **SONG:** Distinctive; a nasal, drawn-out *guuuaaaak* or *ker-gwuuuuaaaa,* lasts 1–1.5 secs. **Status & Distribution** Uncommon. **YEAR-ROUND:** Open or semi-open habitats. **BREEDING:** Highly colonial, traditionally in large marshes in Central Valley, CA; also fields of rice and triticale, and thickets of Himalayan blackberry. **MIGRATION:** After breeding, many move toward the coast or northward, starting in mid-July. **WINTER:** Closely associated with rangeland, dairy operations, parks, and landfills. **RARE STATUS:** Previously rare to casual to WA, OR, and NV, but has now bred in isolated small colonies in those states. Casual to southeastern CA.

Population Endangered. Current population is estimated at 177,000 birds (370,000 in 1994). Historically, a single colony in Glenn Co., CA, held more than 200,000 birds, and many colonies had more than 100,000. Marsh drainage, the huge increase of sterile nut orchards, and the harvesting of agricultural fields with active colonies imperil the population. In 2018, the species was granted special protection in CA, but the USFWS has yet to take action.

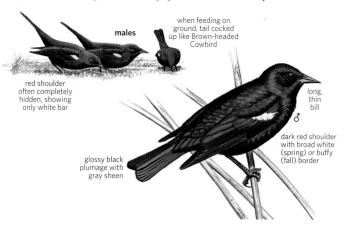

when feeding on ground, tail cocked up like Brown-headed Cowbird

males

red shoulder often completely hidden, showing only white bar

long, thin bill

♂

glossy black plumage with gray sheen

dark red shoulder with broad white (spring) or buffy (fall) border

TAWNY-SHOULDERED BLACKBIRD *Agelaius humeralis* TSBL ▪ 5

The Tawny-shouldered is an accidental stray from the West Indies. Polytypic (2 sspp.; likely *humeralis* from nearby Cuba and northern Haiti). L 8" (20 cm) **Identification** Like a small and slim Red-winged, but plumage entirely black with tawny lesser coverts, median coverts tawny with blended yellowish tips. Bill slim and sharply pointed. Frequently flips tail up. **Similar Species** Red-winged Blackbird is larger and bulkier, with a red epaulet and a wider yellow border. Tawny-shouldered is more arboreal and has a different song.

Voice CALL: A variety of notes, some like other blackbirds. **SONG:** A muffled buzzy drawn-out *zwaaaaaaaa,* lasting just over a second in length is given by both sexes.

Status & Distribution Resident on Cuba and northern Haiti. Arboreal, preferring open woods, edge, and park-like settings, where it forages in noisy groups as they move through trees; also feeds on ground and at feeders. **RARE STATUS:** Accidental Key West Lighthouse, FL (27 Feb. 1936); two were collected from a flock of Red-wingeds. **Population** Nominate subspecies

stable; *scopulus* from Cayo Cantiles, Cuba, considered rare and endangered.

tawny lesser coverts with blended edge

slender, pointed bill

adult ♂

COWBIRDS Genus *Molothrus*

The true cowbirds, of which there are five species, are obligate brood parasites, laying their eggs in the nests of other species. Parental care is performed entirely by the hosts. Unlike many cuckoos, cowbirds are generalists, using various host species. They have stocky bodies and short tails. Males are glossy black; females and juveniles are duller.

SHINY COWBIRD *Molothrus bonariensis* SHCO ▪ 3

In N.A., the Shiny Cowbird is found primarily in FL. It is one of the few brood parasites in N.A. Polytypic (7 sspp.; *minimus* in N.A.). L 7.5" (19 cm) **Identification** Shaped like Brown-headed Cowbird, but longer billed with a longer tail and slimmer body. Legs and bill black, eyes dark. **ADULT**

MALE: Entirely black with violet-blue iridescence on head and anterior part of body. **ADULT FEMALE:** Dull brownish throughout, with paler supercilium and darker wings and tail. **JUVENILE:** Similar to female, but obscurely streaked below.

Similar Species Male Shiny's strong

gloss and violet head separate it from Brown-headed Cowbird. Female is extremely similar to a Brown-headed. Shiny is generally darker, lacking white throat, but with a noticeably paler supercilium, a longer and slimmer body, and a longer black bill. A female Brown-headed shows a paler bill with a horn or yellowish base to lower mandible. On closed wings, the secondaries do not show obvious pale fringes on the Shiny. Finally, Shiny has shorter, more rounded wings; the outermost primary (P9) is equal in length to P7 or P6 but noticeably shorter than P8, the second outermost primary.

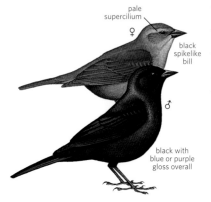

pale supercilium

♀

black spikelike bill

♂

black with blue or purple gloss overall

Voice CALL: A soft *chup*; females give a chatter. **SONG:** Primary song strange sounding, liquid and bubbling, lasting two to three seconds. It begins with several purring bubbly notes and then a screechy series of high-pitched notes, *blurr-glurr-glurr-pt-tcheeeEEE*. Males also give a flight whistle, which may be considered a secondary song rather than a call. It is a more complicated and long series of short whistles, and it shows a great deal of geographic variation. Whistle is given both in flight and while perched.

Status & Distribution Found S.A. and West Indies (occupied during 20th century). Partially migratory. Rare to uncommon in FL. **BREEDING:** Open and edge habitats; agricultural areas for foraging and forest edge habitats for finding host species. **MIGRATION:** Arrivals from south appear Mar.–Apr. **WINTER:** Open areas, often at feeders.

RARE STATUS: Casual to FL and north to ME and NB; also west to AL (many), MS (several), TN (once), OK, LA (several), and TX (12). Most records away from FL have occurred in spring and summer. **Population** First detected in FL in 1985. Many recorded in the late 1980s and early 1990s. Since then, reduced sightings from most regions, and declining numbers in South FL.

BRONZED COWBIRD *Molothrus aeneus* BROC ◼ 1

Larger than the Brown-headed or Shiny Cowbird, the Bronzed Cowbird has a rather sinister appearance due to the male's typically hunchbacked look and blood-red eye. It gives a unique hovering display. Polytypic (4 sspp.; 2 in N.A.). L 8.7" (22 cm)
Identification A thick-set cowbird with a large and deep black bill. Legs black. **ADULT MALE:** Entirely black with bronzed body iridescence, becoming blue-green on wings and tail. Eyes bright red. **ADULT FEMALE:** Varies geographically, see below. More widespread eastern subspecies has a black female plumage, lacking strong gloss, with browner wings and tail.

Eyes red. **JUVENILE:** Similar to female but dark brown.
Geographic Variation The more widespread *aeneus* is found from south-central TX eastward; *loyei* from NM and West TX to southeastern CA. Males are similar, but females blackish in *aeneus* and grayish brown in *loyei*.
Similar Species Bronzed is larger than other cowbirds, and adults show bright red eyes and thick bills. Juveniles could be mistaken for female Brown-headed Cowbirds, but note their larger size, larger bulk, and thick bill.
Voice CALL: A rasping *chuck*. Females give a rattle. **SONG:** A series of odd squeaky gurgles, *gluup-gleeeep-gluup-bloooop*. Flight whistle highly geographically variable, given in flight and while perched. About four seconds long, it is a series of long sliding or vibrating whistles. Three general

both subspecies

loyei

aeneus

dialects in N.A.: AZ (CA to westernmost TX), Big Bend (Big Bend, TX), and south TX (east of Big Bend).
Status & Distribution Found TX and Southwest south to central Panama. In US uncommon to locally common. Southwest *loyei* migratory in US. Open shrubland, forest edge, and agricultural areas (especially those associated with livestock). **BREEDING:** Brood parasite, specializes on sparrows and orioles. **MIGRATION:** Spring movements in TX in Mar., after mid-Apr. in southeast CA. Southbound movements in Sept. Small but increasing numbers winter along Gulf Coast to FL. **WINTER:** In flocks, particularly in agricultural areas. **RARE STATUS:** Casual to coastal Southern CA, southern NV, and CO. Accidental to KS, MO, TN, SC, MD, NY, ME, and NS. **Population** The range and population of this species began a marked expansion in the 1950s. Currently stable. Recently spread to FL and Gulf states.

red eye

ruff often raised, giving thick-necked look

thick bill

highly iridescent plumage

loyei ♂

grayish plumage, unlike female *aeneus*

red eye

loyei ♀

less scaly above with larger bill than juvenile Brown-headed

juvenile *loyei*

blackish plumage

TX ♀ *aeneus*

BROWN-HEADED COWBIRD *Molothrus ater*   BHCO ■ 1

The most widespread brood parasite in N.A. Polytypic (3 sspp.; all in N.A.). L 7.5" (19 cm)

Identification A smallish, compact blackbird with a short and thick-based bill, almost finchlike. Foraging birds are commonly seen on ground with tail cocked. **ADULT MALE:** Glossy black body, with a greenish iridescence, contrasts with a brown head. Eyes are dark; bill and legs are black. **ADULT FEMALE:** Face has a beady-eyed look due to dark eyes; lores are pale; underparts are obscurely streaked. **JUVENILE:** Like female, but is more strongly streaked below, and upperparts are scaly-looking due to pale feather fringes.

Geographic Variation Three subspecies, differing mainly in size and darkness of females, are not field identifiable; western *obscurus* is smallest.

Similar Species Male's glossy black body and brown head diagnostic. Female similar to Shiny Cowbird female. Brown-headed is paler overall, showing a whitish throat and a pale face with a beady-eyed look. Shiny has a more marked dark eye line and paler supercilium, giving it a more striking face pattern. Compared to a Shiny, Brown-headed is more compact, with a shorter tail and bill. Brown-headed's dark bill shows a pale or horn base to lower mandible; on Shiny, bill is shiny black. On closed wings, secondaries show obvious pale fringes on Brown-headed. Streaked juvenile is easily confused with more strongly and darkly streaked female Red-winged Blackbird; note bill shape differences.

Voice CALL: A soft *kek*. Females give a distinctive dry chatter, while males may give a single modulated whistle, particularly just after taking off. **SONG:** Primary song is a series of liquid, purring, gurgles followed by a high whistle. Song of Brown-headed has the highest frequency range of any species in N.A. Males also give a flight whistle, which is a geographically variable series of two to five whistles, often fre-

quency modulated. Flight whistle is given both in flight and while perched. Primary songs appear to be hardwired, while flight whistles are learned; this accounts for why dialects are found in the latter but not in the former.

Status & Distribution Common. **BREEDING:** Open and edge habitats. **MIGRATION:** Northbound mid-Mar.–mid-Apr., southbound late July–Oct.

Population Increased its range and population greatly during the 1800s, when eastern forests were cleared. Brown-headed was restricted to Great Plains, where the buffalo herds were, and spread east and west from there. More recently, numbers have declined.

streaked below and scaly above, can be confused with female Red-winged Blackbird

Yellow Warbler

juvenile

parasitic female Brown-headed Cowbirds lay eggs in nests of other species, so young raised by other parents

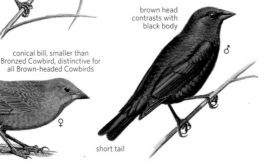

immature ♂ in molt

young males seen in this transitional plumage in late summer and early fall can be confusing

conical bill, smaller than Bronzed Cowbird, distinctive for all Brown-headed Cowbirds

Brown-headed Cowbirds often feed with tail cocked up

♀

short tail

brown head contrasts with black body

♂

BLACKBIRDS Genus *Euphagus*

The two species in *Euphagus* are closely related to the grackles and the Neotropical genus *Dives*. They resemble grackles in having ruff-out displays that accompany the song: the tail is cocked, the wings are drooped, the body is ruffled, and often the pale eyes are prominent. Unlike the grackles, *Euphagus* have standard-shaped tails.

RUSTY BLACKBIRD *Euphagus carolinus*   RUBL ■ 1

The Rusty is a blackbird of swampy forests, adorned with rusty in fall and winter. Polytypic (2 sspp.; both in N.A.). L 9" (23 cm)

Identification Slim with a slender, pointed bill. **SUMMER MALE:** Black. Black bill and legs; bright yellow eyes. **WINTER MALE:** Black with cinnamon, buff, or warm edges. In fall and early winter edging broad, breast and back appear largely rusty. Buff supercilium and malar. Coverts and tertials tipped rusty; rump blackish. As winter progresses, edges wear and plumage blackens. **SUMMER FEMALE:** Blackish gray with darker wings, tail, and lateral throat stripes. Bill black, legs black, and eyes yellow. **WINTER**

FEMALE: Widely edged rusty and buff. Similar to winter male, but paler and even more rusty, and rump grayish. Dark triangular area on face and lores sets off prominent supercilium.

Geographic Variation Subspecies *nigrans*, which breeds in the Maritimes and Newfoundland, is not field separable from widespread nominate.

Similar Species Brewer's is similar but has a shorter bill. In winter, wide rusty edging on Rusty's plumage is distinctive, although immature male Brewer's can show some dull buff on breast and upperparts. Brewer's does not show rusty edges to tertials. Male Rusty in summer like Brewer's but is less glossy; separated by bill shape and habitat. Summer female Rusty more grayish than Brewer's, and shows yellow eye. A few Brewer's females also show pale eyes but are browner than Rusty.

Voice CALL: A *chuck*, not as deep as that of a grackle. **SONG:** A squeaky, sweet, rising *kush-a-lee* or *chuck-la-weeeee*. A secondary song begins with two or three musical notes followed by a harsher long note. Females sing a weaker version.

Status & Distribution Uncommon to locally fairly common. **BREEDING:** Bogs in boreal forest and tundra regions. **MIGRATION:** Diurnal migrant; arrives southern Canada by late Mar., southbound late Sept.–Nov. **WINTER:** Wet open woodlands, or fields near wetlands. **RARE STATUS:** Very rare to casual migrant to much of West including Pacific region from BC to CA (a few winter), also casual to Pribilofs, St. Lawrence I., Russian Far East, and Bermuda. In West, often associates with Brewer's, particularly in winter.

Population Vulnerable. Numbers have declined precipitously since 1960, with some sources estimating a 90 percent drop between the 1960s and 1990s.

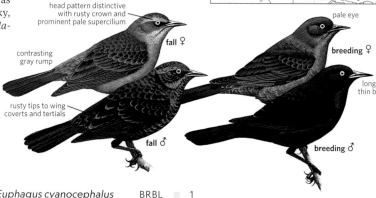

head pattern distinctive with rusty crown and prominent pale supercilium

contrasting gray rump

fall ♀

rusty tips to wing coverts and tertials

fall ♂

pale eye

breeding ♀

long, thin bill

breeding ♂

BREWER'S BLACKBIRD *Euphagus cyanocephalus* BRBL ■ 1

A common and widespread ground-dwelling icterid of the West, the Brewer's in some ways replaces the Common Grackle ecologically. Monotypic. L 9" (23 cm)

Identification Slim but slightly stouter build than Rusty. **MALE:** Basically black with bright yellow eyes. Strongly iridescent, with a bright blue or purplish blue sheen on head, while body is greenish. **FEMALE:** Dull brownish gray and unstreaked; dark eyes. **IMMATURE MALE:** Many young males show some buffy feather tips on breast and warmer buff tipping on back.

Similar Species A summer Rusty Blackbird is similar but has a more slender pointed bill. Male Brewer's is more strongly glossy, showing blue on head and green on body. In fall and winter some young male Brewer's show buff tipping on breast, head, and upperparts, but tertials are nearly always entirely black. Female Rusty in fall is rusty overall with a contrasting dark face, a prominent pale supercilium, and a gray rump. Female Brewer's typically show a dark eye, although a few show pale eyes. Common Grackle has a long, graduated tail often held in a deeply keeled shape. Common Grackles show more complex and brighter iridescence patterns than Brewer's, and in versicolor shows a break in color from head iridescence to that of body.

Voice CALL: A *chak*, or *chuk*, similar to that of Rusty Blackbird. When alarmed, it gives a whistled *teeeuuuu* or *sweeee*. **SONG:** Often gives a faint and unappealing raspy *schlee* or *schrrup* during the ruff-out display; both sexes sing and display.

Status & Distribution Common. **YEAR-ROUND:** Varied habitats, including urban areas, golf courses, agricultural lands, open shrubby areas, forest clear-cuts, and riparian forest edges. Requires open ground for foraging and some dense vegetation or edge for nesting. Farther east, where sympatric with Common Grackle, it takes more open sites than the grackle. **MIGRATION:** Poorly understood, easternmost populations more migratory. Wintering groups somewhat nomadic. **RARE STATUS:** Rare in Midwest and at eastern edge of breeding range. Casual to Northeast, AK, and NT.

Population Declining in parts of range (e.g., CA).

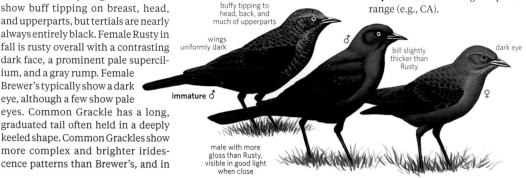

buffy tipping to head, back, and much of upperparts

wings uniformly dark

immature ♂

♂

bill slightly thicker than Rusty

dark eye

♀

male with more gloss than Rusty, visible in good light when close

GRACKLES Genus *Quiscalus*

There are seven grackle species worldwide, one of which is now extinct—the Slender-billed Grackle (*Q. palustris*) of central Mexico was last recorded 1910. The males have a glossy black plumage and often have yellowish eyes. During breeding displays, the strongly graduated tail is held deeply keeled, giving a V-shaped cross section.

BOAT-TAILED GRACKLE *Quiscalus major*   BTGR   ▮   1

Inhabits coastal marshes of the Atlantic and Gulf Coasts. Polytypic (4 sspp.; all in N.A.). L 15–16.5" (38–42 cm)
Identification A large grackle with a moderate-size bill, a rounded crown. **ADULT MALE:** Black with obvious blue iridescence, becoming violet on head. Eyes yellow to dull yellowish or even honey brown; bill and legs black. Long, deeply keeled tail. **ADULT FEMALE:** Smaller than male. Brown above; warm tawny on head and below,

darker on belly and vent. Eyes usually dark. Does not hold tail in deep keel.
Geographic Variation The four subspecies vary primarily in eye color, some having yellow eyes, others brown. Otherwise, they are not field-identifiable other than by range.
Similar Species Very similar to Great-tailed Grackle. Boat-tailed shows a steeper forehead, rounder crown, and smaller bill; this is more obvious on males than on females. The two east-

ern subspecies show dark eyes. Display differs where wings are flipped high over the back for Boat-tailed. Female Boat-tailed lacks or has an indistinct lateral throat stripe.
Voice CALL: A low *clak* or a *kle-teet*. **SONG:** Distinct from Great-tailed. A continuous, long, harsh trilling song interspersed with other notes, *jeeb-jeeb-jeeb tireeet chrr chrr chrr chrr tireet tireet tireet tireet*. Wing-flipping display accompanies lower *chrr* notes.
Status & Distribution Common. **YEAR-ROUND:** Coastal marshes and a variety of nearby habitats. **RARE STATUS:** Casual in northern New England. **Population** Stable, has been expanding its range since the 1890s.

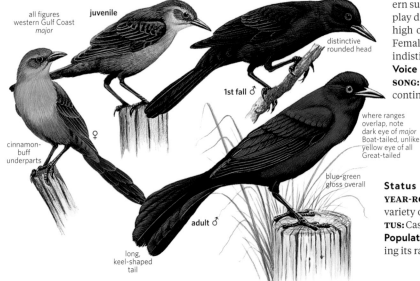

Subspecies of Great-tailed Grackle

Eight Great-tailed Grackle subspecies are recognized, but only three are found in N.A. These northern subspecies are *prosopidicola*, found in the eastern part of the Great-tailed's range west to central TX; *monsoni*, found from central AZ east to West TX; and *nelsoni*, found in CA and western AZ. All three subspecies of the Great-tailed are spreading northward in the US. For the most part, there is little information regarding which subspecies have spread to which areas, therefore the range descriptions given above are tentative.

Great-tailed Grackle, *monsoni*

And some intergradation may be occurring now that these subspecies are coming widely into contact.

The males of these subspecies are similar, differing mainly in size, with *prosopidicola* and *monsoni* being large subspecies while *nelsoni* is noticeably smaller. With regards to plumage, a male *monsoni* shows on average more of a purplish gloss, but this is variable. Differences in plumage are much more marked in females. In general, *monsoni* females are darkest below, *nelsoni* palest, and *prosopidicola* intermediate, although closer to *monsoni*. In fact, *nelsoni* females

GREAT-TAILED GRACKLE   *Quiscalus mexicanus*   GTGR ■ 1

This huge blackbird is hard to ignore due to its raucous song, which can be deafening in large nighttime roosts. Polytypic (8 sspp.; 3 in N.A.). L 15–18" (38–46 cm)

Identification Long, deeply keeled tail. Long bill, with nearly straight culmen; shallow forehead. **ADULT MALE:** Black with violet-blue iridescence. Eyes yellow. **ADULT FEMALE:** Smaller and shorter tail than male with no keel to tail. Brown above with dull iridescence on wings and tail; buffy on head and below, darker on belly and vent. Dark malar usually obvious. Eyes yellow. **JUVENILE:** Like female, but paler and shows diffuse streaking below and darker eyes.
Geographic Variation See sidebar below.
Similar Species The very similar Boat-tailed Grackle overlaps with Great-tailed Grackle in southwestern coastal LA and eastern TX. See Boat-tailed Grackle.
Voice CALL: A low *chut*; males may give a louder clack. Eastern males give a striking ascending whistle *twooo-eeeeeeee!* **SONG:** The eastern bird sings a four-part song beginning with harsh notes similar to the breaking of twigs, then a soft undulating *chewechewe*, and then twig-breaking notes and finally

several loud two-syllable *cha-wee* calls, *crrrk crrrk chewechewe crrk cha-wee cha-weewlii*. Subspecies *nelsoni* sings a repeated series of notes, ending in a more accented note *chk-chk-chk-chap-chap-chap-chap-CHWEEE*, often interspersed with various other repeated notes.
Status & Distribution Fairly uncommon to abundant. Mostly open habitats, from agricultural to urban. **MIGRATION:** Not well understood; more are wintering farther north now. **RARE STATUS:** Casual to the north of its range, from BC east to NS.
Population The species has experienced a great range and population increase in the US, showing a 3.7 percent annual increase from 1966 to 1998.

juvenile

streaked underparts

buffy supercilium

cinnamon-buff underparts

♀

long, rather thin bill

♂

purple gloss with no head and body contrast as in Common Grackle

western ♀
nelsoni

very long, keel-shaped tail

nelsoni is smaller and paler than other subspecies

---

may be pale grayish below with a nearly white throat; this coloration is strikingly different from the buff to warm brown underparts of *prosopidicola* and *monsoni*. The pale plumage combined with the small size sets *nelsoni* well apart from *prosopidicola* and *monsoni*.

Historically, the mountains of central Mexico divided the general population of Great-tailed Grackles into an eastern and central group and a western group. The western Great-tailed Grackles—from *nelsoni* in the north, to coastal forms in west Mexico south to Guerrero—are small,

Great-tailed Grackle, *prosopidicola*

they have a noticeably different song than the more eastern populations, and there are genetic differences. However, with the opening of more grackle-friendly habitats throughout the area due to agricultural and urban development, this previously isolated population has come into contact with eastern Great-tailed Grackles. In AZ intergradation between *nelsoni* and *monsoni* appears to be common, and birds that have *monsoni* mitochondrial DNA are by measurements small and like *nelsoni*.

COMMON GRACKLE *Quiscalus quiscula* COGR ■ 1

The Common Grackle is a common and often urban blackbird of eastern N.A. Highly gregarious in the non-breeding season. Polytypic (3 sspp.; all in N.A.). L 12.6" (32 cm)

Identification A large blackbird with a long, graduated tail that is often held in a deep keeled shape during the breeding season. **ADULT MALE:** Blackish with iridescence in good light. Widespread *versicolor* shows bronze gloss to body, blue head, and purple or blue iridescence on wings and tail. Eyes are bright yellow. **ADULT FEMALE:** Smaller and duller than male and does not hold tail in deep keel shape; in widespread *versicolor* bluish head still sharply contrasts with brown body. **JUVENILE:** Brown, with dark eyes and faintly streaked on breast.

Geographic Variation "Bronzed Grackle" (*versicolor*), found west of the Appalachians and in the Northeast, has bronze iridescence on body, a blue head, and purplish tail and wings. "Purple Grackle," which includes *stonei* found east of the central Appalachians, has a purplish body and head, and a blue or greenish glossed

tail; slightly longer billed *quiscula*, ranging from FL to southern LA and SC, is similar but has olive green gloss on back and underparts.

Similar Species Brewer's and Rusty Blackbirds lack long, graduated tail of Common Grackle, and they never hold it in a keeled shape. Boat-tailed and Great-tailed Grackles are much larger, with even more striking tails. Common (*versicolor*) shows clear, abrupt division between the gloss color of head and body.

Voice CALL: A loud and deep *chuck*. **SONG:** A mechanical, squeaky *readle-eak*. Both sexes sing.

Status & Distribution Common, but declining. **YEAR-ROUND:** Open and edge habitats, urban areas, agricultural lands, golf courses, swamps, and marshes. **MIGRATION:** Diurnal migrant; southern populations resident. Arrive at breeding areas mid-Feb.–mid-Mar.

and early Apr. in northernmost sites. Southward movements mostly Oct.– early Nov. **RARE STATUS:** Very rare or casual in Pacific states, BC, and Bermuda. Casual to far north, AK, YT, NT, NU, and Churchill, MB.

Population Near Threatened. Declines of more than 50 percent between 1970 and 2014.

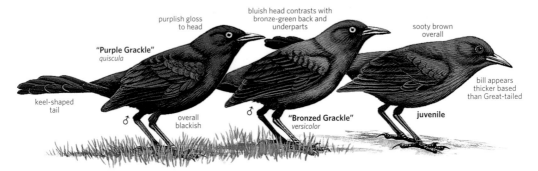

"Purple Grackle" *quiscula*

purplish gloss to head

bluish head contrasts with bronze-green back and underparts

sooty brown overall

keel-shaped tail

♂

overall blackish

♂

"Bronzed Grackle" *versicolor*

juvenile

bill appears thicker based than Great-tailed

WOOD-WARBLERS Family Parulidae

Yellow-rumped Warbler (NJ, Oct.)

Wood-warblers include some of our most colorful birds. Most members of this New World family are Neotropical migrants.

Structure Most are small with variable, colorful plumage in spring; some are duller in fall. All have nine primaries. Some show rictal bristles; some don't.

Plumage Many species have a brighter breeding plumage than nonbreeding plumage. Yellow, green, and olive figure prominently in many plumages and prominent wing bars are found on some species. Many species show extensive white on the outer tail feathers, which may appear as white spots.

Behavior All species in North America are generally insectivorous, but many feed on fruit or nectar in winter. Most species build a cup nest, placed on the ground or low to high in a shrub or tree; only two species nest in cavities. Many species have two song types: an often accented, primary song used for defending territory and an often unaccented, alternate song used in a variety of situations (e.g., near the nest when paired with a female). Songs are quite varied. Many eastern species undertake trans-Gulf migrations in spring and fall; many take a trans-Atlantic fall migration.

Distribution About 108 species breed from Alaska to northern Argentina. Fifty species breed annually in North America (not including Bachman's Warbler, which is almost certainly extinct), with the majority of those in the East; six additional species have been recorded as strays.

Taxonomy Nearly a third of the world's 108 species of wood-warblers are in in the genus *Setophaga*; other important genera include *Myiothlypis* (15 Central and South American species) and *Geothlypis*, *Basileuterus*, and *Myioborus* (each with 12 species). A number of species, many in North America, are classified in their own genus, and their closest relatives within the family are uncertain. Recent molecular work radically changed the taxonomy of wood-warblers. The American Redstart (*Setophaga*) was found to be firmly embedded within genus *Dendroica*, and by the Rule of Priority (which taxon was named first) all *Dendroica* became *Setophaga*. Additional significant changes include: moving most species out of the genera *Vermivora* and *Oporornis* and a completely new linear sequence. The two waterthrushes were found to be very different from the Ovenbird, a fact that ornithologist Kenneth Parkes discovered decades earlier when he discovered that their juvenal plumages were totally different. The new waterthrush genus, *Parkesia*, honors his insights. Two aberrant species formerly considered wood-warblers have been removed to their own families: Yellow-breasted Chat (Icteriidae) and Wrenthrush (Zeledoniidae).

Conservation Wood-warblers have been declining in recent decades. Most North American wood-warblers migrate to the tropics, which puts them at risk not only through potential habitat destruction on both breeding and wintering grounds, but on their migration corridors as well. In particular, the hazardous, difficult trans-Gulf migration often results in high mortality. Some species have restricted and specialized breeding ranges. Brown-headed Cowbird parasitism affects these rare species most dramatically, but as forests become fragmented many species become more vulnerable. Birdlife International codes: 10 NT, 7 VU, 6 EN, 1 CR, and 1 CR (PE).

Genus *Seiurus*

Formerly this genus included the two waterthrushes; now it contains only the Ovenbird and is basal (most primitive) to the wood-warbler family.

OVENBIRD *Seiurus aurocapilla*   OVEN ■ 1

A familiar voice from the forest, the Ovenbird spends much of its time walking on the ground or branches, often with its tail cocked. Gleans insects from forest floor. Its domed, well-hidden nest, with its side entrance, is built on the ground and encloses three to six eggs in June. Polytypic (3 sspp.; all in N.A.). L 6" (15 cm)
Identification ADULT: Olive upperparts, black lateral crown stripes, bold white eye ring, orange crown patch. Underparts white; bold black spots form streaks on breast, sides. **JUVE-** **NILE:** Very different from adult; brown above with indistinct blackish streaking and mottling; buffy below with faint markings; head pattern subdued. **Geographic Variation** Slight variation among three subspecies, usually not evident in the field. Western *cinereus* is grayer and paler on upperparts than eastern nominate; *furvior* (breeds NL) is slightly darker. **Similar Species** Northern and Louisiana Waterthrushes are slimmer and brown backed; they also lack white eye ring. Waterthrushes habitually bob their tails. **Voice CALL:** A loud, sharp *tsick*, often in a rapid series when alarmed. **FLIGHT CALL:** Thin, high *seee*. **SONG:** Primary song a loud, ringing *cher-tee cher-tee cher-tee* or *tea-cher tea-cher tea-cher* in a rising crescendo. Also an elaborate flight song.

Status & Distribution Common in mature forests, rare in the West. Migrates on a broad front in the East. **BREEDING:** Deciduous or mixed forests dominated by deciduous trees. **MIGRATION:** Arrives on the Gulf Coast late Mar.; in much of the South early Apr.,

rufous crown with blackish lateral crown stripes

olive above

bold white eye ring

extensively spotted below

peaks there mid-Apr.–early May; in the upper Midwest peaks mid-May. Departs as early as late July, earliest arrivals in the South early Aug.; peaks in much of the East in mid-Sept., stragglers into Dec. (sometimes in northern states). **WINTER:** Primary and second-growth forests from southern TX and coastal NC to southern FL south through C.A. and the Caribbean. **RARE STATUS:** Rare migrant in the West. Casual to Ecuador, Greenland, and Europe.
Population Recent significant declines likely due to forest fragmentation on breeding grounds.

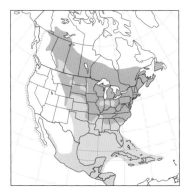

Genus *Helmitheros*

WORM-EATING WARBLER   *Helmitheros vermivorum*   WEWA   ■ 1

This inconspicuous warbler feeds mostly on caterpillars by probing into suspended dead leaves. Its nest is found at the base of a sapling on a hillside or ravine, usually under dead leaves. Monotypic. L 5.3" (13 cm)
Identification Large bill; pinkish-colored legs; short tail. **ADULT:** Rich buffy head with bold black lateral crown and eye stripes, buffy underparts, grayish olive upperparts. **IMMA-TURE:** Very similar to adult, with duller head stripes and rusty tips on tertials,

olive-brown back

boldly striped head

dead leaf specialist

pumpkin buff color on head and underparts

greater and median wing coverts, wearing off by fall.
Similar Species Swainson's Warbler lacks bold head stripes and rich buff coloration.
Voice CALL: Includes a soft *chip*. **FLIGHT CALL:** A doubled *zeet-zeet*, also given at other times. **SONG:** Dry, high-pitched trill similar to Chipping Sparrow, but drier and less staccato, and to Pine Warbler, but less musical. Note habitat of singer.
Status & Distribution Uncommon and often inconspicuous; migrates across a broad front. **BREEDING:** Large tracts where deciduous and mixed forests overlap with moderate to steep slopes. **MIGRATION:** In spring, arrives in FL and Gulf Coast in last half of Mar. continuing to early May; arrives in Midwest mid-Apr.–early-May. In fall, inconspicuous departure from breeding grounds mid-July–Aug. in northern areas, through Sept. in southern areas. **WINTER:** Forest and scrub

habitats of eastern Mexico south to Panama and the northern Caribbean (including the Greater Antilles). **RARE STATUS:** Very rare to casual in migration to northern New England, Maritimes, QC, ON, northern Great Plains, and the West (e.g., nearly 150 records from CA and 35+ from AZ). One record from Venezuela.
Population Probably stable; however, it is difficult to census populations accurately. The species is sensitive to forest fragmentation.

Genus *Parkesia*

Both species in this genus have loud, ringing songs. They have streaked breasts and are terrestrial, walking, not hopping, around wet areas. These two species were formerly placed in the genus *Seiurus* with Ovenbird.

LOUISIANA WATERTHRUSH   *Parkesia motacilla*   LOWA   ■ 1

Larger and longer-billed than the Northern Waterthrush, Louisiana has a distinctive ringing song. Between Apr. and June, it will nest in a hollow of a fallen tree's root system or a streamside bank and lay three to six eggs. Monotypic. L 6" (15 cm)
Identification Bobs tail slowly up and down, with some side-to-side motion. Large bill, bright pink legs. **ADULT:** Olive-brown crown, back, wings, tail. Two-tone supercilium: anterior pale

buff then pure white and broader behind eye. White underparts usually with distinct pinkish buff on flanks, undertail coverts; blackish streaking on breast, sides. Usually very small, indistinct white spots on outer tail feathers. **IMMATURE:** Usually identical to adult; some with buffy or rusty tips on tertials and more pointed tail feathers; lacks any white on outer tail feathers.
Similar Species Compare to Northern Waterthrush. See sidebar opposite.
Voice CALL: Loud, rich *chik* or *chich*, less ringing or metallic than Northern. **FLIGHT CALL:** High *zeet*. **SONG:** Loud, rollicking primary song with clear slurred whistles ending in sputtering notes: *seeeu seeeu seeeu seewit seewit ch-wit it-chu*.
Status & Distribution Uncommon. **BREEDING:** Along clear streams in hilly deciduous forest, in cypress swamps and bot-

tomland forest. **MIGRATION:** Arrives early, typically Gulf Coast in mid-Mar., peaks late Mar.–mid-Apr., reaches Great Lakes mid- to late Apr. Migration generally complete by mid-May. Departs early, mostly by mid- to late July; migrates through eastern N.A. during early to mid-Aug., a very few into early Sept., casually to early Oct., exceptional later. **WINTER:** Rivers and

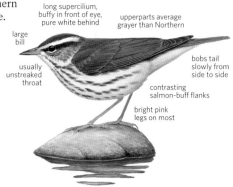

long supercilium, buffy in front of eye, pure white behind

upperparts average grayer than Northern

large bill

usually unstreaked throat

bobs tail slowly from side to side

contrasting salmon-buff flanks

bright pink legs on most

streams in hilly or mountainous areas from coastal northern Mexico through C.A. and Caribbean to extreme northwestern S.A. Very rare southern FL, AZ. **RARE STATUS:** Casually overshoots in spring migration to ND, QC, NS. Casual in West: CA, NM, AZ, NV, CO, Baja California, and Venezuela. Accidental OR, ID, and to Canary Is. and Morocco. **Population** Range is expanding in the Northeast; populations stable or slightly declining elsewhere.

Identification of Waterthrushes

The two species of waterthrush—Northern and Louisiana—are very similar in appearance, habits, and habitat. They can occur together during migration.

Singing birds are easily identified by song and, with experience, by call. See accounts for descriptions of their songs and calls.

The head patterns are distinctive. Northern's supercilium is uniform in width and color (usually yellow, sometimes white). Louisiana has a subtly bicolored supercilium, buffy above the lores and white behind, that broadens noticeably behind the eye.

The waterthrushes' throat pattern show average differences. Both species show a narrow malar streak. Northern also shows small streaks on the throat, whereas

Northern Waterthrush (NJ, Sept.)

Louisiana Waterthrush (CA, Sept.)

Louisiana usually shows a clean white throat or a very few streaks.

The underparts of Northern are boldy streaked and are usually uniformly yellowish, though sometimes may be whitish. The Louisiana shows white underparts, usually with pinkish buff flanks, and less-distinct and sparser streaking.

Louisiana has a distinctly longer and heavier bill but a disproportionately shorter tail.

Leg color is also helpful in identification. Generally, Louisiana's legs are a brighter pink than Northern's legs.

Behavior differs. Louisiana's tail bobbing is more exaggerated, but slower and more circular than Northern's rapid up-and-down bobbing.

NORTHERN WATERTHRUSH *Parkesia noveboracensis* NOWA ▪ 1

The waterthrushes are appropriately named, as they are almost always found near water, and on or near the ground, whether in migration or when breeding in northern forests. Often located by its loud song and calls, the Northern lays one to five eggs in a hollow of a fallen tree's root system or a streamside bank between May and June. It is now regarded as monotypic. L 5.8" (15 cm)

Identification Vigorously bobs tail up and down. Dull pink legs. **ADULT:** Olive-brown crown, back, wings, and tail; yellowish, buffy, or white supercilium (even width throughout); underparts whitish tinged with yellowish, or mainly whitish, with blackish streaks on throat, breast, and sides. Usually very small, indistinct white spots on outer tail feathers. **IMMATURE:** Usually identical to adult; some with buffy or rusty tips on tertials or more pointed tail feathers, lacking any white on outer tail feathers.

Geographic Variation Slight, clinal variation not well correlated with the three formerly recognized subspecies. Yellowest in Northeast, but much variation throughout range.

Similar Species Compare to Louisiana Waterthrush. See sidebar above.

Voice CALL: Sharp, ringing, metallic *chink.* **FLIGHT CALL:** Buzzy *zeet.* **SONG:** Primary song is loud and ringing, rather staccato: *twit twit twit sweet sweet sweet chew chew chew.*

Status & Distribution Common; trans-Gulf, circum-Gulf, trans-Caribbean migrant; migrates on a

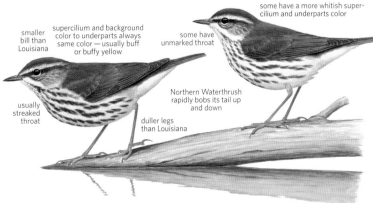

smaller bill than Louisiana

supercilium and background color to underparts always same color — usually buff or buffy yellow

usually streaked throat

some have unmarked throat

duller legs than Louisiana

Northern Waterthrush rapidly bobs its tail up and down

some have a more whitish supercilium and underparts color

broad front through the East. **BREED-ING:** Wooded areas with slow-moving water (e.g., swamps, bogs, margins of lakes). **MIGRATION:** In spring, a slightly more westerly route: arrives on Gulf Coast early Apr., peaks late Apr.–early May; arrives in the Great Lakes and Northeast late Apr., peaks mid-May. In the West, smaller numbers mainly mid-May–early June. In fall, a slightly more easterly route: Departs breeding grounds in July, main arrival early Aug., peaks Sept. in northern US, few linger into Oct. **WINTER:** Mangroves, rainforest, second growth from coastal northern Mexico through C.A., Caribbean, and northern S.A. to northeastern Peru and northern Brazil. A few overwinter in southern FL, along Gulf Coast. **RARE STATUS:** Casual in migration to north slope of AK, Bering Sea islands, Wrangel I., Chukotski Peninsula; also Greenland and Europe.

Population Breeding areas appear stable and secure; wintering habitat (especially mangroves) threatened with development.

Genus *Vermivora*

This genus now comprises a total of three species. They are recognized by their smaller size, shorter tails, and thin, pointed bills. These species have tail spots and lack rictal bristles. Their vocalizations consist primarily of a series of buzzes.

BACHMAN'S WARBLER   *Vermivora bachmanii*   BAWA ■ 6

Almost certainly extinct, this warbler was found in southern swamps. Monotypic. L 4.8" (12 cm)

gray crown; breast gray or yellowish

adult ♀

adult ♂

large black throat patch extending down through breast; black forecrown

Identification SPRING MALE: Thin, long, pointed, and slightly decurved bill. Yellow forehead, supercilium, eye ring, chin, and shoulders. Black crown and bib, gray nape. White spots on outer tail feathers. **SPRING FEMALE:** Duller than male with no black. Grayish on crown, throat, and breast; yellow forehead. **IMMATURE:** Male shows less black on crown and bib than adult male; smaller tail spots. Female duller than spring female.

Similar Species Hooded has yellow undertail coverts and larger white tail spots; bill not decurved. Immature female Bachman's is similar to several species, notably nominate subspecies of Orange-crowned, but distinguished by pointed, decurved bill; complete white eye ring; and whitish undertail coverts. Female Common Yellowthroat and immature Yellow Warbler have also been confused with female Bachman's.

Voice CALL: A buzzy *zip* or *zeep* given by both sexes. Little known. **SONG:** A rapid series of buzzy notes, similar in quality to alternate song of Blue-winged and Golden-winged, delivered rapidly and on one pitch.

Status & Distribution Last confirmed record was spring 1962, near Charleston, SC. Formerly a local breeder from southernmost MO to SC. Most records were migrants from mainland FL, Key West, and southeastern LA. **BREEDING:** Probably canebrakes (bamboo) in swamps and bottomlands. **WINTER:** Cuba and Isle of Pines in semi-deciduous forest and forested urban open space. **RARE STATUS:** Casual to NC, VA.

Population Probably Extinct. Degradation of breeding habitat is the most likely cause of its decline.

GOLDEN-WINGED WARBLER   *Vermivora chrysoptera*   GWWA ■ 2

Boldly patterned with black, white, gray, and yellow, the Golden-winged Warbler is an active and acrobatic forager; often probes into dead leaves. Monotypic. L 4.8" (12 cm)

same pattern as found on male but much duller

yellow wing patch

slender bill

♀

yellow crown and chickadee-like face pattern

♂

Identification ADULT MALE: Gray upperparts with a bright yellow cap. Broad wing bars appear as a single large yellow patch. White underparts. Black cheek and throat contrast with white malar and supercilium. Large white spots on outer tail feathers. **ADULT FEMALE:** Similar to male, but upperparts more olive-gray, cheek and throat gray, duller yellow cap blending in with gray nape; two yellow wing bars usually distinct, not forming a large yellow patch on wing. **IMMATURE:** Duller than adult. Male more olive above, with duller yellow cap; black throat and cheek often slightly fringed with buff. Female more olive above with duller yellow cap contrasting little with grayish olive nape, paler gray throat and cheek patch. **HYBRID:** See Blue-winged Warbler.

Similar Species Throat pattern and foraging behavior suggest Black-capped Chickadee. Look carefully at all characters for backcrossing hybrids with Blue-winged (e.g., some yellow on breast).
Voice CALL: Most often heard call is a gentle *tsip*, similar to Field Sparrow and identical to Blue-winged Warbler. **FLIGHT CALL:** A high, slightly buzzy *tzii*, often doubled. **SONG:** Primary song is a high-pitched, buzzy *zeee bee bee bee*, with a higher-pitched first note. Alternate song is indistinguishable from Blue-winged Warbler, with a stuttering first note and a flatter second note.
Status & Distribution Uncommon to rare and declining; a trans-Gulf migrant. **BREEDING:** Wet shrubby fields, marshes, and bogs on edge of woodlands, most often restricted to early successional habitat. **MIGRATION:** In spring, peaks early to mid-May in MI and PA, to late May in northern WI. In fall, often departs breeding grounds undetected during Aug., with latest dates in early to mid-Oct. **WINTER:** Woodlands and forest borders from southern Mexico to Panama. Casual in northern Colombia, Venezuela, and Caribbean. **RARE STATUS:** Very rare to casual to New England (formerly bred) and casual to Atlantic Provinces. Casual in West, recorded from most western states and provinces; nearly 90 records from CA. Accidental to Greenland, UK (winter), and the Azores.
Population Near Threatened. Population declining as Blue-winged moves into and gradually takes over Golden-winged Warbler territories through hybridization and earlier arrival.

BLUE-WINGED WARBLER *Vermivora cyanoptera* BWWA ■ 1

The Blue-winged Warbler is a bright yellow and has a buzzy song. The feeding behavior is like the closely related Golden-winged Warbler. It nests on or near the ground like Golden-winged. Monotypic. L 4.8" (12 cm)
Identification ADULT MALE: Bright yellow forehead, crown, and underparts contrast with greenish yellow nape and back; white undertail coverts. Bluish gray wings with two distinct white wing bars. Black lores with short black postocular line behind eye. Large white spots on outer tail feathers. **ADULT FEMALE:** Duller than male, with yellow forehead blending into yellowish olive crown, nape, and back; whitish undertail coverts. Wing bars less distinct, sometimes tinged yellow. Lores and eye line slightly duller. **IMMATURE:** Duller, similar to adult female. **HYBRIDS:** Blue-winged x Golden-winged Warbler hybrids are well documented, and two types have been named; "Brewster's" is more frequent, as all first-generation hybrids and some later-generation hybrids result in this type. A few backcrosses result in the rarer "Lawrence's." Some hybrids do not resemble these two types and combine characters of both parents. Hybrids usually sing the songs of one species or the other.
Similar Species Long and well-defined black or dark eye line diagnostic. Note also white undertail coverts and extensive white in tail.
Voice CALL: Most often heard call is a gentle *tsip*, identical to Golden-winged Warbler and similar to Field Sparrow. **FLIGHT CALL:** A high, slightly buzzy *tzii*, often doubled. **SONG:** Primary song is typically a high-pitched, buzzy, inhaled-exhaled *beeee-bzzzz*. Alternate song is somewhat similar, but with a stuttering first note and a flatter second note: *be-ee-ee-ee-bttttt*.
Status & Distribution Fairly common in brushy meadows and second-growth woodland edges; more catholic in its choice of breeding habitats than Golden-winged. A trans-Gulf migrant. **BREEDING:** Overgrown old fields and brushy swamps, usually in early to mid-succession. **MIGRATION:** In spring, early Apr.–early May in TX, late Apr.–mid-May in OH, late Apr.–late May in MN. In fall, usually departs breeding grounds undetected. As early as late July in NJ, peaking in late Aug., a few to late Sept. Occasional Nov. records in Southeast, a few farther north. **WINTER:** Humid evergreen and semi-deciduous forest and edge from southern Mexico to Costa Rica, rarely south to Panama and West Indies. **RARE STATUS:** Casual in West and central Canada, including SK, AB, WY, CO (approximately 40 records), NM (approximately 12 records), AZ (nine records), NV, WA, OR, and CA (nearly 60 records). Rare north to Atlantic Provinces, mostly in early fall; accidental in Iceland, Azores.
Population Expanding northward in Great Lakes into range of Golden-winged Warbler, but declining due to reforestation of open habitats and urban sprawl in Northeast US.

black eye line

slender bill

white undertail coverts

two white wing bars on grayish wings

"Brewster's Warbler"
first-generation hybrids and backcrosses exhibiting genetic dominance have Blue-winged head pattern

Blue-winged x Golden-winged hybrids

"Lawrence's Warbler"
rarer recessive hybrids from backcrosses have Golden-winged head pattern

Genus *Mniotilta*

BLACK-AND-WHITE WARBLER *Mniotilta varia* BAWW ▪ 1

This distinctive black-and-white striped warbler, which creeps on trunks and horizontal branches while foraging, is more like a nuthatch than most other wood-warblers. Between Apr. and June, it nests on the ground against a shrub, tree, stump, rock, or log and produces four to six eggs. This species has an extensive breeding range. Late spring migrants heading well north are still passing through TX after local breeders have fledged young. Monotypic. L 5.3" (13 cm)

Identification Long, slightly decurved bill; long hind claw grips tree trunks. **BREEDING MALE:** Striped black and white overall, with black cheek and throat; black crown with white median stripe. First-spring males show more white, or all-white, on chin and throat. **SPRING FEMALE:** Similar to male, but with grayish buff cheek, narrow black line behind eye, white chin and throat, duller streaks on sides. **WINTER ADULT:** Male with chin and throat mottled with white (or nearly entirely white), smaller black cheek patch. Female very similar to spring, but sometimes with pale buffy wash on underparts. **IMMATURE:** Male similar to adult male, but with clear white cheek; female dullest, with richer buff on flanks and undertail coverts; paler, blurry side streaks; buffy cheek patch.

Similar Species Spring male Blackpolls show white cheeks; Black-throated Grays lack white median crown stripe, have solid gray back, small yellow spot in front of eye.

Voice CALL: A dull *chip* or *tik*. **FLIGHT CALL:** A doubled *seet-seet*. **SONG:** Primary song is a long, slightly variable, thin, rhythmic *weesee weesee weesee weesee weesee weesee*, sometimes described as sounding like a squeaky wheel.

Status & Distribution Common and

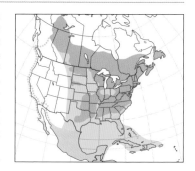

widely distributed; a mid- to long-distance migrant, generally along a broad front. **BREEDING:** A variety of habitats, including deciduous and mixed woodlands, both mature and second growth. **MIGRATION:** In spring, arrives earlier than many warbler species, reaching central FL and TX in early Mar., PA and NJ by mid-Apr., New England by early May. Peaks in Great Lakes in early May. Departs breeding areas in late July; main migration from late Aug. through late Sept.; stragglers into Nov. **WINTER:** A wide variety of habitats, including mature forest, mangroves, and open areas, from coastal NC to FL and southern TX through C.A. and Caribbean to northern S.A., south to Ecuador (rare) and Peru (casual). **RARE STATUS:** Rare migrant through much of the West, also noted rarely in winter. Casual to WA, AK, and in fall to Europe and the Azores.

Population Appears stable, but may be negatively affected by severe forest fragmentation.

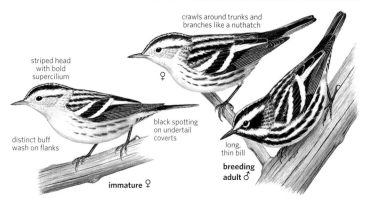

crawls around trunks and branches like a nuthatch

striped head with bold supercilium

♀

black spotting on undertail coverts

distinct buff wash on flanks

long, thin bill

breeding adult ♂

immature ♀

Genus *Protonotaria*

PROTHONOTARY WARBLER *Protonotaria citrea* PROW ▪ 1

A denizen of wooded swamps, this is the only eastern wood-warbler that nests in natural and artificial cavities (three to seven eggs, Apr.–June). Monotypic. L 5.5" (14 cm)

Identification ADULT MALE: Bright golden yellow head, underparts; beady black eye; long black bill; greenish back; bluish gray wings, tail; white undertail coverts; large white tail spots; short tail. Some males and females have orange heads. Pinkish-colored lower mandible in fall. **ADULT FEMALE:** Similar to male, but greenish olive wash on rear of crown, nape; smaller white spots on tail. **IMMATURE:** Male like adult male but with some dark stippling on crown. Female with extensive olive on crown, nape, cheek.

Similar Species Blue-winged shows black eye line, whitish wing bars. The smaller Yellow shows pale edges to wing coverts and tertials and has yellow tail spots and a smaller bill.

Voice CALL: Loud, dry chip note, like Hooded Warbler. **FLIGHT CALL:** Loud *seeep*. **SONG:** Simple, loud, ringing series of notes: *sweet sweet sweet sweet sweet sweet*.

Status & Distribution Fairly common, often local; mainly a medium-distance

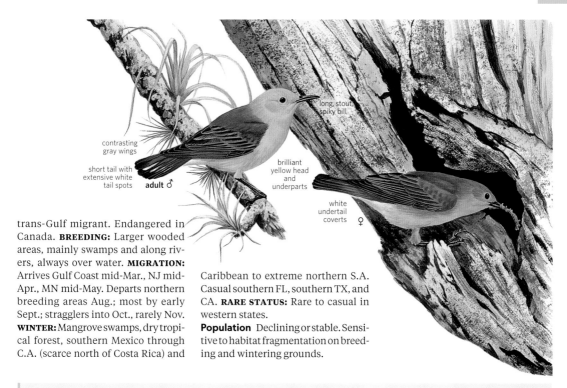

long, stout spiky bill

contrasting gray wings

short tail with extensive white tail spots **adult ♂**

brilliant yellow head and underparts

white undertail coverts ♀

trans-Gulf migrant. Endangered in Canada. **BREEDING:** Larger wooded areas, mainly swamps and along rivers, always over water. **MIGRATION:** Arrives Gulf Coast mid-Mar., NJ mid-Apr., MN mid-May. Departs northern breeding areas Aug.; most by early Sept.; stragglers into Oct., rarely Nov. **WINTER:** Mangrove swamps, dry tropical forest, southern Mexico through C.A. (scarce north of Costa Rica) and Caribbean to extreme northern S.A. Casual southern FL, southern TX, and CA. **RARE STATUS:** Rare to casual in western states.

Population Declining or stable. Sensitive to habitat fragmentation on breeding and wintering grounds.

Genus *Limnothlypis*

SWAINSON'S WARBLER *Limnothlypis swainsonii*   SWWA ■ 2

The skulking Swainson's Warbler is often difficult to observe. It usually forages by shuffling through leaf litter (quivering its tail). Builds nest in dense ground vegetation. Monotypic. L 5.5" (14 cm)

Identification Large; heavy bodied; short tail; flat forehead, long pointed bill with pinkish-colored lower mandible and pinkish legs. **ADULT:** Rufous-brown crown; duller olive-brown above; pale supercilium; dark brown eye line; whitish or pale yellowish underparts, grayish tinge on sides.

Similar Species Worm-eating has conspicuous black stripes on head. **Voice CALL NOTE:** Distinctive chip note; louder, sweeter than Prothonotary Warbler. **FLIGHT CALL:** High, thin, slightly buzzy *swee* notes, sometimes doubled. **SONG:** Somewhat variable; loud, ringing *whee whee whee wee tu weeu*. Like Louisiana Waterthrush, without sputtering notes.

Status & Distribution Uncommon. **BREEDING:** Damp bottomland hardwoods, canebrakes (bamboo), and rhododendron thickets. **MIGRATION:** Medium-distance migrant, mainly across eastern Gulf. Arrives late Mar.–late Apr. Departs between mid-Sept. and mid-Oct., a few as early as mid-Aug. or as late as mid-Nov. **WINTER:** Swamps and river floodplain forests in Yucatán, Belize, Guatemala, and western Caribbean. **RARE STATUS:** Overshoots in spring, where recorded casually or accidentally north to WI, MI, OH, NJ, NY, MA, ON, NS; and west to KS, NE, CO, AZ, and NM.

Population Declines in parts of range, e.g., mid-Atlantic and eastern TX.

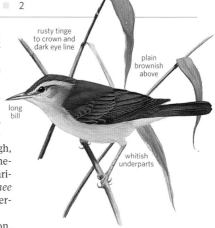

rusty tinge to crown and dark eye line

plain brownish above

long bill

whitish underparts

Genus *Oreothlypis*

This genus formerly held eight N.A. species. Recently, only the brightly colored, tropical Crescent-chested and Flame-throated Warblers, the latter endemic to the mountains of Costa Rica and western Panama, have been retained in *Oreothlypis*; the other, duller species are likely soon to be moved to *Leiothlypis*.

CRESCENT-CHESTED WARBLER *Oreothlypis superciliosa* CCWA ◾ 4

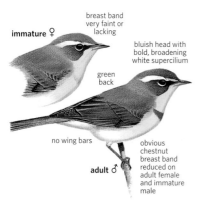

immature ♀

breast band very faint or lacking

bluish head with bold, broadening white supercilium

green back

no wing bars

adult ♂

obvious chestnut breast band reduced on adult female and immature male

This tropical montane species occurs casually north of Mexico and was formerly placed in the genus *Parula* along with Flame-throated (*O. gutturalis*). Polytypic (5 sspp. show slight variation, likely northwest *sodalis* in N.A.). L 4.3" (11 cm)
Identification Blackish upper man-dible; blackish lower mandible with pinkish-colored base. **ADULT MALE:** Gray head with broad white super-cilium and small white arc below eye; green back; yellow throat and breast with distinct crescent-shaped chest-nut band on breast; whitish belly and undertail coverts. Gray wings and tail. **ADULT FEMALE:** Smaller chestnut breast band. **IMMATURE:** Breast band is much reduced or lacking.
Similar Species Both Northern and Tropical Parulas lack white supercil-ium and have white wing bars. Rufous-capped Warbler is rufous on crown and lacks chestnut breast band.
Voice CALL: A soft *sik*, similar to but softer than Orange-crowned Warbler. **FLIGHT CALL:** A high, thin *sip*. **SONG:** A short, flat buzz, reminiscent of a "Bronx cheer": *t-t-t-t-t-t-t-t-t*.
Status & Distribution Uncommon to fairly common in native range. **BREEDING:** Montane pine-oak and cloud forests from northern Mexico to north-central Nicaragua. **RARE STATUS:** Casual stray to AZ (nearly 15 records) in spring, fall, and win-ter; nearly all are from southeast AZ, except for one for Yavapai Co. Most of the records have been from the mountains; some of these birds have joined mixed-species flocks. Two unconfirmed records for TX.
Population Stable.

TENNESSEE WARBLER *Oreothlypis peregrina* TEWA ◾ 1

This rather plain short-tailed warbler is a spruce-budworm specialist and nests on the ground at base of small shrub or tree (three to eight eggs, June). Monotypic. L 4.8" (12 cm)
Identification SPRING MALE: Green upperparts, contrasting gray crown and nape, white supercilium and underparts, blackish eye stripe. Faint whitish wing bar. **SPRING FEMALE:** Duller than male, less contrasting crown and back, tinged yellow below. **FALL ADULT:** Duller than spring adult, usually some pale yellowish below, less contrasting nape and back. **IMMA-TURE:** Bright olive-green upperparts. Supercilium pale yellowish; dusky eye line. Pale to bright yellowish under-parts. Bright white undertail coverts,

slender, straight spike-like bill

bright green back, faint wing bars

fall

pale supercilium in all plumages

breeding ♀

fall

bright green upperparts

whitish undertail coverts on most, sometimes pale yellow

gray cap and dark eye line

white belly

short tail

breeding ♂

sometimes tinged pale yellow.
Similar Species Similar to Orange-crowned, but with straighter bill, shorter tail, and nearly always white undertail coverts. Philadelphia and Warbling Vireos have thicker bills with hooked tips.
Voice CALL: Sharp *tsit*. **FLIGHT CALL:** Thin, clear *see*. **SONG:** Loud, staccato series of chip notes, increasing speed in two or three distinct steps.
Status & Distribution Fairly common; mostly a trans-Gulf migrant. **BREED-ING:** Mainly mixed and coniferous forest and bogs in boreal zone. **MIGRA-TION:** Arrives early Apr. on Gulf Coast. Bulk of migration in Great Lakes dur-ing mid- to late May. Some adults depart breeding grounds very early, often in July. Peaks mid-Sept.–early Oct., a few to late Oct. **WINTER:** South-ern Mexico to Panama and northwest-ern S.A. Very rare in coastal CA. **RARE STATUS:** Rare migrant in CA, very rare elsewhere in the West. Rare in AK (some breed), casual to Greenland, Europe, and the Azores.
Population Declining.

ORANGE-CROWNED WARBLER *Oreothlypis celata* OCWA ■ 1

This rather yellowish green warbler, named for its least conspicuous character, forages lower than many species. Polytypic (4 sspp.; all in N.A.). L 5" (13 cm)

Identification Nominate described. **ADULT MALE:** Olive green; yellowish narrow broken eye ring, indistinct dusky eye line. Greenish yellow underparts with indistinct blurry streaks. Undertail coverts always brighter yellow than belly. **ADULT FEMALE:** Duller and grayer than male. **IMMATURE:** Duller and grayer; eye arcs whitish. **Geographic Variation** Northern nominate *celata* is dullest, West Coast *lutescens* is brightest yellow, Rocky Mts. and Great Basin *orestera* is intermediate,

and immatures resemble *celata* as they have grayish heads; coastal Southern CA *sordida* (mainly Channel Is.) is darkest green.

Similar Species Compare with shorter-tailed Tennessee Warbler. In West, Yellow and Wilson's Warblers are similar to *lutescens*, but both show plain faces with prominent dark eyes. **Voice CALL:** Most frequent call a very distinctive, hard *stick* or *tik*. **FLIGHT CALL:** A high, thin *seet*. **SONG:** A high-pitched loose trill becoming louder and faster in the middle, weaker and slower at the end; faster in *lutescens*.

Status & Distribution Common in West, uncommon in East. **BREEDING:** Brushy deciduous thickets and second growth, from boreal forest in East to a great variety of habitats in West. **MIGRATION:** In spring, northern subspecies takes a more westerly route up the Mississippi River Valley, mid-

Apr.–late May. Pacific *lutescens* moves shorter distances, earlier peaking in late Mar.–early Apr. In fall, northern subspecies move much later than other warblers, peaking in mid-Oct. in much of northern US. Western subspecies moves earlier, mid-Aug.–early Oct. in AZ and CA. **WINTER:** Generally from southeast US and CA, through Mexico to Guatemala. Rarely lingers in northern US. **RARE STATUS:** Nominate *celata* rare in fall to Atlantic Provinces; casual to Bahamas, Cuba, Jamaica, Cayman Is.

Population Stable.

brighter overall and more of a lemon yellow below

lutescens ♂

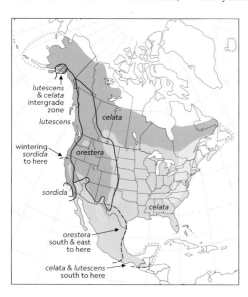

lutescens & celata intergrade zone

lutescens

wintering sordida to here

orestera

celata

sordida

celata

orestera south & east to here

celata & lutescens south to here

immature ♀
celata

grayish head

slender bill slightly decurved

split eye ring

grayish underparts

longer tail than Tennessee

even drab birds have yellowish undertail coverts

blurry streaks on breast

adult ♀
celata

COLIMA WARBLER *Oreothlypis crissalis* COLW ■ 2

The Colima Warbler breeds only in the Chisos Mts. of West TX and adjacent mountain ranges in north-central Mexico. It nests on the ground in grasses and produces three to four

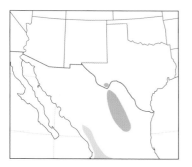

eggs between May and June. Monotypic. L 5.8" (15 cm)

Identification This is the largest and longest-tailed species in the genus *Oreothlypis*; it only rarely wags its tail. **ADULT:** Grayish head, brownish back, and yellowish olive rump. Olive-brown sides and flanks. Undertail coverts orange buff. Bold white eye ring. Partially concealed rufous crown patch. **JUVENILE:** Like adult but lacks rufous crown patch and has pale buff wing bars.

Similar Species Virginia's Warbler is smaller, shorter tailed, and grayer, lacking brown coloring. Adults have a yellow patch on breast. It frequently

brownish wash to back and sides

apricot-colored rump and undertail coverts

long tail

bobs its tail, unlike Colima Warbler. It has likely hybridized with Virginia's in the high Davis Mts. (Mt. Livermore),

TX; there may be a small hybrid population there.

Voice CALL: A loud, sharp *plisk*, similar to calls of Nashville, Virginia's, and Lucy's Warblers. **SONG:** A simple musical trill in which final two notes are down-slurred, similar to song of Orange-crowned Warbler.

Status & Distribution Fairly common but local. **BREEDING:** Montane areas of oak, pinyon, and juniper. **MIGRATION:** Essentially unknown as a migrant. Arrives on breeding grounds as early as mid-Mar., all by late Apr. Departs the Chisos Mts. by mid-Sept. **WINTER:** Brushy areas of humid montane for-

ests in southwestern Mexico, including Colima, hence its English name. **RARE STATUS:** Casual to Davis Mts. Accidental early May to Kenedy Co., TX.

Population Chisos Mts. (TX) population varies, with between 40 and 80 breeding pairs on censuses from 1960s to 1990s (fewer in drought years).

LUCY'S WARBLER   *Oreothlypis luciae*   LUWA ■ 1

A small gray bird of the arid southwest deserts, the Lucy's Warbler is one of two species of cavity-nesting warblers. It makes its nest in a natural cavity, including crevices, under loose bark, or in an abandoned woodpecker hole. Monotypic. L 4.3" (11 cm)

Identification This very small warbler flicks its short tail. **ADULT MALE:** Pale gray upperparts, slightly darker wings and tail, white underparts. Dark eye stands out on pale gray face, which shows white lores and an indistinct, white eye ring. Chestnut crown patch and rump are both often concealed. Outer tail feather shows a small white spot near tip. **ADULT FEMALE:** Similar to male, but paler face, smaller crown

patch, and paler rump. **IMMATURE:** Similar to adult female, with pale buff cast on underparts. When fresh, immatures have two indistinct pale buffy wing bars.

Similar Species Virginia's and Colima Warblers show yellowish on rump and undertail coverts, not chestnut. Bell's Vireo has a thicker bill and wing bars. Juvenile Verdin has a much sharper, thicker-based bill and a longer tail. Females of Yellow Warbler's desert subspecies (*sonorana*) always show some greenish on wings and tail as well as pale yellowish spots on tail.

Voice CALL: A sharp *chink*, similar to the call of Virginia's Warbler. **FLIGHT CALL:** A weak *tsit*. **SONG:** A loud, lively, sweet song somewhat similar to Yellow Warbler's: *tee-tee-tee-tee-tee-sweet-sweet-sweet*.

Status & Distribution Fairly common. **BREEDING:** Dense lowland riparian mesquite, cottonwood, and willow woodlands, mainly in the Sonoran Desert. **MIGRATION:** Earlier in spring than most migrants; arrives in southern AZ in early Mar. In fall, departs breeding grounds early, sometimes beginning as early as late July. Most

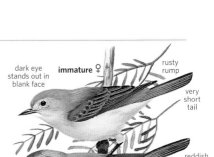

dark eye stands out in blank face immature ♀ rusty rump

very short tail

reddish chestnut rump

reddish chestnut cap ♂

gone from AZ by early to mid-Aug., with latest in Oct. Rare on CA coast late Aug.–late Nov., casual in winter. **WINTER:** Thorn forest and riparian scrub in western Mexico. **RARE STATUS:** Casual in southwestern CO. Casual or accidental in OR, WA, BC, ID, LA, VA, and MA.

Population Nests in unusually high densities for a noncolonial species. Declines due to the loss of riparian habitat from water projects, the cutting of mesquite trees, and increase in thickets of introduced tamarisks.

NASHVILLE WARBLER   *Oreothlypis ruficapilla*   NAWA ■ 1

The Nashville is a small olive-and-yellow warbler with a broad white eye ring and a gray head. Polytypic (2 sspp.; both in N.A.). L 4.8" (12 cm)

Identification ADULT MALE: Gray head contrasts with olive back, wings, and tail; white eye ring; yellow underparts. White area around vent. Rufous crown patch usually hidden. **ADULT FEMALE:** Duller than male, with more olive-gray head. Smaller rufous crown patch. **IMMATURE:** Duller than adult female, with more brownish olive upperparts. Whitish throat. Some are quite dull.

Geographic Variation Compared to eastern nominate, western *ridg-*

wayi ("Calaveras Warbler") is brighter yellow-green on rump and brighter yellow on breast; has more white on vent; and often wags its longer tail; vocalizations differ. Still, field identification of potential strays is problematic. Breeding ranges do not overlap.

Similar Species Connecticut Warbler is much larger, with a gray or brownish hood, including throat and upper breast. Virginia's has less yellow below and is gray (including edges of remiges) above. Nashville is grayer above and has olive-edged flight feathers.

Voice CALL: A distinctive flat *tink*; sharper in *ridgwayi*, like Virginia's. **FLIGHT CALL:** A high, thin *tsip* or *seet*.

SONG: A two-part, loose series of sweet notes; first part bounces, second part is slightly faster: *see-bit, see-bit, see-bit, see see see see see*. Western subspecies has a similar but sweeter series of *see-bit* notes with less pattern.

Status & Distribution Common.

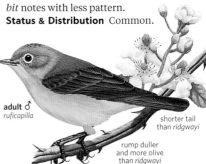

adult ♂ *ruficapilla*

shorter tail than *ridgwayi*

rump duller and more olive than *ridgwayi*

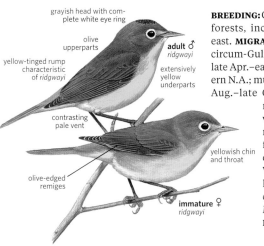

grayish head with complete white eye ring

olive upperparts

yellow-tinged rump characteristic of *ridgwayi*

contrasting pale vent

olive-edged remiges

adult ♂
ridgwayi

extensively yellow underparts

yellowish chin and throat

immature ♀
ridgwayi

BREEDING: Open deciduous or mixed forests, including spruce bogs in east. **MIGRATION:** Eastern birds are circum-Gulf migrants. Spring peak late Apr.–early May in much of eastern N.A.; much of fall migration late Aug.–late Oct. Western *ridgwayi* migrates through Southwest and along coast, most in Apr. in West; in fall, earlier than the eastern *ruficapilla*. **WINTER:** Wooded areas in highlands and coastal areas, coastal CA (rare) through Mexico to Guatemala. **RARE STATUS:** Very rare

in fall and winter in Caribbean. Casual to AK, including Bering Sea region. Accidental to Greenland. **Population** Stable.

ridgwayi

ruficapilla

VIRGINIA'S WARBLER *Oreothlypis virginiae*   VIWA ▪ 1

A tail-bobbing gray, white, and yellow warbler of the central and southern Rockies and the mountains of the Great Basin. Monotypic. L 4.8" (12 cm) **Identification SPRING MALE:** Gray above with yellowish green rump and uppertail coverts. Prominent white eye ring. Whitish below with a yellow patch on breast and yellow undertail coverts. Extensive rufous crown patch often partially concealed. **SPRING FEMALE:** Like male, with less extensive yellow on breast and smaller rufous crown patch. **FALL ADULT:** Browner above, buffy-tinged underparts. Yellow breast patch mixed with gray. Rufous crown

patch obscured with gray. **IMMATURE:** Brownish gray upperparts, buffy tinge below, yellow breast patch usually small or absent. Rufous crown patch very limited or absent.
Similar Species Western subspecies of Nashville Warbler also bobs its slightly shorter tail, but it shows entirely yellow underparts, and its upperparts show at least some greenish (including on remiges), not uniformly gray as Virginia's. Colima Warbler is larger and browner, it lacks yellow on breast, and its undertail coverts are more of an apricot color.
Voice CALL: A loud, sharp *chink*. **FLIGHT CALL:** A very high, clear *seet*. **SONG:** Similar to song of Nashville Warbler's western subspecies, but less structured and on one pitch: *s-weet, s-weet, s-weet, sweet-sweet-sweet.*
Status & Distribution Fairly common. **BREEDING:** Scrubby areas in steep-sloped pinyon-juniper and oak woodlands. **MIGRATION:** In spring, later than most other warblers, arriving in West late Apr.–mid-May, late Apr. in CO, and early May in WY. In fall, departs AZ and NM in early Aug.–Sept. **WINTER:** Arid to semiarid scrub in

highlands in southwest Mexico. Casual in southern TX and Southern CA. **RARE STATUS:** In fall, rare to coastal Southern CA; very rare to coastal northern CA and casual to OR. Casual to accidental nearly throughout eastern N.A., recorded northeast to Goose Bay, Labrador; accidental to Belize, Guatemala, and Grand Bahama I.
Population Apparently stable on breeding grounds, but numbers recorded in coastal CA are now fewer.

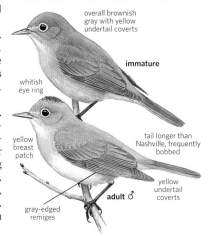

overall brownish gray with yellow undertail coverts

immature

whitish eye ring

yellow breast patch

tail longer than Nashville, frequently bobbed

yellow undertail coverts

adult ♂

gray-edged remiges

Genus *Oporornis*

This genus now contains only the Connecticut Warbler. The other three (Kentucky, Mourning, and MacGillivray's) were moved into *Geothlypis*. Unlike those species the Connecticut has a distinctive walking gait. Surprisingly, its closest relative may be Semper's Warbler (*Leucopeza semperi*), a likely extinct species from St. Lucia (Lesser Antilles).

CONNECTICUT WARBLER *Oporornis agilis*   CONW ▪ 2

This secretive, very terrestrial warbler of northern spruce bogs is one of the most challenging wood-warblers for birders to see. Monotypic. L 5.8" (15 cm)

Identification Habitually walks on or near ground. Large and chunky; short tail projection and long primary extension. Pinkish legs; bill mostly

pinkish with dusky brown culmen and tip. In all plumages, a prominent circular white or whitish eye ring, often with a slight break in the rear.

SPRING MALE: Gray head, throat, and upper breast form a distinct hood, paler on throat; olive above, pale yellow below. **SPRING FEMALE:** Similar to male, but paler hood tinged brownish olive. **FALL ADULT:** Similar to spring, but washed with olive on head. **IMMATURE:** Similar to fall female, some duller; extensively washed with olive-brown on head; throat buffy with brownish olive chest band; buff-tinged eye ring.

Similar Species MacGillivray's and Mourning Warblers are smaller with longer tail projection; show dark lores with broken or absent eye rings; and hop instead of walk. Nashville Warbler is much smaller and shows yellow on throat and breast.

Voice CALL: Loud, nasal *chimp*, rarely

heard. **FLIGHT CALL:** Buzzy *zeet*, also given when perched. **SONG:** Primary song is loud, staccato, and more emphatic toward end (similar in quality to Northern Waterthrush): *chuppa-cheepa chuppa-cheepa chuppa-cheep*.

Status & Distribution Uncommon; long-distance migrant. **BREEDING:** Boreal forest, spruce-tamarack bogs, muskeg, poplar woodlands, and deciduous forest. **MIGRATION:** In spring, through Caribbean (unrecorded Cuba!) to FL, then northwest to breeding grounds; later than most other warblers. Arrives in FL early May, peaks in Great Lakes in late May, stragglers into early June. Much more easterly route in fall, begins in late Aug., peaks in last half of Sept., stragglers to late Oct. on southern Atlantic

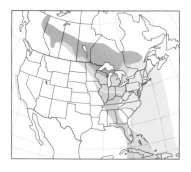

coast. Most fly nonstop across the Atlantic from southern New England or the mid-Atlantic to S.A.; 75 were found on Bermuda (25 Sept. 1987) after the passage of Hurricane Emily. **WINTER:** Recent evidence indicates that most may winter in eastern Bolivia in woodlands, forest edge, and dense second growth. A few recorded south of Amazon in Pantanal of Brazil to Bolivia. **RARE STATUS:** Casual in spring east of Appalachians and to LA, TX, and CA. Very rare to accidental in West, in fall most records along Pacific coast (about 120 records from CA). Rare to casual in Atlantic Canada. Casual to northern Baja California, Honduras, Costa Rica, and western Panama. Accidental Azores (fall).

Population Probably stable.

short tail projection past undertail coverts

brownish hood

large bill

immature

walks rather than hops; habits suggest Ovenbird

adult ♀

white eye ring sometimes slightly broken at rear

gray head

adult ♂

underparts duller yellow than Mourning, especially in immatures

---

Genus *Geothlypis*

Twelve rather skulking species; five in N.A. Vocalizations are loud and rhythmic. This genus was expanded in 2011 to include the Kentucky, Mourning, and MacGillivray's Warblers (all formerly in *Oporornis*). The latter two species fit well genetically into *Geothlypis*, but Kentucky less so. Connecticut Warbler remained in *Oporornis*.

GRAY-CROWNED YELLOWTHROAT *Geothlypis poliocephala* GCYE ■ 4

An aberrant *Geothlypis*. Polytypic (6 sspp.; *ralphi* in N.A.). L 5.5" (14 cm) **Identification** Moderately thick, bicolored bill, curved culmen; long tail often cocked and waved. **ADULT**

MALE: Gray crown; black lores extend below eye; broken, white eye ring; all yellow below, brownish on flanks. **ADULT FEMALE:** Similar to male, but olive-gray crown, slate-gray lores. **IMMATURE:** Similar to adult female.

Similar Species Common Yellowthroat has thinner all-dark bill and lacks strongly curved culmen. A possible Gray-crowned × Common Yellowthroat hybrid was photographed in Zapata Co., TX, in winter of 1995–96. **Voice CALL:** Includes a distinctive, nasal *cheed-l-eet*. **SONG:** A rich, scratchy warble like song of a

thick bill with pinkish base

ralphi ♂

long tail

ralphi ♀

gray head with broken white eye ring and dark lores

Passerina bunting or Blue Grosbeak.

Status & Distribution Historically resident (through 1910) but now casual in extreme southern TX. At least six records in the last 25 years.

Population Stable in C.A. range.

MACGILLIVRAY'S WARBLER *Geothlypis tolmiei* MGWA ■ 1

Closely related to the eastern Mourning Warbler, the more westerly MacGillivray's is a close relative; the two hybridize frequently where their ranges meet in Alberta. MacGillivray's breeds mainly in montane areas. Polytypic (2 sspp.; both in N.A.). L 5.3" (13 cm)
Identification ADULT MALE: Blue-gray head and throat form a hood; black lores; black and gray mottling on upper breast do not form a solid patch.

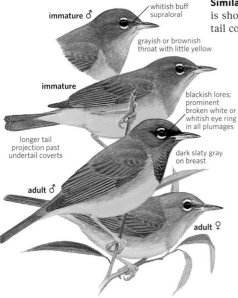

immature ♂
whitish buff supraloral
grayish or brownish throat with little yellow
immature
blackish lores; prominent broken white or whitish eye ring in all plumages
longer tail projection past undertail coverts
dark slaty gray on breast
adult ♂
adult ♀

Bold white crescents above and below eye. Yellow below. **ADULT FEMALE:** Similar to male, but with paler gray hood, paler throat. **IMMATURE:** Similar to adult female, but with more brownish olive hood; throat grayish, rarely yellow tinged.
Geographic Variation The two subspecies are not field-identifiable. Nominate from Pacific coast is more yellow-olive on upperparts and deeper yellow below than interior *monticola*.
Similar Species Mourning Warbler is shorter tailed, has longer undertail coverts; song and calls different. Adult male Mourning usually lacks black lores and usually shows an extensive black breast patch; white eye arcs are absent or very narrow. Immature Mourning Warblers show a yellowish throat and an incomplete breast band; white around eye, when present, is thinner, often forms a nearly complete ring.
Voice CALL: A loud, sharp *tsik*, distinctly different from Mourning. **FLIGHT CALL:** A penetrating *tseep*. **SONG:** Primary two-part song, with a somewhat burry first part and a variable (higher- or lower-pitched) second part: *churry churry churry tree tree tree* or

sweet sweet sweet sweet peachy peachy.
Status & Distribution Fairly common. **BREEDING:** Thickets and more open areas in montane mixed and coniferous forests, usually near rivers and streams. **MIGRATION:** In spring, arrives in southern AZ and Southern CA in early Apr., peaking mid-May over much of the West. In fall, departs breeding areas during Aug., peaking late Aug.–mid-Sept., with some to mid-Oct. **WINTER:** Densely vegetated habitats, from northwestern Mexico to central Panama. Casually in Southern CA, AZ, and TX. **RARE STATUS:** Casual in eastern N.A., where records are widespread in migration and some have wintered.
Population Stable. Benefits from burns and logging.

MOURNING WARBLER *Geothlypis philadelphia* MOWA ■ 1

This skulking warbler arrives on its breeding grounds later than most warblers. Some immatures are very difficult to separate from closely related MacGillivray's Warbler and are best told by call. Monotypic. L 5.3" (13 cm)
Identification Chunky, short tailed. Pinkish-colored legs, mostly pinkish bill somewhat dusky on upper mandible. Plumage somewhat variable. **ADULT MALE:** Blue-gray head and throat form a hood; black bib on upper breast sometimes extends onto throat; lores sometimes black; very rarely with thin, broken eye ring. Olive-green back, wings, and tail. Bright yellow lower breast, belly, and undertail coverts. **ADULT FEMALE:** Similar to male, but with paler gray hood and lacking black bib. Can show nearly complete, very thin eye ring in fall. **IMMATURE:** Similar to adult female, but with more olive-gray hood and

incomplete olive-gray bib; yellowish throat, sometimes quite pale. Shows a thin, broken eye ring (thicker on some); some males may show limited black spotting on bib.
Similar Species Adult male MacGillivray's has different song and especially call, shows black lores and slaty, not black, mottling on throat; all ages and sexes show distinct broadly broken eye arcs, above and below eyes, and longer tail projection. Connecticuts show distinct, complete eye rings and shorter tail projection; they walk instead of hop. Immature and female Common Yellowthroats also hop and are longer tailed, lack gray on crown, have a smaller all-dark bill, and show no yellow on belly.

immature males often with some black stippling on lower throat
immature ♂
dull yellow supraloral
short tail projection past undertail coverts
yellow breaks through breast; extends into throat on most immatures
immature
black chest patch
adult ♂
adult ♀
some problematic immature Mourning or MacGillivray's best separated by call

Voice CALL: A distinctive, sharp, scratchy *chit* or *jip* suggestive of one of Bewick's Wren calls. **FLIGHT CALL:** A thin, sharp *seep*. **SONG:** Variable, most often two-part, with first part louder and slightly burry, second part faster and lower: *curry churry churry chorry chorry*. Also, a one-part song, somewhat similar to some songs of Kentucky Warbler.

Status & Distribution Uncommon to fairly common. **BREEDING:** Second growth in thickets in disturbed areas and along streams. **MIGRATION:** Circum-Gulf migrant. In spring, a few arrive in southern TX as early as very late Apr., peaking in Upper Midwest in late May–early June. Departs breeding grounds in early Aug., peaking late Aug.–early Sept., casually to late Oct., accidentally later. **WINTER:** Dense thickets, overgrown fields, shrublands, and other semi-open areas, often near water; southern Nicaragua to Ecuador. **RARE STATUS:** Very rare in much of Southeast. In West mainly from CA, where very rare: about 170 records, most fall, but including two exceptional coastal winter records. Casual elsewhere in the West. Accidental in fall to AK. Two records from Baja California, three specimens from Greenland.

Population Stable; benefits from openings created by fires, logging and power lines.

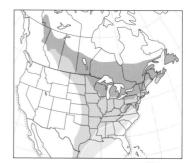

KENTUCKY WARBLER *Geothlypis formosa* KEWA ■ 1

This skulking warbler is more often heard than seen. Monotypic. L 5.3" (13 cm)
Identification Chunky with short tail, olive above and uniform bright yellow below. **ADULT MALE:** Black cap; yellow supercilium forms spectacles; black lores and triangular cheek patch. **ADULT FEMALE:** Similar to male, but reduced black on crown, cheek. **IMMATURE:** Male similar to adult female.

Female dullest; suggestion of face pattern; black replaced by dark olive.
Similar Species Common Yellowthroats lack yellow spectacles and are not solid yellow below.
Voice CALL: Low, distinctive *chup*, similar to Hermit Thrush. **FLIGHT CALL:** Loud, buzzy *zeep*. **SONG:** Series of rich rolling notes: *churree churree churree churree*, suggestive of Carolina Wren.
Status & Distribution Uncommon, primarily a trans-Gulf migrant. **BREEDING:** Dense understory in deciduous or mixed woodlands. **MIGRATION:** Arrives Gulf Coast late Mar.–early Apr., northern extreme of range early May. Departs early Aug., peaks late Aug.–early Sept., stragglers to late Sept. and (exceptional) early Oct. **WINTER:** Forested lowlands, second growth from northeastern Mexico to Panama and extreme northeastern Colombia and northwestern

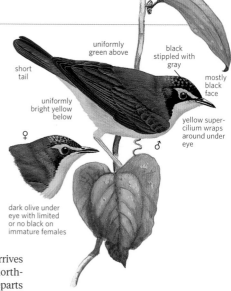

uniformly green above

short tail

uniformly bright yellow below

black stippled with gray

mostly black face

yellow supercilium wraps around under eye

♀

♂

dark olive under eye with limited or no black on immature females

Venezuela. **RARE STATUS:** Spring overshoots are very rare north in eastern N.A. Rare to casual in West. Casual in winter to FL, TX, and CA.
Population Declining, perhaps in part from tropical deforestation.

COMMON YELLOWTHROAT *Geothlypis trichas* COYE ■ 1

The Common Yellowthroat is partial to wetlands over much of range. Polytypic (13 sspp.; 10 in N.A.). L 5" (13 cm)
Identification Short winged and long tailed. Nominate described. **ADULT MALE:** Broad black face mask with grayish border. Bright yellow throat, undertail coverts; paler belly. **ADULT FEMALE:** Face brownish olive; yellowish throat. **IMMATURE:** Somewhat like adult female; many males show some black on face.
Geographic Variation Mask of eastern adult males usually bordered by gray; in West, such as widespread *occidentalis*, mask bordered by whitish. Nominate found in much of East; *chryseola* (southwestern TX to southeastern AZ) entirely yellow below, mask border tinged with yellow. Great Plains *campicola* and northerly *yukonicola* (if recognized) have more restricted yellow throat on mostly white underparts.
Similar Species Kentucky has yellow spectacles and is all-yellow below. Stockier. Mourning and MacGillivray's are shorter tailed and are uniformly yellow from belly to undertail.

Voice **CALL:** Includes husky *tschep*; rapid chatter. **FLIGHT CALL:** Buzzy *dzip*. **SONG:** A variable, loud, rolling *wichity wichity wichity wichity wich.* **Status & Distribution** Common in dense vegetation, usually in or near wetlands. **MIGRATION:** Peaks late Apr.– early May and Sept. in East. In West, extended spring peak; fall Sept.–mid-Oct. Stragglers late Oct. through much of range. **WINTER:** South to extreme northwestern S.A. A few in Great Lakes marshes. **RARE STATUS:** Casual to Arctic Canada, Greenland, Iceland, UK, and Azores. **Population** Stable.

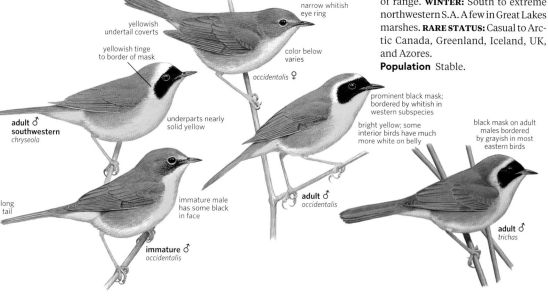

narrow whitish eye ring

yellowish undertail coverts

yellowish tinge to border of mask

color below varies

occidentalis ♀

adult ♂ **southwestern** *chryseola*

underparts nearly solid yellow

prominent black mask; bordered by whitish in western subspecies

bright yellow; some interior birds have much more white on belly

black mask on adult males bordered by grayish in most eastern birds

long tail

immature male has some black in face

adult ♂ *occidentalis*

adult ♂ *trichas*

immature ♂ *occidentalis*

Genus *Setophaga*

Most of these species were formerly placed in the now-defunct genus *Dendroica*. The expanded genus now includes both Northern and Tropical Parulas and Hooded Warbler, for a total of 25 species in N.A.; nine other species are resident in the West Indies for a total of 34 species. Most have pale wing bars or patches and tail spots, and shorter and thicker bills than warblers in *Oreothlypis* and *Vermivora*. Many species have primary songs along with a moderately to distinctly different alternate second song.

HOODED WARBLER *Setophaga citrina*   HOWA ■ 1

The Hooded Warbler is often seen only as a flash in open deciduous woodland. Monotypic. L 5.2" (13 cm)
Identification Opens and closes tail, revealing white outer tail feathers; dark lores; large bill, eye. **ADULT MALE:** Black hood, throat; contrasting bright yellow forehead, cheeks; green above; bright yellow below. **ADULT FEMALE:** Like male, but variable head pattern: most show only narrow black border; a few show nearly complete black hood.

IMMATURE: Male similar to adult male; female dullest; no black on throat and crown; indistinct yellow supercilium, throat; olive cheeks; plumage held for over a year.
Similar Species Smaller, immature female Wilson's Warbler has smaller bill and eye; lacks dark lores, white tail spots. Yellow Warbler has yellow tail spots.

rapidly opens and closes tail

adult ♂

black hood

all Hoodeds have dusky lores

adult ♀

older females have some black to extensive black in hood and necklace

extensive white in outer tail feathers

large dark eye and blank face

immature ♀

Voice **CALL:** Loud *chink*. **SONG:** Clear and whistled, with emphatic ending: *ta-wee ta-wee ta-wee ta-wee tee-too* with variations.
Status & Distribution Fairly common. **BREEDING:** Mixed hardwood forests in north; cypress-gum swamps in south, preferring larger woodlots.

MIGRATION: Primarily trans-Gulf migrant. Arrives Gulf Coast mid-Mar.; Midwest late Apr.; northern range early May. Slightly more easterly route in fall: earliest arrive FL mid-July; latest depart early Sept., small numbers in South into early Oct., casually into Nov. **WINTER:** Males lowland mature forest; females scrub, secondary forest, disturbed habitats. C.A.; very rarely southern FL, northern Colombia, Venezuela. **RARE STATUS:** Rare migrant to Atlantic Provinces and to much of West (where has bred). Casual in winter to southern and Pacific states; accidental UK and Azores. **Population** Stable. May be sensitive to forest fragmentation.

AMERICAN REDSTART *Setophaga ruticilla*   AMRE ▪ 1

This bird is conspicuous in lower levels of vegetation as it frequently fans its tail, showing large tail spots, and as it sallies out to catch insects. It lays one to five eggs in its nest, built against the main trunk of a tree or woody shrub or, sometimes, on a horizontal branch away from the trunk (May–June). Monotypic. L 5.3" (13 cm)
Identification Flat, flycatcher-like bill with prominent rictal bristles. **ADULT MALE:** Largely black with conspicuous orange patches at sides of breast, on bases of wing feathers (forming broad wing stripe), and extensively at bases of outer tail feathers. White belly and undertail coverts. First-spring male sometimes nearly identical to spring female; most often with irregular black blotches on face, head, back, and underparts. **ADULT FEMALE:** Male's orange patches replaced with yellow (orange in some older females), light gray head, olive green back, white underparts. **IMMATURE:** Similar to adult female, often with narrower wing stripe. Sexes often indistinguishable; some males as well as adult females show orange-yellow patch on sides of breast.
Similar Species Slate-throated and Painted Redstarts similar in behavior, but only show white in tail and are markedly different in plumage.
Voice CALL: Thin chip note, like Yellow Warbler. **FLIGHT CALL:** Penetrating, clear *seep*. **SONG:** Significantly variable, high-pitched series, usually with down-slurred, slightly burry final note: *zee zee zee zee zweeah*. Some versions similar to songs of other warbler species.
Status & Distribution Fairly common to common. **BREEDING:** Wide variety of open wooded habitats. **MIGRATION:** Migrates along a broad front Usually arrives Gulf Coast first half of Apr., peaks mid-May through much of northern US. Departs breeding grounds July, peaks Upper Midwest late Aug.–mid-Sept., some into Oct. **WINTER:** Forest, woodland, lower and middle elevations from southern FL (rare in Southern CA) and southern Mexico through Caribbean and in S.A. south to Peru. **RARE STATUS:** Rare migrant in West away from breeding range. Casual to Bering Sea islands, Arctic Canada, Europe, and the Azores. **Population** Declines in parts of range (e.g., New England).

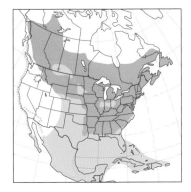

gray head with white eye ring

often spreads and holds open tail

♀

extensive yellow base to outer tail feathers

adult ♂

dark lores

1st spring ♂

1st-spring male usually with some blackish spotting on throat or breast

KIRTLAND'S WARBLER *Setophaga kirtlandii*   KIWA ▪ 2

The rare Kirtland's Warbler has substantially increased in recent decades. It forages lower than most warblers when it persistently pumps its tail. Monotypic. L 5.8" (15 cm)
Identification ADULT MALE: Bluish gray upperparts, with black lores and black below eye; fine black streaks on crown, bolder on back. Yellow underparts with bold black streaks on sides, white undertail coverts. Broken whitish eye ring, indistinct wing bars. **ADULT FEMALE:** Similar to male, but duller, less boldly streaked sides. Upperparts tinged brownish gray. **IMMATURE:** Male similar to adult female. Female more brownish above, more spotted below.
Similar Species Female and immature Magnolias are smaller; show green on back, yellow rump, more prominent wing bars, white band at base of tail, and do not pump tail. Smaller Prairie wags its tail but has a yellow face pattern (except immature female).
Voice CALL: A low, forceful chip note, similar to Prairie Warbler. **FLIGHT CALL:** A thin, high *zeet*. **SONG:** A loud,

deliberate, rich, throaty *chew chew chee chee wee wee*, suggesting Northern Waterthrush.

Status & Distribution Locally fairly common within its limited range and breeding habitat. Rarely observed in migration or on wintering grounds. **BREEDING:** Extensive, dense stands of young jack pine (seven to 15 years old) on sandy soil in central MI with far fewer on Upper Peninsula; a small population of about 50 now breeds in WI, and a few singing males have been noted in southern ON. **MIGRATION:** Rare, but a few reported regularly from Lake Erie region in early to mid-May. Arrives on breeding grounds around May 12 (some earlier). Fall movement as early as late Aug., a few linger as late as early Oct. Fewer reports of fall migrants; some are seen on the spring route, but now annually recorded from central Appalachians (a few from East Coast), perhaps indicating a more easterly route than spring.

WINTER: Dense scrubby undergrowth mainly in central Bahamas; most recently found on Eleuthera. **RARE STATUS:** Casual IN and IL. Accidental MO, NY, and ME.

Population Near Threatened. Removed from the federal Endangered Species List in 2018. Extensive management of breeding habitat (through burning or clear-cuts and replanting of jack pines, as well as, until recently, the removal of parasitic

Brown-headed Cowbirds) has allowed the population to increase from a low of 167 singing males in 1974 to a total population of 5,000 in 2016. Managing for larger tracts of breeding habitat has proven to be the most effective way to increase the population.

female paler yellow below, spotted across breast

grayish lores

immature ♀

black lores and anterior part of face; broken white eye ring

adult ♀

adult ♂

bobs tail like Prairie and Palm Warblers

CAPE MAY WARBLER *Setophaga tigrina*   CMWA ▪ 1

This is one of the most aggressive wood-warblers. Monotypic. L 5" (13 cm) **Identification** Slightly decurved, finely pointed bill. Pale patch on side of neck. **SPRING MALE:** Sides of neck yellow with chestnut cheek patch and a yellow rump. Yellow below with extensive black streaking; whitish undertail coverts, white wing patch. **SPRING FEMALE:** Similar to male, but duller with paler yellow on face and underparts; indistinct wing bars. **FALL ADULT MALE:** Slightly duller. **IMMATURE:** Duller than adult male, lacks chestnut in cheeks, slightly reduced white in wing. Female streaked below; very dull grayish sometimes with some yellow on face; underparts (sometimes tinged yellow) with faint streaks. Note greenish edges on flight feathers and greenish rump.

Similar Species Immature female is similar to "Myrtle" Yellow-rumped and Palm Warblers, which are larger and browner above, with heavier bills and longer tails; they lack yellowish patch on sides of neck. Dull "Myrtle" has a contrasting bright yellow rump. Tail-pumping Palm obscurely streaked and has yellow undertail coverts.

Voice CALL: A high, sharp *tsip*. **FLIGHT CALL:** A soft buzzy *zeet*. **SONG:** Primary song a high-pitched penetrating *seet-seet-seet-seet-seet*. Alternate song lower pitched with several two-syllable notes: *seetee seetee seetee seetee seetee*; similar to Bay-breasted.

Status & Distribution BREEDING: Coniferous forest and bogs with spruce; more in areas with spruce budworm infestations. **MIGRATION:** In spring, more westerly route. Rare on western Gulf Coast. Arrives on FL coast in late Mar., peaks in late Apr. In Midwest, arrives early May, peaks in mid-May. In fall, peaks in Great Lakes and Northeast in mid-Sept., FL in late Sept.–mid-Oct. **WINTER:** Mainly in Caribbean (uncommon in south FL), where it feeds substantially on nectar using its specialized semi-tubular

tongue. **RARE STATUS:** Casual to very rare in West, including AK (has bred), mostly late Sept.–early Oct. In CA, about 250 records. Accidental UK.

Population Variable depending on spruce budworm outbreaks.

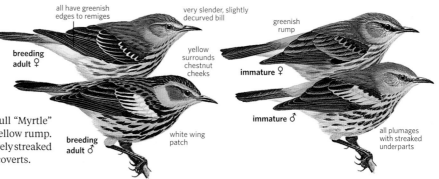

all have greenish edges to remiges

breeding adult ♀

very slender, slightly decurved bill

greenish rump

immature ♀

yellow surrounds chestnut cheeks

immature ♂

breeding adult ♂

white wing patch

all plumages with streaked underparts

CERULEAN WARBLER   *Setophaga cerulea*   CERW ■ 2

This beautiful sky-blue warbler sings, forages, and nests higher in the treetops than most other species. Monotypic. L 4.8" (12 cm)

Identification Long wings with thick white wing bars and short tail. **ADULT MALE:** Cerulean upperparts, brightest on crown; black streaks on back; white underparts; narrow dark blue band across lower throat; bold dark blue streaks on sides. Indistinct supercilium and reduced breast band in first-spring male. **ADULT FEMALE:** Bluish green crown, back, and rump; unstreaked back; whitish underparts variably washed with yellow; gray eye line, broad and long yellowish supercilium; diffuse greenish streaks on sides, no breast band. **IMMATURE:** Similar to adult female with bold supercilium. Male's plumage a mixture of bluish gray and greenish above, some streaking on back, more prominent side streaks. Female dullest; greenish upperparts with little or no bluish tint; more yellow on underparts (especially throat).

Similar Species Immature female Blackburnian is most similar to immature Cerulean, but shows dark auricular patch with broad pale supercilium connected to yellow patch on sides of neck and more olive-gray upperparts with pale streaking on sides of back. Immature Black-throated Gray is similar, but is never bluish or greenish.

Voice CALL: A slurred chip note. **FLIGHT CALL:** A buzzy *zzee*. **SONG:** A variable, buzzy, accelerating, rising series of notes, the last note highest: *zhee zhee zhee zizizizizi zzzziiii*. Similar to some songs of Northern Parula, but the rising buzzy note at end is distinctive.

Status & Distribution Uncommon and declining. **BREEDING:** Mature deciduous woods with open understory. **MIGRATION:** Trans-Gulf migrant. In spring, arrives on Gulf Coast early to mid-Apr.; arrives in Great Lakes early May. In fall, departs breeding grounds early and sometimes arrives on wintering grounds as early as Aug. Some linger in Southeast into Oct. **WINTER:** Canopy and forest borders at middle and lower elevations on east-

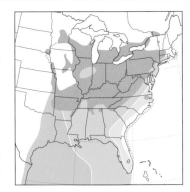

ern Andean slopes from Colombia and Venezuela south to Peru, rarely to Bolivia. **RARE STATUS:** Casual in West: recorded from CO, NM, AZ, NV, and CA (13+ records), as well as Baja California. Also casual in New England, Maritimes, NL, Bermuda, Bahamas, and Greater Antilles. Accidental Iceland and Azores.

Population Vulnerable. Formerly more numerous, but with steep declines in recent decades and recently extirpated as a breeder in several states. Identified causes of declines include deforestation, fragmentation, management of forests for even-aged stands, and loss of some key tree species including oaks, elms, chestnuts, and sycamores.

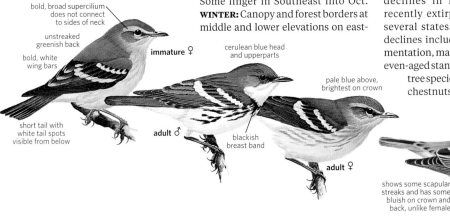

bold, broad supercilium does not connect to sides of neck

unstreaked greenish back

bold, white wing bars

short tail with white tail spots visible from below

immature ♀

cerulean blue head and upperparts

adult ♂

blackish breast band

pale blue above, brightest on crown

adult ♀

shows some scapular streaks and has some bluish on crown and back, unlike female

fall immature ♂

NORTHERN PARULA   *Setophaga americana*   NOPA ■ 1

One of our smallest and shortest-tailed warblers, the Northern Parula inhabits mossy environments and is often found singing high in the treetops. It usually builds its nest in bunches of epiphytes (*Usnea* lichen in north, *Tillandsia* in south) on the end of a branch, ranging from low to high in a tree; it lays between three and five eggs from Apr. to July. Monotypic. L 4.5" (11 cm)

Identification Bicolored bill: black upper mandible and yellow lower mandible. **SPRING MALE:** Blue-gray upperparts, including head and sides of throat, with bronze-green upper

back; two broad, white wing bars; white eye ring broken at front and rear; black lores. Yellow throat and breast with variable reddish and black bands across breast; white belly and undertail coverts. Large white spots on outer tail feathers. **SPRING FEMALE:** Similar to male, but duller blue-gray above; breast bands are much reduced or absent. **IMMATURE:** Similar to spring female, but duller with more greenish upperparts and smaller white tail spots. **HYBRID:** "Sutton's Warbler" (not illustrated) is a very rare hybrid of Northern Parula with Yellow-throated Warbler. Its appearance is similar to

Yellow-throated Warbler, but it has Northern Parula's green back patch, no white ear patch, and fewer side streaks.

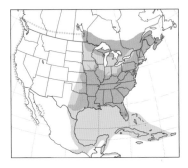

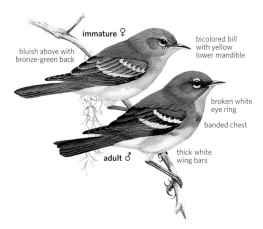

immature ♀

bluish above with
bronze-green back

bicolored bill
with yellow
lower mandible

broken white
eye ring

banded chest

adult ♂

thick white
wing bars

Similar Species Tropical Parula has a black face mask, but it lacks black or chestnut bands on breast; it also lacks white eye ring. Beware of hybrids with intermediate characteristics; they are noted somewhat regularly in southwest portion of Edwards Plateau, TX.

Voice CALL: A sharp *tsip*. **FLIGHT CALL:** A high, weak, descending *tsif*. **SONG:** Primary song a rising trill with a terminal note that drops sharply. The song of western populations has a less emphatic, up-slurred ending, more like Tropical Parula. Alternate song is more complex and wheezy: *b-zee-b-zee-b-zeee-zee-zee-zee-up*.

Status & Distribution Fairly common to common. **BREEDING:** Moist deciduous, coniferous, or mixed woodlands in north; hardwood bottomlands along rivers and swamps. Almost always associated with epiphytic growth. **MIGRATION:** Eastern birds migrate through the Caribbean; western birds migrate through TX and Mexico. Arrives widely in southern US late Mar.–early Apr., in northeastern US in mid-May. Rare over much of the West and occasionally breeds in coastal central and northern CA. In fall, it departs breeding grounds in Aug., with migration peaking Sept.–mid-Oct. **WINTER:** In a wide variety of habitats, preferring undisturbed areas to disturbed areas. In Caribbean, most of Mexico south to El Salvador, rarely from Nicaragua to Panama. Rare in south TX and CA. Casual in LA, NM, and SC. **RARE STATUS:** Rare from Prairie Provinces and the West (scarcer in fall than in spring). Accidental in AK (sight record from Middleton I.). Casual to Greenland, Europe (more than 15 records), and the Azores.

Population Due to habitat requirements, populations are locally variable. Long-term trends are stable, but recent shorter-term trends show declines.

TROPICAL PARULA *Setophaga pitiayumi* TRPA ▪ 3

Tropical Parula, a close relative of the Northern Parula—with which it is sometimes considered conspecific—reaches N.A. in south TX. Polytypic (9 sspp.; at least 1 (*nigrilora*), likely 2, in N.A.). L 4.5" (11 cm)

Identification Black upper mandible, yellow lower mandible. **SPRING MALE:** Bright blue-gray upperparts, including head, with green-bronze upper back; two broad white wing bars, small black face mask. Yellow throat and belly, orange-yellow breast, white undertail coverts. Large square white spots on outer tail feathers. **SPRING FEMALE:** Similar to male, but duller above. Black face mask is lacking or much duller; breast shows little or no orange. **IMMATURE:** Similar to spring female, but more greenish above and duller yellow below.

Geographic Variation The east Mexican subspecies (*nigrilora*) breeds north to south TX. The west Mexican subspecies (*pulchra*) may account for the records in AZ and the one in CA. It differs from the eastern subspecies by its larger size, longer tail, more white on greater wing covert tips, and more cinnamon on flanks. S.A. subspecies differ more subtly from each other.

Similar Species Northern Parula shows a broken white eye ring, and blue-gray coloring extends to sides of throat. It has a white belly and less black on face (lores only). Males show a breast band of black and chestnut. Tropical Parula hybridizes with Northern Parula in TX.

Voice CALL: A thin slurred chip note, similar to Northern Parula's call note. **SONG:** The primary song is a rising trill that ends abruptly with a separate buzzy, down-slurred terminal note. This song is similar to the song of Northern Parula from western populations. The alternate song is quite variable and more complex, but it always has a buzzy quality.

Status & Distribution Found from south TX and northwest Mexico to northern Argentina. **BREEDING:** Mainly in live-oak woodlands with abundant epiphytes, especially Spanish moss. Arrives on nesting grounds in TX by Apr. Has spread north to southwestern Edwards Plateau, TX. **MIGRATION:** In spring, they arrive in south TX in mid-Mar. In fall, they depart south TX by Sept. **WINTER:** A few overwinter in the southern part of the breeding range (south TX) in a variety of woodlands; often joins mixed species feeding flocks. Some postbreeding dispersal to the north and east happens in fall and winter. Winter range of TX breeding birds uncertain. **RARE STATUS:** Rare to West and north TX. Casual or accidental to CA, AZ, LA, MS, CO, and Baja California.

Population The southern TX population declined significantly after a devastating freeze in the winter of 1951. Habitat destruction may have hindered the Tropical Parula's recovery. It is common in most of its Neotropical range.

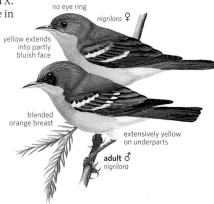

no eye ring

nigrilora ♀

yellow extends
into partly
bluish face

blended
orange breast

extensively yellow
on underparts

adult ♂
nigrilora

MAGNOLIA WARBLER *Setophaga magnolia*   MAWA   ▪ 1

Easily identified in all plumages from below by its unique broad white base to a black tail. Monotypic. L 5" (13 cm) **Identification** Diagnostic tail pattern. **SPRING MALE:** Broad white line separates gray crown from black mask; black back; yellow rump. Yellow underparts with broad black streaks on breast and sides. White lower belly and undertail coverts. White wing patch. Black tail with broad white band near base. **SPRING FEMALE:** Variable; duller than male, with narrower wing bars and breast streaks, grayish mask, some greenish on back, less purely gray crown. **FALL ADULT:** Duller with reduced black. Narrow white eye ring on grayish cheek, white wing bars, and narrower breast streaks; black upper-tail coverts edged grayish. Female duller, with more greenish on back, less breast streaking. **IMMATURE:** Similar to fall adult female, but duller, sometimes nearly lacking streaks on breast and sides (especially female). **Similar Species** Immature Prairie and Kirtland's Warblers are somewhat similar, but Magnolia shows a unique tail pattern and complete white eye ring

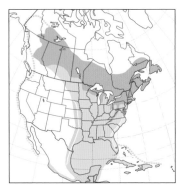

on gray face, and it does not wag its tail like those species. In particular, some first-spring female Magnolias are frequently confused with the larger Kirtland's Warbler, but note Kirtland's nearly continuous tail bobbing and very different tail pattern. Prairie has a distinct subocular and is more extensively yellow below, with less distinct wing bars.
Voice CALL: A unique nasal *enk*, most often heard in fall migration and on wintering grounds. **FLIGHT CALL:** A high, buzzy *zee*. **SONG:** Primary song is a variable short, musical *weeta weeta wit-chew*, accented on the final notes. Alternate song is an unaccented *sing sweet*.
Status & Distribution Fairly common to common. **BREEDING:** Dense young coniferous or mixed woodland, usually fairly low to ground. **MIGRATION:** Mainly a trans-Gulf migrant. In spring, arrives on Gulf Coast in mid-Apr., peaks in northern US in late May, arrives on breeding grounds in late May–early June. In fall, departs breeding grounds mid-Aug., peaks through Sept., and lingers as late as late Oct. (sometimes into Dec. and Jan. in LA and FL). **WINTER:** Shrubby second growth, wooded, and agricultural areas in Caribbean and from southern Mexico to Costa Rica; uncommon south FL. **RARE STATUS:** Very rare spring and rare fall migrant west to Pacific states (casual in winter). Casual to the Azores. Accidental in Bar-

bados, Trinidad, Tobago, Venezuela, northern Colombia, and the UK.
Population Stable or slightly increasing. Uses second growth.

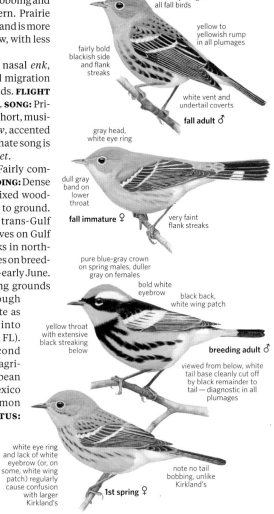

white eye ring on all fall birds

yellow to yellowish rump in all plumages

fairly bold blackish side and flank streaks

white vent and undertail coverts

fall adult ♂

gray head, white eye ring

dull gray band on lower throat

fall immature ♀

very faint flank streaks

pure blue-gray crown on spring males, duller gray on females

bold white eyebrow

black back, white wing patch

yellow throat with extensive black streaking below

breeding adult ♂

viewed from below, white tail base cleanly cut off by black remainder to tail — diagnostic in all plumages

white eye ring and lack of white eyebrow (or, on some, white wing patch) regularly cause confusion with larger Kirtland's

note no tail bobbing, unlike Kirtland's

1st spring ♀

BAY-BREASTED WARBLER *Setophaga castanea*   BBWA   ▪ 1

In fall plumage, very similar to a fall Blackpoll. Monotypic. L 5.5" (14 cm) **Identification SPRING MALE:** Black face; chestnut crown, throat, sides; cream neck patch; two broad white wing bars. **SPRING FEMALE:** Somewhat to much duller than male. **FALL ADULT:** Sexes similar, female slightly duller. Crown, nape, and back are yellowish olive, with indistinct black streaks. Indistinct eye line and super-cilium; breast unstreaked, whitish to pale buff undertail coverts; some or little chestnut on flanks; dark legs and

feet. **IMMATURE:** Unstreaked underparts whitish buff with yellow tinge on breast; chestnut on flanks reduced (male) or absent (female).
Similar Species Pine has unstreaked back; yellow or whitish of throat extends behind auriculars; wing bars contrast less; much longer tail extends well beyond undertail coverts; primary projection shorter. Blackpoll more similar. (See sidebar, p. 702.)
Voice CALL: Loud, slurred *tchip*. **FLIGHT CALL:** Buzzy *zeet*. **SONG:** A variable series of very high-pitched

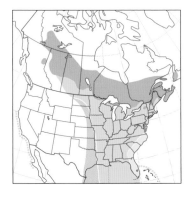

lisping notes: *see-see-swee-see-see-swee-swee-see.*

Status & Distribution Uncommon to fairly common, depending on spruce budworm outbreaks. **BREEDING:** In boreal forest, mainly in mature dense, spruce-fir forests. **MIGRATION:** Mainly a trans-Gulf migrant, rarely through Caribbean. Arrives Gulf Coast mid- to late Apr., peaks Great Lakes last half of May. In fall, more easterly than in spring, departs late July; peaks eastern N.A. late Aug.–late Sept.; stragglers to late Oct., later on Gulf Coast. **WINTER:** Forest edges, second growth from Costa Rica through Panama to northwestern Colombia, northern Venezuela. Feeds mainly on fruit.

RARE STATUS: Casual to accidental in migration to nearly all western states (very rare in CA; casual in winter), including western AK. Casual in Labrador; accidental Greenland and UK.

Population Spraying for spruce budworm has reduced the breeding populations in recent decades.

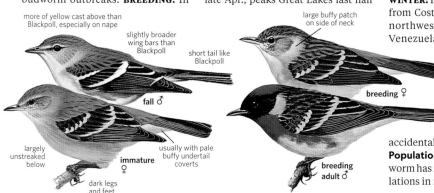

more of yellow cast above than Blackpoll, especially on nape

slightly broader wing bars than Blackpoll

short tail like Blackpoll

fall ♂

largely unstreaked below

immature ♀

usually with pale buffy undertail coverts

dark legs and feet

large buffy patch on side of neck

breeding ♀

breeding adult ♂

BLACKBURNIAN WARBLER   *Setophaga fusca*   BLBW   ▪ 1

The fiery orange throat and head markings of the spring male Blackburnian Warbler are striking. Monotypic. L 5" (13 cm)

Identification Dark cheek patch; a broad, pale supercilium connects to pale sides of neck; pale stripes on sides of back (braces). **SPRING MALE:** Black triangular cheek patch is surrounded by fiery orange. Black crown with orange patch; black back with conspicuous pale yellowish stripes on side of back. Belly pale yellow (white on first spring male); white undertail coverts. White wing patch. Extensive white on outer tail feathers. **SPRING FEMALE:** Similar to male, but duller, grayish brown replaces black on head and back; orange duller and and less white in wing. **FALL ADULT:**

Male with duller orange coloring; its black areas are veiled with olive edges; and it has less white in wing. Female similar but has yellower head markings. **IMMATURE:** Male is similar to fall adult female, but with black eye line; brighter throat and head markings. Female dull, with throat and supercilium colored pale yellow to buffy (sometimes almost whitish) and indistinct streaks on sides.

Similar Species Immature female Cerulean Warbler shows plain greenish back, a less extensive ear patch; supercilium not joined to pale areas on sides of neck; tail shorter. Fall Bay-breasted Warbler and Blackpoll Warbler lack a well-defined supercilium; note plumper bodies and shorter tails.

Voice CALL: A rich *tsip*. **FLIGHT CALL:** A buzzy *zzee*. **SONG:** Primary song is a very high-pitched, ascending series of notes, ending with an almost inaudible trill: *see-see-see-see-ti-ti-ti-siiii.* Alternate song begins at a high pitch and concludes with a lower-pitched ending: *tsee-tsee-tsiii-chi-chi.*

Status & Distribution Fairly common. **BREEDING:** Mature coniferous or mixed forests in Appalachian Mts. **MIGRATION:** Trans-Gulf migrant. In spring, takes a more westerly route, arriving on Gulf Coast in early Apr. and peaking in southern ON in mid-May. In fall, departs breeding grounds early and arrives in the Great Lakes by early Aug., where it peaks in mid-Aug. In the South, peaks during Sept. and on through early Oct. Stragglers can be found into Nov. and Dec. **WINTER:** Most numerous in the montane forests of northern Andes; also in Amazonia and northern S.A. A few winter north to Costa Rica; casual in N.A. **RARE STATUS:** Rare to casual in all western states; more than 500 records from CA. Accidental to Greenland and Europe.

Population Appalachian breeding population is threatened by the loss of hemlock trees killed by introduced insect pests.

breeding ♀

bold white wing patch

breeding adult ♂

fiery orange throat

adult males have tawny yellow belly

dark triangular auricular patch

bold, broad supercilium connects to pale sides of neck

all Blackburnians have pale mantle lines

fall adult ♂

immature ♀

YELLOW WARBLER *Setophaga petechia* YEWA 1

This very familiar and widespread warbler is often associated with willows. It builds its nest in an upright fork of a bush, sapling, or tree and lays four to five eggs (May–June). When its nest is parasitized by a Brown-headed Cowbird, the Yellow Warbler is known to build a new nest on top of the old one. Polytypic (43 sspp.; ±11 in N.A.). L 5" (13 cm)

Identification Only wood-warbler with yellow tail spots. **SPRING MALE:** Entirely yellow; crown sometimes tinged chestnut. Conspicuous black eye on yellow face. Wing feathers broadly edged with yellow. Chestnut streaks on breast and sides. **SPRING FEMALE:** Similar to male, but duller yellow on head and underparts. Breast streaking indistinct or absent. **IMMATURE:** Duller and generally lacking chestnut breast streaks; crown and face more olive-yellow, as on back. Sometimes very olive or grayish overall.

Geographic Variation The 43 subspecies are often arranged into three main groups based on plumage characteristics and geographic distribution: "Northern" (*aestiva*), "Golden" (*petechia*), and "Mangrove" (*erithachorides*) groups. "Northern" group consists of nine subspecies, eight breeding across N.A., they tend to be darker in the north and paler in the south; southwestern subspecies (*sonorana*) is most distinctive, showing minimal chestnut breast streaking. "Golden" group consists of 18 subspecies, mostly in Caribbean. They are represented by a single subspecies (*gundlachi*) in N.A., which breeds in mangroves in south FL and Cuba; it shows an extensively olive crown and shorter wings. "Mangrove" group (16 sspp. found coastally from Mexico to Galápagos) is represented by two to three subspecies in the US, of which *oraria* is resident in coastal southern TX; all adult males in this group (except *aureola* from Galápagos and Isla de Coco) have chestnut heads.

Similar Species Duller immatures of northwestern (*rubiginosa*), AK (*banksi*), and south FL (*gundlachi*) subspecies can be very grayish yellow and can be confused with Orange-crowned or Wilson's. Yellow Warbler can be recognized by its plain, unmarked face with a narrow yellow eye ring and by its yellow tail spots.

Voice CALL: A husky, downslurred *tchip* or a thinner *tsip*. **FLIGHT CALL:** A buzzy *zeet*. **SONG:** Primary song somewhat variable between individuals, usually most similar to *sweet sweet sweet sweeter than sweet*. Alternate songs are longer and more complex, sometimes sounding similar to Chestnut-sided.

banksi
rubiginosa
parkesi
"NORTHERN" group
amnicola
brewsteri
morcomi
aestiva
rhizophorae
sonorana
gundlachi
"GOLDEN" group
castaneiceps
oraria
"MANGROVE" group
dugesi
bryanti
phillipsi
multiple subspecies

Status & Distribution Common in second growth and shrubby areas; a long-distance migrant to C.A. and S.A. **BREEDING:** Wet deciduous thickets, often dominated by willows. **MIGRATION:** In East, mostly circum-Gulf or along western Gulf, arriving in southern US by mid-Apr. and in Great Lakes in late Apr.–early May. In West, variable depending on subspecies, but generally by late Mar. in southern AZ, late Apr. in OR and WA, and late May in AK. In East, one of the earliest departing warblers in fall, with birds leaving Great Lakes beginning in mid-July and rarely later than mid-Sept. (the more northerly eastern *amnicola* departs later). Small numbers along the Gulf Coast into mid-Oct. Later in West, beginning in late July, peaking in late Aug. and early Sept., and extending into mid-Oct. **WINTER:** Mexico to central Peru and northern Brazil. Rare in Southern CA. **RARE STATUS:** "Mangrove" group is accidental to Southern CA, southern AZ, and Rockport, TX. "Northern" group birds are casual in fall to Greenland and Europe.

Population Stable and widespread.

dark eye stands out in blank face
immatures of widespread eastern aestiva typically pale yellow
pale tertial edges
immature ♀ aestiva
adult ♂ aestiva
immature ♀ gundlachi
red streaks on underparts
whitish underparts
immature ♀ amnicola
all Yellow Warblers have pale legs and feet
overall duller and darker than aestiva
faint streaks
short tail with yellow tail spots
♀ aestiva
immature rubiginosa
immatures of northern subspecies are duller and darker and are washed with olive or brownish

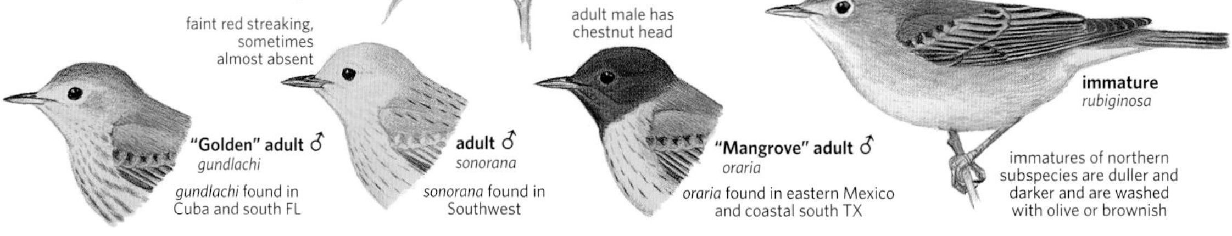
faint red streaking, sometimes almost absent
adult male has chestnut head
"Golden" adult ♂ gundlachi
gundlachi found in Cuba and south FL
adult ♂ sonorana
sonorana found in Southwest
"Mangrove" adult ♂ oraria
oraria found in eastern Mexico and coastal south TX

CHESTNUT-SIDED WARBLER *Setophaga pensylvanica*   CSWA   ■   1

The Chestnut-sided Warbler is commonly seen foraging low in small trees and shrubs with drooped wings and cocked tail. Nests low in shrubby understory. Monotypic. L 5" (13 cm)

Identification Distinct white tail spots in all plumages. **SPRING MALE:** Bright yellow crown; black lores, eye line, and whisker; white cheek and underparts; extensive chestnut on sides. Wings with two pale yellow wing bars. First-spring males usually have a shorter chestnut strip, not reaching the flanks. **SPRING FEMALE:** Similar to male, with greener crown, duller upperparts, less black on face; the blackish in malar region is often broken and does not meet the chestnut on the sides; and less chestnut on the sides. **FALL ADULT:** Distinctive; very different from spring adult. Bright lime green upperparts and yellow wing bars. Pale grayish cheek, throat, and upper breast. Whitish underparts. Conspicuous white eye ring. Chestnut reduced and largely restricted to sides and flanks; more restricted on female. **IMMATURE:** Similar to fall adult, but all females and some males completely lack chestnut on sides. Back streaking is faint on female.

Similar Species Essentially unmistakable in all plumages. Note lime green upperparts, yellow wing bars, and distinct white eye ring on all fall birds.

Voice CALL: A loud, sweet *tchip*, almost identical to Yellow Warbler, but not delivered in a rapid series like a Yellow when agitated. **FLIGHT CALL:** A very burry, slightly musical *breeet*. **SONG:** Primary song is similar in quality to Yellow Warbler, but its phrasing is different, with variations similar to *please, please, pleased to meetcha*, the last note with a distinct drop. Alternate song is rather nondescript and is more similar to Yellow Warbler's song and to some songs of American Redstart, with variations similar to *wee-weewee-wee-chi-tee-wee*.

Status & Distribution Fairly common breeder in early successional second growth. **BREEDING:** Northern hardwood and mixed woodland. **MIGRATION:** Medium-distance migrant to C.A. In spring, arrives in southern US in mid-Apr. and over much of eastern N.A. by mid-May. Common migrant along western Gulf Coast, very rare in Caribbean. In fall, departs breeding grounds in Aug. and Sept.; more easterly migration continues without a clear peak through Sept. and some to mid-Oct. **WINTER:** Southern Mexico through Panama; most numerous in Costa Rica in a variety of forested and second-growth habitats; rare in AZ and CA. **RARE STATUS:** Rare in the West. Accidental in AK. Casual in northern S.A. Accidental to Greenland and UK.

Population Chestnut-sided Warbler has benefited from deforestation, which has opened up breeding habitat. Population declines since the 1960s are possibly related to urbanization as well as to reforestation and maturation of habitat.

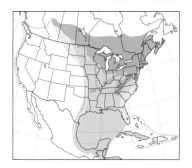

lime green upperparts

yellowish wing bars

gray face with white eye ring

grayish white underparts

immature

usually cocks tail

greenish crown

breeding adult ♀

yellow crown

chestnut sides

breeding adult ♂

BLACKPOLL WARBLER *Setophaga striata*   BLPW   ■   1

The black-and-white males are easily identified in spring; duller fall birds make identification challenging. Monotypic. L 5.5" (14 cm)

Identification SPRING MALE: White cheek; black cap and malar stripe; bold black streaks on sides of white underparts. Two bold white wing bars, orange-yellow legs and feet, yellowish lower mandible. **SPRING FEMALE:** Variable. Olive-gray upperparts, with dark streaks on back. Dark eye line, and streaked malar. Distinct side streaks. Some more yellowish olive above and below. **FALL:** Upperparts olive with dark streaks on back; underparts yellowish with narrow streaks on sides; white undertail coverts. Pale or dark legs, always with yellow on soles of feet. Immatures are very similar to fall adult, but some are yellower below with duller streaking on sides. Sexes indistinguishable.

Similar Species In fall, Pine Warbler is unstreaked above and dark cheek sharply contrasts with throat; it has a much longer tail and shorter primary projection. In fall, Bay-breasted is very similar. (See sidebar, p. 702.) Spring male is superficially similar to Black-and-white; note different foraging style.

Voice CALL: A loud, sharp

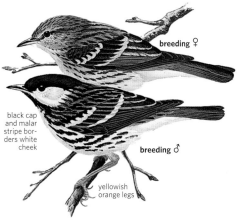

breeding ♀

black cap and malar stripe borders white cheek

breeding ♂

yellowish orange legs

tchip. **FLIGHT CALL:** A loud, sharp, buzzy *zeet.* **SONG:** A series of very high-pitched staccato notes, inaudible to many, usually louder in the middle: *tsit tsit tsit tsit tsit tsit tsit.*

Status & Distribution Common; undertakes extremely long nonstop

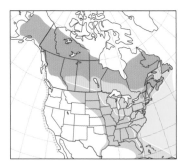

overwater migrations. **BREEDING:** Boreal spruce forest and spruce-alder-willow thickets. **MIGRATION:** In spring, takes a more westerly route, through the Caribbean as far west as coastal TX; most later than most other wood-warblers. Departs wintering areas in Apr. and arrives on breeding grounds mid-May–early June. Peaks in Midwest and mid-Atlantic states mid- to late May, with a few into early June. In fall, a more easterly route, including an overwater flight from northeastern coastal US and Atlantic Canada to northern S.A.; a few south through eastern Caribbean. Peak numbers through northeastern US from mid-Sept. to early Oct., with some into Nov., a very few to mid-Dec. **WINTER:** A variety of wooded habitats

in northern S.A. east of Andes south to northern Bolivia. One valid N.A. winter record from Long I., NY. **RARE STATUS:** Rare fall migrant in West, primarily along Pacific coast; very rare elsewhere in West. Very rare in spring west of Rockies. Very rare in Costa Rica and Panama. Casual in fall to Greenland, Europe, and the Azores. **Population** Near Threatened.

faint streaks on sides of breast

short tail with long undertail coverts and white tail spots

fall

yellow soles to feet and usually to back of legs

Separation of Bay-breasted and Blackpoll Warblers in Fall

These two species define the phrase "confusing fall warblers." The similar Pine Warbler can usually be distinguished by its very different face pattern in all plumages with a darker cheek sharply delineated from the throat, shorter wings, and much longer tail projection past the undertail coverts. Adult Bay-breasteds and Blackpolls appear considerably different in fall than in spring. They and dull fall immatures look quite similar to one another.

Both species show a similar face pattern, with a dark eye line and a pale broken eye ring. The eye line is better defined in Blackpoll.

Both species have greenish olive backs with blackish streaks. Streaks tend to be more conspicuous on Blackpoll and less so on Bay-breasted, but this is somewhat variable. Bay-breasted has a brighter yellowish olive on the crown, nape, and back. When noting wing bars, look for Bay-breasted's often broader white wing bars.

Bay-breasted Warbler, immature (GA)

Blackpoll Warbler, immature (NY)

Bay-breasted tends to be buffier from throat to undertail coverts, often buffiest on vent; it also has whitish undertail coverts and, sometimes, a little to moderate chestnut (males) or richer buff (females) on the flanks. Blackpoll tends to be yellowish with contrastingly white (or occasionally yellow) undertail coverts. Unlike Bay-breasted, it also usually shows narrow dark olive streaks on the sides of the breast and flanks.

Both species have white tail spots and long undertail coverts; however, Blackpoll's undertail coverts tend to be longer and whiter, whereas Bay-breasted's are slightly shorter and pale buff, but can be nearly whitish.

Foot color is also important. Although immatures of both species can show dark legs, the soles of Bay-breasted's feet are bluish gray whereas Blackpoll's are yellowish. Sometimes this yellowish color occurs narrowly up the rear of the tarsus.

BLACK-THROATED BLUE WARBLER *Setophaga caerulescens* BTBW ▪ 1

The sexes of this species are strikingly different in plumage and easy to identify. Polytypic (2 sspp.; both in N.A.). L 5.3" (13 cm)

Identification ADULT MALE: Dark blue above. Black face, throat, and flanks. White underparts. Large white wing patch at base of primaries. **ADULT**

FEMALE: Greenish gray upperparts, buffy underparts. Dusky ear coverts, whitish supercilium, and white crescent under eye. White spot at base of primaries. **IMMATURE:** Similar to respective sex of adult; pale primary spot smaller or absent (some females). Male green-tinged above; black

throat and breast tipped with white. **Geographic Variation** Males, particularly adults, of the southern Appalachians *cairnsi* tend to have more black markings on back, sometimes coalescing into a patch. **Similar Species** Female vaguely suggestive of Orange-crowned Warbler, but head pattern differs and pale patch is at base of primaries, not at bend of the wing on marginal coverts.

Voice CALL: Popping *tuk*. **FLIGHT CALL:** A prolonged *tseet*. **SONG:** Primary song a slow series of buzzy notes, rising at the end: *zhee zhee zhee zeeee*; sometimes faster. Alternate song shorter: *zree zree zhrurrr*.

Status & Distribution Fairly common. **BREEDING:** Deciduous and mixed woodland. **MIGRATION:** Migrates mainly through East Coast and Great Lakes. Arrives in Southeast mid-Apr., peaks in Great Lakes mid-May. Departs breeding grounds

mid-Aug; peaks somewhat later than other warblers (late Sept.–early Oct.) in Great Lakes; a few later. **WINTER:** Forested areas of Greater Antilles; some in Bahamas, a few in Yucatán Peninsula, Belize. **RARE STATUS:** Very rare to the West in fall, casual in spring and winter; accidental AK. Casual to the Azores; accidental to Iceland.

Population Slight declines.

whitish supercilium with dark cheek

whitish eye arc

buffy underparts ♀

most females show whitish patch at base of primaries

dark blue crown

black throat

black stippling on back

caerulescens ♂

white patch larger in adult males

Appalachians ♂
cairnsi

PALM WARBLER *Setophaga palmarum*   PAWA   ▪ 1

The most terrestrial *Setophaga*, the Palm Warbler wags its tail vigorously. Polytypic (2 sspp.; both in N.A.). L 5.5" (14 cm)

Identification Western *palmarum* described. **SPRING MALE:** Chestnut crown, streaked grayish brown back, yellow-olive rump. Yellow supercilium with dark eye line, narrow crescent below eye whitish. Bright yellow throat and undertail coverts contrast with whitish belly; thin dark chestnut malar streak and narrow streaks on breast and sides. **SPRING FEMALE:** Very similar to male; often not distinguishable, sometimes less chestnut in crown, paler yellow areas. **FALL ADULT:** Chestnut on crown usually lacking; dull whitish supercilium, submoustachial, and throat; less distinct breast and side streaks. Bright yellow undertail coverts (typically the only yellow at this season). **IMMATURE:** Very similar, often indistinguishable

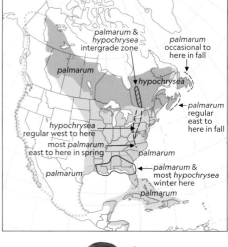

palmarum & hypochrysea intergrade zone

palmarum occasional to here in fall

palmarum

hypochrysea

palmarum regular east to here in fall

hypochrysea regular west to here

most *palmarum* east to here in spring

palmarum

palmarum

palmarum & most *hypochrysea* winter here

palmarum

western breeding *palmarum*

duller midsection contrasts with yellow throat and undertail coverts

both subspecies constantly bob tail

white tail spots

distinctive pale supercilium and dark eye line

western fall *palmarum*

yellow undertail coverts

western fall *palmarum*

some are tinged yellow throughout underparts

more olive above and more uniformly yellow below in all plumages than *palmarum*

rufous cap

eastern breeding *hypochrysea*

all yellow underparts with chestnut streaks on sides of breast

underparts and supercilium yellowish throughout

eastern fall *hypochrysea*

earlier spring and later fall migrant than *palmarum*

from adult. More pointed tail feathers. **Geographic Variation** The more widespread "Western Palm" (*palmarum*) breeds from the Hudson Bay region west. The slightly larger "Yellow" or "Eastern Palm" (*hypochrysea*) breeds from eastern QC to Atlantic Canada and northern New England, and shows entirely yellow underparts (chin to undertail coverts) with little or no contrast; yellowish narrow crescent below eye; yellow-green tinged upperparts; and broader, brighter chestnut breast and side streaks. "Yellow Palm" is duller in fall but still distinguishable. **Similar Species** Prairie and Kirtland's share tail-wagging habit but lack contrasting bright yellow undertail coverts. Similarly, female Cape May and fall Yellow-rumped lack bright yellow under-

tail coverts and do not wag their tails. **Voice CALL:** A distinct, sharp *chick*. **FLIGHT CALL:** A high, light *seet* or *see-seet*. **SONG:** A somewhat buzzy or gravelly series of notes, uttered somewhat unenthusiastically and often more forcefully in the middle: *zwee zwee zwee zwee zwee zwee zwee zwee.* **Status & Distribution** Common. **BREEDING:** Best known from bogs. **MIGRATION:** Timing differs between subspecies. "Eastern" migrates earlier in spring and later in fall than "Western." "Western" migrates through Mississippi Valley to Canada, typically arrives in Upper Midwest mid- to late Apr., peaking in late Apr.–early May; "Yellow" migrates northeast along the Atlantic coast, arrives by mid-Apr. (to NS) to early May (to NL). "Western"

takes a more easterly route than in spring, mainly arriving mid-Sept., peaking late Sept.–early Oct. through much of the East; "Yellow" is very rare west of the Appalachians; casual in West. **WINTER:** A variety of open woodlands, second growth, thickets, and open areas. "Yellow" primarily along Gulf Coast (LA to FL); nominate throughout southeast (TX to VA, rarely farther north), Caribbean, and eastern coastal Mexico to Honduras. **RARE STATUS:** Rare to very rare throughout West, most numerous in fall from along coast (some winter); casual AK (fall). Casual in spring in West. Casual in Costa Rica, Panama, northwestern Colombia, and western Venezuela. Accidental Europe and Azores. **Population** Stable.

PINE WARBLER *Setophaga pinus* PIWA ■ 1

The Pine Warbler appropriately prefers pines. It often forages on the ground in fall and winter with Yellow-rumped Warblers, Chipping Sparrows, and Eastern Bluebirds. Polytypic (4 sspp.; 2 in N.A.). L 5.5" (14 cm)
Identification ADULT MALE: Olive green upperparts. Yellow throat, breast, and belly, extending to rear of olive cheeks; white lower belly and undertail coverts. Dull olive to blackish indistinct

immature ♀
pinus

brownish above

pale wraps around auriculars

whitish below

long tail projection past undertail coverts

immature ♂
pinus

unstreaked back in all plumages, unlike Blackpoll and Bay-breasted

adult ♀
pinus

dark auriculars contrast sharply with throat

adult ♂
pinus

white lower belly and undertail coverts

streaks on sides of breast. Indistinct broken yellow eye ring and supercilium. Two whitish wing bars; large white tail spots on outer tail feathers; tail extends well past the undertail coverts. **ADULT FEMALE:** Similar to male, but paler yellow. **IMMATURE:** Duller than respective of adult; male washed with some brownish above; female quite brownish above and whitish below. **Geographic Variation** Widespread N.A. *pinus* and *florida* from south FL not field identifiable. Two others are resident in West Indies (*achrustera* from the Bahamas and *chrysoleuca* from Hispaniola). **Similar Species** Yellow-throated Vireo surprisingly causes confusion; it is larger and more sluggish and has a thicker hooked bill and conspicuous yellow spectacles. In fall, Bay-breasted and Blackpoll are similar. Pine lacks streaking on upperparts; has yellow or pale extending up behind contrasting auriculars, longer tail, and shorter primary projection. Wing bars contrast less. **Voice CALL:** A slurred *tsup*, similar to call of Yellow-throated Warbler. **SONG:** A musical trill, most similar to Chipping Sparrow or Dark-eyed Junco, but usually softer, more musical, and shorter, varying in speed. Occasionally sings two-parted songs, with the second part being faster and higher pitched. Also similar to songs of Worm-eating and Orange-crowned Warblers.

Status & Distribution Common, occurring in N.A., Bahamas, and Hispaniola year-round, with northern populations migratory. **BREEDING:** A broad range of pine habitats. **MIGRATION:** One of the earliest spring migrant warblers in many areas: begins northward movement in late Feb., arriving by mid-Apr. in southern Great Lakes, mid-Apr. in New England, and late Apr.–early May in northernmost breeding areas. A late fall migrant warbler: departs northernmost breeding areas as early as late Aug. but peaks late Sept.–mid-Oct., some into Nov. and rare into Dec. and Jan. **WINTER:** Pine forests in southeastern US. **RARE STATUS:** Rare, mainly in fall and early winter, in Atlantic Canada. Rare or casual in northern Great Plains and Prairie Provinces. Very rare in West, with most records (150+) from coastal Southern CA in fall and winter. Rare Bermuda. Casual in southern Caribbean in winter; accidental in fall in Greenland. **Population** Stable or increasing.

YELLOW-RUMPED WARBLER *Setophaga coronata*   YRWA   ■ 1

These are probably the best known and most frequently encountered wood-warblers. Although variable, all Yellow-rumped Warblers possess a bright yellow rump, which is shared with only two other species. Yellow-rumped's unique ability to digest the waxes in bayberries allows it to winter farther north than other warblers. Polytypic (4–6 sspp.; 3–5 in N.A.). L 5.5" (14 cm)

Identification SPRING MALE: Crown and back blue-gray streaked with black. Yellow crown patch, distinct rump patch, and patches at sides of breast. White or yellow throat. Black streaks on upper breast and side. White wing bars or more solid white wing patch. White spots in outer tail feathers. **SPRING FEMALE:** Similar to male, but brownish above with smaller tail spots. **FALL ADULT:** Similar to spring adult, but generally browner above in both sexes with less black on breast. **IMMATURE:** Similar to spring female; some immature females very dull with indistinct streaking and much reduced yellow on sides of breast.

Geographic Variation Four to five subspecies are placed into two groups that were formerly considered full species until 1973: "Myrtle" and "Audubon's." They hybridize extensively in portions of BC and AB. "Myrtle" (nominate over most of range and *hooveri*, not recognized by many, in northwest N.A.) shows a white throat, black lores and ear coverts, a white line above lores and eye, two distinct white wing bars, a broken white eye ring, and large white spots on outer three tail feathers. "Audubon's Warbler" (*auduboni* over most of range), larger and darker *nigrifrons* of northwestern Mexico, and largest and darkest *goldmani* (mainly from Guatemala) shows a yellow throat, bluish gray sides of head (including ear coverts), two broad white wing bars often forming a distinct patch, a broken white eye ring, and white spots on outer four or five tail feathers; the latter may be a separate species; songs differ and males in winter resemble breeding birds. Some birds breeding in the southwestern mountains are intermediate between *auduboni* and darker-faced *nigrifrons*. Telling "Myrtle" from "Audubon's" is more difficult in winter, especially the immatures. "Myrtle" is browner, less grayish above with a pale supercilium and a darker cheek. Its whitish throat is angled on the sides. Most immature "Audubon's" have a yellowish throat, but it can be whitish; it is more rounded on the edges. The cheek is paler and face lacks a supercilium. The flatter and lower-pitched call of "Myrtle" is a useful character. Only breeding-plumaged intergrades are likely to be detected.

Subspecies *auduboni*, "Audubon's Warbler" Group

auduboni & coronata intergrade zone

nigrifrons

goldmani

Subspecies *coronata*, "Myrtle Warbler" Group

coronata & auduboni intergrade zone

"Audubon's Warbler"

more extensively black overall

Southwest breeding ♂ intergrade

fall ♀
auduboni

breeding ♂
auduboni

most have pale yellow in rounded throat patch

yellow crown patch

yellow throat

"Myrtle Warbler"

breeding ♀
coronata

angled whitish throat

yellow patches on sides of breast

breeding ♂
coronata

browner above than fall "Audubon's"

yellow rump

distinct whitish supercilium

whitish patch sharply angled on sides of throat

fall ♀
coronata

Similar Species Compare to Magnolia and Cape May Warblers, which have yellow rumps but also yellow underparts. Palm Warbler has yellow undertail coverts.

Voice CALL: "Myrtle" gives a loud, husky, flat *chek*; "Audubon's" gives a loud and richer *chep*. **FLIGHT CALL:** A high, clear *sip*. **SONG:** A variable, loosely structured trill, sometimes with two parts—the first higher pitched and the second lower and trailing off at the end: *chee chee chee chee wee wee wee we.* Louder and richer on breeding grounds than in migration. "Audubon's" song is similar, but it is simpler and weaker.

Status & Distribution Common breeder in coniferous woodlands; very common short- to medium-distance migrant to central US south to Caribbean and central Panama. **BREEDING:** Northern boreal and mixed forest, and montane coniferous woodland. **MIGRATION:** In spring, generally arrives earlier than other warblers, returning to northern breeding areas by late Apr. In fall, generally migrates later than other warblers, peaking in northern portions of nonbreeding range in late Sept.–mid-Oct. **WINTER:** Large numbers of "Myrtles" winter along the East Coast. Small numbers irregularly through northern Caribbean and south in C.A. and western Panama; casual to southern Caribbean and northern S.A. **RARE STATUS:** "Audubon's" is casual in eastern N.A.; a spring Attu I. record may represent a ship assist. "Myrtle" casual to Aleutians (fall), Bering Sea islands, Baffin I., Greenland, Europe, Atlantic islands, and Russian Far East.

Population Breeding range of "Myrtle" has been expanding south in the East.

YELLOW-THROATED WARBLER   *Setophaga dominica*   YTWA   ▪ 1

The foraging behavior of this long-billed warbler, creeping along trunks and branches, is unusual among species in the genus *Setophaga*. Now treated as monotypic. L 5.5" (14 cm)
Identification ADULT MALE: Blackish forehead; faintly marked or plain gray crown and upperparts; yellow throat and breast; black triangular ear patch; bold black streaks on sides; white supercilium, crescent below eye, and patch at sides of neck. **ADULT FEMALE:** Similar to male, but with less black on forehead. **IMMATURE:** Similar to adult, but female shows the least amount of black on forehead; ear patch duller. Belly and undertail coverts washed with buff and brownish tint above.

Geographic Variation Although regarded as monotypic by most, breeding birds west of the Appalachians (formerly *albilora*) usually show a white supraloral. More easterly birds (formerly nominate *dominica*) show yellow supraloral (rarely white). Breeding birds from the Delmarva Peninsula

and the FL Panhandle and adjacent AL (latter area formerly named *stoddardi*) have the longest and most slender bills. The now endemic Bahama Warbler (*S. flavescens*) was regarded until recently as a subspecies of Yellow-throated Warbler. It is resident in Caribbean pine forests on Grand Bahama and Little and Great Abaco in the northern Bahamas and has not been recorded in N.A. It is distinctive and shows a noticeably longer bill with decurved culmen and straight lower mandible, little black on forehead, brownish gray upperparts, much narrower supercilium with yellow to rear of eye, much smaller white spot on side of neck, narrow streaks on whitish undertail coverts, and yellow extending well into the belly.

Similar Species Grace's broad supercilium is more extensively yellow; it has much less black on cheek and face and a different foraging style. Immature male Blackburnian shows streaked and more brownish back with pale braces, and has deep yellow supercilium and on sides of neck.

Voice CALL: A high, soft chip note, similar to Pine Warbler, identical to Grace's. **FLIGHT CALL:** Clear, high *see*. **SONG:** A somewhat variable series of clear, ringing, down-slurred notes, rising and weaker at the end: *tee-ew tee-ew tew tew tew tew wi.* Lacks an alternate song.

Status & Distribution Fairly common woodland species. Most are migratory. **BREEDING:** In South prefers cypress swamps and live-oak stands, especially those with large amounts of Spanish moss. In North, prefers bottomland with large sycamores or dry upland pine-oak forests. **MIGRATION:** Spring migration very early. Migrants reach southern breeding grounds by mid-Mar. and northern breeding grounds second week of Apr. to late Apr. Departs breeding grounds mid-Aug.–late Sept. in northernmost areas. A few remain well north into Dec. and Jan. **WINTER:** Swamps and more open areas in southeastern US and semi-open woodlands, city parks, and gardens in Caribbean and northeastern Mexico to Costa Rica. Over much of range partial to palms. Has attempted to winter as far north as Newfoundland. **RARE STATUS:** Rare in spring migration and sometimes late fall north of breeding range to Great Lakes, New England, and Maritimes; recorded north to James Bay, ON, and NL; very rare to casual in spring and fall in western states. Accidental southeast AK and the Azores.

Population Stable. Long-billed population from northwestern FL and coastal AL (formerly *stoddardi*) is rare and declining.

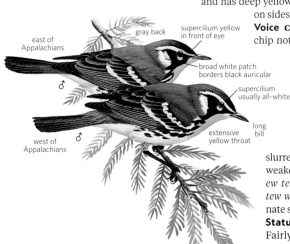

east of
Appalachians

gray back

supercilium yellow
in front of eye

broad white patch
borders black auricular

supercilium
usually all-white

long
bill

extensive
yellow throat

west of
Appalachians

PRAIRIE WARBLER *Setophaga discolor* PRAW ■ 1

This tail-wagging warbler has a distinctive, buzzy song. Often forages low. Polytypic (2 sspp.; both in N.A.). L 4.8" (12 cm)
Identification Distinctive facial pattern: yellow supercilium, dark line through eye, broad pale crescent below eye, and dark lower border to cheek. **ADULT MALE:** Black and yellow face pattern; olive upperparts, usually with reddish spots on back; yellow underparts; bold black streaks on sides; yellowish wing bars. **ADULT FEMALE:** Similar to male, but dark olive and yellow face pattern, sometimes a little black on lower cheek. Less distinct reddish spots on back and black streaks on sides. **IMMATURE:** Duller than adult. Male with very limited or no black on face, narrow side streaks. Female dullest, with indistinct face pattern showing gray, not black, and yellow

replaced by pale whitish, very indistinct side streaks.
Geographic Variation Nominate (most of range) and sedentary *paludicola* (coastal mangroves of FL) are not field-separable.
Similar Species Pine Warbler is larger, shows no black on head or reddish on back, does not wag tail as habitually as Prairie. Immature Magnolia Warbler shows yellow rump, complete white eye ring and gray face, lacking the pale subocular of Prairie, and shows much more white in lower belly to undertail; the tail when viewed from below shows a white base and a broad black tip. It does not wag its tail.
Voice CALL: A smacking *tsip*, or *tchick*, similar to Palm and especially Kirtland's Warbler. **FLIGHT CALL:** A thin *seep*. **SONG:** Primary song a rapid or slower series of buzzy notes evenly ascending in pitch: *zee zee zee zee zee zee zee zee zee.*
Status & Distribution Fairly common,

declining in some areas. **BREEDING:** Not on prairie. Nominate in a variety of shrubby old fields, dunes, pine barrens, early successional habitats. Now rare and very local in the Great Lakes region. **MIGRATION:** Arrives on Gulf Coast of FL mid-Mar.; arrives mid-Apr. to Ohio River Valley, by early May in Great Lakes. In fall, reaches FL and Bahamas mid- to late July. Northern breeders depart early Aug.–early Oct., stragglers through Dec. **WINTER:** Second growth and forest edge, mangroves, and gardens mainly in Caribbean and most of FL, rarely to coastal TX and NC. A few from coastal Yucatán to El Salvador. **RARE STATUS:** Rare to Atlantic Canada in fall; very rare to CA mostly on coast in fall; casual to accidental elsewhere in West in spring and fall (including AK); accidental to the Azores.
Population Declining in parts of breeding range.

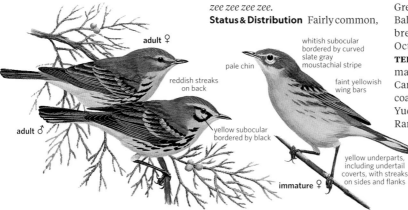

adult ♀
pale chin
reddish streaks on back
adult ♂
yellow subocular bordered by black
whitish subocular bordered by curved slate gray moustachial stripe
faint yellowish wing bars
yellow underparts, including undertail coverts, with streaks on sides and flanks
immature ♀

GRACE'S WARBLER *Setophaga graciae* GRWA ■ 1

This short-billed montane pine specialist is found in the US in Southwest mountains. Polytypic (4 sspp.; nominate in N.A.). L 5" (13 cm)
Identification ADULT MALE: Gray above with black streaks on crown and

back. Yellow throat and breast, white belly and undertail coverts. Black streaks on sides and flanks, yellow supercilium becomes white behind eye, small yellow crescent below eye, black lores and moustachial area. Two white wing bars, large white spots on outer tail feathers. **ADULT FEMALE:** Similar to male, but duller and paler above; finer streaks on crown, back, and sides; gray lores and moustachial area. **IMMATURE:** Male similar to adult female, with upperparts unstreaked and more brownish, duller yellow underparts, little black on crown and face, finer streaks on sides, buff wash on flanks. Female duller than

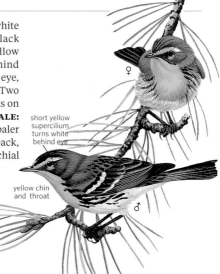

short yellow supercilium turns white behind eye
yellow chin and throat
♀
♂

immature male, lacking black on crown and face; more buff on flanks and belly.

Geographic Variation Slight and clinal, with back color brownest in northernmost subspecies, *graciae* (breeding in N.A.), to more blue-gray in three C.A. subspecies; yellow of throat paler in north and deep orange-yellow in south.

Similar Species Yellow-throated is larger with a bolder face pattern and different foraging style. "Audubon's" Yellow-rumped shows yellow rump and dark cheek patch.

Voice CALL: A soft, slurred chip note, identical to Yellow-throated Warbler. **FLIGHT CALL:** Very high, thin *sip*. **SONG:** Accelerating series of chip notes, most often two-parted: *chew chew chew chew chew chew chee chee chee chee.*

Status & Distribution Uncommon to fairly common. Pine forest to pine-oak woodland from the Southwest to northern Nicaragua. **BREEDING:** Mainly montane open, parklike forests of tall pines. **MIGRATION:** Arrives on breeding grounds early Apr.–early May. Departs breeding grounds as late as late Sept. **RARE STATUS:** Very rare Southern CA, mostly in fall and winter on coast, but also in mountains in late spring and summer (ponderosa pines); casual northern CA. Accidental NE, IL, ON, and NY.

Population Stable.

BLACK-THROATED GRAY WARBLER *Setophaga nigrescens* BTYW ◼ 1

This boldly patterned warbler forms a superspecies group with the Golden-cheeked, Townsend's, Black-throated Green, and Hermit Warblers, sharing call notes and long tails with extensive white in the outer tail feathers. Black-throated Gray is believed to have diverged earlier than the above species. Now generally treated as monotypic (birds from southern part of range, formerly *halseii*, show slight average differences including larger size, paler coloration, more white in tail, and heavier side streaks). L 5" (13 cm)

Identification ADULT MALE: Black-and-white head with broad white supercilium and malar; tiny yellow spot on lores. Gray back with black streaks, two white wing bars. White underparts. **ADULT FEMALE:** Similar to male, but grayer head, white chin, black throat patch mixed with white. **IMMATURE:** Male similar to

adult male but with gray in center of crown; cheek and throat not as black. Female with brownish tinge above, buffy white underparts with fainter streaks.

Similar Species Townsend's has a green back and yellow underparts. Spring male Blackpoll has all-white cheek and shorter tail. Black-and-white has a pale median crown stripe, a striped back, and very different foraging style. Immature male Cerulean shows bluish on upperparts, yellow on breast, and has less solid and dark auricular; its tail is much shorter.

Voice CALL: A dull *tip*. **FLIGHT CALL:** A high, clear *see*. **SONG:** Primary song is a series of two-syllable buzzy notes, the second syllable louder and higher pitched, the final note falling: *buzz see buzz see buzz see buzz see wueeo.* The alternate song is longer, more complex, lacking downward-inflected ending.

Status & Distribution Fairly common. **BREEDING:** Mixed or coniferous

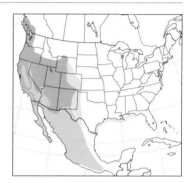

woodland with brushy undergrowth. **MIGRATION:** Medium-distance migrant. CA spring migration peaks mid- to late Apr. in south, mid-Apr.–early May in north. In fall, migration in OR mid-Aug.–mid-Sept., rarely late Oct. In Southern CA, significant movement occurs mid-Oct. **WINTER:** A variety of forest, scrub, and thickets in Mexico from Baja California Sur to central Oaxaca. A few in central CA. Rare in LA, southern FL, southern AZ, southern TX. **RARE STATUS:** Casual to accidental in migration east to NS, MA, NJ, and north to MT.

Population Stable.

gray upperparts

♀

bold white supercilium

yellow supraloral spot in all plumages

broad white submoustachial

black throat

adult ♂

broad white supercilium

upperparts tinged brown

clean white throat

immature ♀

TOWNSEND'S WARBLER *Setophaga townsendi* TOWA ◼ 1

This boldly patterned warbler of the Pacific Northwest gleans insects on its breeding grounds; on its wintering grounds, it will also exploit honey-dew excreted by sap-sucking insects. Monotypic. L 5" (13 cm)

Identification Dark auriculars surrounded by yellow, yellow on breast, extensive white on outer tail feathers, two white wing bars. **ADULT MALE:** Black cheek surrounded by yellow, with small yellow crescent below eye; black crown. Olive-green back streaked black. Black chin, throat, and broad side streaks. Yellow breast and sides; white belly and undertail coverts. **ADULT FEMALE:** Duller than male, with olive-green crown and auriculars; yellow chin and throat with limited black; narrower side streaks. **IMMATURE:** Male very similar to adult female. Female much duller, with no streaking on crown and back, no black on chin and throat, very narrow side streaks. **HYBRID:** Townsend's frequently hybridizes with Hermit where their ranges meet in WA and northern OR. Most hybrids closely resemble Hermit, especially in face, which is often entirely yellow. Crown varies from black to yellow and breast from yellow to white; flank streaks are extensive or absent; back is olive green or gray. Most individuals tend to show a Hermit's face pattern and yellow on the underparts more closely matching Townsend's. Less frequently, hybrids show Townsend's head pattern; a green or gray back with black streaks;

and a breast with little or no yellow.

Similar Species Black-throated Green is superficially similar to adult females and immatures, but it lacks dark auricular and yellow on throat and breast. Immature female Blackburnian shows less green and more streaking on back, including pale streaks on sides of back and a trace of a pale forehead stripe; yellow on underparts paler, sometimes almost whitish by mid- to late fall.

Voice CALL: A high, sharp *tsik*. **FLIGHT CALL:** A high, thin *see*. **SONG:** Primary song a variable, buzzy *weazy weazy weazy dzeee*. Alternate song a buzzy *zi-zi-zi-zi-zi-zi, zwee zwee*.

Status & Distribution Fairly common in coniferous forests of the Pacific Northwest. **BREEDING:** Tall coniferous and mixed woodlands, preferring mature or old-growth forest. **MIGRATION:** A medium- to long-distance migrant. Migrants from C.A. arrive in Southern CA in mid-Apr, peaking late Apr.–mid-May. Arrives as early as late Apr. in AK, with most arriving in May. Departs AK by early Aug., lingering into early Oct. Peaks late Aug.–early Sept. in OR; peaks mid-Aug.–mid-Oct. in CA. More numerous in Rockies and Great Basin in fall. **WINTER:** A variety of habitats in coastal Pacific Northwest south to Southern CA. Also rare in Central Valley, low

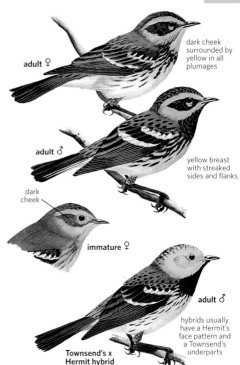

dark cheek surrounded by yellow in all plumages

adult ♀

adult ♂

yellow breast with streaked sides and flanks

dark cheek

immature ♀

adult ♂

hybrids usually have a Hermit's face pattern and a Townsend's underparts

Townsend's x Hermit hybrid

western Sierra, and the Southwest; also winters commonly in montane forests from northwestern Mexico to northern Nicaragua, fewer to westernmost Panama. **RARE STATUS:** More than 100 records nearly throughout eastern N.A.; mostly in fall on East Coast. Casual to accidental in Bermuda, Bahamas, northern AK, western Aleutians (to Shemya I.), and other Bering Sea islands.

Population May be expanding northward in AK. Stable to increasing in rest of range.

HERMIT WARBLER *Setophaga occidentalis* HEWA ■ 1

This striking golden-headed warbler is often difficult to see high in its conifer haunts. It is a sister species (closest relative) to the Townsend's Warbler and frequently hybridizes with it where the

two ranges overlap in the Northwest. It appears that Hermit Warblers are dominated by Townsend's. (See Population, below.) Monotypic. L 5.5" (14 cm)

Identification ADULT MALE: Yellow face, crown; black chin, throat, upper breast, lower nape. Gray back with black streaks. White underparts. Two

white wing bars. Extensive white on outer tail feathers. **ADULT FEMALE:** Duller than male, with lower nape,

grayish upperparts

adult ♀

yellow face

adult ♂

black throat

white underparts with no or limited flank streaks

immature with olive cast to back

yellow face with little dark in auriculars

immature ♀

black on throat much reduced; dull white to light gray sides, smaller tail spots; back without prominent streaks. **IMMATURE:** Male with olive on crown and ear coverts; chin and throat extensively mottled with olive green, yellow, and black. Dull whitish or grayish underparts. Female dullest, with extensive olive on crown and ear coverts; no black on back, nape, or throat; yellow face, throat, and narrow eye ring. Underparts tinged buffy. **Similar Species** Female and immature Olive Warblers are similar to immature female Hermit but show bright white patch at base of primaries and a different face pattern with a hollow

(in center) auricular patch, and have much different voice. Black-throated Green has an outlined greenish auricular patch, a bright green back, and has some yellowish on underparts, including the sides of the vent. See Townsend's for hybrids with Hermit. **Voice CALL:** A flat *tip*, similar to Townsend's. **FLIGHT CALL:** A high, clear *sip*. **SONG:** Primary song a variable *weezy weezy weezy weezy zee* (last note highest). Alternate song is variable and more complex: *che che che che cheeo ze ze ze ze ze ze seet*. **Status & Distribution** Fairly common; partial to conifers. **BREEDING:** Coniferous forest. **MIGRATION:** A medium-

to long-distance migrant. Arrives in southern AZ and Southern CA in mid-Apr., peaking late Apr.–mid-May. Departs breeding grounds early, some by mid-July, peaking mid-Aug.–early Sept., with a few to mid-Oct. **WINTER:** Montane pine-oak and other woodlands from northwestern Mexico to Nicaragua. Also rarely in live oak and coniferous woodland in coastal CA. **RARE STATUS:** Casual to accidental in central and eastern US and Canada. Accidental to western Panama. **Population** Reduced breeding range. Likely declining from competition and hybridization with Townsend's; also from droughts and resultant fires.

GOLDEN-CHEEKED WARBLER *Setophaga chrysoparia* GCWA ■ 2

This aptly named warbler breeds only in central TX, where its nest is constructed from Ashe juniper bark. Monotypic. L 5.5" (14 cm) **Identification** Clear yellow cheeks, black line from eye to bill. **ADULT MALE:** Bright golden yellow sides of

dark eye line

immature ♀

unmarked, bright yellow cheek in all plumages

black back

adult ♀

all plumages with white vent

adult ♂

head. Black crown, back, throat, and breast. White from belly to undertail coverts; white wing bars. **ADULT FEMALE:** Duller than male, black eye line narrower, black-streaked olive-green upperparts; narrower streaking below. **IMMATURE:** Male similar to adult female. Female duller, plainer backed, and with no blackish on throat. **Similar Species** Black-throated Green is green above with a green-outlined auricular and has yellow on side of vent. Immature female Golden-cheeked most similar to Black-throated Green but with stronger eye line and clear yellow cheek that lacks a dark border; lacks yellow below. Female and immature Hermit lacks eye line and is grayer above. **Voice CALL:** A *tsip*, similar to Black-throated Green Warbler. **FLIGHT CALL:** A high, thin *see*. **SONG:** A variable, short buzzy *dzee dzweeee dzeezy see*. Alternate song simpler: *zee zee zee zee see*. **Status & Distribution** Very restricted range. **BREEDING:** Mature juniper-oak woodlands of central TX. Arrives early to mid-Mar.; departs as early as late

June, but typically present to early Aug., stragglers until mid-Aug. **MIGRATION:** Migrates through Sierra Madre Oriental (mainly in oaks to 3,000 ft). **WINTER:** Montane pine-oak forests from southern Chiapas, Mexico, to north-central Nicaragua. **RARE STATUS:** Accidental on upper TX coast, NM, CA, FL, and Virgin Is. (sight record). **Population** Endangered; also listed as Endangered by USFWS with an estimate of 13,500 pairs; recent petitions to delist the species have been declined. Prime habitat is unfragmented, old-growth juniper-oak woodlands, which is being lost to development.

BLACK-THROATED GREEN WARBLER *Setophaga virens* BTNW ■ 1

A familiar bird of mixed and deciduous woodlands, the Black-throated Green Warbler is often detected by its buzzy, cheery song. Now treated as monotypic. Breeding birds (formerly recognized as *waynei*) from coastal Atlantic coast in cypress and other deciduous swamps from southern VA to SC are slightly smaller and have a slightly shorter and more delicate bill. L 5" (13 cm)

Identification Yellow face, plain green back, yellow sides to vent. **ADULT MALE:** Yellow sides of head with olive-yellow auriculars; greenish olive crown, back, and rump. Black chin, throat, breast, and broad side streaks. Two white wing bars, whitish to pale yellowish belly and undertail coverts. **ADULT FEMALE:** Similar to male; chin and throat variable but with

less black, usually with pale chin; back never shows streaks; narrower side streaks. **IMMATURE:** Belly and undertail coverts more strongly tinged with yellow. Male similar to adult female. Female with whitish chin and throat (no black); very narrow side streaks. **Geographic Variation** Smaller overall size and bill length of isolated southeastern coastal subspecies *waynei* not detectable in the field and not recognized by most authorities.

Similar Species Golden-cheeked always shows a pure yellow auricular and a dark line from bill through eye to dark nape. Some hybrids of Townsend's and Hermit can resemble Black-throated Green Warbler, but they lack yellow on the sides of the vent.

Voice CALL: A soft, flat *tsip*. **FLIGHT CALL:** A high, sweet *see*. **SONG:** Primary song is variable with a whistled, buzzy quality; the last note is the highest: *zee-zee-zee-zoo-zee*. Alternate song a

variable, more deliberate *zoo zee zoo zoo zee*.

Status & Distribution Common. **BREEDING:** Mainly coniferous and mixed forest, but sometimes deciduous. Subspecies *waynei* breeds in cypress swamps. **MIGRATION:** Primarily a trans- and circum-Gulf migrant, some move through FL to Caribbean. Arrives in Gulf states in late Mar., in Midwest by late Apr., peaking in mid-May. Fall migration is more protracted than other eastern warblers, as early as late July. Migration peaking mid- to late Sept., many well through Oct.; stragglers widely recorded in North into Nov. **WINTER:** Mature montane forests from northeastern Mexico to Panama; much smaller numbers in Caribbean. Small numbers winter in southern FL, southern TX. Rare in northern Colombia and western Venezuela. **RARE STATUS:** Casual to rare for most western states, mostly in late fall and many (300+) in

CA. Casual in coastal BC, southeastern AK, and NT. Casual to the Azores; accidental in Greenland, Iceland, and Germany (specimen in 1858 from Heligoland, North Sea).

Population Relatively stable.

yellow face with greenish border to auriculars in all plumages

adult ♀

all plumages have a yellow side to the vent

greenish cheek

green back

extensive black throat

adult ♂

immature ♀

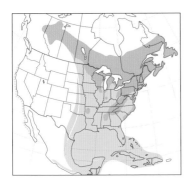

Genus *Basileuterus*

This genus has recently been divided, with 15 primarily S.A. species transferred to *Myiothlypis*. All three species recorded in N.A. remain in *Basileuterus*. Most species of this mainly tropical genus are olive to gray above, yellow to whitish below, with some patterning on the crown or face.

FAN-TAILED WARBLER *Basileuterus lachrymosus* FTWA ■ 4

The Fan-tailed Warbler is a large tropical warbler. It is rather secretive and is usually seen low in vegetation or walking on the ground. Formerly placed in the monotypic genus *Euthlypis*; molecular studies showed it belonged with *Basileuterus*. Now treated as monotypic, although there is a cline of increasing darkness from north to south. Formerly three subspecies were recognized; specimens from AZ and Baja California were assigned to northwestern *tephra*. L 5.8" (15 cm)

Identification Long, graduated white-tipped tail held open, swung up and down and side to side; long pink legs. **ADULT:** Distinct head pattern: blackish head with broken white eye ring, white supraloral spot, and yellow crown patch; gray above, yellow below with tawny-orange wash on breast. Female similar but black areas replaced with slate gray. **IMMATURE:** Face slightly paler.

Similar Species Yellow-breasted Chat has less yellow on underparts; lacks yellow crown patch and white supraloral spot; terrestrial habits and tail spreading and bobbing of Fan-tailed also differ.

Voice CALL: A distinctive, high, thin *tsew* or a penetrating *schree*. **SONG:** Primary song a series of sweet notes ending in a sharp up-slur or down-slur: *che che che a-wee wee che-cheer*; suggestive of Swainson's Warbler.

Status & Distribution Breeds in moist, shady, steep-walled ravines from Pacific slope of northwestern Mexico south to Nicaragua, with a disjunct population in coastal east-central Mexico. **RARE STATUS:** Casual spring and fall to southeastern AZ (10 records); one lingered into late June. Accidental to east-central NM in spring, to West TX (Big Bend NP) in fall, and Baja California Norte in early winter.

Population Stable.

yellow crown patch bordered by black

uniformly gray above

whitish supraloral spot and broken eye ring

tawny wash on breast

spreads open graduated tail with white tips; bobs tail up and down or side to side

RUFOUS-CAPPED WARBLER   *Basileuterus rufifrons*   RCWA   3

The tropical Rufous-capped Warbler, rare in N.A., inhabits dense brush in montane canyons. It is difficult to locate when not vocalizing. Polytypic (8 sspp.; 2 in N.A.). L 5.3" (13 cm)

Identification Short, thick bill. It cocks and waves its long, thin tail like a wren or gnatcatcher. **ADULT:** Distinct head pattern with rufous on crown and cheeks, broad white supercilium, dark lores; grayish olive upperparts; bright yellow throat and breast, white belly, grayish buff sides and flanks.

Geographic Variation Records from TX, substantiated by specimens, are *jouyi*; it is slightly darker and more richly colored than the northwestern and likely subspecies recorded from AZ (*caudatus*). Two subspecies groups: five in the northern *rufifrons* group, which are found south to northern Guatemala, and three in the southern "Chestnut-capped" *delattrii* group from southern Guatemala to northern Colombia and western Venezuela. They differ by plumage (southern group entirely yellow below) and vocalizations, and are perhaps different species.

Voice CALL: A hard *tik*, sometimes doubled or in a rapid series. **SONG:** Primary song is a rapid, variable series of chip notes, chirps, and trills, usually changing in pitch and pace.

Status & Distribution From northern Mexico south to Panama and extreme northwestern S.A. **RARE STATUS:** Casual year-round to western and central TX and southeastern AZ (some involving long-staying individuals and successful breeding). Most recent AZ records have come from French Joe Canyon in the Whetstone Mts., Hunter Canyon in the Huachuca Mts., and Florida Canyon in the Santa Rita Mts. Most recent TX records have been from the southern Edwards Plateau. Recently recorded at Guadalupe Canyon, NM.

Population Stable.

long tail is often cocked

bright rufous crown and auriculars; bold white supercilium

olive-brown above

bright yellow throat and white submoustachial

GOLDEN-CROWNED WARBLER   *Basileuterus culicivorus*   GCRW   4

The most widespread member of the genus, the tropical Golden-crowned occurs very casually north of Mexico. Polytypic (14 sspp.; *brasherii* in N.A.). L 5" (13 cm)

Identification Northeast *brasherii* described. **ADULT:** Olive-gray upperparts; bright yellow underparts, tinged olive on sides and flanks. Distinctive head pattern: yellow central crown patch, sometimes suffused with orange, especially more southern subspecies, a broad dull yellowish olive supercilium extends through the supraloral, broad black lateral stripes, grayish olive auriculars, dull yellowish olive supercilium, gray loral spot, narrow broken yellow eye ring. **IMMATURE:** Very similar to adult; some show slightly less distinct head pattern.

Geographic Variation The 14 subspecies of Golden-crowned Warbler break into four groups (three S.A., one M.A.). The M.A. populations (four subspecies) are olive gray above with a yellowish supercilium; *brasherii* of northeast Mexico has made it to N.A.; *flavescens* from west-central Mexico is genetically distinct.

Similar Species Orange-crowned and Wilson's are superficially similar, lack distinct crown stripes.

Voice CALL: Often repeated, slightly buzzy *tuck*. **SONG:** Several clear whistled notes, ending with a distinct up-slur: *see-whew-whew-wee-see?*

Status & Distribution Disjunct populations from west-central and southeastern Mexico through much of C.A., northern S.A., the Guianas, and much of eastern Brazil to northern Argentina and Uruguay. **RARE STATUS:** Casual, mainly in winter from Nov.–Mar. (one record Oct.; one late Apr. record), to extreme southern TX (20+ records). Accidental in spring to southern coastal TX (Nueces Co.), east-central NM (Roosevelt Co.), and eastern CO (Cheyenne Co.).

Population Stable.

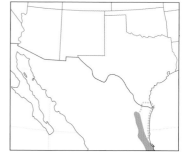

greenish supercilium

yellowish green median crown stripe and blackish lateral crown stripe

plain grayish olive above with yellowish underparts

brasherii

broken eye ring

Genus *Cardellina*

Formerly, the genus was monotypic with only the Red-faced Warbler, but it has been expanded to include the beautiful Red and Pink-headed Warblers (formerly in *Ergaticus*) and Wilson's and Canada Warblers. The latter two and the Hooded Warbler were formerly in the now-defunct *Wilsonia*. All have long tails that are frequently flipped.

WILSON'S WARBLER   *Cardellina pusilla*   WIWA   ■   1

This small warbler feeds very actively, often fly-catching. Polytypic (3 sspp.; all in N.A.). L 4.8" (12 cm)

Identification Flips long tail up and down and in a circular motion. Nominate described. **ADULT MALE:** Olive-green upperparts (including wings, tail, cheeks); solid shiny black cap; bright yellow forehead, lores, eye ring, broad supercilium, entire underparts. **ADULT FEMALE:** Similar to male, but smaller black cap, usually nearly absent or restricted to front half

smaller black cap; absent in immature female

adult ♀ *pusilla*

black cap and olive face

long, plain tail lacks white   *pusilla* ♂

dark eye stands out in plain face

auriculars contrast less with supercilium than *pusilla* and *pileolata*

yellow wash on back

gold forehead

chryseola ♂

brighter yellow below

of crown (often mottled with olive). **IMMATURE:** Similar to adult; male with extensive olive mottling on black cap, female usually with olive crown, forehead; lacks black cap.

Geographic Variation Nominate breeds through much of boreal Canada to NT; *pileolata* (AK, Rocky Mts.) has brighter yellow forehead and underparts. Pacific *chryseola* is brightest overall: yellowish wash on cheeks, orange-tinged forehead and upperparts a lighter and brighter yellow-green color on the upperparts. This subspecies is an earlier spring migrant, arriving in Southwest (southeast AZ and west) by early Mar. The different songs of each may indicate two or even three species.

Similar Species Larger immature female Hooded has larger bill and eye, dusky lores, white in tail, and a distinctly different chip note. Shorter-tailed Orange-crowned is drabber with diffuse breast streaks on most, a dusky eye line, and more pointed bill. Its slightly shorter tail is not flipped around like Wilson's. Immature Yellow Warbler has a shorter tail that is bobbed; also has yellow tail spots and pale-edged wing feathers.

Voice **CALL:** A somewhat nasal *timp*. **FLIGHT CALL:** A sharp, slurred *tsip*. **SONG:** Primary song short and chattering, dropping in pitch toward end: *chi chi chi chi chi chi chet chet*. Songs of *pileolata* and *chryseola* louder and faster and lack the drop-off at the end.

Status & Distribution Uncommon in eastern N.A. to common in West. **BREEDING:** Wet situations with dense

pileolata

pusilla

chryseola

ground cover, low shrubs. **MIGRATION:** Medium- to long-distance migrant. Eastern birds are circum-Gulf migrants, arrive in southern TX late Apr., in Great Lakes late May. Pacific coast birds arrive in southern AZ as early as late Feb., but migration extends to late May (early spring migrants are nearly all *chryseola*); arrives in southwestern BC late Apr. In East, earliest fall migrants mid-Aug., peak late Aug.–mid-Sept., casually into Nov. Pacific birds begin to depart in mid-July. **WINTER:** Wide variety of habitats; northwestern Mexico to central Panama, rarely to extreme southeastern TX, southern LA, coastal Southern CA; very rare FL; casual farther north on Atlantic coast. **RARE STATUS:** Casual in winter to Colombia, northern Bahamas, Cuba, Jamaica. Accidental in Arctic Canada and Greenland, UK.

Population Declining (especially in the West), possibly due to cowbird parasitism and loss of riparian habitats.

CANADA WARBLER   *Cardellina canadensis*   CAWA   ■   1

Sometimes called the "Necklaced" Warbler, the Canada Warbler's behavior is very similar to Wilson's. It generally forages low in shrubs and bushes,

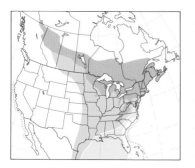

but also feeds higher in trees. When feeding will make short sallies for insects but also gleans from leaves. Monotypic. L 5.3" (13 cm)

Identification Active; often cocks tail, flicks wings. Pink legs. **ADULT MALE:** Plain blue-gray upperparts (including wings, tail); black forehead and forecrown with thin gray tips; center of crown spotted with black; lores and anterior auricular also black; distinct complete white or yellow eye ring, yellow supraloral line, forming spectacles; bright yellow underparts; long rows of

black spots across breast form a necklace; white undertail coverts. Black

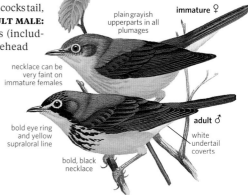

plain grayish upperparts in all plumages

immature ♀

necklace can be very faint on immature females

bold eye ring and yellow supraloral line

bold, black necklace

adult ♂

white undertail coverts

markings reduced in first-spring males; necklace is shorter. **ADULT FEMALE:** Duller than male, lacks black on face; less distinct dark necklace. **IMMATURE:** Male similar to spring female; more olive-gray above. Female duller; indistinct to almost absent necklace formed by grayish olive spots.
Similar Species Kirtland's and Magnolia Warblers both have white in tail. Smaller and shorter-tailed Nashville Warbler lacks breast streaks and is entirely yellow below.
Voice CALL: Sharp *tchup* or *tik*.

FLIGHT CALL: High *zzee*. **SONG:** Primary song variable, staccato series of jumbled notes; usually begins and ends with loud *chip* note, then short pause: *chip . . . chupety swee-ditchety chip*. The song is closest to the song of Magnolia Warbler.
Status & Distribution Fairly common.
BREEDING: Cool, moist mixed forests with dense understory in the boreal and Appalachian regions. **MIGRATION:** Mainly a circum-Gulf migrant; most numerous on immediate TX coast; rare on Gulf Coast east of southwest

LA. Late spring migrant; arrives southern TX after mid-Apr., Upper Midwest mid-May, continues into early June. In fall, peaks in mid- to late Aug., continues into mid-Sept., some until early Oct., casual to early Nov. **WINTER:** Dense montane undergrowth; from northern S.A. in Andes to central Peru. **RARE STATUS:** Very rare to casual in West; casual in Caribbean and to Greenland; accidental Iceland.
Population Declining, likely from forest succession and loss of forested wetlands.

RED-FACED WARBLER   *Cardellina rubrifrons*   RFWA ◼ 2

The Red-faced Warbler, a strikingly plumaged southwestern mountain species, is easily identified. In May–June, it constructs a nest in a depression on the ground, often on a slope at the base of woody plants, and often with an overhang that helps conceal and protect the four to six eggs within. Monotypic. L 5.5" (14 cm)
Identification Short and thick bill with a decurved culmen and prominent rictal bristles. Head appears peaked. Flips long tail around in a similar manner to other *Cardellina*, including Wilson's and Canada. Long, pointed wings. **ADULT MALE:** Bright red face, upper breast, and sides of

neck; black crown and cheeks; white nape spot, rump, and underparts; a single white wing bar on median coverts; gray upperparts. **ADULT FEMALE:** Nearly identical to male, but with slightly duller, more orange-red face. Adults in fresh fall plumage have a pink suffusion across breast, often extending into belly. **IMMATURE:** Similar to respective adults. Immature female is more orange-salmon on head on dullest birds; upperparts tinged brown. Some immature males nearly as bright as adult males.
Similar Species Painted Redstart is the only other warbler with red in plumage (but not on face), and it shows large white patches in wings and tail. Chickadees are similarly gray above, and Red-faced Warbler is sometimes similarly acrobatic, but once the red face is seen there can be no confusion. The white rump on Red-faced is often very visible when it makes short flight sallies for insects. Also note that chickadees do not show white nape spots or white rumps.
Voice CALL: A sharp *chup* or *tchip*, suggesting Black-throated Gray Warbler. **SONG:** Primary song a series of assorted thinner notes

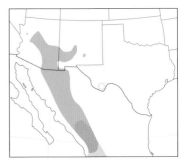

with an emphatic ending: *wi tsi-wi tsi-wi si-wi-si-whichu*.
Status & Distribution Fairly common. **BREEDING:** Montane mixed and deciduous woodland. **MIGRATION:** Rather long-distance migrant. Not often observed in migration; arrives late Apr.–late May. Departs as early as Aug., some into Sept. **WINTER:** Humid montane forest, pine-oak forest, and riparian woodland from northwestern Mexico south to eastern Honduras. **RARE STATUS:** Casual in TX (mostly Big Bend NP) and CA (all but one from Southern CA); accidental in CO, WY, LA, and GA.
Population Difficult to determine; possibly declining slightly. Likely vulnerable to logging and wildfires.

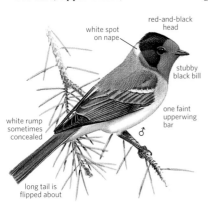

red-and-black head
white spot on nape
stubby black bill
one faint upperwing bar
♂
white rump sometimes concealed
long tail is flipped about

Genus *Myioborus*

When it was first described, the Painted Redstart was thought to be closely related to the American Redstart, which was previously named mainly for its red tail patches. All 12 members of this mainly tropical genus show white tail patches, thus they are sometimes called whitestarts, and only two species show any red in the plumage at all.

PAINTED REDSTART   *Myioborus pictus*   PARE ◼ 2

With its bold black, red, and white coloration and its conspicuous wing- and tail-fanning behavior while creeping along branches, the Painted Redstart

is unlikely to be confused with any other warbler species. It is one of the most visible of our wood-warblers, and detection is made even easier by its

habit of frequently giving its loud and distinctive call note. It makes its nest on an embankment, often near water—well hidden under rocks, grasses, or roots. Polytypic (2 sspp.; nominate in N.A.). L 5.8" (15 cm)

Identification ADULT: Mainly black with large white patches in wing, white on outer three tail feathers, and a small white crescent below eye. Red lower breast and belly (female's red slightly paler); undertail coverts mixed slate and white. **JUVENILE:** Retains juvenal plumage later than most warblers, June–Aug. Similar to adult, but without red. Sooty lower underparts; undertail coverts mottled grayish and white. Red belly is gradually acquired in late summer or early fall.

Similar Species Slate-throated Red-

start lacks white wing patches and eye crescent, shows less white in more graduated tail, and has slaty (not black) upperparts and duller red underparts.

Voice CALL: A unique, scratchy, whistled *sheu* or a richer *cheree*, similar to calls of Pine Siskin. **SONG:** Primary song a somewhat variable series of rich, two-part syllables, ending with one or two inflected notes: *weeta weeta weeta wee.*

Status & Distribution Uncommon to fairly common. **BREEDING:** Most numerous in heavily wooded canyons in mountains with running water. Very uncommon in West TX (primarily from Boot Canyon, Chisos Mts., Big Bend NP). **MIGRATION:** A short-distance migrant. Arrives on breeding grounds mid- to late Mar. The few migrants noted away from the breeding grounds are noted primarily in Apr. and Sept. Departs breeding grounds in Sept., with some lingering into mid-Oct. **WINTER:**

Pine-oak woodlands, sometimes at lower elevations than when breeding; from northwestern Mexico south to northern Nicaragua, where it mixes with sedentary subspecies (*guatemalae* with reduced white in tail and on tertial edges). Rare in winter in breeding range. **RARE STATUS:** Rare to uncommon in fall and winter in Southern CA (more than 100 records). Casual in northern CA, Baja California, southwestern AZ, southwestern UT, southwestern CO, and southern TX. Accidental in BC, MT, MB, WI, MI, southern ON, OH, NY, MA, LA, MS, AL, and GA.

Population Stable.

white crescent under eye — *pictus* — bold white wing patch — red belly — white outer tail feathers — juvenile *pictus* — sooty breast

SLATE-THROATED REDSTART *Myioborus miniatus*   STRE ■ 4

An easily identified warbler, the Slate-throated Redstart occurs casually north of Mexico. Polytypic (12 sspp.; NM specimen is *miniatus*). L 6" (15 cm)

Identification Frequently fans and holds open its long, distinctly graduated tail. **ADULT MALE (*MINIATUS*):** Slate-gray upperparts and wings; a blackish face, throat, and sides; a dark chestnut crown patch; a black tail with large white spots on outer three tail feathers; red breast and belly; undertail coverts mottled white and slate gray. C.A. and S.A. subspecies show orange or yellow breast and belly. **ADULT FEMALE:** Nearly identical to adult male, but with duller red underparts; a slate-

gray face, throat, and sides; and smaller chestnut crown patch. **JUVENILE:** Similar to adult, but lacks red underparts and chestnut crown patch; paler slaty coloring on breast and belly; undertail coverts mottled with cinnamon-brown.

Geographic Variation This species comprises 12 subspecies: Northern *miniatus* is found south to Isthmus, Mexico; another northern subspecies, *malochinus*, from Sierra de Tuxtla of southern Veracruz, is darker above, more richly red below. Farther south, underparts grade from salmon to orange to yellow-orange and to yellow in S.A.

Similar Species See Painted Redstart.

Voice CALL: A single, high *tsip*, similar to chip call of Chipping Sparrow and totally unlike the call of Painted Redstart. **SONG:** Primary song a variable series of *sweet s-wee* notes, often accelerating toward the end of the series. Although the pattern can suggest Painted Redstart, it is higher pitched and thinner, with slurred single notes rather than doubled ones. Some

chestnut crown patch — no white under eye — uniformly dark gray above — face and throat more grayish on female — more graduated tail than Painted with white only in tips

versions recall the songs of Yellow-rumped Warbler.

Status & Distribution Breeds in montane coniferous and mixed forest from northwestern and central Mexico (northernmost populations migratory) south to northern S.A., south in Andes to central Bolivia. **RARE STATUS:** Casual (mostly Apr.–May) to southeastern AZ (about 10 records), including a hybrid pairing with Painted Redstart in the Chiricahua Mts.; southeastern NM (one record) and western and southern TX (more than 10 records), all but two from the Chisos and Davis Mts., West TX, from mid-Apr. into June. The two south TX records involve spring migrants in Mar. and Apr.

Population Stable.

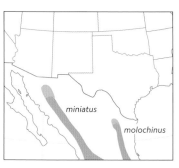

miniatus — *molochinus*

CARDINALS AND ALLIES Family Cardinalidae

Dickcissel, male (OH, July)

I n North America, the Cardinalidae is represented by some of the region's brightest and most striking species, such as the Northern Cardinal; Indigo and Painted Buntings; and Scarlet, Western, and Summer Tanagers. Members of the family have stout to very thick bills. Most are strongly sexually dimorphic, with males typically uniquely colored and unmistakable, but females, and in some cases immature males, being duller and more challenging to identify. Important features used to identify similar females of closely related species include bill size and shape, overall color, and presence or absence of wing bars. In N.A., all breeding species except Pyrrhuloxia and Northern Cardinal are nocturnal Neotropical migrants. The Rose-breasted Grosbeak and Indigo Bunting are among the more common species seen during migratory fallouts along the Gulf Coast, and Western Tanagers are familiar migrants throughout the West.

Structure Considered part of the nine-primaried oscines, members of the Cardinalidae family are variable in size, with cardinals, grosbeaks, and *Piranga* "tanagers" relatively large for passerines and buntings relatively small (about the size of a sparrow). Most have distinctly shaped conical bills adapted for eating seeds and fruit. Some species have disproportionately large bills (Yellow Grosbeak). Species in the genus *Cardinalis* are characterized by a long, pointed crest.

Behavior They mainly feed on a combination of insects (breeding season), seeds, and fruit (fall and winter). Grosbeaks in particular switch to almost exclusively fruit beginning in late summer. Songs are mostly similar across the family: a series of somewhat robin-like phrases, often paired; the Dickcissel's harsher song is an outlier. Calls usually consist of sharp single notes, often metallic in quality and difficult to differentiate within a genus. Members of the genus *Passerina* (buntings and Blue Grosbeak) are known for tail-twitching behavior. Except for the Northern Cardinal, all are medium- to

long-distance nocturnal migrants, wintering from Mexico into northern South America.

Plumage Bright male plumages make these some of the most distinctive and recognizable species in North America. Females of many species are equally famous for their dull brown or green plumages and difficult identifications. Males attain adult plumage by their second winter, normally with a distinct first-summer plumage. Immature males typically are intermediate between adult male and female plumage; some (e.g., Varied Bunting) closely resemble the female. Immature males are known to establish territories and sing. Winter adult males usually have brown edging to plumage, obscuring bright colors.

Distribution The family occurs from Canada south through tropical South America. Members of the genus *Cardinalis* are generally resident, while all other North American cardinalids are migratory. Cardinalids are generally more prevalent south of the US, with no fewer than 30 additional species found exclusively in Central and South America. Fourteen species regularly breed in North America, and four others occur as strays from Mexico. The Crimson-collared Grosbeak and Blue Bunting are casual in the Rio Grande Valley, while the Yellow Grosbeak is casual in southeastern AZ. The Flame-colored Tanager is very rare but nearly annual in southeastern AZ and has bred. Species that breed in North America are found in a variety of habitats, from deciduous forest canopies to oak-conifer woodlands, arid brushy hillsides, and weedy meadows. Several species are associated with riparian woodlands.

Taxonomy The Cardinalidae is closely allied with the Thraupidae and other nine-primaried songbird families; the family includes 48 species in 11 genera. *Piranga* and related genera, long considered members of the tanager family Thraupidae, are clearly cardinalids but retain the English name "tanager." The Dickcissel, in the genus *Spiza*, has sometimes been aligned with the icterids, but genetic studies place it with the cardinalids; the family also includes the three *Granatellus* chats (Mexico to South America) formerly considered wood-warblers. Closely related species in the genus *Pheucticus* (grosbeaks) hybridize when they come in contact, as do Indigo and Lazuli Buntings in the genus *Passerina*.

Conservation As most species are found in a variety of disturbed and semi-disturbed habitats, they are generally considered common and not threatened at present. Pesticide use in the Neotropics may adversely affect Dickcissel populations on the wintering grounds. In the East, Scarlet Tanager populations have declined because of forest fragmentation and negative effects from cowbird parasitism. Western populations of Summer Tanager are declining due to loss of riparian habitats. In addition to habitat issues, many Neotropical species have suffered from exploitation for the cage-bird trade. BirdLife International codes: 4 NT and 1 EN.

Genus *Piranga*

After recent genetic studies the nine *Piranga* "tanagers" (all found in the Americas) were moved out of family Thraupidae (Tanagers) and placed in family Cardinalidae. In the ABA area there are four regular breeding species and one casual visitor; all are Neotropical migrants. Males are mostly bright red in plumage; females are duller, greenish yellow.

HEPATIC TANAGER *Piranga flava*   HETA ■ 2

Uncommon to fairly common in the pine and pine-oak forests of the Southwest. Normally found in different habitat and at higher elevation than the Summer Tanager. Polytypic (15 sspp.; 2 in N.A.). L 8" (20 cm)

Identification Bill heavy and dark gray with "tooth" on upper mandible. **MALE:** Brick-red overall, mixed with gray on auricular, back, and flanks. **FEMALE:** Mostly greenish, with strong gray overtones to back, auricular, and flanks, contrasting with a yellow throat and forehead. A more orange morph is rather rare. **IMMATURE MALE:** Like female. **JUVENILE:** Similar to female, but more streaked above.

Geographic Variation In the US,

the subspecies *hepatica*, from southeastern CA and southern AZ, tends to be larger and duller than *dextra*, from southeastern CO to West TX, with more extensive gray on auriculars and flanks. Three other subspecies are found south to northern Nicaragua. Two other subspecies groups also occur to the south, one "Highland" group (*lutea* group) from Costa Rica and S.A. with multisyllabled calls, and one more southerly S.A. "Lowland" group (*flava* group); these likely reflect additional species.

Similar Species Adult males similar to brighter red Summer Tanager, which lacks gray auriculars and grayish wash on back and flanks and typically has paler bill. Female Hepatic's grayer overall plumage, contrasting with its bright yellow throat, is quite different from more uniform ochre plumage of Summer. Hepatic's call is very different from other *Piranga*.

Voice CALL: A single low sharp *chuck*. **SONG:** Robinlike and clearer than

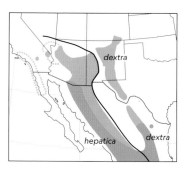

"tooth" on upper mandible

dark bill

grayish cheek

color brightest on forehead and throat in all plumages

dull red plumage

adult ♂ hepatica

hepatica ♀

grayish wash

Western; more similar to the song of Black-headed Grosbeak, with which it overlaps.

Status & Distribution Found south to southern S.A. Uncommon to fairly common in montane Southwest mixed coniferous forests. **MIGRATION:** Rare in lowlands, mainly in riparian areas, along streams. **WINTER:** Rare winter resident in southeast AZ, mainly in the Patagonia-Nogales. **RARE STATUS:** Very rare to coastal CA; casual over most of TX to east of breeding range. Accidental to IL, WY, and LA.

Population US populations are susceptible to long-term drought.

SUMMER TANAGER *Piranga rubra*   SUTA ■ 1

The adult male Summer Tanager is one of the more colorful breeding birds of N.A. Summer Tanager feeds mainly on fruit, except during breeding season. Quite vocal, Summer is often detected by its distinctive call. Polytypic (2 sspp.; both in N.A.). L 7.8" (20 cm)

Identification Sexually dimorphic. The bill is generally bulky and long and ranges from gray to pale horn in color. The head often shows a slight crested appearance. **BREEDING MALE:** Adult male is unmistakable—all bright red—and achieves its brightest plumage by the fall of its second calendar year. **BREEDING FEMALE:** Generally ochre or mustard in color, with greener wings and upperparts. Some (more in nominate subspecies?) have a brighter morph with dull red mixed in. Other females are greener, largely lacking

ochre tones, and more closely resemble a female Scarlet. **FIRST-SPRING MALE:** Mixture of red (except on flight feathers) and greenish yellow, sometimes blotchy, sometimes with entirely red head and breast. First-fall males look like females.

Geographic Variation Adult male eastern *rubra* is deeper red than adult male western *cooperi*. Both sexes of *cooperi* are larger, paler overall, and have larger and paler bills than *rubra*.

Similar Species Adult male Scarlet Tanager has black wings. Male Hepatic Tanager is more brick red in color and has a grayer bill, grayer flanks and cheek; call very different. Female Summers largely lacking ochre tones are more problematic and easily confused with female Scarlet. Note yellow undersurface of tail in Summer, lack-

ing in Scarlet. Summer is also uniformly colored above (wings darker on Scarlet) and has a longer tail and longer bill than Scarlet. Female and immature male Hepatics are less uniform underneath, with brighter yellow throats and crowns contrasting with grayer cheek and back.

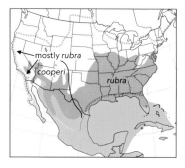

mostly rubra

cooperi

rubra

Voice CALL: A distinctive *pi-tuck* or *pi-ti-tuck* or *ki-ti-tuck*, sometimes extended to several notes. **SONG:** An American Robin–like series of warbling phrases. Easily confused with Rose-breasted or Black-headed Grosbeak. **FLIGHT NOTE:** A soft, wheezy *verree*. **JUVENILE:** Begging note similar to Black-headed Grosbeak's *veeooo*.
Status & Distribution In the East, common in the canopy of oak and pine-oak woodland. In the Southwest, common in cottonwood-willow habitats along permanent streams and rivers. **BREEDING:** Nesting birds arrive mid-Apr.–early May. **MIGRATION:** Nominate *rubra* is mainly a trans-Gulf migrant, common from along upper TX coast to coastal FL. Regular at the Dry Tortugas. **WINTER:** Mainly southern Mexico through C.A.,

uncommonly to northern S.A. Rare in winter in southern US from FL to CA. **RARE STATUS:** Rare (mainly spring) New England and Maritimes, central and northern Great Plains, and much of West (majority are *rubra*). Casual NL. Accidental UK.
Population Western population threatened by loss of riparian habitats.

red head and patches elsewhere

greenish primaries and secondaries

1st spring ♂
rubra

large bill

some females with patchy dull red throughout

red morph ♀
rubra

rubra ♀

overall ochre plumage

overall rosy red

adult ♂
rubra

SCARLET TANAGER *Piranga olivacea* SCTA ■ 1

The breeding male Scarlet Tanager is striking. Often seen in small flocks during spring migration, Scarlet Tanager sings and forages on the breeding grounds high in the canopy. Monotypic. L 7" (18 cm)
Identification Sexually dimorphic. Bill short and stubby; short tailed, whitish wing lining. **BREEDING ADULT MALE:** Brilliant red, with black wings and tail. **FEMALE:** Females are yellow-green with yellower throat and sides, dark wings and tail, a thin eye ring, and wing coverts with greenish edging. Some adult females show weak wing bars. **WINTER ADULT MALE:** Entirely greenish yellow, but retains black scapulars, wings, and tail. Late summer birds can be blotchy red. **FIRST-SPRING MALE:** Resembles adult male with black scapulars and wing coverts, but flight feathers are brownish.

Similar Species Immature females are similar to some Summers that largely lack ochre tones. Note Scarlet's structural differences, darker wings, and gray undersurface to tail, which is yellowish in Summer. A worn late-summer female Western can show reduced wing bars; note grayer back contrasting with yellower rump, and a longer, paler bill.
Voice CALL: A hoarse *chip* or *chip-burr*, unlike other tanagers. **SONG:** Robinlike but raspy *querit queer querry querit queer*, very similar to Western Tanager. **FLIGHT NOTE:** A whistled *puwi*.
Status & Distribution Fairly common. **BREEDING:** Nests in deciduous forests in the eastern half of N.A.; arrives late Apr.–mid-May. **MIGRATION:** Trans-Gulf migrant. Arrives Gulf Coast early Apr., peaks mid- to late Apr. Fall migration more easterly and mostly late (late Sept.–Oct.). **WINTER:** Mainly in Amazonia and the foothills of the Andes. Casual in N.A. **RARE STATUS:** Casual in Atlantic Provinces north of breeding range. Very rare in Southern CA, mainly in Oct. and Nov.; casual elsewhere in West; accidental to AK.

1st spring ♂

small bill

♀

wings contrast darker to olive-green back

brownish primaries

shorter tail than Summer

black contrasts with greenish remiges

1st fall ♂

Casual western Europe and the Azores.
Population Sensitive to forest fragmentation and parasitism by Brown-headed Cowbird.

breeding adult ♂

all adult males have solid black wings and tail

fall adult ♂

WESTERN TANAGER *Piranga ludoviciana*   WETA   1

The striking black-and-yellow Western Tanager, with its bright red head, is one of the more characteristic summer species of western pine forests. Like other *Piranga*, it can be quite inconspicuous when feeding on insects high in the canopy or singing for long periods without moving. During migration, the Western Tanager often feeds on fruit and is more conspicuous. Monotypic. L 7.3" (19 cm)

Identification Sexually dimorphic. Western is the only regular N.A. tanager with wing bars. Note its yellow wing linings in flight. Its bill is small, larger in size than Scarlet, but smaller than other species. **BREEDING ADULT MALE:** Unmistakable. Bright yellow underparts, yellow rump, black back, conspicuous wing bars that contrast with black wings (yellow upper bar, lower white bar), and bright red head. First-summer male is duller with duller red head and brownish flight feathers. **BREEDING FEMALE:** Duller, mostly greenish yellow below, grayish back and wings, with two thinner, pale wing bars. Underparts variable; some have brighter yellow belly and flanks, while others are quite gray. **NONBREEDING ADULT MALE:** Similar to breeding male, but loses red on head; duller red confined in varying amounts to forehead and chin. Black in plumage not as crisp, with some greenish edging to back. **FIRST-FALL MALE:** Generally yellow-green, yellower below with a yellow upperwing bar, and white lower bar. Bill noticeably pale. **FIRST-FALL FEMALE:** Can be very dull grayish yellow, still with two thin whitish wing bars.

Similar Species Adult males unlikely to be confused with other tanagers. In

1st fall ♂

gray "saddle"

gray morph ♀

larger bill than Scarlet

♀

white wing bars and tertial edges

some very dull below

winter adult ♂

reddish head

breeding adult ♂

black "saddle" and yellow rump

southeastern AZ, beware of confusing Western Tanager with Flame-colored Tanager, which has two white wing bars, a striped flame-and-black back, white tips to tertials, and white-tipped tail feathers. Western has hybridized with Flame-colored; offspring have a mixture of Western and Flame-colored characteristics. Female Flame-colored has a streaked back, larger bill, and white tips to both its tertials and tail. Also note that some female Scarlets show pale wing bars, but female Westerns usually show a yellow upper bar. Some worn Westerns show little or no wing bars, but when compared to Scarlet, Western has a grayer back, longer tail, and larger bill.

Voice CALL: A *pit-er-ick*, with a noticeably rising inflection. Very different from both Summer and Scarlet, but indistinguishable from Flame-colored. **SONG:** Similar in tone and pattern to both Scarlet and Flame-colored. **FLIGHT NOTE:** A whistled *howee* or *weet*.

Status & Distribution A common bird of western coniferous forests, although it breeds in deciduous riparian habitats as far north as southeast

AK. **MIGRATION:** Common migrant throughout the lowlands of the West in spring (mid-Apr.–early June) and fall (mid-July–early Oct.). **WINTER:** Found mainly from central Mexico south (rarely) to Costa Rica and Panama. Uncommon to rare in winter in southern coastal CA. **RARE STATUS:** Very rare, mainly in fall and winter, to eastern N.A.

Population Stable.

FLAME-COLORED TANAGER *Piranga bidentata*   FCTA   3

A montane resident of Mexico south through C.A., the Flame-colored Tanager has become a somewhat regular summer visitor to the mountain canyons of southeastern AZ. It behaves much like other *Piranga* tanagers, mainly singing and feeding high up in oaks, sycamores, or conifers with virtually all records from between 5,000 and 7,000 ft in elevation. Adult males

are striking in appearance. In AZ, the Flame-colored Tanager occasionally forms mixed pairs with the similar-sounding Western Tanager. Its alternative English name is Stripe-backed Tanager, reflecting a key field mark. Polytypic (4 sspp.; probably 2 in N.A.). L 7.3" (19 cm)

Identification Sexually dimorphic. **ADULT MALE:** Males are quite variable.

head and below bright yellow with a tinge of orange

1st spring ♂
bidentata

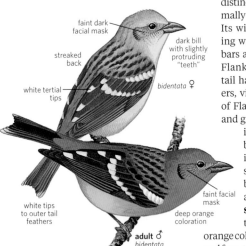

faint dark facial mask

dark bill with slightly protruding "teeth"

streaked back

white tertial tips

bidentata ♀

white tips to outer tail feathers

faint facial mask

deep orange coloration

adult ♂ bidentata

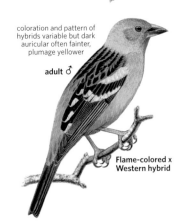

coloration and pattern of hybrids variable but dark auricular often fainter, plumage yellower

adult ♂

Flame-colored x Western hybrid

Some have bright, flame orange-red heads and underparts, whereas others are duller. Characteristic feature of male is orange-red back streaked with black. Rump is also usually orange-red with black streaks. Flame-colored has a distinctive darker cheek patch, normally with a blackish rear border. Its wings are blackish, contrasting with two distinct white wing bars and white tips to all tertials. Flanks are grayer, and blackish tail has white tips to outer feathers, visible from underneath. Bill of Flame-colored is rather heavy and gray. **ADULT FEMALE:** Female is much different in color, being greenish yellow, but is similarly patterned with a streaked back, two white wing bars, and white tips to tertials and outer tail feathers. **FIRST-SPRING MALE:** Yellower than typical adult male, its reddish orange coloring is confined to forehead and face. **IMMATURE FEMALE:** Back has less-distinct streaking.

Geographic Variation Males of the western Mexican subspecies *bidentata* are more orange than the eastern subspecies *sanguinolenta*, which tends to be redder.

Similar Species Flame-colored Tanager overlaps with the similar-sounding Western Tanager in the mountains of southeast AZ, where the species have been known to form mixed pairs. Hybrids are well documented, with some resembling male Flame-coloreds, though they usually have some noticeable Western feature, such as a yellow upperwing bar, a yellow unstreaked rump, or intermediate white spotting on the tertials or tail tips. Hybrid females undoubtedly occur but are more difficult to distinguish.

Voice CALL: A rising *pit-er-ick*, virtually indistinguishable from the rising call of Western; also gives a huskier, low-pitched *prreck*. **SONG:** Similar to Western Tanager.

Status & Distribution BREEDING: From northern Mexico to western Panama. Has nested several times in southeast AZ, mostly mixed pairs with Western Tanager. All nests have been in well-wooded mountain canyons in the Santa Rita, Huachuca, and Chiricahua Mts. of southeastern AZ, usually where sycamores occur. **RARE STATUS:** Rare to casual spring and summer visitor to montane canyons in southeastern AZ. Casual in the Chisos and Davis Mts. of West TX and in the Lower Rio Grande Valley.

Population Stable.

bidentata

flammea

sanguinolenta

bidentata

Genus *Rhodothraupis*

CRIMSON-COLLARED GROSBEAK   *Rhodothraupis celaeno*   CCGR ■ 4

This distinctive species, in its own monotypic genus, is endemic to northeastern Mexico, but wanders casually to south TX. This species often skulks on or near the ground. It prefers to eat green leaves, but also feeds on fruit (e.g., mulberries). The color combinations of Crimson-collared Grosbeak are unique. Monotypic. L 8.5" (22 cm)

Identification Sexually dimorphic; thick, stubby bill with a curved culmen. **ADULT MALE:** Pinkish red underparts and collar contrast with the black hood and bib. Blackish above, with two thin pinkish wing bars. **ADULT FEMALE:** Females have a similar pattern, with pinkish red replaced by greenish yellow. **IMMATURE MALE:** Like immature female. Can show pinkish red blotches.

IMMATURE FEMALE: Like adult female, but reduced black hood and bib. **Similar Species** Adult male unmistakable. Female superficially similar in

on adult female, black hood distinctly defined on breast

yellow-olive coloration on female and immature male

adult ♀

black head

thick stubby bill with curved culmen

intense deep red on nape and below

adult ♂

body, head, and bib color to the adult male Audubon's Oriole, but note the oriole's different shape, particularly bill length and shape, and its white wing bars.

Voice CALL: A penetrating, rising and falling *seeiyu*. **SONG:** A varied warble.

Status & Distribution BREEDING: Endemic to northeastern Mexico. **RARE STATUS:** Casual fall and winter (mostly early Nov.–late Apr.) visitor to southern TX (about 35 records), mainly to well-vegetated parks and refuges (also residential yards in towns) in the Lower Rio Grande Valley. The winter of 2004–2005 was exceptional, with at least 15 individuals found. Some individuals remain until spring. Recorded north on Rio Grande to Webb Co. and on coast to Galveston Co.

Population Stable.

CARDINALS Genus *Cardinalis*

The Northern Cardinal and Pyrrhuloxia are characterized by long, pointed crests and stout, conical bills. They are mainly resident within their home range. Males are brightly plumaged; females and immatures are duller. Cardinals feed mainly on seeds and fruit and commonly come to seed feeders.

NORTHERN CARDINAL *Cardinalis cardinalis* NOCA 1

The red male Northern Cardinal, with its conspicuous crest, is one of the most recognizable birds in N.A. It is found commonly through virtually all of the eastern US in a variety of habitats, including suburban areas; frequents feeders in winter. Males usually sing from exposed perches. Polytypic (19 sspp.; 5 in N.A.). L 8.8" (22 cm)

Identification Sexually dimorphic. Thick cone-shaped bill. **ADULT MALE:** Unmistakable. Males are red with a black face, red bill, and an obvious, pointed crest. **ADULT FEMALE:** Females are buffy brown with a reddish tinge on wings, tail, and crest. **JUVENILE:** Both sexes have dark bills. Like adult female but generally browner overall with less reddish tones in wings and tail; female duller, lacking reddish tones in wings and tail.

Geographic Variation Five subspecies described north of Mexican border. Size and coloration varies clinally from east to west, with the eastern birds (*cardinalis*) being smaller, shorter crested, duller red, and slightly more black on face than the southwestern

superbus. The other three N.A. subspecies, *canicaudus*, *floridanus*, and *magnirostris*, are intermediate.

Similar Species In East, not really confused with any other species. Note that the adult male Summer Tanager is also bright red, but it lacks both crest and black face. In southern TX and the Southwest, Northern Cardinal overlaps with the very similarly shaped Pyrrhuloxia, but note their color differences. Female and immature Pyrrhuloxias are similar to female and juvenile cardinals. Pyrrhuloxia has a noticeably yellow bill that has a distinct downward curve to culmen, whereas the cardinal has a distinctly straighter culmen as well as a pointier, reddish or dark bill. Plumage of the female Pyrrhuloxia is grayer with very little red.

Voice CALL: A sharp, loud chip note. **SONG:** Variable. A liquid, whistling *cue cue cue*, or *cheer cheer cheer*, or *purty purty purty*. Both sexes sing virtually all year, though females sing less frequently than the males do.

Status & Distribution Common. Generally nonmigratory. **BREEDING:**

Throughout the East, found in a variety of habitats, including woodland edges, swamps, streamside thickets, and suburban gardens. In the West, restricted mainly to southwestern TX, southern NM, and southern AZ, where it is common in mesquite-dominated habitats, usually near water. **RARE STATUS:** Casual to Prairie Provinces and MT. Now casual in southeastern CA and likely extirpated along the Colorado River. Casual north to Inyo Co., CA, and southern NV.

Population Range expanded north in eastern N.A. in 20th century.

long crest

black surrounds red bill

straight culmen

cardinalis ♂

dark bill

cardinalis ♀

overall buffy brown and dull red

juvenile ♂ *cardinalis*

longer crest than eastern birds

less black around red bill than eastern birds

male brighter red than eastern birds

southwestern ♂ *superbus*

PYRRHULOXIA *Cardinalis sinuatus*   PYRR ■ 1

This bird's coloration and bill shape and color distinguish it from the Northern Cardinal. Polytypic (3 sspp.; 2 in N.A.). L 8.8" (22 cm)
Identification Sexually dimorphic. **ADULT MALE:** Pearly gray body and blood-red face, bib, center to breast and belly, and tip of crest. Thick, yellow bill and curved culmen. Red-edged primaries and tail feathers. **ADULT FEMALE:** Duller, lacks most of red coloration. **JUVENILE:** Lacks red tones in wings and tail. Bill not as yellow as adult.

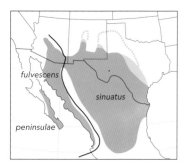

Geographic Variation Nominate subspecies *sinuatus* breeds from south TX to southern NM; *fulvescens* breeds in southern AZ (underparts of adult male are paler orange-red rather than deep red).
Similar Species Females and juveniles can be confused with female and juvenile Northern Cardinal. Distinguished by curved culmen, yellower bill, fewer red tones, and spikier crest.
Voice CALL: A *chink*, more metallic than Northern Cardinal. **SONG:** Reminiscent of, but thinner and shorter than, Northern Cardinal's liquid whistles.
Status & Distribution Pyrrhuloxia is fairly common. **BREEDING:** Thorny brush and mesquite thickets in Southwest lowland desert; more arid environments than cardinal. Often in brushy borders near houses in desert ranchland.

WINTER: Much more widespread; moves north from breeding areas in AZ, NM, and TX (where also moves south and east). **RARE STATUS:** Casual, mainly in fall, winter, and spring to Southern CA, southern NV, and southwestern UT, as well as the Great Plains and Gulf of Mexico (oil rigs). Accidental to OR, WI, and ON.
Population Still common, but potentially threatened by loss of natural desert habitats in the Southwest.

yellowish bill with strongly curved culmen

fulvescens ♀

grayish overall with pale buffy breast

fulvescens ♂

overall gray with patchy bright red

GROSBEAKS Genus *Pheucticus*

Two species breed in N.A.; one is a casual visitor from Mexico. Adult males are brightly colored; females and immature males are duller. First-spring males are separable from older birds. With large, conical bills, they feed on insects during the breeding season and fruit during fall and winter. They are also known to frequent seed feeders.

YELLOW GROSBEAK *Pheucticus chrysopeplus*   YEGR ■ 4

The Yellow Grosbeak is found mainly in western and southern Mexico. In western Mexico, where more numerous in summer, it prefers thorn forest and riparian areas. Polytypic (3 sspp., presumably *dilutus* in N.A.). L 9.3" (24 cm)
Identification ADULT MALE: Bright yellow with black-and-white wings: white median coverts, white-tipped

greater coverts, and white-based primaries form a white patch on folded wing. White tips to tertials. Black tail with white inner webs. **ADULT FEMALE:** Duller yellow and more streaking on upperparts. Wings and tail browner. **IMMATURE MALE:** Like female, with a yellower head. It achieves adult male plumage by its second winter.
Similar Species Female and imma-

overall bright yellow, but female and immatures duller

huge bill

adult ♂

black wings with extensive white

adult ♀

ture male Black-headed Grosbeaks are much more buffy underneath and have a more patterned head and smaller bill.
Voice CALL: *Pheucticus*-type *eek*, like Black-headed. **SONG:** Rich and warbled, also similar to Black-headed.
Status & Distribution Northwest Mexico from central Sonora to Guatemala; northern birds migratory. Casual late spring and early summer to riparian areas in the lowlands and canyons of southeastern AZ. There is a winter record from Albuquerque, NM. Records from IA and eastern CA are questioned on origin.
Population Stable.

ROSE-BREASTED GROSBEAK   *Pheucticus ludovicianus*   RBGR ■ 1

The striking Rose-breasted Grosbeak is found in wooded habitats across much of eastern and midwestern N.A. Monotypic. L 8" (20 cm)

Identification Sexually dimorphic. Takes more than a year to acquire adult plumage. **ADULT MALE:** Black head, back, wings, and tail contrast with gleaming white underparts and rump. Bright rosy pink patch on breast. Rose-red wing linings. Large, pale bill. **ADULT FEMALE:** Streaked brownish upperparts, whitish below with extensive dark streaks. Dark yellow wing linings. **WINTER MALE:** Molts into winter plumage before migrating. Brownedged head and upperparts, barred rump, and dark streaks on sides and flanks. Wings as adult. **FIRST-FALL MALE:** Buffy wash with fine streaks across breast, pink wing linings, usually a few pink feathers on sides of breast. **FIRST-SPRING MALE:** Similar to adult male, but browner, particularly wings and tail.

Similar Species Adult males unmistakable. Compare female and first-fall male Rose-breasted to the comparable

breeding adult ♂

winter adult ♂

rose red breast and wing linings

breeding adult ♂

rich buff chest, often with a few pink feathers; reddish pink wing linings

1st fall ♂

pale bill

1st spring ♂

strong streaking across breast

♀

plumages of the similar Black-headed Grosbeak. First-fall male Rose-breasted is deep buff across breast like Blackheaded, but the color is not quite as buffy orange and the breast is usually a bit more streaked. Rose-breasted has pink wing linings and often one or more pink feathers are visible on breast. Female Rose-breasted is usually broadly and evenly streaked across the breast against a whitish background, and is overall colder in coloration; its squeakier call is diagnostic.

Voice CALL: A sharp *eek*, squeakier than Black-headed. **SONG:** A robinlike series of warbled phrases, but shorter.

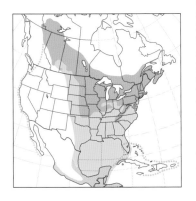

Songs are similar to those of Blackheaded Grosbeak.

Status & Distribution Common. **BREEDING:** Deciduous forest habitats. **WINTER:** Mainly Mexico, C.A., and rarely Cuba; casual or very rare in southern US, including coastal CA. **MIGRATION:** Peak spring migratory period in eastern US mid-Apr.–midMay; peak fall migratory period Sept.–mid-Oct. Often seen in flocks during spring migration fallouts. **RARE STATUS:** Rare during late spring and fall in the Southwest. Casual to Pacific Northwest and southeastern AK. Also casual Europe and the Azores (mainly in Oct.).

Population Possible decline due to forest fragmentation.

BLACK-HEADED GROSBEAK   *Pheucticus melanocephalus*   BHGR ■ 1

The western counterpart of Rosebreasted Grosbeak; the two occasionally hybridize where they come into contact on the western Great Plains. Found in a variety of habitats, from open coniferous forests to riparian areas. Adult males are striking; females and immatures are duller. Polytypic (2 sspp.; 2 in N.A.; subspecies are poorly differentiated; adult male coastal birds, *maculatus*, have a tawny supercilium; adult male *melanocephalus* has a solid black head). L 8.3" (21 cm)

Identification Bill is large and conical, typically bicolored; wing linings are yellow in both sexes. **ADULT MALE:** Cinnamon-colored body including collar, with nearly black head, mostly black back, black wings. Center to lower belly yellow. **ADULT WINTER:** Plumage is acquired on the winter grounds. Similar to adult male, but head less black, supercilium and crown strip cinnamon, and streaks on back more prominent. **FIRSTSUMMER MALE:** Like adult winter male, but wings less black. **FIRST-**

WINTER MALE: Like female, but underparts more cinnamon without streaking, and underside of tail gray. **ADULT FEMALE:** Mainly dull brown, with streaked back, buffy white nape and supercilium. Underparts variable, from buffy to almost white, with varying amount of streaking on sides of breast and flanks, but center of breast is typically unstreaked.

Similar Species Adult male unmis-

takable. Female and first-winter male is easily confused with the female and first-winter male Rose-breasted. Note different amount of streaking on underparts and color of bill. On Black-headed, the color of the breast, especially in fresh plumage in late summer and fall, is a richer buff and the streaking is confined to the sides of the breast. The bill is darker, particularly the upper mandible, and less pinkish red overall. The general coloration is warmer, but the wing linings are slightly paler. The call notes are diagnostic.

Voice CALL: A sharp

eek, similar to the Rose-breasted, but decidedly less squeaky. Juveniles give an incessant whistled *whee-o* begging call once fledged. **SONG:** Robinlike series of warbled phrases, like Rose-breasted; also difficult to separate from Hepatic Tanager's song where the two species overlap in the mountains of the Southwest.

Status & Distribution Common throughout entire West. **BREEDING:** Nests in a variety of habitats, from mixed coniferous forest to montane riparian. Arrives on breeding ground from early Apr. (in south) to mid-May (in north). **MIGRATION:** Migrates in spring singly or in smaller groups, unlike Rose-breasted, which sometimes migrates in flocks. In fall, not uncommon to see several in a fruiting tree. Adults begin migrating south by mid-July. Immatures migrate later, Aug.–mid-Sept., a few into early Oct. **WINTER:** Mainly Mexico. **RARE STATUS:** Casual or accidental wanderer, mainly in fall and winter, to virtually all eastern states and provinces. Casual in migration to AK (including Gambell) and NT.

Population Stable.

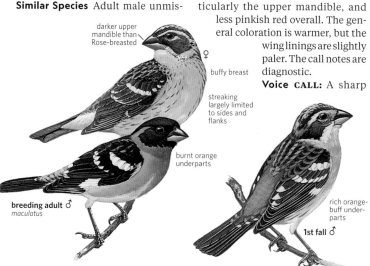

darker upper mandible than Rose-breasted

♀

buffy breast

streaking largely limited to sides and flanks

burnt orange underparts

breeding adult ♂
maculatus

rich orange-buff underparts

1st fall ♂

Genus *Cyanocompsa*

BLUE BUNTING *Cyanocompsa parellina*   BLBU ◼ 4

This M.A. endemic rarely visits south TX. Polytypic (3 sspp.; records from US likely *beneplacita*). L 5.5" (14 cm)
Identification Sexually dimorphic. Stout blackish bill with curved culmen. **ADULT MALE:** Blackish blue body with brighter blue crown, cheeks, shoulder, and rump. Lacks wing bars. **IMMATURE MALE:** Like adult male, but with brownish cast to wings. **FEMALE:** Uniformly rich, buffy brown, lacking streaks or wing bars.
Geographic Variation Eastern *bene-*

placita from northeast Mexico and nominate *parellina* from Veracruz and Yucatán belong to *parellina* group. Female of West Mexico *indigotica*, found from central Sinaloa to southwest Chiapas, is overall paler and grayer and gives a different call note, possibly indicating species-level differences.
Similar Species Female Varied Bunting, also plainly colored, has less rusty brown and is smaller billed, with a bluish tint to primaries and tail.
Voice CALL: A metallic *chink*, simi-

lar to Hooded Warbler. Subspecies *indigotica* from west and southwest Mexico gives a very different call note. **SONG:** A series of high, clear warbled phrases, usually beginning with two separate notes.
Status & Distribution Resident northwest and northeast Mexico to north-central Nicaragua. **RARE STATUS:** Casual to rare, nearly all valid records are in winter (mid-Nov.–mid-Mar.), in brushy areas and feeders in the Rio Grande Valley of south TX.
Population Stable.

overall a rich cinnamon-brown

large dark bill

♀

deep, dark blue overall with paler blue highlights

adult ♂

parellina group

indigotica

BUNTINGS Genus *Passerina*

Adult males are very brightly plumaged; females are very dull brown or green, some with wing bars, some without. All have the characteristic behavior of twitching the tail while perched. They feed mainly on insects during breeding season, seed during fall and winter. They are nocturnal Neotropical migrants.

BLUE GROSBEAK *Passerina caerulea*   BLGR ■ 1

Closely related to the other *Passerina* buntings, the larger Blue Grosbeak is primarily found in the southern US. It occurs in overgrown fields, along brushy roadsides, or in riparian habitats. Although it usually stays low in the brush, it often sings from an exposed perch. Like *Passerina* buntings, it has the distinctive behavior of twitching and spreading its tail. It feeds mainly on seeds. Polytypic (7 sspp.; 3 in N.A.; eastern N.A. birds slightly smaller and larger-billed than western N.A. birds). L 6.8" (17 cm) **Identification** Sexually dimorphic. Note large, massive bill. **ADULT MALE:** Male is deep blue and has bright chestnut wing bars. Also note black in face and chin and indistinct blackish streaking to upperparts. **ADULT FEMALE:** Pale grayish brown body overall with thick, buffy wing bars, lower wing bar paler. Indistinct streaking on back, some blue tinge to scapulars and tail. **FIRST-SUMMER MALE:** A mixture of

male and female plumage, favoring female usually with a mostly blue head, sometimes with blue patches on breast, and blue tinge to primaries and tail. Some have a touch of blue on forehead. **FIRST-WINTER MALE & FEMALE:** Similar. Overall richer rufous brown than adult female, with chestnut wing bars; unstreaked below.

Similar Species Larger and larger-billed than Indigo Bunting. Female and immature Indigo Bunting always with faint to moderate streaking on underparts. Calls and song very different.

Voice CALL: A loud, sharp *chink*. **SONG:** A series of rich, rising and falling warbles.

Status & Distribution Uncommon to fairly common. **BREEDING:** Found in a variety of brushy habitats, often near water, all across

the southern US. **MIGRATION:** Arrives on the southern breeding grounds by early to mid-Apr., northern range by mid-May. Eastern birds are trans-Gulf migrants. Regular Bermuda (more in fall). **WINTER:** Mexico to central Panama. Casual to CA, AZ, and FL. **RARE STATUS:** Rare in migration to New England and Atlantic Provinces. Casual to northern Great Lakes, southwestern AB, and southern BC. Accidental as far north as southeastern AK. Rare to Cuba and Bahamas.

Population Stable.

caerulea ♀

rusty buff
wing bars

thick
bill

plain
rich buff
below

immature
caerulea

chestnut
wing bars

variable
amount
of blue
on head

1st spring ♂
caerulea

breeding adult ♂
caerulea

frequently
twitches tail

LAZULI BUNTING *Passerina amoena*   LAZB ■ 1

A western counterpart to the Indigo Bunting, the Lazuli Bunting is found in a variety of habitats, and its blue, white, and rich buff plumage makes it one of the most attractive songbirds of the West. Its behavior is similar to other *Passerina* buntings. Frequents brushy borders to fields and roads, seen sometimes in single-species flocks, and often twitches and spreads its tail while perched. Occasionally hybridizes with Indigo Bunting. Monotypic. L 5.5" (14 cm) **Identification** Highly sexually dimor-

phic. Achieves adult plumage by second winter. **ADULT MALE:** Distinctive male is bright turquoise blue above and on throat with a rich cinnamon-buff wash across breast; white on belly with a thick white upperwing bar. **WINTER MALE:** Like Indigo, blue plumage obscured by buffy brown edging to feathers, with blue visible on head and throat. **ADULT FEMALE:** Very different, duller, resembling female Indigo. Note drab grayish brown coloration with buffy wash across breast, two narrow white wing bars, and complete lack

of streaking on underparts. **WINTER FEMALE:** Like breeding female but with warmer buff wash on breast, grayer throat, and buffier wing bars. **FIRST-SUMMER MALE:** Like adult male with brown feathers intermixed with blue. **JUVENILE:** Like female, but with distinct fine streaking across breast.

Similar Species Male unmistakable. Females confused with female Indigo and Varied Buntings and with female Blue Grosbeak. Note overall grayer plumage, warm buff wash across breast, lack of streaking on underparts (except in juvenile, which has finer, less blurry streaks), which is present on all female Indigos, and more distinct narrow white wing bars.

Voice CALL: A dry, metallic *pik*, similar to Indigo.

SONG: A series of varied phrases, sometimes paired, like Indigo, but faster and less strident.

Status & Distribution Fairly common. **BREEDING:** Found in open deciduous or mixed woodland and in chaparral, particularly along streams and rivers; also favors burns. **MIGRATION:** Con-

gregates during migration in the Southwest (late July–Oct.), where molting occurs, before continuing migration to Mexico. **WINTER:** Mainly west Mexico. Rare in southeastern AZ. **RARE STATUS:** Casual in migration north to AK (including Gambell) and eastern N.A. **Population** Stable.

winter ♀

rich and extensively buffy below but unstreaked, unlike all female Indigos, which have blurry streaks

buffy orange breast band

rich powder or lazuli blue on head

breeding adult ♀

white patch on median coverts

juvenile

fine dark pinstripes below, the common plumage seen in late Aug. and Sept. in much of West; adults have migrated south earlier

strong white wing bars, upper thicker

breeding adult ♂

buffy breast contrasts with white belly

INDIGO BUNTING *Passerina cyanea*   INBU   ■   1

The male Indigo Bunting is a familiar sight singing on wires in eastern N.A. In the spring, Indigo Bunting may be present in large flocks, particularly during migration fallouts, and is often seen in brushy habitat or along weedy margins of fields and roads. It sometimes hybridizes with the Lazuli Bunting on the Great Plains and in the West. Two genetic studies using mtDNA indicated that Blue Grosbeak, formerly in the monotypic genus *Guiraca*, and Lazuli are sister species (each other's closest relatives). A subsequent study using nuclear DNA not surprisingly showed Lazuli and Indigo Buntings are sister species. Monotypic. L 5.5" (14 cm)

Identification Highly sexually dimor-

phic. **SUMMER ADULT MALE:** Entirely bright blue. **WINTER ADULT MALE:** Blue obscured by brown and buff edging; mottled brown and blue early in winter. **SUMMER FEMALE:** Dull brown, usually with two faint wing bars and indistinct streaking on underparts. Whitish throat, small conical bill with straight culmen, relatively long primary projection. **FIRST-SUMMER MALE:** Patchy dark blue and brown. **IMMATURE & WINTER FEMALE:** More rufescent overall than breeding female, with

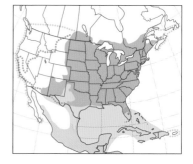

breeding adult ♀

faint wing bars

all females have blurry streaks below

deep blue color

breeding adult ♂

1st spring ♂

variable patchy blue and brown plumage

mostly brown but with some veiled dark blue showing

winter adult ♂

fall ♀

fall females and immatures a richer cinnamon-brown color than breeding females blurry streaking on breast and flanks. **Similar Species** Much smaller and smaller-billed than adult male Blue Grosbeak, which has broad chestnut wing bars, black around the face, and dark streaks on back. Female and immature Indigos are similar to comparable plumages of Lazuli and Varied Buntings. Lazuli and Indigo Buntings are closely related, and immature males resemble older males by the next spring, but with patchy brown. First-spring males of Painted and Varied Buntings (each other's closest relatives) look like females. Female and winter immature male Indigo Buntings always show extensive blurry streaks below, the single best field mark—thus any streaked *Passerina* bunting in winter is an Indigo. Adult female Lazuli Buntings have a buffy wash across the breast but are always unstreaked below; juveniles of both sexes (plumage seen into Oct.) show sharp, well-defined streaks across the breast. Female and immature Indigos are more extensively darker brown than Lazuli, with slightly less distinct wing bars. Varied Buntings (females and immature males) are similar to Indigo but are always unstreaked ventrally and show a slightly more curved culmen. Juvenile Painted is similarly unstreaked and gray overall with some greenish on the back.

Voice CALL: A dry, metallic *pik*, just like Indigo. **SONG:** A series of sweet, varied phrases, usually paired. **Status & Distribution** Common. **BREEDING:** Found in brushy borders to mainly deciduous woodland throughout eastern US. In the Southwest, mainly found in riparian habitats. **MIGRATION:** Arrives on breeding grounds mid-Apr.–early June. In fall, mainly mid-Sept.–mid-Oct. **WINTER:** Mainly Mexico through C.A., rarely to northern S.A. Also on Caribbean islands. Uncommon in Lower Rio Grande Valley, TX. Rare along Gulf Coast. **RARE STATUS:** Rare to Pacific states (hybridizes with Lazuli) and Atlantic Provinces; casual YT and to Europe and the Azores. **Population** Stable.

VARIED BUNTING *Passerina versicolor*   VABU ■ 2

One of the more colorful of the buntings, adult males of this Southwest specialty appear uniformly dark in poor light. Found in more arid habitats than other buntings, it is often seen in desert washes or on cactus-laden hillsides of canyons. Similar in behavior to other *Passerina* buntings, including twitching their tails. An inhabitant of dense mesquite thickets, the Varied Bunting is usually difficult to see, but it often sings from exposed perches. Adult males are stunning in good light; females and immatures are extraordinarily dull. Polytypic (4 sspp.; 2 in N.A.). L 5.5" (14 cm)

Identification Sexually dimorphic. **SUMMER MALE:** Quite colorful, but colors difficult to see. Striking are bright red nape, deep blue head and rump, and dark purplish body. Bill is rather small, with distinctive curved culmen. **SUMMER FEMALE:** Extremely dull and featureless: almost entirely grayish brown, with very faint suggestion of wing bars, but no obvious

pale edging to tertials. Female and juvenile Varied Buntings show no trace of streaking below, a character shared with female and immature Painted Buntings. Note curved culmen. **WINTER MALE:** As in other *Passerina* buntings, its color is obscured by brownish edging to many feathers. **WINTER FEMALE:** Similar to summer female, but more richly colored and slightly more rufescent overall. **FIRST-SPRING MALE:** Like adult female, but with varying amounts (often none) of purple on forehead; sings as ardently as older males and defends breeding territory.

Geographic Variation Nominate subspecies *versicolor* from southeastern NM and southern TX has, on average, duller red nape, purple throat with reddish tinge, and pale blue rump. Subspecies *dickeyae* from southeastern AZ and *pulchra* from Baja California Sur have brighter red nape, purple throat with no red tones, and purplish blue rump. Given the individual

variation, even identification of out-of-range adult males is problematic. Females grayer in AZ *dickeyae*. **Similar Species** Males unmistakable in good light. Females more of a challenge. Lack of distinguishing characteristics, including no trace of ventral streaking, is actually an identifying feature of female Varied, but note distinctive curved culmen. Female Lazuli has distinct pale wing bars and a buff wash across breast. Female Indigo is darker, always with some streaking

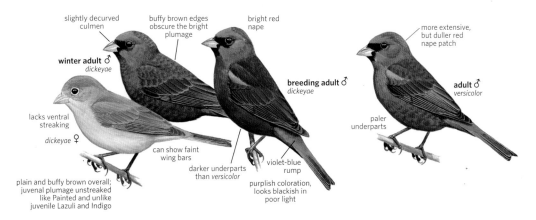

slightly decurved culmen

buffy brown edges obscure the bright plumage

bright red nape

winter adult ♂
dickeyae

lacks ventral streaking

dickeyae ♀

plain and buffy brown overall; juvenal plumage unstreaked like Painted and unlike juvenile Lazuli and Indigo

can show faint wing bars

darker underparts than *versicolor*

violet-blue rump

purplish coloration, looks blackish in poor light

breeding adult ♂
dickeyae

more extensive, but duller red nape patch

adult ♂
versicolor

paler underparts

on underparts. Juvenile Painted Buntings are very drab but always show some greenish on the back.

Voice CALL: A loud, rich chip note, similar to Lazuli and Indigo Buntings. **SONG:** A series of rich, sweet, unrepeated phrases, very similar to Painted Bunting, its closest relative. The two are hard to differentiate by song in areas of overlap in south and West TX and southern NM; however, the call notes of these two species are different.

Unlike Painted, Varied sings through the monsoon in late summer.

Status & Distribution Found south to Guatemala. In U.S., uncommon to fairly common, and local. **BREEDING:** In summer, inhabits dense, thorny thickets in desert washes, canyon hillsides, and sometimes along streams and rivers. Mostly found at lower elevations. **MIGRATION:** Most of the N.A. breeding population migrates south in winter. Population in southwest TX may be

at least partially resident. TX breeding population arrives mid-Apr.; AZ population very late arriving, mostly late May–early June (very rare before mid-May). **WINTER:** Western Mexico, mainly Sonora to Oaxaca. Casual in southwestern TX. **RARE STATUS:** Casual to Southern CA (late fall and winter) and most of TX away from mapped range. Accidental LA (Jul.), PA (spring), and ON (spring). **Population** Stable.

PAINTED BUNTING   *Passerina ciris*   PABU   ▪ 1

Found across the South in two discrete populations, the adult male Painted Bunting, with its multiple colors, can be considered N.A.'s gaudiest songbird. Even so, it can be difficult to see, often singing from inside dense thickets, but at other times sings from telephone lines. Females and juveniles are very different. The Painted regularly comes to seed feeders during the nonbreeding season. Polytypic (2 sspp.; 2 in N.A.). L 5.5" (14 cm)

Identification Sexually dimorphic. **ADULT MALE:** Unmistakable. Bright red underparts and rump, lime-green back, dark wings with green edging, and dark blue head (except for red chin and center to throat), red eye ring, dark tail. **ADULT FEMALE:** Upperparts lime green, dull yellowish below. Greenish edging on wings and tail. **FIRST-WINTER MALE:** Resembles female. **FIRST-SUMMER MALE:** Also like female, but some show scattered blue feathers on head or red on breast. Achieves adult plumage by second winter. **JUVE-NILE:** Very drab, grayish below and on head, back shows some greenish tones. Upperparts are greener than plain underparts. Subsequent molt during the fall results in a greener plumage.

Geographic Variation Although the variation in plumage between the eastern nominate subspecies *ciris* and western *pallidior* is slight, migratory

and molt patterns differ. Eastern birds molt on the breeding grounds and then winter in FL and the Caribbean, while western birds molt at migratory staging grounds or on the winter grounds in Mexico and C.A.

Similar Species Adult male is unmistakable. Green plumage of adult female and older fall immatures of both sexes suggest a miniature female Scarlet Tanager. All other female and immature *Passerina* buntings are brownish. Juveniles are perplexingly plain and grayish; however note the greenish wash on lower back. The lack of streaks below is distinctive from female and immature Indigo and juvenile Lazuli. Female and immature Varieds are also unstreaked, but note that they are brownish, not grayish, and lack any green color.

Voice CALL: A loud chip note, sweeter than other *Passerina* buntings. **SONG:** A rapid series of varied phrases, very similar to Varied.

Status & Distribution Fairly common. **BREEDING:** Often secretive in brushy thickets and woodland borders, often along streams and rivers, along the southeast coast (NC, SC, GA, FL) and much of the south-central US, north to western TN, west to West TX and southeastern NM. **MIGRATION:** Spring migration relatively early (early Apr.–early May). Fall migrations differ by

population; eastern birds mainly late Sept.–Oct. and western birds late July–early Oct.; by mid-Aug. most adults have left N.A. Western birds interrupt their migration to molt out of breeding plumage, mainly in Sonora, Mexico, but some to southeastern AZ. **WINTER:** Mainly Mexico south to Costa Rica and Panama; also central and southern FL, Bahamas, and Cuba. Rare up Atlantic coast to VA; casual farther north. **RARE STATUS:** Rare to casual, primarily in spring and fall north and west of mapped breeding range. This species is kept in captivity, so stray records, especially those of adult males, are sometimes debated in regard to origin.

Population Affected by capture for bird trade. Atlantic subspecies (*ciris*) has small population and is impacted by loss of breeding habitat.

bright green upperparts similar overall in coloration to larger female Scarlet Tanager

yellow eye ring

♀

adult ♂

unmistakable

overall plain and grayish, usually with some greenish on lower back

juvenile

Genus *Spiza*

DICKCISSEL *Spiza americana*   DICK ■ 1

The Dickcissel, a sometimes abundant migrant and breeding bird of the central US, gets its name from the verbal interpretation of the male's song. Year-round usually found in open prairie or weedy agricultural fields, its numbers and breeding distribution varies from year to year. Males often sing while sitting up on a tall weed, shrub, or wire. In migration, the Dickcissel is usually detected when flying over and giving its distinctive, flat call. Large flocks congregate during migration and on the winter grounds. Often found with House Sparrows at feeders in the East in winter. Monotypic. L 6.3" (16 cm)

Identification Sexually dimorphic. Note long, pale supercilium, broad pale submoustachial, and distinct dark malar streak. **ADULT MALE:** Distinctively patterned; grayish brown and sparrow-like. Underparts with a black bib surrounding white chin, yellow breast, whitish lower belly and undertail coverts. Grayish back, boldly streaked black, and gray rump is unstreaked. Wings with bright chestnut shoulder patch. Gray nape. Complex face pattern: gray auriculars; yellow eyebrow above and in front of eye, white behind eye; yellow submoustachial, widening to white neck mark. Bill rather long and conical. **WINTER ADULT MALE:** Similar to breeding male, but browner overall and with a less-distinct black bib. **ADULT FEMALE:** Similar to male but duller and browner. Varying amount of black speckling instead of black bib. Eyebrow duller; buffy behind eye. **IMMATURE:** Fine streaking on breast and flanks and bolder streaking on back; immature male has relatively broad, rusty tips to median coverts, breast with some yellow, lacks bib; immature female is very dull with narrow, buff tips to median coverts.

Similar Species Immatures, which lack chestnut shoulder and have no yellow on underparts, can resemble female or juvenile House Sparrows. Note distinct white upperwing bar and lack of any streaking on underparts of House Sparrow. Note also House Sparrow's plumper shape, shorter bill, and shorter primary projection.

Voice CALL: A flat *bzrrrrt* is the typical flight call; also given when perched; also a husky *check*. **SONG:** A variable, buzzy insectlike *dick dick di'cissel*.

Status & Distribution Common, sometimes locally abundant. Numbers and range fluctuate annually outside core breeding range; in some years, moderate numbers well to the north and east. **BREEDING:** Nests in open, weedy meadows, in agricultural fields, and in prairie habitats. **MIGRATION:** Mainly a nocturnal migrant, but large flocks are seen migrating during the day. Flocks coalesce, sometimes reaching numbers in the thousands. Spring migration mid-Apr.–mid-May; fall migration mid-Aug.–mid-Oct. **WINTER:** Mainly the llanos (seasonally flooded grasslands and savannas) of Venezuela; uncommon and local along the Pacific coast of C.A. In US very rare to Northeast and Atlantic Provinces in winter, usually at feeders and often with House Sparrows. **RARE STATUS:** Rare, mainly in fall, to East Coast; very rare to West Coast; accidental AK. Casual to the Azores; accidental Norway.

Population Large-scale conversion of native grasslands to agriculture in the main breeding areas has likely had a negative impact on population size.

breeding ♀

winter adult ♂

pale supercilium

large bill   immature ♂

broad, pale yellow submoustachial

chestnut median coverts

immature ♀

chestnut lesser and median coverts

black bib and yellow breast

longish wings

breeding ♂

TANAGERS AND ALLIES Family Thraupidae

Red-legged Honeycreeper, breeding male *carneipes* (Costa Rica, July)

Ranging from intensely multicolored to relatively dull brown or gray in plumage, tanagers are common and remarkably diverse through the New World tropics; one of the largest of all bird families, Thraupidae includes some 371 species in 98 genera. Work by taxonomists in recent years has obliterated any former definitions of what constituted a "tanager." Many birders are understandably confused that all North American birds called tanagers are now placed in the cardinal and grosbeak family (Cardinalidae). The only North American species now placed within the Thraupidae, all marginal, are Red-legged Honeycreeper, Morelet's Seedeater, Yellow-faced and Black-faced Grassquits, and Bananaquit. Morelet's Seedeater is now a rare resident in south TX. The other four species are casual visitors to either TX or FL. None of these species have "tanager" in their English names!

Structure Part of the huge radiation of nine-primaried oscines, tanagers are small to moderately large songbirds exhibiting great variations in bill shape including relatively large-billed fruit-eaters, stout-billed seedeaters, and thin-billed insect and nectar feeders. They have a slender to slightly stocky build and medium-length wings and tails.

Behavior Foraging behavior is as diverse as the morphologies shown by the members of this family; frugi-vory (fruit eating) is common, but there are insect gleaners, nectar feeders, flower piercers, and even aerial salliers. Most species are mainly found in pairs or travel in small flocks; many join mixed-species flocks. Tanagers are not gifted vocalists, with most species giving thin chip notes, twitters, squeaks, or scolds; a few groups have more pleasant warbles or high whistles.

Plumage Many tanagers sport brilliant plumage, as in the almost psychedelic genus *Tangara*. In many species the sexes are similar, but in others the females are considerably duller. Recent taxonomic changes have moved many plainly marked genera—with plumages of grays and browns to yellows, and including many streaked, sparrow-like birds—into the tanager family.

Distribution Tanagers are found through much of the New World, from Mexico and the West Indies south to southernmost South America. Most species are sedentary or undertake short altitudinal movements, but the family includes some austral and intratropical migrants. Three enigmatic taxa are endemic to south Atlantic Ocean islands, and the famous "Darwin's Finches" (now usually placed in the Thraupidae) occur on the Galápagos Is. and Cocos I. off the northwest coast of South America.

Taxonomy Tanagers are perhaps most closely related to the cardinals and their relatives (family Cardinalidae) and, in turn, to the New World sparrows (Passerellidae), blackbirds (Icteridae), and wood-warblers (Parulidae). Recent taxonomic upheavals have drastically changed the composition of this family, with many genera transferred to other families (Cardinalidae, Fringillidae). Many more genera have recently been transferred into the Thraupidae (mainly sparrowlike seedeaters from the Emberizidae). Reflecting the diversity within this family, the AOS recognizes nine thraupid subfamilies from Mexico south to Panama.

Conservation Many species have very restricted ranges and are rare within those ranges, such as Cherry-throated Tanager of Brazil and the 15 "Darwin's finches" of the Galápagos. Others suffer from habitat degradation and/or exploitation for the pet trade. Eighty species are at some level of risk; BirdLife International codes: 28 NT, 32 VU, 16 EN, and 4 CR.

HONEYCREEPERS Genus *Cyanerpes*

Four small species with decurved bills that occur from Mexico to S.A. Males are mostly blue with black wings; leg color varies between species, as does the pattern of black. Females are largely green and streaked below.

RED-LEGGED HONEYCREEPER *Cyanerpes cyaneus* RLHO ■ 5

There is one accepted record of the tropical Red-legged Honeycreeper from south TX. Polytypic (11 sspp., no specimen from N.A., but only *carneipes* occurs in M.A.). L 4.5" (11 cm) **Identification** Small with decurved bill. Unusual among tropical birds, males have a postbreeding plumage that resembles female. **BREEDING ADULT MALE:** Plumage held Dec. through Aug., deep blue and black with turquoise crown. Bright yellow underwing coverts and inner webs of flight feathers. Bright red legs.

WINTER ADULT MALE: Greenish, but retains black lores, wings, and tail. **FEMALE AND JUVENILE:** Overall yellow-green with obscure streaks below; legs reddish to dull pink. Juvenile male has complete postjuvenal molt and resembles winter male by early winter.

Similar Species Female and juvenile could be confused with a wood-warbler. Note the decurved bill and leg color. Leg color separates it from other M.A. honeycreepers.

Voice CALL: A mewing *meeah*; also a thin *ssit*, often repeated.

Status & Distribution From northeast Mexico, southeast San Luis Potosí, and northern Veracruz to Ecuador and Brazil; also Cuba (said to have been introduced in 19th century). **RARE STATUS:** Accidental south TX, one at Estero Llano Grande SP, Hidalgo Co.,

juvenile similar to female; young male acquires some black on wings and tail by early winter

pale supercilium

long, slightly decurved bill ♀

faint yellow wing bar

overall greenish coloration

diffuse olive streaking on breast

short tail

shiny blue crown

legs of female and juvenile duller red than male

plumage deep blue and black; in winter, blue parts of plumage replaced by green but black retained

breeding adult ♂

bright red legs year-round

27–29 Nov. 2014. Six recent records from south FL, perhaps from Cuba, but not accepted on origin grounds. **Population** Stable.

Genus *Coereba*

BANANAQUIT *Coereba flaveola* BANA ■ 4

The Bananaquit is a small, short-tailed, black-and-white bird with a yellow breast and rump and a long, thin, decurved bill. Its tongue is specialized for feeding on nectar; it often pierces the bases of flowers. (41 sspp.; *bahamensis* in N.A.). L 4.5" (11 cm) **Identification ADULT:** Mostly black above, with long, curved, white supercilium and yellow rump. White patch at base of primaries and small white tips on outer tail feathers. White below with yellow breast patch. Red

corners at base of bill. **JUVENILE:** Duller than adult, with yellowish supercilium and dull yellow on underparts. Also duller rump patch and duller gape.

Geographical Variation Plumage variable (over 40 subspecies); yellow on underparts and throat color varies geographically. Many subspecies have grayish throats; birds from southern Caribbean are almost black.

Similar Species Adults are distinctive and unlike any other N.A. species. Juveniles may resemble a wood-warbler (*Parulidae*), but are readily distinguished by their decurved bill.

Voice CALL: An unmusical *tsip*. Young birds give a *chit, chit, chit* similar to warblers. **SONG:** Several ticks followed by rapid clicking in *bahamensis*.

Status & Distribution Resident throughout the Caribbean, except Cuba (rare visitor), and from south-

bold white supercilium

juvenile *bahamensis*

yellowish rump

decurved bill with reddish gape

yellow breast

white patch at base of primaries

adult *bahamensis*

white tail spots

ern Mexico through S.A. lowlands to southern Brazil, but mostly absent from Amazon Basin. **BREEDING:** Related to rainfall, peaking in Feb.–Apr. in Bahamas. **RARE STATUS:** Casual to southeast FL, mostly southeast coast and upper FL Keys (60+ records), mainly Jan.–Mar., some into May; a very few in fall. Accidental in west FL. **Population** Stable.

GRASSQUITS AND SEEDEATERS Genera *Tiaris* and *Sporophila*

These Neotropical genera include two *Tiaris* and 40 *Sporophila* species; both genera were formerly placed with New World sparrows (Passerellidae). All are small, with strong sexual dimorphism. *Tiaris* grassquits have a short, conical bill and a short, squared tail; *Sporophila* seedeaters have a distinctive "Roman nose" bill and a fairly long, rounded tail.

YELLOW-FACED GRASSQUIT *Tiaris olivaceus*   YFGR ■ 4

This bird feeds on grass seeds in vacant lots and weedy edges. Polytypic (5 sspp.; 2 in N.A.). L 4.3" (11 cm)
Identification Subspecies *pusillus* described. Very small; short tail; small, conical bill; straight culmen. **ADULT MALE:** Black head, breast, and upper belly; golden yellow eyebrow, throat, and crescent below eye; olive above.
ADULT FEMALE & IMMATURE MALE: Traces of same head pattern; olive above, lacks black below.
Geographic Variation Subspecies *pusillus*, with extensive black below, occurs from Mexico to northwest S.A. and accounts for TX records. FL records pertain to *olivaceus* from

Cuba, including Isla de Juventud, Cayman Is., Jamaica, and Hispaniola, which has more restricted black on face and underparts. Female *olivaceus* entirely lacks black coloration and is very plain overall with a short yellow supercilium. Three other subspecies occur: *bryanti* from Puerto Rico is similar but brighter olive-green above, more yellowish below; *intermedius* from Isla Cozumel, off Yucatán, is blacker below than *olivaceus*; and *ravidus* from Isla de Coiba, off southwest Panama, is even blacker than *pusillus* and black extends down to nape and through belly.
Similar Species Male distinctive; female nondescript with suggestion of the male face pattern. Cuban Grassquit (*T. camorus*), a declining endemic species from Cuba (introduced and established Nassau, New Providence I., Bahamas), is occasionally seen in south FL. It has a yellow frame around a black (male) or chest-

nut (female) face. It is a popular cage bird in Cuba. Cuban populations have significantly declined and FL records are all treated as escapes.
Voice CALL: A high-pitched *sik* or *tsi*. **SONG:** Thin and variable insectlike trills.
Status & Distribution Common in native range. Resident in M.A. and S.A. and West Indies. Introduced to HI. In Mexico, *pusillus* found from central Tamaulipas and Nuevo León south to S.A. where found in western Venezuela, Colombia, and Ecuador; *olivaveus* from the Greater Antilles is absent from the Bahamas and Cayman Is. **RARE STATUS:** Casual in southern FL (*olivaceus*) and south TX (*pusillus*) with about five records from each region. All but one of the TX records are of males; one at Bentsen-Rio Grande Valley SP in June 2002 built a partial nest, but did not attract a female.
Population Stable.

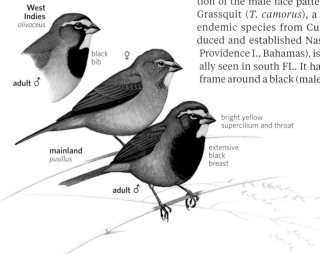

West Indies
olivaceus

black bib ♀

adult ♂

mainland
pusillus

adult ♂

bright yellow supercilium and throat

extensive black breast

BLACK-FACED GRASSQUIT *Tiaris bicolor*   BFGR ■ 4

Behavior is like the Yellow-faced Grassquit. Polytypic (8 sspp.; *bicolor* in N.A.). L 4.5" (11 cm)
Identification Very small; short tail; a small, conical bill and a straight culmen. **ADULT MALE:** Black face and upper breast, dark olive elsewhere; black bill, gape becomes vivid pink during breeding season. **FEMALE:** Nondescript; pale gray below, gray-olive above; pale horn bill. **JUVENILE:** Like female.
Geographic Variation Multiple subspecies in Caribbean and northern S.A. FL records pertain to widespread Baha-

mas (and very locally in Cuba) subspecies, *bicolor*; *omissus*, from Puerto Rico and south through the Lesser Antilles to Tobago and northern Venezuela, is smaller; *marchii*, from Jamaica and Hispaniola, has black confined to throat and breast; *grandior*, from San Andrés Archipelago in southwest Caribbean, like *omissus* but is much larger and brighter above. Three subspecies off northern Venezuela are *johnstonei* on La Blanquilla and Los Hermanos Is., which is the blackest subspecies; *sharpei* from Netherlands Antilles, which is similar but paler and

tortugensis on La Tortuga I., which is much paler, grayish olive above. S.A.

bicolor

marchii   omissus

huliae from Magdalena Valley, central Colombia, is brownish olive above with grayish flanks.

Similar Species Male unmistakable; female is extremely drab, compare with female *olivacea* Yellow-faced Grassquit that has yellow facial markings. Larger female and immature Indigo Buntings are brownish with faint wing bars.

Voice CALL: A lisping, soft and musical *tst*. **SONG:** A buzzing *tik, tik see*, the second part being buzzier.

Status & Distribution Common resident in a variety of open areas with grasses and shrubs, including forest edges and along the sides of roads. Also

found in urban areas in gardens, etc. Widespread in the Caribbean. Found through much of the West Indies, except rare and very local in Cuba (found only in the vicinity of Gibara and on Tío Pepe Cayo) and absent from the Cayman Is. Also found north coastal S.A. and central Colombia. **RARE STATUS:** Casual in southeastern FL (about 10 scattered records). **Population** Stable.

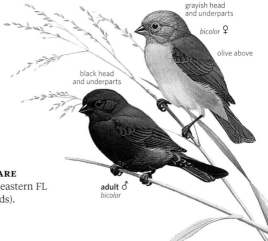

grayish head and underparts

bicolor ♀

olive above

black head and underparts

adult ♂
bicolor

MORELET'S SEEDEATER *Sporophila morelleti* MOSE ▪ 3

During spring, male Morelet's sing from high perches. Females and non-breeders can be secretive as they forage near the ground. Loosely colonial as nesters, this species may form small flocks in the nonbreeding season. In the US, found only in the Rio Grande Valley, south TX. Formerly known as the White-collared Seedeater (*S. torqueola*); the English name change resulted from a species split with the West Mexican endemic *torqueola* subspecies group. The West Mexican taxon is now recognized as a separate polytypic species, Cinnamon-rumped Seedeater (*S. torqueola*). Morelet's Seedeater was named by Charles Lucien Bonaparte (a French naturalist and a nephew of Napoleon Bonaparte) in 1850 after Pierre Marie Arthur Morelet, another 19th-century French naturalist, but he misspelled the scientific name as *morelleti*. The rules of the International Code of Zoological Nomenclature do not allow fixing this type of error, thus the misspelling is fixed for the scientific name, but the English name recognizes the correct spelling. Polytypic

(2 sspp.; *sharpei* in N.A.). L 4.5" (11 cm) **Identification** Subspecies *sharpei* described. Small size, stubby bill with strongly curved culmen, and rounded tail are distinctive. **ADULT MALE:** Black cap and wings, white crescent below eye, an incomplete buffy collar, white throat and breast washed with buff, narrow white wing bars, and white patch at base of primaries. **IMMATURE MALE:** Paler, browner, less distinctly marked. **FEMALE:** Brown overall; darker on wings with paler buffy color on breast; narrow wing bars.

Geographic Variation TX males are the relatively indistinctly marked *sharpei* of northeast Mexico (found south to eastern San Luis Potosí and northern Veracruz). Adult males of *morelleti* from southern Mexico through C.A. more black-and-white, with black breast band; intergrades with *sharpei* from Veracruz through Guatemala.

Similar Species Females can be confused with *Passerina* buntings or female Lesser Goldfinch; note the seedeater's smaller size, strongly curved culmen, more rounded tail, different coloration, and distinctive calls. The Cinnamon-rumped Seedeater (found from central Sinaloa to southern Oaxaca) has occurred in southeast AZ, Southern CA (San Diego area), and south (Rio Grande Valley) and West (El Paso) TX. All are believed to be escapes from captivity. Adult males are distinctive, with a cinnamon rump and buffy cinnamon underparts with a black chest band. Females and immatures are similar to Morelet's but lack wing bars. Morelet's

MORELET'S SEEDEATER

atriceps   *sharpei*

torqueola   *morelleti*

CINNAMON-RUMPED SEEDEATER

and Cinnamon-rumped Seedeaters occur near each other in Oaxaca but are separated by the Sierra Madre Oriental.

Voice CALL: Distinct, high *wink*; loud, whistled *chew*; and husky, rising *che*. **SONG:** A variable, clear-toned *sweet sweet sweet sweet cheer cheer cheer* that is pitched high, then low; recalls song of American Goldfinch.

Status & Distribution Resident in northeast Mexico from Tamaulipas and Nuevo León south through M.A. to west Panama. In US, uncommon and local resident along Rio Grande from southern Val Verde Co. south to northern Starr Co., where it prefers stands of cane. **RARE STATUS:** Formerly occurred farther east along Rio Grande but now casual east to Hidalgo Co. Accidental to Kenedy Co. (five birds, 1997) and Goliad Co. (2009). **Population** Formerly more widespread in south TX, it was greatly reduced during the 1950s and 1960s due to habitat loss and possibly pesticide use. By 1980 only a very small population persisted in Starr Co. Slow recovery in recent decades has resulted in gradual but significant expansion north along Rio Grande. Small extant US population is stable but vulnerable.

sharpei ♀

tiny size

thin, pale wing bars

short, stubby bill with strongly curved culmen

buffy below

1st winter ♂
sharpei

blackish head

pale buffy white collar

distinctly patterned wings

adult ♂
sharpei

Redwing

White Wagtail
alba

Dunlin
schinzii

GREENLAND

Greenland (part of Denmark) is the largest island in the world, most of it lying north of the Arctic Circle, and it forms the most northeastern part of N.A. Greenland is some 200 mi from Baffin I., across the Davis Strait, but is only 10 mi from Ellesmere I. across the Robeson Channel. Iceland lies about 190 mi to the southeast across the Denmark Strait. Greenland is now covered (again) by AOS, but not ABA. More than 240 species have been recorded there and nearly all are substantiated by specimens, most housed at the Zoological Museum in Copenhagen (ZUMC). The definitive ornithological reference is *An Annotated Checklist to the Birds of Greenland* (1994) by David Boertmann (Bioscience 38). Greenland receives limited coverage; it is difficult and expensive to reach. The status and distribution of birds in Greenland offers important clues for eastern Canada and the US, as does more thoroughly covered Iceland. The online Icelandic Birding Pages offer lots of information about birding in Iceland and the status and distribution of its birds, including detailed information on the various rarities that visit, most of which are from Europe or North America.

Boertmann divides Greenland into four regions (North, West, Northeast, and Southeast) and details the bird distribution for each. Greenland's avifauna is a mix of both Palearctic and Nearctic species, many of which are strays respectively from Europe (including Iceland) and mainland N.A. The breeding avifauna includes Pink-footed, Greater White-fronted (the endemic and declining breeding subspecies *flavirostris*), and Barnacle Geese, White-tailed Eagle (fairly common in West Greenland), Fieldfare (now extirpated), Redwing, White Wagtail (nominate *alba*), and Meadow Pipit. The resident subspecies of Mallard is endemic (*conboschas*); it is larger, but with a smaller bill. Three subspecies of Rock Ptarmigan are found in Greenland, two of which (*saturata* and *capta*) are endemic. Two subspecies of Dunlin breed in Greenland, one of which (*arctica*) is an endemic breeder in the northeast; the other (*schinzii*) breeds in southern Greenland as well as northwestern Europe. Merlin specimens from Greenland are of *subaesalon*, an endemic breeder on Iceland; also recorded is the more widespread *aesalon*, breeding mainly in northern Europe. Red Crossbill specimens are of nominate *curvirostra* from the Palearctic, yet a similar number of White-winged Crossbill records are *leucoptera* from N.A. rather than *bifasciata* from the Palearctic. The two Bohemian Waxwing records are of nominate *garrulus* from Europe, and there are multiple records for Common Scoter. Other noteworthy records include Emperor Goose, Eared Grebe, Common Gallinule (six records), American Avocet, Eurasian Woodcock, Crested Auklet, Yellow-nosed and Black-browed Albatrosses, Yellow-breasted Chat (three records), and Yellow-headed Blackbird. Four extant specimens of Eskimo Curlew were taken, all in Aug.–Sept., the last in 1882. The above information and the list below should alert observers to potential visitors to eastern Canada and northeastern US.

The following species are recorded from Greenland but *not* from Canada or the US. If known, the subspecies is given. **Ruddy Shelduck** *Tadorna ferruginea*: Four collected from three different locations in the summer of 1892, an invasion year for the species in northwestern Europe (see p. 52); these records add credibility to the flock of six photographed in Canada on Southampton I. being wild birds. **Velvet Scoter** *Melanitta fusca*: Nine records. **Western Water-Rail** *Rallus aquaticus hibernans*: Four records. The subspecies *hibernans* is endemic to Iceland. **Spotted Crake** *Porzana porzana*: Eleven records, nearly all in fall, from western Greenland. **Oriental Plover** *Charadrius veredus*: One May record (1948) from western Greenland. A remarkable record, as it breeds on the steppes of East Asia and winters in Australia. **Eurasian Spoonbill** *Platalea leucorodia leucorodia*: One Oct. record (1909) for western Greenland. **Rook** *Corvus frugilegus frugilegus*: One Mar. record (1901) for the Southeast. **Hooded Crow** *Corvus cornix* (now split by most authorities from Carrion Crow, *C. corone*): Two spring records (1897, 1907). **Eurasian Blackcap** *Sylvia atricapilla atricapilla*: One Nov. record (1916) from southeast Greenland. **White's Thrush** *Zoothera aurea aurea*: One Oct. record (1954) for northeast Greenland. Following recent treatments, it is now recognized as a separate species from the more southerly breeding Scaly Thrush complex (*Z. dauma*) and the endemic Amami Thrush (*Z. major*), which is resident on Amami Ō Shima, Ryukyu Is., Japan. **Meadow Pipit** *Anthus pratensis pratensis*: Scarce breeder in eastern Greenland. **Lesser Redpoll** *Acanthis cabaret*: One Sept. record (1933).

BERMUDA

Bermuda (part of the UK)—named for its discoverer, Juan de Bermúdez (ca. 1505)—consists of 150 islands and islets (21 sq mi; most islets are uninhabited) and lies 570 mi east of Cape Hatteras, NC.

Bermuda is the only breeding site of the Bermuda Petrel (also called the Cahow, Bermuda's national bird). Discovered early in the 17th century, it was then thought to be extinct soon after its discovery, until it was rediscovered in 1951 when 17 nesting pairs were found on several rocky islets. Before colonization and predation by humans and their introduced predators (rats, cats, dogs, and pigs), the population was perhaps a half million birds. Since 1951 a number of individuals and government agencies have worked diligently to save this species, and slowly the population has increased. In 2016–17 the count was 117 nesting pairs. Efforts included removing Bermuda's only recent record (1987) of a stray Snowy Owl. It was preying on the petrels, killing five adults during its stay (Jan.–Mar. 1987). In addition to restorative landscaping and monitoring, biologists have been hand-fostering chicks. It now nests on six islets: Long Rock, Inner Pear Rock, Horn Rock, Southampton I., Green Island, and best-known Nonsuch I. Bermuda Petrels arrive and court by Nov., the best month to see the species, and leave largely in May. The efforts to save the species were led by David Wingate (since 1951) and chronicled in Elizabeth Gehrman's book *Rare Birds* in 2012.

Bermuda Petrel

The White-tailed Tropicbird (aka "Longtail") reaches its northernmost breeding range here. Between 3,500 and 4,000 pairs nest in Bermuda, representing about 60 percent of the breeding population in the North Atlantic. The only extant, endemic breeding land bird is a resident subspecies of White-eyed Vireo (*bermudianus*). Other native resident species include Common Ground-Dove (*bahamensis*), Common Gallinule, Barn Owl, and Eastern Bluebird. Least Tern, Roseate Tern, and Audubon's Shearwater have been extirpated as breeders. Established exotics not found in N.A. include European Goldfinch, Common Waxbill, and Orange-cheeked Waxbill. Great Kiskadee, introduced in 1951 to control an introduced *Anolis* lizard, is now abundant and a pest, predating on birds.

Fossil and subfossil remains reveal that 10 or more extinct species were once on Bermuda, including the Great Auk and Bermuda Towhee (*Pipilo naufragus*). In addition, fossil remains have revealed an adult and nestling Short-tailed Albatross, suggesting a former pan-oceanic distribution.

White-tailed Tropicbird
catesbyi

Bermuda is well known for its migrants and strays, which make up most of Bermuda's extensive species list—in excess of 390 species. Given its small size and isolated location, it is a beacon for birds, particularly in the fall. Strong cold fronts and hurricanes can produce major fallouts, which reveal the migration routes of some of our trans-Atlantic migrants. Hurricane Emily made a direct hit on Bermuda on 25 Sept. 1987, and the next day 75 Connecticut Warblers and thousands of Bobolinks were counted! Surprising northern species that have occurred include Northern Goshawk, Northern Hawk Owl, Gyrfalcon, Northern Shrike, Bohemian Waxwing, Pine Grosbeak, and White-winged Crossbill. Most of the strays come from N.A. (including the West, e.g., Say's Phoebe, Varied Thrush, and Townsend's Warbler) or Europe, but a few (Large-billed Tern, Fork-tailed Flycatcher, and Yellow-green Vireo) are from S.A. Red-necked Stint, Arctic Warbler, and, even more surprisingly, Dark-sided Flycatcher (a late-Sept. specimen of nominate *sibirica*) are from Asia. Other oddball records of note include multiple records of Black Swift (subspecies uncertain), Common Swift (1986), and Common House-Martin (1960s). Much information on Bermuda bird distribution can be found online in the annual reports of the Bermuda Audubon Society. Recent birding references for Bermuda are *A Guide to the Birds of Bermuda* (1991) by Eric Amos and *A Birdwatching Guide to Bermuda* (2002) by Andrew Dobson.

The following species are recorded from Bermuda but *not* from mainland N.A.: **West Indian Whistling-Duck** *Dendrocygna arborea*: Two records (1907 and 2015). This species is resident on some of the Bahamian islands. (A record from VA is of uncertain origin.) **Ferruginous Duck** *Aythya nyroca*: A winter sight record (1987). **Alpine Swift** *Apus melba*: June record (2015); also six records from the West Indies. **White Tern** *Gygis alba*: A remarkable Dec. record (1972). Photographic evidence indicates that it was not nominate *alba* (tropical South Atlantic) but one of the subspecies from the Pacific! **Striated Heron** *Butorides striata*: One record (1985) of a long-staying bird. **Western Marsh Harrier** *Circus aeruginosus*: Two records of long-staying birds (2015–16). **Booted Eagle** *Hieraaetus pennatus*: A Sept. sight record (1989).

Great Kiskadee

CONTRIBUTORS

The following authors and other contributors formed the talented team that wrote and edited the first edition of *Complete Birds of North America* (2006). They include many of North America's most well-known ornithologists, birders, and birding authors. Under the direction of editor Jonathan Alderfer and associate editor Jon L. Dunn, the first-edition team created a state-of-the-art handbook on the identification of North American birds. The superlative work of map consultant Paul Lehman complemented the written text, and a team of illustrators produced numerous new illustrations. (Illustrations Credits appear on pages 739–740.)

The second edition was published in 2014 with the same editorial team. Writers Edward S. Brinkley, Kimball L. Garrett, and Paul Hess contributed new species and family accounts, and artists Jonathan Alderfer, N. John Schmitt, and Thomas R. Schultz created additional illustrations.

Since the second edition, new identification information has come to light, additional bird species have been recorded in North America, status and distribution have changed for many species, and genetic and observational studies (augmented by the increasingly sophisticated analysis of vocalizations) have had a profound effect on bird taxonomy.

This third edition (2021)—based on the 2018 AOS Supplement—incorporates as much of that new information as we could fit onto 752 pages. The editorial team remains unchanged. Jonathan Alderfer designed the layout, reviewed text, assisted in text editing, and wrote some of the new species accounts. Jon L. Dunn wrote most of the new species accounts and edited the entire text as needed. Paul Lehman reviewed the status and distribution sections of all the species accounts and updated many range maps. Kimball L. Garrett reviewed and updated the taxonomy entries of the family accounts. Sixteen art figures for 13 new species were created for this edition: Jonathan Alderfer (Kermadec, Juan Fernandez, and Tahiti Petrels), David Quinn (Red-backed Shrike, Thick-billed Warbler, River Warbler, European Robin, Pied Wheatear, and Mistle Thrush), N. John Schmitt (Black Kite and Great Black Hawk), and Thomas R. Schultz (Cassia Crossbill and Black-backed Oriole).

Found below are short biographies, listed alphabetically, of editorial and writing contributors for all editions and for artists who painted new illustrations found in this edition. Since new material by other authors and the editors has been woven into the second- and third-edition text, the original author credits found in the first edition are no longer accurate and have been dropped from the family accounts and table of contents. However, at the end of each biography found below is a list of the first-edition bird families that each contributor originally wrote (or co-wrote) or of the editing work they did.

Jonathan Alderfer is a widely published author and field guide illustrator. One of the nation's foremost birding artists, he is well known in the birding community for his expertise as a field ornithologist and for his knowledge of North American birds. With Jon L. Dunn he co-authored the seventh edition of *National Geographic Field Guide to the Birds of North America* and was a contributing artist to the last five editions (2003–2017). In addition to creating field guide illustrations, Alderfer creates paintings and woodblock prints; more of his artwork can be viewed at *www.jonathanalderfer.com*. He lives in Edgecomb, Maine, on the banks of the Sheepscot River. *Editor and contributing artist (all editions).*

Jessie H. Barry works at the Cornell Lab of Ornithology, where she is the project leader for Merlin. Merlin Bird ID is an app designed to help people identify North America's most common birds. Jessie has a passion for bird ID and sharing it with others. She is a member of the Cornell Lab's Team Sapsucker and holds the national record for most species of birds seen in 24 hours, 294 species. *Ducks, Geese, and Swans.*

Edward S. (Ned) Brinkley is a lifelong birder whose interests include migration, seabird identification and distribution, and the effects of weather and climate on birds. He has guided birding tours around the world since 1993 (guiding for Field Guides, Inc.) and served as editor (1996–2016) of *North American Birds,* a quarterly journal of ornithological record. He has published widely in the field, including over 100 articles and several books, notably *The National Wildlife Federation Field Guide to North American Birds* (2007). He died birding in Ecuador in 2020. *New species accounts (second edition).*

Steven W. Cardiff is collection manager for birds and mammals at Louisiana State University Museum of Natural Science (since 1984). Steve also serves as chair of the Louisiana Bird Records Committee and as an eBird reviewer for Louisiana and several counties in far western Texas. His many interests include status and distribution of Louisiana birds, Gulf of Mexico seabirds, identification/plumages/molt of large gulls, and natural history of the Yellow Rail and *Myiarchus* flycatchers. (He co-authored the Ash-throated and Brown-crested Flycatcher accounts in *Birds of North America* with Donna L. Dittmann.) Steve is a lifelong birder who grew up in Southern California, moved to Baton Rouge, Louisiana, to attend LSU in 1979, and has lived in St. Gabriel, Louisiana, since 1990. *Rails, Gallinules, and Coots; Nighthawks and Nightjars; Tyrant Flycatchers.*

Allen T. Chartier picked up a pair of binoculars and a field guide at the age of 11 and has been hooked on birds ever since. His strong interests in participating in scientific projects, writing, and teaching have led him beyond birding, to become a bird bander specializing in hummingbirds, and a freelance environmental educator. He has served on the Michigan Bird Records Committee and the editorial board for Michigan's state bird journal, *Michigan Birds and Natural History*, for many years. Among the many birding publications he has authored or co-authored are *A Birder's Guide to Michigan*, *Michigan's Second Breeding Bird Atlas*, and *The ABA Field Guide to Michigan Birds* (2018). He lives in Inkster, Michigan. *Larks; Swallows; Kinglets; Olive Warbler; Wood-Warblers; Bananaquits.*

Cameron D. Cox has been an avid birder for 27 years who works as a birding and nature photography guide. In 2013, he co-authored the Peterson *Reference Guide to Seawatching: Eastern Waterbirds in Flight,* a book inspired by his years observing waterbirds in Cape May, New Jersey. His primary interest is studying the identification of waterbirds and sharing his enjoyment and knowledge of these birds with other birders. *Ducks, Geese, and Swans.*

Donna L. Dittmann is a former member of the American Birding Association (ABA) Checklist Committee and has been secretary of the Louisiana Bird Records Committee since 1988. Avian interests range from genetics to identification and molt, and she has written a variety of technical and popular articles. She is co-founder of the Yellow Rails and Rice Festival (*yellowrailsandrice.com*). She works at LSU Museum of Natural Science where she is the genetic resources collections manager, permits coordinator, and museum preparator. Donna is a native of San Francisco, California; moved to Louisiana in 1984; and has lived in St. Gabriel, Louisiana, since 1990. Website: *snowyegretenterprises.com. Rails, Gallinules, and Coots; Nighthawks and Nightjars; Tyrant Flycatchers.*

Jon L. Dunn is an expert on the identification and distribution of North American birds. Since 1980 he has served as the chief consultant or lead author for all seven editions of the *National Geographic Field Guide to the Birds of North America.* Among the other books he has written are *Field Guide to the Warblers of North America* and *Birds of Southern California*, both co-authored with Kimball L. Garrett. He has served as a longtime member of the California Bird Records

Committee and is currently on the American Ornithological Society Committee on Classification and Nomenclature and previously was on the American Birding Association Checklist Committee. He has been a senior tour leader for WINGS birding tours for over 40 years. He lives in Bishop, California. *Old World Flycatchers; associate editor (first and second editions); editor and contributing writer (third edition).*

Ted Floyd is the editor of *Birding*, the flagship publication of the American Birding Association. Floyd has served on the boards of directors of Colorado Field Ornithologists and Western Field Ornithologists, and is a frequent speaker at bird festivals and ornithological society meetings. He is the author of five bird books and more than 200 articles and technical papers on birds and other aspects of natural history. Floyd has a special interest in bird vocalizations, especially nocturnal songs and calls. He lives in Lafayette, Colorado. *Wrens; Dippers; Starlings; Old World Sparrows.*

Kimball L. Garrett is a lifelong resident of the Los Angeles area, where he has served as Ornithology Collections Manager at the Natural History Museum of Los Angeles County since 1982. He is past-president of Western Field Ornithologists, a longtime member of the California Bird Records Committee, a former member of the ABA Checklist Committee, and a fellow of the American Ornithological Society. Among the many birding books and articles he has written are several co-authored with this guide's two principal editors. *Tropicbirds; Boobies and Gannets; Pelicans; Cormorants; Darters; Frigatebirds; Swifts; Woodpeckers and Allies; new family accounts (second edition); review of taxonomy entries in family accounts (third edition).*

Matthew T. Heindel has been birding for over 40 years with a focus on identification, migration, and status and distribution. His articles have appeared in *Birding, Birder's Journal, Western Birds, Audubon Field Notes,* and *North American Birds,* and he was editor of the California Christmas Bird Counts for many years. He served several terms on the California Bird Records Committee, including multiple stints as the chairman and vice-chairman. He currently lives near San Antonio, Texas. *Thick-knees; Lapwings and Plovers; Oystercatchers; Stilts and Avocets; Jacanas; Sandpipers, Phalaropes, and Allies; Pratincoles; Trogons; Hoopoes; Vireos; Old World Warblers and Gnatcatchers.*

Paul Hess, a retired newspaper editor, has co-authored or edited nearly two dozen books on birds and natural history for various publishers including National Geographic. He writes the "Frontiers in Ornithology" column for *Birding,* the American Birding Association's magazine; has contributed many articles to the state journal *Pennsylvania Birds;* and has chaired the Pennsylvania Ornithological Records Committee. He has received state and regional awards for outstanding contributions to Pennsylvania ornithology and bird conservation. He lives in Natrona Heights, Pennsylvania. *Shrikes; Chickadees and Titmice; editing (first edition); new species and family accounts (second edition).*

Steve N. G. Howell is an international bird tour leader with WINGS, a research associate at the California Academy of Sciences, and a widely published author. His more recent books include the widely acclaimed *Oceanic Birds of the World: A Photo Guide*, co-authored with Kirk Zufelt (2019, Princeton University Press) and *The Peterson Guide to Bird Identification: In 12 Steps*, co-authored with Brian Sullivan (2018, Houghton Mifflin Harcourt). Beyond birds, Steve's interests include chocolate, flying fish, and tequila. *Loons; Grebes; Cranes; Skuas, Gulls, Terns, and Skimmers; Hummingbirds.*

Marshall Iliff grew up in Annapolis, Maryland, and has been birding for 35 years. He has worked as a tour guide for Victor Emanuel Nature Tours and has served on three state Records Committees. His long interest in helping birders to collect better data led to his current position as eBird project leader. A free, global website, eBird (*www.ebird.org*) allows all birders to keep their personal records, track their lists, and see what others are seeing. The open-access database has become invaluable to the science and conservation communities: it now contains almost 800 million records and has truly revolutionized how birders contribute to the understanding and protection of the birds they love. *Emberizids.*

Alvaro Jaramillo was born in Chile but began birding in Toronto, Canada. He was trained in ecology and evolution with a particular interest in bird behavior. He is the author of the *American Birding Association Field Guide to the Birds of California*, *Birds of Chile*, and *New World Blackbirds: The Icterids*. Alvaro writes the "Identify Yourself" column in *Bird Watcher's Digest*. He co-authored the Emberizids chapter for the *Handbook of the Birds of the World* and is writing a photo guide to the birds and wildlife of Patagonia. He runs Alvaro's Adventures, a birding tour company that also emphasizes natural history and culture while traveling. Alvaro lives with his family in Half Moon Bay, California. *Ducks, Geese, and Swans; Skuas, Gulls, Terns, and Skimmers; Babblers; Accentors; Blackbirds.*

Ian L. Jones has studied seabird behavior, ecology, and conservation since 1982. He received his M.Sc. in zoology researching Ancient Murrelets. His Ph.D. focused on Least Auklets in the Pribilof Islands, Alaska, and his postdocs concerned Crested Auklets at Buldir, Aleutian Islands, Alaska, and founding a seabird demography project at Triangle Island, British Columbia. Ian has led seabird research and conservation projects in Newfoundland and the northwestern Hawaiian Islands. His main focus is auklets (*Aethia* sp.) in the Aleutians sector of the Alaska Maritime National Wildlife Refuge, pursuing conservation projects, especially the impact of introduced Norway rats. Movement and migration of auklets outside the breeding season is currently (2020) his major focus. Ian has been a professor of biology at Memorial University in St. John's, Newfoundland, since 1995 and with collaborators has published more than 100 scientific papers. *Auks, Murres, and Puffins.*

Paul Lehman has birded extensively throughout North America since he was young and has written many papers dealing with the continent's bird distribution and field identification. He has been the principal range-map maker for many of the popular field guides used today, including multiple editions of the *National Geographic Society Field Guide to the Birds of North America*. He is a past editor of *Birding,* the American Birding Association's magazine, and a former geography instructor with special interests in meteorology, biogeography, and the effects of weather on bird migration and vagrancy. He currently lives in San Diego, California, and previously resided in Cape May, New Jersey; Santa Barbara, California; and the New York City area. *Maps (all editions); text editing (first and second editions); review of status and distribution entries (third edition).*

Tony Leukering has published extensively on his primary birding interests: migration, identification, and distribution. He has presented more than 500 online photo quizzes for the American Birding Association and the Colorado Field Ornithologists (CFO). He is the primary author of *In the Scope*, a column for *Colorado Birds* that covers identification issues from a Colorado perspective, and past chair of the Colorado Bird Records Committee. In 2008, he was awarded the CFO's Ron Ryder Award, given for scholarly contributions to Colorado field ornithology. Tony has been a working ornithologist for 30 years, primarily at bird observatories scattered from coast to coast. Currently based in Colorado, he works as a consultant, conducting fieldwork in many interior western states. *Thrushes.*

Mark W. Lockwood is the regional director for Texas State Parks in the Trans-Pecos region in the State Parks Division of the Texas Parks and Wildlife Department. He has a lifelong interest in the status and distribution of birds in the state. Mark is the author or co-author of five books, including *Birds of the Texas Hill Country* and the *TOS Handbook of Texas*

Birds. Mark lives in Alpine, Texas. *Curassows and Guans; New World Quail; Cuckoos, Roadrunners, and Anis; Kingfishers; Penduline Tits and Verdins; Long-tailed Tits and Bushtits.*

Guy McCaskie is a retired engineer with a lifelong interest in birds. He is a past-president of Western Field Ornithologists, a member of the California Bird Records Committee since 1970, and a co-author of the southern California regional report for *North American Birds* and its predecessors since 1963. Guy is one of the co-authors of *Birds of the Salton Sea* (2003, University of California Press) and is keenly interested in that region, birding it far more than anyone else, since the early 1960s. He formerly lived in Great Britain but is a longtime resident of Imperial Beach, California. *Pigeons and Doves.*

Bill Pranty has birded extensively in Florida for more than 40 years. His studies emphasize the documentation of Florida's diverse avifauna, with a focus on its exotic species. His research has added four species to the *ABA Checklist*— Egyptian Goose, Purple Swamphen, Nanday Parakeet, and Common Myna—all of them exotics from Florida. Pranty is a former chairman of the ABA Checklist Committee and a technical reviewer for and frequent contributor to *Birding* magazine. In addition to dozens of peer-reviewed papers, Pranty is the author or co-author of five books, among these the latest book on Florida ornithology. He lives in Bayonet Point, Florida. *Bitterns, Herons, and Allies; Ibises and Spoonbills; Storks; Flamingos; Limpkins; Parakeets, Macaws, and Parrots; Bulbuls; Weavers; Estrildid Finches.*

David E. Quady began birding in graduate school, as something to do while camping. But soon camping became simply something to do to facilitate birding. His fondness for owls began with a close, pre-dawn encounter with a vocalizing Northern Spotted Owl and has grown to the point that he now sometimes wonders why people bother to go birding in the daytime. He has birded widely in North America and abroad, leads field trips and has taught owl classes for Golden Gate Audubon Society, and is a past president of Western Field Ornithologists. He lives in Berkeley, California. *Barn Owls; Typical Owls.*

David Quinn is a British artist, working in a variety of drawing and painting media. A former *British Birds* Bird Illustrator of the Year winner, he has illustrated many ornithology publications dealing with the identification, behavior, and ecology of birds. David contributed artwork to five editions of the *National Geographic Field Guide to the Birds of North America*, specializing in the Old World species that have strayed to North America. He also painted the Bald Eagle for the cover of the seventh edition. *Contributing artist (all editions).*

Kurt Radamaker has been interested in birds and nature most of his life. He grew up in southern California, where he started birding at the age of 8 and by 15 had completed Cornell Laboratory's Seminars in Ornithology. He worked as an ornithology teacher for the University of La Verne in La Verne, California, and has authored or co-authored a handful of books on birds and contributed articles for a number of journals including *Western Birds* and *Birding*. He has a keen interest in birding in Mexico and was editor of the Mexican journal *The Euphonia* for five years. He is a member of the Arizona Bird Committee and past president of Arizona Field Ornithologists. He resides in Cave Creek, Arizona. *Bitterns, Herons, and Allies; Ibises and Spoonbills.*

Gary H. Rosenberg has been a bird-watcher since he was a young boy, when his father introduced him to bird-watching. He studied biology at Arizona State University and then ornithology at Louisiana State University, where he earned a master's degree in zoology, studying birds and specializing on Amazonia river islands in Peru. Since 1986, Gary has been a professional birding tour leader, first working for WINGS and then starting his own company—Avian Journeys. He has led hundreds of birding tours in the neotropics and across North America. Gary has published numerous scientific and popular papers and has been part of the discovery of three birds new to science. He has served as secretary of the Arizona Bird Committee since 1987. He lives in Tucson, Arizona. *Mockingbirds and Thrashers; Wagtails and Pipits; Tanagers; Cardinals, Saltators, and Allies; Fringilline and Cardueline Finches.*

N. John Schmitt is a well-known artist with a lifelong passion for natural history and the study of birds, especially raptors. He has contributed illustrations to numerous books including *Birds Asleep; Cornell Handbook to Bird Biology; Birds of South Asia: The Ripley Guide; A Field Guide to Raptors of Europe, the Middle East, and North Africa; Birds of Prey of the Indian Subcontinent; Birds of Peru; A Field Guide to the Birds of Trinidad and Tobago; Raptors of Mexico and Central America; and National Geographic Field Guide to the Birds of North America* (third to seventh editions). John has also worked as a co-leader in an ecotourism company, participated in numerous bird surveys, and devoted several years to both the Peregrine Falcon and California Condor recovery programs. He is a lifelong resident of California and currently lives in the Kern River Valley. *Contributing artist (all editions).*

Thomas R. Schultz has been drawing and painting birds since the 1970s and began a career in illustration in 1982 with National Geographic's new book, *Field Guide to the Birds of North America*. Since then he has provided additional work for subsequent editions—up through the seventh. He has also worked on numerous other bird books and field guides, including *A Field Guide to Warblers of North America* and *Birds of South Asia*. He is a past president of the Wisconsin Society for Ornithology and has served on the WSO's board of directors for the past 31 years. *Contributing artist (all editions).*

Brian Sullivan has conducted fieldwork on birds for the past 30 years. Research, photography, and field projects have taken him across the planet in search of birds. He has written and consulted on various books and popular and scientific literature on North American birds and is co-author of the forthcoming *Princeton Guide to North American Birds*. He is currently digital publications lead at the Cornell Laboratory of Ornithology, focused mainly on the Birds of the World project, and is also a former eBird project leader. He served as photographic editor for the American Birding Association's journal *North American Birds* from 2005 to 2013. *Albatrosses; Shearwaters and Petrels; Storm-Petrels.*

Clay Taylor started birding in Connecticut in 1975. Viewing and photographing raptors became his prime passion while attending college in Rochester, New York, near the Braddock Bay hawk-watch site. As a birder, tour leader, and hawk bander in the 1980s he started the hawk-banding project at Braddock Bay that developed into Braddock Bay Raptor Research. He joined Swarovski Optik in 1999 and is now the naturalist market manager, finding time to visit many of the top hawk migration sites during his travels. His Corpus Christi, Texas, yard list has 23 species of raptors seen from the property. *New World Vultures; Hawks, Kites, Eagles, and Allies; Caracaras and Falcons.*

Chris Wood is the assistant director of information science at the Cornell Lab of Ornithology. Much of his time is focused on eBird, which has grown from a small citizen science project into one of the largest in the world. Today, eBird is a major source of biodiversity data, increasing our knowledge of the dynamics of species distributions and having a direct impact on the conservation of birds and their habitats. Chris is widely recognized as a leading authority on bird identification and distribution, and has written and consulted on popular bird books and the scientific literature. These activities are all built around Chris's primary interest—developing ways to connect birders with scientists and the conservation community to better understand and conserve birds and their ecosystems. *Partridges, Grouse, and Turkeys; Tyrant Flycatchers; Jays and Crows; Nuthatches; Creepers; Waxwings; Silky-flycatchers.*

ILLUSTRATIONS CREDITS

ART CREDITS

The following artists contributed the illustrations for this volume. Jonathan Alderfer, David Beadle, Peter Burke, Marc R. Hanson, Cynthia J. House, H. Jon Janosik, Donald L. Malick, Killian Mullarney, Michael O'Brien, John P. O'Neill, Kent Pendleton, Diane Pierce, John C. Pitcher, H. Douglas Pratt, David Quinn, Chuck Ripper, N. John Schmitt, Thomas R. Schultz, Daniel S. Smith, and Sophie Webb.

Front cover: Great Egret photographed near Shark River in Everglades National Park, Florida (Carlton Ward); **front flap:** Wilson's Phalarope (Killian Mullarney); **back cover, clockwise from top left:** Calliope Hummingbird (Jonathan Alderfer), Common Eider (Jonathan Alderfer), Northern Harrier (N. John Schmitt), Blue Jay ((N. John Schmitt); **5**—Alderfer, except Green-breasted Mango by Schmitt and Alderfer; **8**—Alderfer, except Common Black Hawk by Schmitt and gull heads by Schultz; **9**—Alderfer, except American Black Duck by House and gull wing by Schultz; **11–13**—House; **14**—Schultz, except Graylag Goose by Alderfer; **15**—Quinn, **16–17**—Mullarney; **18**—Alderfer; **19**—House; **20**—Schmitt, except "Aleutian" and "Cackling" by House; **21**—House, except "Lesser" by Schmitt; **22**—House; **23**—House, except Tundra and Trumpeter Swans (heads) by Mullarney; **24**—House, except Egyptian Goose by Schmitt; **25**—House, except Muscovy Duck (flight) by Schmitt; **26**—House, except Baikal Teal by Mullarney; **27**—House, except Cinnamon Teal (female) by Schmitt; **28**—Mullarney, except Northern Shoveler by House; **29**—House; **30**—House, except wigeon hybrid by Schultz; **31**—House, except Eastern Spot-billed Duck by Alderfer; **32**—House; **33**—House, except American Black Duck (head) by Schmitt and Mottled Duck (*fulvigula*) by Schultz; **34**—House; **35**—House, except Green-winged Teal (female) by Schmitt; **36–40**—House; **41**—House, except Common Eider (heads and flight) by Alderfer; **42**—House, except Harlequin Duck (adult males) by Alderfer; **43**—House, except Labrador Duck by Alderfer; **44**—House, except Common Scoter and White-winged Scoter (*fusca*) by Mullarney and White-winged Scoter (*stejnegeri*) by Alderfer; **45–47**—House; **48**—House, except goldeneye hybrid by Alderfer; **49**—House, except Common Merganser ("Goosander") by Alderfer; **50–52**—House; **53**—Schmitt; **54**—Pendleton; **55**—Pendleton, except (*floridanus* and *taylori*) by Schmitt; **56–57**—Pendleton; **59**—Pendleton, except Chukar (flight) by Schmitt; **60–61**—Pendleton; **62**—Pendleton, except Greater Sage-Grouse (displaying male) and Gunnison Sage-Grouse by Alderfer; **63**—Pendleton; **64**—Pendleton, except Rock Ptarmigan (*evermanni*) by Schmitt; **65–66**—Pendleton; **67**—Pendleton, except Sooty Grouse (*howardi* and *sitkensis*) by Schmitt and Sharp-tailed Grouse (displaying male) by Alderfer; **68–69**—Pendleton; **71**—Alderfer, except Greater Flamingo (standing and swimming) by Pierce; **72**—Alderfer; **73**—Alderfer, except Red-necked Grebe by Janosik; **74**—Alderfer; **75**—Janosik, except Western Grebe (winter adult) by Alderfer; **76**—Janosik, except Clark's Grebe (winter adult) by Alderfer; **77**—Pratt; **78**—Pratt, except Scaly-naped Pigeon by Beadle; **79**—Pratt, except Band-tailed Pigeon (juvenile) by Alderfer; **80**—Pratt, except Oriental Turtle-Dove by Schmitt and Alderfer, and European Turtle-Dove by Quinn; **81–82**—Schmitt and Alderfer; **83**—Alderfer; **84**—Pratt, except Common Ground-Dove by Alderfer; **85**—Alderfer; **86**—Pratt, except Ruddy Quail-Dove by Quinn; **87**—Pratt; **88–89**—Schmitt and Alderfer; **90–91**—Quinn; **92**—Pratt; **93**—Pratt, except Greater Roadrunner by Schmitt; **94**—Pratt; **96**—Schmitt, except Common Nighthawk by Schultz; **97**—Schultz; **98**—Schmitt, except Common Pauraque (tails) by Ripper; **99**—Schmitt, except tails by Ripper; **100**—Schmitt, except tails by Ripper; **101**—Schmitt, except Mexican Whip-poor-will (tail) by Webb and Gray Nightjar by Beadle; **103–106**—Schmitt; **107**—Schmitt, except Antillean Palm-Swift by Beadle; **109**—Schmitt and Alderfer, except "Parts of a Hummingbird" by Alderfer; **110–112**—Schmitt and Alderfer; **113**—Schmitt and Alderfer, except Lucifer Hummingbird by Webb; **114–115**—Schmitt and Alderfer, except spread tails and wing details by Alderfer; **116**—Schmitt and Alderfer, except spread tail and wing detail by Alderfer and Bumblebee Hummingbird by Beadle; **117–120**—Alderfer; **121–122**—Schmitt and Alderfer; **123**—Schmitt and Alderfer, except Cinnamon Hummingbird by Beadle; **124**—Webb; **125**—Hanson, except Yellow Rail (flight) and Black Rail by Schultz; **126–128**—Schultz; **129**—Schultz, except Sora (standing and head) by Hanson; **130**—Schultz, except Paint-billed Crake and Spotted Rail by Beadle; **131**—Quinn, except Purple Gallinule by Hanson; **132**—Schultz, except Common Gallinule (swimming) by Hanson and Common Moorhen by Beadle; **133**—Hanson; **134**—Alderfer; **135**—Hanson, except flight by Pierce; **137**—Pierce, except flight by Schultz; **138**—Pierce, except Common Crane (flight) by Schultz; **139**—Beadle; **140**—Quinn; **141**—Janosik; **142**—Quinn;

143—Janosik, except American Oystercatcher (*frazari*) by Schultz; **145**—Mullarney, except Black-bellied Plover by Alderfer; **146–147**—Alderfer; **148**—Mullarney, except Snowy Plover (standing) by Pitcher and (flight) by Smith; **149**—Quinn, except Wilson's Plover (standing) by Pitcher and (flight) by Smith; **150**—Quinn, except Wilson's Plover (standing) by Pitcher and (flight) by Smith; **151**—Pitcher, except flight by Smith; **152**—Pitcher, except flight by Smith; **153**—Mullarney, except Killdeer (flight) by Smith; **154**—Mullarney, except Mountain Plover (dorsal flight) and Eurasian Dotterel (flight) by Smith; **155**—Janosik; **157**—Mullarney, except flight by Smith; **158**—Schmitt, except flight by Smith; **159**—Smith, except Little Curlew by Schmitt; **160**—Smith; **161**—Smith, except Far Eastern Curlew (standing) by Schmitt; **162**—Quinn, except Eurasian Curlew (standing) by Schmitt and (flight) by Smith; **163**—Alderfer, except Bar-tailed Godwit (flight) by Smith; **164**—Alderfer; **165**—Alderfer, except Marbled Godwit (flight) by Smith and Ruddy Turnstone by Pitcher; **166**—Pitcher, except flight by Alderfer; **167**—Schultz, except Red Knot (flight) by Alderfer; **168**—Pitcher; **169**—Mullarney, except Ruff (flight) by Smith and "*Calidris* Plumage" by Alderfer; **170**—Pitcher, except Sharp-tailed Sandpiper (standing) by Mullarney and (flight) by Smith; **171**—Alderfer, except flight by Smith and Curlew Sandpiper (standing) by Schultz; **172**—Pitcher, except flight by Smith; **173**—Pitcher; **174**—Pitcher, except Sanderling (standing) by Schultz and (flight) by Smith; **175**—Schultz, except flight by Smith; **176–179**—Pitcher, except flight by Smith; **180**—Mullarney, except flight by Smith; **181**—Pitcher, except Sharp-tailed Sandpiper by Mullarney and Semipalmated Sandpiper (breeding female) by Smith; **182**—Pitcher; **183–184**—Alderfer; **185**—Quinn, except American Woodcock by Alderfer; **186**—Alderfer; **187**—Quinn, except Common Snipe by Alderfer; **188**—Alderfer, except Terek Sandpiper by Pitcher; **189–191**—Pitcher; **192**—Pitcher, except Lesser Yellowlegs (flight) by Smith; **193**—O'Brien; **194–195**—Pitcher, except flight by Smith; **196**—Quinn, except Wood Sandpiper by Pitcher; **197**—Quinn; **198–199**—Mullarney; **200**—Quinn; **202–206**—Schultz; **208**—Ripper, except winter (flight) by Alderfer; **209**—Ripper, except swimming by Schmitt and flight by Alderfer; **210**—Alderfer; **211**—Ripper, except Great Auk by Alderfer; **212**—Ripper, except all *mandti* by Alderfer; **213**—Alderfer, except Pigeon Guillemot (swimming) by Ripper; **214–215**—Ripper; **216–217**—Alderfer; **218**—Schmitt and Alderfer; **219**—Schmitt, except Parakeet Auklet by Alderfer; **220–221**—Alderfer; **222**—Alderfer, except Atlantic Puffin by Ripper; **223**—Alderfer, except Atlantic Puffin (swimming) by Ripper; **224**—Alderfer; **226**—Schultz, except "Parts of a Gull" by Alderfer; **227–229**—Schultz; **230**—Schultz, except Gray-hooded Gull by Quinn; **231–260**—Schultz; **261**—Schultz, except Whiskered Tern by Quinn; **262–267**—Schultz; **268**—Janosik, except *dorotheae* head by Alderfer; **269–270**—Janosik; **271–274**—Quinn, **276–293**—Alderfer, except Great-winged Petrel by Beadle; **295**—Alderfer, except Trindade Petrel by O'Brien; **296**—Alderfer; **297**—O'Brien; **298**—O'Brien, except Juan Fernandez Petrel by Alderfer; **299**—Alderfer, except Fea's Petrel by O'Brien; **300–306**—Alderfer; **307**—Alderfer, except Short-tailed Shearwater (darkish underwing) by Hanson; **308**—Alderfer, except Sooty Shearwater (flight) by Hanson; **309–312**—Alderfer; **313–314**—Pierce; **315**—Janosik, except Great and Lesser Frigatebirds by Schultz; **317–325**—Alderfer; **326–328**—Janosik; **329**—Janosik, except *californicus* by Alderfer; **331**—Burke, except Yellow Bittern by Quinn; **332**—Burke, except Bare-throated Tiger-Heron by Alderfer; **333**—Pierce; **334**—Pierce, except Gray Heron by Alderfer and Great Egret (*modesta*) by Quinn; **335–336**—Quinn; **337–338**—Pierce, except flight by Schultz; **339**—Schultz, except Tricolored Heron (wading) by Pierce; **340**—Pierce, except flight and *coromandus* by Schultz; **341**—Pierce, except Chinese Pond-Heron by Quinn and Green Heron (flight) by Schultz; **342–343**—Burke; **344**—Pierce; **345**—Burke, except flight by Pierce; **346**—Pierce; **348–349**—Malick; **350**—Malick, except *ridgwayi* by Schmitt; **352**—Malick, except Hook-billed Kite perched adults by Pendleton and juvenile by Schmitt; **353**—Schmitt, except Hook-billed Kite (black morphs) and Swallow-tailed Kite by Malick; **354**—Malick; **355**—Schmitt; **356**—Malick, except juvenile (flight) by Schmitt; **357**—Malick, except juvenile (flight) by Schmitt; **358**—Malick, except juvenile (flight) by Schmitt; **359**—Schmitt, except Mississippi Kite (perched) by Malick and 1st summer (flight) by Pendleton; **360**—Malick, except 3rd year by Schmitt; **361**—Schmitt; **362**—Pendleton, except Crane Hawk by Schultz and Snail Kite (flight) by Schmitt; **363–365**—Schmitt; **366**—Malick, except flight by Schmitt; **367–368**—Schmitt; **369**—Schmitt, except perched Short-tailed Hawk by Malick; **370**—Malick, except Zone-tailed Hawk (flight) by Schmitt; **371**—Malick, except flight by Schmitt; **372–373**—Schmitt; **374**—Schmitt, except perched adult by Malick; **375**—Malick, except adults (flight) by Schmitt; **376**—Malick; **378**—Malick, except Oriental Scops-Owl by Schultz; **379–384**—Malick; **385**—Malick, except Mottled Owl by Schultz; **386–387**—Malick; **388**—Schmitt, except Long-eared Owl (perched) by Malick and Stygian Owl by Schultz; **389**—Malick, except *domingensis* by Schmitt; 390—Malick, except Northern Saw-whet Owl (*brooksi*) by Schmitt; 391—Quinn, 392—O'Neill, except Elegant Trogon (juvenile) by Schmitt; 393—Quinn; 395—Malick; 396—Alderfer, except Green Kingfisher (perched) by Malick and (flight) by Schmitt; 397—Quinn; 398–403—Malick, 404—Malick, except American Three-toed Woodpecker by Schultz; 405—Malick, except American Three-toed Woodpecker (*bacatus*) by Schultz; 406—Malick, except Great Spotted Woodpecker by Quinn, 407—Malick, except Nuttall's Woodpecker (juvenile head) by Schmitt; 408–410—Malick; 411—Malick, except intergrade (head) by Schultz; 412–413—Malick; 415—Malick, except Collared Forest-Falcon by Schultz; 416—Malick, except dorsal flight by Schmitt; 417—Malick, except Eurasian Kestrel (flight) by Schmitt and Red-footed Falcon by Schultz; 418—Malick, except *suckleyi* and flight by Schmitt; 419—Schmitt, except Aplomado Falcon (perched) by Malick; 420—Malick, except flight by Schmitt; 421—Malick, except flight by Schmitt; 422—Malick, except flight by Schmitt; 423—Schmitt; 424—Schmitt, except Carolina Parakeet by Alderfer; 425–431—Schmitt; 432—Alderfer; 433—Alderfer, except Gray-collared Becard by Schultz; 435—Beadle, except White-crested Elaenia by Schmitt; 437–439—Burke; 440—Schultz, except Great Kiskadee (flight) by Pratt and Social Flycatcher by Alderfer; 441—Schultz, except Piratic Flycatcher by Burke; 442—Schultz, except Variegated Flycatcher by Burke; 443—Alderfer; 444—Alderfer, except Thick-billed Kingbird (spring/summer) by Schultz; 445—Alderfer, except Eastern Kingbird by Schultz; 446—Schultz, except Loggerhead Kingbird by Beadle; 447—Pratt, except Fork-tailed Flycatcher (immature) by Schmitt; 448–456—Beadle; 457—Beadle, except Pine Flycatcher by Schultz; 458–459—Beadle; 460–461—Pratt; 462—Quinn; 463—Quinn, except Loggerhead Shrike (perched) by Pratt and (flight) by Schmitt; 464—Pratt; 465—Pratt, except White-eyed Vireo (immature head) by Schultz; 466—Schultz; 467—Pratt, except Gray Vireo by Schultz, and Hutton's Vireo and Ruby-crowned Kinglet by Schmitt; 468—Pratt, except Cassin's Vireo by Schultz; 469—Schultz; 470—Schultz, except Philadelphia Vireo by Beadle; 471—Beadle, except Red-eyed Vireo by Pratt; 472—Schultz; 473—Beadle; 474–476—Pratt; 477—Pratt, except Blue Jay (flight) by Schmitt; 478–479—Schmitt; 480—Schmitt, except Woodhouse's Scrub-Jay (*texana*) by Schultz and Mexican Jay (juvenile) by Pratt; 481—Pratt; 482—Pratt, except Yellow-billed Magpie (juvenile) by Schmitt; 483–485—Schmitt; 486—Schmitt, except Chihuahuan Raven (flight) and Common Raven (flight and calling) by Alderfer; 488–489—Beadle; 491—Schmitt, except *arboricola* by Schultz; 492—Beadle; 493—Beadle, except Tree Swallow by Pratt; 494—Alderfer, except Violet-green Swallow dorsal flight and perched adult male by Schmitt, and juvenile and ventral flight by Pratt; 495—Pratt, except Northern Rough-winged Swallow by Schmitt; 496—Schmitt; 497—Pratt, except Cave Swallow (*pallida* flight) by Beadle; 498—Pratt; 499—Pratt, except Common House-Martin by Quinn; 501–508—O'Brien; 509–510—O'Neill; 512—Pratt; 513—Schultz; 514—Pratt; 515—Schultz; 517—Schmitt, except Canyon Wren by Pratt; 518—Pratt; 519—Pratt, except Pacific Wren (*pacificus*) by Beadle and Winter Wren by Schmitt; 520—Schmitt; 521—Schmitt, except Carolina Wren by Pratt; 522—Pratt, except *calaphonus* by Schmitt; 523—Schmitt, except Cactus Wren (center) by Pratt; 525—Pratt, except Blue-gray Gnatcatcher (*obscura*) by Schultz; 526–528—Pratt; 529–530—Schmitt; 531—Quinn, except Common Chiffchaff by Alderfer; 532–535—Quinn; 536—O'Brien; 537–545—Quinn; 546—Quinn, except Common Redstart by Alderfer; 547–548—Quinn; 549–551—Pratt; 552—Pratt, except Brown-backed Solitaire by Schultz; 553—Beadle; 554–556—Schultz; 557—Schultz, except Eurasian Blackbird by Quinn; 558–560—Quinn; 561—Pratt, except White-throated Thrush by Burke; 562—Pratt; 563—Schultz, except Varied Thrush by Pratt; 564—Schmitt; 566—Alderfer, except Gray Catbird by Pratt; 567—Schmitt; 568—Schultz; 569–571—Schmitt; 552—Pratt, except Northern Mockingbird (flight) by Schmitt; 574—Pratt; 575—Alderfer, except Hill Myna by Pratt; 577—Pratt, except Bohemian Waxwing (perched adults and flight) and Cedar Waxwing (flight) by Quinn; 579—Alderfer, except Phainopepla by Pratt; 580—Pratt; 581—Quinn; 582–587—Pratt; 589—Pratt; 590—Quinn, except Gray Wagtail by Pratt; 591—Pratt, except Tree Pipit by Quinn; 592–594—Quinn; 596—Quinn, except Brambling by Pierce; 597—Pierce; 598—Quinn, except Common Rosefinch by Schultz; 599—Quinn, except Pine Grosbeak by Pierce; 600—Pierce, except Asian Rosy-Finch by Schmitt; 601—Schmitt, except *littoralis* and male *tephrocotis* by Beadle and *umbrina* by Pierce; 602—Schmitt; 603–605—Pierce; 606—Pierce; 607—Pierce, except Common Redpoll (*rostrata*) by Schmitt; 608—Pierce; 609—Pierce, except Cassia Crossbill by Schultz; 610—Pierce, except Eurasian Siskin and Pine Siskin (green morph) by Quinn; 611–612—Pierce; 614–617—Pierce, except wing figures by Schultz; 618—Pierce, except 1st winter female (flight) by Alderfer; 620–623—Quinn; 625–627—Burke; 628—Quinn; 629–630—Burke; 631–632—Schultz; 633—Pierce; 634–636—Schultz; 637—Pierce, except Worthen's Sparrow by Beadle; 638—Pierce, except Vesper Sparrow by Beadle;

639—Pierce; 640—Pierce, except Bell's Sparrow (three tails) by Schultz; 641—Pierce, except Bell's Sparrow (*canescens*) by Beadle; 642–643—Schultz; 644—Pierce; 645—Pierce, except LeConte's Sparrow by Schmitt; 646–647—Schmitt; 648—Schultz; 650—Pierce, except juvenile by Schultz; 651—Schultz, except Swamp Sparrow by Pierce; 652–653—Pierce; 654—Schultz; 655—Pierce; 656—Pierce, except *shufeldti* by Beadle and flight by Schmitt; 657—Pierce, except *mearnsi* by Beadle; 658—Schultz; 659—Burke, except female *auricollis* by Schultz; 661—Pratt, except Bobolink by Schultz; 662–663—Schultz; 664—Burke, except Orchard Oriole (*fuertesi*) by Schultz; 665–668—Burke; 669—Burke, except Black-backed Oriole by Schultz; 670—Schultz, except Red-winged Blackbird (males) by Pratt; 671—Schultz; 672—Schmitt, except Tawny-shouldered Blackbird by Beadle; 673—Beadle, except Bronzed Cowbird (*aeneus*) by Pratt; 674—Schmitt; 675–679—Pratt; 680—Pratt, except Louisiana Waterthrush by Schultz; 681—Schultz; 682–683—Pratt; 684—Schultz; 685—Pratt, except Crescent-chested Warbler by Beadle; 687–688—Pratt; 689—Schultz; 690—Schultz, except Gray-crowned Yellowthroat by Burke; 691—Schultz; 692—Pratt; 693—Schultz, except Hooded Warbler by Pratt; 694–695—Pratt; 696—Pratt, except immature male by Schultz; 697—Pratt; 698—Schultz; 699—Pratt, except Bay-breasted Warbler (fall male) by Beadle; 700—Schultz; 701–702—Pratt; 703—Pratt, except Palm Warbler by Schultz; 704—Pratt; 705—Pratt, except *coronata* fall female by Schultz; 706–707—Pratt; 708—Pratt, except immature female by Schultz; 709–710—Pratt; 711—Pratt, except Fan-tailed Warbler by Burke; 712—Burke; 713—Schultz, except Canada Warbler by Pratt; 714—Beadle; 715—Pratt; 717–719—Burke; 720—Burke, except tanager hybrid by Schmitt; 721—Pierce, except *superbus* by Beadle; 722–725—Pierce; 726–728—Schultz; 729—Pierce; 731—Schmitt, except Bananaquit by Pratt; 732—Burke; 733—Pratt, except Morelet's Seedeater (adult male) by Burke; 734—Quinn, except White Wagtail by Alderfer and Dunlin by Schultz; 735—O'Brien, except White-tailed Tropicbird by Janosik and Great Kiskadee by Schultz.

PHOTOGRAPHY CREDITS

Most of the photographs in this book were contributed by the following photographers: Richard Crossley, Rob Curtis/The Early Birder, Mike Danzenbaker, Kevin T. Karlson, Maslowski Wildlife Productions, Robert Royse, Larry Sansone, Brian E. Small, and Tom Vezo. Abbreviations: t–top, b–bottom, c–center, l–left, r–right.

Front cover—Carlton Ward; 2–3—Alexander Koenders/Minden Pictures; 10—Crossley; 53—Small; 54—Vezo; 58—James Cumming/NG Your Shot; 70—Karlson; 72—Vezo; 76—Maslowski; 89—Tom J. Ulrich; 95—Glenn Bartley; 102—Glenn Bartley; 108—Small; 124—Vezo; 134—Chris Jimenez; 135—Troy Lim/NG Your Shot; 136—Vezo; 139—Christopher L. Wood; 140—Maslowski; 142—Vezo; 144—Vezo; 155—Crossley; 169—Sansone; 200—Danzenbaker; 201—Small; 207—Crossley; 225—Crossley; 243—Sansone (l) and Danzenbaker (r); 244—Crossley (both); 268—Tom J. Ulrich; 271—Karlson; 275—Danzenbaker; 282—Royse; 285—Tomi Muukonen/Alamy; 292—Danzenbaker; 313—Karlson; 314—Judd Patterson; 316—Glenn Bartley; 321—Crossley; 326—Karlson; 327—Sansone; 330—Karlson; 343—Karlson; 347—Royse; 350—Darran Rickards; 351—Vezo; 356—Karlson; 357—Sansone; 376—Taxi/Getty Images; 377—Small; 391—Dorothy E. Harris/NG Your Shot; 393—Axel Hilger/NG Your Shot; 394—Small; 397—Danny Brown/NG Your Shot; 414—Crossley; 423—Curtis; 430—S. B. Nace; 432—Chris Jimenez; 434—Curtis; 436—Small (all); 462—Maslowski; 464—Small; 473—Daniel Parent/Getty Images; 487—Small; 490—Danzenbaker; 498—Danzenbaker (l) and Crossley (r); 500—Maslowski; 502—Karlson (both); 509—Royse; 510—Roger van Gelder; 511—Maslowski; 515—Maslowski; 516—Small; 524—Matthew Studebaker; 527—Alan Fuchs; 528—Bhalchandra Pujari/Dreamstime.com; 529—Karlson; 530—Matthew Studebaker; 535—Small; 536—Dobyter/Dreamstime.com; 538—Danzenbaker; 540—Mehrdad Sadat; 548—Maslowski; 550—Royse (l and r) and Bob Steele (c); 565—Vezo; 573—Verastuchelova/Dreamstime.com; 576—Small; 578—Maslowski; 580—Royse; 581—Sansone; 582—Tom Gatz/USFWS; 583—Tze-hsin Woo/Getty Images; 584—Kimball L. Garrett; 586—Curtis; 588—Maslowski; 595—Bob Steele; 605—Karlson (l) and Small (c and r); 613—Brian Wolitski/NG Your Shot; 615—Howard Cummings (l), Small (c-l and c-r), and Sansone (r); 619—Ralph Martin/Alamy; 624—Small; 658—Glenn Bartley/Alamy; 659—William Leaman/Alamy; 660—Danzenbaker; 671—Small (l), Vezo (c), and Bob Steele (r); 676—Small; 677—Karlson; 678—Crossley; 681—Karlson (upper) and Small (lower); 702—Giff Beaton (upper) and Vezo (lower); 716—Small; 730—Alessandro Tramonti.

INDEX

ACKNOWLEDGMENTS

The editors, map consultant, and artists wish to thank the following individuals and institutions for their valuable assistance in the preparation of this edition and the two previous editions of this work.

Third edition (2021): Louis R. Bevier, David Boertmann, Daniel D. Gibson, Ian Lewington, Natural History Museum, Tring, UK (Mark Adams, Robert Prys-Jones, and Hein van Grouw), Natural History Museum of Los Angeles County, Los Angeles, CA (Allison J. Shultz and Kimball L. Garrett), Brainard Palmer-Ball, Jr., David E. Quady, Lara Tseng, and World Museum, Liverpool, UK (Clem Fisher and Tony Parker). Special thanks to the National Geographic team behind the book, Susan Blair, Melissa Farris, Bridget Hamilton, Susan Tyler Hitchcock, Judith Klein, Linda Makarov, Nicole Miller, Moriah Petty, and Jennifer Conrad Seidel.

Second edition (2014): Bruce Anderson, David Arbour, Christian Artuso, Pierre Bannon, Louis R. Bevier, Dick Cannings, Russell Cannings, Steve Cardiff, Stephen Dinsmore, Cameron Eckert, Richard Erickson, Michael Force, Daniel D. Gibson, Joe Grzybowski, Peter Hamel, Margo Hearne, Matt Heindel, Steve N. G. Howell, David Irons, Oscar Johnson, Rudolf Koes, Yann Kolbeinsson, Mark Lockwood, Derek Lovitch, Carl Lundblad, Bruce Mactavish, Ron Martin, Chris McCreedy, Chet McGaugh, Robert McKernan, Martin Meyers, Charles Mills, Eric Mills, Steve Mlodinow, Kenny Nichols, Brainard Palmer-Ball, Brian Patteson, Mark Peterson, Bill Pranty, Peter Pyle, David E. Quady, Larry Rosche, Scott Seltman, David Sibley, John Sterling, Peter Taylor, Thede Tobish, Michael Todd, David Vander Pluym, Ron Weeks, Sandy Williams, Alan Wormington, and Andy Wraithmell.

Special thanks to Terry Chesser and Andy Kratter of the AOU Committee on Classification and Nomenclature for providing the accepted AOU English names for all North American bird families; and to the American Birding Association for use of their bird abundance codes.

First edition (2005): George Armistead, Louis R. Bevier, Edward S. Brinkley, Jamie Cameron, Richard A. Erickson, Tom Gatz, Daniel D. Gibson, Shawneen Finnegan, Nancy Gobris, Robert A. Hamilton, Floyd Hayes, Tom and Jo Heindel, Greg Lasley, Paul Lehman, Bruce Mactavish, Curtis Marantz, Ian A. McLaren, Steven Mlodinow, Michael O'Brien, Michael A. Patten, J. Brian Patteson, Grayson Pearce, Peter Pyle, Van Remsen, Bill Schmoker, Willie Sekula, Larry Semo, Debra Shearwater, Douglas Stotz, Giles Timms, Bill Tweit, Phil Unitt, Nick Ward, Angus Wilson, and Sherrie York.

Institutions: Archbold Biological Station, Venus, FL; Burke Museum, University of Washington, Seattle; Cornell Laboratory of Ornithology, Ithaca, NY; Field Museum of Natural History, Chicago, IL; Museum of Natural Science, Louisiana State University, Baton Rouge; Museum of Vertebrate Zoology, University of California, Berkeley; Natural History Museum of Los Angeles County, Los Angeles, CA; PRBO Conservation Science, Stinson Beach, CA; San Diego Natural History Museum, San Diego, CA; and Wings, Tucson, AZ.

Since 1888, the National Geographic Society has funded more than 14,000 research, conservation, education, and storytelling projects around the world. National Geographic Partners distributes a portion of the funds it receives from your purchase to National Geographic Society to support programs including the conservation of animals and their habitats.

Get closer to National Geographic Explorers and photographers, and connect with our global community. Join us today at nationalgeographic.com/join

For rights or permissions inquiries, please contact National Geographic Books Subsidiary Rights: bookrights@natgeo.com

ISBN: 978-1-4262-2188-0

Printed in Hong Kong

21/PPHK/1